LICHENS OF NORTH AMERICA

Lichens
OF NORTH AMERICA

IRWIN M. BRODO

SYLVIA DURAN SHARNOFF

STEPHEN SHARNOFF

with Selected Drawings by Susan Laurie-Bourque

and published in collaboration with the

Canadian Museum of Nature

Yale University Press/New Haven and London

HAZARD WARNING: Some of the chemicals and procedures described in this book for the identification of lichens may be hazardous if used or performed inappropriately. While efforts have been made to indicate the hazards associated with the different reagents and procedures covered in this book, it is the ultimate responsibility of the reader to ensure that safe working practices are used.

Neither the author nor the publisher has any legal responsibility or liability for errors, omissions, outdated material, or the reader's application of the information or testing procedures contained in this book.

Copyright © 2001 by Yale University. All rights reserved. This book may not be reproduced, in whole or in part, including illustrations, in any form (beyond that copying permitted by Sections 107 and 108 of the U.S. Copyright Law and except by reviewers for the public press), without written permission from the publishers.

Designed by Nancy Ovedovitz and set in Minion type by B. Williams & Associates. Printed in China.

Library of Congress Cataloging-in-Publication Data
Brodo, Irwin M.
Lichens of North America / Irwin M. Brodo, Sylvia Duran Sharnoff, and Stephen Sharnoff ; with selected drawings by Susan Laurie-Bourque.
      p.   cm.
Includes bibliographical references (p.).
ISBN 978-0-300-08249-4 (cloth : alk. paper)
1. Lichens — North America — Identification.   2. Lichens — North America — Pictorial works.   I. Sharnoff, Sylvia Duran, 1944–   II. Sharnoff, Stephen, 1944–   III. Title.
QK586.5 .B76 2001
579.7'097 — dc21   00-049541

A catalogue record for this book is available from the British Library.

The paper in this book meets the guidelines for permanence and durability of the Committee on Production Guidelines for Book Longevity of the Council on Library Resources.

10   9   8   7   6

Publication of *Lichens of North America* has been made possible by the generous support of the following benefactors:

Educational Foundation of America

Ms. Susan Gibson and Dr. Allan Stone of the Lichen Foundation

Furthermore, the publication program of The J. M. Kaplan Fund

The Bay Foundation

The Ottawa Field-Naturalists' Club

The British Lichen Society

Dr. Glenna Dean

Averill Babson and Greg Sohns

Dr. Trudy Murphy

B. J. Fontana

An anonymous donor

Fieldwork for *Lichens of North America* was sponsored by the Missouri Botanical Garden, St. Louis.

Support for photographic expenses, especially fieldwork, was provided by:

Educational Foundation of America

National Geographic Society

National Fish and Wildlife Foundation

Frank Weeden Foundation

Norcross Wildlife Foundation

New England Biolabs Foundation

Maki Foundation

Davis Conservation Foundation

California Native Plant Society

Support for Irwin Brodo's fieldwork and travel was provided by:

The Canadian Museum of Nature, Ottawa.

Original artwork was financed with a donation from Elisabeth Lay.

Dedicated to the memory of

MASON E. HALE, JR., who gave North Americans their first popularized guidebook to the lichens of this continent

VICTOR DURAN, Sylvia Sharnoff's father, who inspired the photographic work, for his close focus on nature — especially lichens — and his patient generosity

ELIZABETH DURAN for her sincere wish to be of service to the natural world and its human inhabitants

Also dedicated to

FENJA BRODO, for her love and her unfailing help and encouragement

CONTENTS

Foreword by Peter H. Raven	xvii
Preface	xix
Acknowledgments	xxi

PART ONE / ABOUT THE LICHENS

1 Lichens: The Organism — 3
  What Lichens Are — 3
  What Lichens Are Not — 3
  The Components — 4
    Fungi, 4   Photobionts, 4
  Picking a Partner — 6
  The Lichen Association — 6
  The Mysterious Transformation — 8

2 How Lichens Are Built: Thallus Shapes and Structure — 9
  Growth Forms — 9
    *Foliose Lichens, 13   Fruticose Lichens, 14   Crustose Lichens, 16
    Squamulose Lichens, 17   Combinations of Growth Forms,
    and Intermediates, 18*

Special Features of Lichen Thalli   18
    Cilia, 18 Pruinose and Scabrose Surfaces, 18 Tomentum, 19
    Cortical Hairs, 19 Maculae, 20 Pseudocyphellae, 20
    Cyphellae, 21 Cephalodia, 21

3   How Lichens Reproduce   23

Fruiting Bodies in Lichens   23

Ascomata   23

Asci   24

Ascospores   26

The Tissues of Apothecia   26

Perithecia   28

Mushroom Lichens   29

Pycnidia and Conidia   30

Vegetative Reproductive Structures   30
    Soredia, 30 Isidia, 31 Granules and Blastidia, 32
    Schizidia, 33 Lobules, 33

4   Colors of Lichens   34

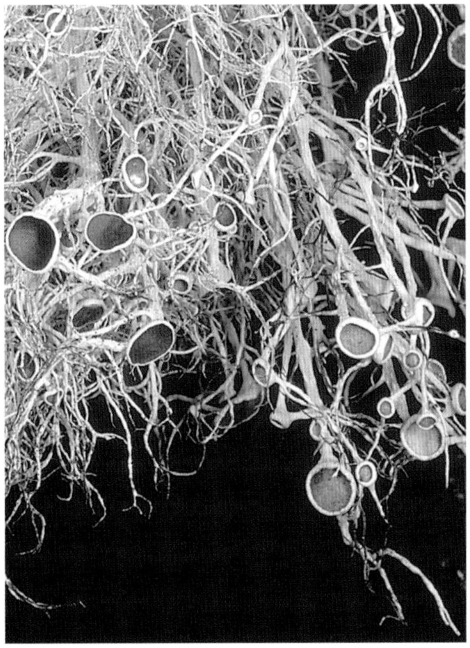

5 How Lichens Live: Physiology and Growth 36
   Producing and Sharing Food: Photosynthesis in Lichens 36
   Using Food for Energy: Respiration 37
   Finding the Balance: Photosynthesis versus Respiration 37
   Water Relations 38
   Temperature 39
   Minerals 39
   Nitrogen 40
   Growth 40
   Putting It All Together: Physiology and Habitat Selection 41

6 Lichen Chemistry 42
   Lichen Substances: What They Are 42
   Where the Substances Occur, and What They Do for the Lichen 42
   Using Chemistry for Identifying and Classifying Lichens 44

7 Where Lichens Grow: Their Substrates 45
   General Characteristics of Substrates 45
   Bark 45
   Wood 46

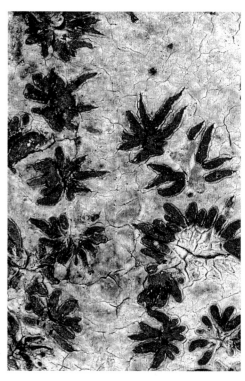

Mosses and Dead Vegetation	46
Rock	48
Soil	50
Leaves	51
Artifacts as Substrates	52
Animals	52

8 Lichens and Ecosystems — 54

Nature's Pioneers — 54

Colonization of Rock: Succession and Soil Formation, 54
Colonization and Stabilization of Soil, 55 Colonization of Trees, 57

Contribution of Lichens to the Nitrogen Cycle — 58

Animals and Lichens — 59

Lichens as Food, 59 Lichens as Nesting Material, 60
Lichens as Camouflage, 61

9 Geographic Distributions in North America — 62

How Have Distribution Patterns Arisen? — 62

Classifying Patterns of Distribution — 63

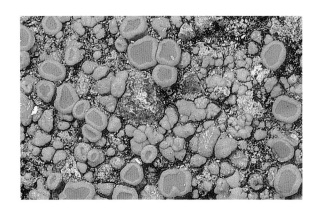

	Arctic-Alpine Element	67
	Boreal Element	67
	Temperate Element	68
	Eastern Sector, 69 Western Sector, 71	
	Western Montane Element	72
	Madrean Element	73
	Tropical Element	74
	Oceanic Element	75
	Maritime Element	77
10	**Lichens and People**	78
	Lichens as Food	78
	Clothing	79
	Dyes	80
	Perfumes	82
	Medicines and Poisons	82
	Models and Decorations	83
	Lichenometry	84

	The Down Side of Lichen Growth: Lichen-Caused Damage	84
	Human Impacts on Lichens	85
11	**Environmental Monitoring with Lichens**	89
	Lichens as Pollution Monitors: Why Does It Work?	89
	Pollution Indexes, Scales, and Maps	89
	Transplants	91
	Lichen-Based Programs of Pollution Monitoring	92
	Prospecting	92
	Ancient Forests	92
12	**Naming and Classifying Lichens**	94
	Lichen Names	94
	Scientific Names and Their Authors	94
	Lichen Classification: Families of Lichens	95
	Scientific versus Vernacular Names	95
13	**Collecting and Studying Lichens**	97
	Conservation	97

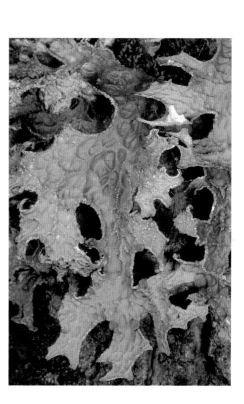

	Tools	97
	Transporting Lichens	98
	Back Home	99
	Preparing Specimens for Study and Storage	99
	Preparing a Label	100
	The Permanent Collection	100
	Studying Lichens	101
	Equipment, 101 Microscopic Study, 101	
	Chemistry	103
	Spot Tests, 103 Crystal Tests, 108 Thin-Layer Chromatography, 108 Ultraviolet Light, 109	
	Photographing Lichens: Techniques and Advice	109
14	**Using This Book to Name a Lichen: Hints and Conventions**	111
	Organizing the Task	111
	Keys	111
	Main Entries	112
	Distribution Maps	113
	Plates	113

PART TWO / GUIDE TO THE LICHENS

Identification Keys to Genera and Major Groups	117
Key to Keys	117
Key A: Fruticose Lichens	118
Key B: Dwarf Fruticose Filamentous Lichens	121
Key C: Sterile Crustose Lichens	122
Key D: Crustose Perithecial Lichens and Lichens with Ascomata Resembling Perithecia	124
Key E: Crustose Script Lichens	126
Key F: Crustose Disk Lichens	126
Key G: Squamulose Lichens	134
Key H: Umbilicate and Fan-Shaped Lichens	136
Key I: Jelly Lichens	137
Key J: Yellow Foliose Lichens	137
Key K: Foliose Lichens That Are not Umbilicate, Jelly-Like, or Yellow	139
Descriptions, Illustrations, Keys to Species, and Maps	145

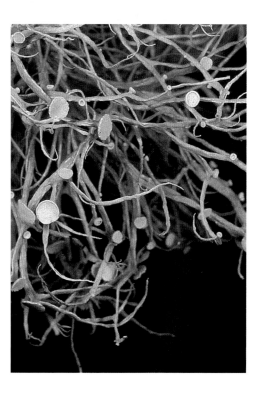

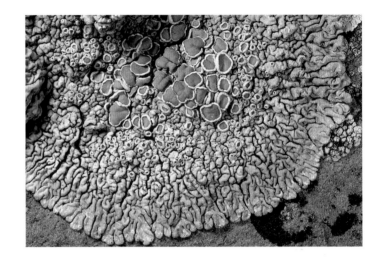

Appendix: Classification of Lichen Fungi	751
Glossary	755
Further Reading and Bibliography	765
Index of Names	771

FOREWORD

In trying to understand and appreciate the biological diversity of this continent, especially in the face of human impact on the environment, it is essential not to overlook any significant part of that biodiversity. This, unfortunately, has been the case with lichens, even though there are over 3,600 species in North America that are beautiful, diverse, interesting, and ecologically important. *Lichens of North America* addresses this oversight, providing for the first time a comprehensive guide to the lichens of North America north of Mexico. Decades of careful study by Irwin Brodo, combined with the photographic talents of Steve and Sylvia Sharnoff, have produced an enormously useful work in which the professional skill and enthusiasm of the authors is evident on every page.

In this book, the fascinating world of lichen biology is introduced in a lengthy opening section covering not only what lichens look like but also how they function, why they are found where they are, and what practical and ecological importance they have. Illustrating nearly a quarter of the lichens of Canada and the United States and mentioning almost half of the recorded species, and providing readily accessible keys for recognizing most of the more abundant or conspicuous lichens of this region, this book is an ideal introduction to the lichens.

Why is this important? Fundamentally, because lichens are fascinating organisms that grow almost everywhere. Anyone who loves the out-of-doors is probably aware of them as orange, yellow, or gray patches on alpine rocks or roadside trees; yellowish green masses festooning the pines and oaks of the Pacific coast; multicolored incrustations on palms in the southeast; or mile after mile of gray reindeer "moss" covering the ground in northern or mountainous regions. Environmentally, lichens play a significant role in ecological succession — the progression by which bare, open soil or rocks are gradually converted to habitats suitable for occupation by vascular plants and the formation of complex communities in ecosystems. As in the great coniferous forests of western Canada and the northwestern United States, lichens in which the photosynthetic partner is a blue-green alga also contribute significant quantities of nitrogen to the ecosystems where they occur, thus playing a role of fundamental importance in creating rich and diverse biological communities. In the region covered by this handsome and useful book, more than half a million species of organisms occur, and fewer than a third are known to science. Despite their importance, lichens are one of the least appreciated of these groups.

The biodiversity of North America is under siege as a result of growth in human numbers, the desire for increased levels of consumption everywhere, and the widespread use of inappropriate technologies. With respect to the latter, lichens are sensitive indicators of air pollution and therefore often absent from urban areas. This quality also makes them valuable for monitoring trends that directly affect the success of agriculture and forestry, as well as human health.

Saddened by the untimely death of Sylvia Sharnoff, whose inspiration this book was, I nonetheless congratulate the Sharnoffs on their accomplishment in illustrating so beautifully and well such a large proportion of the species of lichens of Canada and the United States and so effectively complementing Irwin Brodo's clear and accurate text. The descriptions, keys, maps, and photographs in this book will provide a way for specialists and novices alike to identify the most noticeable and noteworthy species of lichens of its region. There is no longer any reason for the importance of lichens to be overlooked on this continent by anyone who would like to understand them better: forest managers, park personnel, teachers, and air quality researchers. Most laudable of all, however, is the way in which this book opens the world of lichens to appreciation and enjoyment by ordinary people who love the outdoor world. Lichens represent an often colorful part of the world of nature, one that can be appreciated in winter or summer, in the north, in the deserts, in the tropics, along roadsides, or deep in forests — they exist in every nonpolluted habitat on this continent.

This book is thoroughly up-to-date, covering the latest literature and the best contemporary understanding of lichen characteristics, classification, and distribution. The authors have presented us with an outstanding gift as a result of their productive labors.

Peter H. Raven
Missouri Botanical Garden

PREFACE

The tiny moss has been the theme of many a gifted poet; and even the despised mushroom has called forth classic works in its praise. But the Lichens, which stain every rock, and clothe every tree, which form:

    Nature's livery o'er the globe

    where'er her wonders range

have been almost universally neglected, nay despised.

                                                  Lauder Lindsay

One of the ongoing puzzles that have plagued the authors is why so few people, including amateur and professional naturalists, know about — or, indeed, care about — lichens. As a part of nature, lichens seem to have everything going for them. They are colorful, varied, can be found in winter and summer, occur almost everywhere (except in the centers of big cities), are often easy to recognize, and require little equipment to appreciate. Because they prefer unpolluted landscapes, lichens are the essence of wildness. To find them in abundance is to find a corner of the universe where the environment is still pure and unspoiled.

Although some lichens are little more than a smudge on a rock or twig, others form brilliant orange, yellow, or milky white patches the size of your palm, and others drape tree branches in gossamer curtains that can be several meters in length. Are they useless, insignificant ornaments of our forests and fields? Hardly. Lichens form the basic sustenance of huge herds of caribou, the nesting materials of birds and mammals, and the basic organic matter replenishing soil over bare rock and recharging the soil with nitrogen. Peoples of many lands have used lichens as food, medicines, dyestuffs, and clothing material. Lichens are sought out as an early warning system for deteriorating air quality and as indicators of ancient forests.

Why then have North American lichens been so neglected for so long? We believe it is because there has never been an easily accessible, full-color guide to even the most common and conspicuous species. Europeans have produced numerous, well-illustrated books dealing with their lichens, covering the Swedish, Norwegian, British, French, German, and Spanish regions. Yet nothing comprehensive has been available for North Americans.

The idea of preparing a book to fill that gap slowly grew from the conversations Steve and Sylvia Sharnoff and I had on two long field trips we took together, first to the High Sierra of California in 1986 and then to the Queen Charlotte Islands and neighboring parts of British Columbia in 1988. Would it be possible to prepare an up-to-date, illustrated guidebook to the lichens of North America thorough enough to make it truly useful, and yet not so technical as to make the book frustrating for novices? High-quality, color illustrations seemed to be the key. Steve and Sylvia, already accomplished lichen photographers, had been compiling photographs of the lichens of western America and had worked out the techniques of taking close-up lichen portraits in nature with controllable lighting that would reproduce the true color of the lichens. It would be their task to assemble the lichen photographs that would form the core of the book. I had been working at the Canadian Museum of Nature as a lichen taxonomist for 25 years and had field experience with lichens from most parts of the continent. My job would be to identify the voucher specimens collected for each photograph and, building on my experience teaching lichenology and writing popular regional guides and articles, I would write most of the text. A collaboration seemed to be logical and potentially fruitful, and so we made plans for a lichen identification guide that would cover the United States and Canada.

I approached the Canadian Museum of Nature for support in 1992. After my major research commitments had been fulfilled in 1993, I began working almost full time on the project. Meanwhile, Steve and Sylvia began making dozens of grant applications to see if money could be raised to support the three to four years of full-time lichen photography that would be needed to illustrate the book. Without waiting for the financial backing they hoped for, the Sharnoffs sank their savings into a small RV that could serve as home, office, and laboratory, and they set off on a marathon that would take them around the continent three and a half times, from Alaska to Florida and from Nova Scotia to California, a trip that covered 100,000 miles over four and a half years. As the grant money began trickling in and Yale University Press expressed an interest in publishing the book, it seemed that the project would certainly succeed.

The road to a finished manuscript was a rocky one, but obstacles were overcome one by one. Close to 4000 photo voucher specimens were identified, several hundred of them by generous colleagues who named lichens in their areas of expertise. (Another 1500 are still only tentatively identified or had to be left unstudied owing to time constraints.) The Sharnoffs were able to photograph almost 1300 species, and it is from that set that the images for this volume were selected. The selection process was often painful because we were reluctant to exclude even one species. Many of our favorite photographs had to be set aside in order to ensure that the book would be both a reasonable size and an affordable price.

Our book attempts to cover over 1500 common, conspicuous or otherwise important lichens from the arctic to the Mexican border, from coast to coast, with color photos of 804, nearly all taken in their natural settings. Readers will be pleased to see that crustose lichens, traditional stumbling blocks for amateurs and often excluded from popular guidebooks, have not been neglected. Most genera are represented by at least one species and some by many. Except in the case of endangered species, virtually every photograph taken is supported by a voucher specimen identified in the laboratory using modern methods of morphological and chemical analysis. These specimens have been deposited in the National Herbarium of Canada (Canadian Museum of Nature, Ottawa), with many duplicates placed in the U.S. National Museum (Smithsonian Institution, Washington, D.C.), for future reference. The photographs and discussions are organized alphabetically by genus, with frequent cross-references to closely related or superficially similar genera to make comparisons easier. A short description of each species is followed by notes on chemistry and habitat, as well as comments on similar lichens. Finally, extensive notes are given for past and present practical uses of various lichens, often with comments on their ecological roles or uses by wildlife.

A somewhat controversial decision was made to include vernacular names (in English) for most lichens, especially the foliose and fruticose species. Many of these names have been invented by various lichenologists in North America and Europe because no true common names existed. The rationale for this decision is presented in the text.

Identification keys are provided for about 1050 species, the largest coverage of the North American lichen flora since Bruce Fink's classic *Lichen Flora of the United States* (1935). Although this represents only 30 percent of the approximately 3600 species of lichens known for North America, the coverage of the macrolichens is much more complete.

Besides Fink's now out-of-date treatise, other important and useful books have dealt with parts of the North American lichen flora. A list of regional floras and guidebooks is presented in the section entitled "Further Reading." Most notable is Mason Hale's comprehensive guide to the macrolichens of North America, *How to Know the Lichens,* published by Wm. C. Brown in 1969, with a second edition in 1979. The present volume is, in a way, an extension and elaboration of that immensely valuable, groundbreaking work.

One final word. The last stages of preparing this volume for publication were done without benefit of the considerable talents and sage advice of our coauthor, Sylvia Sharnoff, who died before the book was completed. The original inspiration for doing a book such as this was hers, and her contributions to every facet of the book — photographs, text, layout, and overall vision — are incalculable. We hope that this book will be a lasting tribute to her efforts.

IRWIN M. BRODO

ACKNOWLEDGMENTS

These acknowledgments are lengthy for good reason. We have been the happy recipients of countless favors and kindnesses from many people: professional lichenologists, amateur naturalists, nature interpreters, park administrators, teachers, foresters, and others.

The North American Lichen Project, which consisted primarily of the photographic fieldwork for this book, was sponsored by the Missouri Botanical Garden, with generous financial support from the Educational Foundation of America, the National Geographic Society, the National Fish and Wildlife Foundation, the Weeden Foundation, the Norcross Wildlife Foundation, the New England Biolabs Foundation, the Maki Foundation, the Davis Conservation Foundation, and the California Native Plant Society. The Canadian Museum of Nature funded 12 weeks of fieldwork during 1994 and 1995 as well as several trips to California to permit the authors to get together, select the final photographs, and discuss matters related to the text.

Publication of the book was supported by grants to Yale University Press from the Educational Foundation of America, Lichen Foundation, the J. M. Kaplan Fund, the Bay Foundation, the Ottawa Field-Naturalists' Club, the British Lichen Society, and many individuals: Averill Babson, Glenna Dean, Trudy Murphy, B. J. Fontana, Greg Sohns, and an anonymous donor.

Professional and amateur lichenologists, botanists, ecologists, interpretive naturalists, conservationists, ethnobotanists, land administrators, and many other people contributed their expertise and assistance. They identified specimens, shared information, guided us to interesting lichen habitats (often in person), gave us permission to collect specimens or take lichen-related photographs, wrote letters of support, put us in contact with other people whose help was vital to the project, and (along with friends and relatives) extended hospitality as we traveled.

We are especially grateful to Dr. Peter H. Raven, director of the Missouri Botanical Garden, for adopting the photographic component of this project when it was homeless. His mentoring and encouragement made the book possible. At the Garden, we also want to thank Dr. Marshall R. Crosby and Barbara Mack.

Special mention must be made of the following individuals for their critical support: Fenja Brodo, Dory Cameron, Laurie Consaul, Chicita Culberson, William

Culberson, Susan D'Alcamo, Paula DePriest, Chiska Derr, David Ehrenfeld, Linda Geiser, Susan Gibson, Trevor Goward, Richard Harris, Bruce McCune, John Powers, and Allan Stone.

Early drafts of the introductory text were reviewed by Barbara H. Smith, Sheila Thomson, and Fenja Brodo, all of whom made many useful comments. We are deeply indebted to Richard C. Harris for his comments on the entire manuscript and keys, and for his insightful suggestions on the classification section.

A number of hardworking volunteers in Ottawa prepared almost all the maps and helped in many other ways such as incorporating text corrections into early drafts. Fenja Brodo organized and supervised the volunteer crew and did a great deal of the mapping work. The team consisted of Peggy Holton, Sylvia Edlund, Miriam Sussman, Lise Dubé, Hebe Naomi Gouda, and Binson Wei. At the Smithsonian, volunteer Michael Sanders helped us gather specimen data for the maps.

Many lichenological colleagues spent a great deal of time helping with the identification of voucher specimens, adding data to our maps, or checking maps within their areas of expertise: Teuvo Ahti (Cladoniaceae), Vagn Alstrup (lichen parasites), Othmar Breuss (*Catapyrenium*, *Placidium*, and *Dermatocarpon*), Stephen Clayden (*Rhizocarpon* and *Stereocaulon*), Phillipe Clerc (*Usnea*), William L. Culberson (*Haematomma*), Paula DePriest (specimen data from USNM), Theodore Esslinger (*Phaeophyscia*, *Physconia*), Trevor Goward (*Peltigera*, *Hypogymnia*), Corrina Gries (southwestern records), Beatrice Hale (early maps of M. E. Hale), Samuel Hammer (*Cladonia*), Richard Harris (many groups), Josef Hafellner (*Pleopsidium*), Hannes Hertel (*Lecidea*), Jim and Pat Hinds (records from Maine), Doug Ladd (Missouri records), François Lutzoni (*Omphalina*, *Hymenelia*, *Ionaspsis*, and comments on lichen classification), Janet Marsh (*Niebla*), Aino Henssen (Ephebaceae, Lichinaceae), Per Magnus Jørgensen (Pannariaceae), Scott La Greca (*Ramalina*), Roland Moberg (*Physcia*), Louise Lindblom (*Xanthoria*), H. Thorsten Lumbsch (*Diploschistes*), Bruce McCune (Pacific Northwest lichens), Thomas Nash III (*Xanthoparmelia*), Ron Peterson (*Multiclavula*), Roger Rosentreter (vagrant lichens), Claude Roux (*Acarospora* and *Pleopsidium*), Bruce Ryan (*Lecanora*), Steve Selva (Caliciales), John Sheard (*Rinodina*), Ulrich Søchting (*Caloplaca*), Larry St. Clair (Utah lichens), Isabelle Tavares (*Usnea*), John W. Thomson (northeastern lichens), Lief Tibell (Caliciales), Tor Tønsberg (crustose lichens), William A. Weber (southwestern lichens), Clifford Wetmore (Heppiaceae, *Caloplaca*, and records from the University of Minnesota), Pak Yau Wong (*Collema*, *Placopsis*, and others).

We are grateful for the time and effort spent by the following friends and colleagues, who tested early drafts of the identification keys: Alex Ciegler, Oliver Crichton, Lee Crane, Joan Crowe, Robert Egan, Barbara Gaertner, Elisabeth Lay, Rob Lee, Claude Roy, Isabelle Tavares, Sheila Thomson, and Darrell Wright.

The following individuals are also thanked for their help in the field, with grant applications, and with manuscript problems, and for so many other small and large favors: Vernon Ahmadjian, Kat Anderson, Karl Anderson, Daniele Armaleo, Buzz Baker, Doris Baltzo, Rudi Becking, Jayne Belnap, James Bennett, Paula Benshoff, Ken Berger, Anna Bilsky, Meredith Blackwell, Lois Brako, Charis Bratt, Sharon Brodo-Smith, William Buck, Ann Buckley, Don Buso, Karen Casselman, Joe Castalano, Sarah Chaney, Russell Chapman, Daniel Clement, Barbara Conklin, Judith Hazen Connery, John Cooke, Max Copenhagen, Peter Cory, Tom Darden, Jerry Davis, John Davis, Glenna Dean, Ann DeBolt, Frederica deLaguna, William Denison, Barbara Deutsch, Barry Deutsch, Martyn Dibben, Janet Doell, Mary Edwards, Robert Egan, Katherine Enns, Phyllis Faber, Richard Felger, Jonathan Ferabee, David Galloway, Andrea Gargas, Bernard Goffinet, Dana Griffin, Dennis Haddow, Beth Hagan, Neil Hagedorn, Lee Heinmiller, Julie Henderson, Ted Hendrickson, Jaymes Henio, Ann Hitchcock, Phil Hyatt, Alison Isenberg, Anna Jacobson, Steven Jacobson, Hans Martin Jahns, Lawrence Janeway, Marshall Johnston, Elizabeth Kneiper, Craig Kirpatrick, Kurt Kotter, John Krug, Douglas M. Ladd, John Ladd, Donna Lamb, Keith Langdon, Elisabeth Lay, Yin-May Lee, Robin Lesher, Xavier Llimona, Thomas E. Lovejoy, Lynn Lozier, Francois Lutzoni, Bruce Mace, Norma Mark, Jon Martin, Steve McCormick, Cheryl McJannet, Nancy Morin, Patricia Muir, Mary Muller, Robert Munter, Barbara Murray, David Murray, Thomas H. Nash, III, Nancy Naslund, Peter Neitlich, Robert Nelson, Rita O'Clair, Bill Pell, Eric Peterson, Larry Pike, Sherry Pittam, Catherine J. Porter, Bea and George Prehara, Joe Pretti, Richard Reagan, Fred Rhoades, David H. S. Richardson, Jim Riley, David Riskind, Jamie Ross, Bruce Ryan, Steve Selva, David Shaw, Sam Shushan, Harrie Sipman, Martha Sherwood, Bob Simons, Steve Sillett, Janna Six, Clifford Smith, J. Mark Smith, Raul "Sonny" Solis, Larry St. Clair, Soili Stenroos, Tom Stohlgren, Rita Suminski, Lowell Suring,

Chris Swail, Anders Tehler, John W. Thomson, Sheila Thomson, Leif Tibell, Nancy Tileston, Einer Timdal, Christopher Topik, Dale Turner, Nancy J. Turner, Shirley Tucker, Heino Uaenskae, Linda Vorobik, William A. Weber, Dan Wedemeyer, Thomas Wendt, Alix Wennekens, Paul Whitefield, Adam Willett, David Williams, Ronald Wilson, Volkmar Wirth, Darrell Wright, Sylvia Wright, Becky Yahr, and Barbara Zimmer.

The Sharnoffs wish to thank their children, David and Meera, for patiently putting up with long absences from home, camping trips in rainy weather, and the dislocations to their lives that resulted from Steve and Sylvia's enthusiasm for lichens.

Irwin Brodo could not possibly have completed the task without the unwavering support and tireless work of his wife, Fenja, who served as his unpaid assistant for six years and gave him constant encouragement throughout.

Finally, we extend our heartfelt thanks to our editor, Jean E. Thomson Black, who stuck with the project through thick and thin, and whose energy, guidance, and dedication were indispensable. Her unfailing support under circumstances that would have caused many in her position to abandon ship has earned our eternal gratitude. Nancy Moore Brochin, Nancy Ovedovitz, Paul Royster, and the Yale University Press staff applied their considerable talents to assuring that the manuscript, illustrations, and book design were as perfect as possible.

ILLUSTRATIONS

With few exceptions, all the photographs in the book were taken by authors Sylvia and Stephen Sharnoff. Plates 43, 62, 83, 88, and 90 are by Irwin Brodo; plate 6 is by Vernon Ahmadjian; plate 46 is by George Calef; and plate 65 is by Dorothy I. D. Kennedy. Full acknowledgments are made in the captions.

Susan Laurie-Bourque created all but one of the original line drawings and modified some of the previously published drawings. Figure 36b was drawn by Irwin Brodo.

PART ONE

ABOUT THE LICHENS

LICHENS: THE ORGANISM

O broken life! O wretched bits of being,

Unrhythmic, patched, the even and the odd!

But Bradda still has lichens worth the seeing,

And thunder in her caves—thank God! Thank God!

Thomas Edward Brown

WHAT LICHENS ARE

Lichens are unique in the world of vegetation in that they cannot be neatly classified into any of the ordinary categories we think of as "plants." The reason is simple: a lichen is not a single entity, but a composite of a fungus and an organism capable of producing food by photosynthesis. Lichen fungi can associate with green algae or cyanobacteria (the latter also known as *blue-green algae*), or sometimes both, and none of these three groups are plants in the strict, modern sense (which now include mainly mosses and vascular plants). The special biological relationship found in lichens is called symbiosis. The resulting composite of a fungus and its photosynthetic symbiont (photobiont, for short) has been such an evolutionary success that there are close to 14,000 species of lichens in the world, tremendously diverse in size, form, and color. They are found from the poles to the tropics, from the intertidal zones to the peaks of mountains, and on every kind of surface from soil, rocks, and tree bark to the backs of living insects!

WHAT LICHENS ARE NOT

Lichens are often informally grouped with mosses, liverworts, free-living fungi, and algae under the name, *cryptogams* (referring to plants with "hidden marriages," that is, reproducing by spores rather than seeds). Curiously, although lichens consist of fungi and algae, they are often thought of as a kind of "moss." (Reindeer "moss," for example, is a lichen.) Mosses and their close relatives, the liverworts and hornworts (together called bryophytes), are members of the plant kingdom, biologically very different from lichens. Mosses and leafy liverworts have tiny green "leaves" composed of cells that contain chloroplasts, and they are relatively easy to distinguish from lichens. The flat, broadly lobed liverworts are superficially more lichen-like and can be confusable until you learn to recognize them. Liverworts and other bryophytes are generally grassy green, at least when they are wet. They are about the same size as lichens and often join them in the same habitats, which is why so many of our photographs of lichens contain a generous helping of bryophytes.

1. Many lichen fruiting bodies resemble those of unlichenized fungi such as this cup fungus, *Sowerbyella rhenana*. ×1.5

THE COMPONENTS

Fungi

Fungi are numerous and diverse, and they are very different from ordinary plants. It is now clear they have a separate ancestry and must be placed in a kingdom of their own. All fungi are basically built from microscopic threads called hyphae, some of which are modified to give the appearance of round or squarish cells. All lack any chlorophyll, the green pigment capable of taking carbon dioxide and water in the presence of sunlight and converting them into sugars through photosynthesis. Because fungi cannot make their own food, they have to find it just as animals do. They can live on dead organic materials such as dead leaves or wood as saprophytes, or they can live on living organisms as either benign or death-dealing parasites.

The most familiar fungi are those that form mushrooms; there are also a myriad of less conspicuous kinds that are, however, no less important. Think of the fungi that make blue cheese, those we use to make beer and bread (yeast is a fungus), and those that decompose fallen trees, converting them into soil. There are, of course, also the disease-producing fungi: those that make our feet itchy and wither our toenails, and those, such as wheat rust and the potato blight, that devastate our crops.

The fungi that form lichens belong, for the most part, to the Ascomycetes, or "sac fungi," distinguished by producing their microscopic reproductive spores in tiny sacs called asci (described later). There are almost 30,000 species of ascomycetes, almost half of which form lichens (plate 1). A large and diverse variety of ascomycetes, including 13 major groups (orders) out of 43, are represented among the lichens. Only a handful of mushroom-producing fungi and their relatives (the Basidiomycetes) are also lichen-forming. Because such a large number of unrelated fungi are involved in lichen formation, biologists know that lichens have not all evolved from a common ancestor. What links lichens together is their mode of nutrition, not their ancestry. Thus, they cannot be considered collectively as a single branch on the evolutionary tree and therefore do not constitute a taxonomic unit.

Do lichen-forming fungi receive a share of the photosynthetic products of the photobionts in a more or less reciprocal partnership, or are they parasites, exploiting their algal or cyanobacterial hosts? That is not so easy to answer because the relationship is complex and differs from one lichen to another (see the section below entitled "The Lichen Association"). Lichenologist Trevor Goward has described lichens as "fungi that have discovered agriculture," and there is much truth to that.

Photobionts

There are about 25 genera of green algae, a few golden algae, one brown alga, and 12 genera of cyanobacteria (or blue-green algae) that become associated in lichens as photobionts. Another few are algae of other kinds. Only a dozen genera, however, represent the photobionts in the vast majority of lichens. Many photobionts are not easy to identify within the lichen thallus, because they can be substantially changed by the symbiotic state. To be identified with absolute certainty and precision, even to genus, most have to be isolated from the lichen and grown by themselves in culture. The photobionts of only about 2 to 3 percent of all lichens have been identified to the species level. The more common and easily identified genera are illustrated in figure 1 and are described briefly in the glossary. Distinguishing green algae from cyanobacteria, however, is not difficult. The massed green algae generally form a layer in the lichen that is grassy green; cyanobacteria form a dark blue-green or blue-gray layer, or they are in lichens that are almost black and jelly-like when wet.

Green algae (plate 2) are in the kingdom Protoctista, class Chlorophyceae, and, like plants, have chlorophyll-containing bodies called chloroplasts in their photosynthetic cells. They include *Trebouxia*, which is rarely found free-living in nature but is the most common alga found in lichens; *Trentepohlia*, a filament-

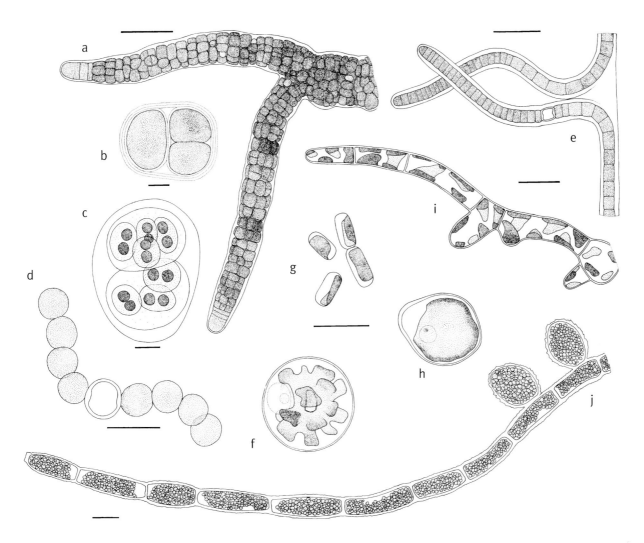

Fig. 1 Photobionts. a–e, cyanobacteria: (a) *Stigonema*, (b) *Chroococcus*, (c) *Gloeocapsa*, (d) *Nostoc*, (e) *Scytonema*; f–j, green algae: (f) *Trebouxia*, (g) *Stichococcus*, (h) *Myrmecia*, (i) *Heterococcus*, (j) *Trentepohlia* (in most lichen thalli, *Trentepohlia* develops as individual cells, not filaments). Scale: a, e = 100 μm; b–d, f–j = 10 μm. (Reproduced courtesy of British Lichen Society, from Purvis et al., 1992, fig. 42.)

producing alga with many free-living species, especially common in tropical crustose lichens; *Coccomyxa*, a common terrestrial alga, occurring in the green species of *Peltigera* and *Solorina* and the mushroom lichens; *Dictyochloropsis*, common in *Lobaria* and *Pseudocyphellaria*; and *Stichococcus*, found in many stubble lichens (*Calicium* and *Chaenotheca*).

Golden algae are classified in the class Tribophyceae (or Xanthophyceae) and are distinguished by the flagella that propel their motile spores through the water and by their pigments. *Heterococcus* (Fig. 1i) is the photobiont of some species of the crustose lichen *Verrucaria*.

A few maritime species of *Verrucaria*, including *V. tavaresiae* on the California coast, are associated with *Petroderma maculiforme*, a crustose species of brown alga (class Fucophyceae or Phaeophyceae). Many brown algae are extremely large, such as the typical brown "seaweeds" and kelp one sees on shoreline rocks or in coastal bays. Some parasitic fungi are constantly associated with certain brown algae, but none form typical lichens.

Cyanobacteria belong to an entirely different kingdom, the Monera, characterized by the lack of a well-defined nucleus. Their chlorophyll occurs throughout the cell fluid rather than in special chloroplasts. All cyanobacterial photobionts can be found free-living in nature. The most common is *Nostoc* (plate 3), found in most jelly lichens (e.g., *Collema* and *Leptogium*) and in many species of *Peltigera*. *Scytonema* and *Stigonema* are filament-forming, branched cyanobacteria that are common on moist rocks and occur in minutely filamentous lichens such as *Ephebe* and *Polychidium*. *Gloeocapsa*, also commonly free-living on moist rocks, is a single-celled to colonial genus that is the photobiont of *Thyrea* and *Pyrenopsis*. All these cyanobacteria can also be in cephalodia (described below).

2. (right) Not all algae live in water. Some, like this colony on an apple tree, are "terrestrial" with the potential of becoming lichenized. This particular species, however, rarely does.

3. (below) Species of *Nostoc*, cyanobacteria frequently found free-living as small, globular colonies, are often co-opted as lichen photobionts. ×2.8

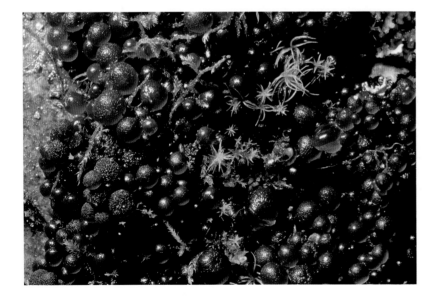

PICKING A PARTNER

A species of lichen collected anywhere in its range has the same lichen-forming fungus and, generally, the same photobiont. (A particular photobiont, on the other hand, may associate with scores of different lichen fungi.) So few photobionts have been identified to species, we are not sure how many exceptions exist to the general specificity of a lichen fungus with a particular species of photobiont. The exceptions we do know about are very interesting. For example, there are many cases of the same lichen fungus associating with different species of algae in different parts of the lichen's geographic range. Recently, in fact, it has been shown that, within its normal life cycle, the same fungus can sometimes associate with a green alga and other times associate with a cyanobacterium. This amazing ability is most common in species of *Peltigera*, *Sticta*, and *Nephroma*. The resultant lichens, one with cyanobacteria and the other with green algae, can sometimes be rather similar (as in the case of *Peltigera britannica*) or quite different (as in some *Sticta* and *Lobaria* species). These observations show that although the fungus controls the appearance of the lichen, the photobiont plays a critical and indispensable role in this transformation. (See also the section below entitled "Mysterious Transformation").

THE LICHEN ASSOCIATION

Despite the fact that lichens have for generations been considered as the most perfect example of a harmonious and mutually beneficial biological partnership called mutualistic symbiosis or, often, simply *symbiosis*, the relationship between the fungus and photobiont (either green alga or cyanobacterium, or both) is far from simple. In fact, the relations between the "partners" in the wide variety of lichens run the gamut between a fairly innocuous, mild parasitism to a rampant, photobiont-destroying disease. There are few if any lichens in which the algae or cyanobacteria are not invaded and killed to some extent by the fungus. In all lichens, of course, the photobiont cells are able to reproduce faster than they are destroyed—otherwise the lichen would "eat itself alive." Is there, then, any reason to continue to regard lichens as examples of a true partnership or mutualism?

In most lichens, the fungus envelops the algal or cyanobacterial cells with tiny branches of its hyphae, the tips slightly expanded and tightly pressed against the photobiont's cell walls. The fungus apparently "recognizes" the right alga by virtue of certain proteins (lectins) on the cell wall. Often, short pegs called haustoria penetrate into the photobiont cells to a greater or lesser extent. The photobiont, being photosynthetic, produces sugars and other carbohydrates. While in a symbiotic state within the lichen thallus, the photobiont is chemically affected by the presence of the fungus in a curious way. The walls of the photobiont cells become relatively permeable to the carbohydrates it produces, and these substances "leak out" and are absorbed by the fungus. Some of the carbohydrate, of course, is retained by the photobiont and provides an energy source for its own needs. (This process is described in more detail in Chapter 5.) The fungus, for its part, protects the photobiont within its tissues and provides a steadier supply of moisture by conducting water within its cell walls. Most common green algal photobionts are damaged by excess light, and fungal tissue in the lichen, together with the pigments it produces, provide the algae with a partial light shield. Perhaps the most important contribution of the lichen fungus is to provide a habitat for the photobiont. At least the green algal photobionts would scarcely be able to survive in a free-living state outside the lichen—certainly not on the rocks, dry tree bark, and other surfaces frequented by lichens. The very shape of the lichen, whether broad and leafy or many-branched, increases the surface area exposed to light, also benefiting the photobiont. So, although strictly speaking the lichen symbiosis is a kind of controlled parasitism of the photobiont by the fungus, the lichen itself retains many aspects of a mutually beneficial partnership.

But the relations between fungi and algae are often diverse and complex. Some seaweeds such as the brown alga, *Ascophyllum,* are almost always found to be infected by certain fungi. Are these lichens? Some mushrooms or bracket fungi are always blanketed with a coating of green algae. Does that make them lichens? Some lichens live on other lichens, sharing or even stealing their photobionts (plate 4). Other lichens like *Diploschistes muscorum* begin development as a lichen parasite, only later becoming independent. Still other lichens, or fungi that are closely related to lichen-forming fungi, are destructive parasites on lichen thalli, not associating with the photobionts at all (for example, *Caloplaca epithallina*). (We are not referring to disease-producing fungi like *Illosporium carneum* (plate 5) that can infect and damage lichens but which are totally unrelated to lichen-forming fungi.)

4. The white lichen, *Acarospora stapfiana*, invades another lichen, the orange *Caloplaca trachyphylla*. ×2.6

5. The pink fungus, *Illosporium carneum*, is a common parasite on *Peltigera didactyla* and other lichens. ×2.8

6. Pure cultures of the components of *Acarospora fuscata* (the green algae on the left and the fungus on the right) look nothing like the intact lichen. (Photo by Vernon Ahmadjian)

What, then, is a lichen? A definition accepted by members of the International Association for Lichenology is "an association of a fungus and a photosynthetic symbiont resulting in a stable vegetative body having a specific structure." This is the definition we use in this book, although we do include some marginal fungus-alga associations (such as *Multiclavula*).

THE MYSTERIOUS TRANSFORMATION

Lichens, being composed of two organisms, can be taken apart, and the components can be grown separately in the laboratory. In culture, the green algal photobionts produce little green colonies (plate 6) composed of cells that often do not closely resemble those found within the intact lichens. Cyanobacteria, by comparison, do not change much in culture. The fungus, in the absence of its photobiont, produces a characterless heap of hyphae not at all like the lichen from which it was isolated (plate 6). When the fungus and photobiont are mixed together under the right conditions, however, they form a new association that takes the form of the parent lichen (plate 92). What, then, triggers this fascinating physical transformation called morphogenesis? The fungus alone seems to contain all the genetic information it needs to create the characteristic form of the lichen, but it cannot do this alone. The alga or cyanobacterium is somehow able to "turn on" the fungal genes that control the morphogenesis, but how is this is done? These important basic biological questions, for which there are still few answers, are relevant to more general studies of cell transformations, even cancer research, and make lichens a potentially significant research tool.

HOW LICHENS ARE BUILT: THALLUS SHAPES AND STRUCTURE

Yet lovely was its pleasant shade;
Lovely the trunk with moss inlaid;
Lovely the long-haired lichens grey;
Lovely its pride and its decay.

Mary Russell Mitford

What defines lichens is not their form or color but their biology; therefore, it is not easy to answer the frequently asked and entirely legitimate question, "What does a lichen look like?" Lichens can stand erect like little shrubs, drape tree limbs like Christmas tinsel, or appear to be little more than a black smudge on a rock. They come in an array of colors from brilliant yellow or red to mundane white or gray. Some are less than a millimeter tall, and others can grow to several meters in length (plates 7 and 8).

To answer the question of what a lichen looks like, we will first try to break down its considerable variation into some general "growth form" types and then examine how these growth forms are constructed.

GROWTH FORMS

It is necessary to organize lichens into recognizable categories in order to deal with their bewildering diversity. In formal classifications, we try to establish categories that are "natural," that is, reflecting some "real" relationship, one that is based on common ancestry (see Chapter 12). It is useful, however, for more casual reference, to classify lichens informally into general growth forms, with no implied natural relationships (Fig. 2). In fact, as we will see, many lichens that look very much alike are actually hardly related. Furthermore, although the growth form categories are convenient ways to refer to lichens while identifying them, many growth forms have intermediate types that defy easy classification. The growth form categories described in the paragraphs that follow are: foliose, fruticose, crustose, and squamulose (plate 9). We deal with intermediates in the keys and descriptions by including them under more than one category if necessary.

In talking about growth forms, we refer to the vegetative body of the lichen, the thallus. In most cases, the bulk of a lichen consists of its thallus (as opposed to the reproductive or fruiting structures). When we say that a lichen is green or yellow, big or small, thread-like or spherical, we are referring to the thallus unless the fruiting bodies dominate the lichen, as in *Caloplaca holocarpa* or *Lecanora dispersa*.[1]

---

[1] Lichens to which references are made, with few exceptions, are pictured in the main section of the book, in Part II, which is organized alphabetically by genus and species.

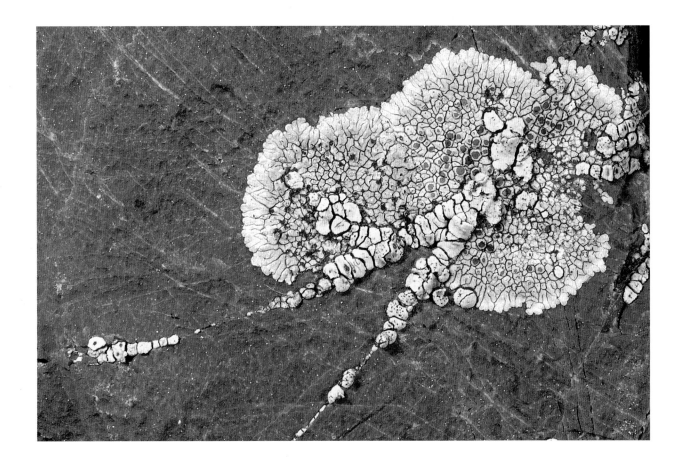

7. (facing page) Strands of *Usnea longissima*, some more than 3 meters long, festoon the branches of trees in the Washington Cascades.

8. (above) Bright white and brilliant yellow lichens like *Dimelaena radiata* and *Pleopsidium* species decorate rocks on the California coast. ×4.8

9. (left) The gray foliose lichen, *Parmotrema hypotropum*, shares a twig in Texas with two fruticose species, *Teloschistes exilis* (orange) and *Usnea cirrosa*. ×1.9

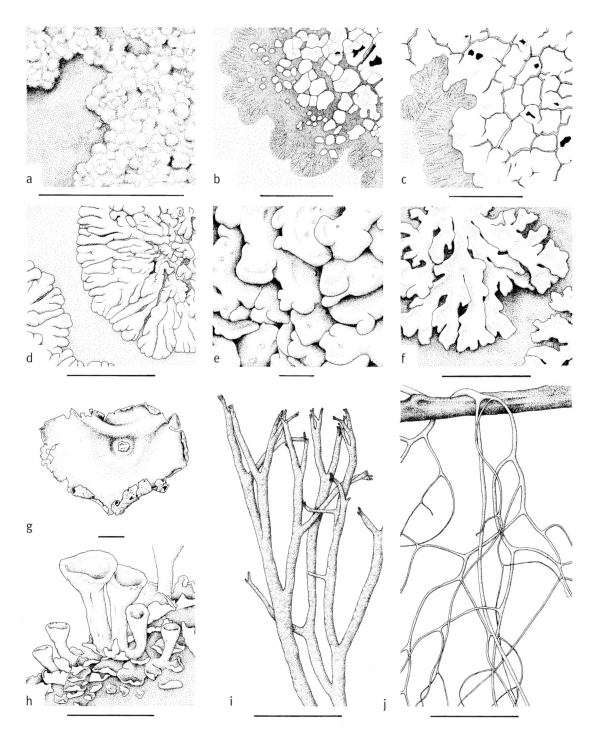

Fig. 2 Thallus growth forms, with most examples from the North American flora. a–d, crustose: (a) leprose (*Lepraria lobificans*), (b) areolate (*Buellia ocellata*: British), (c) rimose (*Buellia disciformis*), (d) lobed (*Caloplaca ignea*); (e) squamulose (*Endocarpon*); (f) foliose (*Physconia muscigena*); (g) foliose, umbilicate, as seen from the underside showing the central holdfast (*Umbilicaria proboscidea*); h–j, fruticose: (h) cup-like (*Cladonia carneola*), (i) erect, shrubby (*Cladonia furcata*), (j) pendent, hair-like (*Bryoria capillaris*). Scales: a–c, e = 1 mm; d, f–j = 5 mm. (Reproduced courtesy of British Lichen Society, from Purvis et al., 1992, fig. 36.)

Foliose Lichens

Lichens that have a more or less flattened thallus with easily distinguished upper and lower surfaces are called foliose. The larger foliose lichens such as *Lobaria pulmonaria* or *Flavoparmelia caperata* fall into this category easily. Foliose species that are very closely attached to the substrate such as *Hyperphyscia syncolla* or *Xanthoria elegans,* or that have almost vertically oriented lobes such as *Cetraria islandica* or *Kaernefeltia merrillii,* are harder to place. Most foliose lichens have rounded or somewhat angular lobes at the margins. The shape of the lobes, their length and width (Fig. 3), and their configuration (e.g., upturned versus curled under, concave versus convex, inflated versus flat) are all important characteristics that distinguish one lichen from another.

Most foliose lichens are built in layers and are said to be stratified. A thin slice of a stratified foliose lichen from the top surface to the bottom will reveal these layers clearly (Fig. 4). It is important to become familiar with these tissues because they are important characters in the identification of lichens.

Because the bulk of most lichens consists of the fungal component (rather than photosynthetic cells), most of the tissues are made up entirely, or mostly, of fungal cells. As we mentioned above, the tissues of most fungi consist basically of cells arranged in threads called hyphae. The hyphae, however, can vary a great deal in length, shape, thickness of their walls, arrangement, branching pattern, and so on. It is this potential for variation that makes possible the tremendous array of tissue types seen in lichen thalli. Fungal tissue composed of uniform cells roughly equal in width and height, usually with thin walls, can be called pseudoparenchyma, or "false parenchyma," referring to its superficial similarity to unrelated tissue found in higher plants (Fig. 5d,f). When the cells are more elongate, typically with the thick cell walls glued together and indistinct, the tissue is called prosoplectenchyma (Fig. 5e).

The upper surface of most foliose lichens has a well-developed cortex usually consisting of thick-walled cells closely packed in a common, gelatinous, often tough matrix. This is thought to serve the lichen as a protective skin. The cortex can be colorless, or it can contain a variety of pigments, giving the lichen its characteristic color (see Chapter 4). The cells of the cortex can be rounded, as in *Peltigera* (Fig. 5f), arranged in vertical rows, as in *Roccella* (Fig. 5a), or fibrous and parallel with the surface, as in *Heterodermia* (Fig. 5b).

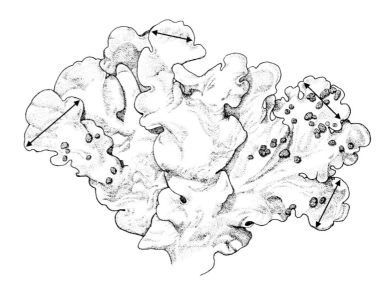

Fig. 3 Lobes, and how they are measured. (Redrawn from Brodo, 1988, fig. 34.)

Below the cortex of stratified foliose lichens is a green layer made up of cells of the photobiont, either green algae or cyanobacteria, enmeshed in a network of fungal hyphae. The photobiont can be single-celled or filamentous. This tissue is called simply the photobiont layer (or "algal" layer), even though fungal hyphae are present as well. The color of the photobiont layer can be bright grass green, orange-green, dark grayish green, or distinctly blue-green, depending on what kind of photobiont is present in the lichen.

Under the photobiont layer is a much more loosely packed medulla, which often makes up the bulk of the lichen thallus. The hyphae in the medulla are generally thin-walled and obviously threadlike and branched. The medulla is usually white, but many species characteristically produce pigments that make the medulla yellow, orange, or pink. The color is often most intense (or only detectable) close to the photobiont layer.

In most foliose lichens, the lower surface is protected by a lower cortex. Its structure is much like that of the upper cortex. It can be white, but most often it is darkly pigmented, even pitch black, and can be smooth, warty, or wrinkled and ridged in various ways that are often characteristic of particular species. In many foliose lichens, however, there is no lower cortex, and the medulla itself (pigmented or not) can be seen on the lower thallus surface. The absence of a lower cortex is, in itself, a distinguishing characteristic for some genera (e.g., *Peltigera*) or species (e.g., within the genus *Heterodermia*).

Foliose lichens are attached to a surface (the substrate) either directly by the hyphae of the lower cortex (if a cortex is present) or medulla (if there is no lower cortex) or by special attachment threads called rhi-

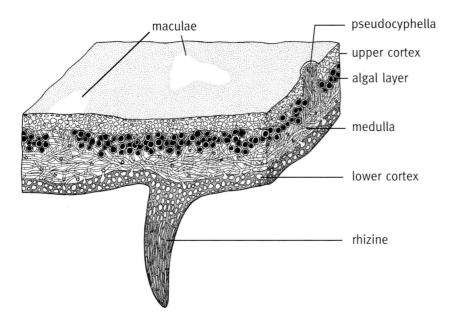

Fig. 4 Stratified foliose lichen thallus, a 3-dimensional view. In the pseudocyphella, the hyphae of the medulla reach the surface through a small hole in the cortex; maculae show as pale spots on the thallus surface because of an interruption in the algal layer just below it.

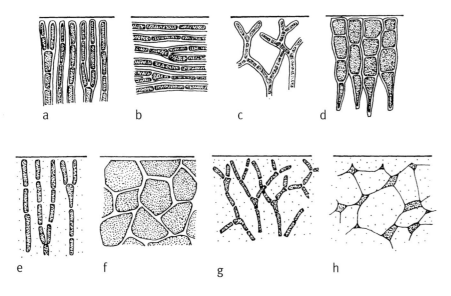

Fig. 5 Tissue types produced by different kinds of hyphal growth: (a) hyphae perpendicular to the surface forming a palisade-like cortex (as in *Roccella*); (b) distinct hyphae lying parallel to the surface (cortex of *Anaptychia*); (c) net-like, branched hyphae (close to surface of *Collema*); (d) pseudoparenchyma formed by palisade-like hyphae (cortex of *Allocetraria*); (e) prosoplectenchyma, palisade-like here, but can also be longitudinally arranged as in the cortex of *Bryoria*; (f) pseudoparenchyma of equal-sized cells (upper cortex of *Peltigera*); (g) net-like hyphae in a type of prosoplectenchyma (part of the cortex of *Ramalina*); (h) small cells in an amorphous, swollen matrix (as in the cortex of some species of *Lecanora*). (Reprinted by permission from Henssen and Jahns, 1974, fig. 3.5.)

zines (Fig. 6). Rhizines can be unbranched or branched into dichotomies (a series of small "Ys"), they can have perpendicular side branches like a bottlebrush (squarrose), or they can simply be bundles of hyphae characterized as tufted or fibrous. Some lichens form a short, hairy nap called a tomentum on the lower surface. In *Anzia*, a hypothallus of anastomosing, darkly pigmented hyphae develops from the lower cortex.

Some foliose lichens are attached to the substrate (almost always rock, in this case) by a single, stout, more or less central peg or holdfast. Because this central attachment is reminiscent of an umbilical cord, such lichens are called umbilicate (Fig. 2g). The umbilicate growth form occurs in a number of quite unrelated lichens, for example, in *Dermatocarpon*, *Rhizoplaca*, and *Omphalora*, as well as in the rock tripes, *Umbilicaria* and *Lasallia*.

So far, we have described mainly stratified foliose lichens, those that have clearly defined algal and medullary layers. In some lichens, especially those containing cyanobacteria rather than green algae, the photobiont is uniformly mixed with the medullary fungal hyphae without forming layers, resulting in unstratified thalli. Unstratified lichens such as *Collema* and *Leptogium* usually exhibit another characteristic that makes them easy to spot: they are gelatinous and translucent when wet, rather like thin sheets of blue-green or black gelatine, giving them the common name "jelly lichens." Some have upper and lower cortices, and some lack them (Fig. 7).

Fruticose Lichens

Lichens that grow erect or that are pendent, whose thalli, even if flattened, have no clearly distinguishable upper and lower surfaces, are called fruticose lichens. This growth form includes lichens that are highly branched and shrubby and those that form unbranched stalks or filaments. We are including here even tiny lichens with cushionlike or filamentous thalli such as *Polychidium* and *Ephebe*, but not the stubble lichens (*Calicium* and *Chaenotheca*, and related genera) in which the stalks of the fruiting bodies are hair-like and only 1 to 3 millimeters high and the thalli are clearly crustose. In some genera such as *Cladonia* and *Pilophorus*, only part of the thallus is fruticose. (These special cases are discussed below under "Combinations of Growth Forms, and Intermediates.")

Fruticose lichens are very much like foliose lichens in having a layered structure, but the thalli are built in three rather than two dimensions. There are no "upper" and "lower" cortices, but rather a single cor-

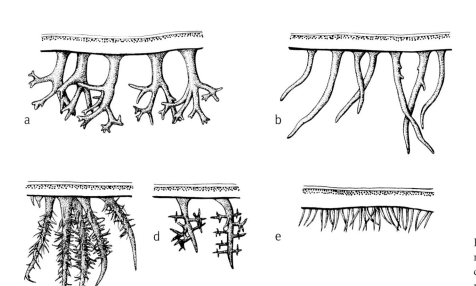

Fig. 6 Rhizines and tomentum: (a) forked or dichotomously branched rhizines (*Hypotrachyna*); (b) unbranched rhizines (*Parmelia saxatilis*); (c) squarrose rhizines (*Physconia detersa*); (d) squarrose rhizines (*Parmelia squarrosa*); (e) tomentum (*Lobaria pulmonaria*); (f) fibrous, tufted rhizines (*Peltigera canina*); (g) hypothallus of *Anzia colpodes*; (h) same, at higher magnification, showing the black anastomosing hyphae. Scale: a–g = 0.5 mm; h = 300 μm. (b and c redrawn from Purvis et al., 1992, fig. 39b,c.)

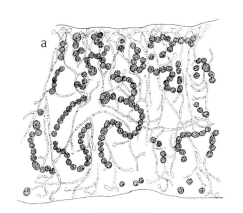
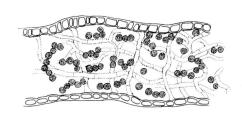

Fig. 7 Thallus structure in the jelly lichens: (a) *Collema* (lacking a cortex); (b) *Leptogium* (upper and lower cortices present). Scale = 30 μm. (Reprinted courtesy of the Canadian Museum of Nature, from Brodo, 1988, figs. 35, 37.)

Fig. 8 Fruticose thallus cross sections, showing location of supporting tissue (stippled) in different genera. a–d, pendent lichens: (a) in cortex, as in *Bryoria* and *Alectoria*; (b) in central cord, as in *Usnea*; (c) in separate strands, as in *Letharia*; (d) with no special supporting tissue, as in *Evernia*. e–g, erect lichens: (e) in cortex, as in *Sphaerophorus*; (f) in cylindrical stereome, as in *Cladonia*; (g) constituting entire stalk to which vegetative phyllocladia are attached, as in *Stereocaulon*.

al, algae; *axis*, central axis; *cor*, cortex; *med*, medulla; *phyl*, phyllocladium; *st*, stereome; *sup*, supporting strands. Schematic, at various magnifications.

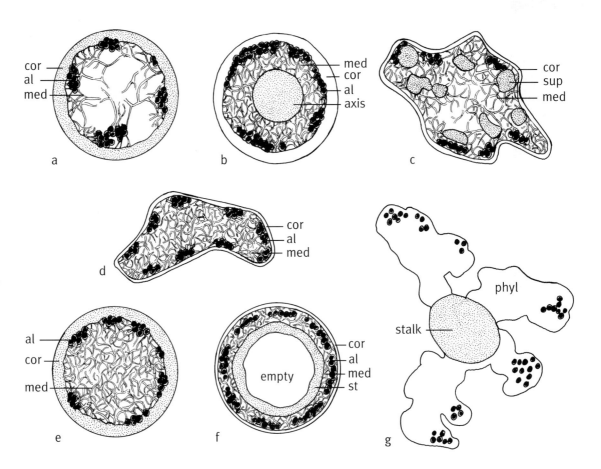

tex that entirely envelops the thallus stalks or filaments, with the photobiont layer just internal to the cortex, followed by some sort of medulla. The central core region can be hollow as in *Cladonia* or *Thamnolia*, or it can be stuffed with loose or dense fungal hyphae as in *Evernia* or *Stereocaulon*, respectively.

The stalks and branches of fruticose lichens are supported in a variety of ways (Fig. 8). Most of them involve the formation of a tough, cartilaginous tissue (prosoplectenchyma) with long, thick-walled cells running parallel to the axis of the branches. These support "strategies" are characteristic of specific genera and are useful in distinguishing one from another. The supporting tissue in the beard lichen, *Usnea*, is a tough, somewhat elastic central cord; in the stalks (podetia) of *Cladonia* species, the innermost portion of their medulla forms a more or less complete cartilaginous cylinder called a stereome; the wolf lichen (*Letharia*) has several narrow strands within the medulla; and the hair lichens (*Alectoria* and *Bryoria*) as well as *Ramalina* and *Niebla* build their supportive tissue right into the tough cortex.

Fruticose lichens are attached to their substrates directly at one point or at relatively few points. Many filamentous forms are at first attached by some sort of holdfast, but can continue growing simply by being draped over branches or even telephone wires. At a distance, pendent lichens can resemble Spanish moss (see plate 7). The superficial resemblance of *Usnea* to this tropical flowering plant (which is in the pineapple family) is reflected in the scientific name of Spanish moss: *Tillandsia usneoides*, meaning "the *Tillandsia* that looks like *Usnea*."

Crustose Lichens

Lichens that simply form crusts over their substrates are called crustose lichens. Their entire lower surface grows on and among the particles of the substrate without an intervening lower cortex (plate 10). Some form very large, brightly colored patches and can be quite thick or rough. Others can barely be detected without the help of a magnifying glass. Crustose lichens cannot be removed from the substrate in one piece. To collect one, we must therefore remove the underlying substrate with the lichen (see Chapter 13).

The structure of the thicker crustose lichens is not unlike that of stratified foliose lichens, with an upper

10. Crustose lichens, like these in central California, occur in many colors and shapes. ×2.4

cortex, photobiont layer, and medulla, but without a lower cortex of any kind (Fig. 9). Thinner crustose lichens may lack an upper cortex or distinguishable medulla, or they can be leprose, composed entirely of powdery particles or granules of various sizes and forms (Fig. 2a).

Crustose thalli can be smooth and unbroken (continuous, e.g., *Buellia stillingiana*), broken up by deep cracks that may close up when the thallus is wet (rimose, e.g., *Haematomma fenzlianum*; Fig. 2c), or divided into irregular or regular, contiguous or dispersed patches called areoles (e.g., *Rhizocarpon geographicum*; Fig. 2b). A crustose lichen broken into patches because of deep cracks in the thallus is sometimes described with the combined term rimose-areolate (e.g., *Lecidea lapicida*). A rough or bumpy thallus can be called verrucose if the bumps resemble warts, or verruculose if the bumps are very small.

Crustose thalli are frequently surrounded by a fringe of unlichenized (i.e., purely fungal) tissue called a prothallus. It is often black but can be white as in *Lecanora thysanophora* or colored like the thallus as in *Caloplaca flavogranulosa*. In lichens such as *Rhizocarpon geographicum*, the prothallus can be seen between the areoles as well as bordering the thallus.

If large areoles become upturned at the edges when well-developed, they can resemble little scales (see *Lecidea atrobrunnea*). This type of thallus intergrades with the final major growth form, the squamulose lichens.

Squamulose Lichens

Squamulose lichens are clearly intermediate between foliose and crustose growth forms. They are composed of a few to hundreds of relatively small scales, most commonly 1–15 mm in diameter, attached by the lower surface or along one edge like tiny shingles (Fig. 2e). Collectively, the scales can form fairly large patches. Examples are species of *Psora* and *Hypocenomyce*. Each scale making up a squamulose lichen has the basic architecture of a foliose lichen, sometimes even producing a lower cortex. Because of the overall small size and stature of squamulose growth forms, however, they are generally grouped together with crustose lichens under the general category microlichens. The fruticose and foliose lichens, then, are collectively called macrolichens.

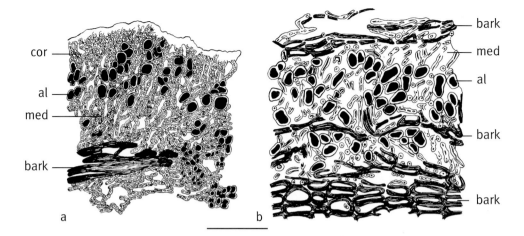

Fig. 9 Crustose lichen structure: (a) on bark surface (epiphloeodal), *Lecanora;* (b) within bark layers (endophloeodal), *Graphis.* Scale = ca. 0.5 mm. (Reprinted by permission, from Ozenda, 1963, figs. 28D, 29A.)

al, algae; *bark*, cells of bark; *cor*, cortex; *med*, medulla.

Combinations of Growth Forms, and Intermediates

As we mentioned at the outset, growth form categories are little more than conveniences and can be somewhat imprecise. One can encounter many intermediates between all these categories, and we take that into consideration in our descriptions and identification keys.

Most species of the common lichen genus *Cladonia*, for example, are combinations of two growth forms: fruticose and squamulose. The horizontal primary thallus, that is, the one that normally appears first, is clearly made up of squamules. From this squamulose thallus, the secondary vertical thallus consisting of "fruticose" stalks develops. The secondary thallus represents the highly developed stalks of fruiting bodies, which often appear at their summits. We treat *Cladonia* species as fruticose lichens here, but some species that rarely produce stalks are also classified as squamulose lichens. The pink earth lichen, *Dibaeis baeomyces,* has a white, crustose, primary thallus together with stalked fruiting bodies that can be considered as a fruticose secondary thallus.

Many species of crustose lichens become conspicuously lobed at the margins, a foliose trait, but remain basically crustose in structure in that they are in intimate contact with the substrate over virtually the entire lower surface (Fig. 2d). Good examples can be found in *Caloplaca, Lecanora,* and *Acarospora.* On the other hand, some foliose lichens such as *Xanthoria elegans* and species of *Hyperphyscia* are so closely appressed and tightly attached to the substrate, that one has to section the thallus and check with a microscope for the presence of a discernible lower cortex to be sure that the lichen is foliose.

SPECIAL FEATURES OF LICHEN THALLI

Besides the basic anatomy of the lichen thallus, a number of structures and anatomical modifications affect the appearance of lichens and can be used in lichen classification and identification. Most of these features play important roles in the biology of the lichen.

Cilia

Many foliose and some fruticose lichens can produce rhizine-like appendages called cilia on the margins of thallus lobes or fruiting bodies (see *Teloschistes chrysophthalmus*). They can be long and evenly branched, as in *Heterodermia echinata* and *Parmotrema perforatum,* slender and pale as in *Physcia adscendens,* or short and stubby with an inflated, bulbous base as in the tiny cilia of *Bulbothrix* (Fig. 10).

Pruinose and Scabrose Surfaces

Clumps of crystals and dead fungal cells deposited on the surface of the thallus sometimes give the surface a powdery, dusted, or frosted appearance. This "frosting" is called pruina, and lichens that have such deposits are said to be pruinose (plate 11). Pruina can be produced on fruiting bodies as well (discussed later), and can be white or colored. For example, white pruina is produced on the lobe tips of species of *Physconia,* and yellow pruina covers the apothecial disks of *Lecanora cupressi.*

A surface that becomes rough and scaly like a bad case of dermatitis is said to be scabrose. This appearance is usually caused by an irregular growth of cortical cells. The thallus surface of *Peltigera scabrosa* is an example. Scabrose fruiting bodies, such as occur in *Ochrolechia upsaliensis,* are the result of both anatomical roughness and heavy, scaly crystal deposits.

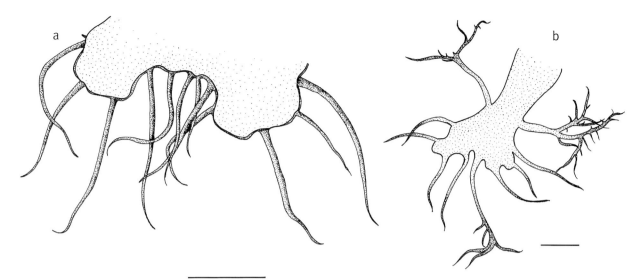
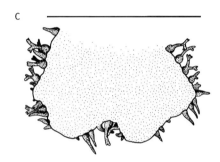

Fig. 10 Types of cilia: (a) unbranched, as in *Parmotrema perforatum*; (b) branched, as in *Heterodermia speciosa*; (c) bulbous, as in *Bulbothrix*. Scale = 1 mm.

In these cases, it is frequently difficult to determine the dividing line between a scabrose and a pruinose condition.

Tomentum
If the cells of a thallus cortex grow out into undivided or branched, colorless hyphal threads, they form a cottony fuzz that we call tomentum. It can be rather thick and erect or thin and lying flat on the thallus surface (plate 12). If it is sparse, tomentum can easily be missed. It is, however, very abundant on the upper surface of *Peltigera rufescens* and *Erioderma*, and on the lower surface of species of *Lobaria*, *Sticta*, and *Pseudocyphellaria* (Fig. 6e).

Cortical Hairs
Intermediate between cilia and tomentum are fine, transparent hairs produced by a number of foliose lichens on lobe surfaces or margins. Examples can be seen in *Phaeophyscia cernohorskyi*, *Leptochidium albociliatum*, and *Lobaria hallii*.

11. White pruina on the lobes of *Physconia detersa*. ×7.4

HOW LICHENS ARE BUILT

12. (top) The fuzzy tomentum on the upper surface of *Peltigera rufescens* is particularly thick. ×8.2

13. (right) Network of white maculae on *Rimelia reticulata*. Unbranched black cilia are on the lobe margins. ×5.4

14. (bottom) White, slightly raised pseudocyphellae on the lower surface of *Pseudocyphellaria anthraspis*. ×2.8

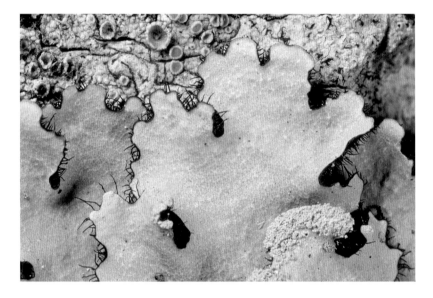

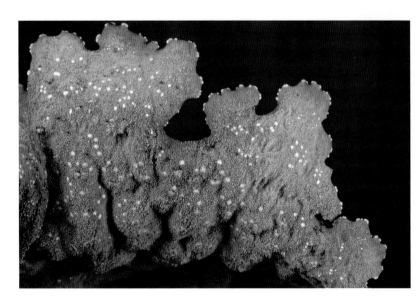

Maculae

The photobiont layer in many lichens is not entirely continuous (Fig. 4). Because the photobiont contributes to the color of the thallus, an interruption of the photobiont layer is visible on the surface of the lichen as a pale (usually irregular) spot or line. These pale areas are called maculae. Species of *Rimelia* have a particularly conspicuous, net-like pattern of maculae on the lobe surface (plate 13), but maculae are common in many genera of stratified foliose and fruticose lichens such as *Parmelia*, *Peltigera*, and *Cladonia*.

Pseudocyphellae

Pseudocyphellae, like maculae, are spots on the thallus, but they are associated with a break in the cortex. As a result, hyphae from the medulla can grow through the photobiont layer and sometimes right through the interruption in the cortex (Figs. 4, 11). Pseudocyphellae can be round or irregular in shape, and can be sunken (as in *Cetraria aculeata*) or result in a little bump (as in *Bryocaulon divergens*). They can be white or colored, depending on the color of the medulla. On some stratified foliose lichens, such as *Cetrelia* and *Punctelia*, they develop on the upper surface; on others, for example *Pseudocyphellaria*, they are on the lower surface (plate 14). Some pseudocyphellae can even become pruinose. Examples of pseudocyphellae can be found among many foliose, fruticose, and even crustose lichens.

Cyphellae

These are very specialized round pits or holes in the cortex that have their own lining of round, thin-walled cells. Unlike *pseudo*cyphellae (false cyphellae), they are not simply breaks in the cortex. Nevertheless, one can find some intermediates in certain lichens (e.g., *Aspicilia hispida*). Cyphellae on the lower surface of the thallus characterize all species of *Sticta*, the only genus north of Mexico to have them (plate 15).

Cephalodia

Lichen fungi can sometimes associate with more than one photobiont in a single thallus (see Chapter 1). The most common three-way symbioses involve lichens with green algae as their main or primary photobiont, together with gall-like growths containing a cyanobacterium as a secondary photobiont. These growths are called cephalodia, which means "resembling heads," and they can assume a variety of forms, from almost spherical and brain-like (as in *Amygdalaria paneola*

15. Cyphellae on the lower surface of the thallus characterize the genus *Sticta*. ×2.8

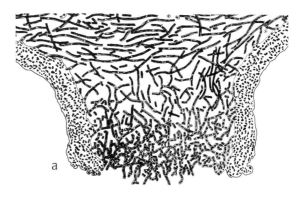

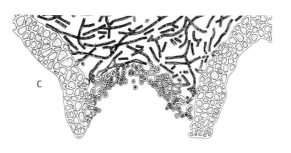

Fig. 11 Cyphellae and pseudocyphellae. a–b, pseudocyphellae: (a) on lower surface of thallus, slightly raised, *Pseudocyphellaria*; (b) on upper surface, depressed, *Bryocaulon divergens*; (c) young cyphella on lower thallus surface, *Sticta*. Magnification unknown. (Reprinted by permission from Ahmadjian and Hale, 1974, figs. 89, 90, 94.)

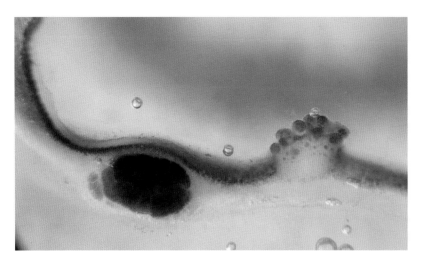

16. In this section of the thallus of *Lobaria pulmonaria*, a cephalodium forms a large, dark lump (on the left) below the green algal layer. Soredia burst through the upper cortex in a soralium (on the right). Approx. ×100

and some species of *Stereocaulon*) to appressed and lobed (as in *Placopsis lambii* or *Peltigera britannica*) or buried entirely within the parent thallus producing only vague bumps on the lower or upper surface (as in *Lobaria pulmonaria* or *Nephroma arcticum*) (plate 16). In *Solorina crocea,* a separate layer of cyanobacteria just below the layer of green algae represents an extreme modification of a cephalodium.

Because cyanobacteria are able to take nitrogen from the air and transform it into a form of nitrogen that the lichen can use for growth ("nitrogen fixation"), cephalodia are of tremendous importance to lichens, especially those living on nitrogen-poor substrates. This is especially true because specialized nitrogen-fixing cells called heterocysts are particularly numerous in the cyanobacteria found within cephalodia, as compared with cyanobacteria of the same species not associated with cephalodia. With cephalodia providing a direct source of nitrogen, it is no surprise that many early colonizers of bare rock or soil, such as *Placopsis, Pilophorus,* and *Stereocaulon,* are lichens with cephalodia. (See also Chapter 5.)

HOW LICHENS REPRODUCE

I find myself inspecting little granules as it were on the bark of trees—little shields or apothecia springing from a thallus—such is the mood of my mind—and I call it studying lichens.

Henry David Thoreau

FRUITING BODIES IN LICHENS

Because lichens consist of two (or three) different kinds of organisms, reproduction can be a little complicated. Each of the individual components can be successful in reproducing itself, but both will have to get together somewhere along the line in order to recreate the lichen. The fungus can produce microscopic spores by the millions, but a new lichen can be formed only if a germinating spore encounters the appropriate photobiont, and, if the combination is to continue to maturity, this must happen in the right habitat. The problem of how a fungus recognizes the appropriate photobiont and initiates a new lichenization is a fascinating subject in itself. Our concern here is how a lichen reproduces sexually.

The fruiting bodies of lichens are actually those of the fungal component. In most cases, the photobionts do not produce recognizable reproductive structures, nor do they reproduce sexually once they are "lichenized" (i.e., in a lichen association).

ASCOMATA

Because almost all lichen fungi are Ascomycetes, their fruiting bodies, called ascomata (singular: ascoma), are usually one of two basic types: apothecia or perithecia. Apothecia are basically disk- or cup-shaped fruiting structures that have an exposed spore-producing layer on the upper surface (Fig. 12). Perithecia are flask-shaped structures enclosing the spore-producing layer and opening by some kind of pore or hole at the top. (These are described in detail later in this chapter.) Unfortunately, there are some intermediates and variations on these themes such as the rather perithecium-like apothecia seen in *Pertusaria, Thelotrema,* and *Diploschistes,* and the elongated, sometimes branched ascomata called lirellae seen in *Graphis* and *Opegrapha.* One can also be misled by ascomata that superficially look like apothecia or perithecia but have an entirely different development and ancestry. For example, the disk-shaped ascomata of some species of *Arthonia* are not true apothecia (as in *Biatora*) although they look like them, and the perithecium-like ascomata of *Arthopyrenia* (not included in this book) or *Thelotrema* are not related by common ancestry to the perithecia of lichens such as *Porina*. These special cases are explained later in the descriptions of those lichens (Part II).

Sectioning various kinds of ascomata from top to bottom and viewing the sections under a microscope will reveal certain similarities and differences. The fea-

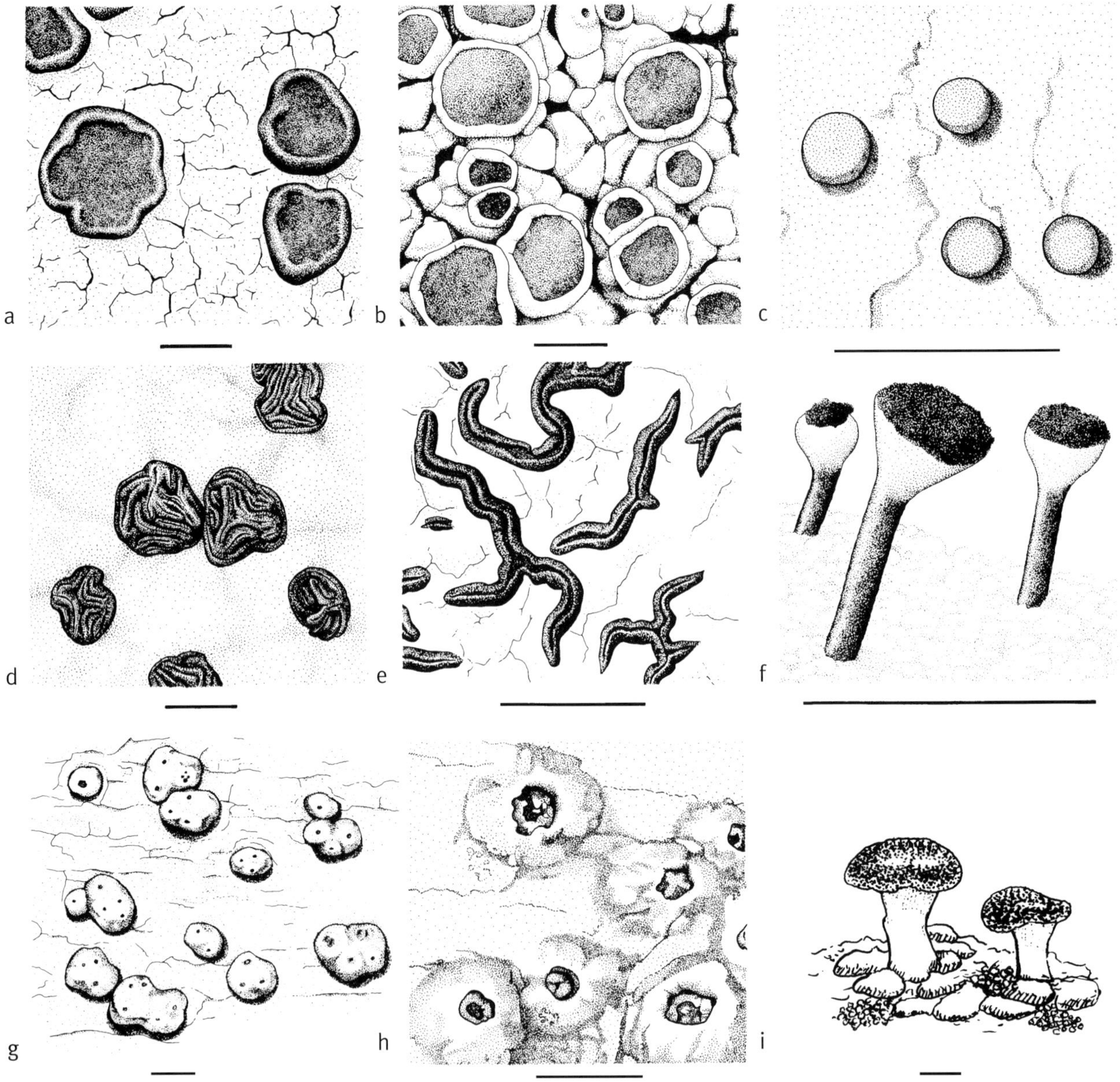

Fig. 12 Types of apothecia: (a) lecideine, lacking a thalline margin (*Porpidia macrocarpa*); (b) lecanorine, with a thalline margin (as in *Lecanora cenisia*); (c) arthonioid (as in *Arthonia patellulata*); (d) gyrose, with concentric sterile ridges (*Umbilicaria hyperborea*); (e) lirellae (as in *Opegrapha* and *Graphis*); (f) stalked mazaedia (*Calicium viride*); (g) pertusarioid (*Pertusaria macounii*); (h) double-walled (*Thelotrema lepadinum*); apothecia raised on podetia (*Baeomyces rufus*). Scales = 1 mm. (a–b and d–f reproduced courtesy of British Lichen Society from Purvis et al., 1992, fig. 28; g and i reprinted courtesy of the Canadian Museum of Nature, from Brodo, 1988, figs. 24 and 68.)

tures of most types of apothecia are illustrated in Fig. 13. (Perithecial sections are shown in Fig. 17.) Because the classification and identification of almost all crustose lichens are based on microscopic features of the fruiting bodies, it is essential to learn to recognize the different tissues, structures, and spores of the ascomata, and to assimilate the necessary terminology. Although we will try to keep the jargon to a minimum, the basics cannot be avoided.

ASCI

All ascomata contain a layer or cluster of cylindrical to club-shaped, sac-like cells called asci (singular: ascus). It is inside the young developing ascus that sexual fusion of cell nuclei occurs, followed by reduction division (meiosis: when the doubled set of chromosomes, one from each parent, is again reduced to one set per cell), with the subsequent development of ascospores. When mature, each ascus typically contains eight ascospores, although there can be as few as one and as many as several hundred.

Asci can have walls consisting of a single layer, but much more commonly in lichen fungi the wall has several layers of different kinds, which are involved in the ejection of the spores. The walls of asci, viewed under a microscope, change color or "stain" with iodine in different ways; these staining patterns correlate

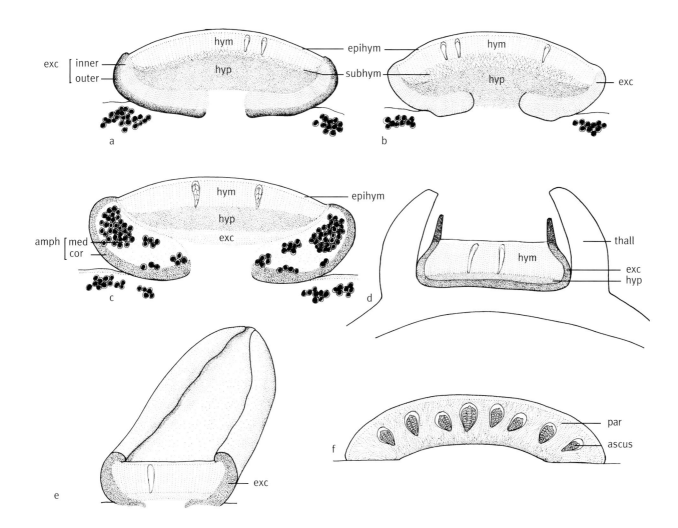

Fig. 13 Apothecia seen in section: (a) lecideine, excicple with a dark outer layer and pale inner layer, lacking algae; (b) biatorine, with a pale and often radiating exciple, lacking algae; (c) lecanorine, with a thin exciple, surrounded by a thalline margin containing algae; (d) double, both the exciple and thalline margins are well developed and distinct; (e) lirella, like an elongate lecideine apothecium, usually lacking algae; (f) arthonioid, asci arising in a rather uniform tissue consisting of paraphysoids. *amph*, amphithecium; *cor*, cortex; *epihym*, epihymenium; *exc*, exciple; *hym*, hymenium (containing asci and paraphyses); *hyp*, hypothecium; *med*, medulla; *par*, paraphysoid tissue; *subhym*, subhymenium; *thall*, thalline margin. Schematic, at various magnifications.

with other morphological features that make ascus morphology useful in classification. In some asci, all or some of the wall layers stain a deep blue; in others, the walls remain entirely unstained (or take on a yellowish color). This subject is discussed in more detail under "Studying Lichens: Microscopic Study," in Chapter 13.

The tips of the asci are greatly thickened in many lichens due to an expansion of the inner ascus wall, giving rise to the *tholus* (not to be confused with *thallus*). Parts of the tholus have different patterns of staining blue with iodine. For the purposes of this book, we refer only to the major types of staining patterns. The ones illustrated (Fig. 14) will help you distinguish many families and genera of lichens, especially those of crustose lichens.

In most kinds of lichen fungi, the ascus is like a tiny gun under pressure. At a certain point of maturity, controlled and aided by the structure of the walls and tip, the ascus "gun" fires its complement of ascospores into the air, where the spores can be carried away by air currents. Some other types of spore dispersal are mentioned in the discussions of special kinds of fruiting bodies.

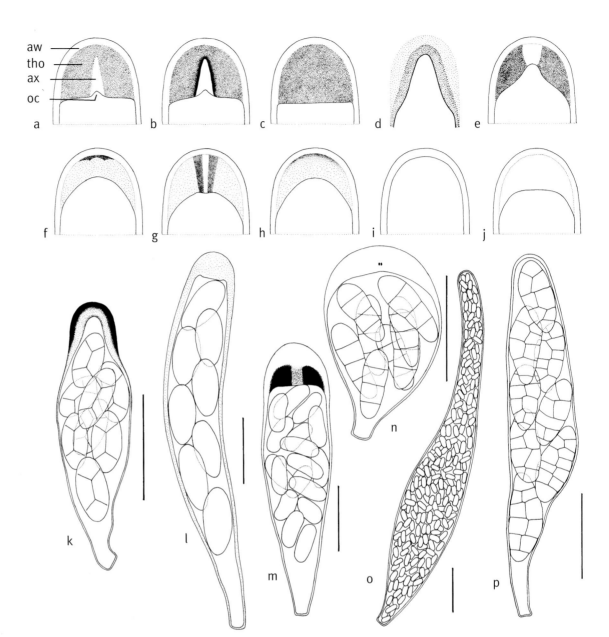

Fig. 14 Ascus types:
(a) *Bacidia*-type;
(b) *Biatora*-type;
(c) *Catillaria*-type;
(d) *Fuscidea intercincta*–type;
(e) *Lecanora*-type;
(f) *Lecidea*-type;
(g) *Porpidia*-type;
(h) *Rhizocarpon*-type;
(i) *Schaereria*-type;
(j) *Tremolecia*-type;
(k) *Xanthoria parietina* (*Teloschistes*-type);
(l) *Trapelia coarctata*;
(m) *Candelariella vitellina*;
(n) *Arthonia radiata*;
(o) *Thelocarpon epibolum* (not included in book);
(p) *Gyalecta truncigena*.
aw, ascus wall; *ax*, axial body; *oc*, ocular chamber; *tho*, tholus.
Scale: k–p = 20 μm (a–c and e–p reproduced courtesy of British Lichen Society from Purvis et al., 1992, fig. 33.)

ASCOSPORES

Ascospores are tremendously diverse in shape and size (Fig. 15). They can be composed of a single cell, or divided into many cells. They can be grain-like, long and tapered, or bulky and potato-like. They vary from less than a few microns long to over 200 μm long and 75 μm wide.[1] The walls of spores can be thin or very thick, smooth or sculptured. The cells of the spore can be cylindrical, lens-shaped, or angular, and the walls either colorless or darkly pigmented. Some spores have a gelatinous outer sheath called a perispore or "halo." Such spores are said to be halonate. All these characteristics are used in distinguishing species, genera, and even families of lichens.

THE TISSUES OF APOTHECIA

Although apothecia vary a great deal in color, shape and size, they all consist of the same basic set of tissues (Fig. 13).

---

[1] "Micron" is the shortened form of "micrometer." It is 1/1000th of a millimeter (1/2500th of an inch) and is abbreviated "μm." The micron is the normal unit of measurement for microscopic objects.

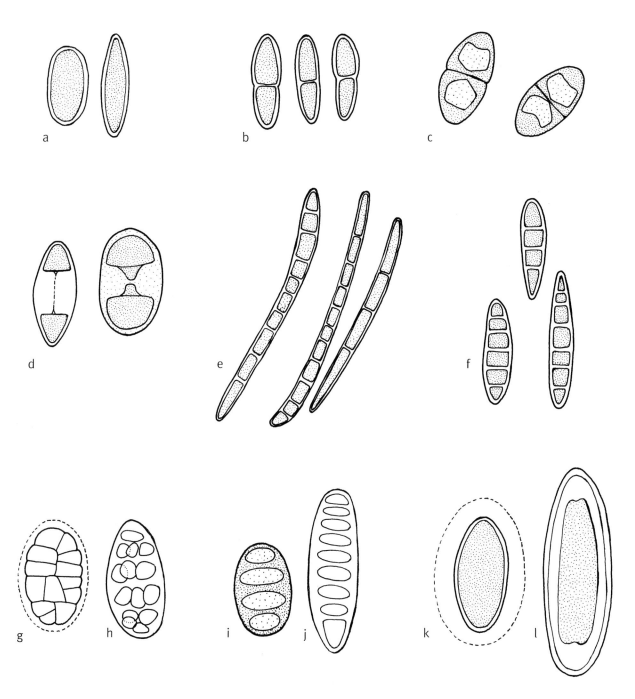

Fig. 15 Spore types. (a) 1-celled: broadly ellipsoid (*Lecanora pulicaris*); narrowly ellipsoid (*L. strobilina*); (b) 2-celled (*Lecania dubitans*); (c) *Physcia*-type: angular locules and unevenly thickened walls (*Rinodina subminuta*); (d) polari-locular: 2-celled, with a thickened septum (*Caloplaca holocarpa* on left; *C. litoricola* on right); (e) needle-shaped, 4- to 11-celled (*Bacidia schweinitzii*); (f) fusiform, cylindrical cells, 4- to 7-celled (*Bacidia sabuletorum*); (g) muriform, thin-walled with gelatinous halo (*Rhizocarpon*); (h) muriform, locules small and walls thick (*Graphina*); (i–j) septate, with lens-shaped locules; (i) brown, 4-celled (*Pyrenula*); (j) colorless, many-celled (*Graphis*); (k) 1-celled, with gelatinous halo (*Porpidia*); (l) 1-celled, with thick, layered walls (*Pertusaria*). Schematic, various magnifications. (c–f redrawn from Brodo, 1988, figs. 30, 15, 11b, 13, respectively.)

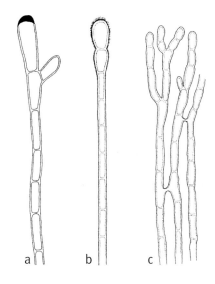

Fig. 16 Paraphyses: (a) almost unbranched, barely expanded at summit, but with pigment within cell at tip; (b) unbranched, with an expanded end cell that is slightly pigmented on the outside; (c) branched and anastomosing paraphyses, not expanded or pigmented at tips. Magnification not known. (Reprinted by permission from Wirth, *Die Flechten*, p. 30, © 1995 by Verlag Eugen Ulmer, Stuttgart.)

A hymenium normally forms the upper surface of the apothecial disk. It consists of hundreds of vertically oriented asci, usually interspersed with some branched or unbranched sterile hyphae that help support the asci. These hyphae within the hymenium are called paraphyses, pseudoparaphyses, or paraphysoids depending on their anatomical origin and mode of development (Fig. 16). For the sake of simplicity, we refer only to paraphyses (the most common type) in the description below.

The upper portion of the hymenium has a character of its own. Above the tips of the asci, the paraphyses often branch, or are swollen or pigmented, forming an epihymenium. The expanded tips of paraphyses often are pigmented, contributing to the color of the apothecial disk. This layer can contain its own pigment outside the tips of the paraphyses, and it can contain granules as well.

Below the hymenium, a tissue of densely branched, randomly oriented hyphae called a hypothecium usually can be distinguished. It can be colored or colorless.

Surrounding all these tissues is the wall of the fruiting body, the exciple, which typically forms the margin of the apothecial disk as seen from above with a hand lens. If the exciple is at least partly blackened and carbon-like (carbonaceous), the apothecium is called lecideine (Figs. 12a, 13a) because it is characteristic of the genus *Lecidea*. If the exciple is colorless or lightly pigmented, the apothecium can be called biatorine (Fig. 13b), being the type found in *Biatora* and similar lichens. Additional thallus-like tissues containing a cortex, photobiont cells, and a medulla (together termed the amphithecium) can develop outside the exciple, forming a thalline margin, as in species of *Lecanora*, *Physcia*, and *Ochrolechia*. An apothecium with a thalline margin is said to be lecanorine (Figs. 12b, 13c).

In *Arthonia* and its relatives, the ascomata may resemble apothecia in being disk-like (Fig. 12c), but they are special in lacking an exciple and having a less structured hymenium. The ascoma begins development as a hemispherical mass consisting of a more or less uniform tissue of highly branched hyphae. The asci grow up into this mass, forming "holes" or locules for each ascus (Fig. 13f). Such ascomata are therefore called ascolocular. The asci of ascolocular lichens have an intriguing "jack-in-the-box" mechanism of breaking open and dispersing the spores that involves the splitting open of an outer wall and extension of the inner wall. In *Graphis*, the structure of the ascoma is basically like that of a lecideine apothecium, but the fruiting body grows at only two points rather than radially and becomes an elongated lirella (Figs. 12e, 13e).

Very specialized apothecia occur in members of the order Caliciales, including *Calicium*, *Cyphelium*, and *Thelomma*. In these fruiting bodies, the asci are slender and thin-walled when young. As the asci mature, their walls disintegrate, releasing the spores into a dry, powdery mass (a mazaedium) within the cup-like exciple (Fig. 12f).

Some lichens such as *Thelotrema* and *Diploschistes* have apothecia in which both the exciple and thalline margins are well developed and often distinct, giving the fruiting bodies a double-margined appearance (Figs. 12h, 13d).

PERITHECIA

Lichens whose ascomata are perithecia can collectively be called pyrenocarpous (with "fruit like a pear," referring to the shape of the perithecium). Examples of pyrenocarpous lichens are *Pyrenula*, *Porina*, and *Verrucaria* among the crusts, and *Dermatocarpon* among the foliose lichens.

Perithecia have a much-reduced hymenium entirely enclosed by the excipular wall, except for a hole called an ostiole usually at or close to the summit (Fig. 17a). The exciple can be colorless, but it is usually pig-

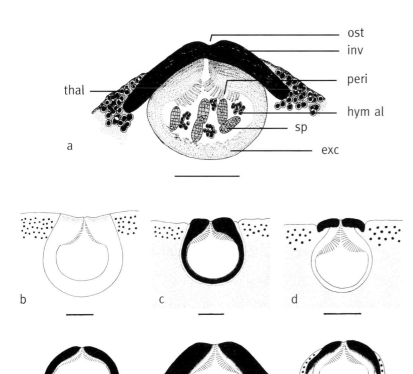

Fig. 17 Perithecia in section: (a) perithecium of *Staurothele fissa*. *exc*, exciple (excipulum); *hym al*, hymenial algae (found only in *Staurothele* and *Endocarpon*); *inv*, involucrellum; *ost*, ostiole; *peri*, periphyses; *sp*, spore; *thal*, thallus tissue. b–g, types of perithecia: (b) immersed in foliose thallus, involucrellum absent, exciple pale (*Dermatocarpon miniatum*); (c) immersed in limestone, involucrellum absent, exciple black and friable (as in *Verrucaria calciseda*); (d) immersed in limestone, involucrellum lid-like, exciple mostly pale (*Verrucaria* sp.); (e) prominent, involucrellum absent, exciple black and friable above, pale at base (*Thelidium* sp.); (f) involucrellum well developed, exciple pale below (as in *Pyrenula pseudobufonia*); (g) immersed in thalline wart, involucrellum present, exciple pale below (as in *Staurothele fissa*). Scale = 200 μm. (a, reproduced from Brodo, 1988, fig. 26; b–g, reproduced courtesy of British Lichen Society from Purvis et al., 1992, fig. 41.)

mented brown to carbon-black (Fig. 17b–g). In some lichens, the exciple is itself partially or entirely enclosed by a second dark layer called an involucrellum.

In most pyrenolichens, the ascomata look like perithecia but develop in a manner much like *Arthonia* (described in the preceding section); they are therefore also ascolocular, and the ascomata should not, strictly speaking, be called perithecia. Nevertheless, we do use the term broadly in this book.

Perithecia can develop on the surface of the thallus, but more frequently they are partly or entirely buried within the thallus with only the ostiole showing at the surface. Certain lichens develop thalline warts in which one or more perithecia are buried. These fertile warts, called pseudostromata (singular: pseudostroma), are composed mainly of fungal tissue but almost always incorporate some material from the substrate (generally bark cells). Stromata are similar warts consisting of only fungal tissue, but these almost never occur in lichens. *Trypethelium virens* and *Laurera megasperma* are examples of lichens with perithecia embedded in pseudostromata. *Glyphis cicatricosa* is a lichen with lirellae rather than perithecia formed on a pseudostroma.

Paraphyses can be present, although in some groups such as the genus *Verrucaria* they fail to develop or they gelatinize by the time the perithecium is mature. In perithecia, the ascospores are released through the ostiole, either directly from the ascus or in a drop of jelly extruded from the perithecial cavity.

MUSHROOM LICHENS

In Basidiomycete lichens such as *Omphalina*, the fruiting bodies are ephemeral mushrooms like those of their unlichenized cousins. The sexual spores are produced at the ends of little club-shaped cells called basidia, which entirely coat the surfaces of the gills of the mushrooms or the outer surface of the club. When mature, the spores drop off or are popped off the basidia and fall between the gills toward the ground, where they are picked up by air currents and dispersed. *Multiclavula* is a basidiomycete always associated with algae, but that does not form a special thallus

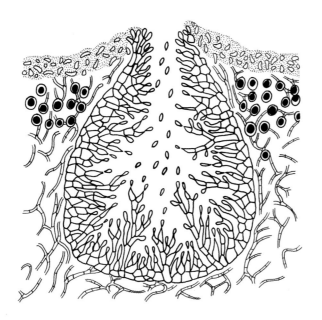

Fig. 18 Pycnidium, producing 1-celled, ellipsoid conidia. (Redrawn from Vobis, *Bau und Entwicklung der Flechten-Pycnidien und ihrer Conidien*, 1980, J. Cramer, Vaduz, fig. 5d.)

and therefore, strictly speaking, is not a lichen. It has club-like fruiting bodies. We have included an example in the book because it is biologically interesting, and lichenologists often collect and study it.

PYCNIDIA AND CONIDIA

Pycnidia are tiny, perithecium-like structures that contain hundreds of spores not associated with asci (Fig. 18). The spores are called conidia. Conidia can germinate like ascospores, thereby providing the lichen fungus with a nonsexual means of reproduction. Many conidia also function as a kind of sperm cell, providing a fertilizing "male" nucleus in the sexual reproduction of the fungus.

In many lichens, pycnidia are black and conspicuous (e.g., along the lobe margins of species of *Cetraria* and *Tuckermannopsis*), but in others they are pale and embedded in the thallus and therefore very inconspicuous (except for their often dark ostioles).

The color, shape, and location of pycnidia are useful characteristics for the identification of lichens, as are the shape, dimensions, and mode of development of the conidia.

VEGETATIVE REPRODUCTIVE STRUCTURES

Because the chances are relatively small that an ascospore or conidium of a lichen fungus will germinate extremely close to just the right photobiont on the right substrate, it is not surprising that various mechanisms for bypassing spore-based reproduction have developed in lichens over the course of evolution. In fact, any fragment of a lichen containing both the fungal and algal components can, theoretically, form a new lichen, although it now seems that most fragments must first "deconstruct," breaking down into undifferentiated fungal and algal cells, before a new lichen can be reassembled. Exceptions would be long fragments of hair lichens or beard lichens (*Bryoria*, *Alectoria*, and *Usnea*), which are easily torn from the parent thallus and can become reestablished elsewhere. Specialized types of thallus fragments that can function as easily dispersed propagules have evolved independently in many lichens (these are described below) (Fig. 19).

Soredia

Soredia are tiny balls of hyphae enveloping a few photobiont cells. They can be extremely fine and powdery, almost like flour (called farinose for that reason), or they can have a more grainy texture like corn meal (called granular or granulose). The dividing line between the two types can be arbitrarily set at about 30 μm in diameter, but no distinct boundary exists between them, and intermediates abound sometimes even on the same lichen. Some soredia (mostly of the farinose type) can occur in granular or irregular agglomerations up to 150 μm in diameter and can be called consoredia.

Soredia are produced at the level of the photobiont layer and therefore have no cortex. They are released through breaks in the thallus cortex as in *Parmelia sulcata* or *Lobaria pulmonaria* (Fig. 19i; plate 16), or they can develop after a portion of the cortex has fallen away or disintegrated (Fig. 19h) as in some species of *Cladonia* (e.g., *C. coniocraea* or *C. deformis*).

Soredia can form on the thallus in discrete structures called soralia or can cover large, irregular areas. Soredia sometimes comprise the entire thallus, in which case we say the thallus is leprose (as in species of *Lepraria*). The shape of a soralium and where it is found are usually species-specific, making them important characteristics in lichen identification (Fig. 20; plate 17). Marginal soralia form along the lobe margins (e.g., *Physcia atrostriata*, *Parmotrema stuppeum*); laminal soralia are on the upper surface of the lobes (e.g., *Dirinaria picta*, *Pseudocyphellaria anomala*); submarginal soralia are usually on the upper surface close to the margin (e.g., *Parmotrema arnoldii*, *P. chinense*); and labriform soralia (meaning "lip-shaped") form at the lobe tips or on special marginal lobes by producing

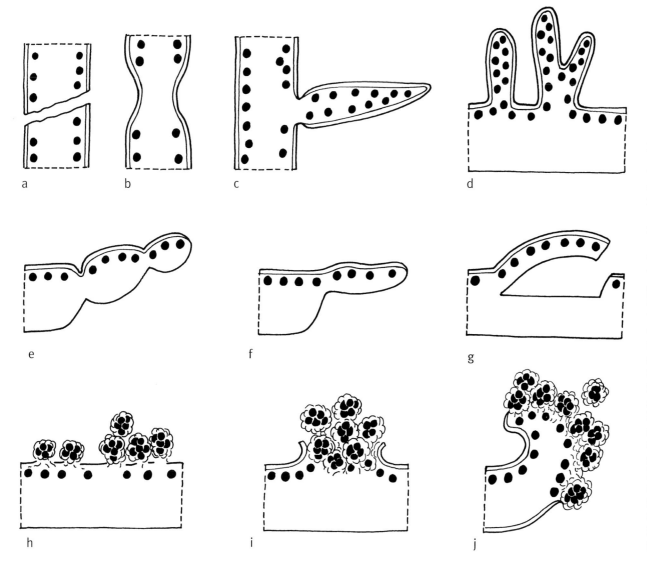

Fig. 19 Vegetative (nonsexual) propagules: (a) thallus fragmentation; (b) weakened segment for fragmentation; (c) lateral spinule, narrowed at base; (d) isidia, with cortex intact; (e) blastidia, budding off granules from thallus margin; (f) lobule; (g) schizidium, a fragment of the upper cortex and algal layer; (h) soredia formed from eroded thallus surface; (i) soredia breaking through cortex and forming a soralium; (j) soredia formed from lower cortex of recurved lobe (labriform soralium). (Redrawn from Purvis et al., 1992, fig. 29.)

soredia on the undersurface of curled back or crescent-shaped areas as in *Xanthoria fallax*, or even by bursting from the interior of an inflated lobe tip as in *Hypogymnia physodes*.

Isidia

Isidia are tiny, cylindrical or granular outgrowths of the thallus that are covered with the same layer of cortex as that of the thallus itself (Fig. 19d). Most isidia are constricted at the base and fragile, and they are therefore easily dispersed. They can be spherical (e.g., *Xanthoparmelia mexicana*), elongated in various ways (*Evernia mesomorpha*, *Parmelia saxatilis*), or even branched (*Loxosporopsis corallifera*), and they are generally easy to recognize (plate 18). In some lichens (e.g., *Melanelia subaurifera* and some forms of *Lobaria pulmonaria*), the isidia themselves break down into soredia, or the cortex is thin and soon lost as in *Parmelia hygrophila*. The boundary between soredia and isidia, in such cases, is rather imprecise. Some lichens that lack a cortex, such as the jelly lichen, *Collema*, can produce granular outgrowths on the thallus surface that resemble isidia. Although these structures do not have a cortex, they retain the color and texture of the thallus surface, so we continue to call them isidia.

Granules and Blastidia

Rounded corticate fragments of the thallus (granules) can develop in a number of ways depending on the lichen, but when they bud off from finely divided lobe margins or squamules they are called blastidia (Fig.

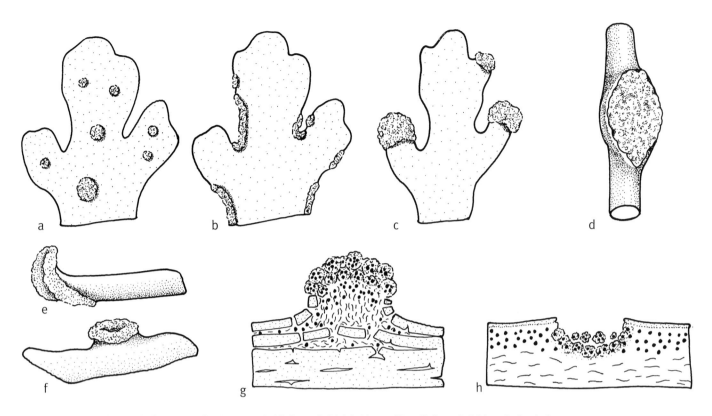

Fig. 20 Soralia: (a) laminal; (b) marginal; (c) terminal; (d) fissural; (e) labriform; (f) cuff-shaped; (g) hemispherical or erumpent; (h) excavate. (a–c, e–h redrawn from Purvis et al., 1992, fig. 43.)

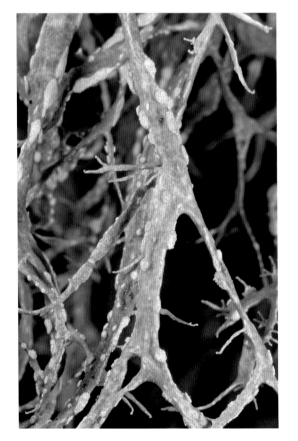

17. (left) Oval soralia filled with powdery soredia occur on the margins and surface of branches of *Ramalina farinacea*.

18. (above) Branched isidia on *Platismatia herrei*.

19e). Very small granules or blastidia are almost indistinguishable from soredia. Lichens with granules or blastidia include *Cladonia chlorophaea* and *Physcia millegrana*, respectively. In both lichens, the particles grade into soredia.

Schizidia
Fragments of the upper layers of a stratified thallus, including portions of the upper cortex and algal layer, can develop from the breakdown of pustule-like protuberances on the thallus surface or simply flake off the thallus surface (Fig. 19g). These fragments are called schizidia (meaning "split off"), and they function as the propagules for many lichens, including *Fulgensia bracteata*, *Hypotrachyna osseoalba*, and *Loxospora pustulata*.

Lobules
Lobules, as the name implies, are minute lobes having the same basic anatomy as the thallus from which they develop (Fig. 19f). They can form on the thallus and apothecial margins as in *Anaptychia palmulata* or *Physconia americana*, or all over the thallus surface as in *Lobaria tenuis* (plate 19). In some species of *Peltigera* (e.g., *P. elisabethae* or *P. pacifica*), lobules develop along cracks or points of injury, as well as on the thallus margins.

19. Scale-like, overlapping lobules are abundant on the upper surface of *Lobaria tenuis*. ×2.8

4. COLORS OF LICHENS

A quaint enclosure fenced from modern times,
And their destructive influences—ground
Held sacred to the past. Its dampest nooks
Are green with moss, and rusty with red gold
Of coloured lichens such as painters love.
Philip Gilbert Hamerton

In the absence of any special pigments in a lichen's upper cortex, the lichen is generally a shade of gray or greenish gray. When brown pigments such as melanin are in or between the cortical cells, the lichen is brown, at least when dry. When a lichen is wet, the cortex is more transparent, and we can see the underlying photobiont layer, making a lichen that is gray or brown when dry turn bright green or olive (plates 20A and 20B). Any of a number of brightly colored pigments can also be deposited in the cortex (see Chapter 6 and plates 10 and 29). The most widespread pigment in lichens is the pale yellow usnic acid, the substance that gives species of *Usnea* and *Xanthoparmelia* their characteristic pale yellowish green or "usnic yellow" color. Other pigments include yellowish xanthones, as in *Buellia halonia;* brilliant yellow pulvinic acid derivatives in species of *Vulpicida, Letharia,* and *Pleopsidium;* and bright yellow, orange, or red anthraquinones, coloring species of *Caloplaca, Xanthoria,* and the bright red fruiting bodies of British soldiers (*Cladonia cristatella*).

Because of differences in the concentration of pigments from one population to another owing to differences in thallus age, exposure to sunlight, genetics, or other factors, colors vary from specimen to specimen, even within the same species. In selecting photographs to illustrate the species, we tried to choose images that display a typical color. The range of color variation is mentioned in the text.

But, as lichenologist Jouko Rikkinen asked in the title of his book on lichen photobiology, "What's behind the pretty colours?" Flowers are colored to attract insects for pollination, and birds are colored to attract a mate. Why, then, are so many lichens brightly pigmented? The lichens that tend to be most deeply colored—whether yellow, orange, or even brown—are those in the most exposed, often dry habitats (plate 10). Clearly, light has something to do with it. We know that the green photobionts of lichens can be damaged by too much light—specifically, ultraviolet light. There is good evidence that the pigments deposited in the cortex of exposed lichens protect the algae from this radiation. Lichens on exposed mountaintops and in the lower arctic latitudes, where radiation levels are particularly high, are darker than those at higher latitudes in the arctic, where the angle of the sun is lower and radiation is not as much of a problem.

Different pigments act in different ways. Usnic acid,

20. *Heterodermia hypoleuca* when dry (A) and wet (B). Strikingly different colors when wet and dry also occur in many other lichens, including species of *Physconia* and *Peltula* and in *Dermatocarpon luridum*. ×1.6

a pale yellow pigment, is most effective against short-wavelength ultraviolet light, whereas pulvinic acid pigments, which are intensely yellow, absorb longer-wavelength ultraviolet light. Brown melanin pigments also absorb and scatter radiation.

Exposed lichens that are whitish because of a thick tomentum or heavy pruina are also protected by the light-scattering properties of these physical features. *Peltigera* species such as *P. rufescens* that are adapted for a life in open fields are much more heavily tomentose than closely related species adapted for the shade such as *P. canina*. (See also Chapter 5, under "Temperature.")

Thallus color may also have a role in regulating temperature within the lichen, at least in part. Arctic and alpine lichens tend to be dark, which would promote the absorption of radiant energy, warming the thallus. Desert lichens are commonly white-pruinose or scabrose, scattering and reflecting the sun's rays, keeping the thallus cooler.

5 HOW LICHENS LIVE: PHYSIOLOGY AND GROWTH

... the lichens, gray, crisp, brittle, and crusted ... deriving their food from certain kinds of small algae which they hold enslaved in their meshes.

W. F. Ganong

A lichen thallus consists mostly of fungal tissue, but because its symbiosis involves photosynthetic algae or cyanobacteria, the lichen functions much like a green plant in that it requires light. On the other hand, its life processes differ in some important ways from those of a typical leafy plant. In order to understand why lichens are found where they are, and how they can survive in some of the earth's most forbidding regions and challenging habitats, we have to consider some aspects of their physiology.

PRODUCING AND SHARING FOOD: PHOTOSYNTHESIS IN LICHENS

Organisms that contain the green pigment chlorophyll can perform what is certainly one of the most amazing and important chemical transformations on earth: photosynthesis. With chlorophyll as a catalyst, plants, algae, and cyanobacteria can combine carbon dioxide from the air with water, in the presence of sunlight, to make carbohydrates such as sugar, transforming the sun's energy into potentially useful chemical energy. These carbohydrates then serve as an energy source for all the organism's other life functions, including growth. In the process, oxygen is released back into the air.

Fungi have no chlorophyll, but lichen fungi have solved that problem by associating with photosynthetic symbionts (photobionts): green algae or cyanobacteria (or sometimes both). The photobionts in ordinary stratified foliose or fruticose lichens are organized within the thallus in a way that maximizes their photosynthetic function. In fact, a foliose lichen resembles the structure of a green leaf in many respects (see Chapter 2). The upper cortex is the equivalent of the leaf's upper epidermis, the algal layer functions like the palisade layer in a leaf, the medulla is loosely packed with air spaces like the leaf's mesophyll, and the lower cortex is protective like a leaf's lower epidermis. Pseudocyphellae can be regarded as the lichen's answer to the leaf's stomates, providing gas exchange to the interior.

In lichens, the photobionts make enough carbohydrate to supply both themselves and their fungal cohabitant. The presence of the fungus, in fact, makes the photobionts relatively "leaky" with respect to the sugars they produce, and a good proportion of the carbohydrate passes into the fungal tissue as either sugar alcohols such as ribitol or sorbitol (in the case of green algae) or glucose (in the case of cyanobacteria) (Fig. 21). These carbohydrates are stored in the fungus as mannitol, another sugar alcohol. When we isolate the algae and grow them apart from the fungus, this flow of carbohydrates out of the photobiont quickly shuts off.

USING FOOD FOR ENERGY: RESPIRATION

Respiration is essentially the reverse of photosynthesis: using carbohydrate (usually with oxygen) to generate chemical energy for growth and repair, reproduction, the manufacture of pigments, and other life processes, while producing carbon dioxide and water as by-products. Because respiration uses up the carbohydrate produced by photosynthesis, in order for a lichen to survive and grow independent of outside sources of carbohydrate, its rate of photosynthesis has to be higher than its rate of respiration. It is the balance sheet of photosynthesis and respiration that determines, to a large degree, where a lichen can grow.

FINDING THE BALANCE: PHOTOSYNTHESIS VERSUS RESPIRATION

Both respiration and photosynthesis are affected by such external factors as temperature and moisture levels. Photosynthesis is also affected by the amount of available light because it occurs only in light; respiration, on the other hand, can proceed in light or in complete darkness. The two processes, however, react somewhat differently to all these factors. The subject is complex, but some generalizations are possible.

In most lichens, photosynthesis functions best when the thallus is about 50–70 percent saturated with water, whereas the rate of respiration is highest when the thallus is entirely or almost entirely saturated. Most lichens, when air-dried, have a moisture content of about 15–30 percent, so they have to absorb more water before either photosynthesis or respiration can begin.

A dry lichen, neither burning nor generating energy, is in a dormant state. Photobiont cells may shrivel a bit, and some changes may occur to their cell walls, but, unlike the cells of most vascular plants, they will not be permanently damaged.

When a dry lichen absorbs water gradually, its rate of photosynthesis will be higher than its respiration rate until the thallus is close to saturation, at which point respiration will exceed photosynthesis.

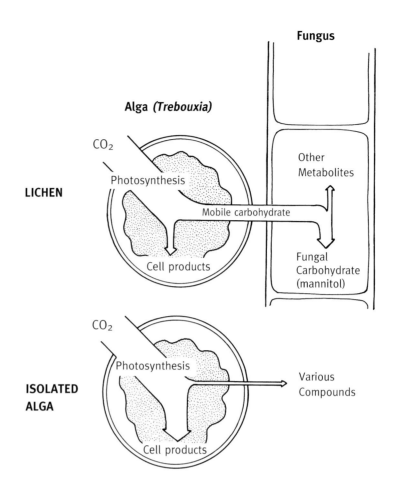

fig. 21 Carbon exchange. In the lichen, most of the carbohydrate manufactured by the alga becomes mobile, passing into the fungus, where it is stored as mannitol. When the alga is separated out of the lichen and isolated, the products of photosynthesis largely remain within the algal cell. (Redrawn from Ahmadjian, 1993, fig. 43.)

The optimal temperature for photosynthesis is lower than that for respiration in most lichens, but different species have different optimal temperature ranges, and most correlate well with their natural habitats. Antarctic lichens, for example, photosynthesize best at 0–10°C, whereas temperate species do better at higher temperatures (10–15°C). Respiration, however, is most efficient at about the same range for most lichens (15–30°C).

The light level at which photosynthesis rates are highest varies from species to species according to habitat and photobiont type. In general, cyanobacterial lichens (such as species of *Nephroma* and *Peltigera*) that live in shaded habitats need less light than do green algal lichens (such as *Ramalina* and *Usnea*) that live in open habitats. Although the green algae of most lichens can actually be damaged by too much light (see the discussion of pigments, Chapter 4), light is essential and determines the habitats of many species.

21. Crustose lichens, adapted for life in this periodic watercourse in Texas, form black patterns on the granite.

22. *Siphula ceratites* is a semiaquatic fruticose lichen that always grows with the basal parts of its stalks in mud or standing water.

WATER RELATIONS

Lichens, unlike leafy plants with their waxy cuticle, have no special water-repellent covering that can help them conserve water. Dry habitat lichens typically develop a thick outer layer of dead, heavily gelatinized cortical material that undoubtedly slows water loss to some extent. Nevertheless, evaporation occurs over the entire thallus surface, and so lichens lose water and dry out very quickly. On the other hand, they also can easily absorb moisture through all surfaces of the thallus. Most lichens are extremely drought-resistant and are able to survive weeks or even months in a perfectly dry state, resuming photosynthesis and respiration soon after being rewetted. *Ramalina maciformis*, a desert lichen, was stored for 12 months at 1 percent moisture content with no apparent ill effects. Aquatic lichens such as *Siphula*, *Hydrothyria*, and some crustose species (plates 21, 22), which live permanently or intermittently submerged in water, succumb to dry conditions in a comparatively short period of time.

One of the things that make lichens special is their ability to absorb moisture directly from the air when the humidity is high enough. Most lichens can survive quite well on fog or dew even where there is hardly any rain, as in the coastal fog deserts of Baja California. In such areas, lichens are tremendously abundant and diverse (plate 23). Powdery lichens such as *Lepraria* or *Chrysothrix*, in fact, cannot absorb liquid water at all (the drops just bead up and roll off), but they do very well even under overhanging rocks where only the humidity in the air can serve as a source of moisture (plate 24).

Most lichens, of course, can absorb rainwater and do so efficiently. There is, however, no conducting tissue for the rapid movement of water from one part of the lichen thallus to another. (The "veins" of species of *Peltigera* are simply thickened areas of medulla.) Lichens such as *Cladina stellaris*, a reindeer lichen, can be crispy dry at the top where they are exposed to the drying sun, and thoroughly wet at the bottom (keeping the soil moist at the same time; see "Nature's Pioneers: Colonization and Stabilization of Soil," in Chapter 8). Rhizines do not function as absorbing organs, although they do help to trap the moisture under a foliose lichen and keep the lichen wet longer than would be the case without rhizines. The highly branched morphology of many lichens promotes moisture retention by creating an area of high humidity around the branched clump, which slows evaporation.

23. Lichen communities in the coastal fog deserts of Baja California are specially adapted to take advantage of humid air, even in the absence of rain.

24. Leprose lichens such as the brilliant yellow *Chrysothrix* grow so thickly on overhanging rocks in humid localities that passersby can carve their initials in the lichen.

TEMPERATURE

Lichens can survive in the coldest as well as the hottest spots on earth. Of course, different species are involved. Indeed, like other organisms, lichens do have their upper and lower limits of temperature tolerance in the natural world.

Laboratory experiments with lichens have shown that certain species can survive extremely high temperatures (up to 90°C) and others can be frozen to the temperature of liquid nitrogen (−196°C), and still resume photosynthesis rather quickly when brought back to room temperature. In nature, the microhabitats that lichens occupy can, in fact, have very extreme temperatures, for example, up to 70°C on exposed soil in desert environments. Lichens can survive such extremes *in the dry state*. If they are moist, they perish essentially by being cooked. Temperate lichens, normally resistant to the coldest weather, can apparently be killed by a particularly cold winter under certain circumstances.

Nevertheless, lichens are the dominant vegetation in the coldest parts of the Antarctic and Arctic, on the highest mountain peaks, and in some of the hottest deserts.

The possible role of color in modifying thallus temperature is mentioned in Chapter 4.

MINERALS

Because lichens can absorb water and dissolved minerals through any part of their thalli, they have no need for roots or other special absorptive structures. The minerals that lichens require are generally supplied by moist air or rain, or in the water that flows over thalli that are still attached to their substrate.

Lichens are very efficient accumulators of minerals, and they can actually concentrate minerals found only in trace amounts in the substrate or atmosphere. This ability has practical implications, positive and negative, both for lichens and for other creatures, including human beings. On the negative side, some of the compounds absorbed by lichens, such as sulphates and metal compounds, are toxic. They can damage and finally kill the photobiont, ultimately causing the death of the lichen. In the days of atmospheric testing of atomic bombs, reindeer lichens proved to be efficient accumulators of radioactive nucleotides such as strontium 90. When caribou ate these lichens, the radioactive material was concentrated further in their

25. A pack rat "latrine" at the base of basaltic columns in central British Columbia is the preferred habitat for these orange lichens requiring high levels of nitrogen.

bones and tissues. This radioactivity, concentrated still more in people who ate contaminated caribou meat, caused increased incidence of bone cancer and leukemia. This same chain of events occurred again after the explosion of the Chernobyl nuclear power plant in Ukraine, necessitating the large-scale destruction of reindeer herds in Lapland.

On the plus side, lichens can be tested for metal concentrations and can give useful information regarding metal-bearing ores, such as copper and iron, in the immediate vicinity. Testing lichen thalli for accumulated pollutants is commonly done in pollution-monitoring programs. Such uses of lichens are discussed further in Chapter 11.

NITROGEN

All organisms require nitrogen for the manufacture of proteins and many other essential organic compounds needed for life. Although nitrogen is abundant in the air, this gaseous form cannot be used by organisms. Lichens often live on nitrogen-poor substrates such as sandy soil or bare rock, and mechanisms for dealing with low levels of nitrogen have evolved in many species. Lichens that contain cyanobacteria, either as the primary photobiont or in cephalodia, have a ready source of usable nitrogen because cyanobacteria are able to convert atmospheric nitrogen into biologically usable compounds by a process called nitrogen fixation. The impact of nitrogen-fixing lichens on the ecosystems they inhabit is discussed more fully in Chapter 8.

Green algal lichens, however, must find nitrogen in the natural substrate or from their source of water. Some habitats are especially rich in nitrogen. For example, on dung-enriched bird perches or animal dwellings, or on trees and rocks around farms, abundant nitrogen is available from urea or its breakdown product, ammonia. Some species of lichen are especially adapted to such habitats and are commonly found there (plate 25). These ornithocoprophilous (literally, "bird dung–loving") lichens may also be responding to the high levels of calcium and phosphorus or the high pH of bird dung. Some examples are *Xanthoria candelaria*, *X. elegans*, *Lecanora muralis*, and *Physcia caesia*.

GROWTH

The symbiosis of the lichen is so well balanced that the two components grow at just the right speed to maintain the partnership. If something decreases the growth rate of the fungus but leaves the growth rate of the alga unimpaired, or vice versa, the lichen will break down. Maintaining this balance has its price, however — namely, overall productivity. Compared with most leafy plants or other fungi, lichens grow very slowly.

Lichen growth can be measured in terms of height or weight or as the increase in the radius of a circular patch: radial growth. The latter is the expression of growth most often used in discussing flat, more or less two-dimensional foliose and crustose lichens. The yearly radial growth of crustose lichens is generally in the range of 0.5–2 mm; for foliose lichens, it is usually 0.5–4 mm (plates 26A and 26B). Most fruticose lichens grow about 1.5–10 mm in height or length each year. Some lichens can grow relatively rapidly, however, with a few *Peltigera* species adding more than an inch (27 mm) every year. The fastest-growing lichens may be the lace lichen (*Ramalina menziesii*) and lettuce lichen (*Lobaria oregana*), which can increase their weight by a third each year. These lichens are found in the foggy or rainy West Coast forests, and they grow both at their tips as well as in the older parts of the thallus. Most foliose lichens, by contrast, grow almost entirely at the tips or lobe margins, with the older, more central parts of the thallus becoming inactive and finally dying, sometimes creating concentric rings, seen clearly in *Arctoparmelia centrifuga*. Crustose lichens grow radially at the margins, but many can continuously replace older tissue to create very old, unbroken patches. Some extremely large patches of the map lichen (*Rhizocarpon geographicum*) may be over 4000 years old, estimated by multiplying its growth rate by its radius. (This technique is the basis of lichenometry, discussed in Chapter 10.) Most unusual is the growth of rock tripes (*Umbilicaria* and *Lasallia*). In these genera, growth is patchy throughout the thallus but highest around the central umbilicus, leaving the oldest, often decaying areas at the margins.

PUTTING IT ALL TOGETHER: PHYSIOLOGY AND HABITAT SELECTION

Most lichens do best in localities that have abundant moisture and light, but moderate temperatures (keeping photosynthesis high and respiration low). Such conditions are found in coastal oceanic regions, the canopies of temperate rain forests and montane tropical cloud forests, and some coastal fog deserts, and it is there that lichens are most numerous and diverse. Even in these habitats, however, lichens are largely restricted to substrates that permit them to compete successfully with faster-growing vascular plants and bryophytes.

Because lichens consist mainly of fungal tissue, which respires but does not photosynthesize, they do best at cool latitudes or in mountains where respiration is held in check. Because photosynthesis can occur at low temperatures as long as there is liquid water and light, lichens can be productive all year long if the temperature is above freezing at the substrate surface or in the thallus, or, in some cases, even a few degrees below freezing. The sun can warm up a rock even if the air temperature is relatively low, and the dark pigmentation of many arctic and alpine lichens (like *Bryoria nitidula*, *Pseudephebe minuscula*, and *Lecidea atrobrunnea*) appears to be an adaptation for absorbing heat. Conversely, during the hot summer, when lichens respire away more carbohydrate than they can manufacture, growth is comparatively slow.

When we consider the highly specific physiological requirements and tolerances of the majority of lichens as well as their rates of growth, it is easier to appreciate why certain lichens flourish in some habitats but are absent in others.

A

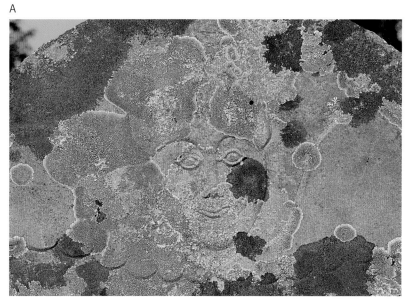

B

26. This gravestone in Cape Cod covered with *Dimelaena oreina* and a species of *Aspicilia* was photographed in 1983 (A) and 1994 (B) and shows the radial growth and decay of the lichens. See Fig. 24 (p. 85) for an analysis of the incremental changes. × 0.42

LICHEN CHEMISTRY

Thou for frozen lands wast meant,
Ere the winter's frost was sent;
And in love he sped thee forth
To thy home, the frozen north,
"Where he bade the rocks produce
Bitter lichens for thy use."
Mary Howitt

LICHEN SUBSTANCES: WHAT THEY ARE

Among the most fascinating aspects of lichen biology is the fact that lichens produce over 600 "secondary" compounds, almost all of them unique to lichens. These are substances not involved directly in the primary metabolism of the lichen. Rather, they represent, in most cases, chemical by-products of the fungal component and are deposited on the outer surface of the hyphae. Although the fungus can manufacture large quantities of some of these substances within the lichen (commonly up to 5 percent of the total weight), the isolated fungus grown apart from its photobiont can produce only minute quantities, if any. A great diversity of compounds is involved — including xanthones, pulvinic acid derivatives, usnic acids, aliphatic ("fatty") acids, and triterpenoids — but by far the most numerous and interesting are weak phenolic acids called depsides and depsidones. They are built from two or three phenolic rings derived from orcinol or ß-orcinol, joined together by ester, ether, or carbon-carbon linkages (Fig. 22). It is beyond our task here to describe the chemical products of lichens in detail, but some aspects of lichen chemistry are important to know.

WHERE THE SUBSTANCES OCCUR, AND WHAT THEY DO FOR THE LICHEN

Lichen substances are not distributed uniformly throughout the thallus and fruiting bodies. Certain substances — including usnic acid, atranorin, and pigments such as xanthones and pulvinic acid derivatives — are found mainly in the thallus cortex, where they seem to function as light screens (protecting the light-sensitive algae) or foul-tasting deterrents to browsing invertebrates. The depsides, depsidones, and fatty acids are most commonly medullary compounds. It is thought that by coating the hyphae of the medulla with an insoluble crystalline material, lichen substances help to repel water and provide air spaces within the lichen. These spaces are critical for the efficient gas exchange needed for photosynthesis.

Certain lichen substances have antibiotic properties, and some can inhibit the growth of soil fungi and even the germination of the seeds of vascular plants. These properties may give the lichens, which are notoriously slow-growing, some competitive advantage in nature.

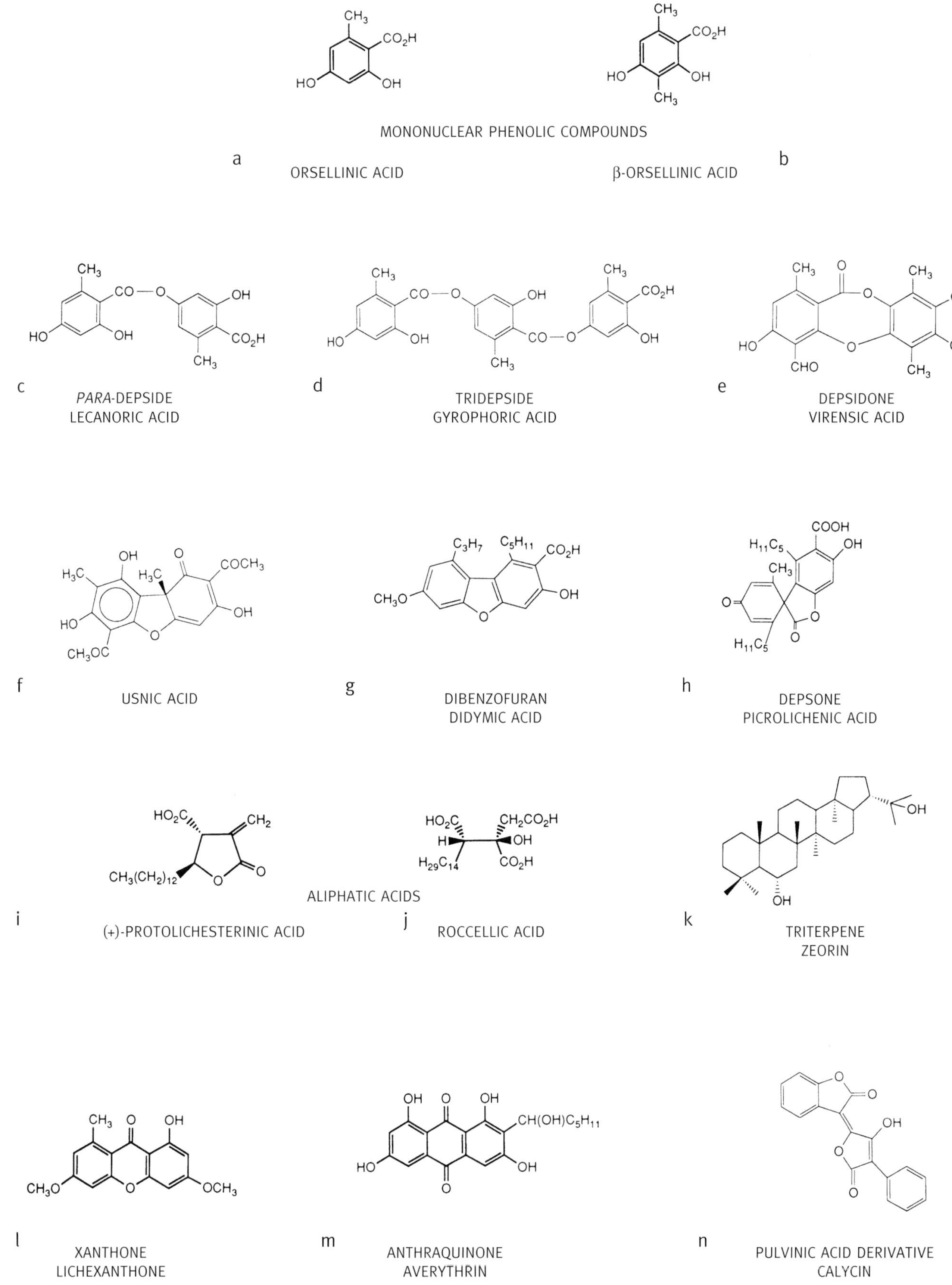

MONONUCLEAR PHENOLIC COMPOUNDS

a ORSELLINIC ACID

b β-ORSELLINIC ACID

c PARA-DEPSIDE
LECANORIC ACID

d TRIDEPSIDE
GYROPHORIC ACID

e DEPSIDONE
VIRENSIC ACID

f USNIC ACID

g DIBENZOFURAN
DIDYMIC ACID

h DEPSONE
PICROLICHENIC ACID

ALIPHATIC ACIDS

i (+)-PROTOLICHESTERINIC ACID

j ROCCELLIC ACID

k TRITERPENE
ZEORIN

l XANTHONE
LICHEXANTHONE

m ANTHRAQUINONE
AVERYTHRIN

n PULVINIC ACID DERIVATIVE
CALYCIN

Fig. 22 Chemical structures of some common lichen compounds: a–b, basic building blocks of many lichen compounds; c–e, main types of medullary compounds, usually giving a positive spot test; f, most common cortical pigment and an effective antibiotic; g–h, less common lichen compounds; i–k, compounds that are negative with PD, K, KC, and C; i–j, common fatty acids; k, the most common triterpene; l, example of a xanthone (some are yellow pigments, most are UV+ yellow to red); m–n, bright yellow, orange, or red pigments, usually in the cortex. (Reprinted by permission from Nash, 1996, *Lichen Biology,* figs. 9.2–9.5.)

USING CHEMISTRY FOR IDENTIFYING AND CLASSIFYING LICHENS

Over a hundred years ago, it was discovered that applying certain common household chemicals such as lye (potassium or sodium hydroxide) or bleach (a solution of calcium or sodium hypochlorite) on various lichens produces color changes that seem to be correlated with morphological differences. In the 1930s, *para*-phenylenediamine was added to the suite of reagents used for making color tests. The substances responsible for most of those color reactions are depsides and depsidones, as well as some organic pigments such as anthraquinones and xanthones.

Because certain lichen substances are characteristic of morphologically recognizable lichens, it is not surprising that the presence or absence of specific compounds was quickly adopted as a taxonomic marker. Morphological characteristics tend to vary depending on the lichen's habitat, age, or health. In such cases, chemistry can be extremely useful in confirming the identity of a specimen for a species whose chemical profile is already known. Chemical products, therefore, provide us with a reliable set of characters to use in conjunction with morphological features for the identification of lichens.

The location of specific compounds in related lichens can also be used as an aid in lichen identification. For example, in certain species of the crustose lichen genus *Ochrolechia,* the depside gyrophoric acid is found in the thallus cortex and epihymenium of the apothecium; in others, it is only in the epihymenium; and in still others, gyrophoric acid is found in both the cortex and medulla as well as in the epihymenium.

The main advantage of using chemistry in the classification of lichens is its predictability. Very often, two populations that appear to be distinguishable only on the basis of their chemistry can later be shown to differ in subtle ways in their morphology or ecology. Examples can be found in *Cladonia, Cetrelia, Lecanora, Ochrolechia,* and many other genera. Furthermore, even higher categories in the classification system such as genera or families can be shown to have their own chemical profiles.

The creation of separate names for chemically distinct but morphologically identical populations of lichens is a subject of continuing debate in lichenology. If it can be shown that a morphologically defined lichen contains one compound in part of its geographic or ecological range and another compound in another part of its range, such "chemical races" or "chemotypes" are commonly given separate names, either at the species level or for designating a subspecies or variety. The more the range of two chemotypes overlaps, the more controversial is the taxonomic recognition of the entities.

Many methods can be used to reveal differences in the chemistry of lichens. In the descriptions and keys found in this book, we frequently refer to color tests or "spot tests," and we indicate the substances responsible for these reactions. These tests, and other methods used to identify the substances in lichens, are described in Chapter 13.

The secondary compounds helpful for identifying lichens are the same ones that make lichens useful to people. They are employed in the manufacture of dyes and antibiotics from lichens and are the basis for the use of lichens as perfumes and poisons. These topics are revisited in Chapter 10.

7

WHERE LICHENS GROW: THEIR SUBSTRATES

Sharing the stillness of the unimpassioned rock, they share also its endurance; and while the winds of departing spring scatter the white hawthorn blossoms like drifted snow, and summer dims on the parched meadow the drooping of its cowslip-gold — far above, among the mountains the silver lichen-spots rest, star-like, on the stone.

John Ruskin

We have already mentioned some aspects of a lichen's substrate, the material upon which it grows. Natural lichen substrates include tree bark, wood, rock, soil, peat, mosses, and other lichens. Lichens, however, can also grow on glass, metal, plastic, and cloth, and they can be found on the shells of living tortoises and the backs of certain insects! Unlike their unlichenized fungal brethren, most lichens derive little from their substrate except a place to grow. There are some interesting exceptions to this general rule, including the lichens that parasitize other lichens, but by and large, lichens do not extract organic nutrients from their substrates. Nevertheless, most lichens are, to a greater or lesser degree, restricted to certain substrate types. Tree lichens are rarely found on rocks; those on limestone are not normally found on granite. Substrate preferences can, in fact, help to characterize individual lichen species.

GENERAL CHARACTERISTICS OF SUBSTRATES

Three features of a substrate are critical in determining whether a particular lichen will find it a happy home: its texture (rough or smooth, stable or unstable); its ability to absorb moisture quickly as well as its ability to retain it; and its chemistry (the organic compounds and minerals it contains, its relative acidity, and its ability to dampen or "buffer" the effects of sudden changes in acidity). Let us look at different types of substrates and see how these three factors vary.

BARK

The bark of one species of tree is not the same as the bark of another with regard to texture, moisture-holding capacity, and chemistry, so it is not surprising that the lichens that live on some kinds of trees never seem to be found on others (plate 27). For example, conifer bark differs from the bark of deciduous trees principally in its chemistry. It contains organic resins and gums normally lacking in hardwoods, has much lower levels of inorganic nutrients (measured as "ash content"), and tends to be quite acidic. In addition, the canopies of many conifers tend to be very dense, allowing very little light to fall on the bark, whereas the canopies of deciduous trees are bare for as much as six months of the year. On conifers such as spruce and fir with branches that slope downward, rain tends to

27. Community of crustose and foliose bark lichens in northern California. (*Lecanora pacifica* surrounded by species of *Physcia*, *Physconia*, *Phaeophyscia*, and *Xanthoria*.) × 2.9

28. (facing page) Brightly colored crustose lichens on slate in the Sierra foothills of central California.

flow directly off the canopy into the soil rather than flowing from small branches to larger branches and finally over the lichens growing on the trunk (carrying nutrients washed from the leaves).

Lichens such as *Imshaugia placorodia* are found almost exclusively on pine bark, which is very acidic, tends to be dry and flakes off easily and often. All bark-dwelling species of hair lichens (*Alectoria* and *Bryoria*) are largely restricted to conifers. Interestingly, two tree types seem to cross over these categories with respect to their lichen flora: birch, which is a deciduous tree, has bark that often supports "conifer lichens"; the northern white cedar (*Thuja occidentalis*), a conifer, is the substrate for lichens more typically found on deciduous trees.

Many lichens, such as *Lecanora allophana*, *Xanthoria fallax*, and certain species of *Caloplaca*, are found most often on the bark of poplars (including aspens and cottonwoods) and elms, which is low in acidity, stable, and fairly absorbent with regard to moisture. Oak, hickory, and linden, with hard, rough, acidic bark, share a distinctive set of lichens. Young maples, which have smooth, hard bark, have a related but slightly different flora. (*Lecanora thysanophora* and *Conotrema urceolatum* are highly specific to sugar maples.) As maples age, the bark becomes softer and more absorbent. Cracks in the bark ooze alkaline, nitrogenous compounds that add to the already nutrient-rich character of the bark, making sugar maples in the east and big-leaf maples in the west extremely favorable habitats for a great diversity of lichens, including lush colonies of *Lobaria* and *Pseudocyphellaria*. Trees with soft, flaky bark, such as white oak and hop hornbeam, support lichens that apparently need a substrate that remains wet after a rain, perhaps because these lichens have high moisture requirements.

A number of lichens, mostly crustose, live only on trees with smooth, living, green bark, such as holly and beech trees, perhaps indicating that they actually derive some nutrition from the green bark cells (although there is still no evidence to prove this). Examples are *Trypethelium virens*, *Pyrenula pseudobufonia*, and *Graphis afzelii*.

The physical relations between lichens and the bark substrate differ from one growth form to another. The rhizines of foliose lichens remain superficial, being barely attached to the outermost cells of the bark. The thalli of crustose lichens such as *Lecanora allophana* partly invade the outer layers of weathered bark in addition to forming a relatively thick layer on its surface. The thalli of *Trypethelium virens*, *Astrothelium versicolor*, and many other tropical and subtropical crustose lichens, however, develop entirely *under* the outermost layers of bark and are therefore *within* the bark rather than on it. Light penetrates through the thin, transparent outer bark cells and allows the lichen's photobiont to photosynthesize. Such lichens are called endophloeodal (*endo* = within; *phloe* = inner bark). (See Fig. 9.)

WOOD

Bark and dead wood (or lignum) are related substrates, but lichens are typically restricted to one or the other. Lichens that can live on both are usually found on the wood of trees whose bark they frequent, almost certainly because of similarities in substrate chemistry. (*Parmeliopsis ambigua* and *Letharia vulpina* grow on both the bark and wood of conifers.) Not only is the origin of the wood important, the degree of decay is also critical because it affects the wood's moisture-retaining ability as well as its chemistry. For example, *Icmadophila ericetorum* always occurs on soft, well-rotted logs or stumps (or peat), but *Xylographa* species grow almost exclusively on hard, weathered wood (plate 29).

MOSSES AND DEAD VEGETATION

Although most lichens grow more slowly than bryophytes (mosses and liverworts), some species can grow

29. (right) Wood-dwelling group of lichens in coastal California dominated by *Thelomma californicum* and species of *Usnea, Xanthoria,* and *Hypogymnia.*

30. (far right) The white band of rock above the water level of this reservoir in Yosemite National Park is where the lichens *are not;* the granite mountains above are covered with a mantle of gray lichens.

31. (facing page) In a boreal woodland north of Quebec City (Les Grands Jardins), granitic boulders are totally covered with an array of lichens, including species of *Stereocaulon, Arctoparmelia,* and *Aspicilia,* as well as some dark patches of moss.

over them, especially when the bryophyte is a bit sick. *Bacidia sabuletorum* is such a lichen, although it can also grow on moss-covered bark or even rock. The minute fruticose lichen *Polychidium muscicola,* as its name implies (*musc* = moss; *cola* = dweller), also develops in clumps of moss, not so much covering the bryophyte as growing among its branches. Moss-inhabiting lichens like *Ochrolechia upsaliensis* are also frequently found growing over decaying clumps of flowering plants, especially heaths, in arctic or alpine tundra communities, or growing over the fern ally, spike moss (*Selaginella*), in dry regions.

ROCK

Different rocks and minerals differ so widely in chemical composition, texture, and even water-holding capacity that most rock-inhabiting lichens have become adapted to living on specific rock types (plate 28). The most significant property of a potential rock substrate, from a lichen's point of view, is its calcium carbonate (lime) content. Rocks rich in this mineral are said to be calcareous (*calcis* = lime). The lichens that inhabit calcareous rocks such as limestone and marble comprise communities that are very different from those growing on rocks low in calcium carbonate but rich in silicates (called siliceous) such as granites or schists. For example, *Placynthium nigrum, Lecanora dispersa,* and *Caloplaca feracissima* are characteristic invaders of limestone as well as of mortar and concrete, which resemble limestone chemically. Lichens restricted to siliceous rocks include most species of *Umbilicaria, Xanthoparmelia,* and *Pseudephebe,* in some cases giving entire mountain ranges their color (plate 30). Granitic boulders can be so totally covered by lichens that the rocks themselves are obscured (plate 31). Rocks with especially high concentrations of metals such as iron and magnesium have their own special lichen flora, including species such as *Stereocaulon pileatum.* Many of these species can deposit iron oxides in their cortex, giving them a rusty color (see *Tremolecia atrata*).

Sandstones have their own particular characteristics. They often have the ability to retain moisture for relatively long periods of time, and their chemistry can vary depending on the material binding the rock particles together. Calcareous sandstone may have a lichen flora much like that of limestone; calcium-poor sandstones support lichens found on siliceous rock.

Although rock is, almost by definition, hard and impenetrable, it is astounding how deeply lichens can penetrate into stone. The hyphae of lichens on coarse-grained granite can grow several millimeters into the rock, something easily seen if you pry up the rock crystals under lichens such as *Polysporina* and *Porpidia* (plate 32). Even lichens growing on glass-like obsidian can take advantage of the slightest cleavage planes and lift up flakes of the mineral, often incorporating them into the lichen thallus. On limestone, which can be dissolved by the carbonic acid produced by the lichen's normal metabolism, the hyphae of some lichens have been shown to extend into the rock up to 16 millimeters! Limestone-dwelling species of *Verrucaria* such as *V. calciseda* dissolve out small depressions under each perithecium, leaving a pitted surface on the rock after the perithecia disintegrate. This ability of lichens to grow between mineral crystals, loosening the particles as the hyphae swell and dry or freeze and thaw, is of obvious relevance with regard to rock breakdown, soil formation, and glass etching (see "The Down Side of Lichen Growth," in Chapter 10). Lichens that grow entirely or largely within the rock substrate, often revealing only the fruiting bodies at the surface, are called endolithic (*endo* = within; *lithos* = stone).

SOIL

Soil, like rock, can be either calcareous or siliceous depending on the chemistry of its parent material, and this influences the lichen flora. *Squamarina lentigera*, *Solorina saccata*, and *Fulgensia bracteata*, for example, are found only on calcareous soils. The texture and stability are obviously also of great importance. Many lichens are well adapted to live on friable, unstable soils such as sand or eroding clay. Examples include *Placynthiella oligotropha* and the pink earth lichen (*Dibaeis baeomyces*) as well as many species of *Cladonia*. Others, like *Stereocaulon tomentosum* and species of earthscales (*Peltula*, *Psora*, *Placidium*, and *Catapyrenium*), need relatively stable, well-packed surfaces (plate 33).

Soil lichens produce bundles of hyphae that grow among the soil particles, binding them together as they penetrate the soil. Lichens, in fact, are important soil stabilizers and make significant contributions to the fertility of the soil in the form of organic matter and fixed nitrogen. Light-colored lichen crusts reflect heat, helping the soil to remain cool and moist. The destruction of soil lichens by overgrazing livestock has

32. White fungal hyphae and the green photobiont of an endolithic crustose lichen can be seen within the upper layers of this section of sandstone. On the rock surface, only the large black apothecia are visible. × 23

33. *Psora decipiens* and a species of *Xanthoparmelia* growing on bare soil in southern Idaho. × 1.4

34. Vagrant lichens including species of *Rhizoplaca* and *Xanthoparmelia* form a diverse soil community in a high valley of southeastern Idaho.

35. *Cladonia grayi* has broad substrate tolerances. This patch is growing luxuriently on a discarded shirt in Martha's Vineyard, Massachusetts. × 2.9

become a serious concern in the West (see "Human Impacts on Lichens," in Chapter 10).

"Vagrant" lichens are specially adapted to living on flat, open areas of bare gravelly or sandy soil in arid or semiarid regions (plate 34). They are not attached to any fixed substrate, often forming relatively tight balls that can roll over the ground. There are vagrant foliose lichens such as *Xanthoparmelia chlorochroa* and vagrant forms of umbilicate lichens including *Dermatocarpon miniatum* and *Rhizoplaca melanophthalma*; squamulose forms of *Cladonia* species like *C. perforata*, *C. prostrata*, and *C. strepsilis*; and fruticose species like *Aspicilia hispida*. Crustose lichens such as *Aspicilia contorta* become vagrant by completely overgrowing small pebbles, which can then roll from place to place in a strong wind or heavy rain.

LEAVES

In the tropics and subtropics, there are lichens that grow on the older leaves of palms and broad-leaved shrubs. A few species in the northern conifer forests can use evergreen needles as a substrate. Leaf-inhabiting lichens are either entirely superficial or grow just below the waxy surface layer of the leaf (the cuticle).

36. Lichens never grow on washed cars, but if a car remains dirty for some years, anything is possible.

37. Bones and antlers are favorite substrates for many lichens, especially species of *Xanthoria*.

They rarely do the leaf much harm. Many leaf lichens have peculiar lobed or helmet-shaped structures, called campylidia, that produce conidia. These can develop on lichens with or without ascomata. An example of a leaf-inhabiting lichen is *Calopadia fusca*.

ARTIFACTS AS SUBSTRATES

The artificial substrates that lichens colonize generally resemble their natural substrates in the essentials of texture, chemistry, and water-retaining properties. Thus, lichens that grow on glass are species associated with hard, siliceous rocks; those on cloth are lichens normally found on soil or dead vegetation; species on leather usually inhabit wood; those on metal or plastic are generally rock lichens or lichens that are not choosy about their substrates (plates 35, 36).

ANIMALS

Although animals and lichens interact in numerous important ways, the idea of an animal as a *substrate* is a bit difficult to comprehend. Lichens have long been known to grow on the decaying horns or bones of mammals (plate 37). Most are species normally found on hard wood or limestone, substrates that resemble bone in texture and chemistry, respectively. Those growing on human skulls had special significance in early medicine (see "Medicines and Poisons," in Chapter 10).

But can lichens develop on *living* animals? Not only can certain animals act as lichen substrates, some, especially insects, are specially adapted for this role. The larvae of the lacewing (*Notida pavida*), for example, excrete a sticky silk on their backs that can "catch" bits of lichen. They then carry the soredia and lichen fragments in packets on their backs (plate 38). When stationary, the larva looks like a bit of lichen. The lichens provide camouflage for the insect, while the insect serves as a dispersal mechanism for the lichens — a kind of double-barreled symbiosis. Although we have no North American examples of lichens on adult insects, several species of weevils in New Guinea have roughened, scaly, and often sticky backs, an adaptation that promotes the growth of tiny foliose lichens, which then serve to camouflage the insect.

A few species of *Pyrenocollema* (not included in this book) live exclusively on the shells of barnacles in the intertidal zones of both the east and west coasts. The only lichen known to use large land animals as a substrate is *Dirinaria picta*, a foliose bark lichen common throughout the tropics including the southeastern United States. This species grows luxuriantly on the carapace of the Galápagos land tortoise, *Geochelone elephantopus*—not quite within our study area, but worth the mention nevertheless.

38. The leprose lichens covering the larva of the lacewing, *Notida pavida*, blend with nearby bark lichens, providing the insect with an effective camouflage. × 3.1

LICHENS AND ECOSYSTEMS

I flushed a Ledge Locust. . . . His mottled wing covers matched the ash-gray of weathered rock as well as the greenish-gray of some lichens and the black of others.
Vincent G. Dethier

Lichens play a significant role in nature almost everywhere they occur. They form the dominant vegetation over about 8 percent of the earth's terrestrial surface, fundamentally influencing the growth and development of other plants and animals sharing the same environment.

NATURE'S PIONEERS

Lichens have been nicknamed "nature's pioneers" because they have the ability to colonize bare rock and are often the first plant-like forms to become established on newly exposed surfaces. Although they share this ability with many kinds of algae, cyanobacteria, and bryophytes, lichens constitute the most abundant and diverse group of primary invaders of rock.

Colonization of Rock: Succession and Soil Formation
We have already described how the fungal hyphae of crustose lichens can grow several millimeters into a rock, developing between the rock crystals and fragments and along cleavage lines, and even etching pits into the rock material itself. We know that certain lichen substances (especially depsides and depsidones) can chemically combine with rock minerals, creating metal complexes that make the rock slightly more soluble, thus speeding the weathering process caused by freezing and thawing, heating, and natural chemical alterations (plate 39). This weathered rock is the first stage of soil formation. It should be emphasized that the process of soil formation is extremely slow, more often measured in centuries than in decades. Lichen substances such as depsides and depsidones are extremely weak acids and are only slightly soluble in water. On the other hand, carbonic acid, an organic acid formed by the lichen's ordinary metabolic activity, is relatively more powerful and actually can eat away at rocks rich in calcium carbonate such as limestone.

The tissues of the crustose, foliose, and fruticose lichens that grow over smooth, inhospitable rock surfaces, however, constitute the most important contribution lichens make to soil formation, although certain hardy mosses also have this ability. When these lichens and bryophytes die, their decayed organic material is added to the minerals of the rock. Even more important, they intercept and trap dust and silt blowing across the rock surface, causing a more rapid accumulation of fine-grained material. Eventually, the spores of other mosses and seeds of hardy vascular plants can begin developing on and among the clumps of lichens and in pockets of soil. A typical succession over siliceous rock in eastern North America might involve an initial colonization by crustose species like *Sarcogyne, Polysporina, Acarospora, Porpidia, Rhizocarpon,* and *Lecidea,* foliose lichens such as *Xanthoparmelia,* and fruticose lichens including *Stereocaulon,* as well as mosses such as *Andreaea* and *Grimmia.* These are followed (or accompanied) by *Phaeophyscia* and *Parmelia* species, which are finally overtaken by *Cladonia* and *Cladina,* as well as grasses, hairy-cap mosses, and herbs (plate 40).

Lichens can function as pioneers because of several quirks of their biology: they can withstand long periods of drought (for example, 62 weeks in the case of some rock tripes!); they are self-sufficient, taking what few minerals they require from the ambient dust and from whatever is dissolved in the moisture they receive; they contain their own suppliers of carbohydrates, the sugar-producing photobionts; and their propagules are extremely tiny, enabling them to become established on all but the smoothest surfaces.

Many pioneer lichens have one more ecological advantage. Lichens such as *Placopsis gelida, Stereocaulon vesuvianum,* and *Pilophorus acicularis* have cephalodia, which contain cyanobacteria and can be thought of as nitrogen-fixing factories. Thus, on nitrogen-poor substrates such as newly exposed rock surfaces, these lichens provide their own source of nitrogen and are in an excellent position to become established.

Colonization and Stabilization of Soil
On soil, lichens are typically poor competitors against grasses, herbaceous plants, or even mosses, which grow much faster. There are, however, vast areas such as deserts and tundra where lichens thrive on soil under conditions that are inhospitable to most flowering plants. Some lichens are important colonizers of disturbed or sandy soil. In such habitats, they can flood an area with propagules such as soredia or fragments of squamules and, even at a growth rate of only a few millimeters per year, can quickly cover bare and even eroding soil with large numbers of individual thalli. Examples are *Dibaeis baeomyces,* species of *Collema, Trapeliopsis granulosa,* and *Peltigera rufescens* over disturbed soil banks in the east; *Cladonia cristatella, C. polycarpoides,* and other small *Cladonia* species over abandoned eastern farmland; and *Placynthiella oligotropha* over sand (plate 41).

The important role of lichens in consolidating and stabilizing soil, and providing organic matter and ni-

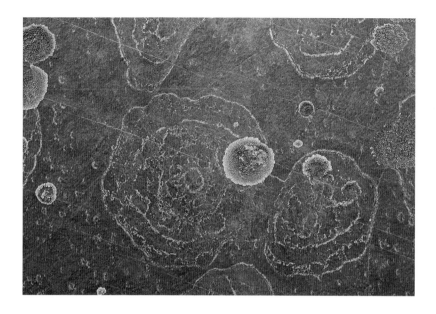

39. Ghostly patterns of lichen patches that had been scraped off this gravestone years before attest to the rock-altering power of some lichens. × 0.45

40. In this oak glade in the Ozark Mountains of Arkansas, one can trace the gradual invasion of lichens and mosses over bare sandstone rock. The succession begins with crustose and foliose lichens, and is followed by mosses and finally reindeer lichens.

41. Five species of *Cladonia* and two of *Cladina* are crowded into a few square centimeters of this sandy field on the north shore of Lake Superior. × 1.0

42. Soil crusts in Utah receive close attention from research ecologist Jayne Belnap of the U.S. Geological Survey.

trogen, has already been mentioned. A soil crust made up of, to a large extent, squamulose and crustose lichens such as *Placidium* and *Psora*, and of nitrogen-fixing lichens such as *Collema tenax* and *Peltula* (see comments below), is of particular importance over rangeland, especially in the Great Basin and arid parts of the prairies, where soil quality tends to be poor (plate 42).

The ability of certain lichens to change the reflectivity of the soil surface (from a heat-absorbing, drought-inducing dark brown to a reflective pale gray) is acknowledged by the eastern forest industry. In Quebec, Canada, for example, species such as *Trapeliopsis granulosa* have been disseminated over large areas of recently burned terrain in an effort to improve the soil and increase the success of new tree plantings.

The boreal forest covers a large part of the northern hemisphere and is the source of a significant amount of our water and wood. There is no doubt but that lichens have a profound effect on the functioning of this ecosystem. Where lichen cover reaches 50 to 90 percent as in the northern boreal forest region, with reindeer lichens and species of *Stereocaulon* forming an almost seamless blanket (plate 43), lichens become critical in preserving soil moisture, reflecting heat, adding organic matter, and trapping seeds. Lichens also influence the ecological dynamics of the boreal forest by intercepting significant amounts of rainfall, thereby affecting soil moisture and the rate of runoff into nearby streams and rivers. On a negative note, some lichen substances can inhibit the development of beneficial soil fungi and lower the germination rate of certain types of seeds.

The boreal lichen woodlands are frequently subjected to fire, which is a major factor in permitting lichens to maintain their dominance in that region. Lichen succession after fire takes between 80 and 120 years to progress from bare, charred soil, through stages dominated by crusts such as *Placynthiella* and then cupped and stalked cladonias, to a mature mat of reindeer lichens.

Colonization of Trees
Plants that live on other plants are known as epiphytes. Although lichens are not, strictly speaking, plants, those that grow on plants are also called epiphytes, together with certain bryophytes, ferns, and flowering plants (such as bromeliads). Outside the tropics and coastal rain forests, lichens are the most important epiphytes of forest trees and shrubs. Even in the tropics and rainy coastal areas where lichens must compete with larger, faster-growing epiphytes including bryophytes, lichens are numerous and conspicuous (plate 44).

Tree bark, generally speaking, is a dry, hard, and seemingly forbidding substrate, but when you consider that lichens are prime colonizers of rock, it is not surprising that lichens do well on bark — for all the same reasons. On bark, however, the succession of species is not the classic "crustose followed by foliose followed by fruticose" pattern regarded by some as typical of colonization on rock. Faster growing foliose or fruticose lichens are often the first to appear on young twigs and young bark, followed later by crustose lichens. As the bark ages, its texture, moisture-holding properties, and even chemistry change, and these changes cause modifications in the lichen communities that inhabit it. Thus, the succession of species of lichens over bark is caused by both the interactions of the newly introduced elements of the lichen community as well as by changes in the bark itself.

The lichens that inhabit forest trees affect several aspects of the forest habitat. They absorb significant amounts of nutrients from rainwater that passes through the canopy over the leaves and differentially absorb minerals flowing down the trunk. The huge amounts of lace lichen (*Ramalina menziesii*) in Coast Range valleys of California have been shown to intercept significant amounts of nutrients dissolved in the ambient fog, thereby influencing the composition and concentration of nutrients in the soil below the trees. By absorbing and then releasing water after a rain, lichens can also influence the level of humidity within a forest ecosystem. Most interesting and potentially important, however, is the lichens' contribution of fixed nitrogen to ecosystems (discussed below).

But are there any *negative* effects of lichens on trees? Can lichens damage the tree that serves as its substrate? Fruit growers are sometimes worried about the "growths" on trees in their orchards, especially because lichens tend to be more abundant on dead or dying trees than on healthy ones.

There have been unconfirmed reports of lichen substances entering the vascular system of trees and harming them, but this is extremely unlikely because these substances are almost insoluble in water. It has recently been suggested that *Evernia prunastri* (oakmoss) may damage a tree by growing deep into its living wood tissue, but even if this observation is proven to be accurate, the situation would clearly be an exception. The most likely reason for the abundance of lichens on sick or dead trees is that the loss of leaves opens the canopy, allowing more light to reach the bark, thereby stimulating more rapid growth of lichens. Furthermore, the bark of dead standing trees is

43. A mat of *Cladina stellaris* (over 20 cm thick in places) breaks into irregular polygons when dry, blanketing the ground in this northern boreal woodland in Labrador. (Photo by I. M. Brodo.)

44. Epiphytic species of *Usnea, Ramalina,* and *Evernia* cover deciduous trees at the edge of a forest in the Willamette Valley of Oregon.

a more stable substrate because it is sloughed off at a slower rate than it is from living trees.

It is true that a thick growth of foliose or fruticose lichens can harbor insects. In most cases, these insects are harmless, but in eastern Canada a moth with a very destructive caterpillar called the hemlock looper (*Lambdina fiscellaria*) lays its eggs almost exclusively on hair lichens such as *Bryoria trichodes*. This moth can defoliate large tracts of fir or hemlock, and the lichen is critical in its life cycle.

CONTRIBUTION OF LICHENS TO THE NITROGEN CYCLE

Anyone who has tried to grow anything — from house plants to vegetables to lawn grass — knows that nitrogen, in a form that can be used by plants, tends to be in short supply and yet is essential for good growth. It is the critical building block of proteins, the very stuff of life. The air we breathe has abundant nitrogen, but nitrogen does not occur naturally in native minerals. The nitrogen of the air must be converted into nitrates or ammonium compounds (a process called nitrogen fixation) before it can contribute to plant growth or, indeed, to the growth of animals, which ultimately depend on plants. This critical step can be accomplished only by a small group of organisms, mainly certain bacteria and cyanobacteria.

We have mentioned before that because many lichens contain cyanobacteria as the primary photobiont or in cephalodia, they can provide themselves with sufficient nitrogen compounds to survive. This nitrogen can become available to plants in the immediate area when the lichens die and decay, or when nitrogen compounds leach from living lichens. Although the amounts of these contributions are still uncertain, it is clear that they are significant in certain ecosystems such as old growth, conifer forests of the Pacific Northwest, where there is an abundance of cyanobacterial lichens such as *Lobaria, Peltigera, Pseudocyphellaria,* and *Nephroma* (plate 45). Lichen contributions of up to 50 percent of the total nitrogen input have been reported, but follow-up studies to confirm these figures still have to be made. Soil-dwelling species of *Stereocaulon, Collema,* and *Peltula* add sizable amounts of nitrogen to the soil in regions where these lichens form an important component of the ground cover. Studies in the Great Basin still need to be conducted to ascertain the precise levels of this contribution; it is not certain how much of the nitrogen

fixed by the cyanobacteria of the lichen crusts remains in the soil before being broken down by denitrifying bacteria (performing the opposite reaction to nitrogen fixation).

ANIMALS AND LICHENS

Many kinds of animals, from mites to musk oxen, use lichens for either food or shelter. In some cases the importance of lichens for the animal is minor, but in others it is critical, indicating once again how intricately interdependent the elements of an ecosystem are.

Lichens as Food
Perhaps the most widely known fact about lichens is that they are eaten by reindeer (hence, the popular but misleading common name, "reindeer moss," for some lichens). Lichens make up about 90 percent of the winter diet of caribou and their European counterparts, reindeer. In addition, many other ungulates including mule deer, black-tailed and white-tailed deer, mountain goats, and moose can sometimes depend on arboreal lichens for food. There is some evidence that pronghorn antelope, a species of the semiarid plains, eats substantial quantities of terrestrial lichens in winter. Fifty-one percent of the stomach contents of one road-killed pronghorn in Idaho consisted of lichens, mostly vagrant species of *Rhizoplaca* and *Xanthoparmelia* (see plate 34).

Most lichens are high in carbohydrates but low in protein. Although they can provide high-energy food all winter to ungulates (with a rumen and a bacterial flora to break down the lichen carbohydrate), these animals, however, still need to browse on grasses and shrubs for a protein supplement. Nevertheless, even during the summer, lichens constitute about 50 percent of the diet of caribou. Surprisingly, large, nitrogen-fixing, foliose lichens that are comparatively rich in protein, such as species of *Lobaria* and *Peltigera*, are generally not eaten. Mountain goats in Alaska, in fact, seem to be the only ungulates that eat substantial quantities of *Lobaria*.

Caribou are lichen specialists, and most herds cannot survive without them (plate 46). A small herd introduced to St. Matthew Island in the Bering Sea increased to about 6000 animals in 20 years, but at that point, almost all the lichens had been eaten. As a result, during the winter of 1963, all but 50 caribou perished from starvation.

In winter, caribou feed on terrestrial lichens such as

45. *Lobaria oregana* grows in profusion on trees in the Columbia River watershed of western Washington, contributing to the nitrogen input of forest ecosystems in the area.

46. These barren-land caribou in Old Crow, Yukon Territory, are digging holes or "craters" in the snow to reach the caribou lichens on the ground. Two caribou have lifted their heads with their ears back, preparing to rear up and fight over a feeding hole. (Photo by George Calef, from *Caribou and the Barren-lands*, 1981, Firefly Books, Ltd., and the Canadian Arctic Resources Committee, Toronto, p. 141.)

47. (right) *Alectoria sarmentosa* and *Bryoria* growing in abundance on a Douglas fir trunk is a common sight in the Cascades in Washington and Oregon.

48. (below) Mule deer in coastal California have created a distinct browse line by eating the lichens from the branches of oaks.

Cladina stellaris and *C. rangiferina* (reindeer lichens), *Stereocaulon,* and *Masonhalea* by kicking holes called "craters" through the snow. In wooded areas, they feed on tree lichens such as *Alectoria* and *Bryoria* species and, to a much lesser extent, on foliose lichens such as *Hypogymnia physodes, Tuckermannopsis ciliaris,* and *Lobaria pulmonaria*. This is especially true of woodland caribou, the subspecies that lives in the boreal forest. Deer, elk, and moose also prefer arboreal lichens, not having developed the cratering technique for reaching terrestrial species. In winters with heavy snow and depleted browse, lichens blown down from tree branches (litterfall) are of great importance for the survival of western ungulates. Wildlife managers sometimes even cut down lichen-laden trees to provide easy access to *Alectoria* and *Bryoria* as food for starving deer (plates 47 and 48).

Some small mammals also eat lichens in large quantities. Species of *Bryoria* and *Usnea* can constitute up to 93 percent of the winter and spring diet of northern flying squirrels in eastern Oregon and adjacent Idaho, and 80 percent of its spring and summer food in the mountains of West Virginia. *Bryoria* species and reindeer lichens are important for rodents such as the California and boreal red-backed voles. These animals are, in turn, the main food source for many other animals and birds such as the boreal owl.

Many tiny terrestrial arthropods such as mites, springtails, bark lice, and silverfish, as well as slugs and snails, are enthusiastic consumers of lichens. Because these organisms are basic items in the food web of many terrestrial ecosystems, lichens can sometimes figure prominently as primary producers (plate 49).

Although birds seldom eat lichens, they by no means ignore them, as we see in the following section.

Lichens as Nesting Material
Birds are well known for their resourcefulness in using appropriate material in constructing their nests. Fibrous materials are particularly favored, so it is not surprising that a large variety of pendent fruticose lichens have been found incorporated into the nests of close to 50 species of birds, including mergansers, hawks, chickadees, thrushes, and a large number of warblers. In the northern part of its range, the northern parula warbler is closely associated with lichens, making its nest as a hollow in a clump of *Usnea* with hardly any additional nest lining. Where *Usnea* is scarce, the bird uses a hanging clump of hemlock or spruce twigs and lines the hollow with bits of *Usnea*.

Some western birds use the lace lichen, *Ramalina menziesii*, in much the same way. Gnatcatchers, flycatchers, and hummingbirds carefully attach lobes of foliose lichens such as *Parmelia sulcata* to the outside of their nests, providing effective camouflage.

The northern flying squirrel, which makes its home in the cavities of snags, builds its nests largely with *Bryoria* species in habitats where *Bryoria* occurs (plate 50). Because the squirrels also eat these lichens, they can munch on their homes whenever they are hungry.

Lichens as Camouflage

In many small animals that live in lichen-covered habitats, mechanisms have evolved that allow them to blend into the lichenous background. The most famous is the English peppered moth, which, in lichen-rich habitats, is most abundant in its pale, mottled color phase, closely resembling the lichen-covered bark. Where the lichens have been killed by air pollution and the tree bark is blackened by soot, the moth occurs primarily in its uniformly smoky-gray phase. Several kinds of tree frog (*Hyla versicolor* and *H. avivoca*) are very effectively camouflaged on lichen-covered trees, and the green salamander (*Aneides aeneus*) blends similarly with the rocks it frequents.

A more direct stratagem has been adopted by a number of insects that are specially adapted to cover their backs with living lichens, completely obscuring them to potential predators. The lacewing larva, described in Chapter 7, is a good example (see plate 38).

49. A snail or slug has chewed its way across this *Pertusaria*, leaving characteristic rasp marks in the lichen. × 3.0

50. Researchers in Idaho examine the *Bryoria* nest of a northern flying squirrel.

LICHENS AND ECOSYSTEMS

GEOGRAPHIC DISTRIBUTIONS IN NORTH AMERICA

Where no tree nor grass can grow,
On a far northern hill,
This humble lichen brought to me,
Doth their place in Nature fill.
Jones Very

Virtually every organism is found in some parts of the world and not in others. The geographic distribution of a species can be extremely broad, including several continents, or it can be limited to a single locality. In this chapter, we examine the geographic distribution of lichens in North America. If you flip through the distribution maps presented with the species descriptions, you will notice that certain distribution patterns recur in only slightly modified form in many species. These recurring patterns can be used as a clue to the likelihood of finding a species in an area from which it is not yet known.

HOW HAVE DISTRIBUTION PATTERNS ARISEN?

For many reasons, species of plants and animals, as well as lichens, are not ubiquitous. First, consider their *biological* requirements and tolerances: certain species are adapted to live in cold climates, others in the tropics; some need to be periodically submerged in salt water, others cannot even tolerate occasional salt spray; many can grow on only one or two species of trees. That is why there are arctic and tropical lichens, maritime species, and species that follow the range of a particular tree.

How, then, do we explain the fact that every tropical lichen species is not found in every tropical region or that every birch lichen is not found everywhere birch occurs? The reasons are usually *historical,* involving the geologic and climatic history of the earth. For example, although most parts of the North American continent were covered by ancient seas or continental glaciers at some time or other over the past 100 million years, the Appalachian region has been available for colonization over that entire period. During the ice ages, many lichen distributions that once were continuous became broken up by expanding continental glaciers. As conditions changed, organisms migrated southward in front of the advancing ice. Only some of them were able to reestablish themselves northward as the ice retreated. Many lichens became stranded as relicts on isolated mountain peaks in southern regions, permanently separated from the main population that closely followed the retreating ice front. Some lichens survived the last ice age in parts of the northern range that were never covered by ice. Such areas are called refugia. Examples include the Alaskan-Yukon corridor and part of Banks Island.

The separation of large continental plates from what once was a huge supercontinent (continental drift) took place between 100 and 200 million years ago, mainly during the Jurassic and Cretaceous periods, after most genera of lichens had evolved. This accounts for the scattered distributions of many lichens. Populations of a species separated by long distances are said to have disjunct distributions, and the separated populations are often referred to as disjuncts. Many disjunctions are found in the temperate zone as a result of the breakup of what was a continuous, deciduous forest during the Tertiary epoch, about 50 million years ago. (This is described later in this chapter, in the section entitled "Temperate Element.")

Geographic factors also help determine a specific pattern of distribution. The existence of a continuous coastal plain, mountain range, or system of deserts that produces corridors along which taxa can easily migrate, and the combination of geological features and local weather patterns that create areas of high or low rainfall or temperature are all important in this regard.

Very limited geographic ranges can result from the evolution of new species that lack either the reproductive means or the time to become more widespread. Characteristic adaptations and genetic uniqueness can slowly evolve in isolated populations. Gradually, the populations diverge from their parent species in morphology, chemistry, or physiology, responding to new selective forces in the isolated regions. These populations can eventually become sufficiently distinct to justify their being recognized as new species. Often, recently evolved species remain endemic to limited areas. We can speak of a species being endemic to North America on a global scale, endemic to the Pacific Northwest or central grasslands on a continental scale, or even endemic to a particular mountain peak on a regional scale.

CLASSIFYING PATTERNS OF DISTRIBUTION

All lichens have certain climatic tolerances and requirements, so it is logical to organize the basic distribution types according to major climatic regimes. Within these climatic categories, geographic features are most important in creating basic distribution patterns, although the biological requirements of each taxon contribute much to the configuration of each lichen's specific range. Species that share a common history and basic climatic requirements and tolerances have patterns of distribution that are roughly congruent. These species, collectively, constitute certain floristic *Elements*. To better understand the factors that have shaped the North American lichen flora, it is useful to relate distribution patterns with geologic history and view them in a worldwide context, and so this also figures into the organization of distributional categories. Here, we cover only the major Elements (based mainly on climate), and then discuss the various distribution types found within them (table 1; Fig. 23).

One must bear in mind that individual species may belong to more than one Element as a result of broad ecological or physiological tolerances or special historical circumstances, and this results in combined distribution patterns. A lichen with its main distribution in the spruce-fir forests of the boreal region may also be abundant in the more temperate, but mainly coniferous, Pacific Northwest. Species with disjunct distributions commonly have patterns of distribution including several distribution types (e.g., *Alectoria sarmentosa* and *Pertusaria amara*). This will become clearer as we describe the individual patterns and give examples of each. The examples we give are those lichens (almost all of them included in this book) that best demonstrate the core pattern — they are not necessarily common or conspicuous species.

Table 1 Distribution Types of North American Lichens

1. Arctic-Alpine Element
    a. Widespread
    b. Beringian
2. Boreal Element
3. Temperate Element
    a. Pan-Temperate
    b. Eastern Sector
        i. East Temperate (widespread)
        ii. Appalachian–Great Lakes
        iii. Coastal Plain
    c. Central Grasslands
    d. Western Sector
        i. West Temperate (widespread)
        ii. Interior Basin
        iii. Pacific Northwest
4. Western Montane Element
    a. Rockies
    b. Cascades
5. Madrean Element
    a. Californian
    b. Southwestern Desert
6. Tropical Element
7. Oceanic Element
8. Maritime Element

Fig. 23 Distribution patterns of lichens in North America. The various distribution types are labeled on the individual maps (a–r).

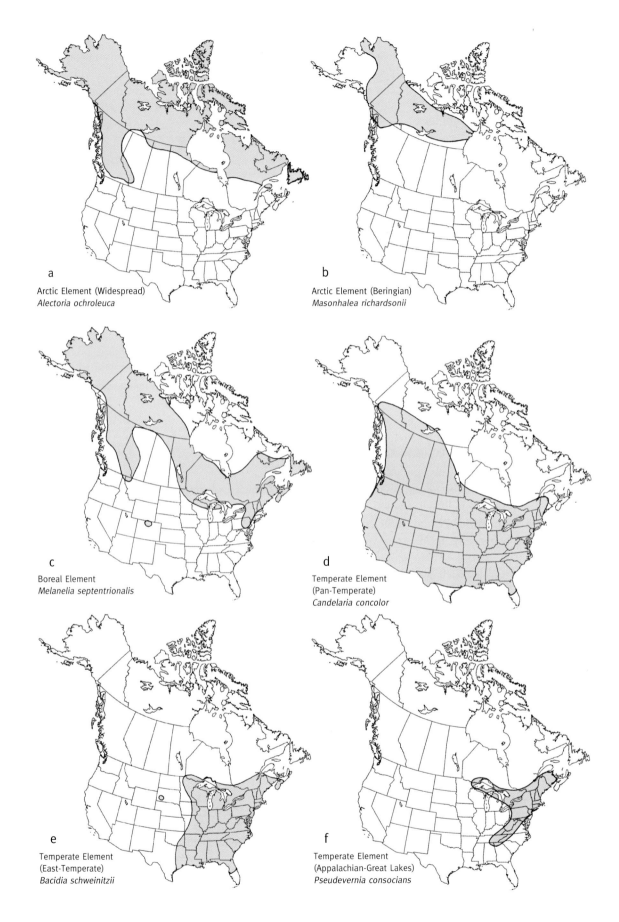

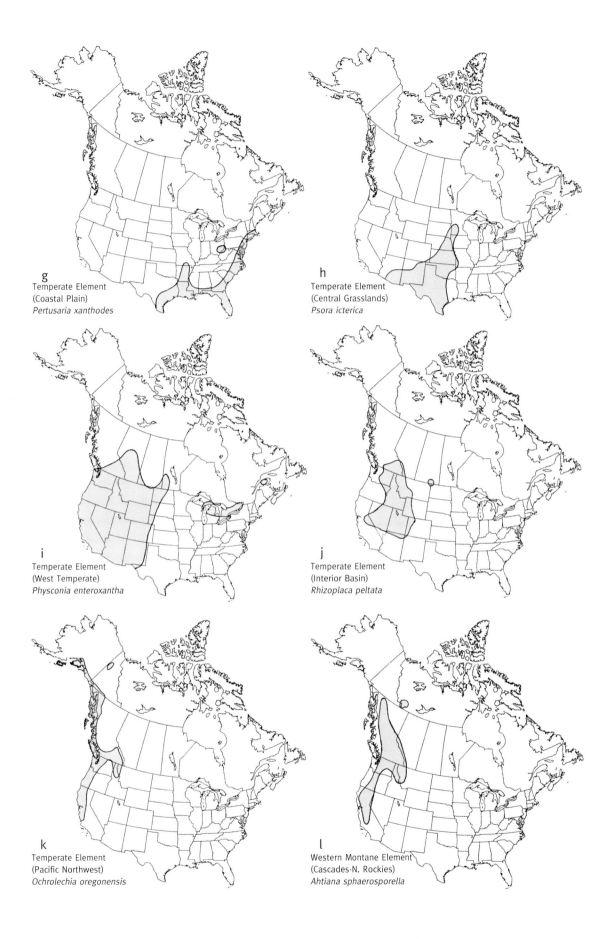

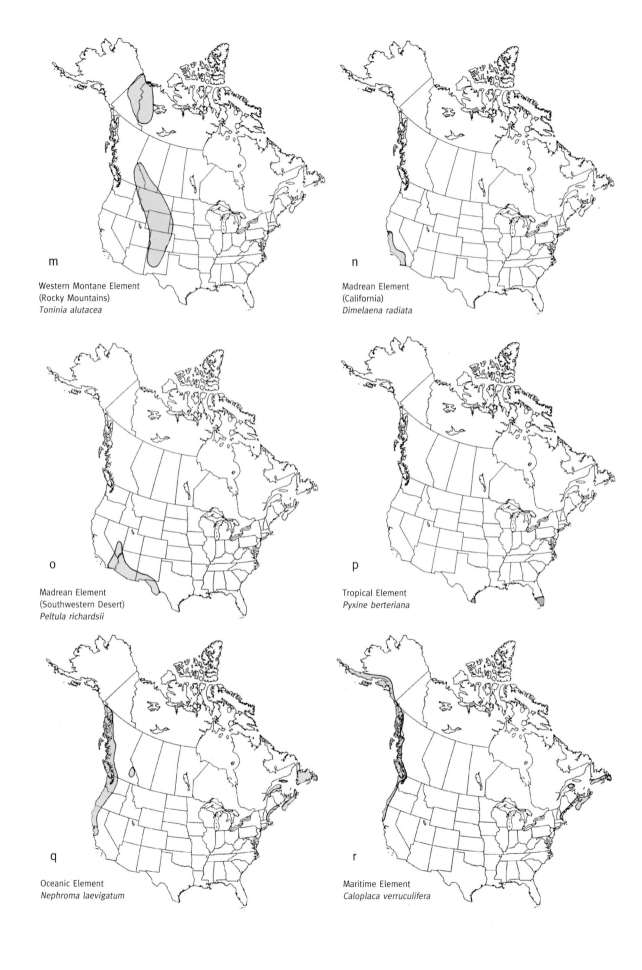

51. White Pass at the Alaska/British Columbia border is a typical alpine-subalpine habitat dominated by tundra lichens.

ARCTIC-ALPINE ELEMENT

We begin our discussion in the far north, even though relatively few people live there. In this land of treeless tundra, lichens form a particularly conspicuous part of the vegetation. The lichens and plants that live in arctic or alpine environments in North America are generally widely distributed not only across this continent but throughout the arctic northern hemisphere. Lichens that can withstand the rigors of the extreme north (a comparatively dry region with little snow, sometimes called the "high arctic") can usually be found in the snowier, more southern "low arctic" as well. Arctic species also frequently occur in the tundra-like alpine zones of the Rocky Mountains and northern Cascades (plate 51). Some widely distributed Arctic-Alpine species are *Vulpicida tilesii*, *Dactylina ramulosa*, *Cetraria islandica* ssp. *islandica*, *Rinodina turfacea*, and *Alectoria ochroleuca*.

Some arctic species are most abundant in Alaska, the Yukon Territory, and the western parts of the Northwest Territories, with only scattered occurrences as far east as Hudson Bay. This *Beringian distribution type* developed from ancient land connections and Ice Age refugia in the region of the present Bering Strait. Most Beringian species also occur frequently on the Siberian side of the strait. *Masonhalea richardsonii* and *Asahinea chrysantha* belong to this group.

There are a few arctic endemics in North America (e.g., *Teloschistes arcticus*), but all are rare species and not included in this book.

BOREAL ELEMENT

The North American boreal forest region extends as a 1000-km-wide belt from west central Alaska, southeasterly to southern Hudson Bay and the north shore of Lake Superior, then eastward to Labrador and the island of Newfoundland. This is the land of spruce and fir, interspersed with patches of larch, white birch, balsam poplar, and pines (plate 52). The ground is covered with lichens in the open, and feather mosses in the shade. And black flies reign supreme!

Where the boreal forest intersects mountain ranges that run north to south, elements of the forest (such as white spruce and trembling aspen) continue southward into the mountains. In the east, the boreal forest merges with the red spruce forest, following the Appalachians as far south as the Great Smoky Mountains.

52. Lichens are diverse and abundant on soil, rocks, and trees in this boreal woodland near Parc des Laurentides, Quebec.

In the far west, the boreal forest region meets the Pacific Northwest coniferous forests that range from Alaska to northern California. The species of trees that dominate montane and west coast coniferous forests are different from those of the Canadian boreal region, but the lichens that associate with them are often the same. These lichens of the coniferous forests constitute the Boreal Element.

The boreal forest region can be divided along climatic lines: cooler and drier in the north and west, and milder and wetter in the south and east. Lichen distributions sometimes reflect these differences, but most boreal lichens are not restricted to one climatic zone. *Bryoria lanestris*, however, is a lichen with a clearly northern boreal distribution pattern, preferring the areas with widely spaced black spruce. In the southern sector of the boreal forest, the climate is more humid and temperature more moderate, resulting in a thicker forest where white spruce is more abundant and black spruce and larch grow mainly in bogs. Lichens following this southern boreal zone include *Bryoria trichodes, B. nadvornikiana, Melanelia septentrionalis,* and *Stereocaulon dactylophyllum*.

Boreal lichens, like those in the Arctic Element, tend to be widely distributed in equivalent forest types throughout the northern hemisphere. For this reason, guidebooks for the lichens of Scandinavia can be useful for studying lichens of the North American boreal region.

TEMPERATE ELEMENT

Most large population centers of the United States and Canada occur in the temperate part of the continent, so this is the region most familiar to us. Roughly speaking, the temperate region begins just north of the Canadian border and includes most of continental United States. The western mountains, together with their rain-shadowed Great Basin, form such a formidable barrier to dispersal of species east and west that it is possible to divide the temperate region into two parts meeting approximately at the eastern edge of the prairies (close to 95°W longitude). The dominant forest trees of the eastern half of temperate North America are deciduous species such as oaks, maples, hickories, elms, and ash. The western half contains a complex of distinct geographic features and vegetation types, many of which are reflected in lichen distributions.

The temperate region contains some of the oldest landmasses on the continent. The southern Appalachians and Ozark Mountains have not been disturbed by glacial ice or inundated by inland seas for over 100 million years. This long period of isolation and stability has allowed the evolution and establishment of many species; this is why the Temperate Element contains most of North America's endemics. Paradoxically, lichens having highly discontinuous (disjunct) distributions are also found mainly in the Temperate Element. For an explanation, we have to look back 37 to 65 million years to the early Tertiary period. At that time, the earth had a wide-ranging, moist, mild climate, and many organisms had broad distributions within a lush, temperate deciduous forest that extended around the world. Later in the Tertiary, parts of the area became dry or hot owing to changes in weather patterns that resulted from mountain-building. Remnants of this Tertiary forest and the lichens that inhabited it became isolated in small pockets that retained this mild, moist climate. Such fragments are now found in the Southern Appalachians, parts of the west coast, and mountains in central Europe and eastern Asia.

Both the eastern and western sectors of the Temperate Element are highly diverse. Based on geographic features, climatic and vegetational differences, the geologic history of certain areas, and dispersal routes, one can recognize a variety of distribution types within each sector. Nevertheless, some lichens seem to have broad tolerances with regard to habitat and climate, and also have particularly effective means of dispersal. These species are, not surprisingly, found all across the continent; they can be called *Pan-Temperate*. Excellent examples are *Candelaria concolor, Physcia aipolia,* and *Phaeophyscia ciliata.*

Eastern Sector
Wide-ranging species of the deciduous forest region (mainly east of the prairies and the Mississippi River), belong to the *East Temperate distribution type* (plate 53). Some of these species clearly have northern or southern concentrations, but they are grouped here if they are about equally represented in the lowlands and mountains. Most East Temperate lichens do not occur on the southeastern coastal plain (discussed below). Typical examples are *Loxospora pustulata, Myelochroa aurulenta,* and *Phaeophyscia rubropulchra.* A few East Temperate lichens have disjunct populations on the west coast (e.g., *Buellia stillingiana, Lecanora hybocarpa,* and *Pertusaria amara*) or southern Arizona and southeastern New Mexico (e.g., *Heterodermia obscurata*), remnants of what once was a continuous distribution before topographic and climatic changes and subsequent vegetational changes occurred in the central parts of the continent.

The Great Lakes region east to the maritime provinces shares many tree species with the Appalachian chain (e.g., sugar maple, beech, and black oak) and has a roughly equivalent climate. Many lichens are more or less restricted to that area, producing a characteristic "prostrate Y" distribution pattern that we call the *Appalachian–Great Lakes distribution type.* Examples are *Anaptychia palmulata, Allocetraria oakesiana, Lobaria quercizans, Pseudevernia consocians,* and *Umbilicaria mammulata.* Some Appalachian–Great Lakes species have disjunct populations on the northwest coast (e.g., *Menegazzia terebrata*); in the southern Rocky Mountains (e.g., *Imshaugia placorodia, Ramalina intermedia,* and *Lasallia papulosa*); or in both areas (e.g., *Hypotrachyna revoluta*). When the pattern also includes the Ozark Mountains, as with *Anzia colpodes,* this category merges with the East Temperate distribution type, but with montane concentrations.

Some Appalachian lichens do not range into the Great Lakes region but simply follow the chain of mountains from New Brunswick to Tennessee (e.g., *Pseudevernia cladonia, Hypogymnia krogiae*). A few, such as *Melanelia culbersonii* and the endangered *Gymnoderma lineare,* are largely or entirely limited to the southern Appalachians. Species occurring in the northern but not the southern Appalachians are almost always found in the Great Lakes region as well.

53. Sugar maples, even those on residential lawns such as this one near St. John, New Brunswick, can be covered with a variety of lichens, many of them endemic to North America.

Spreading to the east and south of the Appalachians is a low, flat region known as the Coastal Plain, dominated by various oaks and pines, bald cypress, and, in the south, live oak, tupelo, and palmetto. Lichens with a *Coastal Plain distribution type* can be found in suitable habitats all along the Atlantic coast as far north as Nova Scotia, and along the Gulf Coast from Texas to Florida (plate 54). Many range northward into the Mississippi valley as well. Some Coastal Plain lichens, however, are restricted to the Gulf Coast and the Florida peninsula. Lichens that have their origins in the tropics (and are therefore part of the Tropical Element) as well as those with temperate ancestors (often from the Appalachians) may have Coastal Plain distributions. It is hard to place them in one Element or the other unless one knows their occurrences outside North America. As a rule, Coastal Plain species that are absent from the southern tip of Florida (e.g., *Ramalina willeyi,* a North American endemic) belong to the Temperate Element, but this is not always the

54. The coastal plain is characterized by low, flat, sandy areas and lichens such as *Cladina evansii*. This colony is growing under palmettos in northern Florida.

case (e.g., *Parmotrema tinctorum* is a widespread tropical lichen). *Bulbothrix confoederata, Parmotrema ultralucens, Pertusaria xanthodes,* and *Physcia atrostriata* are other Coastal Plain lichens.

Lichens that occur predominantly in the grassland areas of the central United States and southern third of the prairie provinces of Canada belong to the *Central Grasslands distribution type*. This is a rather small group, partly because the climate can be rigorous, with cold, dry, windy winters, wet springs, and hot, rather dry, late summers, but also because of the scarceness of ideal lichen habitats in the prairies. Heavy grass and herb growth overshadow lichens on the ground over much of this area. Trees are not abundant, although cottonwoods, elms, willows, and bur oaks grow along water courses or in scattered clumps. The trees usually have a relatively poor lichen flora made up mainly of widespread eastern species, but some bark-dwelling lichens are particularly abundant in the central part of the continent (e.g., *Hyperphyscia syncolla, Phaeophyscia cernohorskyi, Punctelia bolliana,* and *Rinodina populicola*). A few like *Ramalina celastri* are restricted to isolated trees in the southern grasslands or mesquite savannah. *Melanelia albertana*, a North American en-

55. Lichen-covered rocks in central Texas include large white patches of *Haematomma fenzlianum* together with yellow *Acarospora* and orange *Caloplaca* species.

demic, is unusual in being almost entirely confined to the Canadian prairies. Lichens growing on rocks in the prairies (especially calcareous rocks) are generally very widespread, with none showing a Central Grasslands distribution pattern (plate 55). In drier habitats where grasses are sparse, earth scales of different kinds (e.g., *Psora icterica*) and other soil lichens can become abundant.

Western Sector
Lichens that are restricted largely to the western half of the United States and adjacent Canada compose the western sector of the Temperate Element. Those concentrated mainly in the mountains or in southern California belong to the Western Montane or Madrean Elements, respectively, and will be discussed later.

West Temperate lichens, with broad ecological tolerances with regard to climate and substrate, are relatively widespread. They are the western equivalent of the *East Temperate* distribution type mentioned above. West Temperate lichens such as *Melanelia subolivacea, Physconia enteroxantha, Psora globifera,* and *P. nipponica* frequently have disjunct populations in the east, especially in the Great Lakes region.

Lichens with an *Interior Basin distribution type* have ranges centered in and around the dry, sagebrush country of the Great Basin, sometimes with extensions into the warmer southwestern deserts, eastward into parts of the prairies, or northward into eastern Oregon, Washington, and southern Idaho on the intermountain plateau, with some populations occurring as far north as interior British Columbia and southern Yukon Territory. In this region, trees, shrubs, and grasses are sparse, with lichens exploiting the bark and wood of scattered junipers and the many patches of open soil (plate 56). Lichens with this distribution type include *Aspicilia hispida, Psora cerebriformis,* and *Rhizoplaca peltata*.

Lichens having a *Pacific Northwest distribution type* closely follow the ranges of the Douglas fir, western hemlock, and western red cedar from northern California to southeastern Alaska, often showing incursions through the Columbia or Fraser Valleys into the wetter parts of the interior (plate 57). The Pacific Northwest is an area with relatively abundant summer as well as winter precipitation. Along the coastal strip from northern California to southeastern Alaska, temperatures throughout the year are moderated by the proximity of the Pacific Ocean and westerly winds, producing what is called an "oceanic climate." (Lichens belonging to the more widespread Oceanic dis-

56. (above) The interior basin of south central Idaho is rocky and dry and has its own characteristic community of lichens including species of *Rhizoplaca* and *Xanthoparmelia* as well as the widespread *Xanthoria elegans*.

57. (left) Tall Douglas firs in the Siuslaw National Forest, Oregon, support a variety of lichens characteristic of the Pacific Northwest.

58. These boulders on a mountaintop in northern New Mexico have been colonized by the potato chip lichen, *Omphalora arizonica*, and associated species.

gana, Pertusaria subambigens, Pseudocyphellaria anomala, and *Platismatia lacunosa.*

A subcategory of Pacific Northwest lichens comprises species concentrated mainly in the drier parts of the northwest dominated by species of pine, especially ponderosa and lodgepole pines. This pattern generally extends from northern California into the dry, intermountain regions of eastern Oregon, Washington, Idaho, Montana, and British Columbia. *Vulpicida canadensis, Alectoria imshaugii, Bryoria fremontii,* and *Ochrolechia juvenalis* are examples of lichens with this distribution type.

WESTERN MONTANE ELEMENT

Western North America is dominated by several high mountain ranges running in a roughly north-south direction from Alaska to Mexico. The easternmost ranges are the Rocky Mountains. The Cascade and Coastal Ranges, extending from Alaska to California, are farther to the west. The Coast Range is floristically connected to the Cascades by intervening patches of conifer forest. These ranges, being close to the Pacific, receive the greatest amount of rain, although the western slopes of the northern Rockies have a well-developed "wet belt" with some typical coastal conifers such as western hemlock and western red cedar and the lichen communities that are associated with those trees. The driest ranges are the Sierra Nevada in eastern California and the mountains of Arizona and western New Mexico (plate 58).

Geographers have long noted that the altitudinal zones on a mountain mimic the latitudinal zones of the continent in many ways. We have already mentioned that most arctic and boreal lichens range southward into the alpine and coniferous zones of the western mountains. In fact, a large percentage of the montane lichen flora consists of lichens that are included in the Arctic and Boreal Elements. Arctic species are most typically found in the Rockies (e.g., *Cetraria ericetorum*) but others cling closer to the Cascades and Coast Ranges (e.g., *Massalongia carnosa*). Arctic and boreal lichens are less likely to be represented in the drier ranges south of the Great Divide Basin including the Southern Rockies, and there are very few in the Sierra Nevada.

There is, however, an element of the lichen flora entirely or largely confined to the western mountains, and it is this group that comprises the *Western Montane Element*. Interestingly, a large number of these are North American endemics. These species are usually

tribution type are discussed later.) Sitka spruce becomes an additional component of the forest, and an important lichen habitat, near the coast. Redwoods occur in the southernmost part of the Pacific Northwest forest but contribute little to the lichen distribution pattern because the tree is not a particularly good lichen substrate. Conversely, some deciduous trees characteristic of certain habitats within the same forest region (e.g., big-leaf maple, Gary oak, cultivated apple, red alder), although not always particularly abundant, are adopted as substrates by a wide variety of lichens. *Ochrolechia laevigata* and *Hypotrachyna sinuosa* are examples of Pacific Northwest lichens that grow almost exclusively on deciduous tree bark. Some lichens with cyanobacteria as primary or secondary photobionts, including species of *Sticta, Nephroma, Pseudocyphellaria,* and *Lobaria,* are particularly abundant on these trees.

Most Pacific Northwest lichens are endemic to the region, but many also occur in western Europe (*Evernia prunastri* is very common in Europe; *Ochrolechia oregonensis* is very rare). Examples of endemic Pacific Northwest lichens include *Cladonia verruculosa,* many species of *Hypogymnia, Lecanora pacifica, Lobaria ore-*

present in more than one of the mountain ranges, especially in the north, with the pattern bifurcating in southern B.C. (between the Rockies and Cascades) or in central California (between the Coastal Range and Sierra Nevada) (plate 59). Lichens with distributions mainly in the Rockies include *Evernia divaricata, Lecanora novomexicana,* and *Peltigera kristinssonii;* some Cascade and Coastal Montane lichens are *Ahtiana pallidula, Kaernefeltia merrillii, Letharia columbiana,* and *Tuckermannopsis platyphylla.* Although the dominant tree species in the Southern Rockies (e.g., ponderosa pine, Engelmann spruce, and subalpine fir) are also found farther north, some lichens remain concentrated in the southern mountains, often with distributions that extend into the Sierra Nevada through the mountains of southern Arizona and New Mexico. Many of these lichens, like *Candelariella rosulans, Physcia biziana,* and *Physcia callosa,* occur in dry, open habitats of the interior plateaus.

MADREAN ELEMENT

Western lichens with a distinctly southern center of distribution are grouped together as the Madrean Element, taking its name from the Sierra Madre Occidentale of Mexico. These lichens are confined mainly to the extreme southwest, from western Texas to central California. For the most part, they are lichens in habitats subject to long periods of drought, although some of these lichens thrive only in the foggy coastal strip from the southwestern corner of California (including the Channel Islands) into Baja California. Two distribution types can be recognized within the Madrean Element: one associated with the savannahs and coastal areas of California, the other correlated with the warm southwestern deserts.

California, with its two parallel mountain ranges, relatively dry valleys (at least in the south), and foggy coastline, is geographically complex and has correspondingly complex vegetation. The species that center in the California mountains and isolated peaks in the desert are best grouped with other Western Montane lichens, as we have already mentioned. There is, however, a unique *Californian distribution type* especially in the southwestern sector where there is a warm, temperate climate with cool winters and hot, rainless summers — very similar to the climate found in the Mediterranean region. The flora of California reflects this similarity.

Along the windy and misty coastal strip from San Francisco to Baja California, there is a unique lichen

59. The branches of this Sierra juniper in Yosemite National Park are covered with bright yellow wolf lichens (*Letharia*).

community whose distribution forms a special part of the Californian element. Shrubby sage, herbs, and grasses dominate the flowering plant vegetation. The lichens, able to absorb all the water they need directly from the moisture-laden air and benefiting from abundant sunshine and mild temperatures, thrive on the coastal cliffs and the branches of shrubs. Many lichen genera characteristic of the coastal strip are common to California and the Mediterranean region (e.g., *Roccella, Schizopelte,* and *Niebla*), but most of the lichens of coastal California (many ranging into Baja California) are endemic. Examples are *Dendrographa* species, *Buellia halonia, Lecania brunonis,* and *Niebla homalea. Dirina catalinariae* is conspicuous on the Channel Islands.

The foothills and interior valleys are characterized by a savannah type of vegetation: dry grassland with scattered trees. In California, the trees are mainly broadleaf evergreens such as coast live oak, California black oak, and California buckeye, as well as several species of pine (plate 60). The lichen flora, although retaining its Mediterranean character, is composed of more wide ranging genera than those found exclusively along the coast. Some species of this distribu-

60. *Ramalina menziesii* and species of *Usnea* are common on deciduous trees in the coast ranges of California. They are most conspicuous before the trees leaf out, as in this early spring scene.

tion type are also found in southern Europe (e.g., *Dimelaena radiata*, *Melanelia glabra*, and *Parmelina quercina*), but most are endemic (e.g., *Physconia isidiigera* and *Rinodina bolanderi*). *Ramalina menziesii*, although most abundant in the valleys and foothills of California, ranges northward in suitable habitats of coastal forests all the way to southeastern Alaska.

The hot Sonoran, Mohave, and Chihuahuan deserts dominate the southwestern corner of the United States and form the core of the *Southwestern Desert distribution type* (plate 61). Rainfall is extremely sparse, typically in the form of torrential but brief showers that quickly disperse. The rain is of use only to highly specialized desert plants, although it is quickly absorbed by lichens on the ground. Creosote bush, cactus, ocotillo, and yucca dominate the flowering plant vegetation over much of the region. Lichens with a Southwestern Desert distribution often range northward, reaching parts of the cool deserts of the Great Basin (which are sometimes classified as part of the Madrean region), and almost all have distributions that include parts of Mexico. Typical representatives include several species of *Peltula* and *Psora* as well as *Candelina submexicana*, and perhaps *Speerschneidera euploca*. Some Mexican montane lichens such as *Pseudevernia intensa* become abundant in forested, high-elevation areas of both the Chihuahuan Desert and Sonoran Desert, including the Big Bend area of western Texas.

TROPICAL ELEMENT

Lichens of the Tropical Element appear in our flora mainly in southern Florida, but most are more common in the West Indies and Central and South America and are, in fact, widespread in tropical areas throughout the world. In Florida, they typically inhabit stands of evergreen hardwoods called "hammocks," which form islands of dense forest among the expanses of sawgrass or flooded bald cypress swamps (plate 62). Other tropical species are found on oaks in the more open sand pine scrub, and others are epiphyllous on the leaves of palmetto or various laurel-like evergreen shrubs. The truly tropical species like *Pyxine berteriana* and *Sarcographa labyrinthica*—those having no tolerance for frost—extend only as far north as Lake Okeechobee and the southern tip of Texas. Many others, however, have broader climatic tolerance and are better referred to as subtropical.

They can be found on the bark of oak, tupelo, sweetgum, and similar hardwoods in floodplain forests on the Gulf and Atlantic Coastal Plains, often with a distribution pattern almost identical to that of temperate coastal plain species. Subtropical and tropical lichens occurring on the coastal plain include *Cryptothecia rubrocinta, Dirinaria confusa, Leptogium marginellum, Ramalina paludosa,* and *Usnea baileyi.*

OCEANIC ELEMENT

The Oceanic Element comprises lichens that are narrowly restricted to regions having an oceanic climate, where summers are cool, winters are mild, and there is abundant precipitation with frequent fog throughout the year. This climate is called oceanic because it is most frequently encountered in coastal regions (plate 63). This contrasts with a continental climate, generally well inland, characterized by hot summers and clear, very cold winters, often coupled with sporadic or relatively meager precipitation.

In North America, four areas have a pronounced oceanic climate: the extreme west coast from northern California to southeastern Alaska, the coasts of Newfoundland and nearby Nova Scotia and New Brunswick, the north shore of Lake Superior, and the southern Appalachian Mountains. Lichens that belong to the Oceanic Element tend to be very widely distributed, often occurring in at least two (and frequently, all four) of these regions and in one or more climatically similar areas elsewhere in the world. Oceanic lichens of North America often turn up along the west coasts of Ireland, Wales, Norway, and Portugal, some localities in the Alps, parts of eastern Asia, cloud forests of Central America, and even in the southern hemisphere, especially in New Zealand and Tasmania. A likely explanation for this wide-ranging distribution is that Oceanic lichens are ancient relics of the lichen flora that inhabited the temperate forest that was widespread during the Tertiary epoch (see the discussion under "Temperate Element"). Examples of lichens found in at least two or three different oceanic areas in North America (and common in western Europe) are *Pseudocyphellaria crocata, Platismatia norvegica, Nephroma laevigatum,* and *Cavernularia hultenii;* those in only one of the areas (with occurrences outside North America) include *Erioderma sorediatum* (+ Greater Antilles) and *Polychidium contortum* (+ New Zealand).

61. The Sonoran desert, despite its severe heat and lack of moisture, supports its own community of lichens, including many species of *Xanthoparmelia.*

62. Smooth-barked trees in the tropics, like these fig trees near Sarasota, Florida, can be covered with a mosaic of crustose lichens. (Photo by I. M. Brodo)

63. (overleaf) The misty islands of southeast Alaska have tall spruces draped with *Alectoria sarmentosa* growing with many other oceanic lichens.

MARITIME ELEMENT

Lichens that are directly associated with the marine environment — the intertidal zone and the shoreline subjected to salt spray — are grouped within the Maritime Element. Obviously, this categorization cuts across climatic boundaries, but many Maritime species show preferences for one climatic regime or another.

Because the majority of Maritime lichens are rock-dwelling species, this Element is mainly restricted to regions where there are coastal rocks, or at least glacial boulders and pebbles. That is why the Maritime Element is poorly represented on the Atlantic coast south of Long Island, New York. On the Pacific side, Maritime lichens are found from southern California to the Aleutians. There appear to be few Maritime species along arctic shorelines, probably because of ice scouring.

The main factor influencing the local distribution of Maritime lichens is the extent to which they are exposed to salt water. Daily tides and periodic episodes of salt spray create a distinct series of zones along the shore. The color of each zone is determined largely by the presence (or absence) of specific lichen species. Although the lichen species may change in different parts of the world, a black belt dominated by *Verrucaria maura* and related species is almost universally present in the upper intertidal zone. Coastal rocks on the Queen Charlotte Islands display a typical northwestern zonation pattern (plate 64). The white lichen is mainly *Coccotrema maritimum*. Below the black *Verrucaria maura* zone is a bare area free of both lichens and marine algae; still lower, the algae take over. Farther south and along the east coast, species of *Caloplaca* and *Xanthoria* replace *Coccotrema* as the dominant lichen in the lower salt-spray zone, making that belt orange instead of white.

Species restricted to a single coast tend to be North American endemics. Those that are widespread enough to occur on both the Atlantic and Pacific coasts are usually found in Europe and Asia as well. *Coccotrema maritimum* and *Caloplaca flavogranulosa* are examples of endemic Maritime lichens from the northern Pacific coast. *Caloplaca coralloides* and *Lecanora pinguis* are endemic to the coast of California. Species found on both the northwest and northeast coasts, as well as in Europe, include *Lecanora xylophila*, *Verrucaria maura,* and *Caloplaca verruculifera*.

64. A specially adapted community of lichens inhabits many levels of the intertidal zone on this rugged, rocky headland on the Queen Charlotte Islands, British Columbia, each dominant species creating a distinct zone. The white lichen is mainly *Coccotrema maritimum,* and the black one is almost entirely *Verrucaria maura*. Below the *Verrucaria* is a bare zone free of both lichens and marine algae, and, still lower, the algae take over.

10. LICHENS AND PEOPLE

Headstone and half-sunk footstone lean awry,
Wanting the brick-work promised by-and-by;
How the minute grey lichens, plate o'er plate,
Have softened down the crisp-cut name and date!
Robert Browning

Lichens play a significant role in nature, influencing soil fertility, the growth of surrounding plants, and the formation of soil over bare rock or sand, as well as providing food, nesting materials, and shelter for mammals, birds, and invertebrates. Although these effects on the natural world have obvious importance for people, it is not unreasonable to ask if lichens have any commercial or practical value. The answer is, "Yes, indeed." People have used lichens over the centuries as food and decoration, and as a source of dyes, medicines, poisons, and fiber. More recently, they have been employed mainly in the manufacture of perfumes and antibiotics, and as pollution monitors and indicators of old growth forests. The special importance of lichens as environmental indicators is covered in Chapter 11.

LICHENS AS FOOD

Although lichens are not especially tasty, few of them are poisonous, and they have been utilized as foodstuffs in numerous cultures. The nutritional value of lichens for human beings is quite limited because, unlike animals that chew their cud, we lack the rumen with its bacterial flora needed to break down complex lichen carbohydrates into something that can be absorbed. Perhaps for this reason, the partially digested lichen (primarily *Cladina*) from the stomachs of killed caribou is a more successful food than fresh lichen and has been eaten by arctic groups from Alaska to Baffin Island. A mixture of this material and raw fish eggs was regarded as a favorite dish and was called "stomach ice cream."

Horsehair lichens, especially *Bryoria fremontii*, have been eaten by native peoples throughout the northern range of the lichen from Washington, Idaho, and Montana into British Columbia. Although *Bryoria* (usually under the old collective name *Alectoria jubata*) is often mentioned in ethnobotanical accounts as a food eaten only in times of famine, it was a favorite food of the Interior Salish of the Okanagan-Colville language group. They baked the lichen in pits over leaves, sometimes adding wild onions, blue camas bulbs, saskatoon berries, or other flavorings, until it coalesced into a black, gelatinous mass (plate 65). In later times, the baked lichen was served with sugar, raisins, or apples, or recooked with meat in a soup or with flour and butter to make a pudding. Now called "black tree-lichen," it is still being prepared by the Nlaka'pamux people of south central British Columbia (in modern ovens, these days) and served as a kind of taffy candy called *we'ia,* with the texture and appearance, if not the flavor, of licorice.

Probably the most publicized use of edible lichens has been as an emergency food in the far north. Raw lichen, although poor in available starches and protein, does nevertheless fill the stomach and can provide a modicum of nutrition that, in times of starvation, might just make the difference. Rock tripes or *tripes de roches* (species of *Umbilicaria* and *Lasallia*) can be boiled and, after several changes of the water to remove some of the bitter lichen substances, eaten alone or cooked into a soup. The Woods Cree, for example, used *Umbilicaria muehlenbergii* as an ingredient in a thick fish broth. One kind of rock tripe is collected in Japan and prepared in soups and salads as a delicacy under the name *iwatake* (rock mushroom). Members of the ill-fated Sir John Franklin expedition ate rock tripes to keep themselves alive, as did many a stranded pilot lost in the boreal woodland or arctic tundra.

Brødmose (bread moss) is one of the names that Scandinavians give to *Cetraria islandica* (Iceland lichen) because it was ground and added to wheat flour or potatoes in times of famine to stretch their meager supplies. Apparently, Iceland lichen is quite palatable when it has been soaked with lye or soda to remove the bitter acids and then made into a pudding with milk. Interestingly enough, this same species has also been used as a laxative. A variety of lichens, including *Cetraria islandica, Alectoria ochroleuca,* and species of *Cladina,* were used in Russia during World War II to make a kind of molasses. Here also, people resorted to eating lichens only when other food — in this case, potatoes and beet sugar — was not available.

Lichens have been involved in beverage production as well. *Cladina rangiferina* was used in Sweden in the 1880s as a source of sugar for a short-lived brandy-distilling industry. When supplies of the lichen were exhausted (perhaps locally), the distilleries had to close down. Monks in at least one Siberian monastery used *Lobaria pulmonaria* in the seventeenth and eighteenth centuries as a bitter flavoring for beer. Much more recently, the Tarahumara of northern Mexico have used species of *Usnea* as catalysts for making fermented corn beverages.

Legend has it that a desert vagrant lichen in Iran and northern Africa, *Aspicilia esculenta* (closely related to *Aspicilia contorta*), may have been the biblical manna (Exodus 16:31) that was eaten by the Israelites during their 40-year trek through the Sinai wilderness. The lichen forms rather hard, almost spherical growths resembling small pebbles, usually less than a few centimeters in diameter, which lie unattached on the soil. They can be whipped up by a strong wind and blown into heaps where the lumps soften and swell in the heavy morning dew, which is common in the desert. *Aspicilia esculenta* can be abundant enough to be fed to sheep and goats in certain regions. People in west central Asia are known to have eaten it, at least in times of famine. Interestingly, the lichen is not found in the Sinai — at least not now.

CLOTHING

The fibrous lichens *Alectoria* and *Bryoria* have been incorporated into clothing in a number of cultures. Native people of the British Columbia interior, especially the Nlaka'pamux (Thompson Indians), called *Bryoria* "tree hair" in their own language and "black moss," later "black tree lichen" in English. They skillfully wove it together with cedar or silverberry bark fiber to make vests, leggings, and moccasins (plate 67). Because such clothing was not very durable or comfortable, it was used mainly by poorer people who could not acquire skins, or it was used indoors for ritual or ceremonial purposes. (In rainy weather, a garment made of soggy lichens would prove very unsatisfactory!)

65. Nettie Francis (left) and Cecelia Pichette remove pit-cooked black tree lichen (*Bryoria fremontii*) from cooking sacks. Colville Indian Reservation, Washington State. (Photo by Dorothy I. D. Kennedy.)

66. (above) Chilkat Tlingit dancers in Haines, Alaska, still use decorative blankets and costumes colored with lichen dyes, especially those derived from *Letharia vulpina*, a lichen of the drier interior. They obtain the lichen by bartering coastal products such as fish oil.

67. (right) This shirt and pair of leggings, made by the Nlaka'pamux (Thomson River) people, incorporates *Bryoria* as a fiber. Garments collected by J. Teit in 1903 and now in the collection of the American Museum of Natural History.

DYES

Lichens have been used as a source of dyes for thousands of years and in many parts of the world. North American aboriginal peoples from New Mexico to the Arctic used lichen dyes for coloring wool (mainly of mountain goats in the Northwest) and porcupine quills, and even as body paint (plate 66). Lichen dyes remain popular today among weavers and other artisans. The chemicals that make lichens such good sources of dyes are the same unique lichen substances that are used for lichen identification.

There are two types of lichen dyes, distinguished by the kind of lichen substances involved and the techniques used for extracting them. The first, called boiling water dyes, are prepared from colorless compounds that turn red or yellow with K or PD (for definitions, see "Chemistry," in Chapter 13), stictic, salazinic, and norstictic acids, for example, as well as from a few types of yellow pigments such as vulpinic and pinastric acids. The lichens producing such dyes, especially *Parmelia omphalodes* and *P. saxatilis*, and sometimes the dyes themselves, are called "crottle." The beautiful, earthy russets, browns, and yellows used by Scottish craftsmen for Harris Tweed formerly came

80 ABOUT THE LICHENS

mostly from boiling water dyes. The precise methods varied a great deal, but one technique required a large pot in which the wool was layered alternately with a dye-producing lichen such as *Parmelia omphalodes, Lobaria pulmonaria,* or *Letharia vulpina* in about equal quantities. Water was added, and the mixture was boiled for several hours until the color developed in the wool. Although the dye traditionally was used only with animal fibers (wool and silk), it works quite well with a variety of natural and synthetic fibers including cotton, linen, orlon, and rayon, and the dye is relatively color-fast (plate 68).

The fermentation dyes, the second type of colorant, give rich red and purple hues (plate 69). They develop when the appropriate lichen is mixed with a source of ammonia (traditionally stale urine, but currently a solution of pure ammonia) and allowed to stand, or "ferment," for a period of weeks. (The smell of the urine gradually weakens and disappears, and the dye, they say, finally takes on the aroma of violets!) Although the colors produced are very beautiful, they become dulled to a pale brown by direct exposure to sunlight unless fixed with mordants. Fermentation dyes come from lichen substances such as crythrin or lecanoric, gyrophoric, and alectoronic acids, which react C+ red or KC+ red in spot tests. *Ochrolechia tartarea* (a European species) and Mediterranean species of *Roccella* are rich in these compounds, and they were once collected in huge quantities as a source of dyes. "Cudbear" is the dye produced from *Ochrolechia*, and "orchil" refers to the *Roccella* dye. All North American species of *Roccella*, including *R. babingtonii*, are rare lichens and should *never* be collected for dyeing. Rock tripes (*Umbilicaria* and *Lasallia*), *Ochrolechia oregonensis,* or the common *Punctelia rudecta* have the same or related compounds and are much more abundant, making them a better choice as a source of dyes, although any lichen population can be wiped out by overcollecting.

Many books on natural dyes have been written for the benefit of weavers and dyers, and specific recipes and techniques can be found in those references (see the section "Further Reading," at the back of the book). Karen Casselman has developed lichen dyeing techniques that require much smaller quantities of lichens than are generally recommended in traditional recipes. In the interest of lichen conservation, these more frugal (but equally effective) methods should be adopted.

68. The Ramah Navajo of northern New Mexico use all natural materials, including lichens such as *Xanthoparmelia chlorochroa*, to dye their wool blankets.

69. Karen Casselman and Glenna Dean used a variety of boiling water and fermentation methods of dye production to color these beautiful lichen-dyed samples of wool yarn.

LICHENS AND PEOPLE 81

70. A variety of commercial products use lichen extracts, especially usnic acid, as an essential ingredient.

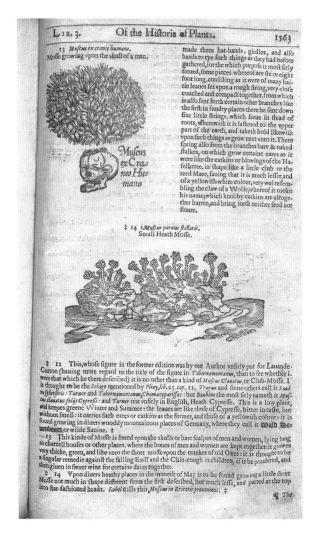

PERFUMES

For decades, thousands of tons of "oakmoss," *Evernia prunastri* and *E. mesomorpha,* and "treemoss," *Pseudevernia furfuracea,* have been collected annually in Europe and North Africa to supply the perfume industry. Certain compounds in these lichens have the ability to "fix" fragrances, making them last longer when applied to the skin, and some, atranorin in particular, may have agreeable aromas of their own. The North American species of *Pseudevernia* (*P. cladonia, P. consocians,* and *P. intensa*) have not been tested for these properties. Although these lichens and both species of *Evernia* are fairly common in North America, they have never been collected commercially on this continent and probably would not survive the levels of harvesting practiced in Europe, even for a few years.

In the seventeenth and eighteenth centuries, oakmoss lichens were ground up and dusted on powdered wigs, again for their ability to retain a fragrance, and as a cleansing agent. Certain lichens are still used in India as ingredients in hair-cleansing preparations. Besides being used in the production of perfumes, lichens are also ingredients (as imported "lichen extracts" of *Usnea* species or unspecified sources) in soaps and deodorants, perhaps in part because of their antibiotic properties (plate 70).

MEDICINES AND POISONS

Fourteenth-century herbalists believed that plants resembling or reminiscent of a part of the body or a disease symptom, are God's sign that the plant is useful in treating afflictions of that part of the body or that condition. This "doctrine of signatures" dictated that a lichen scraped from a human skull could cure epilepsy (and was literally worth its weight in gold) (plate 71). Similarly, it was believed that the dog lichen (*Peltigera*

71. The curative value of certain lichens was extolled in early herbals, such as this one from 1636 by John Gerarde illustrating *Muscus ex Craneo Humano.* The lichen, described as green in this edition (as a true moss?), was said to be "whitish" in a later edition by John Parkinson (*Theatrum Botanicum,* London, 1640). Parkinson also stated that it "cureth wounds" and was particularly valued "because it is rare and hardly gotten." The most desirable "mosse," he reported, "should be taken from the sculls of those that have been hanged or extented for offences" (p. 1313). (Photograph from J. Gerarde, *The Herball or Generall Historie of Plantes,* London, Norton & Whitakers, 1636.)

canina), whose apothecia resemble dogs' teeth, must be effective against rabies; that remedies made from the beard lichen, *Usnea*, would help strengthen the hair; and that the orange *Xanthoria parietina* could be used to treat jaundice. Unfortunately, none of these remedies worked. On the other hand, it was also said that the tree lungwort, *Lobaria pulmonaria*, could be used as a cure for tuberculosis because its pitted surface resembled lung tissue, and, as it turns out, some of the substances found in this lichen are indeed effective against tuberculosis bacteria. Iceland lichen (*Cetraria islandica*), which does not resemble a lung, was also used to treat tuberculosis and other infections even into modern times, and many other lichen extracts have been found to be effective in killing Gram-positive bacteria (so called because of the ability of the bacteria to take up a specific stain).

In the 1940s, large numbers of lichens were screened for antibiotic properties. Usnic acid, the lichen substance found in scores of pale yellowish lichens, especially the genus *Usnea*, was found to be sufficiently active to make it valuable commercially. It is still used in central Europe principally in ointments for superficial infections, either by itself or combined with other antibiotics, although the usefulness of usnic acid is now being questioned because many people are allergic to it. The substance, in fact, can cause serious skin disorders in certain susceptible individuals, especially loggers and others who are exposed to large amounts of lichen-covered wood.

The absorptive as well as antibiotic properties of *Usnea*, *Alectoria*, and *Bryoria* have been recognized independently in many cultures from British Columbia to New Zealand and have led to their use as wound dressings, baby diapers, and feminine sanitary absorbents. Species of *Usnea* have been used by the Greeks and Chinese to treat various gynecological ailments, and a sodium salt of usnic acid is still prescribed in Russia to treat certain infections including those of *Trichomonas*, a protozoan that can cause a disease of the uterine cervix.

Other lichens used as medicines — either internally as laxatives, expectorants, or tonics, or externally as a healing paste or poultice — include *Flavoparmelia caperata*, *Lobaria retigera*, *Parmelia saxatilis*, *Parmotrema perforatum*, *P. chinense*, and *Peltigera aphthosa*. The search for new pharmaceutical uses of lichens is still under way in various parts of the world, especially in Japan, and some promising discoveries have recently been made. For example, some lichen polysaccharides, glucans, and glycoproteins show antitumor activity, and a polysaccharide from the edible rock tripe, *Umbilicaria esculenta*, inhibits the growth of HIV, the virus that causes AIDS.

Very few lichens are truly poisonous. The wolf lichen (*Letharia vulpina*) and species of *Vulpicida* apparently are exceptions. These lichens contain the toxic, bright yellow pigment, vulpinic acid, and were used in Scandinavia to poison wolves. The lichens were added to various baits such as reindeer blood and other meats, and were sometimes mixed with ground glass. The unfortunate wolves who ate such concoctions were reported to succumb in less than 24 hours. Although two species of *Letharia* are common in western North America, there is no record of their ever having been used as wolf poisons on this continent. They were, however, used by the Achomawai people of northern California to poison arrowheads. The arrowheads were soaked in lichen for a year, sometimes with the addition of rattlesnake venom. It is therefore curious that these same wolf lichens were used to treat sores and inflammations by indigenous people in northern California and southern British Columbia, and even taken internally. One lichenologist experienced severe respiratory irritation and even nosebleed from a heavy exposure to *Letharia vulpina* (he was making a mass collection), indicating that the lichen's toxicity should be taken seriously.

MODELS AND DECORATIONS

Many people who have never noticed a lichen in a forest or on a fencepost have seen scores of lichens used to simulate trees and shrubbery on architectural mock-ups or in model train landscapes (plate 72). Reindeer lichens, soaked in a preservative with some glycerine and dyed green, red, or yellow, are used commercially in large quantities to represent miniature trees and shrubs. These lichens, especially *Cladina stellaris*, are also used in commercial floral displays with flowers (dried or fresh) and are gathered by the ton for use as grave decorations in Europe. This latter tradition has not reached this side of the Atlantic to any great extent, although the depletion of *Cladina* mats (known locally as "moss") for use in floral arrangements has become a problem in, for example, the Ouachita National Forest in Arkansas. Certainly, lichens should not be exploited commercially south of the vast lichen woodlands of northern Canada and Alaska. The hairlike appearance of *Usnea* and *Bryoria* has frequently been exploited by artists, both ancient and contemporary (plate 73).

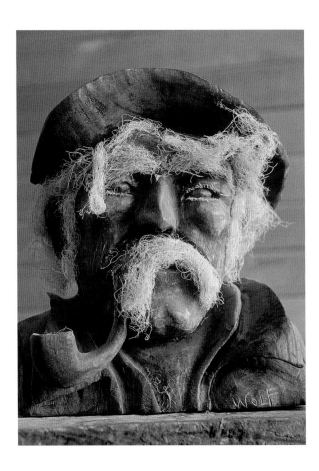

72. Lichen "shrubbery" is remarkably realistic in this HO scale model train set.

73. This modern carving with *Usnea* hair and mustache was seen in Bellingham, Washington.

LICHENOMETRY

Because lichens grow so slowly, a relatively small circular patch can actually be very old. In the arctic, crustose species such as the map lichens (*Rhizocarpon geographicum* and related species) add only a fraction of a millimeter of radial growth each year. If it is possible to reliably estimate the average growth rate of a lichen, for example, by measuring lichens on dated gravestones in the vicinity or by photographing or tracing the same thallus over a period of years (Fig. 24; see plate 26), then one can measure the diameter of the largest lichen thallus on an exposed rock surface and calculate with reasonable accuracy how long that surface was available for colonization by the lichens — that is, the minimum "age" of that surface. This dating technique is called lichenometry, literally, "the measurement of lichens." It has been in use for over 40 years (developed by an Austrian-Canadian lichenologist, Roland Beschel), mainly for dating the age of glacial moraines to estimate the rate of retreat of glaciers in the Alps, Alaska, and Canada. Lichenometry has also been used for dating the age of artifacts, including the ancient stone monuments of Easter Island (which, according to the lichens, are about 450 years old). Estimating the age of ancient rock slides using lichenometry has helped researchers date major earthquakes in the Sierra Nevada. Even experienced practitioners admit, however, that there are limits to the reliability of the method owing to the uncertainties involved in using growth rate figures calculated for lichens that are not present at the study site.

THE DOWN SIDE OF LICHEN GROWTH: LICHEN-CAUSED DAMAGE

Despite their obvious beauty, lichens are not welcome everywhere. Fragile rock carvings and paintings (petroglyphs and pictographs) made by native people centuries ago inevitably become covered with lichens in suitable habitats. To preserve these archeological and cultural treasures, the lichens have to be removed and prevented from reinvading. Unpolished gravestones make excellent lichen substrates, and lichens can obliterate (or "soften down," as Robert Browning put it) the writing on the stones. In well-maintained cemeteries, lichens are routinely scraped or washed off (plate 39). Lichen lovers have a hard time convincing caretakers of rural cemeteries that the lichens themselves are worthy of preservation (plate 74). In cities, monu-

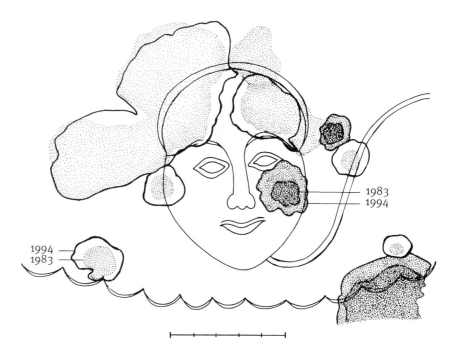

ments are kept free of most lichens by "virtue" of the ambient air pollution, but statues in rural locations can become badly disfigured. Stained glass windows can also be etched and weakened by crustose lichens.

Most lichen growth can be easily controlled by sprays containing copper salts (many commercial products are available), or simply by the installation of a source of copper or zinc above the surface to be protected (plate 76). Scraping does not remove the entire lichen thallus, which can usually penetrate into the rock to some extent (plate 39), and scraping further damages the lichen-weathered surface.

Bark-dwelling lichens do virtually no direct damage to trees and do not have to be removed (see "Colonization of Trees," in Chapter 8).

HUMAN IMPACTS ON LICHENS

If, as we have seen, lichens contribute to the well-being of humankind in numerous ways, and their destructive roles are relatively minor, it is only fair to pose the question the other way around. What impacts are we having on lichens?

Looking first at the positive side, human activities have created many new niches where lichens can grow. Just as some birds take advantage of the undersides of roof eaves to build their nests or barnacles find a home on the pilings of piers, lichens have colonized many constructed surfaces, from old wooden buildings and fences to statues, stone and concrete walls, and sidewalks (plate 75). In a less obvious example, nitrogen-rich dust from the fertilizer used on agricultural fields promotes the growth of such lichens as *Xanthoria fallax* on trees along roads in farming country.

On balance, however, it must be said that the overall human impact on lichens has been severely negative, largely because of air pollution and habitat alteration. Although the effects are most pronounced in urban areas, whole regions of North America have become poor in lichen diversity as a result of airborne contaminants. As with other types of organisms, the most sensitive species have died off, and they have either not been replaced or been supplanted by those few species with a tolerance for dirty air. It is encouraging to note that wherever society manages to reduce airborne pollutants to low levels, many lichens recolonize areas in which they once flourished (see Chapter 11).

The widespread modification of natural habitats is continuing to reduce lichen populations and lichen

Fig. 24 (above right) Tracings from gravestones showing growth increments (compare with plates 26A and 26B). The heavily stippled, dark lichen is *Aspicilia* sp.; the paler one is *Dimelaena oreina*. The inner, more darkly stippled thalli are the lichens in 1983; the outer margins represent the lichens in 1994. Scale: 12-mm increments.

74. (above left) Lichens decorate the robe and hair, but not the smooth face, of this statue of an angel in Manchester, California.

75. A concrete house near the ocean in Martha's Vineyard, Massachusetts, makes an ideal substrate for *Xanthoria parietina*.

76. (near right) A thin copper wire embedded in the beautiful etched outline of a killer whale on this Haida gravestone in Haida, British Columbia, has prevented the establishment of lichens on the surface of the stone.

77. (far right) People encounter lichens in a variety of ways. In this yard in Oregon, children will play under a tree draped with *Usnea*.

species diversity across the continent. This takes its most extreme form in the case of spreading cities and suburbs. Brooklyn, New York, one of the most thoroughly urbanized areas in North America, once was home to a rich lichen flora including many species such as *Lobaria pulmonaria, Pannaria lurida,* and *Rimelia reticulata,* now found no closer to Brooklyn than the most pristine localities of eastern Long Island 100 miles away. A survey conducted on Long Island in the 1960s counted 261 species (including some purely historical records), rather poor for a coastal area 116 miles long and 20 miles wide. One can only guess at how many more species have disappeared under asphalt and buildings over the past 30 years. Similar impoverishment resulting from habitat destruction and pollution has been documented in many parts of the continent, but nowhere with more devastating effects than in the state of California, with its diverse, unique flora. This problem was noted as early as 1936 in California by A. W. C. T. Herre, who commented in *Our Vanishing Lichen Flora,* "The regions where Bolander gathered amazing forms in abundance have long since been devastated by 'real estaters,' while it is now absolutely impossible to collect lichens in the favorite haunts of Dr. Hasse and myself, where hitherto unknown species were brought to light every year.... Miles of terrain are covered with asphalt, concrete and houses.... Their resident lichens have not been merely discouraged, they have been wiped out of existence" (p. 199).

Less extreme in impact than urbanization, but vast in extent, is the conversion of old growth forest ecosystems to intensively managed silvicultural plots. As discussed in Chapter 11, in the section on "Ancient Forests," old forests offer a combination of substrate continuity, habitat variety, and microclimate characteristics that foster a wealth of lichen species. In British Columbia, for example, lichenologist Trevor Goward has studied lichens in wet-zone unlogged forests, where fires are a rare occurrence. These forests are much older than the oldest trees in them; they may represent communities that are continuous back to the last ice age. He has found a very high diversity of lichen species there, including many rare lichens. Forests dedicated to the production of lumber or wood pulp are never allowed to develop beyond a stage of early succession; their simplicity of vascular plant species (and of fauna) is mirrored by a similar impoverishment in the diversity of lichens.

A more subtle but also important alteration has taken place in the arid rangelands and deserts of the West. Before the advent of widespread livestock grazing, most soil surfaces were colonized by a mixture of lichens, free-living algae, and cyanobacteria, the "cryptogamic" or "microbiotic" crusts discussed under "Colonization and Stabilization of Soil," in Chapter 8. They controlled wind and water erosion and contributed nitrogen and organic material to the thin soils in these areas. Where grazing has been heavy, the delicate crusts have all but disappeared, and they are very slow to recover. As ecologist Thomas Fleisher put it in the September 1994 issue of *Conservation Biology,* "If a single footprint can bring a local nitrogen cycle almost to a halt, the impact of a century's work of livestock hoofprints can easily be imagined" (p. 633). The problem is recognized by range managers and U.S. federal agencies, and efforts are being made to control and mitigate the damage. The destruction of soil crusts by off-road vehicles is a relatively new phenomenon, but one which is serious enough to warrant close attention (plate 78).

Sometimes habitat alteration can be an inadvertent by-product of human actions. Along the west coast of Oregon, for example, are stretches of sand dune communities with scattered conifers and an understory of shrubs such as huckleberry and manzanita. The open structure of the community, with its high light levels, abundant moisture, and complex array of substrates, provides a rich, unique habitat for many lichens, including a number of rare species such as *Erioderma sorediatum.* The area, however, is becoming completely overrun by Scotch broom, a weedy European shrub introduced to Vancouver Island in 1850 for its showy yellow flowers. It spreads quickly and forms

78. Soil lichens destroyed by an off-road vehicle in this lichen-juniper sand flat near Valemount, British Columbia, will take many years to recover.

79. Rock climbers at the Enchanted Rock State Natural Area in Texas can have an impact on the rock lichens, and the lichens, especially when wet and slippery, can have an impact on the climbers.

dense thickets, crowding out native shrubs and completely destroying the lichens. Similar invasions of aggressive plants are a serious threat to native species in many areas of North America and around the globe as well.

The increasing popularity of rock climbing as a sport has put some of the rarer cliff-dwelling species of lichens at risk. Park managers and environmentalists have already begun to assess the damage that can be caused by the heavy use of some cliffs by climbers (plate 79).

Although lichen communities may eventually recover if the causes of their destruction are removed, there is an ultimate limit: an extinct species will never return. Some species are approaching local extinction; an example is *Roccella babingtonii*, which is essentially gone from southern California (except for the Channel Islands) but is still fairly common along the west coast of Baja California. Others are globally endangered, such as the two lichens in the United States that are listed under the Federal Endangered Species Act, *Cladonia perforata*, found only on scattered sandy plots in Florida, some threatened by housing developments, and *Gymnoderma lineare*, which is restricted to a few wet cliff faces in the southern Appalachians. Many other rare species probably deserve protected status. Lichens have not been studied as thoroughly as plants, and no one really knows how many species became extinct in North America before they were known to science.

Truly, where lichens flourish, we can be confident of a healthy, stable environment. Where they are absent, one immediately suspects an ecosystem that is too simple, too young, or too polluted to be entirely "natural." Landscapes where lichens typically grow but are now absent look weirdly bare. Future generations that accept this impoverishment as normal will never even be aware of the richness they are missing. Natural communities will regrow if given a chance, but humans will have to take appropriate action to make that possible.

ENVIRONMENTAL MONITORING WITH LICHENS

The quality [of lichens near Manchester] has been much lessened of late years through . . . the influx of factory smoke, which appears to be singularly prejudicial to these lovers of pure atmosphere.

L. H. Grindon (1859)

LICHENS AS POLLUTION MONITORS: WHY DOES IT WORK?

Undoubtedly, the most important modern use of lichens is for monitoring air quality. For over 140 years, lichens have been known to be extremely sensitive to air pollution. This sensitivity derives from their ability to absorb chemicals rapidly from the air and rainwater, and from the delicate balance within the lichen symbiosis between the needs of the fungus and those of the photobiont. If a pollutant even slightly affects the well-being of one component — for example, by damaging the photosynthetic ability of the alga — the partnership quickly breaks down and the lichen dies.

POLLUTION INDEXES, SCALES, AND MAPS

Lichens can be harmed by a variety of pollutants, especially sulphur dioxide, a by-product of the burning of fossil fuel. Sulphuric and nitric acids (components of acid rain), fluorides, ozone, hydrocarbons, and metals such as copper, lead, and zinc are other important pollutants affecting lichens. Some lichens are more sensitive than others to these pollutants, so a survey of the lichens in and around an urban or industrial area can give a good indication of air quality when the survey is combined with a study of the lichens found in more pristine habitats in the same region or historical records of lichens from the area (plates 80A and 80B). Close to a pollution source, lichens are completely absent, creating a "lichen desert." Farther away, pollution-tolerant species appear; farther still, the lichen flora approaches normality in species richness and abundance.

The most pollution-sensitive lichens include the filamentous, fruticose species such as *Usnea*, *Ramalina*, and *Teloschistes* as well as epiphytic lichens containing cyanobacteria as the principal or secondary photobiont — for example, species of *Lobaria*, *Pannaria*, and *Nephroma*. *Usnea longissima*, once fairly common in humid regions of Europe, is now almost extinct on that continent, apparently owing to widespread air pollution. A 10-point scale or zone system that can be directly related to average sulphur dioxide levels was devised by David Hawksworth and Francis Rose based on the differential sensitivity of lichens. It has been used successfully to map pollution levels in England, Wales, and Ireland. The method is so easily taught that schoolchildren have been enlisted in local programs not only to help in the task but also to

80. The casual observer would assume that (A) red alder bark is white, but in fact, the bark is almost completely covered with white crustose lichens. (B) Downwind from a pulpmill in the Tongass National Forest, Alaska, the lichens cannot survive, and so the brown color of the bark becomes apparent.

learn about the effects of air pollution on their local environment.

Lichen surveys can be done as straightforward inventories of species or can involve carefully designed plot sampling and quantitative analysis, depending on the levels of pollution in an area and the need for precision. An "Index of Atmospheric Purity" (IAP) for a particular region can be calculated using a formula based on the occurrence, abundance, and pollution sensitivity of various lichen species. The IAP method, developed by Fabius LeBlanc and Jacques DeSloover in 1970 for a study of the Montreal area, has been used more recently for tracking pollution in industrial areas of Sudbury (Ontario) and Indiana.

A map of the distribution of the lichens or of the IAP values can often provide a clear picture of not only the levels of pollution but also their source and direction of movement (plate 81). When surveys are repeated over time, mapping the location of lichens as they were and as they are can graphically demonstrate the deterioration or improvement of air quality. For example, Clifford Wetmore compared his inventory of the lichens in the Indiana Dunes Recreation Area near Chicago with one made 90 years before and documented the disappearance of 80 percent of the lichen flora. He had similar results in a study of the Cuyahoga Valley Recreation Area in northern Ohio, where 79 percent of the lichen diversity was lost over a period of 70 years. On the other hand, the efficacy of pollution abatement programs has been amply demonstrated. Mark Seaward mapped the reinvasion of *Lecanora muralis* in Yorkshire, England, that followed passage of pollution-reduction laws in 1956 and 1968 (Fig. 25). In southeastern England, pollution-sensitive lichens such as *Usnea* have again become established in areas devoid of lichens only decades before. Lichens started to reinvade the trees in the nickel-smelting region near Sudbury, Ontario, and near a power plant in southeastern Ohio after higher chimneys and new pollution scrubbers were installed that substantially reduced gaseous sulphur dioxide at the ground level.

Because lichens absorb pollutants far more efficiently than most other organisms, they can be analyzed in the laboratory for the polluting compounds (plate 82). Such laboratory analysis of lichen samples has been used to assess environmental contamination by aromatic hydrocarbons and highly toxic polychlorinated biphenyls (PCBs) as well as sulphur dioxide and metal pollution emanating from smelters and other industrial sites. Lichen analysis was used to monitor the deposition of radioactive materials following the Chernobyl incident. Often, floristic surveys and lichen tissue sampling are done together to provide a more complete picture of pollution levels.

TRANSPLANTS

Transplanting healthy lichens into polluted areas or into areas that are undergoing pollution abatement programs can give investigators a good idea of the extent of pollution or the effectiveness of the abatement. One method uses bark disks with foliose lichens such as *Hypogymnia physodes* or *Flavoparmelia caperata* (plate 83). In other studies, lichens, either by themselves or still attached to their original substrate, were fastened in new localities with inert adhesives such as silicone caulk or by tying them in place with monofilament fishing line.

LICHEN-BASED PROGRAMS OF POLLUTION MONITORING

Lichen surveys, analyses, and transplant programs are much cheaper and faster for monitoring air quality than are studies that rely on expensive, sophisticated air analysis equipment. Because lichens respond to changes in microclimate and habitat loss as well as pollution levels, they can also tell us a great deal about general environmental degradation or improvement. The use of lichens as pollution "canaries" is now so firmly a part of environmental assessment that the U.S. Forest Service and National Park Service have es-

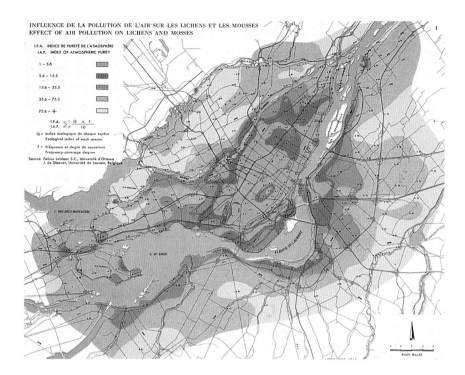

81. Map of the Index of Atmospheric Purity values of lichens and mosses in and around Montreal, Quebec, shows the deterioration of epiphytic vegetation as one approaches the built-up area. Purple = "lichen desert" where no lichens are found; yellow = areas with relatively normal lichen growth; other colors = intermediate conditions. An "island" of improved lichen growth in the city core centers on an isolated mountain (Mount-Royal), reaching above the heavy layer of pollution and providing a refuge for some lichens. (From Leblanc and DeSloover, 1970.)

82. The lichens in these crucibles will be analyzed for accumulated pollutants as part of an integrated investigation of air quality in southeast Alaska.

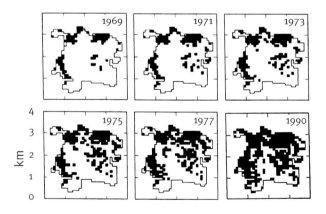

Fig. 25 Reestablishment of *Lecanora muralis* in West Yorkshire, England, after pollution reduction. The black squares indicate 1-km² areas where the lichen was found. (Reproduced by permission from Richardson, 1992, fig. 8.)

83. This bark disk with *Flavoparmelia caperata*, and four other disks, were placed at varying distances from New York City as part of a study to assess the influence of the city climate and pollution on the lichen flora of Long Island. (Photo by I. M. Brodo.)

84. To study lichen growth and diversity in the canopy of an old forest in Washington, researchers use a crane to sample the lichens.

tablished active programs of baseline studies of lichen vegetation in national parks, forests, and recreation areas across the United States and have also conducted detailed investigations of lichen sensitivity and sampling methods (see the "Further Reading" section at the back of the book). The Canadian government, through its Environmental Monitoring and Assessment Network (EMAN), has also begun to investigate the use of lichens for large-scale monitoring programs. Although more research is needed on the use of lichens as environmental indicators, we can expect to see a much wider application of lichen study in this realm in the future.

PROSPECTING

If lichens can absorb and accumulate tiny quantities of metal pollutants, they can also absorb trace amounts of metals from soil dust whipped up from nearby areas. Metal-rich dust can reveal the presence of ore bodies. Lichen analysis has been used in certain regions for mineral prospecting under the assumption that the presence of lichens that contain high levels of copper may indicate copper deposits in the neighborhood. This technique, developed mainly in eastern Europe, has never been used extensively in North America.

ANCIENT FORESTS

Certain lichens have been found to be restricted to forests that have been undisturbed for very long periods of time (e.g., 200–800 years). Such forests contain trees and understory vegetation of mixed age as well as dead standing trees, logs in various stages of decomposition, and scattered openings caused by windthrows. The well-developed canopies and thick, moist soil of old forests contribute to a microclimatic buffering that provides a more uniform moisture regime than is found in younger forests. The variety of unique habitats and microclimatic conditions promotes an increased species diversity, making such forests especially rich. Recently, it has been shown that complex self-contained ecosystems exist high in the canopies of old growth forests, and that lichens are critical components of these habitats, contributing nitrogen and minerals and even carbohydrates to the canopy ecosystem (plate 84).

Certain lichens respond to the special microclimates, substrates, and other rare ecological conditions found only in forests that have remained undisturbed for hundreds of years, and these lichens can be used as indicators of ecological continuity. Examples include many of the stubble lichens (species of *Chaenotheca* and *Calicium*), *Lobaria* species, *Usnea* species, *Pseudocyphellaria rainierensis*, *Nephroma occultum*, and many crustose species. Because many of these lichens are restricted to specific geographic regions or forest types, an indicator species in one part of the continent may not serve this function in another area. Furthermore, a species may thrive in some habitats but be much more vulnerable (and therefore a good indicator) in others. For these reasons, it is necessary to discover and list the indicator species for each major vegetation type and region. Reliable, easily identified indicators of forest continuity become more and more important to forest managers and natural resource departments as more and more old growth forests disappear, and with them, the animals and plants that rely on ancient forests for habitat. Much research is therefore currently being focused on old growth indicators and their ecology.

NAMING AND CLASSIFYING LICHENS

But Nature, ever prone to fling
Some beauty round the rudest thing,
Has clothed the avalanche of stone
With moss and lichens, all her own.

John Critchley Prince

LICHEN NAMES

The appearance of each lichen is determined almost entirely by the genetic information contained in the fungus, which, in most cases, determines the lichen's structure. Evidence of various kinds supports the assumption that, with few exceptions, every recognizable lichen is derived from a different species of lichenized fungus. For this reason, and in agreement with internationally accepted rules of nomenclature, the name we give to a lichen is actually the name of its fungal component. When we say, "This lichen is *Cladonia cristatella*," we mean that the fungus of the lichen is *Cladonia cristatella*; the photobiont has its own name, in this case, the green alga *Trebouxia erici*.

We cannot place intact lichens within the hierarchical systems of biological classification according to categories including kingdom, phylum, class, order, family, genus, and species, because lichens are dual organisms, and each component has its own classification. Relationships among lichens are expressed in the classification of the lichen-forming fungus alone. Many orders, families, and even genera of fungi, in fact, contain both lichen-forming and non-lichen-forming species, and recent work with the DNA (basic hereditary material) of lichen fungi shows that some lichens are more closely related to nonlichenized fungi than to other lichens.

SCIENTIFIC NAMES AND THEIR AUTHORS

Scientific names consist of two elements: the name of the genus (in italics, beginning with a capital letter) followed by the species designation or "epithet" written entirely in lowercase italic letters. If you see two lichens with the same genus name, it is reasonable to expect them to have a lot in common. *Usnea longissima, Usnea baileyi,* and *Usnea subscabrosa* all belong to the genus *Usnea*, sharing a similar thallus structure, spore type, chemistry, and so forth. The species epithets are Latinized descriptive terms of some kind: *longissima* means "very long" in Latin; *subscabrosa* means that it is similar to another species called *Usnea scabrosa*, and *scabrosa* means "rough," referring to the abundant isidia covering the branches. *Usnea baileyi*, "Bailey's *Usnea*," was named in honor of F. M. Bailey, who first collected the lichen (in Australia).

The unabridged scientific name of a species consists not only of the genus epithet and species epithet but also of the name of the individual who first described that species, the author.[1] For example, in 1753, Linnaeus coined the name *rangiferinus* for the reindeer lichen (because reindeer belong to the genus *Rangifer*), and he placed his new species in the genus *Lichen*. The complete scientific name was then written: *Lichen rangiferinus* L., the "L." referring to Linnaeus. (The author's name is commonly abbreviated.) As we learn more about a species, however, we often find that it must be reclassified, which may involve transferring it to a different genus. If this is done, the species epithet has to be combined with a different genus name. The author of this "new combination" is also cited in the scientific name, following the author of the original name, which is now placed in parentheses. Thus, in 1780, when Weber decided that *Lichen rangiferinus* is better classified in the genus *Cladonia*, the name of the lichen became "*Cladonia rangiferina* (L.) Weber" (*rangiferinus* changing to *rangiferina* because of Latin grammar). Still later, in 1867, Nylander placed the species in his new genus *Cladina*, and the name and its authorities changed once again, this time to "*Cladina rangiferina* (L.) Nyl."

Some species are first described as varieties or subspecies and are only later raised to the species level, sometimes not even within the same genus. The authors involved, however, are cited in the same way: first the originator of the species epithet in parentheses, followed by the author who made the currently used combination. For example, "*Lecanora glabrata* (Ach.) Malme" was first described as a variety in 1810 by Acharius: "*Lecanora subfusca* var. *glabrata* Ach." A hundred years later, Malme recognized it as a full species, still within the genus *Lecanora*.

LICHEN CLASSIFICATION: FAMILIES OF LICHENS

The urge to organize diverse things into groups is one of the most basic of human traits. Creating order out of apparent chaos is a deeply satisfying endeavor, whether it involves organizing food in a kitchen, items in a newspaper, or species in the living world. We have already seen that the two-element scientific name of each species reveals the first level of organization: the genus. It is equally useful to group related genera into families, related families into orders, and so on. Knowing the families of living things tells us a great deal about basic relationships. For example, you may see a warning on a can of insecticide that the spray can damage all members of the rose family, which would include strawberries, cherry trees, and wild roses. These plants are superficially different, but the structure of their flowers shows that they share certain evolutionary advancements, and this links them together within the same family. Similarly, we group the foliose lichen genera *Ahtiana*, *Parmelia*, *Hypogymnia*, and *Cetraria* together with *Bryoria*, a fruticose lichen, because similarities in their fruiting structures reveal what appears to be their common ancestry. They are all placed in the family Parmeliaceae. (Families of plants always end with "-aceae.") Members of the Lobariaceae (*Sticta*, *Pseudocyphellaria*, and *Lobaria*) also have many basic characteristics in common such as their photobionts and cephalodia, and how their apothecia develop. To help the reader better appreciate these kinds of relationships among the lichens, we have presented, as an appendix, a classification of most North American lichen fungi including all those mentioned in this book. Although this classification (like all classification systems) is in a state of flux due to ongoing research, especially DNA studies, we believe the outline provides the intellectual framework needed for better understanding the natural relationships among lichens and their diversity.

SCIENTIFIC VERSUS VERNACULAR NAMES

In these introductory chapters, we have referred to lichens largely by their Latin or scientific names, with just a smattering of names in English. For better or worse, most lichens simply do not have truly "common" names, that is, names used in the popular culture. If we want vernacular names now, we must invent most of them. The questions then become, should we do so, and if we coin names, how should we do it?

Scientific names, ideally, are relatively stable and are used consistently throughout the world. The internationally sanctioned Code of Botanical Nomenclature governs how scientific names of plants and fungi can be created, published, and changed, with the aim of making scientific names stable and reliable. Papers published in Chinese, English, Czech, Portuguese, and so on all use the same Latinized scientific name for the same taxon. Common names, by contrast, can vary from culture to culture. For example, species of *Usnea* are commonly called "beard lichens," but some people also use that name to refer to species of *Alectoria* or

---

[1] In this book, the complete scientific name of each lichen, including the authors of the name, can be found in the index, not in the main body of the text.

Evernia. Bryoria is called "black tree moss," "tree hair," "witch's hair," or "horsehair lichen" as well as "we-ia" (Nlaka'pamux), "wa-kamwa" (Oregon), Moosbart [moss-beard] (Germany), and so on.

If the importance of names lies in the information associated with them, then at least in the case of lichens, scientific names have infinitely more value than do vernacular names, especially newly created ones. Knowing the scientific name of a lichen opens the door to a whole body of literature about that lichen, including its description, history, uses, and distribution. You have seen in the section above that using the scientific name allows you to learn something about a lichen's classification (specifically, its genus and related species) at the same time you are learning what to call it.

But scientific names are indeed in a foreign language for most people, and that makes them harder to pronounce, harder to remember, and close to impossible to understand. For those people who encounter lichens only casually—on a guided nature walk in a park or in the course of reading a popular article on natural history—being introduced to a lichen by its unpronounceable Latin name is frustrating. The name is easily forgotten, and so is the lichen. Teachers and interpretive naturalists have therefore been urging lichenologists to introduce vernacular names into their popular articles and books so that lichens can be discussed more easily and effectively with people who are relatively unfamiliar with the natural world. We agree that the need is there.

In this book, we have included vernacular names for most of the macrolichens, at least at the genus level, and also for most genera of crustose lichens. We feel that introducing a "common" name for every lichen in the book is impractical and, indeed, undesirable. Having two names for every lichen can double the job of becoming familiar with a species. In the case of crustose species, one usually needs to look at microscopic features to name the lichen, and so vernacular names, even at the genus level, would be difficult to use reliably in the field.

Vernacular names are not entirely unknown in North American lichenology. Guy Nearing, in his *Lichen Book*, gave English names to every species in the northeastern flora, but few of these names have caught on. We use some of them here, but most seem inappropriate for one reason or another. Trevor Goward, Bruce McCune, and Linda Geiser have introduced English names in their publications covering the lichens of the northwest, and many have been adopted in this volume or are listed as alternatives. An informal committee of lichenologists has attempted to come up with a list of English names, with only partial success. We have used many of the committee's suggestions and translated aboriginal as well as newly created vernacular names from Europe, but have, nevertheless, finally had to invent many of our own names. Our approach has been to use names that are descriptive and easily remembered. If an English translation of the Latin name makes sense, we have used it (e.g., "kidney lichens" for *Nephroma* in reference to the kidney-shaped apothecia, or "bloodspot lichens" for *Haematomma*, describing the color of the apothecia). Although we have tried to avoid references to microscopic characters, sometimes we have found them to be the best choice (e.g., "can-of-worms lichen" for *Conotrema urceolatum*, relating to the long, segmented spores of that species). We also tried to be straightforward rather than overly metaphoric, using "green," for example, instead of "frog-" in reference to color. We hope that those requiring vernacular names find our choices acceptable. If the existence of vernacular names helps popularize lichens, then our purpose will have been served.

COLLECTING AND STUDYING LICHENS

There is a low mist in the wood—
It is a good day to study lichens.

Henry D. Thoreau

Sometimes it may be necessary to collect a lichen that you wish to identify. If you bring the specimen back to your home or office, you can use a microscope to examine features that cannot be seen with a hand lens, and perhaps use some chemical reagents that are not convenient to use in the field.

CONSERVATION

A few words, however, must first be said about conservation and collecting ethics. Lichens grow very slowly, and an entire population can be removed from an area with the slice of a knife or the swing of a hammer. It is a good idea, therefore, not to collect anything unless you are serious about studying it later and intend to add it to your collection, and unless the lichen is abundant enough to permit leaving a portion of the colony behind for regeneration. A good rule to follow is: do not collect more than you need, and never collect every last scrap of anything.

Also, it is very important to obtain permission to collect from property owners, park managers, or local authorities. Be aware that disturbing or collecting living organisms, or even rocks, is illegal in most public parks unless you have a permit. Collecting lichens usually involves some damage to trees or rocks. Property owners or managers should therefore be made aware of what you are about to do so that misunderstandings are avoided. One should strive to minimize the damage to a tree. Never chip lichens off gravestones, and obtain permission to collect from cement walls and stone fences.

TOOLS

To begin with, you need some sort of magnifier. The characteristics you need to see, and sometimes the lichens themselves, are often very small. A hand lens that can magnify about ten times ("10×") is ideal, but anything from 7× to 16× will work. At the lower powers, you may miss some details, and at the upper end your viewing field is small and it is hard to hold anything in focus.

Because lichens grow on a variety of substrates, including tree bark, twigs, wood, soil, and rock, you need a variety of tools to remove them. Most useful is a strong, sharp knife about 12–18 cm (5–7 in) long, carried in a leather sheath. If you use a pocket knife, make sure it has a locking blade. Collecting lichens on hard bark or wood often involves a certain amount of prying, so the knife should be thick enough to withstand that kind of abuse without breaking. It should be sharp so that it cuts into the bark rather than slides over it, risking a bad cut to your hand or, worse, ruining the lichen you're trying to collect. Some collectors find that a broad wooden chisel 3.5–5 cm (1.5–2 in) wide used with a hammer is ideal for collecting on tough bark or wood. Many lichens grow on small branches or twigs, and for these, good-quality pruning shears (from a garden shop) are a great help.

What about lichens that grow on rock? For reasons of bulk and weight, the first choice is to remove rock lichens from the rock surface when possible. This can be done only with foliose or fruticose lichens because crustose lichens cannot be separated from the rock substrate. Foliose lichens, even those that are fairly closely attached to the rock, can be removed in one piece with very little damage by wetting them with water from a small spray bottle and then carefully lifting the lichen from the rock using an artist's palette knife. Such a tool is thin and flexible and can be manipulated along the contours of the rock under the lichen. You can even fold back the detached part of the lichen as you work to get a better view of your progress. A palette knife, of course, works just as well on bark lichens, especially tightly adherent foliose species.

Collecting crustose lichens on rock involves the use of a hammer and cold chisel—there is no other way. Any small, well-balanced hammer will do, such as a ¼–½-kg (16–32-oz) sledge or a geologist's hammer. Cold chisels come in various sizes, but those 12–20 mm (½–¾-in) wide are best. Keeping them reasonably sharp is a chore but worth the effort. Carbide-tipped chisels are expensive but retain their edge very well. Always wear a glove, at least on the chisel hand, as well as protective eyeglasses, when using a cold chisel. With a little experience and patience, you will learn the characteristics of most rock types and will soon be able to chip off nice, specimen-sized fragments with the lichen intact. (A frustration all lichenologists share, however, is that the most interesting lichens always seem to grow on the smoothest, hardest, most inaccessible rock surfaces.)

Soil lichens are usually easy to collect, but we sometimes require special techniques and materials for transporting them (see below). A knife can be used as a narrow trowel for picking up patches of soil lichens, especially crustose or squamulose species. If the soil is sandy or otherwise loose, cut a fairly thick slice of the underlying soil together with the lichen. Fruticose cladonias can usually be simply picked up (together with the squamulose primary thallus), the loose soil brushed away, and placed in a packet or bag.

We have mentioned the need to collect "specimen-sized" samples. But what does that mean? Usually that depends on the lichen. Try to include at least part of the lichen's margin or edge (this area has many important identifying characters), as well as fruiting bodies if they can be found. The specimen, or the portion of substrate in the case of tiny lichens, ideally should be about the size of your palm, enough to fill the 10 × 15 cm (4 × 6 in) packet we describe below. It will often end up as a number of fragments, which is perfectly acceptable as long as they are not so small that they are hard to handle.

TRANSPORTING LICHENS

It is easy to ruin a beautiful lichen specimen between the tree or rock where it was found and the place where it is to be stored and studied. Once collected, lichens can be put either in small paper bags (no. 1 to no. 3 size), or prefolded paper envelopes or packets. Packets can be folded from sheets of ordinary paper or even newspaper (see "Preparing Specimens for Study and Storage," below). Avoid plastic bags because they promote the growth of mold on the lichens. Paper bags hold more than packets and tend to remain closed, but if you are not collecting duplicates for later exchange, packets are usually sufficiently large and will keep the lichens flat. Information on the exact habitat of the lichen (e.g., forest type, tree species, position on tree, proximity of a road or lake, rock type, amount of shade) should be written with a soft pencil or with waterproof ink on the outside of the paper bag or packet. It is best to put only one species in each bag. If you put several species together to save time in the field, numbering and sorting will take longer once you get the specimens home. A compromise is to write how many species are included, or their names if you know them, on the outside of the bag as you put the lichens in.

Some types of specimens require special care. Soil lichens should be carefully wrapped in facial tissue or toilet paper to keep the soil intact before putting the lichens in the bag. Each rock fragment bearing a lichen should be wrapped separately in a small square of newspaper or tissue to prevent the rocks from rubbing against the lichens. Lichens collected on resinous bark such as balsam fir or spruce should be placed in the bag or packet so that the sticky, cut side of each bark fragment is against another cut surface, protecting the lichens from being covered with sap.

The stubble lichens (e.g., *Calicium* and *Chaenotheca*) are extremely delicate. Ideally, each major fragment of bark or wood bearing the lichen should be wedged or glued into a small box to prevent the specimen from being crushed. Another, somewhat less effective way of safely transporting stubble lichens is to wedge the wood or bark pieces into a small, heavy paper bag so that the lichens do not touch other fragments or the sides of the bag.

It is a good idea to use a larger bag of some kind (grocery bag or small cloth sack) for each locality you visit. When you move on to a new locality, use a new bag, indicating the complete locality information on a slip of paper (or an empty collecting bag) thrown into the bag. An alternative is to number each bag with a locality number corresponding to a description of the locality in your field notebook. Then, lichens collected in different localities can be thrown in together, even directly into a backpack. Some collectors carry prenumbered slips of paper and throw one into the specimen bag as each lichen is gathered.

BACK HOME

Once the lichen specimens are safely on the kitchen table or in the lab, it is time to sort and dry them and to number them if you have not done so in the field. Some collectors (like us) number everything they collect, whether or not they keep it. Others wait until they have identified their specimens. There are various systems of numbering, but giving each specimen a unique number associated with the locality and date of collection is very important for future reference. If several lichens have been stored within a single bag, they should be sorted, separated, and given their own numbers. A collection book made from a bound notebook with all your specimens listed one line at a time is very useful. As your specimens are identified, the names, and perhaps the substrate, can be entered, giving you a permanent record of all the species collected at each locality. A computerized collection book is, of course, possible, but the existence of a bound, hard copy is reassuring in these days of constantly changing software, hard drive crashes, and electronically corrupted data.

Even in good weather, many lichens are damp when they are collected. It is very important that any lichen that is not perfectly dry be allowed to dry out quickly. Damp bags or packets can simply be laid out on the floor or table overnight or placed in a plant press with a fan blowing air through the corrugated cardboards. Bulky lichens such as reindeer lichens or species of *Parmotrema* should be pressed lightly before being packeted. If they are too dry and stiff to press, soak them in distilled water (tap water contains chemicals that can cause many lichens to discolor), squeeze out the excess, and then put the specimens into the plant press. Use only a light weight on the press rather than straps so that the lichens are not squeezed into unrecognizable pancakes. *Never* use heat, because many lichens, especially the jelly lichens (e.g., *Collema*) and others with cyanobacteria, can turn black and be destroyed by excessively heated air.

PREPARING SPECIMENS FOR STUDY AND STORAGE

Most lichens can be stored in a paper packet easily folded from ordinary, letter-size paper (21.5 × 28 cm; 8½ × 11 in). You will not need training in origami. An easy method is illustrated in figure 26. Ideally, your packets should be made of durable, acid-free paper to prevent discoloration and deterioration of the lichens inside. Packets about 10 × 15 cm (4 × 6 in) are a good size.

Flat samples of foliose or fruticose lichens can be placed directly in the packets. In many large herbaria including the one at the Canadian Museum of Nature, a piece of stiff cardboard is first placed inside the packet to support the lichen and make it easy to handle. To make an even more elegant preparation, you can place some sort of cushioning material, even a folded facial tissue, under the lichen so that it will not slip to the bottom of the packet when stored vertically, while at the same time giving the specimen some protection against being crushed.

Fragments of rock or bark bearing crustose lichens should be glued to a piece of cardboard. This makes it

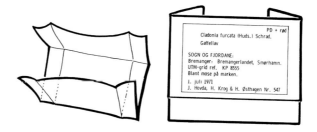

Fig. 26 How to fold a packet. (Reprinted by permission from Krog, Østhagen, and Tønsberg, *Lavflora*, 1980, Universitets forlaget, Oslo, fig. 23.)

```
         LICHENS OF    VERMONT
              Windham County

    Hypogymnia physodes (L.) Nyl.
          on cherry

     South of Wardsboro Center on south
    facing hill north of Rice Mountain, elev.
    1500 ft.  26-27 Aug. 1962.

    Collected by Clifford M. Wetmore    No.  12094
```
a

```
              LICHENS OF ALASKA
    Lecidella stigmatea (Ach.) Hertel & Leuckert


                                    58°37'N 134°56'W
    Juneau: Sunshine Cove, at west end of road, 25 miles west of Auke Bay. Boulder
    beach and rocky cliff at north end of cove. On rocks at upper edge of beach, at
    point.

    NOTES: Thallus C-.

    05 JUN 1988                        I.M. Brodo, no: 26024A
                                       with F. Brodo & R. O'Clair
    CANL 108194                Det:
```
b

Fig. 27 Prepared labels: (a) typed; (b) computer-generated.

very easy to view these specimens and provides a way to indicate (by means of an arrow drawn on the cardboard) which fragment was studied for the identification written on the label. It is, unfortunately, very easy to collect a mixture of superficially similar species, thinking they are the same. The arrow takes the guesswork out of subsequent examinations. Ordinary white casein glue (e.g., Elmer's™), which is water-soluble, works well as an adhesive. Never glue a foliose or fruticose lichen directly to a cardboard or sheet; the substrate is the only thing that the glue should touch.

Soil lichens, especially those growing on loose soil or sand, tend to fall apart in a packet unless the soil mass is stabilized in some way. The lower part of the clump of soil can be dipped in a mixture of one part white glue to one part water, allowing the mixture to be absorbed. Care has to be taken to avoid submerging the lichen itself in the glue. When the glued soil dries, the whole mass can be attached with undiluted glue to the cardboard.

PREPARING A LABEL

No lichen is a useful specimen without some information about where and when the lichen was collected, who collected it, what the collection number is, and something about the habitat and substrate. *Least* important is the name of the lichen. That can always be determined later; the locality data cannot. Labels have to be easy to read, but a clearly handwritten label in waterproof ink is just as good as one prepared with the best laser printer. A simple but adequate label and a fancy computer-generated label are shown in Fig. 27, but many other formats are equally good.

Labels should be permanent; try to use the best-quality, acid-free paper you can get. Attach the labels to the front of the packet with a good-quality glue (white glue is excellent), not with rubber cement, glue sticks, or other adhesives that dry out or become discolored. Only one strip of glue along the upper edge of the label is necessary.

THE PERMANENT COLLECTION

The reason for keeping specimens is to have a reference collection for purposes of comparison. Keeping a private collection can be very useful and satisfying, and we encourage you to do so if the need is there. You may find, however, that a nearby university or natural history museum has its own research collection (herbarium) and would welcome the material you have collected. If your specimens are deposited in such a herbarium, they are likely to be consulted more frequently and thus to contribute to our general knowledge of lichens.

Assuming, however, that you are creating your own herbarium, most of the lichens you have collected should be kept in the packets described and illustrated above. Bulky specimens such as those on unbreakable rock can be stored in oversized packets or in small boxes. Packets can be stored vertically in shoe boxes, which hold about 20–30 specimens (depending on the thickness of the lichens) and can easily be labeled on the outside for later reference.

Lichens can be stored for decades with little or no special care as long as they are kept dry. They are rarely bothered by insect pests such as dermestid beetles, although in damp conditions silverfish can be a problem. Specimens should not be stored with any chemicals such as paradichlorobenzene (PDB, or moth flakes), commonly used in collections of fungi, because they might interfere with later chemical tests.

STUDYING LICHENS

Equipment

A hand lens of about 10× magnification (described earlier) and a few reagents available from the grocery store are all that is needed to use the basic keys to groups and even to species of the macrolichens. To make full use of the information provided in the descriptions and identification keys, however, you will need some additional equipment. This supplementary equipment will simply make your work easier and more enjoyable.

Beyond a hand lens, the most useful aid for the study of lichens is a stereomicroscope, also called a dissecting microscope. The magnifications available are usually 6× to 50×. A good light source is important for the higher magnifications, and one provided with a light blue filter will ensure a white rather than an unnatural yellowish light on the subject. This is critical for judging colors.

A compound microscope with magnifications from 100× to 400× (60× to 1000× is better) is used for examining spores, photobionts, and certain anatomical features of lichens. These examinations are usually necessary for the confident identification of crustose lichens, although many crustose lichens are recognizable with just a hand lens.

Other important tools are: razor blades (single-

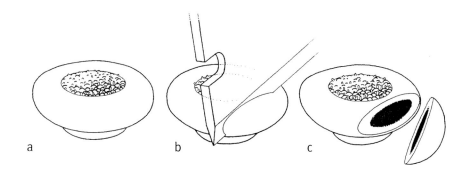

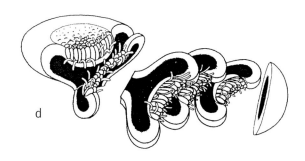

edged are safest), a pair of fine forceps, and some sort of needle for probing and moving things around. A heavy one mounted in a wooden handle is called a dissecting needle. Pins used for mounting insects are ideal for fine work.

Microscopic Study

Examining lichens under a stereomicroscope requires no special preparation. To see the anatomical details of an ascoma or thallus, however, you have to use a compound microscope, which involves making a very thin, almost transparent section of the lichen and placing it in a drop of water on a microscope slide. Sections are made most easily by slicing the material with a new razor blade while viewing it under the dissecting microscope. Sections are always made top to bottom and are most easily and successfully done while the ascoma is still attached to the thallus and substrate. The substrate holds the material in place like a vice. It takes practice to make really thin sections easily, so don't get discouraged when you start out.

For crustose lichens firmly attached to bark or stone, wet the ascoma with a drop of water. After the water has been absorbed and the ascoma is soft, make a series of sections, leaving them attached at the base (Fig. 28). The razor can be rested against a fingernail as a guide. Once the slices have been made, remove the first few thick sections from the outer margin of the

Fig. 28 Technique for sectioning an apothecium (Reprinted by permission of the publisher from W. M. Malcolm and D. J. Galloway, *New Zealand Lichens: Checklist, Key, and Glossary*, Museum of New Zealand Te Papa Tongarewa, Wellington, 1997, p. 185.)

ascoma. (You can save them at the edge of the microscope slide for later chemical tests.) Then cut through the bases of the central sections and move the sections on a corner of the blade over to a small drop of water on the microscope slide. Spread out the sections with a pin or needle and then cover the preparation with a cover slip, starting at one edge to avoid trapping air bubbles.

Ascomata attached to moss or soil should not be moistened before they are sectioned. Instead, wet the edge of the razor blade (for example, in the drop of water on your microscope slide) and then moisten only the portion of the ascoma to be sectioned, not the entire ascoma. This keeps the underlying moss or soil rigid, supporting the material being sectioned.

Sections of foliose thalli are best made with lobes removed from the lichen and placed on a microscope slide in a small drop of water. While being sliced, again, under the dissecting microscope, the lobe has to be held in place with one finger, a dissecting needle, or a cover slip.

Another technique is to embed your specimen (thallus or fruiting body) in a drop of white, water-soluble glue (e.g., Elmer's School Glue™) on a microscope slide, allow it to solidify partially, then cut the sections as usual. The sections (and surrounding glue) are then transferred to a small container of water where the glue dissolves, allowing the thin slices to float free. These can then be transferred to a drop of clear water or mounting medium using a fine, camel's hair paintbrush with all but a few of the bristles removed.

To see spores, we normally begin with sections prepared in the manner just described. More often than not, some ripe spores will be released into the drop of water and can easily be viewed and measured. (Measure only mature spores, the ones with clearly defined walls or cells.) Frequently, however, some spores have to be coaxed out of the asci by gently crushing the sections under the cover slip with a stout dissecting needle or the eraser on a pencil. With the needle, you will be able to crush the asci while watching the results through the compound microscope at 100× magnification. Broken cover slips will be a common result of this procedure, even after you gain some experience, so be patient.

The asci and paraphyses in a hymenium are often stuck together in a kind of gel. To see the asci and spores most easily, it is best (and usually possible) to dissolve the gel using 10% KOH (the same reagent used in color spot tests; see the section on "Chemistry," below). This is done by placing a drop of KOH at one edge of the cover slip and then drawing the liquid under the cover slip by putting a small square of paper towel or filter paper on the opposite edge of the cover slip. Once the KOH has replaced the water, you can try crushing the sections of the ascoma once again. Partially collapsed spores fill out to their normal dimensions in KOH, but be aware that they can sometimes swell abnormally, especially in width. The spores of *Caloplaca* can even change in shape. If you suspect that this has occurred, compare the spores in a water mount. The outer cell walls, cross walls (septa), and cell shapes do not change their shape or thickness in water.

The spores of many lichens have a gelatinous outer envelope (called a perispore or halo) that presents a special problem. Halonate spores, characteristic of certain genera such as *Rhizocarpon* and *Porpidia* (see Fig. 15g,k), are usually difficult to recognize because the halo is transparent and often diffuse. To make the halo visible, add a tiny drop of India ink to one edge of the cover slip, mix it with a drop of water, and draw the mixture under the cover slip as described for adding KOH. The ink particles will flood the section, darkening most of the field but leaving a clear ring or halo around the spores with a gelatinous perispore.

It is sometimes necessary to stain the tissues of the ascoma, asci, or spores with iodine. To do this, a drop of a 1.5% solution of iodine in 10% potassium iodide (Lugol's solution) can be introduced under the cover slip as we described with KOH, or the sections can be placed directly in a drop of Lugol's. (Some lichenologists prefer a weaker solution: 0.5–1.0% iodine; others mix the iodine with lactic acid instead of water, which helps clear the preparation and makes it less prone to drying out.) If the sections are already in KOH, the KOH will have to be rinsed out before introducing the iodine by adding water at one edge of the cover slip and drawing it through with a piece of blotting paper at the opposite side. In fact, staining with iodine after treatment with KOH (the "K/I" stain) is a standard and often important technique for staining ascus tips and for distinguishing one type of hymenial gel from another. When the hymenium turns dark blue with iodine, it is described as amyloid, with a starch-like reaction; if it does not undergo a color change with iodine, the hymenial tissue can be called nonamyloid. If the hymenium is entirely or largely negative, greenish, or orange-red with iodine but turns dark blue with a K/I stain, then it is said to have a hemiamyloid reaction.

Preparing asci for examining the iodine staining

patterns sometimes requires other techniques. Rather than prepare thin sections of the entire ascoma, you can cut out and mount chunks of the hymenium or make somewhat thick sections and then cut away the tissues below and surrounding the hymenium. This leaves the ascus-containing hymenium free of other tissues that might prevent you from squashing it effectively when you press down on the cover slip. You can apply the series of reagents and stains (KOH, water, iodine, water) to these chunks of tissue *before* dropping the cover slip over the preparation, which is much easier than running the series on a squash under the cover slip. (See Fig. 14 for examples of ascus tips stained with iodine.)

CHEMISTRY

Hundreds of the lichens described in this book are recognizable without the use of any chemical tests or analyses. In many cases, however, distinguishing one species from another, or even one genus from another, is much more easily and confidently done with the help of some simple chemical tests. If you ignore the chemical diversity among lichens when identifying specimens, you are working at a significant disadvantage. We therefore list the major chemical products of the species we illustrate. In many cases, this information will help you confirm the identifications you have made using the photographs, keys, and descriptions.

Spot Tests

In all but a very few cases, the only chemical analyses required in the keys are "spot tests." These involve applying tiny amounts of a reagent to a lichen and observing any resulting color change. Of the three or four reagents used in these spot tests, 10% potassium hydroxide (KOH, abbreviated here as K) is probably the most useful. If K is not readily available, dissolving 10 pellets of household lye (sodium hydroxide; NaOH) in about 20 ml (1 oz) of water will give nearly the same results. The second reagent, a solution of sodium hypochlorite (abbreviated C), is even more readily available. It is simply commercial laundry bleach. The bleach is used straight out of the bottle (additives such as "lemon scent" are suspect; avoid them). The last of the routinely used reagents is an alcoholic solution of *para*-phenylenediamine (abbreviated PD). It is made by adding a drop of 70% ethyl alcohol to a few crystals of PD on a microscope slide or in a tiny vial. PD is the most difficult item to obtain. It can be ordered from a chemical supply house, but generally only in large amounts (ca. 250–500 g). So little is needed for spot tests that 10 grams or so should last you more than 10 years. Try to order some together with other lichen aficionados. Iodine, the same 1.5% Lugol's solution used for staining asci, is also used as a spot test reagent for certain lichens. Alkaline iodine (AI) will turn blue or violet on tissues containing stictic acid and is especially useful for distinguishing species of *Parmotrema* and *Xanthoparmelia* that have mixtures of stictic and norstictic acids from those with salazinic acid. (Mix 1.5 ml of 20% Lugol's solution with 18.5 ml of pH 11.0 buffer.) Unfortunately, the mixture lasts only a few days.

In fact, all these reagents deteriorate with time—some faster than others. The alcohol solution of PD is useless after an hour, the bottle of C should be changed every week or two, and the K will last for about six months to a year. Lugol's solution fades with time and is unreliable when it is no longer a deep reddish brown. A relatively stable aqueous solution of PD ("Steiner's solution") can be prepared and is often very useful. Although the colors it produces are sometimes paler than those you get with the alcohol solution, Steiner's solution can last for some months in a dark bottle away from direct light and works well with the "filter paper method" described below. First, make a 10% solution of sodium sulfite (10 g Na_2SO_3 dissolved in 90 ml of water) and add five or six drops of liquid detergent. To this, add 1 g of PD. Stir or swirl the mixture for several minutes until as much of the PD has dissolved into solution as possible. Then filter the saturated solution and discard the undissolved crystals.

CAUTION:

Chemical reagents of any kind, even those you can buy in the corner grocery store, should be handled with care. Potassium hydroxide and lye are extremely caustic. The pellets should never be picked up with the fingers. Even diluted KOH can damage unprotected skin, as can ordinary bleaching solution. Spills should be *carefully* wiped up as soon as they occur. *Para*-phenylenediamine is extremely poisonous, and even the fumes can stain paper, clothing, desktops, and other surfaces. (The aqueous solution is just as nasty as the alcohol solution.) Make sure your desk is protected wherever PD is used, and handle the reagent with respect. I have used PD for 40 years and never stained any clothing, and two inventors and prime users of PD for lichen study, Y. Asahina of Japan and A. W. Evans of

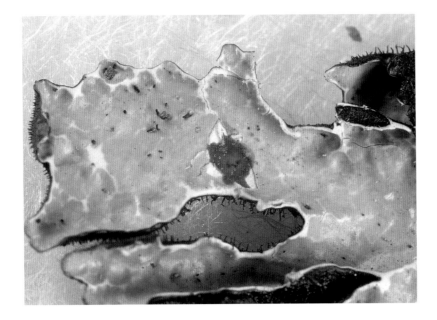

85. The blood red reaction with a tiny drop of 10% KOH ("K") on the medulla of *Parmelia sulcata* reveals the presence of salazinic acid. ×9.7

86. When the medulla of *Parmotrema tinctorum* is moistened with a drop of undiluted bleaching solution ("C"), it turns deep red because of lecanoric acid in the lichen. ×9.8

the United States, lived healthy and active lives well into their nineties.

Reagents (except alcohol preparations of PD) are normally stored in dark dropper bottles, 30–45 ml size (1–1.5 oz). The finer the dropper (for application purposes), the better. A toothpick or very fine paintbrush (prune it down to 3–4 bristles) moistened with the reagent enables more control over the amount dispensed. The best way of applying the reagents, however, is by using a finely drawn-out capillary pipette, like the ones used by nurses to take up blood from your punctured finger for those painful blood tests. Each capillary tube can be heated over a flame and drawn out in the middle, to be broken into two tubes. A very tiny, easily controlled amount of K or C can be taken up into the capillary tube from the dropper in the reagent bottle. The PD, mixed freshly on a microscope slide, is drawn into the tube directly from the slide. The tubes can be marked with a wax pencil for easy recognition (e.g., one stripe for K, two for C, and none for PD). Another tube can be prepared for iodine tests, easily distinguished because of the color of the iodine.

To perform a spot test, you simply apply an appropriate reagent to a reactive part of the lichen and observe the color change, if any. Because different compounds are produced in specific parts of a lichen, references to spot tests usually tell you which reagent to use and which tissue should be tested, as well as what color to expect. For example, "medulla K+ yellow" means that in this species, if you moisten the medulla with KOH it will turn yellow. Different lichen substances result in different colors. In lichens containing salazinic acid (e.g., most species of *Parmelia*), the medulla turns yellow, rapidly changing to blood red, with the application of K (plate 85); moistening a fresh area of medulla with PD produces a deep yellow, almost orange color; applying C does not cause any color change and is said to be "C negative" or simply C–. Moistening the medulla of *Parmotrema tinctorum* with C, however, turns the spot deep red, owing to the presence of lecanoric acid (plate 86), whereas K and PD are ineffective (K–, PD–). In *Ochrolechia oregonensis*, all parts of the lichen turn reddish with C (owing to the presence of gyrophoric acid), but in the somewhat similar *Ochrolechia juvenalis,* only the apothecial disk contains the compound and turns red with C. The situation is more complicated when a lichen contains more than one substance. For example, the medulla of *Cladonia strepsilis* turns PD+ deep yellow owing to the presence of baeomycesic acid; and C+ greenish, K– (no reaction with K), and KC+ blue-green owing to strepsilin. (A "KC" test is performed by first moistening the reactive part of the lichen with K and then, after the K has had a chance to soak in, moistening the *same* spot with C. Table 2 summarizes the reactions to be expected for the most commonly encountered lichen substances.

In general, the most reactive parts of the lichen (the parts of the lichen producing the highest concentrations of lichen substances) are: the medulla, especially close to the photobiont layer; the growing tips; soralia; pseudocyphellae; and the apothecial margin of lecanorine apothecia. Testing a fresh area of medulla exposed with an oblique cut using a razor (Fig. 29) is better than testing medulla exposed by an old break in the thallus. To check the tissue of an apothecium, make thick sections (top to bottom) and place them on a microscope slide without water (or with just enough water to stabilize the sections). Color reactions are best viewed under the dissecting microscope using as little reagent as possible. PD tests should be made on portions of the lichen removed from the thallus and placed on a glass slide. The analyzed fragments are then *carefully* discarded. A PD-moistened fragment left in a packet will soon produce a huge black spot on the packet, often going right through to the label. Removing the unsightly bits of lichen tested with K or C is also a good idea for aesthetic reasons, although they normally will not ruin the packet.

One variation of the spot test works well with slender, darkly pigmented lichens such as *Bryoria*. Squares of filter paper (or white paper towel, or even white bond paper) about 1 cm² are placed on a microscope slide, which, for safety's sake, is itself placed on a glass plate or disposable cardboard. Four or five filaments of the *Bryoria* are placed on the paper square and are then flooded with one or two drops of reagent, either KOH or Steiner's solution (for PD)—enough reagent to leave the entire square wet. The reactive lichen substances will then dissolve in the reagent and flow out onto the paper from the cortex and exposed medulla (e.g., where the filaments were broken), and the color changes will be easily seen as the paper begins to dry (plate 87). Because wetting the filaments of *Bryoria* with Steiner's solution makes them more transparent, reactions in the medullary area become visible even though they are normally obscured by the dark brown cortex. Thus, one can observe under the stereomicroscope any PD+ red reaction that may occur in the

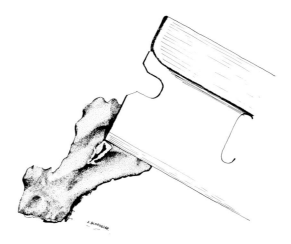

Fig. 29 Slicing method to expose medulla for a spot test. (Drawing by J. Schroeder. Reprinted by permission of The McGraw-Hill Companies, from Hale, *How to Know the Lichens*, 1979, p. 11.)

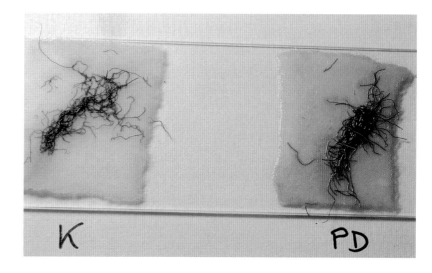

87. To reveal cortical and some medullary color reactions on the slender branches of *Bryoria*, we use the filter paper method. A drop of K floods the sample of *B. capillaris* on the left, and Steiner's solution (PD) is used on the right. The resulting color flows out onto the paper.

Table 2 Commonly encountered lichen substances (or complexes of several related substances) and the reactions they produce with reagents and ultraviolet light as seen in various lichen specimens

Chemical	PD	K	C	KC	AI	UV
Fatty acids						
Caperatic acid	–	–	–	–		–
Protolichesterinic acid	–	–	–	–		–
Roccellic acid	–	–	–	–		–
Orcinol depsides						
Cryptochlorophaeic acid	–	Reddish (slow)	Purple/pink	Purple/pink		Dull whitish
Divaricatic acid	–	–	–	–		White
Erythrin	–	–	Red	Red		White
Evernic acid	–	–	–	–		White
Glomelliferic acid	–	–	–	Red		
Gyrophoric acid	–	–	Pink	Red		Whitish
Homosekikaic acid	–	–	–	–		White
Imbricaric acid	–	–	–	–		
Lecanoric acid	–	–	Red	Red		–
Merochlorophaeic acid	–	–	Pink-violet	Pink-violet		Whitish
Olivetoric acid	–	–	Red	Red		White
Paludosic acid	–	Wine-red (slow)	Purple/pink	Purple/pink		
Perlatolic acid	–	–	–	–		White
Sekikaic acid	–	–	–	–		White
Sphaerophorin	–	–	–	–		White
Stenosporic acid	–	–	–	–		White
Orcinol depsidones						
Alectoronic acid	–	–	–	Red		White
α-Collatolic acid	–	–	–	Pink		White
Grayanic acid	–	–	–	–		White
Lividic acid complex	–	Pink-brown	–	–		
Lobaric acid	–	–	–	Red/violet		White
Norlobaridone	–	–	–	Pink		
Physodic acid	–	–	–	Pink		Whitish
Orcinol depsone						
Picrolichenic acid	–	–	–	Purple		Whitish
ß-Orcinol depsides						
Atranorin	Pale yellow	Pale yellow	–	–		–/Dull whitish
Baeomycesic acid	Deep yellow	Pale yellow	–	–/Yellow		Yellow
Barbatic acid complex	–	–	–/Orange	Pink-orange		Whitish
Diffractaic acid	–	–	–	–		White
Nephroarctin	Deep yellow	–/Yellow	–/Yellow	–/ Yellow		–
Squamatic acid	–	–	–	–		White
Thamnolic acid	Orange	Deep yellow	–	–		–
ß-Orcinol depsidones						
Argopsin	Orange-red	–	–	–		
Eriodermin	Orange	–	–	–		–
Fumarprotocetraric acid	Red	Brownish	–	–	–	–
Galbinic acid	Orange	Orange-red	–	–		
Norstictic acid	Yellow-orange	Dark red	–	N/A	–	–
Pannarin	Orange	–	–	–		–
Physodalic acid	Red-orange	–/Brownish	–	–		–
Protocetraric acid	Red-orange	–	–	Pink	–	–

Chemical	PD	K	C	KC	AI	UV
ß-Orcinol depsidones						
Psoromic acid	Bright yellow	–	–	–		White
Salazinic acid	Orange	Dark red	–	N/A	–	–
Stictic acid complex	Orange	Yellow	–	–	Blue	–
Virensic acid	Orange-red	–	–	–		Yellowish?
ß-Orcinol dibenzyl esters						
Alectorialic acid	Dark yellow	Pale yellow	Red	Red		Yellow
Barbatolic acid	Yellow	Yellow	–	–		–
Dibenzofurans and usnic acids						
Didymic acid	–	–	Green	Green		White
Pannaric acid	–	–	Olive-green	Dark olive-green		
Strepsilin	–	–	Green	Green		–
Usnic acid	–	–	–	Yellow-orange		Absorbs
Chromone						
Siphulin	–	Yellow-brown	Violet (fading)	–		Whitish
Xanthones						
Arthothelin	–	–	Orange	Orange		–
Lichexanthone	–	–	–	–		Yellow
Thiophanic acid	–	–	Orange	Orange		–
Anthraquinones						
Nephromin	–	Red-purple	–	–		–
Parietin	–	Red-purple	–	–		–
Skyrin	–	Red-purple	–	–		–
Triterpenoids						
Zeorin	–	–	–	–		–
Pulvinic acid derivatives						
Calycin	–	–/Pink?	–	–		Dull dark orange
Rhizocarpic acid	–	–	–	–		Orange
Vulpinic acid	–	–	–	–		–

Note: The reactions may be somewhat different with pure substances. PD = *para*-phenylenediamine (alcohol solution); K = 10% potassium hydroxide; C = undiluted household bleach; KC = C applied to a tissue moistened with K; AI = iodine in a pH 11 buffer; UV = longwave ultraviolet light. A slash (/) stands for "or."

medulla or inner part of the cortex, as it does, for example, in the branch tips of *B. trichodes,* although no color flows onto the paper square. Remember to discard the paper squares quickly and safely after they have been used.

A variation of the filter paper test for C and KC tests uses a white porcelain plate with shallow depressions. Place a few strands of a single branch of the *Bryoria* to be tested into the depression and add acetone drop by drop, allowing the white residue to accumulate on the sides of the depression as the acetone evaporates. (CAUTION: Acetone is highly inflammable, and the fumes are unhealthy, so avoid an open flame and work in a well-ventilated room.) After removing the extracted lichen fragments from the depression, allow a drop of K to run down one side of the depression to view the K color reaction on the residue, and do the same with C on the other side of the depression. The KC reaction will be observed where the two drops meet at the bottom of the depression.

Write the results of the spot tests on the label or on a slip of paper placed inside the packet. This will make it unnecessary to repeat the test on the same specimen at a later time.

Variations in the color reactions and problems in interpretation can be caused by (1) variations in the

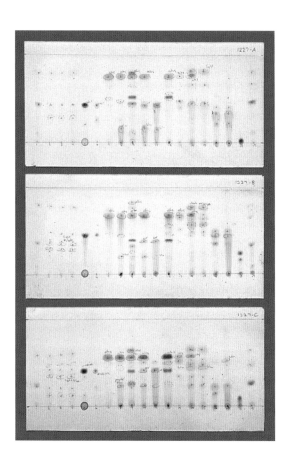

88. The potassium salt of norstictic acid forms characteristic clusters of straight, red needles when K is added to a lichen extract containing the acid. Magnified 400×, in polarized light. (Photo by I. M. Brodo.)

89. Thin-layer chromatography has been used here with a variety of lichen compounds in three solvent systems. A: toluene, dioxane, acetic acid (180:45:5); B: hexane, methyl *tert.*-butyl ether, formic acid (140:72:18); C: toluene, acetic acid (170:30). The relative distance traveled by the substances in the three solvents can be used to identify the compounds.

concentrations of substances found in a particular specimen (old, decaying lichens or those growing under poor conditions will react less vividly); (2) the strength and freshness of the reagents; (3) the amount of reagent applied (flooding a spot will obscure results, and too little will be ineffective); (4) reactions that fade rapidly, such as the C and KC tests; and (5) combinations of compounds within a single specimen (influencing the shade or intensity of color, or masking it).

Crystal Tests

Distinctive, microscopic crystals are formed from lichen substances extracted from lichen fragments with acetone and dissolved in certain reagents. This was once an important way of identifying specific compounds, although the technique is now considered unreliable and will not be described further here. A few compounds, however, form crystals directly from K or PD, and these are worth noting. Norstictic acid produces red needles, singly or clustered, after it is dissolved in K. The presence of norstictic acid in some crustose lichens such as *Buellia stillingiana* and *Bellemerea alpina* is easily detected in ordinary apothecial sections mounted in KOH by the formation of the crystals (plate 88). Variolaric acid, found in certain species of *Ochrolechia,* forms characteristic clusters of colorless needles in KOH preparations. Pannarin can be detected in certain tissues of *Lecanora cinereofusca* and some *Pannaria* species by producing clusters of short, orange crystals in sections of apothecia or thalli mounted in alcohol solutions of PD. Use a small fragment of a broken cover slip (left over from your attempts at crushing apothecial sections) to reduce the amount of PD needed for the test.

Thin-Layer Chromatography

Crystal tests were largely abandoned when thin-layer chromatography (TLC) became popular in the early 1970s. With TLC, all the compounds in a lichen are extracted, usually with acetone. The extracts are concentrated in spots on a glass or aluminum plate coated with silica gel. Then the compounds in the spots are separated by placing the lower edge of the plate into some solvent mixture, allowing the compounds to migrate in the advancing solvent, which migrates much like a drop of liquid on a paper towel. Different compounds migrate at different rates and therefore to different distances from the point of origin. Because the spots are usually colorless, they have to be made visible by either spraying the plates with 10% sulphuric acid followed by a short period of heating or by viewing them under ultraviolet light. When carefully performed using a standarized method developed by Chicita Culberson, TLC is a technique that one can use to accurately identify hundreds of different lichen substances (plate 89). Unfortunately, the technique requires a chemical laboratory, and so we do not describe it in more detail here. The methods are fully discussed in articles and books listed at the back of this book under "Further Reading." Mention of the compounds contained in the lichens illustrated in the book will give those who want to pursue such chemical analyses an idea of what to look for.

Ultraviolet Light

Many lichen substances are fluorescent in longwave ultraviolet (UV) light. It is therefore possible to use UV light to reveal these compounds as an aid in lichen identification. If a compound is expected in the thallus cortex, one can determine its presence or absence simply by shining the UV light on the lichen in the dark (plate 90). Species of *Pyxine*, *Cladonia*, *Hypotrachyna*, *Ochrolechia*, and *Thamnolia* are often distinguished this way. Substances found only in the medulla of the lichen cannot be seen until the lichen's medulla is exposed by removing the thallus cortex.

A well-equipped laboratory will usually have a UV light chamber (a self-contained, light-free viewing box, often with both longwave and shortwave UV light sources). It is much cheaper, however, to obtain a hand-held longwave UV light from a store catering to mineral collectors and hobbyists. A closet or even a cardboard carton will provide enough darkness for viewing the lichens. A worthwhile lichenological adventure can be had—especially in the subtropics, where many lichens contain cortical xanthones that are brightly fluorescent—by taking a portable, battery-equipped UV lamp outside at night and shining it on lichens growing on the trees. Try it!

PHOTOGRAPHING LICHENS: TECHNIQUES AND ADVICE

The techniques used to photograph the subjects in *Lichens of North America* were designed to overcome the most vexing problem inherent in close-up photography: the lack of adequate depth of field, or sharpness of focus from the front to the back of the picture. Depth of field becomes much more limited as the magnification increases, usually forcing the photographer to make critical choices about which parts of the photo will be sharp and which parts out of focus. One can partially overcome this problem by making the lens opening (aperture) smaller, setting the f-stop at f/32 to f/16, but this creates the further problem of not having sufficient light to take the picture. Using a long time exposure can help, but only with a tripod and a subject that is not moving, and long exposures can produce a color shift toward blue in the film. Our solution is to use a twin-flash system.

We use 35mm equipment, usually with a macro lens, separated from the camera body with rings whenever we need a magnification greater than ×1 (life size on the slide). In order to use a very small aperture yet allow enough light to reach the subject, we use a pair of electronic flashes, one on either side of the lens. They are mounted on adjustable, custom-made brackets whose design has evolved from a set invented by Sylvia Sharnoff's father, Victor Duran, a professional scientific photographer for many years. The brackets allow us to position each flash independently at any angle to the subject and at any distance from it. With this system we can take photographs using apertures as small as f/32, the one most commonly employed for this book. This aperture works well to magnifications of about ×1. At higher magnifications, the performance of lenses decreases when used at such small f-stops so, for smaller subjects, we increase the aperture: by the time we are at ×4 we use f/8 or f/11, depending on how flat the subject is.

The total amount of light on the subject is determined by the flash duration, which is controlled by the

90. Lichexanthone in the cortex of this specimen of *Ochrolechia mexicana* fluoresces bright yellow under longwave ultraviolet light. (Photo by I. M. Brodo.)

camera's computer using a "through-the-lens" metering system. We can balance the light between the two flashes by independently adjusting the distance between each flash and the subject, and we can create shadows or eliminate them by changing the flash angle. Flat subjects often need a harsher light to bring out the details; very three-dimensional subjects generally want a soft light, which we can create by using acetate diffusers over each flash head. The dark backgrounds seen in many of the photographs are the result of flashes trying to illuminate empty space.

An effective technique often used for photographing lichens is to spray the thallus with a bit of water, then wait for it to partially dry out. When it is slightly damp but no longer soaking wet, its colors will be at their peak; many of the photographs in the book were taken this way. The technique works much better with some lichens than with others, because different species of lichens vary dramatically in their response to wetting. Cleaning the subject with fine forceps or a small paintbrush (using a hand lens) is also helpful.

14 USING THIS BOOK TO NAME A LICHEN: HINTS AND CONVENTIONS

The habit of looking at things microscopically as the lichens on the trees & rocks really prevents my seeing aught else in a walk.

Henry D. Thoreau

ORGANIZING THE TASK

The elements of this book that enable you to identify a lichen are: the keys, the descriptions, the maps, the photographs, and the glossary. In most cases, proceeding from one element to the next is the best bet, although the temptation to start by leafing through the pretty pictures will be hard to resist. Remember that there are over 800 of them, and they are arranged in alphabetical order according to the genus and then the species, not according to their overall appearance. You therefore need a way to at least reduce the number of pictures that need to be considered, or better yet, we can lead you directly to the most likely candidate. For that, you need the keys.

KEYS

Identification keys are simply a series of choices, each choice leading to another until the name of the lichen is determined. If the choices are reasonably clear (and we've tried to make them that way), the main problems encountered will probably be terminology and interpretation of descriptive remarks. The glossary, with references to illustrations in the introductory chapters, is designed to help remedy the difficulties in learning new terms; experience is the only thing that will help you with interpretation.

The choices are presented as "couplets," two at a time, never as three or more alternatives. The couplets are numbered sequentially, and each choice leads to either the number of another couplet or to some end point, normally a lichen name or the title of an additional key that should be consulted elsewhere in the book.

In using keys, remember to read *both* choices before deciding which lead to follow. (The first may sound good, but the second may be even better.) If you encounter an impasse, that is, a place where neither choice works, then either the lichen is not in the key (our coverage includes only about 30 percent of the North American flora) or you have made an incorrect choice somewhere along the way. Checking the photographs as you proceed will help clarify choices and keep you on track. This is particularly important with regard to color choices. Keep in mind that the color descriptions are based on the lichen in a *dry* state, unless we specifically indicate otherwise.

Sometimes, you may think you know the name of a

lichen (e.g., by recognizing a lichen from the photographs or by consulting another book), but you do not understand how the keys would lead to that name. In such cases, the keys may be used *backwards,* starting with the name (it can be found using the index), and then reading the choices leading to that name. To help you find leads that originate more than two couplets away, the originating lead is given in parentheses. For example, in Key A, couplet 20 originates from a choice in couplet 12, and so "12" is placed in parentheses after number 20. Because couplet 12 comes from couplet 6, the number "6" is in parentheses. Couplet 6 originates from couplet 5 just above, and so no parenthetical reference is needed.

The keys include all 804 illustrated species as well as more than 250 other common lichens of North America, so it is likely the lichen you are trying to identify will be covered. Even if your lichen is not a main entry or in the keys, you may find it mentioned in the text as a similar but rarer species (an additional 500 taxa), so it is worth reading the Comments under lichens that are suggested by the keys and probably resemble yours even though they do not quite fit.

Depending on your level of experience, you may want to begin with the "Key to Keys" section in Part II (p. 117), which clarifies the major growth form divisions, or you might want to jump in at the next level, with the keys to growth form groups (fruticose, foliose, squamulose, and so on) or species groups (for example, lichens with perithecia) found in keys A–K. These group keys either lead you directly to species for certain genera or point to species keys found under each genus entry in the main text. You can even go directly to the genus if you think you recognize it. Similar genera are mentioned in the Comments on each genus, to apprise you of other possibilities. The really confusable genera are generally keyed out together (e.g., all the *Parmelia*-like genera). The coverage is indicated in the title of each key, and cross-references are provided under each genus.

MAIN ENTRIES

Genera: Genus descriptions are provided for most genera with two or more species in North America. To save space, there are a few exceptions with genera having only two or three species. When features of the genus apply to all the species in a genus, they are not repeated.

Names: The names used in this book, with few exceptions, are those listed by Esslinger and Egan in the "official" North American checklist posted on the Internet as of July 1999. Full author citations are supplied in the index of this book. Names in boldface in the species discussions refer to lichens that do not have a main entry. These names may appear under more than one main entry, but each time it will be in boldface to make it easy to find.

Synonyms: In most cases only very recent synonyms are given, except for those names used by Mason Hale in *How to Know the Lichens.* His names, when different from ours, are given as synonyms to help link the two books.

English names: If more than one English name is given, those following the first name are potentially useful names that were introduced by other authors. (See Chapter 12 for a full discussion of this topic.)

Description: As in the keys, colors generally refer to a dry lichen unless otherwise mentioned. This is important because many lichens are pale green or yellowish green when wet and can be confused with the pale yellowish green color (when dry) of lichens containing usnic acid. This color, sometimes referred to as "usnic yellow," will be interpreted by some readers as yellowish and some as simply green. Because no two people see (or understand) colors in the same way, we suggest using the photographs as a guide, using the color descriptions to help visualize the possible variation in a species. Chapter 4 is devoted to a discussion of lichen colors.

The photobiont is generally referred to as "bluegreen" rather than as "cyanobacteria," and "green" rather than as "green algae" unless such a simplification would be misleading. In most cases, the color of the photobiont (dark blue-green or blue-gray for a cyanobacterial lichen and grassy green for a green algal lichen) seen by scraping or cutting away the upper cortex is indicative of the photobiont. In a few cases, a microscopic preparation may be needed.

Lobes are measured as in figure 3. We use metric measurement throughout (25 mm to the inch). Measurements are generally rounded to the nearest half-unit except for extremely narrow or small features (e.g., lobes 0.2–0.6 mm). Dimensions given as, "spores $(5-)7-11(-14) \times (1-)2-4(-6)$ μm" mean: most spores measure 7–11 μm long and 2–4 μm wide, but exceptional spores may be as little as 5 and as much as 14 μm long. The extreme range is given in parentheses. (For an explanation of microns (μm), see footnote 1 in Chapter 3.)

Chemistry: General methods for determining color reactions are found in Chapter 13, although some special techniques may be described under certain genera. In general, only the main constituents, especially the compounds responsible for the spot test reactions, are mentioned. If spot test results are given under the genus, they refer to all species in the genus and are not repeated under each species entry.

Habitat: A brief indication of the substrates and habitat of each lichen is given under "Habitat." It is possible that some lichens normally found on one substrate may be found on a different one under special circumstances. This is fully discussed in Chapter 7. References to "wood" as a substrate do not include bark; if the lichen also occurs on bark, we indicate so. We use "poplar" in a broad sense, including cottonwoods and aspens.

Distribution: A brief description of the general range of the lichen is given only when insufficient reliable information was available for a map (as was the case for 15 species) and when mapping that species would be impractical or misleading. This was the case, for example, for lichens that are particularly inconspicuous and therefore rarely collected, or which have been so frequently misidentified that herbarium specimens and literature reports are unreliable.

Comments: Notes on variation, problems in interpretation, and comparisons with similar species are made in this section.

Importance: If a lichen has some special use by people or wildlife, or has an ecological role worthy of note, it is mentioned here.

We have made a strong effort to avoid unnecessary technical terms and jargon, but we have found that some specialized terminology is essential for precision and conciseness. We encourage readers to use the glossary frequently until the terminology is learned. Entries in the glossary often refer to figures in the introductory chapters where the morphological features are explained in context.

DISTRIBUTION MAPS

These small maps give you an idea of where the lichen is known to occur. If used in conjunction with Chapter 9, they may also give you a hint as to where the species might turn up in the future. The maps are based on the latest studies of individual genera or families, wherever possible. We also gathered locality data from specimens in the Canadian Museum of Nature, the United States National Museum, Duke University, and the New York Botanical Gardens, as well as data from the University of Colorado Museum, University of Minnesota, Arizona State University, and Oregon State University and, in many cases, unpublished records shared with us by other lichenologists. This information was supplemented by reliable records from close to 70 floristic studies found in scholarly books and journals. First, dot maps were prepared with each known locality represented by a black dot on a blank map of North America. These were then translated into sketches showing the outline of the presumed range. We were conservative in sketching ranges based on the dot maps. That is, we often left spots or areas disjunct rather than assuming connections between them. Extensions or adjustments will surely be possible with almost all the maps as the species receive more study, specimens upon which the records were based are checked, or more herbarium collections are consulted.

PLATES

Almost all the photographs of species were taken in nature. Pay particular attention to the magnifications given in the captions for every image. Some photographs are highly magnified and can be misleading. A magnification of "×6.8," for example, means that the printed photograph is 6.8 times larger than the original subject. We have tried to capture every species in its most characteristic, healthy state, but invariably some pictures of young or somewhat atypical subjects had to be used. Such cases are noted in the caption.

PART TWO

GUIDE TO THE LICHENS

IDENTIFICATION KEYS TO GENERA AND MAJOR GROUPS

Readers unfamiliar with identification keys will find it useful to read the section under "Keys" in Chapter 14. There you will find an explanation of how keys work and how to use them most effectively.

Certain conventions and peculiarities of the keys in this book are important to know. (1) All choices are given two at a time; there are no three-choice couplets. (2) The "Key to Keys" leads you to Keys A through K for eleven major groups of lichens (indicated in boldface); these twelve keys precede the general section with descriptions and photographs of the lichens. All the other keys to genera and species are found under the appropriate genus entry following the genus descriptions. (3) Genera or species in square brackets do not have a main entry in the book; they may or may not be discussed further in the text. The index indicates where the species is mentioned. (4) Numbers in parentheses following a couplet number indicate the origin of that choice; these are useful in working backward in the key (see Chapter 14 under "Keys" for guidance on how this works).

KEY TO KEYS

1. Fruiting body a mushroom or fleshy, ephemeral, club-shaped stalk with a mushroom-like consistency 2
1. Fruiting body a perennial ascoma (apothecium, perithecium, or lirella); or thallus without any fruiting bodies 3

2. Fruiting body a pale to dark yellow, gilled mushroom, growing out of a lichenized thallus consisting of either dark green spherical granules or green, lobed squamules ..... *Omphalina*
2. Fruiting body club-shaped, usually unbranched, white to orange, growing out of a gelatinous, dark green, barely lichenized algal film ..... *Multiclavula*

3. Thallus erect, shrubby, tufted, or pendent, at least in part (sometimes with a crust-like or scaly basal thallus as well); branches or stalks round, angular, or flat in cross section; but if flattened, both surfaces similar in color and texture 4
3. Thallus composed entirely of flat lobes, branches, or scales, with clearly different upper and lower surfaces, or crust-like and firmly attached to the substrate over the entire lower surface, or growing within the substrate 5

4. Thallus relatively large, more than 5 mm long or high, with most branches or stalks more than 0.2 mm in diameter (Fig. 2h–j) ...... **Fruticose lichens (Key A)**
4. Thallus forming small tufts or clumps rarely over 5 mm long or high, consisting of fine, hair-like elements rarely more than 0.2 mm in diameter (see, e.g., *Polychidium*, *Ephebe*, or *Coenogonium*) ....
.... **Dwarf fruticose filamentous lichens (Key B)**
5. Thallus consisting of a crust that grows with its entire lower surface in intimate contact with the substrate and cannot be removed from the substrate intact (i.e., without leaving part of the lichen's lower surface behind); thallus can be smooth and continuous, powdery, cracked into angular patches (areoles), extremely thin (even growing within the substrate), or quite thick and warty, sometimes lobed at the margins (Fig. 2a–d) .... 6 (Crustose lichens)
5. Thallus consisting of more or less flat dorsiventral lobes or scales, ascending or closely appressed to the substrate, more or less easily removed from the substrate with the lower surface intact (Fig. 2e–g) 9
6. Fruiting bodies absent ....
.... **Sterile crustose lichens (Key C)**
6. Fruiting bodies (ascomata) present 7
7. Ascomata flask-shaped, perithecia or resembling perithecia (opening by a deeply concave or pit-like ostiole) (Figs. 17, 12h, 13d), superficial or buried in the thallus or in a fertile wart (pseudostroma) with only the ostiole visible. [*Note:* Pycnidia can sometimes resemble perithecia, but they contain hundreds of conidia not associated with asci; see Fig. 18.] ....
.... **Crustose perithecial lichens (Key D)**
7. Ascomata apothecia: (1) disk- or cup-shaped (Figs. 12a–c; 13a–c,f), sometimes raised on a stubby or slender stalk (Fig. 12f); (2) buried in thalline warts (as in *Pertusaria subpertusa* or *P. xanthodes*) (Fig. 12g); or (3) elongated and sometimes branched (lirellae)(Figs. 12e; 13e) 8
8. Ascomata lirellae, short and ellipsoid to quite long, script-like, and sometimes branched ....
.... **Crustose script lichens (Key E)**
8. Ascomata apothecia (disk- or cup-shaped or buried within thalline warts) ... **Crustose disk lichens (Key F)**
9.(5) Thallus composed of separate or overlapping scales (squamules) that are rarely more than 5 mm long or wide (Fig. 2e) .... **Squamulose lichens (Key G)**
9. Thallus composed of narrow or broad, flat lobes; thallus generally exceeding 5 mm long or wide (Fig. 2f–g) .... 10 (Foliose lichens)
10. Thallus attached to the substrate by a holdfast at a single point, either central, giving it an umbrella shape (Fig. 2g), or at one edge, giving it a fan shape. (Lobes sometimes crowded and overlapping, obscuring the attachment.) ....
.... **Umbilicate or fan-shaped lichens (Key H)**
10. Thallus attached to the substrate at numerous points, with hair-like attachment organs (rhizines), a fuzzy tomentum, or directly by the lower surface (Fig. 2f) .. 11
11. Thallus, when wet, translucent and jelly-like, and typically black to very dark gray, brown, or olive (e.g., *Collema*); photobiont blue-green, usually distributed more or less uniformly within the thallus, i.e., not confined to a definite layer (Fig. 7) ... **Jelly lichens (Key I)**
11. Thallus remaining opaque when wet, color various; photobiont green or blue-green, confined to a definite layer within the thallus (Fig. 4) 12
12. Dry thallus orange, yellow, greenish yellow, or yellowish green (see, e.g., *Xanthoria*, *Candelaria*, and *Flavoparmelia*) .... **Yellow foliose lichens (Key J)**
12. Dry thallus shades of white, gray, pale green, olive, brown, or black, without a yellowish tint ....
Foliose lichens that are not umbilicate, jelly-like, or yellow (Key K)

KEY A: FRUTICOSE LICHENS

1. Thallus bright yellow, bright chartreuse, or orange ... 2
1. Thallus pale greenish yellow, yellowish green, green, white, gray, brown, olive, or black 5
2. Branches distinctly flattened; medulla yellow; cortex and medulla K–; on the ground in arctic or alpine sites ....*Vulpicida tilesii*
2. Branches round to angular; medulla white; on trees or, if on the ground, not arctic-alpine 3
3. Thallus bright yellow to chartreuse, K–. Most frequently in inland, montane localities .... *Letharia*
3. Thallus orange, at least on exposed parts, K+ dark red-purple 4
4. Attached directly to rock along seashores in California .... *Caloplaca coralloides*
4. On trees, shrubs, or soil .... *Teloschistes*
5.(1) Thallus pendent or almost pendent (i.e., growing downward or outward); most on trees, shrubs, or vertical surfaces of rocks (includes hair and beard lichens) 6
5. Thallus erect or prostrate (i.e., basically growing upward, at least initially); most on the ground or on horizontal surfaces of rocks (includes lichens with upright stalks or podetia) 28
6. Thallus greenish yellow or yellowish green (containing usnic acid in the cortex) 7

6. Thallus shades of white, gray, brown, olive, or black (lacking usnic acid) 12
7. Branches with a tough, single, central cord .... *Usnea*
7. Branches with a more or less uniform medulla, without a single, central cord 8
8. Pseudocyphellae present 9
8. Pseudocyphellae absent 10
9. Branches flattened in cross section, at least at the base .... *Ramalina*
9. Branches round in cross section, except sometimes at the axils .... *Alectoria*
10. Thallus soft and pliable, with a thin cortex ... *Evernia*
10. Thallus stiff because of a thick, tough cortex 11
11. Pycnidia black, abundant, appearing as black dots on the margins and sometimes the surface of the thallus; restricted to coastal California .... *Niebla*
11. Pycnidia, when present, pale and inconspicuous; widespread .... *Ramalina*
12.(6) Branches distinctly flattened, at least at the base or tips 13
12. Branches round or angular in cross section throughout (sometimes grooved or twisted, or flattened at the axils) 20
13. Thallus brown to greenish black .... *Kaernefeltia*
13. Thallus pale gray to smoky brownish gray 14
14. Cilia present on the branches or apothecia 15
14. Cilia absent 17
15. Branches 10–25 mm long; medulla C+ pink .... *Everniastrum catawbiense*
15. Branches 40–150 mm long; medulla C– 16
16. Cortex K+ yellow; medulla K+ red (atranorin and salazinic acid); sorediate on the lower surface of the lobe tips .... *Heterodermia leucomela*
16. Cortex and medulla K–; lobes without soredia .... *Anaptychia setifera*
17. Fruiting bodies almost spherical, black, powdery masses formed close to the tips of the flattened branches; on mossy trees or rocks in Pacific Northwest coastal forests ...... [*Bunodophoron melanocarpum;* see couplet 34]
17. Fruiting bodies round or elongate apothecia, never powdery; not found in Pacific Northwest coastal forests 18
18. On trees or wood in the southwest or Appalachian–Great Lakes regions; cortex K+ yellow (atranorin); lower surface gray to black in part ..... *Pseudevernia*
18. On rocks and shrubs in coastal California; cortex K–; lower and upper surfaces the same color 19
19. Cortex C+ red, PD– (lecanoric acid) .... *Roccella*
19. Cortex C–, PD+ red (protocetraric acid) .... *Dendrographa*
20.(12) Thallus almost white, sometimes darkening to dark yellowish gray; coastal rocks and shrubs in California; cortex PD+ red .... *Dendrographa*
20. Thallus shades of gray or brown, or almost black; cortex PD– or PD+ yellow (except rarely in some dark species of *Bryoria*); on bark or wood, rarely on rock 21
21. Thallus with thick main branches and much finer secondary and tertiary branches, gray-green, often browned at the tips; fruiting body a spherical mazaedium at the branch tips .... *Sphaerophorus globosus*
21. Thallus gradually tapering, mostly uniform in color; fruiting bodies apothecia 22
22. Thallus shrubby, with divergent, stiff branches, usually with a spiny appearance 23
22. Thallus usually pendent with parallel branches 25
23. Thallus dark greenish black, at least close to the branch tips; apothecia appearing terminal on branches, almost black; epihymenium K+ violet .... *Kaernefeltia californica*
23. Thallus and apothecia brown; epihymenium K– ... 24
24. Thallus distinctly reddish brown; cells of the cortex (viewed at 100–400×) interlocking like a jigsaw puzzle; branches grooved and channeled throughout .... *Nodobryoria abbreviata*
24. Thallus brown without a reddish tint; cells of the cortex elongate, parallel, not interlocking; branches usually uniformly terete except at the axils, where they can be flattened or pitted .... *Bryoria*
25. Strongly depressed pseudocyphellae forming long, deep, linear grooves down the branches, sometimes twisting around the branch; thallus pale, dull, mottled with reddish brown to gray-brown streaks; rare .... *Sulcaria badia*
25. Pseudocyphellae present or absent, linear or ellipsoid, but not forming deep grooves; thallus typically more or less uniform in color 26
26. Thallus gray to pale or dark brown or olive ... *Bryoria*
26. Thallus distinctly reddish brown 27
27. Branches grooved and pitted, brittle; apothecia brown, without pruina; cortical cells in surface view (viewed at 100–400×) resembling pieces of a jigsaw puzzle .... *Nodobryoria oregana*
27. Branches more or less smooth except for the axils; apothecia, if present, with yellow pruina; cortical cells straight, long, and parallel .... *Bryoria fremontii*
28.(5) Stalks or branches solid 29
28. Stalks or branches hollow, at least in part 63

29. Stalks or branches distinctly flattened at least at the base or tips 30
29. Stalks or branches most commonly round or angular in cross section (sometimes flattened at the axils) 40
30. Thallus yellow-green (containing usnic acid) 31
30. Thallus gray, brown, or black (lacking usnic acid) ... 33
31. Pseudocyphellae present, slightly depressed, on the lower surface of the branches; lobes thin, entirely flattened and erect .... *Flavocetraria*
31. Pseudocyphellae absent, although the branches sometimes have raised, white warts; lobes relatively thick 32
32. Lobes irregularly branched, strongly curled and twisted, usually with white warts on the margins or tips of the branches; growing unattached over bare soils in the interior high plateaus .... *Rhizoplaca haydenii*
32. Lobes regularly dichotomous, not curled, generally round, at least toward the base, without white warts; on soil in arctic or alpine sites .... *Allocetraria madreporiformis*
33.(30) Thallus gray or gray-green 34
33. Thallus very dark brown to black 35
34. On limestone, in arid south central U.S.; branches more or less uniform in width and very regularly dichotomous. .... *Speerschneidera euploca*
34. On mossy trees or rocks, in humid coastal forests of the Pacific Northwest; main branches broad, giving rise to clusters of much narrower branches, irregularly dichotomous .... [*Bunodophoron melanocarpum* (see couplet 17)]
35. Directly on rock 36
35. On soil, heath, or mosses (sometimes over rock) ... 38
36. Thallus abundantly branched, forming wool-like cushions .... *Pseudephebe minuscula*
36. Thallus with relatively few strap-shaped branches; forming small clumps 37
37. On limestone; photobiont blue-green; thallus jelly-like when wet .... *Jelly lichens (Key I)*
37. On siliceous rock; photobiont green; thallus unchanged when wet .... *Cornicularia normoerica*
38.(35) Branches very flat, regularly dichotomous, with broad, pruinose patches (pseudocyphellae) on lower surface; growing unattached (vagrant), rolling freely over the soil and heath ...... *Masonhalea richardsonii*
38. Branches usually grooved or inrolled at the margins; pseudocyphellae linear or irregular in shape, not broad and pruinose; growing in more or less fixed clumps, not rolling freely over the soil or heath 39
39. Medulla C+ pink (gyrophoric acid); lobe margins more or less even, without projections .... *Cetrariella delisei*
39. Medulla C–; lobe margins with stalked pycnidia or cilia .... *Cetraria*
40.(29) Thallus yellowish green (containing usnic acid) 41
40. Thallus white, gray, brown, olive, or black (lacking usnic acid) 45
41. Thallus prostrate, with long, slender branches, 0.5–3 mm in diameter, 30–50 mm long, usually irregularly branched and tangled .... *Evernia divaricata*
41. Thallus at least partly erect; branches short, usually under 30 mm high 42
42. Branches very slender, less than 0.3 mm in diameter, covered with cottony granules .... *Leprocaulon*
42. Branches thicker than 0.3 mm; surface relatively smooth, at least at the base 43
43. In arctic or alpine sites in western mountains; not cushion-forming .... *Allocetraria madreporiformis*
43. Not arctic-alpine; forming rounded cushions or clumps 44
44. On seashore rocks in California; branch tips becoming granular or sorediate, lacking white warts; cortex C+ orange, KC+ red-orange (xanthones) .... *Lecanora phryganitis*
44. Rolling unattached over bare soils in the interior high plateaus; branches with white warts but without granules or soredia; cortex C–, KC+ yellow (usnic acid alone) .... *Rhizoplaca haydenii*
45.(40) Branching frequently 46
45. Unbranched, or branching at most once or twice ... 59
46. Thallus black or very dark brown to almost black ... 47
46. Thallus reddish brown, yellowish brown, olive, gray, or white 49
47. Attached directly to rock .... *Pseudephebe*
47. On soil or heath 48
48. Pseudocyphellae pale, raised, usually conspicuous; thallus pale at the base; cortex and usually medulla C+ pink, PD+ yellow .... *Alectoria nigricans*
48. Pseudocyphellae dark, level with surface or depressed, inconspicuous; thallus dark throughout; cortex C–, PD–; medulla PD+ red .... *Bryoria nitidula*
49.(46) Thallus shades of brown or olive 50
49. Thallus white to dark greenish gray 53
50. Photobiont blue-green .. *Dendriscocaulon intricatulum*
50. Photobiont green 51
51. Pseudocyphellae absent .... *Sphaerophorus*
51. Pseudocyphellae white, conspicuous 52
52. Pseudocyphellae usually raised like small bumps, C+ pink; medulla usually C+ pink as well (olivetoric acid) .... *Bryocaulon divergens*

52. Pseudocyphellae depressed, C–; medulla C– .... *Cetraria aculeata*
53.(49) Photobiont blue-green .... *Dendriscocaulon intricatulum*
53. Photobiont green 54
54. Branches very slender, less than 0.3 mm in diameter, covered with cottony granules .... *Leprocaulon*
54. Branches thicker than 0.3 mm; surface relatively smooth, at least at the base 55
55. Thallus growing as prostrate cushions up to 30 mm in diameter on soil in the arid interior of North America; round, depressed pseudocyphellae conspicuous on branches .... *Aspicilia hispida*
55. Thallus colonies usually larger than 30 mm, not on soil in the arid interior of North America; pseudocyphellae, if present, not round and depressed 56
56. Cortex C+ red; on coastal rocks in southern California; fruiting bodies large, lecanorine, with heavily pruinose disks .... *Schizopelte californica*
56. Cortex C–; fruiting bodies not pruinose 57
57. Stalks with a tough, cartilaginous central core, more or less covered with granular, cylindrical, or scaly outgrowths (phyllocladia) (Fig. 8g) .... *Stereocaulon*
57. Stalks filled with white medulla, without a cartilaginous central core; without phyllocladia (Fig. 8e) ... 58
58. Branch tips blackened; pseudocyphellae usually conspicuous .... *Alectoria nigricans*
58. Branch tips browned, not blackened; pseudocyphellae absent .... *Sphaerophorus*
59.(45) On conifer branches in subalpine sites; fruiting body a mazaedium at the tip of a stubby, barrel-shaped stalk .... *Tholurna dissimilis*
59. On soil, rock, or moss; fruiting body not a mazaedium 60
60. Stalks white, smooth or with longitudinal furrows, not growing from a primary thallus, without fruiting bodies of any kind; cortex C+ violet (fading quickly) .... *Siphula ceratites*
60. Stalks almost white to gray or olive-gray, without furrows, developing from a crustose or granular primary thallus; cortex C– 61
61. Cephalodia absent .... *Baeomyces* key (*Baeomyces, Dibaeis*)
61. Lumpy pink to dark gray cephalodia present on primary thallus, sometimes on stalks 62
62. Apothecia usually present, black, cylindrical or almost spherical; stalks without soredia .... *Pilophorus*
62. Apothecia extremely rare, brown; granular soredia at the tips of the stalks .... *Stereocaulon pileatum*
63.(28) Photobiont blue-green; thallus olive-brown; stalks cylindrical, most commonly under 2 mm high; on rock .... *Peltula cylindrica*
63. Photobiont green; thallus not olive-brown; stalks over 2 mm high 64
64. Lacking a basal thallus 65
64. With a scaly or crustose basal thallus 70
65. Stalks abundantly branched 66
65. Stalks unbranched, or branched only once or twice 67
66. Thallus under 20 mm tall; branches relatively stocky, usually violet or pinkish as a result of pruina; interior cavity containing webby hyphae ... *Dactylina ramulosa*
66. Thallus over 20 mm tall, without pruina; interior cavity empty, lined with a cartilaginous stereome (Fig. 8f) ... . *Cladonia* key (*Cladonia, Cladina*)
67. Thallus C+ red (erythrin); stalks inflated, brittle; on coastal rocks in southern California .. *Hubbsia parishii*
67. Thallus C–, or, if C+ pink, then arctic-alpine 68
68. Stalks white, slender, 1–2.5 mm in diameter; usually partly or entirely prostrate ... *Thamnolia vermicularis*
68. Stalks yellowish, brown, or greenish gray; erect 69
69. Stalks inflated, thin-walled and brittle, with no stereome; entirely without squamules ... *Dactylina arctica*
69. Stalks more or less slender, not inflated, stiff, supported by a cartilaginous stereome, sometimes bearing squamules .... *Cladonia*
70.(64) Primary thallus squamulose .... *Cladonia*
70. Primary thallus crustose, granular, not at all squamulose 71
71. Stalks appearing inflated; apothecia brown; cephalodia absent; eastern North America ... *Pycnothelia papillaria*
71. Stalks slender, not inflated; apothecia pitch black; cephalodia present on primary thallus; northwestern North America .... *Pilophorus acicularis*

KEY B: DWARF FRUTICOSE FILAMENTOUS LICHENS

1. Thallus pale: gray, greenish, olive, or orange 2
1. Thallus dark: black, olive, or brown 6
2. Thallus forming cottony tufts of fine, branching filaments or hairs; photobiont green or blue-green 3
2. Thallus with more or less erect, fine stalks; photobiont green 4
3. On subtropical and tropical trees and shrubs; tufts cottony, sometimes flattened, pale green to orange, almost white when old; disk-shaped apothecia common, yellow to orange; photobiont green (*Trentepohlia*) .... *Coenogonium*
3. On twigs and branches on the very humid west coast; thallus forming small, foam-like clumps, pale olive or

olive-gray; photobiont blue-green (*Hyphomorpha*) ... *Polychidium contortum*

4. Stalks branched, covered with cottony granules, giving clumps the appearance of a pale gray to whitish, leprose, crustose lichen .... *Leprocaulon gracilescens*
4. Stalks branched or unbranched, without granules or soredia; thallus thin and membranous 5

5. Stalks like long, slender isidia, often branched, containing green algae; apothecia sometimes produced on basal crust; on bark in the Pacific Northwest .... *Loxosporopsis corallifera*
5. Stalks colorless, without algae, unbranched, tipped by star-shaped, umbrella-like conidia-containing structures; on mosses and tree bases in the southeast .... *Gomphillus americanus*

6.(1) Thallus dark red-brown, with abundant, large, red-brown apothecia; growing in patches of moss over rock; photobiont blue-green (*Nostoc*) .... *Polychidium muscicola*
6. Thallus black to shades of olive or olive-gray; apothecia inconspicuous, rare, or absent; photobiont blue-green or green 7

7. On bark or moss; thallus olive to olive-gray or brownish gray; photobiont blue-green 8
7. On rock; thallus usually black to very dark olive, at least when dry 9

8. On moss or mossy trunks; branches thick, 0.1–0.6 mm in diameter, spiny, dark olive-gray to brownish; photobiont *Nostoc* .... *Dendriscocaulon intricatulum*
8. On twigs and branches in humid localities on the west coast; branches hair-like, less than 0.1 mm in diameter, not spiny; thallus forming small, foam-like clumps, pale olive or olive-gray; photobiont *Scytonema* .... *Polychidium contortum*

9. Thallus like black wool, the hair-like filaments 10–20 μm thick containing the green alga *Trentepohlia*; typically on shaded rock faces 10
9. Thallus forming tufts or clumps of much thicker filaments (40–140 μm) containing cyanobacteria (*Stigonema*); on dry or wet rocks, usually exposed 11

10. Cells enveloping algal filaments long and straight (Fig. 34) .... *Racodium*
10. Cells enveloping algal filaments irregular and knobby .... [*Cystocoleus*]

11. Thallus forming rounded clumps or cushions, 5–15 mm across and 6 mm high, on dry or wet rock; blue-green, rhizine-like hyphae developing at the base of the cushions .... *Spilonema revertens*
11. Thallus forming irregular patches or felt-like mats (when abundant), not rounded cushions; usually on wet rocks near a stream or lake shore; lacking any blue-green, rhizoid-like hyphae at the base of the clumps ... . *Ephebe lanata*

KEY C: STERILE CRUSTOSE LICHENS

1. Soredia and isidia absent 2
1. Soredia or isidia present 10

2. On bark; older parts of thallus covered with hollow pustules that break up into granular or soredia-like schizidia .... *Loxospora pustulata*
2. On wood, soil, mosses, dead vegetation, or rock 3

3. On rock; thallus cortex C+ red (lecanoric acid) .... *Dirina catalinariae*
3. On wood, soil, mosses, or dead vegetation 4

4. Thallus yellow to orange, or greenish yellow to yellow-green 5
4. Thallus without yellow or orange tint 6

5. K– .... *Lecanora*
5. K+ purple; on soil .... *Fulgensia*

6. Thallus pale 7
6. Thallus dark 9

7. Thallus cortex and medulla PD+ red (fumarprotocetraric acid); thallus almost white, with tall verrucae; arctic-alpine .... *Pertusaria dactylina*
7. Thallus cortex and medulla PD– or PD+ yellow (fumarprotocetraric acid absent); arctic-alpine or temperate 8

8. Thallus C+ pink or red, PD– (gyrophoric acid; baeomycesic acid absent) .... *Trapeliopsis granulosa*
8. Thallus C–, PD+ yellow (baeomycesic acid; gyrophoric acid absent) .... *Dibaeis baeomyces*

9.(4) Thallus dark steel gray to dark greenish gray; verrucose .... *Trapeliopsis flexuosa*
9. Thallus dark brown, granular .... *Placynthiella oligotropha*

10.(1) Isidia present 11
10. Isidia absent 15

11. Thallus olive, distinctly lobed; photobiont blue-green; directly on rock in and near southeastern Alaska .... *Vestergrenopsis isidiata*
11. Thallus white to gray-green or reddish brown, not lobed at the edge; photobiont green; on bark, wood, mosses, or dead vegetation, temperate or tropical ...12

12. On exposed wood, thallus reddish brown; boreal to northern Temperate .... [*Placynthiella icmalea*]
12. On bark, mosses, or dead vegetation; thallus pale, white to gray-green 13

13. Thick red or white prothallus present; isidia subspheri-

cal (granular) to cylindrical; southeastern U.S. ....*Cryptothecia*

13. Prothallus absent; isidia cylindrical; arctic-alpine or temperate 14

14. On mosses or dead vegetation in arctic-alpine habitats; thallus white, verrucose; isidia thick and crowded .... *Pertusaria dactylina*

14. On bark in the coastal Pacific Northwest; thallus yellowish gray to greenish gray, more or less smooth; isidia long and slender, often branched .... *Loxosporopsis corallifera*

15.(10) Soredia comprising entire thallus 16
15. Soredia in delimited soralia, or diffuse on older parts of the thallus. 23

16. Thallus pale green, gray, brown, or black, without yellow or orange tint 17
16. Thallus yellowish green, greenish yellow, bright yellow, or orange 19

17. Thallus forming round, shingled lobes with a basal cottony brown to white hypothallus; thallus KC−, PD+ orange or PD− (pannaric and roccellic acids; usnic acid absent) .... *Leproloma membranaceum*
17. Thallus thin, never forming shingled lobes; hypothallus absent, prothallus sometimes present; pannaric acid absent 18

18. Prothallus white, forming a fibrous fringe; thallus KC+ gold, K−, PD− (usnic acid and zeorin present) .... *Lecanora thysanophora*
18. Fibrous white prothallus absent; thallus KC−, K+ yellow or K−, PD+ yellow or red (usnic acid absent) .... *Lepraria*

19. Thallus greenish-yellow or yellowish green 20
19. Thallus yolk-yellow, lemon-yellow, yellow-orange, or orange-yellow 21

20. Prothallus absent, thallus thick, edge indefinite or weakly lobed; thallus PD+ orange, KC− (stictic acid present, usnic acid absent) .... *Lepraria lobificans*
20. Prothallus present, white and fibrous; thallus thin, edge definite; thallus PD−, KC+ gold (stictic acid absent, usnic acid present) .... *Lecanora thysanophora*

21. Soredia K+ purple (anthraquinones present, calycin absent) .... *Caloplaca citrina*
21. Soredia K− (anthraquinones absent, calycin present) 22

22. Thallus uniformly leprose, the soredia not clustered or associated with granules; in shaded, humid habitats .. .. *Chrysothrix candelaris*
22. Thallus usually producing soredia in small clumps, originating from the breakup of tiny areoles, but the soredia can become confluent in well-developed specimens; very common on exposed deciduous trees even close to urban areas .... *Candelariella efflorescens*

23.(15) Thallus distinctly bright yellow or orange, K+ deep purple or K− 24
23. Thallus not distinctly yellow or bright orange, K− or K+ yellow to red 27

24. On soil 25
24. On bark, wood, or rock 26

25. Thallus K+ deep purple .... *Fulgensia*
25. Thallus K− .... *Arthrorhaphis*

26. Thallus margin distinctly lobed, with a lower cortex, producing pustules and granular schizidia .... *Xanthoria sorediata* (see also *Caloplaca*)
26. Thallus not lobed, entirely crustose, lacking pustules or schizidia .... *Caloplaca* key: couplet 21

27.(23) Cephalodia present. 28
27. Cephalodia absent. 29

28. Thallus continuous, flat, with a distinctly lobed margin; cephalodia flat, disk-shaped, on the thallus surface ... . *Placopsis lambii*
28. Thallus verrucose or dispersed areolate, not lobed at the margin; cephalodia convex, cushion-shaped, between the areoles .... *Amygdalaria panaeola*

29. On rock 30
29. On bark, wood, or soil 31

30. Thallus thin, shiny, C−; soralia KC+ purple; common and widespread .... *Pertusaria amara*
30. Thallus thick, dull, C+ pink or red; soralia KC− or KC+ pink or red; restricted to coastal rocks in California .... *Dirina catalinariae* f. *sorediata*

31. On trees in tropical and subtropical regions; prothallus thick and conspicuous 32
31. In temperate to boreal regions; prothallus absent or thin and inconspicuous 33

32. Prothallus red or white; thallus not lobed, more or less smooth, not cottony or webby .... *Cryptothecia*
32. Prothallus black; thallus clearly lobed, cottony or webby .... *Crocynia pyxinoides*

33. Soredia diffuse, not confined to discrete soralia even in young parts of the thallus, arising as schizidia from the breakup of small, hollow pustules 34
33. Soredia at first produced in discrete soralia, or from the disintegration of solid granules or verrucae, sometimes becoming confluent in older parts of the thallus; pustules and schizidia absent 35

34. Thallus PD+ yellow, K+ yellow becoming blood red (norstictic acid) .... *Phlyctis argena*
34. Thallus PD+ orange, K+ bright yellow (thamnolic acid) .... *Loxospora pustulata*

35. Thallus cortex or soredia C+ pink or red, KC+ red (gyrophoric acid) 36
35. Thallus cortex C–, KC– (gyrophoric acid absent); soredia KC– or KC+ purple 37

36. Soredia white or yellowish; thallus continuous, verruculose or more or less smooth .... *Ochrolechia androgyna*
36. Soredia pale greenish to dark green; thallus verrucose, areolate, or granular .... *Trapeliopsis*

37. Thallus greenish or olive to brownish gray to dark greenish gray; soredia green or yellow-green 38
37. Thallus pale to dark slate gray or yellowish white; soredia white, pale to dark gray, or brownish 40

38. Thallus distinctly lobed, appearing crustose but lower cortex present .... *Hyperphyscia adglutinata*
38. Thallus not at all lobed; truly crustose 39

39. Soralia PD+ red (fumarprotocetraric acid); often with brown prothallus; common, boreal to northern temperate .... [*Fuscidea arboricola*]
39. Soralia PD– (contains perlatolic and hyperlatolic acids); with or without a brown or black prothallus; rare, coastal British Columbia .... [*Ropalospora viridis*]

40. Soralia KC+ purple; thallus cortex PD– .... *Pertusaria amara*
40. Soralia medulla KC–; thallus cortex PD+ yellow, orange, or red, or PD– 41

41. Thallus PD–, K–; thallus with conspicuous blue-black or dark gray prothallus. Soredia usually grayish blue, at least in part; all grayish blue tissue turning reddish purple in nitric acid; common in Pacific Northwest on alders .... [*Mycoblastus caesius*]
41. Thallus PD+ yellow or red; prothallus absent or pale 42

42. Soredia olive-black, brown, or dark greenish gray, at least in part, PD+ yellow, K+ red (norstictic acid); on bark and wood on the west and east coasts .... [*Buellia griseovirens*]
42. Soredia white, PD+ red or orange, K– or K+ yellow 43

43. Thallus very thin, whitish; soralia PD+ red, K– or brownish (fumarprotocetraric acid); boreal forest, usually on conifers and birch ... *Pyrrhospora cinnabarina*
43. Thallus thin or more frequently thick and verruculose; soralia (actually, sorediate fruiting warts) PD+ orange, K+ deep yellow (thamnolic acid); temperate eastern regions, mainly on deciduous trees .... *Pertusaria trachythallina*

KEY D: CRUSTOSE PERITHECIAL LICHENS AND LICHENS WITH ASCOMATA RESEMBLING PERITHECIA

1. Spores 1-celled, ellipsoid, and large (mostly 45–150 × 25–60 μm), most commonly with thick walls; ascomata buried in thalline warts 2
1. Spores 1-celled, ellipsoid, and small (less than 45 μm long), with thin walls; or 2- to many-celled with either thin or thick walls; ascomata superficial or more or less buried in warts, pseudostromata, or the thallus 4

2. Spore walls conspicuously thick-walled, often layered (Fig. 15l); exciple poorly defined and always colorless; widespread .... *Pertusaria*
2. Spore walls relatively thin, never layered; exciple colorless but distinct and well-developed; Pacific Northwest 3

3. Ostiole deep and hole-like; hymenium and asci IKI– or orange; thallus cortex and medulla K+ yellow, PD+ orange, UV– (stictic acid); on coastal maritime rocks; locally common .... *Coccotrema maritimum*
3. Ostiole slightly depressed, not forming a hole; hymenium and asci IKI+ blue; thallus area around ostiole C+ pink (gyrophoric acid), medulla UV+ white (alectoronic acid); on maritime or alpine rocks near coast; rare .... [*Ochrolechia subplicans*]

4. Growing on rock, soil, mosses, dead vegetation, barnacle shells, lichens, or leaves 5
4. Growing on bark or wood 17

5. Spores numerous in each ascus; inconspicuous, rather rare, temperate to boreal lichens 6
5. Spores 8 or fewer per ascus 7

6. Ascomata red-brown to black perithecia, without pruina; spores ellipsoid to fusiform, 4-celled; on mosses and plant remains; asci cylindrical or club-shaped .... [*Thelopsis*]
6. Yellow pruina present on the thalline covering of the perithecium-like ascomata; spores ellipsoid, 1-celled; on soil or decaying lichens; asci pear-shaped (Fig. 14 o) .... [*Thelocarpon*]

7. Spores 1-celled 8
7. Spores 2- or more-celled 10

8. Photobiont blue-green; thallus reddish and gelatinous when wet; on rock .... *Pyrenopsis polycocca*
8. Photobiont green; thallus not reddish and gelatinous when wet 9

9. Paraphyses persistent; on soil. Thallus thin, continuous, greenish or greenish-brown; widespread but overlooked .... [*Thrombium epigaeum*]
9. Paraphyses absent, disappearing or not developing; on rock .... *Verrucaria*

10.(7) Spores muriform, many-celled 11
10. Spores transversely septate; rarely submuriform . . . 13

11. Ostioles deeply depressed (hole-like), sometimes encircled with radiating ridges; paraphyses persistent; thallus thick, gray; medulla C+ red . . . . *Diploschistes*
11. Ostioles level with perithecial surface, slightly depressed, or prominent, without radiating ridges; paraphyses disintegrating or not developing; thallus thin or thick, pale gray to dark brown; medulla C– 12

12. Algae present in perithecial cavity . . . . *Staurothele*
12. Algae absent from perithecial cavity . . . . [*Polyblastia*]

13.(10) On mosses, dead vegetation, or leaves; spores 2- to 8-celled, rarely submuriform . . . . *Strigula*
13. On rock or barnacle shells 14

14. Spores 4- to many-celled; paraphyses persistent . . . . *Porina*
14. Spores 2-celled (rarely 4-celled); paraphyses persistent or disappearing 15

15. On intertidal calcareous rocks or barnacle shells; thallus light brown to pale orange, often only a membranous stain; photobiont blue-green; spores constricted at the septa; on Atlantic and Pacific coasts . . . . [*Pyrenocollema halodytes*]
15. On nonmaritime, mostly calcareous, rocks; thallus pale gray or yellowish white; photobiont green; spores not constricted at the septa; northern temperate to arctic-alpine 16

16. Paraphyses disintegrating or not developing; excipulum carbonized or darkly pigmented; periphyses conspicuous; spores narrow, prevalent length to width ratio 2–3:1; mostly arctic-alpine . . . . *Thelidium*
16. Paraphyses persistent; excipulum pale or colorless; periphyses absent; spores broadly ellipsoid, prevalent length to width ratio 2:1 or less; northeastern U.S. and Canada . . . . [*Acrocordia conoidea*]

17.(4) Spores very long and narrow, many-celled, not tapering, constricted at the septa, giving the spores a segmented, worm-like appearance . . . . *Conotrema urceolatum*
17. Spores broadly to narrowly ellipsoid, fusiform, or needle-shaped, often tapering, never worm-like 18

18. Ostiole deep, forming a pit or hole; excipie pale or carbonized, well developed, surrounded by a thalline envelope or covering, giving the ascoma a double-walled appearance, especially when the two walls are not fused 19
18. Ostiole not deep or forming a hole; excipie or excipulum dark or pale, surrounded by thalline tissue or not 21

19. Short, hair-like hyphae resembling periphyses lining the inner face of excipie around the ostiole; ascomata lacking a column of sterile tissue (columella); northeastern and western coasts as well as southeast . . . . *Thelotrema*
19. Hair-like hyphae resembling periphyses absent; many species with a columella developing in the ascomatal cavity; southeastern coastal plain 20

20. Excipie (inner wall of ascoma), and columella when present, black . . . . *Ocellularia*
20. Excipie, and columella when present, pale to reddish brown . . . . *Myriotrema*

21.(18) Spores pale to dark brown when mature 22
21. Spores colorless 23

22. Spores with unevenly thickened walls, central spore locules about equal in size, lens-shaped; spores not constricted at the septa . . . . *Pyrenula*
22. Spores with uniformly thickened walls, spore locules not equal in size; spores constricted at the septa. Thallus pale gray, endophloeodal; perithecia partly immersed, wall dark above and pale below; spores fusiform, with pointed ends, 18–24 × 5–9 μm . . . . [*Eopyrenula intermedia*]

23. Spores 1-celled 24
23. Spores 2- or more-celled 25

24. Asci containing numerous spores; thallus consisting of yellow pruinose warts containing perithecium-like ascomata; on wood; spores smooth; paraphyses branched or unbranched; northern temperate . . . [*Thelocarpon*]
24. Asci with 8 spores; thallus not at all yellow; perithecia black, not in thalline warts; on bark; spores conspicuously warty on the surface; paraphyses branched; southeastern coastal plain . . . . [*Monoblastia*]

25. Asci containing numerous spores . . . . [*Thelopsis* (see couplet 6)]
25. Asci usually 8-spored 26

26. Perithecia buried in a wart-like pseudostroma, often several perithecia per wart . . . . *Trypethelium* key
26. Perithecia discrete, not in a pseudostroma 27

27. Spores with unevenly thickened walls . . . . *Trypethelium* key
27. Spores with uniformly thickened walls 28

28. Spores 6- or more-celled 29
28. Spores 2–4-celled 33

29. Spores 18–27 × 4–7 μm, (5–)6–8(–9)-celled, constricted at the septa; perithecia black or almost black; thallus whitish, within substrate, indistinct; southern coastal plain . . . . [*Polymeridium quinqueseptatum*]
29. Spores 24–125 μm long, 8–14-celled, sometimes only slightly, but usually not, constricted at the septa; perithecia pale to dark brown or black; thallus yellowish

gray to greenish gray, or pale greenish, superficial or within bark 30

30. Thallus and sometimes perithecia with sparse to dense isidia. Spores 8-celled, 35–47(–57) × 5.5–8 µm, tapered, not constricted at septa; southeastern coastal plain ... . [*Porina scabrida*]
30. Thallus and perithecia smooth, not isidiate 31

31. Spores 8-celled, fusiform (widest at or close to the center), often slightly constricted at the septa, 24–42 × 5–7.5 µm; ascus thickened at tip but IKI–, with distinct ocular chamber; northeastern to Great Lakes region .. .. [*Strigula stigmatella*]
31. Spores 8–14-celled, wider at one end than the other, not constricted at septa; ascus not thickened at tip, IKI–, with no ocular chamber 32

32. Spores 60–125 × 9–15 µm; perithecia covered with thalline tissue; inner wall pale brown to yellowish brown; southeastern U.S. ....*Porina heterospora*
32. Spores (32–)38–50 × 5.5–7.5 µm; perithecia exposed, brown to black, wall dark brown to black; widespread, East Temperate .... [*Trichothelium cestrense*]

33.(28) Paraphyses abundantly branched and anastomosing; asci club-shaped; spores narrowly ellipsoid to fusiform, prevalent length to width ratio over 3:1, 2–4-celled, 14–20 × 4.5–6 µm; East Temperate .... [*Anisomeridium polypori* (syn. *A. nyssaegenum*)]
33. Paraphyses unbranched or slightly branched; asci narrowly cylindrical; spores ellipsoid, prevalent length to width ratio under 3:1 34

34. Spores 4-celled, 18–27(–30) × 7–10(–12) µm; perithecia flask-shaped, with distinct, usually off-center to lateral neck; ostioles prominent; involucrellum brown; hymenium IKI+ greenish blue, paraphyses unbranched; northeastern .... [*Lithothelium hyalosporum*]
34. Spores 2-celled, 11–16.5 × 6–9.5 µm; perithecia hemispherical, ostioles level with perithecial surface; involucrellum carbonized; perithecial cavity IKI–; paraphyses slightly branched; north central U.S. .... [*Acrocordia cavata*]

KEY E: CRUSTOSE SCRIPT LICHENS

1. On rock 2
1. On bark or wood 4

2. Spores 1-celled; thallus white, K+ red (norstictic acid); western coastal mountains and Gaspé Peninsula (Quebec); very rare .... [*Lithographa tesserata*]
2. Spores septate 3

3. Hymenial surface ("disks" of ascomata) pruinose; exciple contains crystals .... *Lecanographa*
3. Hymenial surface not pruinose; exciple lacks crystals .... *Opegrapha*

4. Lirellae immersed in wart-like pseudostroma 5
4. Lirellae developing directly on the thallus 6

5. Lirellae black, opening with only a narrow fissure, pseudostroma heavily pruinose; spores narrowly ellipsoid, 4–6-celled when mature .... *Sarcographa*
5. Lirellae brown, flat, with a completely exposed hymenium; pseudostroma without pruina or lightly pruinose; spores fusiform, 6–8-celled ... *Glyphis cicatricosa*

6. Spores 1-celled 7
6. Spores 2- or more-celled 8

7. Lirella wall brown, not carbonized; common and widespread .... *Xylographa*
7. Lirella wall black, carbonized; rare, Pacific Northwest .... [*Ptychographa xylographoides*]

8. Spores muriform 9
8. Spores only transversely septate 11

9. Spores brown to olive .... *Phaeographina*
9. Spores colorless 10

10. Lirellae with well-developed black wall (although sometimes immersed); asci cylindrical to club-shaped .... *Graphina*
10. Lirellae lacking any wall; asci balloon-shaped .... *Arthothelium* (key to *Arthonia*)

11. Spore walls evenly thickened, cells cylindrical (appearing square) 12
11. Spore walls unevenly thickened, cells lens-shaped ... 14

12. Ascomata appearing lecanorine (surrounded with thalline tissue); usually pruinose .... [*Schismatomma*]
12. Ascomata not at all lecanorine in appearance; pruinose or not 13

13. Lirellae with persistent carbonized walls (exciple) .... *Opegrapha*
13. Lirellae without distinct walls (exciple absent) .... *Arthonia*

14. Spores pale to dark brown when mature .... *Phaeographis*
14. Spores colorless .... *Graphis*

KEY F: CRUSTOSE DISK LICHENS

1. Ascomata or pycnidia at the summit of a slender or stout stalk 2
1. Ascomata immersed to superficial, broadly attached or sometimes constricted at the base, but not raised on a conspicuous stalk 5

2. Stalks stout; ascomata more than 0.5 mm in diameter, with flat to convex or hemispherical disks; spores remain within asci at maturity 3
2. Stalks slender, hair- or stubble-like (Fig. 12f); ascomata or pycnidia usually under 0.4 mm in diameter, irregular

in shape or almost spherical; spores massed at the summit, loose or remaining within asci at maturity 4

3. Apothecia broader than stalk; disks pink to brown .... Baeomyces key (*Baeomyces, Dibaeis*)
3. Apothecia immersed in the tip of the stalk; disks black .... *Pertusaria dactylina*

4. On moss; spores (conidia) in a gelatinized mass, threadlike, many-celled .... *Gomphillus americanus*
4. On bark, wood, or rarely rock; spores (ascospores) usually in a dry mass, spherical to ellipsoid, 1–2(–4)-celled .... key to *Calicium* and similar lichens

5.(1) Apothecia deeply concave or opening by a pit-like hole; margin double, with a thalline margin partially or entirely enclosing a well-developed excipulum (Figs. 12h, 13d) .... see Key D
5. Apothecia disk- or cup-shaped, or immersed in the thallus or in thalline warts, convex, flat, or somewhat concave, not opening by a deep, pit-like hole; margins various (Figs. 12a–c, 13a–c) 6

6. Apothecia buried in thalline warts, 1 or more per wart; spores very large and thick-walled, 1–2-celled, often with walls having 2 layers (Fig. 15l) 7
6. Apothecia disk- or cup-shaped, with the hymenium exposed; spores 1- to many-celled, thin- or thick-walled 8

7. Spores 1-celled; apothecia in fruiting warts opening to the surface with one or more small ostioles with the hymenium largely enclosed (Fig. 12g) .... *Pertusaria*
7. Spores 2-celled, constricted at the septum and easily breaking in two; apothecia in fruiting warts broadly opened to the surface, pinkish or yellowish, sometimes with the ascus tips showing as tiny glistening dots. Thallus white, thick or thin, sometimes becoming sorediate; medulla K–, C+ red (lecanoric acid); spores 200–400 × 72–135 µm; on moss, dead vegetation, wood, bark, or stones; arctic-alpine .... [*Varicellaria rhodocarpa*]

8. Apothecia mazaedial: spores lying loose in a powdery mass within the exciple 9
8. Apothecia usually with a well-developed hymenium; spores remaining within the asci until maturity, not forming a loose mass 11

9. Spores 4-celled, with unevenly thickened walls; ascomata with 2 chambers, one above the other, the lower one immersed in the substrate, the upper one cylindrical, about 0.5 mm high; thallus UV+ yellow (lichexanthone), thin, continuous; southeast coastal plain .... [*Pyrgillus javanicus*]
9. Spores 1–2-celled, rarely submuriform; ascomata with one chamber; thallus UV– or UV+ orange from yellow pigments (lichexanthone absent); most commonly northern or western 10

10. Exciple brown to black, well-developed; spores ellipsoid, 2-celled or submuriform in one species (*C. notarisii*); thallus thin; apothecia immersed or superficial, not in thalline warts .... *Cyphelium*
10. Exciple pale and weakly developed; spores globose or broadly ellipsoid, 1–2-celled; thallus thick; apothecia entirely immersed in thalline warts, disk level with surface .... *Thelomma*

11.(8) Apothecia pale yellow, bright lemon- or yolk-yellow, orange, or red 12
11. Apothecia green, gray, olive, brown, pink, or black .. 36

12. Asci containing numerous spherical spores; apothecia tiny, 0.1–0.2 mm in diameter, bright red, K–; rare .... [*Strangospora microhaema*]
12. Asci containing 1–32 spores; apothecia usually larger than 0.2 mm in diameter 13

13. Apothecia blood red to cinnabar or pale red 14
13. Apothecia yellow or orange, not red 17

14. Spores polarilocular; apothecia pale red; California ... . *Caloplaca luteominia* var. *bolanderi*
14. Spores thin-walled, not polarilocular 15

15. Apothecia biatorine; spores ellipsoid, 1-celled .... *Pyrrhospora*
15. Apothecia lecanorine; spores fusiform, 2–8-celled ... 16

16. Southeastern and southwestern U.S.; thallus pale gray to greenish gray (lacking usnic acid) ... *Haematomma*
16. Western and northern North America; thallus distinctly yellowish (usnic acid in cortex) .... *Ophioparma*

17.(13) Pigmented tissues (epihymenium or thallus cortex) K+ deep purple or dark red-purple (anthraquinones) 18
17. Pigmented tissues K– or K+ pinkish (lacking anthraquinones) 22

18. Spores polarilocular, with a thickened septum .... *Caloplaca*
18. Spores 1- to several-celled or muriform, not polarilocular 19

19. On dry soil; spores 1–2-celled, thin-walled ... *Fulgensia*
19. On rocks, bark, or bryophytes; spores 1- to several-celled or muriform 20

20. On rock, especially limestone; apothecia with thin, disappearing margins; spores 1-celled .... *Protoblastenia*
20. On bark, bryophytes, or decaying vegetation; apothecia with prominent, persistent margins; spores transversely septate or muriform 21

21. Thallus olive to orange; spores with unevenly thickened walls, sometimes forming spiraled locules .... *Letrouitia*
21. Thallus white to pale gray; spores thin-walled, muriform, with many cells .... *Brigantiaea*

22.(17) Apothecia biatorine, lacking algae in the margin ... 23
22. Apothecia lecanorine (with algae in the margin) or cryptolecanorine (sunken into the thallus, and lacking a recognizable exciple) ... 28

23. Spores 2- or more-celled ... 24
23. Spores 1-celled ... 27

24. Apothecial margin fuzzy or tomentose (byssoid); paraphyses branched; asci club-shaped ... *Byssoloma*
24. Apothecial margin smooth; paraphyses unbranched, slender; asci narrowly cylindrical ... 25

25. Spores 4–8-celled, 16–48 per ascus ... [*Pachyphiale fagicola*]
25. Spores 2-celled, 8 per ascus ... 26

26. Thallus very thin, not gelatinous; photobiont *Trentepohlia*; hymenium K/I+ blue; apothecia pale orange to pinkish orange, 0.2–1.5 mm in diameter ... *Dimerella*
26. Thallus gelatinous; photobiont Chlorococcoid; hymenium K/I–; apothecia very pale yellowish white, 0.3–0.4 mm in diameter ... [*Absconditella*]

27.(23) Thallus leprose, bright yellow, UV+ orange (rhizocarpic acid); on rock ... *Psilolechia lucida*
27. Thallus smooth to granular, without soredia, greenish yellow, UV– (usnic acid); on bark ... *Lecanora symmicta*

28.(22) Apothecia cryptolecanorine, immersed in thallus, the disk flush with the thallus surface ... 29
28. Apothecia lecanorine, broadly attached or constricted at the base; thallus pale yellow or yellow-green or gray ... 30

29. Spores numerous in each ascus; thallus bright yellow ... *Acarospora* key
29. Spores 8 per ascus; thallus distinctly rusty orange or brownish gray (shade forms) ... *Ionaspis lacustris*

30. Apothecial disks bright yellow or bright yellow-orange ... 31
30. Apothecial disks pale greenish yellow, pale yellow, or pale pinkish orange ... 34

31. Thallus bright yolk-yellow (calycin) ... 32
31. Thallus yellow-green or yellow-gray (usnic acid) ... 33

32. Thallus distinctly lobed; lower surface corticate ... *Candelina*
32. Thallus edge indefinite, never lobed; lower surface without a cortex ... *Candelariella*

33. Apothecial disks not pruinose; apothecial margin flush with disk, becoming thin and disappearing in maturity ... *Lecanora symmicta*
33. Apothecial disks yellow pruinose; apothecial margin prominent ... *Lecanora cupressi*

34.(30) Spores conspicuously thick-walled ... *Pertusaria*
34. Spores thin-walled ... 35

35. Spores 10–20 μm long; paraphyses unbranched; apothecia generally less than 1.5 mm in diameter ... *Lecanora*
35. Spores over 25 μm long; paraphyses highly branched; apothecia frequently over 1.5 mm in diameter ... *Ochrolechia*

36.(11) Apothecia lacking algae in the margin: lecideine, biatorine, or without an exciple; apothecia superficial, or sunken into thallus but retaining a distinct, usually pigmented exciple ... 37
36. Apothecia with algae in the margin or in thalline tissue surrounding the sunken disk: lecanorine (superficial) or cryptolecanorine (sunken into the thallus, usually without a distinct exciple; but see couplets 108–111) ... 98

37. Spores 1-celled, spherical to narrowly ellipsoid (if spores are fusiform with length to width ratio more than 3:1, they may be species that become septate later in development; see, e.g., *Icmadophila*) ... 38
37. Spores septate (2- or more-celled), broadly ellipsoid to thread-shaped ... 62

38. Apothecia biatorine, with a colorless to pigmented exciple that is soft rather than brittle, with a clearly defined and usually radiating cellular structure throughout, or exciple very reduced and indistinct ... 39
38. Apothecia lecideine, with an exciple that is very dark brown to black, usually somewhat carbonized, occasionally brittle, with a poorly defined cellular structure, at least on the outer part ... 58

39. Spores conspicuously thick-walled (more than 2.5 μm thick) ... 40
39. Spores thin-walled relative to the spore size (usually less than 1.5 μm thick) ... 41

40. Apothecial disks black; thallus white to pale greenish gray; usually with a blood-red area below the hypothecium (plate 515); epihymenium green; spores 1 or 2 per ascus; thallus cortex K+ yellow (atranorin) ... *Mycoblastus*
40. Apothecial disks dark reddish brown; thallus brown or pale; tissue below hypothecium not red; epihymenium brown; spores 8 per ascus; thallus cortex K– (atranorin absent) ... *Japewia tornoensis*

41. Directly on rock ... 42
41. On bark, wood, soil, or mosses or dead vegetation ... 43

42. Thallus pale gray, C+ pink (gyrophoric acid); apothecia pinkish or brown, with a rough or ragged margin; asci cylindrical, usually with uniformly K/I+ pale blue walls including the tip (Fig. 14l) ... *Trapelia*
42. Thallus dark, brownish, olive, or dark gray, C–; apothe-

cia dark brown to black, with a smooth margin; ascus club-shaped, the tips staining as pale and dark blue layers in K/I (Fig. 14d) . . . . *Fuscidea*

43. On moss, soil, or dead vegetation 44
43. On bark or wood 50

44. Spores numerous within the ascus, 5–8 × 2–2.3 µm, colorless. Apothecia red-brown, translucent when wet, 0.1–0.5 mm in diameter . . . . . [*Sarcosagium campestre*]
44. Spores 8 per ascus 45

45. Hypothecium dark red-brown, brown, or black . . . 46
45. Hypothecium colorless, yellowish, or very pale brown 48

46. On moss; hypothecium distinct from exciple; paraphyses usually unbranched; thallus pale greenish gray to whitish; asci *Porpidia*-type 47
46. On soil or decayed peat; hypothecium merging with exciple; paraphyses abundantly branched; thallus dark brown or olive-brown; asci K/I– . . . . *Placynthiella*

47. Thallus thick, granular to warty or areolate; tips of paraphyses somewhat expanded and brown; spores 10–16(–19) × 3.5–5(–6) µm, rarely 2-celled . . . . [*"Mycobilimbia" berengeriana*]
47. Thallus thin, membranous; tips of paraphyses not expanded; spores 10–16(–19) × 4.5–6(–7), 2-celled spores often mixed with 1-celled spores . . . . [*"Mycobilimbia" hypnorum*]

48.(45) Thallus cortex and medulla C+ pink, KC+ red (gyrophoric acid); epihymenium shades of olive or green; paraphyses abundantly branched . . . . *Trapeliopsis*
48. Thallus cortex or medulla C– or C+ orange, KC– or KC+ yellow to orange (gyrophoric acid absent); epihymenium shades of yellow or brown, or colorless; paraphyses branched only at tips 49

49. Thallus usually pale green or gray-green; apothecia usually pale to dark yellowish brown, but sometimes dark brown; exciple almost colorless or yellowish internally, sometimes pigmented at outer edge, with slender radiating hyphae . . . . *Biatora*
49. Thallus dark brown or olive, thick, mostly commonly continuous or dispersed areolate; apothecia dark brown to almost black; exciple distinct, dark brown, with large cells . . . . *Lecidoma demissum*

50.(43) Spores numerous within the ascus. Apothecia black, 0.3–0.5 mm, convex; rare and widely distributed in the west to the arctic and Great Lakes . . . . [*Strangospora moriformis*]
50. Spores 8 per ascus 51

51. Thallus leprose, dark yellow, C+ orange (xanthones); epihymenium K+ deep purple (anthraquinones) . . . . *Pyrrhospora quernea*

51. Thallus continuous to areolate or granular, C–, C+ orange, or C+ pink; epihymenium K– 52

52. Spores almost spherical, 5–10(–13) × 4–8 µm; apothecia brown to black; margins PD+ red. Very common on conifers in western mountains . . . [*Lecanora fuscescens*]
52. Spores ellipsoid; apothecial margins PD– or rarely PD+ 53

53. Apothecia greenish to yellow-green; thallus cortex KC+ gold (usnic acid) . . . . *Lecanora symmicta*
53. Apothecia pale beige, pinkish, brown, or black; thallus cortex KC– (usnic acid absent) 54

54. Paraphyses abundantly branched and anastomosing; thallus often C+ pink (gyrophoric acid) 55
54. Paraphyses unbranched except for tips; thallus C– . . . 56

55. Thallus reddish brown to dark brown, granular to isidiate; hypothecium brown, merging with exciple . . . . *Placynthiella*
55. Thallus pale gray to dark greenish gray; hypothecium pale to colorless . . . . *Trapeliopsis*

56. Thallus a mass of isidia or coarse granules (squamulose in younger parts); spores 9–12 × 1.8–2.3 µm . . . . granular forms of *Phyllopsora parvifolia*
56. Thallus continuous, areolate to somewhat granular, never squamulose; spores broader than 2.5 µm 57

57. Thallus KC+ orange (somewhat faint); spores broad (length to width ratio 1.5–2:1). Apothecia dark red-brown, 0.2–0.4 mm in diameter; very common on trees in the east . . . . [*Pyrrhospora varians*]
57. Thallus KC–; spores narrow (length to width ratio 2–3:1) . . . . *Biatora*

58.(38) Spores 8 per ascus 59
58. Spores many more than 8 per ascus 60

59. Spores brown, walls thickened at equator; prothallus thick, black; arctic-alpine . . . . *Orphniospora moriopsis*
59. Spores colorless, walls uniformly thin; prothallus present or absent . . . key to *Lecidea* and *Lecidea*-like crusts

60. Thallus well-developed, often lobed at the margin; medulla C+ pink (gyrophoric acid); apothecia immersed in thallus between the areoles, disk flush with thallus; epihymenium green or brown; spores globose or broadly ellipsoid . . . . *Sporastatia*
60. Thallus almost entirely within the substrate, absent from view, lacking gyrophoric acid; apothecia sessile; epihymenium brown or black; spores narrowly ellipsoid 61

61. Apothecial disks rough, umbonate (with sterile columns in apothecial disk); apothecia often concave or crushed as a result of crowding; apothecial margin rough or minutely fissured; exciple dark brown to carbonized throughout . . . . *Polysporina*

61. Apothecial disks smooth, without umbos; apothecia flat or soon convex; apothecial margin smooth; exciple dark or carbonized only at edge, pale internally . . . . *Sarcogyne*

62.(37) Spores muriform 63
62. Spores only transversely septate 69

63. Apothecia biatorine, concave, pale and waxy-looking, pink, yellow, or pale orange; epihymenium and hypothecium colorless, paraphyses unbranched; asci narrowly cylindrical . . . . *Gyalecta*
63. Apothecia biatorine or lecideine, flat to somewhat convex, brown or black, not waxy; epihymenium pigmented; hypothecium colorless to yellowish, or brown to black; paraphyses unbranched or abundantly branched or difficult to distinguish; asci club-shaped 64

64. Directly on rock; spores halonate or not 65
64. On bark, soil, leaves, or mosses; spores not halonate 66

65. Spores halonate . . . . *Rhizocarpon*
65. Spores not halonate . . . . . [*Buellia* key (*Diplotomma*)]

66. On leaves; apothecia 0.2–0.5 mm in diameter, disks pale to dark brown; apothecial margin even with disk . . . . *Calopadia*
66. On bark, mosses, or soil; apothecia usually more than 0.5 mm in diameter, disks black or almost black; apothecial margin prominent 67

67. Apothecial margin yellow or bright orange, paler than disk; exciple radiate, yellow; epihymenium K+ red or deep purple-red (anthraquinones); on bark in tropical and subtropical regions . . . . *Letrouitia*
67. Apothecial margin black, the same color as the disk; exciple dark brown at edge, pale internally; epihymenium unchanged in K (anthraquinones absent); on bark, soil, or mosses in northern regions 68

68. Spores 1 per ascus, over 50 μm long . . . . *Lopadium*
68. Spores 8 per ascus, less than 30 μm long . . . . [*Buellia* (*Diplotomma*)]

69.(62) Apothecia pink to yellowish pink, 1.5–4 mm in diameter; thallus green to greenish white, continuous, sometimes verrucose; growing over well-rotted wood or peat . . . . *Icmadophila*
69. Apothecia pale brown to black (if pinkish, then under 1 mm in diameter); thallus and habitat various 70

70. Spores 4- or more-celled 71
70. Spores 2-celled 82

71. Photobiont blue-green; thallus with a conspicuous blue-black prothallus . . . . *Placynthium*
71. Photobiont green; lacking a blue-black prothallus . . . 72

72. Spores brown . . . . *Buellia*
72. Spores colorless 73

73. Spores slender and needle-shaped, some often curved or bent; length to width ratio 7:1 or more 74
73. Spores ellipsoid, narrowly ellipsoid, or fusiform, straight; length to width ratio less than 7:1 76

74. Thallus bright yellow; apothecia *Arthonia*-like; apothecial margin absent; growing on soil or lichens . . . . *Arthrorhaphis*
74. Thallus greenish, olive-gray, or brownish; apothecia biatorine, with a persistent margin; on bark, mosses, or dead vegetation 75

75. Spores many per ascus; thallus edge definite, with a brown prothallus; exciple pale internally, dark brown at edge; ascus *Fuscidea*-type, with a layered appearance when stained with K/I . . . . *Ropalospora chlorantha*
75. Spores 8 per ascus; thallus edge usually indefinite; prothallus present or absent; exciple pigmented in various patterns, or sometimes colorless; ascus *Bacidia*-type, with a more or less uniformly dark blue tholus in K/I . . . *Bacidia* key (including *Bacidina* and *Scoliciosporum*)

76.(73) Thallus thick, areolate to squamulose; apothecia black; on soil or rocks, usually in arid or alpine habitats . . . . *Toninia*
76. Thallus usually thin, membranous to leprose or granular; apothecia very pale to black; on bark, moss, or dead vegetation, occasionally on rock or soil 77

77. Exciple absent; asci broad, balloon-shaped, thick at the tips but almost entirely K/I– . . . . *Arthonia*
77. Exciple usually present, although sometimes indistinct, lecideine or biatorine; asci club-shaped, K/I+ or K/I– 78

78. Exciple lecideine, dark brown to black 79
78. Exciple biatorine, usually pale; apothecia not, or lightly, pruinose 80

79. Apothecia usually pruinose or dull; margin not radially fissured . . . . *Lecanactis*
79. Apothecia black, shiny, never pruinose; margin broken up by radial fissures; on moss; arctic-alpine . . . . [*Sagiolechia rhexoblephara*]

80. Apothecial margin prominent or even with disk, fuzzy or tomentose (byssoid); apothecia flat when mature; epihymenium colorless . . . . *Byssoloma*
80. Apothecial margin becoming thin and disappearing in maturity, or absent; apothecia flat to convex or hemispherical; epihymenium shades of yellow, brown, or green 81

81. Paraphyses unbranched or slightly branched at the tips; epihymenium always C– . . . . *Mycobilimbia*
81. Paraphyses highly branched and anastomosing; epihy-

menium C– or often C+ pink (usually disappearing quickly) .... *Micarea*

82.(70) Spores dark brown when mature 83
82. Spores colorless 84

83. Spores halonate; ascus tips K/I essentially negative; on rock .... *Rhizocarpon*
83. Spores not halonate; ascus tips K/I+ dark blue, at least in part; on various substrates ....
    .... *Buellia* key (*Buellia, Amandinea, Diploicia*)

84. On rock 85
84. On bark, wood, moss, leaves, or dead vegetation ... 89

85. Hypothecium yellow-brown; asci balloon-shaped (Fig. 14n); exciple absent (Fig. 13f); on maritime rocks along both coasts .... [*Arthonia phaeobaea*]
85. Hypothecium dark brown to black; asci club-shaped; exciple well-developed, biatorine or lecideine 86

86. Photobiont blue-green; prothallus conspicuous, blue-green; epihymenium emerald green ..... *Placynthium*
86. Photobiont green; prothallus conspicuous or not, but not blue-green; epihymenium brown, reddish, to olive or greenish 87

87. Spores small, 6–15 × 2.5–5 μm; paraphyses expanded and pigmented at tips; widespread, temperate to boreal
    .... [*Catillaria chalybeia*]
87. Spores large, 12–30 × 3–14 μm 88

88. Spores narrow, 3–5 μm wide, not halonate; on rock or soil; epihymenium gray to purplish .... *Toninia*
88. Spores broad, 8–12 μm wide, halonate; on rock; epihymenium brown or greenish, rarely reddish ....
    .... *Rhizocarpon*

89.(84) Apothecial margins fuzzy and tomentose (byssoid); thallus sometimes UV+ orange; on bark or leaves on the southeastern coastal plain .... *Byssoloma*
89. Apothecial margin, if present, smooth, not at all fuzzy; thallus always UV–; most commonly East Temperate, northern, or western, on various substrates 90

90. Hypothecium dark reddish brown to black 91
90. Hypothecium colorless to pale brown 93

91. Paraphyses abundantly branched and anastomosing; exciple absent (Fig. 13f); asci balloon-shaped (Fig. 14n). Spores 8–15 × 2.5–5 μm; apothecia black, without a margin (Fig. 12c); epihymenium K–; forming white patches on poplar bark; common in boreal region from maritimes to B.C. .... [*Arthonia patellulata*]
91. Paraphyses usually unbranched except for tips; exciple lecideine or biatorine; asci usually club-shaped or cylindrical 92

92. Spores 6–12 × 2.5–4 μm; epihymenium brown; hypothecium brown to black; widely distributed on bark of all kinds, but especially conifers ....
    .... [*Catillaria glauconigrans*]

92. Spores 13–18(–24) × 5–7(–8); epihymenium black to greenish; hypothecium red-brown; Appalachian–Great Lakes distribution, on hardwoods, especially beech and maple .... *Megalaria laureri*

93.(90) On leaves of conifers and broadleaf trees and shrubs, rarely on twigs; apothecia tiny, 0.1–0.3 mm in diameter, very pale beige; spores 9.5–14(–16) × 3.5–7 μm, tapered and constricted at the septum; Pacific Northwest and southeastern U.S. ....
    .... [*Fellhanera bouteillei*]
93. On bark, wood, or moss; apothecia (0.1–)0.3–2 mm in diameter, pink to brown, gray, or black 94

94. Paraphyses abundantly branched and anastomosing. On rotting wood and occasionally bark; thallus dark green, granular; exciple poorly differentiated; apothecia pale gray to gray-brown or black; epihymenium K+ violet; spores 1–2-celled; widespread ....
    .... [*Micarea prasina*]
94. Paraphyses most commonly unbranched, except at the tips. 95

95. Spores 2–4 μm wide 96
95. Spores 4–7 μm wide 97

96. Apothecia pruinose, very pale pinkish to black or mottled; paraphyses barely expanded and not pigmented at the tips; hypothecium colorless; spores 8–16 × 2.5–3.5 μm; on bark of all kinds; coastal .....
    .... *Cliostomum griffithii*
96. Apothecia not pruinose, brown to black, not pinkish; paraphyses expanded and pigmented at the tips; hypothecium pale brown; spores 8–10 × 2–4 μm; on bark, especially poplars and white cedar, from North Dakota to southern Ontario and eastern U.S. ....
    .... [*Catillaria nigroclavata*]

97. Apothecia black, usually flat to lightly convex, with a persistent margin; spores all 2-celled, 10–15 × 5–7 μm; on bark of different kinds, especially poplar and cedar; widely distributed, especially in northern temperate to boreal region and California ....
    .... [*Catinaria atropurpurea*]
97. Apothecia yellow- to red-brown, convex; spores 1–2-celled, 11.5–19 × 4–7 μm; on moss, rarely bark; widespread .... *Biatora vernalis*

98.(36) Spores 2- or more-celled 99
98. Spores 1-celled 112

99. Spores pale to dark brown when mature, most commonly 2-celled 100
99. Spores colorless, 2- or more-celled 102

100. Spores 4 per ascus, usually over 35 μm long; cyanobacteria in the continuous crustose thallus, and green algae in lobes forming the apothecial rim; apothecia concave, cup-like, reddish or orange-brown ....
     .... *Solorina spongiosa*

100. Spores 8 or more per ascus, most commonly under 35 μm long; photobiont green with no cephalodia; apothecia flat or soon convex, black to dark brown . . . 101
101. Thallus thin, areolate, forming round, often lobed rosettes on rock; spores 8–12 μm long, with thin, uniformly thickened walls . . . . *Dimelaena*
101. Thallus not forming rosettes; if lobed, then not on rock; on bark, soil, mosses, or rock; spores 11–40 μm long, almost always with unevenly thickened walls at least when young . . . . *Rinodina* key (*Rinodina*, *Phaeorrhiza*, and *Hyperphyscia*)
102.(99) Spores muriform, 1 per ascus; apothecia buried in thallus . . . . *Phlyctis*
102. Spores transversely septate, 2–8-celled, 8 per ascus 103
103. Spores 2-celled, ellipsoid, straight or bean-shaped 104
103. Spores 4–8-celled, fusiform 105
104. Spore walls evenly thickened, thin-walled; apothecial disks pale to dark brown, frequently pruinose . . . . *Lecania*
104. Spore walls with a thickened septum (polarilocular); apothecial disks brown to almost black, sometimes pruinose . . . . *Caloplaca*
105. Hypothecium colorless; apothecia lightly pruinose or not pruinose, brown to black 106
105. Hypothecium brown to black; apothecia white because of a heavy pruina, black beneath. Occurs mainly in coastal California 108
106. Spores straight to slightly curved, with rounded ends, not twisted in the ascus; apothecia pale to very dark brown or black, pruinose or not; on trees or rocks . . . . *Lecania*
106. Spores sinuous, curved and twisted in the ascus; apothecia pinkish brown to red-brown, lightly pruinose or without pruina; on bark in temperate to boreal regions 107
107. Thallus producing long, slender, sometimes branched isidia; thallus K–, PD–; coastal forests of Pacific northwest . . . . *Loxosporopsis corallifera*
107. Thallus lacking isidia; thallus K+ deep yellow, PD+ orange (thamnolic acid); eastern temperate forests to boreal forest . . . . *Loxospora*
108.(105) Paraphyses abundantly branched and anastomosing; spores halonate; apothecia appear lecanorine when immersed in the thallus, but with a dark brown to black exciple; thallus cortex and medulla usually C+ red, KC+ red; on rocks . . . . *Lecanographa*
108. Paraphyses usually unbranched or branched only at the tips; spores not halonate; apothecia truly lecanorine or appear to be lecanorine; thallus cortex and medulla C+ or C–, KC+ or KC–; on rocks or bark 109
109. Hypothecium extending like a peg to the substrate (seen in a section through the center of the apothecium); thallus cortex of North American species C–, KC–; California 110
109. Hypothecium lens-shaped, not extending to the substrate; thallus cortex or medulla C+ red, KC+ red (erythrin, lecanoric acid); California or Florida . . . . *Dirina*
110. Thallus PD–; ascomatal disks irregular in shape, lobed or slightly elongate; on bark or rock . . . . [*Roccellina franciscana*]
110. Thallus PD+ yellow or red-orange; ascomatal disks round; on bark or wood 111
111. Thallus creamy white when fresh, PD+ yellow (psoromic acid); on trees along coast and into chaparral, San Francisco to Los Angeles . . . . [*Sigridea californica* (syn. *Schismatomma californicum*)]
111. Thallus tan to greenish brown, PD+ red-orange (unknown substance); on bark or wood, Santa Catalina Island . . . . *Roccellina conformis*
112.(98) Thallus gelatinous (and reddish) when wet; photobiont blue-green (*Gloeocapsa*); apothecia opening by a deep pore; on streamside or lakeside rocks . . . . *Pyrenopsis*
112. Thallus not gelatinous, most with green photobionts; apothecial disks broad; on various substrates 113
113. Cephalodia conspicuous on the thallus surface, pink to brown, disk-like and more or less lobed; apothecia usually pink . . . . *Placopsis*
113. Cephalodia absent or very inconspicuous; apothecia pink, brown, or black 114
114. Apothecia more or less immersed in thallus, disk flush with thallus surface (cryptolecanorine) 115
114. Apothecia superficial: adnate or constricted at base 125
115. Spores numerous in each ascus . . . . *Acarospora* key (*Acarospora*, *Pleopsidium*)
115. Spores 8 per ascus 116
116. Thallus distinctly lobed at the margin 117
116. Thallus margin without distinct lobes 118
117. Thallus and apothecia heavily pruinose, thin and clearly crustose; apothecia black (beneath the pruina); thallus cortex and medulla K–, PD– . . . . *Aspicilia candida*
117. Thallus and apothecia without pruina, very thick and almost foliose; apothecia dark brown; thallus cortex and medulla K+ yellow becoming blood red, PD+ yellow (norstictic acid) . . . . *Lobothallia*

118. Spores conspicuously thick-walled . . . . *Megaspora verrucosa*
118. Spores thin-walled 119
119. Thallus pale yellow to yellowish white, thick, KC+ gold (usnic acid); epihymenium green . . . . *Lecanora marginata*
119. Thallus not at all yellowish, KC– (lacking usnic acid); epihymenium shades of yellow, brown, or green, or colorless 120
120. Apothecial disks C+ yellow, heavily pruinose; on siliceous rocks . . . . *Lecanora rupicola*
120. Apothecial disks C–, pruinose or not (if pruinose, then on calcareous rock) 121
121. Apothecial disks black; epihymenium shades of green 122
121. Apothecial disks pink to brown or gray, rarely darkening to black; epihymenium shades of gray, yellow, or brown 124
122. Epihymenium yellow-olive to olive-brown, unchanged or intensifying with nitric acid. Ascus K/I–; widespread and common . . . . *Aspicilia*
122. Epihymenium blue-green to olive-green, changing to wine-red with nitric acid, or unchanged 123
123. Mainly arctic-alpine species; thallus thin, continuous to rimose; ascus K/I– . . . . *Ionaspis* key
123. East Temperate; thallus thick, white to pale gray, areolate; ascus with a K/I+ blue tip . . . . *Lecanora oreinoides*
124.(121) Ascus tips with a IKI+ dark blue "plug" in a light blue tholus (*Porpidia*-type); spores always halonate; on dry rocks; apothecial disks very dark brown or red-brown (especially when wet) . . . . *Bellemerea*
124. Ascus tips entirely IKI–; spores not, or sometimes vaguely halonate; on wet or submerged rocks, or on dry rocks; apothecia pinkish to brown . . . *Ionaspis* key
125.(114) Spores conspicuously thick-walled (more than 2.5 μm thick) 126
125. Spores thin-walled (usually less than 1.5 μm) 128
126. Apothecia pale, pruinose or not . . . . *Ochrolechia*
126. Apothecia black, sometimes whitened by pruina 127
127. Epihymenium olive to greenish, K+ more clearly green; spores 30–65 μm long, 8 per ascus; boreal to arctic, on trees, bryophytes, and vegetation . . . . *Megaspora verrucosa*
127. Epihymenium brown, K– or K+ violet; spores usually longer than 60 μm, 1–8 per ascus; widespread on various substrates . . . . *Pertusaria*
128.(125) Spores very numerous in the ascus, 3–6 × 1.5–2.5 μm. Asci similar to *Fuscidea*-type; apothecia dark brown to almost black, 0.4–1 mm in diameter; thallus brownish gray, bumpy; on trees; East Temperate . . . . [*Maronea constans*]
128. Spores 1–32 per ascus, larger than 6 × 2.5 μm 129
129. Hymenium purple; hypothecium brown; apothecia pitch black . . . . *Tephromela atra*
129. Hymenium colorless; hypothecium colorless to yellowish, rarely brown; apothecia various shades of pink and brown, or black 130
130. Spores with rough, bumpy, sculptured walls (clearly seen under 400× magnification); thallus brownish to green, granular to, more commonly, squamulose; on mossy logs, peat, and soil in boreal to arctic regions 131
130. Spore walls smooth; thallus rarely squamulose, pale to dark; on various substrates 132
131. Photobiont blue-green . . . . *Pannaria pezizoides*
131. Photobiont green . . . . *Psoroma hypnorum*
132. Spores over 25 μm long; paraphyses abundantly branched; apothecia pale, usually pinkish to yellow-pink . . . . *Ochrolechia*
132. Spores mostly 10–20 μm long; paraphyses unbranched or sometimes sparsely branched, especially at the tips; apothecia pale to dark brown or black 133
133. On dry soil; thallus clearly lobed at the margins, pale yellowish to chalky white and pruinose, rarely brownish green . . . . *Squamarina*
133. On rock, wood, bark, dead vegetation, or peat; thallus not yellowish white and chalky pruinose 134
134. Thallus pale greenish gray to pinkish white, C+ pink (gyrophoric acid); apothecial margin often with a frayed or ragged appearance; on rock, especially small stones . . . . *Trapelia*
134. Thallus pale or dark, C– in North American species (lacking gyrophoric acid); apothecial margins smooth or bumpy but not ragged; on various substrates . . 135
135. On bark, wood, mosses, or dead vegetation . . . . *Lecanora*
135. On rock 136
136. Thallus distinctly lobed at the margin 137
136. Thallus not lobed 138
137. Thallus cortex K+ yellow becoming blood red (norstictic acid), rarely K–; tips of paraphyses septate and constricted like a string of beads (moniliform); asci K/I– . . . . *Lobothallia*
137. Thallus cortex K– or K+ yellow (norstictic acid absent); tips of paraphyses not appearing "beaded"; asci with K/I+ blue tips . . . . *Lecanora*
138. Ends of spores pointed; thallus grayish brown to reddish brown, shiny; apothecial disks chocolate brown and shiny . . . . *Protoparmelia*

138. Ends of spores rounded; thallus rarely brown and shiny; apothecia various colors .... *Lecanora* key (*Lecanora, Rhizoplaca*)

KEY G: SQUAMULOSE LICHENS

1. Thallus gelatinous when wet 2
1. Thallus dull or shiny, but not gelatinous when wet ... 3
2. Thallus without a cortex .... *Collema*
2. Thallus with a cellular cortex .... *Leptogium*
3. Thallus greenish yellow, yellow-green, yellow, or orange 4
3. Thallus pale green, gray, brown, olive, or black, without a yellowish or orange tint 17
4. Thallus greenish yellow or yellowish green 5
4. Thallus bright yolk-yellow, lemon-yellow, or shades of orange 10
5. Directly on rock 6
5. On soil 8
6. Apothecia entirely immersed in thallus, disk flush with thallus; asci containing numerous spores .... *Acarospora contigua*
6. Apothecia sessile or constricted at base; asci containing 8 spores 7
7. Apothecia constricted at base; disks yellow-orange, lightly pruinose; thallus not forming radiate, lobed rosettes .... *Rhizoplaca subdiscrepans*
7. Apothecia sessile; disks pale greenish yellow, not pruinose; thallus usually forming distinctly lobed rosettes .... *Lecanora muralis*
8. Squamules ascending, with the lower surface easily seen; thallus cortex KC+ yellow, UV– (usnic acid) .... *Cladonia* species, especially *C. robbinsii*
8. Squamules closely appressed or lifting only at the edges; apothecia frequently seen, black, directly on the squamules; thallus cortex KC–, UV+ orange (rhizocarpic acid) 9
9. Squamules flat, often turned up at the edges, sturdy; spores one-celled; southwestern and south central U.S. .... *Psora icterica*
9. Squamules closely appressed, convex, fragile; spores needle-shaped, 9–12-celled; boreal to arctic .... *Arthrorhaphis citrinella*
10.(4) Thallus yellow-orange or orange-yellow; thallus cortex K+ dark purplish (anthraquinones) 11
10. Thallus yolk-yellow or lemon-yellow; thallus cortex K– (anthraquinones absent) 14

11. Squamules ascending, with white lower surface easily seen; on bark or rock .... *Xanthoria*
11. Squamules ascending or closely appressed; lower surface, if visible, yellow; on wood, soil, or rocks 12
12. Squamules thick, yellow to orange, shiny, ascending, with deep yellow lower surface clearly visible; apothecia black, hemispherical and marginless; spores spherical, colorless, 1-celled, 3–4 µm in diameter; Texas to Mexico, on limestone .... [*Xanthopsorella texana*]
12. Squamules usually thin, appressed, lower surface not easily seen; apothecia orange, clearly rimmed 13
13. On soil in the interior of the continent north to the arctic; spores 1–2-celled, with uniformly thin walls ... . *Fulgensia*
13. On rock, often in maritime or nearby coastal localities; spores 2-celled, polarilocular .... *Caloplaca*
14. Squamules ascending, with the pale lower surface easily seen; mainly on bark; thallus thin .... *Candelaria*
14. Squamules closely appressed; on rock or soil; thallus thick 15
15. Apothecial disks black, convex; spores needle-shaped, septate .... *Arthrorhaphis citrinella*
15. Apothecial disks brown or yellow; spores globose or ellipsoid, 1-celled 16
16. Apothecia sessile, lecanorine; apothecial disks yellow; spores 12.5–16.0 × 4.8–6.5 µm, 8 per ascus .... *Candelariella rosulans*
16. Apothecia entirely immersed in thallus, disk level with thallus, apothecial disks dark brown, or reddish brown to orange-brown; spores 3–4 × 2–3 µm, many per ascus .... *Acarospora*
17.(3) Squamules ascending or erect, flat or cylindrical, with the lower surface easily seen 18
17. Squamules closely appressed, or lifting only at the edges 23
18. Thallus dwarf fruticose, consisting of erect cylindrical to somewhat flattened lobes 19
18. Thallus composed of flat, clearly dorsiventral squamules 20
19. Photobiont blue-green; spores globose, 1-celled, more than 8 per ascus .... *Peltula cylindrica*
19. Photobiont green; spores ellipsoid, muriform, 2 per ascus .... *Endocarpon pulvinatum*
20. Squamules elongate and branched; apothecia clustered in groups on the squamule tips; thallus cortex K+ yellow (atranorin); very rare .... *Gymnoderma lineare*
20. Squamules not more than 3 times longer than broad; apothecia occurring singly, thallus cortex K– (atranorin absent); relatively common species 21

21. Squamules lobed or finely divided, apothecia raised on a stalk or stipe; thallus cortex C–, KC– (gyrophoric acid absent as main compound) .... *Cladonia* species (e.g., *C. caespiticia*)
21. Squamules round, or scalloped (with rounded lobes); apothecia sessile; thallus cortex C+ pink or red, KC+ pink or red (gyrophoric acid) 22

22. Thallus pale gray; squamules thick, distinctly lobed, with edges turned down; apothecia flat or convex, pink to lead gray, with a prominent margin .... *Trapeliopsis wallrothii*
22. Thallus olive to greenish brown; squamules thin, round, with edges turned up, apothecia becoming hemispherical, dark brown to black; apothecial margin absent .... *Psora nipponica*

23.(17) Photobiont (and exposed algal layer) blue-green 24
23. Photobiont (and algal layer) green 29

24. Soredia present 25
24. Soredia absent 26

25. Thallus olive brown; on rocks in arid or damp habitats in the western interior; soredia farinose; squamules round, with edges turned down or relatively flat .... *Peltula euploca*
25. Thallus blue-gray to yellowish gray, on bark or sometimes rock in humid, boreal to subarctic habitats; soredia coarsely granular; squamules scalloped (with rounded lobes) or elongate, with edges turned up .... *Pannaria conoplea*

26. Squamules extremely narrow-lobed (less than 0.3 mm wide), building into an areolate crust with abundant isidia; apothecia black, lecideine .... *Placynthium nigrum*
26. Squamules usually wider than 0.3 mm, isidiate or not; apothecia red-brown, lecanorine or biatorine 27

27. Thallus olive or olive-brown; apothecia partly or entirely immersed in the thallus; photobiont *Scytonema* or *Anacystis* 28
27. Thallus brownish or gray-brown; apothecia sessile or constricted at the base; photobiont *Nostoc* .... *Pannaria* key (*Pannaria*, *Fuscopannaria*, *Parmeliella*)

28. Spores ellipsoid, 8 per ascus .... *Heppia conchiloba*
28. Spores globose or broadly ellipsoid, up to 100 per ascus .... *Peltula*

29.(23) Soredia present 30
29. Soredia absent 34

30. On soil or rock, rarely wood; cortex and medulla K+ persistently yellow, PD+ orange (stictic acid) .... *Baeomyces rufus*
30. On bark, wood, mosses, or lichens; thallus cortex and medulla K–, PD– (stictic acid absent) 31

31. Soralia raised and finally cup-like; thallus olive-brown, with round, tiny squamules (less than 0.5 mm in diameter); apothecia blue-gray, biatorine; spores 4-celled, fusiform; California .... [*Waynea californica*]
31. Soralia on lobe surface or on margin of lower surface, not raised or cup-like 32

32. Thallus pale green; squamules thin with a raised, thickened rim, separate, not overlapping; on mosses and lichens (especially cyanobacterial lichens), also bark and wood in humid habitats. Thallus cortex and medulla C–, KC– .... *Normandina pulchella*
32. Thallus olive to brownish or gray-brown; squamules thick, commonly overlapping like shingles; on wood (especially charred wood) or conifer bark 33

33. Thallus olive; cortex and medulla C+ pink, KC+ pink or red, PD– (lecanoric acid) ... *Hypocenomyce scalaris*
33. Thallus greenish brown to gray-brown; cortex and medulla C–, KC–, PD + red (fumarprotocetraric acid) .... *Hypocenomyce anthracophila*

34. (29) Ascomata absent. Squamules green with raised or thickened rims 35
34. Ascomata present, either perithecia or apothecia ... 36

35. Squamules less than 2.5 mm wide; on bark, mosses, and lichens (see couplet 32) .... *Normandina pulchella*
35. Squamules 2–5 mm wide, often lobed, closely appressed to soil or peat; arctic-alpine to boreal. Fruiting bodies, when present, pale yellow mushrooms .... [*Omphalina hudsoniana*]

36. On bark or wood 37
36. On soil, mosses, dead vegetation, or rocks 40

37. Ascomata perithecia buried in the thallus, only the ostioles showing as dark spots at the surface 38
37. Ascomata reddish-brown apothecia on the thallus surface 39

38. Spores muriform, brown, 2 per ascus; squamules thin, closely appressed, relatively flat edges; perithecial cavity containing algae .... *Endocarpon pusillum*
38. Spores 1-celled, colorless, 8 per ascus; squamules thick, free at the margins, with the edges turned down; perithecial cavity containing no algae .... *Placidium tuckermanii*

39. Apothecia biatorine, 0.3–0.5 mm in diameter, flat to convex, margin becoming thin and disappearing in maturity; spores narrowly ellipsoid, 1.8–2.2 µm wide .. .. *Phyllopsora parvifolia*
39. Apothecia lecanorine, 0.5–3 mm in diameter, flat to concave, margin prominent; spores ellipsoid, 6.5–10.5 µm wide .... *Psoroma hypnorum*

KEYS TO GENERA AND MAJOR GROUPS

40.(36) Ascomata perithecia buried in the thallus with only the dot-like ostiole showing 41
40. Ascomata apothecia, superficial or sunken into the thallus; or ascomata absent 42

41. Spores muriform, brown, 2 per ascus; perithecial cavity containing algae . . . . *Endocarpon pusillum*
41. Spores 1-celled, colorless, 8 per ascus; perithecial cavity containing no algae . . . . *Placidium*

42. Apothecia raised on a stalk or stipe . . . . *Baeomyces*
42. Apothecia not raised on a stalk or stipe 43

43. Apothecia lecanorine or cryptolecanorine (with algae in the margin, or immersed in the thallus) 44
43. Apothecia lecideine or biatorine (lacking algae in the margin) 51

44. Spores up to 100 per ascus . . . . *Acarospora*
44. Spores 8 or fewer per ascus 45

45. Spores 2- or more-celled 46
45. Spores 1-celled 48

46. Spores dark brown; growing on calcareous soil, mosses, or dead vegetation; arctic-alpine . . . . *Phaeorrhiza nimbosa*
46. Spores colorless; growing directly on noncalcareous rock 47

47. Spore walls evenly thickened; epihymenium K– (anthraquinones absent); in coastal California . . . . *Lecania brunonis*
47. Spores polarilocular; epihymenium K+ deep purple-red (anthraquinones); in the arid southwest . . . . *Caloplaca pellodella*

48. Apothecia entirely immersed in thallus, disks flush with thallus; apothecial disks black or almost black, heavily pruinose; spores globose or broadly ellipsoid, 15–25 μm wide . . . . *Aspicilia contorta*
48. Apothecia sessile or somewhat raised; apothecial disks brown to yellowish orange, not or lightly pruinose; spores ellipsoid, 4–10 μm wide 49

49. Directly on rocks; thallus shiny . . . . . *Lecanora muralis*
49. On soil, mosses, or dead vegetation; thallus dull . . . 50

50. Thallus thick, yellowish white, pruinose; apothecial margin flush with disk, or becoming thin and disappearing in maturity; on soil in relatively dry habitats . . . . *Squamarina lentigera*
50. Thallus thin, brown to gray-green; apothecial margin prominent; on mossy soil and peat in relatively humid habitats . . . . *Psoroma hypnorum*

51.(43) Spores 2- or more-celled, prevalent spore length to width ratio over 3.0 . . . . *Toninia*
51. Spores 1-celled, prevalent spore length to width ratio 3.0 or less 52

52. Squamules closely appressed 53
52. Squamules lifting, at least at the edges 54

53. On siliceous rock; thallus shiny, with a black prothallus; apothecia lecideine, black; apothecial margin prominent or flush with disk . . . . *Lecidea atrobrunnea*
53. On calcareous soil; thallus dull, without a prothallus; apothecia biatorine, reddish brown or orange-brown; apothecial margin absent . . . . *Gypsoplaca macrophylla*

54. Apothecial disks lead gray, pale brown, or pink . . . . *Trapeliopsis wallrothii*
54. Apothecial disks black or almost black, dark brown, or reddish brown to orange-brown . . . . *Psora* key (*Psora, Psorula*)

KEY H: UMBILICATE AND FAN-SHAPED LICHENS

1. Thallus attached by a holdfast at one edge of the lobe and becoming fan-like 2
1. Thallus attached by a holdfast close to the center of the thallus 5

2. Thallus remaining opaque when wet; with a distinct grass-green algal layer (photobiont a green alga); lower surface white, yellowish, or dark brown; on bark, or soil 3
2. Thallus jelly-like and translucent when wet; lacking a distinct algal layer (photobiont cyanobacteria); lower surface black or gray; on rock 4

3. Thallus yellow-green, rather shiny; lobes with a network of depressions and sharp ridges, wrinkled or bumpy (rugose); lower surface about the same color as the upper surface, with a smooth, more or less uniform cortex; on bark . . . . *Ramalina sinensis*
3. Thallus dark gray-green when dry and grass-green when wet, lobes smooth and even, upper surface dull; lower surface distinctly different from the upper surface, entirely without a cortex, webby or cottony, basically white, with conspicuous dark brown veins; on soil . . . . *Peltigera venosa*

4. On dry calcareous rocks in the open; lower surface more or less uniform; thallus black . . . . . *Lichinella nigritella*
4. Aquatic, on submerged rocks in mountain streams; lower surface with conspicuous veins; thallus mineral gray, brown, or olive, at least when dry . . . . *Hydrothyria venosa*

5.(1) Thallus yellow-green; cortex KC+ orange-yellow (usnic acid) 6
5. Thallus gray, brown, or olive, without a yellow tint; cortex KC–, or KC+ red (lacking usnic acid) 7

6. Rhizines abundant, stubby and peg-like; thalli large (up to 15 cm across); apothecia with pale to dark red-brown disks . . . . *Omphalora arizonica*

6. Rhizines absent; thalli up to 6 cm across; apothecia various colors, but not clear red-brown . . . . *Rhizoplaca*
7. Black dots abundant all over upper surface caused by immersed perithecia . . . . *Dermatocarpon*
7. Black dots absent or rare, and then caused by scattered pycnidia; fruiting bodies, when present, apothecia . . . 8
8. Apothecia brown, sometimes pruinose, broken up into segments by sterile tissue, cryptolecanorine, sunken into thallus; thallus thick, surface areolate or cracked, chalky white to pale gray; spores spherical, 3–4 μm in diameter, many per ascus; rare; in the arid interior or arctic, on limestone . . . . [*Glypholecia scabra*]
8. Apothecia black, usually with concentric or radiating bands of sterile tissue, lecideine, superficial or sunken; thallus thin or thick, very dark brown to gray; spores never spherical, 8 or fewer per ascus; common and widespread, on siliceous rock . . . . *Umbilicaria* key (*Lasallia*, *Umbilicaria*)

KEY I: JELLY LICHENS

1. Thallus crustose 2
1. Thallus foliose or fruticose 3
2. Thallus granular, reddish when wet; photobiont *Gloeocapsa* . . . . *Pyrenopsis*
2. Thallus membranous, olive to black when wet; photobiont *Nostoc* . . . . [*Lempholemma*]
3. Aquatic, on submerged rocks; lower surface with conspicuous veins . . . . *Hydrothyria venosa*
3. Not aquatic (i.e., not growing on submerged rocks); lower surface without veins 4
4. Lower surface tomentose (hairy or furry) . . . . *Leptogium* key (*Leptogium* and *Leptochidium*)
4. Lower surface smooth or wrinkled 5
5. Thallus pitch black when dry, foliose to fruticose, attached by a single point with the lobes fanning out; on limestone; photobiont Chroococcales (e.g., *Gloeocapsa*) 6
5. Thallus shades of gray, brown, or olive, often dark but not pure black; foliose, broadly attached; on various substrates; photobiont *Nostoc* 7
6. Lobes rounded, 0.7–2.5(–3.5) mm wide, ascending, often shell-like or crinkled, frequently with thickened margins, never pruinose, up to 250 μm thick when wet . . . . *Lichinella nigritella*
6. Lobes strap-shaped, 0.3–1.5 mm wide, usually prostrate, more or less flat, rarely thickened at margins, often pruinose, up to 500 μm thick when wet . . . *Thyrea confusa*
7. Upper and lower surfaces with a cortex, usually a single layer of square to roundish cells; thallus often shiny, gray to reddish brown . . . . *Leptogium*
7. Upper and lower surfaces lacking a cortex; thallus always dull; olive to olive-brown or almost black 8
8. Spores transversely septate or muriform; very common lichens . . . . *Collema*
8. Spores 1-celled; uncommon, inconspicuous lichens . . . . [*Lempholemma*]

KEY J: YELLOW FOLIOSE LICHENS

1. Thallus orange to yellow-orange; cortex K+ dark purple (anthraquinones) . . . . *Xanthoria*
1. Thallus deep yolk-yellow, bright yellow, greenish yellow or yellowish green ("usnic-yellow"), yellowish olive, or yellowish gray; cortex K– or K+ yellow (anthraquinones absent) 2
2. Thallus deep yolk-yellow or bright yellow 3
2. Thallus greenish yellow, yellowish green ("usnic-yellow"), yellowish olive, or yellowish gray 5
3. Medulla white; calycin present, pinastric and vulpinic acids absent 4
3. Medulla bright yellow; calycin absent, pinastric and vulpinic acids present . . . . *Vulpicida*
4. Thallus very closely attached, almost crustose (but with a lower cortex), lacking rhizines; on rock . . . . *Candelina submexicana*
4. Thallus clearly foliose, with rhizines; on bark or wood, less frequently on rock . . . . *Candelaria*
5.(2) Rhizines absent; lower surface with or without tomentum (see Fig. 6) 6
5. Rhizines abundant or sparse; lower surface without tomentum 13
6. Thallus attached by a single central point (umbilicate), or not attached to the substrate (vagrant); lower surface without tomentum . . . . *Rhizoplaca*
6. Thallus not umbilicate or vagrant; lower surface with or without tomentum 7
7. On bark 8
7. On rock, soil, or moss 10
8. Photobiont green; upper surface rather shiny; soredia absent but lobules present; medulla PD+ orange . . . . *Lobaria oregana*
8. Photobiont blue-green; upper surface dull or scabrose; soredia present, lobules absent; medulla PD– or PD+ yellow 9
9. Lobes with a network of depressions and sharp ridges; thallus greenish yellow; lower surface smooth or wrinkled, dark in the center, pale to dark brown close to the margin; medulla PD–, K– . . . . *Nephroma occultum*
9. Lobes smooth and even, or with rounded depressions; thallus yellowish olive or pale green; lower surface pale

or dark tomentose interrupted by small or large bald spots; medulla PD+ yellow, K+ yellow to orange (stictic acid with some norstictic acid) . . . *Lobaria scrobiculata*

10. Thallus of erect, elongated lobes, almost fruticose; lower surface yellow; pycnidia abundant and conspicuous as black dots, occurring mostly along the lobe margins . . . . *Flavocetraria*
10. Thallus prostrate; lobes rounded, not erect, closely appressed or loosely attached over entire surface; lower surface pale to dark brown or black; pycnidia absent or sparse and very inconspicuous 11

11. On mossy rock in humid, temperate regions; algal layer dark blue-green (photobiont cyanobacteria); soredia present; lower surface with tomentum interrupted by small or large bald spots . . . . *Lobaria scrobiculata*
11. On the ground or over rocks in alpine or arctic regions; algal layer grass-green (photobiont green algae); soredia absent 12

12. Lobes smooth and even, or with shallow depressions or wrinkles; cephalodia producing broad gray bumps on the thallus surface; pseudocyphellae absent; lower surface with a tomentum; apothecia abundant, produced on the lower surface of lobe margins . . . . *Nephroma arcticum*
12. Lobes with a network of depressions and sharp ridges; cephalodia absent; pseudocyphellae usually abundant and conspicuous; lower surface smooth, lacking tomentum; apothecia rare, on the upper surface of the lobes . . . . *Asahinea chrysantha*

13.(5) Cilia bulbous, common along lobe margins. Small lichens (lobes less than 2 mm wide) of the subtropics and tropics; rare . . . . [*Relicina*] (see comments under *Bulbothrix*)
13. Cilia, if present, not bulbous 14

14. On bark or wood 15
14. On rock or soil 28

15. Soredia present 16
15. Soredia absent 22

16. Rhizines forked in regular dichotomies; soredia on upper surface of lobe tips; medulla PD+ distinct yellow, K+ red (salazinic acid) . . . . *Hypotrachyna sinuosa*
16. Rhizines unbranched; soredia on the lobe margins or on the upper surface; medulla PD– or PD+ red-orange, K– (salazinic acid absent) 17

17. Tiny white dots (pseudocyphellae) present on the upper surface of the lobes; lower surface dark brown to black; medulla C+ red or pink (lecanoric acid) 18
17. Pseudocyphellae absent; lower surface white, brown, or black; medulla C– or C+ red 19

18. Soralia almost entirely marginal; pseudocyphellae sparse and inconspicuous . . . . *Flavopunctelia soredica*
18. Soralia both on the thallus surface and along the margins; pseudocyphellae abundant and conspicuous . . . . *Flavopunctelia flaventior*

19. Lower surface black with a brown, naked zone near margins; rhizines black; medulla PD– or PD+, KC+ pink or red 20
19. Lower surface white to pale brown; rhizines pale or dark brown; medulla PD–, KC– 21

20. Soredia coarsely granular, entirely laminal; medulla PD+ red-orange, KC+ pink, C– (protocetraric acid) . . . . *Flavoparmelia caperata*
20. Soredia both laminal and marginal; medulla PD–, KC+ red, C+ red (lecanoric acid) . . . *Flavopunctelia soredica*

21. Soredia in round soralia on thallus surface; upper surface dull; rhizines abundant, brown; medulla UV+ blue-white (divaricatic acid) . . . . *Parmeliopsis ambigua*
21. Soredia in elongate soralia along the lobe margins; upper surface rather shiny; rhizines sparse, white or very pale tan; medulla UV– (divaricatic acid absent) . . . . *Allocetraria oakesiana*

22.(15) Lower surface dark brown or black 23
22. Lower surface white to pale brown or yellow 26

23. Pseudocyphellae abundant and conspicuous; medulla KC+ red, C+ red (lecanoric acid) . . . . *Flavopunctelia praesignis*
23. Pseudocyphellae absent; medulla KC–, C– 24

24. Thallus closely attached with abundant rhizines almost to the margin; cilia and isidia absent; medulla PD+ red; south Texas, New Mexico, and Missouri . . . . [*Flavoparmelia rutidota*]
24. Thallus loosely attached; rhizines absent from a broad zone at the margin; isidia and cilia present; medulla PD–; southeastern U.S. 25

25. Medulla white; isidia often with short, black cilia growing out of the tips; marginal cilia common and abundant, lower surface smooth; cortex K–, KC+ orange-yellow (usnic acid present, atranorin and vulpinic acid absent) . . . . *Parmotrema xanthinum*
25. Medulla yellow; isidia without cilia, marginal cilia sparse; lower surface wrinkled; cortex K+ yellow, KC– (atranorin and vulpinic acid present, usnic acid absent) . . . . *Parmotrema sulphuratum*

26.(22) Medulla bright yellow (pinastric and vulpinic acids); apothecia along the lobe margins . . . . *Vulpicida viridis*
26. Medulla white or pale yellowish orange; apothecia laminal, not marginal 27

27. On southeastern coastal plain; medulla often becoming yellowish to orange in places, PD+ orange, K+ yellow (stictic acid) . . . . *Pseudoparmelia uleana*

27. Western or Appalachian–Great Lakes region; medulla white, PD–, K– . . . . *Ahtiana*
28.(14) Soredia present 29
28. Soredia absent 31
29. Soredia on thallus surface; lower surface and rhizines black 30
29. Soredia along the lobe margins; lobes elongated; lower surface and rhizines white to pale brown; medulla PD–, KC– . . . . *Allocetraria oakesiana*
30. Widespread temperate; lobes rounded; medulla PD+ red, KC+ pink (protocetraric acid) . . . . *Flavoparmelia caperata*
30. Arctic-alpine; lobes narrow, under 1 mm wide, convex, medulla PD–, KC+ red (alectoronic acid) . . . . [*Arctoparmelia incurva*]
31. Hollow pustules resembling isidia present in clumps on the upper surface, sometimes breaking into granular fragments; medulla PD+ red, KC+ pink, K– (protocetraric acid) . . . . *Flavoparmelia baltimorensis*
31. Hollow, isidia-like pustules absent; solid isidia present or absent; medulla with various reactions 32
32. Lobes 8–12 mm wide; medulla PD–, K–; marginal cilia common and abundant; rhizines sparse, absent from a broad zone close to the margin . . . . *Parmotrema xanthinum*
32. Lobes less than 4 mm wide; marginal cilia absent; rhizines usually abundant to the thallus edge 33
33. Upper surface dull; lower surface white to dull black; medulla K–, PD–, KC+ red (alectoronic acid); arctic-alpine to boreal . . . . *Arctoparmelia*
33. Upper surface usually shiny; lower surface brown to pitch black, shiny; medulla usually PD+ yellow to red, K+ yellow to red; widespread from arctic to southern temperate . . . . *Xanthoparmelia*

KEY K: FOLIOSE LICHENS THAT ARE NOT UMBILICATE, JELLY-LIKE, OR YELLOW

1. Lichens on the ground (soil or mossy turf) 2
1. Lichens directly on bark, wood, or rock 17
2. Thalli ascending, forming almost erect fruticose tufts; pseudocyphellae conspicuous on the lower surface of the lobes or branches . . . . *Cetraria* key (*Cetraria, Cetrariella*)
2. Thalli prostrate, not erect; pseudocyphellae, if present, on upper surface of lobes 3
3. Algal layer blue-green (photobiont cyanobacteria) . . . 4
3. Algal layer green (photobiont green algae) 8
4. Lobes less than 2 mm broad; on mossy or bare soil . . . . *Massalongia carnosa*
4. Lobes usually more than 2 mm wide 5
5. Lower surface covered by a rather thick, blue-black tomentum; subtropics to eastern coastal plain . . . . *Coccocarpia*
5. Lower surface with a pale brown to brown-black tomentum, or veined 6
6. Cyphellae present on lower surface; thallus dark brown; rare, arctic-alpine . . . . [*Sticta arctica*]
6. Lacking cyphellae on lower surface 7
7. Lobes usually under 4 mm wide; apothecia embedded in thallus lobes; lichens mainly of arid localities; spores 1-celled; photobiont *Scytonema* . . . . *Heppia conchiloba*
7. Lobes usually greater than 4 mm wide; apothecia marginal; spores septate; photobiont *Nostoc* . . . . . *Peltigera*
8.(3) Lobes usually less than 3 mm wide 9
8. Lobes usually more than 3 mm wide 12
9. Rhizines absent; lobes generally appearing puffed . . . 10
9. Rhizines present, sparse or abundant; lobes solid and flat, not appearing puffed 11
10. Lobes tube-like, hollow . . . . *Hypogymnia*
10. Lobes solid . . . . *Brodoa oroarctica*
11. Lower surface predominantly pale (but dark in the center) with sparse, unbranched rhizines; lobes 0.2–0.5(–1) mm wide, not pruinose . . . . . . *Phaeophyscia constipata*
11. Lower surface black with a dense mat of squarrose rhizines; lobes 1–3 mm wide, lightly to heavily pruinose . . . . *Physconia muscigena*
12. (8) Lower surface jet black, shiny; thallus lacking cephalodia; upper surface K+ yellow (atranorin); western arctic-boreal 13
12. Lower surface pale tan to black, rarely shiny; cephalodia present as small warts on the lower or upper surface or as internal patches; upper cortex K– (atranorin absent); arctic to temperate 14
13. Rhizines sparse but present, especially on older parts of thallus; isidia absent; rare, Bering Sea coast of Alaska . . . . [*Cetrelia alaskana*]
13. Rhizines entirely absent; isidia present on thallus surface; Alaska interior to Hudson Bay . . . . [*Asahinea scholanderi*]
14. Upper surface of lobes with a network of ridges and depressions; lower surface uniformly pale brown, sparsely to clearly tomentose . . . . *Lobaria linita*
14. Upper surface of lobes smooth or wrinkled but not forming a network of ridges 15
15. Cephalodia appearing as gray to brown scales on the upper surface; apothecia marginal on small lobes . . . . *Peltigera*
15. Cephalodia hidden within the thallus, on the lower

surface, or forming low bumps on the upper surface 16

16. Thallus lobes strongly crinkled and crisped, ascending; apothecia rare, on the lower side of small marginal lobes . . . . *Nephroma expallidum*
16. Thallus more or less smooth, prostrate; apothecia common, immersed in the thallus lobes, often in depressions . . . . *Solorina*

17.(1) Lobes usually under 3 mm wide 18
17. Lobes usually over 3 mm wide 87

18. Lobes inflated and hollow; lower surface lacking rhizines or tomentum . . . . *Hypogymnia* key (*Hypogymnia, Menegazzia*)
18. Lobes solid; rhizines or tomentum present or absent 19

19. Thallus shades of brown, olive, or black 20
19. Thallus shades of gray 45

20. Algal layer blue-green . . . . . . *Pannaria* key (*Pannaria, Parmeliella, Massalongia, Vestergrenopsis*)
20. Algal layer green 21

21. Rhizines and tomentum absent, or thallus so tightly attached that it is hard to tell 22
21. Rhizines or tomentum present, sparse or abundant 27

22. Thallus ascending, very loosely attached at only a few points 23
22. Thallus appressed, closely attached at numerous points 26

23. Lobes strap-shaped, squamule-like, dark olive-green; apothecia spherical; southern Appalachian Mountains, rare and endangered . . . . *Gymnoderma lineare*
23. Lobes almost fruticose, not flat and strap-shaped; apothecia flat; not from southeast 24

24. Branches short, usually up to 15 mm in length; thallus dark greenish brown to black . . . *Kaernefeltia merrillii*
24. Branches quite long, over 20 mm in length; thallus brown to brownish gray, without a greenish tint . . . 25

25. Lobe margins with long, branched, thallus-colored cilia; lower surface white, webby . . . *Anaptychia setifera*
25. Lobe margins often with short, unbranched, black projections but no true cilia; lower surface brown and shiny like the upper surface . . . . *Tuckermannopsis subalpina*

26.(22) Lobes convex, appearing inflated but actually solid; on rock . . . . *Allantoparmelia* key (*Allantoparmelia, Brodoa, Lobothallia*)
26. Lobes flat; on bark . . . *Hyperphyscia* (*Phaeophyscia* key)

27.(21) Cortex K+ yellow (atranorin); thallus mainly brown on the lobe tips, gray in older parts of the thallus . . . 28

27. Cortex K–; thallus uniformly brown, olive, or blackish 29

28. Pseudocyphellae conspicuous, white, round to elliptical; medulla K–, C+ red (gyrophoric acid) . . . . *Punctelia stictica*
28. Pseudocyphellae in a net-like pattern on ridges; medulla K+ red, C– (salazinic acid) . . . . *Parmelia*

29. Lobes distinctly pruinose at least at tips; rhizines usually squarrose . . . . *Physconia*
29. Lobes without pruina 30

30. Soredia present 31
30. Soredia absent 36

31. Soredia mainly in patches (soralia) on the thallus surface 32
31. Soredia mainly at the lobe tips or along the margins 34

32. Medulla C+ pink or red . . . . *Melanelia*
32. Medulla C– 33

33. Rhizines not visible from above; thallus very dark brown, often shiny; spores colorless, 1-celled . . . . *Melanelia*
33. Rhizines often protruding in a fringe around lobes; thallus dark to pale brown or dark greenish gray, not shiny; spores brown, 2-celled . . . . *Phaeophyscia*

34. Medulla K+ red or C+ reddish . . . . *Melanelia*
34. Medulla K–, C– 35

35. Lower surface and rhizines pale brown; rhizines sparse, not extending out from margins and visible from above; lobes chocolate brown to olive-brown, flat, crisped, or undulating . . . *Tuckermannopsis chlorophylla*
35. Lower surface usually black; rhizines most commonly black (often with white tips), abundant and extending out from the margins; lobes flat, not undulating or crisped . . . . *Phaeophyscia*

36.(30) Isidia present 37
36. Isidia absent 39

37. Isidia mainly marginal, especially on young lobes . . 38
37. Isidia mainly on the upper surface of the lobes, even on young lobes . . . *Melanelia* key (*Melanelia, Neofuscelia*)

38. On rock . . . . *Phaeophyscia sciastra*
38. On wood, or sometimes bark . . . . *Tuckermannopsis coralligera*

39. Pycnidia prominent, black, along the lobe margins; pseudocyphellae often conspicuous especially along the lobe margins 40
39. Pycnidia absent or immersed in thallus with only a pale or dark ostiole showing, not largely confined to the margins; pseudocyphellae present or absent, not especially marginal 41

40. On bark or wood .... *Tuckermannopsis*
40. On rock .... *Melanelia hepatizon*

41. Lobes with conspicuous ridges and depressions ......
    .... *Kaernefeltia merrillii*
41. Lobes more or less flat, without ridges and depressions
    42

42. Lower surface and rhizines pale brown 43
42. Lower surface and rhizines usually dark brown or black
    44

43. On rock in arid habitats; lobules, if present, not strap-shaped; medulla K+ yellow, PD+ orange (stictic acid); spores colorless, 1-celled; southwestern U.S. ....
    .... *Neofuscelia atticoides*
43. On rock or bark in mossy, forest habitats; lobules strap-shaped, marginal; medulla K–, PD–; spores brown, 2-celled; eastern North America ....
    .... *Anaptychia palmulata*

44. Apothecia very dark brown to black, dull; spores brown, 2-celled; rhizines usually very abundant, often visible from above as a cilia-like fringe and frequently growing on the apothecial margins ..... *Phaeophyscia*
44. Apothecia yellowish brown to reddish brown, shiny; spores colorless, 1-celled; rhizines abundant or sparse, not forming a fringe visible from above and not growing on the apothecial margins .... *Melanelia*

45.(19) Photobiont blue-green .... *Pannaria* key
45. Photobiont green 46

46. Lower surface white, tan, yellow, or orange 47
46. Lower surface dark brown to black (except sometimes near lobe tips) 59

47. Lobes thick and convex, appearing inflated (although solid); rhizines absent; PD+ yellow; on rock ....
    ... *Allantoparmelia* key (*Allantoparmelia*, *Lobothallia*)
47. Lobes thin, convex or flat, not appearing inflated; rhizines present or absent; on various substrates 48

48. Rhizines and tomentum absent 49
48. Rhizines or tomentum present, sometimes sparse ... 55

49. Thallus very closely attached to substrate over the entire thallus surface (almost crustose in appearance) ...
    .... *Hyperphyscia* (*Phaeophyscia* key)
49. Thallus loosely attached and ascending 50

50. Marginal cilia absent 51
50. Marginal cilia present and conspicuous 54

51. On rock 52
51. On trees 53

52. Lobes convex, 0.1–0.3(–0.5) mm wide; on dry limestone or sandstone, south central U.S. ....
    .... *Speerschneidera euploca*

52. Lobes flat, 0.8–1.3 mm wide, strap-shaped; on wet rock walls in forests of the southern Appalachian Mountains
    .... *Gymnoderma lineare*
53. Lobes smooth to wrinkled, often strongly convex; eastern or southern .... *Pseudevernia*
53. Lobes generally wrinkled and ridged, flat or concave; humid forests on the west coast .... *Platismatia*

54.(50) Soredia produced on the lower surface of the lobe tips; cortex K+ yellow (atranorin) ....
    .... *Heterodermia leucomela*
54. Soredia absent; cortex K– .... *Anaptychia setifera*

55.(48) Lower surface with a short, pale tomentum, sometimes very sparse and limited to the low areas between raised bald spots; cephalodia visible on the lower surface as small bumps that are dark blue-green inside ...
    .... *Lobaria*
55. Lower surface with distinct rhizines; cephalodia absent
    56

56. Soredia present ....
    ...... *Physcia* key (*Heterodermia*, *Physcia*, *Physciella*)
56. Soredia absent 57

57. Black cilia with bulbous bases fringing the lobe margins .... *Bulbothrix*
57. Cilia, if present, not black and bulbous 58

58. Thallus cortex PD+ orange, K+ deep yellow (thamnolic acid); apothecia pale brown, without pruina; spores colorless, 1-celled .... *Imshaugia*
58. Thallus cortex PD– or pale yellow, K+ yellow (atranorin); apothecia, if present, dark brown to black, often pruinose; spores brown, 2-celled ....
    .... *Physcia* key (*Heterodermia*, *Physcia*)

59.(46) Distinct rhizines absent 60
59. Distinct rhizines present 65

60. Lower surface with a spongy black hypothallus consisting of intricately interconnected fibers (Fig. 6g, h) ....
    .... *Anzia colpodes*
60. Lower surface naked, without a spongy hypothallus ..
    61

61. Lower surface pitted with many tiny perforations; thallus small, usually forming rosettes less than 2.5 cm across .... *Cavernularia*
61. Lower surface not perforated or pitted; thalli larger than 3 cm long or wide 62

62. On alpine or arctic rocks; lobes thick, convex ....
    ...... *Allantoparmelia* key (*Allantoparmelia*, *Brodoa*)
62. On bark or wood 63

63. Closely appressed over most of the thallus surface ....
    .... *Dirinaria* (see *Physcia* key)
63. Very loosely attached by relatively few points 64

64. Lobes flat to concave, often wrinkled; pycnidia along the lobe margins; medulla C–; in humid or boreal forests along the west coast . . . . *Platismatia*
64. Lobes convex; pycnidia buried in lobe tips, not along the margins; medulla C+ red (lecanoric acid); montane and interior sites . . . . *Pseudevernia*

65.(59) Pseudocyphellae present on lobe surface 66
65. Pseudocyphellae absent 67

66. Pseudocyphellae dot-like, most easily seen on young lobes; medulla K– . . . . *Punctelia*
66. Pseudocyphellae net-like on reticulate ridges; most species with medulla K+ red (salazinic acid) . . . . *Parmelia*

67. Cilia present on lobe margins or in the axils of the lobes (sometimes very sparse) 68
67. Cilia absent 72

68. True cilia arising from the margins or axils of the lobes
 . . . . *Parmelia* key (*Bulbothrix, Myelochroa, Parmelina, Parmelinopsis, Relicina*)
68. "Cilia" actually a fringe of rhizines extending beyond the lobe margins, not arising from the lobe margins themselves 69

69. Rhizines long and forked, uniformly black . . . .
. . . . *Hypotrachyna* (*Parmelia* key)
69. Rhizines unbranched or squarrose, or frayed to brush-like at the tips, often with white tips 70

70. Cortex K+ yellow 71
70. Cortex K– . . . . *Phaeophyscia*

71. Lobe margins abundantly squamulose . . . .
. . . . *Heterodermia squamulosa*
71. Lobe tips sorediate, not squamulose . . . .
. . . . [*Heterodermia casarettiana*]

72.(67) Rhizines forked in regular dichotomies; thallus without pruina . . . . *Hypotrachyna*
72. Rhizines unbranched, squarrose, or brush-like, rarely forked; thallus with or without pruina 73

73. Thallus very loosely attached, strongly wrinkled, especially on the lower surface; on conifers in the west . . . .
. . . . *Esslingeriana idahoensis*
73. Thallus closely attached over most of its surface; smooth, or only slightly wrinkled, on both surfaces . . .
. 74

74. Cortex K– (without atranorin); thallus usually dark greenish gray 75
74. Cortex K+ yellow (atranorin); thallus usually pale gray
. . . . 77

75. Thallus without pruina . . . . *Phaeophyscia*
75. Thallus pruinose at least at the lobe tips 76

76. Rhizines squarrose (like bottlebrushes); most commonly northern . . . . *Physconia*
76. Rhizines unbranched or rarely forked; northern to subtropical . . . . *Pyxine* (see *Physcia* key)

77.(74) Thallus pruinose, at least at lobe tips . . . .
. . . . *Physcia* key (*Physcia, Pyxine*)
77. Thallus without pruina 78

78. Medulla pale yellow, at least close to algal layer . . . .
. . . . *Myelochroa* (*Parmelia* key)
[Rare nonpruinose specimens of *Pyxine eschweileri* or adnate *Esslingeriana* will key out here]
78. Medulla white 79

79. Soredia absent; squamulose lobules abundant on lobe margins . . . . *Heterodermia squamulosa*
79. Soredia present 80

80. Soredia on or close to the lobe margins; medulla C– . .
. . . . 81
80. Soredia laminal or on the lobe tips 82

81. Medulla PD–, K+ yellow; rhizines sparse and distinct
. . . . *Physcia sorediosa*
81. Medulla PD+ orange, K–; rhizines abundant, forming an intricate mat . . . .
. . . . *Pyxine eschweileri* (nonpruinose morph)

82. Lobes tips often curled into tubes; medulla C+ red (gyrophoric acid) . . . . *Hypotrachyna revoluta*
82. Lobes more or less flat; medulla C– 83

83. Lobes 0.5–2 mm wide 84
83. Lobes 2–4 mm wide 86

84. On rocks; medulla PD+ red-orange (protocetraric acid); southeastern U.S. . . [*Paraparmelia alabamensis*]
84. On bark or wood; medulla PD– or PD+ orange . . . 85

85. Rhizines black; southeastern coastal plain . . . .
. . . . *Pyxine eschweileri*
85. Rhizines pale tan to brown, not black; western and northern . . . . *Parmeliopsis hyperopta*

86. Medulla UV+ white (divaricatic acid); lower surface and rhizines dark brown . . . . *Canoparmelia texana*
86. Medulla UV–; lower surface and rhizines black . . . . . .
. . . . *Myelochroa aurulenta* (rare specimens with a white medulla)

87.(17) Lobes inflated and hollow . . . . *Hypogymnia*
87. Lobes solid 88

88. Thallus brown, brownish green, olive, or black when dry 89
88. Thallus pale gray to greenish gray when dry 103

89. Algal layer blue-green 90
89. Algal layer green (with a lower, secondary, blue-green layer in *Solorina*) 95

90. Lower surface smooth or with a short tomentum, usually pale brown (or gray to bluish in *Pannaria*), lacking veins; usually with a lower cortex; without discrete rhizines 91
90. Lower surface cottony or webby, often with branching or interconnecting veins, lacking a lower cortex; white to brown or black, with discrete or tufted rhizines .... *Peltigera* key (*Peltigera, Erioderma, Leioderma*)
91. Small round holes or pits (cyphellae) on the lower surface (plate 15) .... *Sticta*
91. Cyphellae absent 92
92. Tiny, white or yellow, raised spots (pseudocyphellae) on the lower surface (plate 14) ..... *Pseudocyphellaria*
92. Pseudocyphellae absent 93
93. Tomentum on lower surface most commonly gray to bluish; apothecia on upper surface of lobes .... *Pannaria*
93. Tomentum on lower surface, or lower surface itself, pale brown 94
94. Thallus olive-gray to brownish gray, sorediate; apothecia infrequent, laminal .... *Lobaria*
94. Thallus brown, sorediate or not; apothecia frequent, on lower surface of lobe margins; medulla PD– .... *Nephroma*
95.(89) Lower surface tomentose or veined at least in part; cephalodia present (warts or secondary algal layers containing cyanobacteria) 96
95. Lower surface smooth or wrinkled, not at all tomentose; cephalodia absent 98
96. Lower surface more or less veined or webby, without a cortex; apothecia immersed in thallus, often in depressions .... *Solorina*
96. Lower surface uniform in color, pale brown, partly or entirely tomentose, often with raised, smooth, naked areas, with a cortex; apothecia superficial or raised 97
97. Pseudocyphellae (yellow or white raised spots) on lower surface of lobes .... *Pseudocyphellaria*
97. Pseudocyphellae lacking on lower or upper surface ... . *Lobaria*
98. Thallus loosely attached, ascending 99
98. Thallus more or less closely appressed to substrate, except at lobe tips 100
99. Thallus greenish black or dark olive-brown; apothecia black or very dark brown; pycnidia immersed in thallus; pseudocyphellae very inconspicuous or absent ... . *Kaernefeltia merrillii*
99. Thallus brown to olive-brown; apothecia red-brown; pycnidia black, prominent; pseudocyphellae often conspicuous .... *Tuckermannopsis*

100. Lobe margins fringed with black rhizines extending out from below and appearing like cilia; rhizines forming an interwoven mat below .... *Phaeophyscia hispidula*
100. Lobe margins without cilia-like rhizines; rhizines separate and distinct 101
101. Lower surface and rhizines black; upper surface usually gray in part, K+ yellow (test the parts that remain gray); surface of lobes usually with a network of ridges and depressions, appearing like hammered metal 102
101. Lower surface and rhizines pale to dark brown (rhizines rarely black); upper surface uniformly brown or olive, K–; lobes smooth or rough, but rarely with a network of ridges .... *Melanelia* key (*Melanelia, Neofuscelia, Tuckermannopsis*)
102. Pseudocyphellae milk-white, very conspicuous, round to elongated; medulla K–, C+ red (gyrophoric acid) .... *Punctelia stictica*
102. Pseudocyphellae pale, net-like or irregular in shape, occasionally round; medulla K+ red, C– (salazinic acid) .... *Parmelia*
103.(88) Lower surface tomentose, cottony, or veined, at least in part, rarely naked; discrete rhizines sometimes present; photobiont blue-green or green ... 104
103. Lower surface smooth, wrinkled, or rough, but not tomentose or cottony; discrete rhizines usually well developed; photobiont green 112
104. Pseudocyphellae present on the upper or lower thallus surface 105
104. Pseudocyphellae absent 106
105. Pseudocyphellae on the upper surface of the lobes; medulla C+ red .... *Punctelia subrudecta*
105. Pseudocyphellae on the lower surface of the lobes; medulla C– .... *Pseudocyphellaria*
106. Soredia present 107
106. Soredia absent 109
107. Lobes 2–5 mm wide; soredia marginal, coarsely granular; medulla PD+ orange (pannarin) .... *Pannaria conoplea*
107. Lobes 5–20 mm wide 108
108. Lower surface vaguely veined, cottony (not tomentose); soredia usually marginal ..... *Peltigera collina*
108. Lower surface uniformly brown, tomentose (or with scattered naked areas); soredia on both the margins and upper surface of the lobes .... *Lobaria*
109. Algal layer blue-green 110
109. Algal layer green 111
110. Lower surface with a thick or thin gray to blue-black tomentum ... *Pannaria* key (*Coccocarpia, Pannaria*)

110. Lower surface with distinct or indistinct veins . . . . *Peltigera*

111. Cephalodia in the form of brown or gray warts or lobed squamules on the upper surface of the thallus; lower surface webby or cottony, white at the lobe edge and dark brown to black in the center, sometimes with veins . . . . *Peltigera*

111. Cephalodia in the form of small round warts or galls on the lower surface of the thallus; lower surface without veins, abundantly or sparsely tomentose . . . . *Lobaria*

112.(103) Lower surface entirely white or very pale brown; rhizines white or pale brown 113

112. Lower surface pale to dark brown or black, sometimes blotched with white over small or large areas, but in such cases, always brown to black in the oldest, central area; rhizines black or brown 114

113. White dots (pseudocyphellae) present on upper surface of lobes . . . . *Punctelia*

113. Pseudocyphellae absent . . . . *Physcia biziana*

114. Upper cortex K–, UV–; lobes fringed with black cilia-like rhizines . . . . *Phaeophyscia hispidula*

114. Upper cortex K+ yellow, UV–, or rarely K–, UV+ yellow; with or without cilia or cilia-like rhizines . . . 115

115. Apothecia common, black; spores brown, 2-celled; closely adnate tropical species on bark . . . . *Dirinaria confusa*

115. Apothecia, when present, brown; spores colorless, 1-celled . . . . *Parmelia* key

DESCRIPTIONS, ILLUSTRATIONS, KEYS TO SPECIES, AND MAPS

Nature has a day for each of her creatures—her creations. Today it is an exhibition of lichens at forest Hall—The livid green of some—the fruit of others. They eclipse the trees they cover.

Henry D. Thoreau

Acarospora (77 N. Am. species)
Cobblestone lichens, cracked lichens

DESCRIPTION: Areolate to almost squamulose crustose lichens in various colors, such as white, gray, brown, and brilliant yellow. Photobiont green (unicellular). Apothecia usually immersed in the thalli with round to irregular disks, but occasionally somewhat prominent with lecanorine margins; paraphyses mainly unbranched; asci K/I– including the thickened tip (except K/I+ blue in *A. heppii*); spores colorless, ellipsoid to spherical, dozens to hundreds per ascus. CHEMISTRY: Yellow species containing pulvinic acid derivative pigments such as rhizocarpic acid in the cortex (UV+ bright orange); a few species with gyrophoric or norstictic acids. HABITAT: On rock of various kinds, usually in full sun; a few species on soil. COMMENTS: The brown and gray species of *Acarospora* can resemble some *Aspicilia* species, but these have large spores, 8 per ascus. *Polysporina* and *Sarcogyne* have many-spored asci, but their apothecial margins are lecideine and usually carbon-black. The yellow species of *Acarospora* are very similar to *Pleopsidium*, which differs in details of the ascus. At the present time, the species limits of the North American representatives of *Acarospora*, especially the yellow ones, are poorly understood.

KEY TO SPECIES (including *Pleopsidium*)

1. Thallus distinctly yellow 2
1. Thallus brown or gray, without a yellowish or orange tint 4

2. Growing on soil .... *Acarospora schleicheri*
2. Growing on rocks 3

3. Thallus areolate, dispersed or contiguous, not conspicuously lobed at the margin; asci not thickened at the apex, K/I– .... *Acarospora contigua*
3. Thallus somewhat lobed at the margin, otherwise areolate; asci with a distinctly thickened, partly K/I+ blue tholus .... *Pleopsidium flavum*

4. Thallus brown, not pruinose; thallus cortex usually C+ pink, KC+ pink but sometimes hard to demonstrate (gyrophoric acid present); on noncalcareous rock .... *Acarospora fuscata*
4. Thallus gray to gray-brown, lightly to heavily pruinose; thallus cortex C–, KC– (gyrophoric acid absent); on calcareous rocks 5

5. Thallus usually reduced to narrow, reddish brown,

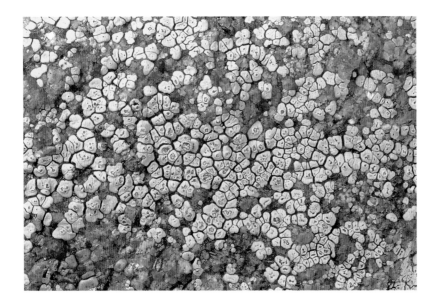

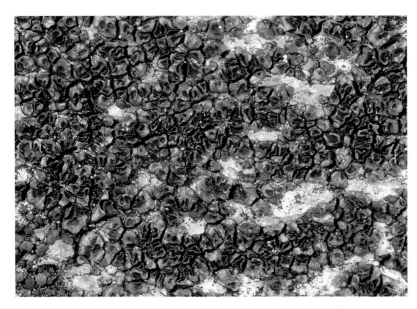

91. *Acarospora contigua*   hill country, central Texas ×2.8

92. *Acarospora fuscata* Lake Superior region, Ontario ×5.4

sometimes pruinose margins around the apothecia, which become *Lecanora*-like; apothecial disks lightly pruinose . . . . *Acarospora glaucocarpa*

5. Thallus well-developed, areolate; one to several apothecia immersed in each areole, never *Lecanora*-like . . . . *Acarospora strigata*

Acarospora contigua
Gold cobblestone lichen

DESCRIPTION: Thallus bright sulphur-yellow to greenish yellow, composed of scattered or contiguous areoles 1–2.5 mm in diameter, convex or flat, with smooth or scalloped margins, sometimes lightly dusted with a white pruina. Apothecia reddish brown to very dark brown, immersed in the areoles, 1–4 per areole, sometimes becoming somewhat prominent with thallus-colored margins; disks often broken up with patches or a network of sterile thallus-colored tissue; spores ca. 3–4 × 2–3 µm, many per ascus. CHEMISTRY: Cortex and medulla PD–, K–, KC–, C– (pulvinic acid pigments). HABITAT: On siliceous rock in open, arid sites. DISTRIBUTION: Probably widespread throughout the arid southwest, but also known from glades in the Ozarks. COMMENTS: The closely related *A. schleicheri* grows on soil and has a more marginally lobed thallus. The taxonomy of this group is still poorly worked out, and so precise distribution maps of *A. contigua* and *A. schleicheri* are impossible at this time.

Acarospora fuscata
Brown cobblestone lichen

DESCRIPTION: Thallus areolate, the areoles often becoming slightly lifted at the edges and becoming squamulose, dispersed or contiguous, 0.5–2(–3) mm in diameter, yellowish brown to deep reddish brown, often shiny, sometimes black at the edge, concave, flat, or convex, generally with 1–5 darker brown, irregular disks of immersed apothecia in the center of each areole; spores 4–6 × 1–1.5 µm. CHEMISTRY: Cortex and sometimes medulla PD–, K–, KC+ red, C+ pink (gyrophoric acid), but the reaction is usually difficult to observe against the brown cortex. (Sections of the thallus should be tested under the stereomicroscope.) HABITAT: On granitic rocks in full or partial sun. COMMENTS: This is by far the most common and widespread of the many brown, dispersed areolate species of *Acarospora*. The C+ pink reaction of the thallus is a good confirming character but cannot always be seen. *Acarospora badiofusca* has a similar thallus, but the apothecia are more prominent and generally have a complete or partial black margin. *Acarospora smaragdula,* a widely distributed brown species with about the same amount of thallus variation as *A. fuscata,* reacts C–, and most specimens are K+ red because of norstictic acid. A few brown species occur on calcareous rocks: *Acarospora heppii* is a widespread but infrequently collected species with small, dispersed areoles, each containing one apothecium. *Acarospora macrospora* is a western species with contiguous areoles having several apothecia and, unusual for the genus, large spores (ca. 7 × 3 μm), only a few dozen per ascus. See also *A. glaucocarpa.*

Acarospora glaucocarpa
Rimmed cobblestone lichen

DESCRIPTION: Thallus consisting of dispersed pale reddish brown to medium reddish brown squamules, each containing a single broad apothecium (up to 3 mm across) with a lightly pruinose, brown disk; thallus tissue usually reduced to a rim around each immersed apothecium, giving the fertile squamule the appearance of a lecanorine apothecium; spores 4–8 × 1.3–3 μm, many per ascus, but often not developing. CHEMISTRY: All reactions negative (no lichen substances). HABITAT: On exposed calcareous rocks and pebbles. COMMENTS: This lichen looks like a broad pruinose *Lecanora* on limestone. Although the multispored asci identify it as an *Acarospora,* the apothecia, unfortunately, often have no mature asci. *Acarospora glaucocarpa* should also be compared with *Sarcogyne regularis,* another pruinose multispored lichen that frequently grows as a neighbor in the same habitat. It has a black (often thin) lecideine apothecial margin, and its thallus is almost entirely endolithic.

Acarospora schleicheri
Soil paint lichen

DESCRIPTION: Thallus pale to bright sulphur yellow, sometimes pruinose, squamulose-areolate, often lobed at the margins, with large red-brown, sometimes scabrose apothecia embedded in areoles; hundreds of spores per ascus. CHEMISTRY: Negative to all reagents but UV+ orange (no lichen substances other than a yellow pigment, rhizocarpic acid, in the cortex). HABI-

93. *Acarospora glaucocarpa* Rocky Mountains, Alberta ×5.6

94. *Acarospora schleicheri* Rocky Mountains, Colorado ×5.4

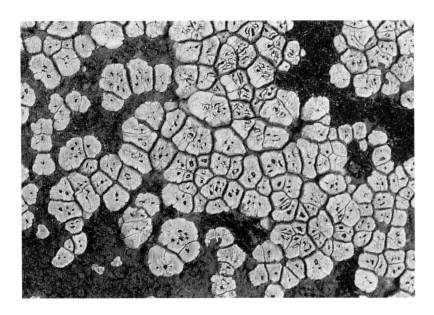

95. *Acarospora strigata* northern New Mexico ×5.1

TAT: On stabilized soil in the open. DISTRIBUTION: Throughout the more arid parts of the western interior from Mexico to southern British Columbia and Saskatchewan. See Distribution under *A. contigua*. COMMENTS: This is the only common yellow *Acarospora* on soil. Because it is extremely variable in thallus development, color, amount of pruinosity, and color of the apothecia, many varieties have been given names. Some lichenologists include saxicolous populations, such as *A. contigua*, within *A. schleicheri*. *Acarospora stapfiana,* a yellow species made chalky white by a heavy pruina, has an areolate thallus with sunken brown apothecia; it parasitizes *Caloplaca trachyphylla*. *Fulgensia* species also produce yellow thalli on arid soil, but they can be identified, even when sterile, by their K+ deep purple reaction.

Acarospora strigata
Hoary cobblestone lichen

DESCRIPTION: Thallus consisting of areoles that are pale gray or yellowish gray because of a heavy pruina (brown where the pruina has been removed), round or angular, 0.5–1.5 mm in diameter, dispersed or growing together in an almost continuous crust, the surface cracked and creased, especially around the apothecia. Apothecia sunken in the areoles, one to several per areole, with lightly pruinose, reddish brown disks, or without pruina; spores ellipsoid, ca. 4–7 × 2–3.5 μm, hundreds per ascus. CHEMISTRY: All reactions negative (no lichen substances). HABITAT: On calcareous rock. COMMENTS: Similar lichens on soil (all in the southwest) include *Acarospora nodulosa,* with a more or less lobed thallus with flat areoles (K+ red) and broad, reddish to dark brown *Lecanora*-like apothecia; and *Acarospora thelococcoides,* with broad, black apothecia immersed in strongly convex areoles (K–), and with perfectly spherical spores, 10–12 μm in diameter, about 24–36 per ascus.

Ahtiana (3 N. Am. species)
Candlewax lichens

DESCRIPTION: Medium-sized, yellowish, closely appressed to ascending foliose lichens without pseudocyphellae or soredia; surface usually wrinkled; cilia sparse or absent; lower surface pale, wrinkled, with fairly abundant, pale rhizines. Photobiont green (*Trebouxia*?). Apothecia lecanorine, with light brown, shiny disks, produced on the upper surface or margins of the lobes; spores colorless, 1-celled, spherical, 8 per ascus; pycnidia black, spherical and prominent, laminal or marginal; conidia dumbbell-shaped. CHEMISTRY: Cortex PD–, K–, KC+ yellow, C– (usnic acid); medulla PD–, K–, KC–, C– (fatty acids, especially caperatic, lichesterinic, and protolichesterinic acids). HABITAT: On bark and wood, usually of conifers. COMMENTS: Except for its color and chemistry, *Ahtiana* is very similar to *Tuckermannopsis*. *Allocetraria* (*A. oakesiana*) is closely related, differing in details of the ascus tip, conidia, and upper cortex. Species of *Flavoparmelia* are generally greener with a black lower surface. Most are sorediate or isidiate and have a more temperate distribution.

KEY TO SPECIES

1. Thallus pruinose, at least at lobe tips; lower surface pale yellow, wrinkled .... *Ahtiana pallidula*
1. Thallus entirely without pruina; lower surface white to pale brown, smooth 2

2. Thallus closely appressed; lower surface white or almost white; western montane .... *Ahtiana sphaerosporella*
2. Thallus loosely attached; lower surface pale brown; in Appalachian–Great Lakes region .... *Ahtiana aurescens*

Ahtiana aurescens (syn. *Cetraria aurescens*)
Eastern candlewax lichen

DESCRIPTION: Thallus appressed to somewhat raised; lobes flat and branched or crowded into overlapping cushions, usually divided into small, round lobules 0.5–2(–3) mm wide, with no soredia or isidia; lower surface pale brown, shiny; rhizines short, pale, sparse to abundant, looking like cilia when growing along the lobe margins. Apothecia abundant, red-brown, 2–7 mm in diameter, somewhat raised, marginal; pycnidia marginal or laminal, very conspicuous. HABITAT: On trees, especially cedars and pines, rarely hardwoods, in the Appalachians. COMMENTS: In the east, this species is most similar to *Allocetraria oakesiana*, but without soredia. It is often seen with *Imshaugia placorodia* on exposed pines.

Ahtiana pallidula (syn. *Cetraria pallidula*)
Pallid candlewax lichen

DESCRIPTION: Thallus loosely attached to ascending, pale yellow or greenish yellow, often pruinose on some lobes, with rounded, sometimes toothed lobes 4–10 mm wide; soredia and isidia absent; lower surface pale yellow, strongly wrinkled with a network of sharp ridges; rhizines abundant, pale, and slender. Apothecia abundant, produced on the margins, up to 10 mm in diameter, disks brown; black pycnidia prominent, marginal or laminal. HABITAT: On conifers such as Douglas fir, ponderosa pine, and western larch at low to medium elevations. COMMENTS: Within its range, *A. pallidula* cannot be confused with any other lichen. *Ahtiana sphaerosporella* has laminal apothecia and grows mainly on trees at subalpine elevations. Species of *Tuckermannopsis* are brown, not yellow-green.

96. *Ahtiana aurescens* Appalachians, North Carolina ×2.7

97. *Ahtiana pallidula* northern Sierra, California ×2.1

98. *Ahtiana sphaerosporella* Coast Mountains, British Columbia ×1.7

Ahtiana sphaerosporella (syn. *Parmelia sphaerosporella*)
Mountain candlewax lichen

DESCRIPTION: Thallus dull, pale yellowish green, often with dark greenish black margins, closely appressed; lobes thin, rounded, about 2–4 mm across, with wrinkles and folds covering the upper surface; without pseudocyphellae, soredia, or isidia; lower surface pale tan, with short, slender, often branched, woolly rhizines. Large, pale brown apothecia common on the upper surface. HABITAT: Closely restricted to whitebark pine and subalpine larch in intermontane, subalpine localities. COMMENTS: This lichen looks like a very appressed *Flavoparmelia caperata* without soredia, but *Flavoparmelia* has a black lower surface and is rarely fertile.

Alectoria (7 N. Am. species)
Witch's hair

DESCRIPTION: Shrubby to pendent, light greenish yellow (rarely gray to black) fruticose lichens with slender, hair-like branches basically round in cross section or sometimes partially flattened, angular, or pitted; supporting tissue in the cortex; medulla dense or cottony; raised, white pseudocyphellae present on the branches of all species; photobiont green (*Trebouxia?*). Apothecia *Parmelia*-like but infrequent, scattered along the branches; spores large, brownish, 1-celled, 2–4 per ascus. CHEMISTRY: Cortex PD–, K–, KC+ yellow, C– (usnic acid) in all species except *A. nigricans;* medulla often with orcinol depsides, reacting KC+ red, CK+ gold, or C+ red. HABITAT: On bark and wood, on the ground, rarely on rock, in well-lit situations. COMMENTS: *Alectoria* can closely resemble the pendent species of *Usnea* (beard lichens) in color and general habit, but *Usnea* has a central supporting cord or "axis." The stretch test distinguishes them: hold a fairly thick branch between your fingers and gently pull it apart, stretching it until it breaks. If it breaks cleanly, it is *Alectoria;* if it has a rubberband-like central axis that persists before breaking (Fig. 36, p. 710), it is *Usnea*. Greenish yellow species of *Alectoria* can be mistaken for *Ramalina* or *Evernia*. *Evernia* is usually soft and pliable and does not have raised white pseudocyphellae. *Ramalina* usually has flattened branches; the species with round, slender branches (like *R. thrausta*) have tips that curl up and develop a few granules, and the pseudocyphellae, if present, are more like dots.

KEY TO SPECIES (including *Bryocaulon, Bryoria, Nodobryoria,* and *Sulcaria*)

1. Thallus pale greenish yellow or yellowish green; cortex KC+ gold (usnic acid); pseudocyphellae conspicuous, usually slightly raised 2
1. Thallus gray, olive, or shades of brown to almost black; if yellowish gray, then cortex KC+ pink, not gold; pseudocyphellae conspicuous or inconspicuous, level with the surface or slightly to deeply depressed 6

2. Thallus forming erect clumps, or sometimes prostrate, on the ground; tips of branches usually becoming black or greenish black .... *Alectoria ochroleuca*
2. Thallus shrubby to pendent, on trees or shrubs, rarely with blackened branch tips 3

3. Thallus forming bushy clumps, usually less than 10 cm long 4
3. Thallus pendent to slightly pendent, usually 8–20 cm long when mature 5

4. Thorny isidia and spinules developing in elongate pseudocyphellae and fissures; apothecia very rare; cortex K+ bright yellow, PD+ orange (thamnolic acid) or K–, PD– (squamatic acid); medulla KC– .... *Alectoria imshaugii*
4. Isidia and spinules absent; apothecia usually present, brown; cortex K–, PD–; medulla KC+ red (alectoronic acid) .... *Alectoria lata*

5. Branches very slender and pale, with tips curled up and granular or sorediate; medulla KC– .... *Ramalina thrausta*

5. Branches slender or thick, not curled up at the tips; without soredia; medulla KC+ red (alectoronic acid), or infrequently KC– (usnic acid alone) . . . . *Alectoria sarmentosa*

6.(1) Thallus forming erect or prostrate clumps on soil, heath, or sometimes rock (rarely on shrubs) 7
6. Thallus bushy to pendent, on trees and shrubs, rarely on rock 10

7. Thallus brown with a distinct reddish tint 8
7. Thallus black to very dark brown, not especially reddish 9

8. Pseudocyphellae dot-like, level to slightly raised, white, C+ red (olivetoric acid) . . . . *Bryocaulon divergens*
8. Pseudocyphellae round to elliptical, depressed, white, C– . . . . *Cetraria aculeata*

9. Thallus dull, gray to yellowish gray at the base with blackened branch tips, occasionally entirely black; pseudocyphellae raised, white; cortex PD+ yellow, KC+ pink (use the filter paper test) . . . . . *Alectoria nigricans*
9. Thallus usually shiny, uniformly dark brown to black; pseudocyphellae level with the surface, brown and very inconspicuous (seen as dull, fusiform areas); outer cortex PD–; medulla and inner cortex PD+ red (fumarprotocetraric acid; diffusing onto filter paper only where the branch is broken) . . . . *Bryoria nitidula*

10.(6) Thallus forming rounded or irregular bushy tufts or clumps with divergent branching 11
10. Thallus pendent or almost pendent when mature . . . 14

11. Branches without soredia or isidia 12
11. Branches with scattered soralia 13

12. Thallus red-brown; apothecia shiny red-brown, decorated with spiny cilia on the margins; epihymenium K– . . . . *Nodobryoria abbreviata*
12. Thallus brown at base, greenish black at tips; apothecia greenish black; epihymenium K+ purple . . . . *Kaernefeltia californica*

13. Thallus shiny, uniformly brown, with fissural soralia that contain tiny, thorn-like isidia; outer cortex and medulla PD+ red, K–, KC– (fumarprotocetraric acid) . . . . *Bryoria furcellata*
13. Thallus dull, gray to pale or dark brown, with round, warty soralia that contain no isidia; cortex PD+ orange-yellow, K+ bright yellow, KC+ pink (alectorialic acid) . . . . *Bryoria nadvornikiana*

14.(10) Branches with soralia 15
14. Branches without soralia 19

15. Soredia bright yellow; thallus with some stout main branches and slender, perpendicular side branches; medulla and soralia PD– 16
15. Soredia white (sometimes flecked with black) 17

16. Pseudocyphellae yellow, twisting around branches . . . . *Bryoria tortuosa* (rare sorediate form)
16. Pseudocyphellae absent (although vague white cracks sometimes look like pseudocyphellae) . . . . *Bryoria fremontii*

17. Pseudocyphellae present, rather long, but sometimes hard to see; cortex PD+ orange-yellow, K+ yellow, KC+ pink (alectorialic acid); main branches with short or long, perpendicular side branches . . . . *Bryoria nadvornikiana*
17. Pseudocyphellae absent; cortex PD– or PD+ red; soredia PD+ red (fumarprotocetraric acid); perpendicular side branches present or absent 18

18. Thallus very dark brown, almost black; branches very slender (less than 0.2 mm in diameter), brittle; soralia typically abundant, often broader than the branch; cortex PD– . . . . *Bryoria lanestris*
18. Thallus dark to medium brown to olive; branches 0.2–0.6 mm in diameter, not very brittle; soralia relatively sparse, usually not broader than the branch; cortex PD+ red or PD– 18a

18a. Thallus dark to medium brown, typically dull, usually paler at the base; angles between the branches (axils) acute, not rounded; soralia tuberculate or fissural; cortex PD+ red or PD– . . . . *Bryoria fuscescens*
18a. Thallus olive to olive-brown, shiny, not paler at base; branch axils broad and rounded; soralia always fissural; cortex PD– . . . . *Bryoria glabra*

19.(14) Branches with deep, longitudinal grooves (special pseudocyphellae), often quite long; branches commonly twisted; cortex K+ yellow, PD+ yellowish or brownish, KC– or KC+ yellow (atranorin); very rare, California and Oregon . . . . *Sulcaria badia*
19. Branches without deep grooves, with or without pseudocyphellae 20

20. Pseudocyphellae absent; medulla PD– 21
20. Pseudocyphellae present, usually conspicuous; medulla PD+ red or PD– 22

21. Thallus uniformly red-brown, with slender, brittle branches 0.1–0.2 mm in diameter; apothecia fairly common, with red-brown, nonpruinose disks; cortex with jigsaw puzzle–shaped cells as viewed at 100–400× magnification . . . . *Nodobryoria oregana*
21. Thallus dark reddish brown to yellowish brown; thallus with stout, twisted and dented main branches 0.4–1.5(–4) mm in diameter, and slender, perpendicular side branches; apothecia uncommon, with yellow-pruinose disks; cortex with elongate cells . . . . *Bryoria fremontii*

22. Pseudocyphellae yellow, twisting around the branch in spirals; thallus reddish brown or often yellowish (vulpinic acid) . . . . *Bryoria tortuosa*

99. *Alectoria imshaugii* Cascades, Oregon ×2.1

100. *Alectoria lata* Coast Range, northern California ×0.96

22. Pseudocyphellae white or pale brown, dot-like, fissural, or twisting; thallus never yellowish 23
23. Thallus not brittle, pale to dark brown or red-brown; outer cortex PD– 24
23. Thallus usually brittle, pale gray to gray brown or dark brown; outer cortex and sometimes medulla PD+ yellow 25
24. Thallus brown, without a reddish tint; pseudocyphellae elongate (fusiform or linear); inner cortex and medulla usually PD+ red (fumarprotocetraric acid; seen through the cortex rendered transparent in the filter paper test), C–; common .... *Bryoria trichodes*
24. Thallus distinctly red-brown; pseudocyphellae dot-like, slightly raised; inner cortex and medulla PD–, C+ red (olivetoric acid); rare, coastal Pacific Northwest ... . [*Bryocaulon pseudosatoanum*]
25. Cortex K+ bright yellow, KC+ red (alectorialic acid); thallus usually pale gray to brownish gray; most commonly in coastal or humid localities .... *Bryoria capillaris*
25. Cortex K+ brownish, KC– (norstictic acid); thallus usually dark brown, less frequently pale; usually in inland, montane forests .... *Bryoria pseudofuscescens*

Alectoria imshaugii
Spiny witch's hair

DESCRIPTION: Thallus shrubby to subpendent, usually about 5–8 cm long, pale greenish yellow, producing isidia from fissures in the cortex that often originate from old pseudocyphellae. CHEMISTRY: Cortex and medulla PD+ orange, K+ yellow, KC–, C– (thamnolic acid), or PD–, K–, KC–, C– (squamatic acid). HABITAT: On exposed pines and other conifers. COMMENTS: *Alectoria imshaugii* is the only *Alectoria* with isidia, and it is the only one with thamnolic or squamatic acids. In general aspect, it resembles some shrubby species of *Usnea*.

Alectoria lata
Flowering witch's hair

DESCRIPTION: Thallus forming shrubby to subpendent tufts, 5–8 (–15) cm long, pale greenish yellow, sometimes with streaks of black at the base, without soredia

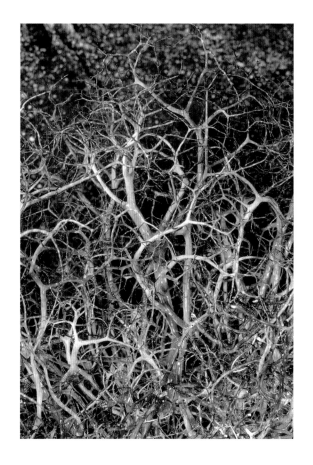

101. *Alectoria nigricans* coastal mountains, Alaska ×1.4

102. *Alectoria ochroleuca* interior Alaska ×1.7

or isidia; branches 0.5–1.5(–2) mm in diameter. Apothecia usually abundant, appearing to be close to the tips of the branches, with brown disks mostly 2–4 mm in diameter. CHEMISTRY: Cortex PD–, K–, KC+ gold, C–; medulla usually KC+ red (alectoronic acid). HABITAT: On conifer branches, especially pine, as well as rock and soil, in montane forests at high altitudes. COMMENTS: This shrubby *Alectoria* is very distinctive when it is abundantly fertile. The less frequent sterile tufts are hard to distinguish from young specimens of *A. sarmentosa,* which is pendent when mature.

Alectoria nigricans
Gray witch's hair

DESCRIPTION: Thallus shrubby, growing erect or tangled on the ground; main branches varying from pale pinkish gray to almost black, but always with black branch tips; surface almost always dull. Large brown apothecia sometimes produced. CHEMISTRY: Cortex and medulla PD+ yellow, K+ yellow, KC+ red, C+ pink (alectorialic acid). In very dark specimens, these reactions are difficult to see, but they can be discerned with the filter paper method described in Chapter 13. HABITAT: Normally a soil lichen in tundra heath, but occasionally growing on the low branches of trees or shrubs. COMMENTS: This is the only *Alectoria* that is grayish instead of yellowish (lacking usnic acid in the cortex). The darkest thalli can be mistaken for *Bryoria nitidula,* but that lichen almost always has some shiny brown parts and its medulla is PD+ red. Even the darkest specimens of *A. ochroleuca* typically have some yellow at the base, but the two species can be very similar otherwise; chemistry will distinguish them.

Alectoria ochroleuca
Green witch's hair

DESCRIPTION: Thallus shrubby, erect, main branches greenish yellow, with the tips darkened greenish to black. CHEMISTRY: Cortex PD–, K–, KC+ yellow, C–; medulla PD–, K–, KC– or yellow, CK+ gold, C– (diffractaic acid). HABITAT: Strictly arctic-alpine, growing on the ground or, rarely, on shrubs. COMMENTS: The chemical reactions distin-

103. (facing page) *Alectoria sarmentosa* Southeast Alaska ×1.1

guish this species from unusually dark specimens of *A. nigricans*. The prostrate, soil-dwelling **A. sarmentosa ssp. vexillifera** has flattened main branches, never becomes erect, and reacts KC+ red in the medulla. IMPORTANCE: In Russia during the 1930s, a method was developed for using *A. ochroleuca* to make a kind of molasses. The process was important because, during World War II, beet sugar was scarce, and potatoes and grain were sent to the military. The lichen reportedly yielded 82 percent of its dry weight in glucose, producing a light yellow syrup.

Alectoria sarmentosa
Witch's hair

DESCRIPTION: Thallus pendent, pale greenish yellow; branches often twisted and somewhat flattened at the axils, with raised, white pseudocyphellae; medulla usually loose and cottony. Round, brown apothecia not uncommon along the branches. CHEMISTRY: Cortex PD–, K–, KC+ gold, C–; medulla KC+ red (alectoronic acid) or KC–. HABITAT: Often draping conifer branches and trunks in cool, coastal areas and some moist inland sites. COMMENTS: Among the tree-dwelling species of *Alectoria*, *A. sarmentosa* is by far the most common. A coarser, somewhat grayer species, **A. vancouverensis,** occurs along the coast from northern California to Vancouver Island; its medulla is compact and dense, and it reacts C+ red and KC+ deep red (olivetoric acid). **Alectoria fallacina** is a knobby, mainly KC–, Appalachian lichen with a very thick cortex. **Alectoria sarmentosa ssp. vexillifera,** found on the ground in the eastern arctic and western alpine tundra, has flattened branches and is often blotched dark greenish black. (Compare *A. sarmentosa* with the descriptions of the shrubbier, western species *A. imshaugii* and *A. lata*.) IMPORTANCE: *Alectoria sarmentosa* is an important food for black-tailed deer, especially in winter when other forage is scarce. Scientists in British Columbia have experimented with reintroducing it after timber harvesting to improve second-growth forests as deer habitat. The Bella Coola Indians of coastal British Columbia used *A. sarmentosa* as artificial hair on dance masks. On Vancouver Island, the Nitinaht used it for making bandages and diapers.

Allantoparmelia (2 N. Am. species)
Rock grub lichens

DESCRIPTION: Very dark olive- to yellow-brown or black foliose lichens with thick, narrow lobes (ca. 0.2–1.5 mm across), closely appressed; the more elongate lobes worm-like, with irregular constrictions (giving the lichen its English name); pseudocyphellae, soredia, and isidia absent; lower surface tan to black, without rhizines but sometimes attached to the substrate with thick, peg-like outgrowths of the lower cortex. Photobiont green (*Trebouxia*?). Apothecia lecanorine; pycnidia common, entirely embedded in the thallus and scattered over the lobes. CHEMISTRY: Cortex PD–, K–, KC–, C– (atranorin lacking); medulla with various compounds. HABITAT: On alpine rocks. COMMENTS: *Hypogymnia*, *Brodoa*, and *Lobothallia* all have thick, convex lobes and can resemble *Allantoparmelia*. *Hypogymnia* is generally larger, with hollow lobes. Most species of *Melanelia* have broader, flatter lobes, often with pseudocyphellae.

KEY TO SPECIES (including *Brodoa* and *Lobothallia*)

1. Lobes long and divergent, not contiguous near the growing margin, gray to dark brown; upper cortex K+ yellow on pale areas (atranorin); medulla PD– or rarely PD+ red .... *Brodoa oroarctica*
1. Lobes long or short, contiguous at the margin of the thallus, not divergent; cortex K– or K+ yellow to red (atranorin absent); medulla PD+ yellow 2
2. Cortex K–; medulla K–, KC+ red (alectorialic acid); apothecia always superficial, not immersed, up to 7 mm across; spores 7.5–10 × 5–7 μm .... *Allantoparmelia alpicola*
2. Cortex and medulla K+ red, KC– (norstictic acid); apothecia at first immersed, later superficial, under 2.5 mm across; spores 10–14 × 6–10 μm 3
3. Thallus gray to gray-brown .... *Lobothallia alphoplaca*
3. Thallus yellow brown to copper brown .... *Lobothallia praeradiosa*

Allantoparmelia alpicola
Rock grubs

DESCRIPTION: Thallus dark brown to almost black, frequently gray in older parts; lobes usually short, up to 1.5 mm across; lower surface pale to dark brown or black. CHEMISTRY: Medulla PD+ deep

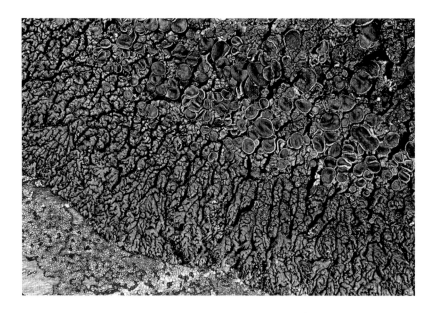

104. *Allantoparmelia alpicola* interior Alaska ×2.1

yellow, K+ pale yellow, KC+ red, C– or C+ pink (alectorialic and barbatolic acids). COMMENTS: This lichen is very similar to *Brodoa* in color and overall appearance, but the lobes are broader and shorter, and the chemistry is different. *Allantoparmelia almquistii* is a rarer arctic-alpine lichen with distinctly flattened, narrower lobes (usually less than 0.5 mm across) becoming entangled and producing slender lobules near the thallus center. In appearance, it is midway between a flattened morphotype of *Pseudephebe minuscula* and *Allantoparmelia alpicola*. *Melanelia stygia* is also similar to species of *Allantoparmelia* but has conspicuous pseudocyphellae and rhizines. *Melanelia panniformis* is thin and flattened at the growing margin, producing rounded, imbricate lobules only on the older thallus portion. Chemistry easily differentiates them: *A. almquistii* has a C+ red, PD– medulla (olivetoric acid); *Pseudephebe* is PD–, C– (no substances); *Melanelia stygia* can be PD+ red (fumarprotocetraric acid) or PD– and is always C–; *M. panniformis* is PD–, C– (perlatolic acid); in *Brodoa*, the cortex is K+ yellow (atranorin), and the medulla is KC+ pink (lobaric acid) and either PD– or PD+ red-orange (protocetraric acid). See also *Protoparmelia*.

Allocetraria (2 N. Am. species)

DESCRIPTION: A genus of mainly foliose, rarely fruticose lichens, yellow to yellowish green or brown; upper cortex usually composed of a palisade-like pseudoparenchyma (Fig. 5d); pseudocyphellae present or absent; rhizines sparse. Photobiont green (*Trebouxia*). Apothecia marginal or laminal but close to the margin, brown, with narrow asci; spores colorless, spherical, 5–10 × 5–8 μm, 8 per ascus arranged in a single row; pycnidia prominent or immersed, distributed usually along the lobe margins on foliose species; conidia thread-like, 10–19 × 0.5–2 μm. CHEMISTRY: Cortex PD–, K–, KC+ gold, C– (usnic acid); medulla of North American species with fatty acids. HABITAT: On the ground at high elevations; one species on trees at low or high elevations. COMMENTS: Most species of *Allocetraria* are found in southeastern Asia. The characteristics that link the two quite dissimilar North American representatives are mainly anatomical features (the cortex, pycnidia, and asci). *Ahtiana* is a closely related genus with dumbbell-shaped conidia and a cortex of pseudoparenchyma (without a palisade arrangement; Fig. 5f).

KEY TO SPECIES: See Keys A and J, and *Dactylina*.

Allocetraria madreporiformis
(syn. *Dactylina madreporiformis*)
V-fingers

DESCRIPTION: Thallus greenish yellow, dull but without pruina, with finger-like stalks branched dichotomously (in regular "Vs") several times, 10–35 mm high, 1–2 mm thick, usually terete and erect, dimpled and ridged, occasionally somewhat flattened and more prostrate; branches filled with a dense, white medulla; cortex composed of palisade pseudoparenchyma. Apothecia very rare, pale brown; conidia thread-shaped, somewhat curved, and slightly broader at one end, 11–18 × 1–1.5 μm. CHEMISTRY: Medulla PD–, K–, KC–, C– (protolichesterinic acid). HABITAT: On calcareous soil in arctic or alpine sites. COMMENTS: This lichen is very similar to *Dactylina*, especially *D. ramulosa*, which often occurs as a neighbor. *Dactylina* species are hollow, have lemon-shaped conidia, and lack fatty acids. In *D. ramulosa*, the medulla is very webby and so the stalks are essentially hollow. In addition, the branches are unevenly divided and usually heavily pruinose at the tips.

105. (above) *Allocetraria madreporiformis* Rocky Mountains, Colorado ×5.0

106. (left) *Allocetraria oakesiana* central Massachusetts ×4.0

Allocetraria oakesiana (syn. *Cetraria oakesiana*)
Yellow ribbon lichen

DESCRIPTION: Thallus yellowish green, uniform, mostly appressed; lobes strap-shaped, crisped at the edges, branched, 1–4 mm wide, lacking pseudocyphellae; soralia yellowish, almost entirely marginal, granular, sometimes developing from coarse pustules; rarely with tiny, almost granular lobules; medulla white to pale orange; lower surface pale brown to almost white, somewhat wrinkled, shiny, with sparse rhizines. Photobiont green (*Trebouxia*). Apothecia rare, marginal or laminal, with pale brown disks; spores spherical, ca. 5 μm in diameter; conidia usually thread-shaped, 8–10 μm long, but rather variable in shape and size. CHEMISTRY: Medulla PD–, K+ yellow, KC+ orange to red-orange, C– (caperatic, protolichesterinic, lichesterinic, and secalonic acids). HABITAT: On bark and wood, usually of conifers and birch; sometimes on rock; typically in shaded forests. COMMENTS: Except for the color, this species is similar to *Tuckermannopsis chlorophylla*. It is also very close to species of *Ahtiana* but has different conidia.

Amandinea (6 N. Am. species)
Button lichens

DESCRIPTION: Crustose lichens with brown to gray or almost white, thick, or, more frequently, very thin thalli. Photobiont green (*Trebouxia*). Apothecia black, lecideine or lecanorine, usually with a persistent margin; hypothecium usually dark brown but sometimes pale; epihymenium brown to greenish; tips of paraphyses enlarged and pigmented; asci variable in K/I reaction, but usually *Bacidia*-type; spores brown, 2-celled, with uniformly thickened walls or with a thickened septum; conidia very slender, long, and curved, mostly about 15–30 μm long. CHEMISTRY: North American species without lichen substances. HABITAT: On rocks, bark, or wood, rarely soil. COMMENTS: The long, arc-shaped conidia separate this genus from the closely related *Buellia*. Thallus and apothecial type vary from species to species, and even the spores are not uniform. In the coastal rock species *A. coniops* (British Columbia, Alaska, and Newfoundland), the septum separating the two cells is conspicuously thickened; in most other species of *Amandinea*, it is not. This species also differs from the others in having a thick, areolate to verrucose, brown to gray-brown thallus and large spores, 12–18 × (6.3–)8–9.5 μm.

KEY TO SPECIES: See *Buellia* and *Rinodina*.

Amandinea punctata (syn. *Buellia punctata*)
Tiny button lichen

DESCRIPTION: Thallus thin and barely perceptible to moderately thick, cracked or areolate, gray to brownish, green when wet. Apothecia 0.2–0.5 mm in diameter, with a thin, black margin that often disappears in maturity; exciple uniformly pigmented dark brown; spores (7–)11–16 × (4–)5–8 μm, not constricted, 8 per ascus. HABITAT: On various types of bark and wood, and occasionally on siliceous rock. COMMENTS: *Amandinea polyspora*, an East Temperate species, is very much like *A. punctata* but has 12–32 spores per ascus, somewhat smaller spores (7.5–11.5 × 3.5–5.5 μm), and an exciple that is pale within. It grows only on bark. Two species of *Amandinea* with incipient to well-developed lecanorine margins are ***A. dakotensis*** (syn. *Rinodina dakotensis*) and ***A. milliaria*** (syn. *R. milliaria*). The latter has apothecia that erupt from the pale gray thallus, sometimes leaving a very thin thalline border outside the lecideine margin. It is almost entire coastal, from Maine to Texas, with a few records from the shores of the Great Lakes. *Amandinea dakotensis* has a dark gray-green thallus, and the apothecia, although also initially erupting, finally have well-developed lecanorine margins as in species of *Rinodina*. It is found throughout the midwest, rarely occurring on the east coast.

Amygdalaria (7 N. Am. species)
Almond lichens

DESCRIPTION: Crustose lichens, most often with a thick, areolate thallus with pinkish or brownish to creamy tones; cephalodia almost always present containing one of several types of cyanobacteria, sometimes on the same thallus; main photobiont green (unicellular). Apothecia with black, immersed disks, like *Aspicilia;* exciple brown-black; spores large, almond-shaped, halonate; asci *Porpidia*-type. CHEMISTRY: Gyrophoric acid and the stictic acid complex are the most common. HABITAT: On siliceous rocks, mainly in humid, oceanic habitats; some are arctic. COMMENTS: *Amygdalaria* is distinguished from *Porpidia* mainly in having cephalodia, larger spores with a compact, conspicuous halo, and more often immersed apothecia. It is also associated with humid, oceanic habitats. Related genera with *Porpidia*-type asci, halonate spores, and immersed apothecia include *Bellemerea* and *Immersaria,* both of which lack cephalodia. *Bellemerea* has brown apothecial disks, a colorless exciple, and smaller spores. *Immersaria carbonoidea* is a rare lichen from arctic Alaska with a dark brown, areolate thallus, black apothecial disks often clustered in groups, a brown epihymenium, and a dark brown hypothecium and exciple. Its spores are 12–13 × 6.5–7.5 μm, and the thallus medulla is PD–, K–, C–, IKI+ blue.

KEY TO SPECIES: See Keys C and F.

Amygdalaria panaeola
Powdery almond lichen

DESCRIPTION: Thallus dispersed areolate to verrucose, pale brownish gray to yellowish pink; areoles dissolve into coarse granular soredia especially at the base or sides; cephalodia dark pink or olive (depending on the photobiont) brain-shaped galls on or between the areoles. CHEMISTRY: Soralia C+ red; cortex PD–, K–, KC+ red, C+ red (rarely C–); medulla PD–, K–, C–, KC– (gyrophoric acid often with confluentic acid and related compounds). HABITAT: On arctic or alpine rocks or, at lower elevations and latitudes, on humid rock faces such as those near waterfalls. COMMENTS: This lichen is almost always sterile

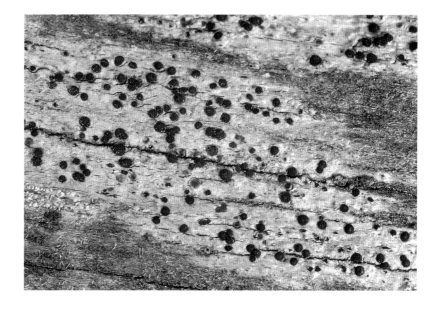

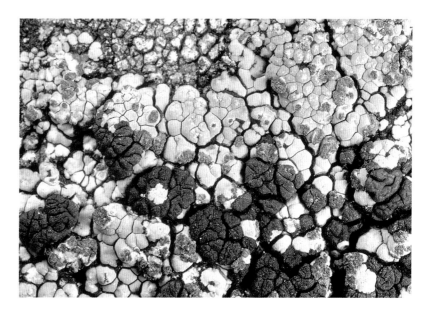

107. *Amandinea punctata* herbarium specimen (Canadian Museum of Nature), Quebec ×5.6

108. *Amygdalaria panaeola*   interior Alaska ×4.7

109. *Anaptychia palmulata* Appalachians, Virginia ×2.2

and is best distinguished from most other sterile crustose lichens on rocks by its combination of cephalodia and soredia, and by its chemistry. ***Coccotrema pocillarium*** is a sorediate, usually sterile crustose lichen on bark in humid coastal forests of the Pacific Northwest. It contains stictic acid (PD+ orange, K+ yellow). Most other species of *Amygdalaria* are relatively rare and are confined to the humid northwest coast from Washington to southeastern Alaska.

Anaptychia (4 N. Am. species)
Fringe lichens

DESCRIPTION: Foliose lichens, loosely attached to ascending, almost fruticose in some species; greenish gray to brown or olive-brown, with narrow, short to extremely long lobes and branches, 0.2–1(–2) mm wide; without soredia or isidia in North American species, but some with narrow lobules; upper surface with a vague combed appearance because of a fibrous upper cortex running the length of the lobes (see Fig. 5b); lower surface pale, most commonly without a cortex, usually with rhizines and (or) marginal cilia, the latter often very long and branched. Photobiont green (*Trebouxia*?). Apothecia lecanorine, with thallus-colored margins; spores brown, 2-celled, most with thin, more or less evenly thickened walls, or with a thickened septum when young. CHEMISTRY: All reactions negative (no lichen substances). HABITAT: On rocks and bark in shade and sun. COMMENTS: These lichens have the general appearance of a *Physconia* or *Phaeophyscia* except for the sometimes fruticose *A. setifera*. The closely related *Heterodermia* has atranorin in the cortex (K+ yellow) and *Physcia*-type spores.

KEY TO SPECIES: See Keys A and K.

Anaptychia palmulata (syn. "*A. palmatula*")
Shaggy-fringe lichen

DESCRIPTION: Thallus greenish gray to brownish, often coarsely white scabrose on the lobe tips, flat to ascending at the margins, forming rosettes 4–8 cm across; lobes rather elongate, often fan-like at the tips, 0.7–1.5(–2) mm wide, the margins covered with tiny, strap-shaped lobules but few cilia; lower surface pale, with pale, unbranched rhizines that later become squarrose or brush-like. Apothecia common, 1–2 mm in diameter, with dark brown, nonpruinose disks and thick, prominent margins that sometimes develop small lobules. HABITAT: On bark of hardwoods and white cedar or on shaded rocks; in forests. COMMENTS: *Physconia subpallida,* an eastern lichen growing in similar habitats, also has a lobulate thallus, but the rhizines are black and squarrose when mature, the apothecia are pruinose, and the spores have unevenly thickened walls (*Physcia*-type). *Physconia americana* is the western equivalent. *Melanelia panniformis* is a saxicolous species with a black lower surface that typically produces masses of overlapping lobules on the central part of the thallus. A rare lobulate *Anaptychia, A. bryorum,* grows on tundra soil and vegetation. It is dark brown with some cilia along the lobe margins, and the lobules are extremely narrow and frequently forked.

Anaptychia setifera
Hanging fringe lichen

DESCRIPTION: Thallus foliose to fruticose, consisting of very long, narrow, abundantly branched lobes in cushions or, more frequently, in pendent clumps up to 8 cm long; pale gray to smoky gray-brown, with a slightly fuzzy surface; lobes 0.3–0.5(–1.5) mm wide with long, tapered, pale gray to black, unbranched or forked cilia all along the margins; soredia and isidia absent; lower surface white, with a fibrous groove or channel (lacking a cortex); rhizines absent. Apothecia frequent, raised, with concave, heavily pruinose disks and abundantly ciliate margins. HABITAT: On limy cliffs, in full sun or partial shade especially overlooking streams or lakes; rarely on shrubs. COMMENTS: *Heterodermia leucomela* has similar long, linear lobes and long, black cilia, but it is usually paler and the lobe tips are sorediate. Its cortex is K+ yellow (atranorin). The rare Californian species *Teloschistes californicus* has fuzzy, gray, subfruticose branches, but they are much broader and irregular and are not ciliate.

Anzia (3 N. Am. species)
Black-foam lichens

DESCRIPTION: Medium-sized to small foliose lichens, gray to greenish; lower surface covered with a thick, spongy, black hypothallus (figs. 6g,h). Photobiont green (*Trebouxia*?). Apothecia lecanorine, with concave, shiny brown disks; spores colorless, tiny (3–6 × 1.2–2 μm), 1-celled, oblong or curved, many per ascus. CHEMISTRY: Cortex PD–, K+ yellow, KC–, C– (atranorin); medulla PD–, K–, KC–, C–, UV+ white (divaricatic acid). HABITAT: On bark, usually in forests. COMMENTS: The black, spongy lower surface is unique among North American lichens.

KEY TO SPECIES: See Key K.

Anzia colpodes
Black-foam lichen

DESCRIPTION: Thallus foliose, forming rosettes of greenish gray, dichotomously branching, convex lobes, 1–2 mm wide, without soredia or isidia, lobes developing small, round lobules. Apothecia common, up to 5 mm in diameter, very concave; pycnidia usually buried in the tips of the lobes, seen as black dots. HABITAT: On hardwoods in deciduous forests. COMMENTS: *Anzia colpodes* at first glance looks like a *Hypogymnia,* especially *H. krogiae,* but the lobes (apart from the hypothallus) are actually thin and solid. The two other North American species of *Anzia* are very rare. *Anzia americana,* a southern

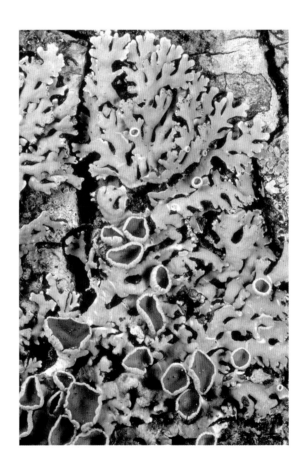

110. *Anaptychia setifera* herbarium specimen (New Brunswick Provincial Museum), New Brunswick ×2.8

111. *Anzia colpodes* Smoky Mountains, Tennessee ×2.4

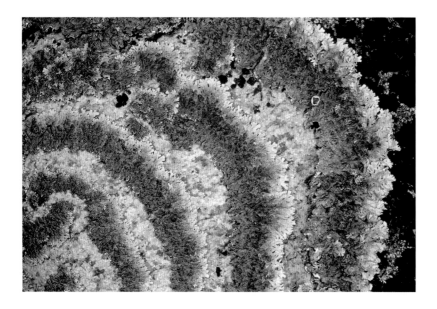

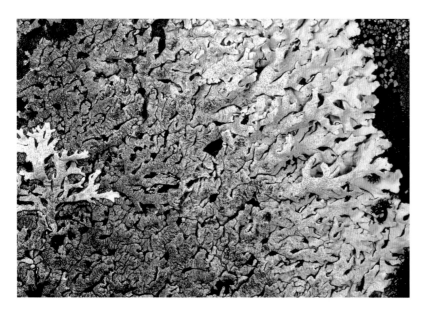

112. *Arctoparmelia centrifuga* Laurentides, Quebec ×0.64

113. *Arctoparmelia separata* interior Alaska ×1.5

Appalachian species, is sorediate; *A. ornata,* on the southern coastal plain, is marginally isidiate.

Arctoparmelia (3 N. Am. species)
Ring lichens

DESCRIPTION: Narrow-lobed (0.3–0.5 mm across), dull greenish yellow foliose lichens, commonly forming concentric rings of radiating lobes, with the thallus dying in the center of the rosette; lower surface with a dull, white or pale tan to very dark gray cortex; rhizines unbranched and scattered. Photobiont green (*Trebouxia*?). Apothecia lecanorine, with brown disks and thallus-colored margins; spores colorless, 1-celled, ellipsoid, 8 per ascus. CHEMISTRY: Cortex PD–, K+ yellow, KC+ gold, C– (usnic acid with atranorin); medulla PD–, K–, KC+ red, C– (alectoronic acid). HABITAT: On siliceous rocks usually in the open; mainly arctic or alpine. COMMENTS: *Xanthoparmelia* superficially resembles *Arctoparmelia* but usually has a shiny upper surface, a brown to black, shiny lower surface with branched rhizines, and an entirely different medullary chemistry.

KEY TO SPECIES: See *Xanthoparmelia*.

Arctoparmelia centrifuga (syn. *Xanthoparmelia centrifuga*)
Concentric ring lichen

DESCRIPTION: Lobes flat to slightly convex, 1–2 mm wide, without soredia or isidia; lower surface white or almost white, with scattered dark rhizines. Apothecia fairly common. COMMENTS: The white lower surface distinguishes this lichen from the much rarer *A. separata* and *A. incurva;* both have a mouse-gray lower surface, and *A. incurva* produces large, irregular soralia on the upper surface. Rarely, *A. centrifuga* lacks usnic acid and is gray instead of yellow-green, with the cortex reacting K+ yellow, KC–. IMPORTANCE: In the Arctic, *A. centrifuga* has been used as a source of red-brown dyes for wool.

Arctoparmelia separata (syn. *Xanthoparmelia separata*)
Rippled ring lichen

DESCRIPTION and COMMENTS: Very similar to the much commoner *A. centrifuga*, but the lower surface is a dull mouse-gray, at least in the center of the thallus and sometimes out to the edge. (The more appressed the thallus, the darker the undersurface.) It tends to be thicker and stiffer than *A. centrifuga*.

Arthonia (110 N. Am. species)
Comma lichens

DESCRIPTION: Crustose lichens, with thalli usually very thin or actually growing beneath the upper layers of the substrate. Photobiont green (*Trentepohlia* or unicellular species similar to *Desmococcus*). Fruiting bodies (ascomata) irregular to elongate and branching, or round and disk-like; asci broadly club-shaped or balloon-shaped with a thickened summit, K/I–, lying in a more or less uniform layer of branched and tangled hyphae making up the hymenium, which is usually K/I+ blue (hemiamyloid); without an exciple (Fig 13f); spores colorless, ellipsoid to fusiform, usually with one end broader than the other, 2–8 cylindrical, thin-walled cells, often with 1 or 2 of the cells conspicuously larger than the others (a distinguishing character). CHEMISTRY: Most species have no lichen substances and react PD–, K–, KC–, C–, but some have orange, K+ red-purple, anthraquinone pigments on the fruiting bodies. HABITAT: On bark, wood, or, less frequently, rocks (see below). Some species are partially or entirely parasitic on other lichens. Most species are from tropical or temperate latitudes. COMMENTS: This very large genus is greatly in need of study. The genus *Arthothelium* is very similar and closely related, but it generally has larger fruiting bodies and larger, muriform spores. The script lichens (*Graphis* and *Phaeographis*) have lirellae with a well-developed exciple, and the locules of the spores are lens-shaped. Most *Arthonia* species grow on tree bark, but some are restricted to rock. ***Arthonia phaeobaea***, for example, forms large colonies on maritime rocks along both coasts. It has a thin, orangish to brown thallus with round, black, convex ascomata resembling apothecia, 0.15–0.3 mm in diameter, and 3–6-celled, somewhat constricted, tapering spores. The few species we include here give only a very slight idea of the diversity of the genus.

KEY TO SPECIES (including *Arthothelium*)

1. Spores muriform; ascomata lobed to star-like or almost round 2
1. Spores only transversely septate; ascomata round to script-like 3

2. Spores (15–)17–24(–26) × 7–9.5(–10.5) μm .... [*Arthothelium ruanum*]
2. Spores 26–36 × 12–15 μm .... [*Arthothelium spectabile*]

3. Ascomata round, with the appearance of a biatorine apothecium 4
3. Ascomata elongate or branched lirellae 6

4. On maritime rocks. Thallus brown, membranous; spores 3–6-celled, 15–23 × 5.5–8 μm; on both Atlantic and Pacific coasts .... [*Arthonia phaeobaea*]
4. On bark 5

5. Ascomata convex, bluish gray owing to a heavy white pruina; spores 4-celled; on a variety of trees in the northeast .... *Arthonia caesia*
5. Ascomata flat to slightly convex, black, not at all pruinose; spores 2-celled, strongly tapered, (8–)10–12 × 3.5–5 μm; found almost exclusively on trembling aspen trees, northern temperate to southern boreal .... [*Arthonia patellulata*]

6.(3) Spores with 4 equal-sized cells .... *Arthonia radiata*
6. Spores (4–)5– to 8-celled, tapered, with at least one end cell larger than the others 7

7. Ascomata ellipsoid to Y-shaped, convex, red pruinose on the margins or over the entire surface (anthraquinones); spores 18–28 × 7–10 μm ... *Arthonia cinnabarina*
7. Ascomata irregularly branched, flat or immersed, not pruinose (anthraquinones absent); spores 26–35 × 10–15 μm 8

8. Ascomata black, flat; spores tapered, with uppermost cell considerably larger than lower cells; coastal Pacific Northwest .... [*Arthonia ilicina*]
8. Ascomata brown to reddish brown, usually immersed and narrowly or widely open, appearing like irregular cracks in the bark; spores only slightly tapered if at all, both end cells larger than middle cells; southeast ..... [*Arthonia rubella*]

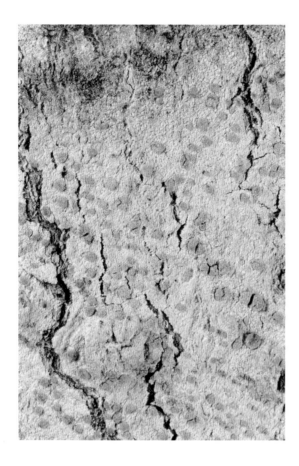

114. *Arthonia caesia* southwestern Quebec ×6.8

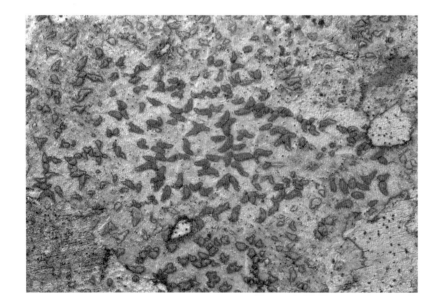

115. *Arthonia cinnabarina* Gulf Coast, Florida ×4.2

Arthonia caesia
Frosted comma lichen

DESCRIPTION: Thallus thin, yellowish white to yellowish green, granular or powdery. Photobiont green (unicellular, not *Trentepohlia*). Ascomata small (0.2–0.4 mm in diameter), heavily blue-gray pruinose, convex, marginless; internal tissues dark brown, spores 15–24 × 4–6 μm, 4-celled, with the upper 2 cells slightly larger or equal to the lower 2, and slightly constricted in the middle. CHEMISTRY: Spot tests unreliable (contains usnic acid and probably triterpenes in the thallus). HABITAT: On bark, mainly deciduous trees and shrubs such as willow, alder, and maple. COMMENTS: The combination of the small, blue-gray, marginless ascomata on a powdery yellowish crust makes this lichen easy to identify. Sterile specimens, however, closely resemble species of *Lepraria* or *Lecanora thysanophora*. Old specimens of *A. caesia* develop a fuzz of colorless needles, indicating the probable presence of triterpenes.

Arthonia cinnabarina
Bloody comma lichen

DESCRIPTION: Thallus largely immersed in the bark, forming a gray to pinkish patch often bordered by a thin, brown prothallus. Ascomata irregularly ellipsoid, rarely branched, 0.3–1 × 0.2–0.5 mm, brown, usually with dark red or red-orange pruina on the edges; surface of "disk" white or red pruinose, or naked; spores mostly 5–6-celled, strongly tapered, with the end cell noticeably larger than others, 18–28 × 7–10 μm. CHEMISTRY: Red pruina is K+ red-purple (several anthraquinones). HABITAT: On bark, usually in shaded woods. COMMENTS: The spores and red, K+ purple pruina on the fruiting bodies are usually enough to distinguish this striking lichen.

Arthonia radiata
Asterisk lichen

DESCRIPTION: Thallus thin, pale to white, in small patches on trees. Ascomata black, elongate and branching, up to 1.5 mm long and 0.1–0.2 mm broad, forming star-like clusters similar to those of a a tiny script lichen; spores 15–21 × 4.5–7 μm, 4-celled, with all the cells approximately the same size (fig. 14n). CHEMISTRY: No lichen substances. HABITAT: On bark of many kinds, especially the smooth bark of hardwoods; typically in forests. COMMENTS: In the east, *Arthonia radiata* is often accompanied by *A. patellulata,* a common species with a hidden thallus that creates pale patches on the bark of poplars and other hardwoods. Its fruiting bodies are black and round, resembling biatorine apothecia without a margin, and the spores are 2-celled.

Arthrorhaphis (4 N. Am. species)
Dot lichens

DESCRIPTION: Thallus consisting of bright yellow or gray granules, convex areoles, or squamules, or growing within the tissues of a host lichen; medulla sometimes containing masses of colorless, irregular crystals of calcium oxalate (insoluble in K, soluble in strong acid). Photobiont green (unicellular). Ascomata pitch black, flat to slightly convex, resembling apothecia, without a distinct margin; exciple poorly developed; paraphyses branched and anastomosing; asci K/I–, barely thickened at the tip; spores colorless, needle-shaped, 9–12-celled, 8 per ascus. CHEMISTRY: PD–, K–, KC–, C–, UV+ orange (rhizocarpic acid). HABITAT: On disturbed soil such as on roadsides and frost boils; boreal to arctic. COMMENTS: Several species in this genus are lichen parasites.

KEY TO GENUS OR SPECIES: See Keys C, F, and G.

Arthrorhaphis citrinella
Golden dot-lichen

DESCRIPTION: Thallus bright yellow to chartreuse, forming strongly convex areoles or squamules, or sometimes granules; medulla lacking crystals of cal-

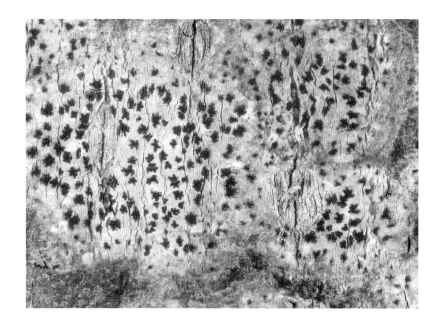

116. *Arthonia radiata* southern Ontario ×4.6

117. *Arthrorhaphis citrinella* Laurentides, Quebec ×5.2

118. *Asahinea chrysantha* interior Alaska ×1.5

cium oxalate; cortex especially brittle so that the areoles break down into a powdery, leprose crust. Ascomata occasionally present; spores 45–65(–110) × 2.5–3.6 μm. HABITAT: On noncalcareous soil. COMMENTS: The very similar *A. alpina* usually has larger, smoother, less fragile areoles and verrucae, contains calcium oxalate crystals in the medulla, has shorter spores (most less than 50 μm long), and occurs on either calcareous or noncalcareous soil; soredia are present or absent. Few other soil crusts are such a bright yellow. The yellow hue of *Fulgensia* species is due to anthraquinones (K+ deep red-purple). *Acarospora schleicheri* is restricted to arid regions of the western interior.

Asahinea (2 N. Am. species)
Ground rag lichens

DESCRIPTION: Rather large, loosely attached foliose lichens; lobes thin, wavy or undulating, often wrinkled, 4–30 mm across, with or without pseudocyphellae; lower surface shiny black, entirely lacking rhizines; pycnidia and apothecia, when present, scattered on the upper surface. Photobiont green (*Trebouxia?*). CHEMISTRY: Cortex containing at least atranorin, sometimes usnic acid; medulla of all species PD–, K–, KC+ red, C– (alectoronic acid). HABITAT: On rock or soil. COMMENTS: At first glance, *Asahinea* resembles a *Platismatia* or large *Parmelia,* but these have rhizines and do not normally grow on soil.

KEY TO SPECIES: See Keys J and K.

Asahinea chrysantha
Arctic rag lichen

DESCRIPTION: Thallus yellow to very pale yellow-brown, with broad, rounded lobes, 10–30 mm wide, usually with a network of sharp ridges and depressions on the upper surface; pseudocyphellae present mainly on the ridges; without soredia or isidia. Apothecia extremely rare. CHEMISTRY: Cortex PD–, K+ yellow (but not detectable on the yellow thallus), KC+ gold, C– (usnic acid and atranorin). HABITAT: On the ground in arctic-alpine sites. COMMENTS: This lichen can resemble a sterile, small specimen of

Nephroma arcticum or *N. expallidum* in having usnic acid in the upper cortex and rather broad, rounded lobes. However, the lower surface of *Nephroma* species is almost always very pale brown at the edges and always has some fuzzy tomentum. ***Cetrelia alaskana,*** a rare species from arctic Alaska, has a gray thallus (lacking usnic acid) and always has at least a few rhizines on the lower surface. The other species of *Asahinea, A. scholanderi,* also grows on exposed rocks and boulders and on decaying vegetation on the tundra. It has a pale yellowish gray thallus, becoming dark brown or black especially toward the exposed margins, with lobes 5–15 mm wide. The surface is covered with abundant, cylindrical to branched, pale or black-tipped isidia. In place of usnic acid in the cortex, it has atranorin (PD–, K+ yellow, KC–, C–). *Asahinea scholanderi* looks much like a ragged, coarse *Platismatia* with black-tipped isidia. *Parmelia saxatilis* can resemble it from above, but it has abundant, black rhizines below and reacts K+ blood red in the medulla.

Aspicilia (61 N. Am. species)
Sunken disk lichens

DESCRIPTION: Crustose or, rarely, fruticose lichens with gray, white, or greenish thalli and immersed, pure black apothecia (sometimes hidden by pruina). Photobiont green (unicellular). Apothecial disks more or less flush with the thallus surface or with a slightly raised, lecanorine rim; hypothecium and exciple colorless; epihymenium pale green to olive-brown; tips of the paraphyses composed of short, rounded cells that look like a string of beads; asci K/I–, the tip only slightly thickened; spores colorless, 1-celled, ellipsoid to subspherical (9–40 × 6–20 μm), not halonate, 4–8 per ascus; conidia long, straight, and thread-like. CHEMISTRY: Often containing β-orcinol depsidones such as norstictic and stictic acids; others with fatty acids or triterpenes. HABITAT: On rocks of all kinds, in sun or shade. COMMENTS: The variability of the species of *Aspicilia* in both thallus morphology and chemistry makes this an especially difficult group to identify. As the taxonomy has never been adequately worked out for North American species, distribution maps are either not presented here or should be regarded as tentative. The pale exciple, large spores, and beaded paraphyses distinguish *Aspicilia* from other genera with immersed, disk-like apothecia, such as *Bellemerea, Lecidea, Amygdalaria,* and *Acarospora.* (See also the notes on *Lecanora oreinoides* under *Buellia spuria.*) *Hymenelia* and *Ionaspis* are closely related but generally have smaller spores, paler apothecia (with different pigments if black), and lack lichen substances in the thallus.

KEY TO SPECIES

1. Thallus fruticose, with clumps of tangled, terete branches; white, depressed pseudocyphellae abundant and conspicuous; apothecia rare ..... *Aspicilia hispida*
1. Thallus crustose, with or without pseudocyphellae; apothecia abundant 2

2. Thallus areolate with areoles dispersed or contiguous, strongly convex or pyramidal, dark olive to dark olive-gray, often with white pseudocyphellae; spores almost spherical, 15–25 μm wide .... *Aspicilia contorta*
2. Thallus usually continuous, smooth or rimose-areolate, white to gray, rarely olive; lacking pseudocyphellae; spores ellipsoid, 6–16 μm wide 3

3. Thallus chalky white, often lobed at the periphery; on calcareous rock; all spot tests negative ....
.... *Aspicilia candida*
3. Thallus creamy white or pale to dark gray, rarely chalky white, never lobed; on noncalcareous rocks 4

4. Paraphyses septate and constricted at the tips, resembling a string of beads; asci K/I–; medulla K+ red or yellow, PD+ yellow or orange (norstictic acid or, occasionally, stictic acid) .... *Aspicilia cinerea*
4. Paraphyses expanded at the tips, not constricted or bead-like; asci with a K/I + blue tip; medulla PD–, K–
.... *Lecanora oreinoides*

Aspicilia candida
Chalky sunken disk lichen

DESCRIPTION: Thallus chalky white, almost blue-white, more or less circular, sometimes producing webby white lobes, or the thallus margin can be edged with a bluish black prothallus. Apothecia sunken or somewhat prominent, under 1 mm in diameter, sometimes scattered, with pruinose-scabrose disks; margins darkish but pruinose; spores 14–24 × 10–16 μm, 4–8 per ascus. CHEMISTRY: All reactions negative (no lichen substances known). HABITAT: On calcareous rocks; subalpine to arctic. COMMENTS: The thallus of this species, like that of many arctic crustose lichens, can become eroded by sand and ice leaving only scattered patches and isolated apothecia.

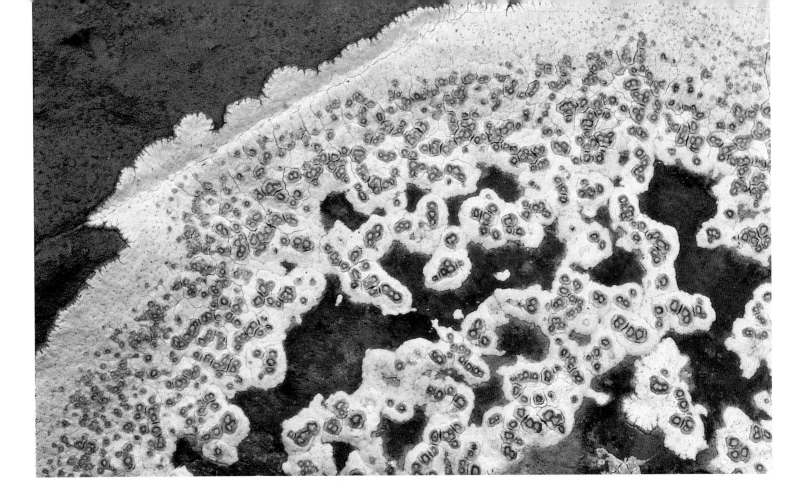

119. *Aspicilia candida* Olympics, Washington ×5.0

120. *Aspicilia cinerea* Appalachians, West Virginia ×3.6

Aspicilia cinerea
Cinder lichen

DESCRIPTION: Thallus thin or thick, continuous and rimose-areolate to somewhat verrucose, very pale to rather dark greenish gray or ashy gray (in some populations, yellowish gray or chalky white). Apothecia black, mostly 0.4–1.2 mm in diameter, disk level with the thallus surface and marginless, or with a thin, slightly prominent margin often darker than the surrounding thallus; spores 12–22 × 6–13 µm, 4–8 per ascus; conidia 11–18(–22) × 1 µm. CHEMISTRY: Medulla PD+ yellow, K+ red, KC–, C– (norstictic acid), or PD+ orange, K+ persistent yellow (stictic acid). HABITAT: On siliceous rocks usually in sun, often covering very large surfaces. COMMENTS: In North America, several species that contain norstictic acid but differ in thallus color and development as well as substrate type can be found in museums filed under "*A. cinerea*." We accept populations with stictic acid and thin, rather smooth to rimose thalli as a variety of *A. cinerea* if all other characters agree. ***Aspicilia verrucigera*** has a rather thick thallus that usually builds into curved and lumpy, sometimes worm-like areoles and contains stictic acid; it is one of the most common species of the genus, especially in eastern North America. *Aspicilia cinerea* generally has a thinner, less bumpy thallus. Like *A. cinerea*, *A. verrucigera* grows on granitic rocks in full sun, and the two species often grow as neighbors. Specimens that are similar to *A. cinerea* but lack lichen substances and have a bluer gray, rimose-areolate thallus are often called **A. caesiocinerea.** Specimens with a K+ red medulla, but growing on limestone, are common in the west. They usually have creamy white thalli and probably represent other species.

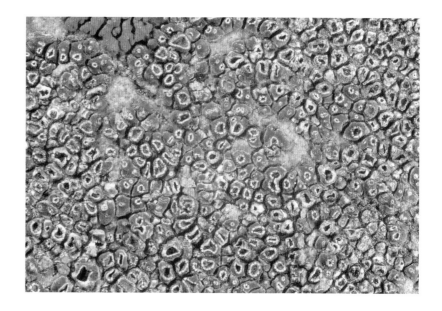

Aspicilia contorta
Chiseled sunken disk lichen

DESCRIPTION: Thallus dark olive to olive-gray, thick, areolate, with hemispherical to "chiseled" areoles that are contiguous or dispersed, often with depressed, white pseudocyphellae. Apothecia deeply immersed, pruinose, usually one per areole; spores broadly ellipsoid to spherical, 20–30 × 15–25 µm, 4 per ascus. CHEMISTRY: All reactions negative (no lichen substances). HABITAT: On noncalcareous or calcareous rock. COMMENTS: The areoles of *A. contorta*, although sometimes contiguous, are not radially oriented or bordered by a distinct prothallus as in *A. calcarea,* a very similar widespread species on limy rocks.

Aspicilia hispida (syn. *Agrestia hispida*)
Vagabond lichen

DESCRIPTION: Thallus fruticose, forming tiny, dull olive-green to blue-green, shrubby clumps up to 3 cm across; branches irregularly divided, terete, 0.5–1 mm in diameter, somewhat flattened in the oldest parts, with conspicuous white, depressed pseudocyphellae, 0.1–0.35 mm in diameter, dotting the branches; medulla white, dense. Photobiont green. Apothecia like those of *Aspicilia contorta* but rare. CHEMISTRY: Medulla PD–, K–, KC–, C– (no lichen substances). HABITAT: Over calcareous soil and pebbles in dry, open prairies, at first attached but soon breaking free and becoming

121. *Aspicilia contorta* herbarium specimen, (Canadian Museum of Nature), Alberta ×5.6

122. *Aspicilia fruticulosa* southeastern Idaho ×2.9

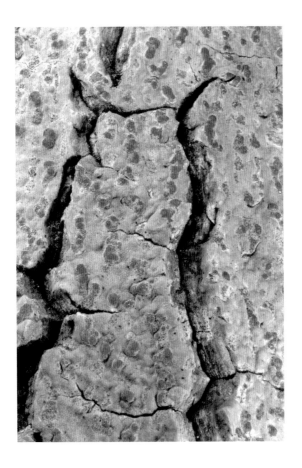

123. *Astrothelium versicolor* northern Florida ×2.9

shaped locules, turning blue with IKI in some species. CHEMISTRY: Thalli often containing lichexanthone (UV+ yellow); warts covered with yellow to orange anthraquinone pigments (K+ red-purple). HABITAT: On bark in tropical or subtropical forests. COMMENTS: Because some species of *Astrothelium* intergrade with species of *Trypethelium,* the boundaries of the two genera are still uncertain. In *Trypethelium,* the perithecia have only one chamber.

KEY TO SPECIES: See *Trypethelium.*

Astrothelium versicolor

DESCRIPTION: Thallus pale, thin, continuous, with clusters of cone-shaped perithecia in warts dusted with deep yellow to orange pruina; spores colorless, 4-celled, 24–30 × 9.5–10.4 µm, IKI–. CHEMISTRY: Thallus UV+ yellow (lichexanthone); pseudostromata K+ red (anthraquinones).

vagrant. COMMENTS: This is a fruticose species within the mainly crustose genus *Aspicilia*. Some lichenologists classify it within its own genus, *Agrestia,* and others regard it simply as a vagrant, fruticose form of *A. calcarea*. *Aspicilia fruticulosa,* shown in plate 122, is a rarer member of this group. It has white pseudocyphellae only on the branch tips, and the tips are blunt rather than pointed. Vagrant species of *Rhizoplaca* and *Xanthoparmelia* are yellow-green (containing usnic acid in the cortex) and usually have flattened lobes, although a form of *Rhizoplaca haydenii* can have terete, blunt-tipped branches.

Astrothelium (6 N. Am. species)
Speckled wart lichens

DESCRIPTION: Tropical crustose lichens, with thin, continuous thalli largely within bark tissue. Photobiont green (*Trentepohlia*); perithecia clustered and buried in yellow or orange warts (pseudostromata) as in *Trypethelium;* perithecia with several chambers but joining at the apex and opening to the surface by a single ostiole; spores colorless, narrowly ellipsoid to fusiform, with gelatinous halo, 4–10-celled with lens-

Bacidia (27 N. Am. species)
Dot lichens

DESCRIPTION: Crustose lichens with thick or thin thalli; soredia and true corticate isidia absent. Photobiont green (unicellular, including *Chlorella* and *Woessia*). Apothecia biatorine, pale pinkish to brown or black, often pruinose; margins prominent or disappearing; hypothecium clearly distinguishable from the exciple, which has rather thick, gelatinized, elongated cells; asci with a thick K/I+ blue tholus (*Bacidia*- or *Biatora*-type); spores fusiform to needle-shaped, straight or curved, 4–28-celled. CHEMISTRY: Many species contain atranorin or zeorin. Spot tests are unreliable, but the pigments in some apothecial tissues change color in K or nitric acid and are useful for distinguishing species. HABITAT: On bark, rocks, and mosses; less frequently on wood, usually in shaded forests. COMMENTS: Many genera are characterized by having biatorine apothecia and slender to fusiform, many-celled spores, including **Bacidina,** *Biatora, Mycobilimbia, Byssoloma, Scoliciosporum,* and *Ropalospora*. Some of these, although similar in many respects, are not even closely related to *Bacidia* if one compares the characteristics of the asci and apothecia. The distinguishing features of these smaller genera tend to be technical and are beyond the scope of this book. Some species of the most common genera are separated in the key. Species of **Bacidina** are most often small and inconspicuous, with mealy or sorediate thalli. The exciple of the apothecium has thin-walled, ellipsoid cells rather than thick-walled, cylindrical cells as in *Bacidia,* and the apothecia are paler. None of the species contain atranorin. Most species of *Bacidina* prefer humid coastal or subtropical habitats.

KEY TO SPECIES (including *Bacidina* and *Scoliciosporum*)

1. On rocks and occasionally wood, rarely tree bases ... 2
1. On bark or mosses, very rarely on rock or wood 4

2. In periodically submerged habitats such as streams and lake shores; epihymenium brown; thallus greenish, usually with a white prothallus. Spores needle-shaped, 4(–8) celled, 24–43 × 2–3 µm; apothecia black when wet, brown when dry; probably widespread temperate to boreal .... [*Bacidina inundata*]
2. In dry habitats; epihymenium greenish or brownish; thallus without a prothallus 3

3. Spores needle-shaped, spirally twisted, (15–)20–30(–40) × 2–3 µm, 4–8-celled. Apothecia red-brown to black; widespread temperate, especially in east .... [*Scoliciosporum umbrinum*]
3. Spores fusiform, not spirally twisted, 5–8 µm wide, 4–6-celled .... "*Bacidia*" *sabuletorum*

4. Apothecial disks black or almost black; epihymenium brown to green 5
4. Apothecial disks pale brown, or reddish brown to orange-brown; epihymenium yellowish or colorless ... 8

5. Hypothecium dark orange-brown to red-brown 6
5. Hypothecium colorless 7

6. On moss on tree bases, rarely directly on bark; apothecia strongly convex to hemispherical; spores fusiform, under 40 µm long, 4–6-celled ....."*Bacidia*" *sabuletorum*
6. On bark; apothecia usually flat; spores needle-shaped, mostly over 40 µm long, mostly 4–9-celled .... *Bacidia schweinitzii*

7. Spores rod- or club-shaped, typically straight, 4–8-celled; thallus gray-green to pale gray, areolate; apothecia usually remaining flat with a persistent margin; spores 11–37(–45) × 1.6–3.7 µm; Pan-Temperate at low elevations .... [*Bacidia circumspecta*]
7. Spores comet-shaped, usually curved and tapering, 5–8-celled; thallus dark green to brownish green, granular; apothecia convex and soon marginless; spores 3–5 µm wide .... *Scoliciosporum chlorococcum*

8.(4) Hypothecium, and exciple below the hypothecium, dark orange-brown or red-brown. Brown pigments in apothecial tissues K+ purple-red; apothecia commonly pruinose at least on margins when young; spores 4–12-celled, 31–74 × 2–5 µm; East Temperate .... [*Bacidia polychroa*]
8. Hypothecium, and exciple below the hypothecium, colorless, pale yellowish, or pale brown 9

9. Thallus consisting of large, round granules; apothecia orange-brown, sometimes pruinose on the margins when young; all apothecial tissues negative or colors intensifying with K .... *Bacidia rubella*
9. Thallus thin or thick, continuous or cracked, not consisting of large, round granules; apothecia pruinose or not; brown pigments in apothecial tissues K+ purple 10

10. Exciple with radiating clusters of crystals; apothecia yellow-brown to purplish brown, rarely red-brown, usually pruinose in part; outermost 4–8-cell layers of exciple very large and distinct from inner cells, which are much narrower; East Temperate .... [*Bacidia suffusa*]
10. Exciple normally without crystals; apothecia mainly orange-brown, not pruinose or pruinose; only outermost 1–2-cell layers of exciple have enlarged cells ... 11

11. Brown pigment of the epihymenium deposited as distinct caps on the tips of the paraphyses (seen best when the hymenium is squashed in K); spores 4–16-celled, 32–67(–73) × 2.5–4.5 µm; on deciduous trees and shrubs; southeastern coastal plain and along the Pacific coast .... [*Bacidia heterochroa*]
11. Brown pigment of the epihymenium distributed uniformly in the upper hymenial jelly; spores 8–29-celled, (50–)57–96(–108) × 2–3.7 µm; on conifers and deciduous trees; Great Lakes to New England, and Pacific Northwest .... [*Bacidia laurocerasi*]

Bacidia rubella (syn. *B. luteola*)
Frosty-rimmed dot lichen

DESCRIPTION: Thallus pale grayish green to dark green, consisting of crowded or scattered round granules. Apothecia pale to dark pinkish brown or yellow-brown, often lightly pruinose on the disks or margins, slightly to strongly convex, 0.6–1.2(–2.0) mm in diameter, with margins that are even with the disk; hypothecium yellowish to very pale orange-brown, usually darker than the exciple, which is colorless at the outer edge and yellowish within; spores straight or curved, needle-shaped, most 4–8-

124. *Bacidia rubella* herbarium specimen (New York Botanical Garden), New York ×6.6

125. *"Bacidia" sabuletorum* herbarium specimen (Canadian Museum of Nature), Ontario ×7.1

celled, (31–)44–63(–104) × 2–3(–4) µm. CHEMISTRY: Contains atranorin. Pigments in the apothecia only intensifying in color with K or nitric acid. HABITAT: In mature forests, on the bark of hardwoods. COMMENTS: The thallus color fades within a few years in a herbarium to a very pale brown. ***Bacidia polychroa*** is a widespread eastern species often mistaken for *B. rubella*. It has a smooth or cracked (not granular) thallus, its apothecia are also pruinose when young, sometimes retaining a pruinose margin, but the hypothecium is darker orange-brown. Most important, in *B. polychroa*, pigmented tissues of the apothecia turn purple-red in K (sometimes pale or faint) when viewed under the microscope. ***Bacidia diffracta*** is like *B. polychroa* in distribution, apothecial pigmentation, and spores, but it has a finely granular thallus more like *B. rubella*, although the granules in *B. rubella* are coarser. Some other common species of *Bacidia* are distinguished in the key.

"Bacidia" sabuletorum (syn. *Mycobilimbia sabuletorum*) *Six-celled moss-dot*

DESCRIPTION: Thallus granular, greenish to greenish gray. Apothecia pale brown (especially when wet) to very dark brown or black, hemispherical, 0.3–1 mm in diameter; margins thin, soon disappearing; epihymenium brown to greenish; hypothecium dark red-brown; spores 4–6-celled, rarely 8-celled, fusiform, 18–30(–40) × 5–8 µm; spore wall minutely warty in some spores (seen best under oil immersion, 1000×). CHEMISTRY: No lichen substances. HABITAT: On moss over rocks, especially calcareous rocks, or on mossy tree bases, and sometimes directly on bark. COMMENTS: A number of superficially similar crustose lichens growing over mossy tree bases have pale thalli and dark apothecia. Among them are three other common species: *Mycobilimbia tetramera* has darker brown or black, flatter apothecia, and has 1–4-celled spores; ***"Mycobilimbia" berengeriana*** has rather flat, red-brown apothecia, often with a persistent and smooth prominent margin, and elongate, ellipsoid, 1-celled spores; "*M.*" *hypnorum* is almost identical to *M. berengeriana* but the paraphyses have narrower tips and the spores are more often 2-celled. The warty outer spore wall of *Bacidia sabuletorum* is not seen in other species of either *Bacidia* or *Mycobilimbia* and, together with some other unusual characteristics, indicates that the species probably should be classified elsewhere.

Bacidia schweinitzii
Surprise lichen

DESCRIPTION: Thallus greenish, varying from granular and almost isidiate to thin and smooth. Apothecia large, pitch black, 0.6–2.0 mm in diameter, somewhat constricted at the base, with prominent rims; epihymenium green; hypothecium very dark reddish brown; exciple merging with the hypothecium internally but almost colorless on the outer edge; spores slender, slightly tapered, straight or curved, mostly 4–9-celled, rarely up to 16-celled, (30–)44–73(–88) × 1.5–3(–4) µm. CHEMISTRY: No lichen substances, or rarely with atranorin. HABITAT: On bark of all kinds, sometimes on moss, usually in shaded forests. COMMENTS: The spores and the surprisingly bright colors of the apothecial tissues in the otherwise pitch black apothecia make *B. schweinitzii* fairly distinctive, although its thallus development is extremely variable. *Lopadium disciforme* grows in similar habitats and also has large, black apothecia, but the thallus is darker, usually olive or brown, and the spores are muriform, 1 per ascus.

Baeomyces (3 N. Am. species)
Beret lichens

DESCRIPTION: Lichens with a crustose primary thallus and fruticose fertile stalks. Primary thallus areolate to squamulose or sometimes lobed and subfoliose, containing green algae (*Coccomyxa* or *Elliptochloris*). Apothecia large, flat to hemispherical, brownish, biatorine, 1–4 mm across, formed at the summits of short, solid, unbranched stalks (podetia) that are 2–4(–8) mm tall, purely fungal or developing thalline tissues (with a cortex and algal layer); spores colorless, fusiform or ellipsoid, 1–4-celled, 8 per ascus; immature specimens without any podetia are very common. CHEMISTRY:

126. *Bacidia schweinitzii*
Ouachitas, Arkansas
×2.9

Thallus PD+ yellow or orange, K+ red or yellow, KC–, C– (norstictic or stictic acid); mature apothecia KC+ red, C+ red (gyrophoric acid). HABITAT: On soil or rocks; less frequently over mosses, wood, or tree bases. COMMENTS: *Dibaeis baeomyces*, as the name suggests, closely resembles species of *Baeomyces*. It has pink rather than brown apothecia and has a white crustose thallus. It also differs chemically and in details of the ascus.

KEY TO SPECIES (including *Dibaeis*)
1. Apothecia pink; thallus almost white; medulla PD+ bright yellow, K– or K+ yellowish (baeomycesic and squamatic acids) . . . . *Dibaeis baeomyces*
1. Apothecia brown; thallus pale green to gray-green or brownish; medulla PD+ orange, K+ yellow (stictic acid) 2
2. Thallus thin, edge indefinite or definite, not lobed at the margin . . . . *Baeomyces rufus*
2. Thallus thick, distinctly lobed at the margin . . . . *Baeomyces placophyllus*

127. *Baeomyces placophyllus* interior Alaska ×2.9

Baeomyces placophyllus
Carpet beret lichen

DESCRIPTION: Thallus greenish white or pinkish gray, browned in parts, thick and crustose, usually with small, rounded, loosely attached squamules (schizidia) on the surface; thallus developing thick, 1–2-mm-wide, rounded lobes at the margins, almost foliose in appearance; podetia up to 8 mm tall, at least partially covered with a smooth, vegetative thallus, each podetium topped with a large, brown, contorted apothecium, 1–4 mm in diameter. CHEMISTRY: Cortex and medulla PD+ orange and K+ deep yellow (stictic acid, with traces of norstictic acid). HABITAT: On disturbed earth such as roadbanks, upturned roots, and old snowbeds, or on mosses. COMMENTS: The thick, lobed thallus is unlike that of any other species of *Baeomyces;* it somewhat resembles a *Squamarina* except for the color and chemistry.

Baeomyces rufus
Brown-beret lichen

DESCRIPTION: Thallus pale green to gray-green, varying from continuous to areolate, minutely verrucose, or clearly squamulose with somewhat overlapping squamules often with loosely attached, round schizidia; some populations becoming abundantly sorediate on the margins of the areoles and squamules, finally coalescing into a sorediate crust, but others virtually without soredia. Podetia short, 1–2(–4) mm high, unbranched, usually without thallus tissue but occasionally with some round thalline squamules; podetia bearing dull brown apothecia at the summits; apothecia solitary or clustered, flat to hemispherical, less than 2 mm in diameter. CHEMISTRY: Thallus PD+ orange, K+ deep yellow, KC–, C– (stictic acid complex and traces of norstictic acid). HABITAT: Often found on mineral soil or shaded rocks, less frequently on wood or the bark of roots. COMMENTS: This is the most common beret lichen. *Baeomyces placophyllus* is rarer and larger, producing a nonsorediate, distinctly lobed thallus. *Baeomyces carneus* resembles *B. rufus* more closely but is even rarer and mainly arctic-alpine. It has smaller squamules and produces abundant norstictic acid with no stictic acid (PD+ yellow, K+ red). *Dibaeis baeomyces* has a whiter, nonsorediate thallus that reacts PD+ yellow, K+ pale yellow (baeomycesic acid); its fruiting bodies are larger and clearly pink, and it grows on soil in sun rather than shade.

Bellemerea (5 N. Am. species)
Brown sunken disk lichens

DESCRIPTION: Gray to creamy-white crustose lichens on rocks. Photobiont green (unicellular). Apothecia brown, either sunken into areoles like an *Aspicilia* or, if there is only one apothecium per areole, appearing more like a *Lecanora;* epihymenium brown, negative with nitric acid; hymenium IKI+ blue; asci *Porpidia*-type; spores colorless, 1-celled, ellipsoid, with a diffuse gelatinous halo that can be seen only in an ink preparation (see Chapter 13), 8 per ascus. CHEMISTRY: Medulla IKI+ blue-black in all species; norstictic acid present in some. HABITAT: On rocks, usually in alpine or arctic sites. COMMENTS: *Bellemerea* resembles a brown-fruited *Aspicilia*, but the latter has black apothecia, its asci are K/I–, and its spores have no halo.

KEY TO GENUS: See Key F.

Bellemerea alpina

DESCRIPTION: Thallus creamy white to very pale brownish, areolate, with contiguous or dispersed areoles, often quite thick, with a well-developed black prothallus; apothecia dark brown to black when dry, brown when wet, not pruinose; embedded in areoles and level with surface, about 0.4–0.8 mm across; spores broadly ellipsoid, $10–20 \times 7–13$ µm. CHEMISTRY: Thallus PD+ deep yellow, K+ yellow turning red, KC–, C– (norstictic acid). COMMENTS: This is a frequent lichen in the western mountains, easily identified by its immersed apothecia like an *Aspicilia* but with reddish brown or pinkish rather than black disks. ***Bellemerea cinereorufescens,*** a western montane to arctic lichen, has an areolate, ashy gray thallus with a conspicuous black prothallus, and the apothecia are darker brown (but not black). It is PD–, K– (lacking norstictic acid). ***Aspicilia myrinii*** is another arctic-alpine species (known also from the Lake Superior region) with an IKI+ blue medulla (sometimes vague) and norstictic acid, but its

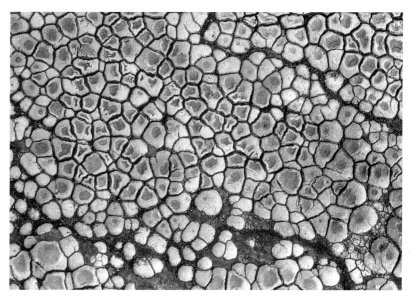

128. (top) *Baeomyces rufus* Cascades, Oregon ×6.1

129. (bottom) *Bellemerea alpina* Selkirks, British Columbia ×4.2

130. *Biatora vernalis* southern New Brunswick ×3.7

epihymenium is greenish or olive (yellowish green with nitric acid), and the spores have no halo. Its apothecia often become superficial with blackish margins. *Lecidea tessellata* has chemical reactions like *B. alpina*, but the apothecia are pitch black and some are prominent; microscopically, they are not at all alike.

Biatora (approximately 14 N. Am. species)
Dot lichens

DESCRIPTION: Crustose lichens with thin or thick, continuous or granular thalli, frequently sorediate. Photobiont green (*Trebouxia*?). Apothecia biatorine, light yellowish brown to dark red-brown without pruina, commonly convex to hemispherical, normally without distinct margins when mature; exciple composed of slender, thin-walled, branching hyphae; hypothecium gelatinized and very distinct, usually colorless or lightly brownish; epihymenium usually colorless or very pale; paraphyses usually unbranched, without expanded or pigmented tips; asci *Biatora*- or *Bacidia*-type; spores colorless, ellipsoid (often narrow), mostly 1–2-celled, sometimes 4-celled. CHEMISTRY: Can contain a variety of compounds including gyrophoric acid, atranorin, and β-orcinol depsidones, especially argopsin (PD+ red, K–). A few species have usnic acid or xanthones. HABITAT: On moss, dead vegetation, wood, and, rarely, bark. COMMENTS: Species of *Mycobilimbia, Bacidia, **Bacidina,*** and *Micarea* have a superficial similarity to *Biatora*. The first three genera differ in the structure and development of the exciple; *Micarea* species have branched paraphyses, a *Lecanora*- or *Cladonia*-type ascus, and a different photobiont (smaller, rather distinctive green algae). *Phyllopsora* is a closely related tropical genus with a thallus of finely divided squamules. *Biatora* is mainly a northern genus, rarely occurring south of 40°N latitude. Although we now know there are 14 species in the west, much taxonomic work remains to be done, especially with eastern taxa.

KEY TO SPECIES: See Key F.

Biatora vernalis

DESCRIPTION: Thallus continuous to wrinkled or granular, pale green or pale to dark greenish gray, without soredia or isidia. Apothecia 0.4–0.8(–1.2) mm in diameter, pale to dark yellowish brown or orange- to red-brown, usually strongly convex and lumpy without visible margins, but sometimes almost flat with persistent, shiny margins slightly paler than the disks; epihymenium, hypothecium, and exciple colorless or pale yellow; hymenium (45–)50–65(–95) μm high; spores mostly 12.5–19 × 4–6(–7) μm, 1- or sometimes 2-celled, rarely 4-celled. CHEMISTRY: Without lichen substances. HABITAT: On mosses over rocks and tree bases, rarely directly on bark, usually in shaded forests. COMMENTS: Most arctic tundra specimens resembling *B. vernalis* are ***B. subduplex,*** a common species with an almost white thallus and darker, orange-brown inner exciple and hypothecium. Specimens on bark are other species; their taxonomy in North America has not been worked out.

Brigantiaea (3 N. Am. species)
Brick-spored firedot lichens

DESCRIPTION: Thallus crustose, white to pale gray, thin or thick, usually continuous. Photobiont green (unicellular). Apothecia biatorine, with yellow, orange, or rusty brown disks and prominent margins; asci with a K/I+ blue tholus and rather thick, K/I– walls; spores colorless, muriform with numerous cells, very large (more than 50 μm long), 1(–2) per ascus. CHEMISTRY: Apothecia containing anthraquinone pigments (K+ deep red-purple); thallus PD–, K–, KC–, C–. HABITAT: On bark in the tropics, or on soil or decaying vegetation in the Arctic. COMMENTS: Although species of *Brigantiaea* resemble *Caloplaca* in general appearance, the muriform spores resemble those of *Letrouitia,* which has different asci and smaller spores, or *Lopadium,* which has black apothecia without any orange anthraquinone pigments.

KEY TO GENUS: See Key F.

131. *Brigantiaea leucoxantha* coastal plain, North Carolina ×3.0

Brigantiaea leucoxantha

DESCRIPTION: Thallus pale greenish gray, continuous, more or less smooth to verruculose. Apothecia mostly 0.5–1.4 mm in diameter; disks orange to olive, with a yellow or orange pruina; margins bright orange and prominent; spores huge, muriform with small cells, 1 spore per ascus. HABITAT: Common on bark in subtropical forests. COMMENTS: This lichen has the same general appearance as several subtropical species of *Letrouitia*. In the field, *Brigantiaea* can be distinguished by its paler, almost white thallus. See Comments under *Letrouitia parabola*. It is somewhat bizarre that this mainly tropical genus should include a species known only from soil and vegetation in alpine and arctic tundra. *Brigantiaea fuscolutea* is a rare but striking tundra species with a white, continuous or verruculose thallus following the contours of the substrate. Its apothecia are dark yellow because of a heavy pruina over the pinkish violet disks. They are 1–3 mm in diameter, round or very irregular in outline, with prominent margins dusted with yellow pruina. Its hymenium is reddish violet in thick section but is otherwise colorless, and the spores measure 50–115 × 24–55 μm.

Many species of *Caloplaca* share the same tundra habitat with *B. fuscolutea,* but they have smaller apothecia and polarilocular (2-celled) spores.

Brodoa (1 N. Am. species)

KEY TO SPECIES: See *Allantoparmelia*.

Brodoa oroarctica (syn. Hypogymnia oroarctica)
Mountain sausage lichen

DESCRIPTION: A foliose lichen, with long, narrow (0.3–2 mm), thick, convex lobes with a dense, solid medulla; lobes constricted at intervals, giving them the appearance of a string of sausages; upper surface dark steel-gray to dark brown, sometimes pale yellowish gray and shiny close to the tips, mottled with an in-

132. *Brodoa oroarctica* Rocky Mountains, Colorado ×4.4

tricate pattern of white spots (maculae); lacking pseudocyphellae, soredia, and isidia; lower surface black or dark gray, often dusted with a white pruina, without rhizines. Photobiont green (*Trebouxia*). Apothecia not common, lecanorine, laminal; spores colorless, 1-celled, ellipsoid, 8 per ascus. CHEMISTRY: Cortex K+ yellow (atranorin); medulla PD– (rarely PD+ red-orange), K–, KC+ red, C– (physodic acid, occasionally with fumarprotocetraric and protocetraric acids). HABITAT: On rocks or soil in alpine or arctic tundra. COMMENTS: *Brodoa* resembles a *Hypogymnia* in general habit, but the lobes that appear to be inflated are actually solid. *Hypogymnia pulverata,* the only North American species of the genus with a solid medulla, has soredia on the upper surface and is known from only a single locality in Oregon and another in boreal Quebec. Smaller specimens of *Brodoa* can resemble species of *Allantoparmelia*, which lack atranorin and have a different type of upper cortex.

Bryocaulon (3 N. Am. species)
Foxhair lichens

DESCRIPTION: Fruticose, pendent or shrubby to prostrate lichens with slender, terete branches (sometimes flattened or channeled in the axils), rather stiff, with the supporting tissue in the cortex (as in Fig. 8a); shiny red-brown, with white pseudocyphellae level with the surface of the branch or slightly raised; medulla very loose and webby. Photobiont green (*Trebouxia?*). Apothecia not common, lecanorine, scattered along the branches; spores colorless, ellipsoid, 1-celled, 8 per ascus. CHEMISTRY: Medulla contains orcinol depsides and depsidones such as olivetoric and physodic acids (KC+ red and often C+ red). HABITAT: On trees or on the ground. COMMENTS: These lichens closely resemble species of *Bryoria* or *Nodobryoria* but differ in the structure of the cortex (two-layered in *Bryocaulon*) and in their chemistry. In the field, they stand out because of their striking red-brown color and raised pseudocyphellae.

KEY TO SPECIES: See *Alectoria.*

Bryocaulon divergens (syn. *Cornicularia divergens*)
Heath foxhair lichen, northern foxhair

DESCRIPTION: Thallus of bushy to sprawling, shiny, red-brown branches mostly 0.3–0.5 mm in diameter, even and round in section or with some depressions. CHEMISTRY: Pseudocyphellae and medulla PD–, K–, KC+ red, C+ pink or red (olivetoric acid), but the reactions, especially on the medulla, are sometimes faint or spotty. HABITAT: On tundra soil and vegetation, often mixed with species of *Bryoria* and *Cetraria*. COMMENTS: *Bryocaulon divergens* resembles *Cetraria aculeata* but is redder and has raised rather than depressed pseudocyphellae. *Bryoria nitidula*, a frequent neighbor of *Bryocaulon divergens*, has similar branching but has dark brown pseudocyphellae, is PD+ red, and is never as reddish. *Bryocaulon pseudosatoanum*, an arboreal member of the genus, is much longer and pendent, resembling a very reddish and shiny *Bryoria* except for the raised pseudocyphellae and C+ red medulla. It often grows on pines in muskeg along the west coast from Vancouver Island to southeastern Alaska.

133. *Bryocaulon divergens* interior Alaska ×3.0

Bryoria (24 N. Am. species)
Horsehair lichens, tree-hair lichens, bear hair

DESCRIPTION: Slender, hair-like, pendent or shrubby, fruticose lichens, dark brown to pale grayish brown or even gray in some species; medulla rather loose; cortex, which constitutes the supporting tissue, composed of a uniform prosoplectenchyma running longitudinally; pseudocyphellae, soredia, and isidia present or absent. Photobiont green (*Trebouxia*?). Apothecia uncommon in most species, lecanorine, with small, colorless, 1-celled, ellipsoid spores, 8 per ascus. CHEMISTRY: Medulla and sometimes the cortex often containing β-orcinol depsidones, especially fumarprotocetraric acid (accompanied by traces of protocetraric acid) or norstictic acid; some species with atranorin, alectorialic and barbatolic acids, or other compounds. To observe some of the critical chemical reactions with PD and K, one must use the filter paper method described in Chapter 13. HABITAT: On trees, almost exclusively conifers, or on tundra soil. COMMENTS: Species of *Bryoria* can be confused with dark brown or black species of *Bryocaulon, Nodobryoria, Pseudephebe, Alectoria, Sulcaria,* or *Cetraria*. IMPORTANCE: The importance of *Bryoria* for wildlife is second only to that of reindeer lichens (*Cladina*). Its uses as food and fiber for humans are mentioned below under *B. fremontii, B. pseudofuscescens,* and *B. tortuosa*.

KEY TO SPECIES: See *Alectoria*.

Bryoria capillaris
Gray horsehair lichen

DESCRIPTION: Thallus usually pale but extremely variable in color, ranging from very pale gray to dark, smoky brown; branches up to 30 cm long, slender, mostly 0.1–0.3 mm thick, rather brittle; pseudocyphellae pale, long and narrow, 0.1–0.25 mm long; soredia absent in North American populations. CHEMISTRY: Cortex and medulla PD+ deep yellow, K+ yellow, KC+ red, C+ pink (alectorialic and barbatolic acids). HABITAT: On forest trees, often well shaded, in the more humid parts of the continent. COMMENTS: *Bryoria capillaris* is among the palest species of the genus. In the west are the much rarer *B. pikei* (shiny and more regularly branched, K– with only alectorialic acid) and *B. pseudocapillaris* (with pseudocyphellae up to 3 mm long, twisting around the branches). The bushier, sorediate, *B. nadvornikiana* is a very similar lichen of eastern forests. (In Europe, *B. capillaris* can also be sorediate; in North America, it never is.) *Bryoria nadvornikiana* and *B. pseudocapillaris* also differ from *B. capillaris* in their short, spine-

134. *Bryoria capillaris* Lake Superior region, Ontario ×1.0

135. *Bryoria fremontii* mountains, central Washington ×2.1

like, perpendicular side branches. See also Comments under *B. pseudofuscescens*.

Bryoria fremontii
Tree-hair lichen, black tree-lichen, edible horsehair

DESCRIPTION: Thallus with very thick, twisted main branches, with dents and ridges and short, slender, perpendicular side branches; thallus commonly 10–30 cm long and sometimes longer; main branches 0.4–1.5(–4) mm thick; color varying from shiny reddish brown (almost like species of *Bryocaulon* or *Nodobryoria*) to very yellowish brown, normally without pseudocyphellae but sometimes having bright yellow, wart-like soralia. Apothecia occasionally present, with bright yellow disks (actually, brown with yellow pruina). CHEMISTRY: Cortex and medulla PD–, K–, KC–, C– (vulpinic acid in the yellow parts). HABITAT: On conifers such as pines, larch, and Douglas fir in open, rather dry stands. COMMENTS: *Bryoria fremontii* is one of the largest species of horsehair lichens. The heavy, twisted main branches with perpendicular side branches are the most reliable characters for field identification. *Bryoria tortuosa*, similar in size and habit, often has a yellowish tint and always has long yellow pseudocyphellae that twist around the branches; it rarely has soralia. Its tissues are PD– (rarely PD+ pale yellow). Most other *Bryoria* species are PD+ red, at least in the soralia. **Bryocaulon pseudosatoanum,** described in the Comments under *Bryocaulon divergens*, should also be considered. IMPORTANCE: *Bryoria fremontii* often grows in dense, abundant clumps, and some native groups developed tasty recipes for it. For example, the Tsimshian, Blackfoot, and Salish peoples bury layers of the lichen with leaves in earth pits and steam it, eating it with onions, fish eggs, or berries (plate 65). It was used by other tribes as a fiber for clothing or shoes. Some Interior Salish people mixed *B. fremontii* with grease and rubbed it on the navel of newborn babies, presumably to prevent infection, and the Nez Perce tribe used it to treat indigestion and diarrhea. According to a colorful Okanagan myth, this lichen owes its origin to Coyote, whose "hair braid" became tangled in a tree. After cutting it loose, he declared, "You shall not be wasted, my valuable hair. After this you shall be gathered by the people. The old women will make you into food."

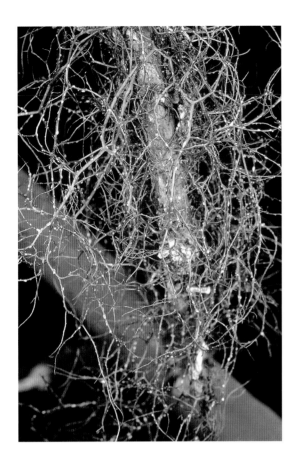

136. *Bryoria furcellata* Oregon coast ×2.3

137. *Bryoria fuscescens* Rocky Mountains, Colorado ×1.4

Bryoria furcellata
Burred horsehair lichen

DESCRIPTION: Thallus forming bushy clumps up to 5 cm across (rarely more) with divergent, short, pointy side branches, giving the lichen a thorny look; branches dark brown, generally shiny, about 0.3–0.5 mm in diameter, with fissural soralia containing farinose soredia as well as short, spiny isidia; pseudocyphellae absent, but young, immature soralia can sometimes resemble them. Apothecia rare. CHEMISTRY: Cortex and medulla PD+ red, K–, KC–, C– (fumarprotocetraric acid). HABITAT: On conifers of all kinds in open, well-lit woodlands. COMMENTS: This is one of the easiest species of *Bryoria* to identify. It is the only one with isidia growing out of broad, fissure-shaped soralia. *Bryoria simplicior* is somewhat similar in general appearance but is much smaller and more strictly northern boreal, and it has no isidia growing out of the soralia, which are greenish black when fresh; its soralia and medulla are PD–.

Bryoria fuscescens
Pale-footed horsehair lichen

DESCRIPTION: Thallus pale to very dark brown (not reddish), sometimes dark olive, the base of the clump almost always paler than the rest of the thallus; mostly 5–15 cm long, with main branches (0.2–)0.3–0.4(–0.6) mm thick, terete or flattened at the axils, occasionally with spiny side branches; producing white, fissural or tuberculate soralia (sometimes sparse); pseudocyphellae normally absent, although an unnamed population on the east coast has white, elliptical pseudocyphellae. CHEMISTRY: Cortex and medulla usually PD– in North America but weakly to strongly PD+ red in some populations; soralia PD+ red; all parts of thallus K–, KC–, C– (fumarprotocetraric acid). HABITAT: A forest species, on spruce in the boreal forest, and pine and Douglas fir in the west. COMMENTS: This horsehair lichen is central to a group of similar, sorediate species with soralia that react PD+ red, K–, KC–, C– (fumarprotocetraric acid).

138. *Bryoria glabra* Olympics, Washington ×2.8

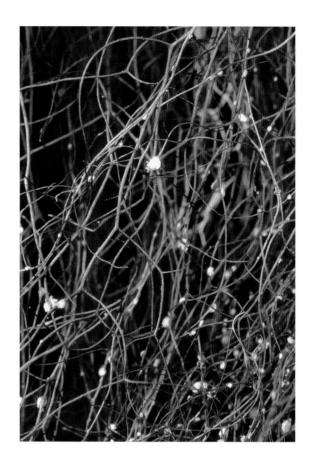

139. *Bryoria lanestris* Coast Mountains, British Columbia ×4.2

Bryoria glabra is smoother and more uniformly olive, with regular dichotomous branching especially close to the base. *Bryoria lanestris* is a northern boreal species that has more slender, blacker branches with soralia that are usually flecked with black. *Bryoria subcana* is a very rare species, pale brown to almost cream-colored, in which the entire thallus, including the outer cortex, reacts strongly PD+ red. It is confined to the west coast in North America (more common in Europe). Very spiny, dark brown to almost olive-black forms of *B. fuscescens* that occur in exposed montane forest habitats in the west from Arizona to Alberta may constitute a different species.

Bryoria glabra
Shiny horsehair lichen

DESCRIPTION: Thallus olive-brown to greenish, shiny, forming pendent clumps 10–15 cm long; branches 0.2–0.4 mm thick, uniform in thickness, branching in wide, equal dichotomies, not becoming twisted or flattened at the axils, with abundant or sparse, purely white, oval, fissural soralia; without spiny side branches or pseudocyphellae. CHEMISTRY: Cortex and medulla PD–, K–, KC–, C–; soralia PD+ red, K– (fumarprotocetraric acid). HABITAT: On conifers of all kinds from sea level to the subalpine, in the open. COMMENTS: This species almost intergrades with *B. fuscescens*, which has more uneven, narrower dichotomies, often twists and becomes flattened at the axils, and frequently has at least a few tuberculate soralia. *Bryoria lanestris* is more slender and uneven, and the soralia are often black-spotted.

Bryoria lanestris
Brittle horsehair lichen

DESCRIPTION: Thallus forming woolly, tangled clumps 5–10 cm long; branches very dark and slender, mostly 0.1–0.25 mm thick, very uneven in diameter and irregular in branching, brittle; abundantly sorediate, with fissural soralia producing little piles of white soredia flecked with black. CHEMISTRY: Cortex PD–, K–, KC–, C–; soralia PD+ red, K–, KC–, C– (fumarprotocetraric acid). HABITAT AND COMMENTS: This species is very common on

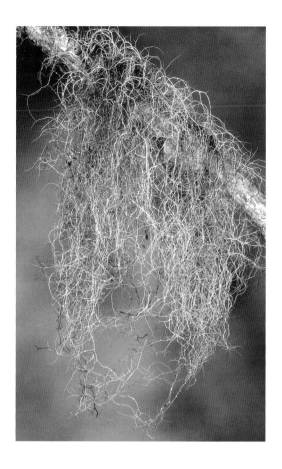
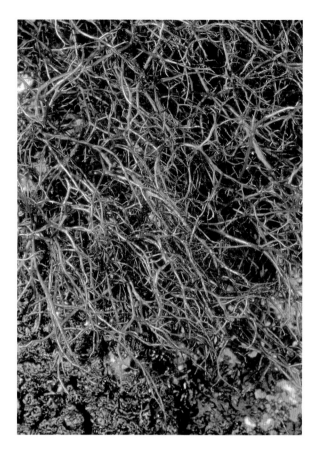

140. *Bryoria nadvornikiana* Lake Superior region, Ontario ×0.74

141. *Bryoria nitidula* interior Alaska ×3.0

branches and twigs of spruce, pine, and larch in boreal forests, where it is usually mixed with **B. simplicior,** a shinier, more divergently branched species with greenish, PD– soralia. See also *B. fuscescens* and *B. glabra.*

Bryoria nadvornikiana
Spiny gray horsehair lichen

DESCRIPTION: Thallus pale gray to grayish brown, rarely dark brown in very sunny habitats, forming shrubby to almost pendent tufts 4–9 cm long, with main branches (0.1–)0.2–0.3(–0.4) mm thick that branch in a fairly regular dichotomous pattern but with many short, pointed, perpendicular side branches; soralia usually common, with piles of white or greenish white soredia, pseudocyphellae white, elliptical to linear, up to 0.5 mm long, sometimes inconspicuous. CHEMISTRY: Cortex, medulla, and soralia PD+ yellow, K+ yellow, KC+ red, C+ pink, or C– (barbatolic acid with small amounts of alectorialic and fumarprotocetraric acids and atranorin). HABITAT: In deeply shaded or open boreal woodlands on conifers and birch; also on rock faces and cliffs, especially in humid sites near waterfalls or lakes. COMMENTS: Typical specimens of this horsehair lichen are easy to name, with their spiny, perpendicular branches, pale grayish brown to mottled surface, and frequent soralia. The closely related *B. capillaris* is more pendent, lacks soralia in North America, and only rarely produces spiny side branches. Unfortunately, one can often find populations of *B. nadvornikiana* that are brown, even dark brown, and sometimes soralia are hard to find. Chemical tests using the filter paper method separate such specimens from *B. fuscescens* or *B. trichodes.*

Bryoria nitidula
Tundra horsehair lichen

DESCRIPTION: Thallus erect and shrubby, 5–8 cm tall, or spreading horizontally over the ground; branches shiny dark brown to almost black, rather thick, about 0.5–0.8 mm in diameter in erect forms but more slender when growing prostrate; lacking soredia or isidia but with dark, elongate pseudocyphellae (often hard to see because of their color).

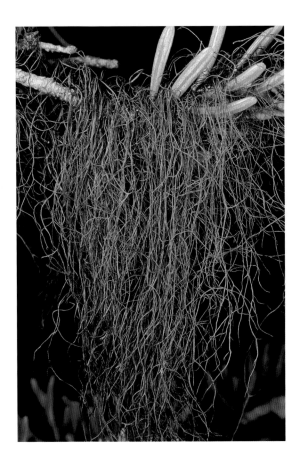

142. *Bryoria pseudofuscescens* Wallowa Mountains, eastern Oregon ×2.2

mm in diameter (reaching 0.8 mm in diameter with contorted, pitted main branches in exceptionally robust specimens); without soredia, but pseudocyphellae usually abundant, white, fusiform to elongate, often spiraling partly around the branches. Apothecia very rare. CHEMISTRY: Cortex and medulla PD+ yellow, K+ yellow changing to dull reddish brown with the filter paper method, KC–, C– (norstictic acid). HABITAT: On conifers, especially subalpine fir and whitebark pine, in montane to subalpine forests, sometimes completely covering the trees with black hair. COMMENTS: Chemistry is the only reliable way to separate several similar and closely related lichens from *B. pseudofuscescens;* in fact, some lichenologists regard all of them as chemical races of a single species, ***Bryoria implexa.*** In the strict sense, *B. implexa* is a rather rare boreal forest species containing psoromic acid (PD+ bright yellow, K–, KC–, C–); *B. friabilis,* a very brittle, often olive-colored, western species has gyrophoric acid (PD–, K–, KC+ pink, C+ pink, but hard to detect); *B. salazinica,* a northeastern endemic, produces salazinic acid (PD+ yellow, K+ red, KC–, C–). ***Bryoria spiralifera*** contains norstictic acid but produces atranorin as well. It is a west coast species with pale, reddish brown branches, very long, spiraled pseudocyphellae (up to 4 mm long), and many spiny, perpendicular side branches. European specimens of *B. pseudofuscescens* are often sorediate, but the species never has soredia in North America. IMPORTANCE: The northern flying squirrel uses *B. pseudofuscescens* and other horsehair lichens as food and as nesting material. Because of its abundance in montane forests, it is probably among the most important arboreal lichens used as a winter food by such ungulates as elk and mule deer.

CHEMISTRY: Cortex PD– or weakly + red, K–, KC–, C–; medulla PD+ deep red, K–, KC–, C– (fumarprotocetraric acid). HABITAT: On tundra vegetation, soil, and rock. COMMENTS: *Alectoria nigricans,* which often grows with *B. nitidula* in the arctic or alpine heath, can also be very dark, almost black, but has raised white pseudocyphellae and generally a paler gray base. Even the blackest thalli of *A. nigricans* give a PD+ yellow reaction with the filter paper method (from the alectorialic acid). *Bryocaulon divergens* is a similar shiny brown (a bit redder) but has slightly raised, white pseudocyphellae that react C+ red; the medulla and cortex are PD–. *Cetraria aculeata* and *C. muricata* are more minutely spiny with depressed white pseudocyphellae, and all reactions are negative.

Bryoria pseudofuscescens
Mountain horsehair lichen

DESCRIPTION: Thallus pendent, mostly 5–10 cm long, very pale grayish brown to almost black, with slender, usually uneven, often twisted branches, 0.15–0.35

Bryoria tortuosa
Yellow-twist horsehair lichen, inedible horsehair

DESCRIPTION: Thallus pendent, up to 30 cm long, dull red-brown to dusky yellow-brown, or occasionally bright greenish yellow because of heavy concentrations of the yellow pigment, vulpinic acid; main branches 0.4–1 mm in diameter, twisted and pitted, with more slender, perpendicular side branches; soralia very rare, bright yellow; long yellow pseudocyphellae usually abundant, typically twisting around the branches in long spirals. Apothecia rare, with yellow pruinose

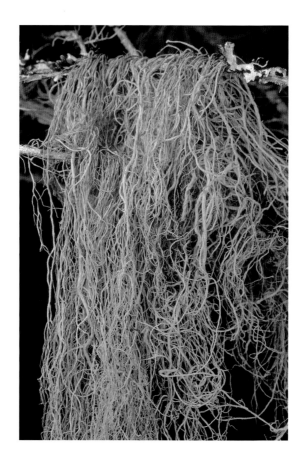

143. *Bryoria tortuosa* northern Idaho ×4.2

144. *Bryoria tortuosa* (yellow form) Washington coast ×2.2

disks. CHEMISTRY: Cortex and medulla PD– (or pale yellow with the filter paper method), K–, KC–, C– (vulpinic acid in yellow parts). HABITAT: On oaks and pines in well-lit forest stands. COMMENTS: This species closely resembles *B. fremontii* morphologically and chemically and has about the same distribution. It has yellow pseudocyphellae, whereas *B. fremontii* has none at all. IMPORTANCE: Distinguishing *B. tortuosa* from *B. fremontii* is not a purely academic exercise: *B. fremontii* is often used as food, and *B. tortuosa* contains relatively high concentrations of vulpinic acid, one of the few poisonous lichen compounds.

Bryoria trichodes
Horsehair lichen

DESCRIPTION: Thallus pendent, mostly 7–15 cm long, pale to dark brown, dull or shiny; branches 0.1–0.4 mm in diameter, even and smooth in ssp. *americana* to uneven in ssp. *trichodes*; pseudocyphellae short, oval, and often raised in ssp. *trichodes*, grading to slender and slightly depressed in ssp. *americana*; soralia very rare except in the extreme northeast where populations have occasional distorted, fissural soralia bending the branches back on themselves. Apothecia frequent, especially in ssp. *trichodes*, scattered along the branches, with pale reddish brown disks. CHEMISTRY: Outer cortex PD–, K–, KC–, C–; medulla and inner layers of cortex PD+ red, K–, KC–, C– (fumarprotocetraric acid, often with atranorin, especially in ssp. *trichodes*). HABITAT: On conifers and birches in forests, in open bogs, and along lake shores; a form of ssp. *americana* can grow on rocks and soil in heathland communities in the east. COMMENTS: Plate 145 shows typical ssp. *americana*. Its distribution is shown in pink on the map. *Bryoria trichodes* ssp. *trichodes* is more abundant in the Great Lakes and Appalachian regions; its range is shown in blue. (Where the ranges overlap, the color is purple.) A spiny, short form of *B. trichodes* is found in open bogs and muskeg, but it intergrades with normal pendent forms in nearby forests. The sorediate form can be mistaken for *B. fuscescens*, which has soralia that never bend the branches in the same way and does not have pseudocyphellae.

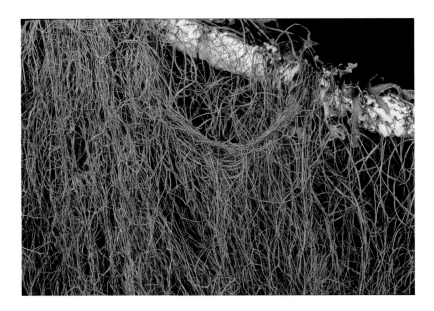

145. *Bryoria trichodes* ssp. *americana* Southeast Alaska ×2.0

Buellia (72 N. Am. species)
Button lichens

DESCRIPTION: Crustose lichens with thin to thick, continuous to rimose or areolate thalli, a few species lobed at the margins; thalli white, gray, brown, or yellowish; prothallus present or absent; some species sorediate. Photobiont green (unicellular). Apothecia lecideine, black (although sometimes pruinose), superficial or immersed in or between thallus areoles, with margins typically black, prominent, and persistent; epihymenium brown to greenish; hypothecium usually brown; exciple usually black to dark brown, at least at the outer edge; hymenium clear or filled with tiny oil droplets (giving it a granular appearance) in some species; paraphyses usually unbranched; asci *Lecanora*-type; spores brown, 2- to rarely 4-celled, often constricted at the septa, with more or less evenly thickened walls (rarely thickened at the septum or tips), (4–)8–16 per ascus. CHEMISTRY: May contain any of a variety of compounds including norstictic acid, atranorin, and xanthones. HABITAT: On bark, wood, rock, or soil; some species parasitic on other lichens. COMMENTS: *Amandinea* is a closely related genus that is sometimes included within *Buellia*. It has very long, thread-like, curved conidia. One example is the very common and widespread *A. punctata*, with small, black, lecideine apothecia, and spores (7–)11–16 × (4–)5–8 μm. Species of *Buellia* with 4-celled to muriform spores, usually with a rather thick and sometimes lobed thallus, are sometimes placed in a separate genus: *Diplotomma*. Except for *Buellia alboatra*, most grow on rock or soil. Species of *Rhizocarpon* have spores with a gelatinous halo, the asci are largely K/I– without a thickened tholus, and the paraphyses are branched.

KEY TO SPECIES (including *Amandinea*, *Catolechia*, and *Diploicia*)

1. Spores 4-celled to muriform 2
1. Spores 2-celled 3

2. Spores muriform, (17–)21–34 × 10–17(–21) μm; hymenium filled with tiny oil drops; on wood and bark along the west coast . . . . [*B. penichra*]
2. Spores 4-celled to few-celled muriform, (11–)15–20(–30) × (5.5–)8–10(–17) μm; hymenium without oil drops; California to northeastern U.S. . . . . [*B. alboatra*]

3. Thallus yellow or with a yellow tint 4
3. Thallus pale gray to greenish gray; cortex C–, KC– . . . 5

4. Thallus greenish yellow; cortex C+ orange, KC+ orange (xanthones); on maritime rocks . . . . *Buellia halonia*
4. Thallus bright lemon-yellow; cortex C–, KC– (containing pulvinic acid-type pigments); on soil and peat. Spores (12–)13–17(–18) × 7–10 μm; rare but widespread arctic-alpine . . . . [*Catolechia wahlenbergii*]

5. Spores with angular locules (*Physcia*-type); thallus thick, distinctly lobed, pruinose and sorediate; California, most commonly coastal, on bark, wood, or rock . . . . *Diploicia canescens*
5. Spores with thin, uniformly thickened walls; thallus thin, not lobed, without pruina or soredia 6

6. Mature apothecia entirely, or almost entirely, immersed in the thallus, the disk level with the thallus surface or slightly depressed; thallus thick, rimose-areolate, with a black prothallus; on rock, southern to eastern U.S. . . . . *B. spuria*
6. Mature apothecia superficial, not sunken into thallus; thallus thick or thin and disappearing, lacking a black prothallus; on bark, wood, or rock 7

7. Thallus composed of scattered or contiguous areoles, sometimes white pruinose; on siliceous rock in California and the southern Rockies . . . . [*B. retrovertens*]
7. Thallus thin, continuous or barely visible, not areolate or pruinose; usually on bark or wood, occasionally on rock 8

8. Thallus almost imperceptible or very thin and dirty greenish gray to pale gray; apothecia less than 0.5 mm in diameter; very common and widespread on wood, but also sometimes on bark or rock . . . . *Amandinea punctata*
8. Thallus thin but clearly visible, pale gray; apothecia 0.4–1 mm in diameter; in temperate to boreal forests, on bark 9

9. Spores mostly 12–17 × 5–8 µm; hymenium yellowish, lacking oil drops; thallus K+ dark yellow to blood red (norstictic acid sparse or abundant) .... *Buellia stillingiana*

9. Spores mostly 18–26(–30) × 6–11 µm; hymenium colorless, but containing abundant oil drops; thallus K+ yellow or K– (although always containing atranorin) .... *Buellia disciformis*

Buellia disciformis
Boreal button lichen

DESCRIPTION: Thallus smooth to rimose-areolate, pale gray. Apothecia black, up to 1 mm in diameter, flat, with prominent margins; hymenium very high, 70–100 µm, filled with oil drops; exciple uniformly black; spores (16–)18–26(–30) × 6–11 µm, ellipsoid or slightly bean-shaped, 8 per ascus. CHEMISTRY: Thallus and apothecial tissues PD–, K–, KC–, C–, but usually contain atranorin. HABITAT: On bark, and occasionally on wood, of deciduous and coniferous trees; most often in forests. COMMENTS: The superficially similar *B. stillingiana* has a lower, yellowish hymenium, and the thallus contains norstictic acid (K+ yellow to red). *Buellia erubescens,* a western species, is distinguished by the lack of oil in the hymenium, and by being found typically on wood rather than bark.

Buellia halonia
Seaside button lichen

DESCRIPTION: Thallus greenish yellow, rimose-areolate or areolate but continuous, in round patches, usually bounded by a thin or broad black prothallus. Apothecia partially immersed at first, but soon superficial, 0.4–1.2 mm in diameter, black with a somewhat bluish pruina, slightly convex; epihymenium brownish olive to green; spores 11–17 × 6–10 µm, dark green to brown, usually with a thickened septum in a water mount, making the spore essentially polarilocular (thickening disappears when spores are mounted in K), 8 per ascus. CHEMISTRY: Cortex PD–, K+ yellow, KC+ red-orange; C+ orange (xanthones); medulla IKI–. HABITAT: On coastal rocks in the open. COMMENTS: In the field, *B. halonia* might be mistaken for one of the yellow, xanthone-containing species of *Lecidella* such as **L. asema,** which sometimes grows in the same habitat. The latter, however, has a more dis-

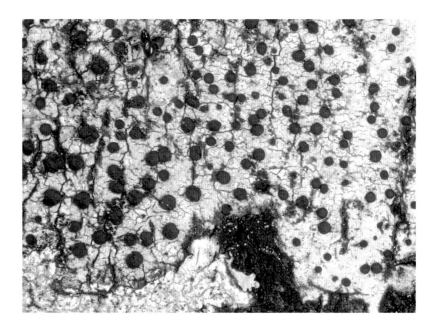

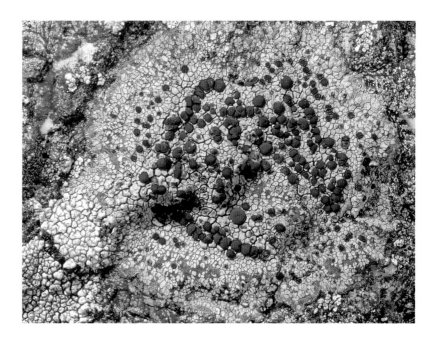

146. *Buellia disciformis*   coastal Alaska ×5.0

147. *Buellia halonia* California coast ×2.0

persed granular appearance with small, convex areoles, and it lacks a black prothallus. Under the microscope, the 1-celled, colorless spores of the *Lecidella* immediately set it apart. No other *Buellia* has the yellowish thallus and green epihymenium of *B. halonia*.

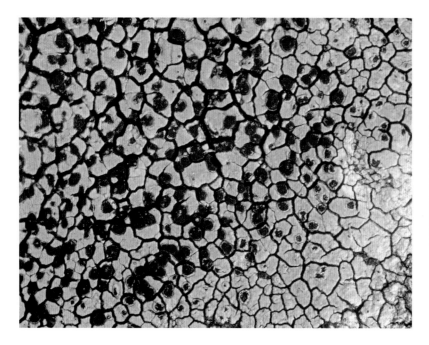

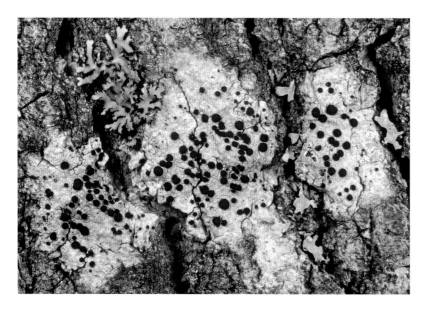

148. *Buellia spuria* herbarium specimen (New York Botanical Garden), Kentucky ×7.5

149. *Buellia stillingiana* Lake Superior region, Ontario ×3.1

Buellia spuria
Sunken button lichen

DESCRIPTION: Thallus gray to yellowish gray, composed of dispersed or contiguous areoles; prothallus black and well-developed. Apothecia at first immersed between the areoles or touching an edge, rarely entirely surrounded by areole tissue, sometimes becoming slightly raised, 0.4–0.7 mm in diameter, with or without thin black margins, round to angular because of pressure from the areoles; epihymenium, hypothecium, and excipl dark brown, although the epihymenium is reported to be greenish; spores ellipsoid, 9–17 × 4–8 µm, 8 per ascus. CHEMISTRY: Medulla PD+ yellow-orange, K+ dark yellow, KC–, C–, IKI+ blue (atranorin and stictic acid complex, with or without a trace of norstictic acid). HABITAT: On siliceous rocks, especially sandstone. COMMENTS: Another common *Buellia* in southeastern U.S., *B. stigmaea*, is very much like *B. spuria* in appearance and also has an IKI+ blue medulla, but it contains abundant norstictic acid without stictic acid (medulla PD+ yellow, K+ red). In the Interior Highlands of Missouri and Arkansas, *B. spuria* is commonly found growing with a superficially similar lichen, *Lecanora oreinoides*, which has small black apothecia (0.25–0.4 mm in diameter) buried within the centers of the areoles rather than at the edges or between them. Aside from the spore and apothecial differences seen under the microscope, the two species differ chemically: in *L. oreinoides*, the medulla is PD–, K–, IKI– (atranorin and confluentic acid).

Buellia stillingiana
Common button lichen

DESCRIPTION: Thallus gray, thin, and continuous. Apothecia flat, pitch black, with prominent black margins, mostly 0.4-0.8 mm in diameter; excipl brownish black to gray-olive at the edge but paler internally; hymenium fairly thin (50–85 µm), without oil drops, yellowish in all but the thinnest sections;

spores mostly 12–17 × 5–8 μm with evenly thickened walls, 8 per ascus. HABITAT: Very common on deciduous trees, but also on conifers in the north. CHEMISTRY: Thallus and apothecial sections produce a deep yellow stain with K, generally developing clusters of red, needle-shaped crystals, sometimes slowly (see Chapter 13); PD+ yellow or PD–, KC–, C– (norstictic and connorstictic acids, atranorin). Northern and western populations produce less norstictic acid and often fail to form red crystals with K, and in some southeastern specimens only atranorin is produced. COMMENTS: *Buellia curtisii* is a superficially identical lichen also common in the eastern U.S., but it has more pointed spores over 17 μm long, the hymenium is colorless, and it usually contains more norstictic acid (quickly producing red crystals in K). *Buellia disciformis* has oil in the hymenium and lacks norstictic acid.

Bulbothrix (5 N. Am. species)
Eyelash lichens

DESCRIPTION: Small, gray, closely adnate foliose lichens with short, stiff, marginal cilia, each with a bulb-like expansion at the base (Fig. 10c); lower surface black or brown, with unbranched or forked rhizines. Photobiont green (*Trebouxia?*). Apothecia lecanorine, under 4 mm in diameter, with brown disks; margins sometimes adorned with black cilia and, in many species, are spotted with black pycnidia; spores colorless, 1-celled, broadly ellipsoid, 8 per ascus; black pycnidia embedded in the thallus or apothecial rims, or emergent; conidia cylindrical, about 6 μm long. CHEMISTRY: Cortex K+ yellow (atranorin); medulla often containing gyrophoric or lecanoric acids (KC+ red, C+ red). One North American species (*B. isidiza*) contains salazinic acid (PD+ yellow, K+ red, KC–, C–) and has abundant isidia and a pale brown lower surface. HABITAT: On bark of trees and shrubs, most commonly in open, tropical to subtropical woodlands. COMMENTS: Species of *Bulbothrix* are often overlooked because of their small size and close attachment to the substrate. The small bulbous cilia are not always easy to find or to recognize; look for them on the margins and in the rounded axils of the lobes. *Relicina* is a very similar genus of the mainly old-world tropics, differing in its yellow-green color caused by usnic acid in the cortex instead of atranorin, and in having dumbbell-shaped conidia 6–10 μm long. In North America, it is represented by two rare species: *R. abstrusa,* from northern Florida and Georgia, with isidia, containing norstictic acid (PD+ yellow, K+ red); and *R. eximbricata,* found in the Everglades and Florida Keys, with smooth lobes, containing fumarprotocetraric acid (PD+ red, K– or brownish).

KEY TO SPECIES (including *Relicina*)

1. Thallus yellowish green ("usnic yellow") . . . . [*Relicina*; see Comments under *Bulbothrix*]
1. Thallus gray 2

2. Isidia abundant on the thallus surface 3
2. Isidia absent; apothecia common . . . . *Bulbothrix confoederata*

3. Lower surface and rhizines black; apothecia abundant; medulla C+ red (lecanoric acid present as main compound) . . . . *Bulbothrix laevigatula*
3. Lower surface and rhizines pale beige to brown; apothecia rare; medulla C+ pink or C–, K+ red or K– 4

4. Medulla C+ pink, K– (gyrophoric acid present as main compound); common throughout southeastern coastal plain . . . . *Bulbothrix goebelii*
4. Medulla C–, K+ red (salazinic acid); common only in Florida . . . . [*Bulbothrix isidiza*]

Bulbothrix confoederata
Smooth eyelash lichen

DESCRIPTION: Thallus with dichotomously branched, short lobes, 0.3–0.8 mm wide, without soredia or isidia; lower surface very dark brown to black, with black, forked rhizines. Apothecia abundant, brown, constricted at base, up to 1.5 mm in diameter, without cilia on the rims. CHEMISTRY: Medulla PD–, K–, KC+ red, C+ red (lecanoric acid). HABITAT: On bark. COMMENTS: This is the most common species of *Bulbothrix* without isidia. *Bulbothrix coronata* is restricted to western Texas. Its medulla contains gyrophoric acid and is C+ pink rather than red as in *B. confoederata*.

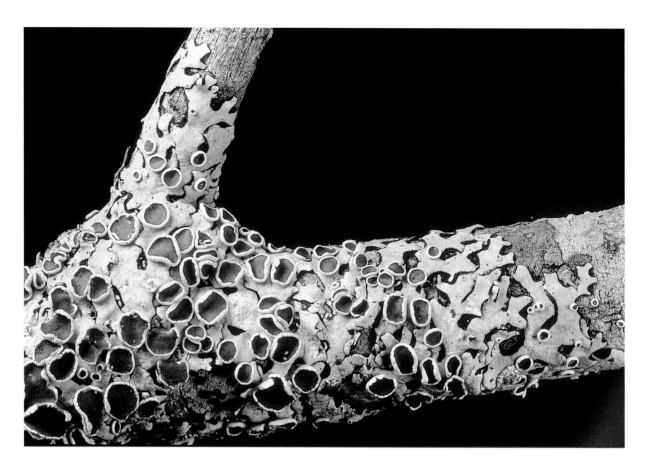

150. *Bulbothrix confoederata* Gulf Coast, Florida ×8.4

Bulbothrix goebelii
Rough eyelash lichen

DESCRIPTION: Lobes dichotomously branched, 0.4–1.3 mm wide, with brown to black, forked cilia; upper surface with scattered or abundant cylindrical, sometimes branched, isidia; lower surface pale to medium brown to almost black in older parts of the thallus, with short, pale, forked rhizines. Apothecia uncommon (although abundant in the specimen shown in plate 151). CHEMISTRY: Medulla PD–, K–, KC+ red, C+ pink (gyrophoric acid). HABITAT: On bark, rarely on rocks. COMMENTS: See *Bulbothrix laevigatula*.

Bulbothrix laevigatula
Matted eyelash lichen

DESCRIPTION, CHEMISTRY, and COMMENTS: *Bulbothrix laevigatula* is extremely similar to *B. goebelii* but with slightly broader lobes (0.7–3 mm) and a black lower surface except for the edge, with a mat of abundantly branched, black rhizines. It is also more commonly fertile than *B. goebelii*, with large brown apothecia and abundant pycnidia. The medulla is C+ blood red in *B. laevigatula* rather than C+ pink because it has lecanoric instead of gyrophoric acid. HABITAT: On the bark of deciduous trees.

Byssoloma (5 N. Am. species)
Fuzzy-rim lichens

DESCRIPTION: Thallus thin, smooth and patchy to granular, white or gray to dark green, often with a white, brown, or black prothallus. Photobiont green (unicellular). Apothecia biatorine, yellowish brown to black, with margins having a cottony or fuzzy surface because the hyphae of the exciple extend to the surface; hypothecium usually dark brown but sometimes pale; epihymenium pigmented or not; asci with a uniformly K/I+ blue ascus tip, or with a more darkly staining tube structure (something like the *Porpidia*-type), without a distinct conical axial body as in *Bacidia*; spores colorless, 2–4-celled, most commonly fusiform, slightly wider at one end, 8 per ascus. CHEMISTRY:

151. *Bulbothrix goebelii* western Mississippi ×1.0

152. *Bulbothrix laevigatula* (cilia in lower right corner) northern Florida ×4.4

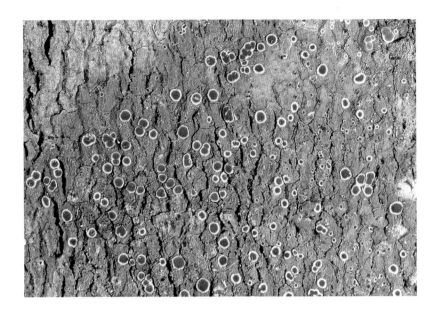

153. *Byssoloma meadii* coastal plain, North Carolina ×5.4

Thallus sometimes contains xanthones (UV+ orange). Apothecial margins K+ yellow (substance unknown) or K–. HABITAT: Most species grow on the leaves of tropical trees and shrubs, a few on bark, and some on both. COMMENTS: The fuzzy margins of the apothecia, as well as K/I staining of the ascus tips, distinguish this genus from *Bacidia* or *Biatora*.

KEY TO GENUS: See Key F.

Byssoloma meadii (syn. *B. pubescens*)

DESCRIPTION: Thallus deep green, granular to verruculose. Apothecia 0.2–0.7 mm in diameter, very flat and thin, with disks varying from pale yellowish to yellowish orange or finally brown-black, and margins white, fuzzy; hypothecium colorless to yellowish or light brown; exciple with hyphae growing out to the surface, forming a fuzz; spores (1–)4-celled, cells not always equal in size, fusiform to tapering, often slightly curved, 10–15 × 2.5–4 μm. CHEMISTRY: Thallus UV+ orange (xanthones). Apothecial margin K–. HABITAT: On bark in shaded forests. COMMENTS: The very similar and equally common *B. leucoblepharum* has a dark brown hypothecium, and the thallus is UV– (lacking xanthones). It grows on leaves as well as bark.

Calicium (14 N. Am. species)
Stubble lichens, pin lichens

DESCRIPTION: Thallus crustose, thin and barely perceptible to granular or powdery. Photobiont green (*Trebouxia* or *Stichococcus*). Apothecia like pinheads consisting of tiny cups (capitula) containing a black mass of ascospores (a mazaedium) at the tips of tiny, erect, unbranched, black stalks typically under 2 mm tall, sometimes dusted with a yellow or white pruina; spores dark brown, 2-celled with a distinct wall dividing the cells, ellipsoid, loose in the mazaedium. CHEMISTRY: Thallus usually PD–, K–, KC–, C–, rarely PD+ yellow, K+ dirty yellow. HABITAT: On wood and bark of different kinds, usually in well-shaded forests. COMMENTS: *Calicium* is distinguished from other stubble lichens by its dark brown, distinctly 2-celled spores. The stalks tend to be stout and relatively short compared with those of other stubble lichens.

KEY TO SPECIES (including *Chaenotheca*, *Chaenothecopsis*, *Mycocalicium*, *Phaeocalicium*, *Sphinctrina*, and *Stenocybe*)

1. Spores ellipsoid, 1–4-celled, brown 2
1. Spores spherical, 1-celled, colorless to brown 10

2. Spores 4-celled. Thallus not lichenized; very common on the trunks and branches of balsam fir .... [*Stenocybe major*]
2. Spores 1–2-celled 3

3. Spores 1-celled. Thallus not lichenized; very common on old stumps in forests .... [*Mycocalicium subtile*]
3. Spores 2-celled 4

4. Spores narrow, 5.5–7.5 × 2–2.3 μm; capitula without pruina, black; thallus not lichenized; widespread boreal .... [*Chaenothecopsis debilis*]
4. Spores broader than 3.5 μm; capitula often pruinose ...5

5. On living or dead twigs and branches of poplars; asci remaining intact when mature; spores 10–13 × 4–6 μm, smooth; thallus not lichenized .... [*Phaeocalicium populneum*]
5. On decaying wood (without bark); asci disintegrating, leaving the spores in a powdery mass; spores ornamented with ridges or warts; usually with algae in the thallus 6

6. Thallus granular, yellowish green; capitulum margin and base without pruina .... *Calicium viride*
6. Thallus whitish or gray, or within substrate and almost absent from view, sometimes forming a whitish stain 7

7. Capitulum with a yellow pruina on the margin and base; common in Great Lakes region, scattered elsewhere .... *Calicium trabinellum*
7. Capitulum with a white or rusty pruina, or not pruinose 8

8. Thallus yellowish white, granular to leprose; stalks thick, unbranched; spores (9–)11–13(–18) × (4.5–)6–7.5(–9.5) μm; capitulum often white pruinose; common along the coast in the Pacific Northwest, occasional elsewhere .... [*Calicium lenticulare*]
8. Thallus almost imperceptible, within the wood; stalks rather slender; spores mostly less than 6 μm wide; capitulum white or brown pruinose; boreal species ... 9

9. Pruina white, on margin of the capitulum .... [*Calicium glaucellum*]
9. Pruina rusty brown, on undersurface of capitulum ... . [*Calicium salicinum*]

10.(1) Parasitic on lichens, especially *Pertusaria*; stalk extremely short or absent, never longer than capitulum, light to dark brown; spores subspherical, 5–7 μm in diameter; widespread .... [*Sphinctrina turbinata*]
10. Not parasitic on lichens; stalks much longer than capitulum, black or almost black; spores less than 5 μm in diameter 11

11. Thallus leprose, with bright yellow-green powdery soredia; stalk covered with yellow pruina; capitulum also yellow pruinose .... *Chaenotheca furfuracea*
11. Thallus within substrate, absent from view, or sometimes consisting of a very thin greenish powder; stalk black, not pruinose; capitulum brown, without pruina .... *Chaenotheca brunneola*

154. *Calicium trabinellum* southern New Brunswick ×9.3

Calicium trabinellum
Yellow collar stubble lichen

DESCRIPTION: Thallus visible only as a pale stain on wood, or occasionally granular; stalks black, relatively short and thick, 0.5–0.9 mm tall, with yellow pruina on the underside of the head-like capitulum; spores 7–12 × 4–6 μm, with rough, sculptured walls. CHEMISTRY: The yellow pruina contains vulpinic acid. HABITAT: On wood. COMMENTS: *Calicium adspersum,* a rare western species, has yellow pruina on the surface of the capitulum, not on the underside. It has a well-developed, gray to whitish thallus that reacts PD+ yellow, K+ dirty yellow. *Calicium glaucellum,* known from British Columbia, the Great Lakes region, and Maine to Nova Scotia, has thick, black stalks and an imperceptible thallus. The capitulum has a thin line of white pruina just on the outer margin of the cup, and the spores are 10–12 × 4–5 μm with sculptured walls. *Calicium abietinum* is similar to *C. glaucellum* and *C. trabinellum* but lacks any pruina on the capitulum. The capitulum of *C. salicinum* is almost spherical, or like a somewhat flattened sphere, and is liberally coated with a rusty reddish brown pruina on the underside. Its spores are 8–11 × 4–5 μm, with walls that have spiraled ridges on the surface. Both *C. abietinum* and *C. salicinum* are widespread and fairly common on wood.

Calicium viride
Green stubble lichen

DESCRIPTION: Thallus well-developed, granular, bright yellowish green; stalks black and smooth throughout, 1.5–2.5 mm tall and about 0.1–0.15 mm thick. Capitula black with no pruina but brown on the underside, 0.15–0.3(–0.6) mm in diameter; spores 11–13.5 × 4–7 μm, constricted at the septum, with coarsely ornamented walls. CHEMISTRY: Thallus PD–, K–, KC–, C–, UV+ orange (rhizocarpic acid and epanorin). HABITAT: Common on conifer

wood and bark in mountain forests but also on deciduous trees. COMMENTS: The vivid yellow-green thallus and relatively long black stalks of this common western stubble lichen are very distinctive. The capitula are much like those of *C. salicinum*. See Comments under *C. trabinellum*.

Calopadia (2 N. Am. species)
Leaf dot lichens

KEY TO GENUS: See Key F.

Calopadia fusca (syn. *Lopadium fuscum*)

DESCRIPTION: Thallus extremely thin, in scattered greenish gray patches. Photobiont green (unicellular). Apothecia biatorine, light brown (as shown in plate 156), more commonly dark to pale grayish brown, 0.2–0.4 mm in diameter, with a thin, slightly paler margin; paraphyses not distinct; exciple colorless; hypothecium yellowish brown; spores colorless, muriform, many-celled, (38–)57–80 × 14–25 µm, 1 per ascus. HABITAT: On palmetto leaves. DISTRIBUTION: Florida and possibly other tropical regions in the east. COMMENTS: The genus *Calopadia* resembles *Lopadium* with regard to spores, but not in features of the asci, paraphyses, hypothecium, and exciple. It comprises small species mainly found on the leaves of tropical or subtropical trees and shrubs.

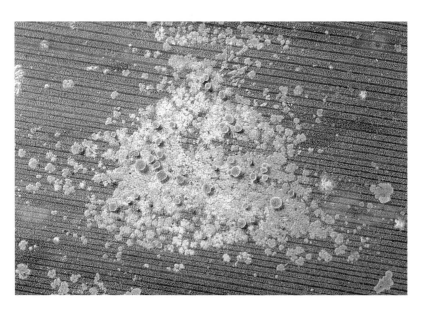

155. *Calicium viride* Rocky Mountains, Montana ×8.0

156. *Calopadia fusca* Gulf Coast, Florida ×5.5

Caloplaca (133 N. Am. species)
Firedot lichens, jewel lichens

DESCRIPTION: Most species bright orange or yellow-orange in either the thallus or the apothecia and often in both; thallus, if not orange or yellow, is white, gray, greenish, or brown. Apothecia usually yellow to orange or rusty red, but dark brown to black in a small group of species; margins usually lecanorine, but some species with biatorine margins; spores 2-celled, colorless, with a conspicuously thickened septum, pushing the cell cavities (locules) to the ends or "poles," therefore called polarilocular. Photobiont green (*Trebouxia*?). CHEMISTRY: Orange pigment in the thallus and apothecia K+ deep purple (one or more anthraquinones). HABITAT: On rocks, bark, and wood of all kinds, as well as on mosses and vegetation, or parasitic on other lichens. COMMENTS: *Caloplaca* species are generally more orange than yellow compared with other yellow lichens such as *Candelariella*, which contain different pigments and are K–. The polarilocular spores are the confirming character, although in a few species, the septum is so thin that the spores resemble ordinary 2-celled spores. Only species of *Caloplaca* and those of related genera such as the foliose *Xanthoria* and the fruticose *Teloschistes* have polarilocular spores. Sulphur yellow lichens on arid or arctic soil are usually species of *Fulgensia* if they react K+ purple. *Acarospora schleicheri* would be K–. The many arctic-alpine species of *Caloplaca* growing on soil or mosses can be identified by using Thomson's *American Arctic Lichens,* volume 2 (see "Further Reading and Bibliography," at the back of this book) and are not included in the key.

KEY TO SPECIES

1. Thallus fruticose, composed of a tangle of cylindrical branches; maritime California coast . . . . *Caloplaca coralloides*
1. Thallus crustose, or squamulose 2

2. Thallus within substrate and absent from view, or very thin and indistinct 3
2. Thallus clearly visible and well-developed 10

3. Parasitic on a variety of saxicolous lichens . . . . *Caloplaca epithallina*
3. Not parasitic 4

4. Growing on bark or wood 5
4. Growing on rock 7

5. In maritime habitats along the west coast, especially on beach logs . . . . *Caloplaca inconspecta*
5. In inland localities throughout the continent, on bark and wood 6

6. Apothecia dark orange to yellowish orange; spores 10–13 × 5–7 μm . . . . *Caloplaca holocarpa*
6. Apothecia rusty red-orange; spores 12–18(–20) × 6–10 (–11) μm . . . . [*Caloplaca ferruginea*]

7.(4) Apothecia biatorine; hymenium containing abundant oil drops . . . . *Caloplaca luteominia*
7. Apothecia lecanorine; hymenium clear, without oil drops 8

8. On shoreline rocks along the west coast; apothecial disks yellow-orange; spores with a broad septum (over 2.5 μm) . . . . *Caloplaca inconspecta*
8. On nonmaritime rocks mostly east of California; apothecial disks dark orange or brownish orange; spores with a very narrow septum (under 2 μm) 9

9. On calcareous rock and concrete; apothecial margin even with disk; epihymenium C– . . . . *Caloplaca feracissima*
9. On noncalcareous rock such as granite or gneiss; apothecial margin prominent; epihymenium C+ purple . . . . *Caloplaca arenaria*

10.(2) Thallus distinctly lobed at the margin 11
10. Thallus not lobed at the margin (but sometimes with lobed squamules) 14

11. Older parts of thallus dissolving into granular isidia or granules, with a ring of narrow lobes at the periphery; in marine habitats (salt-spray zone) . . . . *Caloplaca verruculifera*
11. Older parts of thallus not dissolving into granules; in inland or coastal (but not marine) habitats 12

12. Thallus thin, with rather flat lobes, often broadest at the tips; dark orange to red-orange; found only in southern California . . . . *Caloplaca ignea*
12. Thallus with thick, convex lobes; western to widespread 13

13. Lobes relatively loosely attached; lower surface corticate; very widespread . . . . *Xanthoria elegans*
13. Lobes closely attached; lower surface without a cortex; western, centered in the Great Basin . . . . *Caloplaca trachyphylla*

14.(10) Thallus pale to dark gray or pale brown, not yellowish 15
14. Thallus distinctly yellow, or rusty to bright orange . . . 18

15. Growing on bark; thallus pale gray to blue-gray with similarly gray apothecial margins . . . *Caloplaca cerina*
15. Growing directly on rock 16

16. Thallus pale beige, thin; apothecia biatorine with margins the same color as the disk (orange or red); coastal localities, California to Vancouver Island . . . . *Caloplaca luteominia*
16. Thallus gray; apothecial margins lecanorine, gray to black 17

17. Thallus pale gray, continuous to areolate, not squamulose or lobed; apothecia with a dark, narrow ring of tissue between the disk and the margin; common from the southwest through central U.S. to New England . . . . [*Caloplaca sideritis*]
17. Thallus dark olive-gray or brownish gray, thick, consisting of lobed squamules; lacking a dark ring between the apothecial disk and margin; restricted to the arid southwest . . . . *Caloplaca pellodella*

18.(14) Soredia present 19
18. Soredia absent 22

157. *Caloplaca arenaria* Rocky Mountains, Colorado ×6.2

19. Thallus continuous, with discrete round to irregular soralia on thallus surface; on poplar bark in the boreal region . . . . [*Caloplaca chrysophthalma*]
19. Thallus areolate; on rocks or wood, rarely on bark . . 20
20. Thallus areoles usually on a conspicuous orange prothallus; apothecia common; on rocks in strictly marine habitats (salt-spray zone); areoles often constricted at the base, some dissolving into granular soredia . . . . *Caloplaca flavogranulosa*
20. Prothallus absent; apothecia rare; on rock or wood, rarely bark, marine or not 21
21. Thallus and soredia yellow to orange-yellow; soredia developing first at the edges of the areoles, but frequently taking over the thallus; on rock or wood in maritime or inland localities . . . . *Caloplaca citrina*
21. Thallus and soredia dark orange; soredia usually remaining along the areole margins or in patches on the surface; common on exposed wood or sometimes bark, never rock; from Colorado to the northeast . . . . [*Caloplaca microphyllina*]
22.(18) Thallus consisting of round or squamulose, convex areoles frequently constricted at the base, up to 2 mm in diameter, usually shiny or "waxy"; on rocks, coastal California . . . . *Caloplaca bolacina*
22. Thallus continuous or composed of small areoles (mostly less than 0.6 mm in diameter) or granules; areoles not constricted at the base, not waxy in appearance; on rocks, bark, or wood 23
23. On bark or wood 24
23. On rock, rarely on wood 25
24. Thallus sparse, consisting of scattered small areoles; apothecial margin the same color as the disk; on rocks or wood; maritime, along the west coast . . . . *Caloplaca inconspecta*
24. Thallus continuous and rather smooth; apothecial margin paler than the disk; on bark or sometimes wood; widespread in central and eastern regions . . . . *Caloplaca flavorubescens*
25. Thallus continuous, typically smooth to rimose or verruculose; widespread east of Arizona, Nevada, and Idaho . . . . *Caloplaca flavovirescens*
25. Thallus typically areolate or granular, the areoles contiguous or dispersed, sometimes becoming continuous or thin and rimose 26
26. Most commonly northeastern, not maritime; thallus orange, areolate, the areoles often somewhat lobed; on calcareous rock . . . . [*Caloplaca velana*]
26. Along the extreme west coast, usually maritime; thallus orange to yellowish; areoles lobed or not; on siliceous or calcareous rock 27
27. Thallus well-developed, usually with an orange prothallus (see couplet 20) . . . . . *Caloplaca flavogranulosa*
27. Thallus very thin; prothallus absent 28
28. Apothecia biatorine; apothecial disks dark orange (var. *luteominia*) or red (var. *bolanderi*); hymenium containing abundant oil drops . . . . *Caloplaca luteominia*
28. Apothecia lecanorine; apothecial disks yellow-orange; hymenium clear, without oil drops . . . . *Caloplaca inconspecta*

Caloplaca arenaria
Granite firedot lichen

DESCRIPTION: Thallus growing within the upper layers of rock and barely perceptible. Apothecia abundant and often crowded together, 0.2–0.5 mm in diameter, rusty orange to olive-brown, with thin, rusty margins the same color as the disk or slightly paler; spores narrowly ellipsoid to almost cylindrical, 10–15(–17) × 3.5–5.5 μm, with narrow septa mostly under 2 μm thick. HABITAT: On siliceous rocks (typically granite) in the open. COMMENTS: The spores of this species resemble those of the very similar *C. fraudens,* but the apothecia of the latter have bright orange margins that are much lighter than the dark orange disk; the spore septum is also slightly thicker (2–4 μm). *Caloplaca approximata* is usually found on rocks where birds perch; it has bright red-orange apothecia and smaller spores (8–11.5 × 3.5–4.5 μm).

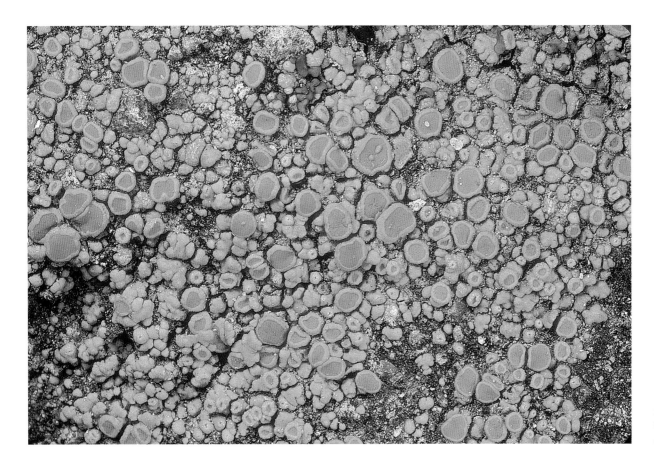

158. *Caloplaca bolacina* Channel Islands, southern California ×4.9

Both species appear to be widespread west of Wisconsin; *C. fraudens* is the more common lichen, especially in the north.

Caloplaca bolacina
Waxy firedot lichen

DESCRIPTION: Thallus yellow-orange, usually somewhat waxy in appearance, composed of rather thick, scattered or contiguous, slightly lobed squamules or convex areoles up to 2 mm across. Apothecia large (0.7–2 mm in diameter), with orange disks and slightly paler, low margins; apothecial tissues (exciple and cortex) composed of elongate, irregularly arranged cells, with few crystals in the medulla; spores 12.5–17.5 × 5.5–8.5 µm, with a septum 3–5.2 µm thick. HABITAT: On coastal rocks of all kinds (rarely on wood or soil) in the salt-spray zone as well as on ridges of up to 1600 m elevation near the coast. COMMENTS: Narrow, elongate cells in the exciple are characteristic of *C. bolacina* and the superficially similar, rarer species, *Caloplaca stantonii,* which is also found on coastal rocks and hills along the west coast. *Caloplaca stantonii* has darker, rustier apothecia, abundant crystals in the medulla, and narrower spore septa (less than 2.5 µm wide). Other confusing west coast maritime species include **C. marina** and *C. flavogranulosa,* which are more minutely areolate to granular. See also *C. verruculifera.*

Caloplaca cerina
Gray-rimmed firedot lichen

DESCRIPTION: Thallus whitish to blue-gray, thin to well-developed, smooth or becoming areolate. Apothecia up to 2 mm in diameter, bright yellow to orange, sometimes slightly pruinose, with a well-developed, raised, gray apothecial margin that is either smooth or somewhat pruinose; spores 10–17 × 6–8.5 µm; septum 3–8 µm thick. HABITAT: Very common on bark, especially poplars and elms, in open woodlands, and on isolated trees. COMMENTS: The only other common, bark-dwelling *Caloplaca* with prominent gray apothecial margins is the very closely related **C. ulmorum,** which has larger apothecia, a heavily yellow-pruinose apothecial disk,

159. *Caloplaca cerina* Lake Superior region, Ontario ×3.8

160. *Caloplaca citrina* east side of Cascades, Oregon ×8.1

and a somewhat squamulose (or at least thick, areolate) thallus. Arctic and alpine populations can be included within *C. cerina* or regarded as a distinct species, *C. stillicidiorum;* these have darker apothecia and grow on dead vegetation and peat in tundra habitats.

Caloplaca citrina
Mealy firedot lichen

DESCRIPTION: Thallus dark yellow to yellow-orange, consisting of irregularly shaped areoles that become granular sorediate starting at the edges, often reducing the entire thallus to a leprose crust. Apothecia rare, with sorediate margins. HABITAT: Widely distributed; on inland to maritime rocks of all kinds, as well as wood and soil. COMMENTS: This is one of the few species of *Caloplaca* that is commonly yolk-yellow like *Candelariella* rather than orange, although many specimens are yellowish orange. The K test quickly settles any question (K+ red-purple in the *Caloplaca*). Most other sorediate, saxicolous *Caloplaca* species are lobed, at least at the edges. Examples include the uncommon, mainly northeastern *C. cirrochroa,* with laminal, excavate, yellowish green soralia, and *C. decipiens,* a southwestern species, which produces piles of granular, orange soredia at the tips of overlapping lobes in older parts of the thallus. *Caloplaca microphyllina* is distinctly orange to red-orange, areolate to squamulose, on bark or wood (frequently seen on fence rails), with soredia produced along the edges of the areoles or squamules. Although it sometimes has abundant apothecia, most specimens are sterile. It ranges from central U.S. to Quebec and New England, but it has been reported from Arizona, and California as well.

Caloplaca coralloides
Coral firedot lichen

DESCRIPTION: Thallus yellow-orange, fruticose, with terete, lumpy branches that fork 1–3 times; branches slender (0.2–0.4 mm) or thicker (up to 1 mm), spreading out onto rock surfaces or growing in tight, erect tufts. Apothecia absent or abundant, forming at the tips of the branches. HABITAT: On seacoast rocks in the salt-spray to upper intertidal zones. COMMENTS: The fruticose habit suggests a species of *Teloschistes,* but most North American representatives of *Teloschistes* grow on trees or on the soil and are much more abundantly branched.

Caloplaca epithallina
Parasitic firedot lichen

DESCRIPTION: Thallus within a host lichen. Apothecia dark rusty red-orange, in clusters; apothecial margins varying from dark orange or reddish to black in places; spores broadly ellipsoid, 8–13 × 5–8 μm, with a septum 2–3.5 μm wide (less than 1/3 the length of the spore). CHEMISTRY: Apothecial disk C–, margin C+ deep red. HABITAT: Parasitic on a variety of arctic and alpine lichens growing on siliceous rocks, especially those frequented by birds. COMMENTS: This is the most common parasitic *Caloplaca*. *Caloplaca castellana* is another common parasitic species on rock lichens in arctic and alpine regions. It generally produces tiny, scattered, orange areoles over the host, together with small apothecia with orange (not rusty) margins that are C– and darker orange to brown disks that can be C+ pink or C–.

Caloplaca feracissima
Sidewalk firedot lichen

DESCRIPTION: Thallus almost imperceptible, sometimes producing a dark gray to dull yellow stain. Apothecia small, 0.2–0.5 mm in diameter, often densely crowded, with dull orange to orange-brown disks and yolk-yellow margins; spores narrowly ellipsoid, 13.5–18 × (5–)6–8(–9) μm, with a narrow septum less than 2 μm thick (barely polarilocular). HABITAT: Extremely common on mortar and cement, especially on less frequented sidewalks; also on natural limestone. COMMENTS: This lichen causes underused sidewalks to become yellow. Although the apothecia are very small, they grow in such profusion that they can color large areas. Other common firedot lichens with disappearing thalli but abundant apothecia include *C. approximata,* with larger apothecia (up to 0.7 mm diameter), and *C. arenaria,* which has a rustier color and shorter spores, and grows mainly on noncalcareous rock.

161. *Caloplaca coralloides* California coast ×2.9

162. *Caloplaca epithallina* (on a lobate species of *Lecanora*) Rocky Mountains, Colorado ×5.4

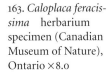

163. *Caloplaca feracissima* herbarium specimen (Canadian Museum of Nature), Ontario ×8.0

164. *Caloplaca flavogranulosa* coast of southern British Columbia ×3.1

Caloplaca flavogranulosa
Grainy seaside firedot lichen

DESCRIPTION: Thallus orange-yellow, squamulose or dispersed areolate with discrete, somewhat lobed areoles, 0.05–0.4 mm in diameter, most commonly constricted at the base or becoming granular; granular sorediate at the edges of the areoles on well-developed specimens, sometimes coalescing into large sorediate patches; usually producing a conspicuous yellow-orange prothallus. Apothecia common, typically 0.5–1.1 mm in diameter, about the same color as the thallus; spores 11–16 × 4.5–7 μm with septa (2–)3–5 μm thick. HABITAT: On maritime rocks in the lower salt-spray zone. COMMENTS: When it lacks one or more characters, this common maritime lichen is easily confused with several others frequently seen on the west coast. When the soredia are absent, it sometimes resembles *C. marina*, which has larger areoles (0.2–0.7 mm in diameter) that are not constricted at the base. Heavily sorediate specimens look like *C. citrina*, which lacks a prothallus and rarely has apothecia. ***Caloplaca rosei*** has a continuous, rimose-areolate thallus as well as a conspicuous orange prothallus. Poorly developed specimens of *C. flavogranulosa* can be confused with *C. inconspecta*, which has no prothallus and almost no thallus areoles, and is never sorediate. ***Caloplaca microthallina*** is an east coast squamulose lichen similar to *C. flavogranulosa* in appearance and habitat.

Caloplaca flavorubescens
Bark sulphur-firedot lichen

DESCRIPTION: Thallus pale sulphur-yellow or almost gray in shade forms (as shown in plate 165), continuous and fairly smooth, without soredia. Apothecia 0.4–1.5 mm in diameter, various shades of orange rarely with a pale yellow or white pruina; margins pale orange, usually with an outer rim of sulphur-yellow thallus-like tissue, sometimes almost white, especially if pruinose; spores 12–19.5 × 7–10.5 μm, with septa over 1/3 the length of the spore. HABITAT: On bark of deciduous trees and conifers, and on wood. COMMENTS: *Caloplaca flavovirescens* is almost the rock-dwelling counterpart of *C.*

flavorubescens, except that its apothecia never become pruinose, the apothecial margins are less thalline in character, and the apothecia are somewhat smaller. The sorediate sister species of *C. flavorubescens* is **C. chrysophthalma,** a common bark-dwelling lichen in the southern boreal forests, particularly on old poplar trees. Its thallus can vary from distinctly sulphur yellow to almost entirely gray, but the bright yellow-orange soralia (usually round, but sometimes irregular) with granular soredia make it rather easy to spot. As in *C. flavorubescens,* the apothecia are bright orange and sometimes have an outer thallus-colored margin besides the orange margin around the disk. The margin of *C. chrysophthalma* contains abundant algae, unlike the otherwise similar **C. discolor,** which has dark, rust-colored apothecia with margins containing no (or few) algae, and is most frequently found along the east coast. *Caloplaca discolor* has coarser soredia as well. Sterile specimens, however, are very hard to distinguish from *C. chrysophthalma*.

Caloplaca flavovirescens
Sulphur-firedot lichen

DESCRIPTION: Thallus cracked-areolate to minutely warty, sulphur-yellow, nonsorediate. Apothecia orange to brownish orange, 0.4–0.8(–1.2) mm in diameter. Apothecial margins almost the same color as the disks or paler, sometimes with an outer thallus-colored layer; spores like those of *C. flavorubescens.* HABITAT: Generally on rocks containing calcium such as limestones and sandstones, and on concrete. COMMENTS: *Caloplaca velana,* a northeastern limestone lichen, has a more distinctly orange, dispersed areolate thallus, the areoles sometimes becoming lobed. See also *C. flavorubescens*.

Caloplaca holocarpa
Firedot lichen

DESCRIPTION: Thallus developing within bark or wood and not visible, or sometimes producing a barely perceptible whitish or gray stain. Apothecia produced in abundance, 0.3–0.5 mm in diameter, with dark to light orange or orangish yellow disks and margins almost the same color or slightly paler; spores small and broad, 10–13 × 5–7 µm, with wide septa (3.5–5 µm). HABITAT: On bark and wood of many types of trees; specimens on rock sometimes called *C.*

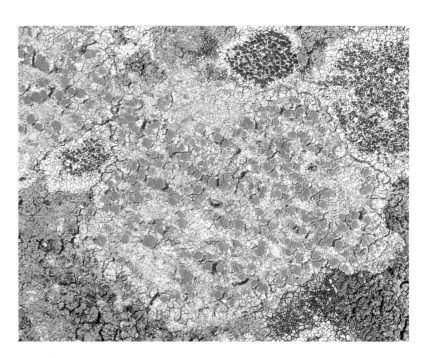

165. *Caloplaca flavorubescens* Channel Islands, southern California ×5.4

166. *Caloplaca flavovirescens* Sonoran Desert, southern Arizona ×3.4

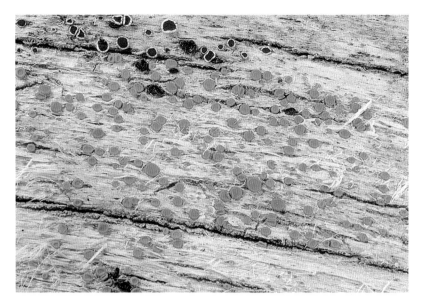

167. *Caloplaca holocarpa* Southeast Alaska coast ×5.6

168. *Caloplaca ignea* Channel Islands, southern California ×4.0

lithophila (see Comments under *C. inconspecta*). DISTRIBUTION: Uncertain, but probably widespread from the northern boreal zone to temperate U.S. COMMENTS: *Caloplaca borealis* is similar to *C. holocarpa*, but at least some apothecia usually have waxy, often grayish margins, and the apothecia are smaller (0.2–0.4 mm in diameter) and scattered. It is very common on a variety of deciduous trees, especially poplars, in temperate to boreal North America and Europe. *Caloplaca ahtii* is much like *C. borealis*, but the apothecial margins are opaque, not waxy. It is actually a sorediate lichen with tiny (less than 0.2 mm diameter), excavate soralia containing blue-green soredia, but the soralia are often so sparse and inconspicuous that they are easily missed.

Caloplaca ignea
Flame firedot lichen

DESCRIPTION: Thallus dark orange to red-orange, lobes narrow, mostly under 0.5 mm wide, flattened to slightly convex, becoming areolate in thallus center. Apothecial disks and margins the same color as the thallus. HABITAT: On dry rocks both along the coast and inland. COMMENTS: This is a common lobed *Caloplaca* in California. Its dark red-orange color and flat lobes make it striking and distinctive. *Caloplaca saxicola*, another common, mainly West Temperate species, is paler orange to yellow-orange with more convex, scabrose lobes (like *C. trachyphylla*, which has broader, contiguous lobes mostly over 0.5 mm wide), and its apothecia are orange, with paler, yellower margins. See also *Xanthoria elegans*.

Caloplaca inconspecta
Seaside firedot lichen

DESCRIPTION: Thallus extremely variable, from virtually absent to yellow or yellow-orange and areolate, with scattered, convex areoles less than 0.2 mm in diameter. Apothecia usually abundant, 0.3–1.1 mm in diameter, most commonly scattered but sometimes in crowded patches, bright yellow to yellow-orange, with thin, even margins; spores small and broad, 10–14 × 3.5–7.5 µm, with septa (2.5–)3–5 µm wide. HABITAT: On rocks and driftwood logs by the sea from the upper intertidal zone to the edge of the

beach. COMMENTS: Specimens with a yellow thallus can resemble *C. flavorubescens* and *C. flavovirescens*, which have much larger spores. However, *C. inconspecta* typically consists of nothing but the bright yellow-orange apothecia and more closely resembles **C. lithophila,** except that this species is found only along the east coast and has slightly smaller apothecia, narrower apothecial margins, and wider septa in the spores. See *C. flavogranulosa*.

Caloplaca luteominia
Red firedot lichen

DESCRIPTION: Thallus often entirely within the rock, or areolate with contiguous or scattered pale brown to pinkish areoles. Apothecia either orange or scarlet, mostly 0.6–1.2 mm in diameter; margin paler than the disk, essentially biatorine (containing no algae), or rarely with a very thin, outer thalline layer; spores 14–20 × 4–7.5(–8.5) μm with septa mostly 2–3 μm wide. HABITAT: On rocks of all kinds, from the upper intertidal zone to shaded sites in coastal mountains. COMMENTS: Specimens of *C. luteominia* with red apothecia (as shown in plate 170) represent var. *bolanderi*. No other firedot lichen has apothecia this color; it is absolutely diagnostic. *Caloplaca luteominia* var. *luteominia* has orange apothecia but is otherwise identical; it is much more common, being found along the coast from Vancouver Island to Baja California. Other coastal species of *Caloplaca* with orange apothecia and inconspicuous thalli such as **C. lithophila** or *C. inconspecta* have algae in the apothecial margins, and the spores are much smaller.

Caloplaca pellodella
Olive firedot lichen

DESCRIPTION: Thallus shiny, dark olive gray or brownish, forming clumps of lobed, convex squamules or simply areolate. Apothecia abundant, 0.2–1 mm in diameter; disks dark orange to brown or almost black, with slightly raised, lecanorine margins the same color as the thallus; spores (10–)11–14(–15) × 5.5–7(–8.5) μm with septa 3–4(–5.5) μm wide. CHEMISTRY: Thallus surface and outer part of exciple turning violet with K, C, or nitric acid. HABITAT: On

169. *Caloplaca inconspecta* California coast ×4.1

170. *Caloplaca luteominia* var. *bolanderi* Coast Range, southern California ×5.8

171. *Caloplaca pellodella* central Arizona ×2.8

siliceous rocks. COMMENTS: With its lobed, olive-gray squamules and dark orange apothecia, this is one of the most distinctive species of *Caloplaca*. Forms with poorly developed thalli sometimes resemble *C. sideritis,* a southwestern to East Temperate species that has a pale brown to pale gray areolate thallus that rarely forms squamules.

Caloplaca trachyphylla
Desert firedot lichen

DESCRIPTION: Thallus orange, with thick, convex, rather lumpy lobes 2–5 mm long and 0.5–1.3 mm wide, typically crowded, parallel, and radiating outward. Apothecia very small, crowded in the center of the thallus. HABITAT: On exposed rocks in dry sites. COMMENTS: *Caloplaca trachyphylla* is very much like *Xanthoria elegans* in general appearance (both have very convex, thick lobes), but the lobes of *C. trachyphylla* cannot be lifted from the substrate with the cortex intact, and the tips are generally rough and scabrose. **Caloplaca saxicola** has much shorter and smoother lobes (1–2 mm long, 0.3–1 mm wide), and *C. ignea* is a redder orange with narrower, flatter lobes. See Comments under *C. ignea*.

172. *Caloplaca trachyphylla* central Utah ×3.3

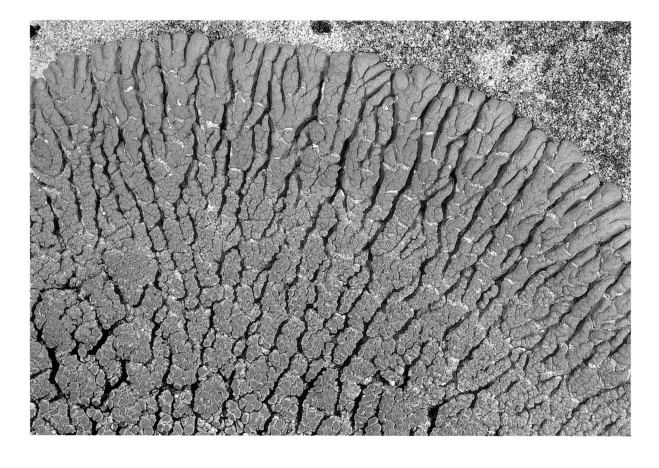

Caloplaca verruculifera
Ringed firedot lichen

DESCRIPTION: Thallus dark yellow to bright orange, in more or less circular patches with narrow, radiating lobes around the periphery; lobes convex, flattening at the tips, 1–5 mm long and (0.2–)0.3–0.8 mm wide, dissolving into coarse granules or granular isidia in the older central portions of the thallus, often leaving a ring of lobes surrounding a largely empty area containing only scattered granules and areoles. Apothecia rare; spores 11–14 × 4.5–7 μm with septa 2.8–4.2 μm wide. HABITAT: On coastal rocks in the salt-spray zone, especially where manured by birds. COMMENTS: *Caloplaca brattiae* is a similar lobed, maritime *Caloplaca* from the west coast with very convex lobes and small, radiating thalli. It grows on rocks from the upper intertidal zone up to the salt-spray zone. *Caloplaca brattiae* has no granular isidia at the center like *C. verruculifera*, producing abundant apothecia instead. The strictly eastern species *C. scopularis* also resembles *C. verruculifera* but has shorter thallus lobes (less than 2 mm long), abundant, somewhat raised apothecia, and no granular isidia. Its spores are larger than those of *C. brattiae* (11–15 × 5–6.5 μm versus 10–13 × 4–5 μm, respectively), and the septa are broader (3.3–5 μm versus ca. 2.5–3.5 μm, respectively). See also *C. flavogranulosa*.

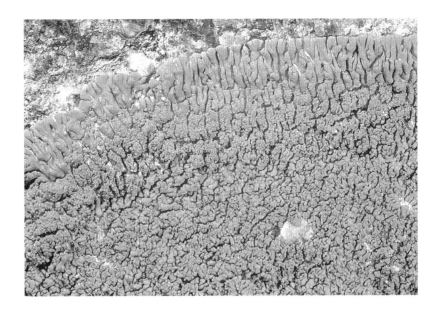

173. *Caloplaca verruculifera* southern New Brunswick ×3.5

lariella and *Candelina*, differing from both in that it has foliose lobes with rhizines. *Candelariella* is entirely crustose; *Candelina*, with at least a partial lower cortex, is attached directly to the substrate and has a growth form much like *Xanthoria elegans*.

KEY TO SPECIES

1. Soredia absent; apothecia abundant .... *Candelaria fibrosa*
1. Soredia present; apothecia rare ... *Candelaria concolor*

Candelaria (2 N. Am. species)
Candleflame lichens

DESCRIPTION: Small, yolk-yellow, foliose lichens with finely divided lobes; lower surface with a well-developed cortex, white or very pale brownish, producing short, unbranched rhizines. Photobiont green (*Trebouxia?*). Apothecia lecanorine, with yellow disks; spores colorless, 1-celled, ellipsoid, 20–50 per ascus. CHEMISTRY: Thallus PD–, K–, KC–, C–. Apothecial disks often K+ pink (yellow pigments are calycin and other compounds related to pulvinic acid). HABITAT: On bark, rarely on rock. COMMENTS: In *Candelaria* the yellow pigments are K– (or light rose-pink), unlike *Caloplaca* and *Xanthoria*, in which the yellows and oranges are due to anthraquinones and turn a deep purple with K. The genus is very closely related to *Cande-*

Candelaria concolor
Candleflame lichen, lemon lichen

DESCRIPTION: Thallus bright yellow, forming small rosettes (less than 1 cm across) of overlapping lobes; lobes only 0.1–0.5 mm across, with lacy margins edged with granules and granular soredia, rarely almost entirely dissolving into granular soredia and essentially becoming leprose. Apothecia uncommon, 0.2–0.7 mm in diameter, having dark yellow to orange-brown disks surrounded by very thin, thallus-colored margins. HABITAT: Extremely common and widespread, especially on nutrient-rich substrates, often forms luxurient colonies along rain tracks on tree trunks or on twigs of trees near farms or towns. COMMENTS: Specimens that consist of only a

few small foliose lobes in a sea of granular soredia are almost indistinguishable from some sorediate species of the crustose genus *Candelariella*, such as *C. efflorescens* or *C. reflexa;* the *Candelariella* species, however, almost always have a few round areoles or granules with sorediate margins. Shade forms of small-lobed, sorediate *Xanthoria* species such as *X. fallax* or *X. candelaria* are sometimes a similar color but are thicker and not so finely divided, and turn dark red-purple with K.

Candelaria fibrosa
Fringed candleflame lichen

DESCRIPTION: Thallus yolk-yellow, brighter in sunny habitats, with very small lobes, 0.3–0.5(–0.7) mm broad, sometimes lobulate or becoming verrucose in the thallus center, without soredia or isidia, rather flat and closely attached to the substrate; lower surface white, with abundant white rhizines sometimes extending out from lobe margins or decorating the apothecial margins. Apothecia common, up to 2 mm in diameter; disks darker yellow than the margins and thallus. HABITAT: On bark or wood. COMMENTS: This species so closely resembles *Xanthoria hasseana* that a K test on the thallus is often needed to distinguish the rare specimens without apothecia. *Candelaria fibrosa* was once much more common than it is now, at least in certain areas. For example, many old specimens from southern Ontario exist in herbaria, but the species has not been collected in that area for almost 100 years. It may be especially sensitive to air pollution or to changes in the climate of the region.

174. *Candelaria concolor*
western Mississippi
×8.6

175. *Candelaria fibrosa*
western Mississippi
×4.2

Candelariella (22 N. Am. species)
Goldspeck lichens, yolk lichens

DESCRIPTION: Yolk-yellow to greenish yellow crustose or squamulose lichens, rarely forming lobed rosettes; thallus sometimes very scanty or growing only within the substrate. Photobiont green (*Trebouxia?*). Apothecia lecanorine with yellow to brownish yellow disks and thallus-colored margins; spores colorless, ellipsoid, 1- or rarely 2-celled (sometimes appearing to be 2-celled because of large oil drops), 8–32 per ascus. CHEMISTRY: PD–, K– or K+ pale rose, KC–, C–, UV+

dull, dark orange (calycin, a yellow pigment). HABITAT: On rocks, bark, and wood of different types; also on soil or tundra peat. COMMENTS: Very few species of *Caloplaca* have the characteristic color of egg yolk seen in *Candelariella*. The spores and the chemical reactions are also entirely different. The southwestern desert genus *Candelina* has the same chemistry and spore type as *Candelariella,* differing in the radiating, convex, almost foliose lobes (but without rhizines). Bright yellow species of *Acarospora* and *Pleopsidium* sometimes resemble *Candelariella,* but their apothecia are usually sunken into the thallus areoles, and their asci generally contain hundreds of spores. Because they contain rhizocarpic acid instead of calycin, their UV fluorescence is much brighter orange.

KEY TO SPECIES

1. Thallus composed of tiny dispersed areoles breaking down into clumps of soredia, sometimes coalescing into a leprose crust . . . . *Candelariella efflorescens*
1. Thallus without soredia 2

2. Growing on soil 3
2. Growing on bark, wood, or rock 4

3. On tundra soil; arctic-alpine; thallus granular to areolate, thin and scanty . . . . *Candelariella terrigena*
3. Typically in sagebrush-juniper habitats; thallus composed of convex, often lobed squamules, relatively thick . . . . *Candelariella rosulans*

4. Spores 16–32 per ascus. On noncalcareous rock or hard, weathered wood . . . . *Candelariella vitellina*
4. Spores 8 per ascus 5

5. Thallus scanty, dispersed areolate; apothecial margin the same color as the disk; on noncalcareous rock, wood, or bark . . . . *Candelariella rosulans*
5. Thallus usually within substrate, absent from view, or poorly developed, granular; apothecial margin paler than the disk; on calcareous rock, occasionally wood . . . . *Candelariella aurella*

176. *Candelariella aurella* herbarium specimen (Canadian Museum of Nature), Ontario ×6.8

Candelariella aurella
Hidden goldspeck lichen

DESCRIPTION: Thallus almost entirely within the substrate, or consisting of scattered, dark yellow granules. Apothecia very abundant, 0.3–1.2 mm in diameter; spores 12–18 × 4.5–7.5 μm, 8 per ascus. HABITAT: On exposed calcareous rock or dust-impregnated wood or bark in areas with lime-rich soil. COMMENTS: This is the most common of the goldspeck lichens on calcareous rock. *Candelariella vitellina* has abundant apothecia, and a poorly developed thallus can resemble *C. aurella*. Its asci, however, contain 16–32 spores, and the species grows on granitic or at least noncalcareous rock.

Candelariella efflorescens
Powdery goldspeck lichen

DESCRIPTION: Thallus consisting of round, flattened areoles less than 0.2 mm in diameter that break down at the edges into fine soredia, creating tiny clusters of soredia that finally coalesce into a yellow, powdery mass. Apothecia uncommon, less than 0.5 mm in diameter; spores 32 per ascus. HABITAT: Very common on bark of all kinds and sometimes wood. COMMENTS: *Candelariella reflexa,* a much less common, mostly western species, is virtually indistinguishable from *C. efflorescens* based on thallus characters alone, but it contains 8 spores per ascus. The thallus of *Candelariella xanthostigma* consists of more evenly distributed, almost spherical, corticate granules (0.05–0.13 mm in diameter) that do not break down into soredia. It is rarely fertile but has many spores per ascus. It grows on bark or wood and has a broad distribution.

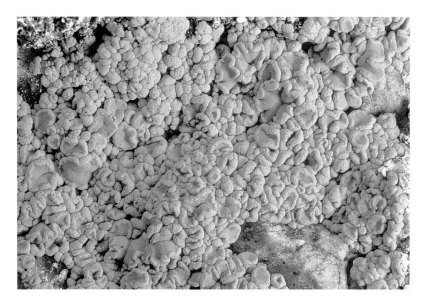

177. *Candelariella efflorescens*   herbarium specimen (Canadian Museum of Nature), Michigan ×8.2

178. *Candelariella rosulans* Mojave Desert, southern California ×5.4

Candelariella rosulans
Sagebrush goldspeck lichen

DESCRIPTION: Thallus dark yolk yellow, forming tiny rosettes of convex to flattened, lobed squamules or squamules reduced to scattered convex areoles. Apothecia 0.4–1.3 mm in diameter, with raised margins; spores 12.5–16 × 4.8–6.5 μm, 8 per ascus. HABITAT: On soil, rock, wood, or bark in juniper, sagebrush, and sometimes canyon habitats. COMMENTS: The lobed areoles and 8 spores per ascus distinguish this species from most other goldspeck lichens. *Candelariella terrigena* is a more alpine species and has smaller areoles. *Candelariella placodizans* has an almost identical thallus but is strictly arctic-alpine on soil, and it produces many spores per ascus. *Candelariella spraguei,* a rare goldspeck lichen on rock from Colorado, superficially looks like *C. rosulans* or *C. terrigena* but has long, noodle-shaped spores 25–46 × 3–5 μm, 1-celled but with the cell contents somewhat discontinuous, making the spores appear septate. The apothecia are large, up to 3 mm across, usually with a rather thick, smooth to verrucose margin.

Candelariella terrigena
Tundra goldspeck lichen

DESCRIPTION: Thallus bright yellow, varying from granular to areolate with slightly lobed areoles, often scanty; thallus frequently sterile. Apothecia 0.4–0.7 mm in diameter with thin or thick, often discontinuous margins; spores 14–18 × 6–8 μm, 8 per ascus. HABITAT: On tundra soil or on moss, lichens, and dead vegetation. COMMENTS: This is the most common arctic and alpine *Candelariella* on soil. Sterile specimens of *C. placodizans* are hard to distinguish from sterile *C. terrigena* and *C. rosulans*. See Comments under *C. rosulans.*

Candelariella vitellina
Common goldspeck lichen

DESCRIPTION: Thallus forming little cushions of flattened granules, very slightly crenulate at the edge in the best-developed specimens, or with scattered yellow areoles; usually fertile, but sterile thalli are also common. Apothecia 0.5–1.5 mm in diameter, often crowded; spores 9–15 × 4–6.5 μm, each commonly containing 2 round oil drops that give the spore the appearance of being 2-celled; 16–32 spores per ascus. HABITAT: Very common on noncalcareous, especially granitic rocks, in sun; also on wood, more rarely on bark. COMMENTS: *Candelariella aurella* is similar but has almost no thallus, is always fertile, has 8 spores per ascus, and is generally found on calcareous rock. *Candelariella placodizans* is a thicker, usually sterile species, with large, lobate areoles or squamules 0.5–1.0 mm across, forming small rosettes on moss or soil in arctic-alpine habitats.

Candelina (2 N. Am. species)
Yolk lichens

DESCRIPTION: Deep yellow, lobed lichens falling somewhere between foliose and crustose in being firmly attached to the substrate but with a lower cortex; thallus areolate in the center, with convex lobes up to 5 mm long and 1 mm broad; medulla white or yellow. Photobiont green (unicellular). Apothecia usually abundant, lecanorine with raised margins, yellow like

179. *Candelariella terrigena* eastern Utah ×3.9

180. *Candelariella vitellina* Cascades, Washington ×6.4

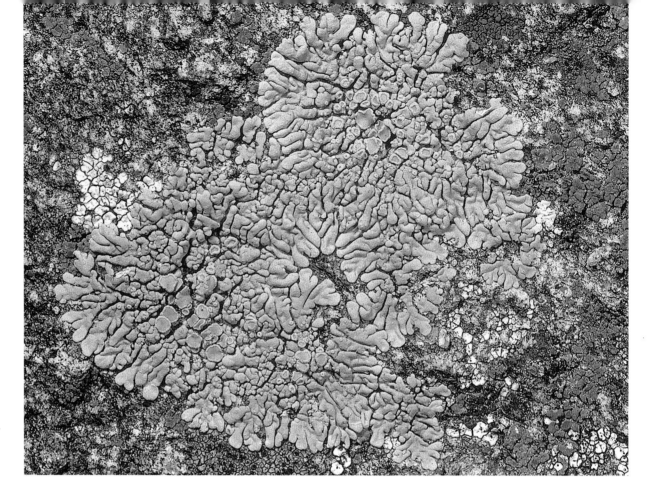

181. *Candelina submexicana* Chisos Mountains, southwestern Texas ×4.1

the thallus or with a slightly more orange or brown disk; spores colorless, 1-celled, elongate ellipsoid, and often slightly curved (bean-shaped), (9–)11–16.5 × 3.5–5 μm, 8 per ascus. CHEMISTRY: Cortex PD–, K– or reddish orange, KC–, C–, UV+ dull, very dark orange (calycin). HABITAT: On exposed granitic outcrops. COMMENTS: *Candelina* is very closely related to *Candelariella*, but even the lobed species of *Candelariella* are entirely crustose. Lobed yellow species of *Acarospora* and *Pleopsidium* might be confused with *Candelina*, but they have a brighter, sulphur-yellow color, sunken apothecia, multispored asci, and UV+ bright orange thalli.

KEY TO SPECIES: See Key J.

Candelina submexicana
Mexican yolk lichen

DESCRIPTION: Medulla white. CHEMISTRY: Medulla PD–, K–, KC–, C–. COMMENTS: *Candelina mexicana* is a less common sister species, even brighter in color, having a deep yellow medulla that is K– or K+ reddish orange.

Canomaculina (syn. *Rimeliella*; 4 N. Am. species)
Mottled ruffle lichens

DESCRIPTION: Gray-green, broad-lobed, foliose lichens with cilia along the margins; maculae on the upper surface scattered randomly, not in a reticulate pattern and not leading to a network of cracks; lower surface uniformly pale to dark brown, not black; rhizines unbranched or forked, some slender and short and others long and stout, developing to the lobe margins. Photobiont green (*Trebouxia?*). Apothecia lecanorine, constricted at the base and usually perforated in the center of the brown disk; spores colorless, 1-celled, ellipsoid, small; conidia thread-like, up to 16 μm long. CHEMISTRY: Cortex K+ yellow (atranorin); medulla with various orcinol and β-orcinol depsidones. HABITAT: On bark or rock, usually in tropical or subtropical areas. COMMENTS: In the closely related *Rimelia*, the maculae are patterned and the lower surface is ebony black except for the marginal area. *Par-*

motrema has a naked zone around the edge of the lower surface, and the conidia are shorter, with a bulge close to one end like a bowling pin.

KEY TO SPECIES: See *Rimelia*.

Canomaculina subtinctoria (syn. *Parmotrema subtinctorium*)
Mottled ruffle lichen

DESCRIPTION: Thallus with broad, round lobes, 5–15 mm across, with black cilia on the margins (often sparse), 0.2–1(–2) mm long; upper surface producing abundant cylindrical to branched isidia. Apothecia not seen. CHEMISTRY: Medulla PD+ yellow, K+ red, KC+ red, C– (salazinic acid and norlobaridone), or less commonly, PD–, K–, KC+ red, C– (norlobaridone alone). HABITAT: On deciduous bark in open woodlands and along roadsides. COMMENTS: *Canomaculina neotropica,* a rather rare southeastern species, is almost identical but contains only salazinic acid. Other broad-lobed, isidiate lichens should be compared, especially *Parmotrema crinitum* and *Rimelia subisidiosa*.

Canoparmelia (7 N. Am. species)
Shield lichens

DESCRIPTION: Medium-sized, pale greenish gray, thin, closely attached, foliose lichens with rounded lobes (mostly 3–5 mm across), without marginal cilia; upper surface often spotted with maculae but without pseudocyphellae; lower surface dark brown to black except for a paler edge; rhizines scattered and unbranched. Photobiont green (*Trebouxia?*). Apothecia laminal or close to the margins, rare in most species; spores ellipsoid, colorless, 1-celled, 10–14 × 6–8 μm; conidia dumbbell-shaped. CHEMISTRY: Cortex K+ yellow (abundant atranorin); a few non–North American species contain usnic acid as well. Medulla may contain orcinol or β-orcinol depsides and (or) β-orcinol depsidones. HABITAT: On bark or rocks. COMMENTS: This is one of many genera of shield lichens found mainly in the southeast. It is characterized by its lack of cilia and pseudocyphellae and its pale, unbranched rhizines.

KEY TO SPECIES: See *Parmelia*.

Canoparmelia caroliniana (syn. *Pseudoparmelia caroliniana*)
Carolina shield lichen

DESCRIPTION: Thallus pale gray, closely adnate, with contiguous, rounded lobes 2–6 mm across; upper surface wrinkled and folded, covered with cylindrical to branched isidia, with many conspicuous maculae close to the lobe tips; lower surface medium to very dark brown, sometimes even brownish black. Apothecia rare. CHEMISTRY: Medulla PD–, K–, KC– or KC+ purplish, C–, UV+ blue-white (perlatolic acid). HABITAT: Common on bark of both hardwoods and conifers in woodland habitats. COMMENTS: *Canoparmelia caroliniana* resembles a very adnate *Punctelia rudecta* (same kind of isidia) but with a network of fine white maculae instead of separate white dots (pseudocyphellae); it also has a darker lower surface and a different chemistry. In *C. amazonica* (syn. *Pseudoparmelia amazonica*), a common lichen in Florida, the medulla reacts PD+ red, K–, KC+ pink, UV– (protocetraric acid). The isidia are not as dense as those of *C. caroliniana,* the lobes are more separated, and the lower surface is jet black. *Canoparmelia salacinifera* (syn. *Pseudoparmelia salacinifera*), another isidiate species found mainly in Florida, has a very uniform upper surface with no maculae and a brown lower surface. It contains salazinic acid (PD+ yellow, K+ red). ***Bulbothrix isidiza*** is

182. *Canomaculina subtinctoria* Appalachians, Virginia ×1.9

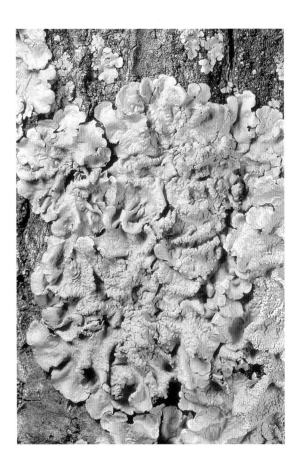

183. *Canoparmelia caroliniana* coastal plain, North Carolina ×3.6

184. *Canoparmelia texana* southeastern Missouri ×1.65

sometimes mistaken for *C. salacinifera* when the bulbous cilia are missed.

Canoparmelia texana (syn. *Pseudoparmelia texana*)
Texas shield lichen

DESCRIPTION: Thallus gray, closely adnate; lobes mostly 2–4 mm wide, with a wrinkled surface and coarse, granular, laminal soredia mainly along the crests of the wrinkles; maculae absent or very inconspicuous; lower surface dull reddish brown darkening to almost black in the center of the thallus. CHEMISTRY: Soralia and medulla PD–, K–, C–, KC– or KC+ faint purple, UV+ bright blue-white (divaricatic acid). HABITAT: On bark of hardwoods and conifers. COMMENTS: *Canoparmelia texana* looks a bit like *Myelochroa aurulenta,* but the medulla is white rather than pale yellow. *Canoparmelia crozalsiana,* a relatively rare species from southeastern U.S. and Arizona, is almost identical to *C. texana* in appearance, but its medulla reacts PD+ orange, K+ yellow, KC– (stictic acid). The only other sorediate *Canoparmelia* is **C. cryptochlorophaea** (syn. *Pseudoparmelia cryptochlorophaea*), a distinctive southern coastal plain lichen that forms almost spherical soralia on the tips of short, erect lobes all over the thallus. The medulla is PD–, K+ dark pink (slowly), KC+ purple, C+ purple, UV± dull blue (cryptochlorophaeic acid). **Paraparmelia alabamensis** (syn. *Pseudoparmelia alabamensis*) is a less common sorediate species, known from northern Alabama and Tennessee, growing on rocks. It has narrow, gray lobes (about 1 mm wide) and unbranched rhizines. The medulla reacts PD+ red, KC+ pink (protocetraric acid). It is much like a gray *Xanthoparmelia* (with atranorin in the cortex instead of usnic acid). IMPORTANCE: *Canoparmelia texana* has been collected and analyzed for pollutants in Brazil, where it was found to be very pollution-tolerant, often profusely covering trees in city parks and along roads.

Cavernularia (2 species worldwide)
Honeycomb lichens, saguaro lichens

DESCRIPTION: Minute, gray-green, foliose lichens with solid lobes 0.2–0.7 mm across, forming small rosettes; lower surface black, without rhizines, covered with minute pits or holes giving it a honeycomb appearance. Photobiont green (*Trebouxia*?). Apothecia lecanorine, raised and cup-shaped, with brown disks; spores colorless, 1-celled, spherical, 8 per ascus. CHEMISTRY: Cortex K+ yellow (atranorin); medulla PD–, K–, KC+ red, C– (physodic acid). HABITAT: On exposed conifer twigs and bark, and on weathered wood. COMMENTS: Several of the genera that resemble *Hypogymnia* have pores of some kind (even though the lobes may not be hollow). *Hypogymnia* is often perforate at the lobe tips, *Menegazzia* has large perforations through the upper surface of the lobes, and *Cavernularia* is abundantly pitted with tiny pores on the lower surface.

KEY TO SPECIES

1. Soredia present on upper surface of lobe tips; apothecia rare . . . . *Cavernularia hultenii*
1. Soredia absent; apothecia abundant, very broad compared to the width of the lobes . . . . *Cavernularia lophyrea*

Cavernularia hultenii
Powdered honeycomb lichen

DESCRIPTION: Thallus with narrow, ascending lobes with rounded piles of greenish soredia on the upper surface of the tips. Apothecia rare. COMMENTS: This honeycomb lichen is a true oceanic species, restricted to wet, forested coasts on both sides of the continent and also in Europe. It forms small, circular thalli resembling a tiny *Hypogymnia tubulosa* because of the patches of soredia on the lobe tips. The abundantly pitted, sieve-like lower surface establishes its identity as a *Cavernularia*. The only other species, *C. lophyrea*, has flatter lobes with no soredia and is almost always fertile.

Cavernularia lophyrea
Fruiting honeycomb lichen

DESCRIPTION and COMMENTS: This is the nonsorediate, abundantly fertile version of *C. hultenii*. The broad, shiny brown apothecia can reach 15 mm in diameter, although the thallus lobes themselves are very narrow (mostly under 0.8 mm wide). The lobes are squarish at the tips and almost always covered with black dots (the buried pycnidia), and the thallus tends to be more closely adnate to the bark than in *C. hultenii*. *Cavernularia lophyrea*, endemic to western North America, is more closely restricted to the coastal forests and is less common.

185. *Cavernularia hultenii* coastal Alaska ×4.0

Cetraria (9 N. Am. species)
Iceland lichens, Icelandmoss, heath lichens

DESCRIPTION: Fruticose, almost fruticose, or erect foliose lichens, yellowish to reddish brown, or less frequently greenish to olive-brown (darker where most exposed to sun); branches flattened or cylindrical, usually with conspicuous pseudocyphellae. Photobiont green (*Trebouxia*?). Apothecia lecanorine, marginal or subterminal, with colorless, ellipsoid, 1-celled spores; asci with a ring structure at the apex that stains darkly in K/I; pycnidia black, buried in conspicuous marginal projections; conidia cylindrical or dumbbell-shaped. CHEMISTRY: All species contain fatty acids, and many have fumarprotocetraric acid as well (PD+ red, K–, KC–, C–). HABITAT: On soil or heath. COMMENTS: *Cetraria* is divided into two groups: (*i*) the Iceland lichens are basically erect foliose lichens taking on a fruticose growth form (e.g., *C. islandica*); (*ii*) the heath lichens consist of species with terete or cylindrical branches that form densely branched cushions (e.g., *C. aculeata*). However, other brown to black, erect, ground-dwelling foliose lichens that closely resemble the Iceland lichens are now excluded from the genus based on features of their asci, thallus anatomy, or chemistry. *Arctocetraria andrejevii* and *A. nigri-*

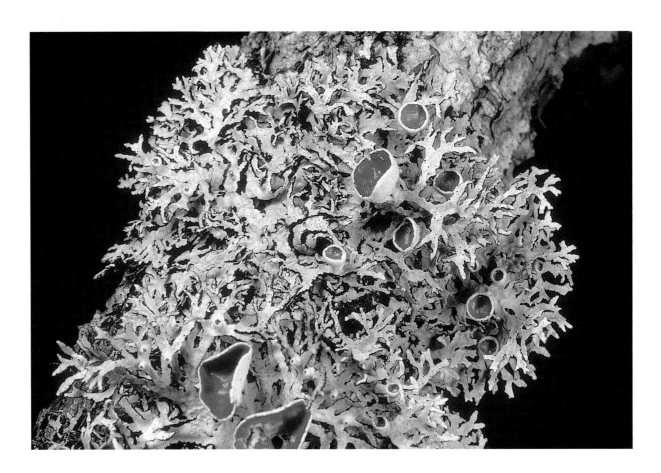

186. *Cavernularia lophyrea* Oregon coast ×4.3

cascens are both arctic species distinguished by the broad bases of their paraphyses. Species of *Cetrariella*, such as *C. delisei*, contain gyrophoric acid in the medulla (KC+ red, C+ pink) and have bottle-shaped conidia. *Flavocetraria* is very similar in growth form but is yellow-green to pale yellow, with usnic acid in the cortex. Other related genera that are more typically foliose include *Tuckermannopsis*, *Melanelia*, and *Esslingeriana*.

KEY TO SPECIES (including *Arctocetraria*, *Cetrariella*, and *Masonhalea*)

1. On trees and shrubs, subalpine; marginal projections very few; medulla PD– .... *Tuckermannopsis subalpina*
1. On the ground, sometimes mixed in moss mats or low heath; medulla PD– or PD+ red (fumarprotocetraric acid) 2

2. Branches round in cross section; thallus entirely fruticose; with round, depressed pseudocyphellae .... *Cetraria aculeata*
2. Branches flat; thallus erect foliose; pseudocyphellae either irregular in shape or linear, depressed or not 3

3. Lower surface pruinose, with an irregular pattern of white and dark areas; medulla UV+ blue-white (alectoronic acid) .... *Masonhalea richardsonii*
3. Lower surface not pruinose; medulla UV– (alectoronic acid absent) 4

4. Medulla KC+ red, C+ red or pink (gyrophoric acid); lobe margins more or less even, not toothed or ciliate .... *Cetrariella delisei*
4. Medulla KC– (gyrophoric acid absent); lobe margins even, or toothed, or ciliate 5

5. Medulla PD+ red 6
5. Medulla PD– 7

6. Pseudocyphellae marginal, forming an almost unbroken line along the lobe margins ..... *Cetraria laevigata*
6. Pseudocyphellae laminal and irregular in shape, but often also along the margins here and there .... *Cetraria islandica*

7. Mainly temperate; thallus gray-olive to olive-brown when dry .... *Cetraria arenaria*
7. Mainly arctic-alpine; thallus yellowish brown to reddish brown when dry 8

8. Pseudocyphellae usually marginal .. *Cetraria ericetorum*
8. Pseudocyphellae most commonly laminal 9

9. Pseudocyphellae broad and conspicuous; margins conspicuously fringed with projections; widespread boreal

to montane; contains protolichesterinic acid and (or) lichesterinic acid . . . .
. . . . rare PD– strain of *Cetraria islandica*

9. Pseudocyphellae small and dark, usually inconspicuous; marginal projections usually very short and tooth-like, rarely forming a noticeable fringe; arctic to northern boreal; contains rangiformic acid . . . .
. . . . [*Arctocetraria andrejevii*]

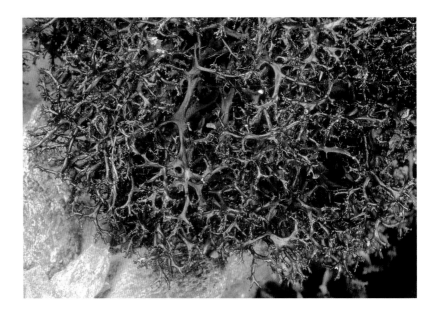

187. *Cetraria aculeata* Olympics, Washington ×2.5

Cetraria aculeata (syn. *Coelocaulon aculeatum*)
Spiny heath lichen

DESCRIPTION: Thallus reddish brown, cushion-forming, very spiny, with irregularly divided, terete to angular branches up to 2 mm wide, longitudinally ridged and grooved, with broad and deep pseudocyphellae (up to 0.6 mm long and 0.3 mm across). CHEMISTRY: Medulla PD–, K–, KC–, C– (protolichesterinic acid group). HABITAT: Growing among heath plants in very exposed, usually alpine and arctic habitats. COMMENTS: The rounded to angular branches distinguish this species (and *C. muricata*) from the true Iceland lichens, all of which have flattened branches. *Cetraria muricata* is extremely close to, and intergrades with, *C. aculeata*, which typically has thicker branches and larger, deeper pseudocyphellae. The branches of *C. muricata* are usually less than 1 mm broad except in the oldest basal parts, and the pseudocyphellae, although sometimes deep, white, and conspicuous, are less abundant and rarely longer than 0.3 mm and wider than 0.2 mm. Also, the branches of *C. aculeata* are more ridged and grooved than those of *C. muricata*. Intermediates are especially abundant in western North America, from Alberta to Alaska. The arctic-alpine *Bryocaulon divergens* is similar in color and branching but has longer, less spiny branches and tends to be more sprawling. The white pseudocyphellae of *Bryocaulon* are slightly raised rather than depressed, and they react C+ red because of olivetoric acid in the medulla.

Cetraria arenaria
Sand-loving Iceland lichen

DESCRIPTION: Thallus gray-olive to olive-brown; lobes rather flat or curled in at the sides, 1–4 mm broad, typically with regularly forked branching, with marginal and rather broad pseudocyphellae or, less frequently, large, irregular, depressed, laminal pseudocyphellae as well; lobes fringed with short, usually unbranched, spiny projections. Apothecia rare, shiny red-brown, on expanded lobe tips. CHEMISTRY: Medulla PD–, K–, KC–, C– (protolichesterinic and lichesterinic acids). HABITAT: Common on sandy soil or on thin soil over bedrock in temperate, lowland areas. COMMENTS: *Cetraria arenaria*, although similar to the more northern species *C. ericetorum* in having marginal pseudocyphellae and a PD– medulla, is more olive in color and has flatter lobes (rarely curled in so strongly that they fuse at the edges).

Cetraria ericetorum
Iceland lichen

DESCRIPTION: Thallus pale to dark brown; lobes narrow, 1–3 mm across, usually curled in and often fused where the edges touch, with spiny marginal projections; long, white pseudocyphellae all along the margins, rarely on the lower surface of the branches as well. Almost all North American populations of *C. ericetorum* have a conspicuously ridged and pitted surface and are called ssp. *reticulata*. CHEMISTRY: Medulla PD–, K–, KC–, C– (lichesterinic acid and two unidentified substances). HABITAT: On the ground with grasses and heath. COMMENTS: To tell one species of Iceland lichen from another, look for the white pseudocyphellae on the branches. In *C. ericetorum*, the pseudocyphellae form

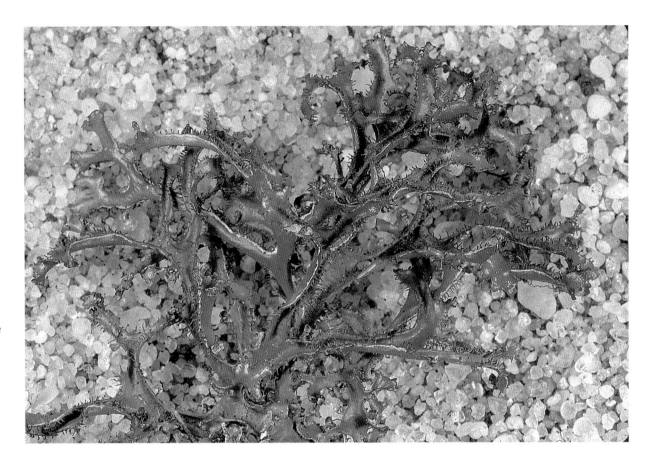

188. *Cetraria arenaria* Cape Cod, Massachusetts ×3.5

189. *Cetraria ericetorum* ssp. *reticulata* southern British Columbia ×2.7

190. (facing page) *Cetraria islandica* ssp. *crispiformis* coastal Alaska ×4.7

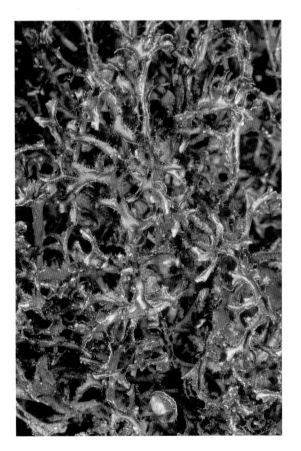

white lines along the lobe margins, sometimes interrupted, sometimes almost continuous. Very few round to irregular pseudocyphellae occur on the surface of the lobes in contrast to *C. islandica*. In addition, the lobes are narrower, on average, and *C. ericetorum* never contains fumarprotocetraric acid. The extent to which the lobes curl inward is another distinguishing characteristic: in *C. ericetorum* and *C. laevigata*, the edges fuse where they meet, almost forming tubes. Some lobes may be flatter, however, as in most subspecies of *C. islandica*. **Cetraria nigricans** is a common PD– species on the arctic tundra. It is quite small, mostly under 2 cm high, very dark (almost black), grows in dense colonies, and has rather long, often branched cilia along the lobe margins.

Cetraria islandica
True Iceland lichen

DESCRIPTION: Lobe width extremely variable, 1–10 mm broad, sometimes more, relatively flat to quite rolled up, almost forming tubes; many irregular, white pseudocyphellae dot the lower

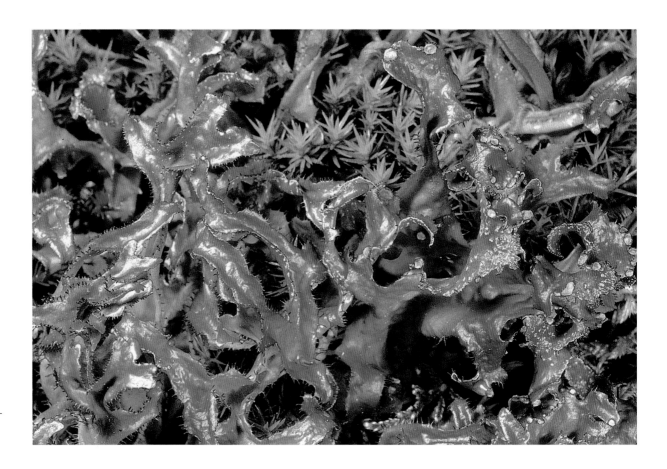

191. *Cetraria islandica* ssp. *islandica* mountains, coastal Alaska ×3.2

surface of the lobes (sometimes also developing along the margins, at least in younger parts); ssp. *crispiformis* differs from ssp. *islandica* in its more irregular (pitted and ridged) surface, especially on the older portions, and in the smaller, less conspicuous pseudocyphellae on the lobe surface. CHEMISTRY: Medulla almost always PD+ red, K–, KC–, C– (fumarprotocetraric acid with protolichesterinic and lichesterinic acids), rarely PD–. HABITAT: Widespread in boreal to arctic heath, and on the forest floor, especially in pine stands. DISTRIBUTION: The map shows the combined distribution of ssp. *islandica* and ssp. *crispiformis*. The ranges overlap over most of North America, but only ssp. *islandica* is found in the Rocky Mountains south of the Canadian border, and only ssp. *crispiformis* extends southward into the mountains of New England. COMMENTS: *Cetraria islandica* ssp. *crispiformis* resembles *C. ericetorum* except for the position of the pseudocyphellae and the PD reaction. IMPORTANCE: Iceland lichen is one of the few lichens that were regularly eaten by northern Europeans, especially in times of food shortages. It was ground and added to wheat flour or mashed potatoes to make bread or porridge. Milk could be added to make easily digested soups as well. A common name for Iceland lichen in Norway is *brødmose* (bread moss).

Cetraria laevigata
Striped Iceland lichen

DESCRIPTION: Lobes narrow, 1–3 mm across, with a smooth and shiny surface; rather broad, white, marginal pseudocyphellae forming a very conspicuous, almost unbroken line from lobe tip to base, with only rare and scattered laminal pseudocyphellae. CHEMISTRY: Medulla PD+ red, K–, KC–, C– (fumarprotocetraric, protolichesterinic, and lichesterinic acids). HABITAT: In tundra heath or on the ground in boreal forest openings. COMMENTS: *Cetraria islandica*, a more broadly lobed PD+ red species, has mainly laminal pseudocyphellae and only interrupted marginal pseudocyphellae (although they are sometimes conspicuous). *Cetraria ericetorum* can be very similar with respect to the pseudocyphellae and lobe width, but it is PD–.

192. *Cetraria laevigata* interior Alaska ×3.3

193. *Cetrariella delisei* mountains, coastal Alaska ×1.0

Cetrariella (2 species worldwide)

KEY TO SPECIES: See *Cetraria* and Key A.

Cetrariella delisei
Snow-bed Iceland lichen

DESCRIPTION: Lobes pale yellowish brown, often mottled with darker brown, much-branched and dissected especially close to the tips, about 1–5 mm wide, not strongly curled into tubes, with few marginal projections or cilia; pseudocyphellae usually laminal but with some interrupted marginal pseudocyphellae as well. Photobiont green (*Trebouxia?*). CHEMISTRY: Medulla and sometimes the cortex and pseudocyphellae PD–, K–, KC+ red, C+ pink (gyrophoric acid). HABITAT: On mineral soil, very often associated with late snow beds in the mountains and arctic regions. COMMENTS: The C+ pink reaction of the medulla distinguishes *C. delisei* from most other Iceland lichens, as does pale color and lack

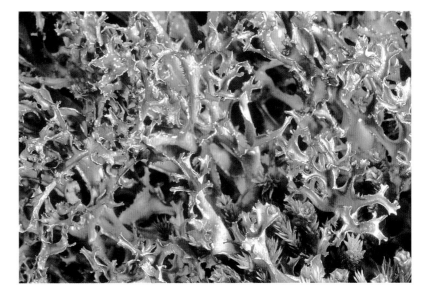

of marginal projections. Most similar is *Cetrariella fastigiata*, which has lobes that are more curled into a tube; its white pseudocyphellae, which are both marginal and laminal, are much less abundant and conspicuous. It is also arctic-alpine.

Cetrelia (4 N. Am. species)
Sea-storm lichens

DESCRIPTION: Broad-lobed (5–18 mm across), greenish gray, foliose lichens with white pseudocyphellae on the upper surface; all species sorediate except *C. alaskana;* lobes without marginal cilia; lower surface black and shiny with brown edges sometimes blotched with white; rhizines sparse, short, and unbranched. Photobiont green (*Trebouxia?*). Apothecia rare in North American species, lecanorine, usually perforate; spores colorless, ellipsoid, 1-celled, 8 per ascus; pycnidia marginal; conidia rod-shaped, 3–6 μm long. CHEMISTRY: Cortex K+ yellow (atranorin); medulla of all species PD–, K–, C– or C+ pink, KC– or KC+ pink (various orcinol depsides). HABITAT: On bark and mossy rock, usually in shaded, rather humid sites. COMMENTS: The sea-storm lichens can resemble species of *Parmotrema*, which never have pseudocyphellae and usually have marginal cilia. Forms of *Platismatia glauca* having only marginal soredia can be similar, but its soredia are coarser, almost isidia-like, and the lobes are more irregular, with ridges and depressions.

KEY TO SPECIES

1. Soredia absent; on tundra in arctic Alaska . . . . [*Cetrelia alaskana*]
1. Soredia present along lobe margins; on trees or rocks 2

2. Medulla C+ red or pink, UV– (olivetoric acid); pseudocyphellae sparse and small (up to 0.3 mm in diameter) . . . . *Cetrelia olivetorum*
2. Medulla C– 3

3. Medulla KC+ pink to red, UV+ blue-white (alectoronic acid); pseudocyphellae abundant and often large (0.15–0.6 mm in diameter) . . . . *Cetrelia chicitae*
3. Medulla KC–, rarely KC+ faint pink (perlatolic or imbricaric acid); pseudocyphellae small and inconspicuous . . . . [*Cetrelia cetrarioides*]

Cetrelia chicitae
Sea-storm lichen

DESCRIPTION: Thallus greenish gray to pale brownish gray, with broad, ascending lobes, undulating and ruffled at the margins, 5–20 mm wide; powdery to coarsely granular white soredia all along the margins of the older lobes, sometimes with round laminal soralia close to the margins; white, irregular pseudocyphellae on the upper surface, mostly 0.15–0.6 mm in diameter; lower surface black in the middle grading to brown, sometimes blotched with ivory at the margins; rhizines sparse, black, absent close to the margins. Apothecia extremely rare. CHEMISTRY: Medulla PD–, K–, KC+ pink, C–, UV+ blue-white (α-collatolic and alectoronic acids). HABITAT: On bark or mossy rocks in forests. COMMENTS: There are three sea-storm lichens with soredia in North America: *Cetrelia cetrarioides, C. chicitae,* and *C. olivetorum.* All three have broad, wavy margins lifting at the edges and fringed with white soredia, looking very much like a frilly petticoat or frothy waves at sea. The three differ in their medullary chemistry and little else. *Cetrelia cetrarioides* is negative with all reagents (or slightly KC+ pink). Populations of *C. cetrarioides* in the Pacific Northwest generally have perlatolic acid; those in the Appalachian–Great Lakes region usually contain imbricaric acid and are sometimes called *C. monachorum. Cetrelia cetrarioides* is the only *Cetrelia* spotted with large white pseudocyphellae on the lower surface as well as the upper surface. It is the least common of the three in North America, although it is very common in Europe. *Cetrelia cetrarioides* resembles *C. chicitae* in its rather conspicuous pseudocyphellae. *Cetrelia olivetorum* has smaller, less abundant pseudocyphellae, and it is the only species that is C+ red. *Cetrelia alaskana* is a tundra species along the west coast of Alaska. It lacks soredia but otherwise closely resembles *C. cetrarioides* (also containing imbricaric acid).

Cetrelia olivetorum
Sea-storm lichen

DESCRIPTION: Much like *C. chicitae* except for the smaller, sparser pseudocyphellae (rarely more than 0.3 mm across) and with more powdery soredia entirely confined to lobe margins.

194. *Cetrelia chicitae* Lake Superior region, Ontario ×2.5

195. *Cetrelia olivetorum* Lake Superior region, Ontario ×1.6

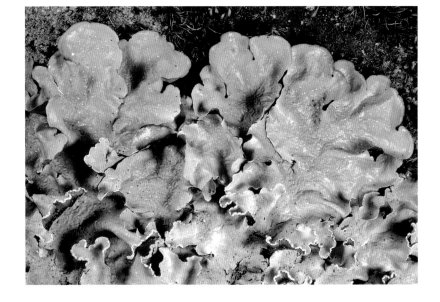

CHEMISTRY: Medulla PD–, K–, KC+ red, C+ red (olivetoric acid). HABITAT: On trees and mossy rocks in shaded forest habitats. COMMENTS: Species of *Punctelia* that have a C+ red medulla can be distinguished from *C. olivetorum* by their abundant rhizines that extend up to the lobe margins. See Comments under *C. chicitae*.

Chaenotheca (16 N. Am. species) (syn. *Coniocybe*)
Stubble lichens, pin lichens, whisker lichens

DESCRIPTION: Crustose lichens with apothecia consisting of minute, cup-like capitula (heads) at the tips of tiny, slender stalks (Fig. 12f); thalli growing within the substrate or forming thin to thick superficial crusts, gray, greenish gray, brown, or bright yellow. Photobiont green (*Stichococcus, Trebouxia, Dictyochloropsis,* or *Trentepohlia*). Asci in the capitula soon break down to produce a brown mass of 1-celled spores (a mazaedium); spores spherical or ellipsoid, brown to almost colorless, with smooth or rough walls. CHEMISTRY: The yellow pigment in the pruina

196. *Chaenotheca brunneola* interior plateau, British Columbia ×11.0

or thallus is vulpinic acid; a few species contain PD+ red depsides or depsidones. HABITAT: Mainly on bark and wood, more rarely on soil or rock in very sheltered areas; most abundant in shaded, old forests. COMMENTS: *Chaenotheca*, one of the most common genera of stubble lichens, is characterized by its slender, usually dark stalks and brown mazaedia with 1-celled, often pale spores. Species of *Cybebe* and *Sclerophora* differ in technical details, and the latter, with pale stalks, colorless spores, and unique spore ornamentation, is probably not even closely related to *Chaenotheca*. *Calicium* usually has thicker, stubbier stalks and dark, 2-celled spores. In *Chaenothecopsis,* the asci persist, so the apothecium is not really a mazaedium. *Mycocalicium* (1-celled, dark brown spores), *Stenocybe* (4–8-celled, dark brown spores), and *Phaeocalicium* (1–2-celled, dark brown spores) are superficially similar to some species of *Chaenotheca* but live as saprophytes or parasites and do not produce a visible thallus. IMPORTANCE: Because many species of *Chaenotheca* and other stubble lichens are excellent indicators of forest continuity, inventories of these tiny lichens are becoming essential to identify old growth forests, which should be conserved for their biological diversity.

KEY TO SPECIES: See Calicium.

Chaenotheca brunneola
Brown-head stubble lichen

DESCRIPTION: Thallus usually within the substrate and not seen, but rarely very thin yet perceptible, powdery, and greenish. Photobiont *Trebouxia* or *Dictyochloropsis*. Fruiting bodies consisting of slender, delicate black stalks, mostly 0.5–1.5 mm tall, with spherical, brown capitula that lack pruina; spores spherical, light brown, 3.3–4.6 µm in diameter. CHEMISTRY: Thallus PD+ orange, K–, KC–, C–, but usually too scanty to test (baeomycesic acid, sometimes with squamatic acid). HABITAT: On hard or well-rotted wood, or on bark (especially of conifers), rarely on bracket fungi. COMMENTS: *Chaenotheca ferruginea* also lacks any pruina but has a well-developed, superficial, pale gray thallus (often with yellow or red spots) and larger spores (5–6.5 µm in diameter). Its capitulum is black and cup-like, and its distribution is similar to that of *C. brunneola*.

Chaenotheca furfuracea
Sulphur stubble lichen

DESCRIPTION: Thallus well-developed, powdery, bright yellow-green or vivid green. Photobiont *Stichococcus*. Stalks 2–4 mm high, very slender, covered with yellow pruina; capitulum almost spherical, mostly 0.2–0.35 mm in diameter, pale brown but usually covered with a thick yellow pruina; spores spherical, very pale brown to almost colorless, 2–4.5 µm in diameter. CHEMISTRY: Vulpinic acid present in the thallus and pruina. HABITAT: Typically on the soil and rootlets of upturned stumps in shaded forests, usually in the most protected recesses; also on wood and stumps, rarely rocks. COMMENTS: Another common yellowish *Chaenotheca* is *Ch. chrysocephala,* often seen on the bark of conifers or birch, as well as on wood, throughout the boreal region. Its thallus is composed of large, bright yellow-green or chartreuse, corticate granules. Yellow pruina is formed on the lower side of the capitulum extending a short way down the stalk but not covering it, and the spores are at least in part ellipsoid, 4.5–6.5 × 2–4 µm.

Chrysothrix (2 N. Am. species)

KEY TO SPECIES: See Key C.

Chrysothrix candelaris
Gold dust lichen

DESCRIPTION: Thallus bright, almost glowing yellow, often flecked with orange, less commonly greenish yellow, composed entirely of powdery soredia (leprose); particles of soredia very tiny (less than 0.1 mm in diameter). Photobiont green. Apothecia not found in North American specimens. CHEMISTRY: PD–, K–, KC–, C– (calycin or pinastric acid, or both; no vulpinic acid). HABITAT: On shaded bark of all kinds and occasionally on rock; widely distributed in rich, old forests, but also on roadside trees. COMMENTS: *Chrysothrix chlorina,* the sulphur dust lichen (plate 24), is the sister species of *Ch. candelaris.* Its color is more like sulphur because of different yellow pigments (calycin and vulpinic acid), although the color is somewhat variable in both species. Its soredia are larger (over 0.1 mm in diameter), and it forms a thicker crust. In addition, *Ch. chlorina* is usually found on shaded rocks rather than trees. Sterile specimens of some stubble lichens (members of the Caliciales) such as *Chaenotheca furfuracea* are easily mistaken for species of *Chrysothrix*, especially *Ch. chlorina. Chaenotheca* species usually have a thinner crust, and a diligent search will usually turn up at least one or two tiny stalks with their pinhead apothecia. *Psilolechia lucida,* which grows on rocks, has a similar, sulphur yellow, leprose thallus, but it usually produces at least a few small, bright yellow, biatorine apothecia. The thallus contains rhizocarpic acid and is therefore UV+ bright orange.

Cladina (14 N. Am. species)
Reindeer lichens, caribou lichens, reindeer moss

DESCRIPTION: Erect fruticose lichens with abundantly branched, hollow podetia lacking an outer cortex but supported by a cartilaginous inner medullary layer (the stereome) that can be translucent or some-

197. *Chaenotheca furfuracea* California coast ×11.4

198. *Chrysothrix candelaris* Coast Range, central California ×1.1

199. *Cladina arbuscula* ssp. *beringiana* Olympics, Washington ×1.8

times becomes blackened; podetia arising from a granular crustose thallus that quickly disappears and is rarely seen. Photobiont green (*Trebouxia*). Apothecia uncommon in most species, small, brown, almost globose, produced at the tips of the podetia; pycnidia common on podetial tips, filled with a clear or reddish jelly. The color of the jelly can be seen by mounting a fresh whole pycnidium in water on a microscope slide, then crushing it under a cover slip with the point of a stiff needle to squeeze out the jelly. The jelly can rarely be seen in specimens more than a year or two old. CHEMISTRY: Frequently with atranorin (K+ yellow), fumarprotocetraric acid (almost always accompanied by protocetraric acid), and some minor PD+ red compounds, some fatty acids, and (or) perlatolic acid as principal products. To get a reliable PD+ reaction, test the tips of the stalks. The K+ yellow reaction of atranorin is faint and hard to interpret in some specimens. HABITAT: On the ground in dry to boggy, usually acidic habitats. COMMENTS: The reindeer lichens differ from most species of *Cladonia* (also stalked, hollow, fruticose lichens) in lacking an outer cortex and any scale-like squamules at the base of the podetia or on the stalks themselves. Because the cortex is absent, the podetia are never shiny and smooth as in the abundantly branched *Cladonia uncialis* and *C. furcata*, which resemble species of *Cladina* in habit. Important characters include the number of branches arising at each axil, whether or not the axils are perforated with an opening into the hollow stem, and how smooth or flocculent (broken into tiny, cottony patches) the surface becomes.

KEY TO SPECIES: See Cladonia.

Cladina arbuscula
Reindeer lichen

DESCRIPTION: Thallus pale yellowish green, with densely branched tips mainly in threes and fours around widely open axils, the tips sometimes bending to one side and looking "combed"; branches short, thick; outer medulla very compact; sometimes resembling an areolate cortex (in ssp. *arbuscula*), or smooth (in the western ssp. *beringiana*, represented in plate 199). CHEMISTRY: Thallus PD+ red, K–, KC+ yellow, C–, UV– (fumarprotocetraric and usnic acids); rarely PD–. HABITAT: Growing abundantly in northern regions over thin soil or rocky ground, among heath plants, or sometimes in wet habitats. COMMENTS: Sometimes slender specimens resemble *C. subtenuis*, in which the branching is in twos, the axils are closed, and the pycnidial jelly is red. Most easily mistaken for *C. arbuscula* is *C. mitis*, which is PD– (lacking fumarprotocetraric acid); in other respects it is very similar, differing mainly in its more open branching at the tips and its generally more slender appearance. IMPORTANCE: The many uses and ecological importance of reindeer lichens are described under *C. rangiferina* and *C. stellaris*.

Cladina evansii
Powder-puff lichen, deer moss

DESCRIPTION: Thallus whitish gray, forming clusters of rounded tufts of densely branched podetia, the tufts 2–3 cm across and 3–6 cm tall; branching mainly in twos and threes, with axils not often perforated. CHEMISTRY: Thallus PD–, K+ yellow, KC–, C–, UV+ blue-white (atranorin and perlatolic acid). HABITAT: On the ground in open or partially shaded, usually sandy areas. COMMENTS: This lichen has the same general growth form as *C. stellaris*, a yellow-green boreal lichen (containing usnic acid) in which the tips usually have 4–5 branches that flare in a star-like cluster around a gaping hole.

200. *Cladina evansii*
northern Florida ×0.86

Cladina mitis
Green reindeer lichen

DESCRIPTION: Thallus pale yellowish gray to yellowish green, the tips only slightly browned; branching tree-like from several main stems, 0.5–0.8 mm thick, commonly in threes but also in twos and fours, the crown typically open (not dense) and divergent (not curved in one direction), although the curvature and density of the branches are variable; stem surface relatively compact and even, not flocculent. CHEMISTRY: Thallus PD–, K–, KC+ yellow, C–, UV– (usnic acid, often with rangiformic acid). HABITAT: Over thin soil, rocks, heath, moss, and grassy areas, usually in full sun. COMMENTS: This species intergrades with forms of *C. arbuscula,* and some lichenologists do not recognize these as separate species. Typically, *C. arbuscula* has denser crowns with more distinctly browned tips that curve in one direction, and it is PD+ red, except in rare populations. *Cladina submitis* is very robust, with stems commonly over 1 mm thick, and with a very thick, compact outer surface. *Cladina portentosa* branches mainly in threes and has a more flocculent surface below the tips; it contains perlatolic acid.

Cladina portentosa ssp. *pacifica*
(syn. *C. pacifica*)
Maritime reindeer lichen

DESCRIPTION: Thallus pale yellowish green (with usnic acid) or pale gray (lacking usnic acid); tips of the branches often conspicuously browned; branching commonly in threes or sometimes in twos; surface tissue (over the inner translucent stereome) very thin, in scattered flocculent patches. CHEMISTRY: Thallus PD–, K–, KC+ yellow or KC–, C–, UV+ blue-white (perlatolic acid with or without usnic acid). HABITAT: In dry or wet habitats, but in the northwest, it is particularly well-developed on hummocks in bogs. COMMENTS: The subspecies *pacifica* of *C. portentosa* is the only subspecies occurring in North America; subspecies *portentosa* is a European lichen. *Cladina mitis* is sometimes quite similar to *C. portentosa* but has numerous branches in fours, less browning on the tips, and a smoother, more uniform surface. *Cladina*

201. *Cladina mitis* Lake Superior region, Ontario ×1.9

202. *Cladina portentosa* ssp. *pacifica* northern California coast ×1.8

203. (facing page) *Cladina rangiferina* Southeast Alaska ×1.9

204. *Cladina rangiferina* (southern form) southeastern Missouri ×0.72

terrae-novae, a northeastern species also common in boggy heath, closely resembles *C. portentosa* but contains atranorin (K+ yellow).

Cladina rangiferina
Gray reindeer lichen

DESCRIPTION: Thallus white to silver-gray, with somewhat browned, deflected or "combed" tips; branching tree-like with main stems and side branches commonly in twos and threes and sometimes in fours; surface dull, fibrous or webby, with scattered, rounded bumps especially on the older parts of the podetia; pycnidia on the branch tips containing a colorless jelly (seen with fresh material in a microscopic squash mounted in water). CHEMISTRY: Tips of the branches PD+ red, K+ pale yellow, KC–, C–, UV– (fumarprotocetraric acid and atranorin). HABITAT: Typically on thin soil, over rock, or in sandy places; also in wet sites especially along the northwest coast; usually in full sun. COMMENTS: Eastern Temperate populations of *C. rangiferina* (as shown in plate 204) are typically darker and more densely branched, with shorter internodes than western populations (plate 203). Most other species of *Cladina* contain usnic acid and are greenish or yellow-green. The color distinction is sometimes difficult because the gray species become much greener when wet. *Cladina stygia* is very similar but has a conspicuously blackened base (and stereome); in *C. rangiferina*, the decaying base remains gray or yellowish brown. In addition, the pycnidial jelly in *C. stygia* is pinkish. *Cladina stygia* is found mainly in boggy areas. The gray form of the western *C. portentosa* sometimes resembles *C. rangiferina* but is more slender and less combed to one side, and reacts PD– and UV+ blue-white. A few species of *Cladonia*, despite their anatomical differences, resemble *C. rangiferina*, the most confusing being **Cladonia wainioi**, a rare northern and west coast lichen. This species has a smoother surface (although in areolate patches over a translucent stereome), lacks the PD+ red fumarprotocetraric acid, and contains merochlorophaeic acid (KC+ pinkish violet, rapidly disappearing). IMPORTANCE: Together with other species of *Cladina*, especially *C. stellaris* and *C. mitis*, *C. rangiferina* constitutes the principal winter food of caribou in North America and reindeer in Europe (see Chapter 8). The Ojibwas have used a decoction of reindeer lichen for bathing newborn babies. Canadian native groups have used it as food, and Canadian woodsmen have cooked it to make a stimulant tea.

Cladina stellaris
Star-tipped reindeer lichen

DESCRIPTION: Thallus pale yellowish green, forming rounded heads 20–40 mm across, consisting of intricately branched podetia with no main stems but with wide-open axils. Podetial tips terminating in star-like clusters of 4–5 branches around a gaping hole; surface uneven and flocculent, especially on older areas. CHEMISTRY: Thallus PD–, K–, KC+ gold, C–, UV+ blue-white (usnic and perlatolic acids); a rare chemical race has psoromic acid in the podetial tips (PD+ bright yellow). HABITAT: On the ground and in heath, sometimes forming extensive mats 10–20 cm thick. COMMENTS: Two North American reindeer lichens form rounded, foam-like tufts: the yellowish green *Cladina stellaris*, which covers thousands of square miles of boreal woodland soil, and the almost white *C. evansii* found on the southeastern coastal plain. *Cladina evansii* contains atranorin instead of usnic acid and is K+ yellow, PD–, C–, and KC–. IMPORTANCE: Although *Cladina rangiferina* is the nominal true reindeer lichen (*Rangifer* is the scientific name of reindeer and caribou), *C. stellaris* is more important as winter food for both wild and domesticated animals and is preferred to all other reindeer lichens. It is also used in the floral industry as a

decoration and as miniature trees and shrubs in architectural models and miniature railroad layouts (plate 72). Tons of *C. stellaris* are used in the pharmaceutical industry in Europe as a source of usnic acid, a mild antibiotic, which is incorporated into topical ointments. One European brand is called "Usno." Finally, the critical role of *C. stellaris* in the boreal ecosystem should not be overlooked (see Chapter 8).

Cladina stygia
Black-footed reindeer lichen

DESCRIPTION, CHEMISTRY, and COMMENTS: *Cladina stygia* is very similar to *C. rangiferina* in form, color, and chemistry, but the surface and stereome at the base of the thallus are dark brown to black. The conclusive characteristic is the pinkish jelly in the pycnidia of freshly collected material (see Comments under *Cladina*). The jelly is colorless in the true reindeer lichen. HABITAT: Most common in boggy or, at least, wet sites in the open.

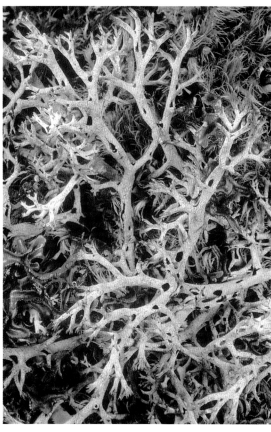

205. *Cladina stellaris* North Woods, Maine ×0.75

206. *Cladina stygia* interior Alaska ×1.9

207. *Cladina submitis* Pine Barrens, New Jersey ×1.7

Cladina submitis
Dune reindeer lichen

DESCRIPTION: Thallus yellowish green, with thick, twisted stems 0.7–2(–3) mm in diameter, sprawling over the sand in tangled mats or growing erect (in more protected sites); older stems can be strongly wrinkled (as shown in plate 207), often sprouting short, isidia-like side branches; surface and medulla compact and thick. CHEMISTRY: Thallus PD–, K–, KC+ yellow, C–, UV– (usnic and pseudonorrangiformic acids). HABITAT: Can be extremely common in sandy areas along the northeast coast. COMMENTS: Erect forms can look much like *C. arbuscula* (PD+ red) or *C. mitis* (also PD–), but only *C. submitis* contains pseudonorrangiformic acid.

Cladina subtenuis
Dixie reindeer lichen

DESCRIPTION: Thallus pale yellowish green to almost gray (in shade forms), rather smooth, with slender branches (0.5–0.7 mm at the base) that fork in regular dichotomies (rarely in threes) and rarely bend in one direction; most axils closed; clumps often forming large mats 5–10 cm or more thick. CHEMISTRY: Thallus PD+ red, K–, KC+ yellowish to gold, C–, UV– (usnic and fumarprotocetraric acids). HABITAT: On open, especially sandy, ground, sometimes mixed with grasses. COMMENTS: *Cladina subtenuis* is the only reindeer lichen centered in the southeast south of Virginia except for *C. evansii*, which is distinctive in color, form, and chemistry; the range of *C. rangiferina* extends into the southern Appalachian Mountains. Its dichotomous branches and closed axils distinguish *C. subtenuis* from most other reindeer lichens in the northeast. **Cladina terraenovae** is superficially similar but has a rougher, more flocculent surface and differs chemically (PD–, K+ yellow, UV+ blue-white, with atranorin and perlatolic acid). It is found along the coast from Newfoundland to New Jersey. IMPORTANCE: *Cladina subtenuis* is used occasionally as packing material by shippers of fragile merchandise in the southeast.

Cladonia (128 N. Am. species)
Cladonia

DESCRIPTION: Species in the genus *Cladonia* are two-part lichens. Development begins with a scaly (squamulose) primary thallus. The squamules making up the primary thallus (therefore called primary squamules) vary greatly in size, the way the margins are subdivided and lobed, and production of soredia, but with rare exceptions they are attached at one edge and have a white lower surface lacking a cortex or rhizines. Erect stalks (podetia) usually develop from the surface or edge of the squamules (Fig. 2h). In some species, the primary squamules disappear, leaving only podetia to constitute the thallus. The podetia are always hollow, but they can be unbranched or highly branched; they can end in a cup-like structure or be blunt or pointed at the tip; they can be covered with squamules, soredia, or flat to rounded granules, or they can be quite smooth. Podetia normally have an outer cortex, an algal layer, and a stiff, cartilaginous, translucent to blackened, supporting layer called a stereome that lines the hollow interior (Fig. 8f). The development of soredia or tiny squamules on the podetia, however, can break up the outer layers of the thallus. Although podetia usually make up the bulk of the vegetative tissue of the lichen, developmentally they are actually a part of the fruiting body. The business end of podetia, accordingly, consists of brown or bright red (or, in a few cases, waxy yellowish beige) biatorine apothecia containing colorless, 1-celled spores, 8 per ascus. The apothecia are generally very convex and cap-like but are sometimes flat. In many species of *Cladonia*, apothecia are rarely produced even though the podetia are well developed. Photobiont green (*Trebouxia* or *Pseudotrebouxia*). CHEMISTRY: *Cladonia* species contain a wide variety of compounds, especially usnic acid (in the cortex) and β-orcinol depsides and depsidones such as barbatic and squamatic acids, atranorin, fumarprotocetraric acid (almost always accompanied by protocetraric acid), and norstictic, psoromic, and thamnolic acids. Perform spot tests on the lower surface of the squamules, sorediate patches, cup margins, or the thickest areas of medulla. HABITAT: On soil, peat, wood, bark, or rock, often mixed with mosses.

208. *Cladina subtenuis* northern Florida ×1.1

COMMENTS: The hollow stalks distinguish *Cladonia* from other fruticose lichens of comparable size, color, and habit (e.g., *Stereocaulon*, *Baeomyces*, *Dibaeis*, *Allocetraria*, and *Pilophorus*); the squamulose primary thallus distinguishes it from *Cladina* and *Dactylina*, which also produce hollow stalks or branches. The primary squamules sometimes resemble other squamulose or minutely foliose lichens.

KEY TO SPECIES (including *Cladina*)

1. Podetia abundantly branched 2
1. Podetia unbranched, or branched only once or twice 25

2. Podetia without a cortex; surface dull and webby 3
2. Podetia with a cortex; surface usually somewhat shiny 13

3. Podetia silver gray to pale greenish gray; thallus KC–, K– or K+ yellow 4
3. Podetia pale yellow-green or greenish yellow (usnic-yellow); thallus KC+ yellow, K– 7

4. Thalli forming tight, rounded tufts without clearly defined main stems; southeastern coastal plain .... *Cladina evansii*
4. Thalli not forming tight, rounded tufts; main stems usually obvious; mainly northern or western 5

5. Branching at tips divergent, not bent in one direction; thallus PD–, K–, UV+ bright blue-white (perlatolic acid) .... *Cladina portentosa*
5. Branching at tips at least partly bent in one direction, giving the thallus a combed appearance; thallus PD+ red, K+ pale yellow, UV– (atranorin and fumarprotocetraric acid) 6

6. Basal part of podetia and stereome blackened; pycnidial jelly red. . . . . *Cladina stygia*
6. Basal part of podetia and stereome pale, more or less the same as upper portions; pycnidial jelly colorless . . . . *Cladina rangiferina*
7.(3) Podetia branching most commonly in twos (dichotomies) 8
7. Podetia branching most commonly in threes and fours 9
8. Podetia thick, 0.7–2 mm wide; podetial axils often open; stereome composed of distinct strands; thallus PD– or PD+ yellow (with or without psoromic acid; fumarprotocetraric acid absent). . . . . *Cladonia pachycladodes*
8. Podetia slender, usually under 0.7 mm wide; podetial axils usually closed; stereome continuous, not broken into strands; thallus PD+ red (fumarprotocetraric acid) . . . . *Cladina subtenuis*
9. Thallus forming tight, rounded tufts without obvious main stems; podetial tips with radiating branches around an open hole. . . . . *Cladina stellaris*
9. Thallus forming loosely organized cushions with clearly recognizable main stems; tips without radiating branches around an open hole 10
10. Surface flocculent (with small, fluffy clumps) except close to the branch tips; thallus UV+ bright blue-white (perlatolic acid). . . . . *Cladina portentosa*
10. Surface compact, smooth or bumpy, not flocculent; thallus UV– (lacking perlatolic acid). 11
11. Thallus PD+ red (fumarprotocetraric acid, lacking fatty acids) . . . . *Cladina arbuscula*
11. Thallus PD– (containing fatty acids) 12
12. Branches very robust, 0.7–2 mm wide, often sprawling and strongly wrinkled; axils broadly open; branching usually in fours; containing pseudonorrangiformic acid; on northeastern coast . . . . *Cladina submitis*
12. Branches usually slender, 0.5–0.8 mm wide, always erect, generally smooth; axils often closed or only slightly open; branching usually in threes, usually containing rangiformic acid; widespread boreal to northern temperate. . . . . *Cladina mitis*
13.(2) Thallus gray to gray-green, often browned in sunny habitats, KC–, K– or K+ yellow (usnic acid absent) 14
13. Thallus yellowish green or greenish yellow, KC+ yellow (usnic acid present) 18
14. Primary squamules large and persistent; podetia up to 25 mm tall; thallus K+ deep yellow (thamnolic acid). . . . . *Cladonia floridana*
14. Primary squamules usually disappearing (although there may be squamules on the podetia); podetia 20–60(–120) mm tall; thallus K– 15
15. Podetia granular sorediate at the tips. . . . . *Cladonia scabriuscula*
15. Podetia without soredia or granules. 16
16. Podetia lacking squamules, surface with flat areoles over a translucent stereome; thallus very *Cladina*-like, PD–, KC+ pinkish violet, rapidly disappearing (merochlorophaeic acid); uncommon, arctic and west coast . . . . . . [*Cladonia wainioi* (syn. *C. pseudorangiformis*)]
16. Podetia usually having at least a few (often many) squamules, with a continuous cortex; thallus stiff, PD+ red, KC– (fumarprotocetraric acid) 17
17. Podetia with unequal branches that are frequently split lengthwise. . . . . uncupped morphotype of *Cladonia multiformis*
17. Podetia with more or less equal, dichotomous branches, split or intact . . . . *Cladonia furcata*
18.(13) Apothecia red, almost spherical . . . . *Cladonia leporina*
18. Apothecia brown or absent 19
19. Stereome rudimentary or absent, not forming a well-defined, cartilaginous cylinder or strands; southeastern coastal plain 20
19. Stereome forming an intact, cartilaginous cylinder or network of cartilaginous cords or strands; southeastern or widespread 21
20. Podetial wall perforated with oval holes; podetial axils open; surface very shiny; very rare . . . . *Cladonia perforata*
20. Podetial wall not perforated or longitudinally split; podetial axils closed; surface smooth but usually dull; common . . . . *Cladonia leporina*
21. Stereome cylindrical, smooth, not broken into strands 22
21. Stereome broken into broad or narrow strands. 23
22. Podetia tall and slender, pointed at the tips or forming narrow but distinct cups; axils often closed or only partially open; barbatic acid present, squamatic acid absent . . . . *Cladonia amaurocraea*
22. Podetia forming densely branched cushions, entirely without cups; axils wide open; barbatic acid absent, with or without squamatic acid. . . . . . *Cladonia uncialis*
23. Podetia slender, 0.5–1.5 mm in diameter, not appearing inflated, usually forming rather flattened mats; stereome cords broad and flat . . . . *Cladonia dimorphoclada*
23. Podetia broad, mostly 3–11 mm in diameter, appearing inflated, usually erect and cushion-forming; stereome with broad or narrow cords. 24
24. On rock, less frequently on soil; stereome barely broken into broad bands; podetial wall intact, not perforated; southeastern U.S. . . . . *Cladonia caroliniana*
24. On sandy soil; stereome broken into a meshwork of

narrow cords; podetial wall perforated and fissured; northeastern coastal plain . . . . *Cladonia boryi*

25.(1) Main thallus consisting of primary squamules; podetia absent or extremely small (under 4 mm high) [*Note:* Many young or poorly developed specimens of *Cladonia* may key out here]. 26

25. Main thallus consisting of many podetia, usually over 4 mm tall; with or without a basal thallus of primary squamules 33

26. Lower surface of squamules pale yellow, upper surface yellow-green to olive; thallus KC+ yellow, or orange (usnic and barbatic acids). . . . . *Cladonia robbinsii*

26. Lower surface of squamules white; upper surface grayish green, brownish, or olive; thallus KC– (usnic and barbatic acids absent). 27

27. Primary thallus forming a lobed, almost foliose, closely attached rosette; medulla K– (atranorin usually absent) . . . . *Cladonia pocillum*

27. Primary thallus composed of discrete squamules, or if forming a rosette, then loosely attached and ascending; medulla K– or K+ pale yellow or red (atranorin, sometimes with norstictic acid). 28

28. Primary squamules extremely large, 10–20 mm long, 2–8 mm wide 29

28. Primary squamules typically less than 8 mm long and 4 mm wide. 30

29. Squamules usually forming a radiating rosette, curled up at the margins when dry; Florida and vicinity. . . . . *Cladonia prostrata*

29. Squamules separate, not forming radiating rosettes; western and northern regions. . . . . *Cladonia macrophyllodes*

30. Thallus C+ green, KC+ green (strepsilin); thallus olive-green, sometimes forming almost spherical, vagrant colonies. . . . . *Cladonia strepsilis*

30. Thallus C–, KC–; thallus gray-green, never forming spherical colonies 31

31. Thallus PD– or PD+ yellow (psoromic or norstictic acid often present) . . . . *Cladonia symphycarpia*

31. Thallus PD+ red (fumarprotocetraric acid). 32

32. Squamules finely divided; apothecia raised on short, noncorticate stalks 1–2 mm high; medulla K–. . . . . *Cladonia caespiticia*

32. Squamules strap-shaped, lobed but not finely divided; apothecia produced directly on basal squamules; medulla K+ yellow (atranorin). . . . . *Cladonia apodocarpa*

33.(25) Most or all podetia with cups or cup-like expansions. 34

33. Most or all podetia lacking cups 71

34. Podetia sorediate, or with spherical, corticate granules 35

34. Podetia without soredia or granules. 50

35. Thallus distinctly yellow-green or greenish yellow, KC+ yellow (containing usnic acid); most species have red apothecia. 36

35. Thallus gray, gray-green, pale green or olive, often becoming brown, KC– or KC+ pinkish orange (usnic acid absent); most species have brown apothecia. 40

36. Podetia mostly over 25 mm tall, with relatively narrow cups; soredia farinose; apothecia red 37

36. Podetia 6–25(–30) mm tall, with narrow or broad cups; soredia farinose or granular; apothecia red or pale brown. 38

37. Cups malformed, often with irregularly dentate margins; podetia typically fissured; medulla (not surface) UV+ blue-white (squamatic acid) . . . . *Cladonia sulphurina*

37. Cups usually well-formed, with even, often dentate or proliferating margins; podetial wall intact or fissured; medulla UV– (zeorin). . . . . *Cladonia deformis*

38. Primary squamules usually sorediate on the lower surface near the margins; cups very narrow; apothecia red; thallus PD+ orange, K+ yellow, UV– (thamnolic acid) or PD–, K–, UV+ blue-white (squamatic acid). . . . . *Cladonia umbricola*

38. Primary squamules never sorediate; cups goblet-shaped, narrow or broad; apothecia red or brown; lacking both thamnolic and squamatic acids, containing zeorin. 39

39. Apothecia and pycnidia bright red; soredia granular; cups relatively broad . . . . *Cladonia pleurota*

39. Apothecia and pycnidia pale brown; soredia farinose; cups broad or narrow . . . . *Cladonia carneola*

40.(35) Cups opening by a gaping hole; cup margins bent inward. . . . . *Cladonia cenotea*

40. Cups closed 41

41. Apothecia or pycnidia red . . . . 42 (also see couplet 37)

41. Apothecia or pycnidia brown or absent 43

42. Primary squamules finely divided, sorediate mostly near margins of lower surface; cups narrow, not toothed; typically coastal . . . . *Cladonia umbricola*

42. Primary squamules large, rounded, lower surface uniformly sorediate; cups trumpet- to goblet-shaped, often toothed at margin; typically inland . . . . *Cladonia digitata*

43. Cups broad, goblet-shaped 44

43. Cups narrow, trumpet-shaped or irregular 45

44. Soredia granular; thallus gray-green or brownish, never gray; lacking fatty acids; very common and widespread . . . . *Cladonia chlorophaea*

44. Soredia farinose or, in part, granular; thallus gray to

gray-green or very pale green; containing fatty acids; west coast . . . . *Cladonia asahinae*

45. Cups proliferating from the center as well as from the margins; podetia blackened at the base with a spotty appearance; soredia coarsely granular . . . . *Cladonia verruculosa*
45. Cups proliferating from the margins (if at all); podetia not blackened at the base. 46

46. Cups well-formed, trumpet-shaped 47
46. Cups flat, shallow or irregular and aborted 48

47. Podetia with only farinose soredia; never with a bluish tint when wet; lacking fatty acids; widespread . . . . *Cladonia fimbriata*
47. Podetia with granular and farinose soredia; often with a bluish tint when wet; containing fatty acids; west coast . . . . *Cladonia asahinae*

48. Cups very narrow and rather flat, with star-like proliferations; podetia brownish to olive, granular sorediate; thallus PD– or PD+ red (homosekikaic acid, with or without fumarprotocetraric acid) . . . . *Cladonia rei*
48. Cups poorly formed and irregular, without regular proliferations on the cup margins; podetia gray-green to pale green or brownish; soredia farinose or granular; thallus PD+ red (fumarprotocetraric acid, lacking homosekikaic acid) 49

49. Soredia produced in irregular patches on the upper ½ to ¾ of the podetia, often becoming continuous at the tip; soredia usually granular; thallus greenish, often browned in part . . . . *Cladonia ochrochlora*
49. Soredia covering podetia, entirely farinose; thallus ashy gray to greenish gray . . . . *Cladonia subulata*

50.(34) Thallus yellowish green or greenish yellow (containing usnic acid); apothecia red 51
50. Thallus not yellowish (usnic acid absent); apothecia brown. 52

51. Podetia and cup interior covered with round areoles, not squamulose; cup margins smooth or proliferating; thallus UV–, PD–, K– (barbatic acid present; squamatic and thamnolic acids absent) . . . . *Cladonia borealis*
51. Podetia and cup margins squamulose; thallus UV+ blue-white, PD–, K–, or UV–, PD+ orange, K+ deep yellow (barbatic acid absent; containing squamatic or thamnolic acid) . . . . *Cladonia bellidiflora*

52. Cups opening by a gaping hole; thallus PD–, UV+ blue-white, or PD+ orange, UV– (with squamatic or thamnolic acid). 53
52. Cups closed or partially perforate; thallus PD+ yellow or red (psoromic or fumarprotocetraric acid). 54

53. Podetial cortex continuous; surface with or without squamules . . . . *Cladonia crispata*

53. Podetial cortex absent; surface covered with large or extremely small squamules . . . . *Cladonia squamosa*

54. Podetia blackened at the base 55
54. Podetial base more or less the same color as upper portions 57

55. Cups proliferating from the margins; widespread . . . . *Cladonia phyllophora*
55. Cups proliferating from the center 56

56. Arctic regions; thallus K+ pale yellow (atranorin); PD+ red (fumarprotocetraric acid); podetia with pale areoles over a black stereome, giving the upper part a mottled appearance. *Cladonia trassii*
56. Eastern and southern coastal plain; thallus K–, PD+ red or yellow (fumarprotocetraric or psoromic acid; atranorin absent); primary squamules gray-green; sometimes blackened at base, but podetia more or less uniform in color, not mottled. . . . . *Cladonia rappii*

57.(54) Cups proliferating from the center. 58
57. Cups not proliferating, or proliferating from the margins. 61

58. Thallus K+ pale yellow (atranorin) . . . . *Cladonia macrophyllodes*
58. Thallus K– (atranorin absent). 59

59. Cups poorly formed and asymmetrical, having both marginal and central proliferations . . . . *Cladonia mateocyatha*
59. Cups round and symmetrical, proliferating almost exclusively from the center 60

60. Cups flaring abruptly, very flat . . . . *Cladonia rappii*
60. Cups tapered, flaring gradually, remaining concave at least at the edges . . . *Cladonia cervicornis* ssp. *verticillata*

61.(57) Primary thallus persistent 62
61. Primary thallus evanescent 67

62. Cups poorly formed and asymmetrical; thallus K– (atranorin absent) 63
62. Cups broad and goblet-shaped, or narrow and trumpet-shaped; thallus K+ yellow or K– 64

63. Podetia 10–60 mm tall; cups with only marginal proliferations, often squamulose at the margins . . . . *Cladonia phyllophora*
63. Podetia under 20 mm tall; cups irregular, also having some central proliferations . . . . *Cladonia mateocyatha*

64. Thallus K+ yellow (atranorin); cups broad or very narrow . . . . *Cladonia ecmocyna*
64. Thallus K– (usually lacking atranorin); cups broad, more or less goblet-shaped 65

65. Primary thallus consisting of thick, radiating, closely appressed, almost foliose lobes; on calcareous soil . . . . *Cladonia pocillum*

65. Primary thallus consisting of distinct, ascending squamules; on acidic soils 66
66. Cortex broken into round areoles that cover the podetia and line the cups; podetia under 30 m tall . . . . *Cladonia pyxidata*
66. Cortex continuous, smooth; podetia 20–50 mm tall . . . . *Cladonia gracilis* ssp. *turbinata*
67.(61) Cups and (or) podetia perforated or fissured, usually irregular in form 68
67. Cups and podetia closed (imperforate) 69
68. Cups with numerous perforations, like a sieve; cortex compact, rather smooth and shiny . . . . *Cladonia multiformis*
68. Cups without sieve-like perforations, but podetia irregularly perforate or fissured; surface dull, almost velvety in texture, especially at tips between areoles . . . . *Cladonia phyllophora*
69. Thallus K+ pale yellow (atranorin) . . . . *Cladonia ecmocyna*
69. Thallus K– (usually lacking atranorin) 70
70. Podetia up to 150 mm tall and 3 mm in diameter . . . . *Cladonia maxima*
70. Podetia 30–80 mm tall and 0.5–1.5 mm in diameter . . . . *Cladonia gracilis* ssp. *gracilis*
71.(33) Podetia sorediate, or with spherical, corticate granules 72
71. Podetia without soredia or granules 85
72. Apothecia and pycnidia red 73
72. Apothecia and pycnidia brown 76
73. Primary squamules with soredia on lower surface; thallus yellow-green or greenish yellow (containing usnic acid) . . . . *Cladonia umbricola*
73. Primary squamules with marginal soredia or lacking soredia; thallus gray, gray-green, or brown (usnic acid lacking) 74
74. Soredia powdery, covering the podetia from top to bottom; apothecia present or very frequently absent . . . . *Cladonia macilenta*
74. Soredia coarsely granular, sometimes sparse; apothecia almost always present 75
75. Primary squamules rounded, only slightly lobed; podetia corticate, at least on the lower half; thallus always PD–, K– (barbatic acid) . . . . *Cladonia floerkeana*
75. Primary squamules finely lobed; podetia entirely without a cortex, revealing the cartilaginous translucent stereome; thallus PD+ orange, K+ yellow (thamnolic acid) or, less commonly, PD–, K– (barbatic acid) . . . . *Cladonia didyma*
76.(72) Soredia coarsely granular, confined to tip of podetia; podetia slender, tall, branched once or twice . . . . *Cladonia scabriuscula*

76. Soredia or granules covering at least the upper half of podetia 77
77. Podetia with a continuous, more or less smooth cortex on the lower ⅓ to ⅔; soredia usually in discrete patches 78
77. Podetial cortex more or less broken up with areoles, granules, or soredia, not smooth and continuous except sometimes very close to the base; soredia, granules or squamules diffuse 80
78. Podetia branched once or twice with predominantly unequal branches, commonly split or fissured; uncommon; northeastern . . . . [*Cladonia farinacea*]
78. Podetia usually unbranched, rarely fissured; common and widespread 79
79. Thallus with gray-green tones predominating; stereome relatively thick and tough . . . . *Cladonia ochrochlora*
79. Thallus typically brownish, especially on upper half; stereome usually relatively thin . . . . . *Cladonia cornuta*
80.(77) Primary thallus disappearing; podetia 30–100 mm tall, often branched once or twice near the tip, covered with powdery soredia . . . . *Cladonia subulata*
80. Primary thallus persistent; podetia usually under 35 mm tall 81
81. Primary squamules very finely divided, usually disintegrating into a granular crust; thallus K+ deep yellow, PD+ orange (thamnolic acid) . . . . . *Cladonia parasitica*
81. Primary squamules remaining discrete, not granular; thallus K–, PD– or PD+ red (thamnolic acid absent) 82
82. Podetia with a mixture of granules (microsquamules) and soredia on the lower half, frequently with brown apothecia at the tips; southeastern coastal plain; thallus PD+ red, UV– (fumarprotocetraric acid) . . . . *Cladonia subradiata*
82. Podetia without a mixture of granules or microsquamules on the lower half, usually without apothecia; mainly absent from coastal plain (but see *C. coniocraea*) 83
83. Thallus greenish gray to brownish, podetia largely without a cortex, 10–40 mm tall, with only scattered granular soredia or microsquamules; thallus PD–, UV+ white (perlatolic acid); boreal to arctic . . . . [*Cladonia decorticata*]
83. Thallus olive or dark green, podetia covered with either cortex or soredia; thallus PD+ red, UV– (fumarprotocetraric acid); temperate to boreal 84
84. Soredia predominantly granular; podetia corticate on lower ¼; squamules finely divided . . . . *Cladonia ochrochlora*
84. Soredia entirely farinose, rarely granular in part, podetia covered with soredia except for a basal area; squa-

mules typically large and rarely lobed . . . .
. . . . *Cladonia coniocraea*

85.(71) Thallus yellow-green or greenish yellow (usnic acid)
. . . . 86

85. Thallus not yellowish (usnic acid absent) 92

86. Apothecia pale brown 87
86. Apothecia red 88

87. Growing on soil or rock; primary squamules dense, crenulate or deeply lobed . . . . *Cladonia robbinsii*
87. Growing on wood; primary squamules scattered, not lobed. . . . . *Cladonia botrytes*

88. Primary squamules sorediate; medulla UV+ white (squamatic acid). 89
88. Primary squamules without soredia; medulla UV– . . .
. . . . 90

89. Soredia on lower surface of squamules; contains didymic acid; common throughout the coastal plain. .
. . . . *Cladonia incrassata*
89. Soredia on margins of squamules; contains grayanic acid; rare, North Carolina to Florida . . . .
. . . . [*Cladonia anitae*]

90. Western and arctic regions; podetial surface areolate, abundantly squamulose. . . . . *Cladonia bellidiflora*
90. Eastern; podetia with a smooth to areolate cortex, squamulose or without any squamules. 91

91. Thallus K–, PD– (barbatic acid); very common, East Temperate to boreal. . . . . *Cladonia cristatella*
91. Thallus K+ deep yellow, PD+ orange (thamnolic acid); Florida. . . . . [*Cladonia abbreviatula*]

92.(85) Primary thallus disappearing 93
92. Primary thallus persistent 99

93. Podetial axils closed 94
93. Podetial axils open 96

94. Thallus K– 95
94. Thallus K+ pale yellow . . . . *Cladonia ecmocyna*

95. Podetia up to 150 mm tall and 3 mm in diameter . . . . .
. . . . *Cladonia maxima*
95. Podetia 30–80 mm tall and 0.5–1.5 mm in diameter . . .
. . . . *Cladonia gracilis* ssp. *gracilis*

96.(93) Podetial axils wide open, forming funnels; thallus PD–, or PD+ orange (squamatic or thamnolic acid; fumarprotocetraric acid absent) . . . . . . *Cladonia crispata*
96. Podetial axils partially open or wide open, not forming funnels; thallus PD+ red (fumarprotocetraric acid). . .
. . . . 97

97. Thallus K+ pale yellow (atranorin); podetia stout, 1–5 mm in diameter . . . . *Cladonia turgida*
97. Thallus K– (atranorin absent); podetia commonly slender, under 2 mm in diameter 98

98. Branching unequal; cortex usually continuous . . . . . .
. . . . *Cladonia multiformis*
98. Branching in more or less equal dichotomies; cortex usually breaking up into small, irregular patches especially on the upper half of the podetia . . . .
. . . . *Cladonia furcata* (western morphotype)

99.(92) Apothecia red; Florida 100
99. Apothecia brown 101

100. Primary squamules small, white below, often dissolving into soredia. . . . . [*Cladonia ravenellii*]
100. Primary squamules large, with a broad yellow band on lower surface, sometimes sorediate on the margins . .
. . . . [*Cladonia hypoxantha*]

101. Podetia mostly 1–3 mm tall, entirely without a cortex; thallus PD+ red, K– (fumarprotocetraric acid) . . . . . .
. . . . *Cladonia caespiticia*
101. Podetia mostly 3–40 mm tall, with or without a cortex
. . . . 102

102. Growing on wood or bark 103
102. Growing on soil, moss, or rock. 105

103. Primary squamules granular at the margins, often reducing the primary thallus to a granular crust; thallus PD+ orange, K+ deep yellow (thamnolic acid) . . . . . .
. . . . *Cladonia parasitica*
103. Primary squamules finely divided but not granular . .
. . . . 104

104. Thallus PD+ yellow, K– (baeomycesic acid); coastal plain . . . . *Cladonia beaumontii*
104. Thallus usually PD–, K–, UV+ white (squamatic acid) or sometimes PD+ orange, K+ deep yellow (thamnolic acid); widespread. . . . . *Cladonia squamosa*

105.(102) Podetia without a cortex 106
105. Podetia at least partly corticate, sometimes broken into areoles or flat squamules. 107

106. Podetia covered with abundant large to tiny, predominantly shingled squamules; containing squamatic or thamnolic acid (see couplet 104); podetial tips blunt, often with a gaping hole; very common and widespread . . . . *Cladonia squamosa*
106. Podetial squamules small and scattered; podetia pointed at tips, sometimes fissured on the sides; PD–, K–, UV+ white (perlatolic acid); boreal to arctic. . . . .
. . . . [*Cladonia decorticata*]

107. Medulla C+ green, KC+ green (strepsilin); thallus distinctly olive in color; podetia more or less inflated . . .
. . . . *Cladonia strepsilis*
107. Medulla C–, KC–, thallus gray-green, without an olive tint; podetia not appearing inflated. 108

108. Thallus K+ yellow, orange, or red 109
108. Thallus K– (or brownish), PD+ red (fumarprotocetraric acid) or PD+ yellow. 112

109. Thallus K+ red, PD+ yellow (norstictic acid) 110
109. Thallus K+ pale yellow, PD– or PD+ yellow or red... 111

110. Podetia rare, stocky, 2–5 mm in diameter, 10–20 mm tall, walls abundantly lacerate and fissured; containing atranorin. .... *Cladonia symphycarpia*
110. Podetia usually abundant, slender, 0.6–2.5 mm in diameter, 8–20 (–30) mm tall, walls intact or sometimes split; lacking atranorin. ...... *Cladonia polycarpoides*

111. Primary squamules large, 2–8 × 1–6 mm; podetia rare, 10–15 mm tall; thallus PD+ yellow (psoromic acid) or PD– (atranorin alone). .... *Cladonia symphycarpia*
111. Primary squamules small, thick, 1–3 × 0.5–2 mm; podetia usually abundant, 10–30 mm tall; thallus PD+ red (fumarprotocetraric acid) or PD– (atranorin alone) .... *Cladonia cariosa*

112.(108) Podetia covered with peltate areoles or squamules (attached in the center); thallus K–, PD+ bright yellow (psoromic acid); primary squamules 3–8 mm in diameter; arctic-alpine. .... [*Cladonia macrophylla*]
112. Podetial squamules and areoles not peltate; thallus PD+ red (fumarprotocetraric acid); squamules less than 3 mm wide; East Temperate 113

113. Apothecia much broader than podetia; podetia minutely warty, slender, often twisted .... *Cladonia peziziformis*
113. Apothecia rarely broader than podetia; podetia smooth, stout, not twisted. ..... *Cladonia sobolescens*

209. *Cladonia amaurocraea* north shore of Lake Superior, Ontario ×2.9

Cladonia amaurocraea
Quill lichen

DESCRIPTION: Primary squamules not seen; thallus consisting of tall, slender, irregularly or dichotomously branched podetia, 15–100 mm high, with a smooth, even surface; yellowish green (grayer green in the shade, as shown in plate 209), often conspicuously mottled with white and green patches. Podetia pointed at the tips or frequently producing narrow cups that are closed or perforated with only a small opening, often proliferating at the margins. CHEMISTRY: Cortex PD–, K–, KC+ gold, C– (usnic acid); medulla PD–, K–, KC–, C–, UV– (barbatic acid). HABITAT: Primarily a boreal forest lichen, most characteristically growing on talus slopes between boulders and on rocky ground, although sometimes in wetter sites as well. COMMENTS: *Cladonia amaurocraea* resembles a slender *C. uncialis* but is generally less branched and almost always has well-developed cups on at least a few of the branch tips. Clumps that have no cups can be distinguished from *C. uncialis* by the sparse branching and chemistry: *C. uncialis* has either squamatic and usnic acids or usnic acid alone, never barbatic acid. Some forms of *C. gracilis* that resemble *C. amaurocraea* are gray-green to brownish, never yellow-green (they lack usnic acid) and react PD+ red (fumarprotocetraric acid). IMPORTANCE: Woodland caribou graze on this lichen in northern Ontario.

Cladonia apodocarpa
Stalkless cladonia

DESCRIPTION: Primary squamules narrow and strap-shaped to branched once or twice, usually with tiny lobules along the margins, 5–12 mm long, 1.5–3 mm wide, grayish green, very smooth on the upper surface; lower surface smooth, slightly concave, pure white, often with a white pruina. Podetia absent, but small brown apothecia can sometimes form on short, naked stalks growing directly on the

210. *Cladonia apodocarpa* eastern Missouri ×1.1

211. *Cladonia asahinae* Coast Range, California ×2.0

squamules. CHEMISTRY: Cortex and especially upper medulla PD+ red, K+ yellow, KC–, C– (atranorin and fumarprotocetraric acid). HABITAT: Growing in clumps on thin or sandy soil or over rock, usually in full sun. COMMENTS: *Cladonia petrophila* is an Appalachian-Ozark species found on shaded woodland rocks and, like *C. apodocarpa* and *C. caespiticia*, produces apothecia almost directly on the squamules. Its squamules are smaller and more divided than those of *C. apodocarpa*, mostly 3–6 mm long and 1–2 mm wide, and its chemistry is unique in the group: PD+ red, K–, UV+ blue-white (sphaerophorin and fumarprotocetraric acid, often with atranorin). When completely lacking apothecia, *C. apodocarpa* is very similar to *C. polycarpia*, *C. symphycarpia*, *C. polycarpoides*, *C. sobolescens*, and so on, which are all cladonias with ascending, strap-shaped squamules that share the same habitat. *Cladonia apodocarpa* differs from these in its chemistry, together with the very smooth upper and lower surfaces of the squamules. *Cladonia peziziformis* and *C. sobolescens*, although containing fumarprotocetraric acid (PD+ red), lack atranorin, and their squamules are smaller. The squamules of *C. apodocarpa* are less divided than those of *C. symphycarpia*, which, in addition, are conspicuously thicker, with a rough, almost scabrose upper surface in older portions. That species contains atranorin and psoromic acid (PD+ bright yellow), or atranorin alone (PD– or pale yellow, K+ yellow). *Cladonia polycarpia* squamules are variable in shape, but they tend to be turned up at the lobe tips and are not concave. This species contains atranorin and norstictic and stictic acids (PD+ orange, K+ red). *Cladonia polycarpoides* has norstictic acid and turns K+ red, PD+ yellow; the latter has atranorin as well.

Cladonia asahinae
Pixie-cup lichen

DESCRIPTION: Primary squamules persistent, minutely lobed, non-sorediate. Podetia 10–25 mm tall, with broad or narrow cups 2–7 mm across, sometimes proliferating at the margins, gray-green to slightly yellowish green or fading to almost white; at least the upper half of the podetia covered with powdery or granular soredia or corticate granules. Apothecia brown, rare. CHEMISTRY: Podetia PD+ red, K–, KC–, C–, UV– (fumarprotocetraric acid and several kinds of fatty acids). HABITAT: On soil or mossy rocks in full sun. COMMENTS: This species often resembles a rather large, unusually gray *C. fimbriata*, which almost always has fine, powdery soredia and narrow, trumpet-shaped cups with few proliferations; it never contains fatty acids.

Cladonia beaumontii
Pale-fruited funnel lichen

DESCRIPTION: Primary squamules finely divided, not sorediate. Podetia gray-green, without cups but often fertile with light brown apothecia, usually slender and hardly branched, 5–20 mm tall, more or less covered with small squamules (or sometimes microsquamules that resemble granules); the area between the squamules either has an intact cortex or is predominantly naked and translucent. CHEMISTRY: Podetia PD+ yellow, K–, KC–, C– (baeomycesic and squamatic acids). HABITAT: On rotting wood, logs, and stumps, rarely soil. COMMENTS: *Cladonia beaumontii* resembles *C. squamosa* but is PD+ yellow, K–. (The thamnolic acid race of *C. squamosa* is PD+ orange, K+ yellow.)

212. *Cladonia beaumontii* central Florida ×2.3

Cladonia bellidiflora
Toy soldiers

DESCRIPTION: Thallus pale yellowish green, with large, deeply lobed, nonsorediate, primary squamules 2–12 × 1–7 mm. Podetia cupped or blunt at the tip, usually abundantly covered with squamules but without soredia, almost always ending in large red apothecia. CHEMISTRY: Cortex KC+ gold (usnic acid); most specimens PD–, K–, UV+ blue-white (squamatic acid). A thamnolic acid–containing race (PD+ orange, K+ deep yellow) occurs from northern Vancouver Island north to Alaska along the coast. HABITAT: On rotting wood and stumps, moss, or soil. COMMENTS: With its abundantly squamulose podetia and bright red fruits, *C. bellidiflora* is one of the most conspicuous western lichens. When the squamules on the podetia are extremely small, almost granular, the lichen can resemble **C. transcendens**, which has granular soredia and is PD+ orange (thamnolic acid). Squamulose forms of the eastern species *C. cristatella* (British soldiers) resemble *C. bellidiflora* but they never have cups, and they have smaller primary squamules and a different chemistry.

Cladonia borealis
Boreal pixie-cup

DESCRIPTION: Thallus pale yellowish green or grayish yellow; primary squamules lobed, 3–10 mm across. Podetia with goblet-shaped cups up to 20 mm high, 4–10 mm across, containing rounded areoles or plates that sometimes occur on the outside of the cups and down the podetium. Apothecia common on the cup margins, bright red. CHEMISTRY: Podetia PD–, K–, KC+ gold, C– (barbatic and usnic acids). HABITAT: Arctic-alpine, on soil or rocks in full sun. COMMENTS: *Cladonia borealis* is about the same size as *C. pleurota* (which contains zeorin), but the latter has granular soredia in and on the cups. The much rarer **C. coccifera** is almost identical to *C. borealis* in appearance and habitat, but it contains zeorin rather than barbatic acid.

Cladonia boryi
Fishnet cladonia

DESCRIPTION: Primary thallus absent. Podetia pale yellowish green, puffed up, contorted, abundantly branched, up to 9 cm tall and 3–11 mm across, eroded here and there leaving irregular perforations, its inner lining (stereome) being reduced to a network of fibrous strands. CHEMISTRY: Thallus PD–, K–, KC+ yellow, C– (usnic acid), or rarely with podetia tips PD+ red-orange (protocetraric acid). HABITAT: On sandy soil and sand dunes, less frequently in open forest glades. COMMENTS: When fully developed, this lichen is unlike any other. Young, more erect specimens, however, look somewhat like *C. caroliniana,* a smaller species with less perforate podetia that grows most commonly over exposed granitic rock.

213. (facing page) *Cladonia bellidiflora* Cascades, Oregon ×5.9

214. (above) *Cladonia borealis* mountains, coastal Alaska ×2.4

215. (below) *Cladonia boryi* Cape Cod, Massachusetts ×2.7

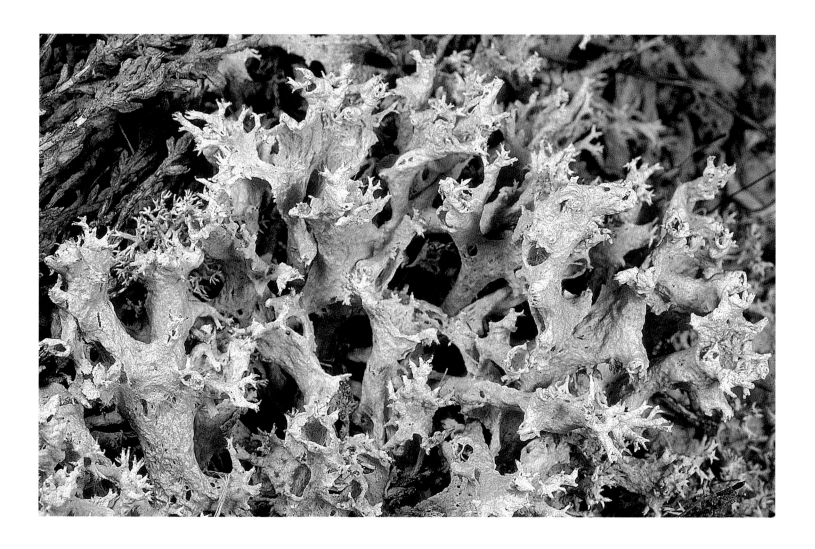

216. *Cladonia botrytes* Laurentides, Quebec ×3.4

217. *Cladonia caespiticia* Adirondacks, New York ×2.9

Cladonia botrytes
Wooden soldiers

DESCRIPTION: Primary squamules tiny and inconspicuous. Podetia yellowish, smooth to bumpy, without squamules or soredia, without cups, mostly 5–10 mm high, topped with large, pale beige apothecia. CHEMISTRY: PD–, K–, KC+ gold, C– (usnic and barbatic acids). HABITAT: On wood, especially stumps and logs, rarely soil. COMMENTS: *Cladonia cristatella* is somewhat similar but has crimson apothecia with only the occasional mutant having waxy yellow fruits (caused by the sporadic loss of the red anthraquinone pigment).

Cladonia caespiticia
Stubby-stalked cladonia

DESCRIPTION: Primary squamules much divided and lacy, without soredia. Extremely short podetia arising from edges of squamules, 2–4 mm high, with little if any cortex. Apothecia large, at the tip of every podetium, yellowish brown to very dark reddish brown. CHEMISTRY: PD+ red, K–, KC–, C– (fumarprotocetraric acid). HABITAT: On mossy rocks, soil banks, and tree bases commonly in partly shaded woodlands. COMMENTS: This lichen looks at first like a very luxuriant but sterile *Cladonia*. The podetia are so short and overshadowed by the large, hemispherical apothecia that they are often not visible from above. *Cladonia apodocarpa* entirely lacks podetia, the squamules are thicker, rounder, and less divided into lobules, and it contains atranorin as well as fumarprotocetraric acid.

Cladonia cariosa
Split-peg lichen, split-peg soldiers

DESCRIPTION: Primary squamules thick, typically tongue-shaped, 1–3 × 0.5–2 mm, sometimes becoming toothed or slightly lobed. Podetia gray to gray-green, torn and fissured, sometimes branched once or twice, without cups, up to 30 mm high; surface areolate, not sorediate. Podetia crowned with large, chocolate-brown apothecia. CHEMISTRY: Two chemical races are common in North America: one has fumarprotocetraric acid and atranorin (PD+ red, K+

yellow on the white lower surface of the squamules); the other has atranorin alone (PD– or pale yellow, K+ yellow). HABITAT: *Cladonia cariosa* grows directly on soil in calcium-rich and calcium-poor sites. COMMENTS: The PD– chemical race is easily mistaken for *C. symphycarpia*, a more temperate species with longer and broader primary squamules that is normally associated with calcium-containing soil. Other cupless cladonias with brown fruiting bodies and similar squamules differ at least chemically. See *C. symphycarpia*, *C. polycarpoides*, *C. peziziformis*, and *C. sobolescens*.

Cladonia carneola
Crowned pixie-cup

DESCRIPTION: Thallus pale greenish yellow; primary squamules deeply divided, not sorediate or (as shown in plate 219) occasionally sorediate on the lower surface close to the margins. Podetia goblet-shaped with jagged, tooth-like cup margins, with broad or narrow cups up to 30 mm high and 5 mm across, covered inside and usually outside with fine powdery soredia. Apothecia and pycnidia often produced on the cup margins, pale waxy-pink to pale brown. CHEMISTRY: Surface PD–, K–, KC± gold, C–, UV– (usnic acid, zeorin, and usually barbatic acid). HABITAT: On rotting wood, soil, or bark, in sun or shade. COMMENTS: Among the yellow-green pixie-cup lichens, this is the only one with pale brown apothecia. Despite the color of its fruits, its chemistry links it with the red-fruited cladonias such as *C. pleurota*: both contain zeorin and develop a crystalline fuzz on the podetial surface after long storage. *Cladonia carneola* is sometimes mistaken for *C. pleurota* when fruiting structures are absent, but the latter usually has a less toothed cup margin and coarser soredia, and it never contains barbatic acid. *Cladonia carneola* is a distinctly northern boreal to arctic-alpine species, whereas *C. pleurota*, although basically boreal, is also common in temperate eastern North America.

Cladonia caroliniana
Granite thorn cladonia

DESCRIPTION: Podetia pale yellowish green, abundantly branched, with pointy side branches, inflated but not perforated, 20–60(–100) mm tall and

218. (top) *Cladonia cariosa* Olympics, Washington ×1.8

219. (overleaf) *Cladonia carneola*   mountains, eastern Oregon ×6.4

220. (bottom) *Cladonia caroliniana* central North Carolina ×1.5

3–6(–10) mm in diameter; inner medulla (stereome) composed of broad, flat cords that coalesce into an almost continuous layer. CHEMISTRY: Cortex PD–, K–, KC+ yellow, C–, UV– (usnic acid); medulla PD–, K–, KC–, C–, UV+ blue-white (squamatic acid, sometimes only in traces). HABITAT: Usually over granitic rock in open areas, less frequently on soil. COMMENTS: *Cladonia caroliniana* is similar to the more northern *C. boryi*, but its podetia are not perforated and lacerated, and not nearly as fibrous. *Cladonia pachycladodes* has a much thicker, dull (soft-looking) wall and is often grayer. *Cladonia dimorphoclada* has more slender branches with more distinct cartilagenous fibers making up the stereome; it usually grows on sandy soil and always lacks squamatic acid. *Cladonia uncialis* has a continuous rather than fibrous stereome, and its branches have a smoother, less inflated appearance.

Cladonia cenotea
Powdered funnel lichen

DESCRIPTION: Primary squamules small, often disappearing. Podetia greenish gray, almost always browned, at least on the lower half, 10–70 mm tall, almost completely covered with powdery soredia, sometimes accompanied on the lower half by very small podetial squamules; the tips flare to form narrow cups that are open to the hollow interior, and the cup margins curl inward as well as proliferate 1–3 times. Apothecia brown, uncommon. CHEMISTRY: Thallus PD–, K–, KC–, C–, UV+ blue-white (squamatic acid). HABITAT: Common boreal forest lichen on wood or earth, usually in shade. COMMENTS: The powdered funnel lichen is the only cupped cladonia with cup margins that curl inward. *Cladonia glauca*, an uncommon boreal lichen, is also covered with fine soredia and contains squamatic acid, but it rarely produces cups. It has tall, slender podetia that typically branch once or twice at the top.

Cladonia cervicornis ssp. *verticillata* (syn. *C. verticillata*)
Ladder lichen

DESCRIPTION and COMMENTS: Podetia smooth or verruculose, nonsorediate, grayish green to olive or brownish, with flaring, closed cups that proliferate from the centers of the cups, not the margins. Subspecies **verticillata** is distinguished from the much rarer **ssp.** ***cervicornis*** by its primary squamules, which are sparse and small, often disappearing, and more frequently proliferating podetia (1–4 times) up to 50 mm tall. Subspecies *cervicornis*, known along the west coast, has persistent primary squamules and podetia usually less than 20 mm tall, proliferating only once. In the east, there is also *C. rappii*, which has more abruptly flaring, flatter cups or tiers. It is usually small, with abundant primary squamules, and typically is associated with boggy areas or other highly acidic habitats on sand. CHEMISTRY: Thallus PD+ red, K–, KC–, C– (fumarprotocetraric acid). HABITAT: Primarily in open, rocky, or grassy areas on thin soil, more rarely on wood.

221. *Cladonia cenotea* interior plateau, British Columbia ×2.2

222. *Cladonia cervicornis* ssp. *verticillata*
Coast Range, Oregon
×2.9

Cladonia chlorophaea
Mealy pixie-cup

DESCRIPTION: Primary squamules abundant, lobed and deeply divided, without soredia. Podetia pale green, gray-green or brownish, up to 35 mm tall, always cupped; cups usually broad and goblet-shaped (young ones sometimes narrow), 2–6 mm across, closed; more or less sorediate inside and outside the cups, the granular soredia extending at least partway down the podetia; soredia sometimes grading to corticate granules, especially inside the cups; podetial squamules not usually abundant, often reduced to almost granule-size. Large brown apothecia frequently produced on short proliferations from the cup margins. CHEMISTRY: PD+ red, K–, KC–, C– (fumarprotocetraric acid). HABITAT: On wood, bark, rock, or soil, in full sun or partial shade. COMMENTS: Unfortunately, many lichens resemble this common and variable species. The true pixie-cup lichen (*C. pyxidata*) has corticate, rounded areoles rather than soredia in and on the edge of the cups. Several lichens are almost indistinguishable from *C. chlorophaea* in appearance but differ chemically. The very common *C. grayi* (see plate 35) contains grayanic acid and sometimes lacks fumarprotocetraric acid; *C. merochlorophaea,* not common but widely distributed, contains merochlorophaeic acid (KC+ pinkish violet, rapidly disappearing). The trumpet lichen, *C. fimbriata,* is more finely sorediate and more narrowly cupped. In *C. conista,* an East Temperate lichen on soil, the farinose soredia are confined mainly to the upper half of the outside of the cup as well as the cup interior, the rest of the podetium retaining the cortex intact. It contains bourgeanic acid, a fatty acid. *Cladonia albonigra* is a pixie-cup lichen found on the oceanic west coast with gray-green to brownish, broad, granular sorediate cups that frequently proliferate from both the margin and center. The bases of the podetia have white areoles or granules over a blackening stereome, giving the lichen a spotty appearance. In *C. albonigra,* grayanic acid usually accompanies the fumarprotocetraric acid.

224. *Cladonia coniocraea* Coast Mountains, British Columbia ×2.6

Cladonia coniocraea
Common powderhorn

DESCRIPTION: Thallus gray-green or olive; primary squamules rather large, up to 6 mm broad, usually rounded, but in some forms divided into lobes; with or without soredia. Podetia arising from the centers of the squamules, strongly tapered (like a powderhorn) or cylindrical, usually unbranched, 10–25 mm tall; very narrow, barely expanding cups occasionally formed at the podetial tips; fine, powdery soredia usually covering the podetia except for the base, sometimes granular in part. Apothecia brown, rare. CHEMISTRY: Thallus PD+ red, K– or brownish, KC–, C– (fumarprotocetraric acid). HABITAT: Very widespread on soil, tree bases, and wood, usually in shade. COMMENTS: Specimens of *C. coniocraea* with incompletely sorediate podetia are sometimes mistaken for *C. ochrochlora,* which has patchier soralia and more granular soredia, and is more often cupped. Small specimens of *C. cornuta* are browner, with podetia that have an intact cortex over more than half their

223. *Cladonia chlorophaea* Lake Superior region, Ontario ×4.5

225. *Cladonia cornuta* ssp. *cornuta* Lake Superior region, Ontario ×1.5

226. *Cladonia crispata* southeast Alaska ×1.8

length; they usually grow in full sun. See also Comments under *C. subradiata* and *C. subulata*.

Cladonia cornuta ssp. *cornuta*
Bighorn cladonia

DESCRIPTION: Primary squamules usually small and relatively inconspicuous. Podetia shiny pale brown to olive-brown, or grayish green in shade forms (as shown in plate 225), tall and slender, typically unbranched, often growing in dense clumps, 25–40 mm tall, with discrete patches of fine soredia on the upper half and fairly smooth elsewhere; tips usually pointed in ssp. *cornuta* but sometimes forming narrow cups, especially in **ssp. *groenlandica*,** in which the cups are often slightly inflated. Apothecia rare, brown. CHEMISTRY: Thallus PD+ red, K– or brownish, KC–, C– (fumarprotocetraric acid). HABITAT: On soil or wood, usually in full sun. COMMENTS: This species, especially ssp. *groenlandica*, is most similar to *C. ochrochlora*, which also has patchy soralia and tall, unbranched podetia but is more yellowish green with cups that do not appear inflated. *Cladonia cornuta* ssp. *groenlandica* is largely restricted to the oceanic coastal parts of the continent. Tall specimens of *C. coniocraea* are distinguished by their olive color and more completely sorediate podetia. *Cladonia farinacea,* an uncommon eastern species, has tall, sparingly branched podetia (mostly 40–60 mm tall), commonly split and perforated, with an intact cortex on the lower half and patches of farinose soredia on the upper half. *Cladonia scabriuscula* has more podetial squamules and produces sparse, granular soredia only at the extreme tips.

Cladonia crispata
Organ-pipe lichen

DESCRIPTION: Primary squamules small and minutely lobed. Podetia 30–50(–80) mm tall, usually brown or olive-brown, rarely greenish, flaring at the tips into narrow funnel-like cups open to the interior, branching in dichotomies from the cup margins; cups sometimes rather oblique and difficult to distinguish from open axils; surface smooth or areolate, without any soredia and with few to many squamules on the podetia.

227. *Cladonia cristatella* (form without squamules) southern Ontario ×3.5

Apothecia common, brown, produced on the cup margins. CHEMISTRY: Two chemical races: (*i*) most common and widely distributed, thallus (especially medulla) PD–, K–, KC–, C–, UV+ white (squamatic acid); (*ii*) mainly on the oceanic west coast with scattered occurrences on the east coast, PD+ orange, K+ deep yellow, KC–, C–, UV– (thamnolic acid). HABITAT: On soil, rocks, or rarely on wood, mainly in open areas, frequently associated with bogs. It is a pioneer species over recently burned land; consequently, its abundance is an indicator of young stands. COMMENTS: *Cladonia atlantica,* found along the southeastern coastal plain, looks very much like *C. crispata* but has baeomycesic and squamatic acids (PD+ yellow, K–) rather than only squamatic acid or thamnolic acid. *Cladonia subfurcata* is a more richly branched, cupless version of *C. crispata* with a conspicuously blackened base; it is a northern boreal species growing on humus. The name *C. carassensis* has been used for eastern specimens of *C. crispata* containing thamnolic acid. See Comments on *C. artuata* and *C. porocypha* under *C. floridana.*

Cladonia cristatella
British soldiers

DESCRIPTION: Primary squamules finely divided and abundant. Podetia typically unbranched, or slightly branched at the tips, without cups (although clusters of short branches at the podetial summit, as shown in plate 227, may give the impression of a cup), usually under 25 mm tall; surface smooth, with a continuous cortex, without soredia; always crowned with one or more large, bright red apothecia. Extremely variable in abundance of podetial squamules and in thallus color; that is, the shadier the habitat, the more squamules occur and the grayer the color. In the open, especially on the ground, podetia have few or no squamules and tend to be very yellowish. CHEMISTRY: PD–, K–, KC+ gold, C– (usnic, barbatic, and didymic acids). HABITAT: Equally at home on wood, soil, or even bark (tree bases). COMMENTS: The British soldiers lichen is perhaps the best known lichen in eastern North America, for three reasons: it is relatively pollution-tolerant and therefore often grows where people live, it is very colorful, and it was given a common name many years ago; people notice it because they have an easy way to refer to it. It is the only northeastern red-fruited cladonia entirely without soredia or granules. *Cladonia incrassata* and *C. anitae* have nonsorediate podetia but sorediate primary squamules; the latter is known from Florida to North Carolina and contains grayanic and squamatic acids. *Cladonia abbreviatula,* also found in Florida, has podetia that broaden toward the apex but are shorter than 10 mm and react PD+ orange, K+ yellow (thamnolic acid). *Cladonia hypoxantha* and *C. ravenelii,* also Florida species with thamnolic acid, are greenish gray, lacking usnic acid. In *C. hypoxantha,* the large primary squamules are sorediate at the margins, with a broad yellow band on the lower surface. *Cladonia ravenelii* has smaller squamules, also sorediate but without the yellow band; its podetia have a rough, verruculose surface sometimes producing patches of soredia. The western counterpart of *C. cristatella* is *C. bellidiflora,* which generally has cups and a patchy cortex over the surface of the stalk, and differs chemically.

228. *Cladonia deformis* Shuswap Highlands, British Columbia ×2.9

Cladonia deformis
Lesser sulphur-cup

DESCRIPTION: Thallus pale greenish yellow; primary squamules small, often disappearing. Podetia 25–85 mm tall, cupped, the cups rather narrow compared to the length of the stalk, covered for most of their length with powdery soredia; podetia almost never with squamules. Red apothecia or pycnidia usually present along the rims of the cups. CHEMISTRY: Podetia PD–, K–, KC+ gold, C–; medulla UV– (usnic acid and zeorin). HABITAT: On rotting wood or soil in full sun. COMMENTS: *Cladonia deformis* most closely resembles *C. pleurota*, which is shorter and has granular rather than farinose soredia, and *C. sulphurina*, which has larger primary squamules and is often taller, with more irregular, split cups. *Cladonia sulphurina* contains squamatic acid, making its medulla UV+ white although its surface may be UV– because the usnic acid in the soredia blocks the squamatic fluorescence; the soredia therefore have to be scraped away before examining the lichens under UV light. Sterile specimens of *C. carneola* are shorter and more goblet-shaped, and they usually have barbatic acid as well as zeorin.

Cladonia didyma
Southern soldiers

DESCRIPTION: Primary squamules lobed, some sorediate or granular. Podetia pale greenish gray, unbranched or branched near the tips, 10–30 mm tall, slender or stout, without cups, with coarsely granular soredia or corticate granules abundant or sparse on a predominantly naked, brownish to translucent stereome. Podetia usually fertile, with red apothecia. CHEMISTRY: var. *didyma*: podetia PD–, K–, KC+ orange, C– (barbatic acid); var. *vulcanica*: podetia PD+ orange, K+ bright yellow, KC–, C– (thamnolic, didymic, and barbatic acids). HABITAT: On sandy soil or rotting wood. COMMENTS: The two chemical races of *C. didyma* are identical morphologically. They differ from *C. macilenta* in being more coarsely granular (*C. macilenta* is almost always farinose sorediate) with more naked stereome showing through. In addition, the primary squamules of *C. macilenta* are less divided. See Comments under *C. parasitica*.

229. *Cladonia didyma* var. *vulcanica* Gulf Coast, Florida ×4.0

230. *Cladonia digitata* southern New Brunswick ×2.2

Cladonia digitata
Finger pixie-cup

DESCRIPTION: Primary squamules very large, 4–10 mm long and broad, rounded or less frequently divided into lobes, often narrowing at the base; abundantly farinose sorediate all over the lower surface, which is predominantly white but becomes orange at the base; upper surface yellowish green to grayish green. Podetia cupped, trumpet- to goblet-shaped, 5–20 mm tall, yellowish green to grayish green; cups 1.5–5 mm across, almost always with tooth-like projections on the cup margins and small, bright red pycnidia or apothecia at the summits of the teeth; cup margins frequently rolled inward to some degree; fine, farinose soredia on the outside of the cups (usually absent from the cup interior), sometimes covering the podetia to the base; podetia arising from the centers of the squamules. CHEMISTRY: Medulla PD+ orange, K+ bright yellow, KC–, C– (thamnolic acid). HABITAT: On well-rotted wood and peat, sometimes mossy tree bases. COMMENTS: The very large, sorediate squamules together with cupped podetia make this a distinctive species. The chemical tests are useful for identifying sterile specimens, which are frequently seen. *Cladonia polydactyla*, except for its much smaller, lobed squamules and grayer color, is very similar to *C. digitata*. *Cladonia incrassata* is another red-fruited cladonia with sorediate primary squamules, but it does not form cups, and its distribution is entirely different. *Cladonia umbricola* has smaller, finely divided squamules, and smaller podetia with more even cups.

Cladonia dimorphoclada
Prostrate thorn cladonia

DESCRIPTION: Primary squamules absent. Podetia highly branched, forming rather flattened colonies with sprawling branches, yellowish green, smooth, shiny in places, somewhat bumpy in older branches, the main branches 0.5–1.5 mm in diameter; inner stereome layer broken into rather broad, flattened cords or barely showing longitudinal striations. CHEMISTRY: Cortex KC+ gold; branch tips and medulla PD–, K–, KC–, C– (containing only usnic acid in

the cortex). HABITAT: On sandy or mossy soil in sun or partial shade. COMMENTS: *Cladonia dimorphoclada* resembles *C. uncialis* but is often prostrate rather than producing upright cushions. In *C. uncialis,* the stereome layer is perfectly uniform, sometimes appearing powdery with white pruina, and the thallus surface is smoother and shinier. Slender specimens of *C. caroliniana* are also sometimes similar to *C. dimorphoclada,* although its branches are usually over 3 mm thick and the stereome is almost uniform, with few of the cords seen in *C. dimorphoclada* or *C. boryi*. *Cladonia caroliniana* usually contains squamatic acid.

Cladonia ecmocyna
Frosted cladonia

DESCRIPTION: Thallus gray to gray-green, usually with a fine white pruina or scaly roughness over the surface of podetial tips; primary squamules rarely seen. Podetia slender and pointed with no podetial squamules, or rather robust, clearly cupped, and covered with podetial squamules; differ-

231. (above right) *Cladonia dimorphoclada* Pine Barrens, New Jersey ×1.3

232. (above left) *Cladonia ecmocyna* ssp. *intermedia* Olympics, Washington ×1.8

233. (left) *Cladonia ecmocyna* ssp. *occidentalis* Cascades, Washington ×0.92

CLADONIA 253

234. *Cladonia fimbriata*
Maine coast ×1.42

ent populations 25–100 mm tall. Cups common in ssp. *intermedia* (plate 232), which is found in the western mountains at high elevations; cups frequently present in the more arctic **ssp. *ecmocyna***. Podetial squamules usually abundant in ssp. *intermedia*, but rare or absent in the others. Subspecies *occidentalis* (plate 233), a slender lichen of lower elevations in the far west, is often quite brown, with the pruina confined to the extreme tips; it sometimes has very narrow cups or is pointed at the tips. CHEMISTRY: Thallus PD+ red, K+ yellow, KC–, C– (abundant atranorin in the cortex; fumarprotocetraric acid in the cortex and medulla). HABITAT: On soil in the open or in partial shade, mainly at high elevations or in the Arctic, except for ssp. *occidentalis* (see above). COMMENTS: *Cladonia gracilis* is somewhat similar to *C. ecmocyna* but is K–, with very little or no atranorin, and it is greener or browner (not at all gray). The map of *C. ecmocyna* shows the combined distributions of all the subspecies. The range of subspecies *occidentalis*, however, is in pink.

Cladonia fimbriata (syn. *C. major*)
Trumpet lichen

DESCRIPTION: Primary squamules lobed, typically without soredia; thallus gray-green to green or olive-gray. Podetia 10–20 mm tall, with relatively narrow, trumpet-shaped cups, usually with even margins except when fruiting; surface covered with fine farinose soredia, rarely with granular soredia. Apothecia brown. CHEMISTRY: Thallus PD+ red, K– or brownish, KC–, C– (fumarprotocetraric acid). HABITAT: On soil or rotting wood, in sun or partial shade. COMMENTS: This lichen closely resembles *C. chlorophaea* but has narrower, less frequently proliferating cups and more powdery soredia. See also *C. asahinae* and Comments under *C. chlorophaea*.

Cladonia floerkeana
Gritty British soldiers

DESCRIPTION: Primary squamules rounded, only slightly lobed, typically without soredia or granules. Podetia greenish gray, without cups, coarsely granular sorediate, with an intact cortex, at least on the lower third to half, unbranched or slightly branched at the tips, 5–20(–40) mm tall, always capped with red apothecia; podetial squamules frequent or sparse. CHEMISTRY: PD–, K–, KC+ orange, C– (barbatic acid, sometimes with usnic and didymic acids). HABITAT: On earth, rotten logs, or soil-covered rock, usually in the open. COMMENTS: This cupless, red-fruited cladonia can be mistaken for *C. cristatella* if one fails to notice the granules or soredia. *Cladonia macilenta* has farinose soredia for almost the entire length of the slender podetia.

Cladonia floridana
Bramble cladonia

DESCRIPTION: Primary squamules rather large, divided into lobes. Podetia slender, branching several times often around a perforated axil (but not forming funnels or cups), 0.5–1.5(–2) mm thick and up to 25 mm tall, pale greenish gray, corticate and smooth, with squamules confined to the lower third;

no soredia on the podetia or squamules. Apothecia brown, sometimes developing in clusters. CHEMISTRY: Thallus PD+ orange, K+ deep yellow, KC–, C– (thamnolic acid). HABITAT: On the ground, usually on sandy soil or under pines. COMMENTS: In the east, there is almost nothing with which *C. floridana* can be confused, but in the west, where *C. floridana* does not occur, several similar lichens contain thamnolic acid. *Cladonia porocypha* is taller (up to 50 mm) and sometimes has podetial squamules, although primary squamules are rarely seen. *Cladonia artuata* has axils that broaden into oblique cups and approaches *C. crispata* in appearance.

Cladonia furcata (syn. *C. subrangiformis* as used in N. Am.) Many-forked cladonia

DESCRIPTION: Thallus greenish gray or olive in the shade, or pale brown in sunnier sites; primary squamules quickly disappearing. Podetia abundantly branched, forming clumps or cushions 2–12 cm high; branches 1–2 mm in diameter, usually bearing many finely lobed podetial squamules that are small or up to 5 mm long, but podetial squamules sometimes absent, especially in western populations; axils wide open, sometimes forming funnels but never cups; in eastern populations cortex fairly smooth and unbroken, but in western populations cortex patchy and almost granular in places; soredia, however, are never present. Apothecia brown, at the tips of the branches. CHEMISTRY: Thallus PD+ red, K– (or brownish), KC–, C–, UV– (fumarprotocetraric acid). HABITAT: This is one of the few branched cladonias found in shaded forests, growing on soil or mossy boulders; also found in sunny sites. COMMENTS: *Cladonia furcata* sometimes resembles a reindeer lichen (*Cladina*), especially if the podetia have few squamules. Reindeer lichens, however, have no cortex. Although *C. furcata* is usually easy to identify, it intergrades with *C. scabriuscula* along the west coast and can be mistaken for certain forms of *C. multiformis* in the north and east. All three species form branched podetia with longitudinally split branches and open axils. *Cladonia multiformis* is the only one of the three forming at least some well-defined cups, each containing one to numerous perforations. *Cladonia scabriuscula* is more sparsely branched and has granular sorediate branch tips. Some western speci-

235. *Cladonia floerkeana* Appalachians, Tennessee ×2.5

236. *Cladonia floridana* Pine Barrens, New Jersey ×2.4

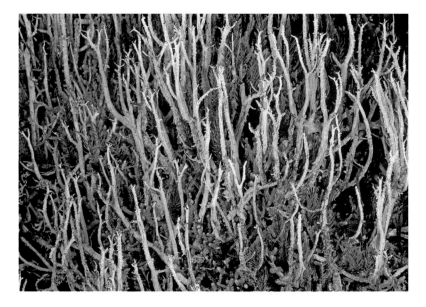

237. *Cladonia furcata* (eastern population, shade form) Ouachitas, Arkansas ×1.5

238. *Cladonia furcata* (western population) Olympics, Washington ×0.96

mens of *C. furcata* with very tiny podetial squamules and an eroded cortex close to the tip mimic this appearance. *Cladonia subfurcata,* an arctic to northern boreal species, is a more sparingly branched version of *C. furcata* that has squamatic acid (PD–, UV+ blue-white) instead of fumarprotocetraric acid. The base of the podetia becomes black, and so the paler, often brown corticate patches are rather conspicuous.

Cladonia gracilis
Smooth cladonia

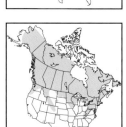

DESCRIPTION: This common and variable cladonia comes in many shapes and sizes; some are named as distinct subspecies, two of which are shown here, in plates 239 and 240. All have podetia with a smooth, unbroken cortex and no soredia, and they are all greenish to olive, becoming quite browned in the sun. If they produce cups, the cups are entirely closed. CHEMISTRY: Thallus PD+ red, K– or brownish, KC–, C– (fumarprotocetraric acid, usually lacking atranorin).

Subspecies *gracilis* (plate 239, top map): DESCRIPTION: Primary squamules disappearing. Podetia slender and usually unbranched, 3–8 cm tall, pointed at the tips and usually without cups (rarely with narrow cups); somewhat squamulose at the base and usually browned at the tips; base yellowish to brown, not blackened. HABITAT: Temperate to southern boreal, usually on rocky ground. COMMENTS: Several subspecies of *C. gracilis* have primarily cupless pointed podetia. *Cladonia gracilis* ssp. **elongata** (formerly ssp. *nigripes*) is found in the arctic to northern boreal regions in a variety of habitats including forest soil and boggy heath. It has slender podetia, mostly 2–5 cm tall, without slits or perforations, not or infrequently branched, with blackening bases and a black stereome. The podetia are almost entirely pointed, or narrow cups sometimes are produced at the tips. In ssp. **vulnerata,** the sides of the podetia are perforated or fissured. It is a common subspecies growing with heath or grass along the rainy northwest coast. It is often over 5 cm long (up to 18 cm) and is slender to robust (0.6–5 mm in diameter). *Cladonia maxima,* which also has tall, largely unbranched podetia, is not black at the base. *Cladonia ecmocyna* can resemble *C. gracilis* when the pruina is sparse, but *C. ecmocyna* always contains abundant atranorin and is generally grayer, at least at the base. The only subspecies of *C. gracilis* regularly containing atranorin (albeit in small quantity) is ssp. *elongata.* IMPORTANCE: Where *C. gracilis* becomes abundant and robust, it forms a part of the winter diet of woodland caribou.

Subspecies *turbinata* (plate 240, bottom map): DESCRIPTION: Primary squamules regularly present. Podetia with well-developed cups, often proliferating at the margins and producing brown apothecia; squamulose or not, generally 2–5 cm tall. HABITAT: Widespread boreal to temperate, normally found on exposed soil but also on wood. COMMENTS: Most other cupped *Cladonia* species have soredia or granules on the surface of the podetia. *Cladonia cervicornis* ssp. *verticillata,* which shares the same habitat, is similar in having a smooth surface and closed cups, but the cups proliferate from the center. Specimens of *C. phyllophora* with well-developed cups can resemble *C. gracilis* ssp. *turbinata,* but these generally have more squamulose cup margins and a duller surface.

Cladonia incrassata
Powder-foot British soldiers

DESCRIPTION and COMMENTS: Like the British soldiers (*C. cristatella*), this cladonia has nonsorediate podetia always topped by a large red apothecium, never a cup. The podetia are short (up to 8 mm) and unbranched. Unlike British soldiers or most cladonias with nonsorediate podetia, its squamules are abundantly sorediate (on the lower surface), often forming an almost leprose crust. See Comments under *C. cristatella.* CHEMISTRY: Cortex PD–, K–, KC+ gold, C– (usnic acid); medulla PD–, K–, KC–, C–, UV+ white (squamatic acid, sometimes with didymic acid). HABITAT: This is a common lichen on well-rotted logs and stumps.

239. *Cladonia gracilis* ssp. *gracilis* White Mountains, New Hampshire ×0.96

240. *Cladonia gracilis* ssp. *turbinata* Cascades, Oregon ×1.3

241. *Cladonia incrassata* Pine Barrens, New Jersey ×3.3

242. *Cladonia leporina* northern Florida ×4.3

Cladonia leporina
Jester lichen

DESCRIPTION: Primary squamules rarely seen. Podetia much branched, forming sprawling cushions commonly up to 10 cm across and 4 cm high, usually yellowish green; branches divergent or curved, smooth to wrinkled, without soredia, 1.5–2(–3) mm in diameter, with a distinct (but usually dull) cortex, thick medulla, and a poorly coalesced, webby stereome; tips of branches often having small, globular, bright red fruiting bodies, much like the pompons on a jester's hat. CHEMISTRY: Cortex PD– or PD+ yellow, K± yellow or K–, KC– or KC+ green, C– or C+ green. Several chemical races are more or less correlated with geographic distribution. Specimens from North Carolina to the Georgia coast usually have baeomycesic acid (PD+ yellow) as well as squamatic, usnic, and low quantities of didymic acids (KC–, C–). Florida specimens frequently lack baeomycesic acid (PD–) and have abundant didymic acid (KC+ green, C+ green, especially close to the algal layer). HABITAT: On sandy soil, rarely on wood. COM-

243. *Cladonia macilenta* Appalachians, Tennessee ×2.8

MENTS: Fruiting specimens cannot be confused with any other lichen, but sterile material sometimes resembles *C. uncialis* or *C. dimorphoclada* (both with a much thinner medulla and always PD–, KC–, C–), *C. pachycladodes* (with at least some open axils and often PD+ yellow), or species of *Cladina* (usually more slender, paler yellowish green or gray, and lacking a cortex). *Cladina submitis*, however, often has much the same prostrate, robust growth form but is confined to the northeast and is always PD–, KC–, and C–.

Cladonia macilenta (syn. *C. bacillaris*)
Lipstick powderhorn, pin lichen

DESCRIPTION: Primary squamules small, less than 2 mm long or broad, thick, relatively undivided. Podetia grayish or greenish, slender, usually unbranched, 10–30 mm tall, often somewhat thicker at the tip, covered with fine soredia with relatively few podetial squamules, ending in blunt or pointed tips frequently with bright red apothecia, without cups. CHEMISTRY: Two main chemical races: PD–, K–, KC+ yellow to orange, C– (barbatic acid with or without squamatic acid), or PD+ orange, K+ yellow, KC–, C– (thamnolic acid with or without barbatic acid). Both chemical races can occasionally contain usnic acid as well. HABITAT: On old wood or soil, occasionally on rocks (especially the thamnolic acid chemical race). COMMENTS: The chemical race with barbatic acid rather than thamnolic acid has been called **C. bacillaris**. *Cladonia macilenta* is similar to *C. umbricola*, but that species is usually yellowish green, frequently has narrow cups, and has large, thin, finely divided primary squamules. ***Cladonia norvegica*** is a rare western species quite similar to the barbatic acid chemical race of *C. macilenta* in that it contains barbatic acid, lacks usnic acid, and has finely sorediate podetia without cups. When fruiting, *C. norvegica* has waxy brown rather than red apothecia. The primary squamules are sometimes sorediate in either species, but more commonly with *C. norvegica*. The podetia of *C. norvegica* are usually more conspicuously tapered from bottom to top, whereas *C. macilenta* can often have a slightly broader (i.e., club-like) tip. See Comments under *C. cristatella* on some similar, southeastern species with thamnolic acid.

Cladonia macrophyllodes
Large-leaved cladonia

DESCRIPTION: Primary squamules very broad and long with lobed margins, reaching 15 mm long and 8 mm wide, ascending, gray-green above with a pure white lower surface lacking a cortex. Podetia gray-green, often slightly pruinose, 7–15 mm tall, with broad, shallow cups proliferating from the center, often with small squamules developing along the rim of the cup; surface of podetia continuous to areolate, without soredia. CHEMISTRY: Lower surface of squamules PD+ red, K+ yellow, KC–, C– (atranorin and fumarprotocetraric acid). HABITAT: On soil and rocks in arctic and alpine sites. COMMENTS: This species resembles the more common *C. cervicornis* (especially ssp. *verticillata*), but the latter is generally browner and lacks atranorin. The primary squamules are among the largest in the genus. *Cladonia subcervicornis* has a dull grayish brown lower surface on the squamules and is known only from the Arctic and western mountains north of Oregon and Montana. *Cladonia schofieldii* is a rare western coastal species with squamules of a similar size (4–50 mm long, 1–5 mm wide) that become partially corticate below; although it has fumar-

244. *Cladonia macrophyllodes* Rocky Mountains, Colorado ×1.8
245. *Cladonia mateocyatha* (squamules) Ozarks, Arkansas ×1.4
246. (right) *Cladonia mateocyatha* (podetia) Ozarks, Oklahoma ×2.1

protocetraric acid like *C. macrophyllodes*, it rarely contains atranorin.

Cladonia mateocyatha
Mixed-up pixie-cup

DESCRIPTION: Primary squamules olive to brownish, very large (1–6 mm wide and up to 12 mm long), the lower surface at first white, commonly darkening to grayish or purplish brown. Podetia with a smooth or areolate cortex, lacking soredia, and with very irregular, somewhat swollen, closed cups 5–20 mm tall, commonly proliferating at the margins and sometimes from the center. Apothecia brown. CHEMISTRY: Thallus PD+ red, K+ brownish, KC–, C– (fumarprotocetraric acid). HABITAT: On thin, sandy soil or rocks, in the open. COMMENTS: *Cladonia mateocyatha* is similar in color to *C. strepsilis* but has much larger squamules and a different chemistry. *Cladonia phyllophora* also has irregular cups, but the cup margins tend to be squamulose, and the podetial base often is blackened or spotted. *Cladonia cervicornis* has more regular cups, and the squamules are smaller.

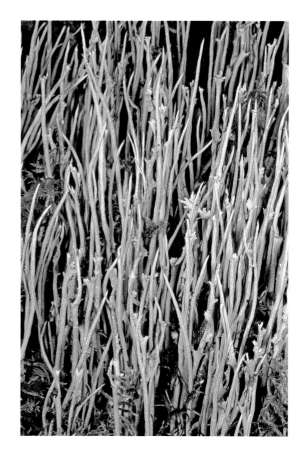

247. *Cladonia maxima*
Southeast Alaska ×0.68

Cladonia maxima
Giant cladonia

DESCRIPTION: Primary squamules never seen. Podetia tall, largely unbranched, grayish green to brownish, very smooth, up to 3 mm wide and more than 15 cm (6 inches!) long, lacking soredia, and with few or no podetial squamules; narrow, closed cups usually present in any clump of *C. maxima*, but they are rare in some populations. CHEMISTRY: Thallus PD+ red, K– or brownish, rarely K+ yellowish, KC–, C– (fumarprotocetraric acid and rarely some atranorin). HABITAT: Usually found in deep moss in forest or boggy sites. COMMENTS: This is the tallest *Cladonia* species on the continent. The sides of the podetia are smooth, only rarely perforated or split, distinguishing it from the very similar *C. gracilis* ssp. *vulnerata*. The latter is usually more slender but occasionally almost as tall. *Cladonia ecmocyna* is generally grayer, somewhat pruinose, and clearly K+ yellow (abundant atranorin in the cortex). IMPORTANCE: Caribou herds feed on *C. maxima* in Newfoundland, where the lichen is particularly abundant.

Cladonia multiformis
Sieve lichen

DESCRIPTION and COMMENTS: Primary squamules usually disappearing. Well-developed podetia are very distinctive, up to 45 mm tall, with cups perforated by numerous holes and with delicate marginal proliferations. It is an extremely variable lichen, as its scientific name indicates. Its cups can be elongated on one side or distorted in other ways, with longitudinally split branches, so that it resembles *C. furcata*. When it bears numerous podetial squamules, it can resemble *C. phyllophora* (with which it often grows), but that species has a softer, duller surface, especially at the tips, and the bases of the podetia are often black-spotted. Specimens with cups having few perforations can be mistaken for *C. gracilis* ssp. *turbinata*, which has cups that are always closed. CHEMISTRY: Thallus PD+ red, K– or brownish, KC–, C– (fumarprotocetraric acid). HABITAT: On thin, mineral soil in sun to partial shade.

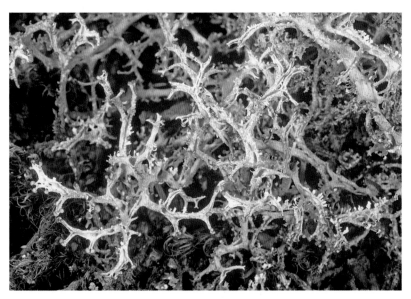

248. (top) *Cladonia multiformis* (perforate cupped form) southern Ontario ×2.4

249. (bottom) *Cladonia multiformis* (cupless form) Maine coast ×1.8

250. (above right) *Cladonia ochrochlora* Southeast Alaska ×1.5

Cladonia ochrochlora
Smooth-footed powderhorn

DESCRIPTION: Primary squamules large, lobed, about 3–6 mm long, 3–7 mm wide. Podetia greenish or olive, rarely brown, unbranched, 10–35 mm tall, with or without very narrow cups, with a continuous cortex on the lower half (more or less), and soredia (usually mealy rather than powdery) on the upper half. CHEMISTRY: Thallus PD+ red, K– or brown, KC–, C– (fumarprotocetraric acid). HABITAT: Almost always on decaying wood, rarely on soil. COMMENTS: The only other tall, narrowly cupped *Cladonia* with soredia more or less confined to the upper half is *C. cornuta*, especially **C. cornuta** ssp. **groenlandica**. The latter is distinctly brown-tinted and frequently found on soil or rock, as well as on wood. The similar *C. coniocraea* is sorediate to the base and rarely cupped. *Cladonia fimbriata* is shorter (usually less than 20 mm tall), has more symmetrical, less frequently proliferating cups, and like *C. coniocraea*, is sorediate to the base. **Cladonia ramulosa** is sometimes similar to cupless specimens of *C. ochrochlora*, but the podetia have less cortex and are more irregular in shape and coarsely sorediate, and the squamules are even finer (sometimes becoming reduced to a granular crust). It is widespread but uncommon.

Cladonia pachycladodes
Thick-walled cladonia

DESCRIPTION: Thallus without primary squamules, very pale yellowish green, often partly grayish. Podetia much branched, with few or many perforate axils, uneven and bumpy but not wrinkled, 25–50 mm long and 0.7–2 mm in diameter, surface dull with a soft appearance owing to the lack of a cortex; forming prostrate cushions, very *Cladina*-like, even with bent-over branch tips, but with very thick walls and a fibrous inner stereome; old specimens frequently covered with fine needle-like crystals (terpenes). Apothecia uncommon, yellowish brown. CHEMISTRY: Podetia PD– or PD+ bright yellow, K–, KC+ yellow, C– (usnic acid alone or, at least in Florida, with psoromic acid in the podetial tips). HABITAT: On sand in the open. COMMENTS: Few other cladonias have such a thick podetial wall and fibrous stereome, and none of the related thorn cladonias (*C. uncialis—boryi* group) have such a grayish hue. The general habit and bumpiness of the podetia of *C. pachycladodes* is similar to *C. dimorphoclada,* which is generally more slender (branches rarely more than 1 mm in diameter) and darker, has a smoother, shiny surface, and never contains psoromic acid. See also *C. leporina.*

Cladonia parasitica
Fence-rail cladonia

DESCRIPTION: Thallus greenish gray; primary squamules finely divided and granular, often forming a thick, granular mat. Podetia often very sparse or absent, short (mostly 3–10 mm tall), without cups, more or less covered with granular areoles or finely divided, coralloid squamules or, less commonly, granular soredia; stereome brownish, translucent, showing through in places with no squamules. Clusters of pale reddish brown apothecia frequently develop on the tips of the podetia. CHEMISTRY: PD+ orange, K+ yellow, KC–, C– (thamnolic acid). HABITAT: On old wood, especially of conifers, less frequently on

251. *Cladonia pachycladodes* central Florida ×1.2

252. *Cladonia parasitica* southern New Brunswick ×1.1

253. *Cladonia perforata* central Florida ×1.3

254. *Cladonia peziziformis* Ouachitas, Arkansas ×2.7

bark. Common on exposed fence rails. COMMENTS: Few other cladonias form a granular crust over rotting wood. The chemistry confirms the identification. *Cladonia macilenta* and *C. didyma* (var. *vulcanica*) have the same chemistry but usually produce at least a few podetia with red apothecia or pycnidia. *Cladonia macilenta* has thick, largely undivided primary squamules; those of *C. didyma* var. *vulcanica* are finely divided but not especially granular.

Cladonia perforata
Perforate cladonia

DESCRIPTION: Primary squamules never seen; thallus composed of tufts of abundantly branched podetia, pale greenish yellow, quite shiny, with a varnished appearance close to the tips. Podetia 30–60 mm long, 1.5–5 mm in diameter, smooth and uniform, frequently perforated with oval holes ca. 1–1.5 mm long; inner medullary surface webby, without a tough stereome. Apothecia unknown. CHEMISTRY: Cortex KC+ gold; medulla PD–, K–, KC–, C–, UV+ white (usnic and squamatic acids). HABITAT: On sandy soil in the open. COMMENTS: This extremely rare lichen is listed as an endangered species in the U.S. It is a large and striking species and is sometimes locally abundant where it is found (Escambia and Highland counties, Florida). *Cladonia perforata* can superficially resemble sterile *C. leporina* (which has more slender branches and different chemistry) and relatives of *C. uncialis*, none of which have round perforations along the sides of the podetia.

Cladonia peziziformis (syn. *C. capitata*)
Turban lichen

DESCRIPTION: Primary squamules small and rounded, rather thick and convex, sometimes coalescing into a crust. Podetia slender, unbranched or once branched, 10–20 mm tall; surface verruculose, not sorediate, split along the sides. Podetia always topped with large, pale brown, turban-like apothecia several times the diameter of the podetia (commonly up to 4 mm in diameter). CHEMISTRY: Medulla and cortex PD+ red, K–, KC–, C– (fumarprotocetraric acid). HABITAT:

On sandy soil in the open; occasionally on wood. COMMENTS: The very large apothecia, rather shell-like squamules, and chemistry distinguish this split-stalked species from related cladonias such as *C. cariosa* and *C. sobolescens*. In *C. sobolescens* and most other peg lichens, the apothecia are barely wider than the podetial diameter, and they sit on the podetial summit more like a skullcap than a turban.

Cladonia phyllophora
Felt cladonia

DESCRIPTION: Primary squamules strap-shaped or rounded, disappearing in most specimens. Podetia extremely variable in form, mostly 10–60 mm tall, 1–4 mm across, with gradually broadening cups having a slightly puffed-up aspect, frequently proliferating from the margins and appearing branched, sometimes with severely altered and aborted cups with perforations or slits; cup margins frequently decorated with small, thick squamules (especially in the northern part of its range); surface dull, appearing soft, especially close to the apex, the rest of the podetial surface smooth to areolate, not sorediate, olive- to gray-green, often browning in the sun; surface near the base of the podetia often breaking up into pale areoles against a dark, sometimes black background. Apothecia brown, on proliferations from the cup margins. CHEMISTRY: Thallus PD+ red, K– or brown, KC–, C– (fumarprotocetraric acid). HABITAT: On soil in the open or in partial shade. COMMENTS: This species is most frequently confused with *C. gracilis* or *C. multiformis*, but neither of these has a puffy, soft surface or spotted base. Some specimens, however, are almost impossible to name with certainty. Populations with squamulose cups are most distinctive. *Cladonia mateocyatha* is a more southern species, with aborted cups, larger squamules, and a more olive color.

Cladonia pleurota
Red-fruited pixie-cup

DESCRIPTION: Thallus pale yellowish green; primary squamules small or rather large, deeply lobed, nonsorediate. Podetia short (6–25 mm high), with broad cups covered with granular soredia; cup margins usually even. Bright red apothecia or pycnidia frequently produced directly on the cup mar-

255. *Cladonia phyllophora* Rocky Mountains, British Columbia ×1.3

256. *Cladonia pleurota* Maine coast ×2.1

257. *Cladonia pocillum* southwestern Yukon Territory ×2.3

258. *Cladonia polycarpoides* Big Thicket, eastern Texas ×1.4

gins or on short proliferations. CHEMISTRY: Thallus PD–, K–, KC+ gold, C–, UV– (usnic acid and zeorin). HABITAT: On wood, bark, or soil. COMMENTS: This resembles a yellowish version of *C. chlorophaea*, but the red apothecia or pycnidia on the cup margins of *C. pleurota* easily distinguish it when fertile. When the soredia are particularly large and partly corticate, *C. pleurota* can intergrade with *C. borealis*, which contains barbatic acid, or with the much rarer, arctic-alpine *C. coccifera*, which also has zeorin. Small specimens of *C. deformis* will key out here, but that species has farinose soredia and more irregular cups. *Cladonia carneola*, also with zeorin, has a more jagged, tooth-like cup margin, pale brown apothecia, and more powdery soredia.

Cladonia pocillum
Rosette pixie-cup, carpet pixie-cup

DESCRIPTION: Primary squamules thick, almost foliose, forming a lobed rosette around the colony. Podetia gray-green to brownish, goblet-shaped, with round areoles dispersed on the inside and outside of the cups; without soredia. Apothecia brown, on the cup margins. CHEMISTRY: Thallus PD+ red, K– or brownish, KC–, C– (fumarprotocetraric acid, very rarely accompanied by atranorin). HABITAT: On lime-containing soil in exposed sites, often on thin soil over rock. COMMENTS: No other pixie-cup has a primary thallus approaching a foliose growth form. In other respects, this species is almost indistinguishable from *C. pyxidata*, which prefers more acidic soil. See also Comments on *C. magyarica* under *C. pyxidata*.

Cladonia polycarpoides
Peg lichen

DESCRIPTION: Primary squamules rather long and strap-shaped, deeply divided into narrow lobes, often dominating the thallus. Podetia usually 10–30 mm high, unbranched except sometimes at the very tip, broader at the top with large, pale to dark brown apothecia often in a cluster. Podetia typically with a continuous cortex and few squamules, sometimes grooved and split; without soredia. CHEMISTRY:

266 CLADONIA

259. *Cladonia prostrata*
Gulf Coast, Florida
panhandle ×2.3

PD+ yellow, K+ red, KC–, C– (norstictic acid; lacking atranorin). HABITAT: On thin, often sandy soil in the open. COMMENTS: *Cladonia polycarpia* closely resembles *C. polycarpoides,* perhaps with a tendency to have shorter (up to 20 mm) podetia, but it contains stictic acid and atranorin as well as norstictic acid. It is widely distributed in the southeast. See Comments under *C. sobolescens.*

Cladonia prostrata
Resurrection cladonia

DESCRIPTION: Thallus consisting entirely of light green to gray-green or olive, radiating rosettes of extremely large squamules, commonly 10–20 mm long and 2–4 mm wide, delicately lobed, often branching in dichotomies, usually lying in one plane but turned up at the margins, revealing the white lower surface; often rolled up and appearing dead when dry and spreading out ("resurrected") when wet. Podetia always stubby and aborted. CHEMISTRY: Undersurface of squamules PD+ red, K+ yellow, KC–, C– (fumarprotocetraric acid and atranorin).

HABITAT: On sandy soil or dunes in the open or under bushes. COMMENTS: The spreading, almost foliose habit of *C. prostrata* is very distinctive. It is loosely attached to the soil, which is different from the foliose lobes of *C. pocillum,* a more northern species on limy soil.

Cladonia pyxidata
Pebbled pixie-cup

DESCRIPTION: Thallus grayish green to olive or brown; primary squamules tongue-shaped, rather thick, up to about 7 mm long and 4 mm wide, divided into rounded lobes. Podetia goblet-shaped, usually under 30 mm tall, with flattened, round areoles or squamules on the inside of the cups and sometimes also scattered over the largely decorticate outer surface of the podetia. Apothecia brown, on short proliferations on the cup margins. CHEMISTRY: Thallus PD+ red, K– or brownish, KC–, C– (fumarprotocetraric acid). HABITAT: Usually on acidic soil or granite. COMMENTS: Among the more common pixie-cups, *C. pocillum* is most like *C. pyxidata,* but it has ra-

diating, almost foliose lobes forming the primary thallus; in *C. pyxidata*, the primary thallus always has distinct squamules. When the flattened areoles of *C. pyxidata* grade into more spherical granules or granular soredia, the lichen can resemble *C. chlorophaea*, although *C. pyxidata* generally has a more discontinuous cortex. The less common, eastern temperate *C. magyarica* is grayer green and slightly pruinose, or olive to brownish in bright sun, and it contains abundant atranorin as well as fumarprotocetraric acid (medulla PD+ red, K+ pale yellow to brownish, KC–, C–). It grows on calcareous soil or over limestone. Its podetia, usually under 25 mm tall, have a slightly more continuous cortex than those of the true pixie-cup, and its cup margins often become quite squamulose.

Cladonia rappii
Slender ladder lichen

DESCRIPTION: Primary squamules usually abundant, elongate and deeply lobed (up to 5 × 1.5 mm), often curling to reveal the lower surface. Podetia gray to gray-green, sometimes becoming browned, 15–40 mm tall, abruptly expanding into flat cups or tiers 2–6 mm across, proliferating or branching up to 4 times from the often slightly raised centers of successive cups; cup margins commonly producing small squamules, dark brown apothecia, or simply tooth-like projections; podetial surface with a continuous cortex, without soredia but sometimes with squamules; base of podetia sometimes blackening. CHEMISTRY: Two chemical races: (*i*) var. *exilior*: PD+ red, K– or brownish, KC–, C– (fumarprotocetraric acid); (*ii*) var. *rappii*: PD+ bright yellow, K–, KC–, C– (psoromic acid). HABITAT: On highly acid soil, less commonly on wood or tree bases. COMMENTS: Some specimens of *C. rappii* var. *exilior* are hard to distinguish from *C. cervicornis* ssp. *verticillata*, which typically has more gradually expanding cups that remain slightly concave, even when proliferating. The latter can be found on drier, less acid soil. *Cladonia rappii* var. *rappii* is distinctive chemically because very few southeastern *Cladonia* species contain psoromic acid. (Note that baeomycesic acid gives almost the same reactions with spot tests; see table 2.) *Cladonia rappii* var. *exilior* has been referred to as ***C. calycantha*** in the old North American literature. True *C. calycantha* is a South American lichen. The map shows the combined range of both varieties.

260. *Cladonia pyxidata* southern Ontario ×1.0

261. *Cladonia rappii* ssp. *exilior* Gulf Coast, Florida ×2.4

Cladonia rei
Wand lichen

DESCRIPTION: Primary squamules small and inconspicuous, often disappearing. Podetia gray-green to olive, sometimes browned, slender, coarsely or finely sorediate on the upper half, with narrow, symmetrical, or, more frequently, lopsided cups, proliferating at the margins to resemble a star or magic wand. Pale to dark brown apothecia common on proliferations along the cup margins. CHEMISTRY: Podetia PD+ red or PD–, K– or brownish, KC–, C– (homosekikaic acid sometimes accompanied by fumarprotocetraric acid). HABITAT: On soil or wood in the open. COMMENTS: *Cladonia rei* resembles *C. gracilis* except for the soredia. *Cladonia coniocraea, C. ochrochlora,* and *C. cornuta* have larger primary squamules and do not normally proliferate at the cup margins; none of these contain homosekikaic acid. In the east, *C. rei* is often a pioneer in old field succession, together with *C. cristatella* and *C. polycarpoides*.

Cladonia robbinsii
Yellow tongue cladonia

DESCRIPTION: Thallus usually without podetia, consisting of nothing but primary squamules, yellowish green or olive, strap-shaped and divided into lobes up to 15 mm long and mostly 1–3 mm wide; lower surface of the squamules distinctly pale yellow or yellowish white. Podetia, when present, up to 5–15(–25) mm tall, without cups but sometimes branched, smooth to areolate, tipped with brown apothecia. CHEMISTRY: Squamules (both surfaces) PD–, K–, KC+ yellow or gold, C– (usnic and barbatic acids). HABITAT: On bare soil, sometimes rocks. COMMENTS: The pale yellow lower surface of the squamules sets this lichen apart from most similar lichens in its range. *C. strepsilis* is another soil cladonia that produces mats of squamules with a yellowish tint, but its chemistry is quite different. The southeastern *C. piedmontensis* also forms mats of yellow-green squamules, but the squamules have a white lower surface. Its podetia are 7–20 mm, without cups, usually unbranched, with pale brown apothecia. The podetial surface is often squamulose but can be continuous to areolate. *Cladonia luteoalba,* a widespread but un-

262. *Cladonia rei* southern Ontario ×2.7

263. *Cladonia robbinsii* Ozarks, Arkansas ×1.5

common lichen in the west, also has squamules with a yellow lower surface and is frequently found without podetia. Its squamules are rounder, more divided at the margins, and curled in. The lower surface is also brighter and more sulphur-like.

Cladonia scabriuscula
Mealy forked cladonia

DESCRIPTION: Primary squamules rarely seen. Podetia gray-green, slender, branched, usually squamulose, up to 100 mm long and 0.5–2 mm wide, with patches of granular soredia at the tips. CHEMISTRY: Podetia PD+ red, K– or brownish, KC–, C– (fumarprotocetraric acid). HABITAT: A woodland species growing over soil and mossy rocks; can tolerate shade, although it can also be found in bogs or on mossy talus slopes. COMMENTS: *Cladonia scabriuscula* is much like *C. furcata* but has granular soredia at the tips of the branches. Western populations of *C. furcata* are particularly confusing because the cortex breaks up into small patches with minute squamules that can appear granular.

264. *Cladonia scabriuscula* Lake Superior region, Ontario ×1.9
265. *Cladonia sobolescens* southeastern Missouri ×2.8

Cladonia sobolescens (syn. C. clavulifera)
Peg lichen

DESCRIPTION: Primary squamules rounded to strap-shaped, grayish green to brown depending on the degree of exposure to sun, sometimes deeply lobed but never frilly, often growing vertically and exposing a white lower surface. Podetia without cups or soredia, short and peg-like, unbranched or rarely branched once or twice near the tips, 5–20 mm tall, topped with brown apothecia only slightly or not at all broader than the podetial tips. CHEMISTRY: Thallus PD+ red, K–, KC–, C– (fumarprotocetraric acid). HABITAT: In very sunny sites on thin or sandy soil. COMMENTS: The peg lichens are a group of similar and presumably closely related species that differ mainly in chemistry. *Cladonia sobolescens* contains fumarprotocetraric acid (PD+ red, K– or brownish); *C. polycarpoides* and one chemical race of *C. symphycarpia* have norstictic acid (PD+ yellow, K+ red); *C. polycarpia* (PD+ orange, K+ red) has both norstictic and stictic acids; *C. brevis*, a northeastern species, and one chemical race of *C. symphy-*

carpia have psoromic acid (PD+ deep yellow, K– or pale yellow), while another chemical race of *C. symphycarpia* (in the west) has only atranorin (PD– or PD+ pale yellow, K+ yellow). *Cladonia peziziformis* is also PD+ red, K– (fumarprotocetraric acid) but has broader, usually paler brown apothecia.

Cladonia squamosa (syn. *C. squamosa* var. *subsquamosa*) Dragon cladonia, dragon funnel

DESCRIPTION: Primary squamules abundant and persistent, very finely divided, nonsorediate. Podetia pale grayish green in the shade to rather browned in sunny sites, short and squat or rather tall and slender (up to 40–50 mm high), abundantly scaly (squamulose), the squamules sometimes extremely small and almost granular but with no true soredia; largely without a cortex between the squamules, leaving a kind of translucent stereome at the surface; tips of the podetia with narrow open cups or cupless and blunt, somewhat branched, with open axils. Brown apothecia sometimes seen. CHEMISTRY: Two chemical races: (*i*) podetia PD–, K–, KC–, C–, UV+ blue-white (squamatic acid); or (*ii*) PD+ deep yellow-orange, K+ deep yellow, KC–, C– (thamnolic acid). HABITAT: Shade-tolerant, most often found on soil or logs in forests, but sometimes in exposed sites. COMMENTS: The thamnolic acid race is indistinguishable in appearance from the more common squamatic acid race, but it is found only along the oceanic west and, occasionally, east coasts. Sometimes the squamules on the podetia of *C. squamosa* are greatly reduced and appear to be granular or even sorediate, but the only sorediate *Cladonia* with open axils and squamatic acid is *C. cenotea*, which has farinose, diffuse soredia and in-rolled cup margins. Most easily mistaken for *C. squamosa* is *C. crispata,* a species with flaring, distinct, funnel-like cups and a more unbroken cortex. **Cladonia singularis** is an uncommon west coast species with pointed podetial tips and sloping, shingle-like podetial squamules. **Cladonia decorticata** is much like *C. singularis* but has more areas without a cortex, and the squamules, which are not at all shingle-like, become reduced to almost granular soredia. *Cladonia decorticata* is a low-arctic to boreal species containing perlatolic acid instead of squamatic acid, although the spot tests are the same (table 2). *Cladonia beaumontii,* a southeastern coastal plain species, resembles *C. squamosa*

266. *Cladonia squamosa* White Mountains, New Hampshire ×1.3

but has better-developed primary squamules and reacts PD+ yellow, K–.

Cladonia strepsilis Olive cladonia

DESCRIPTION: Podetia and squamules with a very distinctive olive or yellowish olive color; primary squamules abundant, with many small rounded lobes, white to slightly grayish below, conspicuously spotted with white maculae, without soredia, sometimes becoming detached from the soil and rounding up into small balls (plate 267). Podetia smoothly corticate and very irregular in form, 5–20 mm high, 1–3 mm wide, swollen and proliferating, often fissured, never forming true cups. CHEMISTRY: Medulla PD+ yellow, K–, KC+ blue-green, C+ green (baeomycesic, squamatic, and often barbatic acids, as well as strepsilin). HABITAT: Common on dry, exposed soil, sometimes becoming vagrant, especially in Ozark glades. COMMENTS: *Cladonia mateocyatha* is most similar to *C. strepsilis* in color and general appearance, but it is PD+ red, KC–, C–, and has some

267. *Cladonia strepsilis* (vagrant clumps of squamules) Ozarks, Arkansas ×2.1

268. *Cladonia strepsilis* (with podetia) Ozarks, Arkansas ×2.3

semblance of cups. The only other cladonias with C+ green, KC+ green reactions are those with abundant didymic acid such as *C. leporina*, a branched, yellow-green, red-fruited species.

Cladonia subradiata
Powdery peg lichen

DESCRIPTION: Primary squamules small, finely divided into lobes, often granular at margins. Podetia cylindrical with pointed or blunt tips, without cups or with very narrow cups, 10–20(–30) mm tall, covered with powdery soredia above, and normally with minute squamules and granules close to the base; with very small dark brown to pale brown apothecia at the podetial tips, single or forming a tiny ring. CHEMISTRY: PD+ red, K–, KC–, C– (fumarprotocetraric acid). HABITAT: Common in open areas on sandy soil; often on wood or bark. COMMENTS: This lichen is somewhat like **C. cylindrica**, a northeastern species that rarely produces apothecia and has coarsely granular soredia especially on the lower half of the podetia. It contains grayanic acid in addition to fumarprotocetraric acid. *Cladonia subradiata* also resembles *C. coniocraea*, but the latter has large, rounder, less divided squamules, its podetia are usually pointed, without apothecia, and it lacks minute squamules mixed with granules at the base. North American lichenologists have used the name *Cladonia balfourii* for *C. subradiata*, but *C. balfourii* is actually synonymous with *C. macilenta*.

Cladonia subulata
Antlered powderhorn

DESCRIPTION: Primary squamules usually disappearing. Podetia ash-gray to greenish gray spotted with brown in places, 30–100 mm tall, slender, frequently with one or two irregular branches at the tip, and sometimes developing very narrow, irregular or aborted cups; fine, powdery soredia covering the podetia from top to bottom. CHEMISTRY: PD+ red, K– or brownish, KC–, C– (fumarprotocetraric acid). HABITAT: On sandy soil, peat, or rotting logs in open boreal forest habitats. COMMENTS: Few, tall, PD+ red cladonias are entirely sorediate. *Cladonia coniocraea* is shorter and more distinctly tapered, with large primary squamules. *Cladonia cornuta* and *C. ochrochlora* are somewhat similar but always have a continuous cortex at the base, usually for at least a third the length of the stalk, and they are rarely branched. *Cladonia farinacea*, in the eastern United States, is a rather tall lichen (up to 100 mm) with open axils and no cups, branching two or three times. It tends to be brownish gray and looks like the dead grass with which it usually grows. *Cladonia ramulosa* (syn. *C. pityrea*) is a widespread, temperate, sorediate species. It is extremely variable but most typically has contorted, bent, more or less pointed podetia covered with coarse, granular soredia. The primary squamules are sometimes also reduced to a granular sorediate crust. *Cladonia rei* is a smaller species covered with granular rather than farinose soredia and contains homosekikaic acid (frequently with fumarprotocetraric acid). *Cladonia glauca* can be similar in color to *C. subulata* but has fewer cup-like tips and is often perforated (being related to funnel lichens such as *C. squamosa*); it is PD–, UV+ blue-white (squamatic acid).

269. *Cladonia subradiata* northern Florida ×2.1

270. *Cladonia subulata* northern British Columbia ×1.7

271. *Cladonia sulphurina* Rocky Mountains, Colorado ×2.7

Cladonia sulphurina (syn. *C. gonecha*)
Greater sulphur-cup

DESCRIPTION: Thallus pale yellow to pale greenish yellow; primary squamules finely divided, up to 8 × 5 mm. Podetia tall, up to 85 mm high and 2–5 mm in diameter, usually very irregular, with or without recognizable, narrow cups, finely sorediate for most of their length, leaving a smooth area with cortex only at the base. CHEMISTRY: Podetia PD–, K–, KC+ gold, C– (usnic acid); medulla UV+ white (squamatic acid). HABITAT: On soil or well-rotted wood such as mossy logs, generally in full sun. COMMENTS: *Cladonia sulphurina* is the tallest of the yellowish, sorediate, red-fruited cup lichens. It is also the most irregular in shape, with the stalks typically split or otherwise misshapen. Although most similar to *C. deformis*, the latter is the less deformed species. Their morphology often overlaps, but chemistry distinguishes them. See Comments under *C. deformis*.

Cladonia symphycarpia
Split-peg lichen, greater thatch soldiers

DESCRIPTION: Primary squamules large (2–8 × 1–6 mm), gray-green, strap-shaped or deeply lobed, curled up when dry, showing the white lower surface. Podetia rarely produced, greenish gray, relatively short (10–15 mm tall), stocky, without cups, unbranched or with a cluster of short branches at the summit, grooved and often perforated or slit; podetial surface usually broken up into small, irregular plates or areoles. Apothecia convex, brown, on branch tips. CHEMISTRY: Medulla (white undersurface of the squamules) KC–, C–, and may be PD–, K+ yellow (atranorin alone), PD+ yellow, K+ red (atranorin with norstictic acid), or PD+ bright yellow, K+ yellow (atranorin with psoromic acid). The psoromic acid chemical race is sometimes called ***C. dahliana***. HABITAT: Common in open areas on thin or sandy soil, especially those areas rich in calcium (e.g., over limestone). COMMENTS: The chemical race with only atranorin can be mistaken for *C. car-*

iosa, a boreal to arctic species with much smaller primary squamules (1–3 × 0.5–2 mm). The other peg lichens are compared under *C. sobolescens*.

Cladonia trassii
Spotted black-foot

DESCRIPTION: Primary squamules round or strap-shaped, 1–5 mm wide, abundant and usually persistent, green or gray-green, often becoming brown in part. Podetia long and slender, 20–80 mm tall, 1–3 mm in diameter, unbranched or with 1–2 branches, usually pointed at the apex but rarely with narrow cups proliferating from the center; surface with greenish areoles, more or less contiguous on the upper third, but scattered over a black stereome over much of the stalk, giving it a spotted appearance; podetial squamules very common, evenly spaced and ear- or wing-like, that is, rounded, sticking out almost at right angles to the podetium. CHEMISTRY: Medulla PD+ red, K+ yellow, KC–, C– (atranorin and fumarprotocetraric acid). HABITAT: On soils rich in humus and in arctic and alpine tundra, usually associated with late snow patches. COMMENTS: This lichen is very distinctive, with its white-on-black spotting and ear-like podetial squamules. *Cladonia stricta,* a similar arctic-alpine species, has a short-lived primary thallus and tall, pointed or narrowly cupped podetia with few podetial squamules; it usually lacks atranorin. *Cladonia phyllophora*, also with a blackened, spotty base, has thicker, stockier podetia with a more continuous cortex and is K– (lacking atranorin).

Cladonia turgida
Crazy-scale lichen

DESCRIPTION: Primary squamules very large, deeply lobed and ascending, up to 10 mm across and 20 mm long. Podetia stout, branched 2–4 times, perforated or with wide-open, gaping axils but never cupped, dark greenish gray or olive, or sometimes almost yellowish green, with a smooth, uniform or strongly maculate, sometimes slightly pruinose surface, without any soredia. Thallus sometimes

272. *Cladonia symphycarpia* southern Ontario ×2.9

273. *Cladonia trassii* coastal Alaska ×2.7

274. *Cladonia turgida* Lake Superior region, Ontario ×0.33

consisting of nothing but a mat of squamules, or, on well-developed old colonies, it can be a tangled mass of podetia with no primary squamules at all! CHEMISTRY: Medulla PD+ red, K+ pale yellow, KC–, C– (atranorin and fumarprotocetraric acid). The K+ yellow reaction is usually obscured by the heavy concentration of fumarprotocetraric acid, which can produce a K+ brown reaction. HABITAT: Shade-tolerant and often growing on forest soil among mosses, but also on thin soils in the open. COMMENTS: Although extremely variable, *C. turgida* is usually easy to recognize because of its large dark green squamules in combination with robust, smooth podetia. Some colonies without squamules somewhat resemble *C. uncialis,* but *C. turgida* is never truly yellowish green, and the medulla is always PD+ red. *Cladonia macrophyllodes* has the same chemistry and also produces large squamules, but the squamules are more crowded and tend to be thicker. The podetia are quite different, with closed cups that proliferate from the center. It is a more arctic-alpine species.

Cladonia umbricola
Shaded cladonia

DESCRIPTION: Primary squamules rather large (1.5–7 × 2–7mm), thin, finely divided, often sorediate; lower surface white to yellowish. Podetia pale yellowish green to pale greenish yellow, rarely gray-green, 6–27 mm high, 0.5–2 mm in diameter, usually unbranched, with or without narrow cups, typically covered with fine, powdery soredia (more rarely granular soredia), cortex disappearing or sometimes persisting at the base of the podetia. Apothecia or pycnidia often present, bright red, at the tips of blunt stalks or on cup margins. CHEMISTRY: This is a chemically diverse species having several chemical races. Usnic acid can be abundant, sparse, or absent (KC+ gold to KC–) and accompanied by either thamnolic acid (PD+ orange, K+ deep yellow, UV–) or squamatic acid (PD–, K–, UV+ blue-white). HABITAT: Mainly in shaded habitats, almost exclusively on rotting wood. COMMENTS: This pretty, red-fruited, sorediate lichen is highly variable in both chemistry and appearance. It

can be abundantly cupped and resemble *C. pleurota* or *C. carneola* (both with broader cups and different chemistry), or it can be entirely cupless and resemble *C. macilenta* except for the primary squamules (small, thick, rounded, and nonsorediate in *C. macilenta*). When it contains usnic acid, *C. umbricola* is quite yellowish; specimens that lack usnic acid are simply pale green or gray-green. **Cladonia polydactyla** is a gray-green, narrowly cupped, sorediate lichen with thamnolic acid (no usnic acid), but it is taller (up to 50 mm), its cup margins commonly proliferate or at least are toothed, and it has more cortex at the base. It is rare in North America. *Cladonia digitata* has similar podetia and chemistry but is distinguished by its very large, almost unlobed primary squamules (up to 15 mm across), which are finely sorediate on the undersurface, especially close to the margins. In **C. transcendens,** the soredia are granular to farinose and occur in patches, and the podetia are ordinarily more squamulose.

Cladonia uncialis
Thorn cladonia

DESCRIPTION: Primary squamules absent. Thallus pale yellow to light greenish yellow, forming cushions of abundantly branched, smooth, thin-walled podetia, 20–60 mm high, with open axils; without soredia or podetial squamules, but surface often mottled with patches of green set off by a white to yellowish network of lines (maculae); inner surface of the hollow podetia (i.e., the stereome) smooth and uniform, usually somewhat pruinose; tips of the branches divergent, pointed and thorn-like, without cups. Podetia variable in size, 0.7–4 mm wide; specimens with the broadest podetia mainly in moist, oceanic regions (plate 277), those with the most slender podetia mainly in dry boreal or eastern temperate regions. Apothecia rare, brown, produced on branch tips. CHEMISTRY: Cortex KC+ gold; medulla PD–, K–, KC–, C–, UV+ white or UV– (usnic acid, with or without squamatic acid). HABITAT: On bare soil or rock, or among mosses. COMMENTS: The smooth inner pode-

275. *Cladonia umbricola* mountains, central Washington ×1.9

276. *Cladonia uncialis* Lake Superior region, Ontario ×0.70

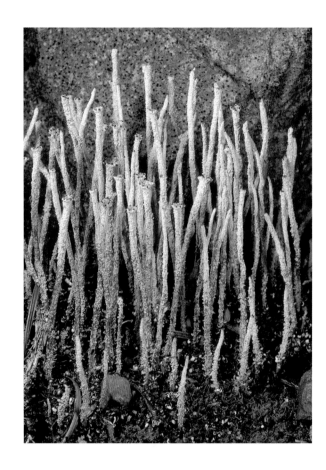

277. *Cladonia uncialis* (closeup) southeast Alaska ×2.0

278. *Cladonia verruculosa* Olympics, Washington ×1.6

tial surface of *C. uncialis* distinguishes it from most of its common relatives, such as *C. dimorphoclada, C. caroliniana, C. boryi,* and *C. pachycladodes*. *Cladonia amaurocraea* is more sparsely branched, and most clumps have at least a few cups; it contains barbatic acid. **Cladonia kanewskii,** a rare northwest coast species, seems closely related to *C. uncialis* but always lacks squamatic acid and has a broadly fibrous stereome and irregular branching more similar to the eastern species, *C. boryi*. It characteristically has a bumpy surface, and old specimens develop a fuzz of fine, colorless crystals at the podetial tips, as does *C. boryi* and its relatives. **Cladonia subsetacea** is a very slender member of the *C. uncialis* group that is found on the southeastern coastal plain. Its tangled branches are mostly under 0.6 mm in diameter, giving it the appearance of a reindeer lichen. However, the shiny, clearly corticate, greenish yellow branches are unlike those of any *Cladina*. Its medulla reacts PD+ yellow (baeomycesic and squamatic acids). IMPORTANCE: Together with reindeer lichens (genus *Cladina*), *Cladonia uncialis* is an important winter food for caribou and was one of the three most preferred lichens in feeding trials in Quebec.

Cladonia verruculosa
Western wand lichen, pebblehorn lichen

DESCRIPTION: Primary squamules small, sometimes disappearing. Podetia brown to olive, tall and slender, 20–70 mm high, ending in pointed tips or with narrow cups that sometimes proliferate from the centers as well as from the margins; granules and granular soredia distributed along the entire podetium or in patches on the upper half; podetial surface, except for the soredia, most commonly without a cortex, revealing the horny stereome, which is dark brown to black toward the base. CHEMISTRY: PD+ red, K–, KC–, C– (fumarprotocetraric acid). HABITAT: On bare soil in full sun. COMMENTS: This lichen resembles a very tall, robust *C. rei*, but in that species, the cups never proliferate from the center, and the podetial base does not become black. In addition, *C. rei* produces homosekikaic acid.

Cliostomum (5 N. Am. species)
Dot lichens

DESCRIPTION: Crustose lichens with thin to moderately thick, continuous to granular thalli, sometimes sorediate, gray to yellow-green. Photobiont green (*Trebouxia?*). Apothecia biatorine; disks yellow to pinkish gray or black; apothecial tissues yellowish or colorless, often containing granules; ascus *Biatora*-type. Spores predominantly 2-celled (occasionally 1–4-celled), colorless, narrowly ellipsoid, 8 per ascus; pycnidia black, with walls turning grayish red or purple in K; conidia almost spherical or ellipsoid. CHEMISTRY: Usually containing atranorin. Some species with usnic, fumarprotocetraric, protocetraric, or divaricatic acids. HABITAT: On bark or wood, usually in oceanic regions. COMMENTS: *Cliostomum* is related to *Biatora*, with similar asci and spores but a different cell structure in the excipulum. The K reaction of the pycnidia characterizes most species of the genus. The yellowish species tend to be accurate indicators of old forests. See Comments under *Psilolechia lucida*.

279. *Cliostomum griffithii* southern New Brunswick ×5.1

KEY TO SPECIES: See Key F.

Cliostomum griffithii
Multicolored dot lichen

DESCRIPTION: Thallus greenish white, thin, granular to verruculose, sometimes located mostly within the substrate (especially when on wood). Apothecia flat, adnate, 0.2–1 mm in diameter, varying (often on the same thallus) from beige or pinkish to dark gray or black, usually lightly pruinose, with thin or prominent biatorine margins; all apothecial tissues colorless, but the epihymenium is filled with granules that dissolve in K with a yellow reaction; spores 8–13(–15) × 2.5–5 μm, mostly 2-celled (occasionally up to 4-celled). CHEMISTRY: PD–, K+ yellowish (seen best in microscopic preparations), KC–, C– (atranorin). HABITAT: On bark and wood in coastal communities. COMMENTS: Its multicolored apothecia on a thin thallus make *C. griffithii* easy to identify in the field.

Coccocarpia (4 N. Am. species)
Shell lichens

DESCRIPTION: Small, closely appressed foliose lichens containing cyanobacteria (*Scytonema*) in a well-defined algal layer with a white medulla; most commonly bluish gray, with rounded, somewhat overlapping lobes or narrow, branching lobes; some species with isidia or lobules but not soredia; cortices present on upper and lower surfaces; lower surface tan to black, with abundant, unbranched, hair-like rhizines that sometimes form a thick hypothallus. Apothecia biatorine, with yellowish to red-brown or black disks and a thin margin disappearing in maturity. Spores colorless, 1-celled, spherical to narrowly ellipsoid. CHEMISTRY: No lichen substances in North American species. HABITAT: On bark or leaves mainly in tropical climates. COMMENTS: *Coccocarpia* resembles several other genera with cyanobacteria, especially **Degelia, Pannaria,** and **Parmeliella**. It is most easily distinguished from these by the biatorine, marginless apothecia, 1-celled spores, or the photobiont. *Coccocarpia* is similar in color to many species of *Leptogium*, but *Leptogium* is a

280. *Coccocarpia erythroxyli* northern Florida ×2.8

281. *Coccocarpia palmicola* hill country, central Texas ×3.5

jelly lichen, becoming translucent when wet. Its photobiont (*Nostoc*) is distributed throughout the thallus, and it therefore lacks a medulla.

KEY TO SPECIES: See *Pannaria*.

Coccocarpia erythroxyli
Fruiting shell lichen

DESCRIPTION: Thallus blue-gray, with rounded, shell-like lobes, mostly 2–7 mm wide; upper surface often having concentric ridges; otherwise smooth and even shiny, without isidia; lower surface pale to dark brown, with a thick blue-green to blue-black tomentum. Brown, convex apothecia, 1–4 mm in diameter, common on the lobe surface. HABITAT: On bark in hardwood hammocks and other tropical or subtropical woodlands; rarely in alpine tundra on the ground. COMMENTS: The two common species of *Coccocarpia* differ in their mode of reproduction. *Coccocarpia erythroxyli* reproduces sexually, and *C. palmicola* reproduces almost exclusively by isidia. *Coccocarpia stellata* is a rare species occurring from South Carolina to Florida, lacking isidia, with very narrow, linear lobes (less than 0.5 mm wide), growing on bark and leaves. *Degelia plumbea* (syn. *Parmeliella plumbea*), although superficially very similar to *C. erythroxyli*, lacks a lower cortex, contains *Nostoc* as its photobiont, has apothecia with distinct pale margins, and has larger spores 17–25 × 7–10 μm). In addition, its upper cortex has cells that are palisade-like rather than running parallel with the surface. *Degelia plumbea* is very rare in North America, known only from a few localities from Maine to Newfoundland.

Coccocarpia palmicola (syn. *C. cronia*)
Salted shell lichen

DESCRIPTION: Thallus dark blue-gray, with rounded lobes 2–5 mm across; cylindrical to globular or lobed isidia abundant or sparse, produced on the thallus surface; lower surface with a thick tomentum varying from pale tan with gray tips, to dark blue-gray, often showing at the edges. Apothecia rare. HABITAT: On tree bark or mossy rock in shaded situations. COMMENTS: *Coccocarpia palmicola* is the isidiate, nonfertile counterpart of *C. erythroxyli*. ***Coccocarpia domingensis*** is a rare tropical species with long, extremely narrow lobes (0.1–0.5 mm wide) and isidia on the lobe margins or upper surface.

Coccotrema (2 N. Am. species)

KEY TO SPECIES: See Key D.

Coccotrema maritimum
Volcano lichen

DESCRIPTION: Thallus creamy white to pinkish gray or yellowish gray, fairly thick, verrucose to areolate, without soredia, with scattered, hemispherical, pinkish to brown-gray cephalodia containing *Calothrix*; no isidia or soredia. Principal photobiont green (*Myrmecia*). Apothecia 0.7–1.5 mm in diameter, embedded in thalline warts with flat summits, opening into the hymenial cavity through a round hole; exciple thick, colorless, distinct; fairly long periphyses around the mouth of the ostiole; paraphyses mainly unbranched; hymenium IKI+ red-orange; asci opening through a slit at the apex, IKI–; spores colorless, 1-celled, mostly 45–65 × 24–32 μm, with thin, uniform walls, (4–)6–8 per ascus. CHEMISTRY: Medulla PD+ orange, K+ yellow to reddish, KC–, C– (stictic and constictic acids, usually with norstictic acid). HABITAT: On maritime rocks in the salt-spray zone, forming a more or less distinct white belt above the black *Verrucaria maura* zone (see Chapter 9, plate 64). COMMENTS: *Coccotrema* has the general appearance of a *Pertusaria* because of the fertile thalline warts, or an *Ochrolechia* because of the pinkish ostioles, which are similar to those of a young *Ochrolechia* apothecium. The double-walled apothecia (with a well-developed exciple surrounded by thalline tissue making up the outer margin) and thin-walled, IKI– asci, however, are unlike those in either of these genera. ***Coccotrema pocillarium*** is a bark-dwelling member of the genus with very similar apothecia, but it produces patches of coarse, granular soredia and contains only stictic and constictic acids. Its thick, creamy thallus can cover fairly large areas of bark or wood. It is restricted to the Pacific Northwest in forests close to the sea.

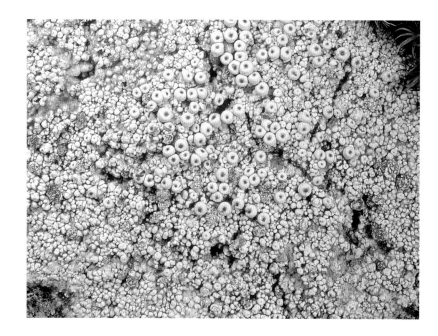

282. *Coccotrema maritimum* Queen Charlotte Islands, British Columbia ×3.1

Coenogonium (7 N. Am. species)
Pixie-hair lichens

DESCRIPTION: Filamentous lichens, usually forming light orange to pale green cottony tufts 10–30 mm across, coalescing into small shelf-like colonies in a few species. Each hair-like element of the thallus consists of a filament of the green alga *Trentepohlia* (or *Physolinum* in ***Coenogonium moniliforme***), enveloped within a network of fungal hyphae. Apothecia biatorine, yellow to pale orange, slightly stalked in some species, disks usually flat, with thin, persistent margins; paraphyses unbranched; spores 1–2-celled, colorless, ellipsoid, 8 per ascus in either a single row or two irregular rows. CHEMISTRY: No known lichen substances. HABITAT: On bark or leaves in shaded tropical or subtropical woodlands. COMMENTS: Most species of *Coenogonium* consist of barely lichenized filaments of *Trentepohlia*. The lichen thalli hardly differ in appearance from the unlichenized tufts of *Trentepohlia*, which are often common in the same habitats. It is therefore essential to find thalli with fruiting bodies, not only to be able to name the species but also to be sure you have a lichen! See also Comments under *Racodium rupestre*.

KEY TO GENUS: See Key B.

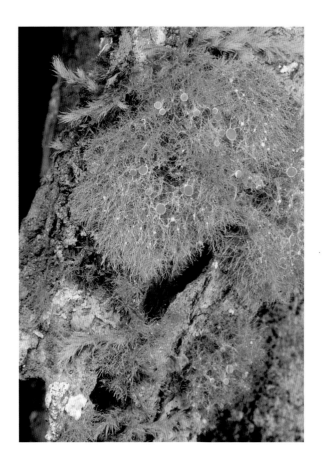

283. *Coenogonium implexum* northern Florida ×1.3

Coenogonium implexum
Pixie-hair

DESCRIPTION: Pale green to orange tufts, lifted slightly from the surface at the lower edge. Apothecia thin, flat, pinkish orange, 0.4–1 mm in diameter, usually appressed to thallus, with a very thin margin; paraphyses thick (resembling the asci in places); spores 2-celled, 7–9 × 2–3 μm, in one row or sometimes doubled at the thickest part of the ascus. COMMENTS: *Coenogonium interplexum* and *C. linkii* have 2-celled spores of about the same size as *C. implexum*. In *C. linkii*, the thalli form shelf-like horizontal patches on the trees with the apothecia on the lower side, and the spores are always in two rows. *Coenogonium interplexum*, like *C. implexum*, has spores usually in one row and does not have shelf-like thalli, but its apothecia are all raised on short stalks and it has slender paraphyses (ca. 1 μm). *Coenogonium interpositum* has 1-celled spores, 6–10 × 2–3 μm in two rows.

Collema (35 N. Am. species)
Jelly lichens, tarpaper lichens

DESCRIPTION: Foliose lichens, black to very dark olive, brownish olive, or dark yellowish brown when dry; when wet, swelling and becoming gelatinous, dark green to blue-green or almost black; thallus sometimes merely a crust-like membrane. Photobiont blue-green (*Nostoc*), not forming a distinct layer within the thallus (i.e., not stratified); thallus without a cellular cortex on either side (Fig. 7a); lower surface much like the top surface, without rhizines but sometimes with tufts of white tomentum. Apothecia lecanorine, developing on the upper surface of the thallus; exciple as seen in a section of the apothecium can be composed of elongate cells (prosoplectenchyma) or rounded cells (pseudoparenchyma); disks brown; spores colorless, 2- to many-celled (muriform), (4–)8 per ascus. CHEMISTRY: No lichen substances. HABITAT: On a wide variety of substrates in wet or dry habitats. COMMENTS: Other unstratified, foliose jelly lichens include *Leptogium*, *Leptochidium*, and *Lempholemma*. The first two differ from *Collema* in having a cortex on at least one surface. Some species of *Leptogium* and *Leptochidium* have white, fuzzy hairs on the thallus or are steel-gray to shiny red-brown when dry, or both. *Collema* species are never hairy and are rarely gray or shiny. **Lempholemma polyanthes** (syn. *L. myriococcum*) forms almost foliose membranes over calcareous rock, soil, or mosses in the northeast. Its spores are 1-celled and almost spherical (9–14 μm in diameter), and the apothecia are clustered on ridges, forming networks on the thallus surface. *Lichinella* is blacker, wet or dry, and is sometimes pruinose; its lobes are more strap-shaped and ascending, almost fruticose in habit, and its photobiont is *Gloeocapsa*, a different cyanobacterium.

KEY TO SPECIES

1. Growing on bark 2
1. Growing on rock, soil, or moss 8
2. Lobes small, under 3 mm wide, crowded into small cushions; spores fusiform, 2-celled, (13–)15–24(–26) × 3–4.5(–6) μm .... [*Collema conglomeratum*]
2. Lobes over 3 mm wide, not cushion-forming; spores fusiform to needle-shaped, 4–16-celled 3

3. Isidia absent 4
3. Isidia present 6
4. Apothecia pruinose .... *Collema pulcellum* (some varieties)
4. Apothecia without pruina 5
5. Tissue below the hypothecium in mature apothecia composed of round to angular cells .... *Collema pulcellum* var. *pulcellum*
5. Tissue below the hypothecium in mature apothecia composed of long, cylindrical cells .... *Collema nigrescens*
6.(3) Isidia cylindrical, unbranched or branched .... *Collema furfuraceum*
6. Isidia spherical or globular 7
7. Thallus relatively flat, although sometimes with folds .... *Collema subflaccidum*
7. Thallus with round to elongate pustules or blisters .... *Collema nigrescens*
8.(1) Lobes more or less uniform in thickness, not thicker at the margins; lobes mostly 2–6(–10) mm wide, with or without erect marginal lobules 9
8. Lobes with distinctly thickened margins, often producing erect, swollen lobules; small species with lobes mostly 0.5–3 mm wide 15
9. Lobes erect and branched, finely divided to lobulate at the margins, under 3 mm wide; spores 4-celled to muriform when mature .... *Collema cristatum*
9. Lobes prostrate or on edge, branched or rounded, often over 3 mm wide; spores muriform or not 10
10. Isidia absent; thallus pustulate; on siliceous rocks; mainly East Temperate .... [*Collema ryssoleum*]
10. Isidia present; thallus pustulate or not; on various kinds of rock 11
11. Isidia flattened like squamules, overlapping .... [*Collema flaccidum*]
11. Isidia cylindrical or globular, not flattened 12
12. Isidia cylindrical on mature specimens, unbranched or branched; lobes strongly ridged and pustulate, 5–10 mm wide .... *Collema furfuraceum*
12. Isidia spherical or globular, rarely cylindrical; lobes more or less smooth or folded, not pustulate, (1–)2–6 mm wide 13
13. On siliceous rock; lobes usually flat, with small, globular isidia (rarely becoming cylindrical); apothecia rare; spores 6–8-celled .... *Collema subflaccidum*
13. On calcareous rock; lobes crowded, often concave, with large globular isidia; apothecia relatively abundant 14
14. Spores fusiform, 4-celled .... *Collema undulatum* var. *granulosum*
14. Spores ellipsoid, muriform ..... [*Collema fuscovirens*]
15.(8) On calcareous rock 16
15. On soil 17
16. Lobes flat, crowded, usually on one edge, without lobules or isidia; apothecia abundant; spores (2–)4-celled, narrow, 6.5–8.5 μm wide .... *Collema polycarpon*
16. Lobes with cylindrical, erect branches; globular isidia sometimes present; spores muriform to 4-celled, ellipsoid, 8–13 μm wide .... *Collema cristatum*
17. Lobes prostrate; lobules, when present, rounded, strap-shaped, or finely divided; globular isidia sometimes present on lobe surface; spores mostly 4-celled and fusiform, sometimes submuriform ..... *Collema tenax*
17. Lobes erect, often with cylindrical lobules; spores 2–4-celled or submuriform 18
18. Spores 2-celled; lobules cylindrical, appearing globular from above, but true isidia absent .... *Collema coccophorum*
18. Spores 4-celled to submuriform; lobules erect, sometimes branched; globular isidia sometimes present .... *Collema cristatum*

Collema coccophorum
Tar jelly lichen

DESCRIPTION: Thallus very dark, with thickened lobe tips that are narrow or rounded, 0.5–3 mm wide, normally without isidia but often developing erect, thickened lobules or cylindrical outgrowths that resemble isidia and can dominate the thallus. Apothecia usually abundant; disks very dark brown, flat, 1–1.7 mm in diameter, with persistent raised margins; spores broadly to narrowly ellipsoid, 2-celled, not constricted in the middle, 11–22 × 5.5–9.5 μm. HABITAT: On dry calcareous soil. COMMENTS: The 2-celled spores and thickened lobe tips distinguish this fairly common lichen from similar soil-dwelling jelly lichens such as *C. tenax* (with 4-celled spores). ***Collema conglomeratum,*** a widely distributed species, is very similar but grows on bark.

284. *Collema coccophorum* (damp) Chiricahua Mountains, southeastern Arizona ×3.4

285. *Collema cristatum* central British Columbia ×3.3

Collema cristatum
Fingered jelly lichen

DESCRIPTION: Thallus very dark olive to black, thin or thick, but more or less uniform in thickness, with relatively narrow, ascending lobes under 3 mm wide, often thickened into cylindrical branches like huge isidia, or with concave, ascending, divided lobes with somewhat thickened margins; globular isidia occasionally produced on the thallus surface. Apothecia numerous or absent, with dark red-brown disks and usually smooth margins; spores muriform or 4-celled, 18–32(–40) × 8–13 μm. HABITAT: On calcareous rocks such as limestone; sometimes on limy soil or among mosses. COMMENTS: This is a common, extremely variable species most easily recognized by its erect, thickened lobules and its habitat. *Collema crispum,* a common species in the west, especially California, has flattened, scale-like isidia or lobules on narrow, somewhat concave, ascending lobes (2–5 mm wide) that are usually crowded and overlapping. It has broader 4-celled spores (never muriform), 26–34 × 13–15 μm. *Collema flaccidum* also has flattened, overlapping isidia but has broad, flat lobes (often over 6 mm wide). It is most common in the east. Its spores are fusiform, 4–6-celled, and 6–8.5 μm wide. See *C. undulatum*.

Collema furfuraceum
Blistered jelly lichen

DESCRIPTION: Thallus lobes broad, 5–10 mm across, with conspicuous ridges and pustules; cylindrical isidia (globular on immature specimens) covering at least the ridges. Apothecia rare; apothecial disks usually without pruina in the north, but a variety in the subtropical southeast has pruinose apothecia. Spores narrowly fusiform to needle-shaped, 40–80 × 3–6.5 μm, 5–6-celled. HABITAT: Usually on trees, especially poplars, also on mossy rock. COMMENTS: This is one of the largest jelly lichens. *Collema nigrescens* also has broad lobes that are pustulate-ridged, but it rarely has isidia, and when present, the isidia are all globular. *Collema subflaccidum* is very similar in size and general appearance but it has no distinct pustules or ridges and its isidia are exclusively globular. Some jelly lichens with flattened isidia are mentioned under *C. cristatum*.

Collema nigrescens
Blistered jelly lichen

DESCRIPTION: Thallus lobes broad (5–10 mm), with conspicuous blister-like pustules and ridges, usually without isidia, or with exclusively globular isidia; dark olive to brownish, or sometimes yellowish brown between the ridges. Apothecia very common, 0.6–1 mm in diameter, without pruina; spores needle-shaped, 50–100 μm, 6–15-celled. HABITAT: Common on the bark of poplars and other trees. COMMENTS: *Collema pulcellum* var. *pulcellum* is virtually indistinguishable from *C. nigrescens* except in details of the anatomy of the apothecium. See Description under *C. pulcellum*. *Collema pulcellum* var. *subnigrescens* and var. *leucopeplum* have moderately to heavily pruinose apothecial disks.

286. *Collema furfuraceum* coastal Alaska ×1.7

287. *Collema nigrescens* (damp) California coast ×3.2

288. *Collema polycarpon* canyon country, Utah ×4.2

289. *Collema pulcellum* var. *subnigrescens* Everglades, Florida ×2.2

Collema polycarpon
Shaly jelly lichen

DESCRIPTION: Thallus small, with thick, deeply divided, erect lobes 1–3 mm wide, conspicuously thickened and black at the edges, yellow-brown to dark olive on the less exposed lobe surfaces, without isidia, but covered with apothecia having dark red-brown to black disks without pruina. Spores narrowly ellipsoid to fusiform, 2–4-celled, mostly 18–28 × 6.5–8.5 μm, but frequently hard to find. HABITAT: On calcareous rocks. COMMENTS: Closely resembles *C. coccophorum*, which has 2-celled spores when mature and grows on soil rather than rock. The erect lobes of *C. cristatum* are cylindrical rather than plate-like.

Collema pulcellum
Blistered jelly lichen

DESCRIPTION and COMMENTS: The variability of *C. pulcellum* is recognized in its three varieties. *Collema pulcellum* var. *pulcellum* is virtually indistinguishable from *Collema nigrescens* except that the exciple (just below the hypothecium) of the latter has long, cylindrical cells, whereas that of *C. pulcellum* has round or angular cells, about equal in length and breadth. *Collema pulcellum* var. *subnigrescens* (without an apothecial cortex) and var. *leucopeplum* (with an apothecial cortex) have apothecial disks that are somewhat to heavily pruinose, unlike *Collema nigrescens*. HABITAT: Common on the bark of poplars and other trees.

Collema subflaccidum
Tree jelly lichen

DESCRIPTION: Lobes large, rounded, dark olive to almost black, mostly 2–6 mm across, with a smooth to folded or puckered surface, and with globular isidia that sometimes become cylindrical or even branched on old lobes. Apothecia rarely seen. HABITAT: On bark of hardwood and sometimes coniferous trees, especially in old forests; also on shaded or mossy rocks. COMMENTS: *Collema furfuraceum* differs from *C. subflaccidum* in having distinct pustules and ridges, with the isidia (which are mainly cylindrical, not globular) confined primarily to the summits of the ridges and blisters. Rare specimens of *C. subflaccidum* have low, poorly defined blisters, but the isidia are globular and are distributed uniformly over the center of the thallus, not just on the ridges.

Collema tenax
Soil jelly lichen, tar-jelly

DESCRIPTION: Thallus black, rather thick, especially at the margins, extremely variable in size (forming colonies 1–10 cm across); lobes closely appressed and almost membranous in part, often with ascending margins, rarely almost cylindrical, variable in width (up to 6 mm) but in most cases under 2 mm across, often forming radiating rosettes; thallus simply folded or isidiate to granular in the center. Apothecia common, with red-brown disks and smooth to verrucose margins, about 0.6–2 mm in diameter; spores predominantly 4-celled and fusiform, rarely submuriform and ellipsoid, 15–25 × 6.5–10 µm. HABITAT: On calcium-containing soil, rarely on calcareous rock. COMMENTS: This is a small, soil-dwelling *Collema* with thickened lobe margins. It is the most common of the jelly lichens on soil, although others also have the swollen and folded lobes characteristic of *C. tenax*. ***Collema bachmanianum***, mainly a western species, has broader, flatter lobes (2–8 mm across), broadly attached to immersed apothecia, and broader submuriform spores (20–35 × 9.5–16 µm). *Collema coccophorum* is a soil-dwelling, mainly western species very similar to *C. tenax*, but with 2-celled, ellipsoid spores. Specimens growing on limestone should be compared with *C. polycarpon*, which has ascending lobes that are generally raised at the edges, forming radiating plates, and 4-celled spores.

Collema undulatum
Jelly flakes

DESCRIPTION: Thallus dark yellowish brown or olive to black, composed of rounded, thin or somewhat thickened, ascending lobes, 1–5 mm across, curled up and contorted like cornflakes, almost always with rather large, globular isidia (0.1–0.2 mm in diameter) on the surface. Apothecia uncom-

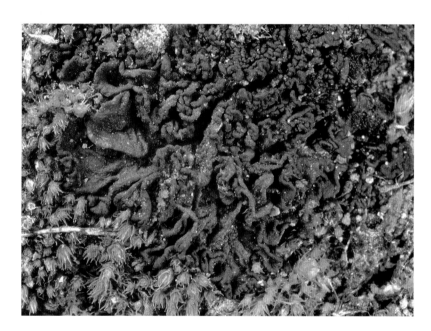

290. *Collema subflaccidum* Appalachians, Tennessee ×3.9

291. *Collema tenax* northeastern Utah ×5.7

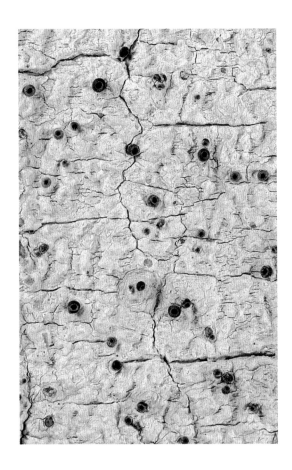

292. *Collema undulatum* var. *granulosum* southwestern Yukon Territory ×3.4

293. *Conotrema urceolatum* North Woods, Maine ×3.6

mon, flat to convex, with red-brown disks and even, persistent margins; spores 4-celled, fusiform, 17–30 × 6.5–9 μm. HABITAT: On lime-containing rocks, less commonly on soil. COMMENTS: The granular isidiate var. *granulosum*, which is shown here, is the most common variety in North America. *Collema undulatum* var. *undulatum* is similar but lacks isidia and usually has abundant apothecia. Sterile specimens of *C. fuscovirens* (syn. *C. tuniforme*) are very similar to *C. undulatum* var. *granulosum*, but the thallus is more often prostrate and the lobes are weakly blistered; fertile specimens have muriform spores.

Conotrema (1 N. Am. species)

KEY TO SPECIES: See Key D.

Conotrema urceolatum
Can-of-worms lichen

DESCRIPTION: Thallus thin, continuous, forming large white patches. Photobiont green (*Trebouxia*). Fruiting bodies are modified apothecia that look like tiny black volcanoes or open barrels, 0.3–0.7 mm in diameter, resembling perithecia but opening to the surface through a deep, wide hole rather than simply with a spot or dimple at the summit; broader apothecia often have a pruinose disk; walls (exciple) pitch black, rarely lightly pruinose, but often partly covered with a thin white thalline layer; spores extremely long and slender (over 100 μm), colorless, with 25–30 short cells constricted at the cross walls, making them look like microscopic earthworms, 8 per ascus. CHEMISTRY: Thallus PD–, K–, KC–, C– (no lichen substances). HABITAT: In North America, almost exclusively on sugar maple bark but sometimes on oak, linden, or a few other deciduous forest species. COMMENTS: *Conotrema* has the general aspect of a very large perithecial lichen, such as an *Eopyrenula*. The broad, deep ostioles of the apothecia, well-developed white thallus, and substrate preference are characters for recognizing the species in the field, and the unique spores will confirm its identity in the lab.

Cornicularia (1 N. Am. species)

KEY TO SPECIES: See Key A.

Cornicularia normoerica
Bootstrap lichen

DESCRIPTION: Small fruticose lichen with tiny tufts of black to very dark red-brown, somewhat flattened, strap-like lobes dichotomously branched, barely 0.5 mm across, and rarely exceeding 10 mm long; cortex thick and tough; medulla dense and solid. Photobiont green (*Trebouxia?*). Often fertile, producing black or dark brown lecanorine apothecia close to the tips of the lobes; spores colorless, 1-celled, ellipsoid, 8 per ascus. CHEMISTRY: All reactions negative (only fatty acids have been reported). HABITAT: On exposed rocks and boulders at high elevations. COMMENTS: *Cornicularia* often grows alongside species of *Pseudephebe*, but even the occasionally flattened *P. minuscula* can be distinguished by having a highly branched mass of much finer branches, more woolly than strap-like. *Lichinella*, a similar black, strap-shaped fruticose lichen on limestone rock, is a nonstratified jelly lichen and contains cyanobacteria.

Crocynia (2 N. Am. species)

KEY TO SPECIES: See Key C.

Crocynia pyxinoides
Lobed cotton lichen

DESCRIPTION: Crustose lichen with a thick, cottony, bluish gray thallus lacking an upper cortex, with narrow, flat, branched lobes at the margins, 0.4–1.3 mm wide,

294. *Cornicularia normoerica* Olympics, Washington ×2.9

295. *Crocynia pyxinoides* south central Florida ×4.0

more or less surrounded by a black prothallus; becoming coarsely sorediate in the older central portions of the thallus. Photobiont green. Apothecia rare. CHEMISTRY: Thallus PD+ orange, K+ yellow, KC–, C– (atranorin, stictic acid, triterpenes, and fatty acids). HABITAT: On bark of hardwoods. COMMENTS: This lichen at first looks like a *Lepraria* or a *Leprolomma,* but it is not leprose except in the old parts of the thallus. The effigurate margin and black prothallus are unique, especially in combination with the chemistry.

Cryptothecia (2 N. Am. species)

DESCRIPTION: Crustose lichens with thick, almost zoned thalli with a conspicuous prothallus developing at the periphery. Photobiont green (*Trentepohlia*). Clearly defined fruiting bodies are not formed. Asci simply arise (sometimes in groups) within the thallus; spores colorless to pale brown, very large, muriform, 1–2 per ascus. CHEMISTRY: Contains *para*-depsides and sometimes chiodectonic acid, a red anthraquinone. HABITAT: On bark in subtropical woodlands. COMMENTS: The two species of *Cryptothecia* are recognized by their thick, round thalli and conspicuous cottony prothallus.

KEY TO GENUS: See Key C.

Cryptothecia rubrocincta
Christmas lichen

DESCRIPTION: Thallus gray-green to mint green with a dull, dusty texture, often aggressively overgrowing other lichens or mosses, forming continuous, rather thick, circular patches with a bright red cottony rim (prothallus); red, spherical to cylindrical isidia-like granules produced on the older central portions of the thallus. Fruiting bodies unknown. CHEMISTRY: Thallus, especially where reddish, PD+ deep reddish purple, K+ very dark purple-red, KC–, C– (chiodectonic and confluentic acids). The reaction with PD is unusual for an anthraquinone. HABITAT: Subtropical to tropical woodlands, most noticeable on bald cypress and oak trees in swamps and hummocks. COMMENTS: The red and green of this unmistakable lichen give it a Christmas wreath look. (After long storage, the green fades to white.) This lichen has never been found fertile, so its classification is based on general thallus structure and chemistry. IMPORTANCE: *Cryptothecia rubrocincta* has been used in Brazil as a source of dye.

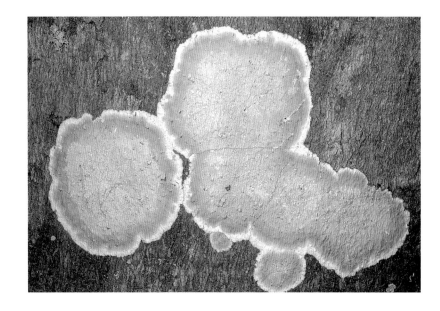

Cryptothecia striata

DESCRIPTION and COMMENTS: *Cryptothecia striata* lacks the crimson pigment of the Christmas lichen, so its conspicuous webby prothalline border is white instead of red. The two species also differ in their chemistry. Spores mostly 55–70 × 23–30 μm, 1(–2) per ascus. This lichen has been called **Chiodecton montagnei,** which is actually a West Indian species with 8 spores per ascus. CHEMISTRY: Thallus PD–, K–, KC+ red, C+ red (gyrophoric acid). HABITAT: Same as *C. rubrocincta.*

296. (facing page) *Cryptothecia rubrocinta* northern Florida ×5.0

297. (above) *Cryptothecia striata* Big Thicket, eastern Texas ×0.70

Cyphelium (11 N. Am. species)
Soot lichens

DESCRIPTION: Crustose lichens with thin to rather thick areolate thalli, gray or yellow. Photobiont green (*Trebouxia*). Apothecia are mazaedia with a well-developed, usually black exciple forming a cup that can be immersed in the areoles or prominent, containing a

298. *Cyphelium inquinans* southern British Columbia ×5.5

loose, sooty mass of spores; spores dark brown to black, 2-celled, or submuriform in *C. notarisii,* often with a rough, sculptured surface. CHEMISTRY: Cortex with atranorin, usnic acid, or yellow pulvinic acid pigments. HABITAT: On hard, weathered conifer wood or bark. COMMENTS: *Thelomma* is a similar, mainly western genus in which the exciple of the apothecium is pale except at the base and the spores tend to have smoother walls.

KEY TO SPECIES

1. Thallus pale to dark gray; apothecium entirely black, without pruina . . . . *Cyphelium inquinans*
1. Thallus bright greenish yellow; margin of apothecium yellow pruinose or not 2
2. Apothecia largely immersed in the thallus with hardly any margin showing; widely distributed . . . . [*Cyphelium tigillare*]
2. Apothecia prominent, with margins clearly visible . . . 3
3. Rim of apothecium yellow pruinose; widely distributed . . . . *Cyphelium lucidum*
3. Apothecium entirely black, without pruina; western montane at high elevations . . . . . . [*Cyphelium pinicola*]

Cyphelium inquinans
Cupped soot lichen

DESCRIPTION: Thallus pale to dark gray or yellowish gray, verruculose to dispersed areolate, often rather thin. Apothecia entirely black, prominent, not at all buried in the thallus, 0.7–1.5 (–2.5) mm in diameter; spores constricted in the middle, about 12–20 × 8–10 μm. CHEMISTRY: Thallus PD– or PD+ pale yellow, K+ yellow or brownish, KC– or KC+ vague orange, C– or (rarely) C+ pale yellow (atranorin, usnic acid, and an unidentified yellow pigment). HABITAT: On bark and wood of conifers, especially in shaded or moist habitats. COMMENTS: This is the most common gray *Cyphelium*. The yellow pigments (including usnic acid) apparently never become sufficiently concentrated to color the thallus, which is almost always some shade of gray. *Cyphelium karelicum,* in the mountains of Alberta and British Columbia, has a darker, greenish to brownish thallus with smaller apothecia (0.4–0.8 mm) and lacks lichen substances (K– thallus). In the rare *Cyphelium trachylioides,* known only from California and northern British Columbia, the apothecia are immersed in thallus warts that are constricted at the base. *Cyphelium chloroconium,* another rare Californian species, has yellow pruina on the cup surface and margin.

Cyphelium lucidum
Yellow soot lichen

DESCRIPTION: Thallus brilliant greenish yellow (chartreuse), verrucose to areolate. Apothecia 0.5–1 mm in diameter, prominent, black, sometimes yellow pruinose, with a black margin that almost always has yellow pruina on the rim. Spores 12–22 × 7–10 μm, with a very rough surface. CHEMISTRY: Cortex PD–, K–, KC–, C– (contains vulpinic acid). HABITAT: On bark of living conifers. COMMENTS: Three North American species of *Cyphelium* have bright yellow or greenish yellow thalli: *C. tigillare, C. pinicola,* and *C. lucidum.* Their distinctions are summarized in the key. The yellow pigment in the thallus of *C. tigillare* and *C. pinicola* is rhizocarpic acid rather than vulpinic acid.

Dactylina (2 N. Am. species)
Finger lichens

DESCRIPTION: Erect, yellowish to pinkish violet fruticose lichens with no primary thallus, with unbranched to abundantly branched stalks filled with a loose, cottony medulla or entirely hollow. Photobiont green (*Trebouxia*?). Apothecia uncommon, lecanorine, with brown disks, on the tips of side branches. CHEMISTRY: Cortex PD–, K–, KC+ gold, C– or C+ pink (usnic acid, sometimes with gyrophoric acid); medulla PD+ red or PD–, K–, KC+ red or KC–, C+ pink or C– (can contain physodalic, physodic, and gyrophoric acids). HABITAT: On the ground in arctic or alpine sites. COMMENTS: The cottony medulla or the absence of squamules distinguishes *Dactylina* from similar species of *Cladonia*.

KEY TO SPECIES

1. Stalks typically unbranched, 20–70 mm tall, 2–6(–14) mm in diameter, entirely hollow . . . . . *Dactylina arctica*
1. Stalks branched, mostly under 20 mm tall and 2 mm in diameter; medulla webby to dense 2
2. Stalks yellowish to pinkish violet, pruinose on young tips, with irregular, divergent branching, almost hollow or partially filled with cobwebby hyphae; medulla KC+ pink, usually PD+ red (physodic acid, usually with physodalic acid) . . . . *Dactylina ramulosa*
2. Stalks greenish yellow, not pruinose, branching in regular dichotomies; medulla densely filled with white hyphae; medulla KC–, PD– (fatty acids) . . . .
. . . . *Allocetraria madreporiformis*

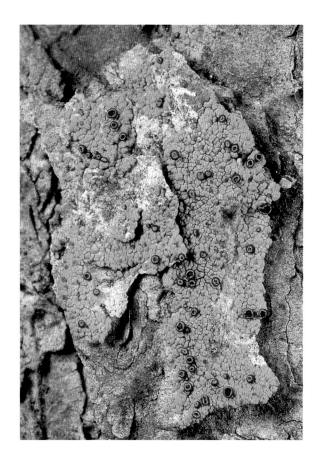

299. *Cyphelium lucidum* Lake Superior region, Ontario ×5.8

Dactylina arctica
Arctic finger lichen

DESCRIPTION: Thallus of finger-like stalks, pale greenish yellow becoming brown at the tips in exposed sites, not pruinose, with a smooth and uniform surface; stalks unbranched or with the occasional side branch, entirely hollow and thin-walled, 20–70 mm tall and 2–6(–14) mm wide. CHEMISTRY: Cortex PD–, KC+ gold, C– or C+ pink; medulla PD– or PD+ red, KC+ red, C+ pink (gyrophoric acid, sometimes with physodalic and physodic acids). HABITAT: In mossy tundra, especially associated with late snowbanks. COMMENTS: The chemical variation in this species is reflected to some extent in geographic distribution. *Dactylina arctica* ssp. *beringica* is PD+ red (physodalic and physodic acids) and is restricted to the extreme northwest (the Beringian area). Both ssp. *arctica* and ssp. *beringica* are C+ pink, KC+ pink owing to gyrophoric acid, but the substance is only in the cortex in ssp. *beringica*, whereas it is confined primarily to the medulla in ssp. *arctica*.

Dactylina ramulosa
Frosted finger lichen

DESCRIPTION: Thallus dull (rarely shiny) straw-colored or brownish to pinkish violet because of a heavy pruina that usually covers at least the newest growth; highly branched stalks with short, divergent branches 0.7–2 mm thick, rarely taller than 20 mm, very brittle, easily breaking open to reveal a webby, almost hollow interior. CHEMISTRY: Medulla of most specimens PD+ red, K–, KC+ pink, C– (physodalic and physodic acids); some lack physodalic acid (PD–, KC+ pink). HABITAT: On calcium-rich soils in arctic or alpine tundra in wet or dry sites. COMMENTS: The pinkish brown pruinose branches distinguish *D. ramulosa* from almost everything else.

300. *Dactylina arctica* interior Alaska ×2.1

301. *Dactylina ramulosa* Rocky Mountains, Alberta ×4.1

When the pruina is sparse or absent, it can closely resemble *Allocetraria madreporiformis*, which has longer branches and more regular branching (internodes up to 11 mm), is not as brittle, has a more densely filled, cottony interior, and is always PD–, KC– (with only protolichesterinic acid, a fatty acid).

Dendriscocaulon (2 N. Am. species)

KEY TO SPECIES: See Keys A and B.

Dendriscocaulon intricatulum
Olive-thorn lichen

302. *Dendriscocaulon intricatulum* Coast Mountains, British Columbia ×3.5

DESCRIPTION: Small, brownish to olive fruticose lichen forming small tufts of abundantly branched stalks, 5–20 mm tall and 0.1–0.6 mm thick (main stalks). Photobiont blue-green (*Nostoc*). West coast material (as in plate 302) uniformly thorny in appearance, with thick main stems and bushy outgrowths, quite white at the base, grading to bluish gray (when dry) and finally brown at the tips; main stems hairy, with hairs made up of rounded, bead-like cells; eastern specimens not hairy and may represent a separate species. CHEMISTRY: No lichen substances. HABITAT: In humid sites, typically on mossy tree trunks in the company of species of *Sticta*, *Pseudocyphellaria*, and *Nephroma*. COMMENTS: *Dendriscocaulon* may actually be formed by the fungus of another lichen. In Europe, it is commonly associated with species of *Lobaria*, especially *L. amplissima*, as a cephalodium. Apparently, the cephalodium sometimes escapes and grows independently. The Western North American lichen forming *Dendriscocaulon* thalli is *Sticta oroborealis*, which is extremely rare in its green-algal form.

Dendrographa (2 N. Am. species)
False orchil

DESCRIPTION: White to pale gray or dark yellowish gray fruticose lichens with abundantly branched, flattened or terete branches growing in tufts 10–90 mm long, often with short, white lateral tubercles that push through the cortex and give the branches a bumpy appearance. Photobiont green (*Trentepohlia*). Fruiting bodies arising along the branches, with the appearance of lecanorine apothecia but actually ascolocular in development; disks black, usually heavily pruinose; thalline margins thin and disappearing; hypothecium brown-black; paraphyses sparsely branched. Spores colorless, 4-celled, fusiform, 8 per ascus. CHEMISTRY: Cortex PD+ orange-red, K– or yellowish brown, KC–, C–; medulla PD+ orange-red, K– or pale yellowish, KC+ pink, C– (protocetraric and usually fumarprotocetraric acids, sometimes also succinoprotocetraric acid). The protocetraric acid is mainly (or entirely) in the medulla. HABITAT: On rocks and shrubs along the foggy California coast. COMMENTS: *Dendrographa* closely resembles, and is related to, *Roccella*, another strap-shaped fruticose lichen of coastal communities with a Mediterranean climate. Chemistry is the surest way of distinguishing them: the cortex is C+ bright red in *Roccella* (lecanoric acid). Also, *Dendrographa* tends to be stiffer and more brittle than *Roccella*. Species of *Ramalina* and *Niebla*, although similar in habit, are distinctly yellowish green because of usnic acid in the cortex.

KEY TO GENUS: See Key A.

303. *Dendrographa leucophaea* f. *minor*
Channel Islands, southern California
×4.6

Dendrographa leucophaea
(syn. *D. minor*)
False orchil

DESCRIPTION: Thallus yellow-gray to gray-brown or almost white, with distinctly flattened branches at least in part (especially at the axils), 0.5–10 mm wide and 15–90 mm long, with a dense but cottony medulla. Fertile form (f. *leucophaea*) rather dark, with smooth branches, even shiny in some cases, with flat to slightly convex, black, pruinose apothecia abundantly produced all along the branches. Sterile form (f. *minor*) with whiter, shorter, more slender, extremely brittle branches, mostly 0.3–1.0 mm wide, up to 30 mm long, round in cross section except for the most basal stems, with short, noncorticate side branches breaking through the cortex and looking like white warts; surface fuzzy throughout; soredia absent, although galls caused by an infecting fungus can resemble soralia. COMMENTS: The sterile *D. leucophaea* f. *minor* (plate 303) is often hard to distinguish from sterile **D. alectoroides** (called f. *parva*). In *D. alectoroides*, all the branches are terete, the cartilaginous medulla makes the thallus stiffer and more brittle, and the thallus is rarely longer than 30 mm. *Dendrographa alectoroides* is restricted largely to the California coast near San Francisco.

Dermatocarpon (10 N. Am. species)
Stippleback lichens, leather lichens

DESCRIPTION: Relatively thick, leathery foliose lichens, typically 1–4 cm across, attached to the substrate by a central holdfast (umbilicate), or sometimes appearing to be broadly squamulose and attached at numerous points; dull gray to brownish when dry; lower surface pale brown to blackish, smooth or ridged; most species lacking rhizines or hairs. Photobiont green (*Myrmecia* and other unicellular algae). Perithecia embedded in the thallus, appearing at the surface as tiny brown or black dots, not containing algae or paraphyses in the hymenium; spores colorless, ellipsoid, 1-celled, 8 per ascus. CHEMISTRY: Cortex and medulla PD–, K–, KC–, C– (no lichen substances). HABITAT: On rocks. COMMENTS: *Dermatocarpon* species have the general aspect of rock tripes (*Umbilicaria*) but produce perithecia instead of apothecia and are therefore not even related. Most species of *Umbilicaria* have a C+ red medulla (gyrophoric acid). *Catapyrenium* is a much smaller but somewhat similar perithecial lichen that has a squamulose thallus with a tomentose lower surface.

304. *Dermatocarpon luridum* Lake Superior region, Ontario ×2.7

KEY TO SPECIES

1. Lower surface covered with short, stubby rhizines..... .... [*Dermatocarpon moulinsii*]
1. Lower surface lacking rhizines 2

2. Thallus entirely without pruinose appearance 3
2. Thallus appearing pruinose, at least at lobe tips...... 4

3. Thallus remaining brown when wet, not grass green; growing on dry rocks........ *Dermatocarpon miniatum*
3. Thallus grass green when wet; growing on at least periodically submerged rocks .... *Dermatocarpon luridum*

4. Lower surface smooth, without papillae, usually brown .... *Dermatocarpon miniatum*
4. Lower surface roughly papillate like coarse sandpaper (fig. 30), usually black ..... *Dermatocarpon reticulatum*

Dermatocarpon luridum (syn. *D. fluviatile*)
Brook lichen, streamside stippleback

DESCRIPTION: Thallus very green and conspicuous when wet but brownish gray when dry, well

305. *Dermatocarpon miniatum* Sonoran Desert, southern Arizona ×4.2

camouflaged among the silt-covered rocks all around it; lobes small (7–20 mm) and overlapping, attached at several points; lower surface smooth or somewhat papillate. HABITAT: Common over a variety of siliceous rocks including granite, in and along streams, and at lake edges. COMMENTS: This is one of the few foliose lichens regularly found in water, although it can grow where it is wet only part of the year. In these respects it is similar to **D. rivulorum,** a rare western species with a strongly wrinkled or veined lower surface. The waterfan *Hydrothyria venosa* is another aquatic foliose lichen, but it is related to the jelly lichens; it has blue-green rather than green algae and lobes that are translucent when wet and are strongly veined. *Hydrothyria* is almost always found submerged. *Dermatocarpon miniatum* sometimes has small overlapping lobes and grows close to the water, but its thallus is usually pruinose when dry, not turning green when wet, and it normally grows over limestone.

Dermatocarpon miniatum
Common stippleback, leather lichen

DESCRIPTION: Thallus variable, often growing as single thalli with few lobes, 15–45 mm in diameter, or developing numerous small lobes especially toward the center of the thallus (polyphyllous); occasionally breaking free of the rock, growing into spherical clumps, and becoming vagrant; upper surface dull, gray, more rarely brown, often appearing pruinose; lower surface pale orange-brown to almost black, perfectly smooth to slightly undulating, occasionally ridged or pitted in a net-like pattern. HABITAT: Usually on limestone rock, but sometimes on noncalcareous rock as well. COMMENTS: *Dermatocarpon miniatum* is the most common limestone-dwelling umbilicate lichen in the east. It is most closely related to *D. luridum,* a semiaquatic lichen with smaller, overlapping lobes. The polyphyllous forms of *D. miniatum* almost intergrade with **D. intestiniforme,** in which the overlapping small lobes are strongly convex throughout, giving the thallus a verrucose appearance; it is more typically found on siliceous rocks. ***Dermatocarpon moulinsii,*** with rhizines on the lower surface, is mainly a western montane species, but rare disjunctions are known from the Thunder Bay district in Ontario and Fundy Park in New Brunswick. See also *Dermatocarpon reticulatum*.

Fig. 30 The lower surface of *Dermatocarpon reticulatum* showing the rough, papillate texture. Scale = 1 mm. (Drawing by Alexander Mikulin. Reprinted by permission from McCune and Goward, 1995, p. 95.)

306. *Dermatocarpon reticulatum* Klamath Range, northern California ×2.8

307. *Dibaeis baeomyces* White Mountains, New Hampshire ×5.6

Dermatocarpon reticulatum
Sandpaper stippleback, Northwest stippleback

DESCRIPTION: Thallus purplish gray to brownish gray, lightly to heavily pruinose, 15–50 mm in diameter, usually somewhat scalloped at the margins with rounded lobes that often become abundant and overlap; polyphyllous forms can become detached from the rock, develop into roundish clumps; lower surface dark brown to black, sometimes developing a network of low, rounded ridges, but always roughened with pyramid-shaped papillae like coarse sandpaper (Fig. 30). HABITAT: On limestone, or rolling freely (vagrant) over exposed soil in the high, dry plateaus of Idaho and Montana. COMMENTS: *Dermatocarpon reticulatum* is the only species of the genus with a rough, black, papillate lower surface. It is more lobed than the typical form of *D. miniatum* and does not develop convex lobes like those of ***D. intestiniforme***. *Dermatocarpon miniatum* can also have reticulate ridges on the lower surface but entirely lacks any roughness below.

Dibaeis (2 N. Am. species)

KEY TO SPECIES: See *Baeomyces*.

Dibaeis baeomyces (syn. *Baeomyces roseus, B. fungoides*)
Pink earth lichen

DESCRIPTION: A two-part lichen with a crustose primary thallus and fruticose apothecial stalks; primary thallus white, continuous, granular to verrucose, with more or less hollow granules or subspherical verrucae produced here and there. Photobiont green (*Coccomyxa*). Turban-like pink apothecia, 1–4 mm across, produced at the summits of short, unbranched, solid stalks 2–6 mm tall; asci thin-walled; spores 1–2-celled,

colorless, fusiform, 8 per ascus. CHEMISTRY: Cortex PD– or PD+ yellow, K–, KC–, C–; medulla PD+ bright yellow, K– or K+ pale yellow, KC–, C–; apothecia PD+ bright yellow, K– or K+ pale yellow, KC+ yellow, C– (baeomycesic and squamatic acids). HABITAT: Very common, often covering large areas of clayey or sandy mineral soil, especially in disturbed areas such as roadsides. COMMENTS: *Dibaeis* resembles *Icmadophila* in some respects and *Baeomyces* in others. Species of *Baeomyces* have stalked apothecia very similar in general habit to those of *Dibaeis baeomyces*, but they all have brown fruits, differ chemically, and have a quite different type of ascus. *Dibaeis absoluta*, a rare East Temperate species on soil and rock, has the same chemistry as *D. baeomyces* but has a thin, continuous thallus without granules or verrucae, and the pink apothecia are virtually stalkless, like those of *Icmadophila*. *Icmadophila ericetorum*, with similar pink apothecia, asci, spores, and photobiont, is more closely related to *Dibaeis* than is *Baeomyces*, although it is entirely crustose. It differs from *D. absoluta* in its thicker, white, granular thallus and chemistry; its distribution and habitat are entirely different.

Dimelaena (6 N. Am. species)
Moonglow lichens

DESCRIPTION: Crustose lichens, greenish yellow, white, gray, or brownish, with rimose-areolate thalli that have radiating, lobed margins. Photobiont green (*Trebouxia?*). Apothecia lecanorine or immersed in the thallus areoles (cryptolecanorine), with black disks, sometimes pruinose; hypothecium pale to dark brown; exciple pale or colorless; spores brown, 2-celled, with thin, uniform walls, 8 per ascus. CHEMISTRY: Cortex with or without usnic acid; may contain pigments that turn C± green, KC± blue-green; medulla may contain gyrophoric acid or various β-orcinol depsidones such as norstictic, stictic, and fumarprotocetraric acids. HABITAT: On rocks in full sun. COMMENTS: The small, black, immersed apothecia resemble those of *Aspicilia* or *Sporastatia*, genera that have colorless, 1-celled spores. *Lecanora oreinoides* can also be very similar. *Buellia*, *Rinodina*, and *Rhizocarpon* sometimes have 2-celled brown spores, but few species of these genera have lobed thalli. In addition, *Buellia* species have a dark hypothecium and exciple, *Rinodina* has spores with unevenly thickened walls, and the spores of *Rhizocarpon* are halonate.

KEY TO SPECIES

1. Thallus greenish yellow (usnic acid); widespread. . . . . *Dimelaena oreina*
1. Thallus white to brown; California to Washington . . . 2
2. Thallus creamy white to brownish gray; medulla PD–, K– . . . . *Dimelaena radiata*
2. Thallus dark brown 3
3. Hypothecium colorless; medulla PD–, K–; common. . . . . [*Dimelaena thysanota*]
3. Hypothecium dark brown; medulla PD+ orange, K+ yellow (stictic acid); rare . . . . . . [*Dimelaena californica*]

Dimelaena oreina
Golden moonglow lichen

DESCRIPTION: Thallus greenish yellow, sometimes blackening at the edges of the lobes or areoles; areoles 0.3–1 mm in diameter; lobes distinct or indistinct, sometimes absent; when well-developed, lobes mostly 0.5–1 mm wide, 1–3 mm long. Apothecia black, normally without pruina, level with the thallus surface, often with a conspicuous thalline margin, or simply immersed in the thallus areoles; spores 8–12 × 5–6.5 μm. CHEMISTRY: Cortex KC+ gold (usnic acid). Several chemical races: medulla PD+ orange to red, K± yellowish, KC–, C– (fumarprotocetraric acid); PD–, K–, KC+ red, C+ red (gyrophoric acid); PD+ orange, K+ yellow to red, KC–, C– (norstictic and (or) stictic acid). Occasionally, gyrophoric acid accompanies one of the PD+ compounds. HABITAT: On sunny siliceous rocks at low and high elevations. COMMENTS: *Dimelaena suboreina* is a very similar southwestern species with heavily pruinose apothecia and elongate lobes that are distinctly pruinose-scabrose on the tips. It sometimes contains stictic acid and (or) gyrophoric acid. Sterile specimens of *D. oreina* can resemble sterile *Lecanora muralis*, which usually has broader, thicker lobes containing only usnic acid and zeorin (medulla PD–, K–, KC–, C–).

Dimelaena radiata
Silver moonglow lichen

DESCRIPTION: Thallus milky white to brownish gray, usually heavily pruinose, or pale yellowish brown when without pruina. Apothecia immersed to sessile,

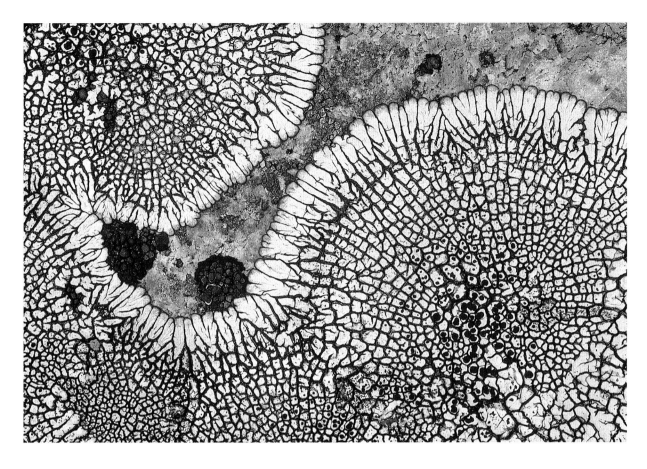

308. *Dimelaena oreina* Lake Superior region, Ontario ×3.6

309. *Dimelaena radiata* Channel Islands, southern California ×4.2

black, often pruinose; hypothecium dark brown; spores 8–12 × 5–8 μm. CHEMISTRY: All reactions negative. HABITAT: On rocks on ridges and in canyons in coastal mountains. COMMENTS: This is the only white *Dimelaena*. *Dimelaena thysanota* has a dark brown thallus (PD–, K–, C–) with black, nonpruinose apothecia and a colorless hypothecium. It is fairly common in the mountains from California to eastern Washington. *Dimelaena californica,* a rare southern California lichen also with a dark brown thallus, has a dark brown hypothecium and contains stictic and norstictic acids (PD+ orange, K+ yellow to red).

Dimerella (2 N. Am. species)
Dimple lichens

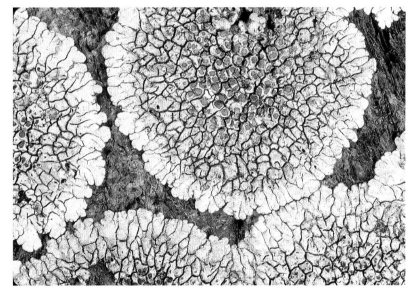

DESCRIPTION: Crustose lichens with very thin, often inconspicuous, greenish gray thalli. Photobiont green (*Trentepohlia*). Apothecia biatorine, pale yellow to pinkish orange; exciple colorless, made up of pseudoparenchyma; paraphyses unbranched, barely expanded at the tips; hypothecium pale; hymenium IKI+ pale blue changing to orange, K/I+ blue; asci thin-walled, slender, cylindrical, not thickened at tip, K/I–; spores

310. *Dimerella lutea* herbarium specimen (Canadian Museum of Nature), Quebec ×6.2

Dimerella lutea
Orange dimple lichen

DESCRIPTION: Thallus membranous, greenish gray. Apothecia flat to slightly convex, 0.3–1.5(–2) mm in diameter, pale orange with a slightly paler margin that is barely or not at all higher than the disk; paraphyses distinctly septate when stained with IKI; spores ellipsoid, 7–12 × 2–3.5 μm; conidia 3–5 μm long, ellipsoid. HABITAT: On shaded mossy bark or rocks, often growing over the moss. COMMENTS: When immature, the apothecia of *D. lutea* can resemble those of *D. pineti*. The apothecia of *D. pineti* are smaller (0.2–0.4 mm in diameter) and noticeably concave, the disks often have a pinkish tint, and the spores are longer (9–14 × 2–3.5 μm). The disjunct southeastern population on the distribution map is somewhat distinctive and probably represents another species.

colorless, elongate ellipsoid, 2-celled, 8 per ascus; conidia ellipsoid to elongate, 1-celled. CHEMISTRY: No lichen substances. HABITAT: On mosses, bark, wood, or rock in shaded, protected, rather humid sites. COMMENTS: *Dimerella* species can look like *Biatora, Bacidia*, or *Cliostomum*, but these all have club-shaped asci with conspicuously thickened, K/I+ blue tips. **Absconditella** is a genus of rarely collected, inconspicuous lichens with apothecia and spores that resemble those of *Dimerella*. Its photobiont (spherical, green algae, not *Trentepohlia*) forms a green gelatinuous film over the substrate (dead peat moss or rotting wood), the hymenium is K/I–, and its slender asci have a thickened tip. Another tiny but interesting lichen in this group is **Pachyphiale fagicola**, which has tiny, partly sunken, concave, orange apothecia, and grows on hardwoods. It contains *Trentepohlia*, and its spores are spindle-shaped with 4–8 cells, many per ascus. It is a northeastern species. See also *Gyalecta*.

KEY TO GENUS: See Key F.

Diploicia (1 N. Am. species)
Lobed button lichens

KEY TO SPECIES: See *Buellia*.

Diploicia canescens

DESCRIPTION: Pale gray, lobed, crustose lichens with folded, almost foliose radiating lobes, mostly 0.7–1.5 mm across, often pruinose at the tips; irregular white to blue-gray soralia on the upper surface; medulla white or pale yellow to mustard-colored; lower surface (where it lifts from the substrate) pale tan to pinkish, without a cortex or rhizines. Photobiont green (*Trebouxia*?). Apothecia rare, black, lecideine; hypothecium brown-black; spores 2-celled, *Physcia*-type, 9–15 × 4.5–7.5 μm, 8 per ascus. CHEMISTRY: Cortex PD– or PD+ pale yellow, K+ yellow, KC–, C–; medulla when white, PD–, K–, KC–, C–; medulla when pigmented, PD–, K+ red, KC+ yellow, C+ yellow (atranorin and diploicin). HABITAT: On bark, wood, or rock near the coast, sometimes in the salt-spray zone. COMMENTS: *Diploicia canescens*, when sterile, can resemble a *Dirinaria*, a *Pyxine*, or a closely adnate *Physcia*, except that it has no lower cortex. *Thelomma californicum* has the same color and lobe type, but it usually has fruiting warts and is not sorediate. The apothecia of *Diploicia* resemble those of *Pyxine,* a genus with some species that also have a pigmented medulla. *Pyxine,* however, always has a black lower surface and rhizines.

Diploschistes (11 N. Am. species)
Crater lichens

DESCRIPTION: Crustose lichens with rather thick thalli, creamy white to gray or yellowish gray, continuous or cracked-areolate; some species heavily pruinose, but none sorediate. Photobiont green (*Trebouxia?*). Apothecia with a well-developed brown to black exciple surrounded by a thalline margin; most common species open by a deep, relatively wide cavity, but some are closed, resembling perithecia; hymenium IKI–; asci cylindrical, thin-walled with a thickening at the apex, IKI–; spores dark brown, muriform, often somewhat shriveled, 4–8 per ascus. CHEMISTRY: All North American species: thallus KC+ red, C+ red (usually lecanoric and diploschistesic acids, rarely gyrophoric acid); some species PD+ yellow, K+ red (norstictic acid). HABITAT: On rock and soil, sometimes on mosses or parasitic on lichens, usually in full sun. COMMENTS: Two of the rarer, perithecium-like species of *Diploschistes* are **D. actinostomus**, with a gray thallus and radiating ostiole area (in eastern U.S.), and **D. aeneus,** with a dark red-brown thallus (in Texas, California, and South Carolina).

KEY TO SPECIES

1. Thallus heavily pruinose, thick and areolate with areoles up to 3 mm across; spores 4–8 per ascus; mainly on bare soil in the arid southwest... *Diploschistes diacapsis*
1. Thallus without or with little pruina, thin or thick, with areoles less than 1.5 mm across; spores 4 or 4–8 per ascus; widely distributed..........................2

2. Spores constantly 4 per ascus; growing on soil, mosses, or lichens.................*Diploschistes muscorum*
2. Spores 4–8 per ascus; growing directly on rock ....
    ....*Diploschistes scruposus*

Diploschistes diacapsis
Desert crater lichen

DESCRIPTION: Thallus gray to chalky white, pruinose, cracked to warty-areolate. Apothecia broad, up to 2.5 mm in diameter, black or pruinose gray; spores 20–38 ×

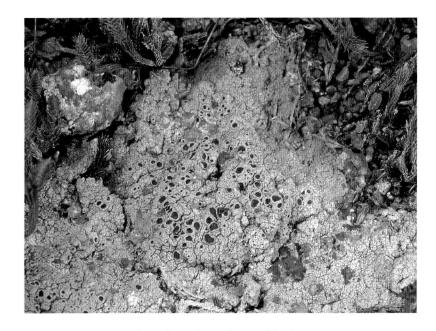

311. *Diploicia canescens* Channel Islands, southern California ×4.2

312. *Diploschistes diacapsis* hill country, central Texas ×1.7

313. *Diploschistes muscorum* southern British Columbia ×2.7

314. *Diploschistes scruposus* White Mountains, eastern Arizona ×5.6

9–17 μm, 4–8 per ascus. CHEMISTRY: PD–, K+ yellow or purple, KC+ red, C+ dark red, or PD–, K+ purple, KC–, C– (lecanoric and diploschistesic acids). HABITAT: Rather common on bare calcareous or noncalcareous soil in arid locations, rarely on calcareous rock. COMMENTS: *Diploschistes muscorum* also grows on soil but never has more than 4 spores per ascus; *D. scruposus* is strictly saxicolous.

Diploschistes muscorum
Cowpie lichen

DESCRIPTION, HABITAT, and COMMENTS: The thallus and apothecial morphology of *D. muscorum* are almost identical to those of *D. scruposus*, but the thallus sometimes becomes pruinose. Unlike *D. scruposus*, it has only 4-spored asci and slightly smaller spores (18–32 × 6–15 μm), and grows only on soil, mosses, wood, and lichens, never rocks. In fact, *D. muscorum* becomes established as a parasite on other lichens (usually on species of *Cladonia*), gradually incorporating the algae of the host, only later becoming more or less free-living on moss and soil (especially calcium-containing). See also *D. diacapsis*. CHEMISTRY: Cortex and often medulla PD–, K+ yellow changing to purple, KC+ red, C+ red (atranorin, lecanoric, and diploschistesic acids).

Diploschistes scruposus
Crater lichen

DESCRIPTION: Thallus forming a thick, greenish gray to almost white crust over rock, verrucose or areolate, without pruina. Black, crater-like apothecia, 0.5–1.5(–3) mm in diameter, embedded in the thallus, often with a double margin: black on the inside and surrounded by a thallus-like outer margin; spores 4–8 per ascus, 25–40 × 10–20 μm. CHEMISTRY: Cortex PD–, K+ yellow to dark red or purple, KC+ red, C+ red (atranorin, lecanoric acid, often with diploschistesic acid). HABITAT: On exposed noncalcareous rock. It is a particularly aggressive crust, often overgrowing everything in its path. COMMENTS: *Diploschistes scruposus* is by far the most common crater lichen in North America. It is extremely variable in the degree to which the fruiting body opens up to reveal the black disk. Often the fruits resemble volcanoes, with a narrow pore at the summit, but sometimes the disks are quite broad, like those of a *Lecanora* or *Aspicilia*. *Diploschistes muscorum* is similar but always begins its development as a parasite and has smaller spores, always 4 per ascus. *Diploschistes actinostomus* is known only from the east and grows on noncalcareous rock. Its apothecia are entirely buried in the thallus, opening to the surface through a narrow ostiole-like hole surrounded by a black, radially cracked ring (the upper part of the exciple). The spores are somewhat shorter than those of *D. scruposus* (16–32 × 10–20 μm).

Dirina (2 N. Am. species)

DESCRIPTION: Crustose lichens, typically creamy white to greenish gray, sometimes browner, usually pruinose, continuous, rimose areolate to verrucose. Photobiont green (*Trentepohlia*). Fruiting bodies resembling lecanorine apothecia, dark gray to white because of a thick pruina; hypothecium brown-black, distinct from the white medulla; paraphysoid threads (like paraphyses) unbranched or sparsely branched, not expanded at the tips; spores colorless, 4-celled, fusiform, often broader at one end, straight or curved, 8 per ascus. CHEMISTRY: All species PD–, K–, KC+ red, C+ red (erythrin and lecanoric acid). HABITAT: On trees and shrubs, or on vertical or overhanging rocks in foggy but otherwise arid coastal sites. COMMENTS: *Dirina* species are prominent members of the California coastal fog zone community, which includes **Schismatomma, Sigridea, Niebla,** and *Dendrographa*. *Roccellina* is a closely related, similar genus common in Baja California. Its apothecia have a black hypothecium usually extending as a "foot" to the base of the thallus or merging with a black medulla.

KEY TO SPECIES AND GENUS: See Keys C and F.

Dirina catalinariae

DESCRIPTION: Thallus variable, milky white to brownish gray, smooth to cracked areolate to verrucose, rarely becoming partially fruticose at the margins of some areoles, usually producing round, hemispherical soralia with coarse, granular soredia (*f. sorediata*; plate 315). Less frequently, it has only pale, lecanorine apothecia (f. *catalinariae*), and even more rarely, both apothecia and soralia. CHEMISTRY: Cortex and medulla C+ red. HABITAT: On coastal rocks and cliffs. COMMENTS: The only non-Californian species of *Dirina* is *D. paradoxa,* a bark-dwelling lichen known from the Florida Keys. Its fruiting bodies are raised somewhat on short stalks, and it is C+ red only in the cortex, not in the medulla.

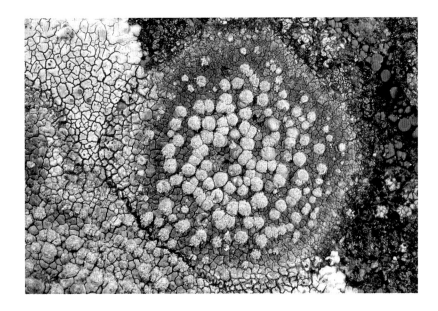

315. *Dirina catalinariae* f. *sorediata* Channel Islands, southern California ×4.5

Dirinaria (10 N. Am. species)
Medallion lichens

DESCRIPTION: Very closely appressed, foliose, sometimes almost crustose in appearance, lobes mostly 1–4 mm wide, pale gray or greenish; lower surface black, corticate, without rhizines; medulla white or, rarely, red. Photobiont green (*Trebouxia?*). Apothecia lecanorine, broadly attached, with black, sometimes pruinose disks; hypothecium brown; spores brown, 2-celled, *Physcia*-type, 8 per ascus. CHEMISTRY: Cortex PD– or pale yellowish, K+ yellow, C–, K–, UV– (atranorin); medulla PD–, K–, KC–, C–, UV+ blue-white (divaricatic or sekikaic acids). HABITAT: On bark or wood, rarely on rock, in tropical and subtropical woodlands. COMMENTS: *Hyperphyscia* species also lack rhizines and are closely appressed, but the upper surface is K– (no atranorin), the hypothecium is pale, and the thick-walled spores have rounded rather than angular locules. Other similar *Physcia*-like genera have abundant rhizines on the lower surface or, like *Diploicia*, are crustose and pale below wherever the lobe margins lift free of the substrate. None of these *Dirinaria* look-alikes have a UV+ blue-white medulla.

316. *Dirinaria aegialita* (damp) herbarium specimen (Canadian Museum of Nature), Florida ×2.9

317. *Dirinaria confusa* South Carolina coast ×2.2

KEY TO SPECIES: See *Physcia*.

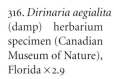

Dirinaria aegialita (syn. *D. aspera*)
Grainy medallion lichen

DESCRIPTION: Thallus pale greenish gray, pruinose, with short lobes at the periphery; upper surface covered with coarse soredia or isidia-like schizidia from the breakdown of small pustules. Apothecia rare. CHEMISTRY: Contains divaricatic acid. HABITAT: On bark of oak, hickory, and other deciduous trees. COMMENTS: See Comments under *D. picta*.

Dirinaria confusa
Medallion lichen

DESCRIPTION: Thallus large, with broad lobes (up to 4 mm across) commonly having narrow folds radiating out to the lobe tips; no soredia or isidia. Apothecia common, with black, nonpruinose disks and rather small, narrow spores (5–8 μm wide). CHEMISTRY: Contains compounds of the sekikaic acid group. HABITAT: On bark. COMMENTS: *Dirinaria confusa* is one of the most common nonsorediate species of *Dirinaria*. The rarer ***D. confluens*** often has smaller lobes and broader spores (usually over 8 μm) and contains divaricatic acid (medulla UV+ blue). *Dirinaria purpurascens* has narrower lobes and apothecia with a purplish pruina.

Dirinaria picta
Powdery medallion lichen

DESCRIPTION: Thallus pale green, thin, flat, with lobes usually under 1.5 mm wide, irregularly pinnately branched and discrete for a good part of the thallus, at least close to the periphery, producing hemispherical mounds of greenish farinose soredia on the thallus surface. CHEMISTRY: Contains divaricatic acid. HABITAT: On bark. COMMENTS: *Dirinaria picta* closely resembles *Physcia americana* but has a black lower surface and different chemistry; small, young thalli can be confused with a *Hyperphyscia*. Distin-

guishing *D. picta* from **D. applanata,** another coastal plain species, is not easy. In *D. applanata,* the lobes grow together in one confluent mass, becoming virtually verrucose-crustose in the center, fanning out at the outer edge but still essentially continuous. The lobes are also often longitudinally folded and somewhat convex. In *D. aegialita,* the "soredia" originate from the breakdown of tiny pustules, much coarser and more thallus-colored than the soredia of *D. picta*. **Dirinaria frostii** resembles *D. picta,* especially in its type of soredia and chemistry, but it grows on rocks rather than on bark. It is a more temperate and widespread eastern species rarely found on the coastal plain.

Dirinaria purpurascens
Purple-eyed medallion lichen

DESCRIPTION: Thallus very pale gray to pale green, lightly pruinose, rather small and tightly appressed, 2–5 cm across, with very narrow, radiating lobes, 0.2–0.7 mm wide. Apothecia usually abundant, 0.5–1(–1.5) mm across, with reddish purple disks caused by a purple pruina that disappears only in very old apothecia, leaving the disks black. CHEMISTRY: Contains divaricatic acid. HABITAT: On bark. COMMENTS: This species is easily identified by the reddish purple pruinose apothecia. See Comments under *D. confusa*.

Endocarpon (8 N. Am. species)
Stippled lichens

DESCRIPTION: Small gray to brown squamulose or dwarf fruticose lichens with perithecia embedded in the thallus; medulla white, with a distinct green algal layer; lower surface sometimes with a pale cortex and attachment hairs. Photobiont green (*Stichococcus*). Perithecia with dark walls; paraphyses not developing, or dissolving; hymenium containing green algal cells (the same as in the thallus but smaller in diameter); spores brown (sometimes pale), muriform with many cells, 2 per ascus. CHEMISTRY: No lichen substances. HABITAT: On rocks and soil, occasionally on tree bases. COMMENTS: *Endocarpon* and *Catapyrenium* are both small, brown, squamulose lichens with

318. *Dirinaria picta* hill country, central Texas ×2.3

319. *Dirinaria purpurascens* Florida Keys ×2.8

320. *Endocarpon pulvinatum* Rocky Mountains, Montana ×5.0

321. *Endocarpon pusillum* herbarium specimen (Canadian Museum of Nature), Ontario ×8.0

perithecia, and both grow on rocks and soil rich in calcium. The perithecia in *Catapyrenium* lack hymenial algae, and the spores are colorless, 1-celled, and 8 per ascus. With regard to other scale lichens, *Peltula* and *Heppia* have blue-green rather than green algae; *Psora* and *Toninia* have apothecia rather than perithecia and are generally thicker. The scattered round areolae of a few species of the crustose lichen *Staurothele* can resemble the squamules of *Endocarpon;* if the edges of the areoles are diffuse and there is no lower cortex or attachment hyphae on the lower surface, it is probably a *Staurothele*.

KEY TO SPECIES: See Key G.

Endocarpon pulvinatum (syn. *E. tortuosum*)
Inflated stippled lichen

DESCRIPTION: Thallus dark reddish brown, dwarf fruticose, consisting of erect, flattened to cylindrical squamules or tiny branches 2–6 mm high; summits with overlapping lobes, or thick, rounded, and recurved, giving the squamules an inflated appearance; lower portions of the squamules or branches often blackened. Perithecia buried in the upper parts of the erect lobes, sometimes with a depressed collar, giving the perithecia slightly prominent summits; spores large, 44–60 × 18–24 μm; hymenial algae long and cylindrical, mostly 2-celled. HABITAT: On rocks in dry or seepage areas. COMMENTS: The erect lobes of *E. pulvinatum* make this a unique species in the genus. It is similar in form to the cyanobacterial lichen *Peltula cylindrica,* but the color and photobiont are quite different.

Endocarpon pusillum
Scaly stippled lichen

DESCRIPTION: Thallus brown to brownish gray when dry, squamulose, each scale 0.5–2 mm across, but sometimes coalescing and almost continuous in the center of the colony and squamulose only around the edges. Perithecia buried within the scales and more or less emergent as black dots or bumps; hymenium containing clusters of small spherical to cylindrical green algal cells; spores 24–38 ×

10–18 μm. HABITAT: Mainly on exposed or shady limestone rock, rarely on tree bases, especially elms. COMMENTS: The presence of algal cells in the perithecium and the muriform spores immediately distinguish *Endocarpon pusillum* from superficially similar species of *Placidium*. The more coalesced specimens of *E. pusillum* can be mistaken for a *Staurothele*, which also has green algae in the perithecia but is truly crustose, having no lower cortex or ascending squamules.

Ephebe (6 N. Am. species)
Shag lichens

DESCRIPTION: Dwarf fruticose lichens, very dark green to olive-black, consisting of branched filaments of *Stigonema* (cyanobacteria), at first lightly lichenized by fungal hyphae that run through its gelatinous sheath (Fig. 31), and which later break up the cyanobacterial filaments with fungal tissue. The lichenized branches grow in tufts or extensive shaggy mats. Pycnidia (and the apothecia that develop from them in these very distinctive fruiting bodies) appear as swellings on the branches and open by means of a narrow ostiole; spores colorless, ellipsoid, 1–3-celled, 8–16 per ascus; conidia ellipsoid, colorless, 1-celled, about 2–4.5 × 1 μm. CHEMISTRY: No lichen substances. HABITAT: On wet, siliceous rocks on lake and stream shores or dripping rock walls. COMMENTS: Colonies of *Stigonema* that are free-living (i.e., not lichenized) have almost the same external appearance as *Ephebe* species but are more blue-green (rather than black). Microscopic preparations stained with lactophenol and cotton blue may be necessary to see the very slender fungal hyphae in the algal sheath. Species of *Spilonema* can be similar to those of *Ephebe* but always have blue-green rhizine-like hyphae growing from the base of the tufts.

KEY TO SPECIES: See Key B.

Ephebe lanata
Rockshag lichen

DESCRIPTION: Filaments intricately and irregularly branched, mostly 70–140 μm in diameter, in tufts 2–3 cm across that lie flat on the rock. Pycnidia and apothecia forming more or less spherical swellings. Spores 8 per ascus. HABITAT: On wet stones in brooks, along lakeshores, and on maritime rocks in the salt-spray zone. COMMENTS: This is by far the most common North American *Ephebe*. It so closely resembles an alga that lichenologists rarely collect it.

Erioderma (3 N. Am. species)
Mouse ears, felt lichens

DESCRIPTION: Small, pale brown to olive-brown, foliose lichens with a fuzzy upper surface caused by erect hairs most easily seen near the lobe margins; lower surface without a cortex, without veins (in North American species), smooth and fibrous, or covered with a hairy tomentum or tufted rhizines. Photobiont blue-green (*Scytonema*). Apothecia lecanorine with the thalline margin often disappearing; spores ellipsoid to elongate, 1-celled, colorless, 8 per ascus. CHEMISTRY: β-orcinol depsidones related to pannarin in a few species. HABITAT: In humid sites on trees and shrubs, usually on mossy branches; best represented in the Tropics of Central and South America, and in oceanic coastal regions of North America. COMMENTS: *Erioderma* closely resembles *Pannaria* and *Leioderma*

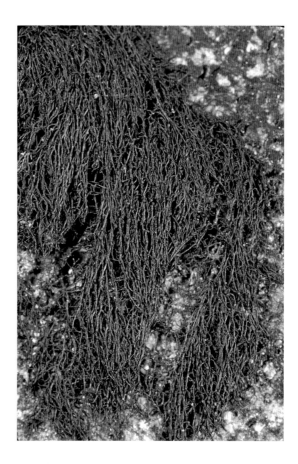

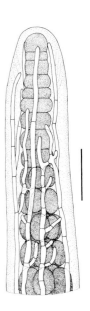

322. *Ephebe lanata* Vancouver Island, British Columbia ×5.3

Fig. 31 A microscopic view of *Ephebe* showing the hyphal sheath over the cyanobacterial filament. Scale = 20 μm. (Reproduced courtesy of the British Lichen Society from Purvis et al., 1992, Fig. 44a.)

323. *Erioderma sorediatum* (damp) Oregon coast ×3.4

PD– medulla. ***Erioderma pedicellatum*** and ***E. mollissimum*** are equally rare species in eastern North America. The latter, found mainly at high elevations in the southern Appalachians, has soredia along the lobe margins, its lower surface has an abundant brownish tomentum, and its medulla is PD–. *Erioderma pedicellatum*, in lichen-rich localities from Newfoundland to Maine, has no soredia and usually produces tiny, hemispherical, brown apothecia on the upper surface. Like *E. sorediatum*, it contains eriodermin (PD+ orange).

and can even be mistaken for a tiny *Peltigera*. The erect hairs on the lobe surface and lack of veins on the lower surface distinguish it from all these. Both *Leioderma* and *Erioderma* are restricted to the most unpolluted habitats in the wettest climates.

KEY TO SPECIES: See *Peltigera*.

Erioderma sorediatum
Mouse ears

DESCRIPTION: Thallus with lobes 3–7 mm across, grayish brown above and yellowish white below, producing abundant blue-gray, granular soredia on the lower surface of curled-back lobe tips; tufted rhizines sparse or abundant. CHEMISTRY: Medulla (and lower surface) PD+ orange, K–, KC–, C– (eriodermin). COMMENTS: This exceedingly rare lichen is included because it is an important indicator of biologically rich, unpolluted sites. The darker color and abundant, erect hairs on the upper surface distinguish this small, shell-shaped, foliose lichen from the very similar, tiny *Leioderma sorediatum*, which has lobes about 3 mm across with bluish soredia along the margins. *Leioderma* is pale gray and has only minute, appressed, woolly hairs on the upper surface and a

Esslingeriana (1 species worldwide)

KEY TO SPECIES: See *Platismatia*.

Esslingeriana idahoensis (syn. *Cetraria idahoensis*)
Tinted rag lichen

DESCRIPTION: Loosely attached foliose lichen, pale gray often with a distinct yellowish tint, smooth to wrinkled, with elongate lobes mostly 1.5–5 mm across; without soredia, isidia, or pseudocyphellae, but sometimes with lobules along the margins; lower surface black, strongly and intricately wrinkled, with sparse, scattered, black, unbranched rhizines. Photobiont green (*Trebouxia?*). Large brown apothecia at the lobe tips are common on better-developed specimens; spores colorless, 1-celled, ellipsoid, 8 per ascus. CHEMISTRY: Cortex PD+ pale yellow, K+ yellow (atranorin), KC–, C–; medulla PD–, K+ purplish pink (test the lobe tips or apothecial margins), KC–, C– (endocrocin). HABITAT: Sometimes abundant on tree branches in open conifer forests. COMMENTS: The tinted rag lichen is very similar to species of *Platismatia*, especially *P. glauca*. That lichen, however, always has at least some isidia, soredia, or other decorations along the margins and is never yellowish. Both species can have depressions and ridges on the upper surface, but they are better developed in *P. glauca*. The yellowish tint of *Esslingeriana* is due to the anthraquinone endocrocin; the yellower the thallus, the deeper the K reaction.

324. *Esslingeriana idahoensis* mountains, northeastern California ×1.8

Evernia (5 N. Am. species)
Oakmoss lichens

DESCRIPTION: Abundantly branched fruticose lichens, pendent on trees or prostrate on the ground, pale yellowish green to straw-colored, with angular and ridged to quite flattened branches; medulla cottony and solid; cortex rather soft and pliable. Photobiont green (*Trebouxia?*). Apothecia rarely seen, lecanorine. CHEMISTRY: Cortex PD–, K–, KC+ gold, C– (usnic acid); medulla PD–, K–, KC–, C– (evernic or divaricatic acids). HABITAT: Usually on trees but a few species ground-dwelling. COMMENTS: Although arboreal species of *Evernia* may at first resemble an *Usnea*, *Alectoria*, *Niebla*, or *Ramalina*, their weak cortex makes them soft and pliable unlike the last three genera, and they have no central cord, unlike *Usnea* (Fig. 8d).

KEY TO SPECIES

1. Branches clearly flattened and dorsiventral throughout, with a pale, almost white lower surface; round soralia on the lobe margins and upper surface, containing coarse, white to blue-gray soredia . . . . *Evernia prunastri*
1. Branches irregular or angular in cross section, usually not flattened or dorsiventral; sorediate or not 2
2. Thallus sorediate on ridges along the branches; bushy to pendent . . . . *Evernia mesomorpha*
2. Thallus without soredia, pendent or prostrate 3
3. Branches stiff and brittle, with a tough, unbroken cortex; medulla dense; prostrate on calcareous soil in western arctic . . . . [*Evernia perfragilis*]
3. Branches usually soft, not brittle; cortex thin and cracked; medulla loose (in pendent specimens on trees) or dense (in prostrate specimens on calcareous soil); Rocky Mountains . . . . *Evernia divaricata*

Evernia divaricata
Mountain oakmoss lichen

DESCRIPTION: Branches regularly or irregularly dichotomous, 0.5–3 mm wide, angular to somewhat pitted or ridged, with no soredia or isidia; pendent specimens on trees up to 30 cm long, making it the longest *Evernia*; specimens on soil have shorter internodes and a more tangled appearance. CHEMISTRY: Contains divaricatic acid. HABITAT: On

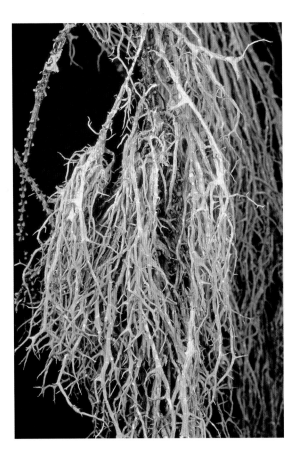

325. *Evernia divaricata* Caribou Mountains, British Columbia ×1.6

326. *Evernia mesomorpha* Lake Superior region, Ontario ×1.2

conifer branches, or on alpine soil (rarely in the Arctic). COMMENTS: Alpine, ground-dwelling forms of *E. divaricata* can be confused with the arctic species *E. perfragilis,* which has stiff branches and a denser medulla. Unlike the common boreal species *E. mesomorpha* or the mainly western *E. prunastri, E. divaricata* has no soredia.

Evernia mesomorpha
Boreal oakmoss lichen

DESCRIPTION: Thallus forming pendent or shrubby tufts (2–)4–8 cm long composed of wrinkled and ridged branches 0.5–1.5 mm thick, irregularly but abundantly divided, with coarse soredia developing on the ridges. CHEMISTRY: Contains divaricatic acid. HABITAT: On branches and trunks of both conifers and hardwoods in sunny sites. It is more pollution-tolerant than *Usnea* and is therefore more frequently found close to urban centers. COMMENTS: See *E. prunastri*.

Evernia prunastri
Oakmoss lichen

DESCRIPTION: Thallus with dichotomously branched, flattened branches 2–4 mm wide, up to 70 mm long, pale greenish yellow above and white on the underside; round soralia frequent or rare along the margins and on the surface of the branches. CHEMISTRY: Contains evernic acid and some atranorin. HABITAT: Growing on trees of all kinds, more rarely on rock walls, in shade or sun. COMMENTS: *Evernia prunastri* resembles some species of *Ramalina*, especially *R. farinacea,* but is much softer and more pliable, and the lower surface is distinctly paler than the upper. Branches of *E. mesomorpha* are angular rather than flattened, and the soredia are more granular and scattered. Although almost entirely restricted to the west coast in North America, *E. prunastri* is quite common throughout Europe. A few good specimens have been seen from the maritime provinces of Canada. Some very old herbarium specimens exist from scattered localities in Ontario close to the

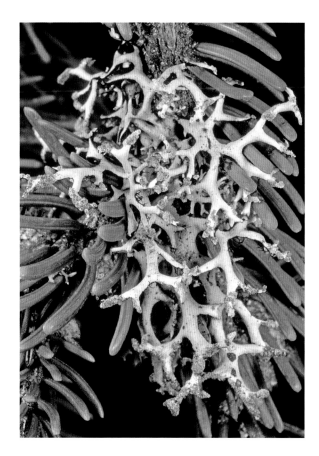

327. *Evernia prunastri* Klamath Range, northwestern California ×1.3

328. *Everniastrum catawbiense* Appalachians, Tennessee ×2.5

Great Lakes, but the species is now almost certainly extinct in that area. IMPORTANCE: In Europe, oakmoss is commercially important in the perfume industry. Its extracts are used as perfume fixatives, and at one time thousands of tons were collected every year in the Slavic republics to supply the cosmetics industry. Large quantities are still shipped from Macedonia to France for that purpose. The lichen is not abundant enough in North America to support commercial exploitation. *Evernia mesomorpha* has sometimes also been used for cosmetics. In Egypt, *E. prunastri* has been used as an additive in bread.

Everniastrum (1 N. Am. species)

KEY TO SPECIES: See Key A.

Everniastrum catawbiense
Powder-tipped antler lichen

DESCRIPTION: Small foliose, almost fruticose lichens forming clumps 2–4 cm across; lobes flattened to strongly convex, 0.8–1.5 mm broad, with a regular dichotomous branching pattern; fine, dark green soredia developing on the upper surface of the lobe tips from an erosion of the cortex; lower surface pitch black and shiny, with black, sometimes branched cilia along the margins, sometimes with black rhizines as well. Photobiont green (*Trebouxia?*). CHEMISTRY: Cortex K+ yellow (atranorin); medulla K–, KC– or KC+ orange, PD–; soralia K+ yellow or –, KC+ pink, C+ pink, sometimes faint (gyrophoric acid). HABITAT: On the branches and wood of conifers, mainly at high elevations. COMMENTS: This species is rare in North America. Like other members of the genus, it is more characteristic of the Tropics. Southern Appalachian specimens of *E. catawbiense* are broader than northern specimens, with more eroded soralia, and they can resemble a small *Pseudevernia*.

Pseudevernia species, however, lack marginal cilia, and none are sorediate. *Hypotrachyna revoluta* also contains gyrophoric acid and has sorediate lobe tips, but it lacks cilia. *Everniastrum catawbiense* is similar to *Hypogymnia tubulosa* in its size, branching pattern, and soralia (at the lobe tips), but the lobes of *H. tubulosa* are clearly inflated and hollow.

Flavocetraria (2 N. Am. species)
Snow lichens

DESCRIPTION: Small to medium-sized, pale greenish yellow to yellow, arctic-alpine foliose lichens; lobes erect; lower surface smooth, the same color as the upper surface, with sparse white pseudocyphellae along the margins; without any rhizines or cilia; medulla white. Photobiont green (*Trebouxia?*). Apothecia rare, marginal, pale brown; spores colorless, 1-celled, ellipsoid, 8 per ascus; pycnidia black, marginal, containing dumbbell-shaped conidia. CHEMISTRY: Cortex KC+ gold (usnic acid); medulla PD–, K–, KC–, C– (sometimes contains protolichesterinic acid). HABITAT: On the ground among mosses and heath. COMMENTS: These lichens resemble *Cetraria*, the Iceland lichens, except for the color, and several species of *Cetraria* grow in the same habitat. *Cetraria* differs in the anatomy of the thallus cortex and in the shape of the conidia (lemon-shaped in *Cetraria*). *Vulpicida tilesii*, another arctic-alpine yellow foliose lichen, is much brighter yellow (because of vulpinic and pinastric acids in the cortex). *Asahinea* species are typically paler and broader than *Flavocetraria* species, and have a black lower surface.

KEY TO SPECIES

1. Lobes flat with a network of depressions and sharp ridges or at least wrinkled; base sometimes turning dark yellow .... *Flavocetraria nivalis*
1. Lobes curled inward, forming a channel, smooth, or undulating and crisped at the margins; base often becoming blotched with red-violet .... *Flavocetraria cucullata*

Flavocetraria cucullata (syn. *Cetraria cucullata*)
Curled snow lichen

DESCRIPTION: Thallus lobes almost vertical, 2–6(–8) mm wide and 25–60(–80) mm high, ruffled at the margins, and curled inward, almost forming a tube (sometimes fusing where the edges touch), often curving back at the tips; margins either smooth or somewhat toothed, with short projections bearing black pycnidia at the tips; base of thallus often becoming red-violet; pseudocyphellae round or linear along the lower surface of the margins. CHEMISTRY: Contains protolichesterinic acid. HABITAT: In open conifer woodlands and tundra, usually at high elevations. COMMENTS: *Flavocetraria nivalis* looks similar but is wrinkled and flat, or at least is never curled into a tube, and the base of the thallus becomes yellow-orange rather than red-violet. IMPORTANCE: Alaskan natives used *F. cucullata* as a flavoring for fish or duck soups.

Flavocetraria nivalis (syn. *Cetraria nivalis*)
Crinkled snow lichen

DESCRIPTION: Thallus lobes erect or prostrate, rather thick, more or less flat with depressions and ridges, not curled inward, divided dichotomously, 1.5–6 mm across, forming tufts or cushions 20–40 mm high; pseudocyphellae round, slightly convex, fairly common on the lower surface; base of thallus often becomes deep yellow or yellow-orange. Black pycnidia on or close to the lobe margins, slightly raised but not on projections. CHEMISTRY: Contains no lichen substances in the medulla. HABITAT: In open heath close to treeline or on tundra soil. COMMENTS: See Comments under *Flavocetraria* and *F. cucullata*. IMPORTANCE: A tea brewed from this lichen has been used as a tonic for altitude sickness and heart problems in the high mountains of midwestern Bolivia by the Qollahuaya Andean people.

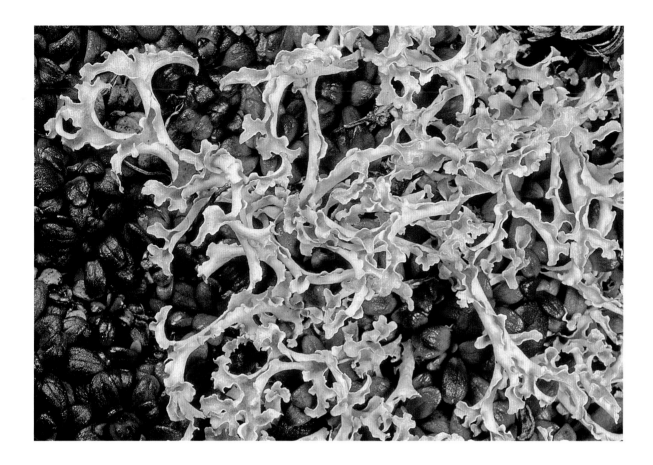

329. *Flavocetraria cucullata* interior Alaska ×2.8

330. *Flavocetraria nivalis* interior Alaska ×1.6

331. *Flavoparmelia baltimorensis* Cape Cod, Massachusetts ×2.3

Flavoparmelia (3 N. Am. species)
Greenshield lichens

DESCRIPTION: Medium-sized, light yellow-green foliose lichens with rounded lobes 2–8 mm wide, forming flat but loosely attached patches 6–20 cm across; upper surface often strongly wrinkled in older parts; lower surface smooth, black except for brown margin, with black, unbranched rhizines. Photobiont green (*Trebouxia*). Apothecia rare. CHEMISTRY: Cortex KC+ gold (usnic acid); medulla PD+ red-orange, K–, KC+ pink, C– (atranorin, protocetraric acid, and the fatty acid caperatic acid). HABITAT: On trees or rocks. COMMENTS: *Flavoparmelia* resembles *Xanthoparmelia*, *Flavopunctelia*, and *Allocetraria oakesiana*. Its broader, rounder lobes and frequently strongly wrinkled thallus surface distinguish rock-dwelling specimens from *Xanthoparmelia*, which is always found on rocks. *Flavopunctelia* usually has conspicuous white pseudocyphellae on the upper surface, and it differs chemically. *Allocetraria oakesiana* has narrow, strap-shaped lobes and marginal soredia. Yellowish green species of *Parmotrema* have a naked marginal zone on the lower surface (i.e., without rhizines) and marginal black cilia. *Ahtiana sphaerosporella* is a superficially similar western montane species that is more closely adnate, without soredia or pustules, and with a different chemistry.

KEY TO SPECIES: See Key J.

Flavoparmelia baltimorensis
(syn. *Pseudoparmelia baltimorensis*)
Rock greenshield lichen

DESCRIPTION: Thallus with globose, pustule-like outgrowths on the upper surface of the lobes, sometimes partially breaking down into granule-sized fragments (schizidia). Apothecia very rare. CHEMISTRY: Medulla PD+ red-orange, K–, KC+ pink, C– (protocetraric acid and, occasionally, traces of gyrophoric acid). HABITAT: Almost exclusively on rocks in sun or shade. COMMENTS: *Flavoparmelia caperata*, which usually grows on bark or wood but can also grow on rocks, has true granular soredia in irregular, laminal soralia that develop from flat pustules, and it never produces gyrophoric acid. *Flavoparmelia baltimorensis* (and rock-dwelling specimens of *F. caperata*) can be mistaken for species of *Xanthoparmelia*, which typically have narrower, more angular or "square" lobes and are rarely wrinkled in the thallus center. Very few *Xanthoparmelia* species have soredia or granules on the upper surface, and most of the common ones contain salazinic, stictic, or norstictic acids (K+ yellow or red). *Xanthoparmelia* also tends to prefer more open, sunlit habitats.

332. *Flavoparmelia caperata* Coast Range, California ×1.1

Flavoparmelia caperata (syn. *Pseudoparmelia caperata*)
Common greenshield lichen

DESCRIPTION: Thallus pale yellow-green when dry, greener when wet; lobes rounded, 3–8 mm wide, smooth or wrinkled, with large or small, irregular patches of coarsely granular soredia from pustules. Apothecia very rare. CHEMISTRY: Cortex PD–, K–, KC+ gold, C–; medulla PD+ red-orange, K–, KC+ pink, C– (usnic, protocetraric, and caperatic acids and atranorin). HABITAT: On bark of all kinds in sun or partial shade, less commonly on rock. COMMENTS: *Flavoparmelia rutidota* (syn. *Pseudoparmelia rutidota*) is an abundantly fertile version of *F. caperata* that produces many dark brown apothecia. It grows on trees in southern Texas. In the east, *F. caperata* can be confused with *Flavopunctelia soredica*, which has few pseudocyphellae. That species, however, has soredia that are primarily marginal, and the medulla reacts C+ red. *Flavopunctelia flaventior* can have some laminal soralia, but the lobes have conspicuous white pseudocyphellae, and it is also C+ red. See also *Flavoparmelia baltimorensis*. IMPORTANCE: The Tarahumar people in Mexico have dried and crushed *F. caperata* and used the powder to treat burns.

uously wrinkled. Photobiont green (*Trebouxia?*). Apothecia lecanorine; spores colorless, ellipsoid, 1-celled, 8 per ascus. Pycnidia black, buried in thallus close to lobe margins; conidia dumbbell-shaped. CHEMISTRY: Cortex KC+ gold (usnic acid); medulla PD–, K–, KC+ red, C+ red (lecanoric acid). HABITAT: Typically on bark and wood. COMMENTS: *Punctelia* differs from *Flavopunctelia* in its gray color (with atranorin rather than usnic acid in the cortex) and its strongly curved conidia. See Comments under *Flavoparmelia*.

KEY TO SPECIES: See Key J.

Flavopunctelia (4 N. Am. species)
Speckled greenshield lichens, green speckleback lichens

DESCRIPTION: Medium-sized, yellow-green, foliose lichens with white spots (pseudocyphellae) on the upper surface of the lobes (sometimes sparse); lower surface brown or black, smooth to minutely wrinkled; rhizines sparse or almost absent, unbranched or forked; upper surface of older portions of the thallus conspic-

Flavopunctelia flaventior (syn. *Parmelia flaventior*)
Speckled greenshield

DESCRIPTION: Lobes rounded, 4–8 mm across, sorediate at the margins and on the upper lobe surface; lower surface black, with a brown edge; rhizines black,

333. *Flavopunctelia flaventior* Coast Range, California ×1.3

334. *Flavopunctelia praesignis* Chisos Mountains, southwestern Texas ×0.88

very sparse. HABITAT: On bark of various trees in fairly open woods or along roadsides. COMMENTS: *Flavopunctelia flaventior* and *F. soredica*, the two yellow-green, foliose lichens with white pseudocyphellae, soredia, and a C+ red medulla, intergrade somewhat and can be difficult to distinguish. *Flavopunctelia flaventior* has more conspicuous pseudocyphellae on the lobe tips, and the soredia are mainly in soralia on the lobe surface, with few on the margins. *Flavopunctelia soredica* is rarer and has few pseudocyphellae, and the soredia are almost all marginal, giving rise to crescent-shaped, powdery lobe tips. *Punctelia subrudecta* can turn a bit yellowish with age and resemble *F. flaventior*, but the *Punctelia* has a pale tan lower surface throughout, and its rhizines are abundant, pale, slender, and hair-like. *Flavoparmelia caperata*, similar in size and general habit, is yellower, has no pseudocyphellae, and has more abundant rhizines; its medulla is PD+ red-orange, C–.

Flavopunctelia praesignis (syn. *Parmelia praesignis*)
Fruiting speckled greenshield

DESCRIPTION: Thallus large; lobes 4–12 mm wide with conspicuous white pseudocyphellae (0.1–0.3 mm across) often irregular in shape; without soredia or isidia; lower surface black except for brown, bare margins; rhizines only in the older parts. Apothecia dark reddish brown, deeply cup-shaped; margins toothed, with many white pseudocyphellae; tiny black dots (pycnidia) peppering the lobe edges. HABITAT: On oaks and conifers in the mountains. COMMENTS: *Flavoparmelia rutidota* is a similar species in the southeast but without white pseudocyphellae and with a C–, PD+ orange medulla (protocetraric acid).

Flavopunctelia soredica (syn. *Parmelia ulophyllodes*)
Powder-edged speckled greenshield

DESCRIPTION and COMMENTS: Very similar to *F. flaventior* except that the soralia are exclusively marginal and the pseudocyphellae are sparse and inconspicuous. HABITAT: On many kinds of bark in open woods.

Fulgensia (4 N. Am. species)
Sulphur lichens

DESCRIPTION: Yellow to orange crustose lichens with areolate to lobed, almost foliose thalli; most species developing hemispherical to scaly schizidia on the upper surface. Photobiont green (*Trebouxia?*). Apothecia lecanorine, dark rusty orange with paler margins; spores 1-celled or, rarely, 2-celled, colorless, with thin walls (not polarilocular), 8 per ascus. CHEMISTRY: Thallus cortex and apothecia K+ deep red-purple (anthraquinones). HABITAT: On lime-rich soil or rock, rarely mosses, in arid regions or tundra. COMMENTS: In color and form, *Fulgensia* closely resembles *Caloplaca*, the firedot lichens. However, the spores of *Fulgensia* are typically 1-celled, and the 2-celled spores do not have the thickened septum characteristic of *Caloplaca*. The production of schizidia and the desert habitat of many species are also characteristic of *Fulgensia*.

KEY TO SPECIES

1. Spores 2-celled . . . . *Fulgensia desertorum*
1. Spores 1-celled 2

2. Thallus areolate, not noticeably lobed at the margin; schizidia often present. . . . . *Fulgensia bracteata*
2. Thallus clearly lobed at the margin; schizidia present or absent 3

3. Lobes long and slightly ascending at the tips; schizidia absent; rare . . . . [*Fulgensia fulgens*]
3. Lobes short, entirely adnate; schizidia abundant on thallus surface; common . . . . . [*Fulgensia subbracteata*]

335. *Flavopunctelia soredica* mountains, central Arizona ×1.44

Fulgensia bracteata
Tundra sulphur lichen

DESCRIPTION: Thallus yellow, almost sulphur-colored at the edges, becoming more orange toward the center, white scabrose or pruinose in part, areolate to verrucose or coarsely granular, without any distinct lobes forming at the periphery. Apothecia abundant, red-orange to brown-orange; spores 1-celled, 9–13 × 4–7 μm. HABITAT: On calcareous soil in tundra or exposed dry sites. COMMENTS: Most North American specimens have been misnamed as ***F. fulgens***, a larger, lobed lichen (see *F. desertorum*). Sterile specimens can be mistaken for ***F. subbracteata***, which has well-developed schizidia on the thallus surface and more lobed squamules. *Fulgensia desertorum* is a southwestern species similar to *F. bracteata*, but with 2-celled spores.

Fulgensia desertorum
Desert sulphur lichen

DESCRIPTION: Thallus yellow, areolate to verruculose, hemispherical warts on the upper surface, with short convex lobes at the periphery, always rough, sometimes whitened with coarse

336. *Fulgensia bracteata* southwestern Yukon Territory ×5.8

337. *Fulgensia desertorum* central Utah ×4.9

pruina on the surface. Apothecia common, 0.8–2 mm in diameter, orange with thallus-colored rims; spores 2-celled, thin-walled (not polarilocular as in *Caloplaca*), 11–15 × 5.5–8 μm. HABITAT: On dry calcareous soil. DISTRIBUTION: In arid parts of the western interior. COMMENTS: The more common *F. subbracteata* has a similar thallus but 1-celled spores. It rarely has apothecia and reproduces with fragments developing from thin and fragile, elevated, scale-like lobules called schizidia that usually cover the upper surface. *Fulgensia fulgens,* a rare species in northern North America, has well-developed lobes often 2–3 mm long, partly lifting from the surface and sometimes appearing almost foliose. It commonly has apothecia, and its spores are 1-celled.

Fuscidea (13 N. Am. species)
Quilt lichens

DESCRIPTION: Crustose lichens, typically brownish or olive, less commonly gray; thallus areolate, verruculose or rimose, usually continuous, with or without soredia; prothallus black or brown, usually forming a distinct line around the small individual thalli, which can grow together, creating a kind of patchwork quilt effect. Photobiont green (unicellular). Apothecia lecideine or biatorine, reddish brown to dark gray or black; epihymenium brown; hypothecium colorless or pale yellowish; exciple usually pale internally, but often brown to black at the outer edge; paraphyses unbranched or branched once or twice, pigmented at the tips or on the walls of the upper third, usually separating easily when squashed in a water mount; asci *Fuscidea*-type (Fig. 14d), with several layers, the innermost one staining most darkly K/I+ blue, surrounded by a weakly stained envelope that sometimes is strongly expanded, and this in turn sometimes has a more darkly stained outer coat; spores 1-celled, ellipsoid to squarish at the ends, straight, or bean-shaped, 8 per ascus, conidia cylindrical or ellipsoid. CHEMISTRY: Commonly contains divaricatic acid; some species have perlatolic, gyrophoric, protocetraric, or fumarprotocetraric acids. Several species have an IKI+ blue medulla, but most are IKI–. HABITAT: Most species on rock; a few on bark. COMMENTS: *Fuscidea* is recognizable in the field by its dark thalli, which often form mosaics on the rock, together with its often brown (rather than black) apothecia. *Ropalospora* differs mainly in having fusiform spores with several cells. *Lecidea* has pitch black apothecia and entirely different asci.

KEY TO SPECIES: See *Lecidea*.

Fuscidea recensa

DESCRIPTION: Thallus dark brownish gray to pale gray, verruculose to areolate, growing in zoned patches one against the other. Apothecia lead-colored, broadly attached (not at all immersed); spores kidney-shaped, slightly curved, rather narrow, 9–11 × 3.5–4.5(–6) μm. CHEMISTRY: Medulla PD–, K–, KC–, C–, UV+ white, IKI– (divaricatic acid). HABITAT: On siliceous rocks in the open; usually in oceanic or at least humid climates. COMMENTS: *Fuscidea recensa* is the most common rock-dwelling species of the genus in North America and is the only one likely to be seen at temperate latitudes in the east. Several relatively rare western species occur along the coast (e.g., *F. intercincta, F. thomsonii*); their apothecia are immersed in the thallus. *Fuscidea arboricola* is a very widely distributed but undercollected sterile species growing on the bark of a variety of trees, especially deciduous species. It is characterized by its dark grayish brown, warty or granular thallus bursting into round, greenish or yellow-green soralia that react PD+ red, K–, KC–, C– (fumarprotocetraric acid). The thallus is often bordered by a brown prothallus.

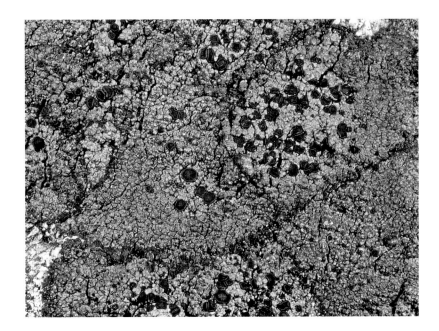

338. *Fuscidea recensa* Cape Cod, Massachusetts ×4.1

Fuscopannaria (about 20 N. Am. species)
Brown shingle lichens, mouse lichens

DESCRIPTION: Thallus squamulose, usually brown, lacking a lower cortex, but often with a blue-black hyphal mat extending around the thallus margin as a prothallus; only a few species with soredia. Photobiont blue-green (*Nostoc*) in a distinct layer. Apothecia with or without a lecanorine margin, sometimes even on the same specimen, mostly 0.5–1 mm in diameter; hymenium hemiamyloid; asci usually with tube structures in the tip that turn dark blue in K/I; spores colorless, ellipsoid, 1-celled, 8 per ascus. CHEMISTRY: Most species lack lichen substances, but some contain fatty acids or triterpenes, sometimes with atranorin. COMMENTS: *Fuscopannaria* has long been included within *Pannaria*, which has a foliose-squamulose thallus usually with gray tones (except for *P. pezizoides*), and often has a true hypothallus. In *Pannaria*, the apothecia have a well-developed lecanorine margin, the asci lack any K/I+ blue stuctures in the tip, and most species contain pannarin. *Parmeliella*, a closely related genus, has apothecia with biatorine instead of lecanorine margins. Only a few North American species of *Fuscopannaria* are sorediate and none of them are common: *F. ahlneri* is discussed with *Pannaria conoplea; F. mediterranea* is a very rare west coast species with small, roundish, closely appressed blue-gray to olive squamules or areoles, granular sorediate on the upturned margins, sometimes reducing the thallus to a sorediate crust.

KEY TO SPECIES: See *Pannaria*.

Fuscopannaria leucophaea (syn. *Pannaria leucophaea*)
Rock shingle lichen

DESCRIPTION: Overlapping, fairly thick squamules, up to 2 mm across, dark grayish brown with no white edges; no isidia or soredia; black prothallus conspicuous or inconspicuous. Apothecia with a persistent thalline margin or no thalline margin at all, or both on the same thallus; disks dark brown to blackish; spores smooth, 13.5–22 × 5–7.5 μm. CHEMISTRY: All reactions negative (no lichen substances). HABITAT: On various kinds of rock, in the shade, especially where there is seepage and the rock is wet. COMMENTS: This widespread but rather nondescript squamulose lichen is

thallus. **Fuscopannaria maritima** has dark yellow-brown to olive-brown squamules with bluish white margins and contains atranorin. It is found on sea-splashed rocks along the northwest coast and is the only *Fuscopannaria* able to tolerate salt water. Compare with *F. praetermissa*.

Fuscopannaria leucosticta (syn. *Pannaria leucosticta*)
Rimmed shingle lichen

DESCRIPTION: Squamules abundant, rounded or crenulate, chestnut-brown, 2–3 mm across, with white felty margins, often on a black prothallus; without soredia or isidia. Lecanorine apothecia with conspicuous white rims usually present; spores 23–27 × 9–11 μm, including a rather thick, clear outer wall often tapering to a fine point at one or both ends. CHEMISTRY: All reactions negative (only triterpenes and fatty acids). HABITAT: On bark or occasionally rocks, often among mosses. COMMENTS: This lichen, with its large, often subfoliose squamules, resembles *Pannaria rubiginosa*,

339. *Fuscopannaria leucophaea* North Woods, Maine ×2.9

notable more for lacking characters (i.e., no white margins, no isidia, etc.) than for having them. Members of the *F. saubinetii* group are west coast species on bark. They have red-brown apothecia with a biatorine margin, more finely divided squamules, and a grayer

340. *Fuscopannaria leucosticta* Ouachitas, Arkansas ×4.3

in which the medulla is PD+ orange (pannarin) and the lobes are somewhat broader (up to 4 mm wide and 8 mm long), without whitened margins. *Fuscopannaria leucostictoides,* a west coast lichen, is grayer, lacks pointed spores, and contains atranorin.

Fuscopannaria leucostictoides
(syn. *Pannaria leucostictoides*)
Petaled shingle lichen

DESCRIPTION: Squamules olive-green but whitish pruinose at the tips, making the thallus blue-gray; lobes flat, radiating, mostly less than 1 mm across and 1–2 mm long, often rough and scaly on the surface, sometimes breaking up into dispersed areoles on a conspicuous, blue-black prothallus; soredia and isidia absent. Apothecia common and abundant, with white, persistent, lecanorine margins; spores ellipsoid, not pointed at the tips, 14–18(–20) × 7–10 μm, with thin, uniform walls. CHEMISTRY: All reactions negative, but contains atranorin. HABITAT: On bark and twigs in coastal forests. COMMENTS: *Parmeliella triptophylla,* also on a blue-black prothallus, has isidia and biatorine, hemispherical apothecia. *Pannaria rubiginosa* has broader marginal lobes (over 1.5 mm) and contains pannarin (PD+ orange) instead of atranorin. *Fuscopannaria leucosticta* is an eastern species with brown, somewhat larger lobes.

Fuscopannaria "praetermissa" (in the broad sense)
Moss shingle lichen

DESCRIPTION: Squamules olive- or red-brown, 0.5–2 mm across, lobed, with gray to blue-white felty tips, usually very abundant and overlapping, forming a continuous crust; usually developing finger-like, erect outgrowths on the lobe margins, giving it a coarsely isidiate or granular appearance; without soredia or prothallus. Apothecia biatorine or lecanorine, but the species is rarely fertile. CHEMISTRY: All reactions negative (steroids). HABITAT: On moss or soil, less commonly on tree bases and rarely on rock, in calcareous habitats in the north. DISTRIBUTION and COMMENTS: In its strict sense, *F. praetermissa* is a common boreal to arctic-alpine lichen. The name has been used, however, for several other recognizable western

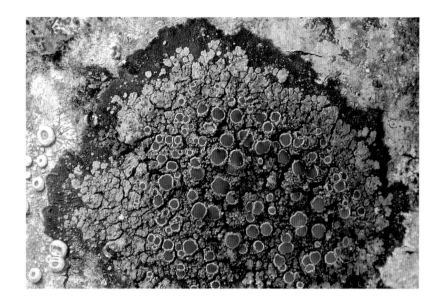

341. *Fuscopannaria leucostictoides* Olympics, Washington ×2.4

342. *Fuscopannaria "praetermissa"* Coast Range, Oregon ×2.9

taxa (one of them represented in plate 342, a photograph taken in Oregon). Distribution maps are still not possible for this confusing species complex, which, fortunately, is under study. The typical *F. praetermissa* can form a granular crust over rather large areas of moss or peat. *Massalongia carnosa* is very similar but has more elongate, strap-shaped lobes, sometimes minutely lobulate but not really felt-tipped, and the spores are narrow and 2-celled.

Fuscopannaria saubinetii group
Pink-eyed shingle lichen

DESCRIPTION: Thallus brownish to gray, squamulose, with overlapping, minutely lobed squamules about 0.5 mm across, occasionally pruinose on the lobe tips; without soredia or isidia, although the tiny lobules can look a bit like isidia; prothallus absent. Apothecia common, biatorine (without a thalline margin of any kind), pale reddish brown, 0.5–0.8 mm in diameter; spores 15–17 × 5–6 µm, but some specimens 18–21 × 7.5–10 µm. CHEMISTRY: All reactions negative (no lichen substances). HABITAT: On conifer branches and twigs in humid coastal forests. DISTRIBUTION: Pacific Northwest. COMMENTS: From the variation in spore size and some other characters, it is clear that what we are calling *F. saubinetii* represents more than one species. Although not common, these lichens can be very abundant in local areas. They are characterized by their overlapping squamules, biatorine apothecia, and bark substrate, and by their lack of a prothallus.

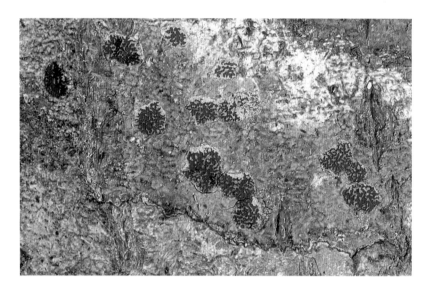

343. *Fuscopannaria "saubinetii"* herbarium specimen (Oregon State University), Oregon ×5.6

344. *Glyphis cicatricosa* southern Louisiana ×4.1

Glyphis (1 N. Am. species)

KEY TO SPECIES: See Key E.

Glyphis cicatricosa
Blistered script lichen

DESCRIPTION: Thallus growing under the outer layers of bark, barely perceptible as an olive stain. Photobiont green (*Trentepohlia*). Fruiting bodies broad, flat, chocolate-brown, script-like lirellae, 0.1–0.3 mm wide, 0.3–2 mm long, marginless, embedded in round, white pruinose, blister-like pseudostromata 1–3 mm across, carbon-black within; lirellae usually abundantly branched, flowing in rounded curves; paraphyses unbranched; spores colorless, (4–) 6-celled, rarely 12-celled, (16–)20–45 × 6.5–11 μm, 4–8 per ascus. CHEMISTRY: All tissues PD–, K–, KC–, C– (no lichen substances). HABITAT: On smooth-barked trees in semitropical forests. COMMENTS: With its cocoa-colored lirellae lying on whitish warts, this lichen can hardly be mistaken for any other. *Sarcographa* species have poorly developed pseudostromata, black or white pruinose lirellae, and shorter spores with fewer cells (usually 4–6); most contain β-orcinol depsidones that react with PD and K.

Gomphillus (1 N. Am. species)

KEY TO SPECIES: See Keys B and F.

Gomphillus americanus
Frazzled dot lichen

DESCRIPTION: Thallus thin and membranous, greenish white. Photobiont green (unicellular). Apothecia biatorine, dark brown to black, convex to hemispherical, extremely small and inconspicuous, 0.25–0.5(–1) mm in diameter, single or clustered in twos or threes, slightly raised on a stubby pedicel when mature; asci long and slender, cylindrical, IKI–; spores colorless, long, thread-like, many-celled but often not maturing. Conidial mass produced at the summit of a colorless stalk 1–2(–3.5) mm long, breaking up at maturity and leaving a ragged fringe of tissues and mass of gelatinized conidia at the summit of the stalks; conidia colorless, thread-like, 180–240 × 1.5–2 μm, with many cells, each cell about 3–4 μm long. CHEMISTRY: No lichen substances. HABITAT: Over mosses and adjacent tree bark in hardwood forests. COMMENTS: This rarely collected lichen is one of the larger species in its family, the Gomphillaceae. Most other members of the family are extremely small species growing on leaves or twigs in tropical forests or in humid, oceanic regions. One example is *Gyalideopsis anastomosans,* which grows on shaded willows and alders in coastal British Columbia and Washington, producing a membranous thallus with tiny spike-like projections resembling isidia under 0.3 mm tall. The apothecia are entirely sessile, and the spores are colorless and muriform.

345. *Gomphillus americanus* herbarium specimen (Duke University), Georgia ×6.4

Graphina (23 N. Am. species)
Script lichens

DESCRIPTION and COMMENTS: Crustose lichens much like *Graphis* (ascomata are elongate lirellae) but with muriform spores; walls of lirellae carbonized or colorless, solid or layered; locules of the spores round and scattered within thickened wall material (see Fig. 15h) or filling the spore like bricks if the walls are thin; 1–8 per ascus. Photobiont green (*Trentepohlia*). CHEMISTRY: Often containing β-orcinol depsidones such as stictic, norstictic, protocetraric, or psoromic acids; lichexanthone sometimes present (UV+ yellow). HABITAT: Usually on bark in tropical or subtropical regions. COMMENTS: Many species of *Graphina* occur in Florida and Louisiana, but only a few of the more common ones are included in the key.

KEY TO SPECIES (based on Harris, *More Florida Lichens,* 1995)

1. Walls of the lirellae black. Spores 1 per ascus, 65–90 × 20–30 μm; lirellae partly immersed, with a narrow disk . . . . [*Graphina xylophaga*]
1. Walls of lirellae pale to colorless. 2

2. Spores 1 or rarely 2 per ascus; lirellae crowded, short and irregular; spores 85–105 × 20–28 μm . . . . [*Graphina cypressi*]
2. Spores 2–8 per ascus 3

3. Spores small, 15–20(–35) × 6–9(–11) μm, 8 spores per ascus. Lirellae appearing like fissures in the thallus, disk not exposed; thallus dark olive to yellow-brown, rather shiny, K–, PD–. . . . . [*Graphina incrustans*]
3. Spores large, over 35 μm long, typically 2–4 spores per

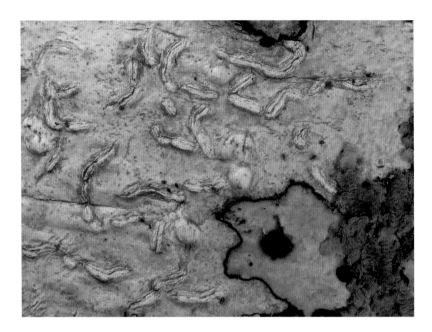

346. *Graphina peplophora* herbarium specimen (Canadian Museum of Nature), Florida × 4.2

ascus; thallus pale gray, K+ or K–, PD+ yellow to red; lirellae white 4

4. Spores ellipsoid, 35–78 × 18–36 μm; walls of lirellae layered and disintegrating; thallus PD+ orange, K+ yellow to red (constictic acid or salazinic acid); rare .... *Graphina peplophora*
4. Spores fusiform, 110–140 × 15–17 μm; walls not disintegrating; thallus PD+ red-orange, K– (protocetraric acid); common in Florida ..... [*Graphina abaphoides*]

Graphina peplophora
Pastry script lichen

DESCRIPTION: Thallus thin, usually within the bark and creating a yellow stain at the surface; lirellae white, pruinose, with crumbling layered walls like a *millefeuille* pastry, 1–5 mm long and 0.5–1 mm broad, usually unbranched; spores colorless, with many cells (10–16 transverse septa and 5–8 longitudinal septa), 2–4 per ascus, 35–78 × 18–36 μm, but they often fail to develop. CHEMISTRY: Tissues of the wall of the fruiting body (in section) PD+ orange, K+ yellow, KC–, C– (constictic acid). HABITAT: On hardwood trees and shrubs.

Graphis (39 N. Am. species)
Script lichens

DESCRIPTION: Crustose lichens, typically with pale, thin thalli growing within upper bark tissues. Photobiont green (*Trentepohlia*). Ascomata elongate lirellae with black or pale walls consisting of the exciple (surrounding the hymenium) and outer thalline tissues often incorporating some bark tissue as well; walls of the lirellae smooth or in layers like a flaky pastry (with a striate appearance), sometimes thin and level with the disk; disk broad and sometimes pruinose or opening by a narrow slit; lirellae of some species buried in the thallus (and bark) and opening only by a slightly raised fissure; hymenium IKI–; spores colorless, fusiform, 4- to many-celled, with lens-shaped locules, usually IKI+ violet; 8 per ascus. CHEMISTRY: Often containing β-orcinol depsidones such as stictic, norstictic, or protocetraric acids; rarely with orcinol depsides (lecanoric acid); cortex containing lichexanthone in some species (thallus UV+ yellow). HABITAT: On bark, commonly in tropical or subtropical woodlands, but a few species in temperate forests. COMMENTS: All but a handful of script lichens are restricted to the southernmost states, especially Florida, where over 25 species are known. Only *G. scripta* is truly widespread, although some southern species are very common within their range; a few are included in our key. Other script lichens are easily mistaken for *Graphis*, especially *Graphina* (with muriform spores) and *Phaeographis* (with brown spores). *Opegrapha* has spores with cylindrical rather than lens-shaped locules, and the hymenium is IKI+ blue-green changing to orange. Species of *Arthonia* with lirellae lack an exciple; that is, the lirellae have no walls at all. Their hymenium is hemiamyloid.

KEY TO SPECIES

1. Lirellae white or thallus-colored; spores 4-celled 2
1. Lirellae black, C–; spores 6–12-celled 3

2. Lirellae prominent, thickly white pruinose, C+ red (lecanoric acid); walls of lirellae black; very common in southeastern coastal plain .... *Graphis afzelii*
2. Lirellae not prominent, seen as raised fissures in the thallus, C–; walls of lirellae colorless or pale; common on trees in southeast and in Pacific Northwest coastal forests .... [*Graphis insidiosa*]

3. Walls of lirellae with long ridges, in layers like a French pastry; Florida 4
3. Walls of lirellae uniform, not ridged or layered 5

4. Spores (6–)9–10-celled, 25–30 × 6–8(–10) μm; thallus UV– (lacking lichexanthone); rare . . *Graphis subelegans*
4. Spores 6–8-celled, 20–30 × 7–9 μm; thallus UV+ yellow (lichexanthone); common. . . . . [*Graphis lucifera*]

5. Spores 6–8-celled; thallus K+ red, PD+ yellow (norstictic acid); common only in Florida and nearby coastal plain . . . . [*Graphis librata*]
5. Spores 8–11-celled 6

6. Lirellae prominent or partly immersed; thallus K–, PD– (no lichen substances); extremely common and widespread, but absent from Florida . . . . *Graphis scripta*
6. Lirellae almost entirely immersed; thallus K+ yellow to red, PD+ orange (stictic acid, sometimes with norstictic acid); common in Florida and Louisiana . . . .
. . . . [*Graphis caesiella*]

347. *Graphis afzelii* northern Florida ×3.5

Graphis afzelii
Powdered script lichen

DESCRIPTION: Thallus within the bark, perceptible only as a vague stain; lirellae white, prominent, straight or curved, usually unbranched, 1–7 mm long, 0.6–0.8 mm wide, opening by a slit; walls carbon-black throughout but hidden by a thick, white, powdery pruina; spores 4-celled, 16–23 × 6–9 μm, IKI+ violet. CHEMISTRY: Surface of lirellae PD–, K–, KC+ red, C+ red (lecanoric acid). HABITAT: Very common on the smooth bark of hardwood trees and shrubs, in open or shaded woods. COMMENTS: This is the easiest script lichen to identify because of its very prominent white lirellae turning C+ deep red. It is the only script lichen with this chemistry.

Graphis scripta
Common script lichen

DESCRIPTION: Thallus within bark tissues and barely visible, or forming circular, yellowish white to greenish gray patches; lirellae black, variable in length and breadth, branching, and shape; mostly 1–7 mm long, 0.15–0.3 mm wide, unbranched (especially on birch bark) or branched once or twice (especially on beech), pointed at the ends; walls thin, black, prominent; disk barely visible under a slit-like opening or relatively broad and lightly pruinose; spores 6–14-celled, 20–70 × 6–10 μm, IKI+ violet. CHEMISTRY: All reactions negative (no lichen substances). HABITAT: On bark of all types of trees, usually in partial shade. COMMENTS: The script lichen is named for its slender, elongate fruiting bodies, which resemble scribbles on the bark, especially in the form having lirellae that branch and curve. On bark with a pronounced grain such as birch, the linear apothecia often follow the bark texture, choosing to "go with the flow." *Graphis scripta* can be confused with *Phaeographis inusta* where their ranges overlap. The spores of *Ph. inusta* are brownish, have 4–5 cells, and the margins of the fruiting bodies are level with the disk. In *G. scripta*, the margins usually stick up a bit above the surface.

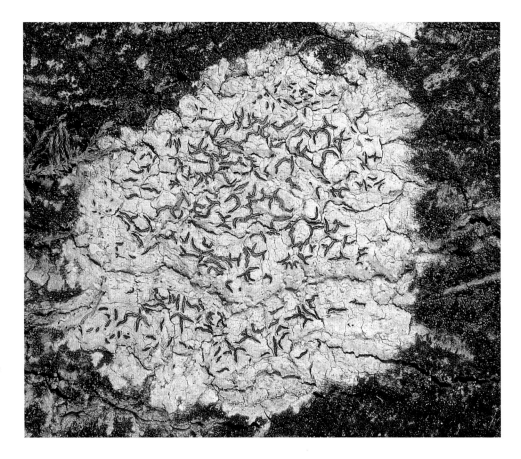

348. *Graphis scripta*
White Mountains, New Hampshire ×3.4

349. *Graphis subelegans*
northern Louisiana ×2.3

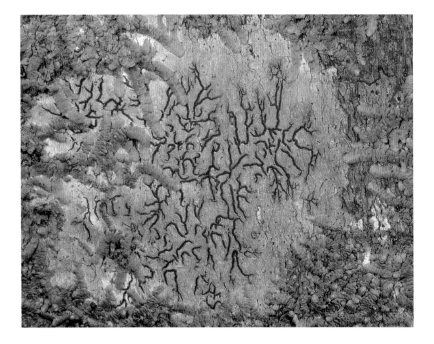

Graphis subelegans
Fluted script lichen

DESCRIPTION: Thallus thin, within bark tissues. Lirellae slender, long, black, with layered, striate walls; inner tissues (e.g., below the hymenium and the inside layer of the walls) somewhat orange; spores (6–)9–10-celled, 25–30 × 6–8(–10) μm. CHEMISTRY: All reactions negative (no lichen substances). HABITAT: On bark, especially on shrubs in open scrub land. COMMENTS: ***Graphis striatula*** resembles *G. subelegans* except that its inner tissues are colorless and its spores are 10–12-celled, 35–45 μm long. The rare **G. elegans,** an oceanic east coast species, has a thicker, partly superficial thallus, and longer spores (10–12-celled, 32–55 × 6–12 μm), and it contains norstictic acid.

Gyalecta (10 N. Am. species)
Dimple lichens

DESCRIPTION: Crustose lichens with thin, pale thalli. Photobiont green (*Trentepohlia*). Apothecia generally biatorine (exciple without algae) but sometimes covered with a thin thalline layer; disks pale yellow to pink or orange, rather waxy-looking, without pruina, typically deeply concave; margins usually prominent; hymenium IKI–, K/I+ blue (hemiamyloid); asci thin-walled, without an apical thickening, K/I+ pale blue; spores colorless, ellipsoid to fusiform, 4–14-celled or muriform, constricted at the septa, giving the spores a bumpy outline, 8 per ascus (Fig. 14p). CHEMISTRY: No lichen substances. HABITAT: On rocks, trees, or soil, especially those rich in calcium. COMMENTS: Several crustose genera resemble *Gyalecta*, with the same sort of pale, pinkish to yellowish, concave, waxy apothecia. *Gyalidea* is most similar but contains a different green alga (*Cystococcus*) and its spores are not constricted at the septa. *Pachyphiale* is similar but very tiny and on bark; its apothecia are often immersed in the bark, and the asci contain 16–48 transversely septate (not muriform) spores that are 3–14-celled. A related lichen, *Petractis farlowii,* is fairly common on limestone in the southeast. The thallus is commonly endolithic, and the apothecia are also immersed, very perithecium-like, opening at the top by radial cracks. Its muriform spores are 18–24(–27) × 9–12(–14) μm. See also *Dimerella*.

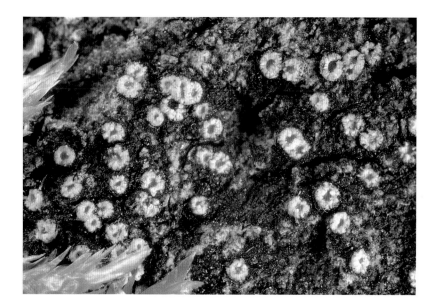

lipsoid spores (12–20 × 5–7 μm). *Gyalecta truncigena,* a bark-dwelling species on a variety of trees, is a real find because it is so small. Its pinkish cup-shaped apothecia, although often very abundant, are usually under 0.25 mm in diameter and the thallus is almost invisible. However, it is probably fairly common in the northern temperate or southern boreal forests and has ellipsoid, muriform spores, 14–28(–31) × 5–9 μm.

350. *Gyalecta jenensis* herbarium specimen (Canadian Museum of Nature), Ireland ×7.7

Gymnoderma (1 species worldwide)

KEY TO GENUS: See Key F.

KEY TO SPECIES: See Key G.

Gyalecta jenensis
Rock dimple lichen

DESCRIPTION: Thallus thin, pale pinkish white but frequently covered with a black crust of cyanobacteria (see plate 350) in particularly humid sites. Apothecia yellowish pink, 0.6–1 mm in diameter; disks flat to deeply concave; margins thick, smooth or cracked into segments; spores ellipsoid, muriform, 13–25 × 7–10 μm. HABITAT: On rocks of many kinds, usually in humid situations. COMMENTS: The superficially similar *G. foveolaris* is frequently seen on tundra soil or peat in the western mountains or in the Arctic; it has 4-celled, transversely septate, el-

Gymnoderma lineare
Rock gnome

DESCRIPTION: Thallus consisting of long, strap-shaped lobes 0.8–1.3 mm wide and 10–25 mm long, bluish green or olive-gray above and white below, blackening at the base, branching in irregular dichotomies, growing in dense, overlapping colonies; cortex on both surfaces fairly thick, making the lobes rather stiff; medulla white, very dense. Photobiont green (*Trebouxia?*). Clusters of almost spherical, dark red-brown to black apothecia, 0.8–1 mm in diameter, develop at the tips of very short, solid podetia (mostly less than 1.5 mm long); spores colorless, el-

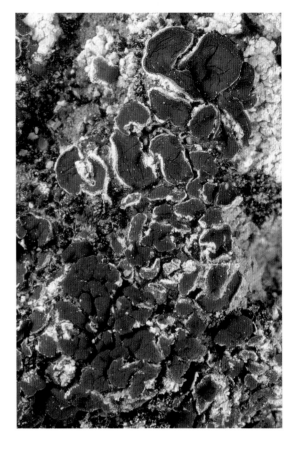

351. *Gymnoderma lineare* (wet) Appalachian Mountains, Tennessee ~×1.5

352. *Gypsoplaca macrophylla* central Utah ×4.1

lipsoid, 1-celled, 9–14 × 4–5 μm, 8 per ascus. CHEMISTRY: Cortex PD–, K+ yellow, KC–, C–; medulla PD–, K–, KC–, C– (atranorin and protolichesterinic acid). HABITAT: On seeping rock walls in deep forests at high elevations. COMMENTS: This rare species is one of the two lichens on the U.S. Endangered Species list (the other being *Cladonia perforata*). It seems to reproduce itself well where it occurs, but it may have extremely narrow environmental requirements that keep it from becoming widespread in the mountains.

Gypsoplaca (1 species worldwide)

KEY TO SPECIES: See Key G.

Gypsoplaca macrophylla
Changing earthscale

DESCRIPTION: Thallus of closely appressed squamules that are rounded to irregular, yellowish brown to olive-brown (sometimes white pruinose at the edges), and 2–9 mm in diameter. Photobiont green (unicellular). Apothecia undelimited, consisting of little more than a transformation of the upper cortex into a fertile layer (hymenium). Apothecia appear as reddish brown swellings on the squamules and can be quite large, sometimes almost taking over an entire squamule (1–3 mm across); hymenium filled with tiny oil drops; spores colorless, 1-celled, ellipsoid to almost spherical, 8 per ascus. CHEMISTRY: Cortex and medulla PD–, K–, KC–, C– (triterpenes). HABITAT: On soil containing gypsum (calcium sulphate) in rather arid sites. COMMENTS: The fruiting bodies of *Gypsoplaca* are unique among the earthscales. In other respects, the lichen resembles a *Psora*.

Haematomma (7 N. Am. species)
Bloodspot lichens

DESCRIPTION: Crustose lichens with thin or thick, whitish to yellowish gray thalli, rarely sorediate. Photobiont green (*Trebouxia?*). Apothecia lecanorine, on the thallus surface or immersed, with blood red to orange-red disks and thallus-colored margins; epihymenium with red pigment but other tissues colorless;

paraphyses somewhat branched and anastomosing; asci *Lecanora*-type; spores colorless, 3–10-celled, oblong and strongly tapered, straight or more frequently curved and twisted, 8 per ascus. CHEMISTRY: Contains a variety of substances including red pigments (russulone or haematommone), atranorin (in all species), sphaerophorin, and placodiolic and pseudoplacodiolic acids. Russulone is detected as follows: drawing K under the cover slip causes a red solution to appear around the pigment, red crystals form in the red solution, and the epihymenium remains red. With haematommone, a violet cloud appears around the red pigment with the addition of K, the epihymenium becomes colorless, and no crystals appear. HABITAT: On bark of trees and shrubs or on rocks, in tropical or subtropical regions, usually in well-lit habitats. COMMENTS: No other crustose lichens have apothecia with blood red disks and white (i.e., thallus-colored) margins. *Ophioparma* has yellowish thalli (containing usnic acid) and is found in arctic or alpine sites. *Pyrrhospora* has almost marginless biatorine apothecia and 1-celled ellipsoid spores. Species of *Caloplaca* are more orange than red (with the exception of *C. luteominia* var. *bolanderi*) and usually have orange margins, and their spores are entirely different (2-celled, polarilocular).

KEY TO SPECIES

1. Growing on rock . . . . *Haematomma fenzlianum*
1. Growing on bark 2

2. Thallus sorediate. . . . . [*Haematomma americanum*]
2. Thallus not sorediate 3

3. Apothecia commonly sunken into thallus; conidia long and curved; epihymenium K+ red . . . . *Haematomma persoonii*
3. Apothecia superficial; conidia rod-shaped or very long 4

4. Spores over 50 μm long; conidia thread-like and twisted; epihymenium K+ red . . . . [*Haematomma rufidulum*]
4. Spores under 55 μm; conidia rod-shaped; epihymenium K+ violet or purple 5

5. Conidia 5–7 × 0.8–1 μm; contains placodiolic acid. . . . . *Haematomma accolens*
5. Conidia 7–8 × 1.5–2 μm; contains isoplacodiolic and isopseudoplacodiolic acids . . . . [*H. flexuosum*]

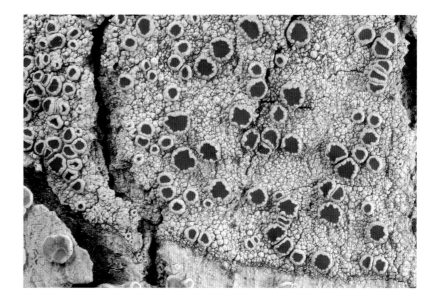

Haematomma accolens
Tree bloodspot

DESCRIPTION: Thallus light to dark greenish gray, rimose to verruculose. Apothecia prominent, broadly attached, 0.8–2 mm in diameter; disk dark red, with a smooth to bumpy or wavy, thallus-colored margin; spores mostly 6–9-celled, oblong, somewhat sinuous, 35–55 × 3–5 μm; conidia 5–7 × 0.8–1 μm. CHEMISTRY: Cortex and medulla PD–, K+ yellow, KC–, C–, UV– (atranorin and placodiolic acid in the medulla); epihymenial pigment haematommone. HABITAT: On bark. COMMENTS: Four other common bark-dwelling bloodspot lichens occur on the southeastern coastal plain. **Haematomma flexuosum**, with spores and apothecia much like those of *H. accolens*, also contains the pigment haematommone but has longer and thicker conidia (7–8 × 1.5–2 μm) and contains isoplacodiolic and isopseudoplacodiolic acids instead of placodiolic acid. The other three all contain russulone. Of these, *H. persoonii* is the most common. Its apothecia are usually sunken, and the disks often fuse. Its spores are 7–8-celled, like those of *H. accolens*, but its thallus contains sphaerophorin (UV+ white). **Haematomma americanum** has 8–14-celled spores (45–60 × 3–5 μm) with superficial apothecia. It is the only sorediate bloodspot lichen in North America (although there is another in the West Indies), and it contains sphaerophorin. **Haematomma rufidulum** has the largest spores, 8–20-celled, 50–80 × 4–6 μm, and contains placodiolic acid.

353. *Haematomma accolens* coastal plain, North Carolina ×4.1

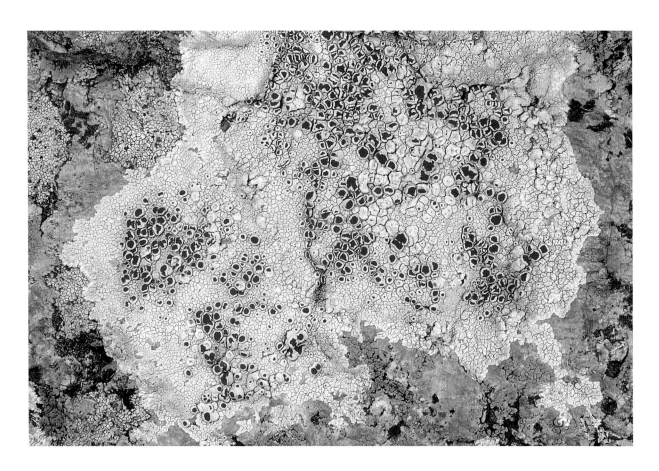

354. *Haematomma fenzlianum* Sonoran Desert, southern Arizona ×2.3

Haematomma fenzlianum (syn. *H. subpuniceum*)
Rock bloodspot

DESCRIPTION: Thallus rimose-areolate, creamy white to brownish gray, without soredia. Apothecia 0.5–1.5 mm in diameter, at first immersed in the thallus but becoming superficial with thick, thalline margins; spores slightly bent to strongly curved or S-shaped and twisted in the ascus, (2–)4–8-celled, 20–30 × 3.5–5 μm; conidia thread-like, curved, 13–18 × 0.9–1 μm. CHEMISTRY: PD– (rarely PD+ yellow), K+ yellow, KC–, C– (atranorin, sphaerophorin, isosphaeric acid, rarely with psoromic acid); red epihymenial pigment russulone. HABITAT: On siliceous rock. COMMENTS: The only other rock-dwelling bloodspot lichen in North America is **H. ochroleucum,** a very rare oceanic species known from the Pacific Northwest. Its thallus is sorediate (almost leprose) and contains zeorin and porphyrilic acid.

Haematomma persoonii
Sunken bloodspot

DESCRIPTION: Thallus creamy white to gray, smooth to rimose or verruculose, without soredia. Apothecia usually sunken (like those of *Aspicilia*), 0.3–1.5 mm in diameter, with scarlet to dark orange-red disks that often grow together; spores 7–8-celled, straight to curved and worm-like, thicker at one end, 30–50(–55) × 3.5–5 μm; conidia thread-like, curved, 16–20 × 1 μm. CHEMISTRY: Thallus PD–, K+ yellow, KC–, C–, UV+ white (atranorin, sphaerophorin); red hymenial pigment russulone. HABITAT: On the bark of a variety of tropical or subtropical deciduous trees and shrubs. COMMENTS: See Comments under *H. accolens*.

Heppia (3 N. Am. species)
Soil ruby lichens, earthscales

DESCRIPTION: Thallus squamulose or rarely granular, dark olive to brownish or gray-brown, most species

with a well-developed upper cortex and some with a loosely organized lower cortex; medullary hyphae arranged vertically, with cylindrical cells expanding to oval cells at the top, like a string of beads. Photobiont blue-green (*Scytonema*) usually in vertical rows within the thallus. Apothecia red-brown (ruby-red when wet), immersed in depressions in the thallus, sometimes very deeply concave; spores colorless, 1-celled, ellipsoid, about 20 × 10 µm, 8 per ascus. CHEMISTRY: All reactions negative (no lichen substances). HABITAT: On bare soil or, less frequently, on rock, in arid sites. COMMENTS: With their deeply sunken, red-brown apothecia, the common species of *Heppia* closely resemble tiny versions of the arctic-alpine genus *Solorina*, which contains green algae as the primary photobiont. *Peltula*, which is very similar to *Heppia*, has numerous tiny spores per ascus. *Placidium* species contain green algae and have black dots on the surface, showing the buried perithecia.

KEY TO SPECIES: See Key G.

Heppia conchiloba
Common soil ruby

DESCRIPTION: Thallus pale brown to olive-brown but usually covered to some extent with a fine gray pruina; it may even be scabrose, making the thallus appear gray; squamules large and foliose, up to 7 mm across, lifting from the soil, often turned down at the margins; upper cortex quite thick (up to 50 µm). Apothecia immersed in thallus, 1–1.5 mm in diameter, 1 to several per lobe; disk level with surface but often at the bottom of a depression; hymenium IKI– or reddish above, but IKI+ blue on the lower third and into the subhymenial tissues. DISTRIBUTION: Endemic to western North America from southern California, Arizona, and Colorado to South Dakota, but precise range unknown. COMMENTS: The only other common *Heppia* in North America is **H. adglutinata**, known from Arizona to Tennessee, a species with small, usually round, more closely attached squamules that remain olive in color, lacking any pruina. Its upper cortex is less well developed, and the hymenium and subhymenial tissues are almost entirely IKI–. *Heppia lutosa* is very rare (only from a few western lo-

355. *Haematomma persoonii* hill country, Texas ×2.5

356. *Heppia conchiloba* (damp) hill country, central Texas ×2.1

357. *Heterodermia albicans* Gulf Coast, Florida ×2.7

calities) and has a black, mainly granular, gelatinous thallus lacking any cortices, resembling a small *Collema* except for the photobiont. Its hymenium and ascus walls turn uniformly blue in IKI.

Heterodermia (25 N. Am. species)
Fringe lichens, centipede lichens

DESCRIPTION: Small to medium-sized, pale gray, foliose lichens with narrow lobes that typically broaden at the tips, almost always fringed with cilia that are unbranched or abundantly branched; upper cortex composed of prosoplectenchyma with elongate cells running parallel to the thallus surface; lower surface white or pigmented, with or without a cortex. Photobiont green (*Trebouxia?*). Apothecia lecanorine, with dark brown disks and prominent margins; spores brown, 8 per ascus, 2-celled with very thick walls, leaving round locules that often bud off to create additional smaller locules. CHEMISTRY: Cortex K+ yellow (atranorin); zeorin is a frequent accessory substance in the medulla, together with other compounds such as norstictic and salazinic acids. HABITAT: On trees and rocks, occasionally on soil. COMMENTS: *Heterodermia* is most similar to *Physcia* in size and color. It differs mainly in the structure of the upper cortex (pseudoparenchyma in *Physcia*), and in spore type (primarily *Physcia*-type spores in *Physcia*). The structure of the upper cortex in *Heterodermia* gives the surface the appearance of flowing toward the lobe tips, whereas the upper surface in *Physcia* has a uniform, unoriented appearance.

KEY TO SPECIES: See *Physcia*.

Heterodermia albicans
White fringe lichen

DESCRIPTION: Lobes 0.5–1.5 mm wide, with coarse white soredia forming all along the lobe margins; lower surface corticate, white to pale tan, smooth and almost shiny, with numerous pale rhizines; marginal cilia (or marginal rhizines) sparse, pale, short, and unbranched. CHEMISTRY: Medulla PD+ yellow-orange, K+ yellow to red, KC–, C– (sala-

358. *Heterodermia appalachensis* Chisos Mountains, southwestern Texas ×2.5

zinic acid). HABITAT: Frequent on hardwood trees, especially oaks; rarely on rock. COMMENTS: *Heterodermia albicans* resembles *H. speciosa* and *H. obscurata* in size and habit, but in the latter two species, the soralia are terminal or crescent-shaped on special lobes, the cilia tend to be more conspicuous, and salazinic acid is absent.

Heterodermia appalachensis
Appalachian fringe lichen

DESCRIPTION: Lobes narrow (0.7–1.5 mm), elongate, partly ascending, with long, pale cilia (darkening at the tips) all along the margins; lobe tips curling back, revealing patches of coarsely granular soredia on the lower surface, which is powdery or webby (without a cortex), almost completely white, but with a pale sulphur-yellow pigment close to tips of many lobes. CHEMISTRY: Medulla and lower surface PD–, K+ yellow, KC–, C– (only atranorin); the yellow pigment is K–. HABITAT: On bark of hardwoods. COMMENTS: The similar *H. leucomela* is K+ red in the medulla (salazinic acid) and has no yellow pigment below. *Heterodermia obscurata* has marginal rhizines (abundantly branched), and the orange lower surface reacts K+ purple.

Heterodermia crocea
Orange-bellied fringe lichen

DESCRIPTION: Closely resembles *H. obscurata*, with a yellow to orange lower surface but with isidia instead of soredia; isidia slender and cylindrical, occasionally branched, produced on the upper surface of the lobes and along the margins; black rhizines predominantly marginal, unbranched or squarrose, sometimes forming a black hypothalline mat (as in *H. obscurata*). CHEMISTRY: Medulla PD–, K+ yellow, KC–, C– (triterpenes); pigment on lower surface K+ red-purple (anthraquinone). HABITAT: On bark. COMMENTS: Another isidiate *Heterodermia*, **H. granulifera,** is common in the southeast. It has somewhat inflated, almost granular isidia on the upper surface of the lobes and a white corticate lower surface; its medulla reacts PD+ orange, K+ red (salazinic acid).

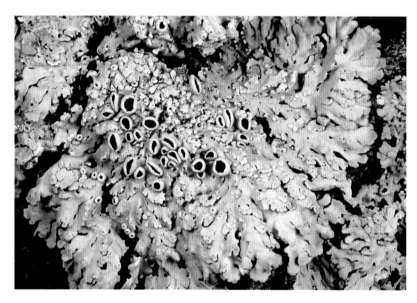

359. *Heterodermia crocea* northern Florida ×2.4

360. *Heterodermia diademata* Chisos Mountains, southwestern Texas ×2.3

Heterodermia diademata
Cupped fringe lichen

DESCRIPTION: Thallus very pale gray to almost white, shiny or with only a light dusting of pruina; lobes mostly 1–2 mm wide, often developing small, round lobules along the margins; cilia sparse; lower surface white to gray, with a cortex; rhizines fairly abundant, usually branched at the tips. Apothecia common, bowl-shaped, 1.5–5 mm in diameter, with dark brown disks and in-turned margins, which sometimes become toothed or lobulate. CHEMISTRY: Medulla PD+ pale yellow, K+ yellow, KC–, C– (zeorin). HABITAT: On hardwoods, especially oak. COMMENTS: *Heterodermia diademata* resembles the eastern species, *H. hypoleuca*, in lacking soredia, but *H. hypoleuca* has no lower cortex, has more abundant cilia, and is more ascending. *Heterodermia rugulosa*, often found in the same habitats with *H. diademata*, is always heavily pruinose, at least on the lobe tips, and has orange spots throughout the medulla.

Heterodermia echinata
Flowering fringe lichen

DESCRIPTION: Thallus whitish gray to gray-green, with ascending lobes forming almost fruticose tufts; lobes 1–3 mm wide, without soredia or isidia, but with abundant, long, unbranched or branched marginal cilia; lower surface white, webby (without a cortex), without rhizines; lobes usually terminating with a large, heavily pruinose apothecium (dark brown beneath the pruina), 1.5–5 mm in diameter, with toothed or ciliate margins. CHEMISTRY: Medulla PD– or pale yellow, K+ yellow, KC–, C– (zeorin). HABITAT: Common on branches of juniper bushes and other trees and shrubs. COMMENTS: The California species *H. erinacea*, also with ascending ciliate lobes, is the most similar to *H. echinata*. Its apothecia are produced on the upper surface of the lobes rather than at the tips. They are heavily pruinose, as in *H. echinata*, but smaller (0.5–2 mm in diameter), and the margins are smooth, without tooth-like lobules or cilia. In the southeast, *H. leucomela* might be mistaken for *H. echinata*, but the lobe tips are sorediate on the lower surface, apothecia are rarely produced, and the medulla contains salazinic acid (K+ red).

361. *Heterodermia echinata* Ouachitas, Arkansas ×3.7

362. *Heterodermia erinacea* Channel Islands, southern California ×2.7

Heterodermia erinacea
Coastal fringe lichen

DESCRIPTION: Thallus with ascending abundantly ciliate branches, 0.5–2 mm wide, without soredia; marginal cilia are very long and hair-like; lower surface white, webby, without a cortex. Apothecia common, slightly raised, laminal (not at the lobe tips); disks black, usually pruinose; margins smooth, almost always lacking cilia. CHEMISTRY: Medulla PD+ yellow or PD–, K+ yellow, KC–, C– (zeorin). HABITAT: On shrubs and rocks along the foggy coast. COMMENTS: *Heterodermia namaquana* is a similar but much rarer species with a sorediate lower surface and somewhat broader lobes. It is also found on the California coast. See Comments under *H. echinata*.

363. *Heterodermia hypoleuca* Appalachians, Virginia ×1.8

Heterodermia hypoleuca
Cupped fringe lichen

DESCRIPTION: Lobes broad, up to 3 mm wide, without soredia or isidia, but with many flat to convex lobules along the margins and apothecial margins; marginal cilia abundant to very sparse (as shown in plate 363); lower surface white, webby (without a cortex), with brown, squarrose rhizines on or close to the margins sometimes coalescing into a brownish mat. Apothecia common, deeply concave and cup-like, with dark brown, nonpruinose disks. CHEMISTRY: Medulla PD– or pale yellow, K+ strong yellow, KC– or KC+ yellow-orange, C– (zeorin). HABITAT: On bark of hardwoods. COMMENTS: This fringe lichen is much like *H. obscurata* and *H. speciosa* but without soredia. Like *H. squamulosa*, it can become somewhat lobulate along the margins, but its lobules are not erect or finely divided as they are in *H. squamulosa*. **Heterodermia rugulosa** is a southwestern species similar to *H. hypoleuca*, but its upper surface can be quite pruinose, and its lower surface is smooth and tan (with a cortex) and produces rhizines. Most important is the presence of orange, K+ deep purple areas in the medulla (sometimes confined to lobe tips). The white parts of the medulla are K–. *Heterodermia diademata* is another nonsorediate, nonisidiate *Heterodermia*. It has a lower cortex similar to that of *H. rugulosa* but lacks the orange areas in the medulla.

Heterodermia leucomela (syn. *H. leucomelaena, H. leucomelos*)
Elegant fringe lichen, elegant centipede

DESCRIPTION: Thallus with long, narrow, branched lobes, 0.5–1 mm wide and 20 mm or more long, lifting from the substrate and creating elegant, fruticose-appearing clumps; granular soredia produced on the lower surface close to the lobe tips; marginal cilia black, 2–5 mm long, often branching once or twice at the tips; lower surface white, cottony or heavily pruinose, without a cortex. CHEMISTRY: Medulla (and soralia) PD+ yellow, K+ red, KC–, C– (salazinic acid). HABITAT: On mossy hardwoods or rock faces in fairly well-lit woodlands. COM-

364. *Heterodermia leucomela* Channel Islands, southern California ×3.0

MENTS: This is not a common lichen, but it is very striking. *Heterodermia appalachensis* is very similar but is pale yellow on the lower surface and lacks salazinic acid (K+ persistent yellow, not turning red). *Heterodermia echinata*, *H. erinacea*, and **H. namaquana** have broader lobes, but only *H. namaquana* has soredia (see Comments under *H. erinacea*). *Anaptychia setifera* can resemble *H. leucomela* in its narrow ascending branches with long marginal cilia, but its branches are dusky brownish gray, more convex, and K– (lacking atranorin), and it does not produce soredia.

cilia and rhizines black, conspicuous, often squarrose, sometimes forming a mat. CHEMISTRY: Medulla PD–, K+ yellow (except for yellow pigment, which is K+ red-purple), KC–, C– (anthraquinone pigments). HABITAT: On bark of hardwoods, occasionally on shaded rocks. COMMENTS: *Heterodermia speciosa* is very similar but has a white corticate lower surface. *Heterodermia crocea* is an isidiate species with an orange lower surface. See Comments under *H. speciosa* and *H. albicans*.

Heterodermia obscurata
Orange-tinted fringe lichen

DESCRIPTION: Thallus with lobes mostly 1–2 mm wide, with gray to blue-gray soralia containing granular soredia on the underside of the lobe tips and on short crescent-shaped side lobes; lower surface yellow or orange, sometimes faint and only in spots, webby (lacking a lower cortex); marginal

Heterodermia speciosa
Powdered fringe lichen, powdered centipede

DESCRIPTION: Lobes slightly broadened at the tips, 0.7–1.5 mm wide; abundant crescent-shaped soralia with powdery soredia produced on the lower surface of lobe tips; abundant, white or tan, branched cilia extending from all lobe margins and sometimes coalescing into a mat; lower surface white to tan, with a cortex (better developed in some spots than in others),

365. *Heterodermia obscurata* (damp) Ozarks, Oklahoma ×2.1

366. *Heterodermia speciosa* (damp) Great Smoky Mountains, Tennessee ×3.2

smooth but slightly fibrous (not webby) in most areas of the lower surface. CHEMISTRY: Medulla PD– or + yellow, K+ yellow, KC± yellow, C– (zeorin). A chemical race containing norstictic acid (PD+ yellow, K+ red) is sometimes called **H. pseudospeciosa**. HABITAT: On trees or rocks, usually in humid or well-established forests. COMMENTS: This is one of the most common of the fringe lichens. In the very similar *H. obscurata*, the cilia are black with white tips, the lower surface is orange, at least at the center, and there is no lower cortex. *Heterodermia albicans* differs chemically (K+ red owing to salazinic acid). Specimens resembling *H. speciosa* but with an obscurely corticate, brownish to purplish black lower surface (except close to the edge) may be **H. casarettiana,** a southeastern species that often contains norstictic acid (medulla K+ red). Specimens like *H. speciosa* but lacking any marginal cilia may be ***Physcia pseudospeciosa***, a rare Appalachian species containing only atranorin. See also *H. albicans*.

Heterodermia squamulosa
Scaly fringe lichen

DESCRIPTION: Lobes 0.8–2 mm wide, overlapping to some extent, the margins and occasionally the upper surface divided into tiny squamules (sometimes so small they look like granules or even soredia); lower surface webby (no cortex), white at the margins but darkening to purplish black in the center in most specimens; rhizines blackening, squarrose to tufted, marginal, fairly abundant or sparse, often forming something resembling a black mat surrounding the lobes. CHEMISTRY: Medulla PD–, K+ yellow, KC– and C– (zeorin). HABITAT: On trees, especially mossy tree bases, in hardwood forests. COMMENTS: The purplish lower surface resembles that of *H. casarettiana*, a southeastern sorediate species that often contains norstictic acid (K+ red).

Hubbsia (1 N. Am. species)

KEY TO SPECIES: See Key A.

Hubbsia parishii (syn. *Reinkella parishii*)
Brittle bag lichen

DESCRIPTION: Thallus fruticose, erect, with creamy white to yellowish gray, hollow, inflated lobes mostly 10–20 mm high, extremely brittle, surface dull, dissolving into granular soredia toward the tips of the branches. Photobiont green (*Trentepohlia*). Ascomata rare, script-like, branched and radiating, 4–5 × 0.1–0.2 mm, commonly immersed in the thallus but later somewhat prominent, retaining the thalline covering on the margins; spores colorless, fusiform, 4–5-celled, 16–17 × 4–5 μm. CHEMISTRY: Cortex PD–, K+ yellowish, KC+ red, C+ red; medulla PD–, K–, KC+ red, C+ red (erythrin and lecanoric acid). HABITAT: On exposed

367. *Heterodermia squamulosa* Appalachians, North Carolina ×2.2

368. *Hubbsia parishii* Channel Islands, southern California ×2.1

369. *Hydrothyria venosa* (photographed underwater) Cascade Range, Oregon ×2.5

rocks near the coast. COMMENTS: This is one of the many wild and wonderful coastal Californian lichens. A second species of *Hubbsia*, *H. californica,* is restricted to the Baja California area and Guadalupe Island. It has solid lobes and more abundant ascomata, and lacks soredia. *Hubbsia parishii* slightly resembles sterile *Schizopelte californica*, a closely related pale gray fruticose lichen from southern California to Baja California, also on coastal rocks and C+ red, but with taller, branched stalks that are solid. Fertile specimens of *Schizopelte* have broad, black to violet pruinose apothecia at the branch tips.

Hydrothyria (1 species worldwide)

KEY TO SPECIES: See Keys H and I, and *Leptogium*.

Hydrothyria venosa
Waterfan

DESCRIPTION: Aquatic jelly lichen with fan-shaped lobes 3–10 mm wide; translucent dark green or brownish when under water, much like a seaweed; dark blue-gray when dry; lower surface of most lobes with smooth, pale, branched veins composed of elongate colorless hyphae; both upper and lower surfaces covered with a colorless cortex of pseudoparenchyma; medulla rather dense and thin. Photobiont blue-green (*Nostoc*). Apothecia common on the upper surface of the lobes, biatorine, orange or red-brown, convex, and without margins when mature. Spores colorless, fusiform, 4-celled, 8 per ascus. CHEMISTRY: Negative to reagents, but Appalachian specimens contain methyl gyrophorate and methyl lecanorate; the western populations lack any lichen substances. HABITAT: Attached to rocks in cool mountain brooks and streams. COMMENTS: *Hydrothyria* must grow entirely submerged in fresh water and so is often overlooked. Another aquatic foliose species, *Dermatocarpon luridum*, can survive quite well on rocks at the stream edge where it is only periodically submerged; it has green algae, and its fruiting bodies are perithecia buried in the thick thallus. *Hydrothyria*, found only in North America, looks like a large blue-gray *Leptogium* when dried, but the thin conspicuous veins on the underside are unique. *Leptogium rivale*, the only other truly aquatic jelly lichen, occurs in the Cascades, Sierra, and Colorado. It has elongate lobes, 0.2–1.5 mm wide, forming small rosettes up to 2 cm across that are tightly attached to rocks in and close to the water.

Hyperphyscia (3 N. Am. species)
Shadow-crust lichens

DESCRIPTION: Small, tightly adnate, foliose lichens, almost crustose in appearance; brownish gray to gray, dull; lower surface pale to black, without (or with few) rhizines. Photobiont green (*Trebouxia?*). Apothecia lecanorine, with dark brown to black nonpruinose disks; spores brown, 2-celled, *Physcia*-type, 8 per ascus. CHEMISTRY: Cortex PD–, K–, KC–, C–; medulla too thin to test (no lichen substances). HABITAT: On bark or wood in well-lit woodlands or on isolated trees. COMMENTS: Superficially similar species of *Phaeophyscia* are rarely as closely attached to the substrate, and they have rhizines.

KEY TO SPECIES: See *Phaeophyscia*.

Hyperphyscia adglutinata
Grainy shadow-crust lichen

DESCRIPTION: Lobes 0.5–1 mm across, confluent, with green granular soredia along the upturned margins of the older central parts of the thallus; lower surface pale. Apothecia rare. HABITAT: On bark of hardwoods. COMMENTS: Closely related *H. minor* has a black prothallus and dark lower surface, and the soredia are laminal.

Hyperphyscia syncolla
Smooth shadow-crust lichen

DESCRIPTION: Thallus forming small patches 1–2 cm across; lobes 0.5–1 mm wide, smooth and nonpruinose; without soredia or isidia; lower surface white. Apothecia abundant, dark brown, 1–2.5 mm in diameter, with prominent margins. HABITAT: On bark, especially roadside trees. COMMENTS: This species resembles a *Lecanora* or *Rinodina* until you notice the well-camouflaged lobes, almost the same color as the bark. To be sure, make a section through the thallus and look for the lower cortex.

370. *Hyperphyscia adglutinata* (damp) central Nebraska ×5.4

Hypocenomyce (7 N. Am. species)
Clam lichens

DESCRIPTION: Small squamulose or dispersed-areolate crustose lichens, with a thallus resembling the squamules of *Cladonia*, usually overlapping but sometimes scattered, often sorediate. Photobiont green (*Trebouxia*). Apothecia biatorine, black or very dark brown, flat to convex; margin thin, sometimes disappearing in mature apothecia; asci with a *Biatora*-like, weakly K/I+ blue tip; spores colorless, 1-celled in North American species, ellipsoid, 8 per ascus. Conidia long or short bacilliform to ellipsoid, 2.5–10 μm long. CHEMistry: Some species contain lecanoric acid, fumarprotocetraric acid, alectorialic acid, or some unidentified compounds that are UV+ white. HABITAT: On the wood or bark of conifers or birch, especially charred logs or stumps. COMMENTS: The overlapping scales of *Hypocenomyce* species closely resemble the primary squamules of *Cladonia*, and if apothecia are absent, as is often the case, the two genera can easily be confused. Chemistry distinguishes *H. scalaris* (no *Cladonia* is C+ pink), and *Cladonia* species rarely have the gray soredia characteristic of *H. anthracophila*.

KEY TO SPECIES: See Key G.

"Hypocenomyce" anthracophila
(syn. *Biatora anthracophila*, *Psora anthracophila*)
Small clam lichen

DESCRIPTION: Thallus consists of greenish brown to brown, rather shiny squamules usually less than 0.8 mm in diameter, scattered or overlapping, flat or concave when young to convex with the edge turned up when mature, revealing very fine gray soredia in a lip-shaped soralium. Apothecia reddish brown, 0.4–0.8 mm in diameter, often absent. CHEMISTRY: Medulla PD+ red, K–, KC–, C–, UV+ white (fumarprotocetraric and protocetraric acids and an unidentified UV+ substance). HABITAT: On conifer wood. COMMENTS: "*Hypocenomyce*" *anthracophila* superficially resembles other species of *Hypocenomyce* but has thread-like conidia, expanded and pigmented paraphyses tips, and a somewhat different ascus tip (almost *Porpidia*-type). It has been placed in *Biatora* by some lichenologists, but its final classification is still to be determined. See also Comments under *H. scalaris*.

Hypocenomyce scalaris (syn. *Psora scalaris*)
Common clam lichen

DESCRIPTION: Thallus composed of overlapping olive-green to brownish, convex squamules 0.4–1.0 mm wide with greenish brown to yellowish powdery so-

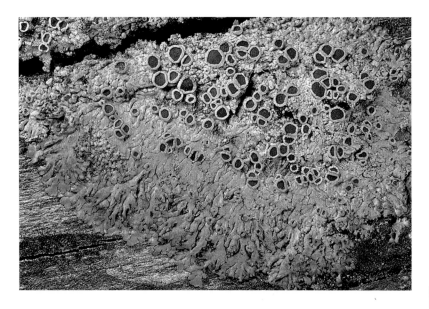

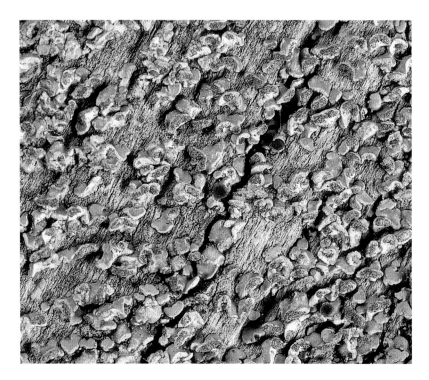

371. *Hyperphyscia syncolla* (Yellowish green color is due to epiphytic algae on the thallus.) western Mississippi ×2.6

372. "*Hypocenomyce*" *anthracophila* herbarium specimen (Canadian Museum of Nature), Michigan ×8.6

redia emerging from the underside of each squamule and fringing the margin. Apothecia black, usually rare (although abundant in the specimen shown in plate 373). CHEMISTRY: Thallus PD–, K–, KC+ red, C+ pink or red (lecanoric acid). HABITAT: Typically on charred wood, but also on conifer bark and other acid-barked trees. COMMENTS: Chemistry distinguishes *H. scalaris* from the smaller, browner, also sorediate *"Hypocenomyce" anthracophila*, which is PD+ red, K–, KC–, C– (fumarprotocetraric acid), or *H. friesii* (syn. *Psora friesii*), a nonsorediate shingle lichen that is PD–, K–, KC+ pink, C– and often produces black apothecia. In the west, there is a brownish shingle lichen *"H." castaneocinerea* that is very similar to *H. anthracophila* but has a PD– medulla and brown-edged, convex squamules; in *H. anthracophila*, the squamules are often white-edged and are flat or slightly concave when young. In *H. anthracophila, H. castaneocinerea,* and *H. friesii* the medulla is brightly fluorescent in UV light owing to unidentified substances. These three species have almost the same habitat requirements as *H. scalaris*. The rare, boreal *H. leucococca* has closely appressed, creamy white areoles that are granular sorediate at the edges and react PD+ yellow, K+ yellow, KC+ red, C+ pink (alectorialic acid). It prefers old birches in partially shaded woodlands. In the arid interior and southwest, one can encounter *H. xanthococca*, a yellowish gray species with convex areoles that are almost squamulose in places, richly fruiting with flat, ebony black apothecia having thin, raised margins; it also contains alectorialic acid.

373. *Hypocenomyce scalaris* (Sorediate squamules in the lower part of the photograph are most typical of the species.) Lake Superior region, Ontario ×5.3

Hypogymnia (23 N. Am. species)
Tube lichens, bone lichens, pillow lichens

DESCRIPTION: Medium to large foliose lichens, usually greenish gray to brownish gray, with more or less inflated lobes that are hollow (tube-like), often perforated with a hole at the lobe tip; appressed or ascending; ceiling of the inside of the tube (the medullary ceiling) dark brown or white (a good character for distinguishing one species from another); lower surface of the lobes black, smooth or wrinkled, without any rhizines. Photobiont green (*Trebouxia?*). Apothecia lecanorine, constricted at the base and often raised or urn-like; disks red-brown, usually concave; spores colorless, 1-celled, ellipsoid, 8 per ascus; pycnidia black, appearing as black dots on the lobe surface and very conspicuous in some species. CHEMISTRY: Cortex K+ yellow (atranorin); medulla K–, C–; most species contain physodic acid (KC+ pink), often with other orcinol and β-orcinol depsidones including protocetraric and physodalic acids (PD+ red). HABITAT: On bark and wood especially of conifers, less frequently on rock or mossy soil. COMMENTS: *Menegazzia* and *Cavernularia* superficially resemble species of *Hypogymnia*: *Menegazzia* has perforations on the upper lobe surface, and *Cavernularia* has pits on the lower surface. *Brodoa* and *Allantoparmelia* can also resemble *Hypogymnia*, but they have solid, not hollow, lobes.

KEY TO SPECIES (including *Menegazzia*)

1. Lobes perforated on the upper surface with round holes; medulla PD+ orange, K+ yellow, AI+ blue (stictic acid) . . . . *Menegazzia terebrata*
1. Lobes perforated at tips, or not perforated at all; medulla PD– or PD+ red, K–, AI– (stictic acid absent) . . . 2

2. Soredia present. 3
2. Soredia absent 7

3. Soredia on expanded, turned-back lobe tips (lip-shaped soralia) 4

3. Soredia on the thallus surface, sometimes close to the lobe tips (but not in lip-shaped soralia)............ 5
4. Medullary ceiling dark brown to grayish black; lobules present on the lobe margins; medulla PD–........... .... *Hypogymnia vittata*
4. Medullary ceiling white; lobules absent; medulla PD+ red (physodalic and protocetraric acids) .... *Hypogymnia physodes*
5. Thallus ascending; lobes elongated, rarely perforated; soredia farinose, at tips of lobes .... *Hypogymnia tubulosa*
5. Thallus closely appressed to substrate; lobes rounded, square, or truncated, perforated at tips; soredia granular, on thallus surface or tips of lobes.............. 6
6. Soredia, for the most part, on the upper surface of the lobe tips .... *Hypogymnia bitteri*
6. Soredia, for the most part, on the older parts of the thallus surface .... *Hypogymnia austerodes*
7.(2) Eastern U.S. and Canada (Appalachians).......... .... *Hypogymnia krogiae*
7. Western or northern North America 8
8. Medullary ceiling white........................ 9
8. Medullary ceiling light to dark brown 13
9. On soil or rock (rarely wood) 10
9. On conifer bark or wood 11
10. Lobules present along the lobe margins; pycnidia sparse or very inconspicuous; in alpine or arctic tundra..... .... *Hypogymnia subobscura*
10. Lobules absent, pycnidia abundant and conspicuous; commonly in the mountains..................... .... *Hypogymnia metaphysodes*
11. Thallus loosely attached, ascending or pendent, sometimes forming round cushions of overlapping lobes; lobes convex, perforated at tips; medulla PD+ red (diffractaic, physodalic, and protocetraric acids present) 12
11. Thallus closely appressed to substrate, forming round, flat rosettes; lobes flat or concave, rarely perforated; medulla PD– or sometimes PD+ red (diffractaic and protocetraric acids absent, physodalic acid sometimes present) .... *Hypogymnia metaphysodes*
12. Lobes forked in rather regular dichotomies, more or less even in width, not constricted at intervals; thallus rather stiff and shrubby; apothecia abundant........ .... *Hypogymnia imshaugii*
12. Lobes branching irregularly, constricted at irregular intervals; thallus usually somewhat pendent, with upturned tips; apothecia rare .... *Hypogymnia duplicata*
13.(8) Lobules present along the lobe margins......... 14
13. Lobules absent............................. 16

14. Medulla PD– (physodalic and protocetraric acids absent); lobes usually less than 3 mm wide .... *Hypogymnia occidentalis*
14. Medulla PD+ red (physodalic and protocetraric acids); lobes usually more than 3 mm wide at the tips 15
15. Lobules strap-shaped; diffractaic acid absent; rare, in coastal forests............ *Hypogymnia heterophylla*
15. Lobules rounded; diffractaic acid present; very common in Pacific Northwest .... *Hypogymnia enteromorpha*
16.(13) Lobes rarely perforated..................... 17
16. Lobes perforated at tips 18
17. Older lobes becoming very wrinkled, often sinuous, convex throughout; apothecia abundant; medulla PD– (hypoprotocetraric acid present).. *Hypogymnia rugosa*
17. Lobes relatively smooth; lobe tips concave; apothecia occasional; medulla PD– or PD+ red (hypoprotocetraric acid absent) .... *Hypogymnia metaphysodes*
18. Lobes 1.5–2 mm wide, more or less even, not constricted at intervals; medulla KC+ red (diffractaic and physodic acids) .... *Hypogymnia inactiva*
18. Lobes 3–4 mm wide, constricted at irregular intervals; medulla KC– (diffractaic and physodic acids absent).. .... *Hypogymnia apinnata*

Hypogymnia apinnata
Beaded tube lichen

DESCRIPTION: Lobes elongate to short, constricted at intervals, appressed to the substrate or loosely attached and partly pendent; lobes 3–4 mm wide, usually perforated with a hole at each lobe tip; lacking soredia and isidia; usually without lobules; medullary ceiling dark. Apothecia common, raised. CHEMISTRY: Medulla PD–, K–, KC–, C– (atranorin alone). HABITAT: On conifers in somewhat exposed to protected sites but not in deep shade. COMMENTS: *Hypogymnia apinnata* is almost identical to *H. enteromorpha*, a species with small rounded lobules along the branches and almost invariably with a PD+ red medulla (physodalic acid). Some specimens with lobules that are PD– may be physodalic acid-deficient specimens of *H. enteromorpha* or lobulate specimens of *H. apinnata*. The two species have almost identical distributions. *Hypogymnia rugosa* can be very similar but has a strongly wrinkled surface on older parts of the thallus, the lobes do not have intermittent constrictions, and the medulla contains hypoprotocetraric and physodic acids (KC+ pink).

374. *Hypogymnia apinnata* Southeast Alaska ×2.0

375. *Hypogymnia austerodes* southwestern Yukon Territory ×3.5

Hypogymnia austerodes
Varnished tube lichen

DESCRIPTION: Thallus pale to dark reddish brown and very shiny, at least at the lobe tips, sometimes pale yellowish gray elsewhere, but strongly mottled with black on the lobe margins, sometimes merging into almost solid black on older portions of the thallus; lobes 1–2.5 mm wide, perforated at the tips; soralia with coarse brownish soredia developing on the central parts of the thallus and sometimes also at the lobe tips, finally coalescing; spherical lobules or tiny isidia-like pustules forming on the upper surface of some specimens. Apothecia rare. CHEMISTRY: Medulla PD–, K–, KC+ pink, C– (oxyphysodic and physodic acids, sometimes with 3-hydroxyphysodic acid). HABITAT: On conifer wood and bark, less frequently on rock or soil, usually at high elevations. COMMENTS: *Hypogymnia oceanica,* an uncommon coastal lowland species on conifer branches, also has coarse soredia on the upper surface of the lobes, but the thallus is very pale and the medulla is PD+ red (physodalic and protocetraric acids).

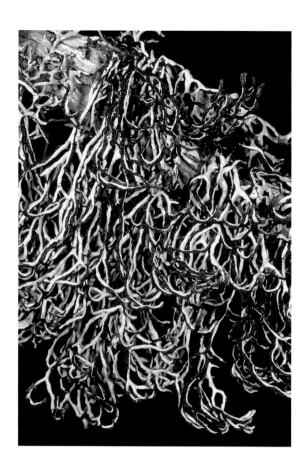

376. *Hypogymnia bitteri* coastal Alaska ×1.7

377. *Hypogymnia duplicata* southeast Alaska ×1.2

Hypogymnia mollis is a very rare lichen restricted to a small area in southern California. It is pale greenish gray and rather adnate, with abundant soredia produced on older portions of the thallus surface; the medulla is PD–. Specimens of *H. austerodes* with sparse soredia can be mistaken for *H. subobscura*, a very similar nonsorediate arctic species that grows on soil, rocks, and moss, and contains physodic and vittitolic acids (PD–, K–, KC+ red, C–). See Comments under *H. bitteri*.

Hypogymnia bitteri
Powdered tube lichen

DESCRIPTION and COMMENTS: *Hypogymnia bitteri* and *H. austerodes* are two similar sorediate species that are often mistaken for each other. Both can be closely appressed, with greenish gray to brown lobes 2–5 mm wide, appearing almost varnished, especially at the tips. *Hypogymnia bitteri* tends to be paler and less browned than *H. austerodes*. Both species have coarse soredia on the upper surface. In *H. bitteri*, the soredia are mainly (sometimes entirely) limited to the lobe tips and are very powdery. In *H. austerodes*, they develop in the oldest parts of the thallus from a disintegration of the upper cortex and only rarely occur on the lobe tips. HABITAT: Normally only on conifer bark and wood at high elevations. (*Hypogymnia austerodes* can occur on rocks or soil as well.) CHEMISTRY: Medulla PD–, K–, KC+ pink, C– (physodic acid). (*Hypogymnia austerodes* contains oxyphysodic acid in addition to physodic acid.)

Hypogymnia duplicata
Ticker-tape lichen

DESCRIPTION: Lobes long and slender, under 1 mm wide, largely pendent, gracefully curving upward at the tips; branching irregular (not strictly dichotomous); medullary ceiling white; without soredia or isidia. Apothecia rare. CHEMISTRY: Medulla PD+ red, K–, KC+ pink, C– (physodalic, protocetraric, and diffractaic acids). HABITAT: Most commonly found on the branches or trunks of conifers in coastal bogs or fens. COMMENTS: The pendent curved branches of *H. duplicata* are very distinctive. The medullary ceiling is also white in *H. imshaugii*, a species with broader, rather stiff, uncurved lobes and regular dichotomous branching.

Hypogymnia enteromorpha
Budding tube lichen, gut lichen

DESCRIPTION: A large lichen with lobes commonly 3–6 mm wide and 5–10 cm long, the lobes relatively short and round on horizontal surfaces but becoming quite elongated on vertical surfaces; irregularly branched; branches alternately constricted and bloated, with numerous tiny round lobules along the margins; medullary ceiling dark. CHEMISTRY: Medulla PD+ red, K–, KC+ pink, C– (protocetraric, physodalic, physodic, and diffractaic acids). HABITAT: On conifer bark or over dry wood in full sun or partial shade. COMMENTS: The budding tube lichen is the most common and conspicuous non-sorediate tube lichen in many parts of the west and also one of the largest. *Hypogymnia imshaugii* is another common western PD+ red tube lichen. It and its PD– counterpart, *H. inactiva*, have neatly forked branches that are usually distinctly ascending, especially when growing on twigs. In addition, *H. imshaugii* has a white medullary ceiling. *Hypogymnia duplicata* is much more slender and pendent, turning up at the tips. Several PD– lichens with irregular branching are similar to *H. enteromorpha*: *H. occidentalis* also has round lobules but is smaller, without constrictions in the branches; *H. apinnata* lacks the small lobules along the margins; and *H. metaphysodes* is much more flattened to the substrate than are the other two species and often has somewhat concave or upturned lobe tips.

Hypogymnia heterophylla
Seaside tube lichen, seaside bone lichen

DESCRIPTION: Thallus fairly large, forming clumps 5–8 cm long; lobes typically variable in width, with very narrow segments (1–2 mm wide) abruptly expanding into flattened lobes up to 6 mm across and then narrowing again, often with long, narrow (not round) lobules, constricted at the base, arising from the branch margins; medullary ceiling dark; soredia and isidia absent. Apothecia common, up to 8 mm across, raised on short, broad stalks. CHEMISTRY: Medulla PD+ red, K–, KC+ red, C– (physodalic, protocetraric, and physodic acids). HABITAT: On bark in open pine stands usually near the ocean. COMMENTS: *Hypogymnia heterophylla* resembles *H. imshaugii* and *H. inactiva* most closely, differing in the irregular branching and uneven lobe width. *Hypogymnia duplicata* can resemble specimens of *H. heterophylla* with especially long branches, but *H. duplicata* is more pendent, with upturned tips, and is PD–. *Hypogymnia enteromorpha* is "fatter," with more bloated (as opposed to flattened), intermittently constricted lobes. IMPORTANCE: *Hypogymnia heterophylla* is one of four lichens officially listed as "rare or endangered" in Canada.

378. *Hypogymnia enteromorpha* Cascades, Oregon ×1.7

379. *Hypogymnia heterophylla* northern California coast ×1.5

380. (facing page) *Hypogymnia imshaugii* Klamath Range, southwestern Oregon ×3.2

Hypogymnia imshaugii
Forked tube lichen

DESCRIPTION: Thallus gray to gray-green, often blackening in highly exposed sites, with relatively slender lobes (1.5–2 mm wide) that are regularly dichotomously branched and ascending; medullary ceiling white. Apothecia common, constricted at the base. CHEMISTRY: Medulla PD+ red, K–, KC+ pink, C– (physodalic, protocetraric, physodic, and diffractaic acids). HABITAT: Common on conifer branches, mainly in moderately dry inland habitats. COMMENTS: *Hypogymnia inactiva* is very similar to *H. imshaugii* but is PD– with a black medullary ceiling. It generally shows more black on the thallus, especially up the sides of the lobes and sometimes partially invading the edges of the upper surface. *Hypogymnia inactiva* is more common along the moister coastal zone than is *H. imshaugii*. *Hypogymnia heterophylla* has very irregular branching and a dark medullary ceiling. *Hypogymnia enteromorpha* is a larger lichen with broader, irregularly branched lobes that are constricted at intervals; its medullary ceiling is dark. See also *H. apinnata*.

Hypogymnia inactiva
Mottled tube lichen

DESCRIPTION and COMMENTS: *Hypogymnia inactiva* closely resembles *H. imshaugii*. The two are compared and described under the latter species. CHEMISTRY: Medulla PD–, K–, KC+ pink, C– (physodic acid). HABITAT: On the bark and branches of conifers in well-lit coastal and inland forests.

Hypogymnia krogiae
Freckled tube lichen

DESCRIPTION: Thalli forming round colonies 3–8 cm across with radiating, slender, overlapping lobes, 0.4–1.5 mm wide, flattened against the substrate but

with upturned tips, usually dichotomously branched, and invariably with holes at the tips; medullary ceiling white to dark brown, often mottled; black pycnidia typically abundant in conspicuous freckle-like clusters on the surface of lobes close to the tips. Apothecia common, raised, often with inflated bases. CHEMISTRY: Medulla PD+ red, K–, KC+ pink, C– (physodalic, protocetraric, and physodic acids). HABITAT: On spruce and fir trees in open or partially shaded forests. COMMENTS: This strictly North American lichen is the only nonsorediate species of the genus from the east, and no other nonsorediate *Hypogymnia* with small lobes grows in overlapping cushions. *Hypogymnia enteromorpha* and *H. apinnata*, both western species, are much larger and have constricted branches. A rare sorediate form of *H. krogiae* also occurs in the Appalachians. The soralia resemble those of *H. physodes*, but the thallus looks like *H. krogiae* in other respects.

381. (left) *Hypogymnia inactiva* coastal Range, northern California ×1.2

382. (below) *Hypogymnia krogiae* coastal Maine ×3.6

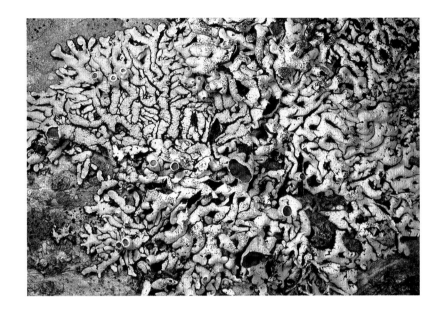

Hypogymnia metaphysodes
Deflated tube lichen

DESCRIPTION: Thallus pale yellowish gray with brown tips, sometimes mottled; lobes flattened (appressed) with slightly upturned or concave lobe tips 1.5–4 mm wide; typically short and divergent but occasionally elongate and irregularly branched; rarely perforated at the lobe tips; without soredia or lobules; medullary ceiling varying from white to very dark brown. CHEMISTRY: Medulla PD– or PD+ red, K–, KC+ pink, C– (physodic acid, sometimes accompanied by physodalic acid). HABITAT: On exposed conifer bark or wood usually at elevations of more than 1000 m, but occasionally found on alder or other trees at lower elevations. COMMENTS: The flattened fanning lobes of this species distinguish it from most other nonsorediate tube lichens. *Hypogymnia occidentalis* has small lobes (mostly under 2.5 mm across) and a dark medullary ceiling, but it produces many tiny lobules along the lobe margins. See also *H. rugosa*.

Hypogymnia occidentalis
Lattice tube lichen

DESCRIPTION: Thallus tends to be appressed to the substrate, with lobes typically under 3 mm wide; without soredia, but with small, rounded lobules sprouting from the sides of the lobes (seen best in the lower left of plate 384), especially close to the tips; upper surface smooth or strongly wrinkled on older lobes; lobe tips usually perforated with round holes; medullary ceiling black. CHEMISTRY: Medulla PD–, K–, KC+ red, C– (physodic acid). HABITAT: Mainly found on trees in open or shaded intermontane forests, rarely along the coast. COMMENTS: This lichen is like a small *H. enteromorpha* except for its chemistry. *Hypogymnia rugosa* has broader lobes, lacks side lobules and perforated tips, and always contains hypoprotocetraric acid.

383. *Hypogymnia metaphysodes* mountains, southern British Columbia ×2.9

384. *Hypogymnia occidentalis* Cascades, Oregon ×2.2

385. *Hypogymnia physodes* Cascades, Oregon ×3.2

Hypogymnia physodes
Monk's-hood lichen, hooded tube lichen, puffed lichen

DESCRIPTION: Thallus extremely variable; usually pale greenish gray and smooth; lobes long or short, appressed or ascending, usually fanning out at the tips; lobe tips mostly 1–2.5 mm wide but can broaden to 5 mm; underside of tips bursting open into lip-shaped soralia containing coarsely granular soredia; medullary ceiling usually white. Apothecia rare. CHEMISTRY: Medulla PD+ red, K–, KC+ pink, C– (physodalic, protocetraric, physodic, and other acids). HABITAT: On bark and wood, primarily of conifers; rarely on moss or soil. COMMENTS: This is one of the most common tree lichens in the north, extending southward in the montane conifer forests. It is usually easy to identify because of its hood-like soralia at the lobe tips. Unfortunately, one often finds thalli that are almost devoid of soredia, but a diligent search usually uncovers one or two sorediate lobes. *Hypogymnia vittata* is a more slender, longer, more irregularly branched species also having lip- or hood-shaped soralia. It has a dark medullary ceiling, black margins, and a PD– medulla. It is much rarer than *H. physodes*. An even rarer sorediate lichen with hooded soralia is a form of *H. krogiae* sometimes found in the Appalachians. Its lobes are 0.4–1.2 mm wide and overlap like those of the non-sorediate forms of *H. krogiae*. *Hypogymnia tubulosa* is a much smaller PD– lichen with soredia produced on the upper surface of the lobe tips. IMPORTANCE: In Europe, *H. physodes* is frequently studied as an important indicator of air quality. Significant concentrations of sulphur dioxide pollution will damage and finally kill it, but it is not so sensitive that it disappears close to built-up areas. The species has been used as a food and medicine as well. The Potawatomi Indians have used it in soups and have eaten it as a cure for constipation. In Scandinavia, *H. physodes* has been reported to yield a brown dye for wool.

Hypogymnia rugosa
Wrinkled tube lichen

DESCRIPTION: Thallus large, appressed, with irregularly branched, sinuous (but not knobby) lobes 3–6 mm wide, strongly wrinkled on the upper surface of older lobes; without soredia or lobules; lobe tips only infrequently perforated. Apothecia common, up to 6 mm across. CHEMISTRY: Medulla PD–, K–, KC+ pink, C– (hypoprotocetraric and physodic acids). HABITAT: On conifers, mainly in intermontane forests at high elevations. COMMENTS: *Hypogymnia rugosa* is difficult to distinguish from other broad-lobed tube lichens in the field. It does not have the constrictions of *H. enteromorpha* or *H. apinnata* and lacks the round marginal lobules of *H. enteromorpha* and *H. occidentalis*. Its rough, wrinkled appearance toward the center of the thallus is distinctive, as are the tortuous short lobes and chemistry; it is the only *Hypogymnia* with hypoprotocetraric acid.

Hypogymnia subobscura
Heath tube lichen

DESCRIPTION: Thallus very much appressed, yellow- to red-brown or very dark brown, sometimes mottled with black, especially along the sides of the lobes, with a shiny, varnished appearance; shaded specimens greenish gray, mottled with black and with shiny red-brown tips; lobes normally 1–3 mm wide; without soredia, but very often producing tiny spherical to club-shaped isidia-like lobules on the upper surface and sides of the lobes; tips rarely

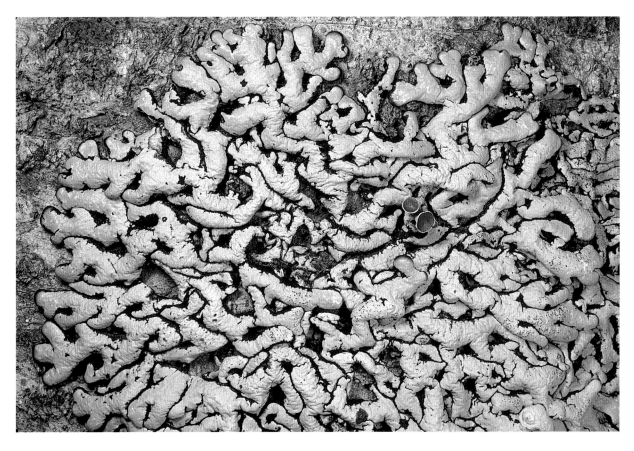

386. *Hypogymnia rugosa* Cascades, Oregon ×1.6

387. *Hypogymnia subobscura* northern British Columbia ×2.9

perforated; medullary ceiling white. Apothecia extremely rare, suggesting that the lobules act as propagules. CHEMISTRY: Medulla PD–, K–, KC+ pink, C– (physodic, 3-hydroxyphysodic, and paraphysodic acids). HABITAT: Over mossy heath, rock, or soil in tundra habitats, less commonly on wood. COMMENTS: The varnished lobes resemble those of the sorediate species *H. bitteri* and *H. austerodes*. Lobules, however, do not normally develop on these species.

Hypogymnia tubulosa
Powder-headed tube lichen

DESCRIPTION: Thallus small, usually under 4 cm in diameter, greenish gray, with forked, divergent, stiff, ascending lobes 0.5–3 mm wide; white, powdery soredia (often flecked with brown or black) on the upper surface of the lobe tips in oval soralia; tips not perforated; medullary ceiling white or pale brown. Apothecia extremely rare. CHEMISTRY: Medulla and soralia PD–, K– or yellow turning brownish, KC+ red, C– (physodic acid and related compounds). HABITAT: Frequent on conifer twigs in

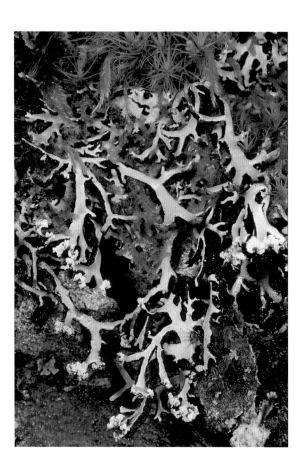

388. *Hypogymnia tubulosa* Klamath Range, northwestern California ×1.4

389. *Hypogymnia vittata* Coast Mountains, British Columbia ×2.1

the eastern and western boreal forests, also on birch or alder bark; usually mixed with the much more abundant *H. physodes*. COMMENTS: This is one of the smallest of the tube lichens and is among the few with powdery soredia on the upper surface of the lobe tips (not bursting from the lobe interiors, as in *H. physodes* or *H. vittata*). The strictly coastal western species *H. oceanica* has similar soralia but is PD+ red; it is larger and more appressed, with marginal lobules.

Hypogymnia vittata
Brownish monk's-hood lichen

DESCRIPTION: Thallus greenish gray to red-brown, at least at the lobe tips, often shiny, appressed or ascending and even somewhat pendent; lobes short to elongate with very irregular branching, 1–2(–3) mm wide; generally producing tiny lobules constricted at the base; lobe tips bursting into lip-shaped soralia; often perforate with small to rather large holes at the lobe tips or on the lower surface close to the tips; medullary ceiling dark brown to black. Apothecia rare. CHEMISTRY: Medulla PD–, K–, KC+ red, C– (physodic and 3-hydroxyphysodic acids). HABITAT: On conifers and mossy or damp rocks, mostly in forests; occasionally on soil or heath. COMMENTS: Only two tube lichens produce soredia from burst lobe tips: *H. vittata* and *H. physodes*. *Hypogymnia physodes* has a PD+ red medulla and a white medullary ceiling, lacks lobules, and has more regular branching. Unfortunately, because of the great variability in *H. physodes*, it is usually necessary to verify the chemistry before making a final determination. *Hypogymnia vittata* is a much rarer species. See Comments under *H. krogiae*.

Hypotrachyna (25 N. Am. species)
Loop lichens

DESCRIPTION: Small to medium, gray or greenish yellow foliose lichens with squared-off lobe tips and broadly rounded axils, causing the divergent lobe tips to overlap and create a loop-like, circular space; pseudocyphellae and marginal cilia absent; soredia, schizidia, or isidia sometimes present; lower surface always black; rhizines black, forked (dichotomously branched), abundant, often extending beyond the lobe margins and resembling cilia. Photobiont green (*Tre-*

bouxia?). Apothecia lecanorine, laminal, with red-brown disks; spores colorless, ellipsoid, 1-celled, 10–18 × 6–12 μm, 8 per ascus. CHEMISTRY: Cortex K+ yellow, KC–, UV– (atranorin), or K–, KC+ gold, UV– (usnic acid), or K–, KC–, UV+ yellow (lichexanthone). The medullary chemistry is extremely varied. HABITAT: On bark and rocks in warm climates. COMMENTS: *Hypotrachyna* is best recognized by its forked rhizines, squared lobes, and round axils between the lobes. Because the species are often defined on the basis of chemistry, we illustrate here only five of the many species in the genus.

KEY TO SPECIES

1. Soredia, or soredia-like fragments originating from the breakdown of pustules or hollow warts (schizidia), present on lobe surface 2
1. Soredia or soredia-like fragments absent 6

2. Rhizines largely unbranched, slender, and short; soredia usually powdery, but sometimes from schizidia, on upper surface of lobe tips; medulla C+ red. . . . . *Hypotrachyna revoluta*
2. Rhizines dichotomously branched, sometimes more than once, robust; soredia powdery or from schizidia; medulla C+ red or C– 3

3. Soredia powdery to granular, not originating from pustules; medulla C+ red (lecanoric and evernic acids); rare, Appalachian, mostly on rocks . . . . [*Hypotrachyna rockii*]
3. Soredia originating from pustules (schizidia); medulla C+ red or C–; on bark. 4

4. Cortex K–, UV+ yellow (lichexanthone); medulla K+ brown-red, PD– (lividic acid). . *Hypotrachyna osseoalba*
4. Cortex K+ yellow, UV– (atranorin); medulla K–, PD+ or PD–. 5

5. Medulla white with orange patches, especially under the pustules, PD+ red-orange, K– where white and K+ purple where pigmented (protocetraric acid and anthraquinone) . . . . *Hypotrachyna croceopustulata*
5. Medulla white throughout, PD–, K–; medulla C+ pink to red (lecanoric and evernic acids); fairly common, Appalachians . . . . [*Hypotrachyna taylorensis*]

6.(1) Isidia present (sometimes sparse), apothecia infrequent; Appalachian 7
6. Isidia absent; apothecia abundant and conspicuous . . 8

7. Isidia cylindrical, laminal; medulla C– or C+ pale orange (barbatic acid complex) . . . . [*Hypotrachyna imbricatula*]

7. Isidia flattened like lobules, on the lobe margins and sometimes on the surface; medulla C+ red (anziaic acid) . . . . [*Hypotrachyna prolongata*]

8. Thallus spotted with maculae; medulla K–, KC+ red, C+ red (evernic and lecanoric acids); southwestern U.S. . . . . *Hypotrachyna pulvinata*
8. Thallus lacking maculae; medulla K+ pinkish brown, KC+ purple-brown, C– (lividic acid); eastern U.S. . . . . *Hypotrachyna livida*

390. *Hypotrachyna croceopustulata* Appalachians, Tennessee × 2.1

Hypotrachyna croceopustulata
Yellow-cored loop lichen

DESCRIPTION: Thallus light gray, closely attached to the substrate; lobes 1–3.5 mm wide, radiating out or somewhat overlapping, with coarse, irregular pustules on the upper surface that break down into soredia-like fragments (schizidia); medulla mostly white, but becoming dark yellow or reddish orange, at least under the pustules. Apothecia rarely seen. CHEMISTRY: Cortex PD–, K+ yellow, KC–, C–, UV– (atranorin); medulla PD+ red-orange, K– (or K+ red-violet where pigmented), KC+ pink, C– (protocetraric acid and rhodophyscin, an anthraquinone pigment). HABITAT: On conifers in open, mostly high-elevation forests, or on hardwoods at lower elevations. COMMENTS: See Comments under *H. osseoalba*.

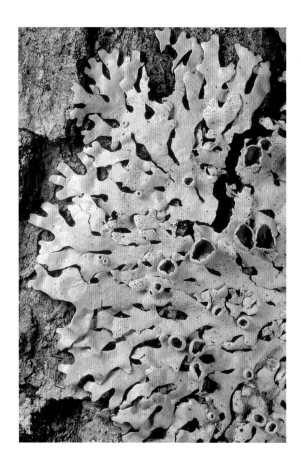

391. *Hypotrachyna livida* central Alabama ×2.7

392. *Hypotrachyna osseoalba* central Florida ×2.9

Hypotrachyna livida
Wrinkled loop lichen

DESCRIPTION: Lobes light gray, strongly wrinkled on the upper surface of older parts of the thallus, often spotted with black pycnidia, 1–4 mm wide, without soredia or isidia. Apothecia abundant, concave, 3–7 mm in diameter, with toothed margins. CHEMISTRY: Cortex K+ yellow, KC–, UV– (atranorin); medulla PD–, K+ pinkish brown, KC+ purple-brown, C– (lividic acid complex). HABITAT: Very common on the bark of deciduous trees. COMMENTS: This species superficially resembles *Myelochroa galbina* but differs in the branching of the rhizines, the white rather than yellowish medulla, and the chemistry. *Hypotrachyna pulvinata* is a fertile, nonsorediate species from the southwest, and its medulla is C+ red (lecanoric acid).

Hypotrachyna osseoalba (syn. *H. formosana*)
Grainy loop lichen

DESCRIPTION: Thallus small, pale gray-green, lobes 1–3 mm wide, with black, forked rhizines that form a mat under and around the lobes, often resembling cilia; coarse soredia or schizidia forming from irregular pustules on the upper surface. Apothecia rare. CHEMISTRY: Cortex K–, KC–, UV+ bright yellow (lichexanthone); medulla PD–, K+ brown-red, KC+ deep brown-red, C– (lividic acid complex). HABITAT: On bark of hardwood trees in open woodlands, rarely on rocks. COMMENTS: A number of *Hypotrachyna* species have coarsely granular soredia developing from pustules and scabs on the upper surface. In *H. croceopustulata,* a dark yellow to reddish orange pigment is often seen in and below the pustules; the cortex has atranorin (K+ yellow, UV–), and the medulla contains protocetraric acid (PD+ red-orange, K–, KC+ pink, C–). ***Hypotrachyna pustulifera,*** common in southcentral U.S., and **H. showmanii,** a rare species from the Ohio Valley, also have atranorin rather than lichexanthone in the cortex. The medullary chemistry of *H. pustulifera* is the same as that of *H. osseoalba;* in

H. showmanii, the medulla is KC+ red owing to unknown substances. **Hypotrachyna laevigata,** a sorediate lichen, produces barbatic, 4-*O*-demethylbarbatic, obtusatic, and norobtusatic acids (KC+ red-orange, C+ orange). Its soredia develop on the upper surface of the lobe tips rather than bursting from pustules. Although rarely collected in North America (only from the southwest), it is common in Europe and the Central American and South American tropics.

Hypotrachyna pulvinata
Smooth loop lichen

DESCRIPTION: Thallus pale gray, with lobes 2–6 mm wide, lacking any soredia or isidia but with white maculae sometimes present on the upper surface; rhizines forming a mat and extending out at the margins like cilia. Apothecia common, up to 9 mm across. CHEMISTRY: Cortex K+ yellow, KC–, UV– (atranorin); medulla PD–, K–, KC+ pink, C+ pink (lecanoric and evernic acids). HABITAT: Common on oaks and pines in dry oak-pine forests. COMMENTS: See Comments under *H. livida.*

Hypotrachyna revoluta
Powdered loop lichen

DESCRIPTION: Thallus pale gray or greenish gray; lobes relatively short, 1–4 mm wide, the margins curled downward (revolute), sometimes almost forming tubes at the lobe tips; granular soredia produced near the lobe tips and sometimes on the lobe surface either from coarse pustules or simply by the erosion of the upper thallus layers; rhizines sparse or abundant, short or long, usually unbranched (those that are branched are forked). CHEMISTRY: Cortex K+ yellow, KC–, UV– (atranorin); medulla PD–, K–, KC+ red, C+ pink (gyrophoric and 4,5-di-*O*-methylhiascic acids). HABITAT: On trees, rocks, and soil. COMMENTS: The size and abundance of the rhizines and the way the soredia are formed are variable in this species. The curled-over lobe tips, the chemistry, and

393. *Hypotrachyna pulvinata* mountains, southern Arizona ×1.7

394. *Hypotrachyna revoluta* Channel Islands, southern California ×2.5

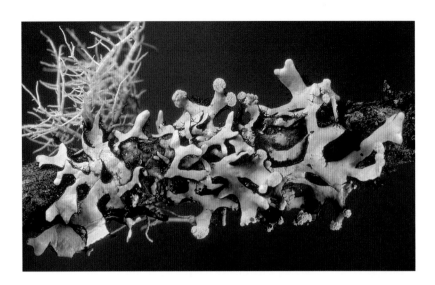

395. *Hypotrachyna sinuosa* Cascades, Washington ×3.7

the sparse branching in the rhizines are the most reliable characters. An Appalachian species on bark, *H. taylorensis*, has pustules that break up into schizidia or granular soredia on the thallus surface or close to the tips, and the rhizines are abundant and richly dichotomously branched. Its medulla reacts PD–, K–, KC+ red, C+ pink (evernic and lecanoric acids). *Hypotrachyna rockii*, also in the Appalachians, closely resembles *H. taylorensis* (and has the same chemistry) but is only found on rock and has a mixture of granular pustules and true soredia on the lobe tips. It is a rare species in North America. *Hypotrachyna laevigata* can have soralia similar to those of *H. revoluta*, but it does not have revolute lobe tips and the medulla is C+ pastel orange, not pink. See Comments under *H. osseoalba* and *Everniastrum catawbiense*.

Hypotrachyna sinuosa
Green loop lichen

DESCRIPTION: Thallus small (lobes rarely wider than 1.5 mm or longer than 3 mm), dichotomously branched, yellowish to yellow-green; fringes of black, forked rhizines confined mainly to the marginal areas but without true cilia; puffs of powdery soredia form on the older lobe tips. CHEMISTRY: Cortex K–, KC+ gold, UV– (usnic acid); medulla PD+ yellow, K+ yellow becoming red, KC–, C– (salazinic acid). HABITAT: Often seen on twigs and small branches of a variety of trees in humid but open forests. COMMENTS: The green loop lichen is unmistakable because of its color, size, forked cilia-like rhizines, and chemistry. This is the only North American *Hypotrachyna* containing usnic acid. *Parmeliopsis ambigua* is somewhat similar but has narrower lobes and unbranched rhizines, and the medulla is K–.

Icmadophila (1 N. Am. species)

KEY TO SPECIES: See Key F.

Icmadophila ericetorum
Candy lichen, spraypaint

DESCRIPTION: Thallus pale green to blue green when fresh, crustose, continuous, fairly thick, smooth to granular or more or less covered with small, spherical, often hollow warts that occasionally break apart into irregular fragments. Photobiont green (*Coccomyxa*). Apothecia abundant, 1.5–4 mm in diameter, biatorine, slightly raised on short stalks; disks flat to slightly convex, pink or pinkish orange with paler, generally wavy margins; asci cylindrical, largely K/I–; spores colorless, fusiform, typically 2–4-celled, 8 per ascus. CHEMISTRY: Apothecia and thallus PD+ orange, K+ deep yellow, KC–, C–, UV+ white (thamnolic and perlatolic acids). HABITAT: Aggressively overgrowing mosses, well-rotted wood, or peat. COMMENTS: With its mint green crustose thallus dotted with bright pink disks, the colorful candy lichen is unlike any other. It is related to the pink earth lichens *Dibaeis baeomyces*, and ***D. absoluta*** looks a bit like them. In *D. baeomyces*, the fruiting bodies are hemispherical and are raised on short, stout stalks; *D. absoluta*, which has sessile apothecia, has a varnish-like, membranous thallus. Although the apothecia of *Icmadophila* are sometimes slightly raised, their stalks are so short that they are virtually invisible. The habitat and distribution of *Icmadophila* and *Dibaeis* are entirely different (see maps).

Imshaugia (2 species worldwide)
Starburst lichens

DESCRIPTION: Small, pale gray foliose lichens; thallus forming appressed rosettes; lobes mostly 1–2 mm wide; pseudocyphellae and soredia absent, but the thallus can be isidiate; lower surface very pale tan with

a well-developed cortex and short, unbranched rhizines. Photobiont green (*Trebouxia?*). Apothecia lecanorine with large, concave, brown disks; asci *Lecanora*-type; spores 1-celled, ellipsoid, colorless, 8 per ascus; pycnidia black, prominent, laminal or marginal. Conidia short bacilliform with a swelling close to one end, or lemon-shaped, 3–4.5 μm long. CHEMISTRY: PD+ orange, K+ deep yellow, KC–, C– (thamnolic acid). HABITAT: On conifer bark and wood in open, well-lit woodlands. COMMENTS: *Imshaugia* resembles rosette lichens (*Physcia*), which have dark brown apothecia, immersed pycnidia, and a different chemistry.

KEY TO SPECIES

1. Isidia absent; pycnidia abundant and conspicuous, scattered on lobe surface; apothecia abundant . . . . *Imshaugia placorodia*
1. Isidia abundant on thallus surface; pycnidia sparse and inconspicuous; apothecia rare . . . . . *Imshaugia aleurites*

Imshaugia aleurites (syn. *Parmeliopsis aleurites*)
Salted starburst lichen

DESCRIPTION: Thallus white to pale gray, shiny; lobe edges sometimes browned; forming more or less circular patches with radiating lobes 0.5–1.2 mm wide; covered with short or long cylindrical isidia except for extreme lobe tips; isidia usually with brownish tips; lower surface tan or almost white, shiny, with fairly abundant short brown rhizines. Apothecia and pycnidia rare. HABITAT: On conifer bark or wood in well-lit conifer forests. COMMENTS: *Imshaugia aleurites* has the general aspect of *Parmeliopsis hyperopta* but has isidia rather than soredia and a different chemistry. (*Parmeliopsis hyperopta* and species of *Physcia* are K+ pale yellow because of atranorin.) No species of *Physcia* has cylindrical isidia. The southeastern **Heterodermia granulifera** is similar but has rounder, more granular isidia and grows on deciduous trees. It reacts PD+ yellow, K+ red owing to salazinic acid in the medulla.

396. *Icmadophila ericetorum* Shuswap Highlands, British Columbia ×3.0

397. *Imshaugia aleurites* Appalachians, Virginia ×2.2

398. *Imshaugia placorodia* mountains, southern Arizona ×3.7

Imshaugia placorodia (syn. *Parmeliopsis placorodia*)
American starburst lichen

DESCRIPTION: Thallus pale gray, narrow-lobed (0.5–1.5 mm), fairly closely appressed at the margins; no soredia or isidia, but with abundant pale brown, somewhat raised, flat to deeply concave apothecia crowded in the center of the thallus; lower surface creamy white to very pale or dark brown, shiny, with short brown rhizines. Apothecia 2–7 mm in diameter with irregular toothed margins; pycnidia black, almost spherical, superficial to slightly raised, common on the thallus surface (especially the older parts). HABITAT: Almost restricted to the bark of pine trees: pitch pine, jack pine, and Virginia pine in the east and ponderosa pine (and Douglas fir) in the southwest; also on wood. COMMENTS: *Imshaugia placorodia* looks something like a big *Physcia* with its narrow lobes and white undersurface, but it is usually abundantly fertile with large pale brown apothecia. In *Physcia* species, the apothecia are always very dark brown to black and generally smaller, and the spores are brown and 2-celled.

Ionaspis (6 N. Am. species)
Watercolor lichens

DESCRIPTION: Crustose lichens usually with pink to orange thalli, sometimes brownish or gray-olive, continuous to rimose-areolate. Photobiont green (*Trentepohlia* or *Trebouxia* in the broad sense). Apothecia round and disk-shaped, sunken into the thallus (cryptolecanorine); disks pale pink or orange to brown, less frequently dark brown or gray, without pruina; margins usually absent or barely perceptible but becoming prominent in a few species; exciple and epihymenium colorless or pigmented with various brown, green, or orange substances turning orange, blue-green, or violet in nitric acid or K; tiny granules typically deposited in a thin amorphous layer over the hymenial surface; paraphyses with one or several rounded cells at the summit; hymenium IKI–; asci IKI– including the ascus tip; spores colorless, ellipsoid, 1-celled, 8 per ascus. CHEMISTRY: All reactions negative (no lichen substances). HABITAT: On siliceous rocks typically in or at the edge of mountain brooks or lakes. One species (*I. alba*) grows on dry, shaded rocks in northeastern deciduous forests. COMMENTS: The splashes of color so commonly seen on rocks in mountain streams are mostly due to thalli of *Ionaspis*. Other common aquatic crustose lichens include **Bacidina inundata,** with greenish thalli and needle-shaped, mostly 4-celled spores; **Rhizocarpon lavatum,** with orangish thalli and large muriform spores; and species of *Verrucaria* and *Staurothele,* perithecial lichens with pale to very dark brown thalli. The genus **Hymenelia** is closely related to *Ionaspis* and comprises mainly tiny arctic crusts on dry rock, both limy and siliceous types; they can have pale pinkish to blue-black disks and have pigments different from those of *Ionaspis*. For example, the dark greenish epihymenial pigments in species of *Hymenelia* turn pinkish violet in nitric acid, whereas in *Ionaspis,* greenish pigments either are unreactive or intensify the green color in nitric acid. No epihymenial pigments in *Hymenelia* turn violet in K, as some do in *Ionaspis*. **Eiglera,** represented in North America by the arctic species **E. flavida,** differs from *Hymenelia* only in having a K/I+ blue ascus tip.

KEY TO SPECIES (including *Hymenelia* and *Eiglera*)
1. Apothecial disks black; epihymenium shades of green, changing to wine-red in nitric acid 2
1. Apothecia pink to brown or gray, rarely darkening to black; epihymenium shades of gray, yellow, or brown, negative or changing to orange-yellow in nitric acid . . 3

2. Ascus tip with a K/I+ blue tholus; thallus pale orange or yellowish when fresh; northern arctic . . . .
. . . . [*Eiglera flavida*]
2. Ascus tip K/I–; thallus whitish to pale yellow-white, gray, or pinkish, often endolithic; arctic-alpine to temperate. . . . . [*Hymenelia*]

3. Thallus pale orange to pale brown; common, Appalachian–Great Lakes–Pacific Northwest distribution . . .
. . . . *Ionaspis lacustris*
3. Thallus pale pinkish (when fresh) to yellowish white; mostly arctic-alpine or boreal 4

4. Apothecia pale pinkish (when fresh); spores 12.5–17 × 7–10 μm; usually on calcareous (rarely siliceous) wet rocks . . . . [*Hymenelia epulotica*]
4. Apothecia smoky gray-brown to brown; spores 8–15 × 5–8 μm; on siliceous rocks in mountain streams; western mountains. . . . . *Ionaspis lavata*

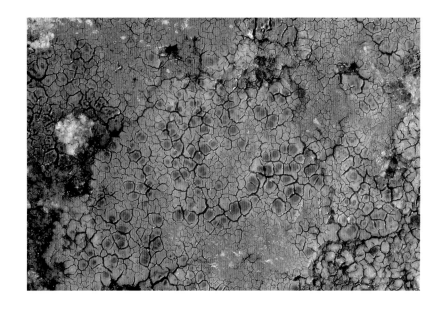

399. *Ionaspis lacustris* herbarium specimen (Canadian Museum of Nature), Quebec ×5.6

Ionaspis lacustris (syn. *Hymenelia lacustris, Lecanora lacustris*)
Rusty brook lichen

DESCRIPTION: Thallus thin, continuous, smooth to cracked, pale orange to tan, highly variable. Photobiont usually *Trebouxia,* rarely *Trentepohlia,* or a mixture of both. Apothecia immersed, 0.2–0.4(–0.7) mm in diameter, with a slightly depressed pale to dark orange disk; margin absent or slightly raised; amorphous layer above the epihymenium containing fine orange granules; spores often with a thin halo, 12–17 × 7.5–8.5 μm. HABITAT: On siliceous rock that is periodically covered with fresh water in streams or at lake edges. COMMENTS: Rusty orange patches on streamside rocks are almost always this lichen; see Comments under *Ionaspis*.

Ionaspis lavata
Mountain pink lichen

DESCRIPTION: Thallus forming round, pink patches 1–2 cm across; thin, continuous to cracked in places, sometimes rimose-areolate. Photobiont *Trentepohlia*. Apothecia 0.15–0.4 mm in diameter, immersed, sometimes coalescing; disks pale to dark grayish brown, slightly concave; epihymenium orange to brown, turning orange with nitric acid

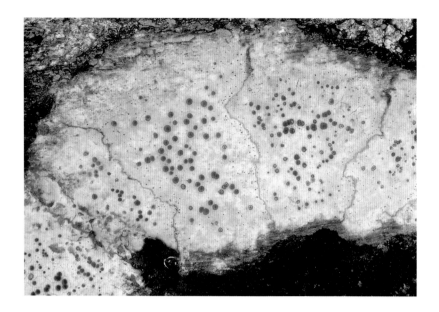

and dark violet with K; spores mostly (8–)10–15 × 5–8 µm, without a halo. HABITAT: On siliceous rocks submerged in clear water, usually in mountain brooks. COMMENTS: The distinctly pink patches "painting" the rocks in streams high in the coastal mountains are an extraordinary sight, especially when part of a quilt-like mosaic with darker aquatic lichens such as *Bacidina inundata, I. odora,* and the orange *I. lacustris. Ionaspis odora* is not uncommon in the same habitats as *I. lavata* in the coastal mountains. It has a darker tan to olive thallus and pale pink to brown apothecia. The thallus tissues and especially the exciple of the apothecia of *I. odora* are K+ violet; the spores are 6.5–9 × 4–6.5 µm.

Japewia (3 N. Am. species)
Dot lichens

DESCRIPTION: Crustose lichens with thin, brown or white thalli. Photobiont green (unicellular). Apothecia biatorine, reddish brown; margins persistent or disappearing; paraphyses slender, branched, and anastomosing, with expanded brown tips; hypothecium colorless; exciple colorless internally but brown at the edge; asci broadly club-shaped with a thick K/I+ blue tholus and K/I– walls (much like the *Lecanora* type); spores 1-celled, ellipsoid, very thick-walled, 8 per ascus. CHEMISTRY: Sometimes containing pigments or lobaric acid. HABITAT: On tree bark and moss in arctic to boreal or oceanic regions. COMMENTS: *Mycoblastus* also has 1-celled, thick-walled spores, but the apothecia are pitch black, and the thallus always contains atranorin (usually with other substances). *Micarea* species do not have thick-walled spores, and the paraphyses are different.

400. *Ionaspis lavata* (photographed underwater) coastal mountains, Alaska ×4.0

401. *Japewia tornoensis* coastal Alaska ×5.6

KEY TO SPECIES: See Key F.

Japewia tornoensis
Hidden dot lichen

DESCRIPTION: Thallus of brown to olive-brown, dispersed, membranous areoles, sometimes difficult to see (e.g., on peat); without soredia or isidia. Apothecia flat to convex, 0.5–1 mm in diameter, irregularly round, marginless; spores broadly ellipsoid to almost spherical, 15–24 × 8–15 µm; walls 2–3.5 µm thick. CHEMISTRY: Thallus and apothecial tissues negative with all reagents (no lichen substances). HABITAT: On bark, mosses, and peat in boreal and arctic regions. COMMENTS: Although very inconspicuous because of its bark-like color, *J. tornoensis* is recognizable in the field because of its brown areolate thallus and red-brown, closely adnate apothecia. Other dot lichens such as species of *Biatora*, *Mycobilimbia*, or *Micarea* rarely have such purely brown fruiting bodies and thallus. The thick-walled spores confirm its identity. The relatively rare sorediate species, *J. carrollii*, is sometimes seen on bark in the Pacific Northwest and southern Appalachians. Its soralia consist of piles of soredia that are brown on the outside and yellow deeper in the soredial mass, resulting from a pigment. The lichen contains small amounts of lobaric acid (KC+ pink).

Kaernefeltia (2 N. Am. species)
Thornbush lichens

DESCRIPTION: Dark brown to greenish black, fruticose or foliose lichens usually forming stiff, erect tufts 5–30 mm high; lacking soredia or isidia, although often with lobules or spiny side branches; pseudocyphellae immersed and extremely inconspicuous or absent. Photobiont green (*Trebouxia*?). Apothecia lecanorine, with dark brown to black disks and thin, bumpy margins, laminal although sometimes appearing to be at the branch tips; asci *Lecanora*-type with a broad axial body at the summit; paraphyses slightly expanded at the tips, not pigmented; spores colorless, 1-celled, ellipsoid, 8 per ascus; pycnidia black, usually immersed and laminal; conidia barbell-shaped (with thickenings close to the ends but with pointed tips), 5–7 μm long. CHEMISTRY: Medulla PD–, K–, KC–, C– (with fatty acids: lichesterinic and protolichesterinic acids and two unnamed fatty acids). HABITAT: On tree branches and wood, mainly of conifers, in open woodlands. COMMENTS: *Kaernefeltia* is one of the *Cetraria*-like lichens, most closely resembling species of *Tuckermannopsis*, especially the *T. fendleri* group. It differs from these in color, features of the cortex, conidial shape, spore shape, and ascus type. *Cornicularia* is similar in its color and general habit, but it grows only on rock, its apothecia are truly at the tips of the branches (not just appearing to be), and its paraphyses have darkly pigmented caps.

KEY TO SPECIES: See Keys A and K.

402. *Kaernefeltia californica* Coast Range, northern California ×2.7

Kaernefeltia californica (syn. *Cetraria californica*, *Cornicularia californica*)
Coastal thornbush lichen

DESCRIPTION: Thallus fruticose, with terete branches often flattened at the axils or with grooves and depressions, dark brown to olive- or greenish black, dull, almost dusty; branches mostly 0.5–1.5 mm in diameter, broadly divergent, with a stiff, spiny or thorny appearance. Apothecia dark brown to black, arising along the branches or close to the tips with the remaining portion of the branch folded back, making the apothecium appear to be at the branch tip; epihymenium turning grayish purple or reddish purple in K; spores ellipsoid, 2.5–4 μm wide. HABITAT: On pine branches and conifer wood in open pine stands in coastal localities at low elevations. COMMENTS: The close resemblance of *K. californica* to *Nodobryoria abbreviata* can cause confusion, although the two lichens have different ranges. The thallus and apothecia of *Nodobryoria* are always red-brown, never greenish black or dusky brown, and the cortex is entirely different. *Nodobryoria* species contain no lichen substances. See Comments under *K. merrillii*.

403. *Kaernefeltia merrillii* Klamath Range, northwestern California ×2.3

Kaernefeltia merrillii (syn. *Cetraria merrillii*)
Flattened thornbush lichen

DESCRIPTION and COMMENTS: The most common form of *K. merrillii* has short (under 15 mm long and 0.4 mm wide), dichotomously branched, flattened but almost fruticose lobes with apothecia (up to 6 mm wide) and pycnidia produced on the upper surface; rhizines are very sparse. Another, much less common form occurs at high elevations on pines and firs. It is broadly foliose (at least in part), has a very wrinkled surface covered with pycnidia, and produces many pale, unbranched or forked rhizines. The extreme fruticose morphotype virtually intergrades with *K. californica*, a less common species restricted to the coastal region. In both, the lobes are abundantly branched and ascending, and both species can become very strongly ridged, like the veins and tendons of a muscular arm. In *K. merrillii*, however, flattened lobes predominate, especially on older parts of the thallus, and some apothecia are always produced on the upper lobe surface. Even the somewhat flattened lobes of *K. californica* look alike on both surfaces, with most branches being round or almost round in cross section, and the apothecia are usually produced close to the lobe tips. In addition, the epihymenium of *K. merrillii* does not change color in K, and the spores are broader (usually over 4 μm wide). *Tuckermannopsis orbata* can be similar but has globose spores and predominantly marginal, prominent pycnidia. HABITAT: On twigs and branches or pines and firs at mid- to high elevations, often in the company of *Nodobryoria* species.

Lasallia (3 N. Am. species)
Toadskin lichens

DESCRIPTION: Dark brown to almost black foliose lichens attached to the rock by a central point (umbilicate); thalli mostly 4–15 cm in diameter, although some can reach 25 cm, with blister-like pustules on the upper surface, each with a corresponding depression on the lower surface; lower surface brown to black, smooth to rough (like coarse sandpaper) or papillate, without rhizines. Photobiont green (*Pseudotrebouxia*). Apothecia black, lecideine, sometimes with a reddish pruina, single or forming dense clusters of mini-apothecia, often with sterile tissue forming columns ("buttons") in the disk; margin thick, black, and persistent; spores muriform, colorless or brown, 1–2 per ascus. CHEMISTRY: Medulla PD–, K–, C+ red, KC+ red (gyrophoric acid); pruina K+ purple (anthraquinones). HABITAT: On noncalcareous rock, especially on boulders and cliff faces in sun or shade. COMMENTS: The toadskin lichens are related to the rock tripes (*Umbilicaria*) and can be confused with some species with warty upper surfaces. Only *Lasallia*, however, has pustules or warts with corresponding depressions on the lower surface. They are also the only umbilicate lichens with one or sometimes two huge muriform spores per ascus.

KEY TO SPECIES: See *Umbilicaria*.

Lasallia papulosa
Common toadskin

DESCRIPTION and CHEMISTRY: Thallus dark brown, upper surface dull to pruinose-scabrose, sometimes covered with an abundant rusty red pruina (anthraquinone pigments: K+ purple); pustules extremely abundant, almost covering the thallus; lower surface pale to medium brown, relatively smooth to rough and grainy. COMMENTS: *La-*

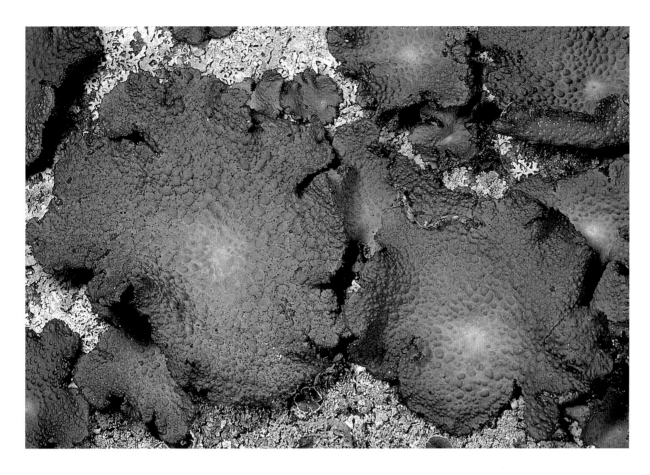

404. *Lasallia papulosa*
central Massachusetts
×1.6

405. *Lasallia papulosa*
(red pruinose form)
Appalachians, Virginia
×1.9

sallia pustulata, a very common lichen in Europe, has been reported from North America but probably does not occur here. It produces stalked and richly branched isidia on the upper surface of the thallus. See *L. pensylvanica.*

Lasallia pensylvanica
Blackened toadskin

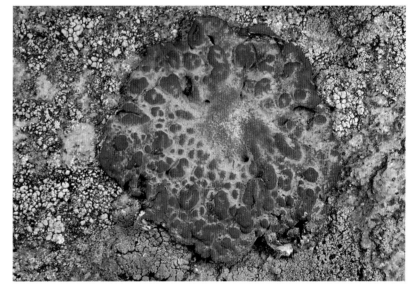

DESCRIPTION and COMMENTS: This species differs from *L. papulosa* in having a smoother thallus with shallower pustules that have sloping sides; in *L. papulosa,* the pustules are steep to vertically sided. Most important, the lower surface of *L. pensylvanica* is pitch black and coarsely papillate, unlike that of *L. papulosa,* which is tan to pale brown and smooth to slightly roughened. The upper surface of *L. pensylvanica* can be smoky brown, especially in the east, or brownish black and often paler around the umbilicus, even becoming white pruinose, especially in the northwest. Thalli of northern populations tend to be much smaller than the eastern ones, rarely exceeding 5 cm across.

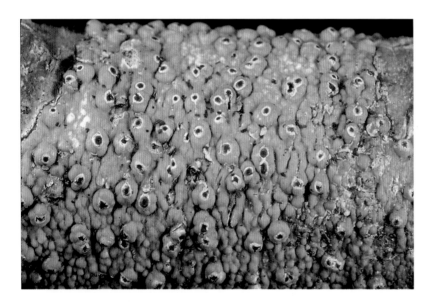

406. *Lasallia pensylvanica* central Massachusetts ×1.9

407. *Laurera megasperma* Gulf Coast, Florida ×3.8

Laurera (2 N. Am. species)
Bark blister lichens

DESCRIPTION: Tropical crustose lichens on bark, dull brownish to olive-brown or greenish yellow, with fertile warts (pseudostromata) containing one to several perithecia; perithecia with black walls; spores colorless, muriform, 2–8 per ascus. Photobiont green (*Trentepohlia*). CHEMISTRY: No lichen substances except for pigments. HABITAT: On bark in tropical and subtropical woodlands. COMMENTS: *Bathelium madreporiforme* used to be included within *Laurera* but differs in its warty brown pseudostromata containing a pigmented, K+ red-purple layer. The pseudostromata are strongly constricted at the base, enclosing 2–8 (or more) perithecia. It has a thin, brownish or olive thallus and muriform spores, 40–50 × 12–17 μm, 8 per ascus. *Trypethelium* is a related genus having spores with only transverse septa and lens-shaped locules.

KEY TO SPECIES: See *Trypethelium*.

Laurera megasperma

DESCRIPTION: Thallus thin, pale sulphur-yellow or yellowish green; perithecia embedded in pseudostromata, usually one perithecium per wart, with a white medulla clearly distinct from the black perithecial wall; from above, the pseudostromata have a broad black ostiole, with a white ring around the opening; spores huge, over 200 μm long and 25 μm broad, 4 per ascus. COMMENTS: *Laurera subdisjuncta* is a less distinctive species with an olive-brown thallus and poorly defined pseudostromata, each containing one to several black-walled perithecia. The muriform spores are 90–130 × 20–30 μm, 8 per ascus.

Lecanactis (5 N. Am. species)
Old-wood lichens

DESCRIPTION: Crustose lichens with thalli that are usually pale gray to almost white, thin, areolate or granular to leprose. Photobiont green (*Trentepohlia*). Apothecia lecideine or somewhat elongate, basically black but often pruinose; margins dull and usually pruinose; hypothecium and exciple very dark brown to black; epihymenium brown; paraphyses branched with expanded tips; asci K/I+ pale blue with a K/I+ blue ring structure in the tip; spores colorless, 4- to many-celled, fusiform, thin-walled; pycnidia black, prominent; conidia rod-shaped. CHEMISTRY: Often contains orcinol depsides such as erythrin and lecanoric and confluentic acids as well as schizopeltic and lepraric acids. HABITAT: On shaded bark, wood, and rock in old humid forests. COMMENTS: *Lecanactis* and *Cresponea* species often give the impression of being unlichenized fungi, especially when the thalli are particularly thin or dispersed. They are generally found only in rich old growth forests or in humid coastal regions. *Cresponea* differs from *Lecanactis* in that the apothecial margins are smooth, often shiny, and carbonaceous, its spores are thick-walled, its pycnidia are entirely or partially immersed in the thallus, and it almost always lacks lichen substances. *Schismatomma* and *Lecanographa* are related genera with lecanorine apothecia, and *Opegrapha*, with clearly elongate lirellae, is also very similar in structure and chemistry.

KEY TO GENUS: See Key F.

408. *Lecanactis abietina* southern New Brunswick ×6.6

Lecanactis abietina
Old-wood lichen

DESCRIPTION: Thallus forming a thin, powdery, yellowish white crust. Apothecia 0.4–0.8 mm in diameter, heavily yellow-pruinose with white pruinose margins; upper part of the hypothecium turning yellowish green in K; spores 4-celled, narrowly fusiform, 28–44 × 3–7 µm, straight or slightly curved or sinuous, without a halo, 8 per ascus; pycnidia stalked but stubby, covered with a white pruina; usually abundant; conidia 10–17 × 2–4 µm, 1-celled, colorless, straight or slightly curved. CHEMISTRY: Thallus and apothecia PD–, K–, KC–, C–; pycnidial pruina PD–, K–, C+ red (lecanoric and schizopeltic acids). COMMENTS: Western populations that lack C+ red compounds in the pycnidia and in which the apothecial margins are rarely white pruinose have been treated as a different species (*L. megaspora*) by some lichenologists. *Cresponia chlorocronia* (syn. *Lecanactis chlorocronia*), although usually black with a black margin, can sometimes have disks with a thin, white or greenish pruina, but the spores are much shorter (11–22 µm) and the apothecial margin is never white pruinose. It is found in the northeastern U.S. and Canada and southern California.

Lecania (25 N. Am. species)
Rim-lichens

DESCRIPTION: Crustose lichens with very thick or extremely thin thalli, pale gray to dark brown. Photobiont green (unicellular). Apothecia lecanorine or biatorine, usually with brown disks, frequently pruinose; exciple usually well-developed even in lecanorine apothecia, with forked hyphae that become gradually thicker at the ends; paraphyses unbranched, easily separated in K, expanded and usually pigmented at the tips; hypothecium colorless; epihymenial pigments typically uneven in distribution, giving the disks a spotty appearance when wet; asci *Bacidia*- or *Biatora*-type; spores 2–4-celled (rarely more), colorless, ellipsoid or fusiform, often bent and bean-shaped, 8–16 per ascus; conidia usually curved. CHEMISTRY: Relatively poor in compounds, but atranorin present in some species. HABITAT: On bark, especially of neutral-barked trees such as poplar and elm; also on calcium-containing or, rarely, siliceous rock. COMMENTS: A

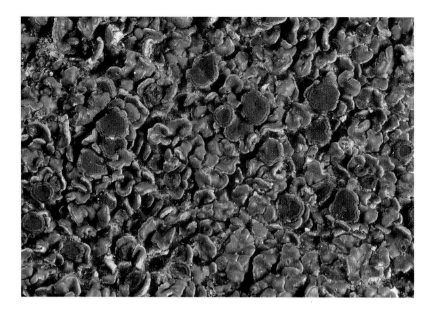

409. *Lecania brunonis* (wet; dry thallus is browner) Channel Islands, southern California ×5.6

great deal of work remains to be done on the North American species of *Lecania*. Many of the specimens in North American herbaria are incorrectly named (e.g., *L. erysibe;* see Comments under *L. brunonis*). *Solenopsora* is another genus with lecanorine apothecia and mainly 1-septate spores; its asci are of the *Catillaria*-type (Fig. 14c).

KEY TO SPECIES (including *Solenopsora*)

1. On rock. 2
1. On bark 6
2. Spores (2–)4-celled, 12–16(–18) × 4–6 µm; apothecia dark brown to black, always pruinose; on limestone, widespread . . . . [*Lecania nylanderiana*]
2. Spores 2-celled; apothecia pale to dark brown or black, sometimes pruinose. 3
3. Thallus areolate to almost squamulose, or clearly lobed at the margins; on calcareous or siliceous rocks; California 4
3. Thallus rimose-areolate, not at all squamulose or lobed, pale greenish to gray; on calcareous rocks. 5
4. Thallus brown, areolate, almost squamulose; apothecia dark brown, not pruinose; on siliceous rocks . . . . *Lecania brunonis*
4. Thallus white, lobed at the margins, not truly squamulose; apothecia almost black, lightly pruinose; usually on calcareous rocks . . . . [*Solenopsora candicans*]
5. Thallus not at all sorediate; mostly in east . . . . [*Lecania perproxima*]
5. Thallus granular to sorediate; mostly in west and north . . . . [*Lecania erysibe*]

6.(1) Spores 4-celled, 12–23 × 4–6 µm, straight or slightly bent, 8–16 per ascus; central parts of continent . . . . [*Lecania fuscella*]
6. Spores 2-celled; widespread 7
7. Spores straight . . . . [*Lecania cyrtella*]
7. Spores bent or bean-shaped . . . . *Lecania dubitans*

Lecania brunonis

DESCRIPTION: Thallus areolate to squamulose, brown or gray-brown when dry, without soredia. Apothecia dark brown with a wavy lecanorine margin that becomes thin in older apothecia and sometimes disappears; spores 2-celled, ellipsoid, 12–20 × 4–7 µm, 8 per ascus. CHEMISTRY: No lichen substances. HABITAT: On siliceous rocks, especially sandstone. COMMENTS: Another somewhat darker brown squamulose *Lecania*, *L. hassei,* is found in California. Its squamules are convex, and its thallus becomes lobed at the circumference. *Lecania perproxima* is a more common eastern *Lecania* on calcareous rock and is often mistakenly named *L. erysibe*. It has a greenish or brownish gray areolate to rimose thallus (without granules or soredia, as one finds in *L. erysibe*), and brown, flat to somewhat convex, non-pruinose lecanorine apothecia (the margins becoming thin), 0.3–0.8 mm in diameter. The spores are 2-celled, ellipsoid, 9–16 × 3–7 µm.

Lecania dubitans (syn. *L. dimera*)
Bean-spored rim-lichen

DESCRIPTION: Thallus chalky white, with a powdery (but not sorediate) texture, forming circular patches. Apothecia very small, 0.2–0.4 mm in diameter, very dark brown to black, flat to convex, scattered over thallus, not crowded, the margin very thin and rarely seen; spores 2-celled, slightly bent or curved, 12–17 × 4–6 µm. CHEMISTRY: No lichen substances. HABITAT: Virtually restricted to poplar bark. COMMENTS: *Lecania cyrtella* is another common, bark-dwelling, northern temperate species known from British Columbia and California to Newfoundland. It has slightly smaller, ellipsoid spores, 10–15 × 3.5–4.8 µm, that are almost always straight. The apothecia are 0.25–0.6 mm in diameter, pale reddish brown, and always crowded together, with pale, thin but persistent margins. *Lecania cyrtella* usually grows on hardwoods such as ash, alder, or maple.

Lecanographa (3 or 4 N. Am. species)

DESCRIPTION: Crustose lichens very similar to *Lecanactis* but with ascus walls largely K/I– with no special structures in the tip (lacking an obvious K/I+ blue ring). Photobiont green (*Trentepohlia*). Apothecia round and disk-like or elongated and script-like; usually heavily pruinose; exciple brown-black; paraphysoids branched and anastomosing; spores colorless, 19–26 (–30) × 4–6 μm, fusiform to oblong, 4–8-celled, halonate, 8 per ascus. CHEMISTRY: Cortex and medulla of North American specimens usually PD–, K–, KC+ red, C+ red (often containing orcinol depsidones such as erythrin and lecanoric or gyrophoric acids). HABITAT: On rocks in the foggy coastal region of California. COMMENTS: This genus has only recently been distinguished from *Lecanactis* and *Opegrapha*, mainly on the basis of ascus type and halonate spores. ***Lecanographa amylacea,*** wide-ranging in shaded old growth forests in Europe, is extremely rare in North America, if present at all. It closely resembles a bark-dwelling species of *Lecanactis* but has *Lecanographa*-type asci and halonate spores.

KEY TO GENUS: See Keys E and F.

Lecanographa hypothallina (syn. *Opegrapha hypothallina, Schismatomma hypothallinum, Lecanactis nashii*)
California chalk-crust

DESCRIPTION: Thallus thick, chalky white with a heavy pruina, rimose-areolate to verrucose. Apothecia round to somewhat elongate, largely immersed in the thallus areoles and creating a kind of lecanorine margin; disk black but covered with creamy white pruina like the thallus; exciple and hypothecium dark red-brown; spores fusiform, straight or slightly curved, 7–8-celled, 19–30 × 4–6 μm. CHEMISTRY: Medulla PD–, K–, KC+ red, C+ red (lecanoric acid and (or) gyrophoric acid). HABITAT: On coastal sandstone rocks. DISTRIBUTION: Coastal southern California. COMMENTS: Species of *Dirina* and *Roccellina* grow in the same habitat (sometimes also on bark) and often have similar pruinose white thalli and black pruinose fruit-

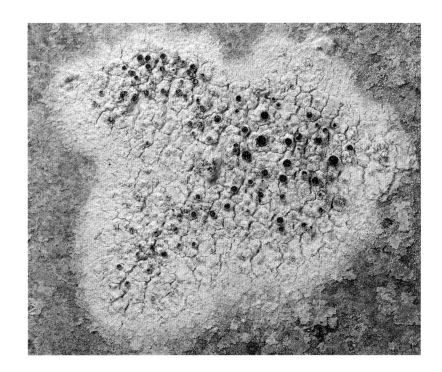

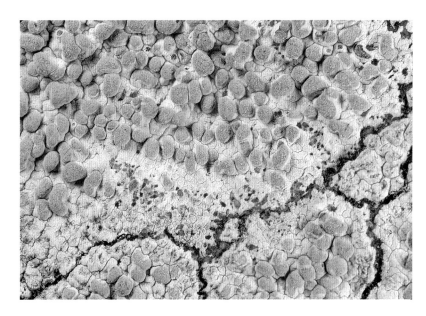

410. *Lecania dubitans* Shuswap Highlands, British Columbia ×4.3

411. *Lecanographa hypothallina* Channel Islands, southern California ×5.1

ing bodies resembling lecanorine apothecia. They have similar chemistry (frequently C+ red), but the spores are 4-celled.

Lecanora (171 N. Am. species)
Rim-lichens

DESCRIPTION: Crustose lichens (rarely fruticose) with thalli varying from very thin and barely perceptible to very thick and lobate; yellowish, gray, brown, or greenish; often sorediate. Photobiont green (*Pseudotrebouxia*). Apothecia yellowish, green, brown, or black, often pruinose, usually with lecanorine margins the same color and texture as the thallus, but sometimes with biatorine or rarely lecideine margins; thalline tissue of the margin (the amphithecium) often containing small or large crystals of calcium oxalate, or more or less filled with algae; paraphyses largely unbranched except at the tips, expanded or not; epihymenium brown, green, or colorless, granular or clear; hypothecium colorless to yellow; asci *Lecanora*-type (Fig. 14e) with the thickened ascus tips (tholus) staining dark blue in K/I, and with a distinct indentation from below (ocular chamber); spores 1-celled, ellipsoid to spherical, ca. 6–20 × 3–12 μm, colorless, 8 per ascus in most species, up to 32 per ascus in a few. CHEMISTRY: The chemistry is extremely varied. Usnic acid, atranorin, or xanthones are commonly present in the cortex, and a variety of compounds can occur in the medulla. HABITAT: On substrates of all kinds, from the Tropics to the Arctic. COMMENTS: The genus *Lecanora* is extremely species-rich and includes a wide variety of growth forms and colors. It is best defined by its crustose habit, *Lecanora*-type asci, and 1-celled spores, but most species have lecanorine apothecia with prominent, thallus-colored apothecial margins that rim the disks (hence the English name). In the field, *Rhizoplaca, Lecania, Protoparmelia, Rinodina,* and *Tephromela* are often confused with *Lecanora;* the apothecia and spores must be examined under the microscope to sort them all out. Identifying species within *Lecanora* often involves determining if atranorin is present in the cortex. Although atranorin turns K+ yellow, this test is not always reliable with some of the smaller species, even when testing the apothecial margin. (K sometimes turns the margin grass-green because of large numbers of algal cells in the amphithecium.) Try testing a part of the thallus or apothecium where the algal cells have been lost or killed because of shading (not disease!). The presence or absence of crystals or granules in tissues of the apothecium (as seen in sections under the compound microscope) is also often important. Tiny oil droplets or air bubbles sometimes look like granules or crystals, so distinguishing these inclusions is critical. This can easily be done using two polarizing filters, one between the light source and the microscope slide (as in the microscope's filter holder), and one between the microscope slide and your eye. By slowly rotating either of the filters to the point at which most light is cut out (the extinction point), the crystalline material will suddenly appear to be brightly lit. Any polarizing lenses will work, even the lenses from an old pair of polarized sunglasses. Two types of granules can be deposited in or on the epihymenium. Some granules are rather coarse (you can see their shape under 400× magnification) and form a layer on top of the epihymenium; they are soluble in concentrated nitric acid but insoluble in K. These granules are found in *L. cenisia* and *L. chlarotera*. The other granules are much smaller (brown particles too small to make out under 400×) and lie between the tips of the paraphyses as well as on the surface; they are insoluble in nitric acid but dissolve in K. These granules are found in *L. circumborealis, L. hybocarpa,* and **L. pulicaris**.

KEY TO SPECIES

1. Growing on bark, wood, mosses, or dead vegetation . . 2
1. Growing directly on rock 24

2. Thallus within substrate, absent from view, or indistinct 3
2. Thallus clearly visible 6

3. Spores ellipsoid to broadly ellipsoid, ratio of length to width 1.8–2.2:1; apothecial disks pruinose; amphithecium filled with opaque crystals . . . . . . *Lecanora hagenii*
3. Spores narrowly ellipsoid, ratio of length to width 2.2–3.0:1; apothecial disks pruinose or not; amphithecium without crystals 4

4. Epihymenium clear red-brown, not at all granular . . . . *Lecanora zosterae*
4. Epihymenium with granules on the surface or between the tips of the paraphyses 5

5. Apothecia pale yellowish brown or reddish brown to black; margin paler than disk, rough; algae filling amphithecial medulla . . . . *Lecanora piniperda*
5. Apothecia yellow; margin disk-colored, smooth; algae very sparse in the apothecial margin, sometimes essentially absent . . . . *Lecanora symmicta*

6.(2) Thallus leprose, greenish yellow or yellow-green, with a well-developed white, fibrous prothallus; apothecia rare . . . . *Lecanora thysanophora*

6. Thallus not leprose, yellowish or gray, lacking a fibrous prothallus; apothecia produced on all mature thalli . . 7

7. Apothecia essentially biatorine, almost without algae in margins (only a few in apothecial base). 8

7. Apothecia lecanorine, with algae relatively abundant in margins 9

8. Apothecia yellowish green to vivid yellow (containing usnic acid); spores narrowly ellipsoid; apothecial margin PD–; widespread on bark or wood of all kinds, boreal to temperate. . . . . *Lecanora symmicta*

8. Apothecia pale to very dark brown to black; spores almost spherical, 6–8 × 5.5–8 μm; apothecial margin PD+ red (fumarprotocetraric acid); common on conifers in western mountains. . . [*Lecanora fuscescens*]

9. Thallus and (or) apothecial margin yellowish, KC+ gold; cortex K– (atranorin absent); spores narrowly ellipsoid, length to width ratio 2.2–3.2:1 10

9. Thallus and apothecial margin cortex KC– or KC+ yellow, K+ yellow (atranorin); spores usually ellipsoid to broadly ellipsoid or spherical, length to width ratio 1.0–2.2:1 13

10. Apothecial disks with a light to heavy yellow pruina; apothecial margin prominent, flexuose . . . . *Lecanora cupressi*

10. Apothecial disks not pruinose or lightly gray pruinose; apothecial margin flush with disk or becoming thin and disappearing in maturity, not flexuose 11

11. Apothecia yellow-brown to black, never yellowish green or greenish yellow, thallus very thin, pale yellowish gray; apothecial margin more or less smooth, not sorediate; amphithecium filled with algae . . . . *Lecanora piniperda*

11. Apothecia usually yellow-green or greenish yellow; thallus pale yellowish green; apothecial margin sorediate or granular; algae sparse or relatively abundant in amphithecium. 12

12. Apothecial margin smooth, containing very few algae, never sorediate or granular . . . . *Lecanora symmicta*

12. Apothecial margin verruculose, granular or sorediate, with algae abundant at least where thalline tissues of the margin are well developed. . . . . *Lecanora strobilina*

13.(9) Apothecia heavily pruinose, making them pinkish or violet; amphithecial cortex lacking, with medullary hyphae growing out to the margin; amphithecial medulla PD+ yellow or red, rarely PD– (protocetraric or virensic acid) . . . . *Lecanora caesiorubella*

13. Apothecia usually yellowish brown to reddish brown without pruina, but if pruinose, then not pinkish or violet; amphithecial cortex present, indistinctly or distinctly delimited from the medulla; amphithecial medulla PD– or PD+ pale yellow. 14

14. Amphithecium containing numerous small crystals; epihymenium red-brown, not at all granular. 15

14. Amphithecium containing a few large, irregular crystals; epihymenium granular on the surface or between the tips of the paraphyses. 16

15. Growing on logs and fences near the sea; amphithecial cortex distinct from the medulla, which is filled with crystals. . . . . *Lecanora xylophila*

15. Growing on bark; amphithecial cortex indistinctly delimited from medulla, with crystals extending from one into the other . . . . *Lecanora allophana*

16. Tiny brown granules, insoluble in nitric acid, distributed in upper third of hymenium; apothecia pruinose or not pruinose 17

16. Coarse granules, soluble in nitric acid, limited to the surface of the epihymenium, not extending into the hymenium; apothecia often pruinose. 18

17. Spores 10–13 × 6–7.5 μm; amphithecial cortex not much wider at the sides than at the base of the amphithecium; apothecial disks pale yellow-brown to reddish brown; eastern temperate deciduous forests . . . . *Lecanora hybocarpa*

17. Spores 13–17.5 × 8–11 μm; amphithecial cortex much wider at the base than at the sides of the amphithecium; apothecial disks dark reddish brown to black; boreal forests . . . . *Lecanora circumborealis*

18. Apothecial disks C+ orange; southeast coastal plain . . . . [*Lecanora louisianae*]

18. Apothecial disks C– 19

19. Apothecia immersed in the thallus for a long time, finally superficial 20

19. Apothecia superficial with well-developed margins 21

20. Spores narrowly ellipsoid, 9–13 × 5–7 μm; apothecia small and very crowded, 0.3–0.8 mm in diameter; southeast coastal plain . . . . [*Lecanora leprosa*]

20. Spores broadly ellipsoid, mostly 10–14.5 × 7–8.5 μm, apothecia not usually crowded together, 0.7–1.5 mm in diameter . . . . *Lecanora cinereofusca*

21. Apothecial margin cortex less than 15 μm thick; apothecial margin strongly verrucose or "beaded"; apothecia never pruinose; epihymenium PD+ orange (with orange crystals slowly developing, as seen under the microscope). . . . . *Lecanora cinereofusca*

21. Apothecial margin cortex 20–50 μm thick; apothecial margin even or slightly bumpy; apothecia often pruinose; epihymenium PD– 22

22. Thallus very thin; apothecia thin and flat; Pacific Northwest. . . . . *Lecanora pacifica*
22. Thallus thick, verrucose to granular; apothecia thick, widespread 23

23. On bark, rarely wood. . . . . [*Lecanora rugosella*]
23. On hard weathered wood, never bark . . . . *Lecanora cenisia*

24.(1) Thallus endolithic, within the rock and out of sight, with only the apothecia on the surface. 25
24. Thallus clearly visible. 26

25. On calcareous rock (bubbling with HCl); apothecial margins white, K–, KC– . . . . *Lecanora dispersa*
25. On siliceous rock (unreactive with HCl); apothecial margins yellowish, K–, KC+ gold (usnic acid) . . . . *Lecanora polytropa*

26. On shoreline rocks; thallus yellowish, often lobed at the margin, cortex C+ orange, KC+ orange (xanthones).27
26. On nonmaritime rocks; thallus yellowish or not, lobed or not, cortex C–, KC– or KC+ gold (xanthones absent) 28

27. In California; apothecia yellowish, heavily pruinose . . . . *Lecanora pinguis*
27. On northern Pacific coast; apothecia red-brown, not or slightly pruinose . . . . [*Lecanora straminea*]

28. Thallus distinctly lobed 29
28. Thallus not lobed 36

29. Thallus cortex K+ blood red (norstictic acid) . . . . *Lobothallia*
29. Thallus cortex K–, or K+ persistently yellow 30

30. Apothecia bright red-brown; epihymenium entirely without granules; thallus pale yellow to gray. . . . . *Lecanora argopholis*
30. Apothecia yellow-brown, greenish, or black; epihymenium usually granular on surface; thallus yellow-green, greenish yellow, or honey-brown 31

31. Thallus honey-brown, often shiny; apothecia without pruina but rarely seen. 32
31. Thallus yellowish green or greenish yellow; apothecia with or without pruina, usually present and abundant 33

32. Low-elevation woodlands of California; contains rangiformic acid; upper cortex containing dead algal cells . . . . [*Lecanora mellea*]
32. High elevations in California, and scrublands in Pacific Northwest and arid interior; contains protolichesterinic acid; upper cortex purely fungal, lacking dead algal cells . . . . [*Lecanora pseudomellea*]

33. Thallus lobes predominantly flat 34
33. Thallus lobes usually somewhat to strongly convex; western. 35

34. Apothecia not pruinose; lobes usually with thickened or slightly raised margins, making the lobe tips somewhat concave; lobe margins usually whitish or pruinose; very widespread, especially on rocks frequented by birds; medulla PD–. . . . . *Lecanora muralis*
34. Apothecia heavily yellowish pruinose; lobes without raised margins, not at all concave; lobe margins not white or especially pruinose; western; medulla PD+ yellow (psoromic acid) . . . . [*Lecanora bipruinosa*]

35. Apothecia black to yellowish green, typically heavily yellowish pruinose but sometimes without pruina; lobes convex, but usually not strongly sinuous or folded, rarely pruinose . . . . *Lecanora novomexicana*
35. Apothecia yellowish to yellow-brown, not pruinose; thallus verrucose or rugose, or with sinuous folds, often pruinose . . . . *Lecanora garovaglii*

36.(28) Ends of spores pointed; thallus grayish brown to yellow-brown, shiny; apothecia shiny dark red-brown . . . . *Protoparmelia*
36. Ends of spores rounded 37

37. On calcareous rock. 38
37. On noncalcareous rock 40

38. Apothecial disks heavily pruinose; apothecia concave; ascus tip K/I– . . . . *Aspicilia candida*
38. Apothecial disks not, or lightly, pruinose; apothecia flat, convex, or becoming hemispherical; ascus tip K/I+ dark blue (*Lecanora*-type) 39

39. Apothecial disks black or almost black; epihymenium green. . . . . *Lecanora marginata*
39. Apothecial disks red-brown; epihymenium brown . . . . *Lecanora argopholis*

40.(37) Thallus leprose, with a conspicuous white, fibrous prothallus; apothecia very rare . . . . *Lecanora thysanophora*
40. Thallus continuous or dispersed areolate, not leprose; prothallus absent; apothecia abundant. 41

41. Apothecial disks shiny red-brown, never pruinose; epihymenium clear red-brown, not at all granular 42
41. Apothecia yellowish to brown or black, dull, pruinose or not; epihymenium granular on the surface or not granular. 43

42. Thallus thick, areolate; apothecia 1–3 mm in diameter, constricted at the base; western . . . *Lecanora argopholis*
42. Thallus thin, continuous; apothecia less than 1 mm in diameter, more or less immersed in the thallus; East Temperate. . . . . [*Lecanora subimmergens*]

43. Thallus white to pale gray or greenish gray (lacking usnic acid); apothecia typically pruinose.......... 44
43. Thallus greenish yellow or yellowish green (containing usnic acid); apothecia with or without pruina 46

44. Apothecial disks C+ yellow, heavily white pruinose .. .. *Lecanora rupicola*
44. Apothecial disks C–, pruinose or not 45

45. Apothecia superficial with well-developed margins; disks brown to black, often pruinose .... *Lecanora cenisia*
45. Apothecia immersed in thallus (cryptolecanorine), lacking margins; disks black, never pruinose......... .... *Lecanora oreinoides*

46.(43) Apothecia 2–7 mm in diameter, constricted at the base; disks shades of orange or pink, lightly pruinose; thallus well-developed, dispersed areolate to almost squamulose .... *Rhizoplaca subdiscrepans*
46. Apothecia 0.3–0.8(–1.3) mm in diameter, sessile; disks pale yellow to pale orange, not pruinose; thallus composed of scattered, very small, flat, adnate areoles .... *Lecanora polytropa*

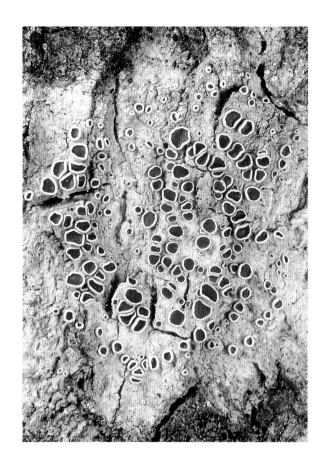

412. *Lecanora allophana* coastal Alaska ×2.5

Lecanora allophana
Brown-eyed rim-lichen

DESCRIPTION and COMMENTS: This is one of the more easily recognized species of *Lecanora*, with its large (up to 2 mm wide), clear, red-brown apothecia constricted at the base, on a white to pale gray, smooth to somewhat rough thallus. A section of the apothecia reveals a clear, red-brown pigment in the epihymenium with no granules; a thick, gelatinous cortex (commonly up to 75 µm wide at the base) with small, angular crystals invading from the medulla; and large spores, (10–)13–19(–21) × 6–11 µm. *Lecanora epibryon* is almost indistinguishable except for its much thicker thallus and its arctic-alpine habitat (on dead moss and vegetation). *Lecanora glabrata* is similar, but its apothecia have thin margins and are not constricted at the base, and it is smaller in all dimensions (apothecial diameter, spore size, etc.). It is common on hardwoods, especially maples, in the northeast. *Lecanora horiza* is sometimes mistaken for *L. allophana* in southern California; it has a clearly defined, clear cortex because the amphithecial crystals are confined to the medulla. CHEMISTRY: Thallus PD– or pale yellow, K+ yellow, KC–, C– (atranorin and triterpenes). HABITAT: On bark, especially of poplars and ash.

Lecanora argopholis
Varying rim-lichen

DESCRIPTION: Thallus variable in both color (distinctly pale yellow to gray) and degree of lobing (entirely verrucose-areolate or with well-developed lobes 0.6–1.5 mm wide). Apothecia clear red-brown, 1–3 mm in diameter, with well-developed margins, especially in the verrucose-areolate forms; epihymenium without any granules or crystals; amphithecium filled with small crystals that are insoluble in K; spores 10–18 × 5–9 µm. CHEMISTRY: Cortex PD+ pale yellow or PD–, K+ yellow, KC+ yellow, C– (with varying concentrations of the yellow xanthone, epanorin); medulla usually PD–, K– (or pale yellow), KC–, C– (atranorin, sometimes with zeorin, gangaleoidin, or fatty acids). HABITAT: Growing directly on rock, both calcareous and noncalcareous types, and occasionally on mosses or plant remains, in very exposed sites such as arctic and alpine areas, boreal outcrops, prairie boulders, or lakeshores. COMMENTS: The varying color of this species (yellow or gray) causes most of the problems with its identification. In

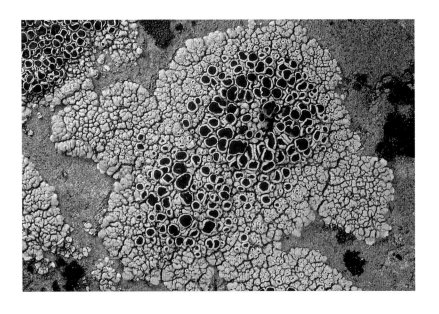
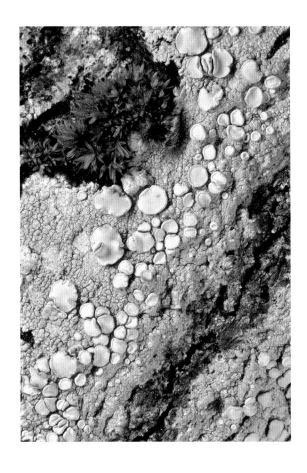

413. (top left) *Lecanora argopholis* canyon country, eastern Utah ×2.0

414. (top right) *Lecanora caesiorubella* ssp. *merrillii* California coast ×2.9

415. (right) *Lecanora caesiorubella* ssp. *prolifera* Ozarks, Arkansas ×2.5

gray forms, the red-brown apothecia give the species the appearance of a member of the *Lecanora subfusca* group such as *L. xylophila* (strictly coastal, with triterpenes) or *L. cenisia* (usually with darker, dull, somewhat pruinose apothecia, large crystals in the amphithecium, and a granular epihymenium).

Lecanora caesiorubella
Frosted rim-lichen

DESCRIPTION: Thallus gray, continuous, smooth to verruculose. Apothecia round to irregular, reaching 3 mm in diameter, appearing pinkish or even lavender because of a heavy white pruina; usually slightly constricted at the base; margins usually thick and prominent; small crystals of calcium oxalate fill the amphithecial medulla, and because the margin has no cortex, the hyphae of the amphithecium grow out to the edge; spores (8–)10–14 × 6–9 μm. CHEMISTRY: To see the chemical reactions most clearly, make thick sections of the apothecia, place them on a microscope slide, then apply the reagents with a fine tube. Apothecial sections PD+ red or rarely yellow or PD–, K+ yellow or becoming red, KC–, C– (atranorin, usually with protocetraric or virensic acids, rarely with psoromic acid, sometimes with norstictic acid). Apothecial disks C– or C+ yellow (xanthones). See Comments below. HABITAT: Most subspecies found on hardwood bark (infrequently on conifers). COMMENTS: This species is one of the largest in the *Lecanora albella* group, which consists of rim-lichens with heavily pruinose apothecial disks, oxalate crystals in the amphithecium, and atranorin in the thallus and apothecial margins (at least K+ yellow). In most species, there is no translucent gelatinous cortex on the apothecial margin, which gives the margins a rather dull surface. *Lecanora caesiorubella* is a widely distributed, basically tropical lichen with several subspecies distinguishable by chemistry and slight morphological differences. Most of the subspecies are PD+ red-orange in the apothecial margins caused by virensic acid (**ssp. caesiorubella,** rarely in ssp. *merrillii*) or protocetraric acid (ssp. *merrillii* and **ssp. glaucomodes**). Subspecies *caesiorubella* has an Appalachian–Great Lakes distribution, and ssp. *glaucomodes* is confined to the southeastern coastal plain. **Subspecies merrillii** (plate 414) is mainly coastal from southern California to Oregon (shown in blue on the map), most commonly on oaks, and contains norstictic acid (K+ red) in the central part of the apothecial base; the apothecial disk is negative with C. **Subspecies prolifera** (plate 415), on the eastern coastal plain (in pink on the map), grows on deciduous tree bark, has norstictic acid in the apothecial margins, not in the base, and the disks turn C+ yellow (due to a xanthone). It is classified at the species level as **L. subpallens** by some lichenologists. One can find **ssp. saximontana** on wood in the Rocky Mountains. It is PD+ yellow, K+ red (norstictic acid, but without protocetraric or virensic acid). Subpecies *caesiorubella* is common on trees in the Great Lakes region south into the Appalachians and, like the subtropical ssp. *glaucomodes,* contains no norstictic acid, and the apothecial disks are C–. *Lecanora albella* is very similar, also with frosted apothecial disks and a PD+ red-orange medulla (protocetraric acid), but the apothecia tend to be smaller (generally under 1.5 mm), with narrower margins, always with C– disks, and it is usually found on conifers. One variety in the Great Lakes region contains norstictic acid in addition to the protocetraric acid. *Lecanora cateilea,* on bark on the west coast and Great Lakes region, has PD+ deep yellow, K+ yellow apothecial sections (psoromic acid), with apothecia up to 1 mm in diameter, C– disks, and 12 spores per ascus instead of the usual 8.

Lecanora cenisia
Smoky rim-lichen

DESCRIPTION: Thallus usually thick and verrucose, very pale gray, almost white. Apothecia 0.5–2 mm in diameter, pale brown in shaded habitats to almost black in the sunniest places, typically with a light pruina; amphithecium containing large irregular crystals; epihymenium pigmented brown or greenish with a superficial layer of coarse granules; spores mostly 11–16 × 7–8 μm. CHEMISTRY: Thallus PD– or pale yellow, K+ yellow, KC–, C– (atranorin and fatty acids, especially roccellic acid; rarely with gangaleoidin). HABITAT: On exposed siliceous rocks or hard, weathered wood. COMMENTS: *Lecanora cenisia* is the most commonly encountered rock-dwelling member of the *L. subfusca* group (brown apothecia, gray K+ yellow thallus, and crystals in the apothecial margin). Its pruinose apothecia and fatty acid production are additional identifying characteristics. *Lecanora argentea* is another rock species sometimes seen in the northern forest region. It resembles *L. cenisia,* but the apothecia are a dark sooty brown to almost black with no trace of pruina. The thallus is thinner than that of *L. cenisia* and contains gangaleoidin, not fatty acids. *Lecanora californica* looks like a very large, extremely pruinose form of *L. cenisia,* with raised apothecia constricted at the base and wavy margins. Its range centers around the central California coast, and it has a unique chemistry: norgangaleoidin and a fatty acid called nephrosteranic acid.

Lecanora cinereofusca
Beaded rim-lichen

DESCRIPTION: Thallus pale gray to almost white, smooth, cracked, or verruculose, rarely producing patches of soredia. Apothecia at first immersed and level with the thallus, finally emerging with a clearly verrucose, almost discontinuous margin having a beaded appearance; disks orange-brown to red-brown, without pruina; cortex thin (under 15 μm);

416. *Lecanora cenisia* mountains, northeastern California ×3.6

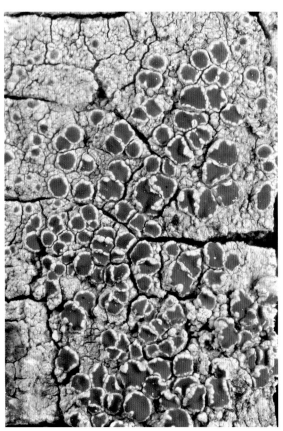

417. *Lecanora cinereofusca* Appalachians, North Carolina ×5.5

large crystals in amphithecium, but sometimes very sparse; epihymenium red-brown, granular; spores (7.5–)10–14.5 × (6–)7–8.5(–9.5) μm. CHEMISTRY: Epihymenium and often apothecial margin PD+ orange (pannarin); roccellic acid and placodiolic acid usually present, especially in eastern specimens. HABITAT: Common on deciduous trees, rare on conifers. A variety in the Appalachian-Ozark region (**var. *appalachensis***) grows on shaded rocks. COMMENTS: The almost immersed apothecia with beaded margins, together with the rather characteristic orangish color of the disks, make this species easy to recognize in the field. The pannarin in the epihymenium can be revealed by putting several apothecial sections under a fragment of a cover slip and introducing PD dissolved in alcohol (not Steiner's solution). The short orange crystals produced by pannarin will flow out from the epihymenium after a minute or two.

Lecanora circumborealis
Black-eyed rim-lichen

DESCRIPTION: Thallus thin and continuous or less commonly dispersed areolate, pale gray, often with a black prothallus. Apothecia mostly 0.4–0.8 mm in diameter, very dark brown to black (rarely medium brown), dull, with a prominent gray margin; amphithecium containing large crystals (often sparse); amphithecial cortex much thicker at the base than at the sides (35–90 μm versus 22–38 μm); epihymenium dark brown to olive-brown with tiny brown granules embedded between the tips of the paraphyses in vertical rows; spores broadly ellipsoid, mostly 13–17.5 × 8–11 μm. CHEMISTRY: Apothecial sections PD–, K+ yellow, KC–, C– (atranorin and roccellic acid). HABITAT: On the bark or wood of conifers, alder, birch, and willow. COMMENTS: *Lecanora pulicaris* is a more temperate species (but also grows in the boreal regions and western mountains). It resembles *L. circumborealis* but has less extreme dimensions, with spores 10.5–15 × 7.5–9.5 μm, cortex 25–65 μm at the base. Its apothecial disk is usually paler, and its medulla is PD+ red (fumarprotocetraric acid). See also *L. hybocarpa*.

Lecanora cupressi
Cypress rim-lichen

DESCRIPTION: Thallus pale yellowish gray to pale sulphur-yellow, rough and cracked, often delimited by a thin black prothallus. Apothecia brownish yellow with a bright lemon-yellow pruina; margins thick and prominent, very bumpy and uneven; spores narrowly ellipsoid, 10.5–15 × 3.5–5 μm. CHEMISTRY: Thallus and apothecia PD–, K–, KC+ gold, C– (usnic acid and zeorin). HABITAT: Frequent on bark (especially bald cypress) and wood. COMMENTS: The chemistry and narrow spores indicate that *L. cupressi* is part of the *L. varia* group, which includes other yellowish rim-lichens such as **L. varia**, **L. symmicta**, and *L. strobilina*. There is, however, no other *Lecanora* or *Lecanora*-like lichen that has such bright yellow, pruinose apothecia.

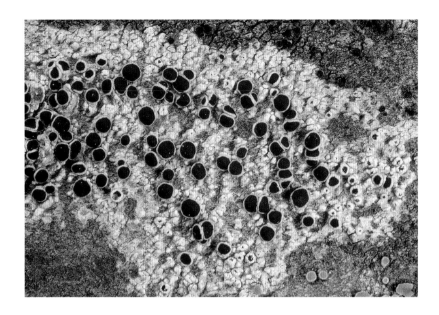

418. *Lecanora circumborealis* Teton Range, Wyoming ×5.6

419. *Lecanora cupressi* Pine Barrens, New Jersey ×8.6

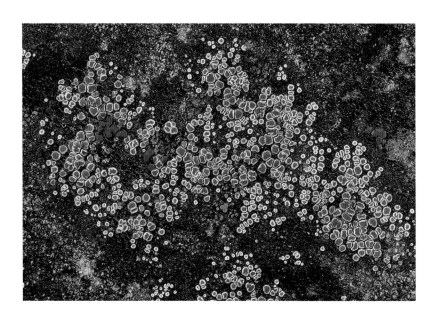

420. *Lecanora dispersa* northern Arizona ×2.8

Lecanora dispersa
Mortar rim-lichen

DESCRIPTION: Thallus usually growing between the rock crystals and absent from view. Apothecia 0.4–1.2 mm in diameter, round or somewhat angular, crowded or dispersed, pale to dark yellow-brown or pinkish brown, without pruina when mature, usually with prominent white margins; spores ellipsoid, 8.5–14 × 3.5–7 µm. CHEMISTRY: PD–, K–, KC–, C– (occasional specimens contain xanthones and sometimes pannarin). HABITAT: On calcareous rocks, often seen on mortar or concrete, even in cities. COMMENTS: This lichen is a survivor. In New York City, it is the only one to persist in central Brooklyn (on concrete fences in Prospect Park). A major factor enabling it to tolerate urban environments is its ability to inhabit rock rich in calcium, which buffers the effects of acid rain and sulphur dioxide pollution, but the lichen must have additional survival mechanisms. Several other species of *Lecanora* resemble *L. dispersa* and grow on limy rock. ***Lecanora crenulata***, like *L. dispersa*, has almost no visible thallus. Its apothecial margins become cracked into disconnected segments, and the disks are sometimes pruinose. ***Lecanora albescens*** forms small round patches of thick white thallus, sometimes lobed at the margins, and the disks have very little or no pruina.

Lecanora garovaglii
Sagebrush rim-lichen

DESCRIPTION: Thallus yellow-green, or pale gray-green because of a thick white pruina, often blackening at the lobe tips; lobes thick, folded and sinuous, quite convex except at extreme tips. Apothecia pale yellowish to dark yellowish brown, becoming greenish black in some forms, dull but without pruina; 0.8–2 mm in diameter; spores ellipsoid, 9–15 × 4–6 µm. CHEMISTRY: Cortex KC+ gold (usnic and isousnic acids); medulla PD–, K–, KC–, C– (zeorin) or rarely PD+ yellow, K– (psoromic acid); also roccellic acid. HABITAT: On rocks, especially sandstone, in arid sites such as sagebrush and juniper scrub communities. COMMENTS: *Lecanora garovaglii* is a common lichen in semidesert regions of central North America. It resembles *L. muralis* but with a much thicker, folded thallus. The color variation in the apothecial disks is remarkable: all intermediates between green-black and pale yellow can be found. ***Lecanora sierrae*** and ***L. pseudomellea*** are closely related lichens that, unlike *L. garovaglii*, have shiny apothecial disks and produce fatty acids rather than zeorin (or other triterpenes). *Lecanora sierrae* has a yellowish to pale blue-green thallus with usnic acid in the cortex; the medulla often contains pannarin or psoromic acid (PD+ orange or yellow) as well as rangiformic acid. Its distribution centers in the Sierra Nevada range of California. *Lecanora pseudomellea*, growing mainly at low to moderate elevations in the Pacific Northwest, has a brownish thallus lacking usnic acid; it contains protolichesterinic acid (a fatty acid). ***Lecanora phaedrophthalma,*** a species common in arid Arizona, often has a pruinose thallus; it contains placodiolic acid and triterpenes. See also *L. novomexicana*.

Lecanora hagenii
Hagen's rim-lichen

DESCRIPTION: Thallus usually thin or even absent from view. Apothecia brown to greenish, distinctly pruinose, 0.4–0.7 mm in diameter; amphithecial cortex thick (especially at the base), up to 45 µm, and gelatinous (i.e., clear); spores ellipsoid but variable in size, 7–14 × 4.5–7.5 µm. CHEMISTRY: Apothecial sections PD–, K–, KC–, C– (no lichen substances). HABITAT: On bark or wood, less frequently peat. DISTRIBUTION: Probably widespread from the arctic to

421. *Lecanora garovaglii* canyon country, northeastern Utah ×2.9

422. *Lecanora hagenii* western Colorado ×5.6

temperate regions. COMMENTS: Our understanding of this species is still incomplete. The name is applied to small, bark-dwelling rim-lichens with a K–, disappearing thallus, and pruinose apothecia. It is very close to *L. dispersa*, a saxicolous lichen with largely non-pruinose apothecia, and **L. umbrina,** usually on wood, which has extremely crowded apothecia with thin, often disappearing margins, and disks that are sometimes pruinose and sometimes not. **Lecanora sambuci** is another similar rim-lichen on bark, especially poplars. It has very small apothecia (0.2– 0.5 mm in diameter) and, most important, 12–32 spores per ascus. See also *L. zosterae*.

Lecanora hybocarpa
Bumpy rim-lichen

DESCRIPTION: Thallus pale gray, continuous, smooth or more commonly bumpy (verruculose). Apothecia mostly 0.5–1 mm in diameter; disks pale or dark orange-brown to reddish brown, dull, usually with prominent, bumpy margins (but sometimes almost smooth); cortex usually not more

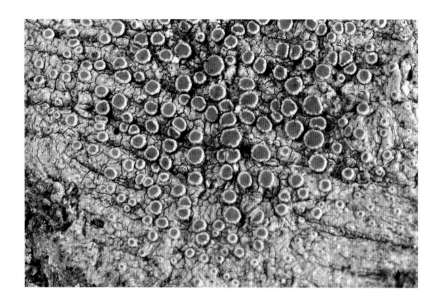

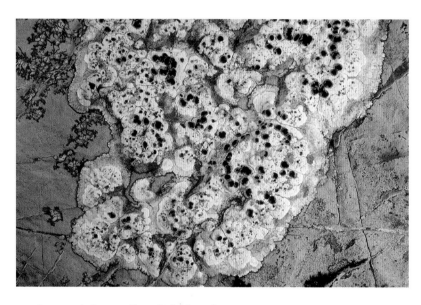

423. *Lecanora hybocarpa* Cape Cod, Massachusetts ×5.6

424. *Lecanora marginata* Rocky Mountains, Alberta ×2.0

than 1.5 times thicker at the base than at the sides; large crystals in the amphithecium; tiny brown granules between the paraphyses tips; spores mostly 10–13 × 6–7.5 μm. CHEMISTRY: Cortex and apothecial sections PD–, K+ yellow, KC–, C– (atranorin and often roccellic acid). HABITAT: On bark of hardwoods, rarely on conifers, in well-lit woodlands or on isolated trees. COMMENTS: This is the most common member of the *L. subfusca* group in the east. The disk color is extremely variable, and pale-fruited specimens sometimes grow beside those with dark apothecia. Although the apothecial margins are typically verruculose, almost "beaded" in the extreme forms, they can also be quite smooth and thin. *Lecanora pulicaris,* a more northerly relative with a thicker cortex noticeably expanded at the base, contains fumarprotocetraric acid (PD+ red). In **L. chlarotera** the epihymenium has a superficial layer of coarse granules rather than fine granules within the epihymenium.

Lecanora marginata

DESCRIPTION: Thallus chalky white to pale sulphur-yellow, less commonly dark yellow; rimose-areolate or sometimes only scattered areoles on a blue-gray prothallus (as shown in plate 424). Apothecia lecideine or partly lecanorine (with a disappearing thalline margin), flat to convex, sometimes becoming hemispherical, 0.6–2.5 mm in diameter; disk black; margin thin, black, usually with a whitish pruina, but typically obscured by the expanding disk; internal apothecial tissues all colorless, but the edge of the exciple and epihymenium are dark blue-green (turning red in nitric acid); spores narrowly to broadly ellipsoid, 8–15 × 4–7.5 μm. CHEMISTRY: Cortex PD–, K+ pale yellow, KC+ yellow, C–; medulla PD–, K– or pale yellow, KC–, C– (atranorin and usnic acid, both variable in concentration). HABITAT: On limestone and other calcareous rocks in exposed alpine or arctic sites. COMMENTS: The apothecia of this species look so much like those of a *Lecidea* that *L. marginata* was classified with that genus until it was realized that its asci and conidia are those of a typical *Lecanora*. The apothecial margin nevertheless lacks algae except in rare specimens.

425. *Lecanora muralis* (Farther inland, this species is typically greener. Red mites are a common sight on lichens.) British Columbia coast ×5.7

Lecanora muralis
Stonewall rim-lichen

DESCRIPTION: Thallus waxy, light yellowish green with little or no pruina; coastal forms (as shown in plate 425) often very pale; areolate in the center of the thallus, becoming lobed at the margins, sometimes squamulose; marginal lobes very flat, closely appressed, 0.3–0.6 mm wide; lobe edges often pale but can also be blackish and somewhat raised; medulla loose and cottony rather than dense and solid. Apothecia yellow- to red-brown, not pruinose, with rather thick, persistent, thallus-colored margins, 0.5–2.3 mm in diameter; spores 9–15 × 4.5–7 µm. CHEMISTRY: Cortex KC+ gold (usnic acid); medulla PD–, K–, C–, KC– (zeorin). HABITAT: On all kinds of rock including cement and mortar, rock walls, and stone fences; on soil in the Great Basin. It is often found on rocks frequented, and fertilized, by birds. COMMENTS: This is one of the most common and widespread lobed species of *Lecanora*. **Lecanora valesiaca** is very similar, but the thallus is heavily pruinose. It is widespread throughout North America. **Lecanora bipruinosa** is a southwestern desert species with patches of pruina on the lobes and pruinose apothecia; it often contains psoromic acid (PD+ yellow). **Lecanora phaedrophthalma,** another yellowish green, lobed *Lecanora* in dry country, has narrow, sometimes pruinose lobes free at the margins, with abundant red- to yellow-brown *L. subfusca*-like apothecia without pruina. The medulla contains placodiolic acid and triterpenes. There are many other lobed species of *Lecanora,* especially in the west. Some of them contain xanthones in the cortex and either react C+ orange or are fluorescent in long-wave ultraviolet light. One example is **L. straminea,** which often accompanies *L. muralis* on bird rocks along the west coast. It has large shiny brown apothecia crowding the center of the thalli, and the marginal lobes are thick and almost foliose. See also Comments under *L. garovaglii,* especially those pertaining to **L. sierrae.**

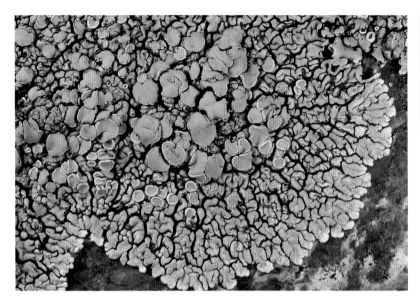

Lecanora novomexicana
New Mexico rim-lichen

DESCRIPTION: Thallus yellowish green, lobed, the marginal lobes thick, convex, and somewhat irregular; medulla thick and very dense, sometimes with scattered rose-colored spots. Apothecia 0.7–2 mm in diameter, yellowish to greenish black or pure black, with a heavy creamy yellow pruina in the typical form but sometimes entirely without pruina; margins prominent. CHEMISTRY: Cortex KC+ gold (usnic acid); medulla PD+ yellow, K–, KC–, C– (psoromic acid), PD–, K–, KC–, C– (only fatty acids), or rarely PD–, K–, KC+ red, C+ red (lecanoric acid). HABITAT: Common on rocks in dry localities at high elevations. COMMENTS: *Lecanora muralis* has a similar aspect but has flat lobes and nonpruinose apothecia. Some forms of *Rhizoplaca melanophthalma* are extremely similar to *L. novomexicana* in thallus color and apothecial characteristics, and a common chemical race also contains psoromic acid; the *Rhizoplaca*, however, is normally umbilicate (attached in the center), not crustose. See also Comments under *L. garovaglii*.

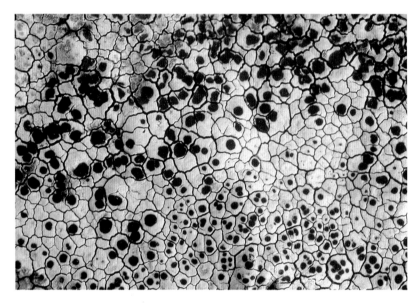

Lecanora oreinoides (syn. *Lecidea tesselina*, *Lecidea oreinodes* [sic])
Sunken rim-lichen

DESCRIPTION: Thallus thick, rimose-areolate, creamy white to pale gray. Apothecia black, entirely immersed in the areoles, one to several per areole, 0.25–0.4(–0.7) mm in diameter, with the disk usually slightly depressed and without any margin; hypothecium colorless; epihymenium greenish to olive, turning red in nitric acid; paraphyses predominantly unbranched, expanded at the tips; spores ellipsoid, (7–)8–14 × 4–6.5 μm. CHEMISTRY: Medulla PD–, K–, KC–, C–, IKI– (atranorin and confluentic acid). HABITAT: On noncalcareous rocks, especially sandstone. COMMENTS: In outward appearance, *L. oreinoides* looks like a typical *Aspicilia*, but its small spores, unbeaded paraphyses, and *Lecanora*-type ascus are different. It is related to the arctic-alpine species, *L. marginata*, which also has a greenish epihymenium. The distinctions of *L. oreinoides* from the very similar *Buellia spuria* are described under the latter species.

426. *Lecanora novomexicana* Rocky Mountains, Colorado ×2.8

427. *Lecanora oreinoides* herbarium specimen (New York Botanical Garden), Missouri ×6.8

Lecanora pacifica
Multicolored rim-lichen

DESCRIPTION: Thallus very thin to slightly verruculose, very pale yellowish gray, often with a thin, blue-black prothallus. Apothecia large and flat, mostly 0.7–1.2 mm in diameter; disks sometimes yellow, sometimes almost black, and sometimes multicolored on the same disk, frequently "frosted" with a light pruina; rims white, prominent, and fairly smooth; amphithecium containing large, irregular crystals; epihymenium coarsely granular on the surface; cortex distinct and relatively uniform in thickness; spores broadly ellipsoid, mostly 12–17 × 7.5–9.5 µm. CHEMISTRY: Apothecial sections PD–, K+ yellow, C–, KC– (atranorin and roccellic acid). HABITAT: Mainly on the smooth bark of deciduous trees such as alder, willow, maple, dogwood, and ash. COMMENTS: *Lecanora pacifica* is probably the most common species of *Lecanora* on the west coast. Spore size and chemistry distinguish *L. pacifica* from other rim-lichens with yellow apothecia such as *L. symmicta* and **L. confusa,** both of which have narrowly ellipsoid spores (3–5 µm wide) and contain usnic acid and zeorin.

Lecanora phryganitis
Shrubby rim-lichen

DESCRIPTION: A small, pale yellowish green fruticose lichen forming small round cushions 10–20 mm high, 15–30 mm across, with intricately branched stems 0.6–1.4 mm in diameter; surface very rough, eroding into granules or granular soredia at the tips; medulla loose with the algae forming a narrow, well-defined layer. Large apothecia often produced along the stems; disks pinkish brown and pruinose, 1–2.5 mm across, with thick thallus-colored margins; spores narrowly ellipsoid, 14–16 × 5–6 µm. CHEMISTRY: Cortex PD–, K+ yellowish, KC+ red-orange, C+ orange (xanthones including thiophanic acid, zeorin, usnic acid). HABITAT: On rocks and soil on bluffs near the sea. COMMENTS: This remarkable species, although belonging to a crustose genus, develops a fruticose thallus. A variety of unrelated lichens including *Caloplaca coralloides, Aspicilia hispida,* **Lecidea ramulosa,** and *Ochrolechia frigida* also show this ability. **Cladidium bolanderi** is superficially similar to *L. phryganitis* but has a tough cartilaginous medulla

428. *Lecanora pacifica* Washington coast ×3.4

429. *Lecanora phryganitis* California coast ×2.5

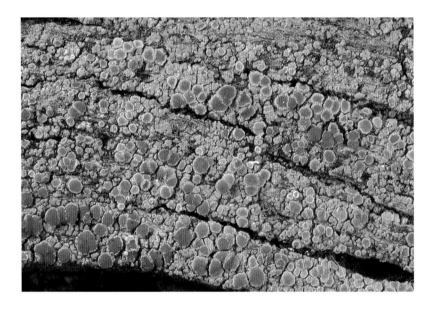

430. *Lecanora pinguis* California coast ×2.2

431. *Lecanora piniperda* Rocky Mountains, Colorado ×5.4

more or less filled with algae, the apothecia are at the branch tips, and the asci are somewhat different. The branches are very stiff and only scabrose, not disintegrating into granules. The cortex is KC+ yellow, C– (usnic acid alone). Like *L. phryganitis*, it grows on rocks along the California coast.

Lecanora pinguis
Seaside sulphur-rim lichen

DESCRIPTION: Thallus very thick with convex folded verrucae, slightly lobed at the margin; pale yellowish green, often with an extremely rough pruinose or scabrose surface and sometimes with irregular patches of granules or granular soredia. Apothecia broad, closely attached, with yellowish, heavily pruinose disks. CHEMISTRY: Cortex PD–, K+ yellow, KC+ red-orange, C+ orange; medulla PD–, K+ yellow, KC+ orange, C+ deep yellow (xanthones including thiophanic acid, zeorin, usnic acid). HABITAT: On coastal rocks. COMMENTS: Although limited in range, this lichen is often a very conspicuous member of the California maritime rock community.

Lecanora piniperda
Wood-spot rim-lichen

DESCRIPTION: Thallus virtually absent or granular to dispersed areolate, yellowish. Apothecia 0.2–0.5 mm in diameter, flat, pale yellowish brown to almost black, sometimes with a thin pruina; margins yellowish, generally thin but persistent; amphithecial cortex thin and indistinct; epihymenium coarsely granular; spores narrowly ellipsoid, 6–14 × 3–5 μm. CHEMISTRY: PD–, K–, KC+ yellow, C– (isousnic acid). HABITAT: On wood and bark of hardwoods and conifers in the open. COMMENTS: Thallus development and apothecial color are extremely variable in *L. piniperda*. *Lecanora saligna,* which particularly resembles pale-fruited forms of *L. piniperda,* can have larger apothecia that are never darker than brown and rarely pruinose when mature; it also has broader spores, 7–13 × 4–7 μm, never narrowly ellipsoid. It has a broad temperate to boreal range. ***Lecanora subintricata*** is another small wood-dwelling rim-lichen with narrowly ellipsoid spores and a granular epihymenium. Its apothecial margins are very thin and disappear in mature apothecia, but the amphithecium nevertheless has a gelatinous cortex. *Lecanora mughicola,* not uncommon on conifers in the western mountains, has large, crowded, greenish black apothecia and large spores (14–18 × 4–5 μm).

Lecanora polytropa
Granite-speck rim-lichen

DESCRIPTION: Thallus scanty or absent from view, sometimes remaining as tiny, scattered, pale yellowish areoles. Apothecia 0.3–0.8(–1.3) mm in diameter, flat, scattered or clustered; disks waxy yellow to pale orange, without pruina; margins smooth and not very prominent, thin or thick, paler than the disk; spores ellipsoid, 8–15 × 5–7 μm. CHEMISTRY: Apothecia PD–, K–, KC+ yellow, C– (usnic acid, zeorin, and fatty acids). HABITAT: On siliceous rocks, especially granite, in full sun. COMMENTS: *Lecanora polytropa* is a very common lichen on exposed granite outcrops and boulders, although it is often missed because of its size. Most similar is the arctic-alpine *L. intricata,* which has darker apothecia (greenish brown to almost black) and a more extensive, yellowish green, areolate thallus. Its apothecia are somewhat sunken, making the disks more or less level with the thallus surface. In *L. polytropa,* the apothecia remain pale and are rarely immersed. The chemistry of the two species is essentially the same.

Lecanora rupicola
White rim-lichen

DESCRIPTION: Thallus usually thick, rimose-areolate, yellowish white or, less frequently, greenish gray. Apothecia mostly 0.5–2 mm in diameter, at first immersed in the areoles, sometimes remaining so but often becoming superficial and somewhat convex with persistent thallus-colored or blackened margins; disks dark brown to black but always covered with a thick creamy white to blue-gray pruina; spores 8–15 × 5.5–7.5 μm. CHEMISTRY: Cortex and medulla PD–, K+ yellow, KC–, C– (atranorin and roccellic acid). Apothecial disk C+ yellow (xanthones). HABITAT: On sunny or shaded rock surfaces, especially granites, mostly at high elevations. COMMENTS: Few other lichens form a creamy white crust with frosted lecanorine apothecia that are sunken into the thick thallus. To be sure of its identity, test the disk with C.

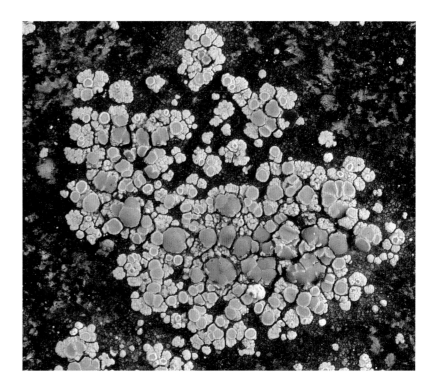

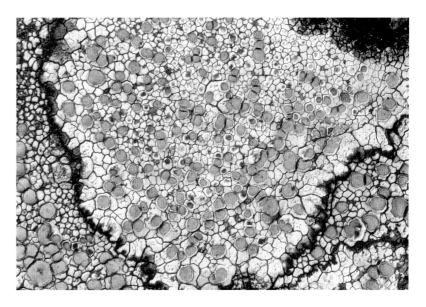

432. *Lecanora polytropa* Lake Superior region, Ontario ×6.9

433. *Lecanora rupicola* (*Rimularia,* a parasite, can be seen invading the thallus of the *Lecanora* in the upper right-hand corner.) Olympics, Washington ×4.2

Plate 433 shows the invasion of a parasitic lichen, *Rimularia insularis,* which forms its own dark brown lichen thallus (with tiny black apothecia) right on the *Lecanora*. The parasite steals the alga of the host to use in its own thallus and forms its own distinctive chemical products, in this case, gyrophoric acid, making the medulla of the parasite C+ red.

Lecanora strobilina
Mealy rim-lichen

DESCRIPTION: Thallus thin, pale yellowish green or rarely gray-green, finely cracked and becoming granular. Apothecia small, 0.4–0.9 mm in diameter, flat to slightly convex, waxy light yellow, with a persistent granular or sorediate margin the same color as the thallus; spores narrowly ellipsoid, 10–15 × 3.5–5 μm. CHEMISTRY: Thallus and apothecial section PD–, K–, KC+ yellow, C– (usnic acid and zeorin). HABITAT: On bark and wood of many kinds in sunny sites. COMMENTS: After long storage, specimens of *L. strobilina* develop a fine white fuzz of tiny, straight, colorless crystals characteristic of lichens containing zeorin or other triterpenes. (See Comments under *Cladonia carneola*.) The most similar lichen is *L. symmicta*, which has smooth, biatorine apothecial margins that, in mature apothecia, soon become level with the disk and then are lost. Specimens of *L. strobilina* that have very thin, almost disappearing margins can therefore be hard to distinguish from *L. symmicta*. **Lecanora conizaeoides** forms a pale yellowish green to grayish green, fairly thick, granular or powdery crust over urban trees on the northeast and northwest coasts. It has broad waxy yellow to pale brown apothecia with sorediate margins. Both the thallus and apothecial margin are PD+ red, K–, KC– because of fumarprotocetraric acid. *Lecanora conizaeoides* is so tolerant of pollution that its presence is regarded as an early indicator of deteriorating air quality.

Lecanora symmicta
Fused rim-lichen

DESCRIPTION: Thallus pale greenish yellow to gray-green or pale green, thin and granular, but sometimes barely developing. Apothecia 0.3–0.8 mm in diame-

434. *Lecanora strobilina* Adirondacks, New York ×6.2

435. *Lecanora symmicta* Lake Superior region, Ontario ×5.8

ter, waxy yellow or vivid yellow to pale yellow-brown, rarely darker, lacking pruina, essentially biatorine, at first flat but becoming convex, often fusing with neighboring apothecia; margins thin, smooth, eventually pushed out of sight by the expanding disks; spores 7–16 × 3–6 μm. CHEMISTRY: Thallus PD–, K–, KC+ yellow, C–; disks sometimes KC+ orange, C+ orange (usnic acid and zeorin, sometimes accompanied by thiophanic acid, a xanthone, in the apothecia). HABITAT: On bark and wood of both conifers and hardwoods, usually in open habitats. COMMENTS: Plate 435 shows a form with exceptionally bright yellow apothecia. Most specimens of *L. symmicta* have paler disks. ***Lecanora confusa*** has similar apothecia and spores, although the apothecial margins tend to be more persistent, but it has a well-developed, yellower, non-sorediate thallus that is KC+ orange, C+ orange. It is fairly common on trees in the Pacific Northwest. In *L. cupressi*, the bright color of the yellow apothecia is due to a pruina. See Comments under *L. strobilina*.

Lecanora thysanophora
Mapledust lichen

DESCRIPTION: Thallus pale green to yellowish green, consisting of a thin patchy layer of granular soredia, almost leprose but with a white, usually conspicuous, webby or fibrous prothallus developing at the thallus margins. Small, yellowish brown to yellowish grey, lightly pruinose apothecia occasionally seen. CHEMISTRY: Thallus PD–, K+ yellow, KC+ gold, C– (usnic acid, atranorin, and zeorin, frequently with porphyrilic acid, as well as a few species-specific triterpenes). HABITAT: A very common lichen on sugar maple bark, but also growing on beech, oak, basswood, or other trees, or even on rocks, in shaded or partly shaded forests. COMMENTS: This powdery lichen looks like a yellowish green *Lepraria*, but a close look will reveal that it is not entirely reduced to soredia. One of the most reliable characters is the white, fibrous margin of the thallus (as shown in plate 436), although in many specimens, it is patchy and not very well developed.

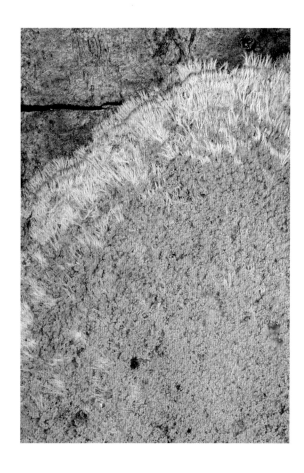

436. *Lecanora thysanophora* Adirondacks, New York ×5.6

Lecanora xylophila (syn. *L. grantii*)
Driftwood rim-lichen

DESCRIPTION: Thallus almost white to yellowish gray, very thin to rather thick and rough, continuous, often covering large areas. Apothecia dark red-brown, shiny, scattered or crowded together, 0.6–1.8(–3) mm in diameter; margins thallus-colored, thick or thin, usually smooth; amphithecium containing abundant small crystals; cortex distinct, typically 20–60 μm thick, about 2 times thicker at the base of the apothecium than at the sides; epihymenium clear red-brown, without any granules or crystals; spores mostly 13–17 × 6–8.5 μm. CHEMISTRY: Thallus and apothecial sections PD–, K+ yellow, KC–, C– (atranorin and characteristic triterpenes). HABITAT: Found on wood (or on rocks along the northeast coast) very close to the maritime shoreline, such as on logs at the upper edge of the beach. COMMENTS: The unique habitat of this lichen together with its striking dark red-brown apothecia make it fairly easy to recognize. *Lecanora allophana* has a thicker, less granular, and less defined cor-

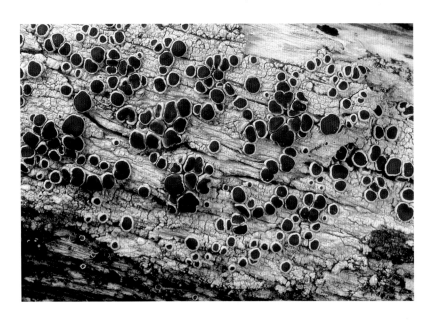

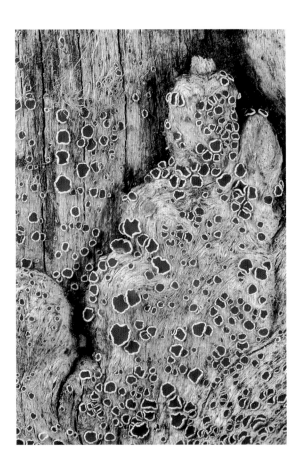

437. *Lecanora xylophila* Southeast Alaska ×2.4

438. *Lecanora zosterae* Southeast Alaska ×4.0

tex, with the amphithecial crystals radiating into the cortex; it is typically found on bark.

Lecanora zosterae
Flat-fruited rim-lichen

DESCRIPTION: Thallus growing within the substrate and virtually invisible at the surface. Apothecia broad (0.4–1.6 mm in diameter), thin, and flat, usually crowded and even overlapping, often undulating or folded, dull red-brown to smoky brown or almost black, sometimes lightly pruinose, with a thin, whitish to pale brown margin; cortex very thick, perfectly clear and gelatinous, 2–3 times thicker at the base of the apothecium than at the sides; spores ellipsoid to narrowly ellipsoid, 10–18 × 4.5–6.5 µm. CHEMISTRY: All reactions negative (no lichen substances). HABITAT: Common on lignum and dead vegetation in the north. COMMENTS: *Lecanora beringii* is a very closely related lichen found on calcareous rock, bones, and antlers in the Arctic. It never has pruina, and its amphithecial cortex is only slighter broader at the base than at the sides. The boundary between these two species is still uncertain. *Lecanora hagenii* has thicker, more heavily pruinose apothecia than does *L. zosterae*, and the apothecial margin is thicker. It grows on bark or, in the Arctic, on dead vegetation and wood.

Lecidea (136 N. Am. species)
Disk lichens, tile lichens

DESCRIPTION: Crustose lichens with black or dark brown, superficial to sunken, lecideine apothecia usually rimmed with disk-colored persistent margins. In section, most apothecial margins are black at the outer edge and paler inside; hypothecium usually dark (less frequently colorless); epihymenium brown or greenish (and then turning red with nitric acid); asci largely K/I– but with a thin K/I+ blue cap at the very tip (*Lecidea*-type; Fig. 14f); spores colorless, 1-celled, 5–20 × 2.5–10 µm, 8 per ascus. Thallus gray, brown, or a rusty orange caused by iron in the substrate; thick rimose to areolate or squamulose to almost lobate, or so reduced that it develops only within the substrate. Photobiont green (usually *Trebouxia* or *Chlorosarcinopsis*). CHEMISTRY: Medulla IKI– or IKI+ blue. Producing a variety of compounds including norstictic, stictic, confluentic, and planaic acids. HABITAT: In the strict sense, *Lecidea* consists only of rock-dwelling lichens. COMMENTS: The genus *Lecidea* has yet to be studied in its entirety, certainly in North America. Some taxa on bark, wood, moss, or soil are included that, without a doubt, will eventually find their home in other genera. Ideally, only lichens with *Lecidea*-type asci should probably be included. At the moment, other lichens with 1-celled spores and lecideine apothecia are classified as *Lecidea* despite their substrate and ascus type. Taken in the broad sense, the disk lichens include more than 25 small genera defined on the basis of various microscopic characters of the fruiting bodies (especially the asci), as well as on chemistry. Although the currently accepted scientific names are used here, all the lichens with lecideine or biatorine margins and colorless, 1-celled spores are included in the key below to make comparisons easier. Only 3 of the 136 known North American species are illustrated in this book because, in the field, most simply look like black dots, and because much work remains to be done on their taxonomy.

KEY TO SPECIES (including *Carbonea*, *Fuscidea*, *Lecidella*, *Schaereria*, and *Tremolecia*)

1. On bark; ascus *Lecanora*-type. 2
1. On rock, wood, soil, or peat; asci various types 3

2. Paraphyses easily separating when mounted in water; hypothecium yellow-brown to colorless, clearly distinguished from exciple, which is dark at outer edge; widespread arctic to temperate, from Lake Superior and James Bay westward. . . . . [*Lecidella euphorea*]
2. Paraphyses sticking together when mounted in water; hypothecium dark purplish brown to black, merging with exciple, which is dark internally and almost colorless externally; boreal forest region and west coast. . . . . ["*Lecidea*" *albofuscescens*]

3. On wood, soil, or mosses 4
3. Directly on rock. 6

4. Thallus dark brown or olive-brown, granular to almost isidiate, K–, PD–, C+ pink (gyrophoric acid) or C–; apothecia dull red-brown to black, flat to strongly convex; paraphyses branched; hypothecium merging with exciple, dark brown; ascus K/I– (*Trapelia*-type). . . . . *Placynthiella*
4. Thallus whitish, extremely thin and often within substrate, or more or less verrucose, K+ yellow, C– (atranorin); apothecia black, margins thin to disappearing; paraphyses unbranched; hypothecium colorless or brown; ascus K/I+ blue (*Lecanora*-type) 5

5. On wood, temperate (mostly northeastern, Black Hills, Colorado); apothecia shiny, flat to convex; exciple dark brown at edge, pale internally; hypothecium pale yellowish brown or colorless; spores narrow, 8–10(–12) × 3–4 µm; thallus PD+ red (fumarprotocetraric acid). . . . . [*Pyrrhospora elabens*]
5. On soil and moss, arctic and alpine tundra; apothecia dull, soon convex; exciple brown or greenish at edge, brown internally; hypothecium red-brown; spores broad, 10–16 × 7–8 µm; thallus PD– . . . . [*Lecidella wulfenii*]

6. Cephalodia present. *Amygdalaria*
6. Cephalodia absent 7

7. Thallus distinctly rusty orange or bright orange. 8
7. Thallus not bright orange 10

8. Apothecia concave, immersed in thallus, usually less than 0.5 mm in diameter . . . . *Tremolecia atrata*
8. Apothecia flat when mature, sessile, usually more than 0.5 mm in diameter 9

9. Spores 14–24 × 6–11 µm, halonate (visible at least in ink preparation); epihymenium olive-brown; paraphyses abundantly branched; medulla IKI– . . . . *Porpidia flavocaerulescens*
9. Spores 9–15 × 4.5–8 µm, without a halo; epihymenium green; paraphyses predominantly unbranched except for tips; medulla IKI+ blue . . . . orange form of *Lecidea lapicida*

10.(7) Thallus red-brown to dark yellow (not rusty orange), shiny, thick, areolate to dispersed areolate over the black prothallus, sometimes appearing almost squamulose 11
10. Thallus gray to white or greenish gray to brownish gray, continuous or areolate, not appearing squamulose, sometimes disappearing (endolithic) 12

11. Thallus reddish brown; medulla IKI+ blue; apothecia immersed between the areoles; ascus tips K/I– (*Lecidea*-type). . . . . *Lecidea atrobrunnea*
11. Thallus dark yellow to bronze; medulla IKI–; apothecia superficial, not between areoles; ascus tips K/I+ blue (*Bacidia*-type). . . . . *Tephromela armeniaca*

12. Hypothecium brown to black, sometimes greenish black. 13
12. Hypothecium colorless to yellowish 18

13. Spores under 3.5 µm wide; spores not halonate. 14
13. Spores over 3.5 µm wide; spores halonate or not 15

14. Apothecia 1–2.5 mm in diameter; thallus usually endolithic and not visible; hymenium under 50 µm high; on alpine and arctic rocks . . . . [*Lecidea auriculata*]
14. Apothecia usually under 0.5 mm in diameter; thallus thin and indistinct or rimose areolate and dark green-

439. *Lecidea atrobrunnea* Rocky Mountains, Colorado ×4.2

ish gray; hymenium 50–60(–80) μm; typically on boulders, small stones, and pebbles; widespread, East Temperate . . . . [*Micarea erratica*]

15. Spores narrow, 8.5–15 × 3.5–5.5 μm. Thallus thin, rimose, or endolithic; apothecia flat, 0.3–1 mm in diameter, with thin, prominent rims; epihymenium and upper part of hymenium emerald green; hypothecium and exciple black, not distinguishable; mostly western arctic and Rocky Mountains . . . . [*Carbonea vorticosa*]
15. Spores more than 5 μm wide 16

16. Paraphyses branched and anastomosing, coherent in a water mount; spores halonate, (10–)13–24 × 6–12 μm . . . . *Porpidia* key
16. Paraphyses predominantly unbranched, easily separated in a water mount; spores not halonate, (8–)10–16 × 6–9 μm 17

17. Thallus white, areolate to verrucose, K+ yellow, C– (atranorin); widespread . . . . [*Lecidella carpathica*]
17. Thallus pale yellowish to greenish yellow, rimose to areolate, C+ orange (xanthones); on rocks along the Pacific coast . . . . [*Lecidella elaeochromoides*]

18.(12) Epihymenium and exciple shades of brown; asci *Fuscidea*-type (Fig. 14d); thallus gray-brown or dark gray, often forming quilt-like patches outlined by black . . . . *Fuscidea*
18. Epihymenium and often the outer layer of the exciple shades of green; asci not *Fuscidea*-type; thallus pale to dark gray, not brownish 19

19. Paraphyses easily separating in water or K; spores 5–9 (–10) μm wide 20
19. Paraphyses coherent in water and K; spores less than 7.5 μm wide 22

20. Apothecia remaining immersed between thallus areoles for a long time; thallus C+ pink (gyrophoric acid); asci cylindrical, thin-walled including tip, K/I– (Fig. 14i) . . . . [*Schaereria fuscocinerea*]
20. Apothecia superficial, not immersed; thallus C– or C+ orange; asci club-shaped with a thick tip that is K/I+ dark blue 21

21. Hymenium containing abundant, tiny oil drops; western montane to arctic, on siliceous rocks. . . . . [*Lecidella patavina*]
21. Hymenium without oil drops; widespread temperate to arctic, usually on calcareous rocks . . . . *Lecidella stigmatea*

22.(19) Prothallus lacking or inconspicuous; thallus gray to orange or mottled with orange areas, continuous or, in part, cracked into areoles . . . . *Lecidea lapicida*
22. Prothallus black, conspicuous 23

23. Thallus white to pale gray, tile-like (areolate) with the black prothallus showing at the margin; medulla IKI+ blue, K–; asci K/I– (*Lecidea*-type) . . . . *Lecidea tessellata*
23. Thallus whitish to pale yellow, continuous to areolate, but not tile-like; medulla IKI–, K+ yellow (atranorin and sometimes usnic acid); asci K/I+ blue at tips (*Lecanora*-type) . . . . *Lecanora marginata*

Lecidea atrobrunnea
Brown tile lichen

DESCRIPTION: Thallus shiny reddish brown, areolate to almost squamulose, the individual roundish areoles 0.4–1.5(–3) mm in diameter, flat to convex, usually on a conspicuous black prothallus. Apothecia pure black, flat to convex, with thin, raised, persistent margins; epihymenium green, hypothecium pale brown, hymenium low (40–60 μm); spores 5–12 × 3–4 μm. CHEMISTRY: Medulla IKI+ blue-black, PD–, K–, KC–, C– (2'-O-methylperlatolic acid). HABITAT: On exposed siliceous rocks. COMMENTS: **Lecidea brunneofusca,** a similar northeastern species with the same medullary reactions, has a duller, more grayish brown thallus. **Lecidea syncarpa,** known from the Rockies, is much like *L. atrobrunnea* but has norstictic acid in the thallus cortex (K+ red) and paler brown areoles. **Lecidea fuscoatra,** a widespread species with pale brown to gray-brown areoles, has an IKI– medulla and a C+ pink cortex (gyrophoric acid). Several other brown areolate-squamulose *Lecidea* species, especially in the western mountains, differ in their thallus and hypothecium color, chemistry, and other features of the apothecium. See also *Tephromela armeniaca*.

Lecidea lapicida
Gray-orange disk lichen

DESCRIPTION: Thallus gray to rusty orange, sometimes with patches of both colors, rimose-areolate, or largely disappearing (endolithic), without a prothallus. Apothecia pure black, broadly attached on the thallus surface or somewhat immersed, with distinct black margins; epihymenium dark green; hypothecium colorless or brownish; exciple almost colorless inside but with a very dark green-black edge; spores 9–15 × 4.5–8 μm. CHEMISTRY: Medulla PD+ orange or yellow, K+ yellow or red, KC–, C–, IKI+ blue-black (stictic or norstictic acid). HABITAT: On siliceous rock such as granite in open areas. COMMENTS: The orange form depicted in plate 440 is the same color as **Rhizocarpon oederi,** *Tremolecia atrata, Porpidia flavocaerulescens,* and forms of *P. macrocarpa.* All these are negative to reagents and can be distinguished by their spores. **Lecidea plana** has a moderately thick, gray thallus and flat, scalloped apothecia with thin, raised margins. It is very close to *L. lapicida* but is K– (containing planaic acid), and the spores are somewhat smaller.

Lecidea tessellata
Tile lichen

DESCRIPTION: Thallus chalky white to blue-gray, forming round patches consisting of well-defined, tile-like areoles bordered by a conspicuous charcoal gray to black prothallus. Apothecia black, sunken between the areoles, or in some forms, prominent; hypothecium and internal parts of exciple almost colorless; epihymenium greenish black; spores small, ellipsoid, 6–13 × 3.5–7 μm. CHEMISTRY: Thallus surface and medulla PD–, K–, KC–, C–, IKI+ blue (confluentic acid). HABITAT: Usually on noncalcareous rock but occasionally on other rock types; in full sun. COMMENTS: The tile lichen can be recognized most easily by its pale, tile-like thallus and black sunken apothecia together with the IKI+ blue reaction. *Lecidea lapicida* is very similar but has slightly larger spores (9–15 μm long) with thinner walls, and it

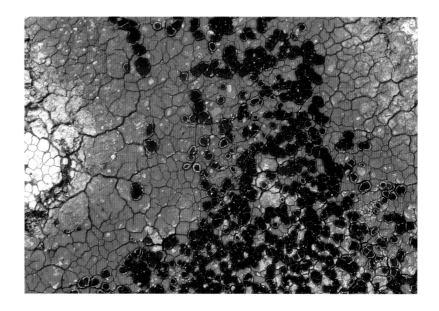

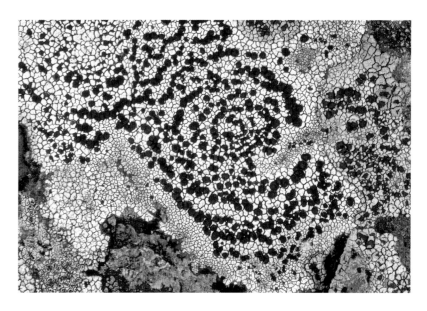

440. *Lecidea lapicida* Olympics, Washington ×4.2

441. *Lecidea tessellata* Lake Superior region, Ontario ×1.7

442. *Lecidella stigmatea* Lake Superior region, Ontario ×5.4

contains at least traces of stictic acid. The apothecia of *L. lapicida* typically have more persistent margins. ***Lecidea confluens*** is much like *L. tesselata* (it also contains confluentic acid), but the hypothecium is pale to dark brown and is easily distinguished from the exciple. In addition, its apothecia are not immersed at maturity and are occasionally pruinose, and its spores are a bit larger (9.5–15 × 4–6 μm).

Lecidella (21 N. Am. species)
Disk lichens

DESCRIPTION: Crustose lichens, most with well-developed, gray thalli but some with imperceptible thalli. Photobiont green (unicellular, including *Chlorella*). Apothecia lecideine, pitch black and often shiny, with black margins level with the disk or prominent; epihymenium and outer edge of exciple usually blue-green or at least olive, rarely brown; hypothecium colorless to brown; paraphyses usually separate easily even when mounted in water (i.e., not glued together in a mass, as in most lichens), unbranched, hardly thickened at the tips; hymenium sometimes contains tiny oil droplets; asci with an IKI+ dark blue tholus (*Lecanora*-type); spores typically broadly ellipsoid, often reaching 9 μm wide, with a distinct wall. CHEMISTRY: Cortex containing atranorin or any of a variety of xanthones that are KC+ orange and C+ orange or that fluoresce in long-wave UV light. HABITAT: On rocks (especially those containing calcium), bark, wood, or soil. COMMENTS: In the field, *Lecidella* may be hard to distinguish from other black disk lichens such as *Lecidea, Porpidia,* or *Buellia*. They are, however, fairly easy to recognize under the microscope by the combination of their easily separating paraphyses, greenish tissues, and broad spores. The darkly staining ascus tip can be used as a confirmation.

KEY TO SPECIES: See *Lecidea*.

Lecidella stigmatea
Disk lichen

DESCRIPTION: Thallus extremely variable, thin, growing within the rock, or fairly well developed and rimose-areolate to verruculose as in the picture, dirty gray to yellowish white. Apothecia mostly 0.4–1.2 mm in diameter, flat or convex; margins barely visible or prominent; epihymenium greenish; hypothecium colorless or yellowish (never brown); exciple almost colorless within and greenish black at the edge, without crystals; spores 10–16 × 6–9 μm. CHEMISTRY: Thallus PD–, K+ yellow, KC–, C– (zeorin and atranorin or, rarely, lichexanthone). HABITAT: Usually on calcium-containing rocks, especially sandstone, but also on granite. COMMENTS: This is by far the most common rock-dwelling *Lecidella*. The widespread, also saxicolous *L. carpathica* differs mainly in its dark yellowish brown hypothecium and darker exciple; it contains atranorin and diploicin (an orcinol depsidone). ***Lecidella patavina*** (syn. *L. spitsbergensis*), common from the Rockies to the Arctic, has the same chemistry as *L. stigmatea*, but its hymenium is filled with oil drops, its exciple contains crystals, and the thallus tends to be thicker. ***Lecidella elaeochromoides,*** on exposed rocks overlooking the Pacific, has a thick, yellowish thallus, with rounded, convex areoles. The yellowish color is due to xanthones (arthothelin) that react C+ orange and KC+ red-orange. Its apothecia are large (up to 1.2 mm), flat, pure black to almost bluish black, with raised, thin, black rims. The hypothecium is dark yellowish brown. Another western species, *L. asema*, is extremely similar to *L. elaeochromoides*. Its thallus is a bit thinner and is also C+ orange, but it contains different xanthones (asemone and thiophanic acid). ***Lecidella euphorea,*** a fairly common and widespread species growing on bark, has a gray to yellowish gray, minutely areolate thallus that reacts PD–, K–, KC– or KC+ pale orange, C–

(xanthones, not arthothelin), and its hypothecium is yellow-brown. Its distinction from *L. elaeochroma,* which differs only in its C+ orange thallus (arthothelin), is not accepted by everyone. *Lecidella wulfenii* is a very common member of the arctic-alpine lichen community that inhabits decaying vegetation and peat and is easily recognized by its pitch black apothecia, which stand out against an almost pure white thallus.

Lecidoma (1 species worldwide)

KEY TO SPECIES: See Key F.

Lecidoma demissum
Brown earth-crust

DESCRIPTION: Thick, dark brown to dark grayish brown crustose lichen, with a smooth or areolate thallus that follows the contours of the substrate, occasionally forming tiny rosettes, and often developing a thickened and slightly lobed margin; lower surface black, attached by black hyphae. Photobiont green (unicellular). Apothecia broad, 0.5–2(–4) mm in diameter, irregularly shaped, dark brown, thin, biatorine, following the contours of the thallus. Epihymenium brown; exciple dark brown; hypothecium colorless; paraphyses forked only at the tips, which are expanded and pigmented; asci K/I+ light blue, with a darkly staining short tube structure in the lower half of the tholus (like an abbreviated *Porpidia*-type structure); spores colorless, 1-celled, ellipsoid, 8 per ascus, 12–16 × 5.5–7 μm. CHEMISTRY: All reactions negative (no lichen substances). HABITAT: On soil and decaying vegetation in alpine to subalpine habitats and in the Arctic. COMMENTS: This species sometimes resembles a *Toninia* but has 1-celled spores, which would be unusual for a *Toninia*. The ascus of *Toninia* has a darkly staining *Bacidia*-type tholus. *Mycobilimbia* species, especially *M. lobulata* (syn. *Toninia lobulata*), are also similar, but these never have a dark lower cortex, and the spores of *M. lobulata* are 2–4-celled. If the thallus of the *Lecidoma* breaks up into small patches, it can have a squamulose appearance and resemble a *Psora*. *Psora*, however, has a different thallus anatomy, discrete round apothecia that do not fuse into irregular shapes, and a more deeply staining ascus tip.

443. *Lecidoma demissum* Coast Mountains, British Columbia ×4.9

Leioderma (1 N. Am. species)

KEY TO SPECIES: See *Peltigera*.

Leioderma sorediatum
Treepelt lichen, mouse ears

DESCRIPTION: Small gray foliose lichens (up to 3 cm across) containing cyanobacteria (*Scytonema*) in a distinct layer; lobes scalloped, flat to slightly concave, ca. 2–4 mm wide; blue-gray granular soredia developing along the lobe margins; medulla white; lower surface webby, lacking a cortex or veins, with scattered tufts of rhizines; upper surface felt-like with tiny, very appressed hairs. Apothecia absent. CHEMISTRY: Upper cortex K+ orange; medulla (and lower surface) PD–, K+ slowly turning pale orange, KC–, C– (ursolic acid, a triterpene). HABITAT: On shrubs (e.g., huckleberry and manzanita) and mossy conifer branches in humid coastal forests. COMMENTS: This extremely rare species is known from only two localities in North America. It looks very much like a tiny *Peltigera* but has no veins below. It is probably most similar to *Erioderma sorediatum* but has longer lobes (less shell-like), is paler gray, and has many more tufted rhizines on the lower surface. The tomentum on the upper surface of *Erioderma* is long, erect, and tufted. In addition, *E. sorediatum* is PD+ red-orange (and K– in the cortex).

444. *Leioderma sorediatum* Oregon coast ×2.4

Lepraria (12 N. Am. species)
Dust lichens

DESCRIPTION: Sterile crustose lichens consisting of nothing but a thick or thin, continuous layer of soredia or sometimes granule-like soredia (actually, consoredia, which are spherical, often fuzzy aggregations of a few individual soredial particles and look like tiny cotton balls). Thallus margins indistinct or, less commonly, clearly defined (but without ascending lobes); thallus blue-gray, gray, green, or yellowish green. Photobiont green (unicellular). Apothecia not present. CHEMISTRY: A variety of compounds including atranorin, stictic acid, and several fatty acids. HABITAT: On bark, wood, soil, or rock in humid, shady sites where they derive all their moisture from the air. In fact, most species of *Lepraria*, like some other leprose lichens, cannot absorb liquid water at all. Raindrops simply bead on the powdery lichen, which is unable to absorb the water because of surface tension. This is undoubtedly why the dust lichens are largely restricted to humid, rocky overhangs and tree bases. COMMENTS: Species of *Lepraria* are distinguished on the basis of subtle differences in color, thallus thickness, substrate, and especially chemistry. Unfortunately, the genus has not yet been studied systematically in North America, although we do know that many European species also occur here. Species of *Leproloma* closely resemble *Lepraria* but often form rounded, membranous, ascending lobes at the thallus margin (as in *L. membranaceum*) and have a distinct cottony hypothallus that is white or brown. They are chemically distinct as well. Other common leprose lichens that are frequently or always sterile include the brilliantly yellow species of *Chrysothrix*, the pale yellow-green *Lecanora thysanophora*, and the creamy yellow *Pyrrhospora quernea*.

KEY TO SPECIES (including *Leproloma*)
1. Thallus forming round, adnate, or somewhat ascending lobes 3–6 mm across; brownish or white cottony hypothallus present; containing pannaric and roccellic acids; on shaded rock faces . . . . . . *Leproloma membranaceum*
1. Thallus not forming well-developed, ascending, round lobes; hypothallus present or absent; pannaric acid absent; on rock or other substrates 2

2. Thallus coarsely granular sorediate, not cottony; margins irregular or, when on rocks, forming distinct, often concentric rings about 20–40 mm in diameter, pale to dark gray; hypothallus entirely absent; on siliceous rocks or moss; thallus K–, PD+ yellow or red (alectorialic, fumarprotocetraric, or psoromic acid) . . . . *Lepraria neglecta*
2. Thallus thick and cottony, never forming circular patches, but occasionally lobed, usually pale green to yellowish green, rarely greenish gray; brownish hypothallus frequently present; on shaded rock, bark, or moss; thallus K+ yellow, PD+ orange (atranorin, stictic acid, and zeorin) . . . . *Lepraria lobificans*

Lepraria lobificans (syn. *L. finkii*)
Fluffy dust lichen

DESCRIPTION: Thallus yellowish green to pale mint green, thick and cottony, without well-defined ascending lobes (as in *Leproloma*); hypothallus absent or brownish and fairly distinct. CHEMISTRY: Thallus PD+ orange, K+ yellow, KC–, C– (atranorin, stictic and constictic acids, and zeorin). HABITAT: Common on tree bases, shaded rocks, and mosses. COMMENTS: *Lepraria lobificans* is one of the most common dust lichens in North America. Its thick, almost yellow-green thallus is fairly characteristic. *Lepraria incana*, an-

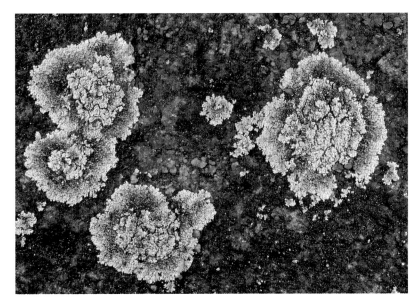

other dust lichen occurring on both rocks and bark, has a thinner bluish gray thallus and contains zeorin and divaricatic acid (UV+ bright white). Its distribution in North America is still uncertain. An apparently unnamed *Lepraria* with a thin, distinctly blue-gray thallus has often been called *L. incana*, but it contains atranorin and zeorin. It is very common in the northeast on both rocks and tree bases.

Lepraria neglecta (syn. *L. zonata*)
Zoned dust lichen

DESCRIPTION: Thallus blue-gray, coarsely granular sorediate, often forming distinctive rings or concentrically arranged "zones" when growing on rock. These rings can rarely be seen on specimens growing over moss. CHEMISTRY: Usually PD+ deep yellow, K–, KC+ red, C+ pink (alectorialic acid); occasionally PD+ red, K–, KC–, C– (fumarprotocetraric acid); rarely PD+ yellow, K–, KC–, C– (psoromic acid). HABITAT: Very common on partially shaded or exposed granitic rocks; perhaps also on mosses in arctic and alpine sites. COMMENTS: The presence of rather large, blue-gray granules forming a kind of super-leprose thallus (usually in circular rosettes) is the best distinguishing feature of this species. Saxicolous material with fumarprotocetraric acid can be called **L. caesioalba,** but we consider it to be a chemical race of *L. neglecta*. A lichen that contains fumarprotocetraric acid and closely resembles *L. neglecta* but grows on tree bark may or may not be the same species; it is not included in our distribution map. A common arctic-alpine species in this group (with alectorialic acid) grows on mosses and peat in the tundra and may be included in *L. neglecta* until the genus is fully studied. However, it too is not included in our map.

Leprocaulon (5 N. Am. species)
Cottonthread lichens

DESCRIPTION: Dwarf fruticose lichens, yellowish to gray, with very slender, erect, solid stalks under 20 mm tall and 0.3 mm thick, bearing cottony granules. Primary thallus either absent or indistinct. Photobiont green (*Trebouxia*). Apothecia not developing. CHEMISTRY: A variety of compounds including atranorin,

445. *Lepraria lobificans* (damp) southern Ontario ×3.8

446. *Lepraria neglecta* Lake Superior region, Ontario ×2.7

447. *Leprocaulon gracilescens* White Mountains, eastern Arizona ×5.4

fatty acids, and many β-orcinol depsidones and orcinol depsides. HABITAT: On soil and rock, often among mosses. COMMENTS: Colonies of *Leprocaulon* resemble thick patches of *Lepraria* or other leprose crustose lichens except for the slender stalks. The species are defined largely on the basis of chemistry.

KEY TO SPECIES: See Key B.

Leprocaulon gracilescens

DESCRIPTION: Thallus pale gray to white, with slender (0.15–0.3 mm thick) stalks usually under 10 mm tall, sparsely or abundantly branched, the stalks bearing at the top and on the sides spherical, fuzzy granules (lacking a cortex); bases of the stems slightly browned. CHEMISTRY: PD–, K+ yellowish (often pale), KC–, C– (atranorin and rangiformic acid). HABITAT: On soil between rocks at high elevations (over 1500 m). COMMENTS: This lichen looks like a cross between a *Lepraria* (granules only, no stalks) and a *Stereocaulon* (stiff and corticate stalks and more flattened or lobed phyllocladia, normally with a cortex). The latter usually also has cephalodia. The rare **L. albicans** (known from coastal Southeast Alaska) is PD+ yellow, containing baeomycesic and squamatic acids. In fact, *L. gracilescens* is scarcely more than a chemical race of that species. *Leprocaulon subalbicans* is fairly common in the mountains in parts of British Columbia and Alaska south to Colorado. Its stalks lack a cartilaginous central core, unlike those of *L. gracilescens* or *L. albicans,* and they are barely more than vertical piles of fluffy granules. *Leprocaulon subalbicans* has several chemical races and can produce psoromic, baeomycesic, or thamnolic acids as well as atranorin. The race with thamnolic acid (PD+ orange, K+ deep yellow) is the most common and widespread. *Leprocaulon microscopicum,* a species seen mainly along the Oregon and Washington coasts, is yellowish green with slender, stiff, almost unbranched stalks 2–6 mm tall, covered with fuzzy or mealy granules. It is the only yellowish *Leprocaulon*. Its chemistry and fuzzy granules also make it unique. The thallus is PD–, K+ yellow, KC+ yellow, and C– (usnic acid, atranorin, zeorin, sometimes rangiformic acid and other fatty acids).

Leproloma (4 N. Am. species)
Lobed dust lichens

DESCRIPTION: Leprose crustose lichens, yellowish green to gray or blue-gray, forming thin or thick heaps of noncorticate soredia or granules that sometimes coalesce into rounded, bracket-like lobes; developing a thin or thick, white to dark brown cottony hypothallus below the layer of soredia or granules. Photobiont green (unicellular). Apothecia unknown. CHEMISTRY: Characteristically with dibenzofuranes such as pannaric, oxypannaric, or porphyrillic acids; some species with roccellic or, rarely, gyrophoric acids. Dibenzofuranes are absent in a few species. HABITAT: In humid habitats such as shaded rock walls, and on mossy rock and peat in arctic or alpine sites. COMMENTS: This character-poor genus is defined mainly by its leprose thallus, cottony hypothallus, and chemistry. See Comments under *Lepraria*.

KEY TO SPECIES: See *Lepraria*.

Leproloma membranaceum

DESCRIPTION: Thallus yellowish green to greenish white, forming small rounded lobes 2–5 mm across, lifting at the lower edge and becoming shingle-like; a white to brownish gray cottony hypothallus always present (but sometimes thin) under the layer of soredia. CHEMISTRY: Thallus PD– or pale orange, K– or yellowish, KC–, C– (pannaric and roccellic acids, and atranorin). HABITAT: On shaded, humid rock walls in forests. DISTRIBUTION: Probably from the southern boreal to temperate regions; appears to be mainly eastern. COMMENTS: No other completely leprose lichen develops slightly ascending lobes. Some North American material has a distinctly bluish gray thallus but is otherwise similar to *L. membranaceum* in appearance. However, it contains protocetraric or fumarprotocetraric acid instead of pannaric acid and may represent another species of *Leproloma* or even a *Lepraria*. **Leproloma vouauxii** resembles *L. membranaceum* in color but lacks well-developed marginal lobes. It is PD+ reddish orange and K+ yellow, with pannaric acid, pannaric acid-6-methylester, and some unknown compounds.

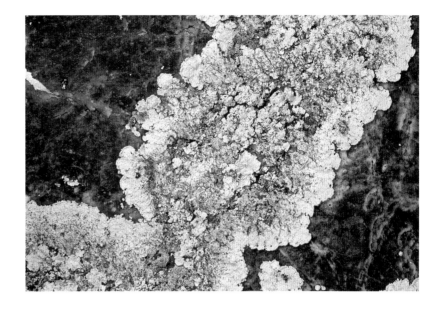

Leptochidium (1 N. Am. species)

KEY TO SPECIES: See *Leptogium*.

Leptochidium albociliatum
Whiskered jelly lichen

DESCRIPTION: Thallus foliose, blackish brown or olive when dry, becoming dark blue-green to black and very jelly-like when wet, like *Collema* and *Leptogium*; lobes 2–7 mm across with stiff, tiny, colorless hairs all along the margins; upper surface with isidia that become flattened and lobe-like when older (often bearing the little white hairs); both surfaces with cell-like cortices; photobiont more or less confined to clearly defined, thick layers just inside the cortices on both sides of the thallus, leaving a thin but distinct white medullary layer in between; lower surface with a thick tomentum of white or brown hairs. Photobiont blue-green (reported to be *Scytonema* but closely resembling *Nostoc*). Red-brown lecanorine apothecia often produced on the lobe surface; apothecial margins usually fringed with white hairs. Spores colorless, 2-celled, ellipsoid, 8 per ascus. CHEMISTRY: All reactions negative (no lichen substances). HABITAT: On mossy rocks, rarely on soil,

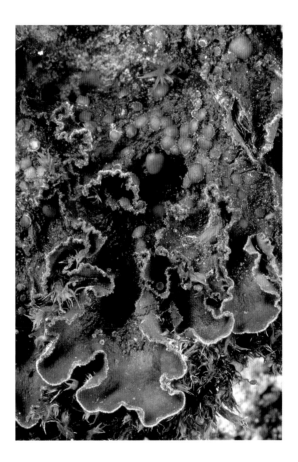

448. *Leproloma membranaceum* Ouachitas, Arkansas ×1.6

449. *Leptochidium albociliatum* Klamath Mountains, northwestern California ×3.0

in dry, open forests or in protected, shady woods. COMMENTS: *Leptochidium* superficially resembles some of the isidiate *Leptogium* species such as *L. saturninum* and **L. burnetiae** that have abundant white hairs on the lower surface; the stiff white "whiskers" on the lobe margins of *Leptochidium* distinguish them. The presence of definite algal and medullary layers in the thallus is also diagnostic for the genus.

Leptogium (53 N. Am. species)
Jellyskin lichens, vinyl lichens

DESCRIPTION: Thin foliose jelly lichens, very small and inconspicuous to medium-sized; lobes 0.1–6 mm wide, bluish gray to olive or brown and often shiny when dry, with a cellular upper and usually lower cortex (Fig. 7b), and some species with cellular fungal tissue (pseudoparenchyma) throughout; lower surface smooth or hairy-tomentose, the same color as the upper surface or somewhat darker. Photobiont blue-green (*Nostoc*), not confined to a distinct layer within the thallus. Lecanorine apothecia common on many species; spores colorless, 4-celled to muriform, 4–8 per ascus. CHEMISTRY: All reactions negative (no lichen substances). HABITAT: On bark, soil, rock, or mosses in very wet or dry habitats. COMMENTS: *Leptogium* and *Collema* have much in common and are sometimes difficult to distinguish in the field, especially when they are moist and swollen (as in many of the photographs shown here). On the whole, species of *Leptogium* are grayer (often steel-gray) or more reddish brown when dry, with a smoother, more skin-like texture. (Colors in the *Leptogium* descriptions normally refer to dry thalli.) *Collema* is never shiny, probably because it has no upper cortex. *Leptochidium* is very much like certain species of *Leptogium*. See Comments under *Leptochidium*.

KEY TO SPECIES (including *Leptochidium* and *Hydrothyria*)

1. Aquatic, on submerged rocks in the western mountains 2
1. In dry or moist habitats, not submerged in water 3
2. Lobes 3–10 mm wide, ascending and fan-like . . . . *Hydrothyria venosa*
2. Lobes 0.2–1.5 mm wide, closely attached to rocks . . . . [*Leptogium rivale*]
3. Lower surface tomentose with a mat of white hairs . . . 4
3. Lower surface smooth, or wrinkled, not hairy (except for scattered tufts in *Leptogium platynum*) 11
4. Isidia and lobules absent; Arizona and New Mexico . . 5
4. Isidia or lobules present on upper surface or margins of lobes 6
5. Surface conspicuously wrinkled . . . *Leptogium rugosum*
5. Surface smooth . . . . [*Leptogium burgessii*]
6. Thallus lobulate, not isidiate 7
6. Thallus isidiate, but isidia can be somewhat flattened in some species 8
7. Lobules confined to the lobe margins; tomentum white, conspicuous, well over 100 μm long, composed of threads with long, cylindrical cells; southwestern . . . . [*Leptogium burgessii*]
7. Lobules on lobe surface as well as margins; tomentum consisting of a faint fuzz made up of threads less than 100 μm long with spherical cells; Appalachians to Ontario and New Brunswick . . . . [*L. laceroides*]
8. Fine, stiff, colorless hairs on the tips or upper surface of the lobes; isidia flattened and lobe-like when older, often bearing fine hairs . . . . *Leptochidium albociliatum*
8. Fine hairs absent or sparse and confined to lobe surface; isidia cylindrical to granular, never flattened or hairy 9
9. Surface of lobes smooth, not wrinkled; isidia granular or cylindrical and branched 10
9. Surface of lobes clearly wrinkled; isidia cylindrical or slightly thicker at tips . . . *Leptogium pseudofurfuraceum*
10. Thallus dark olive-brown or olive-gray when dry, never with fine, colorless hairs; isidia predominantly granular; common . . . . *Leptogium saturninum*
10. Thallus slate gray, without an olive tint when dry, sometimes with scattered, fine hairs on the surface; isidia predominantly cylindrical; uncommon . . . . [*Leptogium burnetiae*]
11.(3) Thallus slate gray 12
11. Thallus brown at least when dry, or olive, black, or greenish gray 22
12. Isidia present, cylindrical or, in part, flattened, resembling narrow lobules; apothecia abundant or rare . . . 13
12. Isidia absent (lobules present or absent); apothecia usually present and abundant 15
13. Upper surface of lobes smooth and even . . . . *Leptogium cyanescens*
13. Upper surface of lobes wrinkled 14
14. Southeastern; isidia flattened to cylindrical; thallus less than 200 μm thick . . . . . *Leptogium austroamericanum*
14. Western montane; isidia cylindrical; thallus more than 200 μm thick . . . . [*Leptogium arsenei*]

15.(12) Lobes curled inward and erect, forming tube-like tips . . . . *Leptogium corniculatum*
15. Lobes flat, convex, concave, or undulating and crisped at the margins, not forming tubes. 16
16. Apothecial margins with abundant lobules; apothecia restricted to the thallus margins . . . . *Leptogium marginellum*
16. Apothecial margins smooth and even or thickly wrinkled, with or without lobules; apothecia on the lobe surface. 17
17. Apothecia constricted at the base and somewhat raised 18
17. Apothecia broadly attached, adnate or sunken into depressions. 19
18. Lobes strongly wrinkled . . . . *Leptogium corticola*
18. Lobes not wrinkled . . . . *Leptogium azureum*
19. Apothecia 2–7 mm in diameter, margins thickly wrinkled; southeastern U.S. . . . . . *Leptogium phyllocarpum*
19. Apothecia 0.2–0.7 mm in diameter, margins smooth and even; western U.S. and Canada 20
20. Rounded, overlapping lobules on thallus surface; thallus 200–500 μm thick, often attached to substrate by tufts of pale hairs . . . . *Leptogium platynum*
20. Finely divided lobules on the lobe margins, or margins fairly even; thallus thin and membranous, less than 200 μm thick, attached directly to substrate, lacking hairs 21
21. Ascospores 4 per ascus; on bark . . . . *Leptogium polycarpum*
21. Ascospores 8 per ascus; on mossy rock, soil, or bark . . . . *Leptogium gelatinosum*
22.(11) Isidia present. 23
22. Isidia absent 24
23. On mossy calcareous rock; lobe margins finely divided, resembling isidia (Fig. 32); apothecia often abundant . . . . *Leptogium lichenoides*
23. On bark; lobe margins scalloped; granular to cylindrical isidia on lobe margins and ridges; apothecia infrequent. . . . . *Leptogium milligranum*
24. Lobes curled inward, forming a tube . . . . *Leptogium corniculatum*
24. Lobes flat, concave, or undulating and crisped at the margins 25
25. Apothecia raised, especially on the crests of ridges and folds; mainly in the southeastern coastal plain . . . . *Leptogium chloromelum*
25. Apothecia adnate or sunken into depressions; mainly western and northern 26
26. Rounded, overlapping lobules on thallus surface; thallus over 200 μm thick, often attached to substrate by tufts of hairs . . . . *Leptogium platynum*
26. Lobules, if present, marginal; thallus thin and membranous, less than 200 μm thick, lacking hairs on the lower surface 27
27. Margins very finely divided (Fig. 32); thallus cushion forming; widespread, especially among mosses over limestone . . . . *Leptogium lichenoides*
27. Margins usually smooth and even, or somewhat lobulate; Pacific Northwest. 28
28. Ascospores 4 per ascus; on bark . . . . *Leptogium polycarpum*
28. Ascospores 8 per ascus; on mossy rock, soil, or bark . . . . *Leptogium gelatinosum*

Leptogium austroamericanum
Dixie jellyskin

DESCRIPTION: Thallus steel-gray with round lobes 2–5(–10) mm across; surface of lobes conspicuously but minutely wrinkled and covered with cylindrical to flattened isidia, which are also present on the lobe margins. In many specimens, isidia broaden into overlapping lobules; lower surface the same as the upper in color. Apothecia rare. HABITAT: On hardwood trees. COMMENTS: This species closely resembles *L. cyanescens* except for the minutely wrinkled thallus.

Leptogium azureum
Blue jellyskin

DESCRIPTION: Thallus blue-gray, with no isidia; lobes thin, folded, and rounded at the margins, 2–7 mm across; lobe surface, although not distinctly wrinkled, sometimes uneven or roughened in the older parts of the thallus. Apothecia commonly produced on the thallus surface, cup-like and constricted at the base, with red-brown disks and thin brownish to thallus-colored margins. HABITAT: On bark of hardwood trees. COMMENTS: *Leptogium cyanescens* is about the same size and color but has marginal and laminal isidia. *Leptogium austroamericanum* and *L. corticola*, often found in the same habitats, have distinctly wrinkled lobes.

450. *Leptogium austro-americanum* western Mississippi ×4.0

451. *Leptogium azureum* northern Florida ×3.8

Leptogium chloromelum
Ruffled jellyskin

DESCRIPTION: Thallus olive to yellowish brown or partly grayish (not blue-gray), with broad, closely appressed to erect and anatomosing lobes, very heavily wrinkled, giving the thallus a ruffled appearance, without isidia or granules. Apothecia abundant, raised or on the crests of folded lobes, with thick, often wrinkled margins; spores partly muriform, 4–7-celled, 20–30 × 9–12 μm. HABITAT: On trees, especially oaks, and sometimes on rocks, in shaded forests. COMMENTS: *Leptogium chloromelum* closely resembles *L. phyllocarpum*, a southeastern species that is distinctly blue-gray, with apothecial margins that are often lobulate. ***Leptogium floridanum*** is a similar species almost entirely restricted to Florida, where it is rather common. It has gray, more prostrate anastomosing lobes with thin, sharp wrinkles running the length of the lobes, building into warty, ruffled knots here and there. The abundant, flat apothecia usually have thick, frilled or ruffled margins. ***Lep-***

togium arsenei is a southwestern species with heavily wrinkled lobes. It is, however, granular isidiate, gray to brownish above and pale below. See also *L. millegranum*.

Leptogium corniculatum (syn. *L. palmatum*)
Antlered jellyskin

DESCRIPTION: Thallus large, usually very red-brown and shiny, at least on the sun-exposed lobe tips, but steel-gray at the base and in shade forms; lobes 1–6 mm wide, curled up into horn-like, erect, branched tubes when dry, with wrinkles running the length of the lobes but without soredia or isidia. Apothecia with red-brown disks, common. HABITAT: Forms luxuriant cushions on and between mossy rocks or directly on the soil in coastal regions. COMMENTS: No other jelly lichen, either *Leptogium* or *Collema*, forms such curled, erect lobes.

452. *Leptogium chloromelum* hill country, central Texas ×2.2

453. *Leptogium corniculatum* Klamath Range, northwestern California ×2.6

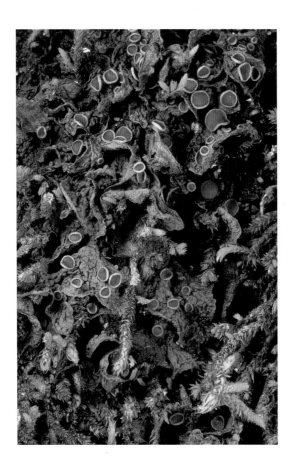

454. *Leptogium corticola* Ouachitas, Arkansas ×2.2

455. *Leptogium cyanescens* coastal Maine ×4.4

Leptogium corticola
Blistered jellyskin

DESCRIPTION: Thallus with broad rounded lobes 2–5 mm across with a distinctly wrinkled surface and puffy blister-like bumps, but without isidia or lobules. Apothecia common, yellow- to red-brown, concave with thin, smooth margins. HABITAT: On hardwoods or occasionally white cedar in the north; sometimes on mossy rocks. COMMENTS: This species closely resembles *L. azureum* except for the heavily wrinkled surface. The lobes never become ascending and anastomosing as in *L. chloromelum* or *L. phyllocarpum*.

Leptogium cyanescens
Blue jellyskin, blue oilskin

DESCRIPTION: Thallus blue-gray, with thin, smooth, spreading or folded lobes 2–4 mm across; lobe margins rounded or somewhat toothed or lobulate;

isidia cylindrical (often branched) to flattened and lobule-like, usually abundant on the lobe margins and upper surface; lower surface of lobes much like the upper surface, occasionally producing tufts of attachment hairs. HABITAT: On bark of all kinds, especially on tree bases, and on logs or mossy, shaded rocks. COMMENTS: This is the most common *Leptogium* in North America. Misidentifications are usually due to the varying quantity of isidia produced on the lobes, or to the slight roughness of the surface of older lobes being interpreted as wrinkles. Species often mistaken for *L. cyanescens* include *L. corticola, L. austroamericanum,* and *L. azureum*. **Leptogium denticulatum** is like *L. cyanescens* but with thin, scale-like lobes instead of cylindrical isidia occurring on the lobe surface and along the margins. It is a tropical species found mainly on rocks in the southwestern U.S. **Leptogium dactylinum** is a small, almost squamulose version of *L. cyanescens* with olive to gray lobes usually under 3 mm wide. Its isidia are long and commonly branched, developing on the lobe surface and margins. Small yellow- to reddish brown apothecia are abundantly produced among the isidia. It is an East Temperate lichen usually found on shaded, mossy limestone, but occasionally it grows on tree bases as well. **Leptogium subaridum** is brown or blackish rather than blue-gray. It is not uncommon in the northwest on dry or mossy rocks. It has distinctly cylindrical to somewhat flattened isidia on the lobe surfaces (sometimes both upper and lower) and finely dissected, lacy margins. The lobes are smooth, unlike the similarly lacy *L. lichenoides,* which has wrinkled lobes and no true isidia on the lobe surface.

456. *Leptogium gelatinosum* (damp) Klamath Range, northwestern California ×2.1

Leptogium gelatinosum (syn. *L. sinuatum*)
Petalled jellyskin

DESCRIPTION: Thallus rather variable, blue-gray to reddish brown where exposed to the sun; lobes 1–4 mm wide, appressed or more commonly ascending, sometimes forming cushions; surface slightly to strongly wrinkled but without isidia; margins typically rounded but sometimes toothed or divided into lobules. Apothecia very common on the lobe surface, 0.35–0.7 mm in diameter, at first partly immersed, later raised, becoming constricted at the base; spores muriform, 25–35 × 12–14 μm. HABITAT: Usually on mossy rocks and soil, rarely on trees. COMMENTS: When the lobes are somewhat indented, *L. gelatinosum* can resemble **L. californicum,** a species with narrower, more finely divided lobes (0.5–3 mm across). Both species are common along the west coast.

Leptogium lichenoides
Tattered jellyskin

DESCRIPTION: Thallus forming small, dark brown or red-brown cushions; lobes fairly wide (1–4 mm), wrinkled, upright, and ornamented on the edges with finely divided, almost cylindrical isidia-like outgrowths (Fig. 32). Apothecia fairly common on the lobe surface, red-brown, concave, 0.2–0.7 mm in diameter. HABITAT: On mossy calcareous rock. COMMENTS: This small, inconspicuous lichen is, nevertheless, widespread and common. The degree to which its marginal outgrowths resemble isidia varies, leading to problems in identification. **Leptogium californicum** is very similar to *L. lichenoides* in having more or less flat lobes standing on edge, with a slightly wrinkled surface and tiny finely divided lobules on the margins of the lobes. In general it is broader than *L. lichenoides,* and the wrinkles are less pronounced. Because intermediates are common, perhaps the species should not be recognized as distinct. **Leptogium subaridum** is a somewhat similar lichen with laminal isidia (see Comments under *L. cyanescens*). Forms of *L. lichenoides* that have particularly long and fine divisions on the lobe margins resemble some extremely minute species with lobes that are under 1 mm

457. *Leptogium lichenoides* (damp) southern British Columbia ×4.2

Fig. 32 *Leptogium lichenoides*: (a) dry thallus; (b) moistened and flattened lobes showing the finely divided, isidia-like margins. Scale = 1 mm. (From Brodo, 1988, fig. 39.)

458. *Leptogium marginellum* northern Florida ×4.1

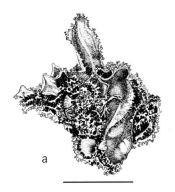

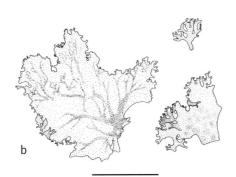

a   b

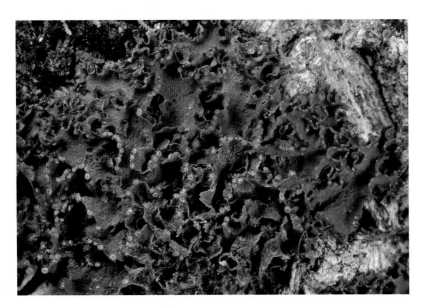

wide from top to bottom. These species differ from most *Leptogium* species in that their thallus is made up of pseudoparenchyma from top to bottom rather than only on the cortical surfaces. The most widespread of this group is **L. tenuissimum,** which has abundant concave aopthecia often fringed with erect, cylindrical, isidia-like branches 0.1–0.2 mm wide. It usually grows on sandy soil, less frequently on sandstone or bark.

Leptogium marginellum
Edge-fruiting jellyskin

DESCRIPTION: Thallus bluish gray; surface deeply wrinkled; lobes rounded, 1–5 mm wide; lobe margins smooth or becoming richly endowed with tiny round lobules and small spherical apothecia (mostly 0.1–0.5 mm in diameter) with pale brown disks and, in older apothecia, lobulate margins; lower surface similar to the upper. HABITAT: On bark, especially in swampy hardwood forests and hammocks. COMMENTS: This is an easily identified species because of its wrinkled thallus and peculiar little globular apothecia with lobulate rims distributed abundantly along its lobe margins. The lobules, which can be cylindrical as well as flattened, especially on the apothecial margins, rarely occur on the lobe surface.

Leptogium milligranum
Stretched jellyskin

DESCRIPTION: Thallus dark olive to olive-brown or partly gray; lobes 2–5 mm across, with wrinkles usually extending lengthwise along the lobes as if caused by stretching; isidia commonly spherical and granular, some cylindrical and even slightly branched, mainly marginal and along ridges (although they can spill over onto the lobe surface); lobes generally anastomosing, at least in the thallus center. Apothecia rare. HABITAT: On bark, especially of oaks, and on wood. COMMENTS: In this large *Leptogium,* the greenish brown color dominates, although the flatter, less isidiate lobes are gray and heavily wrinkled. Because of its dark olive color, this lichen resembles a *Collema*. A thin section through the thallus, however, reveals a single layer of cortical cells on both the upper and lower surfaces. Several isidiate jellyskin lichens with a wrinkled surface occur in the southeast and can be confused with *L. milligranum*. Most of the others tend to be bluish gray rather than olive. *Leptogium austroamericanum* has rather thick isidia, sometimes lobulate, and most of the isidia are on the lobe surface. In **L. isidiosellum,** the lobes remain discrete (not anastomosing). It has very sharp, elevated wrinkles forming a net-like pattern or running longitudinally with the lobes, and it is isidiate along the edges of the wrinkles. In *L. marginellum*, the "isidia" are more like lobules and decorate only the margins of tiny, almost globular apothecia that are distributed along the edges of the thallus lobes. *Leptogium chloromelum* is most similar to *L. milligranum* in color but lacks isidia. See Comments under *L. chloromelum*.

Leptogium phyllocarpum
Frilly jellyskin

DESCRIPTION: Thallus steel-gray, with a very thick, deeply wrinkled appearance; lobes 2–3 mm across, not erect, lying more or less in a plane and conspicuously anastomosing, sometimes crinkled and frilly at the margins; without isidia or lobules. Apothecia broad (2–7 mm in diameter), red-brown, more or less sunken into the thallus or finally superficial with thick, wrinkled margins. HABITAT: Growing on hardwoods in subtropical forests. COMMENTS: *Leptogium chloromelum* is similar to *L. phyllocarpum*, differing mainly in color. In addition, the lobes of *L. chloromelum* usually stand on edge rather than lying flat, and the apothecia are constricted at the base, never immersed in the thallus. **Leptogium floridanum** differs from *L. phyllocarpum* mainly in its conspicuously knotty or warty growth form. See Comments under *L. chloromelum*.

Leptogium platynum
Large-spored jellyskin, wrinkled oilskin

DESCRIPTION: Thallus blue-gray, sometimes browning; lobes rather broad (1–6 mm), without isidia but with a minutely wrinkled surface; often lobulate on parts of the thallus with small, rounded, overlapping lobes (very sparse in plate 461); attached to the substrate by tufts of white hairs. Apothecia abundant but

459. *Leptogium milligranum* hill country, central Texas ×3.9

460. *Leptogium phyllocarpum* (damp) central Florida ×2.5

461. *Leptogium platynum* Coast Range, northern California ×3.8

small (0.2–0.5 mm in diameter), immersed or broadly attached; spores muriform and very large (35–49 × 9–16 μm). HABITAT: On mossy soil or rocks. COMMENTS: *Leptogium platynum* is 200–500 μm thick when wet, which is thicker than **L. californicum** or *L. polycarpum*, two related western species. *Leptogium corticola* is similar but has larger, sessile apothecia containing shorter, more ellipsoid spores (16–26 × 10–13 μm).

Leptogium polycarpum
Four-spored jellyskin, peacock oilskin

DESCRIPTION and COMMENTS: This is a rather rare, western American endemic, much like *L. gelatinosum*, but the asci consistently contain only 4 spores and the apothecia remain half-sunken at maturity. It also resembles **L. californicum**, but in that species, the apothecia are not as crowded, and they are not sunken; their asci contain 8 spores. *Leptogium californicum* grows mainly on mossy rocks and soil. See also *L. pla-* *tynum*. HABITAT: On the bark of deciduous trees in coastal forests.

Leptogium pseudofurfuraceum
Dimpled jellyskin

DESCRIPTION: Thallus broad-lobed (3–8 mm across), reddish brown to brown-gray; surface conspicuously wrinkled with fine, often concentric folds; upper surface more or less covered with cylindrical, partly collapsed isidia, sometimes broadened at the tips; lower surface with a thick white tomentum. Small yellow-brown apothecia sometimes produced on the thallus surface; spores muriform, ellipsoid, 23–29 × 7–10 μm. HABITAT: On bark of various kinds, rarely on rock. COMMENTS: Virtually all North American specimens that have been named as **L. furfuraceum** are actually *L. pseudofurfuraceum*. Several species of *Leptogium* have a hairy, white lower surface. *Leptogium pseudofurfuraceum* is the most common example in the southwest, but there are several others. *Leptogium rugosum* is very similar to *L. pseudofur-*

462. *Leptogium polycarpum* Klamath Range, northwestern California ×6.3

463. *Leptogium pseudofurfuraceum* (damp) Klamath Range, northwestern California ×3.5

464. *Leptogium rugosum* mountains, southern Arizona ×1.9

465. The same thallus of *Leptogium rugosum* when damp.

furaceum but lacks isidia and has fusiform (not muriform), 4–5-celled spores. **Leptogium burgessii,** another nonisidiate species from that area, produces thin lobules on the thallus margins, and the tomentum on the lower surface is very short, with round rather than elongated cells. **Leptogium saturninum,** a more boreal species, is dark olive or olive-gray, has few wrinkles, and the isidia are granular. **Leptogium burnetiae,** mainly an East Temperate–Rocky Mountain species, is the most distinctly gray species of this group. It has short, cylindrical, sometimes branched isidia on the lobe surface and usually has erect white hairs on the upper surface of the lobes close to the margins. **Leptogium laceroides,** found in the Appalachians north to New Brunswick and southern Ontario, is a distinctive brownish to gray lichen covered with isidia on the lobe surface, and with finely dissected, lobulate margins. The lower surface has an extremely short nap of white hairs composed of spherical cells. See also *Leptochidium*.

Leptogium rugosum
Rough bearded jellyskin

DESCRIPTION: Thallus dark gray to olivaceous, with broad lobes 2–8 mm across; surface covered with deep wrinkles, lacking isidia or lobules; lower surface thickly white tomentose. Apothecia common on the lobe surface, 0.5–3 mm in diameter, flat to slightly convex, red-brown, with smooth to wrinkled margins; spores fusiform (not muriform), 4–5-celled, 30–40 × 6–8 μm. HABITAT: On bark in montane woodlands. COMMENTS: This lichen closely resembles *L. pseudofurfuraceum* except for its fusiform spores and the lack of isidia. See Comments under that species. **Leptogium burgessii,** growing on trees or mossy rocks in southern Arizona and New Mexico, has a smooth thallus and abundant lobules along the lobe margins but is otherwise similar to *L. rugosum*.

466. *Leptogium saturninum* Cascades, Washington ×3.0

Leptogium saturninum
Bearded jellyskin

DESCRIPTION: Thallus dark olive to black when dry, with broad round lobes 3–10 mm wide, curling under slightly at the edges; upper surface fairly smooth (i.e., not wrinkled) but covered with granular isidia; lower surface with a blanket of dense white hairs, giving it a bearded look. HABITAT: On bark of a variety of trees, especially poplars and willows, sometimes on mossy rocks. COMMENTS: Two smooth, bearded jellyskin lichens have isidia, *L. saturninum* and **L. burnetiae**. In *L. saturninum,* the isidia are granular or globular; in the rarer *L. burnetiae,* they are cylindrical or branched. The whiskered jelly lichen (*Leptochidium*) has flattened isidia, and the margins are bewhiskered with tiny, stiff white hairs. Two bark-dwelling species, *L. pseudofurfuraceum* (with isidia) and *L. rugosum* (without isidia), differ from the above three in having a rough or wrinkled surface. See Comments under *L. pseudofurfuraceum.*

Letharia (2 species worldwide, both in N. Am.)
Wolf lichens

DESCRIPTION: Brilliant yellow to chartreuse fruticose lichens, forming tufts 3–15 cm across, abundantly branched and rather stiff, the branches ridged and pitted, 0.5–3 mm in diameter; medulla loose, containing strands of tough, fibrous tissue (Fig. 8c). Photobiont green (*Trebouxia*). Apothecia lecanorine, with dark brown disks; spores colorless, 1-celled, 8 per ascus. CHEMISTRY: Cortex and medulla PD–, K–, KC–, C– (the yellow pigment vulpinic acid, and atranorin; norstictic acid often in the hymenium). HABITAT: On conifers and wood, rarely on rocks, especially in dry sites; sometimes lavish on pines east of the Cascades and on red firs in the Sierra Nevada. COMMENTS: The two species of wolf lichen are very similar in size, color, and habitat. On pines, they often grow intermingled with the bright yellow foliose lichen *Vulpicida canadensis*. IMPORTANCE: Native Americans in California have used wolf lichens as an arrow poison, sometimes mixed with snake venom. The Blackfoot and Okanagan-Colville Indians have used them as an external medicine for sores, or even as an internal

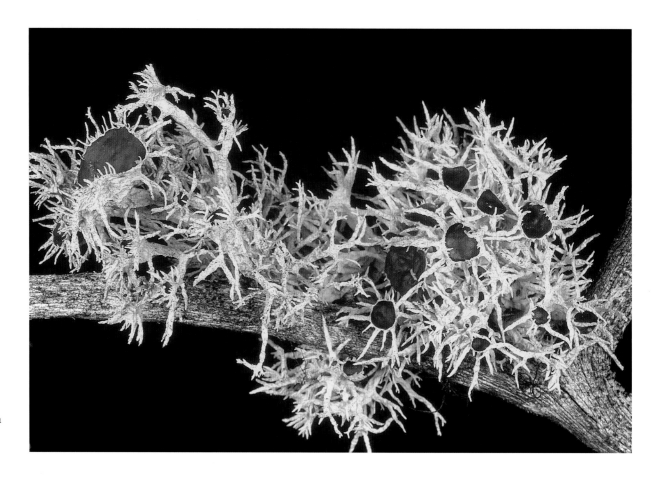

467. *Letharia columbiana* central British Columbia ×2.2

468. (facing page) *Letharia vulpina* Sierra Nevada, California ×2.1

medicine for stomach disorders, which is interesting considering the lichens' reported poisonous properties. The most common traditional use of *Letharia*, however, has been as the source of a bright yellow boiling water dye (see Chapter 10).

KEY TO SPECIES

1. Branches granular sorediate; apothecia rare.......... .... *Letharia vulpina*
1. Branches without soredia; apothecia abundant .... *Letharia columbiana*

Letharia columbiana
Brown-eyed wolf lichen

DESCRIPTION: Almost always fertile, with large, dark brown apothecia up to 15 mm across, fringed with spiny branchlets; soredia and isidia absent; black pycnidia often abundant. COMMENTS: *Letharia columbiana* is the fertile sister species of the sorediate *L. vulpina*, which, because of its thinner cortex, is not as stiff.

Letharia vulpina
Wolf lichen

DESCRIPTION: With coarse granular soredia grading into short, cylindrical isidia on the ridges of the angular branches. Apothecia very rare. COMMENTS: See *Letharia columbiana*. IMPORTANCE: *Letharia vulpina* is also found in northern Europe, where it was once used to poison foxes and wolves (the basis for its scientific and common names).

Letrouitia (3 N. Am. species)
Spiral-spored lichens

DESCRIPTION: Crustose lichens with rather dark greenish, olive, or orange, thin, continuous thalli. Photobiont green (unicellular). Apothecia biatorine, but appearing lecanorine, dark rusty brown to black, with prominent, bright orange margins; exciple biatorine, but sometimes with algae in the medulla at the base of the apothecium; hypothecium pale brown to almost colorless; asci with a darkly K/I staining cap

469. *Letrouitia parabola* Big Thicket, eastern Texas ×2.4

and an expanded K/I+ blue tholus; spores colorless, large, with unevenly thickened walls giving rise to lens-shaped or spiraled locules that, in some species, develop cross walls and become muriform; 2–8 spores per ascus, usually under 50 μm long; conidia rod-shaped. CHEMISTRY: Orange anthraquinones in the epihymenium, outer edge of the exciple, and often in the thallus. HABITAT: Typically on bark in tropical or subtropical regions. COMMENTS: *Letrouitia* species are most similar to *Brigantiaea leucoxantha*, another anthraquinone-containing crustose lichen common in the Tropics. The spores of the *Brigantiaea* are always large (over 50 μm long) and muriform, 1 per ascus, and the asci lack the K/I+ dark blue outer cap.

KEY TO GENUS: See Key F.

Letrouitia parabola

DESCRIPTION: Thallus dull orange to greenish yellow, continuous. Apothecia commonly biatorine (with some algae in the base), up to 1.5 mm in diameter, round or uneven in outline, with dark red-brown to black disks and a thick, smooth, orange margin; spores colorless, narrowly ellipsoid to fusiform, 25–35 × 11–18 μm, with a clearly spiraled locule (4–6 "turns"), subdivided here and there, resulting in a muriform structure; 6–8 spores per ascus. CHEMISTRY: Orange tissues of the thallus and apothecia are K+ red-purple (anthraquinones). HABITAT: On bark in shaded, subtropical woodlands. COMMENTS: All three species of *Letrouitia* have large black apothecia with bright orange margins and are a common sight in shaded southeastern forests. In **L. vulpina** the spores are many-celled with no spiral arrangement at all, even in young spores; 25–35 × 12–17 μm, 2 per ascus. *Letrouitia domingensis* has narrower spores, 9–13 μm wide, with 6–8 lens-shaped locules that are not subdivided or muriform. Most asci contain 8 spores. See also Comments under *Letrouitia*.

Lichinella (6 N. Am. species) (syn. *Gonohymenia*)
Rock licorice

DESCRIPTION: Black, nonstratified or stratified jelly lichens, foliose or fruticose, with clumps of ascending lobes, usually rounded to occasionally strap-shaped, without a cellular cortex. Photobiont blue-green (*Gloeocapsa*-like), cells occurring singly or in small clumps of 1–4. Ascomata are special structures called thallinocarps, consisting of asci developing in groups within small warts on the thallus; extremely inconspicuous and therefore almost never noticed. CHEMISTRY: No lichen substances. HABITAT: On calcareous or noncalcareous substrates. COMMENTS: *Collema*, a similar jelly lichen containing chains of *Nostoc* as a photobiont, is rarely as black as *Lichinella*. Species of *Thyrea* are even more similar (with regard to color and general habit). They have apothecia buried in the thallus that open by pit-like ostioles that look like pinholes.

KEY TO SPECIES: See Keys A and I.

Lichinella nigritella (syn. *Gonohymenia nigritella*)
Rock licorice

DESCRIPTION: Thallus pitch black, shiny or sometimes dull but never pruinose, developing in crowded, frequently umbilicate clumps; lobes thin, 130–250 μm when fully saturated with water, rounded foliose to almost fruticose and strap-shaped, usually with slightly thickened margins, ca. 0.7–2.5 mm across or broadening at the tips to 3.5 mm; surface often covered with mealy, spherical granules (sometimes seen only under high magnification); cyanobacteria 9–14 μm in diameter, with thin sheaths, almost entirely concentrated near the lobe surfaces, leaving a purely fungal medulla in the center. HABITAT: Only on limestone rock. COMMENTS: *Thyrea confusa* is somewhat thicker (230–500 μm), more contorted and deeply lobed, lacking thickened margins, and it is always dull, sometimes becoming bluish with pruina. *Lichinella cribellifera,* known from the central parts of the continent, has rounded, convex lobes, 1–4 mm wide, building into small prostrate, umbilicate cushions, with no isidia or granules on the surface. It grows on siliceous rock.

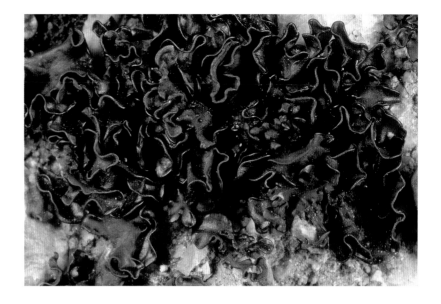

470. *Lichinella nigritella* Coast Range, southern California ×5.3

Lobaria (11 N. Am. species)
Lungworts, lung lichens

DESCRIPTION: Large, usually broad-lobed foliose lichens, typically with rather squarish lobes; lower surface pale brown, with a cortex, and with abundant to sparse, pale brown, fuzzy tomentum (black only in *L. retigera*), often with somewhat raised bald spots, but in some specimens, almost naked. Photobiont either blue-green (*Nostoc* or *Scytonema*) or green (*Dictyochloropsis* or *Trebouxia*); small, wart-like cephalodia occur on the lower surfaces of species with green algae. Apothecia, when present, produced on the lobe surface or along the margins, lecanorine, with brown disks; spores colorless or rarely brownish, fusiform, mostly 2–4-celled, 8 per ascus. CHEMISTRY: Frequently containing β-orcinol depsidones (PD+ yellow to red, K+ yellow to red) or orcinol depsides (KC+ red, C+ pink). COMMENTS: *Sticta* and *Pseudocyphellaria* are most similar to *Lobaria* in size, color, and lower surface, but have cyphellae or pseudocyphellae, respectively, on the lower surface. Among the other broad-lobed lichens with cyanobacteria, *Nephroma* has apothecia on the lower surface of the lobe margins, and *Peltigera,* also with marginal apothecia, lacks a lower cortex.

KEY TO SPECIES

1. Photobiont blue-green; cephalodia absent. 2
1. Photobiont green; cephalodia present 3
2. Stiff, tiny, colorless hairs on the lobe tips or on the upper surface of the lobes; rhizines rope-like, covered with perpendicular hairs; medulla PD–, K– (norstictic and stictic acids absent); infrequent, in Pacific Northwest. . . . . *Lobaria hallii*
2. Stiff, tiny, colorless hairs absent; rhizines tufted; medulla PD+ yellow, K+ red (norstictic and stictic acids present); widespread and frequent. . . . .
. . . . *Lobaria scrobiculata*
3. Lobes with a network of depressions and ridges; thallus loosely attached; medulla KC–, C– (gyrophoric acid absent) 4
3. Lobes relatively smooth and even, without a network of ridges and depressions; thallus closely appressed to substrate; medulla KC+ red, C+ pink (gyrophoric acid) . . . 5
4. Soredia, often mixed with isidia, present on ridges and lobe margins; medulla PD+ yellow to orange, K+ yellow to red (norstictic and stictic acids). . . . .
. . . . *Lobaria pulmonaria*
4. Soredia and isidia absent; medulla PD–, K– . . . .
. . . . *Lobaria linita*
5. Lobes 5–20 mm wide; Appalachian–Great Lakes region . . . . *Lobaria quercizans*
5. Lobes 2–4(–6) mm wide; southeastern coastal plain . . 6
6. Lobules abundantly produced on upper surface and margins of lobes; cephalodia always small and inconspicuous warts, most easily seen on lower thallus surface; pycnidia and apothecia rare . . . . *Lobaria tenuis*
6. Lobules rarely present; cephalodia sometimes forming small, branched outgrowths on the upper thallus surface; pycnidia appearing as abundant black dots on the lobe surface; apothecia abundant . . . . *Lobaria ravenelii*

471. *Lobaria hallii*
Coast Range, northern California ×1.0

472. *Lobaria linita*
(coastal form, wet)
Southeast Alaska ×0.41

Lobaria hallii
Gray lungwort

DESCRIPTION: Thallus light gray, containing cyanobacteria; lobes broad, up to 40 mm across, with fine, colorless hairs on the upper surface of the younger lobe tips (the hairs are sometimes very hard to find); soralia sometimes ring-like, with brown to gray soredia on lobe margins and surface. CHEMISTRY: Cortex K+ yellow (unknown substance); medulla and soralia PD–, K–, KC–, C–. HABITAT: Mainly on cottonwood trees or other kinds of poplar, but also on maples and, occasionally, conifers. COMMENTS: *Lobaria hallii* very closely resembles *L. scrobiculata* except for color and the presence of hairs on the lobe tips. *Lobaria hallii* is grayer than *L. scrobiculata*, which tends to have a yellowish tint. The medulla of *L. scrobiculata* is PD+ orange.

Lobaria linita
Cabbage lungwort

DESCRIPTION and COMMENTS: This species is almost identical to *L. pulmonaria* except for the absence of soredia or isidia and the PD–, K– reaction of the medulla. It is grassy-green when wet. *Lobaria linita* can have abundant apothecia when growing at low elevations along the west coast (plate 472), but arctic and alpine specimens are usually sterile (plate 473). CHEMISTRY: Medulla PD–, K–, KC–, C– (tenuorin). HABITAT: On soil and mossy vegetation in arctic or alpine sites, or on trees, especially tree bases, on the coast. The specimens on trees tend to be much larger than those on the ground. IMPORTANCE: *Lobaria linita* is eaten by mountain goats in southeastern Alaska and perhaps elsewhere.

Lobaria oregana
Lettuce lichen, lettuce lung

DESCRIPTION: Thallus large, leafy, pale yellowish green, containing green algae; lobes commonly 10–30 mm across; margins usually richly decorated with tiny lobules that can resemble isidia, some occurring on the surface of the lobes as well; surface with a network of sharp ridges; lower surface pale tan, tomentose except close to the margins, sometimes with raised, almost white areas. Cephalodia forming very small warts on the lower or sometimes the upper thallus surface. CHEMISTRY: Cortex KC+ yellow (usnic acid); medulla PD+ orange, K+ yellow darkening to red, KC–, C– (stictic, constictic, cryptostictic, and norstictic acids). HABITAT: Characteristically on the upper limbs of trees in northwestern old growth forests, and on the trunks of conifers closer to the coast. COMMENTS: The color of *L. oregana* in the dry state is unlike that of any other lungwort. IMPORTANCE: In the interior, the lettuce lichen can be used as an indicator of old growth forests. Because of its great abundance in some northwestern forests (often more than 1 ton per hectare), *L. oregana* contributes a significant amount of nitrogen to these forest ecosystems, thanks to the nitrogen-fixing cyanobacteria found in its cephalodia. The lichen is also a minor component of the fall and winter diet of the Columbia black-tailed deer on Vancouver Island, British Columbia.

Lobaria pulmonaria
Lungwort, lung lichen

DESCRIPTION: Thallus pale brown to olive-brown when dry and quite green when wet, containing green algae (*Dictyochloropsis*), with a strongly ridged and pitted surface that gives the lichen the appearance of lung tissue; lobes 8–30 mm wide, up to 7 cm long, branching in dichotomies and trichotomies; soralia developing on the lobe margins and along the thallus ridges, often with isidia emerging among the soredia. Tiny, wart-like cephalodia, 0.5–1.5 mm in diameter, common or sparse on the lower surface: cut one open to see the dark blue-green cyanobacteria inside, quite different from the grass-green layer in the main part of the thallus. Apothecia infrequent, usually on or near the lobe margins or along ridges on the upper surface. CHEMISTRY: Medulla PD+ orange, K+ yellow to red, KC–, C– (stictic and norstictic acids), or PD+ yellow, K+ red, KC–, C– (norstictic acid alone). HABITAT: On trees, mossy rocks, and wood in mature forests, usually in the shade. COMMENTS: *Lobaria pulmonaria* is the most widely distributed and common *Lobaria* in North America. In western North America, several lichens with reticulate ridged, lung-like lobes are easily confusable. *Lobaria linita* is PD– and lacks any isidia or soredia. *Pseudocyphellaria anthraspis* and *P. anomala*, besides having pseudocyphellae, are darker brown and contain cyanobacteria as the principal photobiont. ***Lobaria pseudopulmonaria*** resembles *L. pulmonaria* but has cyanobacteria as the main photobiont and lacks soredia and isidia. It is found in many parts of Alaska. The rare, west coast species ***L. retigera*** also contains cyanobacteria. It has isidia, the tomentum on the lower surface is black, and the medulla reacts K–, PD–. In the eastern forests, nothing resembles *L. pulmonaria*. IMPORTANCE: All species of *Lobaria*, but especially *L. retigera*, are good indicators of rich, unpolluted, and often very old forests. Despite its diminishing abundance, *L. pulmonaria* has long been prized as an important source of boiling water dyes. Herbalists have recommended *L. pulmonaria* as a remedy for tuberculosis because of its resemblance to lung tissue, and in India, it has been used to treat lung diseases, asthma, hemorrhages, and even eczema on the head. Lungwort has also been used for brewing in India and Europe. It is apparently a favorite food of moose in the northeast.

473. *Lobaria linita* (tundra form, wet) interior Alaska ×0.70

474. *Lobaria oregana* Olympics, Washington ×1.3

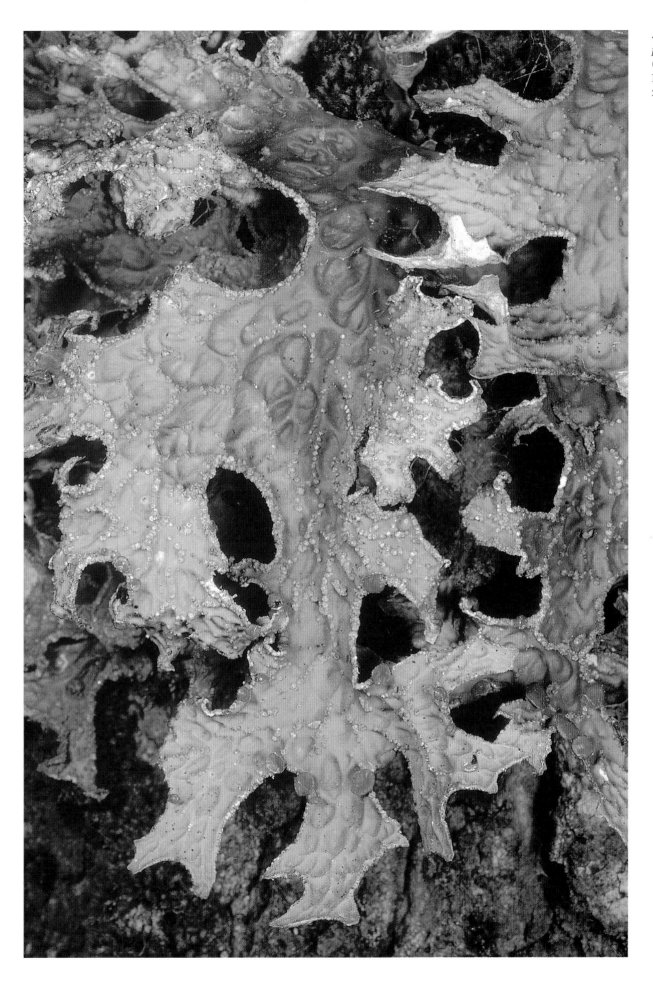

475. *Lobaria pulmonaria* (damp) southern New Brunswick ×2.3

476. *Lobaria quercizans* coastal Maine ×0.90

477. *Lobaria ravenelii* Ouachitas, Arkansas ×1.0

Lobaria quercizans
Smooth lungwort

DESCRIPTION: A large, gray, *Parmelia*-like lungwort, unusual for the genus in having a smooth surface rather than reticulate ridges and depressions, but becoming heavily wrinkled in older parts of the thallus; lobes 5–20 mm wide; soredia and isidia absent. Photobiont green, with infrequent, internal cephalodia that show up on the lower surface as small bumps. Large, red-brown apothecia, up to 4.5 mm across, without marginal lobules, often produced on the thallus surface. CHEMISTRY: Upper cortex K+ yellow (atranorin); medulla PD–, K+ orange, KC+ red, C+ pink (gyrophoric and usually 4–O-methylgyrophoric acids). HABITAT: On bark of deciduous trees, especially sugar maple, sometimes on mossy rock. COMMENTS: This species looks like a large shield lichen (e.g., *Hypotrachyna*), but it has a pale, tomentose lower surface and produces cephalodia. Of the three smooth-lobed, C+ red lungworts with green algae in North America (*L. quercizans*, *L. ravenelii*, and *L. tenuis*), *L. quercizans* is the largest and is the only one that often has prominent, black pycnidia. The latter two are restricted to the southeastern coastal plain. IMPORTANCE: This lichen was once used as both a food and traditional medicine by the Menomini people of Wisconsin.

Lobaria ravenelii
Dixie lungwort

DESCRIPTION: Thallus pale greenish gray or brown to olive-brown, containing green algae; lobes 2.5–6 mm across, smooth or with some depressions, without soredia, isidia, or lobules; cephalodia small and round, or growing out on the upper surface of the thallus and becoming branched. Apothecia generally common, up to 4.5 mm in diameter, often with lobules on the margins. Pycnidia common, entirely buried in thallus. CHEMISTRY: Upper cortex usually K–; medulla PD–, K–, KC+ red, C+ pink (gyrophoric and 4-O-methylgyrophoric acids, rarely with atranorin). HABITAT: On bark. COMMENTS: *Lobaria quercizans* is grayer, has broader lobes (5–10 mm) and larger apothecia that lack lobules, and cephalodia can always be found on the lower surface. See also *L. tenuis*.

478. *Lobaria scrobiculata* Cascades, Washington ×1.4

Lobaria scrobiculata
Textured lungwort

DESCRIPTION: Thallus yellowish green to pale green or olive, sometimes yellowish gray, often brownish at the lobe margins, containing cyanobacteria; lobes 10–20 mm across, usually with depressions and ridges (scrobiculate), but sometimes smooth, without hairs; blue-gray granular soredia produced on the lobe margins and in round to irregular soralia on the upper surface. CHEMISTRY: Cortex K–, KC+ yellow (often vague); medulla and soralia PD+ orange, K+ yellow to orange, KC– or KC+ pink, C– (stictic, constictic, and norstictic acids, scrobiculin, and usnic acid). HABITAT: On a variety of trees or mossy rocks, especially in moist regions. COMMENTS: This is the most common lungwort containing cyanobacteria rather than green algae as a photobiont. *Lobaria hallii* is closely related and very similar but has hairy lobe tips and a PD– medulla. The very rare *L. retigera,* also with cyanobacteria, looks more like *L. pulmonaria* but has only isidia, is dark brown, and has a dark lower surface with black tomentum. IMPORTANCE: This lichen can apparently be eaten. According to a Yup'ik friend from Kwethluk, Alaska, *Lobaria scrobiculata* (or *L. hallii?*) "can be eaten plain, right from the tree." The Yup'ik name for this lichen is Qelquaq.

Lobaria tenuis
Slender lungwort

DESCRIPTION: Thallus gray-green, containing green algae, with flat, appressed lobes 2–4 mm wide; lower surface pale brown, tomentose; soredia and isidia absent, but tiny, rounded to elongate, sometimes branched lobules develop on the surface of older parts of the thallus and along the margins. Cephalodia buried within the thallus, visible as small bumps on the lower surface. Apothecia rare. CHEMISTRY: Upper cortex usually K+ yellow (atranorin); medulla PD–, K–, KC+ red, C+ pink (gyrophoric and 4-O-methylgyrophoric acids). HABITAT: On bark. COMMENTS: *Lobaria tenuis* looks a bit like a *Parmelina* but has a typical tomentose lower surface. *Lobaria ravenelii* is very similar but has few (if any) lobules and more abundant pycnidia and apothecia.

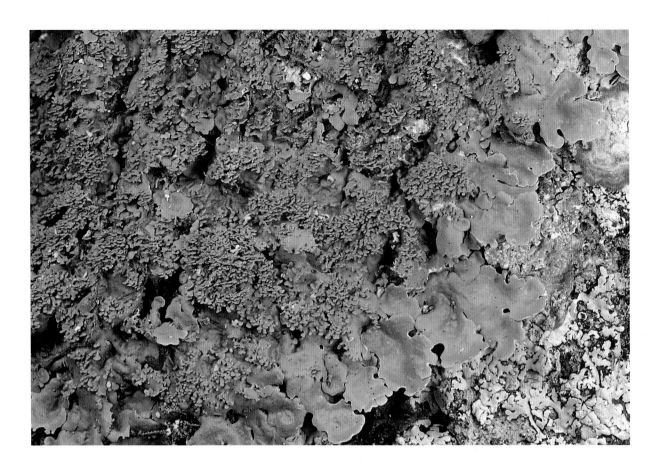

479. *Lobaria tenuis* (*Cryptothecia rubrocinta* on upper right) northern Florida ×3.6

Lobothallia (4 N. Am. species)
Puffed sunken-disk lichens

DESCRIPTION: Very thick, lobed, tightly adnate foliose lichens (becoming areolate and essentially crustose in the center, but with a lower cortex); lobes flat to strongly convex, 0.5–1.0 mm across. Photobiont green (*Trebouxia*). Apothecia sunken or superficial, dark brown to almost black, up to 2.5 mm in diameter, resembling those of *Aspicilia*; paraphyses up to 5 μm thick, septate, constricted like a string of beads at the tips; spores colorless, broadly ellipsoid, 1-celled, 8 per ascus. CHEMISTRY: Cortex and usually medulla PD+ yellow, K+ red, KC–, C– (norstictic acid), or without lichen substances. HABITAT: On noncalcareous rock of different kinds, especially in montane and high desert habitats. COMMENTS: Some specimens are much like *Protoparmelia*, and well-developed specimens can resemble foliose lichens such as *Allantoparmelia*. *Lobothallia* differs, however, in color, chemistry, and characteristics of the apothecia and spores. The genus *Lobothallia* is included within *Aspicilia* by some lichenologists.

KEY TO SPECIES
1. Thallus yellowish to copper brown; lobes flat and appressed to rock . . . . *Lobothallia praeradiosa*
1. Thallus shades of gray, or, less frequently, light brown; lobes very convex, almost foliose in appearance. . . . . . .
    . . . . *Lobothallia alphoplaca*

Lobothallia alphoplaca (syn. *Aspicilia alphoplaca*)

DESCRIPTION: Gray to brown thallus (both illustrated here) with radiating, very convex lobes 0.5–1.5 mm wide, rather foliose at the margins; lower surface pale. Apothecia abundant, superficial to immersed, dark red-brown; spores 10–16 × 6–10 μm. HABITAT: Growing on granite and sandstone. COMMENTS: This species is usually grayer and more easily removed from the substrate than is *L. praeradiosa*, which is typically yellowish brown and has flat lobes closely attached to the substrate. However, as can be seen from plates 480 and 481, the color is variable. Growth habit is therefore more reliable for distinguishing the two. Both are common throughout the southwest, especially in the southern Rockies. *Lobothallia melanaspis* is very similar to *L. alphoplaca* but is entirely K– (lacking norstictic acid) and frequently is submerged in mountain streams for part of the year.

Lobothallia praeradiosa (syn. *Aspicilia praeradiosa*)

DESCRIPTION and COMMENTS: Thallus yellowish to copper-brown and closely appressed to substrate (i.e., more crustose in character), but otherwise much like *L. alphoplaca* in appearance, chemistry, and habitat. Pale forms such as the specimen shown in plate 482 can be mistaken for a lobed *Lecanora* such as *L. muralis*.

Lopadium (5 N. Am. species)
Urn-disk lichens

DESCRIPTION: Crustose lichens with thin, granular to almost squamulose, olive-brown thalli. Photobiont green (unicellular). Apothecia lecideine or biatorine, 0.6–1.3 mm in diameter, black, with a dark brown exciple merging with a brown-black hypothecium; paraphyses branched only at the summit, with expanded, abruptly pigmented tips; asci thick-walled, K/I+ blue; spores very large, muriform with numerous cells, colorless or yellowish, 1 per ascus. CHEMISTRY: All reactions negative. HABITAT: On bark of conifers in shaded forests, or on decaying vegetation and mosses in tundra habitats. COMMENTS: *Schadonia, Brigantiaea,* and *Letrouitia* are all crustose lichens with large muriform spores and biatorine apothecia. The latter two usually contain yellow or orange anthraquinone pigments and are mainly tropical lichens. *Letrouitia* is more closely related to *Caloplaca* in some features of its ascus and spores. *Schadonia* is mentioned in Comments under *L. disciforme*.

KEY TO GENUS: See Key F.

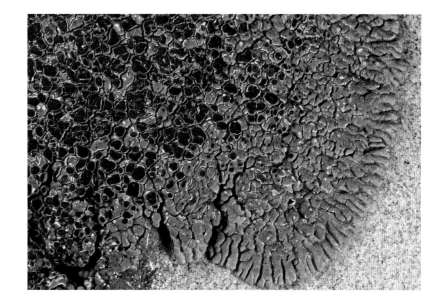

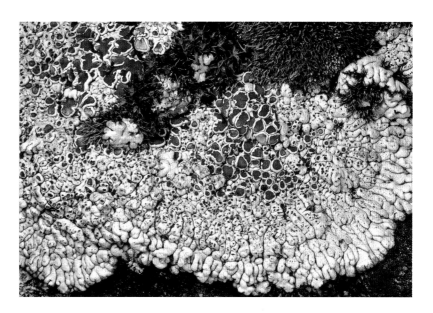

480. *Lobothallia alphoplaca* central Utah ×2.6

481. *Lobothallia alphoplaca* (gray form, wet) Rocky Mountains, southern Colorado ×1.5

Lopadium disciforme

DESCRIPTION: Thallus dark olive to brownish (green when wet, as shown in plate 483), areolate to almost squamulose; hypothecium and epihymenium very dark brown. Apothecia large (0.5–1.5 mm in diameter), black, thick-margined, distinctly constricted at the base; spores huge, colorless, muriform with hundreds of cells, $40–105 \times 14–40$ μm (specimens with larger spores are mostly in the west), 1 spore per ascus. HABITAT: On conifer bark in shaded forests. COMMENTS: *Lopadium pezizoideum* is an extremely similar species growing on moss and dead vegetation in arctic and alpine habitats. *Schadonia*, another black-fruited lichen growing on tundra peat, has branched paraphyses and spores usually less than 40 μm long, 2–8 per ascus. Both *S. alpina* (8-spored) and *S. fecunda* (2–4-spored) are quite rare arctic species, although *S. fecunda* is also known from the mountains of coastal British Columbia. See also *Brigantiaea* and *Letrouitia*.

482. *Lobothallia praeradiosa* southern Arizona ×2.5
483. *Lopadium disciforme* southern New Brunswick ×4.2

Loxospora (4 N. Am. species)
Ragged-rim lichens

DESCRIPTION: Crustose lichens with a thin to thick, yellowish gray thallus. Photobiont green (unicellular). Apothecia lecanorine, with brown disks, sometimes lavender because of a heavy pruina; apothecial margins thin, almost disk-colored, but usually surrounded by a secondary lecanorine margin that is broken up, giving it a ragged appearance; paraphyses slightly branched; asci with a somewhat thickened K/I+ blue tip and a broad axial body; spores colorless, fusiform, somewhat curved or twisted, $30–55 \times 4–5.5$ μm, 4–8-celled, but cross walls (septa) often hard to see, 8 per ascus; several species always or almost always sterile. CHEMISTRY: Thallus PD+ orange, K+ deep yellow, KC–, C– (thamnolic acid). HABITAT: Mostly on bark. COMMENTS: *Loxospora* superficially resembles brown-fruited *Lecanora* or *Lecania* species, but the spores and apothecial tissues are quite different, as is the chemistry. The spores are similar to those of *Haematomma*. Thamnolic acid is found in very few other crustose genera, one of which is *Pertusaria*, which has large, 1-celled, thick-walled spores.

KEY TO SPECIES

1. Apothecia unknown; thallus producing masses of small, hollow pustules that break down into granular schizidia or soredia .... *Loxospora pustulata*
1. Apothecia common, lecanorine, often with a torn or ragged margin; thallus without hollow pustules or soredia 2

2. Apothecial disks red-brown, not pruinose or lightly pruinose; thallus rather thick and irregular; apothecia constricted at base............. *Loxospora ochrophaea*
2. Apothecial disks purplish, heavily pruinose; thallus very thin, more or less smooth; apothecia sessile .... *Loxospora cismonica*

Loxospora cismonica
Frosted ragged-rim lichen

DESCRIPTION: Thallus pale greenish gray, very thin and continuous, without a white prothallus. Apothecia small (0.4–0.7 mm diameter), flat and closely appressed, with purplish, heavily pruinose disks, and usually a ragged, toothed apothecial margin. CHEMISTRY: Thamnolic acid alone. HABITAT: On bark of conifers, especially balsam fir and spruce, and occasionally old birch trees.

Loxospora ochrophaea
Eastern ragged-rim lichen

DESCRIPTION: Thallus usually thick and verrucose, often with a white, webby prothallus. Apothecia large (up to 1.5 mm in diameter) and somewhat raised; apothecial margins prominent, usually ragged, disk-colored to distinctly lecanorine and more like the thallus in color and texture; disks red-brown, without pruina or lightly pruinose, rarely purplish. CHEMISTRY: Zeorin and thamnolic acid. HABITAT: On conifer bark. COMMENTS: This lichen is often mistaken for members of the *Lecanora subfusca* group, but the spores and chemistry easily distinguish them.

484. *Loxospora cismonica* Adirondacks, New York ×5.4

485. *Loxospora ochrophaea* Cape Cod, Massachusetts ×6.1

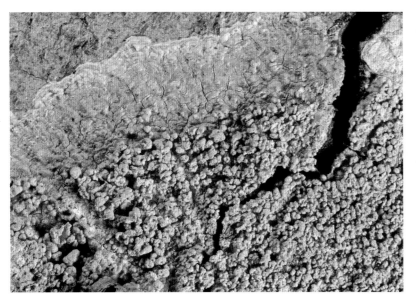

486. *Loxospora pustulata* central North Carolina ×5.2
487. *Loxosporopsis corallifera* Southeast Alaska ×5.6

Loxospora pustulata
Pustule crust lichen

DESCRIPTION: Thallus pale yellowish gray (turning pinkish in old collections), smooth and continuous at the edge but developing large, inflated, hollow warts or pustules (0.3–1 mm in diameter) in older parts that break down into irregular fragments (schizidia), giving the thallus a granular or even sorediate appearance. Apothecia unknown. CHEMISTRY: Usually thamnolic acid alone. HABITAT: Almost exclusively on bark of deciduous trees, especially red maples or oaks, in open or shaded woodlands, very rarely on conifers. COMMENTS: This lichen is very closely related to **L. elatina**, which develops true soredia in greenish soralia rather than having only the whitish pink, granular fragments of the pustules. The habitats and substrates also differ: *L. elatina* is usually found on conifers in boreal woodlands.

Loxosporopsis (1 species worldwide)

KEY TO SPECIES: See Keys B, C, and F.

Loxosporopsis corallifera
Tiny tree-coral lichen

DESCRIPTION: A thin, yellowish white crustose lichen, often covering large areas of bark, with long, slender, branched or unbranched isidia, 0.5–2 mm long (not always as dense as shown in plate 487), without a prothallus. Photobiont green (unicellular). Apothecia infrequent, dark to pale brown, 0.5–1 mm in diameter, irregular in shape, with a thin, thallus-colored margin; asci with a K/I+ blue tholus having an ocular chamber but no axial body; spores slender, pointed at both ends, slightly twisted and curved, 35–50(–65) × 5.5–7.5 μm, with 1–5 cross walls that are difficult to discern; 8 spores per ascus. CHEMISTRY: Thallus PD–, K–, KC–, C–, UV+ blue-white (divaricatic acid); apothecial disk C+ pink (gyrophoric acid). HABITAT: On conifer bark in well-lit Douglas fir or western white pine stands. COMMENTS: This is a fairly common lichen along the west coast

488. *Masonhalea richardsonii* interior Alaska ×1.4

that is rarely collected because it is usually sterile. There is no other crustose lichen on bark with such long, slender isidia.

Masonhalea (1 species worldwide)

KEY TO SPECIES: See Key A and *Cetraria*.

Masonhalea richardsonii
Arctic tumbleweed

DESCRIPTION: A fruticose-foliose lichen with smooth, dark brown, flattened branches, curled up into loose balls when dry; lobes mostly 2–7 mm wide, dividing both regularly in dichotomies and irregularly; lower surface of branches with broad, irregular, whitish pruinose patches without a cortex, and functionally equivalent to pseudocyphellae; lacking cilia or rhizines. Photobiont green (*Trebouxia*). Apothecia extremely rare. CHEMISTRY: Medulla PD–, K–, KC+ pink, C–, UV+ blue-white (alectoronic acid). HABITAT: Rolling freely over the ground in subalpine or alpine heath. COMMENTS: *Masonhalea* bears some resemblance to species of *Cetraria* but has a tumbleweed habit, more dichotomous mode of branching, broad pruinose patches on the lower surface, and different chemistry. IMPORTANCE: Alaskan native people have used *Masonhalea* to prime wood fires.

Massalongia (2 N. Am. species)

KEY TO SPECIES: See *Pannaria*.

Massalongia carnosa
Rockmoss rosette lichen

DESCRIPTION: Small chocolate-brown foliose lichen; lobes 0.5–1.5 mm wide, mostly over 2 mm long, spreading out or overlapping, with tiny lobules or corticate isidia all along the lobe mar-

489. *Massalongia carnosa* Cascades, Oregon ×4.4

gins. Photobiont blue-green (*Nostoc*); clumps of cyanobacteria forming a discrete layer in the thallus. Apothecia frequent, with red-brown disks and paler, biatorine margins; spores colorless, 2-celled, 8 per ascus. CHEMISTRY: No lichen substances. HABITAT: Growing on mosses or mossy rock. COMMENTS: The septate spores, thallus color, and lack of a hypothallus or prothallus distinguish *Massalongia* from similar species of *Fuscopannaria*, *Pannaria*, or *Parmeliella*. *Massalongia microphylliza* is a rare lichen of arid soils and has shorter lobes (less than 2 mm long).

Megalaria (6 N. Am. species)
Dot lichens

DESCRIPTION: Crustose lichens with gray to blue-green, continuous to granular thalli. Photobiont green (*Dictyochloropsis* and possibly other unicellular genera). Apothecia biatorine, black, relatively large (up to 2 mm in diameter); epihymenium greenish to dark purple, turning red in nitric acid; other apothecial tissues often pigmented orange-brown to reddish brown or purplish; exciple with radiating, rather thick hyphae (up to 4 μm wide) not thickened at the tips; paraphyses branched only at the tips, which are not pigmented; ascus tip with a distinct ocular chamber and axial body, *Biatora*-type, *Bacidia*-type, or *Lecanora*-type; spores colorless, 2-celled, ellipsoid with rather thick, smooth walls, 8 per ascus. CHEMISTRY: Atranorin, zeorin, or fumarprotocetraric acid in some species, but most species contain no substances. HABITAT: Generally on bark in humid or old, shaded forests. COMMENTS: *Megalaria* belongs to a group of three closely related genera with colorless, 2-celled spores and black lecideine or biatorine apothecia, the other two being **Catillaria** and **Catinaria**. These differ in lacking an axial body in the ascus tip and in having a brown epihymenium, among other things. In *Catillaria*, the spores have thin, smooth walls, the ascus tips have no ocular chamber (Fig. 14c), and the paraphyses have darkly pigmented cells at the tip; species grow on rock as well as on wood or bark. In *Catinaria*, the spores have roughened, ornamented walls, an ocular chamber is present in the ascus tips, and the paraphyses tips are barely expanded and not pigmented; most species are bark-dwelling. Species of *Rhizocarpon* with colorless, 2-celled spores (like *R. hochstetteri*) can be distinguished from rock-dwelling species of *Catillaria* by their halonate spores and branched paraphyses.

KEY TO SPECIES: See Key F.

Megalaria laureri
(syn. *Catinaria laureri*)

DESCRIPTION: Thallus white to very pale gray, thin, continuous to slightly rimose, conforming to the shape of the bark substrate. Apothecia black, flat to slightly convex, dull or often shiny, 0.4–0.8(–1.1) mm in diameter, broadly attached; margin black, thick or thin, usually not prominent in mature apothecia and soon hardly visible; epihymenium black to greenish black, K+ green or violet; hymenium sometimes reddish purple, K+ green, 45–60 μm high; hypothecium red-brown to orange below, red-purple above, all of it turning purple in K; exciple greenish, but much paler at the outer half, unchanged or turning violet in K; spores ellipsoid to fusiform, 13–18(–24) × 5–7(–8) μm. CHEMISTRY: Thallus PD–, K+ pale yellow, KC–, C– (atranorin). HABITAT: On the bark of a variety of deciduous trees, especially maple, beech, and ash, usually in old, undisturbed forests. COMMENTS: This rather featureless lichen comes alive when its superficially unremarkable black apothecia are sectioned and examined under the microscope. There, tissues of many colors are revealed, with the colors intensifying or changing with the addition of K. *Megalaria grossa,* also found mainly in the northeast, has larger apothecia (0.5–1.2 mm in diameter), larger spores (23–30 × 10–15 μm), a higher hymenium (90–130 μm), and an exciple that is darkly pigmented at the outer edge (usually green) and pale internally; it lacks atranorin (K–). Several other species of *Megalaria* occur in the Pacific Northwest. One of them, *M. brodoana,* has a bluish-gray granular thallus, the blue-green pigment turning red in nitric acid when viewed under a microscope. Its spores are very broadly ellipsoid, (15–)17–21(–26) × (7.5–)9–11(–13.5) μm. It grows on alders and shrubs along the very humid coast from Oregon to the Queen Charlotte Islands.

Megaspora (1 species worldwide)

KEY TO SPECIES: See Key F.

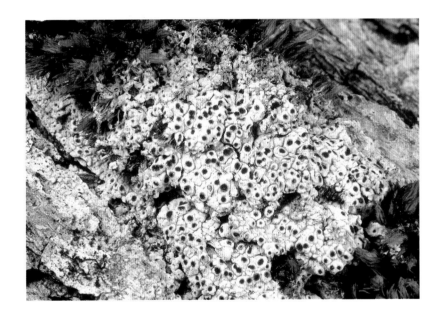

Megaspora verrucosa
(syn. *Pachyospora verrucosa*)
False sunken disk lichen

DESCRIPTION: A gray to white, fairly thick, crustose lichen. Photobiont green (*Trebouxia*). Apothecia with black, mostly sunken disks, sometimes lecanorine with a thick margin; epihymenium olive, turning yellow-green in K; hypothecium yellowish to brown; paraphyses abundantly branched, slender, not expanded at the tips; asci club-shaped, with a K/I+ pale blue tip; spores colorless, 1-celled, broadly ellipsoid, thick-

490. *Megalaria laureri* herbarium specimen (New York Botanical Garden), Michigan ×8.6

491. *Megaspora verrucosa* Coast Range, northern California ×2.9

walled, 30–65 × 16–36 μm, 8 spores per ascus in two rows. CHEMISTRY: Cortex and medulla PD–, K–, KC–, C–. HABITAT: On soil, dead moss, alpine vegetation, or bark, especially poplars; sometimes quite common. COMMENTS: This lichen differs from *Aspicilia* in spore wall thickness, abundantly branched paraphyses, and habitat (all *Aspicilia* species are on rock, rarely on mineral soil). In similar species of *Pertusaria*, the dark epihymenium is often K+ purple, and the asci turn dark blue with K/I.

Melanelia (27 N. Am. species)
Camouflage lichens, brown lichens

DESCRIPTION: Small to medium-sized, dark brown to olive foliose lichens, often with conspicuous pseudocyphellae on the upper surface, shiny to dull, usually closely attached to the substrate; lower surface pale brown to blackish, with unbranched rhizines. Photobiont green (*Trebouxia*). Apothecia lecanorine, usually laminal, with colorless, 1-celled, ellipsoid to spherical spores, 8(–32) per ascus; conidia usually dumbbell-shaped. CHEMISTRY: With various compounds including lecanoric, fumarprotocetraric, stictic, and perlatolic acids, rarely with atranorin. HABITAT: On bark or rock. COMMENTS: *Melanelia* species are the most common brown to olive foliose lichens. *Neofuscelia* is closely related but lacks pseudocyphellae and has an upper cortex with a different chemistry (turning blue-green with nitric acid). Species of *Tuckermannopsis* can be very similar but are generally more ascending, with marginal pycnidia and apothecia and few pseudocyphellae. Species of *Parmelia* that become brown when old or in sunny locations can usually be distinguished by the network of ridges on the upper surface and the K+ red medulla (although a few lack the ridges and may be K–). *Physconia* species are pruinose on the lobe tips and usually have a mat of squarrose rhizines (although a few have unbranched rhizines); *Phaeophyscia* species are generally smaller, and the spores are brown and 2-celled, with round or angular locules.

KEY TO SPECIES (including *Neofuscelia*)

1. Soredia present. 2
1. Soredia absent 9

2. Medulla C+ red or pink 3
2. Medulla C– 6

3. Pseudocyphellae abundant and conspicuous; lobes elongated; gyrophoric acid present; white or brown soralia laminal or partly marginal; soredia granular . . . . *Melanelia tominii*
3. Pseudocyphellae absent; lobes rounded; lecanoric acid present. 4

4. Soredia at least partly on thallus surface (laminal); soredia granular, often mixed with or arising from isidia. 5
4. Soredia entirely marginal, farinose, forming powdery crescents; isidia absent . . . . *Melanelia albertana*

5. Soredia entirely laminal, arising from a disintegration of the cortex or isidia, leaving yellowish green patches; thallus surface not pruinose, without cortical hairs . . . . *Melanelia subaurifera*
5. Soredia both laminal and marginal, arising from pustules or breakdown of the cortex, brown to whitish; thallus surface commonly pruinose, and often with a fuzz of minute, colorless hairs on the lobe tips . . . . *Melanelia subargentifera*

6.(2) Soredia predominantly on upper surface of lobes, dark brown (or whitish where abraded); rhizines abundant, brown or black; lower surface dark brown or black; medulla UV+ blue-white (perlatolic acid) 7
6. Soredia predominantly marginal; rhizines sparse, very pale beige or tan; lower surface pale brown; medulla UV– (perlatolic acid absent). 8

7. Soredia fine, in rounded mounds usually near the lobe tips . . . . *Melanelia sorediata*
7. Soredia coarse, granular, in irregular patches on the older parts of the thallus . . . . *Melanelia disjuncta*

8. On bark and wood, rarely rock, in western North America; medulla K–, PD– (fatty acids); lobes concave, undulating or crisped at the margins. . . . . *Tuckermannopsis chlorophylla*
8. On rock in the Appalachian region; medulla K+ red, PD+ yellow (norstictic and stictic acids); lobes flat . . . . *Melanelia culbersonii*

9.(1) Isidia present 10
9. Isidia absent. 18

10. Lobes 0.4–0.7 mm wide; rhizines predominantly marginal; on wood or bark . . . *Tuckermannopsis coralligera*
10. Lobes usually more than 1 mm wide; rhizines all over lower surface; on rock, wood, or bark 11

11. Isidia spherical or globular. 12
11. Isidia cylindrical, flattened, club-shaped, or conical, unbranched or branched 14

12. On rock; isidia hollow. 13
12. On bark or wood, rarely rock; isidia solid . . . . *Melanelia elegantula*

13. Medulla K–, PD–, KC+ pinkish violet (perlatolic acid) .... *Neofuscelia loxodes*
13. Medulla K+ yellow or red, PD+ orange, KC– (unidentified substance) ..... *Neofuscelia subhosseana*
14.(11) Medulla C+ red or pink (lecanoric acid) 15
14. Medulla C– (lecanoric acid absent) 16
15. Isidia unbranched, mostly shorter than 0.2 mm; abrading to leave yellowish green patches .... *Melanelia subaurifera*
15. Isidia frequently branched and longer than 0.2 mm; not forming yellowish patches where abraded .... *Melanelia fuliginosa*
16. Isidia cylindrical or conical, unbranched; upper surface dull .... *Melanelia elegantula*
16. Isidia flattened or club-shaped; upper surface rather shiny 17
17. Isidia solid; lower surface black; growing on rock; medulla UV+ blue-white (perlatolic acid) .... *Melanelia panniformis*
17. Isidia hollow; lower surface pale to dark brown; growing on bark, rarely wood or rock; medulla UV– .... *Melanelia exasperatula*
18.(9) Growing on bark 19
18. Growing on rock............................... 25
19. Medulla C+ red, KC+ red (lecanoric acid); California .... *Melanelia glabra*
19. Medulla C–, KC– 20
20. Rhizines sparse, predominantly marginal and resembling cilia; lobes 0.5–1.6 mm wide .... *Tuckermannopsis fendleri*
20. Rhizines abundant all over lower surface, never resembling cilia; lobes (0.5–)1–5 mm wide 21
21. Medulla PD+ red (fumarprotocetraric acid) 22
21. Medulla PD– 23
22. Pseudocyphellae sparse or inconspicuous on the thallus and apothecial margins; lobes smooth and even... .... *Melanelia septentrionalis*
22. Pseudocyphellae abundant and conspicuous especially on the apothecial margins; lobes wrinkled or bumpy (rugose) .... *Melanelia olivacea*
23. Spores 12–32 per ascus; lobes often with lobules at least in center of thallus .... *Melanelia multispora*
23. Spores 8 per ascus; lobules absent................ 24
24. Pseudocyphellae absent or rare; western .... *Melanelia subolivacea*
24. Pseudocyphellae on apothecial margins, lobes, and on thallus warts; East Temperate .. [*Melanelia exasperata*]
25.(18) Pseudocyphellae abundant and conspicuous; rhizines mostly on or close to the margins 26

25. Pseudocyphellae absent or inconspicuous; rhizines more or less uniformly distributed............... 28
26. Pseudocyphellae marginal; pycnidia prominent, on lobe margins; medulla PD+ orange, K+ yellow (stictic acid)....................... *Melanelia hepatizon*
26. Pseudocyphellae laminal; pycnidia scattered on lobe surface, entirely buried in thallus with just the ostiole showing; medulla PD– or PD+ red, K– (stictic acid absent)....................................... 27
27. Medulla C+ pink, KC+ red, PD– (gyrophoric acid); thallus thick and stiff; pycnidia sparse or very inconspicuous; conidia rod-shaped ...... *Melanelia tominii*
27. Medulla C–, KC–, PD+ red or PD– (gyrophoric acid absent; fumarprotocetraric acid sometimes present); thallus moderately thick; pycnidia abundant and conspicuous; conidia dumbbell-shaped .. *Melanelia stygia*
28.(25) Thallus brown to olive-brown, usually producing abundant, overlapping, flattened lobules; lower surface wrinkled; conidia rod-shaped .. *Melanelia panniformis*
28. Thallus reddish brown or yellowish brown, without abundant, overlapping lobules; lower surface smooth; conidia dumbbell-shaped 29
29. Lower surface and rhizines dark brown or black; pycnidia sparse or very inconspicuous; medulla K–, PD–, KC+ pinkish violet (perlatolic acid)................ .... *Neofuscelia loxodes*
29. Lower surface and rhizines pale brown; pycnidia abundant and conspicuous; medulla K+ yellow or red-orange, PD+ orange, KC– (stictic and sometimes norstictic acids)................. *Neofuscelia atticoides*

Melanelia albertana (syn. *Parmelia albertana*)
Powder-rimmed camouflage lichen

DESCRIPTION: Thallus olive-brown to red-brown, with numerous marginal or sometimes lip-shaped soralia producing granular soredia; lobes rounded or somewhat elongate, 2–4 mm wide; pseudocyphellae absent; lower surface black in center. CHEMISTRY: Medulla PD–, K–, KC+ red, C+ red (lecanoric acid). HABITAT: On poplars and other broad-leaved trees in interior grasslands. COMMENTS: This is one of the few bark-dwelling, camouflage lichens with soredia, and the only one with marginal soralia. It is similar to *Tuckermannopsis chlorophylla*, which has shinier, more elongate lobes, and a pale lower surface with very few rhizines, and has a C– medulla. In *M. subargentifera*,

492. *Melanelia albertana* Columbia River, Oregon ×4.0

493. *Melanelia culbersonii* Appalachians, Virginia ×2.9

which is also C+ red, the soralia are on the lobe surface as well as the margins. The soredia of *M. subaurifera* are yellowish when abraded and are laminal. The Appalachian rock species *M. culbersonii* also has marginal soralia, but the medulla is K+ red (norstictic acid).

Melanelia culbersonii (syn. *Cetraria culbersonii*)
Appalachian camouflage lichen

DESCRIPTION: Thallus dark olive-brown to blackish brown; lobes under 1 mm wide, with elliptical white soralia, flecked with black, distributed along the lobe margins or less frequently on the thallus surface; pseudocyphellae absent; lower surface pale to dark brown, with brown rhizines originating mainly close to the margins; lobe margins often with black, raised pycnidia. CHEMISTRY: Medulla PD+ yellow, K+ yellow turning red, KC–, C– (stictic and norstictic acids). HABITAT: On rocks in the open, especially on talus. COMMENTS: The absence of pseudocyphellae and presence of soralia distinguish this species from *M. hepatizon*, which can be superficially similar and contains stictic acid. The western *M. tominii,* also sorediate and growing on rock, has a C+ red medulla.

Melanelia disjuncta (syn. *Parmelia disjuncta*)
Mealy camouflage lichen

DESCRIPTION: Thallus very dark brown; lobes 0.8–1.5 mm wide, with coarse, very dark, almost isidia-like soredia in irregular soralia often covering the surface of the older parts of the thallus; lobes flat at the growing edge, but usually convex in older parts of the thallus; pseudocyphellae either small and inconspicuous or quite conspicuous, close to the lobe margins; lower surface black, with many black rhizines. Apothecia rarely seen. CHEMISTRY: Medulla PD–, K–, KC–, C–, UV+ white (perlatolic and stenosporic acids). HABITAT: On exposed, granitic rocks. COMMENTS: Two sorediate camouflage lichens on rock are C–: *Melanelia disjuncta* and *M. sorediata*. The latter has fine soredia in round soralia, predominantly on the margins or tips of the lobes. In *M. tominii,* the medulla and soralia are C+ red. *Neofuscelia loxodes* has coarse, pustular soralia, and the medulla reacts KC+ reddish violet.

Melanelia elegantula (syn. *Parmelia elegantula*)
Elegant camouflage lichen

DESCRIPTION: Thallus with flat, dull, brown to olive lobes, 1–3 (–7) mm across, often lightly to heavily pruinose, with conical to finally cylindrical isidia, many also more or less globular, some appearing to be somewhat inflated but never spoon-shaped or flattened in any way (Fig. 33a); pseudocyphellae indistinct or absent. Apothecia rare. CHEMISTRY: Medulla PD–, K–, KC–, C– (no substances). HABITAT: Mostly on trees and shrubs, sometimes on mossy rock. COMMENTS: There are several olive to brown camouflage lichens of moderate size (lobes 1–4 mm wide) with cylindrical isidia, no obvious pseudocyphellae, and with a PD–, C– medulla. Unfortunately, they are difficult to distinguish, even with experience. *Melanelia elegantula* has small granular to cylindrical isidia (not inflated or flattened) arising from small conical to hemispherical warts; in *M. infumata,* the isidia develop from globular warts constricted at the base; *M. subelegantula* has

494. *Melanelia disjuncta* Columbia River, Washington ×3.2

495. *Melanelia elegantula* Rocky Mountains, Alberta ×6.3

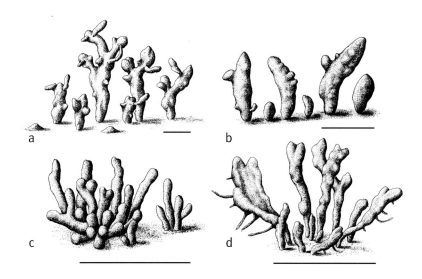

Fig. 33 The isidia of *Melanelia* species: (a) *M. elegantula*; (b) *M. exasperatula*; (c) *M. fuliginosa*; (d) *M. subelegantula*. Scale = 1 mm. (Drawing by Alexander Mikulin. Reprinted by permission of the USDA—Forest Service, from McCune and Geiser, 1997, pp. 159–161, 164.)

496. *Melanelia exasperatula* Rocky Mountains, Alberta ×4.2

slightly flattened but not hollow isidia (Fig. 33d), and the lobes are often somewhat pruinose; *M. exasperatula* has hollow, rather flattened isidia (Fig. 33b), and the thallus is shiny, not pruinose. *Melanelia infumata* is the only strictly rock-dwelling lichen of this group, the others being found mainly on bark. *Melanelia subelegantula* is never on rock.

Melanelia exasperatula (syn. *Parmelia exasperatula*)
Lustrous camouflage lichen

DESCRIPTION: Thallus very appressed, olive-brown, shiny, lobes 2–5 mm wide, without pseudocyphellae or soredia, but with hollow isidia that are constricted at the base, at first round or barrel-shaped, soon flattened and almost lobulate in appearance (Fig. 33b). Apothecia uncommon. CHEMISTRY: Medulla PD–, K–, KC–, C–; containing no lichen substances. HABITAT: Mostly on bark, rarely wood or rock. COMMENTS: See *M. elegantula*. The Latin name *exasperatula* has nothing to do with the frustration we sometimes experience in identifying these confusing isidiate camouflage lichens; it refers to the roughness of the thallus surface (*exasperatis*).

Melanelia fuliginosa (syn. *M. glabratula*, *Parmelia glabratula*)
Shiny camouflage lichen

DESCRIPTION: Thallus olive-green to dark brown, thin, with branched cylindrical isidia (Fig. 33c), 0.2–0.8 mm long; lobes mostly 1–3 mm wide, usually shiny, tending to be pitted or wrinkled, without pseudocyphellae; medulla white, but with patches of orange. Apothecia frequently seen, up to 6 mm in diameter. CHEMISTRY: Medulla PD–, K– (or K+ violet in orange pigmented spots), KC+ red, C+ red (lecanoric acid, and sometimes the anthraquinone rhodophyscin). HABITAT: On bark of coniferous or deciduous trees, or acid rock. COMMENTS: *Melanelia subaurifera*, also C+ red with isidia, is duller and browner, and the unbranched isidia (shorter than 0.3 mm) are easily rubbed off, leaving yellowish patches of soredia behind.

Melanelia glabra (syn. *Parmelia glabra*)
California camouflage lichen

DESCRIPTION: Thallus dark green to olive, becoming brown, lobes 3–4 mm across, thick, becoming wrinkled in the center, with no soredia or isidia; often with a mat of fine, extremely minute, colorless hairs forming a faint fuzz on the lobe and apothecial margins, sometimes spreading to the lobe surface, although other specimens may be perfectly smooth and even shiny; pseudocyphellae only on thallus warts or on the extreme outer edge of the lobes and on apothecial margins; lower surface varying from pale brown at the edges to black in the center, with abundant rhizines. Apothecia common, concave, constricted at the base, 1–5 mm in diameter; margins warty or smooth. CHEMISTRY: Medulla PD–, K–, KC+ red, C+ red (lecanoric acid). HABITAT: On bark of deciduous trees. COMMENTS: This is the only North American, bark-dwelling, camouflage lichen that reacts C+ red and lacks any soredia or isidia. *Melanelia glabroides* is a similar southwestern species that has no cortical hairs, is somewhat shinier, and always grows on rock.

Melanelia hepatizon (syn. *Cetraria hepatizon*)
Rimmed camouflage lichen

DESCRIPTION: Thallus dark brown, usually shiny; lobes 0.4–1.5(–2.5) mm wide, the lobe edges distinctly thickened; soredia and isidia absent; white pseudocyphellae conspicuous along the lobe margins and especially on the margins of the apothecia; lower surface black in center, dark brown at edge; rhizines very sparse and almost entirely marginal. Pycnidia black, cylindrical, usually conspicuous on or near the lobe margins. Apothecia common, red-brown to dark brown, up to 5 mm in diameter; conidia dumbbell-shaped, 4–6 μm long. CHEMISTRY: Medulla PD+ orange, K+ deep yellow, KC–, C– (stictic acid). HABITAT: Growing on exposed, noncalcareous rock. COMMENTS: The closely related **M. commixta,** which shares the same type of habitat, has a pale or brown lower surface (at least not black), and the medulla is PD–, K–. Its conidia are cylindrical or ellipsoid, not thicker at the ends. See Comments under *M. stygia*.

497. *Melanelia fuliginosa* coastal Washington ×2.8

498. *Melanelia glabra* mountains, southern California ×1.2

499. *Melanelia hepatizon* Lake Superior region, Ontario ×2.8

500. *Melanelia multispora* Cascades, Oregon ×2.7

Melanelia multispora (syn. *Parmelia multispora*)
Many-spored camouflage lichen

DESCRIPTION: Thallus brown; lobes 1–3 mm wide, with no soredia or isidia, although often producing small, rounded lobules on the older thallus areas; without pseudocyphellae, except on the apothecial margins; lower surface dark brown to black, usually with abundant rhizines. Apothecia common, 1.3–3 mm in diameter; spores almost spherical, 12–32 per ascus. CHEMISTRY: Medulla PD–, K–, KC–, C– (no lichen substances). HABITAT: On the bark of deciduous trees mainly in humid or mountainous localities. COMMENTS: *Melanelia subolivacea*, a similar western camouflage lichen with no soredia, has 8 spores per ascus and occurs in more inland, drier areas. The southern boreal species **M. trabeculata** also has 16 spores per ascus, but the medulla is PD+ yellow, K+ red (norstictic acid) and the lower surface has overlapping plates of tissue rather than being simply wrinkled as in *M. multispora*.

Melanelia olivacea (syn. *Parmelia olivacea*)
Spotted camouflage lichen

DESCRIPTION: Thallus dark olive to olive-brown; lobes 2–5 mm across, often heavily wrinkled and somewhat ascending, without isidia or soredia; lower surface black with many rhizines; pseudocyphellae conspicuous on the apothecial margins and usually on the lobe tips, although those on the lobes are sometimes hard to find. Apothecia usually abundant, 1.5–5 mm in diameter, somewhat elevated and concave, with thick, bumpy margins, at least when young; spores 12–15 × 7–8.5 μm; pycnidia dotting the thallus, usually abundant. CHEMISTRY: Medulla PD+ red, K–, KC–, C– (fumarprotocetraric acid). HABITAT: On bark, especially birch, in boreal forests. COMMENTS: *Melanelia septentrionalis,* also with a PD+ red medulla, is a smaller lichen with lobes 1–3 mm wide, apothecia rarely more than 1.5 mm in diameter with smoother margins that have fewer pseudocyphellae. *Melanelia halei,* an Appalachian species, is similar to *M. olivacea* but entirely lacks pseudocyphellae, is somewhat paler, develops lobules in the older parts of the thallus, and contains atranorin in addition to fumarprotocetraric acid (which may be responsible for making the medulla react K+ yellowish to reddish brown). It also has larger spores (17–19 × 9–9.5 μm). IMPORTANCE: *Melanelia olivacea* yields a brown dye for wool. See *M. subolivacea.*

Melanelia panniformis (syn. *Parmelia panniformis*)
Shingled camouflage lichen, lattice brown

DESCRIPTION: Thallus dark brown to olive-brown, with narrow lobes (0.3–1.5 mm wide) and even narrower lobules on the thallus surface that are short, rounded, and overlapping like shingles, sometimes resembling flattened isidia; no soredia or true isidia; pseudocyphellae absent or faint; lower surface brown to black. CHEMISTRY: Medulla PD–, K–, KC+ pink or KC–, C–, UV+ white (perlatolic and stenosporic acids). HABITAT: On noncalcareous rock. COMMENTS: This lichen is very distinctive because of its shingled, flattened lob-

501. *Melanelia olivacea* Lake Superior region, Ontario ×2.1

502. *Melanelia panniformis* North Woods, Maine ×2.9

503. *Melanelia septentrionalis* Laurentides, Quebec ×2.7

504. *Melanelia sorediata* Lake Superior region, Ontario ×4.5

ules. *Melanelia exasperatula* and **M. subelegantula** can have somewhat flattened isidia, but the medulla is KC–, UV– (no substances), and they usually grow on tree bark. In *M. subelegantula*, the isidia are initially cylindrical, becoming flattened into lobules (even developing rhizines) only as they become older.

Melanelia septentrionalis (syn. *Parmelia septentrionalis*) Northern camouflage lichen, northern brown

DESCRIPTION: Thallus closely appressed, brown; lobes 0.5–3 mm wide; without soredia or isidia; pseudocyphellae not conspicuous; lower surface brown to black, smooth, with black rhizines. Apothecia broad and flat, shiny yellow-brown, often covering the thallus; spores 8 per ascus, 10–13 × 6–7.5 μm. CHEMISTRY: Medulla PD+ red, K–, KC–, C– (fumarprotocetraric acid). HABITAT: On bark of deciduous and coniferous trees. COMMENTS: *Melanelia trabeculata* is superficially similar but contains norstictic acid, has overlapping plates of tissue on the lower surface, and has spherical spores, 12–16 per ascus. See also Comments under *Melanelia olivacea*.

Melanelia sorediata (syn. *Parmelia sorediosa*) Powdered camouflage lichen

DESCRIPTION: Thallus dark brown, lobes mostly 0.4–1.5 mm wide, flat to slightly concave; surface dull, lacking pseudocyphellae; soralia round, usually terminal on lobe tips or marginal; soredia usually fine and powdery, but occasionally coarse. Apothecia rare. CHEMISTRY: Medulla PD–, K–, KC– or pinkish, C–, UV+ white (perlatolic and stenosporic acids). HABITAT: On acid rock such as granite in open habitats. COMMENTS: This closely resembles *M. disjuncta* but has finer soredia and no pseudocyphellae. See also *M. culbersonii* and *M. tominii*.

Melanelia stygia (syn. *Parmelia stygia*)
Alpine camouflage lichen

DESCRIPTION: Thallus dark brown; lobes thick, flat or convex to almost terete, 0.5–2(–3) mm wide, with a black lower surface and scattered rhizines; without soredia or isidia; pseudocyphellae pale to dark but usually conspicuous, over the lobe surface (not especially marginal). Apothecia common, dark brown to black, constricted at the base, 1–5(–8) mm in diameter; pycnidia common, immersed in lobes; conidia dumbbell-shaped, 3.5–5.5 μm long. CHEMISTRY: Medulla PD+ red, K+ brown, KC–, C– (fumarprotocetraric acid), or PD–, K–, KC–, C– (caperatic acid or no substances). HABITAT: This is a common lichen over noncalcareous rocks at high elevations or in the Arctic. COMMENTS: The most distinctive characters of this species are the conspicuous pseudocyphellae on the thallus surface and the usually convex lobes. The PD– chemical race can be confused with *Pseudephebe minuscula* or ***Melanelia commixta***, both of which are shiny dark brown and grow on exposed rocks in arctic-alpine sites. *Pseudephebe minuscula* has narrower branches and lacks obvious pseudocyphellae, although the pits formed by sunken pycnidia resemble them. In *M. commixta* the pseudocyphellae are on the margins of the lobes rather than on the surface. IMPORTANCE: Like *M. olivacea*, this lichen yields a brown dye for wool.

Melanelia subargentifera (syn. *Parmelia subargentifera*)
Whiskered camouflage lichen

DESCRIPTION: Thallus medium brown to olive, often lightly pruinose or with a fuzz of minute, colorless hairs; surface rough and pitted, especially on older portions; lobes 1–4 mm across, with small, round, brownish gray soralia on the surface of the lobes and along the margins, coalescing into extensive sorediate patches in older parts of the thallus; soredia very coarse and granular; pseudocyphellae absent. CHEMISTRY: PD–, K–, KC+ red, C+ red (lecanoric acid). HABITAT: On bark of all kinds as well as mossy rock. COMMENTS: This species is easily recognized by its laminal and marginal soralia and C+ red reaction. Most similar is *M. albertana*, which has strictly marginal, lip-shaped soralia. *Melanelia subaurifera* has diffuse soredia or isidia that rub off and leave irregular yellowish patches. The rock-dwelling *M. tominii* has a much darker brown thallus and pseudocyphellae on the lobe tips. It contains gyrophoric rather than lecanoric acid.

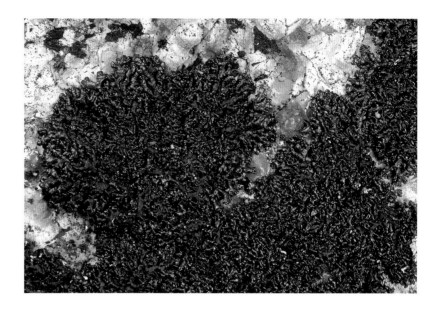

505. *Melanelia stygia* coastal Maine ×2.9

Melanelia subaurifera (syn. *Parmelia subaurifera*)
Abraded camouflage lichen

DESCRIPTION: Thallus olive to chocolate brown, usually dull, but occasionally shiny especially at the edge, not pruinose; lobes rounded, 1–4(–6) mm wide, flat, sorediate or isidiate, usually both, with short, cylindrical, unbranched isidia (mostly less than 0.2 mm long) breaking down into granular soredia on the thallus surface, leaving yellowish patches where they are rubbed off; pseudocyphellae absent or very inconspicuous. Apothecia uncommon. CHEMISTRY: Medulla PD–, K–, KC+ red, C+ red (lecanoric acid). HABITAT: On bark of all kinds, sometimes wood, rarely rock. COMMENTS: *Melanelia subaurifera* is the most common and widespread of the camouflage lichens in eastern North America. Specimens that are mostly isidiate rather than sorediate can resemble *M. fuliginosa*, a species that never becomes sorediate and has longer, usually branched isidia (up to 0.8 mm long). *Melanelia subargentifera* has mainly marginal soredia, without true isidia.

506. *Melanelia subargentifera* Rocky Mountains, southern Colorado ×4.2

507. *Melanelia subaurifera* Lake Superior region, Ontario ×3.1

Melanelia subolivacea (syn. *Parmelia subolivacea*)
Brown-eyed camouflage lichen

DESCRIPTION: Thallus thin, appressed, lacking isidia and soredia, brown to olive-brown, shiny or very dull, sometimes lightly pruinose; lobes 1–4 mm wide; lower surface smooth, light to dark brown, rather shiny, with many rhizines. Usually richly fertile with round, flat, red-brown apothecia having thin margins with few or no pseudocyphellae. Asci contain 8 spores. CHEMISTRY: Medulla and cortex PD–, K–, KC–, C– (no lichen substances). HABITAT: Common on broadleaf trees in dry areas. COMMENTS: *Melanelia multispora* is a very similar western camouflage lichen, but the asci contain 12–32 spores instead of 8, and it often produces small lobules. When *M. subolivacea* bears some marginal tubercules, which it does occasionally, it can resemble the truly isidiate *M. elegantula*. *Melanelia olivacea* has obvious pseudocyphellae, especially on the apothecial margins, the apothecia are more prominent, and the medulla is PD+ red. **Melanelia exasperata** is an East Temperate bark lichen that also resembles *M. subolivacea* but has a dark, rather warty thallus. The small, rounded papillae or warts that cover the thallus have pseudocyphellae at the summits. Pseudocyphellae are also conspicuous on the apothecial margins and lobe tips. Like *M. subolivacea*, *M. exasperata* is negative to all reagents.

Melanelia tominii (syn. *Parmelia substygia*)
Dimpled camouflage lichen

DESCRIPTION: Thallus usually shiny, yellowish brown to almost black; lobes 1–3 mm wide, almost always at least sparsely sorediate on the lobe surface and margins, and usually with conspicuous pale to dark pseudocyphellae on the lobe surface; lower surface black. CHEMISTRY: Medulla PD–, K–, KC+ red, C+ red (gyrophoric acid). HABITAT: On noncalcareous rocks, usually in open, dry sites. COMMENTS: This is the only western rock-dwelling, sorediate camouflage lichen that has a C+ red reaction in the medulla. It is superficially similar to *M. disjuncta* and *M. sorediata*, both C–, as well as *M. culbersonii*, an eastern species.

Menegazzia (1 N. Am. species)

KEY TO SPECIES: See *Hypogymnia*.

Menegazzia terebrata
Treeflute

DESCRIPTION: Thallus greenish gray, paler toward the center, smooth; lobes 1–2 mm wide, puffed and hollow, with large, round perforations in the upper surface; usually with laminal, cuff-shaped soralia (Fig. 20f, p. 32), containing powdery soredia; lower surface black, wrinkled, without rhizines. Photobiont green (*Trebouxia*?). Apothecia very rare. CHEMISTRY: Cortex PD–, K+ yellow, KC–, C– (atranorin); medulla PD+ orange, K+ dark yellow, KC–, C– (stictic, menegazziaic, and constictic acids). HABITAT: On bark, usually deciduous trees, in damp forests. COMMENTS: *Menegazzia terebrata* resembles a *Hypogymnia*, except that its perforations are on the upper surface of the lobes. Perforations in *Hypogym-*

508. *Melanelia subolivacea* Kaibob Plateau, northern Arizona ×3.2

509. *Melanelia tominii* central Utah ×3.7

510. *Menegazzia terebrata* coastal Washington ×5.7

nia, when present, are on the lobe tips or axils, and no species of *Hypogymnia* contains stictic acid. *Cavernularia* species are much smaller lichens and have pits on the lower surface.

Micarea (30 N. Am. species)
Dot lichens

DESCRIPTION: Thallus smooth to granular, generally thin, gray, olive, brownish, or often dark green; growing within the substrate when not visible. Photobiont small, unicellular green algae (including *Elliptochloris* and *Pseudochlorella*); small cephalodia containing *Nostoc* or *Stigonema* present in a few species. Apothecia biatorine, small, usually under 0.6 mm in diameter, convex, often coalescing; disks white, gray, brown, or black, sometimes varying in the same species from almost white to black, depending on the amount of light the lichen receives; exciple with radiating, thick-walled hyphae, very poorly or well developed but always tucked under the main body of the apothecium, rarely forming a visible margin of any kind; paraphyses always branched; asci with a K/I+ blue tholus and a more lightly staining axial body often with a darkly stained lining resembling a tube structure as in the *Porpidia*-type ascus; spores colorless, ellipsoid, and 1-celled to rather long and narrow, with many cells, 8 per ascus. CHEMISTRY: Often containing gyrophoric acid, especially in the apothecium. (Test sections of the apothecium under the microscope by introducing C under the cover slip while observing the sections. The C+ pink reaction in the epihymenium or exciple is usually obvious but rapidly disappears.) A number of species are PD+ red (argopsin), but most species lack lichen substances. HABITAT: Typically in shaded, humid habitats, on wood, bark, rock, or mosses. COMMENTS: The best field characters for recognizing a *Micarea* are the small hemispherical, marginless, often coalescing apothecia. *Bacidia, Bacidina, Catillaria,* and *Mycobilimbia* are similar, but the asci stain differently with K/I. *Arthonia* species have abundantly branched, anastomosing paraphysoids and no exciple, and the asci are K/I– (or turn red-orange).

KEY TO SPECIES: See Key F.

511. *Micarea peliocarpa* herbarium specimen (Canadian Museum of Nature), Ontario ×14

Micarea peliocarpa
Shadow dot lichen

DESCRIPTION: Thallus consisting of pale gray, round granules or small areoles, less than 0.2 mm in diameter. Apothecia hemispherical, very dark brown to black except in deep shade, where they can be almost white, single or coalescing to some degree, 0.2–0.5 mm in diameter; epihymenium greenish; hypothecium colorless; exciple colorless and well developed; paraphyses branching; spores fusiform, 15–23 × 3–6 µm, 2–6 cells, but usually 4-celled, 8 per ascus. CHEMISTRY: Thallus and apothecia PD–, K–, KC+ red, C+ red (gyrophoric and 5-O-methylhiascic acids). HABITAT: A very common and widespread species on shaded wood, bark, bryophytes, or moist, noncalcareous rock. COMMENTS: Other commonly encountered, broadly distributed species of *Micarea* include: *M. melaena,* also with mainly 4-celled spores, 12–21 × 4–5.5 µm, sometimes showing a C+ pink reaction in the thallus but always with black apothecia

512. (facing page) *Multiclavula corynoides* (damp; with *Peltigera venosa* and *Psoroma hypnorum*) Strawberry Range, eastern Oregon ×0.58

that are entirely C–, and a dark brown to purplish hypothecium (often K+ purple); and **M. prasina,** with mainly 2-celled spores, 8–14(–17) × 2.5–4(–5) µm, pale to dark gray or almost black apothecia, and a pale hypothecium.

Multiclavula (5 N. Am. species)
Club-mushroom "lichens"

DESCRIPTION: Thallus consisting of a film of green algae (*Coccomyxa*) over the substrate, through which club-shaped, pale yellow to orange fruiting bodies emerge, 5–15 mm tall, 1–2 mm thick. The fruiting structure is an ephemeral basidiocarp, relating it to mushroom-forming fungi (Basidiomycetes) rather than to cup fungi (Ascomycetes), as in most lichens. CHEMISTRY: No lichen substances. HABITAT: On rotting logs, sandy soil, or peat in exposed or shaded habitats. COMMENTS: *Multiclavula* does not really qualify as a lichen because its association with the algae, although apparently obligate, does not produce a special thallus structure of any kind. The species are distinguished based on characters of the fruiting bodies.

KEY TO GENUS: See Key to keys.

Multiclavula corynoides

DESCRIPTION: Fruiting body very pale yellow; basidiospores 5–8 × 2.0–3.5 µm. HABITAT: On sandy soil, usually with moss, especially road banks. DISTRIBUTION: Pacific Northwest. COMMENTS: It is difficult to distinguish *M. corynoides* from the more common **M. vernalis.** The microcopic structure of the basidiocarps is almost identical, although the spores can reach 12 µm long in *M. vernalis*. The basidiocarps of *M. corynoides* are rather hard and brittle when dry, usually somewhat translucent toward the base, at least when fresh, while those of *M. vernalis* are stouter, fleshier, opaque throughout, and usually larger. *Multiclavula vernalis* is known only from the East. The most common *Multiclavula* is **M. mucida,** an eastern to boreal species on shaded rotten logs, with dark yellow to orange basidiocarps.

Mycobilimbia (7 N. Am. species)
Dot lichens

DESCRIPTION: Thallus ranging from thin and membranous to rather thick, granular, warty, or areolate-squamulose, sometimes sorediate. Photobiont green (unicellular). Apothecia biatorine, usually convex, pale to dark brown or almost black; margins soon disappearing; paraphyses predominantly unbranched except close to the tips; asci similar to those of *Biatora*, with an IKI+ dark blue lining to the cone-shaped indentation or tube in the IKI+ light blue tip (Fig. 14b); spores 1–10-celled, colorless, ellipsoid to fusiform, the long spores often with one end broader than the other; 8 per ascus. CHEMISTRY: All reactions negative (a few species contain zeorin). HABITAT: Commonly growing on mosses, mossy bark, or rock in shaded habitats, a few on tundra peat. COMMENTS: The circumscription of *Mycobilimbia* with respect to other members of its family, the Bacidiaceae, is still unsettled. Some species in the genus are sometimes included within *Biatora*, but the development of the apothecia is different. In general, *Biatora* species differ from most *Mycobilimbia* species in having predominantly 1–2-celled spores; *Micarea* usually has branching paraphyses. *Mycobilimbia* (and *Biatora*) differ from *Bacidia* and *Bacidina* in the branching and orientation of the hyphae in the exciple. The asci of *Biatora*, *Bacidia*, and *Mycobilimbia* can resemble a typical *Bacidia*-type or *Biatora*-type (Fig. 14a,b), depending on the extent of staining around the conical axial body. A few species (**M. berengeriana** and **M. hypnorum**) have a *Porpidia*-type ascus and 1–2-celled spores; they are included in this genus only tentatively.

KEY TO SPECIES: See Key F.

Mycobilimbia tetramera (syn. *M. obscurata, M. fusca, Bacidia obscurata, B. fusca*)
Four-celled moss-dot

DESCRIPTION: Thallus thin, granular to verruculose, sometimes simply forming an irregular membrane over the moss, pale green to greenish gray. Apothecia typically dark brown to black, or paler reddish brown in shade, 0.5–1.4 mm

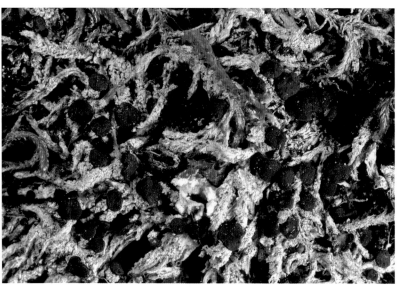

513. *Mycobilimbia tetramera* herbarium specimen (Canadian Museum of Nature), British Columbia ×5.2

514. *Mycoblastus sanguinarius* Klamath Range, northwestern California ×2.8

in diameter, at first rather flat with a thin margin, then becoming convex and losing its margin, often bumpy and irregular when old; epihymenium almost colorless to brownish; hypothecium pale brown or colorless; exciple brownish with radiating hyphae; spores fusiform, 1- to mostly 4-celled, (14–)16–26(–30) × 5.5–8 μm. HABITAT: On bryophytes or mossy bark, sometimes peaty soil. COMMENTS: See Comments under *Bacidia sabuletorum*.

Mycoblastus (6 N. Am. species)

DESCRIPTION: Thallus grayish white or greenish gray, crustose, thin or rather thick, continuous, sometimes sorediate; prothallus frequently present, brownish to blue-green. Photobiont green (*Trebouxia*). Apothecia ebony-black, shiny, hemispherical, tightly appressed, without margins; exciple almost nonexistent; paraphyses abundantly branched and anastomosing; spores colorless, 1-celled, with thick walls, 1–2 per ascus. CHEMISTRY: Thallus K+ yellow (atranorin) in most species; also caperatic, perlatolic, planaic, usnic, and fumarprotocetraric acids. HABITAT: On bark, rarely rocks, mostly boreal. COMMENTS: The fertile species are conspicuous and easily identified (see *M. sanguinarius* below). *Mycoblastus caesius* and *M. fucatus* are two normally sterile species with thin, continuous to areolate gray thalli and discrete patches of granular soredia. They are probably common in more humid boreal forests. Both have blue-green soredia and blue-green prothalli that turn violet with nitric acid. *Mycoblastus caesius* is PD–, K–, KC–, C–, UV+ white (perlatolic acid), and *M. fucatus* is PD+ red, K+ yellow, KC–, C–, UV– (fumarprotocetraric and protocetraric acids and atranorin).

KEY TO SPECIES AND GENUS: See Keys C and F, respectively.

Mycoblastus sanguinarius
Bloody-heart lichen

DESCRIPTION: Thallus white, thick, continuous to verruculose, aggressively overgrowing other lichens and mosses. Apothecia up to 2.5 mm in diameter; upper part of hymenium green, containing violet granules, the color sometimes visible in thin sections with a hand lens; asci containing a single, 1-celled, thick-walled spore, 70–100 × 25–45 μm. A slice through the apothecium reveals its "bloody heart," a bright red zone just below the brown hypothecium. CHEMISTRY: Thallus PD–, K+ yellow, KC–, C– (atranorin and caperatic acids); red pigment K+ brilliant red-orange. HABITAT: On bark and wood of conifers and birch. COMMENTS: The bloody-heart lichen is a very common crustose lichen throughout the northern coniferous forests. *Mycoblastus affinis* is a sister species, without the red pigment in the apothecial base, and with smaller (usually under 70 μm) spores, (1–)2 per ascus. It is a western species. *Mycoblastus alpinus* is a relatively rare boreal *Mycoblastus* similar to *M. affinis,* but instead of apothecia, it usually produces yellow soralia filled with granular soredia. The yellow is due to usnic acid, although both *M. affinis* and *M. alpinus* contain planaic acid as well.

Myelochroa (4 N. Am. species)
Axil-bristle lichens

DESCRIPTION: Small to medium-sized, pale gray to blue-gray foliose lichens; without maculae or pseudocyphellae; medulla pale yellowish, at least under the apothecia or close to the algal layer (sometimes difficult to see); lower surface black, with fairly abundant black, usually unbranched rhizines; short, black, typically unbranched cilia in lobe axils and sometimes on the margins. Photobiont green (*Trebouxia?*). Apothecia lecanorine, reddish brown, not perforate; spores broadly ellipsoid, colorless, 1-celled, 7–15 × 5–10 μm. CHEMISTRY: Cortex K+ yellow (atranorin); medulla with various compounds including zeorin and leucotylic acid (triterpenes) and secalonic acid A (a yellow pigment that reacts C+ yellow, K+ yellow). COMMENTS: This genus is mainly distinguished by its chemistry, including the medullary pigment, and its marginal cilia. Species of *Parmelinopsis* and *Parmelina* also develop marginal cilia but have a white medulla that in most species reacts C+ pink or red.

KEY TO SPECIES: See *Parmelia.*

515. An apothecium of *Mycoblastus sanguinarius* cut open to show blood red tissue of apothecial base. ≈ ×16

Myelochroa aurulenta (syn. *Parmelina aurulenta*)
Powdery axil-bristle lichen

DESCRIPTION: Lobes 2–4 mm wide, rounded to somewhat elongate; cilia often sparse, mostly confined to axils; coarsely granular soredia in irregular soralia, developing from a breakdown of the cortex or burst pustules; maculae absent; yellow medullary pigment mostly in and around soralia. Apothecia rare. CHEMISTRY: Medulla negative to reagents except when yellow (see above); contains leucotylic acid. HABITAT: On bark of deciduous trees, especially maples and oaks in moderately shaded woodlands. COMMENTS: This very common lichen is easily identifiable by its yellow medulla and coarse soredia. *Canoparmelia crozalsiana,* a similar southeastern lichen, has reticulate ridges and depressions on the lobes and reacts PD+ orange, K+ yellow (stictic acid).

Myelochroa galbina (syn. *Parmelina galbina*)
Smooth axil-bristle lichen

DESCRIPTION: Thallus with small, flat lobes (0.8–2 mm), lacking soredia or isidia; usually fertile, with reddish brown apothecia, 2–5 mm in diameter. CHEMISTRY: Medulla PD+ orange, K+ yellow to red (galbinic acid). HABITAT: On bark of deciduous trees,

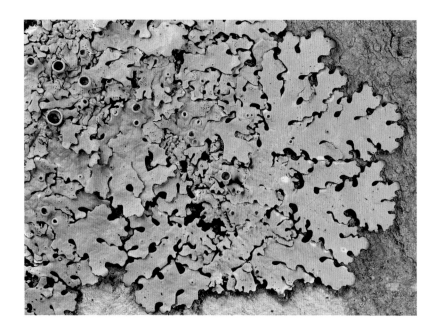

516. (top right) *Myelochroa aurulenta* northern Wisconsin ×2.9

517. (top left) *Myelochroa galbina* Great Smoky Mountains, Tennessee ×3.3

518. (bottom) *Myelochroa obsessa* Appalachians, Virginia ×3.0

rarely on rock. COMMENTS: Two nonsorediate shield lichens might be mistaken for *M. galbina*: *Hypotrachyna livida*, with forked rhizines, broader lobes, a white medulla, and different chemistry; and *Parmelina quercina*, a California species with a white, C+ red medulla.

Myelochroa obsessa (syn. *Parmelina obsessa*)
Rock axil-bristle lichen

DESCRIPTION: Lobes 1–2 mm wide, rather shiny, but largely covered with cylindrical to branched isidia. CHEMISTRY: Medulla PD+ orange, K+ orange to red, KC– or KC+ orange, C– (galbinic acid). HABITAT: On noncalcareous rock, especially sandstone, in oak forests. COMMENTS: This is essentially an isidiate version of *M. galbina* growing on rock. *Imshaugia aleurites*, which resembles *M. obsessa* in size and isidial type, has a very pale lower surface and a white medulla.

Myriotrema (14 N. Am. species)
Volcano lichens

DESCRIPTION: Tropical to subtropical crustose lichens, with thick, continuous to warty thalli. Photobiont green (*Trentepohlia*). Apothecia buried in the thallus, with sunken disks opening to the surface by a pit-like hole, and with a double wall as in *Thelotrema* and *Diploschistes*. The exciple, which constitutes the inner wall, and any sterile column (*columella*) that occurs in the apothecium are colorless or pale reddish brown; spores colorless, IKI+ violet, thick-walled with 3 to many roundish cells either in a single line or irregular, creating a muriform arrangement (see Fig. 15h), 1–8 per ascus. CHEMISTRY: β-orcinol depsidones: psoromic, stictic, norstictic, or protocetraric acids. HABITAT: On bark. COMMENTS: *Thelotrema* and *Ocellularia* are related genera, but *Thelotrema* has no columella in the apothecium, and in *Ocellularia* the columella and exciple are pitch black. *Diploschistes* has a black exciple and dark brown spores, and grows only on soil or rock or over mosses and other lichens.

KEY TO SPECIES (based on Harris, *More Florida Lichens*, 1995)
1. Spores colorless, muriform or transversely septate .... *Myriotrema rugiferum*
1. Spores brown, muriform 2

2. Spores 45–55 × 10–13 μm, many-celled. Ascomata round, not crowded; medulla PD+ orange, K+ yellow (stictic acid complex) .... [*Myriotrema subcompunctum*]
2. Spores under 30 μm long, few-celled 3

3. Ascomata round, with pore-like openings; thallus with a cortex; spores 14–28 × 9–14 μm; medulla PD–, K–, but containing pockets of red crystalline pigment (K+ purple) .... [*Myriotrema wightii*]
3. Ascomata angular, with fissure-like openings, often in whitish areas without a cortex; spores 13–16 × 7–10 μm; medulla PD+ orange, K+ yellow (stictic acid complex) .... [*Myriotrema glaucescens*]

Myriotrema rugiferum

DESCRIPTION: Thallus thick, gray, with tiny, buried apothecia, 0.3–0.4 mm in diameter, opening to the surface by sunken pits; spores colorless, submuriform, 9–20 × 4–6 μm. CHEMISTRY: Thallus PD+ yellow, K+ pale yellow, KC–, C– (psoromic acid).

519. *Myriotrema rugiferum* northern Florida ×4.3

Neofuscelia (10 N. Am. species)
Camouflage lichens

DESCRIPTION: Dark brown foliose lichens with lobes about 0.5–3 mm wide, often elongate, closely or loosely attached to the substrate; lower surface black to pale brown, with short, unbranched rhizines; without pseudocyphellae or soredia, but sometimes with isidia that disintegrate into soredia-like fragments (schizidia). Photobiont green (*Trebouxia*?). Apothecia *Parmelia*-type, brown, nonpruinose, laminal; spores colorless, ellipsoid, 8 per ascus. Pycnidia laminal, immersed and inconspicuous. CHEMISTRY: Upper cortex turns blue-green with nitric acid. (Slice a thin sheet of tissue, place it on a microscope slide in a small drop of water, and test it under the microscope by introducing concentrated nitric acid under the cover slip.) HABITAT: On rock, mostly in arid regions. COMMENTS: This genus is closely related to *Melanelia*; they differ in the reaction of the cortex with nitric acid and in the absence of pseudocyphellae in *Neofuscelia*.

KEY TO SPECIES: See *Melanelia*.

520. *Neofuscelia atticoides* White Mountains, eastern Arizona ×2.0

521. *Neofuscelia loxodes* southern British Columbia ×3.4

Neofuscelia atticoides

DESCRIPTION: Thallus reddish brown to dark olive brown, closely appressed, smooth, or rough and wrinkled; lobes 0.5–2 mm wide, short or elongate, with no isidia but sometimes with lobules on the thallus surface; lower surface pale brown. Apothecia common, 2–4 (–7) mm in diameter with mostly smooth margins. CHEMISTRY: Medulla PD+ orange, K+ yellow or turning red-orange, KC–, C– (stictic and sometimes norstictic acids). COMMENTS: This is the only nonisidiate *Neofuscelia* containing stictic acid. In **N. occidentalis,** the medulla is PD+ red (fumarprotocetraric acid), and in **N. ahtii, N. brunella,** and **N. infrapallida,** all much rarer, the medulla is PD–, K–, containing only fatty acids of different kinds. They differ morphologically from one another in very subtle ways. *Melanelia hepatizon,* which also contains stictic acid, has conspicuous pseudocyphellae on thickened lobe margins. In *M. stygia,* the medulla is PD+ red or PD– and pseudocyphellae are also conspicuous.

Neofuscelia loxodes
Blistered camouflage lichen

DESCRIPTION: Thallus yellowish brown to reddish brown, with lobes mostly 1–3 mm wide, producing small, hollow, globular pustules that resemble globular isidia. These pustules (each larger than 0.2 mm in diameter) break up into soredia-like schizidia. Apothecia uncommon. CHEMISTRY: Medulla PD–, K–, KC+ reddish violet, C– (glomellic, glomelliferic, and perlatolic acids); rarely C+ pink (gyrophoric acid). COMMENTS: The similar *N. subhosseana* has smaller pustules and is PD+ orange and K+ deep yellow. **Neofuscelia verruculifera** has abundant granular isidia, each pustular granule less than 0.2 mm in diameter, forming patches on the upper surface, soon breaking down into schizidia; the medulla is PD–, K–, KC– or + pink, C– or + pink (divaricatic and sometimes gyrophoric acids).

Neofuscelia subhosseana
Erupted camouflage lichen

DESCRIPTION and COMMENTS: Similar to *N. loxodes* but with small, disintegrating pustules like those of *N. verruculifera*, giving it a more sorediate appearance. Its chemistry distinguishes it from both *N. loxodes* and *N. verruculifera*. Apothecia unknown. CHEMISTRY: Medulla PD+ orange (slowly), K+ deep yellow turning red, KC–, C– (unidentified substance).

Nephroma (10 N. Am. species)
Kidney lichens, paw lichens

DESCRIPTION: Medium-sized to large foliose lichens, either dark brown (containing cyanobacteria: *Nostoc*) or green to yellow-green (containing green algae: *Coccomyxa*), except in *N. occultum* (see below). Apothecia kidney-shaped (giving the lichen its name), produced on the lower surface of the lobe tips, which often lift up from the surface and curl back, making the apothecial disk visible from above; lower surface smooth and shiny, or fuzzy with a thin tomentum, without rhizines. Species with green algae develop internal cephalodia that appear as low hills or bumps on the upper or lower surface of the lobes. CHEMISTRY: Except for one species with a yellow medulla and the species containing green algae, all species of *Nephroma* are negative in the medulla with PD, K, KC, and C. They usually contain zeorin and other triterpenes; rarely with depsides nephroarctin and phenarctin. COMMENTS: Specimens with apothecia cannot be confused with any other lichen, but sterile specimens can be similar to some species of *Melanelia*. Other species resemble *Peltigera*, *Platismatia*, *Asahinea*, or other broad-lobed foliose lichens. The color of the algal layer and type of lower surface are helpful in distinguishing these genera.

KEY TO SPECIES

1. Photobiont green; cephalodia present; on the ground and mossy rocks and logs mostly in northern boreal, arctic, and alpine habitats 2
1. Photobiont blue-green; cephalodia absent; on mossy bark or rock 3
2. Thallus pale green, becoming browned especially at the margins; cortex KC– (usnic acid absent); lobes undulating or crisped at the margins; cephalodia forming small, inconspicuous warts on lower thallus surface; lobules present . . . . *Nephroma expallidum*
2. Thallus yellow-green; cortex KC+ orange-yellow (usnic acid); lobes flat; cephalodia forming internal swellings visible externally as rounded, dark bumps on the thallus surface; lobules absent . . . . *Nephroma arcticum*
3. Soredia present 4
3. Soredia absent 5
4. Thallus brown; soredia mainly on the lobe margins and sometimes on the surface; lobes mainly smooth; widespread, common . . . . *Nephroma parile*
4. Thallus yellow-green (with usnic acid); soredia mainly on the thallus surface; lobes with a network of depressions and sharp ridges; western, rare . . . . *Nephroma occultum*
5. Isidia present 6
5. Isidia absent 7
6. Isidia cylindrical, unbranched or branched; lobes with a network of depressions and sharp ridges, or wrinkled; lobules absent; apothecia absent . . . . *Nephroma isidiosum*
6. Isidia flattened; lobes smooth and even; lobules present; apothecia typically abundant. . . . *Nephroma helveticum*
7. Medulla yellow, K+ red or deep purple (anthraquinones) . . . . *Nephroma laevigatum*
7. Medulla white, K– (anthraquinones absent). 8
8. Lower surface tomentose, with scattered, pale bumps. . . . . *Nephroma resupinatum*
8. Lower surface smooth . . . . *Nephroma bellum*

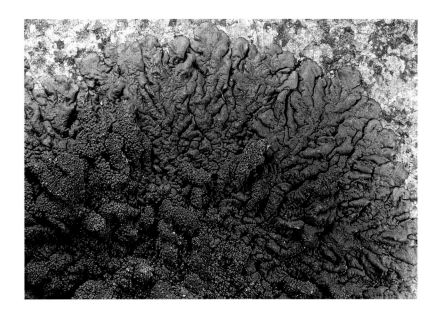

522. *Neofuscelia subhosseana* southern British Columbia ×2.9

523. *Nephroma arcticum* (wet) Southeast Alaska ×1.0

Nephroma arcticum
Arctic kidney lichen, green light

DESCRIPTION: Thallus yellowish green, containing green algae; lobes very broad (up to 30 mm wide), flattened to irregularly wrinkled, without soredia or isidia; lower surface very pale tan at the edge, grading to brown-black in the center; internal cephalodia below the green algal layer, creating broad, flat, gray bumps on the upper surface of the thallus. CHEMISTRY: Cortex KC+ gold (usnic acid); medulla PD+ deep yellow, K– or K+ pale yellow, KC–, C– or C+ pale yellow (zeorin, nephroarctin, and phenarctin). HABITAT: On the ground, usually among mosses; very common throughout subarctic to arctic regions and on mountaintops in the Canadian Rockies. COMMENTS: *Nephroma arcticum* and *N. expallidum* are the only two kidney lichens with green algae. *Nephroma expallidum* is a smaller, less common species with lobes rarely wider than 15 mm; the thallus lacks usnic acid and tends to be brownish, especially close to the margins. IMPORTANCE: Some native groups in Alaska eat the arctic kidney lichen by cooking it with crushed fish eggs. They also cook it alone as a kind of strength-promoting tonic.

Nephroma bellum
Naked kidney lichen

DESCRIPTION: Thallus dull, brown, containing cyanobacteria; lobes 4–10 mm wide, with a white medulla; without isidia or soredia (rarely lobulate); lower surface light brown, almost always smooth and perfectly naked (without tomentum). Apothecia usually abundant and conspicuous. CHEMISTRY: All reactions negative (contains mainly zeorin). HABITAT: On branches and twigs of trees, especially conifers; and also on mossy rocks in humid forests. COMMENTS: *Nephroma helveticum* is much more lobulate and usually has a woolly lower surface. *Nephroma laevigatum*, a bicoastal species, is also naked below, but has a yellow medulla and is more closely appressed.

Nephroma expallidum
Alpine kidney lichen

DESCRIPTION: Thallus pale green to somewhat yellowish green, becoming browned, especially in sunlit habitats, containing green algae; upper surface shiny or pruinose, sometimes fuzzy; lobes crisped, crinkled, crowded, and ascending, up to 18 mm wide, without soredia or isidia but sometimes with lobules; lower surface darkening toward the center, more or less tomentose; cephalodia broad, internal, visible as bumps on the lower surface. Apothecia not common, with pale brown disks. CHEMISTRY: Cortex and medulla PD+ deep yellow, K–, KC–, C– (nephroarctin, zeorin and other triterpenes, lacking usnic acid). HABITAT: Growing among mosses in arctic tundra and northern boreal woodlands. COMMENTS: *Nephroma expallidum* is one of only two kidney lichens with green algae, the other being *N. arcticum*. *Asahinea chrysantha* is yellower (containing usnic acid), has a shiny black lower surface, and has pseudocyphellae on a network of ridges. See Comments under *N. arcticum*.

524. *Nephroma bellum* Southeast Alaska ×2.4

525. *Nephroma expallidum* interior Alaska ×1.8

526. *Nephroma helveticum* ssp. *sipeanum*
Cascades, Washington
×2.9

Nephroma helveticum
Fringed kidney lichen

DESCRIPTION: Dark brown, often shiny, containing cyanobacteria; lobes relatively small, up to 5 mm wide, fringed with lobules and flat isidia, which sometimes also occur on the upper surface of the thallus; lower surface dark brown, naked in part, or more usually with a thin covering of fine woolly hair (not densely woolly). Apothecia common, with dark brown disks and lobulate margins. CHEMISTRY: All reactions negative (only triterpenes). HABITAT: On mossy rocks and bark, especially in humid forests, but with a stronger tolerance for dryness than most other kidney lichens. COMMENTS: The only other isidiate *Nephroma* is the much less common *N. isidiosum*. That species has clusters of true cylindrical or branched isidia on the thallus surface and an upper surface with a network of ridges and depressions. *Nephroma resupinatum* and *N. laevigatum* can sometimes also have lobulate-isidiate margins, but the former has distinctive white bumps or warts on the lower surface, and the latter has a yellow medulla. The thallus of *N. helveticum* is thicker in western populations than it is in the east, and the fuzz on the lower surface is somewhat longer, which is the basis for recognizing the western material as **var. *sipeanum*** (shown in plate 526; depicted in pink on the map). Eastern populations represent **var. *helveticum*** (in blue on the map).

Nephroma isidiosum
Peppered kidney lichen

DESCRIPTION: Thallus brown, with cyanobacteria; lobes up to 10 mm wide, with a network of low ridges and depressions; clustered, cylindrical isidia developing along cracks and ridges of the upper surface and, to a lesser extent, along the lobe margins; lower surface with a thin tomentum, often with scattered, stubby black rhizines. Apothecia unknown. CHEMISTRY: All reactions negative (only triterpenes). HABITAT: On twigs and bark in mature, humid forests. COMMENTS: *Nephroma isidiosum* can resemble *N. parile*, but the latter has a naked lower surface and its propagules, although occasionally resembling isidia, are not corticate and are more like granular soredia. The isidia of *N. helveticum* are clearly flattened. IMPORTANCE: The presence of *N. isidiosum* is a good indicator of the old age and pristine condition of the forest it inhabits.

Nephroma laevigatum
Mustard kidney lichen

DESCRIPTION: Thallus dark brown, with cyanobacteria; lobes very thin and closely appressed to the substrate, 3-9 mm wide, often with lobules along the margins; medulla dark mustard yellow; lower surface smooth, pale brown. Apothecia common. CHEMISTRY: Medulla K+ reddish purple (anthraquinone). HABITAT: On trees and rocks in oceanic forests. COMMENTS: *Nephroma laevigatum* is the only *Nephroma* with a pigmented medulla. The only other cyanobacterial lichens in North America with a pigmented medulla are species of *Pseudocyphellaria*, which have yellow pseudocyphellae on their tomentose lower surface, and an unnamed Appalachian species of *Sticta* related to **S. weigelii,** which has conspicuous cyphellae on the lower surface and marginal lobules.

Nephroma occultum
Cryptic kidney lichen, cryptic paw

DESCRIPTION: Thallus yellowish green or yellow-gray, usually with only the lobe edges becoming brown, containing cyanobacteria; lobes 4–12 mm wide, with net-like ridges and depressions on the thallus surface; quite large granules or granular soredia on the margins and along the thallus ridges; lower surface smooth, creamy yellow to light brown at the thallus edge, darker in the center. CHEMISTRY: Cortex KC+ yellow (usnic acid); other tissues negative to all reagents. HABITAT: On conifers in open, old growth forests. It was first discovered on branches in the canopies of old Douglas fir trees. COMMENTS: Among the sorediate cyanobacterial lichens, *N. parile* has true soredia (although they can be coarse and isidia-like), is brown rather than yellowish, and is not reticulately ridged; *Lobaria scrobiculata* can be somewhat yellowish, but the soredia are finer and mainly on the thallus surface in round patches, and the lower surface is tomentose. IMPORTANCE: *Nephroma occultum* is an important indicator of old growth forests and has already influenced some forest management

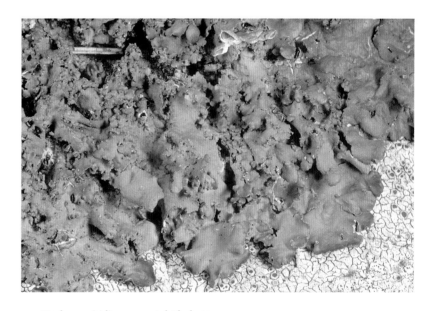

527. *Nephroma isidiosum* coastal Alaska ×2.5

528. *Nephroma laevigatum* British Columbia coast ×2.7

529. *Nephroma occultum* Southeast Alaska ×1.9

decisions. Because of logging pressure, the Committee on the Status of Endangered Wildlife in Canada has listed it as "vulnerable."

Nephroma parile
Powdery kidney lichen

DESCRIPTION: Thallus chocolate brown, rarely brownish gray, becoming olive when wet (as shown in plate 530), usually dull, containing cyanobacteria; lobes mostly 4–8 mm wide, with abundant brown or bluish gray granular soredia along the lobe margins and often in patches on the surface; lower surface pale brown, smooth, and naked (without a tomentum). Apothecia rare. CHEMISTRY: All reactions negative (zeorin in many specimens). HABITAT: On trees and mossy rocks in shaded forests. COMMENTS: This is the only kidney lichen with marginal soredia. When the soredia are large and almost corticate, *N. parile* can be mistaken for *N. isidiosum*. In that species, the lower surface has a thin tomentum, and the isidia are cylindrical (rather than globular), occurring in clusters along cracks and ridges on the thallus

530. *Nephroma parile* coastal Alaska ×3.2

surface. The somewhat similar *Peltigera collina* lacks a lower cortex, so that the lower surface is webby, not smooth. Sorediate species of *Sticta* and *Pseudocyphellaria* have tomentum and either cyphellae or pseudocyphellae on the lower surface.

Nephroma resupinatum
Pimpled kidney lichen

DESCRIPTION: Thallus brown to grayish brown, dull, sometimes short, tomentose especially at lobe tips, fairly thick, ascending; containing cyanobacteria; lobes 5–10 mm wide, with abundant lobules often produced along the margins and thallus cracks; lower surface fuzzy, with scattered pale, naked, pimple-like bumps visible amid the tomentum, often in clumps. Apothecia common. CHEMISTRY: All reactions negative (no lichen substances). HABITAT: On trees and mossy rocks in humid forests. COMMENTS: *Nephroma resupinatum* resembles *N. helveticum* in its typically tiny lobules on the margins, but the latter never has white warts on the lower surface. In addition, the backs of the apothecia of *N. resupinatum* (i.e., the upper surface of the lobe tips) are often hairy rather than rough and wrinkled as in *N. helveticum*.

Niebla (11 N. Am. species) (syn. *Vermilacinia*)
Fog lichens

DESCRIPTION: Pale yellow-green fruticose lichens forming clumps of stiff, rounded to flattened branches; cortex usually tough and thick, composed of cells perpendicular to the surface; medulla cottony and uniform, or dense and containing strands of supportive tissue. Photobiont green (*Trebouxia?*). Apothecia lecanorine, disks usually yellowish or pale pinkish orange to tan, pruinose, containing 2-celled spores. Pycnidia black, buried almost entirely in the thallus, especially along the margins of the branches, but frequently producing abundant and conspicuous black spots over much of the thallus. CHEMISTRY: Cortex KC+ gold (usnic acid); medulla commonly with triterpenes such as zeorin, fatty acids such as bourgeanic acid, and depsides such as divaricatic, evernic, and barbatic acids; β-orcinol depsidones present in species from Baja California (Mexico). HABITAT: On rocks, trees, and shrubs along the foggy coasts of California and Baja California, where on-shore fogs are frequent and levels of light are high. COMMENTS: *Niebla* differs from the closely related *Ramalina*, having a thicker cortex of a different type, and black conspicuous pycnidia in almost all species. *Niebla* also has much more restricted habitat requirements. *Niebla* can be divided into two genera: *Niebla* (in the strict sense), containing strands of supportive tissue in the medulla as described for *Letharia* (see fig. 8c), with a chemistry rich in depsides and depsidones; and *Vermilacinia*, lacking such strands, and containing mainly triterpenes and fatty acids. The chemical and morphological diversity in *Niebla* (in the broad sense) is so great that many distinguishable entities have been given species names. A conservative approach is taken here because studies of this group are still in progress.

531. *Nephroma resupinatum* Cascades, Washington ×1.5

KEY TO SPECIES: See *Ramalina*.

Niebla cephalota (syn. *Vermilacinia cephalota*)
Powdery fog lichen

DESCRIPTION: Thallus very pale yellow to greenish yellow; branches round in section, mostly 20–25 mm long, 0.4–1.5 mm in diameter, with irregular depressions and wrinkles; large, convex soralia containing bluish gray soredia produced along the branches (soralia becoming white-woolly in old specimens). Apothecia not produced. CHEMISTRY: Medulla PD–, K–, KC–, C– (triterpenes including zeorin). Rare populations with salazinic or norstictic acid (K+ red)

532. (top left) *Niebla cephalota* northern California coast ×3.1

533. (facing page) *Niebla combeoides* California coast ×5.5

534. (top right) *Niebla homalea* California coast ×1.38

also occur. HABITAT: On shrubs and trees, sometimes rock. COMMENTS: This small, pale yellowish fruticose lichen resembles a *Ramalina* or *Evernia* but has angular to almost rounded branches and a soft, cottony interior. This is the only *Niebla* that is sorediate, and its range extends farther north than any other *Niebla* species. A lichen called **Niebla ceruchis** by most North American lichenologists is generally similar but is non-sorediate, stiffer, and somewhat larger, with branches often exceeding 30 mm in length and 4 mm in width. It is usually fertile, with one to several apothecia on each branch tip, and it grows exclusively on bark.

Niebla combeoides (syn. Vermilacinia combeoides)
Bouquet fog lichen

DESCRIPTION: Thallus pale yellowish green, shiny or dull, bases gray or blackening; branches forming tight, rounded clumps 20–30 (–60) mm long; branches sparsely divided, round in section, 1–2.5 mm in diameter, with depressions and ridges; tips blunt or with 2–5 apothecia, up to 7 mm in diameter; medulla dense. CHEMISTRY: Medulla PD–, K–, KC–, C– (containing bourgeanic acid and zeorin). HABITAT: On rocky cliffs and boulders. COMMENTS: This is the most common of the clumped fog lichens with terete branches and can be recognized by the apothecia at the branch tips. **Niebla ceruchoides** (syn. *Vermilacinia ceruchoides*) is a much smaller species, forming rounded mounds of slender (less than 0.5 mm in diameter), terete, smooth, usually sterile branches that divide typically by regular dichotomies and have black pycnidia at the tips. **Niebla procera** (syn. *Vermilacinia procera*), also with cylindrical branches, has a much smoother, more even surface commonly splotched with black, and is more abundantly branched.

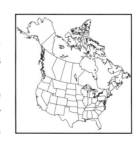

Niebla homalea
Armored fog lichen

DESCRIPTION: Thallus pale yellowish green, rather shiny, with pale brown to blackened bases; branches growing in thick clumps, flattened or strongly angular, with depressions and sharp, low ridges defining "plates" on the surface; branches 1–5(–8) mm

535. *Niebla laevigata* Channel Islands, southern California ×1.5

wide, irregular in width, 20–70 mm long; medulla dense, containing narrow, translucent, cord-like strands of tissue (examine the thicker, more angular branches). Sterile branches terminating in a point, but fertile branches developing lateral or sometimes terminal apothecia, 1–5 mm in diameter. CHEMISTRY: Medulla PD–, K–, KC–, C– (contains either sekikaic, divaricatic, or barbatic acids, often with triterpenes). HABITAT: On rocks or soil. COMMENTS: The flattened branches with armor-like plates make this a distinctive lichen. *Niebla laevigata* is most similar but is greener, has a smooth surface and broader, more flattened, blade-like branches lacking cords in the medulla, often has whitish apothecia on the branch tips, and contains triterpenes.

Niebla laevigata (syn. *Vermilacinia laevigata*)
Black-footed fog lichen

DESCRIPTION: Thallus pale yellowish green; base distinctly blackened (not visible in plate 535); branches flat, sparsely divided, up to 50 mm long, 2–8(–25) mm wide, growing in clumps; surface usually smooth, not ridged or composed of "plates"; medulla thick and cottony, without strands of tissue. Apothecia common, usually terminal, up to 8 mm in diameter. CHEMISTRY: Medulla PD–, K–, KC–, C– (contains bourgeanic acid and triterpenes including zeorin). HABITAT: On cliffs and boulders. COMMENTS: See *N. homalea* above.

Nodobryoria (3 N. Am. species)
Foxtail lichens, red horsehair lichens, chestnut beard

DESCRIPTION: Shrubby or pendent fruticose lichens, distinctly reddish brown, dull, with irregular, ridged, flattened to terete branches without pseudocyphellae, soredia, or isidia; surface of cortex, as viewed at 100–400× magnification, composed of interlocking cells like the pieces of a jigsaw puzzle. (Make a very thin slice of the cortex parallel to the surface, and mount it in water on a microscope slide for examination with a compound microscope.) Photobiont green (*Trebouxia?*). Apothecia lecanorine, red-brown; spores small, ellipsoid, colorless, nonseptate, 8 per ascus. CHEMISTRY: All reactions negative (no lichen substances). HABITAT: On conifer trees. COMMENTS: The *Bryoria*-like habit, dull reddish color, and characteristic cortex make this a unique genus. Only *Polychidium*, a minute fruticose lichen containing cyanobacteria, has a similar cortex. *Bryocaulon* has a similar color, but the branches are shiny and smooth with white pseudocyphellae, and the medulla is usually C+ red.

KEY TO SPECIES: See *Alectoria*.

Nodobryoria abbreviata (syn. *Bryoria abbreviata*)
Tufted foxtail lichen

DESCRIPTION: Thallus shrubby, up to 2.5 cm long, main branches about 0.2–0.4 mm wide, angular and pitted, with a spiny appearance. Apothecia very common, 2–3 mm wide, at least some arising at or close to the branch tips, with spiny cilia on their margins. HABITAT: Mainly on ponderosa pine and Douglas fir in dry forests. COMMENTS: *Nodobryoria abbreviata*, as the name implies, is relatively short and has a remarkable superficial resemblance to *Kaernefeltia californica*. Both are short and shrubby, with deeply channeled and dented branches, producing many apothecia with cilia. *Kaernefeltia californica*, however, is blackish green to dark olive-brown, never red-brown, and the disks are black, giving a K+ purple reaction in the epihymenium when viewed under the microscope. It is a lichen of coastal pine forests, whereas *N. abbreviata* is a more inland species. See Comments under *Nodobryoria oregana*.

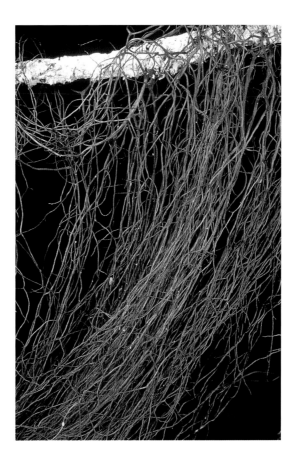

536. *Nodobryoria abbreviata* Wallowa Mountains, eastern Washington ×2.5

537. *Nodobryoria oregana* Coast Mountains, British Columbia ×2.5

Nodobryoria oregana (syn. *Bryoria oregana*)
Pendent foxtail lichen

DESCRIPTION: Thallus pendent, up to 17 cm long; branches slender, 0.1–0.2 mm wide except at the base, with channels and furrows running along the length of the branches. Apothecia infrequent, up to 2 mm wide, arising along the branches (not terminal), margins smooth or sometimes with short cilia. HABITAT: On conifer branches of all kinds, often at high elevations. COMMENTS: The most reliable characters for recognizing *N. abbreviata* in the field are the dull, red-brown color, the irregular branching, and the longitudinal ridges and furrows. Most similar is *Bryoria fremontii*, which is larger and usually shinier and has perpendicular side branches on the thicker stems and furrows and pits mostly confined to the thickest branches and axils. *Nodobryoria abbreviata* is very similar but almost always fertile, with large red-brown apothecia with cilia-like fibrils on the margins. It is usually much shorter and bushier in habit, although some intermediates occasionally occur.

Normandina (1 species worldwide)

KEY TO SPECIES: See Key G.

Normandina pulchella
Elf-ear lichen

DESCRIPTION: Small, green to slightly bluish green squamulose lichen, with very thin squamules, 0.7–2.5 mm across, becoming lobed in well-developed specimens, the edge of each squamule thickened into a thin, uniform, raised rim; often becoming sorediate around the margins (as in lower left-hand corner of plate 538), or on the upper surface; lower surface of squamules white, sometimes tomentose, lacking a cortex. Photobiont green, unicellular (*Nannochloris*). Fruiting bodies perithecia, but exceedingly rare. CHEMISTRY: All reactions negative, but contains zeorin. HABITAT: Commonly growing on mosses and other lichens (especially those containing cyanobacteria such as species of *Pannaria*, *Fuscopannaria*, or *Peltigera*) in humid forests. COMMENTS: This lichen resembles the squa-

538. *Normandina pulchella* (damp) Appalachians, North Carolina ×12.1

mules of *Cladonia*, but the squamules of *Normandina* are more scattered and rounded, with thickened margins, and a bluer green color.

Ocellularia (12 N. Am. species)
Volcano lichens

DESCRIPTION: Crustose lichens that grow on the surface of the bark producing a continuous, rimose to verruculose thallus, or that grow within the outer bark tissues. Photobiont green (*Trentepohlia*). Ascomata buried in hemispherical fruiting warts that closely resemble a *Pertusaria* but often show a rounded black "button" in the ostiole caused by a carbonized mass in the center of the hymenium called a columella, most visible when a section is prepared; periphyses (sterile hairs on the inner face of the exciple wall near the ostiole) absent; hymenium IKI–; paraphyses largely unbranched; exciple black, at least at the top; spores colorless or brown, 4-celled to muriform, IKI+ violet, 1–8 spores per ascus. CHEMISTRY: Frequently with β-orcinol depsidones, especially psoromic and hypoprotocetraric acids, sometimes with depsides (gyrophoric acid). HABITAT: On bark in tropical regions. COMMENTS: See Comments under *Myriotrema*.

KEY TO GENUS: See Key D.

Ocellularia granulosa
Frosted volcano lichen

DESCRIPTION: Thallus thin, continuous, greenish gray, very rough to almost granular in appearance (not actually granular) with a white medulla. Apothecia immersed in the thallus, the openings black, deep, resembling ostioles, or broad and disk-like, usually pruinose; exciple black, forming the inner wall of the fruiting wart; spores colorless, 6–8-celled, with lens-shaped locules, 17–24 × 6–7 μm, 8 per ascus. CHEMISTRY: Medulla PD+ deep yellow to yellow-orange, K–, KC–, C– (psoromic acid). HABITAT: On bark (especially bald cypress). COMMENTS: *Ocellularia americana* is a common species on oak with a pink medulla that reacts C+ red (gyrophoric acid); *O. sandfordiana* has a rough, warty thallus containing no substances, and huge, colorless muriform spores, 1 per ascus.

Ochrolechia (33 N. Am. species)
Saucer lichens, cudbear

DESCRIPTION: Crustose lichens with thin to very thick thalli, smooth and continuous to thick and verrucose, white to ashy gray. Photobiont green (*Trebouxia?*). Apothecia large, generally pink or orange, having thick, lecanorine margins. In several species, apothecia have a thin ring of tissue between the disk and the lecanorine margin, creating a kind of double margin. This tissue is actually the expanded edge of the exciple and is called an excipular ring. Spores over 20 μm long, colorless, 1-celled, thin-walled, 4–8 per ascus; paraphyses very slender, branched and anastomosing; ascus walls generally thick, K/I+ blue-black. CHEMISTRY: Usually with orcinol depsides that react C+ pink or red at least on the disk of the apothecia and often in the thallus cortex or medulla or both. It is important to distinguish between reactions in the cortex and those in the medulla because any differences are useful in identifying the species. To do this, make a thick section of the apothecium from top to bottom and place the section on a microscope slide. Under the dissecting microscope, add fresh C using a fine capillary tube, first to the outer edge (the cortex) and then to the white tissue inside the algal layer (the medulla). To test the reaction of the apothecial disk without making a section of the apothecium, first scratch the disk to penetrate the dead, unreactive layer on the surface of the hymenium, then add the C. The KC test will result in the same reaction, but will give a stronger red. In tissues containing abundant variolaric acid, there will be a C+ yellow reaction. PD and K reactions are always negative or weak. The thallus can be UV+ yellow (lichexanthone) in some species or UV– (or whitish because of gyrophoric acid) in most others. HABITAT: On trees, mosses, rocks, and wood. COMMENTS: Some species of *Ochrolechia* look like large species of *Lecanora*, and others resemble *Pertusaria*. *Lecanora* always has smaller spores, has mainly unbranched paraphyses, and rarely has a C+ red reaction (in North American species). *Pertusaria* is much more closely related, with the same kind of paraphyses; however, the spores are thick-walled, the asci are thin-walled, and the apothecial disks are generally much darker, even black. IMPORTANCE: In Scotland, species of *Ochro-*

539. *Ocellularia granulosa* south central Florida ×7.6

lechia, especially **O. tartarea** (extremely rare in North America) were used as commercial dyestuffs under the name "cudbear" from 1758 until the early nineteenth century. They were scraped from rocks in huge quantities and processed using a secret formula long guarded by the Gordon (née Cuthbert) family. Lichens grow slowly, and the resource was depleted in little more than 50 years. *Ochrolechia* yields a purple dye by the fermentation method because it contains gyrophoric acid. The same lichen substance is produced by more easily collected and more common lichens, especially *Umbilicaria* (rock tripes), which soon replaced cudbear as a source of purple dyes in the north. (In the Mediterranean region, *Roccella,* or orchil, was the source of lichen purple.) In North America, where *O. tartarea* is rarely found, *O. oregonensis* can be used as a substitute dye source, but the lichen should never be collected from standing trees in large quantity, for the sake of both the trees and the lichens.

KEY TO SPECIES
[Note: all spot tests with C must be done on thick sections of the apothecia or thallus, not on the thallus surface.]

1. Growing on rock, soil, peat, or moss 2
1. Growing on bark or wood 5
2. On rock in temperate eastern United States; thallus thick, with thick isidia .... *Ochrolechia yasudae*
2. On soil, peat, or moss; western or arctic-alpine; isidia absent 3
3. Thallus cortex and medulla C– (lacking gyrophoric acid); apothecia abundant, disks coarsely pruinose-scabrose, C– or C+ yellow .... *Ochrolechia upsaliensis*
3. Thallus cortex and often medulla C+ pink (gyrophoric acid); apothecia abundant or rare, disks smooth or cracked, not pruinose or scabrose, C+ pink 4
4. Thallus with round to irregular patches of yellowish granular soredia; apothecia extremely rare .... *Ochrolechia androgyna*
4. Thallus without soredia, or rarely with irregular patches of whitish soredia; apothecia frequent .... *Ochrolechia frigida*
5.(1) Thallus sorediate .... *Ochrolechia androgyna*
5. Thallus without soredia 6
6. Eastern, but some species occasionally in the southwest 7
6. Western, California to Alaska, absent in the southwest 11
7. Thallus thick, more or less covered with thick isidia .. .. *Ochrolechia yasudae*
7. Thallus thin or thick, smooth to verrucose, but not isidiate 8
8. Thallus and medulla C–; on conifer bark and wood; apothecia large and flat, 1.3–5 mm in diameter, with pruinose or rough, scabrose disks; very common .... [*Ochrolechia pseudopallescens*]
8. Thallus cortex or medulla C+ pink or red; mostly on deciduous trees; apothecia 0.6–2(–3) mm in diameter, pruinose or not, rough or smooth 9
9. Cortex of apothecial margin C–, medulla C+ red; apothecia usually pruinose; apothecial margins smooth and even .... *Ochrolechia africana*
9. Cortex of apothecial margin C+ pink, medulla C–; apothecia with or without pruina; apothecial margins smooth or rough.............................. 10
10. Thallus UV– (lacking lichexanthone); apothecia usually without pruina; apothecial margins verrucose.... .... *Ochrolechia trochophora*
10. Thallus UV+ yellow (lichexanthone); apothecia pruinose or without pruina; apothecial margins usually smooth .... [*Ochrolechia mexicana*]
11.(6) Cortex of thallus and apothecial margin C– (containing variolaric acid) 12
11. Cortex of thallus and apothecial margin C+ pink to red (lacking variolaric acid) 13
12. Apothecial disks C+ pink (scratch the surface of the disk before testing), remaining small and pore-like except in oldest apothecia, which become broad; thallus thin or well developed and verruculose .... *Ochrolechia juvenalis*
12. Apothecial disks C–, broadening early in their development; thallus extremely thin .... [*Ochrolechia szatalaënsis*]
13. Thallus very thin and smooth; apothecial margins smooth, lacking a pinkish ring next to the disk, containing few algae in the amphithecium; on deciduous trees, especially alder .... *Ochrolechia laevigata*
13. Thallus thick and often verrucose, or rather thin and verruculose, rarely smooth; apothecial margins smooth to verrucose, often or always with an inner ring of pinkish tissue next to the disk 14
14. Algae abundant in the apothecial margin, but absent or spotty below the hypothecium; inner ring always present on apothecial disks; hymenium 320–410 μm high; on conifer bark and wood .... *Ochrolechia oregonensis*
14. Algae forming a continuous layer under the hypothecium, and also present in the apothecial margin; inner ring sometimes present on apothecial disks; hymenium 180–280 μm high; commonly on deciduous trees, rarely on conifers ... [*Ochrolechia subpallescens*]

Ochrolechia africana
Frosty saucer lichen

DESCRIPTION: Thallus thin or thick, usually bumpy, yellowish gray, with a white, shiny prothallus. Apothecia mostly under 1.5 mm across, usually white pruinose although a form without pruina is known; margins thick and even, usually prominent; algal layer continuous below the hypothecium, or discontinuous; spores 43–67 × 18–30 μm. CHEMISTRY: Thallus UV+ yellow (lichexanthone) or UV–. Apothecial disk C+ red, KC+ red; cortex of the thallus and apothecial margin C–, KC–, but medulla C+ red, KC+ red (gyrophoric, 4-*O*-methylhiascic acids, and sometimes 4,5-di-*O*-methylhiascic acid) and always contains lichexanthone in the cortex (UV+ yellow; see plate 90, p. 109). HABITAT: On bark of deciduous trees, bald cypress, palms, and wood. COMMENTS: This is the most common *Ochrolechia* in the southeast. The form without pruina can look very much like ***O. mexicana***, which reacts C+ red in both the cortex and medulla (4,5-di-*O*-methylhiascic acid in addition to gyrophoric acid). Its apothecia have an excipular ring (described above under the genus). Despite its name, *O. mexicana* is not uncommon in the Great Lakes region, as well as in the southeast from the Ozarks into Texas and Mexico. In *O. trochophora*, the cortex is C+ red and the medulla is C–.

Ochrolechia androgyna
Powdery saucer lichen

DESCRIPTION: Thallus thin to thick, pale gray, with granular, generally yellowish soredia in soralia that are round and hemispherical when young but coalesce into broader, irregular sorediate patches in older parts of the thallus. Apothecia not seen in North America. CHEMISTRY: Thallus cortex and medulla C+ pink, soralia C+ red; thallus and soralia UV– or whitish (gyrophoric acid, rarely with variolaric acid and unknowns). HABITAT: On bark and wood in temperate to boreal regions, and on moss and dead vegetation in arctic and alpine habitats. COMMENTS: This is a variable lichen in form and chemistry, perhaps representing several species. When on bark, it is commonly mistaken for *O. arborea*, which also has C+ red soralia. That species generally has a much thinner thallus, often in small, circular patches. The soralia are small and greenish rather than yellowish, as in *O. androgyna*. In addition, *O. arborea* has a UV+ yellow reaction in the soralia and sometimes the thallus because of lichexanthone. Tundra specimens of *O. androgyna* can resemble the uncommon sorediate morph of *O. frigida*.

Ochrolechia frigida
Arctic saucer lichen

DESCRIPTION: Thallus relatively thick, creamy white, warty to smooth; aggressively covering the substrate, often producing elongate or even fringed, fruticose extensions at the thallus margins; normally without soredia, but occasionally forming granular sorediate patches. Apothecia common, up to 5 mm across with thick margins. CHEMISTRY: Cortex and usually medulla C+ pink to red, UV– (gyrophoric acid). HABITAT: On tundra soil, dead vegetation, and moss, less frequently wood. COMMENTS: No other crustose lichen on the ground forms cilia-like mar-

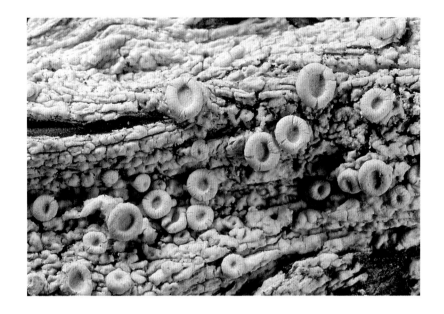

540. *Ochrolechia africana* herbarium specimen (Canadian Museum of Nature), Mexico ×3.9

541. *Ochrolechia androgyna* herbarium specimen (Canadian Museum of Nature), New Brunswick ×7.8

542. *Ochrolechia frigida* interior Alaska ×3.2

543. *Ochrolechia juvenalis* herbarium specimen (Canadian Museum of Nature), British Columbia ×7.7

ginal outgrowths. The sorediate form can resemble the more common *O. androgyna*, which has finer, yellower soredia.

Ochrolechia juvenalis
Juvenile saucer lichen

DESCRIPTION: Thallus pale gray, thin and continuous at the edge, becoming verruculose toward the center. Apothecia 0.8–1.5(–2.0) mm in diameter, with disks that remain very small and pore-like for a long time, finally expanding as normal apothecia, yellowish pink, heavily pruinose and scabrose; margins thick and even, normally without an inner excipular ring; algal layer mainly confined to the margin, with only clumps below the hymenium. CHEMISTRY: Disk K+ yellow, KC+ red, C+ pink (underneath the pruina); cortex K–, KC–, C+ yellow or C–, UV–; medulla K–, KC–, C– (gyrophoric acid in the epihymenium; variolaric and lichesterinic acids in most tissues, but lacking protolichesterinic acid). HABITAT: Most commonly on conifer bark in pine and Douglas fir stands. COMMENTS: Only a few species of *Ochrolechia* react C+ red in the apothecia and are entirely C– elsewhere. Besides *O. juvenalis*, the most common of these is *O. pseudopallescens*, found growing on conifer bark and wood in the Appalachian–Great Lakes region and parts of southern Arizona and New Mexico. It is similar to *O. juvenalis* but has very broad apothecia (1.3–5 mm in diameter) and a fairly thick, rough thallus. It produces protolichesterinic acid in addition to the other compounds. *Ochrolechia subathallina* is an uncommon western species found mainly on hard wood. Its thallus is usually reduced to little more than a gray or white stain.

Ochrolechia laevigata
Smooth saucer lichen

DESCRIPTION: Thallus very thin, smooth and continuous, white to pale yellowish gray or pinkish gray. Apothecia up to 3 mm in diameter, light orange, without pruina; margins thick and smooth, without an excipular ring of pink tissue; algae in margin very sparse and scattered. CHEMISTRY: Apothecial disk and margin cortex, and part of the medulla C+ dark red, UV– (gyrophoric and olivetoric acids). HABITAT: Abundant on alder trees (especially red alder) and other deciduous trees, very rarely on conifers. COMMENTS: Three similar western species, *O. laevigata, O. oregonensis,* and *O. subpallescens,* are compared under *O. oregonensis*.

Ochrolechia oregonensis
Double-rim saucer lichen

DESCRIPTION: Thallus thick, rough to verrucose, yellowish gray to creamy white, with large apothecia (up to 4 mm in diameter) having double margins: an outer, often discontinuous, warty, thalline margin, and an inner, pink or disk-colored, excipular ring around the disk; algae relatively abundant in the margins but not normally forming a continuous layer below the hymenium. CHEMISTRY: Apothecial disk and thick cortex of the thallus and apothecial margin C+ red, UV–; medulla C– (gyrophoric acid alone). HABITAT: On conifer bark and wood, particularly common on Douglas fir. COMMENTS: Most saucer lichens with a thick, C+ red thallus growing on conifer bark in the west are *O. oregonensis,* a very common species. ***Ochrolechia subpallescens*** is another saucer lichen with a strong C+ red reaction in both the apothecial and thallus cortex. (Most specimens contain olivetoric acid as well as gyrophoric acid.) It lacks the double margin of *O. oregonensis* and has a thicker, more verruculose thallus than *O. laevigata*. Both *O. subpallescens* and *O. laevigata* are almost exclusively found on deciduous trees, unlike *O. oregonensis*. However, the algal layer in the apothecia is the best morphological character to distinguish the confusing western C+ red *Ochrolechia* species: in *O. subpallescens,* the algae form an unbroken layer under the hymenium; in *O. oregonensis,* the layer is spotty and broken into clumps; in *O. laevigata,* the thick margin has almost no algae at all.

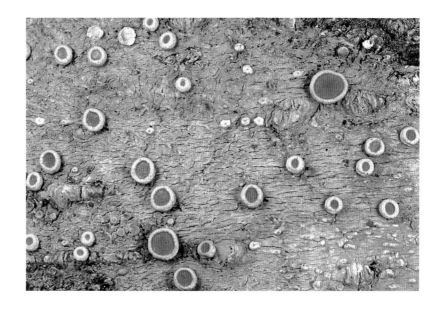

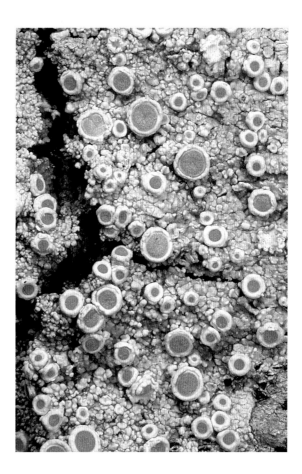

544. *Ochrolechia laevigata* Cascades, Oregon ×2.9

545. *Ochrolechia oregonensis* Coast Range, Oregon ×2.9

546. *Ochrolechia trochophora* Appalachians, North Carolina ×4.2

Ochrolechia trochophora
Rosy saucer lichen

DESCRIPTION: Thallus thin when young, becoming thick and verrucose when well-developed, pale or, less frequently, dark yellowish gray, often shiny. Apothecia rosy or yellowish pink to light orange, usually without pruina, 0.8–3 mm in diameter, with thick, prominent, sometimes verrucose margins. Apothecial disks occasionally developing radiating bands of sterile tissue. Apothecial cortex rather thin and uniform in thickness; algal layer just internal to the cortex and partially continuous under the hymenium. CHEMISTRY: Thallus and apothecial margin cortex PD–, K–, KC+ red, C+ pink, UV–; medulla PD–, K–, KC–, C– (gyrophoric acid, rarely with variolaric acid or atranorin). HABITAT: On bark of deciduous trees and white cedar (*Thuja*), rarely on rocks. COMMENTS: Pastel-colored apothecia on a pale gray thallus make this one of the prettiest saucer lichens. The most similar species is **O. mexicana,** which has more saucer-like apothecia and a thinner thallus, differs chemically, and most frequently grows on conifer bark. *Ochrolechia mexicana* is described and discussed under *O. africana*, another somewhat similar species. Specimens of *O. trochophora* on rock sometimes become thick and produce columnar warts that somewhat resemble the isidia of *O. yasudae*.

Ochrolechia upsaliensis
Tundra saucer lichen

DESCRIPTION: Thallus rather thick, granular, white to yellowish white. Apothecia abundantly produced, mostly 0.6–2 mm in diameter; margins prominent; disks pale yellow, sometimes darkening to a dark reddish purple, very rough, scabrose to coarsely pruinose. CHEMISTRY: Thallus C–, KC–, UV–; apothecial margin cortex and medulla, and apothecial disks C– or yellow, KC– or yellow. Apothecial disks UV– or UV+ yellow (often on the same thallus) (variolaric, murolic, and hydromurolic acids). HABITAT: On soil, moss, and vegetation in arctic and alpine tundra, or in the dry interior. COMMENTS: The tundra saucer lichen is the only *Ochrolechia* growing in both arctic and alpine tundra as well as in some exposed drier localities at lower elevations. In all other *Ochrolechia* species except the following two, at least the apothecial disk turns a shade of pink or red because of gyrophoric acid. These two related but uncommon western species both grow on bark: *O. szatalaënsis,* with a very thin thallus, and *O. farinacea* (especially on Garry oak), with a thick, chalky, sometimes powdery sorediate thallus.

Ochrolechia yasudae
Coral saucer lichen

DESCRIPTION: Thallus thick, continuous, pale ashy gray, with a distinct, often zoned border, covered with thick, cylindrical isidia (rarely sparse). Apothecia usually absent but can be abundant in some specimens, thick, 1–3 mm across, with thick, isidiate to warty margins; disks yellowish pink, without pruina. CHEMISTRY: Thallus cortex and apothecial disk C+ pink, KC+ red, medulla usually C–, KC– (gyrophoric acid); all tissues UV–. HABITAT: On granitic rock or bark (usually oak) in partly shaded, deciduous forests. COMMENTS: This is the only truly isidiate, saxicolous crustose lichen in the east that re-

547. *Ochrolechia upsaliensis* Olympics, Washington ×4.3

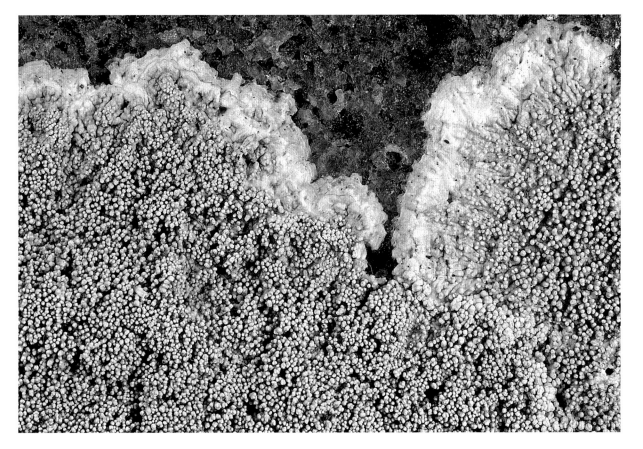

548. *Ochrolechia yasudae* Appalachians, Virginia ×8.1

549. *Omphalina umbellifera* coastal Maine ×3.4

acts C+ red. **Pertusaria globularis** is a somewhat similar, isidiate Appalachian crust growing on mossy rocks and tree bases, but the isidia are much shorter and finer, and the dark greenish gray thallus is C–, KC– (containing 2-*O*-methylperlatolic acid). Most difficult to identify are the occasional rock-dwelling specimens of *O. trochophora*, which can develop small, columnar verrucae and appear almost isidiate in places. **Ochrolechia subisidiata** is a southwestern species with granular or knobby isidia and a C+ pink medulla and C– cortex. It has pruinose apothecia and contains variolaric acid.

Omphalina (5 N. Am. species)
Mushroom lichens

DESCRIPTION: A genus of mushroom-forming fungi with small, pale buff, yellow, or orange mushrooms with thick, waxy gills and smooth stems, without a veil or basal cup; lichenized thallus a mass of green granules, or pale green, round or lobed squamules, around the base of the mushroom (not on the mushroom itself). The green granules are tiny fungal envelopes, 1 cell thick, each envelope packed with hundreds of green algal cells (*Coccomyxa*). CHEMISTRY: No lichen substances. HABITAT: On well-rotted mossy stumps and logs, peat, or mossy soil in shaded boreal or open tundra habitats. COMMENTS: Only three lichen genera in North America have a basidiomycete as the mycobiont. The mushroom-forming *Omphalina* and club-fungus *Multiclavula*, which both associate with green algae, have northern distributions. ***Dictyonema***, a fibrous to somewhat lobed fungus that associates with cyanobacteria (*Scytonema*), is found mainly in tropical or oceanic regions. All three are rarely recognized as lichens and are generally overlooked by lichenologists.

KEY TO GENUS AND SPECIES: See Key to keys, and Key G.

Omphalina umbellifera (syn. *Omphalina ericetorum*)
Greenpea mushroom lichen

DESCRIPTION: Thallus consists of minute, bright green globules 50–300 μm in diameter. Fruiting body a small, pale yellowish mushroom, up to about 4 cm tall, with prominent waxy gills. HABITAT: On rotting wood and peat. COMMENTS: This is an inconspicuous but common lichen in the humid western and boreal forests. Because the mushrooms are produced only at certain seasons, the dark green granular crust is usually the only sign of the lichen. Another common mushroom lichen, ***Omphalina hudsoniana*** (syn. *Coriscium viride*), grows on arctic or alpine peat and soil (rarely boreal), forming pale green, slightly concave squamules up to 5 mm wide much like a *Cladonia* but sometimes develops a more elongate, lobed thallus. Similar, but much smaller squamules occur in *Normandina*, an unrelated lichen with an entirely different habitat.

Omphalora (1 species worldwide)

KEY TO SPECIES: See Key H.

550. *Omphalora arizonica* mountains, central New Mexico ×1.7

Omphalora arizonica (syn. *Omphalodium arizonicum*)
Potato-chip lichen

DESCRIPTION: A very large, greenish yellow umbilicate lichen, up to 15 cm across, very thick and stiff, with a deeply ridged and wrinkled upper surface; lower surface dark brown to black, more or less covered with short, stubby but branched peg-like rhizines, often white at the tips. Photobiont green (unicellular). Lecanorine apothecia common on the upper surface of the thalli, 1–4 mm in diameter, with clear red-brown to very dark brown, nonpruinose disks. CHEMISTRY: Cortex KC+ gold (usnic acid); medulla PD–, K–, KC–, C–. HABITAT: On rocks and cliffs, shaded or exposed. COMMENTS: This spectacular but rather rare lichen can hardly be confused with anything else. Other large umbilicate lichens, such as *Umbilicaria* and *Dermatocarpon*, are brown to gray, never greenish yellow. Some *Rhizoplaca* species are somewhat similar in color but are never more than 1 or 2 cm in diameter, and none have purely red-brown apothecia, a wrinkled upper surface, or rhizines below.

Opegrapha (36 N. Am. species)
Scribble lichens

DESCRIPTION: Crustose lichens with thin or thick thalli, sometimes largely within the substrate; some species sorediate. Photobiont green (*Trentepohlia*). Fruiting bodies short or long lirellae, sometimes branched, with carbon-black walls and base, opening through a narrow or fairly broad slit; paraphyses branched and anastomosing; hymenium usually IKI+ bluish green, changing to reddish orange; asci broadly club-shaped, thick-walled, and largely negative with K/I; spores usually fusiform, often broader at one end than the other, with several square cells and uniformly thickened, rather thin walls, usually with a gelatinous halo visible, at least in ink preparations (see Studying Lichens, in Chapter 13). CHEMISTRY: Typically without lichen substances, but a few species contain gyrophoric acid (PD–, K–, KC+ red, C+ pink). HABITAT: Mostly on bark, especially that of deciduous trees; also on wood and rock in shaded, fairly humid situations. Some species are parasitic on other lichens. COMMENTS: Other script lichens (*Graphis*, *Phaeographis*, *Xylographa*, etc.) have largely unbranched paraphyses; some have thick-walled spores with lens-

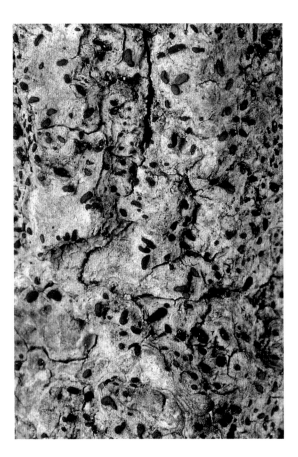

551. *Opegrapha varia* Lake Superior region, Ontario ×5.4

shaped locules, thin-walled cylindrical asci, or fruiting bodies with thin, pale to brown (uncarbonized) walls. Like most lichens containing *Trentepohlia*, *Opegrapha* species seem to prefer shaded forest habitats. Although most species of *Opegrapha* have a thin thallus and grow on bark or wood, some have thick thalli and several species grow on rock. The unlichenized fungus *Hysterium* has black lirellae with thick black walls and can be mistaken for an *Opegrapha*. It is commonly seen on wood but has no trace of algae below the fruiting bodies, and the spores are brown and multicelled.

KEY TO SPECIES

1. On calcareous rock. Spores 4-celled, 22–29 × 5.5–8 μm; thallus thin or endolithic, often parasitic on other lichens, C–; California, Alaska, and Newfoundland . . . . [*Opegrapha rupestris*]
1. On bark or wood 2
2. Spores 4-celled 3
2. Spores 6- or more celled 5
3. Spores (17.5–)19–25 × 5.5–10 μm. Lirellae fusiform, straight or curved, sometimes pruinose; thallus creamy white to pale brown, forming small, discrete patches; western, on bark and wood of various kinds . . . . [*Opegrapha herbarum*]
3. Spores under 5 μm wide; widely distributed 4
4. Thallus white in delimited patches; lirellae long and narrow with thick walls, opening by a narrow slit; spores 13–18(–20) μm long. . . . . [*Opegrapha atra*]
4. Thallus brownish and indistinct, mostly within the bark; lirellae short, often branched once or twice or star-shaped, scattered, with thin walls, opening broadly; spores 15–27 μm long . . . . [*Opegrapha rufescens*]
5.(2) Spores 12–16-celled, 25–60 × 6–9 μm. Lirellae short and broad, elliptical, rarely forked; widely distributed . . . . [*Opegrapha viridis*]
5. Spores 6–8-celled. 6
6. Lirellae with a red-brown surface, long, branched, opening broadly with a very thin black margin that is sometimes hard to see; spores 16–21 × 4–6 μm; southeastern coastal plain. . . . . [*Opegrapha longissima*]
6. Lirellae entirely black, sometimes pruinose, long or short; spores 20–40 μm long. 7
7. Spores 2.5–4.5 μm wide; lirellae long, very narrow (less than 0.25 mm wide), opening by a slit, often branched; eastern . . . . [*Opegrapha vulgata* (syn. *O. cinerea*)]
7. Spores 6–9 μm wide; lirellae fusiform, sometimes forked, opening broadly; very widely distributed . . . . *Opegrapha varia*

Opegrapha varia (syn. *O. pulicaris*, *O. lichenoides*) Scribble lichen

DESCRIPTION: Thallus essentially within bark with little showing at the surface. Photobiont *Trentepohlia*. Fruiting bodies elongated, cigar-shaped, curved, or, infrequently, branched into a Y shape, 0.7–2.5 mm long, up to 0.5 mm wide; margins thick in young lirellae but later become thin and rarely disappear altogether; disk at first only a slit but broadening in maturity and sometimes gray or greenish pruinose; spores colorless, 5–7-celled, and usually with one or two cells in the center larger than the rest, tapering more at one end than the other, mostly 20–37 × 6–9 μm. CHEMISTRY: All reactions negative (no lichen substances). HABITAT: On bark and wood mainly of deciduous trees and eastern white cedar. COMMENTS: A number of bark-dwelling species of *Opegrapha* are widely distributed in temperate North America. Several of the most common ones are separated in the key.

Ophioparma (3 N. Am. species)
Bloodspot lichens

DESCRIPTION: Crustose lichens with yellow or yellow-green, thin or thick thalli. Photobiont green (*Trebouxia*). Apothecia large, dark red, with thick, disk-colored biatorine margins surrounded by a thin, sometimes disappearing, thallus-colored lecanorine rim; asci with a K/I+ blue, uniform tholus (similar to the *Catillaria*-type; Fig. 14c); paraphyses unbranched or slightly branched, hardly or not expanded at the tips; hymenium pale orange, pigments K+ blue to purple (haemoventosin); hypothecium colorless to pinkish; spores colorless, thin-walled, narrowly ellipsoid to fusiform, 1–8-celled. CHEMISTRY: Cortex KC+ gold (usnic acid); medulla in North American populations PD–, K–, KC–, C–, UV+ white (divaricatic acid), rarely PD+ yellow or orange, K+ deep yellow to red, KC–, C– (β-orcinol depsides and depsidones not including thamnolic acid). HABITAT: On noncalcareous rock or bark. COMMENTS: Other smaller bloodspot lichens have whitish to gray thalli (lacking usnic acid) and belong to the genus *Haematomma*; they are found mainly in the southwest and southeast. See Comments under *Haematomma*.

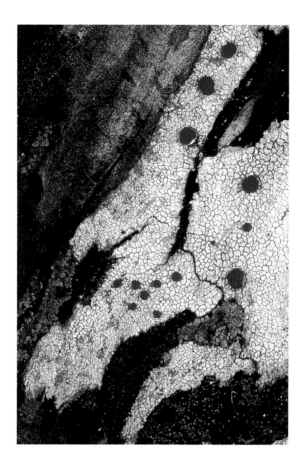

552. *Ophioparma rubricosa* Klamath Range, southwestern Oregon ×2.0

KEY TO SPECIES

1. On bark and wood, rarely rocks, in the Pacific Northwest south to California; spores straight . . . . *Ophioparma rubricosa*
1. On rocks in arctic-alpine regions 2

2. Spores curved, more than 30 μm long . . . . *Ophioparma ventosa*
2. Spores straight, less than 30 μm long . . . . [*Ophioparma lapponica*]

Ophioparma rubricosa (syn. *Haematomma californicum, H. pacificum, O. herrei*)
Pacific bloodspot

DESCRIPTION: Thallus yellowish, thin or thick, granular, the granules or areoles sometimes becoming tall and resembling isidia. Apothecia 1–3 mm in diameter; broadly attached, with a thin margin that disappears in some old apothecia; disks scarlet red without pruina; spores 4-celled, straight, 28–45 × 3–4 μm. CHEMISTRY: Medulla PD–, K–, KC–, C–, UV+ blue white (divaricatic acid). HABITAT: On wood or conifer bark; very rarely on siliceous rock. COMMENTS: The substrate and narrow 4-celled spores are sufficient to separate this species from other species of *Ophioparma*.

Ophioparma ventosa
Alpine bloodspot

DESCRIPTION: Thallus very thick, verrucose or areolate, greenish yellow. Apothecia lecanorine but with very few algae in the margin, up to 3 mm in diameter, round to irregular; disk orange-red to blood-red; margin thick, thallus-colored, becoming very thin to whitish; spores colorless, 4–8-celled, slender and curved, tapering at one end, 35–65 (–70) × 3.5–6 μm, 8 per ascus. CHEMISTRY: Medulla usually PD–, K–, KC– or KC+ pinkish orange, C–, UV+ white (divaricatic acid). Another common chemical race contains hypothamnolic acid (K+ pinkish violet to reddish brown). Rare specimens contain psoromic, stictic, or norstictic acids. In Europe, the most common chemical race contains thamnolic acid

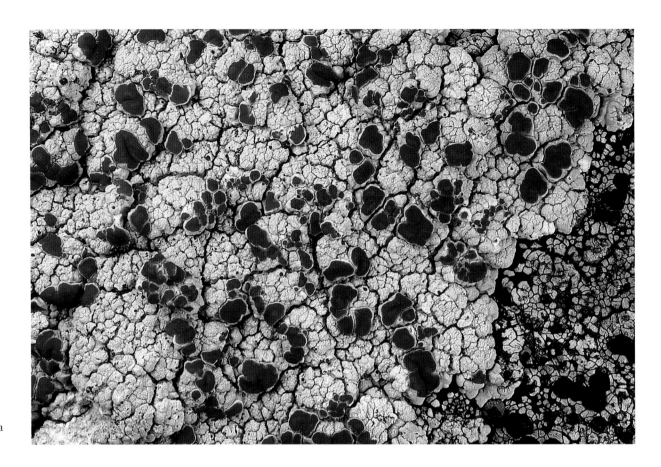

553. *Ophioparma ventosa* interior Alaska ×4.3

(PD+ orange, K+ deep yellow). HABITAT: On non-calcareous rock in open arctic and alpine habitats. COMMENTS: With its thick, yellowish crust and blood-red, rimmed apothecia, the alpine bloodspot is a particularly conspicuous crustose lichen. It is superficially almost indistinguishable from its strictly arctic sister species, **O. lapponica** which has straight, cylindrical to elongate ellipsoid spores, 1–2-celled, 12–25 × 4–6.7 µm. With rare exceptions, all specimens contain divaricatic acid. *Ophioparma rubricosa* is a similar bloodspot lichen in the west that grows on wood and bark. IMPORTANCE: *Ophioparma ventosa* and *O. lapponica* are used as a source of purple-red to magenta dyes; divaricatic acid may be the active ingredient.

Orphniospora (1 N. Am. species)

KEY TO SPECIES: See Key F.

Orphniospora moriopsis
Black-on-black lichen

DESCRIPTION: An arctic-alpine crustose lichen dark in almost all its features: thallus, apothecia, and spores. Thallus areolate and usually rough, blackish to dark gray, usually with a well-developed black prothallus. Photobiont green (unicellular). Apothecia lecideine, black, not pruinose, with or without prominent margins, 0.3–0.8(–1) mm in diameter, either sunken into the thallus areoles or superficial; hypothecium and exciple dark brown; epihymenium greenish, turning red in nitric acid; paraphyses unbranched, expanded at the tips; asci with a thick, almost gelatinous tip (not a tholus) turning blue in K/I; spores brown, 1-celled but with a dark band and a slight thickening on the inner surface of the wall at the equator that make it sometimes appear 2-celled, ellipsoid, 12–18 × 8–10 µm, 8 per ascus. CHEMISTRY: Medulla PD– or PD+ yellow or red, K– or K+ red, KC–, C– (usually with norstictic acid). HABITAT: On exposed rocks in arctic and alpine sites. COMMENTS:

The undivided spores and ascus distinguish this lichen from *Rhizocarpon* or *Buellia* species, which it superficially resembles.

Pannaria (8 N. Am. species)
Shingle lichens, mouse lichens

DESCRIPTION: Small foliose to squamulose lichens containing cyanobacteria as the principal photobiont (*Nostoc* or rarely *Scytonema*). Thallus in most species gray, brownish in a few, never gelatinous; cyanobacteria in a distinct layer; thallus lacking a well-developed lower cortex but often with a blue-black, felty hypothallus. Apothecia with well-developed lecanorine margins; hymenium IKI+ blue, at least near the asci, but the asci entirely negative; spores colorless, ellipsoid, 1-celled, 8 per ascus. CHEMISTRY: Most species contain pannarin. COMMENTS: See Comments under *Fuscopannaria*.

554. *Orphniospora moriopsis* herbarium specimen (New Brunswick Provincial Museum), Newfoundland ×4.2

KEY TO SPECIES (including *Coccocarpia, Degelia, Fuscopannaria, Massalongia, Parmeliella, Psoroma,* and *Vestergrenopsis*)

1. Soredia present along the lobe margins . . . . *Pannaria conoplea*
1. Soredia absent 2
2. Thallus foliose, either minute (with lobes less than 1 mm wide), or relatively broad, generally spreading to form rosettes, not shingled at the thallus margins . . . 3
2. Thallus squamulose, with lobes about as long as they are broad, under 2 mm wide, commonly overlapping, at least in part 14
3. Thallus olive to brown or yellow-brown; lower surface smooth 4
3. Thallus gray to blue-gray or yellowish gray; lower surface blue-black, tomentose or with a conspicuous prothallus 7
4. Lobes 0.1–1.5 mm wide, elongate or branching 5
4. Lobes 2–4(–5) mm wide . . . . *Nephroma helveticum*
5. Thallus chocolate brown, with an upper cortex of pseudoparenchyma; lobes 0.5–1.5 mm wide; on moss or mossy rock; apothecia frequent, biatorine, red-brown; photobiont *Nostoc*. Humid western forests and northern boreal region . . . . *Massalongia carnosa*
5. Thallus olive to yellowish brown, lacking an upper cortex; lobes under 0.5 mm wide; on bare rock; apothecia rare; photobiont *Scytonema* 6
6. Cylindrical isidia on thallus surface; on wet rocks in coastal mountains and foot of glaciers . . . . *Vestergrenopsis isidiata*
6. Cylindrical isidia rare, but margins divided into flattened lobules; on dry rocks mainly in southwestern U.S., just reaching southern British Columbia; rare . . . . [*Koerberia sonomensis*]
7.(3) Isidia or lobules present 8
7. Isidia absent 11
8. Isidia on thallus surface; photobiont *Scytonema*; lower surface with a cortex, at least in part . . . . *Coccocarpia palmicola*
8. Isidia or lobules marginal; photobiont *Nostoc*; lower surface entirely without a cortex 9
9. Prothallus very thick, brown-black, projecting like a fringe all around the thallus; lobes 0.7–2 mm wide; coastal plain, Florida to North Carolina . . . . [*Parmeliella pannosa*]
9. Prothallus thin, conspicuous or not, or prothallus absent; mostly north of the southeastern coastal plain 10
10. Lobes 0.3–1 mm wide, thin, flat, with a conspicuous blue-black prothallus; apothecia rare; medulla PD– . . . . *Parmeliella triptophylla*
10. Lobes 1–3(–4) mm wide, thick, slightly ascending, without a prothallus; apothecia abundant; cortex and medulla PD+ orange (pannarin) . . . . . . *Pannaria tavaresii*
11.(7) Apothecia biatorine, convex with thin or disappearing margins 12
11. Apothecia lecanorine, more or less flat or concave, with persistent raised margins 13

12. Lower surface pale brown to black with a cortex; lobes smooth or with concentric ridges; spores (6–)7–14(–16) × (2–)3–5 μm; mainly southern coastal plain; common . . . . *Coccocarpia erythroxyli*
12. Lower surface blue to black, lacking a cortex; lobes smooth with longitudinal radiate ridges; spores 17–25 × 7–10 μm; northeastern maritime coast; rare . . . . [*Degelia plumbea*]
13. Lobes mostly 0.7–2 mm wide, more or less smooth; apothecial margins smooth and even, or toothed. . . . . *Pannaria rubiginosa*
13. Lobes mostly 2–6 mm wide, with branching veins or ridges on the upper surface; apothecial margins bumpy . . . . *Pannaria lurida*
14.(2) Thallus grass-green to brownish green when wet; photobiont green. . . . . *Psoroma hypnorum*
14. Thallus dark gray to brownish gray when wet; photobiont blue-green 15
15. Thallus with isidia or cylindrical, isidia-like outgrowths or lobules 16
15. Thallus without isidia or isidia-like outgrowths 17
16. On wood or bark, rarely rock; black prothallus well-developed; edges of squamules not white or felty; isidia slender, on the lobe margins and sometimes almost covering the entire thallus . . . . *Parmeliella triptophylla*
16. On soil or moss, rarely rocks; prothallus absent; edges of squamules gray to bluish white and felty; thick, cylindrical to granular outgrowths on the lobe margins . . . . *Fuscopannaria "praetermissa"*
17. Growing directly on rock, especially moist rock walls; edges of squamules not white . . . . *Fuscopannaria leucophaea*
17. Growing on bark, moss, or soil (sometimes on moss over rock) 18
18. Growing on moss, mossy rock or tree bases, or soil; spore walls conspicuously rough or sculptured; prothallus absent; apothecia broad, lecanorine. . . . . *Pannaria pezizoides*
18. Growing directly on bark; spore walls smooth 19
19. Apothecia biatorine with smooth, pale brown margins; prothallus absent; squamules under 1 mm across, edges not or only slightly whitened . . . . *Fuscopannaria "saubinetii"*
19. Apothecia lecanorine with a bumpy, whitish margin; black prothallus usually conspicuous; squamules 0.6–3 mm across, edges white and felty 20
20. Spores tapering to a point at one or both ends; thallus reddish brown; without atranorin . . . . *Fuscopannaria leucosticta*
20. Spores ellipsoid, not pointed at the ends; thallus blue-gray to olive green; contains atranorin (not detectable with a K test) . . . . *Fuscopannaria leucostictoides*

Pannaria conoplea
Mealy-rimmed shingle lichen

DESCRIPTION: Thallus blue-gray to yellowish gray, lobes fairly broad, 2–5 mm across, with upturned and concave tips; non-corticate isidia-like granules or masses of coarse soredia form from the breakdown of the granules (as shown in plate 555) along lobe margins; blue-black hypothallus usually conspicuous at the thallus edge. Apothecia absent. CHEMISTRY: Cortex and medulla PD+ pale orange, K–, KC–, C– (pannarin). HABITAT: On bark, less frequently on rocks. COMMENTS: Few species of *Pannaria*, *Fuscopannaria*, or *Parmeliella* have marginal soredia or isidia. *Pannaria tavaresii* is more distinctly isidiate, never becoming truly sorediate. *Fuscopannaria ahlneri*, a rare lichen of the oceanic west and east coasts, is yellowish brown, with rounded lobes 1–3 mm across and conspicuous heaps of blue-gray, granular soredia produced on the margins, at first developing on the lower surface of the lobe edges. It lacks pannarin and so is PD–. See also *Erioderma* and *Leioderma*.

Pannaria lurida
Veined shingle lichen

DESCRIPTION: Thallus brown or gray-brown when dry, olive-brown or dark green when wet; lobes 2–6(–10) mm across; lower surface pale tan, covered with a dense, brownish to gray tomentum; upper surface dull, often gray pruinose and (or) scabrose especially close to the margins, and generally somewhat wrinkled like the raised veins on the back of a hand; without soredia or isidia. Apothecia abundant, with red-brown disks and persistent thalline margins. CHEMISTRY: PD+ orange, especially at the lobe margins where the lobes curl over, K–, KC–, C– (pannarin), or occasionally PD– (lacking pannarin). HABITAT: On bark or mossy rock. COMMENTS: In general form and chemistry, *P. lurida* resembles *P. rubiginosa*. It is distinguished by its broader, thicker lobes, and especially the wrinkled or veined upper surface. In addi-

555. *Pannaria conoplea* Lake Superior region, Ontario ×0.39

556. *Pannaria lurida* (damp) Ouachitas, Arkansas ×1.6

tion, *P. lurida* sometimes swells when wet, much like a *Collema*; this is never seen in *P. rubiginosa*. Both species are abundantly fertile.

Pannaria pezizoides
Brown-gray moss-shingle

DESCRIPTION: Thallus brownish to gray-brown, remaining brown when wet, consisting of small, lobed squamules 0.5–1 mm across, flat or sometimes overlapping. Apothecia abundant, with flat, brown disks up to 2 mm across, and with thin but prominent, verruculose (bumpy), lecanorine margins; spores 25–30 × 9–12 µm, with conspicuously rough, sculptured walls. CHEMISTRY: All reactions negative (no lichen substances). HABITAT: On soil and moss in exposed or shaded habitats where the substrate remains somewhat damp. COMMENTS: This is the only *Pannaria* or *Fuscopannaria* with spores that have sculptured walls. *Psoroma hypnorum* has comparable spores and is the most similar lichen, but it contains green algae instead of cyanobacteria as the principal photobiont and is grass-green when wet.

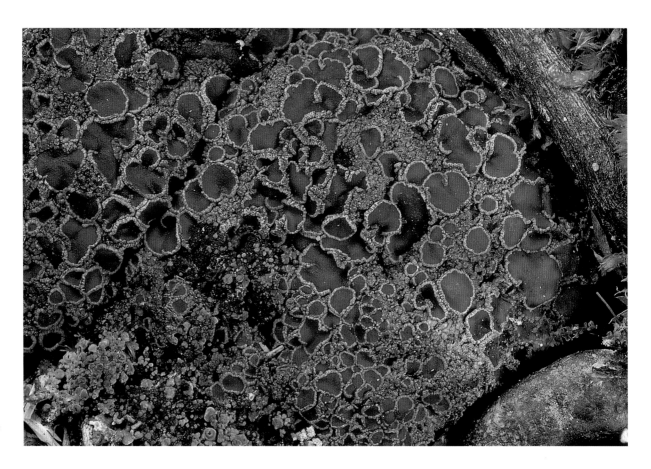

557. *Pannaria pezizoides* interior Alaska ×4.4

558. *Pannaria rubiginosa* Ouachitas, Arkansas ×2.3

Pannaria rubiginosa
Brown-eyed shingle lichen

DESCRIPTION: Thallus blue-gray with more or less discrete ascending lobes, 0.7–2(–4) mm across, without soredia or isidia, but frequently producing round lobules on the margins; lower surface pale with a thick blue-black tomentum. Apothecia abundant, bright red-brown (especially when wet), 0.5–1.5 mm in diameter, with conspicuous, often scalloped or toothed margins. CHEMISTRY: Medulla PD+ orange (often very hard to detect), K–, KC–, C–, sometimes PD– (pannarin usually present). HABITAT: On bark in shaded forests. COMMENTS: *Pannaria lurida* has broader lobes (mostly 2–6 mm) that have a veined appearance on the upper surface. *Fuscopannaria leucosticta* is a definite brownish rather than bluish gray, and it is smaller, with overlapping squamules or lobes (under 1 mm wide and 3 mm long) that are whitish tomentose at the tips; it never contains pannarin. See also *Pannaria tavaresii*.

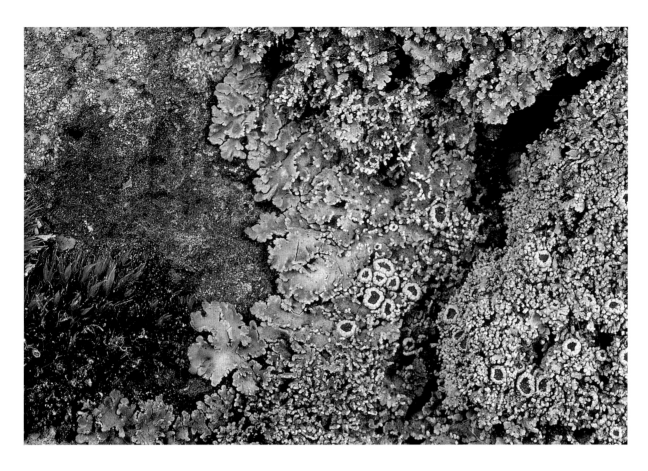

559. *Pannaria tavaresii*
mountains, central
New Mexico ×4.4

Pannaria tavaresii
Coral-rimmed shingle lichen

DESCRIPTION: Essentially like *P. rubiginosa* (see above), but lobe margins with abundant, finger-like isidia. CHEMISTRY: Medulla PD+ orange, K–, KC–, C– (pannarin). HABITAT: Usually on bark, but occasionally on rock, especially in the western mountains. COMMENTS: This lichen can resemble *P. conoplea*, which has at least some true soredia (i.e., without a cortex) on the margins and never has cylindrical isidia.

Parmelia (10 N. Am. species)
Shield lichens

DESCRIPTION: Thallus foliose, with rather squared-off lobes mostly 2–6 mm wide, lacking cilia; pale bluish gray to brown; upper surface usually with a network of white pseudocyphellae or maculae, with many species having shallow ridges and depressions, giving the thallus the appearance of hammered metal; lower surface black, with black rhizines. Photobiont green (*Trebouxia*). Apothecia lecanorine, on the upper surface of the lobes; spores 1-celled, ellipsoid, colorless, 8 per ascus; conidia cylindrical to slightly bone-shaped. CHEMISTRY: Cortex PD–, K+ yellow, KC–, C– (atranorin; usnic acid absent except in the soredia of *P. fraudans*); medulla usually containing β-orcinol depsidones, especially salazinic acid. HABITAT: On bark, wood, rock, or soil. COMMENTS: The name *Parmelia* was used very broadly until 1974, when the genus was broken up into smaller units, including *Bulbothrix*, *Hypotrachyna*, *Parmelina*, *Parmotrema*, and *Xanthoparmelia*. Additional segregations have been made since that time, with the recognition of *Arctoparmelia*, *Ahtiana*, *Canomaculina*, *Canoparmelia*, *Flavoparmelia*, *Flavopunctelia*, *Parmelinopsis*, *Punctelia*, and *Rimelia*. With few exceptions, we call them all shield lichens, although we accept the new classification. The generic distinctions are based mainly on features of the thallus, conidial type, chemistry, and distribution. IMPORTANCE: See notes under *P. saxatilis*.

KEY TO SPECIES (including *Canoparmelia, Myelochroa, Paraparmelia, Parmelina, Parmelinopsis, Pseudoparmelia*, and leads to *Bulbothrix, Canomaculina, Cetrelia, Hypotrachyna, Parmotrema, Platismatia*, and *Rimelia*)

1. Marginal cilia or cilia-like rhizines present on lobe margins or in axils of lobes, often sparse and visible only with a hand lens (Fig. 10) 2
1. Marginal cilia or cilia-like rhizines absent 14

2. Lobes broad, 4–20 mm wide; thallus usually loosely attached over entire surface, ascending 3
2. Lobes narrow to moderate, mostly under 5 mm wide; thallus closely appressed to substrate 4

3. Rhizines abundant to the edge of the lobes; white maculae usually conspicuous, at least on lobe tips in a reticulate pattern; medulla PD+ yellow, K+ red (salazinic acid) .... *Rimelia* key (*Rimelia* and *Canomaculina*)
3. Rhizines absent in a broad zone close to the lobe margin; maculae not common and usually inconspicuous, never reticulate .... *Parmotrema*

4. Cilia bulbous at base, very short and stiff (noticeable only with a hand lens) .... *Bulbothrix*
4. Cilia not bulbous at base, short or long 5

5. Isidia present. 6
5. Isidia absent. 9

6. Rhizines regularly and abundantly forked........... .... *Hypotrachyna*
6. Rhizines unbranched 7

7. Isidia cylindrical to flattened, with black cilia at the tips; medulla C– or C+ pink, KC+ purplish pink, PD– (contains mainly derivatives of gyrophoric and hiascic acids) .... *Parmelinopsis horrescens*
7. Isidia cylindrical, without black cilia growing from the tips. 8

8. Medulla C+ pink to red, KC+ red, PD–, K– (contains mainly gyrophoric acid); medulla white or rarely slightly yellowish. .... *Parmelinopsis minarum*
8. Medulla C–, KC– or KC+ orange-yellow, PD+ orange, K+ yellow to red; medulla pale yellow (secalonic and galbinic acids) .... *Myelochroa obsessa*

9.(5) Rhizines abundantly forked .... *Hypotrachyna*
9. Rhizines not branched 10

10. Soredia or schizidia present; medulla pale yellow, at least close to algal layer or below pustules; apothecia absent or rare.................................. 11
10. Soredia or schizidia absent; medulla white or pale yellow; apothecia abundant 12

11. Medulla C+ pink, KC+ red (gyrophoric acid); lobes square or truncated.......... *Parmelinopsis spumosa*
11. Medulla C–, KC– or C+ yellow, KC+ yellow (secalonic acid derivatives); lobes rounded .... *Myelochroa aurulenta*

12. Rhizines absent from a broad or narrow zone close to the margin; medulla PD+ red (protocetraric acid).... .... *Parmotrema michauxianum*
12. Rhizines more or less uniformly distributed; medulla PD– or PD+ orange 13

13. More or less restricted to California; medulla white, PD–, K–, C+ red (lecanoric acid) .... *Parmelina quercina*
13. Eastern U.S. and adjacent Canada; medulla pale yellow, PD+ orange, K+ yellow or red, C+ yellow (galbinic acid) .... *Myelochroa galbina*

14.(1) Rhizines absent or sparse, at least in a broad or narrow zone close to the lobe margin; thallus usually loosely attached, and (or) with lobes 4–20 mm wide 15
14. Rhizines abundant, uniformly distributed; thallus usually broadly attached, with lobes mostly under 4 mm wide 18

15. Pseudocyphellae conspicuous on the lobe surface, appearing as white dots 16
15. Pseudocyphellae absent or inconspicuous 17

16. Thallus with marginal soredia; lobes with a smooth, even surface; medulla KC+ red and (or) UV+ white .. .. *Cetrelia*
16. Thallus lacking soredia or isidia; lobes with a network of ridges and depressions; medulla KC–, UV– .... *Platismatia*

17. Thallus surface usually very uneven, with ridges and depressions or otherwise wrinkled; rhizines sparse throughout; pycnidia along the lobe margins .... *Platismatia*
17. Thallus generally smooth and at most folded, without sharp ridges and depressions; rhizines absent from a zone near the margin but usually abundant in the thallus center; pycnidia on the lobe surface... *Parmotrema*

18.(14) Pseudocyphellae abundant and conspicuous, or sparse. 19
18. Pseudocyphellae absent 25

19. Pseudocyphellae round or irregular, not forming a reticulate pattern of white markings; medulla K–, often C+ red or pink .... *Punctelia*
19. Pseudocyphellae irregular, forming a reticulate pattern of white markings; medulla usually K+ red, never C+ pink or red 20

20. Isidia present................................. 21
20. Isidia absent................................. 23

21. Isidia globular, dull, with very little cortex, sometimes resembling soredia . . . . *Parmelia hygrophila*
21. Isidia cylindrical or flattened, unbranched or branched, shiny, with a continuous cortex. 22

22. Rhizines slender, squarrose except close to lobe margins; isidia especially abundant along margins but also laminal. . . . . *Parmelia squarrosa*
22. Rhizines thick, unbranched or rarely forked; isidia predominantly laminal . . . . *Parmelia saxatilis*

23.(20) Soredia absent; on rock; montane, boreal and arctic . . . . *Parmelia omphalodes*
23. Soredia present. 24

24. Soredia powdery, on lobe surface and margins; mature rhizines squarrose; on bark, wood, rock, and sometimes soil; extremely common and widely distributed; lacking usnic acid. . . . . *Parmelia sulcata*
24. Soredia coarsely granular, mostly limited to lobe margins; mature rhizines unbranched or sometimes forked; on rock; uncommon, boreal to arctic; contains usnic acid in soralia . . . . [*Parmelia fraudens*]

25.(18) Lower surface white or pale to dark brown. 26
25. Lower surface black. 29

26. Medulla pale lemon-yellow; thallus without soredia, isidia, or lobules; cortex and medulla KC+ orange (secalonic acid). . . . . *Pseudoparmelia uleana*
26. Medulla white, cortex KC– or difficult to interpret because of strong K+ yellow reaction. 27

27. Lower surface white or pale brown; cortex and medulla PD+ orange, K+ deep yellow (thamnolic acid). . . . . *Imshaugia*
27. Lower surface medium to dark brown; cortex PD–, K+ pale yellow (atranorin); medulla PD–, K– 28

28. Thallus surface with isidia; lobe tips with a network of white maculae . . . . *Canoparmelia caroliniana*
28. Thallus surface without isidia, but with granular soredia; maculae sparse and inconspicuous . . . . *Canoparmelia texana*

29.(25) Rhizines dichotomously branched; medulla white . . . . *Hypotrachyna*
29. Rhizines unbranched or sometimes squarrose; medulla white or pale yellow 30

30. Soredia or fragments originating from the breakdown of pustules or hollow warts present 31
30. Soredia and soredia-like fragments absent. 34

31. Medulla pale yellow, at least close to algal layer, K– or K+ yellow, PD–, UV–; common, East Temperate . . . . *Myelochroa aurulenta*
31. Medulla entirely white; southeastern U.S. 32

32. On rocks; lobes 1–2 mm wide, smooth; soralia discrete, in hemispherical mounds, containing dark gray granular soredia; medulla PD+ red-orange, KC+ pink (protocetraric acid). . . . . [*Paraparmelia alabamensis*]
32. On bark of hardwoods; lobes 2–5 mm wide, wrinkled or ridged; soralia discrete or running together, especially along ridges, remaining pale; medulla PD+ orange or PD–, KC– or KC+ faint purple. 33

33. Medulla K+ yellow, PD+ orange, UV– (stictic acid). . . . . [*Canoparmelia crozalsiana*]
33. Medulla K–, PD–, UV+ blue-white (divaricatic acid). . . . . *Canoparmelia texana*

34.(30) Isidia absent; medulla pale yellow, at least near algal layer. . . . . *Myelochroa galbina*
34. Isidia present; medulla white or pale yellow. 35

35. Lobes with a network of whitish maculae, wrinkled and rough; medulla white, PD–, K–, UV+ white (perlatolic acid). . . . . *Canoparmelia caroliniana*
35. Lobes without maculae, smooth and uniform (where there are no isidia); medulla pale yellow, at least close to algal layer, PD+ orange, K+ orange to red (galbinic acid). . . . . *Myelochroa obsessa*

Parmelia hygrophila
Western shield lichen

DESCRIPTION: Thallus bluish gray to pale green; lobes 3–5 mm wide; white pseudocyphellae forming a network extending to the lobe margins; with dull, weakly corticate isidia (often appearing like large granular soredia) on the upper surface; rhizines unbranched or occasionally forked. Apothecia rare. CHEMISTRY: Medulla PD+ yellow, K+ yellow changing to red, KC–, C– (salazinic acid). HABITAT: On bark (rarely rock) in rainy, oceanic areas. COMMENTS: This is a fairly common lichen along the west coast. It resembles the more common *P. sulcata,* except it has unbranched to dichotomously branched rhizines rather than bottlebrush (squarrose) rhizines, and coarsely granular, dull isidia rather than powdery soredia. *Parmelia saxatilis* has unbranched rhizines, like *P. hygrophila,* but its isidia are clearly smooth and shiny (often with brown tips) rather than "soft" and soredia-like. *Parmelia fraudans,* a northern species on rocks, has very coarse soredia that sometimes resemble weakly corticate isidia, but they are yellowish (caused by small amounts of usnic acid) and predominantly marginal. See also *P. squarrosa.*

560. *Parmelia hygrophila* (damp) northern Sierra Nevada, California ×3.6

561. *Parmelia omphalodes* interior Alaska ×2.0

Parmelia omphalodes
Smoky crottle

DESCRIPTION: Thallus bluish gray to brownish gray or dark brown, sometimes almost black, rather shiny; lobes mostly 2–4 mm wide, but some forms have abundant, overlapping lobules only 1–2 mm wide; pseudocyphellae predominantly marginal, or forming a weak reticulation on the lobe surface, less commonly with a network of sharp ridges; without soredia or isidia; rhizines unbranched or forked. CHEMISTRY: Medulla PD+ yellow to orange, K+ yellow turning red, KC–, C–, UV– (salazinic acid, usually with lobaric acid). HABITAT: On rocks in exposed habitats, especially at high altitudes or latitudes. COMMENTS: This species is extremely variable in color and the degree to which a network of ridges and depressions develops. It looks like a nonisidiate *P. saxatilis* except for the more frequent development of lobules. *Melanelia panniformis* can resemble particularly lobulate, brown forms of *P. omphalodes*, but the *Melanelia* has much narrower lobules (actually flattened isidia)

562. *Parmelia saxatilis* Columbia River, Washington ×3.0

and contains perlatolic acid (medulla PD–, K–, KC– or KC+ pink, UV+ white). IMPORTANCE: See *Parmelia saxatilis*.

Parmelia saxatilis
Salted shield lichen, crottle

DESCRIPTION: Thallus greenish gray to bluish gray, turning brown in exposed habitats; lobes 2–4 mm wide, with a conspicuous network of pseudocyphellae, ridges, and depressions; cylindrical to slightly branched isidia forming on the thallus surface, especially the ridges, shiny and often browned at the tips; rhizines unbranched or forked (Fig. 6b). Apothecia fairly common, with isidiate margins. HABITAT: On rock in exposed or shaded habitats, more rarely on bark or wood. CHEMISTRY: Medulla PD+ yellow to orange, K+ yellow turning red, KC–, C– (salazinic acid, usually with lobaric acid; with or without fatty acids). COMMENTS: *Parmelia saxatilis* is the most common isidiate *Parmelia* with unbranched or forked (not squarrose) rhizines. ***Parmelia pseudosulcata*** (which has been called *P. kerguelensis* in North America) is a much rarer, western species with narrow, dichotomously branched lobes that have only weak ridges. It contains protocetraric acid (medulla PD+ red-orange, K–) instead of salazinic acid. In the similar *P. squarrosa* (with squarrose rhizinae), the isidia are usually most abundant on the lobe margins rather than on the upper surface, and the preferred substrate is bark. In *P. hygrophila*, a western species, the isidia are much softer, granular, and more like soredia. IMPORTANCE: This lichen (together with its nonisidiate counterpart, *P. omphalodes*) was traditionally gathered in large quantities in the Hebrides of Scotland as a source of boiling water dyes (deep red-browns and rusty orange). Both these lichens were called crottle by the weavers and dyers. Any of the species of *Parmelia* containing salazinic acid will do the job, but few are as abundant and easily collected as *P. saxatilis* and *P. omphalodes*. A saxicolous species, quite possibly *P. saxatilis*, was made into a medicinal tea by the Maidu people of southern California, and other species of *Parmelia* have been used as traditional medicines in India and China.

563. *Parmelia squarrosa* Oregon coast ×4.2

Parmelia squarrosa
Bottlebrush shield lichen

DESCRIPTION: Thallus pale gray, with lobes 1–5 mm wide, often overlapping, with conspicuous reticulate ridges and pseudocyphellae; isidia commonly cylindrical and shiny, but sometimes becoming almost squamulose or fat and dull, marginal as well as on the lobe surface; rhizines slender, at first unbranched, then squarrose (Fig. 6d). Apothecia rather common. CHEMISTRY: Medulla PD+ yellow, K+ yellow turning blood-red (salazinic and consalazinic acids). HABITAT: On bark or mossy rock, mostly in shaded, humid habitats. COMMENTS: Although the rhizines on young lobes of *P. squarrosa* can be unbranched, they are never thick and forked as in *P. saxatilis* or *P. hygrophila*. See also Comments under *P. saxatilis*.

Parmelia sulcata
Hammered shield lichen

DESCRIPTION: Thallus blue-gray and often browned at the edges, or entirely brownish when in exposed habitats; lobes 2–5 mm wide, with a network of sharp ridges and depressions and whitish pseudocyphellae; powdery soredia along the ridges and lobe margins where the cortex develops cracks; rhizines densely squarrose when fully developed, but on young lobes slender and unbranched. Apothecia rare. CHEMISTRY: Medulla PD+ yellow to orange, K+ yellow turning blood-red, KC–, C– (salazinic and sometimes lobaric acids). HABITAT: Mostly on bark, but also on mossy rocks, wood, and even soil in both shade and sun. COMMENTS: This is an extremely widespread, even weedy species in the north and west. As it is one of the first lichens to invade trees and picnic benches in suburban areas, *P. sulcata* is the lichen most familiar to casual observers in many parts of the continent. It is, unfortunately, also quite variable. For example, soredia can be abundant or hardly produced at all. If soredia are lacking, it can be mistaken for *P. fertilis,* a small, more abundantly fertile species with less conspicuous ridges and pseudocyphellae, known in North America only from Nova Scotia, New Brunswick, Maine, and adjacent Quebec, although common in Asia. In the west, *P. sulcata* should be compared with the more coarsely sorediate *P. hygrophila*. **Parmelia fraudans,** a mainly boreal saxicolous lichen, has unbranched to forked rhizines and yellowish soredia. IMPORTANCE: Like other shield lichens containing salazinic acid, *P. sulcata* can be used for dyeing wool.

Parmeliella (7 N. Am. species)
Shingle lichens

DESCRIPTION: Small brown to gray-brown squamulose lichens. Thallus without a lower cortex and attached to the substrate by masses of hyphae forming a hypothallus, often blue-black. Photobiont blue-green (*Nostoc*). Apothecia brown, biatorine; spores colorless, 1-celled, ellipsoid. CHEMISTRY: Containing no lichen substances. HABITAT: Usually on wood, mossy bark, or rock in shaded, somewhat humid habitats. COMMENTS: *Parmeliella* is extremely close to *Fuscopannaria,* differing mainly in the biatorine rather than lecanorine apothecial margins and the absence of lichen substances. Because some species of *Fuscopannaria* have apothecia with an ephemeral or partial thalline margin surrounding a basically biatorine margin, the boundaries of the genera are rather fuzzy.

KEY TO SPECIES: See *Pannaria.*

564. *Parmelia sulcata* western Idaho ×3.2

Parmeliella triptophylla
Black-bordered shingle lichen

DESCRIPTION: Thallus composed of very flat, gray lobes or squamules, 0.3–1 mm wide, sitting on a thin or thick, conspicuous blue-black prothallus, with long, branching, cylindrical (or sometimes squamulose) isidia along the lobe margins. Sometimes the isidia are so abundant that the entire lichen becomes a mass of isidia. HABITAT: Frequent in cool, humid, coniferous forests, especially on bark or wood, occasionally on rock. COMMENTS: In North America, the name *P. triptophylla* has been used in a broad sense to include both the typical form with cylindrical isidia found in the northern part of the range (shown in plate 565), and a lobulate-isidiate form found in the Appalachians. The western *Fuscopannaria leucostictoides* is similar to *P. triptophylla* when sterile; although it can be somewhat lobulate, it never forms distinct isidia.

565. *Parmeliella triptophylla* Lake Superior region, Ontario ×3.4

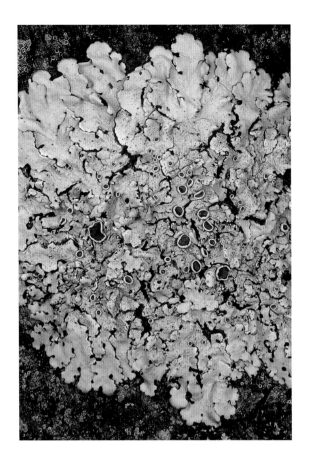

566. *Parmelina quercina* Klamath Range, northwestern California ×1.4

Parmelina (1 N. Am. species)
Shield lichens

KEY TO SPECIES: See *Parmelia*

Parmelina quercina
Fringed shield lichen

DESCRIPTION: Thallus very pale gray to olive-gray, closely appressed to substrate; lobes mostly 2–3 mm wide, often dichotomous, smooth and shiny to dull and almost pruinose, lacking pseudocyphellae, but with white maculae (often faint); producing no soredia, isidia, or lobules, but with black cilia, 0.2–1 mm long, on lobe margins (sometimes sparse); lower surface black, with abundant, rather long unbranched rhizines. The abundant black rhizines merge with the cilia and protrude at the lobe margins, giving the lichen a fringed appearance. Photobiont green (*Trebouxia?*). Apothecia common, 1.5–5 mm across, somewhat raised, with brown disks, often with rhizines developing on the lower side; spores colorless, ellipsoid, 1-celled, 8 per ascus; conidia cylindrical. CHEMISTRY: Cortex K+ yellow (atranorin); medulla PD–, K–, KC+ red, C+ red (lecanoric acid). HABITAT: On bark, especially black oak. COMMENTS: *Parmelina* differs from other shield lichens mainly by its combination of characters: lacking usnic acid and pseudocyphellae while, at the same time, producing cilia and abundant unbranched rhizines. The genus most resembles *Hypotrachyna*, *Myelochroa*, and *Parmelinopsis*, all much more common in the southeast. No other foliose lichens in California have a gray thallus, cilia, no soredia or isidia, and a C+ red medulla. *Hypotrachyna revoluta* is also C+ red but has soredia.

Parmelinopsis (5 N. Am. species)
Shield lichens

DESCRIPTION: Pale gray foliose lichens; lobes 0.5–6 mm wide; cilia present, at least in lobe axils, black, not bulbous at the base, often sparse; no pseudocyphellae or maculae; lower surface uniformly black, with unbranched rhizines; medulla white or slightly yellowish. Photobiont green (*Trebouxia?*). Apothecia uncommon. CHEMISTRY: Cortex K+ yellow (atranorin); medulla at least KC+ red, usually C+ red or pink (orcinol tridepsides, especially gyrophoric acid). COMMENTS: Unlike similar shield lichens, especially *Myelochroa* and *Parmelina*, *Parmelinopsis* has predominantly unbranched rhizines, cylindrical to bone-shaped conidia, 3–5 µm (i.e., very short), occasional to abundant short, black, unbranched cilia, a white medulla, and orcinol tridepsides such as gyrophoric acid. For distinctions among the various genera of shield lichens that have a gray thallus; see the key to *Parmelia*.

KEY TO SPECIES: See *Parmelia*.

Parmelinopsis horrescens (syn. *Parmelina horrescens*)
Hairy-spined shield lichen

DESCRIPTION: Thallus lobes very small, 0.5–2.5 mm wide, shiny, pale gray-green with brownish edges, abundantly covered with cylindrical to flattened isidia, many of them browned or with black cilia at the tip. CHEMISTRY: Medulla PD–, K–, KC+ purplish pink, C– or sometimes C+ pink (mainly 3-methoxy-2,4-di-*O*-

567. *Parmelinopsis horrescens* Great Smoky Mountains, Tennessee ×3.7

methylgyrophoric acid, with hiascic acids). HABITAT: On bark of all kinds, especially in deciduous forests. COMMENTS: *Parmelinopsis minarum* is a very similar isidiate lichen, with cylindrical isidia lacking apical cilia. Its medulla reacts C+ pink (gyrophoric acid).

Parmelinopsis minarum (syn. *Parmelina dissecta*)
Hairless-spined shield lichen

DESCRIPTION: Thallus pale gray, lobes small, 1–3 mm wide, smooth above, covered with cylindrical isidia; unbranched black rhizines sometimes protrude at the margins and look like cilia (which also occur). CHEMISTRY: Medulla PD–, K–, KC+ red, C+ pink (gyrophoric acid, usually with 5-O-methylhiascic acid) and often a UV+ white unknown. In occasional specimens, the medulla is slightly yellowish and K+ pink. HABITAT: On trees and sometimes rocks in deciduous forests. COMMENTS: This species is similar to *P. horrescens*, but no tiny black cilia grow out of the cylindrical (not flattened) isidia, and its medulla is always C+ pink. Compare with *Myelochroa obsessa*, *Bulbothrix goebelii*, and *Canoparmelia caroliniana*, all small, gray, isidiate, foliose lichens in the southeast.

Parmelinopsis spumosa
Pustuled shield lichen

DESCRIPTION: Thallus pale gray-green, somewhat shiny; lobes 0.7–2 mm wide, with abundant, irregular pustules all over upper surface disintegrating into coarse, granular particles (schizidia) resembling soredia; cilia very short, often sparse; medulla faintly yellowish, especially close to algal layer. CHEMISTRY: Medulla PD–, K–, KC+ red, C+ pink (gyrophoric acid, an unidentified pigment, and an unknown substance producing a white fluorescent spot with thin layer chromatography). HABITAT: On bark. COMMENTS: *Parmelinopsis cryptochlora* is a similar but rare lichen also containing gyrophoric acid but producing farinose soredia that do not develop from pustules. It is known from the West Indies and a few places in Florida. The *Hypotrachyna osseoalba* group, also with pustules, has forked rhizines, and the medulla is KC+ purple rather than KC+ red.

568. *Parmelinopsis minarum* Great Smoky Mountains, Tennessee ×2.8

569. *Parmelinopsis spumosa* western Mississippi ×3.6

Parmeliopsis (3 N. Am. species)
Starburst lichens

DESCRIPTION: Small, closely appressed, gray or yellowish green, foliose lichens with very narrow, radiating lobes, 0.5–2 mm wide; sorediate; lower surface almost white to very dark brown, with similarly colored, unbranched rhizines. Photobiont green (*Trebouxia*?). Apothecia uncommon, with brown disks, containing colorless, 1-celled, slightly curved spores; conidia sickle-shaped. CHEMISTRY: Cortex with atranorin or usnic acid; medulla PD–, K–, KC–, C–, UV+ blue-white (divaricatic acid). HABITAT: On bark and wood in exposed habitats. COMMENTS: Species of *Parmeliopsis* most closely resemble *Physcia* species in size and habit, but the spores, chemistry, and often the color are different.

KEY TO SPECIES: See Keys J and K.

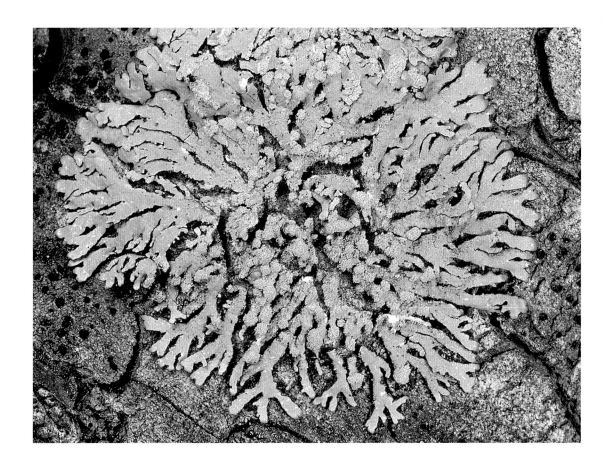

570. *Parmeliopsis ambigua* central Idaho ×4.2

Parmeliopsis ambigua
Green starburst lichen

DESCRIPTION: Thallus pale yellowish green; lower surface dark brown to almost black; powdery to granular soredia in irregular patches on the lobe surface; soralia flat to slightly rounded, not hemispherical, usually laminal; lobe tips often not sorediate. CHEMISTRY: Cortex PD–, K–, KC+ gold, C– (usnic acid). HABITAT: On conifer stumps, logs, and bark in full sun. COMMENTS: Specimens with soredia in hemispherical mounds mainly on the lobe tips can be called *P. capitata*. In that species, which occurs between Lake Superior and the maritime provinces, the lower surface tends to be very pale brown, although it can occasionally become dark brown and shiny as in *P. ambigua*. *Parmeliopsis subambigua,* common along the southeastern coastal plain from Massachusetts to Texas, is very similar to *P. ambigua* except for its pale yellowish white lower surface and more diffuse soredia, which develop from pustules on the upper lobe surface.

Parmeliopsis hyperopta
Gray starburst lichen

DESCRIPTION, CHEMISTRY, HABITAT, and COMMENTS: *Parmeliopsis hyperopta* is virtually identical to *P. ambigua*, except it is pale gray instead of yellowish green and has atranorin rather than usnic acid in the cortex, which is PD+ light yellow, K+ yellow, KC–, C–. The two species often grow side by side on weathered old wood, especially conifer wood and bark. All the northern species of *Physcia* that might be mistaken for *Parmeliopsis hyperopta* have a white lower surface.

Parmotrema (32 N. Am. species)
Ruffle lichens, scatter-rug lichens

DESCRIPTION: Foliose lichens with broad lobes, 4–20 mm across, pale gray or gray-green to yellow-green, most species loosely attached, with ruffled margins; usually with cilia on the lobe margins; lower surface black, at least in the center, dark brown elsewhere

571. *Parmeliopsis hyperopta* Cascades, Oregon ×3.5

except for irregular blotches of ivory white in some species; rhizines usually unbranched, abundant in older parts of the thallus, although usually absent from a broad zone close to the lobe margin; pseudocyphellae absent. Photobiont green (*Trebouxia?*). Apothecia prominent, very large, disks dark to light brown, sometimes with an irregular hole through the center of the disk; spores colorless, 1-celled, ellipsoid; conidia dumbbell-shaped or thread-like. CHEMISTRY: Cortex usually K+ yellow, KC– (atranorin), rarely K–, KC+ gold (usnic acid); medulla with a variety of β-orcinol depsidones especially norstictic, stictic, and protocetraric acids, or depsides (e.g., alectoronic acid), less frequently with fatty acids or orcinol depsides such as lecanoric or gyrophoric acids. In this genus, the alkaline iodine (AI) spot test for stictic acid is particularly useful (see Chapter 13, "Chemistry"). HABITAT: On bark and twigs of all types of trees, less frequently on rock. The genus is best represented in the southeastern U.S., especially along the coastal plain. COMMENTS: *Parmotrema* is most often mistaken for *Rimelia* and *Canomaculina* (with a fine network of maculae and cracks on the upper surface), *Cetrelia* (with conspicuous laminal pseudocyphellae), and *Platismatia* (with sparse rhizines and marginal to almost marginal apothecia and pycnidia). IMPORTANCE: Species of *Parmotrema* are mentioned in the *Indian Materia Medica* (K. M. Nadkarni, ed., 1976) as useful for treating a number of ailments (see *P. chinense*). They are collected in large quantities as a food supplement in India.

KEY TO SPECIES

1. Marginal cilia or cilia-like rhizines present (although sometimes sparse), on lobe margins or in axils of lobes 2
1. Marginal cilia or cilia-like rhizines absent 17

2. Soredia absent. 3
2. Soredia present. 10

3. Isidia present. 4
3. Isidia absent 8

4. Medulla bright yellow. ..... *Parmotrema sulphuratum*
4. Medulla white. 5

5. Thallus yellowish green; cortex K–, KC+ yellow (usnic acid); medulla K–, PD– ...... *Parmotrema xanthinum*
5. Thallus gray to yellowish gray; cortex K+ yellow, KC–

(atranorin); medulla K+ yellow or red, PD+ yellow or orange, or K–, PD– 6

6. Isidia without cilia; maculae present; medulla K+ red, UV+ yellow-orange (mainly in lower half of medulla close to lower cortex) (salazinic acid and lichexanthone)................... *Parmotrema ultralucens*

6. Isidia often with short black cilia growing out of the tips; maculae absent; medulla K– or K+ yellow, UV+ white or UV– (lichexanthone absent) 7

7. Isidia not breaking down into soredia; medulla K+ yellow, KC–, UV– (stictic acid) .... *Parmotrema crinitum*

7. Some or most isidia breaking down into granular soredia; medulla K–, KC+ red, UV+ white (alectoronic acid)....................... [*Parmotrema mellissii*]

8.(3) Apothecial disks not perforated; maculae absent; lower surface brown, never white at edge; medulla PD+ red, K–, KC+ red (protocetraric acid) ....
.... *Parmotrema michauxianum*

8. Apothecial disks perforated with an irregular hole through the center; maculae present on upper surface; lower surface uniform or splotched with white; medulla PD+ orange, K+ red, KC– 9

9. Lower surface usually splotched with white near the margins; contains norstictic acid; mainly southeastern coastal plain and Appalachians...................
.... *Parmotrema perforatum*

9. Lower surface pale to dark brown, not splotched with white; contains salazinic acid; arid central states, Texas to North Dakota .... [*Parmotrema eurysacum*]

10.(2) Medulla K–, PD–, KC+ red (alectoronic acid).... 11

10. Medulla K+ yellow, orange, or red, PD+ yellow, orange, or red, KC– or KC+ pink (alectoronic acid absent) 13

11. Soredia developing from short isidia, many of which are ciliate (with short black hairs) ....
.... [*Parmotrema mellissii*]

11. Soredia developing from breakdown of thallus cortex, not isidia 12

12. Soredia along the lobe margins; southeastern coastal plain................... *Parmotrema rampoddense*

12. Soredia on the upper surface of lobe tips; west coast, Appalachian and Great Lakes regions ....
.... *Parmotrema arnoldii*

13.(10) Cilia short, sparse, and usually confined to lobe axils; medulla PD+ red-orange (protocetraric acid) ..
.... *Parmotrema dilatatum*

13. Cilia long or short, often abundant, scattered along lobe margins; medulla PD+ yellow or orange 14

14. Medulla K+ yellow or orange (stictic acid) ....
.... *Parmotrema chinense*

14. Medulla K+ red (salazinic or norstictic acid) 15

15. Lower surface with blotches of ivory white close to the margins; medulla PD+ yellow (norstictic acid).......
.... *Parmotrema hypotropum*

15. Lower surface uniformly black to dark brown, rarely with white areas; medulla PD+ orange (salazinic acid)
.... 16

16. Soredia on or close to the margins; a network of white maculae and often white cracks usually present, at least on the lobe tips (rarely absent); rhizines abundant ...
.... *Rimelia reticulata*

16. Soredia narrowly restricted to the lobe margins; maculae absent; rhizines sparse ..... *Parmotrema stuppeum*

17.(1) Soredia absent............................... 18

17. Soredia present................................ 20

18. Medulla pale orange-yellow, K+ yellow, C+ pink (gyrophoric acid) .... *Parmotrema endosulphureum*

18. Medulla white, K–, C+ red or C– (gyrophoric acid absent) 19

19. Upper surface dull; medulla C+ red (lecanoric acid)...
.... *Parmotrema tinctorum*

19. Upper surface rather shiny; medulla C–... *Platismatia*

20.(17) Medulla K+ yellow or red..................... 21

20. Medulla K– 22

21. Soredia granular; medulla PD+ red, K+ yellow (protocetraric and echinocarpic acids)...................
.... *Parmotrema dilatatum*

21. Soredia farinose; medulla PD+ orange, K+ red (salazinic acid)................. *Parmotrema cristiferum*

22. Soralia on the thallus surface; thallus closely appressed to substrate, lobes flat; lower surface more or less uniform in color; medulla UV+ blue-white (divaricatic acid)....................... *Canoparmelia texana*

22. Soralia marginal; thallus loosely attached over entire surface, ascending; lobes undulating or crisped at the margins; lower surface pale to dark brown or splotched with white close to the margin; medulla UV– 23

23. Medulla C+ red or pink (lecanoric acid)............
.... *Parmotrema austrosinense*

23. Medulla C–................................. 24

24. Soredia in crescent-shaped soralia on older lobes; on Gulf coastal plain .... *Parmotrema praesorediosum*

24. Soredia scattered over lobe margins, not in crescent-shaped soralia; not on Gulf coastal plain............
.... *Platismatia glauca*

572. *Parmotrema arnoldii* British Columbia coast ×2.4

Parmotrema arnoldii
Powdered ruffle lichen

DESCRIPTION: Thallus pale gray, without maculae; lobes 6–15 mm wide, somewhat divided; soralia on the upper surface of lobe tips close to the margins, making the tips turn downward; with abundant, long cilia especially on nonsorediate lobes; lower surface black to brown with a broad naked zone. Apothecia not produced. CHEMISTRY: Cortex K+ yellow (atranorin); medulla PD–, K–, KC+ reddish, C–, UV+ blue-white (alectoronic acid). HABITAT: On bark of all kinds in open forests, and occasionally on rock. COMMENTS: Only three common, sorediate species of *Parmotrema* occur along the west coast (all less frequent in the Appalachian–Great Lakes region): *P. arnoldii*, *P. stuppeum*, and *P. chinense*; only *P. arnoldii* has a K–, KC+ pink, UV+ medullary reaction. ***Parmotrema margaritatum***, a relatively rare lichen found mainly in the Great Lakes region, is identical to *P. arnoldii* in morphology but contains salazinic acid (PD+ orange, K+ yellow turning red).

Parmotrema austrosinense
Unwhiskered ruffle lichen

DESCRIPTION: Thallus pale gray, with very broad (5–12 mm wide), rounded, smooth lobes, with maculae, wavy and ruffled at the edges, sorediate with very fine and compact soredia on and near the margins of older lobes, but without cilia; lower surface black in the center, but predominantly pale to dark brown, with a broad, naked, often white zone at the edge. CHEMISTRY: Cortex K+ yellow (atranorin); medulla PD–, K–, KC+ red, C+ red (lecanoric acid). HABITAT: On bark and twigs in open situations. COMMENTS: This is an infrequently collected species. It is similar to *P. hypotropum* except for the lack of cilia, the somewhat broader and smoother lobes, and the chemistry. *Parmotrema arnoldii* has more dissected lobes and long cilia, and it is C–.

Parmotrema chinense (syn. P. perlatum)
Powdered ruffle lichen

DESCRIPTION: Thallus pale gray or yellowish gray; lobes 3–15 mm wide, very ruffled, with pillow-shaped marginal or almost marginal soralia on the somewhat downturned lobe tips; cilia mostly under 3 mm, often sparse; upper surface very uniform, with few maculae; lower surface black at the center, with abundant rhizines, brown at the edges. CHEMISTRY: Cortex K+ yellow (atranorin); medulla PD+ orange, K+ yellow or orange, KC–, C– (stictic acid complex, trace of norstictic acid). HABITAT: On bark of all kinds; occasionally on rock in shade. COMMENTS: *Parmotrema chinense* is one of the smaller ruffle lichens. *Parmotrema arnoldii*, another sorediate *Parmotrema* in the west, has cilia up to 6 mm long, darker soredia, a K–, KC+ pink, UV+ medulla, and broader lobes. **Parmotrema hypoleucinum** has the same chemistry as *P. chinense*, but the upper surface of the lobes has an intricate pattern of very minute white maculae, and like *P. hypotropum*, it has a broad white zone on the edge of the lower sur-

573. *Parmotrema austrosinense* hill country, central Texas ×1.5

574. *Parmotrema chinense* Channel Islands, southern California ×2.6

575. *Parmotrema crinitum* Washington coast ×3.6

face. **Parmotrema margaritatum** and *Rimelia reticulata* both lack any white areas on the lower surface, and the K+ red medullary reaction is due to salazinic acid. IMPORTANCE: *Parmotrema chinense* and *P. perforatum* have had medicinal uses in India as a diuretic, headache remedy, sedative, and antibiotic for wounds.

Parmotrema crinitum
Salted ruffle lichen

DESCRIPTION: Thallus pale to greenish gray; lobes 4–12 mm wide, often dissected, isidiate, and fringed with cilia; upper surface more or less covered with dense, cylindrical to branched isidia, the tips of which commonly sprout tiny cilia; lower surface black to brown, with a broad naked zone at the margin. CHEMISTRY: Cortex K+ yellow (atranorin); medulla PD+ orange, K+ yellow, KC–, C– (stictic acid complex). HABITAT: On bark of deciduous trees or white cedar, less frequently on other conifers; also on mossy rocks. COMMENTS: This is the most common and widespread ruffle lichen with distinct isidia and the only common isidiate *Parmotrema* containing stictic acid. *Parmotrema ultralucens* is a similar southeastern species with salazinic acid, and its isidia do not become ciliate. The rare, southeastern *P. internexum* is a small version of *P. crinitum* containing norlobaridone in addition to stictic acid. **Parmotrema mellissii,** another southeastern species, is very similar to *P. crinitum* but has isidia that break down into granular soredia, and it differs chemically (medulla PD–, K–, KC+ pink, C–, UV+ blue-white, because of alectoronic acid). In *Canomaculina subtinctoria*, the lower surface is uniformly brown with rhizines growing right up to the margin, and the medulla contains salazinic acid and norlobaridone (PD+ orange, K+ yellow to red).

Parmotrema cristiferum
Unwhiskered ruffle lichen

DESCRIPTION: Thallus very pale gray to greenish gray, without maculae, but becoming finely cracked on the surface of older central portions; lobes broad and rounded, 10–20 mm wide, with fine soralia along the crinkled margins of the older lobes; lower surface black, with a very broad, brown, naked edge, without white splotches; marginal cilia absent or sparse. CHEMISTRY: Cortex K+ yellow (atranorin); medulla PD+ orange, K+ yellow turning red, KC–, C– (salazinic acid). HABITAT: On sunlit oaks and palms. COMMENTS: A number of sorediate ruffle lichens lack (or usually lack) cilia, but *P. cristiferum* is by far the most common. **Parmotrema rubifaciens** is virtually identical (except, perhaps, in having coarser soredia) but contains norstictic acid; **P. dominicanum, P. dilatatum,** and a chemical variant, **P. gardneri,** are also extremely similar to *P. cristiferum,* but all contain protocetraric acid (PD+ red-orange, K–). *Parmotrema dominicanum* has yellowish rather than white soredia caused by usnic acid, and *P. dilatatum* also contains echinocarpic acid. *Parmotrema praesorediosum* has narrower lobes and is PD–, containing only fatty acids. All these lichens grow together in Florida and nearby parts of the southeastern coastal plain. See also *P. dilatatum.*

lobes, and differs in chemistry. See Comments under *P. cristiferum.*

576. *Parmotrema cristiferum* Gulf Coast, southern Florida ×1.2

Parmotrema dilatatum
Cracked ruffle lichen

DESCRIPTION: Thallus pale gray to greenish gray; older surface often wrinkled and cracked, leaving a network of white lines; lobes 5–20 mm wide, developing ruffled, marginally sorediate lobes on the older parts of the thallus; marginal cilia absent, or short and sparse; lower surface black with abundant rhizines in the center, brown and naked at the edge. Apothecia rare. CHEMISTRY: Cortex K+ yellow (atranorin); medulla PD+ red or red-orange, K+ pale yellow, KC+ pink, C– (protocetraric and echinocarpic acids, with traces of usnic acid). HABITAT: Common on bark in open woods. COMMENTS: **Parmotrema gardneri** is a slightly smaller (lobes mostly under 6 mm wide) and thinner, nonciliate ruffle lichen from Florida and neighboring states. Its marginal soralia are sometimes almost globose. *Parmotrema gardneri* also contains protocetraric acid but lacks echinocarpic acid and has a K– medulla. *Parmotrema praesorediosum* has much the same appearance but contains only fatty acids (PD–, K–, KC–, C–). The ciliate form of *P. dilatatum* can be mistaken for *P. rampoddense,* which has long cilia and narrow, marginal soralia on the young

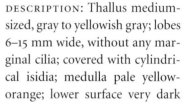

Parmotrema endosulphureum
Yellow-cored ruffle lichen

DESCRIPTION: Thallus medium-sized, gray to yellowish gray; lobes 6–15 mm wide, without any marginal cilia; covered with cylindrical isidia; medulla pale yellow-orange; lower surface very dark brown to black in the center, brown and naked at the margins. CHEMISTRY: Cortex K+ yellow (atranorin); medulla PD–, K+ yellow, KC+ red-orange and C+ red or red-orange close to the cortex and where the isidia have broken off (gyrophoric and often echinocarpic acids, with yellow pigments). HABITAT: On bark in open woods. COMMENTS: *Parmotrema sulphuratum* also has a yellow medulla (more lemon-yellow) but has conspicuous cilia. Except for the yellow medulla, *P. endosulphureum* looks like a small-lobed *P. tinctorum,* which also differs in its pale brown lower surface and C+ blood-red medulla (lecanoric acid).

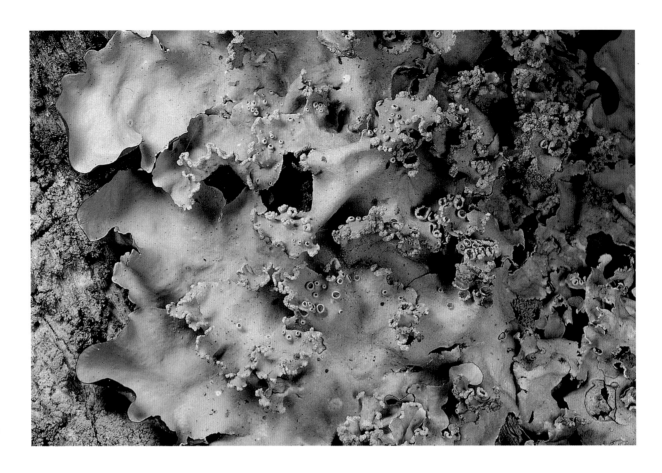

577. *Parmotrema dilatatum* Gulf Coast, Florida ×3.6

578. *Parmotrema endosulphureum* Gulf Coast, southern Florida ×1.9

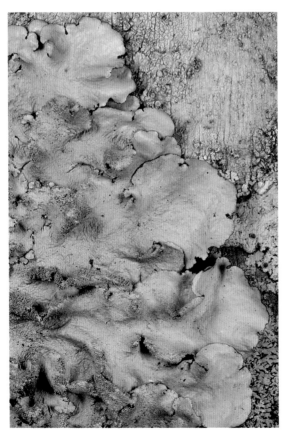

Parmotrema hypotropum
Powdered ruffle lichen

DESCRIPTION: Thallus pale greenish gray, with vague or distinct white maculae on the upper surface; lobes 3–15 mm wide, ascending and often curled back showing ivory-white patches or a continuous white zone at the margins of the lower surface, which is black in the center; patchy soralia on or close to the margins of the lobes; cilia long and usually abundant. Apothecia extremely rare, perforate. CHEMISTRY: Cortex K+ yellow (atranorin); medulla PD+ deep yellow, K+ yellow turning red, KC–, C–, AI– (norstictic acid). HABITAT: On tree trunks and branches, and on shrubs. COMMENTS: **Parmotrema hypoleucinum** is virtually identical, but the medulla is PD+ orange, K+ yellow to orange, AI+ blue, with stictic acid as its main chemical product (in addition to norstictic acid), and it is confined to the southeastern coastal plain and California. Another chemical variant, *P. louisianae* (medulla PD–, K–, KC+ pink; alectoronic acid), is known only from a few localities in the southeast. *Parmotrema stuppeum*, with salazinic

579. *Parmotrema hypotropum* central North Carolina ×2.5

580. *Parmotrema michauxianum* Gulf Coast, Florida ×2.0

acid (PD+ orange, K+ red), has narrow, marginal soralia and lacks any white zone on the lower surface. See also *P. chinense*.

Parmotrema michauxianum
Unperforated ruffle lichen

DESCRIPTION: Thallus uniformly pale gray; lobes 2–8 mm wide, sometimes divided into narrow lobes, with scattered marginal cilia; no soredia, isidia, or maculae, but usually heavily spotted with black pycnidia on the lobe tips; lower surface black with abundant rhizines in center, brown and naked at the edges, not splotched with white areas. Apothecia common, not perforate, flat to bowl-shaped, up to 8 mm wide; conidia straight, stick-like, 5.5–6.7 µm long. CHEMISTRY: Cortex K+ yellow (atranorin); medulla PD+ red-orange, K–, KC+ pink, C– (protocetraric acid). HABITAT: On bark in open woods and along roadsides. COMMENTS: This lichen looks like a small-lobed *P. perforatum* but lacks white blotches on the lower surface, maculae on the surface, and holes through the apothecia. *Parmotrema perforatum* also differs chemically.

Parmotrema perforatum
Perforated ruffle lichen

DESCRIPTION: Thallus pale greenish gray, spotted with white maculae; lobes broad, ascending, often ruffled, 10–20 mm wide, sometimes divided into narrow lobes, with long black cilia; lower surface jet black, with a wide, white margin. Apothecia large (up to 20 mm across), raised, almost invariably perforate with a single, irregular hole, giving the lichen its name. CHEMISTRY: Cortex K+ yellow (atranorin); medulla PD+ orange, K+ red, KC–, C– (norstictic acid). HABITAT: On trees and shrubs in sunny, open sites. COMMENTS: *Parmotrema rigidum* differs from *P. perforatum* in chemistry (typically PD–, K–, KC+ pink, UV+ blue-white: alectoronic acid; occasionally with norstictic acid as well) and in its tendency to have narrower lobes. Another chemically defined relative is *P. preperforatum,* common in east Texas and Louisiana. It contains the stictic acid complex as well as some norstictic acid (medulla AI+ blue). See also *P. michauxianum.* IMPORTANCE: See *P. chinense.*

Parmotrema praesorediosum
Powder-crown ruffle lichen

DESCRIPTION: Thallus pale greenish gray; lobes relatively small, 3–7(–10) mm wide, rounded, not ciliate; producing crescent-shaped soralia on the older margins in the center of the thallus; lower surface dark brown to black in the center and brown at the margins, rarely splotched with white on the oldest, sorediate lobes; rhizines sparse. Apothecia not produced. CHEMISTRY: Cortex K+ yellow (atranorin); medulla: all reactions negative (containing caperatic acid). HABITAT: On tree trunks, especially oaks, in full sun. COMMENTS: *Parmotrema gardneri* and *P. dilatatum* both react PD+ red-orange, KC+ pink (protocetraric acid), but are otherwise very similar to *P. praesorediosum.*

Parmotrema rampoddense
Long-whiskered ruffle lichen

DESCRIPTION: Thallus very pale greenish gray, without maculae; lobes broad, 10–30 mm wide; marginal cilia scattered, very long (up to 5 mm); marginal soralia on the main lobes; lower surface black, rarely blotched with white, almost devoid of rhizines. CHEMISTRY: Cortex K+ yellow (atranorin); medulla PD–, K–, KC+ pink to orange-pink, C–, UV+ blue-white (alectoronic acid). HABITAT: A common lichen on oak and palm bark. COMMENTS: *Parmotrema arnoldii* (with soredia on the lobe tips) and *P. rigidum* (with no soredia) also contain alectoronic acid.

581. (facing page) *Parmotrema perforatum* northern Florida ×3.9

582. (above) *Parmotrema praesorediosum* southern Louisiana ×1.1

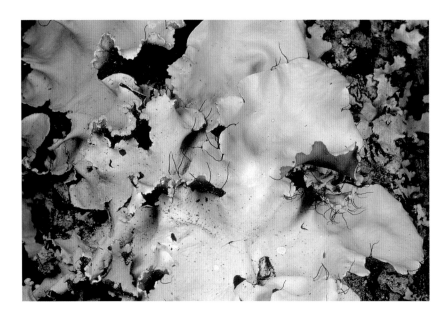

Parmotrema stuppeum
Powder-edged ruffle lichen

DESCRIPTION: Thallus uniformly pale gray, without maculae; lobes rounded and very broad (10–15 mm) with rather erect, ruffled edges having narrowly marginal soralia; cilia often sparse, especially in heavily sorediate specimens; lower surface black in the center, shiny brown and naked at the edges. CHEMISTRY: Cortex K+ yellow (atranorin); medulla PD+ deep yellow, K+ blood-red, KC–, C– (salazinic acid). HABITAT: On bark of deciduous trees, rarely on conifers or rocks. COMMENTS: *Rimelia reticulata* has an identical chemistry, but the surface is conspicuously spotted and cracked with a network of white maculae, which is absent in *P. stuppeum*, and the soredia in the *Rimelia* are coarser and close to the margins, not restricted to the margin edges. ***Parmotrema***

583. *Parmotrema rampoddense* coastal plain, North Carolina ×2.1

584. *Parmotrema stuppeum* Channel Islands, southern California ×3.6

margaritatum, with soralia on short marginal lobes rather than all along the margins, also contains salazinic acid.

Parmotrema sulphuratum
Sulphur ruffle lichen

DESCRIPTION: Thallus yellowish gray to pale yellowish green; lobes 3–8 mm wide, becoming strongly wrinkled and cracked, sparsely to abundantly ciliate with long, slender cilia; surface and lobe margins with cylindrical to slightly branched isidia; medulla bright yellow; lower surface black, although dark brown at the margins. CHEMISTRY: Cortex K+ yellow (atranorin); medulla PD–, K–, KC–, C– (with the yellow pigment, vulpinic acid). HABITAT: On bark of deciduous trees. COMMENTS: The yellow medulla sets this species apart from most other ruffle lichens. *Parmotrema endosulphureum* is another, also with isidia, but it lacks marginal cilia and has a C+ pink medulla (gyrophoric acid).

585. *Parmotrema sulphuratum* herbarium specimen (Canadian Museum of Nature), Florida ×1.8

Parmotrema tinctorum
Palm ruffle lichen

DESCRIPTION: Thallus very pale gray to yellowish gray, dull; lobes broad, 8–15(–20) mm wide, without marginal cilia, covered with granular to cylindrical, or sometimes branched isidia; lower surface black in the center, with a very broad, brown, naked zone at the margin. CHEMISTRY: Cortex K+ yellow (atranorin); medulla PD–, K–, KC+ red, C+ blood-red (lecanoric acid). HABITAT: A very common lichen on bark of all kinds, including palms; occasionally on rock. COMMENTS: This lichen is easily identified by the broad, pale, isidiate lobes and C+ red reaction. It tends to be more closely appressed than most species of *Parmotrema*.

Parmotrema ultralucens
Spotted ruffled lichen

DESCRIPTION: Thallus pale gray, with a vague to conspicuous network of white spots or maculae; lobes 5–15 mm wide with ciliate margins; abundant isidia on upper surface, papillate to slender, cylindrical, often branched, without cilia at the tips; lower surface black (except for brown edges), with a naked zone up to 5 mm from the margin. CHEMISTRY: Cortex K+ yellow (atranorin); medulla PD+ orange, K+ blood-red, KC–, C– (salazinic acid), UV+ bright yellow (lichexanthone). HABITAT: On bark in open woods. COMMENTS: Other southeastern gray ruffle lichens with cilia, isidia, and salazinic acid include *Canomaculina subtinctoria* and **C. neotropica,** both with a brown rather than black lower surface, rhizines reaching almost to the margins, and a UV– medulla. *Canomaculina subtinctoria* also contains norlobaridone (KC+ red, C–). *Parmotrema crinitum* has a K+ yellow medulla (stictic acid) and, like **P. mellissii,** has tiny cilia growing out of the isidia.

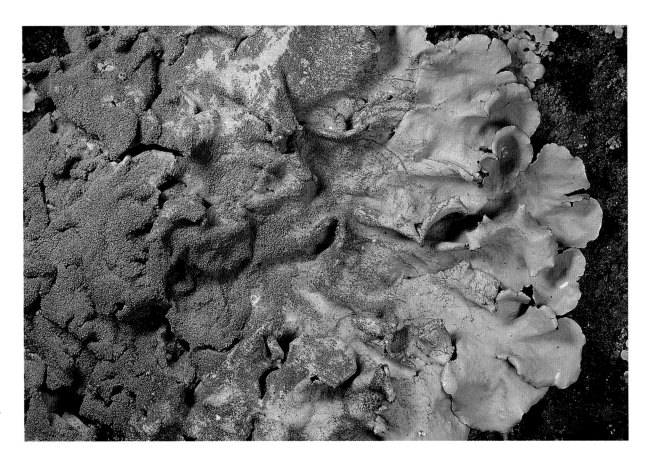

586. *Parmotrema tinctorum* mountains, northern Georgia ×1.9

587. *Parmotrema ultralucens* coastal plain, North Carolina ×1.9

Parmotrema xanthinum
Green ruffle lichen

DESCRIPTION: Thallus pale yellowish green, dull; lobes rounded but often minutely dissected and toothed, 8–12 mm wide, ciliate, densely isidiate on the surface and margins with cylindrical to branched isidia up to 1 mm long, sometimes with cilia growing out of the tips; medulla white; lower surface black with a broad, shiny brown, naked zone at the edge. CHEMISTRY: Cortex K–, KC+ gold (usnic acid); all medullary reactions negative (protolichesterinic acid). HABITAT: On rock outcrops and bark. COMMENTS: **Parmotrema madagascariaceum,** a yellowish ruffle lichen with the same geographic range, differs only in containing gyrophoric acid (medulla C+ pink). *Parmotrema sulphuratum* has a grayer color (without usnic acid in the cortex) and a bright yellow medulla. Other yellow-green foliose lichens with isidia are mainly species of *Xanthoparmelia*, which have much narrower lobes, grow strictly on rock, lack cilia, and have rhizines over the entire lower surface.

Peltigera (28 N. Am. species)
Pelt lichens, dog-lichens

DESCRIPTION: Foliose lichens, most with very broad lobes (up to 40 mm wide); upper surface somewhat tomentose (the "dog-lichens"), or smooth and shiny to scabrose (the "pelt lichens"); lower surface felty, lacking a lower cortex, but most species with raised or flat, white to black "veins" (actually nonconductive thickenings of the medulla); rhizines generally present, either rope-like or tufted (Fig. 6f). Thallus dark gray or brown, photobiont blue-green (*Nostoc*); or thallus brownish green to grass-green, especially when wet, photobiont green (*Coccomyxa*). Cephalodia containing *Nostoc* associated with green algal species. Apothecia on the lobe margins, with red-brown to black disks, often recurved and saddle-shaped; spores 4- to many-celled, colorless, fusiform to needle-shaped. CHEMISTRY: All reactions negative, or rarely KC+ pink, although many species contain small amounts of tridepsides, such as gyrophoric acid and related compounds, and a variety of triterpenes. HABITAT: Most species grow on soil or mossy rock, some on tree bases or bark (especially in humid habitats). COMMENTS: In size, color, and habitat, pelt lichens are most similar to species of *Nephroma*, which have a smooth lower thallus surface (with a cortex). Their apothecia, although marginal as in *Peltigera*, face down; that is, the apothecial disk opens to the lower surface. For all the *Peltigera* species illustrated and discussed below, assume that the photobiont is *Nostoc* unless we indicate otherwise. IMPORTANCE: Some species of *Peltigera*, abundant in humid forests in the Northwest, contribute significant amounts of nitrogen to these ecosystems. They have also been used as traditional medicines in British Columbia and India.

KEY TO SPECIES (including *Erioderma* and *Leioderma*)

1. Photobiont green; cephalodia present 2
1. Photobiont blue-green; cephalodia absent 5

2. Thallus attached by a single point at one edge, fanning out; rhizines absent; cephalodia forming tiny nodular lobules on the lower surface, usually on the veins; disks more or less flat . . . . *Peltigera venosa*
2. Thallus attached by numerous rhizines except at the lobe margin; cephalodia forming scales on the upper surface of the lobes; disks saddle-shaped. 3

3. Cephalodia in the form of lobed scales loosely attached to the upper thallus surface, easily detached. . . . . *Peltigera britannica*
3. Cephalodia in the form of round to slightly scalloped scales closely appressed, firmly fixed to the thallus surface 4

4. Lower surface with conspicuous veins; lobes crisped and undulating at the margins; rhizines separate and distinct; lower surface of apothecia with green, scale-like, discontinuous patches . . . . . . *Peltigera leucophlebia*
4. Lower surface black in the center, changing abruptly to white at the margins, or with a regular pattern of white and dark areas near the margins; lobes more or less flat, not crisped or undulating at the margins; rhizines forming an intricately branched and anastomosing mat; lower surface of apothecia with a continuous, bumpy cortex . . . . *Peltigera aphthosa*

5.(1) Soredia present 6
5. Soredia absent 9

6. Soredia in irregular gray patches on thallus surface. . . . . *Peltigera didactyla*
6. Soredia marginal or on the lower surface of the lobe tips 7

7. Upper surface smooth to scabrous, not tomentose; lower surface with conspicuous veins; photobiont *Nostoc* . . . . *Peltigera collina*
7. Upper surface tomentose, at least near lobe margins; lower surface more or less uniform; photobiont *Scytonema* 8

8. Tomentum on upper surface composed of erect hairs; medulla PD+ orange (eriodermin) . . . . *Erioderma sorediatum*

588. *Parmotrema xanthinum* Appalachians, North Carolina ×2.2

8. Tomentum on upper surface webby; medulla PD– . . . . *Leioderma sorediatum*

9.(5) Isidia present on the thallus surface. 10
9. Isidia absent; lobules present or absent. 11

10. Lobes mostly 5–10 mm wide, concave; isidia flat and scale-like, peltate . . . . *Peltigera lepidophora*
10. Lobes 7–25 mm wide, turned down at the margins; isidia granular to erect and flattened . . . . *Peltigera evansiana*

11. Lobules present on margins and along cracks 12
11. Lobules absent. 14

12. Upper surface dull or tomentose, at least at lobe tips. . . . . *Peltigera praetextata*
12. Upper surface rather shiny. 13

13. Lower surface with conspicuous veins; rhizines distributed along the veins; disks saddle-shaped . . . . *Peltigera pacifica*
13. Lower surface with a regular pattern of white and dark areas; rhizines arising in more or less concentric bands; disks rather flat . . . . *Peltigera elisabethae*

14.(11) Lower surface without distinct veins or patterning 15
14. Lower surface with distinct veins, or with a regular pattern of white and dark areas; rhizines abundant 16

15. Lower surface black in the center grading to white at the margins; rhizines sparse; thallus thick, dark green when wet . . . . *Peltigera malacea*
15. Lower surface pale; thallus thin, not turning dark green when wet; east coast . . . . [*Peltigera hymenina*]

16. Upper surface rather shiny. 17
16. Upper surface dull, scabrose, or tomentose 21

17. Apothecial disks black; maculae present on thallus surface . . . . *Peltigera neckeri*
17. Apothecial disks brown; maculae absent 18

18. Veins pale, narrow, and conspicuously raised, distinctly fuzzy with an erect tomentum; thallus thin and membranous, often with a blistered appearance . . . . *Peltigera membranacea*
18. Veins dark (at least in older parts of thallus), broad or narrow, relatively flat, fuzzy or smooth; thallus not membranous and relatively smooth 19

19. Rhizines arising in more or less concentric bands; apothecia flat; spores mostly under 45 µm long . . . . *Peltigera horizontalis*
19. Rhizines distributed along the veins; apothecia saddle-shaped; spores mostly over 45 µm long 20

20. Lobes 7–15(–20) mm across, with crisped margins; upper surface shiny throughout . . . . *Peltigera polydactylon*

20. Lobes 20–40 mm across, round and not crisped; surface rather dull and often pruinose . . . . *Peltigera neopolydactyla*

21.(16) Lobes concave, or undulating at the margins, upper surface slightly to heavily tomentose 22
21. Lobes usually flat, not undulating at the margins; upper surface tomentose, smooth, or scabrose. 24

22. Lobe margins smooth and even; lobes strongly erect. . . . . *Peltigera didactyla*
22. Lobe margins scalloped; lobes mostly adnate 23

23. Rhizines tufted, running together; lobe surface heavily tomentose . . . . *Peltigera rufescens*
23. Rhizines unbranched or brush-like, separate and distinct; lobe surface heavily tomentose or dull but almost without tomentum . . . . *Peltigera ponojensis*

24.(21) Rhizines rope-like, with a fuzzy surface, unbranched (see couplet 18); tomentum on upper surface often confined to extreme lobe tips with the remainder of the surface rather shiny . . . . *Peltigera membranacea*
24. Rhizines fibrous and smooth, tufted, or brush-like; upper surface tomentose, dull, or scabrose 25

25. Surface more or less smooth, not scabrose or pruinose, always tomentose, at least near the lobe tips; lower surface with raised, pale veins, sometimes dark in the center of the thallus. *Peltigera canina*
25. Surface scabrose or pruinose in part, with or without tomentum; lower surface with dark veins, distinct or indistinct. 26

26. Upper surface dull, sometimes pruinose, but not scabrose; lobes very broad (20–40 mm). . . . . *Peltigera neopolydactyla*
26. Upper surface scabrose. 27

27. Lobes 20–40 mm across; veins distinct, more or less smooth; cortex usually C+ pink (gyrophoric acid and tenuiorin). . . . . *Peltigera scabrosa*
27. Lobes 7–15 mm across; veins very dark, contrasting with white spaces between them, distinctly fuzzy with an erect tomentum; cortex always C– (gyrophoric acid and tenuiorin absent) . . . . *Peltigera kristinssonii*

Peltigera aphthosa
Common freckle pelt, felt lichen

DESCRIPTION: Thallus grass-green when wet, gray-green to various shades of brown when dry (the more sunshine it gets, the browner the color); photobiont green; lobes very broad, up to 40 mm across, smooth except for scattered, grayish brown cephalodia that are 0.5–2 mm in diameter, flat

589. *Peltigera aphthosa* (wet) Cascades, Oregon ×1.1

to convex, tightly attached, warty to scale-like, irregular in shape but generally not lobed; lower surface of thallus uniformly black to dark brown, abruptly turning white close to the margins, much as in *P. malacea* (plate 605). Apothecia 7–15 mm in diameter; disks red-brown, with the back surface covered with a uniform to warty cortex. CHEMISTRY: Tenuiorin, methyl gyrophorate, gyrophoric acid, and triterpenes. HABITAT: On mossy ground, rocks, or tree bases. COMMENTS: This is the largest of the pelt lichens with green algae. The common name is based on the speckled (or freckled) appearance of the upper surface caused by the scattered cephalodia. Only three species of *Peltigera* besides *P. aphthosa* are green: *P. leucophlebia*, with blacker, more distinct veins, ruffled margins, and a patchy rather than complete cortex on the back surface of the fruiting bodies; *P. britannica*, a coastal species with large, lobed, easily detached cephalodia on the upper surface; and the much smaller *P. venosa*, with scattered, roundish lobes and round apothecia that are black when dry (red-brown when wet). IMPORTANCE: All species of *Peltigera* are used for boiling water dyes, usually browns, even though they do not contain lichen substances usually thought to yield these colors (see Chapter 10). In the fifteenth century, when medicinal plants were identified by their resemblance to the symptoms of diseases, the cephalodia of *P. aphthosa* were thought to be similar to thrush eruptions in children's mouths. At the time of Linnaeus (mid-eighteenth century), Swedish mothers still boiled *P. aphthosa* in milk as a remedy for thrush. *Peltigera aphthosa* (and the closely related *P. britannica*) were chewed as a remedy for tuberculosis by the Nitinaht people of Vancouver Island and made into poultices for burns and scalds by the Tlingit of coastal Alaska. Because the cyanobacteria in the cephalodia can fix nitrogen from the air into nitrates that the lichen can use, lichens with cephalodia are an important source of nitrogen in the places they inhabit.

Peltigera britannica
Flaky freckle pelt

DESCRIPTION and COMMENTS: Very similar to *P. aphthosa* except that its cephalodia are lobed and easily detached. Each tiny, lobed cephalodium grows larger and

590. *Peltigera britannica* (with all stages of blue-green and green morphs) Cascades, Washington ×3.2

larger, finally breaking off and growing as an independent lichen. This new lichen (containing the same fungus as the parent but with cyanobacteria instead of green algae) can grow to several centimeters in diameter and looks quite different from the thallus containing green algae, even in general shape. (The blue-green morphs have narrower, crisped lobes, the upper surface often spotted with white maculae, and the lower surface is pale and lacks veins.) This cephalodial lichen then captures the same type of green algae as is found in the parent lichen and develops tiny green lobes, which then grow out into new thalli of *P. britannica*. All these stages are shown in plate 590. Because the cyanobacterial thalli can fix nitrogen, this alternation of green and blue-green forms of the lichen could well be an adaptation for ensuring adequate nitrogen supply to the young thalli under conditions of high rainfall and resulting low nutrients. Besides the difference in cephalodia, *P. britannica* can have slightly better developed veins on the lower surface than does *P. aphthosa*. See also Comments under *P. malacea*. CHEMISTRY: Essentially the same as *P. aphthosa*. HABITAT: Most characteristically on tree trunks and rock faces in extremely humid habitats, also on mossy soil.

Peltigera canina
Dog-lichen

DESCRIPTION: Thallus brown or brownish gray when dry, with a fuzzy tomentum on the upper surface, especially close to lobe margins; lobes mostly 10–25 mm wide, flat and smooth to somewhat blistered, with margins slightly curled downward; lower surface white, or brownish toward the center; veins conspicuous, flat to rounded and raised, mostly white; rhizines tufted and fibrous or brush-like, sometimes running together. CHEMISTRY: No lichen substances. HABITAT: On soil, generally in woodlands, fields, or sandy areas, less commonly on tree bases or mossy rocks. COMMENTS: *Peltigera canina* is similar to several other species with a fuzzy upper surface. *Peltigera rufescens,* common on dry, exposed soil, has narrower lobes that are turned up and ruffled at the edges. Its upper surface is almost white with a thick tomentum, and its rhizines coalesce to form an almost continuous mat below (except at the edges). The western *P. membranacea* is thinner, broader, and less hairy above, and its rhizines are distinctly rope-like rather than tufted. *Peltigera praetextata*, with numerous tiny lobes at the margins and along cracks and breaks in the thallus, has rhizines that are thinner and less fibrous, and veins that are generally lower and less distinct. Because the name *P. canina* has been used so broadly in the past, our distribution map represents only reliable literature records and verified herbarium specimens. The true range of the species is almost certainly broader. See also Comments under *P. scabrosa*. IMPORTANCE: The dog-lichen, as the name implies, was used for treating rabies in medieval Europe based on the "doctrine of signatures" (see Chapter 10). Presumably, the erect apothecia resembled the teeth (or ears?) of dogs. The Nitinaht people on the west coast of Vancouver Island used "*P. canina*" (or more likely *P. membranacea* or *P. aphthosa*) to treat a man who could not urinate.

Peltigera collina
Tree pelt

DESCRIPTION: Thallus gray to brownish gray or dark brown; surface smooth or slightly scabrose (rough and crusty), less frequently pruinose; lobes 5–10 mm wide, the margins usually covered with coarse, bluish gray soredia; lower surface pale with inconspicuous veins and slender or tufted rhizines. Apothecia produced infrequently; when fertile, *P. collina* tends to be less sorediate. CHEMISTRY: Tenuiorin, methyl gyrophorate, gyrophoric acid, and triterpenes. HABITAT: Most common on tree trunks and branches, especially among mosses, less frequently on mossy rocks, rarely on soil. COMMENTS: This is the only *Peltigera* with marginal soredia. The browner forms look a bit like *Nephroma parile*, which it sometimes accompanies on trees in the Northwest, but the lower surfaces are quite different: felt-like and veined in *P. collina* and smooth in the *Nephroma*.

591. *Peltigera canina* Lake Superior region, Ontario ×0.6

592. lower surface of *Peltigera canina* ×3.0

Peltigera didactyla (syn. *P. spuria*)
Alternating dog-lichen

DESCRIPTION: Thallus brownish gray to brown, with small, concave lobes up to about 10 mm across, bearing round patches of blue-gray, granular soredia on

593. *Peltigera collina* (damp) Coast Range, California ×1.5

the upper surface when the lichen is young. As the lichen matures, the patches close over and the lichen develops many red-brown, saddle-shaped fruiting bodies on erect marginal lobes. Lower surface very pale, with low, whitish to pale brown veins; rhizines whitish to brown, fibrous, discrete or tufted. CHEMISTRY: No lichen substances, or with gyrophoric acid and methyl gyrophorate. HABITAT: On soil, especially disturbed soil, or among rocks. COMMENTS: A leopard may not be able to change its spots, but this lichen can! The spots on the upper surface are patches of granular soredia, but they are there only when the lichen is young and disappear when it is mature. This lichen can therefore be sorediate and sterile or nonsorediate and fertile depending on its age. It is the only *Peltigera* with true soredia on the upper surface. The boreal to arctic *P. lepidophora* has tiny flattened granules (isidia), each attached at the center (peltate) on the lobe surface. In *P. collina*, the soredia are marginal. Mature specimens with abundant apothecia and no soredia can be confused with *P. rufescens*. The lobes of *P. didactyla* var. *didactyla* tend to be more separate and erect than those of *P. rufescens*, and the rhizines are less fluffy and confluent; the rhizines of **var. *extenuata*,**

594. *Peltigera didactyla* (damp) Lake Superior region, Ontario ×2.9

however, are like those of *P. rufescens*. In addition, *P. didactyla* is rarely as heavily tomentose as *P. rufescens*, and its apothecia are smaller (mostly 3–5 mm versus 4–8 mm long). *Peltigera didactyla* var. *extenuata*, a common boreal lichen, differs from the var. *didactyla* in having broader, flatter lobes (to 15 mm) and more fibrous, confluent rhizines. Its medullary chemistry is unusual for a tomentose pelt lichen, containing methyl gyrophorate and often traces of gyrophoric acid (but usually reacting C–, KC–).

Peltigera elisabethae
Concentric pelt

DESCRIPTION and COMMENTS: Like *P. horizontalis* except it has abundant isidia-like lobules along the margins and thallus cracks; lower surface predominantly black except at the edge, with small, white, round interspaces instead of distinct veins. In *P. horizontalis,* the veins are more distinct or the interspaces are somewhat elongate. *Peltigera praetextata* also has a ruffled margin and frequent marginal and stress-crack lobules, but the surface is slightly tomentose especially close to the lobe tips, and it has a rather distinct network of pale veins. CHEMISTRY: Tenuiorin, methyl gyrophorate, gyrophoric acid, and triterpenes. HABITAT: On soil or mossy rock faces, typically in forests.

Peltigera evansiana
Peppered pelt

DESCRIPTION: Thallus pale to dark brown when dry, dull and rough; lobes 7–25 mm wide, slightly downturned at the margins, more or less covered with tiny granular to cylindrical or flattened, sometimes branched isidia; finely divided overlapping lobules are also common; lower surface with low but distinct veins, pale at the thallus edge and dark toward the center; rhizines unbranched or slightly brush-like at the tips. CHEMISTRY: No lichen substances. HABITAT: On soil in open or forested habitats. COMMENTS: *Peltigera lepidophora* also has isidia on the upper surface, but they are flat and scale-like, attached at their edge or center, rather than being short-cylindrical or granular. The entire lichen is much smaller, rarely having lobes broader than 10 mm.

595. *Peltigera elisabethae* Lake Superior region, Ontario ×1.3

596. *Peltigera evansiana* southern Wisconsin ×1.6

597. *Peltigera horizontalis* Lake Superior region, Ontario ×1.0

598. lower surface of *Peltigera horizontalis* ×3.0

Peltigera horizontalis
Flat-fruited pelt

DESCRIPTION: Thallus gray to chestnut brown, rather shiny, without any tomentum or pruina on the upper surface; lobes 10–20 (–30) mm across, often with depressions on the upper surface corresponding to the rhizines below; lower surface dark brown with elongate, depressed white spots, as opposed to distinct veins; rhizines brush-like, arising in concentric rows (as shown in the lower left of plate 598). Apothecia flat and rather round; spores 4-celled, 25–40 × 4–6 μm. CHEMISTRY: Tenuiorin, methyl gyrophorate, gyrophoric acid, and triterpenes. HABITAT: On mossy soil, logs, and rocks in forests. COMMENTS: The saddle-shaped apothecia and needle-like spores of *P. polydactylon* easily distinguish this lichen from *P. horizontalis* when they are fertile. Sterile specimens of *P. polydactylon* differ in the scattered arrangement of the rhizines and in the lack of depressions on the upper surface. *Peltigera elisabethae* has abundant marginal lobules but is otherwise almost identical to *P. horizontalis*.

Peltigera kristinssonii
Dark-veined pelt, black-and-white pelt

DESCRIPTION: Thallus light brown to pale gray-brown; lobes 7–15 mm wide, the tips distinctly scabrose and sometimes woolly (on protected lobes), but central portions of the thallus can be smooth and shiny; lower surface with raised, very dark veins contrasting sharply with white interspaces; rhizines short, dark, and tufted, sometimes continuous along the veins. CHEMISTRY: No lichen substances. HABITAT: On mossy soil or bare earth, mostly in lowland, boreal forests. COMMENTS: Because of the scabrose lobe tips, *P. kristinssonii* can resemble *P. scabrosa*, which has lower, less distinct veins and at least some distinct rhizines, never forming single or continuous tufts as in *P. kristinssonii* or the related species *P. rufescens*. The cortex of fresh specimens of *P. scabrosa* reacts C+ pink and KC+ pink because of gyrophoric acid. *Peltigera rufescens* is generally much more heavily tomentose and less scabrose on the lobe tips.

599. *Peltigera kristinssonii* central Idaho ×1.1

600. lower surface of *Peltigera kristinssonii* ~×2.0

Peltigera lepidophora
Scaly pelt

DESCRIPTION: Thallus pale to dark brown, consisting of small, rounded, concave lobes 5–10 (rarely to 20) mm across, with tiny, scale-like, peltate isidia on the upper surface closely resembling cephalodia and, in fact, probably developing in much the same way; lower surface with pale, low, indistinct veins and typically unbranched rhizines. CHEMISTRY: No lichen substances. HABITAT: On exposed soil such as road cuts or trail banks, usually in open, dry habitats. COMMENTS: *Peltigera evansiana* also has isidia scattered on the upper surface, but it has predominantly granular to cylindrical isidia and is a much larger lichen. Young *P. didactyla* has the same general size and appearance but develops patches of gray soredia instead of isidia.

601. *Peltigera lepidophora* canyon country, Utah ×3.3

PELTIGERA 511

602. *Peltigera leucophlebia* Rocky Mountains, British Columbia ×6.4

Peltigera leucophlebia
Ruffled freckle pelt

DESCRIPTION: Thallus grass-green when wet (photobiont green), gray to brownish when dry (the more direct sun exposure it has, the browner it becomes), with scattered dark gray, convex cephalodia tightly attached to the upper thallus surface; lobes rather broad, up to 30 mm across, usually wavy or ruffled at the margins; lower surface with distinct, dark brown veins; rhizines unbranched, tufted, or brush-like, pale or dark. Apothecia with discontinuous patches of cortex on the lower surface (as shown in plate 602). CHEMISTRY: Tenuiorin, methyl gyrophorate, and usually gyrophoric acid; with triterpenes. HABITAT: On moist or dry mossy soil, logs, or rock, especially in somewhat calcareous habitats. COMMENTS: The combination of veined lower surface, patchy cortex on the apothecia, ruffled lobe margins, and tightly attached, convex cephalodia is usually enough to distinguish most specimens of *P. leucophlebia* from its close relatives, *P. aphthosa* and *P. britannica*. Unfortunately, intermediates can occasionally be found between all these species.

Peltigera malacea
Veinless pelt

DESCRIPTION: Thallus rather thick and stiff, dark green when wet (although the photobiont is blue-green), brown when dry; upper surface dull at the lobe tips with erect tomentum, becoming smooth and shiny toward the center; lobes 10–20 mm wide, often overlapping one another like shingles; lower surface pale brown at the margin, abruptly grading into a uniform brown-black (without distinct veins); rhizines sparse, tufted. Apothecia saddle-shaped, on erect lobes. CHEMISTRY: Gyrophoric acid, methyl gyrophorate, tenuiorin, triterpenes. HABITAT: On acidic soil and over moss, usually in forests but also in open tundra. COMMENTS: This lichen is easily recognized by its black, veinless lower surface, thick lobes, erect tomentum, and deep green color when wet. The cyanobacterial form of *P. britannica* also lacks veins, but the thallus is much thinner and smaller, the lower surface is paler, and the upper surface is often mottled with maculae. See also notes on *P. hymenina* under Comments for *P. polydactylon*.

603. lower surface of *Peltigera leucophlebia* ×3.3

Peltigera membranacea
Membranous dog-lichen

DESCRIPTION: Thallus gray to gray-brown when dry, with broad, rounded lobes commonly 20–30 mm across, very thin and membranous (70–100 μm thick); upper surface smooth and rather shiny except for the lobe tips, which are thinly tomentose; lower surface very pale, with a network of narrow, raised, tomentose veins; rhizines long and rope-like, tomentose like the veins. Apothecia red-brown, saddle-shaped, rather frequent. CHEMISTRY: No lichen substances. HABITAT: On almost any type of mossy surface (e.g., rocks, logs, soil), usually in humid habitats. COMMENTS: This very broad-lobed, extremely thin lichen with slender, rope-like rhizines is hard to mistake for anything else, except perhaps *P. canina*. In the latter, the rhizines are more tufted, and the thallus is thicker (300–500 μm thick) and more conspicuously fuzzy, at least on the margins of young lobes. IMPORTANCE: The Kwakiutl tribe of northwest British Columbia used *P. membranacea* as a love charm, but it is not clear how (or if) it worked.

604. *Peltigera malacea*
Shuswap Highlands, British Columbia
×1.3

605. lower surface of *Peltigera malacea*
×1.8

Peltigera neckeri
Black saddle lichen

DESCRIPTION: Thallus shiny, dark brownish gray to bluish gray, lightly pruinose on the lobe tips, otherwise smooth; lobes relatively narrow, 7–10(–15) mm broad; lower surface with broad, flat veins that can be pale to dark brown, sometimes appearing to be white spots on a dark, veinless surface; rhizines fibrous (tufted). Apothecia usually present, saddle-shaped, ebony black, erect on upturned lobes (as shown in the lower left of plate 607). CHEMISTRY: Tenuiorin, methyl gyrophorate, gyrophoric acid, triterpenes. HABITAT: On mossy logs, soil, and tree bases, especially in wet habitats such as lowland forests. COMMENTS: No other *Peltigera* has black apothecia together with a shiny, upper surface and pruinose lobe tips. *Peltigera neckeri* is smaller than its close relatives in the *P. polydactylon* group.

606. *Peltigera membranacea* Cascades, Oregon ×1.3

607. *Peltigera neckeri* southwestern Quebec ×1.5

608. lower surface of *Peltigera neckeri* ×3.1

609. *Peltigera neopolydactyla* coastal Alaska ×0.86

610. lower surface of *Peltigera neopolydactyla* ×2.1

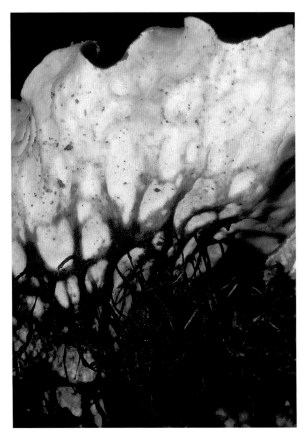

Peltigera neopolydactyla
Carpet pelt

DESCRIPTION: Thallus light brown to greenish gray when dry, smooth but dull, occasionally pruinose; lobes very broad and round, 20–40 mm across; lower surface with a network of pale brown to almost black, low, broad veins interspersed with white, oval spaces, often coalescing and becoming indistinct in the center of the thallus; rhizines commonly fibrous and tufted, commonly 7 mm or more long. Apothecia strongly recurved, red-brown; spores 50–90(–100) μm long. CHEMISTRY: Tenuiorin, methyl gyrophorate, gyrophoric acid, triterpenes. HABITAT: Over moss on the ground and tree bases in mainly coniferous forests. COMMENTS: This broad-lobed member of the *P. polydactylon* group has red-brown apothecia protruding from the margins on the short lobes. The apothecia look very small in comparison to the lobes. *Peltigera polydactylon*, the most similar, is entirely shiny, the lobes rarely exceed 20 mm across, and its spores are rarely longer than 70 μm.

Peltigera pacifica
Fringed pelt

DESCRIPTION: Thallus blue-gray to brown, shiny above, without pruina or tomentum; lobes crisped and divided at the edges, 7–10(–15) mm across, becoming lobulate, with tiny lobes also developing along thallus cracks; lower surface very pale to dark at the center; veins pale to dark, broad, low to somewhat raised (in the thallus center), often indistinct; rhizines 3–7(–9) mm long and rather thick on older portions, dark, discrete to somewhat tufted and brush-like at the tips. Apothecia red-brown, saddle-shaped. CHEMISTRY: Tenuiorin, methyl gyrophorate, gyrophoric acid, triterpenes. HABITAT: Over mossy soil and logs mainly in coastal forests. COMMENTS: This is essentially a *P. polydactylon* that develops squamulose lobules along the lobe margins and on stress cracks. In *P. polydactylon* the veins are more abruptly darkened, with more conspicuous white spaces between them; the rhizines are shorter (2–3(–5) mm long) and are often paler. In *P. elisabethae*, another lobulate species, the veins are hardly distinguishable from the white, oval spaces, the rhizines are in concentric rows, and the apothecia are flat. See also *P. praetextata*.

Peltigera polydactylon
Many-fruited pelt

DESCRIPTION: Thallus greenish gray or more commonly brown, very shiny above, without pruina or tomentum; lobes 7–10(–15) mm across, with crisped margins but not lobulate; lower surface pale at the edge, rather abruptly becoming dark brown, with a network of low, broad veins and white, oval spaces; rhizines 2–4 mm long, pale to dark brown, tufted, or, less commonly, single, not in rows. Apothecia red-brown, saddle-shaped, on upturned lobes; spores mostly 55–70 × 3–4 µm. CHEMISTRY: Tenuiorin, methyl gyrophorate, gyrophoric acid, triterpenes. HABITAT: On soil, moss, or mossy rock in forest habitats. COMMENTS: This is a small pelt lichen with rather crisped lobe margins, short rhizines, and low, broad veins. **Peltigera degenii,** another species with a very shiny upper surface, differs from *P. polydactylon*

611. *Peltigera pacifica* Oregon coast ×1.3

612. lower surface of *Peltigera pacifica* ×1.4

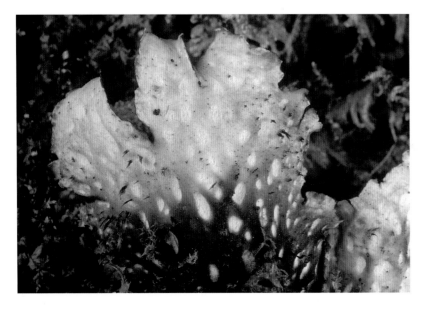

613. *Peltigera polydactylon* Lake Superior region, Ontario ×2.0

614. lower surface of *Peltigera polydactylon* ×2.8

and related species in having raised, rather narrow veins (pale or becoming dark brown) with discrete, often ropy rhizines. *Peltigera degenii* lacks lichen substances and is a relatively rare forest species. **Peltigera hymenina,** mostly on the oceanic east coast (rare in the west) is a similar but taxonomically difficult species with indistinct veins. It can generally be recognized by the uniformly pale lower surface and thin thallus. See also *P. horizontalis, P. neopolydactyla,* and *P. pacifica.*

Peltigera ponojensis
Pale-bellied dog-lichen

DESCRIPTION: Thallus gray to dark brown, surface usually densely tomentose, but can be fairly smooth, although not shiny; lobes 5–15 mm wide, upturned at edges; lower surface rather pale, with somewhat raised, light brown veins bearing predominantly pale, unbranched to fibrous, discrete rhizines. Apothecia produced infrequently, roundish, and flat. CHEMISTRY: No lichen substances. HABITAT: On bare soil or less frequently on moss, in exposed, dry sites. COMMENTS: Superficially very similar to *P. rufescens,* which has dark veins and short, tufted rhizines that run together into a fuzzy mass. *Peltigera rufescens* is always heavily tomentose and also grows in dry, open habitats.

Peltigera praetextata
Scaly dog-lichen, born-again pelt

DESCRIPTION: Thallus light to dark brown, smooth, rather dull, with thin tomentum near the lobe tips; lobes 7–20 mm across, sometimes "crisped" in older parts (as shown in plate 617), developing abundant tiny lobules along the lobe margins and thallus cracks; lower surface rather pale throughout, with light brown, somewhat raised veins sometimes becoming dark in the center of the thallus; rhizines usually discrete, fibrous to ropy, rather long. Apothecia not common. CHEMISTRY: No lichen substances. HABITAT: In forested and open sites on rock, soil, or logs. COMMENTS: The combination of lobulate cracks and a tomentose lobe margin is usually enough to distinguish this species of *Peltigera* from others.

615. (left) *Peltigera ponojensis* Lake Superior region, Ontario ×1.7

616. (below left) lower surface of *Peltigera ponojensis* ×1.5

617. (below right) *Peltigera praetextata* coastal Alaska ×1.4

618. *Peltigera rufescens* mountains, eastern Oregon ×2.2

619. lower surface of *Peltigera rufescens* ×2.9

Except for the tiny lobules, it closely resembles *P. canina*. In addition, its veins are generally less raised and the rhizines are more discrete, never joining at the bases as in *P. canina*. As it is not very tomentose on the surface, it is necessary to examine the lobe tips carefully. (The lobulate species *Peltigera elisabethae* and *P. pacifica* both lack any tomentum on the upper surface.) *Peltigera cinnamomea,* a species of western montane forests, is often mistaken for *P. praetextata,* but it is more tomentose, it does not have squamulose lobules, and the smooth, partly raised veins have a striking cinnamon color.

Peltigera rufescens
Field dog-lichen

DESCRIPTION: Thallus gray to brown, usually with a heavy tomentum on the upper surface; lobes 5–10(–15) mm across, with strongly upturned margins; lower surface with distinct raised veins, pale at the margins but dark elsewhere; rhizines little more than thick tufts of hyphae, coalescing toward the center of the thallus into an almost continuous mat, especially along the veins. Apothecia common, dark red-brown, saddle-shaped, on upright lobes. CHEMISTRY: No lichen substances. HABITAT: On dry, sandy, usually calcareous soil, especially on roadsides or in open fields, always in full sun. COMMENTS: Its relatively narrow lobes, strongly upturned margins, and heavy tomentum distinguish *P. rufescens* from *P. canina*. Fertile specimens of *P. didactyla* are sometimes similar in size and type of apothecia, but the rhizines are separate and usually not tufted. See also *P. ponojensis*.

Peltigera scabrosa
Scabby pelt

DESCRIPTION: Thallus pale green to brownish, sometimes with a bluish or yellowish tint; lobes broad and round, 25–40 mm across, rather flat, conspicuously scabrose (i.e., with a crusty roughness on surface); lower surface predominantly dark, with a network of low, broad, often barely distinguishable dark brown veins and white, oval interspaces; rhizines dark brown, tufted and fibrous, up to 4 mm long. Apothecia infrequently seen. CHEMISTRY: Tenuiorin, methyl gyrophorate, gyrophoric acid. HABITAT: On mossy soil or rock, or on tree bases, usually in open sites. COMMENTS: The rough, crusty surface and broad, rounded lobes of *P. scabrosa* characterize this species. Other pelt lichens with a scabrose surface include *P. kristinssonii*, with well-defined veins; *P. neopolydactyla* (the form with a rough surface), with single rather than tufted rhizines; the rare **P. scabrosella,** which is smaller, with crisped lobe margins and discrete rhizines; and *P. collina,* which is sorediate. **Peltigera retifoveata** is tomentose rather than scabrose, but it is sometimes mistaken for *P. scabrosa* because of the appearance of its lower surface. Except for **P. didactyla var. extenuata,** a much smaller lichen, *P. retifoveata* is the only distinctly tomentose *Peltigera* that contains depsides such as gyrophoric acid and tenuiorin. It is a species of the western boreal region.

Peltigera venosa
Fan lichen

DESCRIPTION: Thallus dark gray-green when dry, deep green when wet. Photobiont green. Lobes fan-shaped to rounded, 10–15 mm wide; upper surface smooth, without tomentum or pruina; lower surface basically white but with raised black veins radiating from the base of each lobe in a striking pattern, bearing granular cephalodia (see Comments below); rhizines absent except for the single attachment point at the base of each lobe. Apothecia almost always present, round and flat, dark red-brown to black when dry, red-brown when wet. CHEMISTRY: Triterpenes (including zeorin) and tenuiorin. HABITAT: On bare mineral soil in moist, shaded nooks and crannies such as along the banks of creeks or roads, generally in regions with high rainfall; the wetter the climate, the more exposure it can tolerate. COMMENTS: With its fan-shaped green lobes, round black apothecia (red-brown when wet, as shown in plate 621), and very conspicuous black veins, this small *Peltigera* cannot be mistaken for anything else. As in all pelt lichens with green algae, *P. venosa* has cephalodia. In this species, the cephalodia are granules or tiny scales on the lower surface of the lobes among the veins, unlike the little galls formed on the upper surface in *P. aphthosa* and *P. leucophlebia*. The cephalodia can become detached and grow independently, as

620. *Peltigera scabrosa* mountains, coastal Alaska ×1.6

621. *Peltigera venosa*
northern Sierra
Nevada, California
×2.6

they do in *P. britannica*, but they remain extremely small, like minute lobes of some tiny *Leptogium*, and they seem to be restricted to the area under and immediately around the parent *P. venosa* thallus.

Peltula (14 N. Am. species)
Rock-olive lichens

DESCRIPTION: Dark olive to olive-brown squamulose to minutely fruticose lichens with flattened to erect, rather round lobes attached to the substrate at the center by a clump of rhizines or single holdfast. Photobiont blue-green (*Anacystis*). Apothecia sunken into the thallus lobes and level with the surface or slightly concave, opening to the surface by a narrow pore or with a broad red-brown disk (looking like a pimento in an olive); asci containing many tiny, colorless, 1-celled spores (up to 100 per ascus). CHEMISTRY: No lichen substances in the thallus, but in some species, the epihymenium gives a K+ red-violet reaction like that of yellow or orange anthraquinone pigments. HABITAT: On rock or soil, usually in dry sites. COMMENTS: *Peltula* superficially resembles the earth or rock scale lichens (*Psora*, *Placidium*, *Psorula*, all containing green algae) and *Heppia*. The latter also contains cyanobacteria, but from a different genus (*Scytonema*), and the asci contain only 8 spores.

KEY TO SPECIES (including *Heppia*)

1. Sorediate . . . . *Peltula euploca*
1. Not sorediate 2

2. Thallus with erect cylindrical lobes . . . *Peltula cylindrica*
2. Thallus squamulose 3

3. Spores ellipsoid, 8 per ascus . . . . *Heppia conchiloba*
3. Spores globose or broadly ellipsoid, many per ascus
   2

4. Spores broadly ellipsoid; growing directly on rock or soil . . . . *Peltula obscurans*
4. Spores globose; growing on soil. Squamules round; epihymenium turning red to violet or purple with K 5

5. Squamules with edges turned down, upper surface rough (rugose), lower surface blackish; apothecial margin usually absent . . . . *Peltula richardsii*
5. Squamules with edges turned up, upper surface smooth, lower surface pale brown; apothecial margin prominent and usually persistent . . . . *Peltula patellata*

622. *Peltula cylindrica* central North Carolina ×8.2

Peltula cylindrica
Cylindrical rock-olive

DESCRIPTION: Thallus consists of crowds of erect, olive-brown, cylindrical lobes, mostly under 2 mm tall, that expand into solid, somewhat spherical areoles, 0.3–1 mm in diameter, flattened at the summits, often having embedded apothecia, 1–3 per lobe, with tiny openings that resemble the ostioles of perithecia; spores globose, 3.5–4.5 μm in diameter. HABITAT: On exposed granitic outcrops along drainage paths and in shallow depressions subject to periodic wetting. COMMENTS: *Peltula tortuosa,* another fruticose *Peltula* that grows on granite, has more flattened, less cylindrical lobes, and the fertile branches are larger than the sterile ones. The lobes are the same size as in *P. cylindrica*. *Peltula clavata* is somewhat isidiate, and most of the globular lobes are hollow or have a very loose medulla. *Peltula zahlbruckneri* is also similar to *P. cylindrica* but has larger, slightly more elliptical spores (4.5–7.5 × 3–4.5 μm) and is largely confined to southern California. All other *Peltula* species grow flat against the substrate.

Peltula euploca
Powdery rock-olive

DESCRIPTION: Olive-brown, rounded to somewhat lobed, umbilicate squamules, 1.5–3(–10) mm in diameter, with the margins dissolved into very fine, gray soredia (with some laminal soralia as well). HABITAT: On noncalcareous rock in open and very dry, or shaded and somewhat damp, situations. COMMENTS: *Peltula bolanderi* is a smaller sorediate *Peltula*, with squamules mostly 1–2 mm in diameter and with more undulating margins. All other *Peltula* species lack soredia.

Peltula obscurans
Common rock-olive

DESCRIPTION: Thallus dark brown when dry, olive when wet, with round to abundantly lobed, umbilicate and somewhat raised squamules, convex or flat, 0.5–2 mm across. Apothecia disk-like,

623. (top left) *Peltula euploca* central Arizona ×3.2

624. (top right) *Peltula obscurans* var. *hassei* Coast Range, southern California ×6.2

625. (right) *Peltula obscurans* var. *hassei* (wet) Sonoran Desert, southern Arizona ×3.0

red-brown (especially when wet), immersed in the center of the squamules; spores broadly ellipsoid, (4–)6–8 × 3–5 μm. HABITAT: Mainly on rock, but also commonly on soil. COMMENTS: This variable species includes three named varieties in North America: *P. obscurans* var. *hassei* (plates 624 and 625 and map) has lobed squamules (the others are generally unlobed), and the epihymenium does not turn red-purple with K as it does in var. *obscurans;* in ***P. obscurans* var. *deserticola***, the apothecia fill the squamules and appear to have a lecanorine margin. The rock-dwelling *P. michoacanensis* has concave squamules. *Peltula richardsii* and *P. patellata* differ, having globose spores.

Peltula patellata (syn. *P. polyspora*)
Stuffed rock-olive

DESCRIPTION: Thallus dark brownish olive, consisting of rather thin, rounded to somewhat lobed squamules mostly 1–3 mm in diameter, usually with a raised margin or rim. Apothecia orange-tan to dark red, 0.2–1.3 mm in diameter, 1 to several per squamule, immersed but with a thin, prominent margin, or sometimes emerging and almost sessile; epihymenium K+ reddish violet; spores spherical, mostly 5–7 μm in diameter. HABITAT: On soil in open, arid sites, widespread in the west. COMMENTS: This species and *P. euploca* are the only two species of *Peltula* that reach as far north as Canada. All the others, except *P. tortuosa* and *P. cylindrica,* are strictly southwestern.

Peltula richardsii
Giant rock-olive

DESCRIPTION: Squamules brownish olive, very large, 2–6(–10) mm in diameter, round, mostly concave to flat, smooth and shiny or rough and fissured, downrolled at the margin. Apothecia large, orange to red-brown, often almost filling the squamule, 1 per squamule; epihymenium K+ red-violet; spores spherical, 6.5–8.5 mm. HABITAT: On limy soil in dry sites. COMMENTS: Because of its size, this *Peltula* looks a lot like *Heppia conchiloba* but has many spores per ascus, and the epihymenium reacts with K.

Pertusaria (75 N. Am. species)
Wart lichens

DESCRIPTION: A large and diverse genus of crustose lichens with thick to very thin thalli. Photobiont green (*Trebouxia*). Fruiting bodies are modified apothecia buried in thallus warts, opening by one or more ostiole-like pores, or broadly open and appearing like ordinary lecanorine apothecia, sometimes breaking down at the summit or rim and appearing coarsely sorediate; thalline warts with vertical or sloping sides, crowded or scattered; epihymenium brown or greenish to pale, sometimes K+ purple; paraphyses very slender, branched, and anastomosing; asci cylindrical, with IKI+ dark blue walls; spores colorless or rarely light brown, 1-celled, large (up to 300 μm long), usually thick-walled, often with clearly distinguishable layers, with the inner wall "trimmed" or squared at the ends (Fig. 15l), 1–8 spores per ascus. CHEMISTRY: Contains a variety of depsides and depsidones in the medulla and often has xanthones in the cortex. HABITAT: On bark, wood, rock, moss, or soil from the Arctic to the Tropics. COMMENTS: The large spores, with their thick and sometimes layered walls, and the unique fruiting warts set this genus apart. Species of

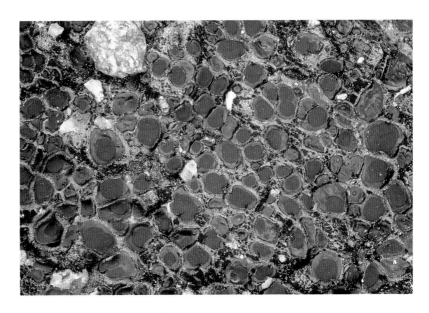

626. *Peltula patellata* Sonoran Desert, southern Arizona ×3.0

627. *Peltula richardsii* Sonoran Desert, southern Arizona ×2.2

Pertusaria with sunken apothecia opening by a pore or ostiole can be roughly divided into two groups: those with fruiting verrucae that have steep, almost vertical sides (the mesa warts, technically called pertusariate) as in *P. subpertusa,* and those with sloping, volcano-like sides (called ampliariate) as in *P. xanthodes*. However, as in most growth-form classifications, there are the inevitable intermediates. Some species of *Ochrolechia* resemble *Pertusaria,* but they have thin-walled spores and thick-walled asci and lack β-orcinol depsidones. *Phlyctis* has low, indistinct and often hidden fruiting warts with asci containing single, muriform spores. **Varicellaria** is also similar but it has huge, 2-celled spores (up to 400 μm long), strongly constricted in the middle.

KEY TO SPECIES

1. Apothecia buried in verrucae with straight or sloping sides, or that are slightly constricted at the base, having small ostiole-like openings 2
1. Apothecia lecanorine, with narrow or broad disks, or becoming partly or entirely sorediate or pruinose at the summit 16

2. Spores (4–)8 per ascus 3
2. Spores 2–4 per ascus 6

3. Thallus yellowish green or sometimes yellow-gray; cortex C+ yellow-orange, KC+ orange, UV+ dark orange (thiophanic acid) 4
3. Thallus greenish gray, not yellowish; cortex C–, KC–, UV– or UV+ yellow 5

4. Ostiole area depressed, dark brown to black, often pruinose, C–; epihymenium K+ violet; medulla K+ red (norstictic acid); Appalachian–Great Lakes distribution . . . . [*Pertusaria rubefacta*]
4. Ostiole area raised and yellow, C+ orange; epihymenium K–; medulla K+ yellow (stictic acid); southeastern . . . . *Pertusaria texana*

5. Spores lined up in a single row within the ascus; thallus and verrucae UV+ yellow (lichexanthone), smooth or rough and warty; ostioles often with a whitish border; throughout the southeast . . . . *Pertusaria paratuberculifera*
5. Spores in two irregular rows within the ascus; thallus and verrucae UV– or orange-pink, rather smooth; ostioles dot-like; Appalachian–Ozark distribution. . . . . [*Pertusaria ostiolata*]

6.(2) Growing on bark, peat, or soil 7
6. Growing directly on rock 15

7. On soil or peat, mostly arctic-alpine (rarely on hardwood bark on east coast). Spores 2 per ascus; abundant black ostioles in separate or frequently confluent verrucae; epihymenium K+ violet; medulla K+ yellow (stictic acid) . . . . [*Pertusaria subobducens*]
7. On bark 8

8. Spores usually 4 per ascus 9
8. Spores 2(–3) per ascus 11

9. Epihymenium K+ violet, fertile verrucae usually with steep sides; southeastern coastal plain . . . . [*Pertusaria sinusmexicani*]
9. Epihymenium K–; fertile verrucae typically with sloping sides 10

10. Spores 75–130 μm long, with rough inner walls; East Temperate . . . . [*Pertusaria tetrathalamia*]
10. Spores 50–100 μm long, with smooth inner walls; East Temperate and west coast . . . . [*Pertusaria leioplaca* (syn. *P. leucostoma*)]

11.(8) Thallus pale greenish yellow to yellow-gray or greenish gray; cortex C+ yellow-orange, KC+ orange, UV+ orange-red; thallus medulla PD+ orange, K+ persistently yellow (stictic acid) 12
11. Thallus gray; cortex C–, KC–, UV– or UV+ yellow or pink 13

12. Inner spore walls rough and grooved when mature; ostioles pale brown, typically separate; mainly on coastal plain . . . . *Pertusaria xanthodes*
12. Inner spore walls smooth; ostioles black, sometimes clustered or fused; East Temperate . . . . [*Pertusaria pustulata*]

13. Ostioles black; epihymenium K+ violet. Medulla K+ yellow, PD+ orange (stictic acid); Great Lakes to Maritime Provinces . . . . [*Pertusaria consocians*]
13. Ostioles pale to dark, sometimes black; epihymenium K– 14

14. Inner wall of spores with radiating channels; spores sometimes brownish and K+ dull violet; medulla K+ yellow, PD+ orange (stictic acid); Appalachian–Great Lakes distribution . . . . *Pertusaria macounii*
14. Inner wall of spores smooth, not channeled, always colorless and K–; medulla PD+ red, K– or brownish (fumarprotocetraric acid); southeastern . . . . *Pertusaria subpertusa*

15.(6) Thallus gray, thin; spores narrowly ellipsoid (75–)85–150(–245) × 25–70 μm, with rough inner walls; thallus cortex C–, KC–; medulla K+ red (norstictic acid); eastern North America . . . . *Pertusaria plittiana*
15. Thallus sulphur yellow, thick; spores 60–90 × 30–50 μm, with smooth walls; thallus cortex C+ orange, KC+ orange (thiophanic acid); medulla K–; southern California . . . . *Pertusaria flavicunda*

16.(1) Growing on mosses or dead vegetation 17
16. Growing on bark or rock 18

17. Thallus white to bluish gray, verrucose, but verrucae not resembling isidia; black, pruinose apothecia immersed in the verrucae; all spot tests negative. . . . . *Pertusaria panyrga*
17. Thallus composed of white, cylindrical isidia with black, sometimes lightly pruinose apothecia at the summits; thallus cortex and medulla PD+ red, K+ brownish (fumarprotocetraric acid) . . . . *Pertusaria dactylina*

18. On rock; thallus sulphur yellow, cortex C+ yellow-orange; spores 2(–4) per ascus (see couplet 15) . . . . *Pertusaria flavicunda*
18. Usually on bark; thallus gray to yellowish green, not sulphur yellow, cortex C–, C+ yellow-orange, or C+ red; spores 1–8 per ascus. 19

19. Thallus UV+ yellow (lichexanthone); spores 1(–2) per ascus, but often absent; eastern. 20
19. Thallus UV–; spores 1, 2, or 8 per ascus 22

20. Verrucae becoming sorediate at the summit; medulla KC+ pink becoming violet, PD– (hypothamnolic acid); throughout southeast . . . . [*Pertusaria hypothamnolica*]
20. Verrucae not becoming sorediate, summit small or broad and lecanorine, white pruinose 21

21. Medulla C–, KC–, PD+ yellow to orange (haemathamnolic acid); verrucae somewhat raised, not expanded; southern coastal plain. . . . . [*Pertusaria copiosa*]
21. Medulla C+ red, KC+ red, PD– (lecanoric acid); verrucae usually expanded into lecanorine apothecia; East Temperate . . . . *Pertusaria velata*

22.(19) Apothecial margins smooth; disk pinkish orange, heavily pruinose but not at all sorediate; thallus cortex and medulla C+ red, KC+ red (lecanoric acid). . . . . *Pertusaria velata*
22. Apothecial margins or the entire verrucae rough or sorediate; thallus cortex C–, medulla KC– or KC+ purple (lecanoric acid absent) 23

23. Soralia or sorediate verrucae KC+ purple (picrolichenic acid); spores 1 per ascus but rarely found . . . . *Pertusaria amara*
23. Soralia (if present), verrucae, and thallus medulla KC–; spores 1–8 per ascus 24

24. Thallus medulla K+ bright yellow, PD+ orange (thamnolic acid); spores 2 per ascus but sometimes hard to find. . . . . *Pertusaria trachythallina*
24. Thallus medulla K–, PD– or PD+ red; spores 1 or 8 per ascus. 25

628. *Pertusaria amara* Lake Superior region, Ontario ×2.7

25. Thallus medulla PD–, K–. Spores 1 per ascus; widely distributed . . . . *Pertusaria ophthalmiza*
25. Thallus medulla PD+ red, K+ brown (fumarprotocetraric acid) 26

26. Apothecial disks usually distinct, various colors from yellow to green or black, usually pruinose; apothecial margins ragged, often in concentric layers; spores 8 per ascus; Pacific Northwest . . . . . . *Pertusaria subambigens*
26. Apothecial disks usually obscured by soredia and aborted, rarely black and pruinose with sorediate margins; spores 1 per ascus; Appalachian–Great Lakes region . . . . [*Pertusaria multipunctoides*]

Pertusaria amara
Bitter wart lichen

DESCRIPTION: Thallus dark gray or greenish gray, often dotted with white pseudocyphellae near the margin (as shown in plate 628), usually rather thick with a clearly delimited border, smooth and shiny with large soralia (actually, fruiting warts that have become sorediate and are rarely fertile) con-

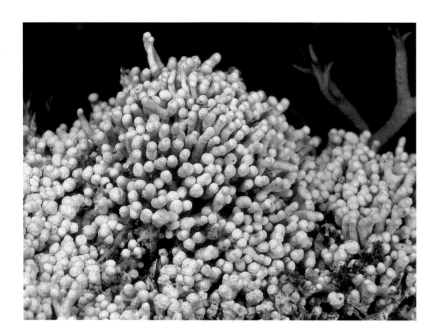

629. *Pertusaria dactylina* (sterile) interior Alaska ×3.4

taining coarse, granular, snow-white soredia that contrast sharply with the thallus. CHEMISTRY: Soralia K–, KC+ purple, C– (picrolichenic acid). *Pertusaria amara* has two chemical races: one PD+ red (protocetraric acid), the other PD–. Both races are equally common along the west coast; in eastern North America, the PD+ race is rare and is found mostly near the Great Lakes. HABITAT: Usually on the bark of hardwoods, also on conifers or rocks. COMMENTS: The bitter wart lichen is aptly named (both in Latin and English). A single sorediate bit, when chewed and allowed to remain in the mouth for a few seconds, produces a very bitter taste that lasts a lot longer than one would like. It is an excellent identifying character if you lack the KC reagent that makes the picrolichenic acid turn a bright purple. Fortunately, the substance is not poisonous! A well-developed dark specimen with pseudocyphellae is not hard to identify, even without a taste test, but paler specimens can be mistaken for a number of sterile, sorediate crustose lichens. One generally has to resort to chemistry to sort them out. See Key C, Sterile Crusts. IMPORTANCE: The bitter taste of *P. amara* (like quinine) apparently was the basis for its use in controlling high fever.

Pertusaria dactylina
Finger wart lichen

DESCRIPTION: Thallus consisting of white, coral-like isidia, 0.3–0.8 mm thick, 0.5–2(–7) mm long, constricted at the bases, forming small, dense patches. Apothecia frequently produced at the tips of the isidia-like stalks, with pale to dark brown or black, flat to slightly convex, lightly pruinose disks; spores 150–340 × (45–)60–90(–120) μm, 1 per ascus. The specimen shown in plate 629 has no apothecia. CHEMISTRY: Cortex PD–, K–, KC–, C–, UV–; medulla PD+ red, K+ brownish, KC–, C– (fumarprotocetraric acid). HABITAT: Growing on arctic or alpine turf. COMMENTS: This lichen is characterized by its erect, thick, white isidia and PD+ red reaction. *Pertusaria panyrga*, another common arctic-alpine lichen, has a whitish thallus with somewhat tall warts and broad, black, pruinose apothecial disks reminiscent of *P. dactylina*, but the thallus never has narrow, columnar isidia and is PD–.

Pertusaria flavicunda
Sulphur wart lichen

DESCRIPTION: Thallus sulphur-yellow, areolate, fairly thick, occasionally with patches of soredia. Apothecia with broad lecanorine disks, black with a coarse yellowish pruina, embedded in crowded angular areoles; spores with very thick smooth walls, mostly 60–90 × 30–50 μm, 2(–3) per ascus. CHEMISTRY: Cortex PD–, K– or K+ yellow; KC+ distinct orange, C+ pale orange, UV+ orange-red (thiophaninic acid); medulla PD– or PD+ faint orange, K– or K+ faint yellow, KC– or KC+ orange in spots, C– (small quantities of stictic acid with a sporadic carry-over of thiophaninic acid). HABITAT: On rocks, especially sandstone, on exposed sites usually not far from the ocean. COMMENTS: This yellow sorediate *Pertusaria* with crowded lecanorine apothecia can hardly be mistaken for anything else. *Pertusaria arizonica* has a similar substrate, chemistry, and color, but fertile specimens have rounded fruiting warts with very small disk-like openings and 4-spored asci. This species never produces soredia and is restricted to Arizona and Texas.

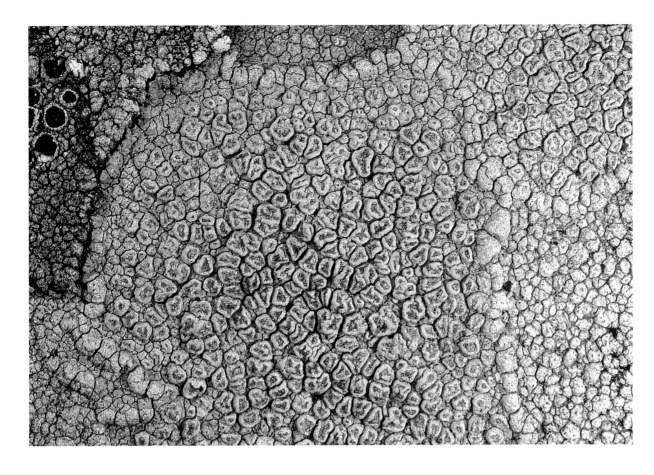

630. *Pertusaria flavicunda* Channel Islands, southern California ×4.4

Pertusaria macounii
Macoun's wart lichen

DESCRIPTION: Thallus shades of gray, thin or thick and verruculose. Fruiting warts steep-sided (mesa-like), containing several apothecia revealed by 2–4(–7) rather pale ostioles per wart; epihymenium pale or dark, K–; spores 2 per ascus, mostly 100–200 × 30–65 µm, with double walls, the inner one with radiating channels; spores occasionally brownish and then turning K+ dull purple. CHEMISTRY: Cortex UV– or UV+ pink-orange (2,7-dichloroxanthone); medulla PD+ orange, K+ yellow, KC–, C– (stictic acid complex). HABITAT: On the bark of hardwoods, rarely conifers. COMMENTS: This is the most common of the 2-spored mesa wart lichens. Its occasionally brownish, K+ purple spores are rather distinctive within the genus. Related species, several of which also have ornamented and sometimes radiately grooved spore walls, are mentioned under *P. subpertusa*.

631. *Pertusaria macounii* Lake Superior region, Ontario ×3.3

632. *Pertusaria ophthalmiza* Lake Superior region, Ontario ×3.9

Pertusaria ophthalmiza
Ragged wart lichen

DESCRIPTION: Thallus usually thin and smooth, pale to dark gray. Fruiting warts with broad, black, coarsely pruinose disks, usually 1 per wart, with margins that are soon torn and fragmented giving the warts a sorediate appearance; epihymenium K–; spores 75–200 × 30–70 μm, 1 per ascus. CHEMISTRY: All reactions negative (no lichen substances). HABITAT: On bark of many kinds. COMMENTS: This is among the most common and widespread of a group of *Pertusaria* species on bark that have ragged, almost sorediate fruiting warts and 1 spore per ascus. *Pertusaria ophthalmiza* is the only member of the group that has entirely negative spot tests. Most similar are *P. amara*, with granular soredia on the fruiting warts, which are KC+ purple, *P. subambigens* (8-spored, PD+ red), and *P. trachythallina* (2-spored, PD+ deep yellow, K+ yellow), all illustrated and described here. ***Pertusaria multipunctoides*** is a similar eastern species with small fruiting warts that react PD+ red (fumarprotocetraric and succinoprotocetraric acids).

Pertusaria panyrga
Arctic wart lichen

DESCRIPTION: Thallus white to bluish gray, with hemispherical to tall, cylindrical fruiting warts, 0.4–1 mm in diameter. Apothecia black with a gray pruina, produced at the summits of the erect warts and rarely broader than their diameter; epihymenium K–; spores mostly 150–200 × 40–85 μm, 1 per ascus. CHEMISTRY: All reactions negative (no lichen substances). HABITAT: Common on dead mosses and vegetation in arctic and alpine tundra. COMMENTS: Some arctic-alpine forms of *Megaspora verrucosa* resemble *P. panyrga*, but that species has 8 spores per ascus. See also *P. dactylina* and *P. ophthalmiza*.

Pertusaria paratuberculifera
Spotted wart lichen

DESCRIPTION: Thallus pale greenish gray, often shiny, rather thick and irregular, spotted with white maculae; large-fruited (fertile warts 1–2.5 mm in diameter) with one to several ostioles per wart; asci narrowly cylindrical, containing 8 spores lined up in a single row; spores 50–130 × 25–45 μm with smooth walls. CHEMISTRY: Cortex UV+ bright yellow (lichexanthone); all other reactions negative (2'-*O*-methylperlatolic acid and sometimes planaic acid). HABITAT: On the bark of deciduous trees, rarely conifers and rocks. COMMENTS: The large, crowded warts, 8-spored asci, and negative chemical reactions (except the UV+ thallus) characterize this common coastal plain species. The somewhat similar *P. tetrathalamia* has 4 spores per ascus, and the thallus is UV–.

Pertusaria plittiana
Rock wart lichen

DESCRIPTION: Thallus gray, thin or thick, cracked into areoles. Fruiting warts usually flat on top with vertical or sloping sides, 0.4–2.5 mm in diameter, with several white-bordered ostioles. Spores 2 per ascus but often aborted, (75–)85–150(–245) × 25–70 μm, with roughened, double walls. CHEMISTRY: Cortex PD–, K–, KC–, C–, UV–; medulla PD+ yellow, K+ red, KC–, C– (norstictic, perlatolic,

633. (left) *Pertusaria panyrga* interior Alaska ×4.4

634. (below left) *Pertusaria paratuberculifera* Cape Cod, Massachusetts ×4.4

635. (below right) *Pertusaria plittiana* (with *Ochrolechia yasudae* on far right) Appalachians, Virginia ×3.0

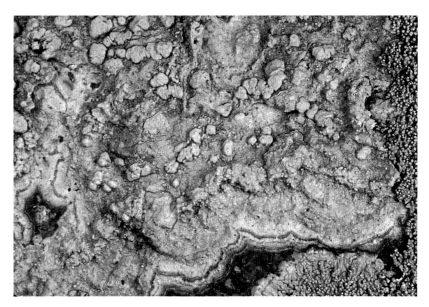

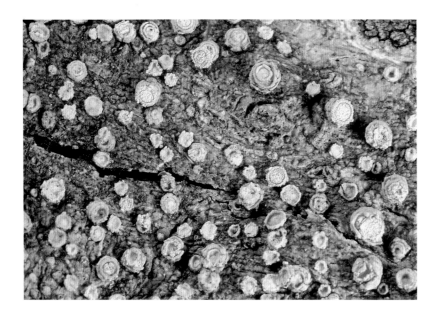

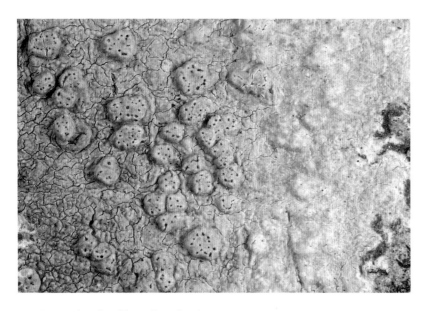

636. *Pertusaria subambigens* Cascades, Oregon ×4.4

637. *Pertusaria subpertusa* Cape Cod, Massachusetts ×5.4

and stenosporic acids), rarely PD–, K– in specimens that have very little norstictic acid. HABITAT: On siliceous rocks, often in shaded forests, where it can be very abundant. COMMENTS: This is the only fertile, rock-dwelling *Pertusaria* in the east. It resembles the corticolous eastern species, *P. neoscotica,* both in appearance and chemistry, although *P. neoscotica* contains planaic acid rather than perlatolic and stenosporic acids in addition to norstictic acid. Specimens of *P. plittii* at the southern edge of the range often have little norstictic acid and do not give positive spot tests.

Pertusaria subambigens
Frosted wart lichen

DESCRIPTION: Thallus ashy gray to greenish gray, shiny, fairly thick, delimited at the margin. Fruiting warts 0.4–3 mm in diameter, each containing a single apothecium having a broad, heavily pruinose disk that can be pale yellowish, pink, greenish, or black, surrounded by a rough margin consisting of concentric rings of tissue; epihymenium K–; spores 9–28 × 5–18 µm, 8 per ascus in a single row. CHEMISTRY: Cortex UV–; apothecial disk and medulla PD+ red, K– or brownish, KC–, C– (fumarprotocetraric and succinoprotocetraric acids). HABITAT: On bark, mainly conifers. COMMENTS: This North American endemic is a common bark lichen along the west coast and is the only one with broad, heavily pruinose disks and ragged, concentrically arranged marginal tissue. The PD+ red reaction of the disk and margin is a confirming character. The somewhat similar but smaller *P. ophthalmiza* gives no positive spot tests.

Pertusaria subpertusa
Mesa wart lichen

DESCRIPTION: Thallus thin to moderately thick, smooth to rough, pale gray to yellowish gray, with scattered or sometimes clustered, steep-sided or basally constricted, flat-topped, mesa-shaped fruiting warts. Fruiting warts mostly 0.4–2 mm in diameter, each containing several apothecia that open to the surface by means of pale to dark ostioles that are not sunken; epihymenium K– or light reddish violet; spores 170–350 × 40–65 µm with thick, layered walls,

170–350 × 40–65 μm with thick, layered walls, 2 spores per ascus. CHEMISTRY: Cortex UV– or UV+ dull orange; medulla PD+ red, K+ yellow, becoming deep red-brown, C–, KC– (fumarprotocetraric and protocetraric acids). HABITAT: On the bark of hardwoods in fields or forests. COMMENTS: *Pertusaria subpertusa*, although usually having steep-sided fruiting warts, can also have some warts with sloping sides. No other *Pertusaria* with apothecia embedded in warts contains fumarprotocetraric and protocetraric acids. There are several North American endemic, mesa wart lichens in the east with 2-spored asci. The most common are *P. macounii*, **P. consocians**, and *P. plittiana* (all distinguished in the key). **Pertusaria neoscotica** is a rarer bark lichen in this group and contains norstictic acid.

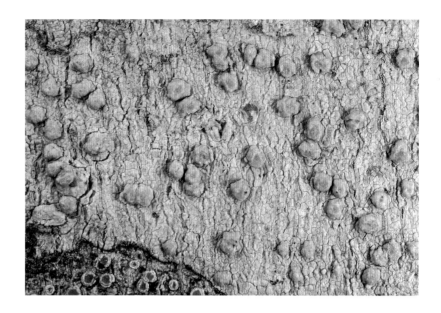

Pertusaria texana
Texas wart lichen

DESCRIPTION: Thallus thin to moderately thick, pale yellowish green or less commonly yellow-gray, smooth or slightly rough; fruiting warts with sloping or straight sides, or even constricted at the base (shown in plate 638), containing 1 to several apothecia with lemon-yellow, somewhat raised ostioles; epihymenium usually K–; spores mostly 30–90 × 20–45 μm, (4–)8 per ascus, usually not arranged in a single row. CHEMISTRY: Cortex and ostioles PD–, K+ yellowish, KC+ yellow-orange, C+ deep yellow to orange, UV+ orange-red (thiophanic acid); medulla PD+ orange, K+ deep yellow (stictic acid complex) or less frequently PD–, K– (without detectable stictic acid). HABITAT: On the bark of hardwoods, rarely on conifers or rock. COMMENTS: *Pertusaria xanthodes*, a common coastal plain species on trees, is identical chemically but has smaller, smoother, more uniformly sloping fruiting warts and 2 spores per ascus; the ostioles are pale and C–. **Pertusaria epixantha** is a rare species from Florida that resembles *P. texana* but usually contains variolaric acid (producing microscopic needles in K) instead of stictic acid, and the ostioles are depressed rather than raised.

Pertusaria trachythallina
Powdered wart lichen

DESCRIPTION: Thallus dark gray to almost white, thin to quite thick, shiny or dull, smooth (especially at margins) or becoming rough (verrucose). Apothecia disklike, mostly 1 per wart, breaking open unevenly leaving a ragged margin, disks black beneath a heavy white pruina, often appearing sorediate; epihymenium K–; spores mostly 50–150 × 20–45 μm, asci containing 2 spores but often immature. CHEMISTRY: Thallus UV–; cortex and fruiting warts PD+ orange, K+ bright yellow, KC–, C– (thamnolic acid). HABITAT: Mostly on bark of deciduous trees, rarely conifers. COMMENTS: This lichen can be distinguished from *P. ophthalmiza* and similar lichens with disk-like, sorediate warts by its chemistry and its 2-spored asci. The apothecia of *P. trachythallina* often appear whiter and more sorediate than those in the specimen illustrated in plate 639, which has newly expanded disks.

638. *Pertusaria texana* coastal plain, North Carolina ×4.2

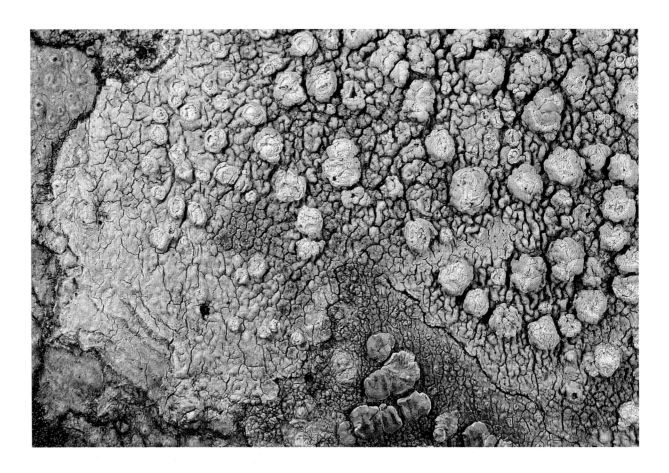

639. *Pertusaria trachythallina* Cape Cod, Massachusetts ×6.0

640. *Pertusaria velata* Appalachians, Tennessee ×4.4

Pertusaria velata
Rimmed wart lichen

DESCRIPTION: Thallus pale gray to yellowish gray, thick and verrucose in the center but becoming smooth and thin at the margins. Apothecia *Lecanora*-like, mostly 0.5–1 mm in diameter, with thick, prominent margins and a pinkish or pale orange disk thickly covered with a coarse white pruina; epihymenium K–; spores 150–300 × 35–95 μm, 1 per ascus. CHEMISTRY: Cortex PD–, K–, C–, KC–, UV– or UV+ yellow (sometimes with lichexanthone); medulla and apothecial disk PD–, K–, KC+ red, C+ deep red, UV– (lecanoric acid). HABITAT: On the bark of either hardwoods or conifers and on mossy rock, usually in shaded, mature forests, mostly sugar maple stands in the northern part of its range and oak-hickory woods to the south. COMMENTS: With its broad, pink, pruinose disks and C+ red reaction, this lichen resembles a species of *Ochrolechia*, for example, *O. trochophora*. However, the huge, thick-walled spores, 1 per ascus, immediately establish the identity of *P. velata*. It is a good indicator of rich, relatively undisturbed forests, often in the company of *Lobaria* species, *Bacidia schweinitzii,* and *Pertusaria amara.*

Pertusaria xanthodes
Volcano wart lichen

DESCRIPTION: Thallus pale greenish yellow to yellow-gray, thin to moderately thick, smooth or rough and cracked, continuous. Fruiting warts typically with sloping sides, sometimes fusing with one another, with pale ostioles at the summits, rarely with a fairly broad opening; epihymenium K–; spores mostly 50–150 × 30–50 μm with thick, layered walls, the inner wall being conspicuously roughened; 2 spores per ascus. CHEMISTRY: Cortex KC+ orange, UV+ orange (thiophanic acid); medulla PD+ orange, K+ yellow (stictic acid complex). HABITAT: On bark of deciduous trees in well-lit forests or in the open. COMMENTS: Of the many other *Pertusaria* species with warts that have sloping sides and 2 spores per ascus, *P. pustulata* is the most similar. However, its thallus is rarely yellow, it has dark ostioles (usually showing a K+ violet epihymenium), and it lacks the roughened spore walls. *Pertusaria texana* tends to have flatter fruiting warts and yellower ostioles.

Phaeographina (4 N. Am. species)
Brick-spored script lichens

DESCRIPTION: Crustose lichens with a thin thallus often developing within bark tissue. Photobiont green (*Trentepohlia*). Apothecia elongate lirellae, often curved and branched; exciple partly or entirely carbonized or colorless; spores muriform or partially muriform, brown to olive. CHEMISTRY: Most species have no lichen substances, although a few have norstictic acid and are K+ red. HABITAT: All species on bark in tropical to subtropical parts of the continent. COMMENTS: *Phaeographis* is a sister genus that has transversely septate rather than muriform spores. *Graphis* and *Graphina* are related genera with colorless spores. However, the young spores of *Phaeographina* can also be colorless.

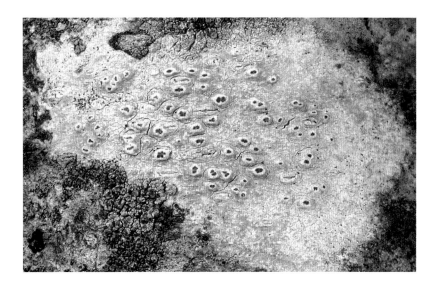

KEY TO SPECIES
1. Lirellae branched; walls of lirellae entirely carbonized . . . . *Phaeographina quassiaecola*
1. Lirellae unbranched; walls of lirellae partially carbonized . . . . *Phaeographina caesiopruinosa*

641. *Pertusaria xanthodes* Cape Cod, Massachusetts ×4.5

Phaeographina caesiopruinosa

DESCRIPTION: Thallus thin, entirely within the bark but perceptible as a yellowish olive or green stain. Lirellae breaking through the bark tissues as an expanding fissure, partly immersed in the thallus or raised above it, commonly 5–10 mm long and up to 0.5 mm wide, bluish pruinose, bordered by a thin black margin; at least upper part of exciple carbon black; hymenium filled with tiny oil drops. Spores dark olive, multiseptate-muriform, 52–75 × 15–23 μm, 8 per ascus. CHEMISTRY: No lichen substances. HABITAT: On hardwood bark.

Phaeographina quassiaecola

DESCRIPTION: Thallus yellow-olive, thin, smooth, with the tissues within the bark. Lirellae strongly raised, emerging from splits in the thallus; walls of the lirellae carbonized, but covered with a thin whitish film; hymenium containing abundant oil droplets; spores 92–110 × 19–27 μm, colorless to brown when mature, muri-

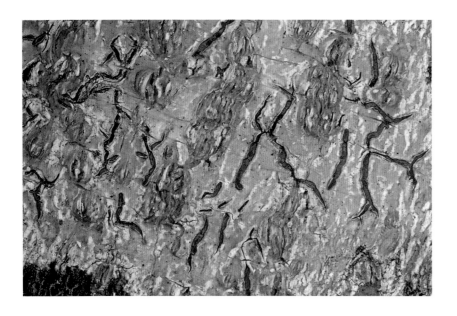

642. *Phaeographina caesiopruinosa* northern Florida ×2.9

643. *Phaeographina quassiaecola* southern Louisiana ×2.4

form and many-celled, 8 per ascus. CHEMISTRY: No lichen substances. HABITAT: On bark of hardwoods. COMMENTS: Few other script lichens have such prominent lirellae.

Phaeographis (14 N. Am. species)
Dark-spored script lichens

DESCRIPTION: Crustose lichens with a thin thallus (often within the bark). Photobiont green (*Trentepohlia*). Apothecia rather broad lirellae; exciple carbonized or colorless; spores brownish with 4–10 lens-shaped locules per spore, 8 per ascus. CHEMISTRY: Some species have lichexanthone in the cortex (UV+ yellow); some have norstictic or stictic acids (PD+ yellow, K+ red, KC–, C–). HABITAT: On bark in temperate and tropical regions. COMMENTS: *Phaeographis* is similar to *Graphis*, but with pigmented spores. *Sarcographa* is a closely related genus that intergrades with *Phaeographis;* it has short lirellae embedded in a pseudostroma.

KEY TO GENUS: See Key E.

Phaeographis inusta

DESCRIPTION: Thallus almost entirely within the bark, appearing as a dark to pale olive-brown stain. Lirellae broad (0.2–0.3 mm wide, 1–2 mm long), radiate, star-like; disk brown to gray (if lightly pruinose); margins not at all prominent (i.e., not visible from surface); hymenium containing oil droplets. Spores 4–6- or rarely 8-celled, pale grayish brown. CHEMISTRY: All reactions negative (no lichen substances). HABITAT: On the bark of deciduous trees and shrubs. COMMENTS: In the east, specimens of *Graphis scripta* with unusually broad lirellae can easily be mistaken for this species. The pigmented spores (usually with only 4 or 5 cells) and the oil droplets in the hymenium are the best characters for distinguishing them. In addition, the spores of *Graphis* turn IKI+ violet; those of *Phaeographis* are IKI– or only weakly purplish.

644. *Phaeographis inusta* coastal New Jersey ×12.2

Phaeophyscia (20 N. Am. species)
Shadow lichens

DESCRIPTION: Rather small foliose lichens with olive to brown or, less frequently, gray thalli that have radiating lobes mostly under 1.5 mm across, a black lower surface (except in *Ph. constipata*), and a lower cortex composed of pseudoparenchyma. Photobiont green (*Trebouxia?*). Apothecia dark brown to almost black, often producing a fringe of rhizines on the lower part of the margins; spores dark brown with thickened walls creating angular or round locules; conidia ellipsoid. CHEMISTRY: Cortex PD–, K–, KC–, C–; medulla also negative with all reagents unless it is pigmented orange or red with anthraquinones giving a K+ red-purple reaction. Thallus either without substances or with zeorin, fatty acids, or orange to yellow anthraquinone pigments; cortex lacking atranorin. HABITAT: On bark, wood, or rocks of all types. COMMENTS: The closely related genera *Physcia*, *Physconia*, and *Physciella* can often be confused with *Phaeophyscia*. Species of *Physcia* all contain atranorin in the upper cortex (K+ yellow) and, except for a few subtropical species, lack rhizines on the apothecial margins. The lower surface is usually white to pale brown but can be black in a few species. *Physconia* can be dark brown like *Phaeophyscia* and the cortex is also K–. However, the lobe tips are almost always heavily pruinose, and the rhizines are branched. *Physciella* is intermediate between *Physcia* and *Phaeophyscia*; it has a pale lower surface with a lower cortex composed of prosoplectenchyma as in most species of *Physcia* but with a K– upper cortex and short, ellipsoid conidia as in *Phaeophyscia*.

KEY TO SPECIES (including *Hyperphyscia* and *Physciella*)

1. Soredia absent. 2
1. Soredia present. 7

2. Rhizines absent; thallus extremely closely appressed, almost crustose (but with a lower cortex) . . . . *Hyperphyscia syncolla*
2. Rhizines present, sparse or abundant; thallus easily detached from substrate, clearly foliose 3

3. Thallus very loosely attached, growing over soil and moss; lower surface white to pale brown; rhizines sparse, white to very pale brown. . . . . *Phaeophyscia constipata*
3. Thallus closely to somewhat loosely attached, growing on bark or rock; lower surface black; rhizines abundant, black. 4

4. Lobes 1–3(–4) mm across; cilia-like rhizines long. . . . . *Phaeophyscia hispidula*
4. Lobes 0.2–1 mm across; cilia-like rhizines very short or absent. 5

5. Growing on bark, moss, or wood 6
5. Growing on rock . . . . *Phaeophyscia decolor*

6. Erect lobules abundant on lobe margins . . . . [*Phaeophyscia squarrosa*]
6. Lobules absent. . . . . *Phaeophyscia ciliata*

7.(1) Rhizines absent; thallus attached directly to substrate, almost crustose . . . . *Hyperphyscia adglutinata*
7. Rhizines sparse or abundant; thallus attached by rhizines, clearly foliose 8

8. Lower surface and rhizines white to pale brown. 9
8. Lower surface and rhizines black 10

9. Cortex K–, atranorin absent; rhizines abundant; conidia ellipsoid. . . . . *Physciella chloantha*
9. Cortex K+ yellow, atranorin present; rhizines sparse; conidia rod-shaped . . . . *Physcia dubia*

10. Medulla red-orange, sometimes only in spots . . . . *Phaeophyscia rubropulchra*
10. Medulla white 11

11. Stiff, tiny, colorless hairs on the upper surface of the lobes . . . . *Phaeophyscia cernohorskyi*
11. Stiff, tiny, colorless hairs absent. 12

12. Soredia black, very coarse (almost isidia), along the lobe margins; on rock. . . . . *Phaeophyscia sciastra*
12. Soredia greenish, granular to farinose, on the margins and tips of the lobes; on bark or rock 13

13. Lobes 1–3(–4) mm across; cilia-like rhizines long. . . . . *Phaeophyscia hispidula*
13. Lobes 0.5–1.5(–2) mm across; cilia-like rhizines very short. 14

14. Soredia on the upper surface of the lobe tips, or on expanded, turned-back lobe tips (lip-shaped soralia). . . 15
14. Soredia on the lobe margins, close to the margins, or on thallus surface. 16

15. Soredia farinose, forming small hemispherical greenish mounds on the upper surface of the lobe tips; lobes mostly less than 0.7 mm wide; on bark, or rarely on calcareous rock (e.g., limestone) . . . . *Phaeophyscia pusilloides*
15. Soredia granular, usually in lip-shaped soralia at the lobe tips; lobes 0.5–1(–2) mm wide; on siliceous rock (e.g., granite), or occasionally on bark. . . . . *Phaeophyscia adiastola*

16. Soredia farinose, mainly on lobe surface in depressed patches, but some marginal ... *Phaeophyscia orbicularis*
16. Soredia granular, mainly on lobe tips, occasionally on surface in mounds .... *Phaeophyscia adiastola*

Phaeophyscia adiastola
Powder-tipped shadow lichen

DESCRIPTION: Thallus dark greenish gray, gray-brown, or brown, lobes 0.5–1(–2) mm wide, with very coarse, almost isidia-like soredia along the margins and at the tips of the lobes, rarely also on the upper surface; lobules occasionally produced at the lobe tips as well; lower surface black with abundant and conspicuous black rhizines with white tips. Apothecia rare. CHEMISTRY: No lichen substances. HABITAT: Quite common on shaded, mossy granitic rocks as well as on bark of different kinds, especially hardwoods. COMMENTS: *Phaeophyscia rubropulchra* can also have coarse soredia but has a red medulla. In *Ph. orbicularis*, the soredia are mainly laminal, and the species is less common in the east. Small specimens of *Ph. hispidula* can be similar but lack lobules, and the soredia are usually laminal, not marginal.

Phaeophyscia cernohorskyi
Hairy shadow lichen

DESCRIPTION: Thallus dark grayish brown, usually blotched with conspicuous maculae; lobes short and narrow, mostly 0.5–1 mm wide, with fine, greenish soredia along the margins, and fine, stiff, colorless, almost transparent hairs on the top surface and edges of the lobe tips (visible on lobes in the lower right of plate 646). Apothecia rare, bearing colorless hairs on the margins. CHEMISTRY: Negative with all reagents (no lichen substances). HABITAT: On the bark of hardwoods and on rock of different kinds. COMMENTS: The stiff hairs of the lobe tips in combination with marginal soredia characterize this inconspicuous lichen. *Phaeophyscia hirtella* (East Temperate) and *Ph. hirsuta* (mostly from the southwest) have similar hairs, but the former has no soredia, and the latter produces its soredia on the lobe tips and lacks the

645. *Phaeophyscia adiastola* (damp) Appalachians, Virginia ×3.4

646. *Phaeophyscia cernohorskyi* (damp) central Nebraska ×5.4

647. *Phaeophyscia ciliata* western Mississippi ×5.0

maculae of *Ph. cernohorskyi*. **Phaeophysica kairamoi,** a species on rock or bark in the western mountains, has cortical hairs on the dark, marginal isidia.

Phaeophyscia ciliata
Smooth shadow lichen

DESCRIPTION: Thallus pale gray to brown, with radiating, narrow, flat lobes 0.5–1.5(–2) mm wide; without soredia, isidia, or erect, colorless hairs; lower surface black, with abundant black rhizines (often with white tips) that can extend out at the lobe margins and resemble cilia. Apothecia usually abundant, with dark brown, almost black disks and prominent margins, 0.5–1.5 mm in diameter, frequently with short rhizines growing from the lower part of the margin. CHEMISTRY: Negative with all reagents (no lichen substances). HABITAT: On deciduous trees and wood; occasionally on mosses over rock, but not directly on rock. COMMENTS: This is the most common nonsorediate species of *Phaeophyscia* on bark. *Phaeophyscia decolor,* a similar species found strictly on rock, has narrower, somewhat more convex lobes and tends to be much darker. **Phaeophysia hirtella,** a nonsorediate species on bark, has stiff hairs on the upper surface, as does *Ph. cernohorskyi*. *Hyperphyscia syncolla* can be similar but is very tightly appressed to the substrate and does not have rhizines growing from the apothecial margins, and its conidia are long (up to 10 μm) and slender.

Phaeophyscia constipata
Pincushion shadow lichen

DESCRIPTION: Thallus pale to dark brown, green when moist (as shown in plate 648) without maculae, narrowly foliose to almost fruticose, with prostrate to ascending main lobes up to 1 mm across (but usually narrower), and even more slender, ascending, elongate but often knobby side lobes approximately 0.15–0.4 mm across; without soredia or isidia, but with marginal cilia, usually pale, unbranched and tapering, rather hair-like; lower surface white to brownish, rarely becoming blackened in older portions; rhizines sparse. Without apothecia. CHEMISTRY: Negative with all reagents (no lichen substances) except for occasional areas of yellow on the lower surface that react K+ purple (an anthraquinone). HABITAT: On soil and moss, often in dry habitats, at low or high elevations. COMMENTS: This remarkable soil-dwelling *Phaeophyscia* is distinctive because of its habitat. **Anaptychia ulotrichoides** is a much rarer but outwardly similar foliose lichen known from Utah. It grows on rocks or soil in arid regions. It has a much duller (but nonpruinose) surface and develops characteristic fine, colorless hairs on the lobe tips. *Speerschneidera euploca,* a somewhat similar fruticose lichen found on dry limestone, has convex lobes and no cilia.

Phaeophyscia decolor
Starburst shadow lichen

DESCRIPTION: Thallus pale to dark gray-brown or brown, becoming pale olive-brown in shade forms, sometimes with a faint pruina, with narrow, flat to somewhat convex lobes, 0.2–0.5 (–1) mm across; without soredia, isidia or cilia; medulla white. Apothecia very common, with red-brown

648. *Phaeophyscia constipata* (damp) Olympics, Washington ×2.8

649. *Phaeophyscia decolor* Rocky Mountains, Colorado ×4.3

650. *Phaeophyscia hispidula* ssp. *limbata* White Mountains, eastern Arizona ×2.4

on the lobe surface (ssp. *hispidula*), but soralia sometimes very sparse or even absent, as shown in plate 650; black rhizines particularly abundant and long, unbranched or squarrose at the base, often coalescing into a mat, appearing as a fringe of "whiskers" when viewed from above. Apothecia uncommon, 1–3 mm in diameter; disks very dark brown to black, with margins that can have lobules as well as a fringe of black rhizines. CHEMISTRY: All reactions negative (no lichen substances). HABITAT: On bark, rock, and moss. COMMENTS: This is one of the largest species of *Phaeophyscia*, distinguishing it, for example, from *Ph. orbicularis*, which can also have soredia on the upper lobe surface, but has finer soredia with lobes that tend to be convex at the tips. Specimens of *Ph. hispidula* that have marginal isidia-like soredia that approach being lobulate can resemble **Ph. squarrosa** (what many have called **Ph. imbricata**), a distinctive eastern species on bark or over mosses on rock that has erect lobules along the lobe margins. That species has narrower lobes (less than 1.5 mm wide) and tends to be dull gray, almost like a *Physcia* (although K– and lacking atranorin). Specimens of *Ph. hispidula* that lack any soredia might key to *Ph. ciliata*, a species with much narrower, more closely appressed lobes.

to black disks, under 1 mm in diameter; spores with angular locules (*Physcia*-type). CHEMISTRY: All reactions negative; contains zeorin. HABITAT: On rock in exposed situations. COMMENTS: This is one of the few rock-dwelling species of *Phaeophyscia* that has no soredia or isidia. The lobes are usually very narrow, often no more than 0.3 mm across. **Phaeophyscia endococcinodes,** rather rare in the southwest and in the southern boreal region, has wider lobes and a red-orange medulla. Its spores are football-shaped with thick walls and round to elliptical (not angular) locules.

Phaeophyscia hispidula
Whiskered shadow lichen

DESCRIPTION: Thallus pale greenish gray or brownish, at least in part; lobes relatively broad, 1–4 (–6) mm wide, with rather concave tips and a white medulla, without colorless cortical hairs; coarse soredia (sometimes almost isidia-like) produced mainly on the margins (ssp. *limbata*), or mainly

Phaeophyscia orbicularis
Mealy shadow lichen

DESCRIPTION: Thallus dark greenish gray to brownish gray, medulla white, with dark gray or greenish, powdery soredia forming irregular soralia on the upper lobe surface often close to the tips (only rarely marginal); lobes 0.5–1.5(–2) mm across, flat to convex at the tips. Apothecia uncommon. CHEMISTRY: All reactions negative (no lichen substances). HABITAT: On rock or bark. COMMENTS: The soralia on the lobe surface rather than along the margins or at the tips characterize this species. The convex lobes, when present, are also not typical of a *Phaeophyscia*. Occasional specimens with a pale to dark brown lower surface can be mistaken for **Ph. insignis,** a rare species known from eastern U.S. Its lobes are narrower (up to 0.5 mm wide) and the laminal soralia are often rounded and hemispherical. **Physciella melanchra,** also with a pale lower surface, has a lower cortex with a different type of tissue (prosoplectenchyma).

Phaeophyscia pusilloides
Pompon shadow lichen

DESCRIPTION: Thallus small, dark greenish gray to brown; lobes less than 1 mm wide, with small, globose, yellowish green soralia produced on upturned lobe tips, giving the soralia an elevated look like a pompon; lobes often overlapping one another like shingles; black rhizines abundant, usually extending out on all sides, giving the lichen a fringed appearance. Apothecia uncommon. CHEMISTRY: All tests negative (no lichen substances). HABITAT: On bark of all kinds; rarely on calcareous rock. COMMENTS: Some forms of *Ph. rubropulchra* are superficially similar, but they have longer, more marginal soralia and a red-orange medulla.

Phaeophyscia rubropulchra
Orange-cored shadow lichen

DESCRIPTION: Thallus extremely variable, pale to dark green or greenish gray to dark brown; medulla red-orange, at least in older parts of the thallus, noticeable in the field where insects have eaten away the cortex, as in the center of the specimen shown in plate 653; lobes 0.5–1.2 mm across, slightly turned up at the tips; soredia usually rather coarse, mostly marginal, but also laminal in part; rhizines short, black with white tips. Apothecia common, mostly under 1 mm in diameter, containing thick-walled spores with round to ellipsoid locules. CHEMISTRY: Reactions negative except for K+ purple medulla (anthraquinone and an unidentified fatty acid). HABITAT: On bark of deciduous trees, rarely on conifers, mosses, or rock, typically in shaded, forest habitats. COMMENTS: The bright orange medulla and presence of soredia distinguish this species from any other lichen. Specimens having only scattered patches of red medulla are sometimes encountered, however, and these can be mistaken for *Ph. orbicularis,* a mainly western species with flat to convex lobes and fine, laminal soredia, or *Ph. adiastola,* which has soredia only on the lobe tips and along the margins and is usually found on rock. Both *Ph. adiastola* and *Ph. orbicularis* have *Physcia*-type spores with angular locules. See also *Ph. pusilloides.*

651. *Phaeophyscia orbicularis* Maine coast ×2.3

652. *Phaeophyscia pusilloides* Lake Superior region, Ontario ×3.7

653. *Phaeophyscia rubropulchra* (damp; showing red medulla) Lake Superior region, Ontario ×4.8

654. *Phaeophyscia sciastra* southern British Columbia ×2.5

Phaeophyscia sciastra
Dark shadow lichen, five o'clock shadow

DESCRIPTION: Thallus very dark greenish gray, appearing almost black; lobes very narrow, 0.15–0.5 (–0.8) mm across, with coarse, black, granular to rarely cylindrical isidia (or isidia-like soredia) along the lobe margins, especially in older parts of the thallus. Apothecia rare. CHEMISTRY: All tests negative (no lichen substances). HABITAT: Forms neat rosettes on exposed rocks (especially sandstone). COMMENTS: This species is the closest thing to a truly isidiate *Phaeophyscia*. Some western populations are virtually nonisidiate (or sorediate) and have somewhat elongated lobes under 0.5 mm. These resemble *Ph. decolor*. ***Phaeophyscia kairamoi*** is another rock-dwelling, somewhat isidiate shadow lichen, but it can be distinguished by the broader lobes (up to 2 mm), more granular isidia, and colorless hairs on the isidia and lobe tips; it is western montane.

Phaeorrhiza (2 N. Am. species)
Brown-fuzz lichens

DESCRIPTION: Dark brown to yellow-brown or yellow-olive, frequently pruinose, crustose to squamulose lichens, with rather thick thalli having lobed margins. Lower cortex indistinct but producing abundant, brown rhizine-like tomentum. Photobiont green (unicellular). Apothecia immersed in the thallus but finally becoming superficial, either lecanorine or lecideine; disks dark brown to almost black; asci *Lecanora*-type; spores brown, 2-celled, with more or less uniform, thin walls, 17–24 × 7–12 μm, 8 per ascus. CHEMISTRY: No lichen substances other than an unidentified pigment on the lower surface of **Ph. sareptana**. (See Comments under *Ph. nimbosa*). HABITAT: Growing on lime-rich soil, mosses, and peat. COMMENTS: The asci, spores, and thallus color place this genus close to *Buellia* and *Rinodina*, which are truly crustose lichens lacking any lower cortex or rhizine-like hyphae below.

KEY TO SPECIES: See Key G and *Rinodina*.

Phaeorrhiza nimbosa

DESCRIPTION: Thallus squamulose or almost continuous, with lobed margins, sometimes overlapping. Apothecia extremely abundant and crowded, immersed when young, becoming lecanorine with flat, dark brown to black disks and thick margins when mature, 0.4–1.5(–2.5) mm in diameter. CHEMISTRY: All reactions negative (no lichen substances). HABITAT: On calcareous soil and peat in arctic and alpine sites. COMMENTS: The only other species of *Phaeorrhiza* in North America is *Ph. sareptana,* a lichen found mainly in the mountains and prairies of the dry interior. Its thallus is distinctly squamulose with black, convex, marginless lecideine apothecia and a K+ violet reaction on the lower surface of the thallus.

Phlyctis (4 N. Am. species)
Whitewash lichens

DESCRIPTION: Thallus crustose, pale yellowish gray to almost white, smooth or rough and coarsely granular with patches of soredia. Photobiont green (*Dictyochloropsis*). Apothecia immersed in the thallus, often very inconspicuous, irregular in shape, with dark disks and disintegrating thalline margins; spores large, colorless, many-celled, muriform, 1–2(–4) per ascus. CHEMISTRY: β-orcinol depsidones. HABITAT: Mostly on bark, rarely rock. COMMENTS: With or without fruiting bodies, most specimens of *Phlyctis* look like sterile crusts, and one must search for the apothecial mounds. The presence of norstictic acid in the two most common species is helpful in their identification. The thalli sometimes cover large areas of bark and resemble white paint.

KEY TO SPECIES: Keys C and F.

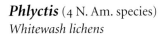

Phlyctis argena

DESCRIPTION: Thallus smooth to very rough, verruculose to granular, the tiny warts breaking down into coarse granular soredia. Apothecia uncommon, the disks irregular, pruinose, with tat-

655. *Phaeorrhiza nimbosa* Rocky Mountains, Colorado ×5.6

656. *Phlyctis argena* Lake Superior region, Ontario ×3.0

657. *Phyllopsora parvifolia* northern Louisiana ×3.1

tered margins; spores huge, 1 per ascus, mostly over 100 μm long. CHEMISTRY: Thallus PD+ deep yellow, K+ yellow changing to blood-red, KC–, C– (norstictic acid). HABITAT: On the bark of deciduous trees and occasionally conifers or rocks. COMMENTS: *Phlyctis agelaea* is a less common species that never produces soredia; it occurs in the eastern boreal region. Its spores are mostly 45–80 μm long, 2 per ascus.

Phyllopsora (9 N. Am. species)
Lace-scale lichens

DESCRIPTION: Thallus squamulose or crustose with lobed, often overlapping margins, sometimes reduced to a mass of isidia or granules, pale greenish gray or darker; white to red prothallus present. Photobiont green (*Pseudochlorella*). Apothecia biatorine, reddish brown; asci *Bacidia*-type; spores colorless, narrowly ellipsoid to fusiform, usually 1-celled, or sometimes appearing 2-celled, 8 per ascus. CHEMISTRY: Various substances including atranorin, argopsin, pannarin, and zeorin. HABITAT: Usually on bark, but occasionally on leaves or rock. COMMENTS: This is a genus of small tropical lichens related to *Bacidia*, but with a squamulose thallus. Some resemble a *Parmeliella*, but that genus has cyanobacteria as its photobiont and has ellipsoid spores.

KEY TO SPECIES: See Keys F and G.

Phyllopsora parvifolia

DESCRIPTION: Thallus gray-green to olive-gray, sometimes brownish at the edge, squamulose to subfoliose, with rounded lobes 0.3–0.5(–1) mm wide, becoming finely divided, reducing the margins to masses of lobed isidia, finally transforming the older parts of the lichen into a coarsely granular crust; lower surface of the squamules composed of a thick, white, fuzzy tomentum, protruding in some places as a white prothallus. Apothecia flat to convex, with thin, disappearing margins, 0.3–0.5 mm in diameter; spores narrowly ellipsoid, 1-celled, colorless, 9–12 × 1.8–2.2 μm. CHEMISTRY: All reactions negative (no lichen substances). HABITAT: On bark in shaded forests.

Physcia (36 N. Am. species)
Rosette lichens

DESCRIPTION: Rather small to medium-sized foliose lichens, pale greenish gray to almost white, frequently spotted with maculae, infrequently pruinose; lower surface usually pale but sometimes brown or even black; lower cortex usually composed of prosoplectenchyma, rarely pseudoparenchyma; rhizines sparse or abundant, usually almost white, tending to be formed close to the margins of the lobes. Photobiont green (*Trebouxia*?). Apothecia lecanorine with dark brown to almost black disks; spores dark brown, 2-celled, thick-walled, with angular or oval locules; conidia stick-like, 4–7 μm long. CHEMISTRY: Upper cortex PD– or pale yellow, K+ yellow, KC–, C–; medulla PD– or PD+ pale yellow, K– or K+ yellow, KC–, C– (atranorin, often with triterpenes such as zeorin). HABITAT: Various substrates. COMMENTS: Most small, light gray foliose lichens with a pale undersurface and K+ yellow upper surface are species of *Physcia*. There are, however, many look-alikes. The key is helpful in sorting them out. See especially *Parmeliopsis hyperopta* and species of *Phaeophyscia, Physciella, Hyperphyscia, Dirinaria, Physconia, Imshaugia, Heterodermia,* and *Anaptychia*.

KEY TO SPECIES (including *Dirinaria, Heterodermia, Imshaugia,* and *Pyxine*)
1. Soredia absent 2
1. Soredia present 23

2. Isidia present 3
2. Isidia absent 5
3. Rhizines more or less uniformly distributed, short and usually unbranched; cortex and medulla K+ deep yellow, PD+ orange (thamnolic acid); on conifers and birch in northern and montane regions .... *Imshaugia aleurites*
3. Rhizines mostly on or close to the margins; cortex K+ yellow; medulla PD– or PD+ yellow (lacking thamnolic acid); on deciduous trees in southeastern U.S. 4
4. Lower surface yellow or orange, webby; medulla K+ yellow (or deep purple where pigmented) .... *Heterodermia crocea*
4. Lower surface white to tan, smooth; medulla K+ red (salazinic acid) .... [*Heterodermia granulifera*]
5.(2) Thallus attached by only a few points, almost fruticose, with long, narrow lobes 6
5. Thallus closely or loosely attached; lobes rarely long and narrow 7
6. Apothecial margins smooth and even; southwestern U.S. .... *Heterodermia erinacea*
6. Apothecial margins toothed or lobulate; south central or southeastern U.S. .... *Heterodermia echinata*
7. Medulla yellow; apothecia lecideine, black; cortex UV+ yellow-orange (lichexanthone) ..... *Pyxine berteriana*
7. Medulla white; apothecia lecanorine, brown to black; cortex UV– (lichexanthone absent) 8
8. Cortex K– (atranorin absent) .. *Phaeophyscia hispidula*
8. Cortex K+ yellow (atranorin, except for *Imshaugia*) 9
9. Medulla with patches of orange-yellow pigment (anthraquinones), K+ yellow (atranorin), or K+ purple where pigmented; thallus heavily white pruinose. Uncommon, in extreme southwest .... [*Heterodermia rugulosa*]
9. Medulla entirely white, K+ yellow (atranorin) or K–; thallus usually without pruina 10
10. Lobules present, at least on the inrolled apothecial margins and usually the lobe margins, sometimes also on the thallus surface (occasionally absent in *H. hypoleuca*, which has brown squarrose rhizines) 11
10. Lobules absent; rhizines never squarrose 13
11. Lobules finely divided, on margins and upper surface of lobes; lower surface becoming purplish black in the center .... *Heterodermia squamulosa*
11. Lobules rounded or strap-shaped, on the margins of the lobes and apothecia; lower surface white to gray throughout 12
12. Lower surface cottony, entirely without a cortex; rhizines squarrose, well developed and often forming a mat; thallus not pruinose; mainly eastern, although also in southwestern U.S. .... *Heterodermia hypoleuca*
12. Lower surface smooth, with a cortex, at least in part; rhizines unbranched or forked, occasionally squarrose, not forming a mat; thallus somewhat pruinose on lobe tips; southwestern U.S. .... *Heterodermia diademata*
13.(10) On rock 14
13. On bark or wood 16
14. Thallus heavily pruinose .... *Physcia biziana*
14. Thallus entirely without pruina 15
15. Lobes 0.7–1.5(–2.5) mm wide; maculae present on upper surface; zeorin present .... *Physcia phaea*
15. Lobes 0.3–0.5 mm wide; maculae absent; zeorin absent .... *Physcia halei*
16.(13) Lower surface black; very tightly attached to substrate (almost crustose in appearance) 17
16. Lower surface white to pale brown; thallus closely or loosely attached to substrate 18
17. Lobes 1–2(–4) mm wide; apothecial disks black, without pruina .... *Dirinaria confusa*
17. Lobes 0.2–0.7 mm wide; apothecial disks reddish purple caused by a purple pruina .... *Dirinaria purpurascens*
18. Spores colorless, single-celled; cortex and medulla PD+ orange (thamnolic acid; atranorin absent) .... *Imshaugia placorodia*
18. Spores brown, 2-celled; cortex and medulla PD– (atranorin present, thamnolic acid absent) 19
19. Medulla K+ yellow (atranorin) 20
19. Medulla K– 22
20. Spores with rounded locules (*Pachysporaria*-type); apothecial margins sometimes with a fringe of rhizines on the lower side; zeorin absent. .... *Physcia neogaea*
20. Spores with angular locules (*Physcia*-type); apothecial margins without rhizines on lower side; zeorin present 21
21. Lobes convex, 0.3–0.5 mm wide ..... *Physcia pumilior*
21. Lobes flat or somewhat concave at the tips, 1–2(–3) mm wide .... *Physcia aipolia*
22.(19) Thallus pruinose .... *Physcia biziana*
22. Thallus entirely without pruina ...... *Physcia stellaris*
23.(1) Lower surface black or purplish 24
23. Lower surface white to pale brown, gray, yellow, or orange 34
24. Rhizines absent, thallus attached directly to substrate; medulla UV+ blue-white (divaricatic acid) 25
24. Rhizines sparse or abundant; medulla UV– or UV+ yellow-orange or red (divaricatic acid absent) 27

25. Soredia granular, originating from the breakdown of pustules or hollow warts, covering the thallus surface . . . . *Dirinaria aegialita*
25. Soredia farinose, in hemispherical mounds. 26

26. On bark, southeastern coastal plain . . . *Dirinaria picta*
26. On rock, East Temperate distribution . . . . [*Dirinaria frostii*]

27.(24) Medulla yellow. 28
27. Medulla white. 30

28. Soredia mostly laminal (but see *P. eschweileri* below), farinose, greenish white or green; thallus closely appressed; southeastern. 29
28. Soredia entirely marginal, granular, bluish gray; thallus with ascending lobe tips; widely distributed, East Temperate; cortex K± yellow, UV– (atranorin) . . . . *Pyxine sorediata*

29. Thallus UV+ yellow, K– (lichexanthone); medulla PD–, never with marginal schizidia . . . . [*Pyxine albovirens*]
29. Thallus UV–, K± yellow (atranorin); medulla PD+ orange; some forms with marginal isidia-like granules (schizidia) from the breakdown of pustules . . . . *Pyxine eschweileri*

30.(27) Thallus with crowded, overlapping lobes; cilia-like rhizines abundant along the margins. . . . . *Phaeophyscia hispidula*
30. Thallus forming flat rosettes; marginal cilia or cilia-like rhizines present or absent. 31

31. Cilia-like rhizines present along the lobe margins . . . . [*Heterodermia casarettiana*]
31. Cilia-like rhizines or cilia absent 32

32. Medulla K+ yellow; rhizines sparse; zeorin present . . . . *Physcia sorediosa*
32. Medulla K–; rhizines abundant; zeorin absent 33

33. Cortex K+ yellow, UV– (atranorin); medulla PD+ orange (testacein); spores 4-celled . . . . *Pyxine eschweileri*
33. Cortex K–, UV+ yellow-orange (lichexanthone); medulla PD–; spores 2-celled. . . . . *Pyxine cocoes*

34.(23) Cortex K– (atranorin absent) 35
34. Cortex K+ yellow (atranorin) 36

35. Lobes without pruina; East Temperate distribution . . . . *Physciella chloantha*
35. Lobes pruinose, at least at the tips; western. . . . . *Physconia thomsonii*

36. Soredia laminal or on the upper surface of the lobe tips 37
36. Soredia mostly marginal, on the lower surface of the lobe tips, or on expanded lobe tips. 39

37. Marginal cilia common and abundant; thallus loosely attached and often ascending; lobes 0.2–0.5(–1) mm wide . . . . *Physcia tenella*
37. Marginal cilia absent; thallus closely attached to substrate; lobes mostly 0.5–1.5(–2.5) mm wide 38

38. Lower surface and rhizines pale brown; soredia blue-gray to white; thallus moderately thick; mainly on rock, rarely wood or bark . . . . *Physcia caesia*
38. Lower surface and rhizines white or almost white; soredia greenish or white; thallus thin and membranous; mainly on bark of deciduous trees, rarely on limestone . . . . *Physcia americana*

39.(36) Marginal cilia or cilia-like rhizines common and abundant. 40
39. Marginal cilia or cilia-like rhizines absent or sparse 45

40. Thalli closely appressed to substrate; soralia crescent-shaped; marginal cilia or cilia-like rhizines branched (Fig. 10b) 41
40. Thalli loosely attached over entire substrate, or ascending and almost fruticose; soralia not crescent-shaped; cilia branched or unbranched. 42

41. Lower surface white to pale brown, K– or K+ yellow, smooth to fibrous (with fibers running in one direction), with a cortex, at least in part. . . . . *Heterodermia speciosa*
41. Lower surface pale orange, at least in spots (white to pale brown elsewhere), K+ purple where pigmented (anthraquinones), entirely without a cortex . . . . *Heterodermia obscurata*

42. Cilia short, 1–2 mm long, unbranched; lobes relatively short, forming crowded clumps; lower surface smooth, with a cortex; soralia on expansions of lobe tips. 43
42. Cilia 2–5 mm long, branched or unbranched; lobes very long and narrow; lower surface cottony or webby, often pruinose, lacking a cortex; soredia on lower surface of lobe tips, which do not widen. 44

43. Soredia within hood-like expansions of lobe tips (between upper and lower cortices) . . . *Physcia adscendens*
43. Soredia on widened, turned-back lobe tips (lip-shaped soralia) . . . . *Physcia tenella*

44. Lower surface usually pale sulphur yellow, at least close to lobe tips (K+ yellow, not an anthraquinone); cilia white or brown, usually unbranched; medulla PD–, K+ yellow (atranorin) . . . . *Heterodermia appalachensis*
44. Lower surface white; cilia branched once or twice; medulla PD+ yellow, K+ yellow changing to red (salazinic acid) . . . . *Heterodermia leucomela*

45.(39) Soredia produced all along lobe margins 46

45. Soredia confined to lobe tips or on crescent-shaped marginal lobes 48
46. Lower surface fibrous, more or less streaked with black or dark gray; thallus pruinose, at least at lobe tips; rhizines brown to black; medulla K+ yellow .... *Physcia atrostriata*
46. Lower surface smooth, more or less uniformly white; thallus entirely without pruina; rhizines white to pale brown, sometimes darkening at the tips; medulla K– or K+ red 47
47. Lobes rounded, thin, finely divided, the margins dissolving into granules (blastidia: Fig. 19e) or soredia; marginal cilia absent; medulla K– .... *Physcia millegrana*
47. Lobes elongated, moderately thick, not finely divided, with white granular soredia (not blastidia) along the margins; marginal cilia sometimes present; medulla K+ yellow changing to red (salazinic acid) .... *Heterodermia albicans*
48.(45) Rhizines black; upper cortex prosoplectenchyma (Fig. 5e, but hyphae longitudinal)................ 49
48. Rhizines white or pale tan, sometimes darkening at the tips, or rarely dark gray; upper cortex pseudoparenchyma (Fig. 5f) 50
49. Lower surface white to pale brown, smooth to fibrous, with a cortex, at least in part . . . *Heterodermia speciosa*
49. Lower surface pale orange, at least in part (K+ purple, anthraquinones), webby, entirely without a cortex.... .... *Heterodermia obscurata*
50. Lobes very narrow, 0.2–0.5 mm wide, hardly widening at the tips; soredia and blastidia very coarse, forming under and at lobe tips; on siliceous rock .... *Physcia subtilis*
50. Lobes 0.3–1.2(–3) mm, usually widening or fanning out at the tips; soredia fine or coarse; on various substrates 51
51. Lobes distinctly upturned with soredia forming in lip-shaped soralia; soredia fine or coarse 52
51. Lobe tips usually flat or down-turned; soralia not lip-shaped; soredia coarsely granular 53
52. Soredia very fine, greenish; on rocks, wood, or sometimes bark, exposed or partly shaded; widespread and very common...................... *Physcia dubia*
52. Soredia coarsely granular, blue-gray to white; on mossy rocks in forest habitats; Appalachian Mountains .... [*Physcia pseudospeciosa*]
53. Lobes flat, finely divided and lacy, dissolving into granules (blastidia: Fig. 19e) or soredia; eastern temperate region; mostly on bark of various kinds, rarely on rock .... *Physcia millegrana*
53. Lobes convex, clearly downturned at the tips, with granular soredia forming on the lower surface of the lobe tips; western; on rock, especially limestone and sandstone .... *Physcia callosa*

Physcia adscendens
Hooded rosette lichen

DESCRIPTION: Thallus pale gray to pale greenish gray, spotted with white maculae and sometimes slightly pruinose, forming small clusters of ascending lobes, many of which expand at the tip to produce inflated and hollow, helmet-shaped soralia, which are formed from a separation of the upper and lower cortices and contain greenish, granular soredia; lobes mostly under 1 mm wide except for the helmets, which can be up to 2 mm across; long, white, usually unbranched cilia (some with darkened tips) grow from lobe margins and tips; lower surface white. Apothecia not common. CHEMISTRY: Medulla K–. HABITAT: On bark, twigs, and wood of a variety of trees, less frequently on rock, in well-lit or slightly shaded sites. COMMENTS: This is the only clearly "hooded" *Physcia*. *Physcia tenella*, a common bark-dwelling, ciliate lichen along both coasts, has lip-shaped soralia on the lobe tips but does not form hollow hoods or helmets. *Physcia leptalea* (syn. *Ph. semipinnata*), a rare, nonsorediate species along the west coast, also has long, pale, marginal cilia.

Physcia aipolia
Hoary rosette lichen

DESCRIPTION: Thallus pale to dark gray, conspicuously spotted with white maculae; lobes narrow and radiating, flat to slightly concave or upturned at the tips, sometimes overlapping, 1–2(–3) mm across, without any soredia or isidia; lower surface white to pale brown with many pale rhizines. Apothecia very common, 1–2(–3) mm in diameter, very dark brown but typically heavily white pruinose (giving the lichen its English name); spores (16–)18–25 × 7–12 µm, with angular cells. CHEMISTRY: Medulla K+ yellow (atranorin and zeorin). HABITAT: On bark and wood of different kinds of trees in open habitats. COMMENTS: *Physcia stellaris* is a very similar species with about the same range as *Ph. aipolia*, but it lacks con-

658. *Physcia adscendens* southern Ontario ×4.3

659. *Physcia aipolia* White Mountains, eastern Arizona ×4.0

spicuous maculae, has flat or somewhat convex lobes, and has a K– reaction in the medulla. *Physcia pumilior* is a small version of *Ph. aipolia* found in the southeast. Its spores are 15–17 × 6–7.5 μm, and the lobes are often convex. *Physcia neogaea* is also very similar but is not heavily white-spotted, the spores have oval to round locules, it sometimes develops rhizines on the apothecial margin, and it lacks zeorin.

Physcia americana
Powdery rosette lichen

DESCRIPTION: Thallus pale greenish gray (fading to white after storage), growing flat against the substrate, without white maculae except on a few of the oldest lobes, sometimes lightly pruinose, with round, laminal soralia containing greenish or whitish, granular soredia; lobes 0.6–1.5(–2.5) mm across; lower surface white, with white rhizines. Apothecia rare. CHEMISTRY: Medulla K+ yellow (atranorin and unknown triterpenes). HABITAT: Mainly on the bark of deciduous trees, occasionally on limestone. COM-

MENTS: Except for the white lower surface, *Ph. americana* looks a bit like *Dirinaria picta*. *Physcia atrostriata* often grows with *Ph. americana,* and the two can be mistaken for each other except that *Ph. atrostriata* has marginal soralia and pruinose lobe tips. **Physcia clementii,** a similar but uncommon species from California, has coarse isidiate soredia occurring over large portions of the thallus surface instead of soredia in round mounds. Specimens of *Ph. americana* with few soredia could be mistaken for *Ph. neogaea*.

Physcia atrostriata
Streaked rosette lichen

DESCRIPTION: Thallus pale green to greenish gray; lobes closely appressed or somewhat overlapping, with pruinose lobe tips and margins, 1–2(–3) mm across; abundant greenish soredia produced along the lobe margins, not on the upper surface; lower surface with a fibrous texture and appearing to lack a cortex, with fibrous black lines fanning out toward the lobe tips, sometimes coalescing to form

660. *Physcia americana* North Carolina coast ×3.5

661. *Physcia atrostriata* (damp) Ouachitas, Arkansas ×3.6

662. *Physcia biziana*
Sangre de Cristo Range, New Mexico
×4.2

an almost black lower surface in the oldest parts of the thallus. Apothecia rare. CHEMISTRY: Medulla K+ yellow (atranorin, zeorin, and an unidentified triterpene). HABITAT: On bark. COMMENTS: The unusual lower surface of *Ph. atrostriata* together with the marginal pruina and soredia make this lichen very distinctive. *Physcia crispa* has a uniformly smooth (not fibrose) white lower surface and broader, more "crisped" lobes, and it is never pruinose. *Physcia undulata* is a rare Florida species similar to *Ph. crispa* except for the K+ yellow medulla and presence of zeorin. *Physcia dimidiata,* a lichen of the arid western interior, has abundant, granular soredia along the margins, but is heavily white pruinose on the upper surface of the thallus. In *Ph. sorediosa,* the lower surface is entirely black, and the soredia are coarser than those of *Ph. atrostriata.* Some species of *Heterodermia* with marginal soredia can be similar, but they generally have cilia.

Physcia biziana
Frosted rosette lichen

DESCRIPTION: Thallus pale to dark gray, usually covered with a coarse pruina over the entire thallus; lobes loosely attached, overlapping, width highly variable (1–5 mm); lobe tips flat to convex, often turned down. Apothecia abundant, 1.5–5 mm in diameter, often raised or at least constricted at the base, disks dark brown but heavily white pruinose; margins prominent and often curled inward; lower surface and rhizines white. CHEMISTRY: Medulla K– (atranorin). HABITAT: On bark or calcareous rocks in open, dry areas. COMMENTS: Of the very few heavily pruinose species of *Physcia,* this is the only one with a K– medulla. For example, **Ph. mexicana,** known from Texas on bark, is K+ yellow. *Physcia cascadensis* is another loosely attached, western species without maculae (on rock or bark) that can become lightly pruinose. The medulla, however, is K+ yellow. Species of *Physconia,* which are also frosted on the lobe tips, have branched, dark rhizines and tend to be brown; they

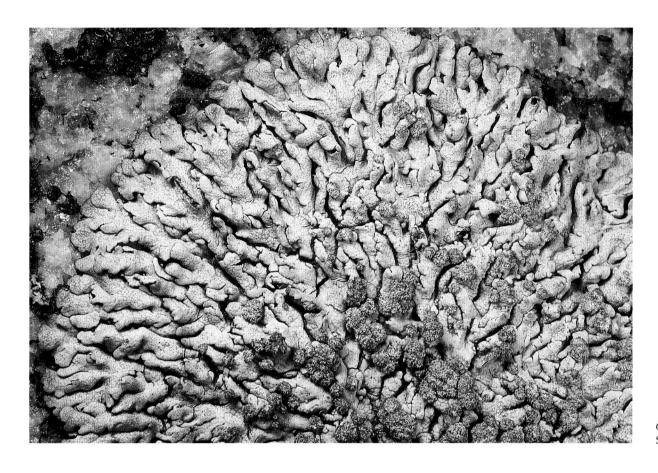

663. *Physcia caesia*
Southeast Alaska ×4.3

have a K– cortex. Exceptionally large specimens of *Ph. stellaris* can resemble *Ph. biziana* but never become as pruinose and the spores are somewhat larger (16–24 × 7–10 μm versus 15–18 × 5–8 μm). See also *Ph. callosa*.

Physcia caesia
Blue-gray rosette lichen, powderback rosette

DESCRIPTION: Thallus pale gray, darker in the center of the thallus, spotted with maculae (although not always conspicuous); lobes 0.5–1(–2) mm wide, convex; hemispherical piles of blue-gray, granular soredia on the upper surface of the lobes; lower surface pale tan to brown, with short, dark rhizines. Apothecia rare. CHEMISTRY: Medulla K+ yellow (atranorin and zeorin). HABITAT: On exposed rock and wood, rarely on bark. It is particularly characteristic of bird rocks, apparently thriving on the highly fertilized surface. COMMENTS: The white rosettes formed on rocks, with round mounds of blue-gray soredia, make this lichen unmistakable. *Physcia phaea* is a fertile counterpart of *Ph. caesia*, with apothecia taking the place of soralia.

Physcia callosa
Beaded rosette lichen

DESCRIPTION: Thallus a clear, almost shiny, uniform gray to almost white, with no white spotting; lobes downturned, 0.4–2 mm wide, becoming divided into lobules at the tips; coarse dark gray soredia produced on the undersides and margins of the lobe tips; lower surface white to pale tan, with abundant pale rhizines. Apothecia constricted at the base; disks dark brown to black, lightly pruinose. CHEMISTRY: Medulla K–. HABITAT: On calcareous rock in open or semishaded sites. COMMENTS: In size, color, habit, and habitat, *Ph. callosa* resembles *Ph. biziana,* which lacks soredia and is heavily pruinose. It looks like a very large version of *Ph. millegrana* (2–3 times the size), an eastern species mainly on bark. *Physcia dubia* also has soredia on the undersides of lobe tips, but these soralia are finer and are produced on distinctly upturned lobes. Both species can be lightly pruinose.

664. *Physcia callosa* Coast Range, California ×4.7

665. *Physcia dubia* eastern Arizona ×3.2

Physcia dubia
Powder-tipped rosette lichen

DESCRIPTION: Thallus pale to dark ashy gray or sometimes yellowish gray, surface spotted with white maculae or almost uniform in color, sometimes with a light frosting of pruina on the lobe tips; lobes 0.3–1.2(–3) mm wide, convex, fanning out at the tips, with abundant gray to greenish soredia produced under the turned-up lobe tips (lip-shaped soralia) clearly visible from above. Apothecia rare. CHEMISTRY: Medulla K–. HABITAT: On rocks of all kinds, wood, and occasionally bark in open or partially shaded sites. COMMENTS: *Physcia callosa* is a larger species on rock with very granular soredia produced under down-rolled rather than upturned lobe tips. In *Ph. subtilis*, the soredia are not produced in masses, and the lobes are uniformly narrow, rarely exceeding 0.5 mm across. See also *Physciella chloantha*.

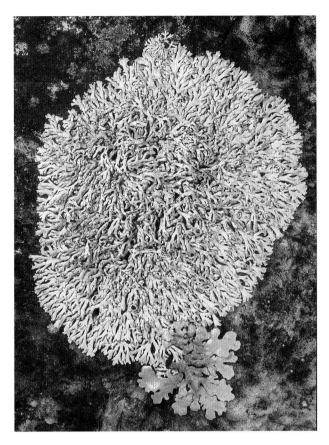

Physcia halei
Granite rosette lichen

DESCRIPTION: Thallus very pale gray, without white maculae or pruina, with closely attached, narrow lobes, approximately 0.3–0.5 mm across; without any soredia, granules, isidia, or lobules; lower surface white, with pale rhizines. Apothecia common. CHEMISTRY: Medulla K+ yellow (atranorin). HABITAT: On granite or sandstone in open sites. COMMENTS: This is the only very narrow-lobed rosette lichen on noncalcareous rock with no marginal granules or soredia. It is very similar to *Ph. subtilis,* an even narrower species with granular soredia on the lobe tips and sometimes on the margins.

Physcia millegrana
Mealy rosette lichen

DESCRIPTION: Thallus pale gray, spotted with white maculae; lobes thin, appressed to somewhat ascending, 0.3–1(–2) mm wide, margins (especially the tips) finely divided and finally dissolving into granular soredia; lower surface white, with pale rhizines. Apothecia very common, under 1 mm in diameter, dark brown, often lightly pruinose. CHEMISTRY: Medulla K– (atranorin). HABITAT: On bark, especially deciduous trees, as well as wood and occasionally granitic rock. COMMENTS: This is among the most common bark-dwelling lichens in eastern North America, even occurring close to urban areas on cultivated as well as wild deciduous trees. Narrow-lobed specimens of *Ph. millegrana* on rock can be confused with *Ph. subtilis,* which has soredia confined mainly to the lobe tips (but which can also be marginal), and has lobes that tend to be long and narrow. In *Physcia dubia,* the soredia are in lip-shaped soralia at the lobe tips, and the margins are never finely divided.

666. *Physcia halei* Ouachitas, Arkansas ×1.8

667. *Physcia millegrana* Cape Cod, Massachusetts ×3.7

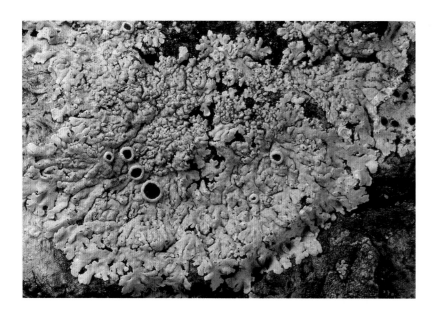

668. *Physcia neogaea*
Atlantic coast, Florida
×4.3

convex, typically crowded and overlapping, 0.7–1.5 (–2.5) mm across; lower surface pale tan to brownish, with abundant or sparse pale to brown rhizines. Apothecia almost always present, perfectly round and somewhat raised, 0.3–1.2(–2) mm in diameter, dark brown to black, rarely with a white pruina. CHEMISTRY: Medulla K+ yellow (atranorin and zeorin). HABITAT: On granitic (less commonly, calcareous) rocks, mostly in the open. COMMENTS: This beautiful rosette lichen resembles *Ph. aipolia*, a bark lichen, but it is found exclusively on rock, and its black apothecia lack the heavy frosting characteristic of *Ph. aipolia*. *Physcia cascadensis* is a rare, rock-dwelling rosette lichen in the northwest that, like *Ph. phaea*, lacks soredia and has a K+ yellow medulla, but it is very pale gray, has no white spotting (maculae), and has pruinose apothecia.

Physcia neogaea
Dwarf rosette lichen

DESCRIPTION: Thallus pale gray, without maculae on lobe tips but often conspicuously white-spotted and even bumpy on older parts of the thallus; lobes very narrow (less than 1 mm and often no more than 0.5 mm), spread out or overlapping; without any soredia, isidia, or lobules. Apothecia common, occasionally with rhizines growing from the lower margin; disks sometimes pruinose; spores 20–25 × 8–11 μm, thick-walled, with oval or round cells. CHEMISTRY: Medulla K+ yellow (atranorin and an unidentified triterpene; no zeorin). HABITAT: On bark. COMMENTS: This small lichen closely resembles *Ph. americana* except it has no soredia. It can also resemble a diminutive *Ph. aipolia*, with thin, flat lobes but without maculae on the lobe tips. In the otherwise similar *Ph. pumilior*, the spore locules are angular (*Physcia*-type). In the Florida populations, the lobes of *Ph. neogaea* tend to be flatter and more finely divided than those in populations elsewhere in its range.

Physcia phaea
Black-eyed rosette lichen

DESCRIPTION: Thallus steel gray, usually with conspicuous white-spotting caused by maculae, but without pruina, soredia, isidia, or lobules; lobes flat to somewhat

Physcia pumilior
Spotted rosette lichen

DESCRIPTION: Thallus pale gray, with conspicuous white spots on all parts of the thallus; lobes narrow, 0.3–0.5 mm across, irregularly branched, bumpy and generally convex; lower surface almost white, with pale rhizines. Apothecia common, round, 1–2 mm in diameter, disks pruinose or not; spores 15–17 × 6–7.5 μm, with angular cells. CHEMISTRY: Medulla K+ yellow, but usually too scanty to test (atranorin, zeorin, triterpenes). HABITAT: On bark and wood. COMMENTS: This southeastern lichen looks like a small *Ph. aipolia* with narrower and more convex lobes, and much smaller spores. See Comments under *Ph. neogaea*.

Physcia sorediosa
Black-bottomed rosette lichen

DESCRIPTION: Thallus pale gray to greenish gray, closely appressed, smooth or sometimes slightly pruinose at the lobe tips; pale greenish, almost white soredia abundantly produced, first as round, very convex mounds along the margins of the older parts of the thallus, but encroaching on the upper surface and appearing to be laminal as the lichen ages; lobes 0.5–1 mm wide; lower surface black except for a white border; rhizines predominantly

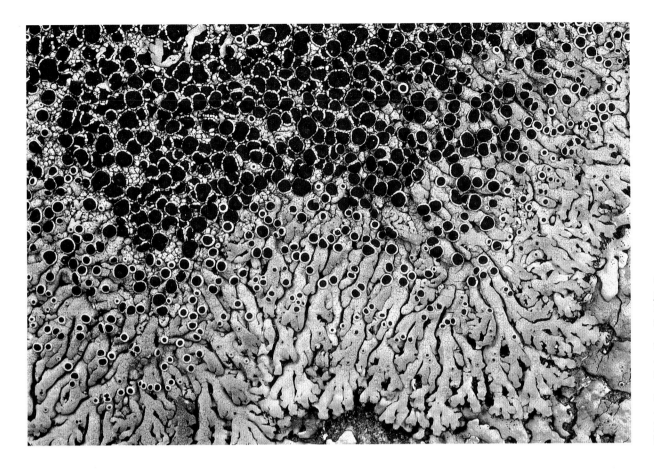

669. (left) *Physcia phaea* Sangre de Cristo Range, New Mexico ×4.4

670. (below left) *Physcia pumilior* western Mississippi ×3.2

671. (below right) *Physcia sorediosa* Appalachians, Virginia ×3.3

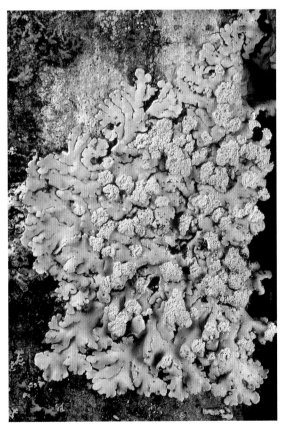

672. *Physcia stellaris* central Massachusetts ×2.0

black. CHEMISTRY: Medulla K+ yellow (atranorin, zeorin, and other triterpenes). HABITAT: On bark of hardwoods. COMMENTS: *Physcia sorediosa* is the only North American *Physcia* with a truly black undersurface, making it easily mistaken for species of *Dirinaria*, especially *D. picta* or **D. applanata,** or for species of *Pyxine* (especially *P. eschweileri*) or *Phaeophyscia*. The *Dirinaria* species have laminal rather than marginal soralia, are greener, and contain divaricatic acid (medulla UV+ blue-white). The lobes tend to be encrusted with a heavy pruina and are almost scabrose, especially in *D. applanata*. Species of *Phaeophyscia* have a K– upper cortex (lacking atranorin). *Physcia atrostriata* can also be dark below but is streaked with dark gray to black radiating lines rather than being uniformly black. *Physcia americana* differs in having truly laminal soredia as well as a white lower surface.

Physcia stellaris
Star rosette lichen

DESCRIPTION: Thallus pale gray, darker gray in the center, more or less smooth and uniform although sometimes with some white spotting, especially on older parts of the thallus; lacking pruina, soredia, isidia, or lobules; lobes radiating out (like a star) but sometimes crowded, flat to convex, 0.5–1.5(–3) mm wide; lower surface white to light brown, with fairly abundant pale to dark rhizines. Apothecia common, 0.7–3 mm in diameter, dark brown, often pruinose; spores 16–22 × 7–10 μm, with angular locules. CHEMISTRY: Medulla K– (atranorin). HABITAT: On bark of many kinds, but especially poplars, alders, and elms; rarely on wood or rock. COMMENTS: The negative K reaction of the medulla is the most reliable way to distinguish *Ph. stellaris* from *Ph. aipolia*, although the convex lobes and lack of conspicuous maculae can also be helpful. The pale rhizines that occasionally protrude from below the lobe margins of *Ph. stellaris* can resemble the ciliate lobes of the rare, mostly coastal species **Ph. leptalea,** but the latter is heavily spotted with white maculae and occurs mainly on conifers.

Physcia subtilis
Slender rosette lichen

DESCRIPTION: Thallus pale to dark gray, more or less uniform with no maculae, very narrow, branching lobes, 0.2–0.5 mm across, spreading or growing over one another, with granular fragments or granular soredia formed from a disintegration of the tissue just under the lobe tips or along the margins; lower surface white to pale tan, with abundant or sparse, pale rhizines. Apothecia uncommon, with very dark brown, nonpruinose disks. CHEMISTRY: Medulla K–, but lobes usually too thin to test. HABITAT: Exclusively on granitic or at least noncalcareous rock, usually in sunny sites. COMMENTS: *Physcia millegrana* closely resembles *Ph. subtilis*, and both are endemic to North America. Although *Ph. millegrana* usually grows on bark, it is occasionally found on rocks. It has broader (up to 2 mm) lobes, usually fanning out, whereas *Ph. subtilis* has almost linear lobe tips that expand very little. *Physcia dubia* has finer soredia produced in more distinctly lip-shaped soralia at the lobe tips, and the lobes are commonly 0.4–1.0 mm across. *Physcia halei*, a mainly southeastern rock species, is similar to *Ph. subtilis* but lacks soredia.

Physcia tenella
Fringed rosette lichen

DESCRIPTION: Thallus very pale gray to greenish gray, yellowish gray, or dark ashy gray (especially on coastal rocks); surface usually uniform, but sometimes spotted with maculae; lobes ascending and often tangled, usually narrow, elongate, and branched, 0.2–0.5(–1) mm across; fine or coarse, greenish soredia produced on upturned lobe tips, or rarely on the lobe surface, sometimes forming neat, round, somewhat excavate soralia; lower surface white to pale tan, with abundant pale to fairly dark brown cilia-like rhizines as well as marginal cilia. Apothecia occasional, somewhat raised, under 1 mm in diameter, with dark red-brown disks. CHEMISTRY: Medulla K– (atranorin).

673. *Physcia subtilis*
Appalachians, Virginia
×4.4

674. *Physcia tenella*
Washington coast
×5.9

675. *Physciella chloantha* (damp) southern Texas ×3.0

HABITAT: On twigs, bark, and rock, mostly in coastal sites in North America. COMMENTS: See Comments under *Ph. adscendens* for similarities between these two species.

Physciella (3 N. Am. species)
Cryptic rosette lichens

DESCRIPTION: Small to medium-sized gray to brownish gray foliose lichens, lobes generally less than 2 mm wide; lower surface pale; lower cortex composed of prosoplectenchyma. Photobiont green (*Trebouxia?*). Apothecia and spores as in *Physcia;* conidia ellipsoid, 2.5–4 × 1–1.5 μm. CHEMISTRY: All reactions negative (no lichen substances). HABITAT: Various substrates. COMMENTS: This genus has an enigmatic mixture of features, in part like *Physcia* (pale lower surface made up of prosoplectenchyma), and in part like *Phaeophyscia* (small ellipsoid conidia and lack of atranorin).

KEY TO SPECIES: See *Phaeophyscia*.

Physciella chloantha (syn. *Physcia chloantha*)
Cryptic rosette lichen

DESCRIPTION: Thallus pale to dark greenish or brownish gray, uniform in color, lacking pruina and usually lacking white maculae; lobes 0.3–1(–2) mm across often ascending; small, lip-shaped soralia on the lobe margins and tips, containing greenish granular soredia; lower surface pale white to pale tan, smooth, with thick, pale, unbranched rhizines, often developing into marginal cilia. Apothecia uncommon, disks very dark brown, not pruinose. HABITAT: Usually on hardwoods; occasionally on rock, especially limestone. COMMENTS: The sister species of *Ph. chloantha* is **Ph. melanchra** (syn. *Physcia melanchra*), which has soralia bursting through the top surface of the lobes as well as on the lobe margins. Marginal cilia (or marginal rhizines that resemble cilia) are less abundant on *Ph. melanchra*. *Physciella chloantha* also resembles *Physcia dubia,* with its brownish gray color and pale lower surface, but the *Physciella* has a K– upper cortex. *Physciella melanchra,* on the other hand, is almost identical to *Phaeophyscia orbicularis* except for the pale rather than black lower surface.

Physconia (11 N. Am. species)
Frost lichens

DESCRIPTION: Moderate-sized foliose lichens, gray-green (usually with brownish tones) to brown, usually green when wet, always with a light to heavy "frosting" (pruina), at least on the lobe tips; medulla white or pale yellow; lower surface generally black, less frequently pale; rhizines abundant, usually thickly branched, squarrose (resembling tiny bottlebrushes), rarely unbranched or forked, often extending out from under the lichen to create a fuzzy fringe. Photobiont green (*Trebouxia?*). Apothecia lecanorine with dark brown disks that are usually whitened with pruina; spores brown, 2-celled, usually with thin, more or less uniform walls except for a thickened septum in young spores. CHEMISTRY: Upper cortex PD–, K–, KC–, C– (no lichen substances); medulla and (or) soredia sometimes containing yellow pigments that are K+ deep yellow and KC+ yellow to orange; medulla C– or rarely C+ pink (gyrophoric acid). HABITAT: On a variety of substrates but most frequently bark. COMMENTS: Because of their brownish color, *Physconia* species can closely resemble large *Phaeophyscia* species except for the pruina and squarrose rhizines. They can be confused with other brownish, K–, foliose lichens including some species of *Anaptychia*. *Physconia* species are the only common brown pruinose lichens in northern regions; farther south, species of *Pyxine* can be common and mistaken for *Physconia,* and in the arid west, several species of *Physcia* become heavily pruinose. Both *Physcia* and *Pyxine,* however, are much grayer and have K+ yellow or, in some *Pyxine* species, UV+ yellow reactions in the cortex.

KEY TO SPECIES

1. Soredia absent; apothecia common 2
1. Soredia present; apothecia rare 4

2. On the ground or over rocks in arctic or alpine sites; thallus loosely attached over entire surface . . . . *Physconia muscigena*
2. On bark or rocks at low elevations; thallus closely appressed to substrate 3

3. Isidia present, usually cylindrical but sometimes flattened. . . . . *Physconia elegantula*
3. Isidia absent, but often with lobules, especially on the apothecial margins . . . . *Physconia americana*

4.(1) Rhizines unbranched or dichotomously branched, pale tan to brown, predominantly on or close to the margins; lobes wrinkled or bumpy (rugose). . . . . *Physconia thomsonii*
4. Rhizines squarrose, black, more or less uniformly distributed; lobes smooth and even 5

5. Soralia discrete, crescent-shaped or lip-shaped on the lobe margins, containing very coarse soredia; lobe margins turned down . . . . *Physconia perisidiosa*
5. Soredia more or less continuous along the lobe margins; lobe margins turned up 6

6. Medulla pale yellow, K+ yellow, KC+ orange-yellow (secalonic acid). . . . . *Physconia enteroxantha*
6. Medulla white, K–, KC– (secalonic acid absent) 7

7. Upper cortex with thick-walled cells (scleroplectenchyma); mostly eastern North America. . . . . *Physconia detersa*
7. Upper cortex with thin-walled cells (pseudoparenchyma); western North America . . . *Physconia isidiigera*

Physconia americana
Fancy frost lichen

DESCRIPTION: Thallus gray-brown to brown, abundantly pruinose all over the upper surface; sometimes the pruina is spotty, leaving olive-brown or brown patches; lobes rather elongate and separate, or sometimes overlapping, 1–2 mm wide; without soredia or isidia but sometimes developing tiny lobules in the older parts of the thallus, and especially on the apothecial margins, giving the lichen a very decorative appearance; lower surface pale brown to black, with a thick mat of dark, entangled, squarrose rhizines. Apothecia almost always present, 1–3 mm in diameter. CHEMISTRY: Medulla PD–, K–, KC–, C– (no lichen substances). HABITAT: On bark or rock. COMMENTS: *Physconia americana* is one of the few North American species of *Physconia* without soredia. Another is the arctic-alpine *Ph. muscigena*. In plate 676, the lobules are particularly clear, but most specimens are more pruinose.

Physconia detersa
Bottlebrush frost lichen

DESCRIPTION: Thallus pale gray-brown to dark red-brown, uniform and often shiny except for the white pruina on the lobes (sometimes confined to the tips); lobes flat to slightly upturned and partly overlapping, (0.6–)1–2(–3) mm wide, bordered with marginal soralia containing brown to gray-green soredia (except for the growing tips); medulla white; lower surface black, producing a thick mat of black, squarrose rhizines. Apothecia rare. CHEMISTRY: Medulla and soredia PD–, K–, KC–, C– (no lichen substances). HABITAT: On bark of various kinds, occasionally on wood or rock. Often on roadside trees, even close to urban areas. COMMENTS: *Physconia detersa* is very common in eastern North America, where its distribution overlaps that of two other, equally common, marginally sorediate frost lichens: *Ph. enteroxantha* and **Ph. leucoleiptes**. *Physconia enteroxantha* is almost identical to *Ph. detersa,* but its medulla is distinctly yellowish (at least in spots) and turns K+ yellow, KC+ orange (secalonic acid). In *Ph. leucoleiptes,* some of the marginal soralia can become crisped and broken up into individual, more or less lip-shaped soralia, and the soredia (not the medulla) are K+ yellow, KC+ yellow-orange (also because of secalonic acid). See also *Ph. perisidiosa* and *Ph. isidiigera*.

676. (facing page) *Physconia americana* (damp) mountains, northeastern California ×5.6

677. (left) *Physconia detersa* Lake Superior region, Ontario ×3.9

678. *Physconia elegantula* White Mountains, eastern Arizona ×2.9

Physconia elegantula
Elegant frost lichen

DESCRIPTION: Thallus brown-gray to yellowish brown, heavily pruinose; lobes rather thick, becoming convex, 0.6–2.5 mm across; true, cylindrical isidia on the margins and upper surface of the lobes often branching and occasionally becoming flattened into tiny lobules; medulla white; lower surface pale brown; rhizines pale to dark, usually squarrose, occasionally unbranched. Apothecia common, up to 3 mm in diameter, with isidiate to lobed margins. CHEMISTRY: Medulla K–, KC– (no lichen substances). HABITAT: On bark in open, fairly arid sites. COMMENTS: No other *Physconia* has true isidia, although *Ph. isidiigera*, with marginal soralia, can have partially corticate soredia that resemble isidia. Isidiate species of *Melanelia* can occasionally become lightly pruinose, but the lobes are not as thick, the isidia are more slender, and the lower surface is generally dark brown to black with unbranched or forked rhizines.

Physconia enteroxantha
Yellow-edged frost lichen

DESCRIPTION: Thallus brownish gray to dark brown, with a white pruina over at least the tips of the lobes; medulla pale yellow or yellowish white, rarely white; lobes 0.6–2(–3) mm wide, with long, continuous, marginal soralia containing yellowish green, granular soredia; lower surface pale at the edge, becoming brown to black in the center, with black, squarrose rhizines. Apothecia rare. CHEMISTRY: Medulla K+ yellow, KC+ yellow-orange, sometimes only in parts of the thallus, C– (secalonic acid A). HABITAT: On bark, wood, and occasionally rock. COMMENTS: This species closely resembles *Ph. detersa*, having the same type of tissue in the upper cortex and the same kind of soralia. They are distinguished by the medullary chemical reaction. ***Physconia leucoleiptes,*** a common eastern species, has the yellowish pigment in the soralia rather than in the medulla, and has some terminal, almost lip-shaped soralia as well as marginal soralia. See also *Ph. isidiigera*.

Physconia isidiigera
Bottlebrush frost lichen

DESCRIPTION: Thallus closely resembles *Ph. detersa* (which is not found on the west coast), but with a different cortical anatomy (thin-walled cells in *Ph. isidiigera* and very thick-walled cells in *Ph. detersa*) and usually with more dissected lobes; soralia long, marginal, sometimes recurved but not lip-shaped; soredia finely granular to distinctly corticate and thus resembling isidia in some forms; lower surface pale brown to black, at least in the center of the thallus. CHEMISTRY: Medulla K–, KC– (no lichen substances). HABITAT: On bark. COMMENTS: *Physconia enteroxantha* is similar and also found in the west, but its medulla is pale yellow and K+ yellow, KC+ yellow-orange. See also *Ph. perisidiosa*.

679. *Physconia enteroxantha* Cascades, Oregon ×4.7

Physconia muscigena
Ground frost lichen

DESCRIPTION: Thallus variable, pale grayish brown to olive-brown or dark brown, lightly to heavily pruinose; lobes elongate to short, dissected, and densely overlapping, often somewhat concave, 1–3 mm wide; without soredia or isidia; lower surface black with a dense mat of black, squarrose rhizines. Apothecia common, up to 5 mm in diameter, with smooth to lobulate margins. CHEMISTRY: Medulla K–, KC– (no lichen substances). HABITAT: On the ground, on mosses, vegetation, or rocks, especially those frequented by birds, usually in lime-rich tundra habitats. COMMENTS: The only other *Physconia* without soredia or isidia is *Ph. americana*, a western bark-dwelling species that generally has a paler lower surface and flatter lobes. *Physconia muscigena* should be compared with *Parmelia omphalodes*, which can be brown and may occur on rocks or tundra soil.

Physconia perisidiosa
Crescent frost lichen

DESCRIPTION: Thallus gray-brown to dark brown, usually heavily pruinose, at least in patches; lobes mostly 0.5–1.5 mm across, relatively short, commonly overlapping like shingles, with small crescent- or lip-shaped soralia containing coarse soredia or isidia-like granules developing along the lobe margins; sometimes with tiny convex lobules at the thallus center; lower surface black except for extreme lobe tips or almost entirely pale, with squarrose rhizines. Apothecia rare. CHEMISTRY: Medulla K–, KC– (no lichen substances). HABITAT: On bark, less frequently on rock or soil. COMMENTS: The discrete, sorediate, marginal lobes distinguish this species from the more common *Ph. detersa, Ph. enteroxantha,* and *Ph. isidiigera,* in which the soredia are strictly marginal, rarely occurring on upturned, labriform lobe tips. In **Ph. leucoleiptes,** one finds somewhat crescent-shaped soralia containing finer soredia that are yellowish and react K+ yellow, KC+ yellow-orange (secalonic acid).

680. *Physconia isidiigera* Channel Islands, southern California ×4.4

681. *Physconia muscigena* southern British Columbia ×3.0

682. *Physconia perisidiosa* southern British Columbia ×4.1

683. *Physconia thomsonii* canyon country, northeastern Utah ×3.0

Physconia thomsonii
Hairy frost lichen

DESCRIPTION: Thallus gray-brown, with abundant white pruina over much of the thallus; lobes 1–2 mm wide, with soralia along the margins (sometimes extending around the lobe tips), containing coarsely granular, usually dark soredia; extremely tiny (10–15 μm) colorless hairs sometimes occur on the tips and edges of the lobes, mixed in with the pruina and therefore very inconspicuous. Medulla white; lower surface white to pale tan, with unbranched or forked (never squarrose) rhizines. Apothecia unknown. CHEMISTRY: Medulla K–, KC– (contains only an unidentified triterpene). HABITAT: On rock, especially in sagebrush juniper communities. COMMENTS: This relatively rare lichen is similar to *Ph. detersa,* but like the exclusively European **Ph. grisea,** it has a pale lower surface and unbranched rather than squarrose rhizines. It differs from *Ph. grisea* in the anatomy of its upper cortex. No other North American *Physconia* has pale cortical hairs, although they also occur on **Anaptychia ulotrichoides,** a nonsorediate lichen found in similar habitats in Utah, and on species of *Phaeophyscia*.

Pilophorus (8 N. Am. species) (syn. *Pilophoron*)
Matchstick lichens

DESCRIPTION: Primary thallus crustose, with an erect, fruticose secondary thallus with solid stalks in most species, (1–)5–15(–25) mm tall, both elements containing vegetative tissue with a green photobiont (*Trebouxia*). Crustose thallus usually well developed, gray-green to olive or brownish, continuous or areolate, producing small hemispherical, pink to brown or black cephalodia containing cyanobacteria (*Nostoc* or *Stigonema*). Tips of stalks usually with black, spherical to elongate apothecia; spores ellipsoid to elongate ellipsoid, colorless, 1-celled. CHEMISTRY: All species contain at least atranorin and zeorin, frequently with other substances. HABITAT: On noncalcareous rock in open or shaded sites. COMMENTS: Species of *Cladonia* can resemble *Pilophorus* in size and color but have a squamulose primary thallus without cephalodia, and the stalks are hollow. *Stereocaulon* has flat, red-brown apothecia, and the cephalodia are usually on the stalks,

684. *Pilophorus acicularis* (wet) coastal Oregon ×4.4

not the primary thallus. Species of *Dibaeis* and *Baeomyces* have pale pink to brown apothecia and lack cephalodia, and the stalks are rarely more than 5 mm high.

KEY TO SPECIES

1. Stalks sorediate at the tip, under 5 mm tall . . . . *Pilophorus cereolus*
1. Stalks not sorediate, over 5 mm tall 2

2. Apothecia elongate, cylindrical; thallus olive to brown . . . . *Pilophorus clavatus*
2. Apothecia spherical; thallus gray-green. . . . . *Pilophorus acicularis*

Pilophorus acicularis (syn. *Pilophoron aciculare*) Devil's matchstick

DESCRIPTION: Primary thallus pale gray-green, granular to verruculose; erect stalks unbranched or occasionally forked once, 5–25 mm tall, becoming hollow when mature, each stalk terminating in a small black, spherical apothecium; surface of the stalks coarsely verrucose, often with black areas showing through; areoles on the stalks closely resembling those of the primary thallus. Cephalodia very large, pinkish to brown, convex to hemispherical, on the primary crustose thallus and sometimes also on the base of the stalks. CHEMISTRY: Thallus and stalks PD–, K+ yellow, KC–, C–. HABITAT: On rock in open sites. COMMENTS: This is the most common species of *Pilophorus*, a true pioneer on newly exposed rock surfaces, often growing with *Placopsis*. Like other pioneer lichens such as *Placopsis* and *Stereocaulon*, *P. acicularis* has cyanobacteria-containing cephalodia, an ecological advantage discussed in Chapter 8 ("Colonization of Rock"). The only other tall, gray *Pilophorus* is ***P. robustus,*** a very rare species that has abundantly branched stalks and grows on tundra soil. ***Pilophorus nigricaulis,*** another relatively rare matchstick lichen with black, spherical apothecia, is restricted to the extreme west coast of British Columbia and Alaska. Its blackened, horny inner medulla of the stalk distinguishes it from other species, except perhaps the even rarer *P. vegae*, which has a continuous, relatively smooth surface with frequently branched stalks lacking apothecia. ***Pilophorus***

685. *Pilophorus cereolus* (damp) Southeast Alaska ×2.9

686. *Pilophorus clavatus* Cascade Range, Oregon ×2.9

fibula is a rare species of the northern Appalachians that has extremely short (under 1 mm) stalks and an areolate to squamulose primary thallus. The apothecia are black, spherical, and about 1 mm in diameter.

Pilophorus cereolus (syn. *Pilophoron cereolus*)
Powdered matchstick

DESCRIPTION: Primary thallus gray-green, areolate, partly sorediate, with small pinkish brown to black, hemispherical cephalodia; stalks short (mostly under 4 mm), unbranched, farinose sorediate, at least on the upper half. Apothecia rarely produced, black, spherical. CHEMISTRY: Cortex and soralia PD–, K+ yellow, KC–, C–. HABITAT: On rock, usually in shaded, rather humid sites. COMMENTS: No other species of *Pilophorus* is sorediate, but the often sterile *Stereocaulon pileatum* can resemble it closely. The latter, however, has stalks that can be somewhat branched and are often smooth at the base rather than areolate as in the *Pilophorus*. Its primary thallus is more granular than areolate, and the soralia are KC+ pinkish purple (lobaric acid).

Pilophorus clavatus (syn. *Pilophoron hallii*)
Tapered matchstick

DESCRIPTION: Thallus rather dark, typically olive to brownish, less frequently dark gray-green, composed of separate to contiguous areoles or corticate, flat granules, with scattered brownish black, lumpy cephalodia; stalks the same color as the primary thallus, with an areolate surface, unbranched, 5–15(–20) mm tall, terminated by an elongated black apothecium, often slightly broader at the top. CHEMISTRY: Cortex PD–, K+ yellow, KC–, C–; medulla PD–, K–, KC–, C–. HABITAT: On rock, often in shaded forest sites, but also in exposed areas. COMMENTS: This is the only *Pilophorus* with elongate apothecia. It often grows with *P. acicularis*, which has a much paler, gray thallus and spherical apothecia.

Placidium (9 N. Am. species)
Stipplescale lichens, earthscales

DESCRIPTION: Gray to dark brown squamulose to almost foliose lichens, dark green when wet; upper cortex composed of large, distinct cells; attached to the substrate with loose bundles of fungal hyphae and sometimes discrete rhizines. Photobiont green (*Myrmecia,* up to 15 µm in diameter). Fruiting bodies perithecia, almost entirely buried in the thallus (rarely between the squamules) with only the mouth appearing at the surface as a black dot; perithecia without algae in the hymenium; paraphyses absent; spores colorless, 1-celled, ellipsoid to almost globose, arranged in a single row in the asci; pycnidia marginal or laminal. CHEMISTRY: No lichen substances. HABITAT: All but a few species grow on bare earth or on rocks, usually in dry, sunny sites. COMMENTS: The closely related genus *Catapyrenium,* taken in the narrow sense as we do here, is distinguished from *Placidium* by its upper cortex, which is composed of very small, poorly defined cells, and by its smaller photobiont cells (up to 10 µm in diameter). *Heteroplacidium* includes mostly saxicolous species and differs in details of the anatomy and conidia. An example is *H. acarosporoides,* an interesting southwestern species, which has small (up to 1.5 mm), shiny red-brown convex squamules and grows on granite and sandstone in sunny habitats. *Endocarpon* is a genus of similar lichens, mostly on rock, but with large muriform spores and with algae within the perithecium. To identify species of *Placidium* or *Catapyrenium* with confidence, prepare thin vertical sections of the thalli and examine them under the microscope. Details of the rhizines and wefts of attaching hyphae as well as features of the upper and lower cortex, structure of the medulla, and characters of the perithecia are all important. To examine the rhizines and attachment hyphae, clean the soil off the lower surface while the thallus is submerged in water in a small dish. As some of these characters are difficult to discern and interpret, only a few of the more common and distinctive species are described here.

KEY TO SPECIES: See Key G.

Placidium lacinulatum (syn. *Catapyrenium lacinulatum*)
Brown stipplescale

DESCRIPTION: Thallus consisting of thick red-brown squamules, green when wet (as shown in plate 688), without any pruina, dispersed or contiguous but rarely overlapping, with black dots (revealing the buried perithecia) scattered over the surface; squamules ap-

687. *Placidium lacinulatum* Sonoran Desert, southern Arizona ×3.3

688. *Placidium lacinulatum* (wet) Sonoran Desert, southern Arizona ×3.2

689. *Placidium tuckermanii* (wet) Ouachitas, Arkansas ×4.3

proximately 2–3 mm across, some lifting at the edges; lower surface pale or light brown; cells of lower cortex irregularly arranged; pale rhizines and clumps of hyphae attaching the thallus to the substrate; walls of the perithecia (the excipulum) colorless; spores 12–17 × 5.5–7.5 μm. HABITAT: Common on soil in dry areas. COMMENTS: This is one of the more common brown *Placidium* species in North America, but it is difficult to distinguish from *P. squamulosum* and *P. lachneum,* all of which have been more or less lumped under "*Dermatocarpon hepaticum*" or *D. lachneum* in the past. *Placidium squamulosum,* widely distributed but the only species common in the east, has much sparser clumps of attachment hyphae and lacks anything that could be called rhizines; *P. lachneum,* a less common but widely distibuted species, has a black lower surface and lower cortex composed of vertically arranged columns of cells. *Catapyrenium cinereum* (syn. *Dermatocarpon hepaticum* in the strict sense), with brown to gray-pruinose, notched squamules 1–3 mm wide, is common in the Rocky Mountains. It has a black lower cortex and dark attachment hyphae. What was called *Catapyrenium plumbeum* in some older literature is, for the most part, *Verrucaria inficiens,* a common lichen in dry montane habitats through the western interior from Arizona and New Mexico to Saskatchewan. It forms a thick, bluish gray, areolate crust on dry calcareous rocks, but begins its development as a parasite on species of *Staurothele*, especially *S. areolata*. The areoles of *V. inficiens* are mostly 0.4–0.9 mm in diameter, somewhat constricted at the base, and very rough-textured (more scabrose than pruinose). The black perithecia are partly emergent, one to several per areole, and the spores are 12–16 × 7.5–10 μm.

Placidium tuckermanii (syn. *Catapyrenium tuckermanii, Dermatocarpon tuckermanii*) *Tree stipplescale*

DESCRIPTION: Thallus consisting of rounded brown lobes (bright green when wet), 2–5 mm wide, dotted with dark brown perithecia; lower surface pale, with tufts of attachment hyphae. CHEMISTRY: All reactions negative. HABITAT:

690. *Placopsis lambii* (wet; soralia sparse, on right side of thallus) coastal Oregon ×3.6

On bark of hardwoods, especially oaks, usually at the mossy base; rarely on limestone. COMMENTS: This is the only common *Placidium* on bark; the others grow on soil or rock. *Catapyrenium psoromoides,* a rare bark-dwelling *Catapyrenium* found in California, has gray, somewhat pruinose lobes. *Endocarpon pusillum* can sometimes occur on bark, but the squamules are smaller (under 3 mm), and the spores are muriform and brown, 2 per ascus.

Placopsis (3 N. Am. species)
Bull's-eye lichens

DESCRIPTION: Thallus crustose, usually with a lobed margin; pinkish white to pinkish or yellowish brown, often turning very green or greenish white when wet; large, pink to brown, strongly or weakly lobed cephalodia almost always present in the common species, located at the center of each round thallus. Primary photobiont green (*Chlorella?*); cephalodia containing *Stigonema* or *Scytonema*. Apothecia common, lecanorine, disks pink to brown; asci cylindrical, slightly thickened at the tip, K/I– or very pale blue; spores colorless, ellipsoid, 8 per ascus, 1-celled, 13–18 × (6–)7–9 (–12) μm. CHEMISTRY: Thallus PD–, K–, KC+ red, C+ pink (gyrophoric acid, sometimes with 5-*O*-methylhiascic acid). HABITAT: On noncalcareous rock in open sites, often occurring as a primary invader of newly exposed surfaces. COMMENTS: The two common North American species *P. gelida* and *P. lambii,* as well as the rare, west coast, isidiate species *P. cribillans,* all have the round, lobed thalli and centrally located, lobed cephalodia typical of the genus. *Placopsis roseonigra,* however, is an exception, with a thallus consisting of pinkish white areoles dispersed on a black bed of cyanobacteria (functioning as a cephalodium). This rather rare species occurs on wet rocks along the British Columbia coast north to southeastern Alaska. The bull's eye lichen gets its name from the large cephalodium that usually sits right in the center of the thallus. In view of this position, it is likely that the lichen establishes itself first as a cephalodium packed with nitrogen-fixing cyanobacteria, perhaps giving the young lichen a boost of nitrogen on the stark new surfaces it tends to invade.

691. *Placynthiella oligotropha* (distant view) Lake Superior region, Ontario

KEY TO SPECIES: See Keys C and F.

Placopsis lambii
Pink bull's-eye lichen

DESCRIPTION: Thallus usually pale pinkish white when dry, pale greenish when wet, but varying to a pinkish brown, somewhat shiny at the lobe tips; cephalodia thick, round, pinkish to yellowish brown, lobed only at the edge (although frequently with fissures that extend to the center); soralia are hemispherical mounds of greenish soredia, less frequently dark gray to brown at the surface, often arranged in concentric rings, sometimes sparse (as shown in plate 690). Apothecia common, disks pink, usually pruinose. CHEMISTRY: Containing 5-O-methylhiascic acid and gyrophoric acid. HABITAT: On exposed rocks. COMMENTS: Although several characters distinguish *P. lambii* from *P. gelida,* which has a similar distribution, each species is variable enough to cause problems in identification. *Placopsis gelida* is almost always relatively dark, often rosy brown; the cephalodial lobes are deep, extending right to the center of each cephalodium; the thallus (especially the lobe tips) is dull, almost pruinose; the soralia are irregular in shape, are clearly excavate (never in a mound), not in concentric rings, and contain dark gray soredia (when they are fresh); the thallus contains only gyrophoric acid. The map shows the combined ranges of both species.

Placynthiella (4 N. Am. species)
Tar-spot lichens

DESCRIPTION: Dark olive-brown to red-brown or almost black crustose lichens with granular to isidiate thalli. Photobiont green (*Chlorella?*). Apothecia lecideine, dark brown to black, convex, with thin to disappearing margins, internal tissues including exciple and hypothecium usually dark red-brown; paraphyses abundantly branched and darkly pigmented at the tips; asci K/I– or pale, even at the tip, *Trapelia*-type (see Fig. 14l); spores colorless, ellipsoid, 1-celled, thin-walled, often containing a few large oil drops, 8 spores per ascus. CHEMISTRY: Frequently with abundant, or at least traces of, gyrophoric acid (PD–, K–, KC+ pink, C+ pink), or lacking lichen substances (all reactions negative). HABITAT: On soil, especially sandy soil, or on rotting wood, rarely rock. COMMENTS: The K/I– ascus tips and pigmented paraphyses tips distinguish *Placynthiella* from most other crustose lichens with biatorine apothecia and 1-celled spores, such as *Micarea* and *Biatora*.

KEY TO SPECIES: See Keys C and F.

Placynthiella oligotropha

DESCRIPTION: Thallus dark olive- to red-brown, composed of granules 100–300 μm (0.1–0.3 mm) in diameter massed into an almost continuous crust. Apothecia very dark reddish brown, almost black, 0.2–0.5 mm in diameter, flat to convex, with a rough surface and a thin, disk-colored margin that disappears in older apothecia; spores 10–15 × 4–7 μm. CHEMISTRY: No lichen substances. HABITAT: On bare peaty or sandy soil, or on well-rotted lignum, usually in the open or in partial shade. COMMENTS: *Placynthiella uliginosa* closely resembles *P. oligotropha* except the thallus granules are much smaller (less than 150 μm in diameter). These tar-spot lichens can form large patches over disturbed, sandy soil and are probably important in helping to stabilize the soil and prevent erosion. They occur less frequently on wood. *Placynthiella icmalea* is restricted to wood and has a thallus of reddish brown, closely packed isidia that react C+ red (under the microscope) because of gyrophoric acid. It is very rarely fertile, whereas *P. oligotropha* and *P. uliginosa* are usually fertile if the thallus is old enough. Dark forms of *Trapeliopsis flexuosa* can resemble *Placynthiella* on wood, but the *Trapeliopsis* has a greenish gray thallus, flat, lead-gray apothecia, and smaller spores (7–9 × 2.5–4 μm).

Placynthium (8 N. Am. species)
Ink lichens

DESCRIPTION: Very dark olive to blackish crustose lichens, often lobed at the margins; lower surface or basal tissue almost always with a blue-green pigment, sometimes with a blue-black prothallus. Photobiont blue-green (usually *Dichothrix*). Apothecia lecideine (lecanorine in *P. stenophyllum*), black, flat, usually shiny; epihymenium blue-green, hypothecium reddish brown, paraphyses usually unbranched except at the tip; spores colorless, mostly 4-celled but can be 2–8-celled, ellipsoid. CHEMISTRY: No lichen substances. HABITAT: On rocks of different kinds. COMMENTS: Sterile specimens of some lobed species can be superficially similar to species of *Vestergrenopsis*, *Koerberia*, or even the foliose *Massalongia*, but all these are generally paler and have pale lower surfaces, and the apothecia are brown.

KEY TO SPECIES: See Keys F and G.

Placynthium nigrum

DESCRIPTION: Thallus very dark olive to black, composed of tiny, minutely lobed squamules that sometimes become densely covered with flattened, granular, or prostrate cylindrical isidia; thallus often cracked into irregular, rather thick areoles; thallus in patches surrounded by a distinctive blue-black, fibrous prothallus. Apothecia dark red-brown to pure black, flat or slightly convex, 0.3–0.8 mm in diameter, with a low, black lecideine margin disappearing only in the oldest apothecia; spores mostly 2–4-celled, narrowly ellipsoid or, rarely, ellipsoid, mostly 10–17 × 3.5–6 μm. HABITAT: On limestone, frequently on old concrete. COMMENTS: Other species of *Placynthium* have no prothallus and are more distinctly lobate at the margins, or are entirely subfoliose. *Placynthium asperellum* is one of the more common examples of these, with dark olive, slender, branched, marginal lobes, 0.5–2 mm long and 0.06–0.2 mm across, flat, convex, or slightly grooved along their length, developing slender, rather long and branched isidia on the older portions, sometimes entirely isidiate, crowded into irregular areoles in the center of the old thallus. The lower surface of the thallus is blue-green (purely coincidental with the blue-green photobiont) and is made up of rectangular cells, often with blue-green hyphae growing out of the lower cortex. *Placynthium flabellosum* has flatter, broader lobes (0.2–0.4 mm) bearing flattened isidia. Both have broad distributions in North America: *P. flabellosum* is southern boreal–montane, on siliceous rocks periodically covered with water, beside streams and lakes; *P. asperellum* is boreal to arctic-alpine, mainly on limy rocks, wet or dry. The relatively rare (or overlooked) *P. stenophyllum* has brownish thalli only 1 cm in diameter with extremely slender, branched, cylindrical lobes less than 3 mm long and 0.1 mm across. It is unusual because it has a pale lower surface and rhizines. It also has a broad montane distribution.

692. *Placynthium nigrum* Lake Superior region, Ontario ×5.1

Platismatia (6 N. Am. species)
Rag lichens

DESCRIPTION: Pale gray-green to slightly brownish, medium to rather large foliose lichens with broad, rather crinkled, ascending lobes, sometimes with pseudocyphellae, commonly with isidia, rarely with soredia; lower surface brown to black, often broadly splotched with white patches close to the margins, with few rhizines, pale or black, usually unbranched. Photobiont green (*Trebouxia*). Apothecia uncommon except in a few species, mostly 4–15 mm in diameter, on or close to the lobe margins, with yellow-brown to red-brown disks; spores colorless, 1-celled, 5–10 × 3–5 µm; pycnidia common in only one or two species, black, immersed in the thallus and appearing as black dots, mainly along or close to the lobe margins. CHEMISTRY: Cortex PD–, K+ yellow, KC–, C–; medulla with all reactions negative (caperatic acid and atranorin in all species) except for *P. lacunosa*, which has fumarprotocetraric acid in addition. HABITAT: Mostly forest species on the trunks, branches, and twigs of conifer trees; some species on rock. COMMENTS: These rather large, "crumpled" lichens resemble several other genera in the large and diverse family Parmeliaceae, especially *Parmotrema* (lacking rhizines at the lobe margins, but usually with marginal cilia), *Cetrelia* (always with pseudocyphellae, usually sorediate, and usually with C+ or KC+ pink depsides), and *Asahinea* (an arctic-alpine genus lacking rhizines and found on rock and turf). All these lichens have broad, round lobes usually with ascending, wavy margins.

KEY TO SPECIES (including *Esslingeriana*)

1. Lobes 0.5–5 mm wide, often elongate or finely divided and sometimes pendent 2
1. Lobes (3–)5–20 mm wide, rarely pendent 4

2. Pycnidia on the thallus surface, common; thallus with a yellowish tint; medulla yellowish and KC+ pink, at least on lobe tips and apothecial margins .... *Esslingeriana idahoensis*
2. Pycnidia, when present, along the lobe margins; thallus without a yellowish cast, KC– 3

3. Isidia absent .... *Platismatia stenophylla*
3. Isidia present along lobe margins .... *Platismatia herrei*

4.(1) Soredia and isidia absent 5
4. Soredia or isidia present 6

5. Northeastern U.S., and southeastern Canada; medulla PD– .... *Platismatia tuckermanii*

5. Western; medulla PD+ red (fumarprotocetraric acid)... .... *Platismatia lacunosa*

6. Isidia and (or) soredia present on lobe margins; lobe surface relatively smooth, without pseudocyphellae ... . *Platismatia glauca*
6. Isidia laminal; surface with a network of sharp ridges and depressions, with white pseudocyphellae .... *Platismatia norvegica*

Platismatia glauca
Varied rag lichen, ragbag

DESCRIPTION: Thallus pale greenish gray, often browning at the edges, uniform, with no pseudocyphellae and few maculae; lobes (3–)5–20 mm wide, ascending and irregular, margins often divided into small rounded to angular lobes with abundant granular soredia-isidia mixtures, only isidia (as shown in plate 693), only soredia (plate 694), or large, branched subfruticose outgrowths, often with some laminal isidia as well; lower surface brown and shiny at the edge, black in the center, but with scattered to continuous patches of ivory-white close to the edge; rhizines sparse. Apothecia and pycnidia very rare. HABITAT: Extremely common in spruce, fir, or Douglas fir forests, especially on branches. COMMENTS: *Platismatia glauca* is the only rag lichen with marginal soredia. The species varies, however, between purely sorediate and almost entirely isidiate, with all intermediates. The lobes can be quite broad, or they can be relatively narrow, resembling the less common *P. herrei*.

Platismatia herrei
Tattered rag lichen

DESCRIPTION: Thallus pale greenish gray, often splotched with black, but without pseudocyphellae; lobes narrow, 0.5–3(–4) mm wide, flat to curled inward, forming channels; margins becoming finely divided into cylindrical or branched isidia, with clumps of isidia also occurring to some extent on the surface of the lobes; lower surface black to brown, but with small to large white patches close to the tips; rhizines rarely produced. Apothecia and pycnidia uncommon. HABITAT: On branches of conifers in coastal forests. COMMENTS: Much like the more common *P. glauca*, but the lobes rarely exceed 3 mm in width, and

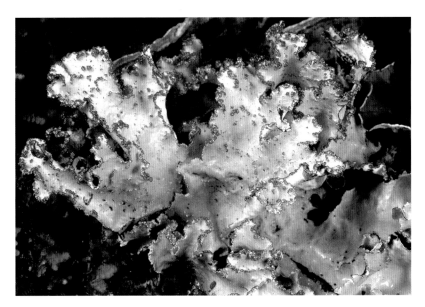
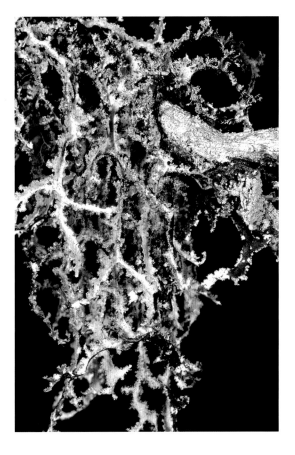

693. (left) *Platismatia glauca* (isidiate form) Columbia River, Washington ×2.0

694. (below left) *Platismatia glauca* (sorediate form) Klamath Range, northwestern California ×1.9

695. (below right) *Platismatia herrei* Oregon coast ×1.7

696. *Platismatia lacunosa* Southeast Alaska ×2.6

the margins are never sorediate, only isidiate. *Platismatia stenophylla* has similar narrow lobes, but without isidia.

Platismatia lacunosa
Crinkled rag lichen

DESCRIPTION: Thallus very pale greenish gray, almost white, although often appearing darkened and soiled with epiphytic algae in the deep pits between the sharp ridges that cover the upper surface; lobes 6–16 mm wide, relatively closely attached to the substrate compared with other species of *Platismatia*; pseudocyphellae sometimes present on the crests of the ridges, but almost invisible against the pale thallus surface; without soredia or isidia, but black, embedded pycnidia often visible along the thallus margins; lower surface dark brown or black to white, often mottled; rhizines sparse to abundant. Apothecia occasional, marginal, with large, folded, brown disks, 4–20 mm in diameter. CHEMISTRY: Medulla PD+ red, K± brown, KC–, C– (fumarprotocetraric acid in addition to caperatic acid and atranorin). HABITAT: On bark of many kinds as well as directly on rock in coastal forests. COMMENTS: The PD+ red medulla distinguishes *P. lacunosa* from all other species of the genus. *Platismatia norvegica* is very similar in appearance but has fairly abundant isidia over the thallus surface, especially on the ridges and wrinkles, and the pits and ridges are shallower than those of *P. lacunosa*. The eastern species *P. tuckermanii*, also with pits and ridges and lacking isidia, has a more crumpled appearance.

Platismatia norvegica
Oldgrowth rag lichen

DESCRIPTION: Thallus pale to dark greenish gray, browned at the edges, with sharp ridges and moderately deep depressions between them; lobes round, 5–25 mm wide; pseudocyphellae easily seen, especially on the ridges; finely isidiate on the ridges and along some margins. Apothecia and pycnidia rare. HABITAT: On bark and rocks in moist localities along both coasts, in old growth forests farther inland. COMMENTS: This closely resembles *P. lacunosa*, which has more pronounced ridges and deeper depressions, lacks isidia, and has a PD+ red medulla.

Platismatia stenophylla
Ribbon rag lichen

DESCRIPTION: Thallus pale gray to greenish gray, often brown at the edges, uniform or spotted with white maculae, but without pseudocyphellae; forming rounded cushions of narrow, branched lobes (0.5–)1–4 mm wide, often strongly curved inward; lower surface brown to white at the growing tips and black in the center, without rhizines. Apothecia occasional, at the lobe tips, red-brown and shiny, 4–10 mm in diameter; black pycnidia embedded in the lobe margins. HABITAT: On trees, mostly in humid coastal forests. COMMENTS: This rag lichen has the general aspect of the more common *P. herrei*, but has no isidia or soredia and is somewhat darker.

697. *Platismatia norvegica* Southeast Alaska ×2.0

698. *Platismatia stenophylla* (wet) Coast Range, Oregon ×1.4

Platismatia tuckermanii
Crumpled rag lichen

DESCRIPTION: Thallus pale gray to greenish gray, often soiled with algae growing on the surface; lobe edges brown to black; spotted with irregular white maculae and sometimes small, round pseudocyphellae; lobes 4–20 mm wide, often giving a rounded, "crumpled" appearance, with fairly deep depressions and sharp ridges; lower surface usually with large, irregular white blotches, or uniformly shiny brown. Apothecia common, along the margins or close to them, shiny red-brown, 2–10 mm in diameter; black pycnidia embedded in the lobe margins. CHEMISTRY: Caperatic acid and atranorin. HABITAT: On bark and wood, mainly of conifers. COMMENTS: Except for the oceanic coastal region, only two common rag lichens are found in the east, one with marginal isidia or soredia (*P. glauca*) and the other with apothecia (*P. tuckermanii*). Small, browned specimens of *P. tuckermanii* might be mistaken for species of *Tuckermannopsis*, but the marginal pycnidia of the latter genus protrude as little cylinders.

699. *Platismatia tuckermanii* Appalachians, North Carolina ×2.2

Pleopsidium (2 N. Am. species)
Gold cobblestone lichens

DESCRIPTION: Bright yellow to chartreuse crustose lichens with areolate thalli, sometimes lobed at the periphery; without soredia or isidia; upper cortex composed of short-celled prosoplectenchyma. Photobiont green (unicellular). Apothecia immersed in the thallus areoles (cryptolecanorine) but can appear to be lecanorine if there is one apothecium per areole; disks yellow to brown; asci K/I+ blue, at least in the lower half of the expanded tholus, similar to the *Lecanora*-type; tiny ellipsoid spores, more than 8 per ascus. CHEMISTRY: All reactions on the cortex and medulla negative (often containing fatty acids); bright yellow pulvinic acid pigments in the cortex. HABITAT: On rocks in open, dry, often hot sites. COMMENTS: *Pleopsidium* differs from yellow species of *Acarospora* (e.g., *A. schleicheri*) in features of the ascus and thallus cortex. The ascus tip in *Acarospora* is only moderately thickened and is K/I–. The cortical differences are difficult to see without thin, stained sections. Yellow *Rhizocarpon* species have black apothecial disks and contain at most 8 spores per ascus. *Caloplaca* species are generally more orange, the apothecia are rarely immersed, and the cortex reacts K+ dark red-purple; they also have only 8 spores per ascus.

KEY TO SPECIES: See *Acarospora*.

Pleopsidium flavum (syn. *Pleopsidium oxytonum*)
Gold cobblestone lichen

DESCRIPTION: Thallus brilliant yellow, with radiating lobes 0.5–2 mm at tips, convex and folded, often quite rough on the surface, becoming areolate in the older central portions, with broad, pale to dark yellowish brown apothecia having persistent thalline margins. DISTRIBUTION: Widespread in the west from California to the dry interior, probably as far north as the Canadian prairies. COMMENTS: *Pleopsidium chlorophanum* (syn. *Acarospora chlorophana*), also found in the western and central region, is similar in form but has a smooth thallus, more convex apothecia with thalline margins that disappear in older parts of the thallus, and the disks are yellower (i.e., not as brown).

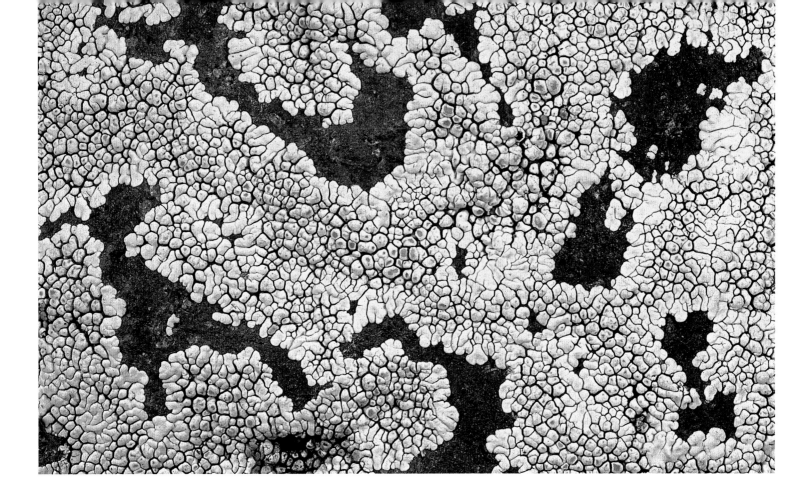

Polychidium (3 N. Am. species)
Woollybear lichens

DESCRIPTION: Pale green to dark brown subfruticose lichens with cyanobacteria (*Scytonema* or *Nostoc*) as the photobiont, forming tiny cushions consisting of very slender, dichotomously branched filaments. The cyanobacteria form the core of the branches, with the fungal component forming a kind of cellular cortex often composed of cells shaped like pieces of a jigsaw puzzle. Apothecia biatorine, pale to dark brown, very common on some species and rare on others; spores 1- or 2-celled, colorless, ellipsoid to elongate, 8 per ascus. CHEMISTRY: No known lichen substances. HABITAT: On bark, mosses, and sometimes rocks in rather humid sites. COMMENTS: *Polychidium* species can resemble other dwarf, branched, cushion-forming lichens with cyanobacteria, such as *Spilonema* or *Dendriscocaulon*. *Ephebe* is generally found only on wet rocks or wood, has longer branches, and is more prostrate.

KEY TO SPECIES: See Key B.

Polychidium contortum
Woollybear lichen

DESCRIPTION: Thallus pale gray-olive to blue-green, sometimes with a violet caste, forming woolly cushions of intricately branched filaments 50–100 μm thick, containing a core of undulating filaments of the cyanobacterium *Scytonema;* cortical cells strongly lobed, shaped like the pieces of a jigsaw puzzle. Apothecia pale brown, arising from pycnidia. HABITAT: On the twigs and branches of trees, especially conifers, in very humid coastal forests; perhaps common but easily overlooked. The cushions are especially inconspicuous because they often grow on the undersides of the tree branches and among the needles of the twigs. DISTRIBUTION: Oregon to southeastern Alaska, mainly along the humid coast. COMMENTS: *Polychidium dendriscum,* known from southeastern Alaska, is extremely similar but appears to be very rare. The species can be distinguished only by making sections of their tiny filaments. *Polychidium contortum* has twisted cyanobacterial filaments and a loose, irregular medulla in the thick branches; *P. dendriscum* has straight cyanobacterial filaments and dense, straight hyphae in the branches.

700. *Pleopsidium flavum* northern Arizona ×5.1

701. *Polychidium contortum* Oregon coast ×5.8

702. *Polychidium muscicola* (damp) northern Sierra Nevada, California ×4.4

Polychidium muscicola
Moss-thorns

DESCRIPTION: Thallus forming tiny cushions of spiny, round, shiny dark brown branches less than 4 mm long and 0.2 mm thick, containing *Nostoc* as its photobiont. Usually fertile; apothecia dark red-brown, up to 2 mm across, with persistent margins; spores 2-celled, ellipsoid to fusiform, usually pointed at the tips, 20–28.5 × 5.5–7(–8) μm (in the west and north) and (12–)14–22 × 4.5–6.5 μm (mainly in the east). HABITAT: Nestled among the mosses of exposed or shaded rocks. COMMENTS: This inconspicuous fruticose lichen looks just like a miniature *Cetraria aculeata*. It also can resemble *Pseudephebe pubescens* and *P. minuscula*, lichens that grow on bare, exposed rock rather than mossy rock. A broken branch of the *Polychidium*, however, will reveal that it contains blue-green cyanobacteria rather than green algae. The tiny, branched, and almost terete lobes of **Leptogium tenuissimum** form small rosettes only a few millimeters across, but in general aspect it resembles *Polychidium muscicola*. Their spores, however, are muriform. It is a widely distributed but extremely inconspicuous lichen. Isidiate species of *Leptogium*, such as *L. lichenoides*, grow in similar habitats and contain *Nostoc*, but close examination shows that these lichens are clearly foliose, at least at the base, and they are jelly-like when wet.

Polysporina (3 N. Am. species)
Coal-dust lichens

DESCRIPTION and HABITAT: Inconspicuous crustose lichens with the thalli growing within the upper layers of the rock substrate between the crystals, and seen only by prying up rock particles to reveal the mixture of white fungal and grass-green algal tissues. Photobiont *Myrmecia*. Apothecia abundant and the only parts of the lichen clearly visible on the rock surface. Apothecia black, lecideine, with a black, rough, brittle, carbonized margin and epihymenium, and a colorless hypothecium; hymenium IKI+ reddish orange, K/I+ blue (hemiamyloid); paraphyses somewhat branched and anastomosing, not expanded at the tips; ascus walls K/I–; spores tiny, elliptical to cylindrical, 3–5 × 1.5–2 μm, hundreds per ascus. CHEMISTRY: Probably

703. *Polysporina simplex*, central Massachusetts ×5.5

no lichen substances. COMMENTS: *Polysporina* grows within rock (endolithic), as can species of *Lecidea, Sarcogyne, Porpidia,* **Clauzadea,** and even *Caloplaca*. Of these, however, only *Sarcogyne* also has black apothecia with many-spored asci. In *Sarcogyne*, the epihymenium is a clear red-brown and does not become rough and blackened with carbonized tissue, the hymenium is IKI+ blue (amyloid), the ascus walls are light blue in K/I, and the paraphyses are unbranched and commonly expanded at the tips into a little cap.

KEY TO GENUS: See Key F.

Polysporina simplex
Common coal-dust lichen

DESCRIPTION: Virtually no thallus showing on the rock surface. Apothecia lecideine, crowded together in groups following cracks in the rock, mostly 0.3–1 mm in diameter, angular in outline and often appearing as if they are crushed from the sides; disks entirely obscured by irregular clumps of carbonized sterile tissue that emerge through the hymenium from below. HABITAT: On siliceous rock, especially sandstone, usually in the open. DISTRIBUTION: Probably widespread in the boreal and temperate regions; also reported from the Arctic. COMMENTS: *Sarcogyne privigna* is often mistaken for *Polysporina simplex*, although the former has apothecia with red-brown disks (when wet) with only scattered bits of carbonized tissue, if any, on the surface. Other distinctions between *Sarcogyne* and *Polysporina* are mentioned in the Comments under the genus. ***Polysporina urceolata*** is a related species on limestone, with tiny (under 0.4 mm in diameter) black apothecia having very thick, carbonaceous margins, usually cracked in several places, and a very small disk. The spores are ellipsoid to spherical, mostly 3–4 × 2–3 μm. Some unusually large specimens of *P. simplex* can resemble **Sarcogyne bicolor,** a southwestern species with a white, areolate thallus and smaller spores (2–3 × 1 μm). See Comments under *Sarcogyne regularis* for notes on other *Sarcogyne* species that can be confused with *Polysporina*.

704. *Porina heterospora* coastal plain, North Carolina ×8.5

and rock, differs in its asci (more cylindrical, with a conspicuously thickened tip); it has 2–10-celled spores. See Comments under *Porina heterospora*.

KEY TO SPECIES: See Key D.

Porina heterospora

DESCRIPTION: Thallus greenish gray to olive, thin, smooth to minutely granular; perithecia prominent, hemispherical, thallus-covered, 0.4–0.7 mm in diameter; with pale yellowish ostioles; walls of perithecia contain large colorless crystals; spores strongly tapered, 10–14-celled, 60–125 × 9–15 μm, with a gelatinous sheath. HABITAT: On bark of deciduous trees. COMMENTS: This is a common, coastal plain lichen on bark. The large, tapering spores and prominent thallus-colored perithecia make it rather distinctive. **Strigula stigmatella,** a fairly common northeastern species, may also key out here. Its thallus forms a smooth, dark greenish covering over bark and mosses, with low black bumps scattered here and there revealing the buried perithecia. Its perithecial walls are pale, but with a black ostiole area. The paraphyses are slender, unbranched to sparsely branched, and persistent. Its asci are cylindrical, containing mostly 8-celled, colorless, fusiform spores, 24–41 × 5.5–7.5 μm. It is commonly found on the bark of white cedar (*Thuja occidentalis*) in lowland forests and wooded bogs, usually in shade. *Porina heterospora* is included within *P. guaranitica* by some lichenologists.

Porpidia (23 N. Am. species)
Boulder lichens

DESCRIPTION: Crustose lichens on rock with thalli that are thick to thin, continuous or composed of dispersed areoles, often endolithic, white to ashy gray, orange or becoming orange because of iron compounds in the rock substrate. Photobiont green (*Trebouxia*). Apothecia lecideine; disks black or very dark brown, lightly or heavily coated with white pruina in some taxa; paraphyses branched and net-like, not conspicuously expanded at the tips; hymenium rather high in most species, 60–130 μm; epihymenium brown, olive,

Porina (11 N. Am. species)
Pimple lichens

DESCRIPTION: Crustose lichens, with thin thalli immersed in the substrate, or superficial and sometimes moderately thick; isidia present in some species. Photobiont green (*Trentepohlia*). Fruiting bodies perithecia, less than 1 mm in diameter, pale brown to black, more or less covered with thalline tissue or naked; involucrellum well developed or absent; excipulum generally pale; paraphyses predominantly unbranched; asci narrow at the apex, with only a slightly thickened tip; spores fusiform to long and slender, colorless, 4- to many-celled, 8 per ascus; a few species have spores with a gelatinous halo. CHEMISTRY: Unknown; presumably no lichen substances. HABITAT: On trees, rocks, mosses, and soil. COMMENTS: Among the perithecial lichens with unbranched paraphyses and many-celled, needle-shaped spores, *Porina* is the best known. Most species are very small, inconspicuous lichens on leaves, bark, or rock and are rarely collected, but a few are encountered often enough to merit inclusion in our keys. **Strigula,** represented by tropical species growing on leaves, and others on bark or green; hypothecium dark brown but often with a colorless layer between it and the hymenium; exciple uniformly brown-black, or grading from black to pale brown from the outside to the inside; asci with a distinctive tube or cylinder in the tip that turns dark K/I+ blue, the remainder of the wall remaining pale blue (see Fig. 14g); spores colorless, 1-celled, large, with a gelatinous halo that is visible, at least in ink preparations (see Fig. 15k), 8 per ascus. CHEMISTRY: Containing a variety of compounds including depsides such as confluentic acid, 2'-*O*-methylsuperphyllinic acid, and 2'-*O*-methylperlatolic acid, as well as β-orcinol depsidones such as norstictic and stictic acids. HABITAT: On siliceous or rarely calcareous rocks. COMMENTS: A large percentage of crustose lichens on rock with large black apothecia (more than 0.75 mm in diameter) belong to this important genus. Other crustose lichens that are superficially similar with a hand lens include species of *Lecidea, Rhizocarpon, Sarcogyne,* and *Buellia.* The black apothecium, distinctive ascus, and large halonate spores set this genus apart, although some less common genera in the same family share these characteristics. Perhaps the most isolated species in the genus is *P. speirea,* with a chalky white thallus, IKI+ blue medulla (rare in the genus), and sunken apothecia; it grows on limestone rather than siliceous rocks. Other genera with lecideine apothecia and the distinctive *Porpidia*-type ascus include *Amygdalaria* (with cephalodia) and *Clauzadea,* a genus of rare arctic-alpine lichens such as *C. monticola,* with an endolithic thallus and very small red-brown to black apothecia. It has small spores and is found on calcareous rocks. *Clauzadea* resembles *Protoblastenia* in some respects and may be closely related. *Melanolecia transitoria* (syn. *Farnoldia jurana*) is also in this group but has larger apothecia and spores. Its distinctions are given in the key.

KEY TO SPECIES (including *Clauzadea* and *Melanolecia*)

1. Thallus orange . . . . *Porpidia flavocaerulescens*
1. Thallus gray 2

2. On limestone; arctic-alpine, rather rare; epihymenium brown or green. 3
2. On noncalcareous rock; temperate to boreal, common; epihymenium brown to olive-brown. 4

3. Apothecia red-brown when wet, very small (under 0.3 mm in diameter), flat, not pruinose; margins sometimes prominent; epihymenium brown; hypothecium and exciple dark brown; spores 6.5–12(–14) × 3.5–7 μm . . . . [*Clauzadea monticola*]
3. Apothecia black when wet, 0.4–1 mm in diameter, flat, prominent margins; disk sometimes lightly pruinose; epihymenium green to blue-green, at least in part; hypothecium and exciple coal-black, not distinguishable; spores mostly 13–28 × 7–14 μm . . . . [*Melanolecia transitoria*]

4. Apothecial disks gray because of a light or heavy pruina; apothecial margins black, contrasting with the disk . . . . *Porpidia albocaerulescens*
4. Apothecial disks black, not pruinose; apothecial margin the same color as the disk 5

5. Apothecial margin brittle and radially cracked. Thallus endolithic; spores 12–18 × 6–8 μm; Appalachian–Great Lakes–Ozarks distribution. . . . . [*Porpidia tahawasiana*]
5. Apothecial margin not brittle or radially cracked 6

6. Apothecia usually less than 1.2 mm in diameter; hymenium 60–75(–100) μm high; spores 10–17 × 5–9 μm; cells in the exciple about 5–8 μm in diameter (seen only in thin sections) . . . . *Porpidia crustulata*
6. Apothecia commonly 1–2.5 mm in diameter; hymenium 80–120 μm high; spores mostly 13–23 × 7–10 μm; cells in the exciple mostly 3–6 μm in diameter . . . . [*Porpidia macrocarpa*]

Porpidia albocaerulescens
Smoky-eye boulder lichen

DESCRIPTION: Thallus pale to dark greenish gray or creamy gray, continuous and fairly smooth or with fine cracks in the thicker portions. Apothecia 0.8–2 mm in diameter, with dark gray to black persistent margins and heavily pruinose disks (without pruina only in old, unhealthy specimens); spores mostly 18–21 × 8–12 μm. CHEMISTRY: Thallus PD+ yellow to orange, K+ yellow or red, KC–, C–; two chemical races: the more common one with stictic and cryptostictic acids, and the other with norstictic and connorstictic acids. HABITAT: On siliceous rocks and boulders in shaded woods. COMMENTS: In the east, this is one of the most commonly seen saxicolous, crustose lichens in shaded habitats. Its gray pruinose apothecial disks set off by a black margin make it very distinctive. *Porpidia carlottiana,* found on exposed rocks along the very humid west coast from Oregon to Alaska, is almost identical in appearance but is PD–, K– (containing 2'-*O*-methylsuperphyllinic and glaucophaeic acids).

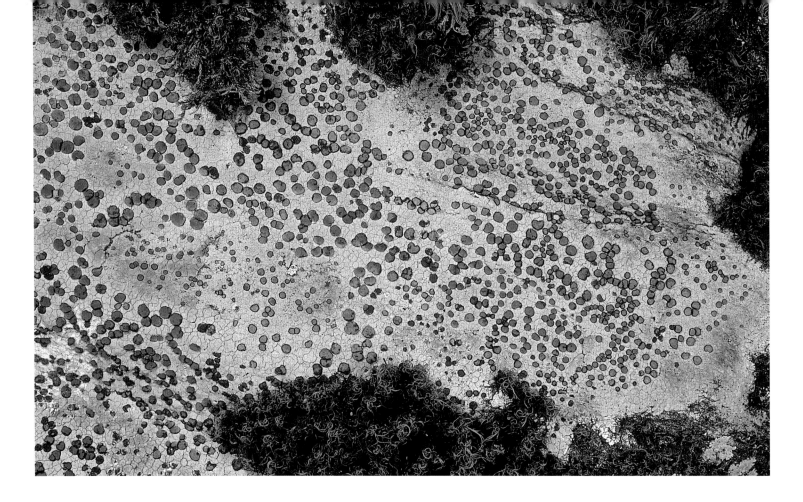

705. *Porpidia albocaerulescens* Appalachians, North Carolina ×1.7

Porpidia crustulata
Concentric boulder lichen

DESCRIPTION: Thallus pale greenish gray, thin but always visible, continuous to somewhat cracked. Apothecia 0.3–1.5 mm in diameter, commonly forming concentric rings (as shown in plate 706); hymenium mostly 60–90 μm high; excipile dark brown but not black, formed from rather large cells approximately 5–8 μm in diameter; spores 10–17 × 5–9 μm. CHEMISTRY: Usually PD– or PD+ vague orange, K– or K+ yellowish, KC–, C–. Spot tests are unreliable, although the thallus contains stictic acid in small amounts. HABITAT: On siliceous rocks, especially pebbles and small boulders, in full sun or shaded woods. COMMENTS: This, together with the very similar *P. macrocarpa,* make up the bulk of the nonpruinose species of *Porpidia* encountered in eastern North America. In *P. macrocarpa,* the apothecia tend to be larger (up to 3.5 mm in diameter), the spores are larger (13–23 × 7–10 μm), the hymenium is higher (80–120 μm), and the cells of the excipile are smaller (mostly 3–6 μm in diameter). *Porpidia thomsonii* is a closely related and similar western to arctic boulder lichen, differing from both these species in having more elongate cells in the excipile and almost always lacking a visible thallus. *Porpidia tahawasiana,* an Appalachian–Great Lakes species, has an imperceptible, endolithic thallus, a carbon-black, brittle and often cracked excipile, and rather narrow spores (12–18 × 6–8 μm). *Rhizocarpon concentricum* is a common, mainly northeastern lichen with concentrically arranged apothecia and can therefore be mistaken for *P. crustulata* in the field. It can be distinguished by its large, colorless, muriform spores.

Porpidia flavocaerulescens
Orange boulder lichen

DESCRIPTION: Thallus bright orange, sometimes with gray patches, rather thick (up to 1 mm), continuous but generally cracked into irregular areoles when dry, without soredia. Apothecia black, the disk (not the margin) usually frosted with white pruina, 0.7–2.5 mm in diameter; spores 14–24 × 6–11 μm. CHEMISTRY: All reactions negative

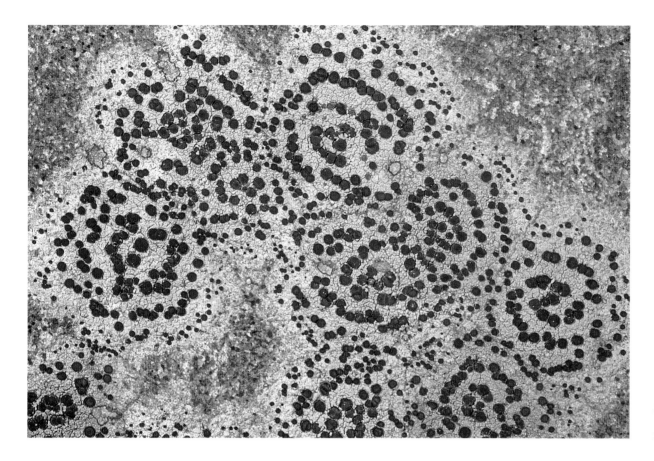

706. *Porpidia crustulata*
Ouachitas, Arkansas
×3.2

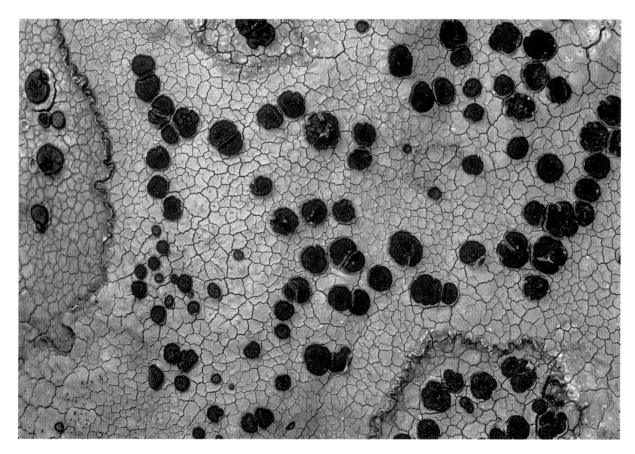

707. *Porpidia flavo-caerulescens* (Two genetically distinct thalli, one darker than the other, are competing for space on this rock, creating a "battle zone" consisting of darkly pigmented hyphae where they meet. This is commonly seen in many species of crustose lichens.)
interior Alaska ×4.4

708. *Protoblastenia rupestris* northeastern Iowa ×4.5

(contains confluentic acid as its main product; one race has norstictic acid and is PD+ yellow, K+ red). HABITAT: Exposed rocks mostly in arctic or alpine tundra. COMMENTS: *Porpidia melinodes,* an uncommon arctic species, is the sorediate counterpart of *P. flavocaerulescens,* producing patches of gray soredia rather than apothecia. *Porpidia macrocarpa* can become quite orange on iron-rich rocks, but its apothecia are not pruinose and it lacks confluentic acid. Some arctic-alpine species of *Lecidea,* such as *L. lapicida,* commonly take on a rusty color because of the chemistry of the rock, but the exciple, spores, and asci separate them. See also *Tremolecia atrata.*

Protoblastenia (4 N. Am. species)
Orange dot lichens

DESCRIPTION: Thallus superficial or buried within the rock, pale to dark yellowish gray, or pale brown, with green algae. Photobiont green (unicellular). Apothecia biatorine, orange to orange-brown, flat to hemispherical, margins soon disappearing; asci *Porpidia*-type; spores 1-celled, colorless, 8 per ascus. CHEMISTRY: Thallus PD–, K–, KC–, C–; apothecia K+ dark red-purple (anthraquinones). HABITAT: On calcareous soil or rock. COMMENTS: This is the only genus of crustose lichens with orange-yellow anthraquinone pigments in the apothecia, and 1-celled spores. Species of *Protoblastenia* look much like a *Caloplaca* without margins on the apothecia. The spores of *Caloplaca* are 2-celled and polarilocular. See also Comments under *Porpidia* and *Pyrrhospora.*

KEY TO GENUS: See Key F.

Protoblastenia rupestris

DESCRIPTION: Thallus dirty white to pale buff, very rough and cracked into tiny areoles. Apothecia flat to strongly convex, round or irregular in shape, sometimes coalescing, sitting on the thallus or slightly sunken, 0.4–1.2 mm in diameter; spores 8–17 × 5–8 μm. HABITAT: On limestone and calcareous sandstone. COMMENTS: *Protoblastenia rupestris* is the only orange dot lichen found outside of the Arctic and mountains. Specimens with especially thin thalli can resemble *P. incrustans,* an arctic-alpine species on limestone. That species produces only a gray stain on the rock because the thallus is mostly endolithic, and its apothecia are sunken into the rock, forming shallow pits.

Protoparmelia (5 N. Am. species)
Chocolate rim-lichens

DESCRIPTION: Crustose lichens with brownish thalli, thick, areolate to verrucose (species on rock) or granular (species on bark or wood). Photobiont green (*Trebouxia*?). Apothecia lecanorine with thick margins, or sunken into the thallus, with dark brown disks; spores colorless, 1-celled, 8 per ascus. CHEMISTRY: Commonly with lobaric acid; sometimes with stictic or norstictic acids, lacking atranorin. HABITAT: Usually on granitic rocks, in well-lit sites, but a few grow on wood or bark such as *P. ochrococca,* not uncommon on conifers in California and Oregon. COMMENTS: The shiny red-brown to dark brown apothecia and well developed chocolate- or red-brown thallus granules or verrucae are the most distinctive features of the genus in the field. Some brown species of *Lecanora* and *Acarospora* are superficially similar.

KEY TO GENUS: See Key F.

Protoparmelia badia

DESCRIPTION: Thallus pale to deep grayish brown to yellowish brown, areolate to verrucose, shiny. Apothecia lecanorine, rather large (mostly 0.7–1.5 mm in diameter) with dark, usually shiny red-brown disks and paler, persistent, smooth margins; spores spindle-shaped (i.e., with pointed ends), 10–16 × 3–6(–7) μm. CHEMISTRY: Cortex and medulla PD–, K– or + yellowish, KC+ red-violet (fleeting), C– (lobaric acid). COMMENTS: This is a very common lichen on granitic rocks in alpine and arctic sites. The pointy spores are unique. Members of the *Lecanora subfusca* group, also with brown fruiting bodies, never have such brown thalli.

Pseudephebe (2 N. Am. species)
Rockwool

DESCRIPTION: Small, very dark brown to almost black, shiny fruticose lichens with thalli resembling coarse, tangled hair, but sometimes with flattened branches at the periphery that are almost foliose. Photobiont green (*Trebouxia?*). Thalli without pseudocyphellae, soredia, or isidia. Apothecia lecanorine, with colorless, 1-celled spores, 7–12 × 6–8 μm; pycnidia abundant, causing slight swellings in the branches with a hole at the top, containing bone-shaped conidia approximately 5–7 × 1 μm. CHEMISTRY: All reactions negative (no lichen substances). HABITAT: On siliceous rocks, usually in open arctic or alpine sites. COMMENTS: Other hair-like, black, fruticose lichens such as *Bryoria* are generally larger, with longer or thicker branches, and they usually contain lichen substances, at least in the soralia. Similar species of narrow-lobed foliose lichens belonging to, for example, *Melanelia*, are discussed below. See Comments under *Polychidium muscicola*.

KEY TO SPECIES

1. Branches very slender and entirely terete, 0.1–0.2 mm in diameter, not flattening at the tips; distance between axils usually 1–3 mm .... *Pseudephebe pubescens*
1. Branches 0.2–0.5 mm in diameter, usually becoming somewhat flattened at the tips; distance between axils 0.2–0.5(–1) mm............. *Pseudephebe minuscula*

Pseudephebe minuscula
Coarse rockwool

DESCRIPTION: Thallus very dark brown to almost black, mainly filamentous, with cylindrical branches 0.2–0.5 mm thick (tapering to 0.1 mm at the tips), but closely appressed to the rock and often with flattened, almost foliose branches up to 1 mm wide at the growing edge of the clump, sometimes breaking up into tangled areoles and creating a crustose appearance in the central portions of old thalli; distance between the axils of the branches is short, 0.2–0.5(–1.0) mm. Apothecia common, up to 3 mm in diameter, with dark red-brown to black disks. HABITAT: Common on granitic or somewhat limy rock or pebbles in relatively dry, windswept arctic and alpine sites, especially in inland areas having a continental climate. COMMENTS: *Pseudephebe minuscula* is extremely variable in form. When its branches are slender, it resembles *P. pubescens*, differing mainly in the shorter, slightly flattened branch elements. When most of the branches are broad, it looks very much like the PD– chemical race of *Melanelia stygia*. *Pseudephebe*, however, has almost no pseudocyphellae and almost always has at least a few knobby, filament-like branches. It contains no lichen substances at all, whereas the PD– *Melanelia stygia* contains some fatty acids. Unfortunately, the conidia are virtually identical.

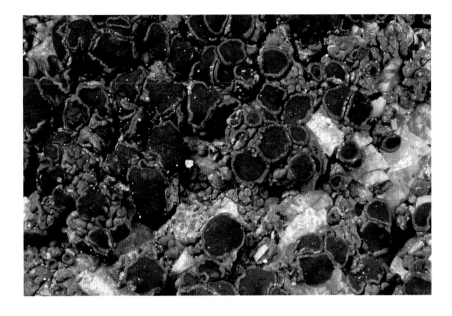

709. *Protoparmelia badia* Rocky Mountains, Colorado ×5.6

Pseudephebe pubescens
Fine rockwool

DESCRIPTION and COMMENTS: *Pseudephebe pubescens* looks very much like a patch of black steel wool. It is very similar to *P. minuscula* and sometimes intergrades with it. *Pseudephebe pubescens*, however, has finer, longer, rounder branches, 0.1–0.2 mm thick on the main stems and less than 0.1

710. *Pseudephebe minuscula* interior Alaska ×3.2

711. *Pseudephebe pubescens* interior Alaska ×2.4

mm at the tips, and apothecia are less common. The distance between the axils (the internodes) is much longer in *P. pubescens*, usually 1–3 mm. HABITAT: Common on granitic rock in arctic and alpine localities that are more humid than those of *P. minuscula*, such as the coastal mountain ranges.

Pseudevernia (3 N. Am. species)
Antler lichens

DESCRIPTION: Very pale to medium gray lichens, fruticose in growth habit (growing out from a single point) but foliose in having flattened branches with distinguishable upper and lower surfaces, although without rhizines or cilia. Photobiont green (*Trebouxia?*). Some species with isidia but none with soredia; lacking pseudocyphellae; lower surface of branches white to dark gray or black. Apothecia common in one species, lecanorine, with broad brown disks; spores colorless, 1-celled, 8 per ascus. CHEMISTRY: All North American species: cortex PD– or PD+ pale yellow, K+ yellow, KC–, C– (atranorin); medulla PD–, K–, KC+

712. *Pseudevernia cladonia* White Mountains, New Hampshire ×1.8

red, C+ pink to red (lecanoric acid). HABITAT: On the branches and bark of conifers. COMMENTS: The forked, antler-like branches forming compact cushions give the lichen its fruticose aspect, even though the branches are flattened almost to the tips in most species, with a pale upper surface and a mousy gray to black lower surface, like a foliose lichen. *Evernia*, especially *E. prunastri*, is similar in some respects but contains usnic acid in the cortex, making it yellowish green, and the lower surface is never black. Narrow-lobed species of *Platismatia* do not have regular, forked branching and are never C+ red.

KEY TO SPECIES

1. Thallus abundantly isidiate on the upper surface. . . . . . .
   . . . . *Pseudevernia consocians*
1. Thallus not isidiate 2

2. Lobes mostly under 1 mm wide; lower surface white except at the base; in the Appalachian mountains . . . .
   . . . . *Pseudevernia cladonia*
2. Lobes 1–3 mm wide; lower surface dark gray to black except at the lobe tips; southwestern . . . .
   . . . . *Pseudevernia intensa*

Pseudevernia cladonia
Ghost antler lichen

DESCRIPTION: Thallus almost white to very pale greenish gray, smooth, lacking isidia or soredia, sometimes slightly pruinose at the lobe tips; lobes branching in regular dichotomies, mostly less than 1 mm wide except at base, almost round in cross section close to the tips, but flattening in older parts; lower surface white to patchy gray, and becoming black at the base. Apothecia rare. HABITAT: On conifers at high elevations in Appalachian forests. COMMENTS: The rare *Everniastrum catawbiense*, also in the Appalachians, is somewhat similar, with evenly forked, ascending branches, appearing almost fruticose, and a C+ pink medulla (gyrophoric acid). However, its lower surface is entirely black, the margins are fringed with black branching rhizines, and the upper surface has patches of soredia.

Pseudevernia consocians
Common antler lichen

DESCRIPTION: Thallus very pale to medium gray; lobes mostly 1–1.5 mm wide close to the tips, with regular to irregular forked branching; cylindrical to branched isidia abundant on the lobe surface and margins, sometimes together with marginal lobules; lower surface pale at lobe tips, darkening to purplish black, becoming black at center. Apothecia rare. HABITAT: On conifers, mainly in forests. COMMENTS: This is the only *Pseudevernia* in North America with isidia. It is larger and less evenly branched than its eastern neighbor, *P. cladonia,* and can be found at lower elevations. IMPORTANCE: The closely related European species, *P. furfuracea,* is an important source of fragrances and fixatives in the perfume and cosmetic industry in France, and more is collected for that purpose than even *Evernia prunastri.* (See Importance under *Evernia prunastri.*)

Pseudevernia intensa
Western antler lichen

DESCRIPTION: Thallus pale gray, smooth or strongly wrinkled; dull and almost pruinose in part, with long or short branched lobes, (1–)1.5–3 mm wide close to the tips, thickened and turned down at the edges, creating channels below; without isidia; lower surface dull, predominantly violet-gray to bluish black, white only at the lobe tips. Apothecia 3–15 mm in diameter, brown, very common on the upper surface of the lobes. Pycnidia forming crowds of black dots on the lobe tips. HABITAT: On conifers in open woodlands, typically at high elevations. COMMENTS: The production of both pycnidia and apothecia on this species suggests that it is the sexual ancestor (primary species) of other species of *Pseudevernia* that reproduce only by fragmentation or isidia. It is the only western member of the genus, and the only one that grows in relatively open habitats. In the southwest, *P. intensa* seems to be the ecological equivalent of *Hypogymnia* species in California and northward, especially species like *H. imshaugii* and *H. inactiva,* which it resembles superficially.

Pseudocyphellaria (5 N. Am. species)
Specklebelly lichens

DESCRIPTION: Medium to large foliose lichens, loosely attached and ascending, greenish gray to dark brown, with conspicuous, often raised, white or yellow dots (pseudocyphellae) on the lower surface of the thallus. Photobiont blue-green (*Nostoc*) or green (*Dictyochloropsis*). Medulla white or bright yellow; lower surface very pale buff to light brown with a fuzzy tomentum instead of rhizines (rarely, the tomentum is sparse). Apothecia on the upper thallus surface or along the margins; spores colorless, 2- to many-celled, with pointed ends, 8 per ascus. CHEMISTRY: Cortex PD–, K–, KC–, C– (no lichen substances); medulla often containing stictic acid or bright yellow, pulvinic acid–related pigments. HABITAT: On both deciduous and coniferous trees, usually in rather humid sites, rarely on mossy rocks; more or less confined to oceanic parts of the world, from the subtropics to the boreal zones. COMMENTS: *Pseudocyphellaria* is closely related to *Sticta* and *Lobaria,* and the species resemble each other closely (as the Comments under the species treated below explain). The principal distinguishing

713. (facing page) *Pseudevernia consocians* Great Smoky Mountains, Tennessee ×4.0

714. (above left) *Pseudevernia intensa* Chisos Mountains, southwestern Texas ×0.32

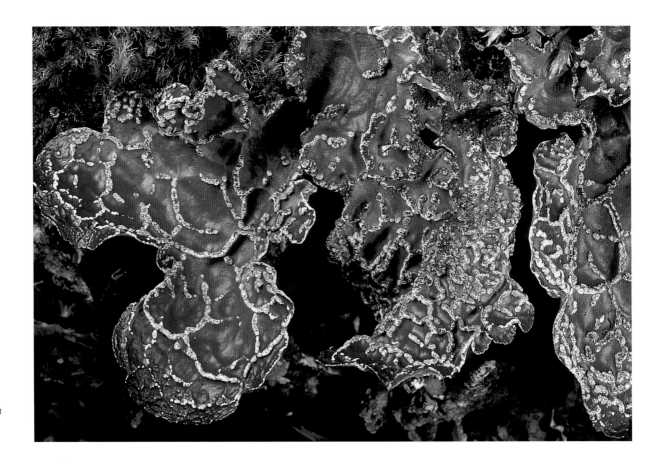

715. *Pseudocyphellaria anomala* Cascades, Oregon ×2.5

character, unique to the genus, is the presence of conspicuous pseudocyphellae on the lower surface of the lobes. Sterile specimens of *Nephroma* also sometimes resemble *Pseudocyphellaria* from above. IMPORTANCE: Because *Pseudocyphellaria* species are often associated with old, humid forests in undisturbed settings, they are often used as indicators of valuable old growth forests.

KEY TO SPECIES

1. Pseudocyphellae and soralia yellow (calycin) 2
1. Pseudocyphellae and soralia (if present) white (calycin absent) 3

2. Photobiont blue-green; medulla predominantly white, except yellow near soralia and pseudocyphellae; lobes with a network of depressions and ridges; medulla PD+ orange, K+ yellow (stictic acid) . . . . *Pseudocyphellaria crocata*
2. Photobiont green; medulla dark yellow throughout; lobes smooth and even; medulla PD–, K– . . . . *Pseudocyphellaria aurata*

3. Thallus greenish gray or pale brown, dull or scabrose; lobes smooth and even; algal layer grass-green; isidia and lobules present along the lobe margins; medulla PD–, K–. . . . . *Pseudocyphellaria rainierensis*
3. Thallus dark reddish brown, rather shiny; lobes with a network of depressions and ridges; algal layer dark blue-green; isidia and lobules absent; medulla PD+ orange, K+ yellow (stictic acid) 4

4. Soredia absent; apothecia abundant . . . . *Pseudocyphellaria anthraspis*
4. Soredia present, mainly on ridges; apothecia rare . . . . *Pseudocyphellaria anomala*

Pseudocyphellaria anomala
Netted specklebelly

DESCRIPTION: Thallus medium to dark chocolate- or reddish brown, usually dull, forming a network of ridges and depressions, the ridges set off by white to gray soredia, although round to irregular soralia can be present between the ridges as well; lobes 10–30 mm across, round or angular. Photobiont blue-green. Apothecia uncommon. CHEMISTRY: Medulla PD+ orange, K+ yellow, KC–, C– (stictic acid complex and triterpenes). HABITAT: On

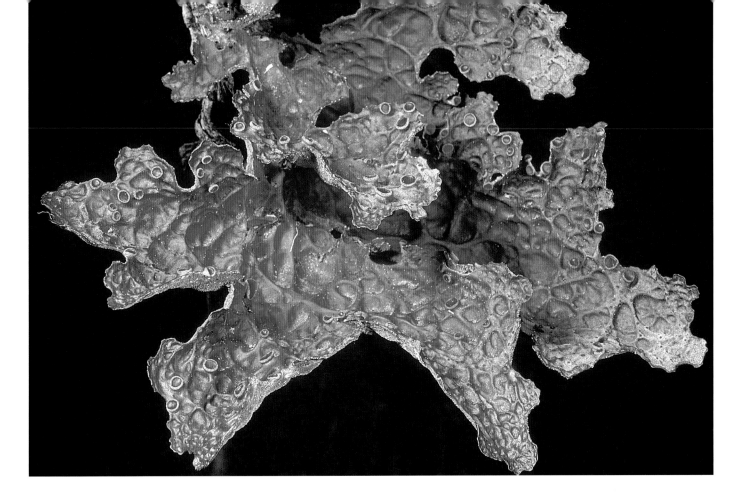

trees of all kinds, especially on mossy branches. COMMENTS: The general aspect of both *P. anomala* and its fertile sister species *P. anthraspis* is much like that of *Lobaria pulmonaria* (olive or greenish brown to yellowish brown, with green algae) or *L. retigera* (dark brown, with cyanobacteria), but *Lobaria* species do not have pseudocyphellae on the lower surface.

Pseudocyphellaria anthraspis
Dimpled specklebelly

DESCRIPTION and COMMENTS: This is the fertile counterpart of the sorediate *P. anomala* and has the same chemistry. The dark brown apothecia, 3–6 mm in diameter, are produced in abundance on the upper surface of the lobes. Both are North American endemics found only along the west coast. It is much darker brown than *Lobaria pulmonaria* and related species, which it resembles in the pitted and ridged surface of the thallus. HABITAT: Same as *P. anomala*.

Pseudocyphellaria aurata
Green specklebelly

DESCRIPTION: Thallus greenish gray to olive or brown when dry, grass-green when wet, with powdery, yellow soredia occurring in marginal soralia (none on the thallus surface). Photobiont green. Lobes 5–10 mm wide; medulla dark yellow (papaya-yellow); lower surface very pale tan to yellowish, with numerous round to irregular bright yellow pseudocyphellae; small cephalodia containing clumps of cyanobacteria are buried within the thallus, producing low bumps on the upper thallus surface. CHEMISTRY: Medulla PD–, K–, KC–, C– (the yellow pigment calycin and triterpenes). HABITAT: On trees at low elevations. COMMENTS: *Pseudocyphellaria crocata* is the only other North American *Pseudocyphellaria* with bright yellow soredia, but it is brown, with coarse soredia or isidiate soralia on both the margins and upper surface of the thallus, the medulla is usually white and reacts PD+ orange, and, most important, the photobiont is blue-green.

716. *Pseudocyphellaria anthraspis* Cascades, Washington ×1.7

Pseudocyphellaria crocata
Yellow specklebelly

DESCRIPTION: Thallus pale to dark chocolate-brown, rarely gray-brown, with bright yellow soralia (often flecked with black) on the low ridges, in the shallow depressions between them, or along the lobe margins; medulla predominantly white, but bright yellow in part, especially in and under the soralia and close to the pseudocyphellae; lobes 5–20 mm wide; lower surface with pale to dark brown tomentum interspersed with round, bright yellow pseudocyphellae, sometimes large and conspicuous and sometimes very sparse. Photobiont blue-green. Apothecia rarely seen. CHEMISTRY: Medulla PD+ orange, K+ yellow, KC–, C– (stictic acid and associated compounds, and calycin, a yellow pigment). HABITAT: On bark, shrubs, and sometimes mossy rock in humid sites. COMMENTS: *Pseudocyphellaria crocata* is morphologically variable but easy to recognize nevertheless. Even when a specimen has very few spots on the lower surface, the combination of brilliant yellow soredia, mostly white medulla, and blue-green photo-

717. *Pseudocyphellaria aurata* Great Smoky Mountains, Tennessee ×2.2

718. *Pseudocyphellaria crocata* Oregon coast ×2.9

biont is unique in North America. The southeastern *P. aurata*, also a yellow sorediate lichen, has green algae and a yellow rather than white medulla, and reacts PD–, K– in the medulla. IMPORTANCE: The yellow pigment in *P. crocata* has made it useful as a source of yellow dyes in Europe, but the lichen is not abundant enough to serve as a dyestuff in North America and should never be collected in large quantity.

Pseudocyphellaria rainierensis
Oldgrowth specklebelly

DESCRIPTION: Thallus greenish or bluish gray to pale brownish, smooth, dull, sometimes scabrose on the lobe tips; medulla white; lobes 6–30 mm across, fringed with lobules and branched isidia along the margins and sometimes in clusters on the thallus surface; lower surface very pale brown to dark, with large, raised, white pseudocyphellae. Photobiont green, but warty cephalodia containing cyanobacteria visible on the lower surface, also seen as low bumps on the upper surface. Apothecia rare. CHEMISTRY: Cortex PD–, K+ yellow, KC–, C–; medulla PD–, K–, KC–, C– (unidentified substance). HABITAT: On the branches of trees and shrubs in coastal forests, especially very old forests, often only in the canopies. COMMENTS: *Lobaria oregana* is similar but is more yellow because of usnic acid in the cortex and lacks white dots on the lower surface. Its medulla reacts PD+ orange and K+ yellow. *Pseudocyphellaria rainierensis* is the only west coast *Pseudocyphellaria* with green algae, and the only one with isidia. IMPORTANCE: *Pseudocyphellaria rainierensis* is a rare lichen and has proved to be a reliable indicator of old growth forests. The Committee on the Status of Endangered Wildlife in Canada (COSEWIC) lists it as "vulnerable."

719. *Pseudocyphellaria rainierensis* Cascades, Washington ×0.58

Pseudoparmelia (3 N. Am. species)

KEY TO SPECIES: See *Parmelia*.

Pseudoparmelia uleana
Lemon-lime lichen

DESCRIPTION: Thallus foliose, yellowish gray to almost bluish green, usually smooth and shiny, rather closely appressed and flat or developing crowded, overlapping lobes in the center. Lobes mostly 2–6 mm wide, without cilia, soredia, or isidia, although sometimes spotted with maculae; medulla pale lemon-yellow, especially close to the algal layer; lower surface pale brown to a smoky or olive-brown, with many short, unbranched rhizines. Photobiont green (*Trebouxia?*). Usually fertile, with many red-brown, bowl-shaped apothecia 1–4 mm in diameter; spores colorless, almost spherical, 5–9 × 4–7 μm, 8 per ascus; conidia slightly bone-shaped, 6–10 × 1 μm. CHEMISTRY: Cortex PD–, K+ yellow, KC+ orange, C–; medulla PD–, K–, KC+ orange, C– or C+ pale yellow-orange (secalonic acid, atranorin at least in traces). HABITAT: Common on the bark of hardwoods and cypress. COMMENTS: The color of this lichen closely re-

sembles that of some *Punctelia* species, but *Pseudoparmelia* species have no white pseudocyphellae. The yellowish tint is caused by the medullary pigments rather than usnic acid in the cortex, as in yellowish green lichens such as *Flavoparmelia* and *Ahtiana*. *Pseudoparmelia* is closely related to the genus *Canoparmelia* (e.g., they have the same cell wall compound, isolichenan), but the latter has abundant atranorin in the cortex, and it generally has a white medulla and black lower surface. The two other North American species of *Pseudoparmelia*, all confined to the extreme southeast, differ from *P. uleana* mainly in chemistry: *P. cubensis* contains compounds in the stictic acid complex (PD+ orange, K+ yellow), and the very rare *P. floridensis* has salazinic acid (PD+ yellow-orange, K+ red). *Pseudoparmelia sphaerospora,* a name long used for the North American species, applies to an African lichen with a thicker, more brittle thallus and somewhat different chemistry.

720. *Pseudoparmelia uleana* northern Florida ×2.3

721. *Psilolechia lucida* Cape Cod, Massachusetts ×8.3

Psilolechia (2 N. Am. species)

DESCRIPTION: Yellowish gray to bright sulphur yellow, powdery or granular crustose lichens. Photobiont green (*Trebouxia* or *Stichococcus*). Apothecia tiny, hemispherical, biatorine, usually pale beige or yellow; paraphyses unbranched to sparsely branched; ascus similar to *Porpidia*-type; spores colorless, 1-celled, slightly broader at one end (teardrop-shaped). HABITAT: On rock, soil, roots, wood, and rarely bark in protected, humid sites. COMMENTS: The genus is characterized by its asci and teardrop-shaped spores, its leprose to granular thallus, and its humid habitat. The thalli of sterile specimens are similar to those of some yellow stubble lichens (e.g., *Chaenotheca furfuracea*), *Chrysothrix,* and yellowish species of *Cliostomum*.

KEY TO SPECIES: See Key F.

Psilolechia lucida
Sulphur dust lichen

DESCRIPTION: Thallus sulphur-yellow to yellow-green (greenish when moist, as shown in plate 721), composed entirely of powdery soredia (leprose), thin or thick, sometimes broken into irregular areoles. Apothecia lemon-yellow, hemispherical to irregular and lumpy, without margins, usually under 0.3 mm in diameter; spores 4–7 × 1–2.5 µm, 8 per ascus. CHEMISTRY: Thallus PD–, K–, KC–, C–, UV+ orange (rhizocarpic acid). HABITAT: On the protected sides of overhanging rocks or shaded boulders or in the crevices of old stone walls, more rarely on wood. COMMENTS: This widely distributed crustose lichen is easy to identify when it is fertile, although yellowish species of *Cliostomum,* such as *C. leprosum* and *C. vitellinum,* can be superficially similar. *Cliostomum* has 2-celled spores and *Lecanora*-type asci, and the species typically grow on bark.

Psora (17 N. Am. species)
Scale lichens

DESCRIPTION: Lichens composed of relatively large, thick squamules (scales), yellow- to red-brown, often pruinose, mostly 2–6 mm across, attached to the substrate by a hairy tomentum; upper cortex with a rather thick, amorphous, gelatinized layer containing old, dead algal cells; lower surface pale, usually with a recognizable cortex; calcium oxalate crystals (insoluble in K) often present in the medulla or lower cortex. Photobiont green (*Myrmecia*). Apothecia biatorine, convex to almost spherical, reddish brown to brown-black, produced either on the top surface of the squamules (laminal) or at the edges of the squamules (marginal); margins usually disappearing in mature apothecia; epihymenium and often exciple containing trace amounts of anthraquinone crystals (sometimes visible only with polarized light); hypothecium pale brown or colorless, containing small crystals that dissolve in acid; hymenium K/I+ blue (hemiamyloid); paraphyses somewhat branched and anastomosing; asci *Porpidia*-type; spores colorless, 1-celled, ellipsoid, 7–18 × 5–9 µm, 8 per ascus. CHEMISTRY: Anthraquinones in the apothecia (epihymenium usually K+ red or violet), and with a variety of medullary compounds. HABITAT: On lime-rich soil or rock, especially in dry sites. COMMENTS: Species of *Psora,* especially when sterile, can be confused with various earth scales such as *Placidium* (with perithecia), *Heppia* and *Peltula* (with cyanobacteria), and some species of *Toninia* (with no lichen substances; spores 2–8-celled; asci *Bacidia*-type). Some rock-dwelling species of *Pannaria* and *Parmeliella,* both with cyanobacteria, can also be mistaken for *Psora* species. *Psorula* has a dark green lower surface and grows over the cyanobacterial lichen *Spilonema*. In *Psora,* the position of the apothecia on the squamules (along the margins or on the upper surface) is critical but sometimes difficult to ascertain, especially when the squamules are overlapping or crowded, as they often are. By moistening a clump and isolating individual squamules, you can better interpret this feature. The amount of pruina is also very variable, even within a small population. Generally, the more protected the squamule, the less pruina will be produced.

KEY TO SPECIES

1. Apothecia mainly at the margins of the squamules; squamules pruinose, at least in part. 2
1. Apothecia mainly on the surface of the squamules; squamules pruinose or not. 4
2. Squamules strongly convex and fissured, yellow-brown to olive-brown . . . . *Psora cerebriformis*
2. Squamules more or less flat, smooth or slightly fissured, reddish or orange-brown 3
3. Squamules brick-red to orange-brown, flat or slightly convex, with edges turned up . . . . *Psora decipiens*
3. Squamules pink to pinkish orange, depressed in the center, with edges turned down . . . . *Psora crenata*
4.(1) Thallus bright yellow or yellow-green (rhizocarpic acid in the cortex) . . . . *Psora icterica*
4. Thallus brown, lacking yellow pigments in the cortex . 5
5. Squamules 0.4–1.5(–3) mm wide 6
5. Squamules 2 mm wide or more 9
6. Parasitic on the cyanobacterial lichen, *Spilonema*; lower surface and margins of squamules blackish or blue-green . . . . *Psorula rufonigra*
6. Not parasitic; lower surface and margins of squamules white or pale brown, not blackish green 7
7. Squamules not lobulate, the edges conspicuously white pruinose; medulla C– . . . . *Psora himalayana*
7. Squamule margins divided into tiny round lobules, sometimes lightly pruinose on the surface, but not along the margins; medulla C+ pink (gyrophoric acid) 8
8. Squamules loosely attached, often ascending; California . . . . *Psora pacifica*
8. Squamules closely attached and flat against the substrate; mostly high elevations in the Rocky Mountains . . . . [*Psora montana*]
9.(5) Squamules pruinose or whitened on the edges and sometimes the surface 10
9. Squamules almost entirely without pruina 13

722. *Psora cerebriformis* canyon country, northeastern Utah ×3.0

10. Medulla K+ red (norstictic acid). Squamules usually concave; southwestern U.S., on soil ... [*Psora russellii*]
10. Medulla K–. 11

11. Found in central and eastern temperate North America on rock; apothecia rusty brown, typically with persistent margins. .... *Psora pseudorussellii*
11. Found in western or northern North America on rock or soil; apothecia red-brown to dark brown or almost black, soon becoming convex and marginless 12

12. Squamules pale yellowish brown to pale reddish brown, 2–5 mm wide, scattered or contiguous and crowded, occasionally overlapping, lightly pruinose on the surface and margins. .... *Psora tuckermanii*
12. Squamules dark reddish brown, 1–3 mm wide, overlapping like shingles, only the edges conspicuously white pruinose .... *Psora himalayana*

13.(9) Medulla of thallus C+ pink or red 14
13. Medulla of thallus C–. 16

14. Squamules pale gray, thick and somewhat convex .... *Trapeliopsis wallrothii*
14. Squamules greenish brown or red-brown, thin or thick 15

15. Squamules olive to greenish brown, often white on the margin, thin, flat or curled inward and cup-like when dry, ascending or standing on edge, (2–)3–6(–10) mm wide .... *Psora nipponica*
15. Squamules (including the margin) red-brown, folded or flat, often shingled, adnate or ascending, 2–4(–6) mm wide. Apothecia black or very dark brown; on soil or rock in California .... [*Psora californica*]

16.(13) Squamules uniformly reddish brown, shiny, smooth or fissured; apothecia very dark brown to black .... *Psora globifera*
16. Squamules pale to medium brown usually with whitish margins, dull, somewhat fissured; apothecia typically red-brown 17

17. Western, on rock or soil; apothecia soon convex and marginless..................... *Psora tuckermanii*
17. Central and eastern regions, on rock; apothecia usually flat, with long persistent margins, finally convex and marginless.................... *Psora pseudorussellii*

Psora cerebriformis
Brain scale

DESCRIPTION: Thallus consisting of pale yellowish brown to olive-brown, strongly convex or, less commonly, flat squamules, (2–)3–5(–8) mm in diameter, usually deeply fissured and brain-like, or resembling a soccer ball, often heavily coated with a white pruina on the surface (not so much on the edges), as shown in plate 722. Apothecia black, hemispherical, up to 2 mm in diameter, restricted to the margins of the squamules. CHEMISTRY: Cortex or medulla PD–, K+ yellow or sometimes K+ red, KC–, C– (atranorin and occasionally traces of norstictic acid). HABITAT: On soil in arid regions. COMMENTS: *Psora decipiens* and the more southern *P. crenata* also produce apothecia on the margins of the squamules. *Psora decipiens* has bright red-brown to almost pink squamules, is generally less pruinose, and usually lacks lichen substances. *Psora crenata*, a subtropical species of deserts and semideserts, has larger, downturned squamules and always contains norstictic acid.

Psora crenata
Brick scale

DESCRIPTION: Squamules pink to pinkish orange, pale gray-green when damp, up to 10 mm in diameter, depressed in the center and turned down at the margins, sometimes fissured, often completely covered with a heavy white pruina. Apothecia black, hemispherical, up to 2 mm in diameter, distributed along or just in from the squamule margins. CHEMISTRY: Medulla (close to algal layer) PD+ deep yellow, K+ red, KC–, C– (norstictic acid), although rare specimens with no reactions or chemical products are known. HABITAT: On soil in arid sites. COMMENTS: See Comments under *P. cerebriformis*.

Psora decipiens
Blushing scale

DESCRIPTION: Squamules bright, brick-red to orange-brown, sometimes pinkish, usually at least partially pruinose on the upper surface and margins, 1–6 mm in diameter, flat to slightly upturned at the edges, which are blanched white and frayed (as if they were old and worn). Apothecia hemispherical, black, 0.7–2 mm in diameter, along the margins of some squamules. CHEMISTRY: Mostly PD–, K–, KC–, C– (no lichen substances), but a K+ red chemical race containing norstictic acid frequently occurs in the Yukon and Alaska. An even rarer race with hyposalazinic acid is also known. HABITAT: Mostly on soil, rarely rock. COMMENTS: *Psora decipiens* is relatively easy to recognize because of its color, white-edged frayed squamules, and marginal apothecia. *Psora himalayana* also has white margins, but the apothecia are laminal. See also *P. cerebriformis* and *P. crenata*.

Psora globifera
Blackberry scale

DESCRIPTION: Squamules shiny, reddish brown or less commonly yellow-brown to greenish brown, almost always without any pruina; frequently with fissures in older squamules; squamules 2–5 mm across, crowded and overlapping one another, ascending, with somewhat upturned margins. Apothecia dark (rarely pale) red-brown to almost black, rarely with a faint yellow or green pruina, slightly convex to hemispherical, 0.7–2 mm in diameter, single or clustered and crowded on the surface, not the margins, of the squamules. CHEMISTRY: All reactions negative (no lichen substances). HABITAT: Mainly on rock but occasionally on soil. COMMENTS: *Psora californica* is a very similar species found only in California, but common there on both soil and rock. Its squamules are usually turned down at the margins and are largely without fissures, and the medulla reacts C+ pink (gyrophoric acid and other compounds). *Psora luridella,* a much rarer species on soil in the southwest, has smaller, more closely attached squamules (see Comments under *P. pacifica*).

723. *Psora crenata* Chisos Mountains, southwestern Texas ×1.7

Psora himalayana
Mountain scale

DESCRIPTION: Squamules medium to dark reddish brown, greenish when wet, mostly 1–3 mm wide, overlapping like shingles or ascending, with rounded lobes and conspicuously pruinose margins, as if the edges of the squamules were dipped in frosting sugar. Apothecia dark brown to dark red-brown, strongly convex to hemispherical, on the squamule surface. CHEMISTRY: All reactions negative (no lichen substances). HABITAT: On rock or soil. COMMENTS:

724. (right) *Psora decipiens* southwestern Yukon Territory ×4.6

725. (below left) *Psora globifera* Coast Range, southern California ×3.8

726. (below right) *Psora himalayana* western Colorado ×4.2

727. Psora icterica Sonoran Desert, southern Arizona ×4.3

Psora tuckermanii is very similar, but the squamules are not as regularly shingle-like. A thin section of a squamule shows that the oxalate crystals in *P. himalayana* are present mainly in the lower part of the medulla, including the lower cortex; in *P. tuckermanii*, the crystals are closer to the algal layer and are absent from the lower cortex.

Psora icterica
Yellow scale

DESCRIPTION: Squamules greenish yellow, smooth or lightly pruinose, rounded, with slightly raised margins that are brighter yellow, 1–5 mm across, rather adnate to soil. Apothecia convex to hemispherical, dark red-brown to black, on surface of squamules. CHEMISTRY: Thallus PD–, K–, KC–, C–, UV+ dull orange (rhizocarpic acid). HABITAT: On soil in arid regions. COMMENTS: This is the only *Psora* that contains bright yellow pigments. **Psora rubiformis,** an uncommon, strictly arctic or alpine species, has yellowish brown squamules with white margins and contains usnic acid in the cortex (UV–). Most specimens contain gyrophoric acid as well (C+ pink).

Psora nipponica (syn. *Psora novomexicana*)
Butterfly scale

DESCRIPTION: Squamules olive to greenish brown, without pruina, often fissured, very broad, (2–)3–6(–10) mm across, relatively thin, round and even, all more or less on edge and curled inward with the pale lower surface exposed; margins pale brown to almost white. Apothecia strongly convex to almost spherical, dark brown to black, 1–2.5 mm in diameter, laminal. CHEMISTRY: Cortex and medulla PD–, K–, KC+ red, C+ pink (gyrophoric acid). HABITAT: On soil or rock, especially mixed with mosses. COMMENTS: This is a very common western lichen. Other western species of *Psora* that react C+ pink or red are much smaller, with squamules rarely more than 3 mm across. These include **P. californica,** with red-brown squamules having downturned edges, and so far known only from Cali-

728. *Psora nipponica*
Rocky Mountains, southern Colorado
×5.2

fornia; a few very small lobate species (see *P. pacifica*); and *Psora rubiformis,* a yellow-brown species containing usnic acid in the cortex, found only above timberline.

Psora pacifica
Pacific scale

DESCRIPTION: Thallus composed of small reddish brown, loosely attached and often ascending squamules 0.4–1.3(–2) mm across, developing small round lobules on the margins, usually with the more erect, emergent lobes becoming covered with a white pruina. Apothecia in the centers of the squamules, up to 1.3 mm in diameter, bright to dark red-brown, flat to slightly convex, without margins. CHEMISTRY: Medulla PD–, K–, KC+ red, C+ pink (gyrophoric, usually with another depside, probably 5-*O*-methylhiascic acid). HABITAT: On bare soil. COMMENTS: There are three very similar western species of *Psora* with small (mostly under 3 mm wide), lobulate squamules: *P. pacifica, P. montana,* and *P. luridella.* **Psora luridella** is the most distinctive, being the largest (squamules 1.5–3 mm wide) and most often pruinose, and lacking gyrophoric acid (medulla C–). It is a fairly rare soil-dwelling species from New Mexico and Colorado. *Psora montana* is a high-elevation, Rocky Mountain species extremely similar to *P. pacifica* but with closely attached squamules and almost black apothecia; it lacks the extra depside found in *P. pacifica.*

Psora pseudorussellii
Bordered scale

DESCRIPTION: Squamules yellow- to red-brown when dry, green when wet, dull or shiny, with or without a light frosting of pruina on the upper surface but with white pruinose margins, flat to slightly concave, rather thick, mostly 2–4 mm wide. Apothecia laminal, rusty brown, sometimes lightly pruinose, 0.3–1.5 mm in diameter, flat to slightly convex (not hemispherical), with long, persistent, red-brown margins, but becoming convex and marginless in maturity. CHEMISTRY: All reactions negative (no lichen substances). HABITAT: On rock. COMMENTS:

Similar to *P. tuckermanii,* but with flatter squamules and redder apothecia that retain the margins much longer. ***Psora russellii,*** mainly from the southwest (California to central Texas and Oklahoma), contains norstictic acid (medulla K+ red), has rounder squamules depressed in the center, and produces apothecia closer to the squamule margins.

Psora tuckermanii
Brown-eyed scale

DESCRIPTION: Squamules pale yellowish brown to chocolate-brown, smooth and shiny or pruinose on the surface and margins, crowded or scattered, sometimes overlapping, mostly 2–5 mm in diameter. Apothecia laminal, reddish brown, or, less frequently, dark brown to black, convex to hemispherical, usually marginless, 0.5–2.5 mm in diameter. CHEMISTRY: All reactions negative (no lichen substances). HABITAT: On soil and rock, especially sandstone. COMMENTS: This scale lichen is common throughout western North America. Its red-brown apothecia are distinctive but are most similar to those

729. *Psora pacifica* Channel Islands, southern California ×5.6

730. *Psora pseudorussellii* (damp) hill country, central Texas ×4.0

731. *Psora tuckermanii* northern Arizona ×3.3

of *P. pseudorussellii*. The latter has flatter apothecia with long, persistent margins and is a bit smaller. *Psora tuckermanii*, like most common lichens, is variable; for example, the apothecia can be quite dark in specimens from California. See *P. himalayana*.

Psoroma (2 N. Am. species)
Green moss-shingle lichens

DESCRIPTION: North American species are brown to green squamulose to granular lichens with finely divided lobes; lower surface pale, attached by tufts of pale hyphae, without rhizines. Photobiont green (*Myrmecia*); cephalodia containing *Nostoc* are gray and lumpy, scattered on the thallus. Apothecia lecanorine, with colorless, 1-celled, ellipsoid spores. CHEMISTRY: All reactions negative (no lichen substances), or rarely PD+ orange (pannaric acid). HABITAT: In humid habitats, mainly among mosses, boreal to arctic-alpine. COMMENTS: *Psoroma hypnorum* is the only common species of *Psoroma* on this continent. The genus is very closely related to *Pannaria pezizoides*, which has cyanobacteria as its principal photobiont.

KEY TO SPECIES: See Key G and *Pannaria*.

Psoroma hypnorum
Green moss-shingle

DESCRIPTION: Thallus squamulose, yellowish to greenish brown when dry, brownish green to grass-green when wet, squamules finely lobed, sometimes becoming almost granular, 0.2–0.5 mm in diameter; cephalodia darker brown, warty to squamulose, mixed with other scales of the thallus. Apothecia abundant, 1–5 mm in diameter, with flat or concave, red-brown disks and a well-developed lecanorine margin with tiny round lobules, giving it a beaded appearance; spores ellipsoid, 19–28 × 8–10 μm, with rough, sculptured walls easily seen under 400× magnification. HABITAT: Among mosses on soil, wood, peat, rock, and sometimes bark. COMMENTS: *Pannaria pezizoides* is an extremely similar scaly crustose lichen growing on mosses and bark in boreal and arctic localities, but its main photobiont is blue-green. When wet, *Psoroma hypnorum* is bright green, whereas *P. pezizoides* remains brown or dark olive. Both species have bumpy spore walls, and the spores are about the same size. *Psoroma tenue,* a very rare species known from Banff National Park, Alberta, has a granular, red-brown thallus and contains pannarin (PD+ orange).

Psorula (1 N. Am. species)

KEY TO SPECIES: See *Psora*.

Psorula rufonigra (syn. *Psora rufonigra*)
Blue-edged scale lichen

DESCRIPTION: Squamules very dark yellow-brown to almost olive-black, usually with a blue-black edge, 1–2 mm long and almost as wide, often with lobed margins; lower surface usually dark green to black, rarely pale close to the edge, scattered singly or more often in a shingle-like arrangement; photobiont green (unicellular). Apothecia black, flat to hemispherical, with a black margin flush with the disk; hymenium K/I–; epihymenium blue-green, hypothecium colorless; exciple dark green to black; spores colorless, 1-

732. *Psoroma hypnorum* Southeast Alaska ×4.3

733. *Psorula rufonigra* (growing over *Spilonema revertens*, visible in lower right among the *Psorula* squamules) hill country, central Texas ×3.0

celled, ellipsoid, approximately 12 × 6 μm. CHEMISTRY: All reactions negative but reported to have atranorin. HABITAT: On granitic rocks, growing over (and possibly parasitic on) olive-brown cushions of the cyanobacterial lichen *Spilonema revertens*. COMMENTS: All specimens of *Psorula* seem to be associated with cushions of the minutely filamentous lichen *Spilonema*. We did not include the localities of *Spilonema* in our map of *Psorula* because we could not check this out in every case. In species of *Psora*, which can be very similar to *Psorula*, the epihymenium reacts K+ red-purple (anthraquinones); in *Psorula*, it is K–, although it turns deep purple with concentrated nitric acid (as green lichen pigments often do). *Psora* species do not parasitize *Spilonema*, and the squamules have a pale lower surface.

Punctelia (12 N. Am. species)
Speckled shield lichens, speckleback lichens

DESCRIPTION: Medium-sized, gray, foliose lichens with conspicuous white pseudocyphellae on the upper surface; lobes mostly 3–10 mm across; medulla white;

734. *Punctelia appalachensis* (damp) Appalachians, Tennessee ×2.0

lower surface pale to black, covered with unbranched rhizines to the lobe edge. Photobiont green (*Trebouxia?*). Apothecia lecanorine, with brown disks; spores colorless, 1-celled, ellipsoid, 8 per ascus. CHEMISTRY: Cortex K+ yellow (atranorin); medulla PD–, K–, KC+ red or KC–, C+ red or C– (often containing orcinol depsides or fatty acids). HABITAT: On bark, wood, or rocks. COMMENTS: *Flavopunctelia* is a closely related genus that is more yellow, containing usnic acid in the upper cortex. Species of *Cetrelia* also have white pseudocyphellae on the lobe surface and often react C+ or KC+ pink or red; they are larger lichens, with lobes 10–20 mm wide, and have a black to dark brown lower surface with few rhizines in a broad zone close to the margin.

KEY TO SPECIES

1. Lower surface and rhizines predominantly black. 2
1. Lower surface and rhizines pale to dark brown 3

2. Growing on rock; lobules absent; medulla KC+ red, C+ pink (gyrophoric acid) . . . . *Punctelia stictica*
2. Growing on bark or, rarely, rock; lobules covering lobe surface; medulla KC–, C–. . . . . . *Punctelia appalachensis*

3. Soredia present along the lobe margins and on the surface, lobules absent . . . . *Punctelia subrudecta*
3. Soredia absent 4

4. Isidia absent; lobules common; pycnidia and apothecia common; medulla C+ red, or C– 5
4. Isidia present, cylindrical, branched or unbranched; lobules absent; pycnidia and apothecia uncommon; medulla C+ red (lecanoric acid) . . . . . *Punctelia rudecta*

5. Medulla C+ red, KC+ red (lecanoric acid); lichesterinic and protolichesterinic acids absent; conidia thread-shaped; southwestern U.S. . . . . *Punctelia hypoleucites*
5. Medulla C–, KC–, lichesterinic and protolichesterinic acids present; conidia rod-shaped; central and eastern U.S. and adjacent Canada. . . . . *Punctelia bolliana*

Punctelia appalachensis (syn. *Parmelia appalachensis*) Appalachian speckled shield lichen

DESCRIPTION: Thallus greenish gray, shiny, spotted with maculae and white pseudocyphellae; lobes mostly 3–6 mm wide; finely divided and often branched lobules cover parts of the thallus, and sometimes the margins; lower surface pitch-black, at least in the center, pale brown at the edges, with sparse to abundant rhizines. CHEMISTRY: Medulla PD–, K–, KC–, C– (protolichesterinic acid). HABITAT: On hardwoods in deciduous forests, occasionally on rock. COMMENTS: This closely resembles *P. bolliana* (with the same chemistry) and looks a bit like *P. rudecta* (with a C+ red medulla), but the lower surface of both these lichens is entirely pale tan. *Punctelia subpraesignis*, another lobulate species with a black lower surface, has a C+ red medulla (gyrophoric acid) and is found in Texas and Mexico. See Comments under *P. hypoleucites*.

Punctelia bolliana (syn. *Parmelia bolliana*) Eastern speckled shield lichen

DESCRIPTION: Thallus blue-gray, lobes 2–6(–10) mm wide, becoming heavily wrinkled and folded in older parts, with many small lobules developing on the thallus surface and edges; lower surface pale tan throughout, with pale rhizines. Apothecia usually abundant, with broad, brown disks, 3–15 mm in diameter, concave to convoluted. Pycnidia abundant, appearing as pale brown to black dots on the thallus surface. CHEMISTRY: Medulla PD–, K–, KC–, C– (protolichesterinic and lichesterinic acids). HABITAT: On the bark of deciduous trees in open woodlands, in fields, and on roadsides. COMMENTS: *Punctelia hypoleucites* and *P. semansiana* are very similar in appearance but are C+ red (lecanoric acid). See also *P. appalachensis*.

Punctelia hypoleucites
Southwestern speckled shield lichen

DESCRIPTION: Thallus very much like *P. bolliana*, but generally less lobulate and with more conspicuous pseudocyphellae; lower surface normally pale brown but can vary to dark brown; conidia 10–13 μm long, threadlike, straight. CHEMISTRY: Medulla PD–, K–, KC+ red, C+ dark red (lecanoric acid). HABITAT: On bark or wood of hardwoods. COMMENTS: *Punctelia semansiana* is almost identical but has conidia only half as long. It is widespread in the central U.S. east to the Appalachians, on both rocks and bark, whereas *P. hypoleucites* is strictly southwestern and corticolous. Specimens of *P. semansiana* on rocks tend to be smaller, with flatter, less wrinkled thalli. *Punctelia subpraesignis* is somewhat similar to *P. hypoleucites* but has a darker lower surface (dark brown to almost black) and contains gyrophoric acid as the main product instead of lecanoric acid. It is a relatively rare lichen, restricted to Texas and Mexico. *Flavopunctelia praesignis* has the same white dots and general appearance as *P. hypo-*

735. *Punctelia bolliana* Appalachians, Virginia

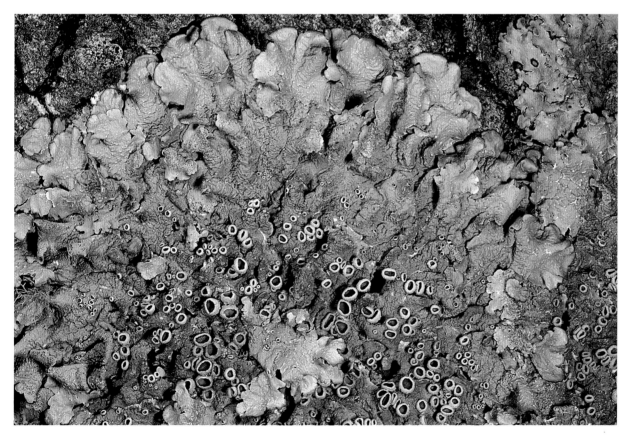

736. *Punctelia hypoleucites* White Mountains, southeastern Arizona ×1.6

737. *Punctelia rudecta* coastal Maine ×2.3

leucites (with abundant apothecia), but is distinctly yellowish green rather than gray (containing usnic acid) and has a pure black rather than pale to dark brown lower surface.

Punctelia rudecta (syn. *Parmelia rudecta*)
Rough speckled shield lichen

DESCRIPTION: Thallus dark greenish gray to almost blue-gray, lobes mostly 3–8 mm wide, more or less covered with cylindrical to branched isidia; white pseudocyphellae usually prominent on lobe tips; lower surface pale tan, with pale rhizines. Apothecia uncommon. CHEMISTRY: Medulla PD–, K–, KC+ red, C+ red (lecanoric acid). HABITAT: On bark of all kinds or shaded rocks. COMMENTS: This is one of the most common eastern isidiate foliose lichens; it is fairly tolerant of pollution, making it familar to city dwellers. Specimens with very few isidia can be mistaken for *P. hypoleucites*.

Punctelia stictica (syn. *Parmelia stictica*)
Rock speckled shield lichen, seaside speckleback

DESCRIPTION: Thallus greenish gray with brown edges, often entirely brown, lobes 1.5–3(–5) mm wide; large, white pseudocyphellae irregular in shape, raised and usually having brown rims, sometimes developing into laminal soralia with coarsely granular soredia; lower surface entirely black except for a narrow brown margin. Apothecia rare. CHEMISTRY: Medulla PD–, K–, KC+ red, C+ pink (gyrophoric acid). HABITAT: On siliceous rocks in exposed situations. COMMENTS: No other small-lobed, rock-dwelling, white-spotted foliose lichen has a black lower surface; the C+ pink medulla confirms the identification. **Punctelia semansiana** is also mostly saxicolous, but the lower surface is pale brown, and the pseudocyphellae are smaller and not raised.

Punctelia subrudecta (syn. *Parmelia subrudecta*)
Powdered speckled shield lichen, forest speckleback

DESCRIPTION: Thallus greenish gray, often browned at the edges; lobes 2–6 mm wide, rather thick and wrinkled in the oldest parts of the thallus; pseudocyphellae small, sparse, and occasionally almost absent; soralia bursting from the lobe margins but also on the top surface, with white to ashy gray, powdery soredia; lower surface pale to dark brown, with abundant, short, slender rhizines, almost like a tomentum. Apothecia rare. CHEMISTRY: Medulla PD–, K–, KC+ red, C+ red (lecanoric acid). HABITAT: Widespread and common on bark of different kinds, and occasionally siliceous rock, usually in open woods and savannas. COMMENTS: *Punctelia punctilla* (syn. *P. missouriensis*) closely resembles *P. subrudecta*, but its soralia, which are almost always associated with pseudocyphellae or cracks in the upper cortex, contain relatively few soredia, and the soredia are coarsely granular to partly corticate or lobule-like. It is a lichen of the Interior Highland, tallgrass prairie

738. *Punctelia stictica* (nonsorediate form; a *Xanthoparmelia* on left) coastal northern California ×2.7

739. *Punctelia subrudecta* mountains, southeastern Arizona ×3.2

740. *Pycnothelia papillaria* coastal Maine ×3.8

region centered in Missouri, and is common on roadside oaks and isolated trees on agricultural land but sometimes grows on shaded rocks as well. Also similar is ***P. perreticulata,*** a much rarer southern interior species found on old pines and junipers. This lichen has fairly narrow lobes (less than 2 mm) and marginal soralia. Its lobes are very rough, with ridges and depressions resembling those of some *Platismatia* species (which have much broader lobes). The pseudocyphellae of *P. perreticulata* are rather sparse and sometimes are difficult to find. ***Punctelia borreri*** is a rare sorediate species from Ohio, West Virginia, and the Pacific Northwest with a black lower surface; it contains gyrophoric acid rather than lecanoric acid.

Pycnothelia (1 N. Am. species)

KEY TO SPECIES: See Key A.

Pycnothelia papillaria
Nipple lichen, gnome fingers

DESCRIPTION: A minute, pale greenish gray fruticose lichen with a persistent, almost white, granular primary thallus, which gives rise to inflated, elongated, unbranched or branched, hollow stalks (podetia) with thin, brittle walls; stalks and branches constricted at the base, 2–15 mm tall, 0.5–2 mm wide, often with brownish, nipple-like tips, or bearing clusters of small, brown apothecia; without soredia; spores 1- or 2-celled, ellipsoid, colorless. Photobiont green (*Trebouxia?*). CHEMISTRY: Thallus PD– or PD+ pale yellow, K+ yellow, KC–, C– (atranorin and protolichesterinic acid). HABITAT: On sandy or gravelly soil on roadsides and open areas. COMMENTS: Although about the same size as many species of *Cladonia* (a genus also characterized by hollow stalks), only *Pycnothelia* has a persistent, granular, primary thallus. In *Cladina*, the primary thallus is granular but quickly disappears; *Dibaeis* and *Baeomyces* have a granular thallus, but the stalks are solid.

Pyrenopsis (12 N. Am. species)
Red water-crust

DESCRIPTION: Granular to areolate crustose lichens, black when dry but distinctly reddish and gelatinous when wet. Photobiont blue-green (usually *Gloeocapsa*, with reddish gelatinous envelopes that turn K+ violet). Apothecia generally open by a small hole (and resemble perithecia); epihymenium pale brown, hypothecium colorless, paraphyses branched or unbranched; asci complex, with K/I– and K/I+ blue zones; spores 1-celled, colorless, ellipsoid, 8 to many per ascus. CHEMISTRY: No lichen substances. HABITAT: On wet siliceous rocks, often on seepages or along streams or lakeshores, rarely on soil. COMMENTS: This genus is poorly understood and needs study. Identifications of North American specimens are suspect. Even the name given to the example in this book should be regarded as tentative; the distribution map is approximate. The genus as a whole is widely distributed from the Arctic to temperate regions. Although species of *Euopsis* are very similar, their fruiting bodies have a broad, disk-like opening (like a lecanorine apothecium), and their asci are somewhat different, with a K/I– or light blue wall and K/I+ blue tholus. *Psorotichia schaereri*, a similar widespread lichen with a black, rimose-areolate thallus found on dry or moist limestone, has *Xanthocapsa*-like cyanobacteria (yellow-brown gelatinous envelopes, K–) and immersed, disk-shaped, black apothecia. Its asci are entirely K/I–, without a thickened tip. In the southwest, some small, black lichens form tiny cushions of stalked granules over soil or calcareous rock. Usually, these are species of *Synalissa*, (with *Gloeocapsa*) or *Peccania* (with *Xanthocapsa*), but they are usually found without fruiting bodies and are difficult for even experienced lichenologists to identify.

KEY TO SPECIES: See Keys D, F, and I.

Pyrenopsis polycocca

DESCRIPTION: Thallus black to reddish black when dry, rusty orange-red when wet, granular with interspersed smooth areas, bearing clusters of perithecium-like apothecia, 0.1–0.2 mm in diameter, with deeply sunken disks that look like holes in the summits; spores 11.5–13 × 7.5–8(–10) μm. HABITAT: In water, semiaquatic; on granite along lakeshores. DISTRIBUTION: Uncertain, but apparently not uncommon in the Appalachian–Great Lakes region. COMMENTS: This and similar species of *Pyrenopsis*, although rarely collected, often form dark reddish incrustations on lakeshore rocks. The photobiont *Gloeocapsa* is common in such habitats in a free-living (nonlichenized) state and looks very much like the lichen. To be sure of what you have, search for the apothecia and ultimately make a microscopic examination of the thallus.

741. *Pyrenopsis polycocca* Lake Superior region, Ontario ×4.7

Pyrenula (48 N. Am. species)
Rash lichens, pox lichens

DESCRIPTION: Crustose lichens with a very thin thallus, usually within the outer tissues of its bark substrate, producing a discoloration of the bark; olive, yellowish, or brownish, or sometimes bright red because of anthraquinones. Photobiont green (*Trentepohlia*). Fruiting bodies fairly large black perithecia, rarely grouped together within wart-like verrucae; perithecial wall carbon-black; hymenial jelly IKI+ greenish blue; paraphyses both branched and unbranched; spores brown, 4-celled to muriform, with unevenly thickened walls and spherical to lens-shaped cells;

spores 8 per ascus; conidia thread-like. CHEMISTRY: Some species produce orange to red anthraquinone pigments (K+ deep red-purple) on the surface of the thallus or perithecia, and some contain lichexanthone and fluoresce UV+ yellow-orange in long-wave UV light. HABITAT: On the bark of various trees, especially those with smooth, living bark tissues; most species tropical or subtropical. COMMENTS: The genus is interpreted broadly here, following the work of Richard Harris, to include species with muriform spores. *Trypethelium* and related genera are superficially similar, but they usually have several perithecia grouped into verrucae or pseudostromata, the spores are colorless, the hymenial jelly is IKI–, and their conidia are rod-shaped and much shorter. Like *Pyrenula*, *Eopyrenula* has brown, septate spores, but the cells are thin-walled and usually not equal in size. **Eopyrenula intermedia** is a fairly common, although inconspicuous eastern lichen that forms a thin, whitish thallus just under the bark surface. Its black perithecia are 0.15–0.4 mm in diameter, with about half the perithecium protruding and thinly covered with a whitish pruina. The spores are 4–7-celled, slightly constricted at the septa, with the largest cells in the middle, 18–24 × 5–9 μm. **Lithothelium** (syn. *Plagiocarpa*) is a genus much like *Pyrenula*, but with ostioles at the sides rather than at the summits of the perithecia, and often with a longish neck. The spores can be colorless or brown. **Lithothelium septemseptatum** is a common brown-spored species in the northeast (spores 8-celled, 30–45 × 12–15 μm); *L. hyalosporum* is another eastern species, but with colorless spores (see Key J).

KEY TO SPECIES (based on Harris, *More Florida Lichens*, 1995). All species from the southeastern coastal plain, except as noted.

1. Spores muriform 2
1. Spores usually transversely septate (occasionally with a few cells divided longitudinally) 6
2. Spores (2–)4(–6) per ascus, 100–150(–190) × 30–41 μm; ostioles off-center (not at the summits), sometimes fused . . . . [*Pyrenula falsaria*]
2. Spores 8 per ascus; ostioles at summit or off-center . . . 3
3. Thallus orange or yellow, K+ purple (anthraquinones); spores 23–35 × 11–17 μm . . . . . . *Pyrenula ochraceoflavens*
3. Thallus not orange or yellow; K– 4
4. Perithecia fused in groups by their off-center or lateral ostioles; spores 45–65 × 18–27 μm . . [*Pyrenula ravenellii*]
4. Perithecia solitary, ostioles at summit 5
5. Spores 45–60 × 16–22(–25) μm . . [*Pyrenula leucostoma*]
5. Spores 30–40(–53) × 11–15 μm . . . . [*Pyrenula thelomorpha*]
6.(1) Spores 27–38 × 12–18 μm 7
6. Spores less than 25 μm long 9
7. Thallus and perithecial warts dark red (the pigment sometimes confined to the perithecial warts) . . . . *Pyrenula cruenta*
7. Thallus and perithecial warts greenish when fresh, whitish to brownish in herbarium 8
8. Overmature old spores containing a red, oily substance; spores 4–6-celled . . . . . [*Pyrenula concatervans*]
8. Overmature old spores empty, collapsing; spores 4-celled . . . . [*Pyrenula punctella*]
9. Spore walls at the tips of the spores thin 10
9. Spore walls at the tips of the spores obviously thick 11
10. East Temperate; no orange pigment on summits of the perithecia; thallus UV+ yellow (lichexanthone) . . . . *Pyrenula pseudobufonia*
10. Pacific Northwest near coast; scant or abundant orange pigment on summits of the perithecia; thallus UV– or dull whitish . . . . [*Pyrenula occidentalis*]
11. Perithecia fused into an extensive raised pseudostroma; spores 17–21 × 7–9 μm . . . . [*Pyrenula anomala*]
11. Perithecia solitary, not in a pseudostroma 12
12. Thallus scurfy, whitish, without a cortex; spores 19–25 × 8–12 μm; perithecia 0.3–0.4 mm in diameter; hemispherical, relatively thin-walled . . . . [*Pyrenula microcarpa* (syn. *P. cinerea*)]
12. Thallus smooth, with a cortex; spores 15–21 × 5.5–8; perithecia 1–1.5 mm in diameter, flattened-conical, thick-walled . . . . [*Pyrenula mamillana* (syn. *P. marginata*)]

Pyrenula cruenta (syn. *Melanotheca cruenta*)
Bark rash lichen

DESCRIPTION: Thallus and perithecial warts crimson, either in patches or over the entire thallus (shade forms are red only around the perithecia); without pseudocyphellae; perithecia 0.5–1 mm in diameter, sometimes clustered and partially fused together; abundant oil drops among the paraphyses; spores 4-celled, 24–35 × 13–17 μm, 8 per ascus. CHEMISTRY: Thallus UV+ reddish; the red pigment is K+ deep red-purple (anthraquinones). HABITAT: On broad-leaved trees in wood-

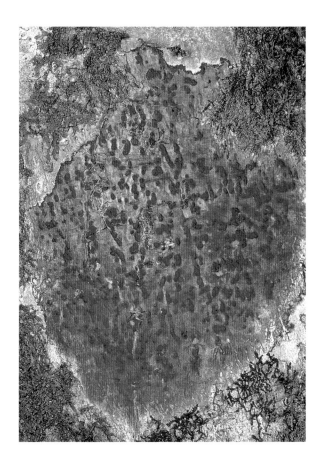

lands and hammocks. COMMENTS: *Pyrenula cruentata,* also with a bright red thallus, is distinguished by its muriform spores, 30–42 × 15–20 μm; it is found only in southern Florida. *Pyrenula cerina,* with a bright orange thallus dotted with pseudocyphellae, is another brightly colored member of the genus with 4-celled spores found in tropical Florida. *Pyrenula ochraceoflavens* and *P. ochraceoflava* both have yellow thalli and muriform spores.

Pyrenula ochraceoflavens
Yellow pox lichen

DESCRIPTION: Thallus dark yellow or orange-yellow, sometimes mainly whitish, with the pigment only over the perithecia; perithecia 0.4–0.6 mm in diameter, almost entirely covered by yellow bark-thallus material, but with an exposed, often nipple-like black ostiole; hymenium without any oil drops; spores muriform, 23–35 × (9.5–)11–17 μm, with spherical cells, 5–7 rows high, and several cells across. CHEMISTRY: Thallus UV+ reddish; pigments of thallus K+ red-purple (anthraquinones). HABITAT: Tropical woodlands, on bark. COMMENTS: Similar in name as well as appearance, *P. ochraceoflava* can be distinguished by its smaller perithecia (0.2–0.4 mm in diameter) and smaller spores (13–25 × 8–13 μm, with spherical cells consistently stacked in four rows).

Pyrenula pseudobufonia
Eastern pox lichen

DESCRIPTION: Thallus thin and endophloeodal (within the bark tissues), causing a grayish olive to olive-brown discoloration of the bark, uniform in color, without pseudocyphellae; perithecia abundant or sparse, 0.4–0.7(–1) mm in diameter, at first covered with a thin coating of bark-thallus tissue in young thalli but emerging and coal-black when mature, scattered rather than aggregated, round to slightly pointed at the summit; hymenium filled with oil drops; spores 4-celled, with lens-shaped cells, the end cells filling the tips of the spores, 13–22(–24) × (7–)8–11(–12) μm. CHEMISTRY: Thallus UV+ orange-yellow (lichexanthone); K– (without anthraquinone pigments). HABITAT: On bark of broad-leaved trees,

742. *Pyrenula cruenta* Big Thicket, eastern Texas ×1.2

743. *Pyrenula ochraceoflavens* Florida Keys ×3.0

744. *Pyrenula pseudobufonia* northern Florida ×2.6

especially beech, holly, and oak, in shaded forests. COMMENTS: This is one of the most common bark-dwelling lichens with perithecia. In the northern part of the range, the perithecia tend to be small and sparse. *Pyrenula occidentalis* is a species from the Pacific Northwest that resembles *P. pseudobufonia*, but the thallus lacks lichexanthone (UV– or + pale beige), and it usually has a thin coating of orange pigment (K+ red-purple) on the summits of some perithecia. Spores are 13.5–19 × 7–9 μm, with walls so thick that the cell cavities often do not reach the outer edge of the spores.

Pyrrhospora (6 N. Am. species)
Crimson dot lichens

DESCRIPTION: Thallus crustose, yellowish to greenish gray, smooth to powdery or with discrete soralia. Photobiont green (*Trebouxia?*). Apothecia biatorine, usually brilliant orange-red, black, or red-brown, with a thin, soon disappearing margin; asci with a broad, K/I+ dark blue tholus (*Lecanora*-type); paraphyses largely unbranched except for tips; spores colorless, 1-celled, broadly to narrowly ellipsoid, 8 per ascus. CHEMISTRY: Frequently with xanthones in the thallus (KC+ orange, C+ orange), and red anthraquinone pigments in the epihymenium and exciple (K+ red-purple); sometimes with β-orcinol depsides and depsidones in the thallus. HABITAT: On bark or wood, rarely on rock. COMMENTS: *Pyrrhospora* is interpreted broadly here to include *P. varians*, a species that does not contain anthraquinones in the apothecium. See comments under *P. quernea*. Other genera with red-orange apothecia can be confused with *Pyrrhospora* species. Species of *Caloplaca* have orange rather than red apothecia (with rare exceptions), and the spores and asci are quite different. Species of *Protoblastenia*, which also have 1-celled spores, grow on rocks or soil and have *Porpidia*-type asci. Red-fruited **Biatorella** species have many spores per ascus; *Haematomma* has lecanorine apothecia and long, many-celled spores. *Pyrrhospora elabens*, with its shiny, ebony-black apothecia, resembles some species of *Lecidella* but has a brown rather than greenish epihymenium and much narrower spores (6–10 × 3–4.5 μm). It has a rather thick, creamy white thallus that contains atranorin and fumarprotocetraric acid (PD+ red, K+ yellow, KC–, C–) and grows on hard wood.

KEY TO SPECIES: See Keys C and F.

Pyrrhospora cinnabarina
Northern crimson dot lichen

DESCRIPTION: Thallus yellowish gray, pale green, or whitish, thin and continuous, often with scattered, small, rounded soralia that contain heaps of granular soredia. Apothecia 0.5–1(–1.5) mm in diameter, bright or dark orange-red, usually abundant (especially in the west), but sometimes entirely lacking, leaving a sterile, sorediate thallus; disks flat, irregular in shape, with thin, disappearing margins the same color as the disks; spores narrow, 8–12 × 2–3(–5) μm. CHEMISTRY: Thallus PD+ red, K– or K+ yellowish, KC–, C–; soralia PD+ red, K+ yellow-brown, KC–, C–, UV– or pale whitish (fumarprotocetraric acid and atranorin); pigmented apothecial tissues K+ red-purple (anthraquinones). HABITAT: On bark and wood, especially of conifers and birch, in shaded or open forests. COMMENTS: Plate 745 shows a richly fertile but nonsorediate specimen. Although easily identified when it has brilliant, orange-red apothecia with thin margins, *P. cinnabarina* is easily confused with several other PD+ red sterile crusts when sterile and sorediate. *Pertusaria pupillaris* and *P. borealis*, for example, also have scattered soralia. These soralia are generally excavate, and their thalli are usually thinner and within the bark, lacking atranorin. In the Great Lakes–Appalachian region, one can encounter sterile specimens of **Megalospora porphyrites** with scattered white soralia; it contains pannarin (PD+ dark orange, K–) and zeorin. Fertile specimens have dark brown biatorine apothecia with 4–5-celled or up to 7-celled, colorless, cylindrical spores, 50–70 × 21–24 μm. It typically grows on white cedar (*Thuja*).

Pyrrhospora quernea
Sulphured crimson dot lichen

DESCRIPTION: Thallus straw-colored to dull sulphur-yellow, consisting entirely of powdery to granular soredia, often sterile, but sometimes producing marginless, black to dark rusty red apothecia, 0.3–0.7(–1) mm in diameter; spores mostly 8–12 × 6–7 μm. CHEMISTRY: Thallus PD–, K+ yellow (difficult to see against a yellow thallus), KC+ orange, C+ orange, UV+ orange (mainly isoarthothelin and thiophanic acid); anthraquinones in the pigmented apothecial tissues (K+ red-purple). HABITAT: On wood and sometimes bark or rock in open areas, in oceanic regions. COMMENTS: *Pyrrhospora varians* is a common northeastern species that also has a thallus that usually reacts KC+ orange because of xanthones. However, it is dull greenish gray and, rather than sorediate, it is granular to continuous, usually in discrete round patches with a black prothallus at the thallus margin. Its apothecia are red-brown to almost black (without anthraquinones), flat or slightly convex, without margins, often clustered, 0.2–0.4 mm in diameter, with spores (8–)10–13 × 5.5–7 μm. It grows on the bark of deciduous trees and shrubs in shady or open sites. Because species of *Lecidella* and *Lecanora* also may contain C+ orange xanthones in the thallus, sterile thalli can be confusing. *Lecanora expallens* and related species are especially problematic when sterile, but their thalli have patchier soralia and contain usnic acid as well as xanthones. *Lecidella scabra* is a rarer species, with a dark greenish gray, granular, sorediate thallus that contains atranorin in addition to xanthones.

Pyrrhospora russula
Southern crimson dot lichen

DESCRIPTION: Thallus dark green to greenish gray, continuous and rough (verruculose), without soredia, sometimes showing a white prothallus. Apothecia red-orange, flat, round to irregular in

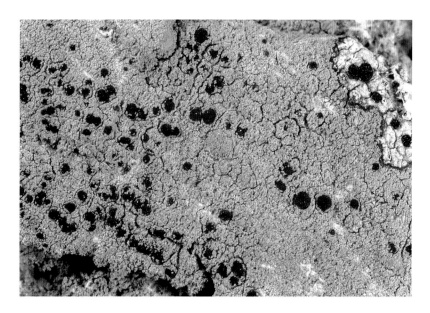

745. *Pyrrhospora cinnabarina* Rocky Mountains, Montana ×6.1

746. *Pyrrhospora quernea* Channel Islands, southern California ×5.7

747. *Pyrrhospora russula* coastal plain, North Carolina ×4.3

shape, with a thin margin the same color as the disk, soon disappearing; spores 8–12 × 3–4 μm. CHEMISTRY: Thallus PD+ red, K– (rarely PD+ yellow, K+ red), KC–, C–, UV+ yellow-orange (fumarprotocetraric acid or rarely norstictic acid, lichexanthone); pigmented apothecial tissues K+ red-purple (russulone). HABITAT: On bark of hardwoods. COMMENTS: Nonsorediate specimens of *P. cinnabarina* have a thinner, whiter thallus that contains atranorin instead of lichexanthone. The ranges do not overlap. The spores distinguish it from similar species of *Caloplaca*.

Pyxine (10 N. Am. species)
Buttoned rosette lichens

DESCRIPTION: Small to medium-sized foliose lichens, usually closely appressed to the substrate, pale gray to greenish gray, smooth or pruinose, frequently with soredia or isidia; lower surface black, with slender, unbranched to forked black rhizines. Photobiont green (unicellular). Apothecia basically lecanorine, although the margins are without algae in most species and resemble lecideine apothecia; disks black, with thick, black margins; spores 2–4-celled, brown, with angular cells and thickened walls (*Physcia*-type). CHEMISTRY: Cortex contains lichexanthone (UV+ yellow-orange) and (or) atranorin (K+ yellow); medulla with various reactions, sometimes containing β-orcinol depsidones and pigments. The presence of triterpenes in the thallus is sometimes revealed by a fuzz of fine, colorless crystals on the thallus surface in old specimens. HABITAT: Usually on bark, but occasionally saxicolous, in tropical or subtropical forests except for *P. sorediata*, which is temperate. COMMENTS: Species of *Pyxine* often resemble those of other members of the family Physciaceae, such as *Dirinaria*, *Hyperphyscia*, *Heterodermia*, *Physconia*, *Physcia*, or *Phaeophyscia*. *Dirinaria*, which is also subtropical to tropical, closely resembles *Pyxine* but lacks rhizines and never has a UV+ thallus. The other genera are mostly more northern. See Comments under *Pyxine sorediata*.

KEY TO SPECIES: See *Physcia*.

Pyxine berteriana
Buttoned rosette lichen

DESCRIPTION: Thallus pale green to grayish green, appressed, often lightly pruinose but with no soredia or isidia; medulla yellow. Apothecia black, up to 2 mm across, having equally black, prominent margins (appearing lecideine); spores 15–22 × 6–8 μm. CHEMISTRY: Cortex K–, UV+ bright yellow (lichexanthone; no atranorin); medulla PD– or PD+ orange, K–, KC–, C– (unidentified pigment, sometimes testacein, triterpenes). HABITAT: On bark, usually in the shade. COMMENTS: Several species of *Pyxine* also have yellow or orange medullary tissue, but they have soredia or isidia. *Pyxine albovirens*, which is also UV+ yellow, has patches of soredia on the lobe surface; *P. caesiopruinosa* has fragments of burst pustules (schizidia), which can resemble soredia, along the margins. See also *P. sorediata*.

Pyxine cocoes
Buttoned rosette lichen

DESCRIPTION: Thallus pale greenish gray, usually with patches of pruina on the thallus surface, with radiating lobes mostly under 1 mm wide, closely appressed, with granular soredia bursting through the cortex, forming small, irregularly shaped, laminal soralia; medulla white. Apothecia black with black margins when mature; young apothecia can have remnants of thalline tissue on the outside of the margin; internal tissues of the base of the apothecium (the stipe) are brownish red and turn deep red with K; epihymenium and external parts of the cortex are K+

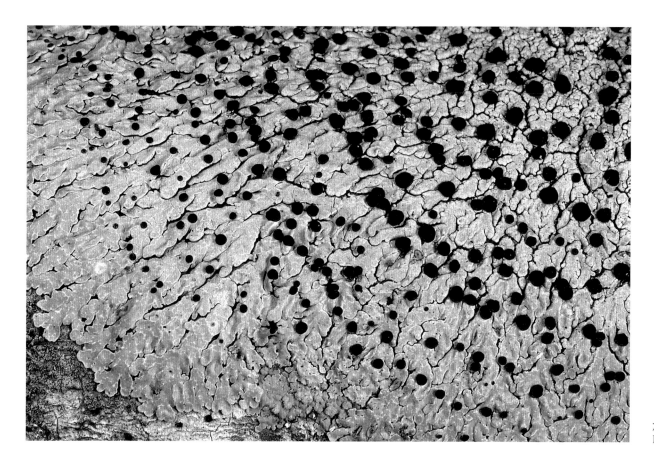

748. *Pyxine berteriana*
Florida Keys ×4.3

reddish purple. CHEMISTRY: Cortex PD–, K–, UV+ yellow-orange (lichexanthone); medulla PD–, K–, KC–, C– (triterpenes). HABITAT: On bark and occasionally rocks. COMMENTS: This species resembles *P. berteriana* but is sorediate and has a white medulla. *Pyxine eschweileri* and *Dirinaria picta* are also closely appressed, pale, foliose lichens on bark with laminal soralia and a white medulla, but the former has 4-celled spores and is PD+ orange in the medulla (testacein), and the latter lacks rhizines. Both have a K+ yellow, UV– cortex (atranorin). *Pyxine subcinerea* is almost identical to *P. cocoes* except the medulla is yellow.

Pyxine eschweileri

DESCRIPTION: Thallus pale gray to greenish gray, forming round rosettes, closely appressed; upper surface smooth, often shiny or somewhat pruinose at the lobe tips; lobes 0.5–1.7 mm across, flat to slightly concave; medulla white or pigmented pale to dark yellow; thallus either forming marginal pustules that break down into soredialike fragments (schizidia), or with laminal soralia containing granular soredia; lower surface with abundant black rhizines. Apothecia occasionally seen, black; spores 4-celled, (17–)21–24(–29) × (7–)8–11 μm. CHEMISTRY: Cortex K+ yellow, UV– (atranorin); medulla PD+ orange, K–, KC–, C– (testacein and triterpenes). HABITAT: On bark. COMMENTS: The two morphotypes of *P. eschweileri* probably represent different species. The morphotype with schizidia represents *P. eschweileri* in the strict sense. Except for *P. sorediata*, a larger more northern lichen, other species of *Pyxine* with a yellow medulla have a K–, UV+ yellow upper surface caused by lichexanthone. See Comments under *P. berteriana*.

Pyxine sorediata
Mustard lichen

DESCRIPTION: Thallus dull bluish gray to green-gray, pruinose at the lobe tips; lobes 1–2.5 mm wide, usually turned up at the tips, often divided into small, round lobules, with blue-gray

749. *Pyxine cocoes*
Florida Keys ×7.3

750. *Pyxine eschweileri* central Florida ×3.9

751. *Pyxine sorediata* Appalachians, Virginia ×2.9

Fig. 34 *Cystocoleus* compared with *Racodium* under the microscope. (a) *C. ebeneus*; (b) *R. rupestre*. Scale = 10 μm. (Reproduced courtesy of the British Lichen Society from Purvis et al., 1992, Fig. 13.)

granular soredia produced abundantly along the lobe margins; medulla mustard-yellow; lower surface coal-black, with abundant rhizines that are slender, forked to tufted, black, but often with pale tips. CHEMISTRY: Cortex K+ yellow (often vague), UV– (atranorin); medulla PD–, K–, KC± reddish slowly, C–. HABITAT: On the bark of hardwoods and white cedar (*Thuja*) in mature forests. COMMENTS: This is the largest *Pyxine* and the only one that is widely distributed in the northeast but absent on the southeastern coastal plain. Other common foliose lichens with a pigmented medulla in the same habitats include *Myelochroa aurulenta*, which has a paler yellow medulla and laminal soredia, and *Phaeophyscia rubropulchra*, a much smaller lichen with a bright orange-red medulla, green soredia, and a K– cortex.

Racodium (1 species worldwide)

KEY TO SPECIES: See Key B.

Racodium rupestre
Rock hair

DESCRIPTION: Hair-like, filamentous fruticose lichen, pure black. The rather stiff and tangled hairs are approximately 10–20 μm thick and consist of filaments of the green alga *Trentepohlia* enveloped by a sheath of pigmented fungal hyphae with cylindrical cells lined up in straight rows (Fig. 34b). CHEMISTRY: No lichen substances. HABITAT: On shady rock faces. DISTRIBUTION: Probably widespread in northern and montane regions west to British Columbia, perhaps rare in the east, but infrequently collected. COMMENTS: This lichen can be distinguished from **Cystocoleus ebeneus** only microscopically (Fig. 34a). The cells that envelop the *Trentepohlia* filaments of *Cystocoleus* are knobby and irregular, not in straight rows. *Cystocoleus* is known from the Rocky Mountains (Colorado to British Columbia) as well as the eastern U.S. Other black filamentous lichens, such as *Pseudephebe* and *Ephebe*, have considerably thicker branches.

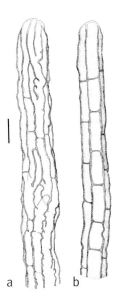

752. *Racodium rupestre* southern New Brunswick ×4.9

Ramalina (44 N. Am. species)
Ramalina

DESCRIPTION: Pendent or shrubby fruticose lichens with long, more or less flattened (rarely terete) branches, broad or slender, usually rather stiff, solid throughout or hollow close to the base, yellowish green (usnic yellow) to yellowish tan, often with pseudocyphellae and (or) soralia; cortex with two distinguishable layers of thick-walled prosoplectenchyma. Photobiont green (*Trebouxia*?). Apothecia lecanorine, with pale yellowish, often pruinose disks; spores colorless, 2-celled, ellipsoid, 8 per ascus. Pycnidia not abundant, pale. CHEMISTRY: Cortex PD–, K–, KC+ dark yellow, C– (usnic acid); medulla and soralia with various reactions (often containing β-orcinol depsidones, and orcinol or β-orcinol depsides). HABITAT: On branches, twigs, and trunks of various types of trees and shrubs, and on siliceous rocks. COMMENTS: Species with flattened, rather stiff branches are easily named as *Ramalina* or its cousin, *Niebla,* but slender, almost filamentous species can be mistaken for *Usnea* or *Alectoria* (see Comments under *R. thrausta*). Species of *Evernia* are much softer, as their cortex contains no stiffening prosoplectenchyma (Fig. 8). All species of *Usnea* contain a central, supportive cord. *Niebla* species, restricted mainly to the California coast (rarely north to Oregon), are extremely stiff, with a much-thickened cortex composed of cells perpendicular to the surface and often have additional cartilage-like strands in the medulla. Almost all species of *Niebla* have abundant, black pycnidia. IMPORTANCE: Species of *Ramalina* have been used to make dye and perfume in Europe and for dye and foodstuffs in India. Some species are used as nesting material by birds such as the Northern Parula Warbler, especially where *Ramalina* replaces *Usnea* as the dominant pendent epiphyte.

KEY TO SPECIES (including *Niebla*)

1. Thallus pendent, 5–100 cm (2–40 in) long, often with clustered or indistinct attachment points 2
1. Thallus growing in bushy or almost pendent tufts from a single point, generally less than 6 cm long 6

2. Branches round in cross section, very fine (mostly under 0.5 mm wide), the tips curled up, often producing a few granules or soredia . . . . *Ramalina thrausta*
2. Branches or stalks distinctly flattened, over 0.5 mm wide 3

3. Branches with expanded, net-like tips . . . . *Ramalina menziesii*
3. Branches not producing net-like tips 4

4. Soredia abundant, in marginal and often laminal soralia . . . . *Ramalina subleptocarpha*
4. Soredia absent 5

5. Branches 3–10 cm long, 2–6 mm wide, more or less flat; western . . . . *Ramalina leptocarpa*
5. Branches 30–50 cm long, 0.2–1.3(–2) mm wide, twisted, with long white pseudocyphellae, giving the branches a striped appearance; subtropical, Florida, Texas, and California . . . . [*Ramalina usnea*]

6.(1) Soredia present; apothecia rare 7
6. Soredia absent; apothecia common 15

7. Soredia white to blue-gray; soralia in rounded mounds, becoming woolly in old specimens; branches mostly round in cross section, often spotted with black bands or dots. On coastal shrubs, trees, and rarely rocks along the west coast . . . . *Niebla cephalota*
7. Soredia yellowish green; soralia of various shapes, never becoming woolly; branches usually flattened, sometimes round in cross section, never spotted with black bands or dots 8

8. Branches at least partly hollow; medulla loose and webby 9

8. Branches solid throughout; medulla dense and compact 10
9. Branches with round to oval perforations, smooth .... *Ramalina roesleri*
9. Branches not perforate or lacerate, but with depressions or ridges .... *Ramalina obtusata*
10. Soralia very irregular in shape, at or close to the lobe tips, which often appear torn, frayed, or hood-shaped; evernic acid present 11
10. Soralia round, elliptical, or elongate; branch tips tapering, sometimes finely divided but not generally torn, frayed, or expanded; evernic acid absent 12
11. Branches narrow or broad, 0.5–3 mm wide; soredia mainly on the lower surface of the branch tips, not developing within hood-shaped expansions .... *Ramalina pollinaria*
11. Branches broad, 1.5–4 mm wide; soredia developing in inflated, hood-shaped expansions of the branch tips.. .... *Ramalina obtusata*
12. Soralia often producing tiny, isidia-like branchlets; branches angular or round in section, or flattened in part; subtropical to tropical regions .... *Ramalina peruviana*
12. Soralia lacking isidia-like branchlets; branches usually flattened throughout; not southern 13
13. Soredia granular, usually concentrated at the branch tips; mainly on rocks in forest habitats; sekikaic acid present .... *Ramalina intermedia*
13. Soredia farinose, marginal or laminal, not concentrated at the branch tips; mostly on trees; sekikaic acid absent 14
14. Soralia mainly linear, along the branch margins, but also elliptical on the branch surface; branches (1–)2–4 (–10) mm wide; soralia always PD–, K–; zeorin present .... *Ramalina subleptocarpha*
14. Soralia mainly elliptical, on the branch margins, and occasionally on the surface; branches mostly 0.5–3 mm wide; soralia PD– or PD+ yellow or red, K– or K+ red; zeorin absent .... *Ramalina farinacea*
15.(6) On coastal rocks in California; pycnidia abundant and conspicuous, black 16
15. On trees, or rarely rocks in interior regions; pycnidia inconspicuous, brown 17
16. Branches distinctly flattened or angular in cross section; slender, cartilaginous cords present in medulla .. .. *Niebla homalea*
16. Branches round in cross section; lacking cartilaginous cords in medulla .... *Niebla combeoides*
17. Branches at least partly hollow, with many perforations; medulla loose and webby .. *Ramalina dilacerata*
17. Branches solid, without perforations; medulla dense and compact 18
18. Branches smooth or with elongate striations....... 19
18. Branches with depressions and ridges or long grooves, or with verrucae, tubercles, papillae 20
19. Branches flattened throughout; spores fusiform, 16–25 (–31) × 3–5 μm .... *Ramalina stenospora*
19. Branches almost round in cross section except for the base; spores elongate ellipsoid to almost fusiform, 11.5–22 × 3–5 μm .... *Ramalina montagnei*
20. Apothecia sometimes on the branch surface or along margins 21
20. Apothecia typically on or close to the tips of the branches 23
21. Warts or tubercles abundant on the branches; divaricatic acid present .... *Ramalina complanata*
21. Warts and tubercles absent; divaricatic acid absent 22
22. Western North America; branches more or less even in width, tapering only at the tip .. *Ramalina leptocarpha*
22. Southcentral U.S.; branches tapering at the base and at the tips .... *Ramalina celastri*
23.(20) Branches with small white tubercles or papillae; medulla K+ red or purplish red 24
23. Branches with depressions and ridges, or with long grooves, but without tubercles; medulla K–........ 25
24. Spiny, perpendicular branches usually present, although sparse; spores ellipsoid, medulla PD+ red or yellow, C– (salazinic, norstictic, or protocetraric acids) .... *Ramalina willeyi*
24. Spiny, perpendicular branches absent; spores fusiform; medulla PD–, C+ pink to red or violet (rapidly disappearing) (cryptochlorophaeic and paludosic acids) ... . *Ramalina paludosa*
25. Branches broad and fan-shaped; pseudocyphellae depressed, between the vein-like ridges on the lower surface; southwestern or north central U.S. .... *Ramalina sinensis*
25. Branches divided and relatively narrow; pseudocyphellae raised or level with surface; south central or eastern U.S. and adjacent Canada .... *Ramalina americana*

Ramalina americana
Sinewed ramalina

DESCRIPTION: Thallus shrubby, in tufts 1–3(–4) cm long; branches with strong ridges and channels, (0.2)0.5–3 mm wide, generally with white pseudocyphellae, solid to the base, frequently fertile with

RAMALINA 621

753. *Ramalina americana* North Woods, Maine ×2.9

754. *Ramalina celastri* hill country, central Texas ×1.3

large flat to contorted apothecia at or close to the branch tips; disks yellow and pruinose; spores 12–13.5 × 4.5–6 μm, straight to slightly bean-shaped. CHEMISTRY: Medulla PD–, K–, KC–, C– (lacking medullary substances). HABITAT: On twigs and branches of various trees, usually in full sun. COMMENTS: Specimens from southeastern U.S. often contain lichen substances in the medulla (e.g., divaricatic, norbarbatic, and 4-*O*-demethylhypoprotocetraric acids; or stenosporic and glomelliferic acids), and these can be called **R. culbersoniorum**. Its range is pink on the distribution map. Tufts of *R. dilacerata* can resemble *R. americana*, but at least the basal branches are clearly hollow and perforated. See *R. sinensis* and *R. paludosa*.

Ramalina celastri
Palmetto lichen

DESCRIPTION: Thallus large, 4–8 cm long with broad branches 3–12(–20) mm across, diverging from a single point like a palmetto leaf (palmate), usually tapering at both ends, creating a blade-like shape; young thalli with few branches, but

side branches developing in older thalli; branches longitudinally ridged, sometimes developing holes and slits, but not conspicuously warty. Apothecia with pale yellowish, pruinose disks, abundantly produced on the lobe surface and along the edges of the branches; spores ellipsoid, ca. 11–14 × 4–6 μm, straight. CHEMISTRY: All medullary reactions negative (only usnic acid). HABITAT: Common on trees, shrubs, and occasionally rocks. COMMENTS: This *Ramalina*, with broad, doubly tapered branches and abundant apothecia, is very distinctive. North American specimens formerly named as *R. ecklonii* belong here.

Ramalina complanata
Bumpy ramalina

DESCRIPTION: Thallus short (2–3 cm long), broad-lobed (1–7 mm wide), bushy, with the branches longitudinally ridged and covered with whitish warts or tubercles; pseudocyphellae at the summits of the warts. Apothecia large (up to 4 mm across), yellowish or pinkish orange, common at the lobe tips or on the lobe surface; spores 9–15(–18) × 4–6 μm, straight. CHEMISTRY: Medulla typically PD–, K–, KC–, C– (divaricatic acid); rarely with protocetraric acid or salazinic acid. HABITAT: On broad-leaved trees and shrubs in the open. COMMENTS: *Ramalina americana* is about the same size, has the same overall habit, and also lacks soredia; it differs in lacking the whitish warts. In the southern species *R. denticulata,* which also has pseudocyphellae raised on little warts, the branches are longer, more strap-shaped, and not as heavily ridged. The apothecia are frequently laminal, marginal, and close to the tips, and the medulla reacts PD+ yellow, K+ red (salazinic acid).

Ramalina dilacerata (syn. "*Fistulariella dilacerata*")
Punctured ramalina

DESCRIPTION: Thallus pale, short and densely shrubby, with a relatively thin cortex and loose medulla; branches 1–2(–3) cm long, 0.4–1.3 mm wide, rather smooth, always more or less hollow with many perforations into the medulla, without pseudocyphellae or soredia. Apothecia abundant, marginal, or mainly at or close to

755. *Ramalina complanata*   hill country, central Texas ×2.0

756. *Ramalina dilacerata* Southeast Alaska ×2.0

757. *Ramalina farinacea* Coast Range, California ×1.2

the branch tips, with pale yellow, pruinose disks. CHEMISTRY: Medulla PD–, K–, KC–, C– (divaricatic acid). HABITAT: On twigs and branches of various trees in the open, typically on lakeshores or maritime coasts. COMMENTS: The thickness of the cortex varies: some thalli are thin and almost translucent, and others are tough and shiny. The larger, more robust west coast populations (including the specimen shown in plate 756) may be more closely related to the southern hemisphere species **R. geniculata**. *Ramalina americana* has a similar aspect but is not hollow or perforate and usually has pseudocyphellae.

Ramalina farinacea
Dotted ramalina, the dotted line

DESCRIPTION: This is an extremely variable species, both morphologically and chemically. Thallus pale to dark yellowish green, almost pendent to bushy, with broad to narrow branches, 0.5–3(–4) mm wide, 3–7 cm long, usually strongly flattened but sometimes almost terete; numerous discrete, elliptical soralia containing greenish, powdery soredia, dotted along the branch margins, but occasionally on the surface as well. Apothecia very rare. CHEMISTRY: Medulla and soralia PD–, K–, KC–, C– (hypoprotocetraric acid or without lichen substances); or PD+ red-orange, K–, KC+ pink, C– (protocetraric acid); or PD+ yellow, K+ red, KC–, C– (salazinic and (or) norstictic acids). HABITAT: On all kinds of trees and shrubs, rarely on rock, in regions with a mild, humid climate. COMMENTS: *Ramalina subleptocarpha* is a similar species along the west coast, but it has long, mostly uninterrupted soralia along the branch margins and the branches tend to be broader; it contains only zeorin besides usnic acid (PD–, K–). Some specimens of *R. intermedia,* an eastern and southern Rocky Mountain species on rock, can resemble *R. farinacea* in morphology but contain sekikaic acid (PD–, K–). Two subtropical sorediate species may also key out here: *R. peruviana* and **R. dendriscoides**. They are compared under *Ramalina peruviana*.

Ramalina intermedia
Rock ramalina

DESCRIPTION: Thallus pale yellow-green, with finely divided, flat branches 1–3 cm long, tapering from 1(–2) mm wide at the base to tips less than 0.2 mm wide and typically curled; soralia concentrated at the branch tips but sometimes marginal, more or less round, containing coarsely granular soredia; pseudocyphellae not abundant or conspicuous. Apothecia rarely seen. CHEMISTRY: Medulla and soralia PD–, K–, KC–, C– (sekikaic acid). HABITAT: On rock faces and boulders in forests, rarely on bark. COMMENTS: *Ramalina pollinaria* and *R. obtusata* are two other sorediate ramalinas that can grow on rock. Both contain evernic acid. *Ramalina pollinaria* has broader, mainly terminal soralia that appear to be produced only on the lower surface; *R. obtusata* has hollow, helmet-shaped soralia at the branch tips. An uncommon Appalachian chemical variant of *R. intermedia* contains protocetraric acid (PD+ red-orange, K–, KC+ pink) and can be called **Ramalina petrina**. The occasional patch of *R. farinacea* on rock can generally be distinguished by its clearly delimited, oval soralia along the branch margins, as well as its chemistry.

758. *Ramalina intermedia* Lake Superior region, Ontario ×2.5

759. *Ramalina leptocarpha* Coast Range, California ×1.7

Ramalina leptocarpha
Western strap lichen

DESCRIPTION: Thallus almost pendent or in tufts, 3–10 cm long; branches rather broad (2–6 mm), strap-shaped, with shallow longitudinal ridges and depressions, without soredia. Apothecia flat to concave, on the margins and sometimes the surface of the branches; white pseudocyphellae on the lower surface of the apothecia. CHEMISTRY: Medulla is PD–, K–, KC–, C– (no lichen substances other than usnic acid). HABITAT: On broad-leaf trees, rarely conifers, in open woodlands. COMMENTS: The lace lichen (*R. menziesii*), which also can have broad branches, is invariably net-like, at least at the tips. The branches of *R. celastri* taper at both ends, and the apothecia tend to be more laminal. The subtropical **R. usnea** is much longer and narrower (see key). See also *R. subleptocarpha*.

Ramalina menziesii
Lace lichen, fishnet

DESCRIPTION: Thallus yellowish to grayish green, pendent, up to a meter long, with fine to extremely broad branches, 1–10(–30) mm wide, the main branches more or less strap-shaped with raised ridges forming striations along their length, but the tips and side branches end in reticulate, net-like expansions; without soredia; elongate pseudocyphellae on the branch margins. Apothecia occasional, on the branch surface. CHEMISTRY: Medulla PD–, K–, KC–, C– (usnic acid alone). HABITAT: On the branches of trees, especially oaks, but also conifers, in full sun or partial shade. COMMENTS: The lace lichen is aptly named, as one can see from plate 760. No other lichen has the lace- or net-like branches. In coastal populations, the nets are limited to tiny expansions at the very tips of slender branches. Farther inland, the branches and the nets are much wider. The thallus can become so slender and finely branched along the northwestern coast that it resembles *Alectoria sarmen-*

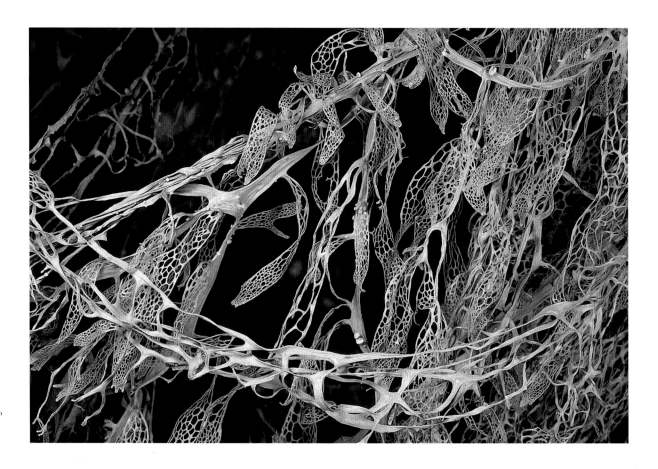

760. *Ramalina menziesii* Coast Range, northern California ×1.5

tosa or *R. thrausta*. In the latter, instead of having nets, the branch tips curl up and become almost granular. The branches are always round in cross section in *Alectoria*; there is no curling and there are no nets formed at the tips. *Ramalina leptocarpha* is very similar to the broader forms of *R. menziesii*; both species have bean-shaped, ellipsoid to narrowly ellipsoid spores, but only *R. menziesii* has net-like branches. **Ramalina usnea,** rare in California, is another long, pendent species that lacks nets (see key). IMPORTANCE: Because of the abundance of this relatively fast-growing lichen and its palatability to deer and sheep, *R. menziesii* is an important source of forage in parts of California. In some coastal valleys of central California, it completely covers the tree branches. Some mule deer have even been known to fight over it. Birds such as the common bushtit, Hutton's vireo, and the house finch build their nests from *R. menziesii* on the California coast.

Ramalina montagnei
Striped ramalina

DESCRIPTION: Thallus in tufts 2–6 cm long, branches more or less round in cross section except for the basal parts, mostly 0.3–0.8 (–1) mm wide, but up to 2 mm wide at the base, with low, white striations and slightly elongate pseudocyphellae. Apothecia abundant, 0.7–2 mm in diameter, pale yellowish orange, pruinose; spores narrowly ellipsoid, straight, 11.5–22 × 3–5 μm. CHEMISTRY: Medulla PD–, K+ slowly turning pink-violet, KC–, C– (sekikaic acid complex) or K– (stenosporic or divaricatic acid). HABITAT: On the branches of trees and shrubs in open woodlands and roadsides. COMMENTS: This species intergrades with *R. stenospora*, which has persistently flattened branches, is more strongly striate, and has somewhat longer spores. *Ramalina stenospora*, however, usually has perlatolic acid, although there is a chemotype with stenosporic acid. *Ramalina montagnei* can resemble *R. willeyi*, another *Ramalina* with rounded branches, but that species has small white bumps (tuberculae)

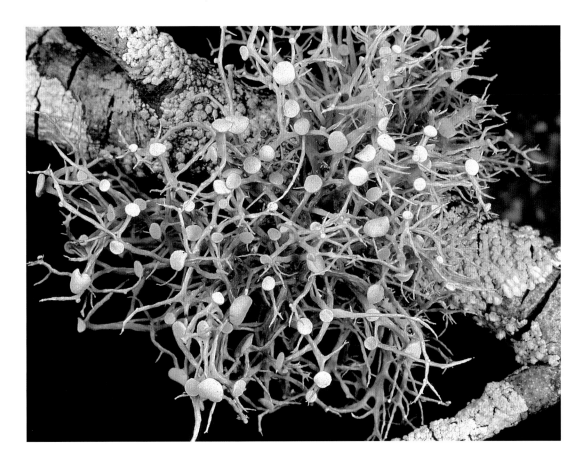

761. *Ramalina montagnei* Florida Keys ×3.9

762. *Ramalina obtusata* White Mountains, eastern Arizona ×2.3

rather than striations, and the medulla is K+ red (salazinic acid). See also *R. paludosa*.

Ramalina obtusata
Hooded ramalina

DESCRIPTION: Branches broad and short, about 1–2.5 cm long and 1.5–3(–4) mm wide, often partly hollow, sometimes almost foliose; soredia produced on the inside surface of inflated, hood-shaped soralia that appear as if they have exploded, leaving soredia on the lower surface. HABITAT: On trees and rocks. CHEMISTRY: Medulla PD–, K–, KC–, C– (evernic and obtusatic acids). COMMENTS: *Ramalina pollinaria* is most similar but does not develop inflated tips. *Ramalina intermedia* is much more slender, dissolving into irregular sorediate patches near the tips, and is almost exclusively on rock (sekikaic acid complex). ***Ramalina canariensis***, which contains divaricatic acid, is similar to *R. obtusata* in the way the soredia form on an interior surface, but in *R. canariensis*, the lobes split at the edges or form irregular holes,

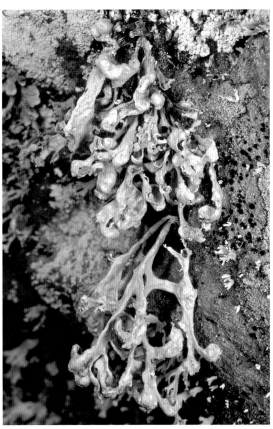

763. *Ramalina paludosa* northern Florida ×4.2

764. *Ramalina peruviana* eastern Texas ×2.1

exposing masses of fine soredia. This ramalina forms small cushions or shrubby masses of short, rather broad lobes (up to 5 mm) on shrubs and trees close to the California coast. The lobes can develop longitudinal ridges or wrinkles but not the reticulate, rather sharp ridges seen in the somewhat similar *R. lacera* (syn. *R. duriaei, R. evernioides*), another California lichen. *Ramalina lacera* has soredia in patches over the branch surface or sometimes on the margins, but contains bourgeanic acid.

Ramalina paludosa
Warty ramalina

DESCRIPTION: Thallus small, shrubby, 1–3 cm long, branches flattened and somewhat twisted, 0.3–1.2 mm wide, with tips almost round in cross section, without soredia but covered with small white bumps (papillae). Apothecia common, almost at the tips of the branches, flat, 1–3 mm in diameter, containing slightly or barely curved, fusiform spores, 9–12 × 3–5 μm. CHEMISTRY: Medulla PD–, K+ wine-red (very slowly), KC+ dark violet-pink; C+ pale violet-pink (cryptochlorophaeic and paludosic acids). HABITAT: On exposed trees and shrubs. COMMENTS: This is the only small, shrubby *Ramalina* with a C+, KC+ medulla, and one of the few with white papillae. *Ramalina willeyi* has similar white papillae, but the branches are more terete, and the chemistry is different. *Ramalina montagnei* has branches that are largely terete and has much longer spores. See Comments under *R. complanata*.

Ramalina peruviana

DESCRIPTION: Thallus shrubby to subpendent with narrow, more or less flattened to almost cylindrical branches, 2–4 cm long, mostly 0.1–0.8 mm wide, usually striate; small round or elliptical soralia produced along the branch margins or frequently on the surface, sometimes giving rise to irregular outgrowths resembling isidia; branches smooth, without tubercles. Apothecia rare. CHEMISTRY: Medulla PD–, K– or slowly pinkish violet, KC–, C– (sekekaic acid complex). HABITAT: On trees and shrubs. COMMENTS: The very similar *R. dendriscoides* contains salazinic acid (K+ red) and has soralia more confined to the branch tips. *Ramalina farinacea*, which has both flat-lobed and almost terete morphotypes, has entirely elliptical, marginal soralia, and never contains sekikaic acid. The ranges of the two species do not overlap.

Ramalina pollinaria
Chalky ramalina

DESCRIPTION: Thallus forming tufts 1–2.5 cm long with flat branches variable in width: broad (1–3 mm), fairly short and shrubby, or slender (0.5–1.0 mm) and branched; powdery, granular soredia mainly on one surface, at or close to branch tips; soralia very irregular in size and shape, appearing where the branch tips seem torn apart. CHEMISTRY: Medulla PD–, K–, KC–, C– (evernic and obtusatic acids). HABITAT: On bark of various trees, especially old ones, or on shaded rocks. COMMENTS: Similar species are mentioned under *R. intermedia*, *R. obtusata*, and *R. farinacea*.

Ramalina roesleri (syn. *Fistulariella roesleri*)
Frayed ramalina

DESCRIPTION: Thallus small, shrubby, 1–4 cm long, with delicate, perforated main stems, 0.1–1(–2) mm wide; lobe tips almost round in cross section, much branched, curled and disintegrating into granules and coarse soredia. Apothecia absent. CHEMISTRY: Medulla PD–, K–, KC–, C– (sekikaic acid complex). HABITAT: On twigs and branches of trees and shrubs in open, humid sites; rarely on wood or shaded rock. COMMENTS: No other lichen forms tiny cushions of perforated, finely divided branches ending in granular soredia. Other than *R. roesleri,* the most common, small *Ramalina* with perforated main stems is *R. dilacerata*. It has no soredia and is usually covered with round, pale yellow apothecia.

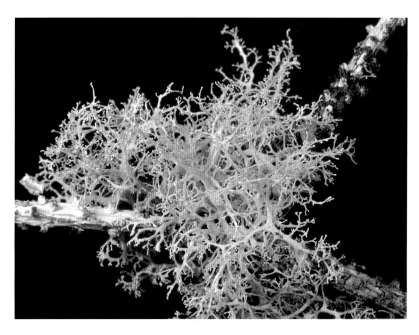

765. *Ramalina pollinaria* California coast ×2.3

766. *Ramalina roesleri* Washington coast ×1.2

767. *Ramalina sinensis* Kaibab plateau, northern Arizona ×1.6

Ramalina stenospora
Southern strap lichen

DESCRIPTION: Thallus somewhat shrubby, 2–4 cm long; branches narrow, mostly 0.3–1(–1.5) mm wide, strap-shaped, entirely flattened, frequently streaked with elongate, white pseudocyphellae; lacking soredia, papillae, or warts. Apothecia common, usually along the margins of the branches, flat to convex when mature, pinkish, pruinose, 1.5–3.5 mm in diameter; spores long and straight, 16–25(–31) × 3–5 μm. CHEMISTRY: Medulla PD–, K–, KC– or KC+ pinkish, C– (stenosporic or perlatolic acids). HABITAT: On the bark of hardwoods in open woodlands. COMMENTS: This species should be compared with *R. montagnei*, with which it sometimes intergrades.

Ramalina sinensis
Fan ramalina, burning bush

DESCRIPTION: Thallus short and broad, 1–3.5(–5.5) cm long, 5–20(–30) mm wide, fan-shaped, undivided and almost foliose to dissected and branched, with apothecia (2–5 mm in diameter) produced at the margins; no soredia, papillae, or warts, but with strong longitudinal, often reticulate ridges; spores ellipsoid, often bean-shaped, but sometimes predominantly straight. CHEMISTRY: Medulla PD–, K–, KC–, C– (only usnic acid). HABITAT: On twigs and branches of various trees and shrubs in the open. COMMENTS: Small specimens of *R. sinensis* are difficult to separate from *R. americana*. The latter species is more abundantly branched and never produces fan-shaped thalli.

Ramalina subleptocarpha
Slit-rimmed ramalina

DESCRIPTION: Thallus pendent or almost so, 3–10 cm long, with flat, strap-like branches, 2–4(–10) mm wide; narrow, long soralia follow the margins of the branches but can be laminal as well. Apothecia unknown. CHEMISTRY: Medulla PD–, K–, KC–, C– (zeorin). HABITAT: On bark and wood in open woodlands. COMMENTS: Small specimens of *R. subleptocarpha* can be mistaken for *R. farinacea*, which has discrete, elliptical soralia that are largely restricted to the margins, and differs chemically. *Ramalina leptocarpha* is the fertile, nonsorediate counterpart of *R. subleptocarpha*.

Ramalina thrausta
Angel's hair

DESCRIPTION: Thallus pendent and hair-like, up to 30 cm long, with extremely slender branches, 0.1–0.4(–0.7) mm wide, that are flattened only at the very base; very pale in color; cortex thin and translucent, smooth and even; branches taper to slender filaments that curl up at the very tips, sometimes ending with a few granules or granular soredia; pseudocyphellae point-like or somewhat elongated, slightly raised. Apothecia very rare. CHEMISTRY: Me-

768. *Ramalina stenospora* Atlantic Coast, northern Florida ×3.7

dulla PD–, K–, KC–, C–, but generally too thin to test reliably (contains only usnic acid). HABITAT: On trees, especially conifers, in forest habitats with high humidity; rarely on rock walls or on tundra soil. COMMENTS: *Ramalina thrausta* closely resembles witch's hair (*Alectoria sarmentosa*), but the extreme tips of the branches always turn up and usually form a little curl, which is the basis of its English name (angels have naturally curly hair). The branches tend to be somewhat more translucent than those of *Alectoria*, and the pseudocyphellae are more dot-like and less ridged, although there is some overlap in this character. The medulla of *Alectoria sarmentosa* is almost always KC+ red. The slender coastal morphs of *R. menziesii* can be confused with *R. thrausta* until a few net-like expansions of the branch tips are found.

769. *Ramalina subleptocarpha* herbarium specimen (Oregon State University), Oregon ×3.0

770. *Ramalina thrausta* Shuswap Highlands, British Columbia ×2.4

Ramalina willeyi
Thorny ramalina

DESCRIPTION: Thallus shrubby, 1–3 cm long, with branches that are flattened only at the base and are almost round in cross section elsewhere, 0.3–1(–2) mm wide, with some short side branches that are stiff and pointed like spines; pseudocyphellae round, white, slightly raised or on sparse to abundant bumpy warts; soredia absent. Apothecia common, 1–4 mm in diameter, flat, appearing terminal; spores straight, ellipsoid, 11–13 × (4–)5–6 μm. CHEMISTRY: Medulla PD+ yellow, K+ red, KC–, C– (salazinic acid, or rarely norstictic acid) or PD+ red-orange, K–, KC+ pink, C– (rare specimens with protocetraric acid). HABITAT: On shrubs and trees in the open. See Comments under *R. montagnei*.

Rhizocarpon (62 N. Am. species)
Map lichens

DESCRIPTION: Crustose lichens, usually with well-developed thalli, continuous to rimose or, more commonly, areolate. Photobiont green (unicellular). Thallus yellow, white, gray, brown, rusty orange, or some combination; typically with a visible black prothallus at the thallus edge and (or) between the thallus areoles. Apothecia lecideine, black; paraphyses branched and anastomosing, not expanded at the tips; exciple black or brown, at least at the outer edge; asci K/I– except for a thin, K/I+ dark blue cap at the ascus summit;

spores colorless, brown, or greenish, 2 or more cells, usually muriform, with a gelatinous outer sheath (halo), 1–8 spores per ascus. CHEMISTRY: Yellow species contain rhizocarpic acid in the cortex; frequently contain β-orcinol depsidones, especially stictic, norstictic, or psoromic acids, and less frequently depsides such as gyrophoric acid. Medulla IKI+ blue (most yellow species) or IKI–. HABITAT: On rock, usually noncalcareous. COMMENTS: Species of *Rhizocarpon* with gray or brownish thalli are hard to separate from other saxicolous crustose lichens with black apothecia, such as *Porpidia*, *Lecidea*, and *Buellia*, without examining the apothecia and spores under the microscope. The gelatinous halo around the spores helps distinguish species with 2-celled spores from similar species of *Buellia* or **Catillaria**. *Porpidia* and *Amygdalaria* have 1-celled spores and a different ascus type. *Lecidea* species have much smaller, 1-celled spores, and their paraphyses are mostly unbranched. Species of *Aspicilia* can resemble those of *Rhizocarpon* with immersed apothecia, but the exciple in *Aspicilia* is always colorless and poorly developed. The genus *Rhizocarpon* is so large and diverse, this small sample of species gives only a brief idea of the range of variation. However, several other common species are included in the key.

771. *Ramalina willeyi*
North Carolina coast
×1.7

KEY TO SPECIES

1. Thallus brown, gray, or white; medulla IKI– or rarely IKI+ blue 2
1. Thallus distinctly yellow; medulla IKI+ blue 13

2. Spores 2-celled 3
2. Spores 4-celled to muriform 5

3. Spores becoming dark brown or greenish. Thallus thick, areolate, pale to dark brown or gray-brown; widespread boreal . . . . [*R. badioatrum*]
3. Spores colorless 4

4. Thallus brownish, usually thin . . . . *Rhizocarpon hochstetteri*
4. Thallus thick, chalk-white, thick or thin; Rocky Mountains to arctic . . . . *Rhizocarpon chioneum*

5.(2) Spores colorless 6
5. Spores dark brown or dark green when mature 9

6. Thallus rusty orange; spores 4-celled or with the occasional longitudinal septum; mostly eastern montane . . . . [*Rhizocarpon oederi*]
6. Thallus white, gray, or brownish, not rusty; spores muriform 7

7. Thallus dark gray to gray-brown, thin, verruculose; no lichen substances; very common and widespread. . . . . *Rhizocarpon obscuratum*
7. Thallus pale brownish gray to gray; containing lichen substances; Great Lakes region to east coast 8

8. Thallus continuous, rimose; medulla K+ yellow or K–, PD+ orange or PD– (stictic acid, sometimes in traces) . . . . [*Rhizocarpon concentricum*]
8. Thallus contiguous or dispersed areolate; medulla K+ red, PD+ yellow (norstictic acid) . . . . [*Rhizocarpon rubescens* (syn. ?*R. plicatile*)]

9.(5) Spores 1–2 per ascus; thallus thick, areolate; medulla IKI– 10
9. Spores 8 per ascus; thallus areolate, often scattered over a black prothallus; medulla IKI+ blue. 12

10. Areoles peltate (free at the margins), brown; mainly western coastal mountains . . . [*Rhizocarpon bolanderi*]
10. Areoles convex, contiguous or dispersed, pinkish brown to gray. 11

11. Spores mostly 1 per ascus; temperate to boreal . . . . *Rhizocarpon disporum*
11. Spores mostly 2 per ascus; widespread, temperate to arctic . . . . [*Rhizocarpon geminatum*]

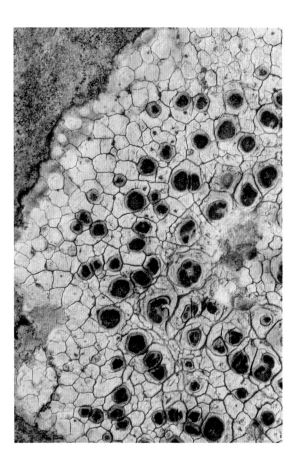

772. *Rhizocarpon chioneum* Rocky Mountains, Alberta ×4.0

12.(9) Thallus brown to gray-brown; medulla or cortex C+ pink, K+ yellow or K– (gyrophoric acid and sometimes stictic acid); very common and widespread . . . . [*Rhizocarpon grande*]
12. Thallus pale brownish gray; medulla and cortex C–; medulla K+ red (norstictic acid); boreal-montane and arctic . . . . [*Rhizocarpon eupetraeum*]
13.(1) Spores 2-celled, dark, 18–32 × 10–15 μm; epihymenium intense blue-green; arctic to eastern montane . . . . [*Rhizocarpon eupetraeoides*]
13. Spores muriform; epihymenium brown to olive-brown 14
14. Thallus greenish yellow, with crescent-shaped areoles surrounding some apothecia; epihymenium unchanged or becoming more intensely green with K . . . . *Rhizocarpon lecanorinum*
14. Thallus lemon-yellow, with apothecia sunken between the areoles; epihymenium becoming violet or purple with K. 15
15. Spores 28–60(–70) × 15–25 μm; hymenium greenish . . . . *Rhizocarpon macrosporum*
15. Spores 24–40 × 11–16 μm; hymenium colorless . . . . *Rhizocarpon geographicum*

Rhizocarpon chioneum
Snowy map lichen

DESCRIPTION: Thallus snow-white or with a bluish tint, thick, continuous or areolate; prothallus black or gray. Apothecia black, somewhat convex, 0.8–2 mm in diameter, with thick black margins that become thin and pruinose, giving the apothecia a lecanorine appearance; epihymenium red-violet to brown, unchanged with K or K+ deeper red-violet; spores 2-celled, colorless, 16–23 × 8–12 μm. CHEMISTRY: Medulla PD–, K– or K+ yellow close to algal layer, KC–, C–, IKI– (small quantities of stictic acid, especially in the apothecia; pigments in the epihymenium). HABITAT: On limestone at high elevations, or in the arctic. COMMENTS: There is no other white *Rhizocarpon* in North America with 2-celled spores. The mainly arctic species *R. umbilicatum* has colorless, muriform spores. *Porpidia speirea* is remarkably similar to *R. chioneum* in its external features, but the thallus is IKI+ blue and it has 1-celled spores. All these white species grow on limestone, which is an unusual substrate for both *Rhizocarpon* and *Porpidia*.

Rhizocarpon disporum
Single-spored map lichen

DESCRIPTION: Thallus usually pinkish brown to brownish gray, sometimes pale gray, areolate, with areoles 0.2–0.6 mm in diameter, usually convex, round or angular by pressure, contiguous or dispersed, occasionally flat; black prothallus usually conspicuous. Apothecia black, flat to convex, round with persistent margins to angular (constricted to that shape between the areoles) with hardly any margin, 0.4–0.7 (–1) mm in diameter; hymenium greenish; epihymenium violet brown, turning reddish purple in K; spores green to brown, muriform with many cells, 48–78 × 18–33 μm, 1 per ascus. CHEMISTRY: Medulla PD– or PD+ yellow, K– or K+ red, KC–, C–, IKI– (sometimes containing norstictic acid). HABITAT: On exposed siliceous rocks. COMMENTS: ***Rhizocarpon geminatum*** looks virtually identical but has 2 spores per ascus. The Latin name *disporum* (meaning 2-spored) is unfortunate, and clearly inaccurate, for this lichen is almost invariably 1-spored. *Geminatum* (referring to the "twins" in the ascus, i.e., 2 spores) is an appropriate name for the 2-spored taxon. *Rhizocarpon geminatum* is more widely distributed than is *R. disporum* and seems to prefer moist habitats, even maritime rocks. **Rhizocarpon grande**, with its brown, rounded areoles, can be quite similar but has larger areoles (0.6–2 mm in diameter) and 8 spores per ascus. This common, mainly boreal to temperate species contains gyrophoric acid (often with other compounds) and reacts C+ red in the medulla.

Rhizocarpon geographicum
Yellow map lichen

DESCRIPTION: Thallus of different shades of yellow; areoles round to angular, flat to slightly convex, either contiguous or dispersed on a black prothallus. Apothecia black, tucked between the thallus areoles and sometimes barely distinguishable from the black prothallus that surrounds and intermingles with the thallus elements; epihymenium reddish (a deeper red-violet in K); spores dark brown, muriform, 5–10 cells per spore, 24–40 × 11–16 μm, 8 per ascus. CHEMISTRY: Cortex UV+ orange (rhizocarpic acid); medulla PD+ yellow or PD–, K–, KC–, C– (psoromic acid present or absent), IKI+ blue. HABITAT: On exposed siliceous rocks, especially in alpine or arctic regions. COMMENTS: *Rhizocarpon geographicum* is the most common and widespread of the yellow map lichens. This group includes more than a dozen species in North America, two of which, like *R. geographicum*, have brown, muriform spores and an IKI+ blue medulla and are described here: *R. lecanorinum* and *R. macrosporum*. Others have 2-celled spores and an IKI– medulla, such as arctic-western montane *R. superficiale* (with spores under 18 μm long) and the strictly arctic *R. inarense* (with spores over 18 μm long). **Rhizocarpon eupetraeoides**, with spores 18–32 × 10–15 μm, has an IKI+ blue medulla. Like many common lichens, *R. geographicum* shows a great deal of variation. For example, the spores can have few or many cells. Thallus color, and areole size and density, can also vary, even in thalli adjacent to one another.

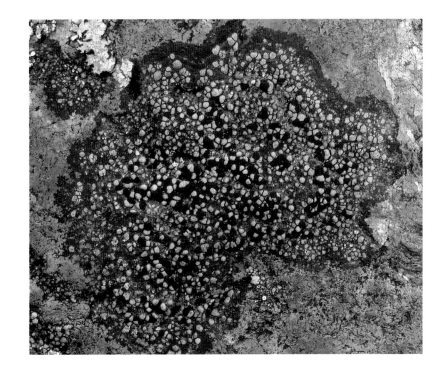

773. *Rhizocarpon disporum* Lake Superior region, Ontario ×3.0

774. *Rhizocarpon geographicum* Rocky Mountains, Colorado ×5.6

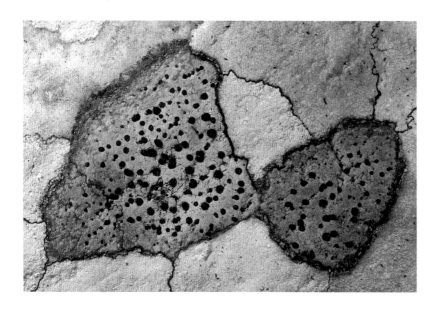

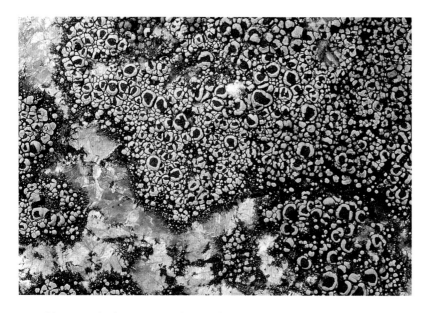

775. *Rhizocarpon hochstetteri* central Massachusetts ×2.8
776. *Rhizocarpon lecanorinum* coastal Maine ×3.0

Rhizocarpon hochstetteri
Smooth map lichen

DESCRIPTION: Thallus pale to dark brown, very thin, usually smooth and continuous. Apothecia 0.9–1.5 mm in diameter, broadly attached and flat to convex, black to dark brown with thin, usually persistent margins the same color as the disk; epihymenium brown to olive; hypothecium and exciple dark brown; spores colorless, 2-celled, ellipsoid, (11–)17–29 × 8–14(–17) µm, 8 per ascus. CHEMISTRY: Spot tests unreliable because thallus is so thin; contains stictic acid complex, often with traces of norstictic acid. HABITAT: On siliceous rocks, usually in damp or shaded sites. COMMENTS: The 2-celled spores superficially resemble those of *Catillaria,* but in *Rh. hochstetteri* the spores are halonate (visible in an ink preparation). The unusually large variation in spore size, epihymenial color, and the tissues of the exciple indicate that several species are covered under this name; fortunately, the problem is under study. *Rhizocarpon cinereovirens,* a fairly common eastern and southern boreal species, is similar to *R. hochstetteri* but contains abundant norstictic acid (K+ red) and has a more areolate thallus.

Rhizocarpon lecanorinum
Crescent map lichen

DESCRIPTION: Thallus yellow-green to greenish yellow, with convex, often curved areoles, 0.5–1.5 mm across. Apothecia partially or entirely enclosed within crescent-shaped areoles; epihymenium greenish brown, K– or greenish; spores dark brown, many-celled, muriform, 29–40 × 13–18 µm, 8 per ascus. CHEMISTRY: Cortex UV+ orange (rhizocarpic acid); medulla PD+ orange, K+ deep yellow, KC+ red or KC–, C+ pink or C–, IKI+ blue (stictic acid, sometimes with gyrophoric acid). HABITAT: On exposed siliceous rocks. COMMENTS: The distinctly greenish tint of the thallus and the crescent-shaped areoles make this one of the easiest yellow map lichens to identify using only a hand lens. It is the most temperate of the yellow species of *Rhizocarpon*.

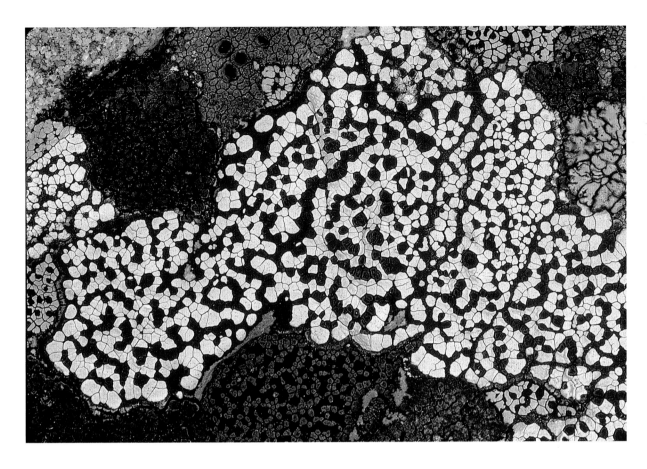

777. *Rhizocarpon macrosporum* Rocky Mountains, Montana ×3.6

Rhizocarpon macrosporum
Lemon map lichen

DESCRIPTION: Thallus bright lemon-yellow, consisting of areoles that are flat and at least partially contiguous, 0.4–1(–1.5) mm across, creating a patchy to continuous rimose-areolate crust surrounded by a broad black prothallus and interrupted by solitary or clustered black apothecia that are sunken between the areoles. Apothecia 0.3–0.8 mm in diameter, black margins persistent or sometimes disappearing; hymenium slightly greenish or brown, K–; spores multicellular, with over 15 cells per spore the norm, 28–60(–70) × 15–25 µm, 8 per ascus. CHEMISTRY: Cortex UV+ orange (rhizocarpic acid); medulla PD–, K–, KC–, C–, IKI+ blue-black (no additional substances), although stictic acid (PD+ orange, K+ yellow) and psoromic acid (PD+ yellow, K–) have been reported. HABITAT: On siliceous rock in open, somewhat humid sites. COMMENTS: This is one of the most brightly colored yellow map lichens. *Rhizocarpon geographicum* has a reddish violet epihymenium and spores with fewer cells (typically 5–9), and it usually contains psoromic acid. The thallus has smaller areoles that are somewhat more golden than lemon-yellow. However, variability is rampant in this group, and exceptions abound.

Rhizocarpon obscuratum
Dusky map lichen

DESCRIPTION: Thallus grayish brown to reddish brown, less commonly dark gray, usually rimose-areolate and continuous but sometimes breaking up and becoming dispersed areolate; black prothallus usually conspicuous. Apothecia somewhat immersed and almost level with the thallus surface, with persistent black margins, but occasionally surrounded by a rim of thallus material and appearing lecanorine; epihymenium olive green to brown, K–; spores colorless, muriform with 3–4(–7) transverse septa and 1(–2) lengthwise septa, 8 per ascus, 20–32 × 9–15 µm. CHEMISTRY: Medulla PD–, K–, KC–, C–, IKI– (no lichen substances), or rarely PD+ orange, K+

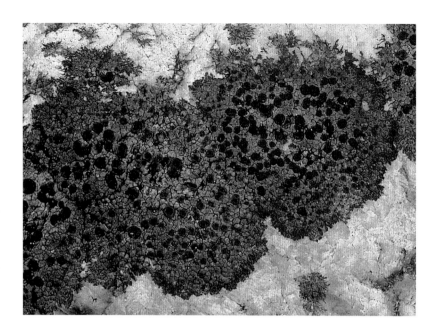

778. *Rhizocarpon obscuratum* Lake Superior region, Ontario ×5.2

yellow (stictic acid). HABITAT: Common on siliceous stones and pebbles as well as outcrops in open or shaded sites. COMMENTS: This is a notoriously variable species. It can be confused with a number of other mainly eastern or boreal map lichens with colorless, muriform spores, and a few are separated in the key. *Rhizocarpon lavatum* can intergrade with *R. obscuratum*; it has a gray to rusty orange thallus and usually grows near running water, and its spores have more numerous cells, with 5–7(–9) transverse septa and 1–2 lengthwise septa.

Rhizoplaca (7 N. Am. species)
Rock-posy lichens, rockbright

DESCRIPTION: Most frequently yellowish green to yellow-gray, but sometimes gray, foliose, umbilicate lichens, rarely crustose or fruticose. Thalli usually rounded and often lumpy, 10–30 mm in diameter (but see *R. haydenii*); lower surface pale to black, without rhizines. Photobiont green (*Trebouxia?*). Apothecia lecanorine, large, orange-pink to dark green or almost black; spores colorless, 1-celled, ellipsoid, small. CHEMISTRY: Cortex KC+ yellow-orange (usnic acid), rarely KC– (lacking usnic acid). Medulla often contains placodiolic or pseudoplacodiolic acid and other depsides, depsidones, and triterpenes in various combinations. HABITAT: On siliceous or calcareous rock, or unattached (vagrant) on soil, in open, especially dry sites.

COMMENTS: Typical umbilicate, yellow-green species of *Rhizoplaca* can hardly be mistaken for anything else, except perhaps the much larger *Omphalora arizonica*, which has wrinkles on the upper thallus surface and rhizines developing on the black to dark brown lower surface. Usnic acid is lacking in the rare, gray, California species, ***R. marginalis,*** with black, pruinose, margined apothecia produced close to the lobe edge, and ***R. glaucophana,*** with brownish, marginless, laminal apothecia. See Comments under *R. subdiscrepans* regarding *Rhizoplaca* species that are not umbilicate. IMPORTANCE: In east central Idaho, vagrant species of *Rhizoplaca* (and *Xanthoparmelia*) sometimes accumulate in sufficient quantity to become an important food source for pronghorn antelope, especially in winter.

KEY TO SPECIES

1. Thallus attached directly to rock 2
1. Thallus growing on soil, unattached (vagrant) 5
2. Thallus crustose, dispersed areolate with very convex areoles; apothecia orange, pruinose . . . . *Rhizoplaca subdiscrepans*
2. Thallus attached by a central holdfast (umbilicate) . . . 3
3. Apothecial disks orange, pruinose . . . . *Rhizoplaca chrysoleuca*
3. Apothecial disks yellow, brown, olive, or black, with or without pruina 4
4. Lower surface rough, broken into areoles with white cracks; apothecial disks yellowish brown, not pruinose; medulla PD+ orange (pannarin) . . . . *Rhizoplaca peltata*
4. Lower surface smooth; apothecial disks yellow-brown to greenish or black, pruinose; medulla PD+ bright yellow, or less frequently PD– (with or without psoromic acid) . . . . *Rhizoplaca melanophthalma*
5.(1) Apothecia usually abundant, yellow-brown to greenish black; lobes lacking whitish warts; medulla usually PD+ yellow (psoromic acid) . . . . *Rhizoplaca melanophthalma*
5. Apothecia lacking; small white warts produced especially on the margins or tips of the lobes; medulla PD– . . . . *Rhizoplaca haydenii*

Rhizoplaca chrysoleuca
Orange rock-posy

DESCRIPTION: Thallus pale yellowish green to yellow-gray, dull, round to irregular, rather closely appressed, often thick and lumpy, covered with a complex system of warts and lobules, less frequently smooth; lower surface pale brown, at least in the center, sometimes greenish black at outer edge, without rhizines but attached by a stout, central holdfast. Apothecia abundant on upper thallus surface, 2–7 mm in diameter, lecanorine with thin or thick margins; disks pale to dark orange or orange-pink, lightly pruinose. CHEMISTRY: Cortex KC+ yellow-orange (usnic acid). Many chemical races exist, but 90 percent of the time, the thallus reacts PD–, K–, KC–, C– (placodiolic or pseudoplacodiolic acid, or usnic acid alone). Other specimens can contain psoromic acid with or without placodiolic or pseudoplacodiolic acid (medulla PD+ yellow, K–, KC–, C–), or rarely lecanoric and pseudoplacodiolic acids (PD–, K–, KC+ red, C+ red). Specimens containing only usnic acid are also common. HABITAT: On exposed granitic rocks. COMMENTS: Because it is closely attached, the foliose thallus can appear to be crustose, but it is easily removed at its holdfast to reveal its umbilicate structure. *Rhizoplaca subdiscrepens* has very similar apothecia but is truly crustose (dispersed areolate to almost squamulose).

Rhizoplaca haydenii
Wanderlust lichen

DESCRIPTION: Thallus yellow-green, dull, extremely variable in form: flattened with the lobes curled under, as shown in plate 780, or branching into almost spherical fruticose clumps with angular branches; producing small white warts on the lobe surface close to the margins or on the tips of branches. Apothecia unknown. CHEMISTRY: Cortex KC+ yellow (usnic acid); medulla PD–, K–, KC–, C– (no additional substances). HABITAT: Growing unattached on exposed soil and calcareous gravels as a vagrant lichen at fairly high altitudes on barren, windswept terraces. COMMENTS: *Rhizoplaca melanophthalma*, when growing as a vagrant lichen, can be very similar. However, it lacks whitish, marginal warts,

779. *Rhizoplaca chrysoleuca* central Arizona ×5.8

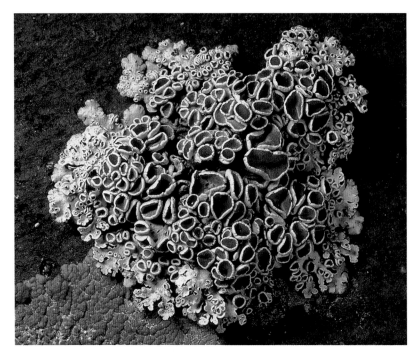

780. *Rhizoplaca haydenii* (form with flattened lobes) southeastern Idaho ×3.2

781. *Rhizoplaca melanophthalma* southern Idaho ×2.2

usually contains at least psoromic acid (PD+ yellow), and is frequently fertile. Apparently, there are several unnamed vagrant species of *Rhizoplaca;* they may accompany *R. haydenii* and *R. melanophthalma* in the same habitat.

Rhizoplaca melanophthalma
Green rock-posy

DESCRIPTION: Thallus yellowish green to dark yellow, often greenish black at the margins, frequently shiny, 7–25 mm in diameter, usually somewhat dissected and lobed, although forming small umbilicate units or becoming detached from the rock and growing into spherical, vagrant forms up to 25 mm in diameter, often abundantly fruiting; lower surface pale to dark brown, often greenish black near the margins, without white cracks. Apothecia abundant and crowded, 2–7 mm in diameter, varying in color from pale yellow-brown through greenish to black, but almost always pruinose; margins thick, often lobed. CHEMISTRY: Cortex KC+ yellow (usnic acid); medulla PD+ yellow or PD–, K–, KC–, C– (psoromic acid present or absent; rarely with placodiolic acid); very rarely PD+ yellow, K–, KC+ red, C+ red (lecanoric acid in addition to psoromic acid). HABITAT: On either siliceous or calcareous rocks, or vagrant on bare soil, in open, usually arid sites. COMMENTS: This species is extremely variable in many respects. It can grow closely appressed to the rock and appear almost crustose, or it can develop into spherical masses that roll over the soil. The variable color of the apothecia is sometimes confusing. Specimens with pale brownish apothecia can be mistaken for *R. peltata*, a much larger lichen with nonpruinose apothecial disks and a different chemistry. The apothecia of *R. chrysoleuca* are always tinted with orange or pink. When disguised as a crustose lichen, *R. melanophthalma* can closely resemble *Lecanora novomexicana*.

Rhizoplaca peltata
Brown rock-posy

DESCRIPTION: Thallus very thick and stiff, pale yellowish green, round to irregular, mostly 20–60 mm across, the surface layer broken into irregular areoles outlined in black; medulla thick, chalky white; lower surface puffy, almost tomentose in places, generally black to greenish black, at least at the margins, and pale to very dark brown in the center, rough, fracturing into areoles leaving a characteristic, net-like pattern of white cracks. Apothecia broad (commonly 3–4 mm in diameter) with waxy yellowish brown, nonpruinose disks surrounded by thick, cracked and split margins. CHEMISTRY: Medulla PD+ orange, K–, KC–, C– (zeorin and pannarin), with usnic acid in the cortex. Pannarin is absent in rare specimens. HABITAT: On calcium-rich rocks in the open. COMMENTS: This large rock-posy is exceptional in many respects other than size. No other *Rhizoplaca* lacks pruina on the apothecia, contains pannarin and zeorin, or has white cracks on the lower surface. It is also unusual in preferring calcareous rocks (although *R. melanophthalma* is also sometimes found on limy rocks, and *R. haydenii* can develop on calcareous gravel).

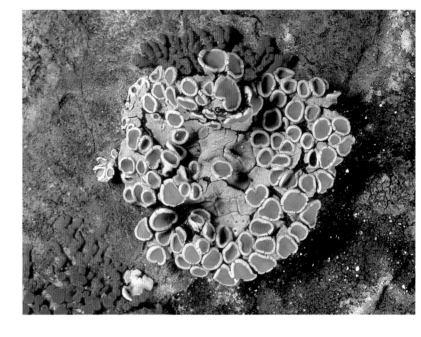

Rhizoplaca subdiscrepans
Scattered rock-posy

DESCRIPTION: Thallus pale green or yellowish green, or rarely gray-green, dull; consisting of scattered round areoles, 0.5–1.2 mm in diameter, usually strongly convex and constricted at the base, less commonly flat and lobed, rarely coalescing into a thick mass attached at the center. Apothecia abundant, pale orange to pinkish orange, pruinose, with raised margins, 0.5–2 mm in diameter. CHEMISTRY: Cortex KC+ yellow (usnic acid); medulla PD–, K–, KC–, C– (usually with pseudoplacodiolic acid). HABITAT: On exposed, granitic rocks. COMMENTS: This species closely resembles *R. chrysoleuca* but has a squamulose-crustose, not umbilicate, thallus, although intergrades can occasionally be found. **Lecanora opiniconensis** is often a neighbor on lakeshore cliffs. Its waxy yellow to yellow-brown, nonpruinose apothecia and shiny, dark greenish yellow, lobed thallus distinguish it from the *Rhizoplaca*.

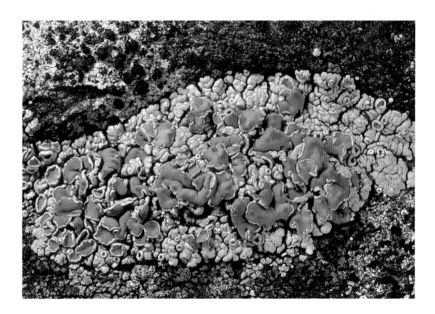

782. *Rhizoplaca peltata* southern Idaho ×3.5

783. *Rhizoplaca subdiscrepans* Appalachians, Tennessee ×2.6

784. *Rimelia cetrata* western Mississippi ×1.0

KEY TO SPECIES (including *Canomaculina*)

1. Maculae scattered, not reticulate or associated with a network of cracks; isidia cylindrical, developing on the lobe surface; medulla KC+ red (norlobaridone) .... *Canomaculina subtinctoria*
1. Maculae creating a network of fine white cracks, at least on the lobe tips 2

2. Soredia and isidia absent; pycnidia and apothecia abundant. .... *Rimelia cetrata*
2. Isidia or soredia present 3

3. Lobes with irregular, sometimes granular isidia, mainly developing close to the margins .... *Rimelia subisidiosa*
3. Lobes with soredia on the upper surface, especially on the lobe tips and margins .... *Rimelia reticulata*

Rimelia (6 N. Am. species)
Cracked ruffle lichens

DESCRIPTION: Medium to large, gray-green foliose lichens; surface with a reticulate pattern of maculae that develops into cracks; isidia or soredia present in some species. Photobiont green (*Trebouxia?*). Lobe margins almost always with unbranched, black cilia; lower surface black in the center, brown close to the margins, with unbranched to squarrose rhizines developing to the thallus edge. Apothecia lecanorine, with brown, sometimes perforate disks; spores colorless, small, 1-celled, 8 per ascus; conidia thread-like to cylindrical, 9–16 × 1 μm. CHEMISTRY: Cortex usually PD–, K+ yellow (atranorin), but rarely UV+ orange-yellow (lichexanthone); medulla with various reactions; can contain orcinol and β-orcinol depsides and depsidones and fatty acids. COMMENTS: Species of *Parmotrema* are about the same size and also have ciliate lobe margins, but they do not develop reticulate cracks in the surface, the rhizines are unbranched, never squarrose, and the conidia are much shorter and of a different shape.

Rimelia cetrata (syn. *Parmotrema cetratum*)

DESCRIPTION: Thallus smooth, maculae often difficult to see but present on at least some lobes; without soredia or isidia; lobes rounded, 4–12 mm across, but sometimes divided into much narrower lobes only 1–2 mm across; cilia 1–2 mm long, sometimes sparse; rhizines unbranched. Apothecia common, with concave, brown disks often perforated with an irregular hole in the center. CHEMISTRY: Cortex K+ yellow, UV–; medulla PD+ orange, K+ blood red, KC–, C– (salazinic acid). HABITAT: On the bark of deciduous trees in open woodlands and fields. COMMENTS: This species superficially resembles *Parmotrema perforatum*, even with perforate apothecia, but has rhizines that extend to the margin, the lower surface has no white splotches, and the chemistry is different.

Rimelia reticulata (syn. *Parmotrema reticulatum*)

DESCRIPTION: Thallus flat, or more commonly folded and curled under, giving a ruffled appearance; lobes 4–15 mm across; maculae and cracks usually conspicuous; soralia mostly on or close to the margins, occasionally at the tips of narrow, elongated lobes, with coarse soredia; lower surface black, with a brown or, rarely, white-splotched marginal zone; rhizines up to or close to the edge, usually unbranched, but sometimes squarrose as well; cilia stout or slender, abundant or sparse, 0.2–1.5 mm long.

785. *Rimelia reticulata* western Mississippi ×3.1

Apothecia rare. CHEMISTRY: Cortex K+ yellow, UV–; medulla PD+ orange, K+ red, KC–, C– (salazinic acid). HABITAT: On bark in well-lit woods. COMMENTS: *Rimelia reticulata* closely resembles a *Parmotrema* (e.g., *P. hypotropum* or *P. margaritatum*) except for the minute cracks on the upper surface. Two other *Rimelia* species from the southern Appalachians have similar soralia: ***R. diffractaica*** (syn. *Parmotrema diffractaicum*), with lichexanthone and diffractaic acid in the medulla (medulla PD–, K–, CK+ orange, KC–, C–, UV+ yellow); and ***R. simulans*** (syn. *Parmotrema simulans*), with caperatic acid (medulla PD–, K–, CK–, KC–, C–, UV–).

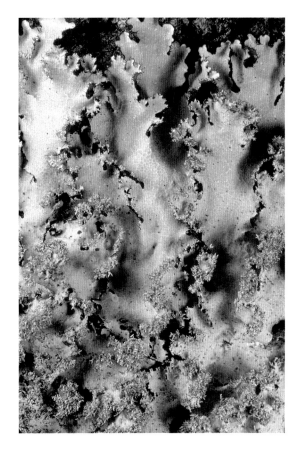

786. *Rimelia subisidiosa* coastal plain, North Carolina ×2.8

Rimelia subisidiosa (syn. *Parmotrema subisidiosum*)

DESCRIPTION: Thallus pale greenish gray, loosely attached; lobes 3–12 mm across, rounded to squarish, with unbranched cilia on the margins (sometimes sparse); reticulate maculae and cracks conspicuous; unbranched or branched isidia (sometimes intergrading with soredia) on thallus surface and sometimes along the margins; isidia occasionally becoming ciliate at the tips; lower surface black except for a brown marginal zone, with black, largely unbranched rhizines to the edge (a few rhizines can be squarrose). CHEMISTRY: Cortex K+ yellow, UV–; medulla PD+ yellow–orange, K+ red, KC–, C– (salazinic acid). HABITAT: On tree bark and sometimes rocks in open woodlands. COMMENTS: The combination of reticulate maculae and isidia make this lichen distinctive. The maculae of *Canomaculina subtinctoria* are not in a reticulate pattern, and the lower surface is uniformly pale to dark brown, not black. Superficially, *R. subisidiosa* resembles a number of other isidiate species such as *Parmotrema crinitum* and *Parmelinopsis horrescens.*

Rinodina (80 N. Am. species)
Pepper-spore lichens

DESCRIPTION: Crustose lichens with thick or thin thalli, gray to brownish or olive. Photobiont green (*Trebouxia?*). Apothecia lecanorine with dark brown to black disks; epihymenium usually brown; hypothecium colorless; asci *Lecanora*-type; spores brown, 2-celled (or rarely 4-celled, as in the west coast species *R. conradii,* an odd species with reddish brown, crowded, convex apothecia), with unevenly thickened walls leaving locules that are angular or round in shape, 8–32 per ascus; conidia rod-shaped. CHEMISTRY: Some species contain atranorin, pannarin, zeorin, or other compounds; many have no lichen substances. HABITAT: On bark, wood, rock, soil, or dead vegetation. COMMENTS: Species of *Rinodina* can be characterized as relatively inconspicuous crustose lichens generally with a greenish or brownish thallus and very dark brown *Lecanora*-type apothecia. It is necessary to check the details of the apothecia and spores in order to confirm the identification. Genera with similar apothecia and spores include *Dimelaena, Amandinea,* and *Phaeorrhiza.* Species of *Hyperphyscia* can also be mistaken for *Rinodina* because their thallus is so closely appressed to the substrate, it can appear to be crustose. *Buellia* is a closely related genus typically with adnate, broadly attached, lecideine apothecia and thin-walled spores, but some species have immersed apothecia and can resemble *Rinodina. Rinodina* is such a large and diverse genus, with species distinguished mainly by microscopic characters, that we can illustrate only a very small sample of representative species, although a number of common species have been added to the key.

KEY TO SPECIES

1. On soil, mosses, or dead vegetation; arctic-alpine 2
1. On bark, wood, or rock; mostly temperate (but see couplet 17) 4
2. Thallus lobed at margins; spores with uniformly thin walls at maturity, mostly smaller than 23 × 11 µm; thallus yellowish, particularly at margins .... *Phaeorrhiza nimbosa*
2. Thallus granular to verrucose or disappearing, not lobed; spores with unevenly thickened walls (*Physcia*-type); thallus, if present, typically reddish brown, never yellow 3
3. Spores 22–35 µm long; apothecia 0.6–1.5 mm in diameter, margins typically prominent, with an expanded lower cortex; thallus granular to verrucose; sphaerophorin present, zeorin absent .... *Rinodina turfacea*
3. Spores 17–26 µm long; apothecia 0.4–0.8 mm in diameter, margins even with disk or becoming thin or disappearing in maturity, without an expanded lower cortex; thallus generally indistinct; sphaerophorin absent, zeorin present .... [*Rinodina archaea*]
4. On rock 5
4. On bark or wood 10
5. Maritime, mostly on shoreline rocks in salt-spray zone; spores inflated at septum, more so in K .... [*Rinodina gennarii*]
5. On nonmaritime rocks 6
6. Spores with a conspicuous pigmented band around the middle near septum, with walls thin at tips, 15–20 × 9–13 µm; on calcareous rock; widespread from arctic southwards .... [*Rinodina bischoffii*]
6. Spores lacking a pigmented band, with walls thickened at tips; on siliceous rock 7
7. Thallus brownish to olive, cortex K– (atranorin absent) 8
7. Thallus creamy white to pale brownish gray, dispersed areolate to verrucose, cortex K+ yellow (atranorin)... 9
8. Spores 17–24 × 8–14 µm, locules becoming round; thallus dull, verruculose to verrucose or rimose-areolate; common, central plains to eastern seaboard, in moist habitats .... [*Rinodina tephraspis*]
8. Spores 23–41 × 11–17 µm, locules angular, walls remaining thickened only at cell tips; thallus more or less smooth, shiny; rare on rock (see couplet 18); northeastern .... *Rinodina ascociscana*
9. Spores 17–22 × 8–11 µm, *Physcia*-type, retaining thick, angular end walls with age; apothecia not pruinose; western mountains, Arizona and California to Alberta .... [*Rinodina confragosa*]
9. Spores 19–31 × 9–16 µm, with a thick septum, locules becoming round in older spores; apothecia often prui-

nose; west coast, California to Vancouver Island. . . . . *Rinodina bolanderi*

10.(4) Thallus edge distinctly lobed; lower cortex present . . . . *Hyperphyscia syncolla*

10. Thallus edge indefinite or definite, not lobed; lower cortex absent 11

11. Spores 12–32 per ascus . . . . *Rinodina populicola*

11. Spores 8 per ascus. 12

12. Spores thin-walled, walls even in thickness 13

12. Spores conspicuously thick-walled, walls unevenly thickened 14

13. Spores even in outline, not constricted; thallus rugose, rimose, or more or less smooth; coastal plain, Massachusetts to Gulf of Mexico . . . . [*Amandinea milliaria* (syn. *Rinodina milliaria*)]

13. Spores becoming slightly constricted at septa; thallus verruculose or areolate; north central to northeastern North America, not coastal plain . . . . [*Amandinea dakotensis* (syn. *Rinodina dakotensis*)]

14. Thallus white or creamy, well-developed, K+ yellow (atranorin); maritime California to southern British Columbia. 15

14. Thallus brown or olive, thin or thick, K– (lacking atranorin); not along west coast. 16

15. Apothecia 0.4–0.7 mm in diameter; spores 15–20 × 8–11 µm; thallus thin, areolate to coarsely granular; pannarin present in epihymenium (PD+ orange crystals) . . . . [*Rinodina marysvillensis*]

15. Apothecia 0.6–1.5 mm in diameter; spores 19–31 × 9–16 µm; thallus very thick, verrucose to areolate; pannarin absent . . . . *Rinodina bolanderi*

16. Arctic to boreal and montane. . . . 17 (see also couplet 3)

16. Temperate (except for *R. archaea*). 18

17. Thallus granular to verrucose; spores 22–35 µm long; sphaerophorin present, zeorin absent . . . . *Rinodina turfacea*

17. Thallus largely within substrate and not visible; spores 17–26 µm long; sphaerophorin absent, zeorin present . . . . [*Rinodina archaea*]

18. Spores 30–40 × 11–16 µm . . . . *Rinodina ascociscana*

18. Spores 12–25 × 6–10 µm 19

19. Thallus membranous, very thin and indistinct, or mostly within substrate 20

19. Thallus clearly visible, granular to verrucose or areolate; southern boreal. 22

20. Arctic to boreal and northwestern montane; apothecia superficial with well-developed thalline margins, at least when young; zeorin present . . . . [*Rinodina archaea*]

20. Temperate or southern boreal, apothecia broadly attached or immersed, typically with poorly developed thalline margins when young 21

21. Spores with 2 round to almost triangular lumina creating an hourglass shape; zeorin absent; coastal plain, Massachusetts to Texas . . . . [*R. applanata*]

21. Spores with angular lumina, at least when young, *Physcia*-type; zeorin present; Appalachian–Great Lakes distribution. On deciduous trees, especially sugar maples . . . . [*Rinodina subminuta*]

22. Thallus bluish gray, dispersed areolate, or with darker granules; apothecial disks pitch-black; epihymenium gray-blue, turning red with nitric acid . . . . [*Rinodina colobina*]

22. Thallus pale greenish gray to brown, not bluish gray; apothecial disks brown to black; epihymenium red-brown, unreactive with nitric acid . . [*Rinodina glauca*]

Rinodina ascociscana
Scaly pepper-spore lichen

DESCRIPTION: Thallus shiny, olive, fairly dark, membranous to almost squamulose in places, often forming squamulose areoles. Apothecia 0.5–0.8 mm in diameter, with a thick, persistent margin; lower cortex of apothecium composed of a mass of gelatinized hyphae, not expanded (as in *R. turfacea*); spores very large for a *Rinodina*, (23–)30–41 × 11–17 µm. CHEMISTRY: Thallus PD–, K–, KC–, C– (no lichen substances). HABITAT: On bark and occasionally rock, often overgrowing bryophytes. COMMENTS: *Rinodina adirondacki* is similar but with pannarin (PD+ orange). It is not common, but occurs from Minnesota to New England, as well as in the mountains of North Carolina.

Rinodina bolanderi
Creamy pepper-spore lichen

DESCRIPTION: Thallus thick, creamy white to yellowish gray, granular, verrucose, or areolate, sometimes almost squamulose. Apothecia 0.7–2 mm in diameter, with thick, persistent margins; disks dark brown to black, with or without white pruina; spores 19–31 × (9–)10–16 µm long, with rounded locules when fully mature. CHEMISTRY: Thallus PD+ pale yellow, K+ yellow, KC–, C– (atranorin, sometimes

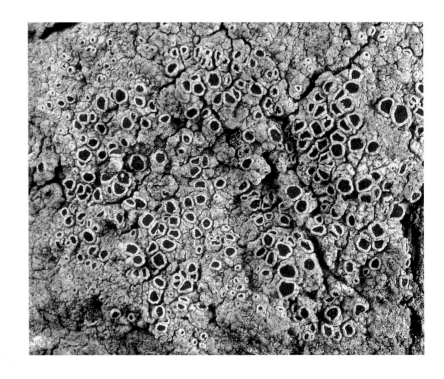

787. (top left) *Rinodina ascociscana* North Woods, Maine ×4.2
788. (top right) *Rinodina bolanderi* Channel Islands, southern California ×4.8
789. (bottom) *Rinodina populicola* Lake Superior region, Ontario ×5.6

zeorin). HABITAT: On soil, rock, or moss, rarely bark, in the open. COMMENTS: *Rinodina bolanderi* is part of the California coastal soil and rock flora. Pruinose specimens can resemble the arctic-alpine species *R. roscida*, which lacks atranorin and has larger spores (over 25 μm long). *Rinodina confragosa* is a similar, more common saxicolous species with smaller spores (less than 22 μm long) that, even in maturity, retain their angular locules.

Rinodina populicola
Poplar pepper-spore lichen

DESCRIPTION: Thallus olive-brown to brownish gray, composed of dispersed to contiguous tiny areoles or verrucae, often on a dark brown prothallus. Apothecia abundant and crowded, irregular in shape because of the crowding, 0.25–0.5 mm in diameter, with very thick, wavy margins and dark brown disks; apothecial cortex thick and conspicuous; spores with round locules and, when mature, evenly thickened walls except for a slightly thicker, darker septum, 11–13(–15) × 6–8 μm, 12–32 per ascus. CHEMistry: Thallus PD–, K–, KC–, C– (no lichen substances). HABITAT: On bark, especially poplars and elms. COMMENTS: This is one of the very few species of *Rinodina* with more than 8 spores per ascus. Another, but rarer, species is *R. polyspora,* with a very thin, whitish thallus and 12–16 spores per ascus. It is known from California and Washington to the maritime provinces and New York. The apothecia are convex with thin margins that are lost in maturity, and a thin, barely discernible apothecial margin cortex. Its spores have distinctly uneven walls.

790. *Rinodina turfacea* herbarium specimen (Canadian Museum of Nature), Ontario ×3.8

Rinodina turfacea
Tundra pepper-spore lichen

DESCRIPTION: Thallus reddish brown or, less commonly, brownish or greenish gray, granular to verrucose, usually almost obscured by the abundance of apothecia. Apothecia mostly 0.5–1.5 mm in diameter, disks black or almost so, margins prominent but thin in very mature apothecia; apothecial margin cortex distinct and very thick, especially at the base of the apothecium; spores (21–)25–36 × (9–)10–14 μm. CHEMISTRY: All reactions negative; contains sphaerophorin. HABITAT: On soil, dead vegetation, peat, and wood in arctic or alpine sites. COMMENTS: Two other common arctic-alpine *Rinodina* species on soil and peat are *R. roscida,* with a creamy white thallus, large black or blue-black apothecia (1–2 mm in diameter) with a heavy white pruina over the disks and margin; and *R. mniaraea,* much like *R. turfacea* in color but with more convex apothecia (mostly under 1 mm in diameter) with or without pruina, and a thin margin with an indistinct cortex. See also *Phaeorrhiza nimbosa.*

Roccella (9 N. Am. species)
Orchil lichens, Canary weed

DESCRIPTION: Pendent to shrubby fruticose lichens with flattened or cylindrical branches, pale brownish gray to violet-gray; medulla solid; cortex usually composed of cells in a palisade arrangement (perpendicular to the surface; Fig. 5a). Photobiont green (*Trentepohlia*). Apothecia appearing lecanorine, black or brown, with a black hypothecium; spores colorless, usually 4-celled, fusiform, straight or curved. CHEMISTRY: Cortex and (or) medulla C+ red in most species (*para*-depsides such as erythrin and lecanoric acid as well as roccellic acid). HABITAT: On coastal rocks and bark, in full sun. COMMENTS: Species of this genus become common only south of the U.S. border, especially in Baja California (Mexico). All the others are quite rare, if still present in North America, having been eliminated from most parts of mainland California by urban development. IMPORTANCE: Mediterranean, Canary Island, and South American species of *Roccella* were collected by the ton for producing the purple dyes called orchil (see Chapter 10). Litmus, a pigment that turns red or blue in acidic or alkaline solutions, respectively, is manufactured from European species of *Roccella.*

KEY TO GENUS: See Key A.

Roccella babingtonii
Orchil lichen

DESCRIPTION: A very pale yellowish gray to almost white fruticose lichen with flattened, wrinkled branches, 1–4 mm wide, 4–12 cm long, usually divided close to the base, with abundant, round, raised soralia filled with white to gray, coarsely granular soredia. Apothecia rare. CHEMISTRY: Cortex and soredia PD–, K–, KC+ red, C+ red; medulla PD–, K–, KC–, C– (lecanoric acid). HABITAT: On coastal trees, shrubs, and rocks, rarely inland. COMMENTS: This

791. *Roccella babingtonii* Channel Islands, southern California ×1.1

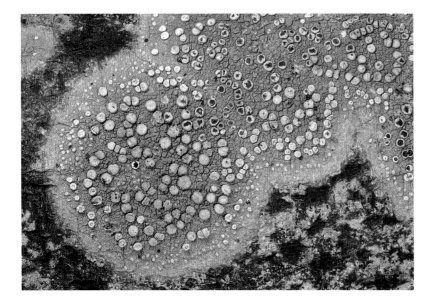

792. *Roccellina conformis* Channel Islands, southern California ×2.4

conspicuous but rare coastal lichen can hardly be mistaken for anything else, except perhaps *Dendrographa* species, which are C– and not sorediate. Another North American *Roccella* occurring north of Mexico is **R. fimbriata,** the nonsorediate equivalent of *R. babingtonii*.

Roccellina (2 N. Am. species)

KEY TO SPECIES: See Key F.

Roccellina conformis
Peg-legged rim lichen

DESCRIPTION: A crustose lichen with a yellowish tan to greenish brown, fairly rough thallus. Prothallus often conspicuous, fibrous, pale to dark brown. Photobiont green (*Trentepohlia*). Apothecia lecanorine, abundant, black disks with a heavy white pruina, constricted at the base, 0.8–1.2 mm in diameter; hymenium IKI+ blue in part; epihymenium brown; paraphysoid hyphae in hymenium usually unbranched; hypothecium black, extending down to the substrate like a peg (visible in vertical sections through the center of the apothecia); spores colorless, 4-celled, fusiform, pointed at the ends, 25–35 × 5–6 µm. CHEMISTRY: Cortex and medulla PD+ red-orange, K– or K+ yellowish to red, KC–, C–, IKI– (unknown substance). HABITAT: On bark of trees and shrubs, narrowly confined to the coastal fog deserts of southern California. COMMENTS: The extension of the black hypothecium down to the substrate distinguishes *Roccellina* from similar species of *Dirina* (e.g., **D. approximata**), which have no such extension. *Dirina* species are usually C+ red in the cortex or medulla. The only other North American species of *Roccellina* is **R. franciscana,** which is on bark or rock and is also on the California coast. It has sessile or immersed apothecia and shorter spores (22–25 µm long) and its thallus is grayer and PD–.

Ropalospora (5 N. Am. species)

DESCRIPTION: Crustose lichens with green to dark gray-brown verruculose thalli, characteristically surrounded by a dark brown to black prothallus. Photobiont green (unicellular). Apothecia biatorine, 0.3–1 mm in diameter, black to dark brown, with pale to brown internal tissues; asci with a layered tholus, K/I+ dark blue zone closest to the spores, surrounded by a K/I+ paler blue layer, which often shows an outer K/I+ dark blue coating similar in most respects to the *Fuscidea*-type (Fig. 14d); hymenial jelly and paraphyses IKI–; spores colorless, fusiform but conspicuously tapered at one end, 4–8-celled, 8 to many per ascus; conidia rod-shaped. CHEMISTRY: Commonly containing perlatolic or gyrophoric acids or atranorin. HABITAT: On a variety of substrates; mostly boreal to arctic-alpine. COMMENTS: The genus *Ropalospora* is characterized by its brownish black, biatorine apothecia, a pale exciple, many-celled, narrow, tapering spores, and asci with rather peculiar staining properties in K/I. Thallus color and ascus characters link this genus to *Fuscidea,* and sterile specimens are difficult to distinguish from it.

KEY TO SPECIES: See Keys C and F.

Ropalospora chlorantha
Comet-spored lichen

DESCRIPTION: Thallus olive-green, usually consisting of discrete granules, less frequently verruculose, in circular patches, often surrounded by a black prothallus. Apothecia lead-black to very dark brown, rather constricted at the base; margins low but discernible, often wavy; exciple pale internally but dark brown at the edge; hymenium filled with oil droplets; hypothecium colorless; spores needle-shaped, indistinctly 4–6-celled, straight or curved, tapered, 16–55 × 1.2–3.1 µm, 30–50 per ascus, but spores often not well developed; pycnidia usually abundant at the thallus edge, dark brown, containing bacilliform conidia, 2.4–4 µm long. CHEMISTRY: Cortex and medulla PD–, K–, KC–, C–, IKI– (perlatolic acid). HABITAT: On the bark of a variety of deciduous and coniferous trees. COMMENTS: ***Ropalospora viridis,*** a mainly sterile species, is only known in North America from coastal Vancouver Island, B.C. It is very similar to *R. chlorantha* but has abundant soredia in the thallus center; *R. chlorantha* rarely produces soredia. The saxicolous, arctic-alpine species **R. lugubris** has a continuous, areolate, brownish gray thallus and 8-celled, strongly tapering spores, 8 in an ascus; it lacks perlatolic acid.

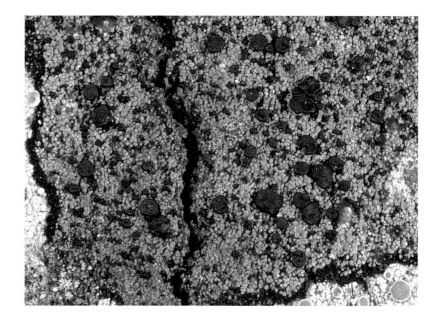

793. *Ropalospora chlorantha* Adirondacks, New York ×6.0

Sarcographa (4 N. Am. species)
Warty script lichens

DESCRIPTION: Crustose lichens with thin thalli growing within bark tissue. Photobiont green (*Trentepohlia?*). Ascomata are branched, script-like lirellae usually embedded in or crowded on the surface of a whitish bump or wart (a pseudostroma) containing carbonized hyphae; spores 4–6-celled, brown, with lens-shaped locules. CHEMISTRY: Some species with stictic or norstictic acid. HABITAT: On the bark of trees in tropical to subtropical woodlands. COMMENTS: The presence of poorly developed, whitish pseudostromata theoretically distinguishes *Sarcographa* from the closely related genus *Phaeographis* but, unfortunately, intergrades are common.

KEY TO GENUS: See Key E.

794. *Sarcographa labyrinthica* southern Florida ×5.1

Sarcogyne (14 N. Am. species)
Grain-spored lichens

DESCRIPTION: Thallus usually within rock visible only as a stain, rarely grayish white and rimose-areolate (e.g., in the southern California species *S. bicolor*). Photobiont green (*Dictyochloropsis* and *Myrmecia*). Apothecia lecideine, black, pruinose or not, with prominent black margins. Ascus walls IKI–, but K/I+ light blue (hemiamyloid), tip thick when immature, becoming much thinner in mature asci, K/I–; paraphyses unbranched, very slightly expanded at the tips; epihymenium usually brown, very rarely carbonaceous (only in *S. bicolor*); hypothecium colorless to brownish; exciple black and often carbonaceous on the outer edge; spores tiny and grain-like, colorless, 100 or more per ascus, approximately 3–6 μm long. CHEMISTRY: No lichen substances. HABITAT: On calcareous or noncalcareous rocks in the open. Some species are widespread, but several are found only in the southwest. COMMENTS: Species of *Polysporina* are easily mistaken for *Sarcogyne*. *Polysporina* is also many-spored but has convoluted, brittle, carbon-black apothecial margins and sterile, black ridges or bumps on the disks. The paraphyses are branched, and the asci are more cylindrical (and K/I–).

Sarcographa labyrinthica

DESCRIPTION: Thallus mostly within the bark; lirellae immersed in dispersed round or irregular white pseudostromata (low, flattish bumps) 3–8 mm in diameter; hymenium containing tiny oil drops; spores colorless, 1–2-celled when young, brown and 4-celled when mature, 15–21 × 5.5–7.5 μm. CHEMISTRY: Sections of pseudostromata: PD+ orange, K+ yellow, KC–, C– (stictic acid). HABITAT: On bark of hardwoods. COMMENTS: Species of *Sarcographa* are not commonly seen, but they are striking in appearance. All North American species are found in Florida. **Sarcographa tricosa** resembles *S. labyrinthica* but has broader spores and lacks lichen substances; *S. intricans* contains norstictic acid, and the hymenium lacks oil drops; *S. medusulina* has 4–6-celled spores and contains stictic acid.

KEY TO GENUS: See Key F.

Sarcogyne regularis
Frosted grain-spored lichen

DESCRIPTION: Thallus crustose, largely within the upper layers of rock, creating a thin white or pale gray to brownish gray stain, continuous and rimose. Apothecia lecideine, quite round, 0.5–1.5(–2) mm in diameter, flat to slightly convex, sometimes partially immersed in the rock; disks reddish brown (especially when wet), lightly to heavily white pruinose; margin black (sometimes white with pruina), thin, even, raised and persistent, or occasionally disappearing; epihymenium brown; hypothecium colorless; exciple black at edge but almost colorless within; spores colorless, narrowly ellipsoid, 3–5(–6) × 1.5–2 μm, more than 100 per ascus. CHEMISTRY: No lichen substances. HABITAT: On limy rocks, pebbles, cement, and mortar, in exposed or somewhat shaded sites. COMMENTS: **Sarcogyne clavus** and *S. similis* also have broad disks and entirely endolithic thalli, but both have nonpruinose apothecia 1–3 mm across, and they grow on siliceous rather than calcareous rocks. They are common species, especially in the east (although also reported from California). In *S. clavus*, the apothecial margin is wavy or lobed and rather thick, and the exciple has a black, carbonized outer layer; the hypothecium is brown. *Sarcogyne similis* has more even apothecial margins, and the cells in the outer layer of the exciple are brown, not carbonized. ***Sarcogyne privigna*** is very similar to *S. clavus* and also grows on granitic rocks. Its apothecia are generally under 1 mm in diameter and are often irregular in shape, the hypothecium is colorless, and the hymenium is somewhat lower on average (65–95 μm versus 85–115 μm).

Schaereria (4 N. Am. species)
False map lichens

DESCRIPTION: Crustose lichens, with well-developed areolate to squamulose thalli, dark gray-brown to steel gray, rarely pale, with a well-developed black prothallus. Photobiont green (*Trebouxia*). Apothecia lecideine, black, with persistent black margins; exciple brown to black in section; epihymenium bright green, sometimes partially violet and turning K+ green; hypothecium predominantly brown; paraphyses with some branching at the tips, easily separable when mounted in K and often even water; asci cylindrical, thin-walled, tips not thickened, K/I– or very faint blue; spores colorless, broadly ellipsoid to almost globose; conidia bacilliform. CHEMISTRY: Often contains gyrophoric acid (C+ pink). HABITAT: On siliceous rocks, mostly in arctic and alpine sites. COMMENTS: The cylindrical, K/I– asci (unique among the *Lecidea*-like crusts), together with the C+ pink medulla, make *Schaereria* distinctive.

KEY TO SPECIES: See *Lecidea*.

Schaereria fuscocinerea (syn. *S. tenebrosa*)

DESCRIPTION: Thallus dark gray (rarely paler or brownish gray), minutely areolate, with rounded to flat areoles in patches surrounded by a conspicuous black prothallus. Apothecia immersed

795. *Sarcogyne regularis*   herbarium specimen (Canadian Museum of Nature), Ontario ×5.8

796. *Schaereria fuscocinerea*   herbarium specimen (Canadian Museum of Nature), Ontario ×11.0

797. *Schizopelte californica* Channel Islands, southern California ×3.6

between or within the areoles when young, becoming superficial, 0.3–1 mm in diameter; spores broadly ellipsoid, 10–16 × 5–8 µm. CHEMISTRY: Medulla PD–, K–, KC+ red, C+ pink, IKI– (gyrophoric acid). COMMENTS: The immersed, black apothecia and areolate thallus give this species the look of a *Rhizocarpon* in the field, but the anatomical characters and spores quickly establish its identity. ***Schaereria cinereorufa***, a rarer species with a similar distribution, has a browner, squamulose thallus and spherical spores, mostly 6–9 µm in diameter.

Schizopelte (syn. *Combea*; 1 species worldwide)

KEY TO SPECIES: see Key A.

Schizopelte californica (syn. *Combea californica*)
Fog fingers

DESCRIPTION: Fruticose lichen forming cushions of cylindrical, branched stalks 1–3 mm in diameter and 10–30 mm high, with a dense, solid medulla; thallus dull, sometimes lightly pruinose, tan to light greenish gray; soredia and isidia absent. Photobiont green (*Trentepohlia*, retaining short filaments in the thallus). Branch tips commonly expanding into broad, irregular, black to pale violet, pruinose apothecia, up to 20 mm across, that appear to have thalline margins; epithecium and hypothecium dark brown; spores 4–8-celled, brown, with cylindrical or somewhat rounded locules, 19–31 × 5.2–8 µm, 8 per ascus. CHEMISTRY: Cortex PD–, K–, KC+ red, C+ red; medulla PD–, K–, KC–, C– (erythrin, lecanoric and schizopeltic acids). HABITAT: *Schizopelte* is a rare lichen, growing on coastal rocks, shrubs, and trees in the fog zone of southern California. COMMENTS: This unusual lichen superficially resembles *Niebla* (with large lecanorine apothecia at the branch tips) but is not at all yellowish green. *Schizopelte* is actually more closely allied to *Roccella* but has fruiting bodies that develop differently. Species of *Roccella* have longer and generally flatter branches, with smaller, marginal or laminal apothecia, and paler, 4-celled spores. See also *Hubbsia parishii*.

Scoliciosporum (4 N. Am. species)

DESCRIPTION: Crustose lichens, most with thin dark green to gray finely granular thalli, sometimes sorediate. Photobiont green (unicellular, up to 16 μm in diameter). Apothecia biatorine, with thin, disappearing margins; epihymenium greenish or brown; paraphyses branched and anastomosing much like the hyphae of the exciple; hypothecium colorless; asci *Lecanora*-type; spores transversely septate, 4- to many-celled, usually curved or spiraled, 8 per ascus. CHEMISTRY: Usually without lichen substances, rarely with traces of gyrophoric acid. HABITAT: On bark, wood, or rocks, often in deep shade. COMMENTS: Species of *Scoliciosporum* often resemble a *Micarea,* but the asci and algae are different and *Micarea* usually does not have curved spores. *Bacidia* and *Bacidina* also have different asci, and their paraphyses are predominantly unbranched.

KEY TO SPECIES: See *Bacidia*.

Scoliciosporum chlorococcum
(syn. *Bacidia chlorococca*)
City dot lichen

DESCRIPTION: Thallus dark green, granular to verruculose, not sorediate. Apothecia very dark brown to black, usually shiny, convex to hemispherical, 0.15–0.4 mm in diameter; epihymenium greenish; hypothecium colorless or very pale brown; spores curved and tapering, with one end distinctly fatter than the other, 18–35 (–40) × 3–5 μm, 5–8-celled. CHEMISTRY: No lichen substances. HABITAT: On wood and bark of all kinds, but most typically on conifers or birches, especially old branches without bark; in shaded forests, but frequently on trees in or close to towns; probably one of the most pollution-tolerant lichens. COMMENTS: The powdery green coating seen on tree trunks and branches, even in cities, is usually the ubiquitous green alga, *Desmococcus viridis* (syn. *Protococcus viridis*).

798. *Scoliciosporum chlorococcum* herbarium specimen (Canadian Museum of Nature), Quebec ×11.3

Sometimes, however, there are patches of darker green granules and tiny verrucae bearing hemispherical black apothecia mixed with this algal cover. These patches almost always turn out to be *Scoliciosporum chlorococcum*. The curved, comet-shaped spores and greenish epihymenium confirm the identification. *Scoliciosporum sarcothamni* is another pollution-tolerant species that is apparently very common in the humid coastal areas of the Pacific Northwest growing on deciduous trees. It differs from *S. chlorococcum* in having scattered yellowish soralia, and much narrower spores (up to 2 μm wide), and in producing gyrophoric acid.

Siphula (1 N. Am. species)

KEY TO SPECIES: See Key A.

Siphula ceratites
Waterfingers

DESCRIPTION: A fruticose lichen with erect, somewhat branched or unbranched white stalks, in dense clumps or more or less scattered, up to 7 cm tall and 1–2 mm thick; stalks blunt at the tips, usually with long furrows, sometimes lumpy, solid, with a dense medulla; without soredia, isidia, or apo-

799. *Siphula ceratites* Queen Charlotte Islands, British Columbia ×1.3

thecia. Photobiont green (*Trebouxia*?). CHEMISTRY: Cortex PD–, K+ yellow to brownish, KC–, C+ violet (color quickly fading), IKI+ blue; medulla PD–, K+ yellow to brownish, KC+ orange to gold, C+ violet (fading), IKI+ blue (siphulin). HABITAT: On soil or mud, sometimes submerged in water at the base, often mixed with mosses. COMMENTS: Only *Thamnolia* species could possibly be mistaken for *Siphula* within its range, but they have hollow, pointed stalks, are usually more prostrate, and have a different chemistry.

Solorina (5 N. Am. species)
Chocolate chip lichens, owl lichens

DESCRIPTION: Mainly foliose lichens (one crustose) with green algae (*Coccomyxa*) as the primary photobiont, and cyanobacteria (*Nostoc*) in cephalodia or forming a secondary layer below the layer of green algae; rarely containing only cyanobacteria; lower surface without or with a poorly developed cortex, sometimes weakly veined, tomentose, with tufts of rhizines. Apothecia large, round, red-brown disks immersed like chocolate chips in the thallus, often in concave depressions; with rounded lobes, or reduced to squamules around the apothecia; spores brown, 2-celled, ellipsoid, very large (30–160 μm long), with sculptured or ornamented walls, 1–8 per ascus. CHEMISTRY: Most species with no lichen substances; depsides and the orange anthraquinone solorinic acid in one species. HABITAT: On soil and rocks, usually rich in calcium; boreal forests to arctic or alpine tundra. COMMENTS: *Solorina*, to some extent, resembles green algal species of *Nephroma* and *Peltigera*. The large brown apothecia in thallus depressions and the large brown spores, together with the 2-photobiont thalli, set it apart from all similar lichens. Most species of *Solorina* are defined on the basis of the number of spores per ascus (2, 4, or 8), with correlating thallus development: scanty in the 2-spored *S. bispora* and 8-spored *S. octospora,* and clearly foliose in the 4-spored *S. saccata*.

KEY TO SPECIES

1. Thallus consisting of two parts: a brownish to gelatinous crustose layer containing cyanobacteria (blue-green in section), and scattered red-brown apothecia with squamulose margins containing green algae . . . . *Solorina spongiosa*

800. *Solorina crocea* (wet) Olympics, Washington ×2.4

1. Thallus foliose, with cephalodia in the form of a discontinuous blue-green layer of cyanobacteria below the green algal layer, or warts on the lower surface. 2
2. Lower surface and medulla bright orange; upper surface not pruinose; arctic-alpine . . . . *Solorina crocea*
2. Lower surface pale brown; medulla white; upper surface white pruinose or scabrose, at least at the lobe margins; widespread . . . . *Solorina saccata*

Solorina crocea
Orange chocolate chip lichen

DESCRIPTION: Thallus olive-brown or olive-gray to red-brown when dry, green when moist (as shown in plate 800), irregular to rounded lobes, 3–15 mm wide; medulla and lower surface bright orange. Photobionts in two layers: a more or less continuous green layer above a patchy blue-green layer of cyanobacteria (constituting the cephalodia); lower surface apparently without a cortex, weakly to strongly veined, the veins sometimes dark brown, with scattered rhizines. Apothecia 2–6(–10) mm in diameter, brown to red-brown, level with thallus or even somewhat convex, not in deep depressions; spores 30–45 × 10–15 μm, 6–8 per ascus. CHEMISTRY: Medulla PD–, K+ deep red-purple, KC and C reactions masked by pigment (solorinic acid, an anthraquinone; occasionally with methyl gyrophorate and gyrophoric acid). HABITAT: On soil, usually in moist spots under late snow patches or seepage areas in arctic or alpine sites. COMMENTS: This chocolate chip lichen is unmistakable, with its bright orange medulla and lower surface.

Solorina saccata
Common chocolate chip lichen

DESCRIPTION: Thallus forming rosettes of broad, round lobes, 10–20 mm across, brown when dry but grass-green when wet, usually coarsely pruinose or scabrose, at least close to the margins; cephalodia forming small blue-green warty patches on the lower surface of the thallus or slightly embedded; lower surface pale brown, webby, sometimes with weak veins; rhizines scattered. Apothecia red-brown,

801. *Solorina saccata* (wet) southwestern Yukon ×2.7

802. *Solorina spongiosa* herbarium specimen (Toby Spribille), Montana ×4.5

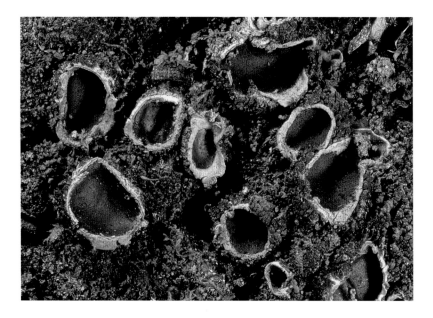

deeply sunken into concave depressions on the upper surface close to the margins, 2–5 mm in diameter; spores 4 per ascus. CHEMISTRY: No lichen substances. HABITAT: On soil or over rocks, especially those rich in calcium, in moist situations on tundra hummocks, cliff seepages, or lakeside rocks. COMMENTS: This is the most widespread species of *Solorina*, and the one with the best-developed thallus, but *S. bispora* and *S. octospora* are both common in the alpine zone of the southern Rockies. *Solorina bispora* has 2 spores per ascus (as the name implies). The spores are so large (up to 0.1 mm long), they can be seen on the apothecial disk and surrounding thallus using only a hand lens. The apothecia are sunken into deep depressions in the relatively small thalli creating round, cup-like squamules and lobes, 6–10 mm across, often pruinose. In the 8-spored *S. octospora*, the thallus can be well developed (and without pruina), or it can be reduced to a single lobe surrounding the apothecia. This species is the only *Solorina* north of Mexico that has a KC+ red, C+ pink medulla (methyl gyrophorate). The tropical montane species, *S. simensis*, has the same chemical reactions (methyl gyrophorate and tenuorin) and occurs in Mexico; it may be discovered in the Southern Rockies some day. This species, which contains only cyanobacteria, has 4 spores per ascus, and its apothecia are flush with the thallus surface.

Solorina spongiosa
Fringed chocolate chip lichen

DESCRIPTION: Thallus with green alga–containing tissue restricted to a relatively narrow, tattered collar around the large, red-brown apothecia, the remainder of the thallus consisting of a mat of cyanobacteria, sometimes brownish and areolate and sometimes gelatinous and barely lichenized; lower surface of green algal thallus pale or dark, without a cortex. Apothecia so deeply concave that they are often cup-like, 1.5–4 mm in diameter; spores 4 per ascus. CHEMISTRY: No lichen substances. HABITAT: Arctic and alpine tundra, rarely in shaded boreal habitats, on soil. COMMENTS: This is the most crustose of the *Solorina* species, with only the apothecial rim being at all expanded.

Speerschneidera (1 species worldwide)

KEY TO SPECIES: See Keys A and K.

Speerschneidera euploca
Pale rockwool

DESCRIPTION: Thallus prostrate fruticose or narrowly foliose, dark to pale greenish gray when dry and bright green when wet, with somewhat flattened, stiff, regularly dichotomous branches, 0.15–0.3(–0.5) mm wide, forming small rosettes, 3–12 cm across; without soredia or isidia; lower surface almost white, without rhizines. Photobiont green (unicellular). Apothecia common, lecanorine, 0.5–2 mm in diameter; disks grayish or reddish brown to almost black, with pale, even, smooth margins; asci *Lecanora*-type; paraphyses unbranched, expanded and pigmented at the tips; spores narrowly ellipsoid, 9–15 × 3–5 μm, colorless, with thin, uniform walls, 2–4-celled when mature, although young spores are 1-celled. CHEMISTRY: Cortex and medulla PD–, K–, KC–, C– (no lichen substances). HABITAT: On limestone and sandstone. COMMENTS: This inconspicuous, delicate lichen, endemic to the south central United States and central Mexico, at first looks like a *Phaeophyscia* or *Physcia*, but its anatomy places it close to the crustose genus *Lecania*.

Sphaerophorus (2 N. Am. species)
Coral lichens, tree coral

DESCRIPTION: Greenish or pale gray to brown fruticose lichens, shrubby to prostrate, with much-divided, cylindrical branches, lacking pseudocyphellae, soredia, or isidia; medulla dense, solid; cortex well-developed, stiff, often shiny. Photobiont green (*Cystococcus*). Apothecia spherical, at the tips of branches, covered with a thallus-colored cortex when young, then bursting open to reveal a black mass of spores (a mazaedium) formed after the breakdown of the ascus walls; spores spherical, brown, 1-celled. CHEMISTRY: Medulla with various reactions caused by a rich variety of compounds including β-orcinol depsides, but always UV+ blue-white (sphaerophorin); medulla IKI+ blue or IKI–. HABITAT: On soil, rocks, bark, moss, and peat. COMMENTS: Fertile specimens can be mistaken for *Bunodophoron melanocarpum*, which has distinctly flattened main branches splaying out from mossy tree trunks in humid coastal forests, or the rare, saxicolous, alpine lichen *Acrocyphus sphaerophoroides,* which has yellowish gray, very stiff, cylindrical branches, 0.8–2 mm thick, with an orange core (K+ deep red-purple) and black mazaedia embedded in the branch tips. *Tholurna dissimilis* has stubby, unbranched stalks on subalpine spruce and fir branches. *Pilophorus* and *Stereocaulon* resemble *Sphaerophorus* in having solid branches and grayish thalli, but both have their supporting tissue in the center of the stalks, not in the cortex.

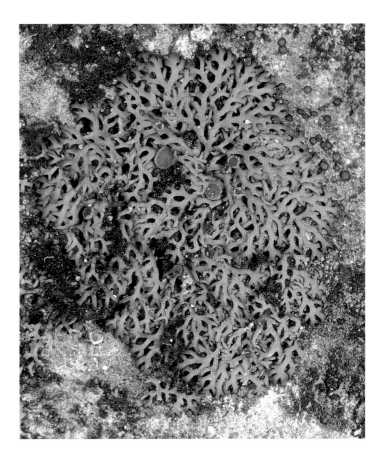

803. *Speerschneidera euploca* southern Texas ×3.9

KEY TO SPECIES
1. Thallus 30–80 mm long, with a stout main stem and finer side branches; fruiting bodies common; medulla IKI+ blue. . . . . *Sphaerophorus globosus*
1. Thallus usually under 20 mm long, dichotomously branched; fruiting bodies rare; medulla IKI–. . . . . *Sphaerophorus fragilis*

blotched with maculae; side branches sometimes slightly constricted at the base. Apothecia spherical, common on branch tips, 1–2.5 mm in diameter. CHEMISTRY: Medulla always IKI+ blue; PD–, K–, KC–, C–; or PD+ yellow, K+ yellow, KC–, C– (thamnolic acid); or PD–, K+ slowly becoming purple, KC+ red or violet-red (fading), C– (hypothamnolic acid); squamatic acid also sometimes present. HABITAT: On trees, especially mossy trunks and branches, in humid coastal forests, exposed or shaded, or on the ground on soil or mosses in arctic or alpine tundra. COMMENTS: The bewildering diversity in form and chemistry of this species has led to its division into more than one species. Because the finely branched and coarser morphotypes have intermediates, only one broadly defined species is recognized at present.

804. *Sphaerophorus fragilis* interior Alaska ×3.0

Sphaerophorus fragilis
Fragile coral lichen

DESCRIPTION: Thallus greenish gray at the base, becoming a shiny, yellow- or red-brown at the tips where exposed to sun; forming dense clumps of stiff and brittle, dichotomously branched stalks, mostly under 20 mm high and 0.3–0.8 mm in diameter. Apothecia rare. CHEMISTRY: Medulla IKI–, usually PD–, K–, KC–, C–, or rarely K+ purple, KC+ red (hypothamnolic acid). HABITAT: On soil or moss, and often on rocks, in arctic and alpine sites. COMMENTS: *Sphaerophorus fragilis* is much smaller, less branched, and less conspicuous than is *S. globosus*, which can also grow in tundra habitats. In cases where identification is problematic, the iodine reaction of the medulla is helpful (IKI+ blue for *S. globosus*).

Sphaerophorus globosus
Coral lichen

DESCRIPTION: Thallus extremely variable in form and color: sometimes with a few, stout main branches (up to 2 mm thick) and tufts of fine side branches (the form shown in plate 805), or sometimes dichotomously branched, gradually tapering at the branch tips; green to pale greenish gray, or quite brown, generally shiny and

Spilonema (3 N. Am. species)

DESCRIPTION: Minutely filamentous fruticose lichens consisting of branching strands of cyanobacteria (either *Stigonema* or *Hyphomorpha*) enveloped in long fungal hyphae, producing blue-green tufts of attachment hyphae at the base of the branches. Apothecia biatorine, brown to black, with a thin, disappearing exciple; epihymenial or hymenial tissue greenish or violet (turning red with concentrated nitric acid); spores ellipsoid, colorless, 1-celled, 8 per ascus. CHEMISTRY: No lichen substances. HABITAT: Most species grow on siliceous rock in the open, usually in moist situations either inland or maritime; rarely on moss or bark. COMMENTS: *Ephebe* has coarser, longer branches and lacks the blue-green attachment hyphae. *Thermutis velutina* is a very rare filamentous cyanobacterial lichen on moist rock that contains *Scytonema* as the photobiont. Its exciple is distinct and persistent, and the brownish hymenial pigments are negative with nitric acid. Species of *Polychidium* have branches enveloped in cellular rather than filamentous fungal hyphae.

KEY TO SPECIES: See Key B.

805. *Sphaerophorus globosus* Cascades, Washington ×1.8

Spilonema revertens
Rock hairball lichen

(Illustrated in plate 733, with *Psorula rufonigra*)

DESCRIPTION: Thallus forming small patches or cushions 5–15 mm across consisting of reddish brown to black (greenish black when wet), spiny, filamentous branches containing *Stigonema*; marginal, prostrate branches radiating from the cushions can be somewhat flattened and lobed; soredia and isidia absent but short branches in the central parts of the cushions can look like isidia. Apothecia uncommon, 0.2–0.5 mm in diameter, buried among the branches; spores 7–9 × 3.5 µm. HABITAT: On moist rocks, especially granitic outcrops, or on coastal rocks in the salt-spray zone. COMMENTS: This species is probably more common than the number of collections would indicate. Because it is often found as a substrate for the parasitic lichen *Psorula rufonigra*, the map of that species should be consulted for possible additional localities. ***Spilonema paradoxum*** is a rare rock species that differs in not forming cushions. See Comments under *Spilonema* for other similar species.

Sporastatia (2 N. Am. species)

DESCRIPTION: Crustose lichens with a well-developed areolate thallus, sometimes lobed at the edge. Photobiont green (unicellular). Apothecia lecideine, black, between the areoles, with thin, slightly raised, black margins on dry specimens; exciple brown and thin; epihymenium brown to green, turning red with nitric acid; hypothecium colorless to brownish; spores colorless, 1-celled, ellipsoid to almost spherical, ca. 2.5–4.5 × 2–3.5 µm, several hundred per ascus. CHEMISTRY: Cortex and medulla PD–, K–, KC+ pink, C+ pink, IKI– (gyrophoric acid). HABITAT: On rocks of all types in open arctic-alpine situations. COMMENTS: Except for the thallus color, *Sporastatia* has the aspect of *Dimelaena*, with black apothecia between the areoles. In *Dimelaena*, however, the apothecia are lecanorine. Other crustose lichens with numerous spores per ascus have quite different thalli and apothecia. See *Polysporina*, *Sarcogyne*, and *Acarospora*.

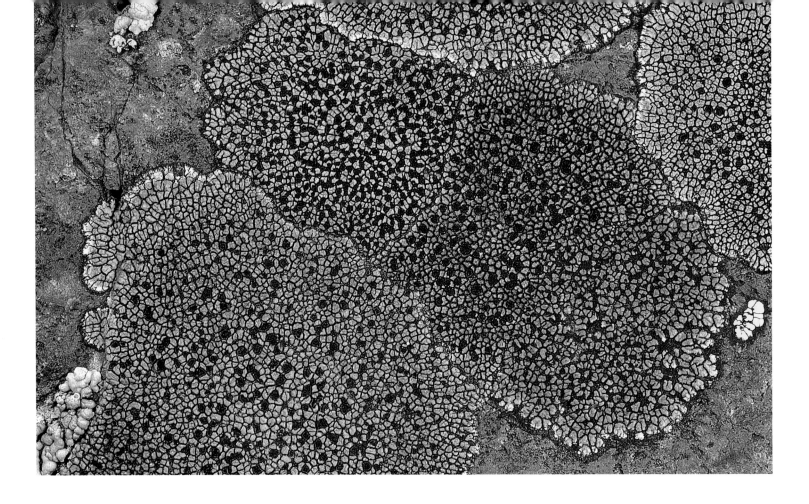

806. *Sporastatia testudinea* Rocky Mountains, Colorado ×3.8

KEY TO GENUS: See Key F.

Sporastatia testudinea
Copper patch lichen

DESCRIPTION: Thallus copper-brown, areolate, usually with distinctly radiating lobes at the margins and with a thick black prothallus. Apothecia 0.2–0.6 mm in diameter; upper part of hymenium and epihymenium green; hypothecium colorless. COMMENTS: The beautifully symmetrical, copper-colored rosettes of *Sporastatia testudinea* with delicately lobed margins are a common feature of arctic and alpine rocks. It resembles species of the *Lecidea atrobrunnea* group or *Dimelaena* but has many-spored asci. *Sporastatia polyspora* is an arctic species (also reported from Colorado) that has a bluish gray to brownish gray thallus and lacks the marginal lobes.

Squamarina (3 N. Am. species)

DESCRIPTION: Lobed crustose or squamulose lichens, forming patches of radiating lobes or overlapping squamules, yellowish white, pale yellow, yellow-green to brownish green, often pruinose, with a well-developed upper cortex and thick white medulla; soredia and isidia absent. Photobiont green (unicellular). Apothecia lecanorine, but margin becoming thin and finally disappearing as the apothecia mature; paraphyses usually unbranched; epihymenium granular; hypothecium colorless; asci *Bacidia*-type; spores colorless, 1-celled, ellipsoid, ca. 10–15 × 4–6 μm. CHEMISTRY: Contains usnic acid in the cortex, and, often, psoromic acid in the medulla. HABITAT: On calcareous soil and rock. COMMENTS: *Squamarina* is very much like a lobed species of *Lecanora* growing on soil but has different asci.

KEY TO SPECIES: See Key G.

Squamarina lentigera
White-rim lichen

DESCRIPTION: Thallus very pale yellowish green to yellowish white or chalky, rarely somewhat brownish green, with white pruina; marginal lobes rather flat, 0.8–2.5 mm wide, becoming slightly concave at the tips; center of the thallus becoming cracked, usually covered with crowded yellow-brown, sometimes pruinose, flat apothecia with thin, pale and often pruinose margins. CHEMISTRY: Cortex K+ pale yellow, KC+ yellow (usnic acid and atranorin); medulla PD–, K–, KC–, C–. (Psoromic acid has been reported in some European populations.) HABITAT: On soil. COMMENTS: No other lobed soil lichen has this combination of yellowish white lobes and brown, lecanorine apothecia. *Squamarina cartilaginea*, which is rare in North America, has very little pruina and contains psoromic acid (medulla PD+ yellow, K–). *Squamarina degelii* forms small, richly fruiting rosettes on rock and has much smaller lobes, usually less than 1 mm across. The lichen itself rarely grows larger than 3 or 4 cm across.

Staurothele (17 N. Am. species)
Rock pimples

DESCRIPTION: Thallus crustose, well developed in most species but scanty and within the rock in a few, typically brown but sometimes gray. Photobiont green (*Stichococcus* or *Desmococcus*). Perithecia embedded in the thallus or prominent; paraphyses absent; hymenium containing spherical or cylindrical green algal cells; spores colorless or brown, muriform, 1–8 per ascus. CHEMISTRY: No lichen substances. HABITAT: On dry or wet calcareous or siliceous rocks. COMMENTS: Species of *Endocarpon*, which also contain algae in the perithecial cavity, are squamulose lichens with a well-developed upper cortex. Hymenial algae are absent in *Polyblastia*, a related genus with muriform spores. Species of *Verrucaria* usually have smaller perithecia. Their spores are colorless and 1-celled.

KEY TO SPECIES

1. Spores 8 per ascus. . . . . *Staurothele diffractella*
1. Spores 2 per ascus 2
2. Thallus continuous, rimose-areolate, without any lobes; perithecia buried in thallus, producing low or high bumps; prothallus absent . . . . *Staurothele fissa*
2. Thallus areolate, the perithecia buried in separate areoles 3
3. On wet or, more rarely, dry rock; areoles containing perithecia are larger than areoles that are sterile; thallus usually with a lobed or fibrous prothallus. . . . . *Staurothele drummondii*
3. On dry rock; fertile and sterile areoles about the same size; thallus lacking a fibrous or lobed prothallus. . . . . [*Staurothele areolata*]

807. *Squamarina lentigera* southwestern Yukon Territory ×2.7

Staurothele diffractella

DESCRIPTION: Thallus pale brown to pale grayish brown, usually continuous, rimose-areolate, and rather smooth. Perithecia embedded in thallus except for the conspicuous black summits that are visible at the surface as black dots with a clearly indented ostiole; involucrellum spreading and distinct from exciple (as shown in Fig. 17a); abundant algal cells (usually spherical) in the hymenial cavity; spores colorless, (4–)8 per ascus, 15–28 × 9–12 μm. HABITAT: Typically on lime-containing rocks, but sometimes on gneiss or even granite; usually in shaded habitats. COMMENTS: This is by far the most common 8-spored *Staurothele*, but the spores are often hard to count because finding a full ascus is not easy. Look for immature asci with small, developing spores.

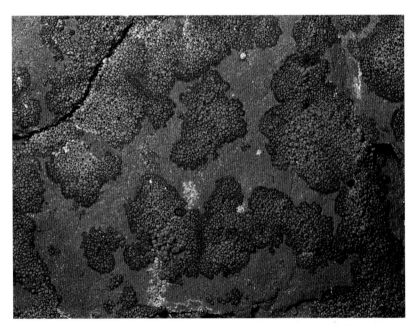

808. *Staurothele diffractella* herbarium specimen (Canadian Museum of Nature), Vermont ×5.7

809. *Staurothele drummondii* (photographed in a stream) Strawberry Mountains, eastern Oregon ×3.6

Staurothele drummondii (syn. *S. fuscocuprea*)

DESCRIPTION: Thallus composed of small, rounded, dark brown to brownish gray contiguous areoles, 0.2–0.5(–0.7) mm in diameter when containing perithecia and notably smaller when sterile; areoles usually narrow; thallus sometimes radiating at the margin, creating more or less distinct lobes; a fibrous black to dark brown prothallus is usually visible, especially on smooth rocks. Perithecia with a conspicuous black involucrellum showing around the ostioles; hymenial algae usually cylindrical, but can be spherical in some populations; spores brown, muriform, 2 per ascus, 24–50 × 11–24 µm. HABITAT: Mainly on at least periodically innundated rocks (e.g., lakeshores and streams), but sometimes on dry limestone. COMMENTS: *Staurothele areolata* is a similar, widespread, 2-spored species, but it prefers dry rocks and lacks marginal lobes. The fertile and sterile areoles tend to be the same size, and they can be dispersed as well as contiguous. The hymenial algae are mainly cylindrical. *Staurothele drummondii* has been called *S. clopima* by some lichenologists, but the application of that name is very unclear.

Staurothele fissa
Lakezone lichen

DESCRIPTION: Thallus dark brown to almost black, continuous but sometimes cracked into small areoles; perithecia 0.3–0.6 mm in diameter, partially buried in the thallus but prominent enough to be visible as bumps with black ostioles; excipie pale, partly covered with a black involucrellum; hymenial algae spherical; spores brown, muriform, many-celled, 27–50 × 14–25 µm, 2 per ascus. HABITAT: On siliceous rocks along lake- and stream shores, forming a conspicuous black zone along the lake edge where the rocks are at least periodically submerged. COMMENTS: A black zone caused by this aquatic lichen is a common sight along northern lakeshores bordered by granite. Other perithecial lichens can contribute to the black zone, especially species of *Verrucaria*, but the dominant member of the community is usually this *Staurothele*. *Staurothele clopimoides* occupies similar habitats in the western mountains (rare in the Great Lakes region). Its perithecia are entirely immersed in the thallus areoles, showing only the black ostiole at the surface.

Stereocaulon (34 N. Am. species)
Foam lichens, Easter lichens

DESCRIPTION: Pale gray to white fruticose (rarely crustose) lichens, with a granular to verrucose primary thallus (disappearing in most species) and a secondary thallus of erect, branched stalks generally in thick clumps or tight cushions, less frequently scattered; stalks supported by a central, cartilaginous column, bearing phyllocladia (granules, verrucae, coralloid branches, or lobed squamules) containing green algae (*Trebouxia*); cephalodia present in most species, generally on the stalks, wart-like, granular or filamentous, containing cyanobacteria (*Nostoc* or *Stigonema*). Apothecia biatorine, brown or rarely black; asci *Porpidia*-type; spores colorless, ellipsoid to needle-shaped, 2–14-celled, 8 per ascus. CHEMISTRY: Cortex K+ yellow (atranorin) in all species; PD+ orange (often faint), KC–, C– (stictic acid), or PD–, KC+ pink or violet (often too faint to detect), C– (lobaric acid) in most species; PD–, KC–, C– (no substances, porphyrilic acid or fatty acids in some species). HABITAT: On soil or rock, commonly mixed with mosses. COMMENTS: Whereas most species of *Stereocaulon* lose their granular or verrucose primary thallus, a few common species, such as **S. condensatum** and **S. glareosum,** do not. *Stereocaulon condensatum,* in the northeast, can cover large areas of sandy soil. It has short, largely unbranched stalks less than 2 cm tall with large apothecia at the summits. Its cephalodia are granular-fibrous and blackish brown. *Stereocaulon glareosum,* a western montane species, has tomentose stalks (1–2 cm tall), also with terminal apothecia, with smooth reddish brown cephalodia. *Stereocaulon* and *Pilophorus* both produce solid stalks, have cephalodia, and have similar asci. *Pilophorus,* however, has 1-celled spores and terminal, almost spherical apothecia; the stalks are almost unbranched, and the cephalodia tend to be on the primary thallus rather than on the stalks. IMPORTANCE: *Stereocaulon* replaces reindeer lichens (*Cladina*) as the dominant ground cover in some parts of the boreal forest. In these regions, it becomes an important component of the caribou winter diet. Although *Stereocaulon* is not as palatable as reindeer lichens, the caribou prefer it to many other kinds of lichens. Some species of *Stereocaulon* have been used as medicine in China and India.

810. *Staurothele fissa* herbarium specimen (Canadian Museum of Nature), Quebec ×8.4

KEY TO SPECIES

1. Stalks sorediate. 2
1. Stalks without soredia 4

2. Stalks 20–40 mm tall; side branches and phyllocladia sorediate; cephalodia on the stalks. . . . . *Stereocaulon coniophyllum*
2. Stalks under 5 mm tall; soredia at stalk summit; cephalodia on crustose primary thallus. 3

3. Soredial mass KC+ pink to violet (lobaric acid); primary thallus usually granular . . . *Stereocaulon pileatum*
3. Soredial mass KC–; primary thallus areolate . . . . *Pilophorus cereolus*

4.(1) Growing attached directly to rock 5
4. Growing on soil (sometimes soil over rock) 10

5. Phyllocladia consisting of convex to flat, more or less round squamules, with dark green centers and pale margins. . . . . *Stereocaulon vesuvianum*
5. Phyllocladia uniform in color 6

6. Tomentum on stalks thick or thin, usually pinkish; cephalodia abundant, consisting of large spherical granules more or less buried in the tomentum, or lumpy galls; phyllocladia squamulose, often deeply lobed . . . . rare, saxicolous forms of *Stereocaulon grande*
6. Tomentum usually thin, or if thick, then gray, not pinkish; cephalodia common, sparse, or absent 7

811. *Stereocaulon coniophyllum* (growing on rock recently uncovered by a retreating glacier) Southeast Alaska ×1.4

7. Phyllocladia flattened and lobed, rarely coralloid; thallus prostrate and clearly dorsiventral, at least at the margins of the colony; cephalodia absent; containing lobaric acid................... *Stereocaulon saxatile*
7. Phyllocladia predominantly coralloid or granular, not flattened; thallus prostrate or erect, dorsiventral or not; cephalodia forming grape-like clusters when present 8

8. Phyllocladia usually in warty or granular clusters at the stalk tips like cauliflower; cephalodia rare; contains porphyrilic acid; arctic-alpine, especially in Alaska ... . [*Stereocaulon botryosum*]
8. Phyllocladia usually coralloid, not especially concentrated at stalk tips; lacking porphyrilic acid......... 9

9. Thallus dorsiventral in part; thallus PD+ orange (stictic acid); cephalodia usually sparse or absent........... .... *Stereocaulon dactylophyllum*
9. Thallus more or less erect, not dorsiventral; thallus PD– (lobaric acid); cephalodia relatively abundant ... . *Stereocaulon tennesseense*

10.(4) Stalks usually under 2 cm high; phyllocladia warty, sometimes lobed 11
10. Stalks mostly 2–8 cm long, erect or mat-forming, with distinct main stems; stems not woody or brittle; phyllocladia granular or squamulose, rarely coralloid..... 13

11. Primary thallus disappearing; stalks mat-forming, without main stems; stems woody, brittle; arctic-alpine .... *Stereocaulon rivulorum*
11. Primary thallus persistent, granular to squamulose; stalks usually erect and distinct.................. 12

12. Mostly northeastern, Great Lakes to east coast, with scattered boreal localities farther north and west; cephalodia black, with a fuzzy or grainy surface, containing *Stigonema* (Fig. 1a) .... [*Stereocaulon condensatum*]
12. Western montane and arctic; cephalodia brown, smooth and often fissured, containing *Nostoc* (Fig. 1d) .... [*Stereocaulon glareosum*]

13.(10) Phyllocladia predominantly granular, clustered; cephalodia in fibrous tufts, dark brown or olive-black, abundant .... *Stereocaulon paschale*
13. Phyllocladia predominantly flat, deeply or shallowly lobed; cephalodia not fibrous, often sparse 14

14. Tomentum thick or sometimes thin, often pinkish; cephalodia common and conspicuous, pinkish brown and lumpy or, rarely, blue-green and granular .... *Stereocaulon grande*
14. Tomentum thick, puffy, creamy white; cephalodia tiny, dark blue-green granules buried in the tomentum 15

15. Thallus PD+ orange (containing stictic acid); thallus erect or prostrate; widespread boreal to North Temperate .... *Stereocaulon tomentosum*
15. Thallus PD– or PD+ yellow (containing lobaric acid); thallus prostrate; mainly western .... *Stereocaulon sasakii*

Stereocaulon coniophyllum
Powdered foam lichen

DESCRIPTION: Stalks erect, up to 4 cm tall, branching, with side branches either dissolving entirely into masses of soredia or becoming flattened and squamulose, with the lower surface and margins sorediate; stalks woody, without tomentum; cephalodia abundant on the stalks (visible on the branch at the bottom of plate 811), pale pinkish brown, lumpy (like a sack of potatoes), 1–2 mm in diameter. Apothecia frequently produced, 1–4 mm in diameter, brown, flat to convex. CHEMISTRY: Contains lobaric acid. HABITAT: On rocks or rarely soil, in moist situations. COMMENTS: Very few foam lichens are sorediate. The most common one is *S. pileatum,* a tiny boreal species with unbranched stalks rarely taller than 5 mm, and with a persistent, granular primary thallus. ***Stereocaulon spathuliferum*** is a rare, granular sorediate species from the Pacific Northwest growing directly on rock. It has horny stems and no tomentum and contains stictic and norstictic acids (PD+ yellow-orange). The phyllocladia are verrucose to granular or grainy in appearance, rarely flattened. Sometimes the soredia, which are formed on the undersides of the tips, are hard to distinguish from the smaller, granular phyllocladia. It has abundant cephalodia resembling clustered grapes.

Stereocaulon dactylophyllum
Finger-scale foam lichen

DESCRIPTION: Thallus prostrate or erect, attached directly to rock; phyllocladia usually elongate and finger-like (coralloid), but can occasionally be flattened and even verrucose; stems without tomentum; cephalodia gray, hemispherical, grape-like, visible from above, but rare or absent in most specimens. Apothecia usually abundant, at the tips of branches, flat to convex. CHEMISTRY: Thallus PD+ orange, K+ yellow, KC–, C– (stictic acid complex, sometimes norstictic acid). HABITAT: On siliceous rocks in the open. COMMENTS: *Stereocaulon dactylophyllum* is very similar to *S. saxatile:* it is directly attached to rock, grows in small rosettes, and is often more or less dorsiventral. The phyllocladia, however, are almost entirely coralloid, and the thallus is PD+ orange. See also *S. tennesseense*.

Stereocaulon grande
Grand foam lichen

DESCRIPTION: Thallus erect, branched, with a rather thick (less commonly thin), white to pinkish tomentum on the stems; phyllocladia somewhat flattened and lobed, often becoming somewhat cylindrical and coralloid; cephalodia lumpy to brain-like, large, ca. 0.5–1.5 mm in diameter, conspicuous on the stems, typically brownish pink, or rarely forming clusters of dark blue-green, tomentose granules. Apothecia frequently produced, terminal or lateral, typically clustered at the branch tips, often broken into segments, dark brown. CHEMISTRY: Contains lobaric acid. HABITAT: On soil or mossy rocks (but rarely attached directly to the rock) in the northern boreal forest zone. COMMENTS: *Stereocaulon grande* intergrades with *S. alpinum*, an arctic-alpine, tomentose species.

812. *Stereocaulon dactylophyllum* Appalachians, Tennessee ×2.6

813. *Stereocaulon grande* Cascades, Washington ×2.4

In its most typical state, *S. alpinum* has verrucose phyllocladia, pinkish stems, smaller, almost granular cephalodia, and mainly terminal apothecia. It can be either prostrate and dorsiventral or entirely erect. The amount of tomentum is extremely variable, as it is in *S. grande*. *Stereocaulon tomentosum* (with stictic acid) and *S. sasakii* (with lobaric acid) are always heavily tomentose and have very inconspicuous cephalodia hidden in the tomentum. Their apothecia are on lateral branches.

814. *Stereocaulon paschale* (unusually fertile specimen) Southeast Alaska ×2.1

815. *Stereocaulon pileatum* coastal Maine ×4.3

Stereocaulon paschale
Easter lichen

DESCRIPTION: Stalks erect, 2–6 cm tall, with rather slender, branched, woody stems, almost devoid of a fuzzy tomentum, or the tomentum can be rather thick; phyllocladia typically clusters of subspherical or granular verrucae, sometimes somewhat elongate and almost coralloid, rarely expanding into finger-like lobes; cephalodia abundant, fibrous and irregular, rarely granular, brown-black (containing *Stigonema*). Apothecia uncommon. CHEMISTRY: Containing lobaric acid. HABITAT: On soil. COMMENTS: This is a common foam lichen on soil and mossy rocks, best characterized by its fibrous cephalodia and granular phyllocladia. It can resemble *S. alpinum* except for the cephalodia.

Stereocaulon pileatum
Pixie foam lichen

DESCRIPTION: Primary thallus persistent, consisting of granular to cylindrical verrucae interspersed with reddish to gray-brown, granular or brain-like hemispherical cephalodia; vertical stalks short and stubby, mostly under 3 mm tall, unbranched or branched once, covered with granular phyllocladia and mealy soredia at the summits. Apothecia very rare. CHEMISTRY: Sorediate tips of stalks KC+ violet, C– (lobaric acid). HABITAT: On siliceous rocks in sunny or partially shaded locations, often near lakeshores or waterfalls. COMMENTS: This miniature foam lichen can be confused with *Pilophorus cereolus*, which has an areolate thallus and KC– soredia.

Stereocaulon rivulorum
Snow foam lichen

DESCRIPTION: Thallus forming low clumps (up to 3 cm high) of thick, contorted, branching stalks, rather woody, with a thin or rarely thick tomentum; main stems not discernible; phyllocladia thick, usually convex and warty to coarsely granular, but often lobed to cylindrical and branched, never flattened; stalks very brittle; cephalodia usually sparse, granular or hemispherical, brownish, scattered on the stem surface. CHEMISTRY: Usually contains lobaric acid; some populations with perlatolic and anzaic acid and some with only atranorin. HABITAT: On moist soil, often associated with late snowbanks, mostly arctic and alpine. COMMENTS: In some forms, the phyllocladia coalesce, giving the thallus the appearance of a crustose lichen when viewed from above. Its fruticose habit is obvious, however, when the colony is broken apart. **Stereocaulon incrustatum** is quite similar but has more distinct, erect stems, more conspicuous, brownish, spherical cephalodia, and normally contains only atranorin.

Stereocaulon sasakii
Woolly foam lichen

DESCRIPTION: Thallus prostrate and forming dorsiventral mats, less frequently erect; phyllocladia flattened, deeply lobed squamules, distributed all along the tall stalks, which can reach 8 cm long; stalks normally covered with puffy clumps of white or creamy tomentum, but in some specimens, tomentum is thin; cephalodia granular, dark blue-green, buried in tomentum. Apothecia common, less than 1 mm in diameter, brown, on the tips of the stalks or on side branches. CHEMISTRY: Tomentum and medulla PD– or PD+ pale yellow (atranorin and lobaric acid). HABITAT: On soil or mossy rock, rarely attached to rock. COMMENTS: This is essentially a lobaric acid-containing west coast variety or subspecies of *S. tomentosum*. In *S. tomentosum*, the tomentum and medulla are PD+ orange (stictic acid), but the reaction can be seen only if the concentration of stictic acid is high. It also has a tendency to be more erect and less dorsiventral than *S. sasakii*, and the squamulose phyllocladia are generally smaller.

816. *Stereocaulon rivulorum* Rocky Mountains, British Columbia ×1.7

Stereocaulon saxatile
Rock foam lichen

DESCRIPTION: Thallus forming centrally or broadly attached mats up to 10 cm across, always dorsiventral, at least at the periphery, but sometimes more or less erect in the center of a colony, 3–6 cm tall; cephalodia absent or extremely rare; stalks usually thinly gray tomentose and rather woody at the base; more rarely, heavily tomentose or entirely without tomentum; phyllocladia squamulose and lobed, sometimes coralloid, covering most of the stalks and sometimes becoming confluent, growing into a somewhat lobed sheet. CHEMISTRY: Contains lobaric acid. HABITAT: Directly attached to siliceous rock, usually in full sun. COMMENTS: *Stereocaulon saxatile* is an extremely polymorphic species, especially in its orientation to the surface, the amount of tomentum, and the shape of its phyllocladia. The similar *S. dactylophyllum* rarely has confluent phyllocladia and contains stictic rather than lobaric acid. Heavily tomentose specimens can be distinguished from *S. tomentosum* and *S. sasakii* (which usually grow on soil) by the lack of cephalodia. Both **S. intermedium** and **S. sterile** contain lobaric acid and are fixed to the rock by a relatively small holdfast rather than a broad area. *Stereocaulon intermedium* usually has branched cylindrical (i.e., coralloid) phyllocladia and almost always has some well-developed, lumpy, gray to blue-green cephalodia (0.5–2 mm in

817. *Stereocaulon sasakii* (damp) Olympics, Washington ×3.3

diameter). *Stereocaulon sterile,* a very common foam lichen in the Pacific Northwest, lacks tomentum on the stalk and has granular to coralloid or lobed squamulose phyllocladia; its cephalodia are not abundant and can be inconspicuous. ***Stereocaulon subcoralloides*** is another species in this group with lobaric acid. Unlike *S. intermedium,* it is more erect and less dorsiventral and frequently has some lobed squamulose as well as coralloid phyllocladia. *Stereocaulon subcoralloides,* however, is similar to *S. dactylophyllum* except in chemistry and somewhat thinner phyllocladia, and so we have come full circle.

Stereocaulon tennesseense
Bony foam lichen

DESCRIPTION: Thallus in coarse tufts firmly attached to the rock, with erect, woody stalks up to 5 cm tall; phyllocladia are all long, branched and cylindrical, like gnarled, bony fingers, up to 2 mm long; cephalodia common, pinkish brown, in grape-like clusters. Apothecia not abundant. CHEMISTRY: Contains lobaric acid. HABITAT: Attached directly to rock, usually at a single point. COMMENTS: The phyllocladia of this species resemble those of *S. dactylophyllum,* which is more dorsiventral and contains stictic acid.

Stereocaulon tomentosum
Woolly foam lichen, eyed foam lichen

DESCRIPTION: Thallus erect to prostrate, often somewhat dorsiventral; stalks covered with puffy clumps of white tomentum; phyllocladia lobed, squamulose, warty or flattened; cephalodia granular, very dark blue-green (*Nostoc*), buried in cephalodia; apothecia common, on lateral branches. CHEMISTRY: Tomentum and medulla usually PD+ orange (stictic acid), but the reaction is often faint. HABITAT: On soil. COMMENTS: Except for the different chemistry, this species is almost identical to the exclusively western *S. sasakii.* (See Comments under that species.) *Stereocaulon paschale* has more granular phyllocladia, and the cephalodia, which are fibrous and conspicuous, contain *Stigonema*.

818. *Stereocaulon saxatile* Lake Superior region, Ontario ×1.9

819. *Stereocaulon tennesseense* Appalachians, Tennessee ×2.9

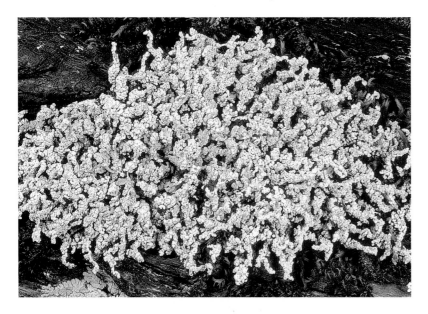

820. *Stereocaulon tomentosum* Laurentides, southern Quebec ~×1.5

821. *Stereocaulon vesuvianum* mountains, coastal Alaska ×2.8

Stereocaulon vesuvianum
Variegated foam lichen

DESCRIPTION: Thallus forming low, loose or fairly firm cushions of rather woody, nodular branches, 1–4(–7) cm high; phyllocladia consisting of lumpy verrucae, at first convex, then flattened and finally developing a depressed, dark olive-green center (relatively broad or almost point-like, most visible on the more shaded branches on the lower part of the stalks); cephalodia usually conspicuous, dark brown or pinkish gray, lumpy or clustered like grapes. Apothecia rare. CHEMISTRY: PD+ orange, often difficult to detect (stictic and usually norstictic acids). HABITAT: Attached directly to exposed rocks; often an early colonizer of newly exposed rock surfaces. COMMENTS: *Stereocaulon arenarium,* also an arctic species, has similar phyllocladia, but the branches grow in extremely tight, mat-forming clumps. It is PD– and contains porphyrilic acid. ***Stereocaulon groenlandicum*** is a coarse-looking saxicolous foam lichen from central Alaska. It has the same nodular appearance as some forms of *S. vesuvianum,* but the lumpy, verrucose phyllocladia do not have dark centers, even in the shaded parts of the thallus, and the cephalodia are more inconspicuous and grain-like. The stalks have no tomentum, and the verrucose phyllocladia often sit on bare areas of the stalk. Chemically, *S. groenlandicum* is very distinctive, containing a mixture of perlatolic, anziatic, miriquidic, and porphyrilic acids.

Sticta (7 N. Am. species)
Moon lichens, crater lichens

DESCRIPTION: Medium to large foliose lichens, mostly dark brown to gray-brown (rarely green), loosely attached, with broad lobes sometimes bearing soredia or isidia; lower surface corticate, fuzzy tomentose (rarely naked), with crater-like cyphellae. Photobiont blue-green (*Nostoc*), or very rarely green (see Comments below). Small cephalodia develop internally or on the lower surface of species with green algae. Apothecia almost never produced. CHEMISTRY: No lichen substances in most species; anthraquinones in the medulla of a few. HABITAT: On mossy rocks or bark, usually in humid sites; one species (*S. arctica*) on tundra soil and peat. COMMENTS: The presence of cyphellae

822. *Sticta beauvoisii* Appalachians, North Carolina ×2.3

on the lower surface immediately distinguishes *Sticta* from the closely related *Lobaria* and *Pseudocyphellaria*. *Sticta wrightii*, a rare western species, is the only North American *Sticta* with a green photobiont. Like many stratified foliose lichens containing cyanobacteria, wet specimens of *Sticta* often have a disagreeable, fishy odor.

KEY TO SPECIES

1. Soredia produced on or close to the lobe margins . . . . *Sticta limbata*
1. Soredia absent, isidia present 2

2. Isidia along the lobe margins or cracks in the thallus; surface of the lobes smooth and often shiny . . . . *Sticta beauvoisii*
2. Isidia covering the thallus surface, which is dull and granular in appearance . . . . *Sticta fuliginosa*

Sticta beauvoisii
Fringed moon lichen

DESCRIPTION: Thallus shades of brown, often dark; lobes round or squarish, 6–10 mm across, smooth and often shiny, with granular to flattened, branched or lobed isidia along the margins and cracks in the upper surface of the thallus; lower surface with a brown, velvety tomentum made up of short, dark brown rhizines and individual hyphae that have brush-like tips. Apothecia extremely rare. HABITAT: On mossy rocks and bark in shaded habitats. COMMENTS: This species has long been confused with *S. weigelii,* a tropical species apparently absent from North America that is characterized best by its yellow medulla reacting K+ purple (anthraquinones). Its tomentum consists of hyphae with clusters of spherical cells at the tips. An unnamed species from the southern Appalachians has the same chemistry as *S. weigelii* but produces lobules rather than isidia. Populations often named as *S. weigelii* occur along the west coast from California to southeastern Alaska, in the south-

823. *Sticta fuliginosa* (damp) Klamath Range, northwestern California ×1.8

ern Rockies, and along the southeastern coastal plain, but their identity or relationship to *S. beauvoisii* is still uncertain. Our map indicates *S. beauvoisii* in a rather strict sense.

Sticta fuliginosa
Peppered moon lichen

DESCRIPTION: Thallus dark brown to olive-gray; lobes broad and rounded, 7–15(–20) mm wide; upper surface somewhat wrinkled, covered with tiny cylindrical to coralloid isidia, giving the whole lichen a dull, granular appearance; lower surface pale brown, covered with an almost white tomentum. HABITAT: On mossy bark, rarely mossy rock. COMMENTS: At least two other isidiate species of *Sticta* occur in North America, but they have isidia or flattened lobules concentrated along the margins: *Sticta beauvoisii*, and what has been called *S. weigelii* (see Comments under *S. beauvoisii*). A third, *S. sylvatica,* is an exceedingly rare western lichen (but common in Europe) that has flattened, lobulate isidia mainly along ridges and cracks on the upper surface.

Sticta limbata
Powdered moon lichen

DESCRIPTION: Thallus chocolate-brown to almost gray-brown, generally remaining shiny; lobes rounded to somewhat divided and ruffled, 6–15 mm wide, with blue-gray granular soredia produced on and close to the margins; lower surface pale brown to yellow, covered with a tomentum of buff to gray, very short, brush-like rhizines. CHEMISTRY: All reactions usually negative, but some specimens have an almost lemon-yellow lower surface that turns red in K (anthraquinones). HABITAT: On mossy rock and bark, especially in coastal forests. COMMENTS: The first thing to do when identifying the brown, cyanobacteria-containing, sorediate foliose lichens in the west is to turn them upside down. *Sticta limbata* has round holes (cyphellae); *Peltigera collina* has veins; *Pseudocyphellaria anomala* has raised white spots (pseudocyphellae); and *Nephroma parile* is smooth and tan. All these species are common in lowland, humid forests, especially on deciduous trees such as alders and maples.

Sulcaria (2 N. Am. species)
Grooved horsehair lichens

DESCRIPTION: Pendent fruticose lichens, pale brown to gray in North America, with a conspicuous, deep, longitudinal groove (a specialized pseudocyphella) running along the length of the branches, often in a spiral. Photobiont green (*Trebouxia?*). Apothecia not observed in North American species, but spores are yellowish to brown, 2–4-celled. CHEMISTRY: Usually with abundant atranorin; sometimes containing β-orcinol depsidones. HABITAT: On deciduous trees and shrubs. COMMENTS: The deep, long pseudocyphellae of species of *Sulcaria* characterize the genus. Some species of *Bryoria* also have long, spiraling pseudocyphellae (see Comments below), and in the absence of apothecia for checking spores, the two genera can be confused.

KEY TO SPECIES: See *Alectoria*.

Sulcaria badia
Bay horsehair lichen

DESCRIPTION: Thallus 20–50 cm long, soft and pliant, pale reddish brown to dusky brown and gray, streaked and mottled, dull; main branching in dichotomies, but with perpendicular side branches; larger branches flattened and deeply grooved with straight or spiraling pseudocyphellae; main branches 0.2–0.4(–1) mm wide; without soredia or isidia. Apothecia and pycnidia unknown. CHEMISTRY: Thallus PD+ pale yellow or brownish, K+ yellow, KC± yellow to pink, C– (atranorin and an unidentified compound). HABITAT: Usually on deciduous trees such as apple, oak, and maple in lowland woodlands and pastures; very rare. COMMENTS: The mottled chestnut color of *Sulcaria badia* is unique among the hair lichens. Although it can form large, robust clumps, it is a very rare species and should be regarded as threatened because its known localities are all fairly close to populated agricultural lands. It is similar to some coastal species of *Bryoria* having long pseudocyphellae, such as **B. pseudocapillaris** and **B. spiralifera**. They

824. *Sticta limbata*
Cascades, Washington
×1.9

825. *Sulcaria badia*
(detail) Coast Range,
northern California
×4.7

826. *Teloschistes chrysophthalmus* Channel Islands, southern California ×2.9

are smaller lichens and differ chemically: *B. pseudocapillaris* is PD+ deep yellow, K+ yellow, KC+ pink, C+ pink (alectorialic and barbatolic acids); *B. spiralifera* is PD+ yellow, K+ red, KC–, C– (norstictic acid and atranorin). The other North American *Sulcaria*, *S. isidiifera*, is even rarer, known only from San Luis Obispo County in California. It is a shorter, shrubbier species producing abundant isidia, and it is PD+ red-orange (protocetraric acid).

Teloschistes (6 N. Am. species)
Orange bush lichens

DESCRIPTION: Bright orange to gray, shrubby or tufted fruticose lichens, with large, orange apothecia when fertile. Photobiont green (*Trebouxia*). Spores polarilocular, 8 per ascus. CHEMISTRY: Orange portions of cortex and (or) apothecial disks PD–, K+ deep red-purple, KC–, C– (yellow or orange anthraquinones); gray areas negative to reagents. HABITAT: On the branches of trees and shrubs, on rocks, and on the ground. COMMENTS: A few species of *Teloschistes* are entirely gray except for the apothecial disks. For example, *T. californicus,* a rare species from coastal southern California into Baja California, looks like some gray aberration of *Ramalina* or *Evernia*, especially because it is almost always sterile. However, it has fuzzy, flattened branches with a reticulate, sinewed surface, lacks soredia, and is entirely negative to spot tests, making it unlike any other lichen.

KEY TO SPECIES

1. Branches producing granular soredia; apothecia rare... .... *Teloschistes flavicans*
1. Branches without soredia; apothecia common....... 2

2. Thallus in tufts up to 2 cm high; apothecial margins abundantly ciliate........ *Teloschistes chrysophthalmus*
2. Thallus in tufts 3–7 cm high; apothecial margins without cilia......................... *Teloschistes exilis*

Teloschistes chrysophthalmus
Gold-eye lichen

DESCRIPTION: Thallus short, consisting of shrubby tufts of rather broad, flattened, but strongly ridged branches, 0.5–1.5 mm wide, usually pale to white below, but deep orange above, sometimes mottled with gray; branches without soredia or isidia, but tips often end in slender, unbranched cilia; pseudocyphellae rare or absent. Large apothecia (1–4 mm across) fringed with cilia almost always on some branch tips, giving the lichen the appearance of a bouquet. HABITAT: On the branches of trees and shrubs. COMMENTS: *Teloschistes exilis*, which also lacks soredia, has rounder, narrower branches and smaller, nonciliate apothecia. In the interior high plateau area of eastern Oregon, Washington, and southern Idaho, *T. contortuplicatus* can be found on moss, soil, or rock in calcareous habitats. It also has flattened, ciliate branches forming small, dense tufts. The tundra-dwelling *T. arcticus,* confined to a small area on and around Banks Island, has a small, shrubby, orange thallus with hollow branches and is always sterile. The distributions depicted in the range maps of *T. chrysophthalmus* and *T. flavicans* are based on old as well as modern collections. The species have been eliminated over parts of their former range, especially in the northeast, because of habitat destruction and air pollution.

827. *Teloschistes exilis* hill country, central Texas ×1.7

Teloschistes exilis
Slender orange bush lichen

DESCRIPTION: Thallus light to dark orange, forming rounded tufts up to 7 cm across; branches partly angular but basically terete, 0.2–0.5(–0.8) mm wide, ending in long, fine, sometimes reddened cilia; pale, raised pseudocyphellae often present but lacking soredia or isidia. Apothecia scattered along branches, 0.8–3 mm in diameter, with thin, nonciliate margins. HABITAT: On the branches of trees in open woodlands. COMMENTS: *Teloschistes flavicans*, a sorediate species, is similar in habit but lacks cilia. The branches of *T. chrysophthalmus* are broader, flatter (up to 1.5 mm wide), and paler (sometimes almost entirely grayish), and terminate in large, ciliate apothecia.

Teloschistes flavicans
Powdered orange bush lichen

DESCRIPTION: Thallus bright orange, forming tufts 2–5 cm across; branches slender (under 0.6 mm wide), mostly terete or angular, with abundant yellow soralia breaking through the cortex. Apothecia rarely seen. HABITAT: On the branches of trees and shrubs, and often on the ground, in open areas, especially near the coast. COMMENTS: This is the only North American sorediate species of *Teloschistes*. It has become extremely rare along the east coast but can still be found in the west and south.

Tephromela (4 N. Am. species)

DESCRIPTION: Crustose lichens, varied in appearance but with well-developed thalli. Photobiont green (unicellular). Apothecia lecanorine or lecideine, immersed in the thallus or superficial; asci *Bacidia*-type with a

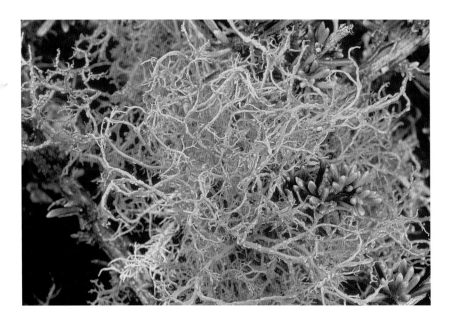
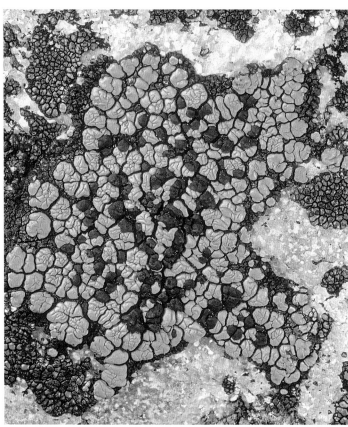

828. *Teloschistes flavicans* Channel Islands, southern California ×3.0

829. *Tephromela armeniaca* Rocky Mountains, Colorado ×3.1

K/I+ blue tholus; spores colorless, 1-celled, ellipsoid; conidia elongate to thread-like, produced from chain-like basal cells. CHEMISTRY: A variety of compounds including atranorin (in most species), usnic, alectoronic, and α-collatolic acids. HABITAT: Most species on siliceous rocks; one also on wood and bark. COMMENTS: This genus is defined mainly on microscopic anatomical details, especially the structure of the ascus, paraphyses, and conidia-bearing tissue. Some species resemble *Lecanora*; others, *Lecidea*.

KEY TO SPECIES: See Key F.

Tephromela armeniaca

DESCRIPTION: Thallus thick, dark yellow, shiny, areolate to almost squamulose, the areoles folded and cracked in various ways; prothallus conspicuous, black. Photobiont green (unicellular). Apothecia flat, 0.7–2(–4.5) mm in diameter, almost marginless, pitch-black, rather thin; epihymenium green; hypothecium colorless or yellowish; spores small, ellipsoid, 9–12 × 3.5–5 µm. CHEMISTRY: Cortex PD–, K+ red-orange, KC+ red-orange, C– or C+ pink; medulla PD–, K–, KC–, C–, IKI–, or sometimes PD+ yellow close to the algal layer (alectorialic acid). HABITAT: On rock, in arctic-alpine sites. COMMENTS: This species looks like a member of the *Lecidea atrobrunnea* group but with a more yellow-brown (rather than red-brown) thallus. In *Lecidea atrobrunnea*, the medulla is IKI+ blue and the ascus tips are largely K/I–.

Tephromela atra
Black-eye lichen

DESCRIPTION: Thallus creamy white to yellowish gray, often shiny, very thick and areolate when on rock, with smaller areoles when on wood, and relatively thin and verruculose when on bark; without soredia or isidia; prothallus inconspicuous or absent. Apothecia pitch-black, shiny, with a thick or thin margin, thallus-colored or partially blackened like the disk, 0.7–3 mm in diameter; hymenium violet; hypothecium yellowish brown to golden yellow; spores broadly ellipsoid, 10–15 × 5–8 μm. CHEMISTRY: Cortex PD–, K+ yellow, KC+ yellow, C–, UV–; medulla PD–, K–, KC+ pink-violet, C–, UV+ white (atranorin, α-collatolic acid, sometimes alectoronic acid). HABITAT: On siliceous or nutrient-rich rocks, on wood, and on bark. COMMENTS: It would seem that this lichen has an extraordinary breadth of ecological tolerance, growing on bark, wood, or rock from the Arctic to the Tropics. Some recent research shows that more than one species may be involved, especially in the Subtropics and Tropics. Despite the variation in thallus morphology and apothecial size, the purple hymenium, golden yellow hypothecium, and broadly ellipsoid spores make it easy to identify when viewed under the microscope. In the field, however, it can be mistaken for some black-fruited species of *Lecanora* such as ***L. gangaleoides*** and *L. cenisia*. Specimens on bark, as shown in plate 830, have a thinner thallus than do those growing on rock.

Thamnolia (1 species worldwide)

KEY TO SPECIES: See Key A.

Thamnolia vermicularis
Whiteworm lichen

DESCRIPTION: Fruticose lichen, with prostrate or erect, ivory-white, hollow stalks, unbranched or with few branches, pointed at the tips, 2–7 cm long, 1–2.5 mm in diameter, growing in tufts or singly, lacking soredia, isidia, and fruiting bodies (reproducing by fragmentation alone). Photobiont green (*Trebouxia?*). CHEMISTRY: Two chemical variants: (1) cortex PD+ yellow-orange to red-orange, K+ bright yellow, KC–, C–, UV– (thamnolic acid); (2) cortex PD+ deep cadmium-yellow, K+ pale yellow, KC–, C–, UV+ yellow (baeomycesic and squamatic acids). Both chemotypes can be somewhat UV+ white in the inner tissues. HABITAT: On exposed gravelly soil, among mosses and heath, and on rocks, mainly in alpine and arctic sites, but also on windswept slopes close to sea level on the northwest coast. COMMENTS: Plate 831 shows the thamnolic acid chemotype, representing *Th. vermicularis* in the strict sense. The baeomycesic acid chemotype, called **var. *subuliformis***, is identical in morphology. The distributions largely overlap in the American Arctic, but the thamnolic acid chemotype (var. *vermicularis*) predominates in the coastal mountains, whereas var. *subuliformis* is the variant encountered in the Rocky Mountains and northern Appalachians. IMPORTANCE: The golden plover uses *T. vermicularis* as nesting material.

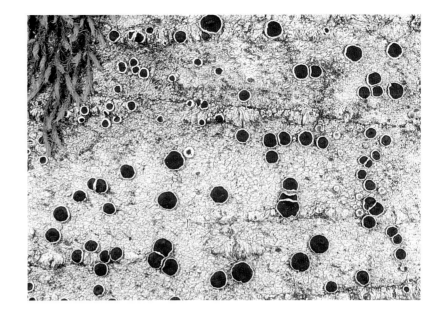

830. *Tephromela atra* (Specimens on bark, as in the photograph, have a thinner thallus than those on rock.) Cascades, Oregon ×2.8

Thelidium (16 N. Am. species)
Speck lichens

DESCRIPTION: Crustose lichens with perithecia and colorless, septate spores (1–4 cells), otherwise very much like *Verrucaria* (1-celled spores) or ***Polyblastia*** (muriform spores); paraphyses absent; excipulum generally dark brown; involucrellum present or absent; thalli well developed or almost imperceptible. Photobiont green (*Protococcus*). CHEMISTRY: No lichen substances. HABITAT: On rock, mostly calcareous types; some species on soil or peat. COMMENTS: The taxonomy of this genus is still poorly worked out, especially in North America. Only a single species is presented here, although many are common limestone lichens in temperate as well as arctic latitudes. To distinguish the species, one must prepare perithecial sections carefully to see the excipulum and involucrellum. Unfortunately, very few reliably identified specimens occur in North American herbaria for comparison.

KEY TO GENUS: See Key D.

831. *Thamnolia vermicularis* Olympics, Washington ×1.9

832. *Thelidium pyrenophorum* Rocky Mountains, Alberta ×5.6

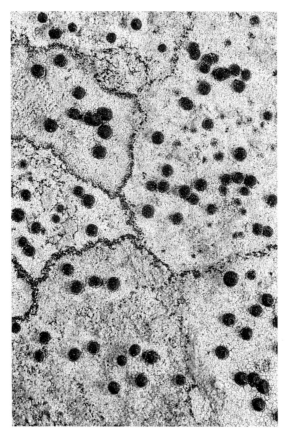

Thelidium pyrenophorum

DESCRIPTION: Thallus very thin, gray, or almost entirely endolithic; perithecia partially embedded in the rock or superficial; approximately 0.3–0.6 mm in diameter; excipulum entirely brown to black, fused to black involucrellum on upper half; spores colorless, 2-celled, approximately 18–30 × 9–14 μm; 8 per ascus. HABITAT: On limy rocks. DISTRIBUTION: Mainly in arctic Canada and Alaska and in alpine sites farther south. COMMENTS: The identity of the material shown in plate 832 is uncertain, but it conforms with published descriptions.

Thelomma (6 N. Am. species)
Nipple lichens

DESCRIPTION: Crustose lichens with thick, areolate, or lobed thalli, gray to yellowish or tan. Photobiont green (*Trebouxia*). Ascomata embedded in thallus warts or areoles, consisting of a black mass of spores (mazaedium); excipy difficult to distinguish from the dark brown hypothecium; spores 1- or 2-celled, brown-black, ellipsoid, with a thick, smooth, or somewhat

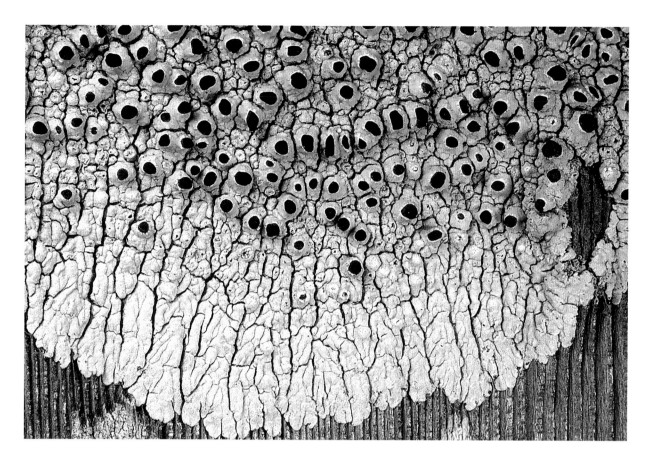

833. *Thelomma californicum* California coast ×3.4

ridged wall. CHEMISTRY: A variety of compounds including divaricatic, variolaric, norstictic, and salazinic acids, the yellow pigments vulpinic and rhizocarpic acids, and skyrin. HABITAT: Most species on exposed rock, but a few on wood, mostly in coastal, Mediterranean climates. COMMENTS: This genus is similar to *Cyphelium* in many ways, but the exciple of *Thelomma* is not nearly as well-developed, the apothecia are always immersed, and the spores are often smoother.

KEY TO SPECIES

1. Thallus areolate to verrucose, not lobed, shiny, never pruinose; spores 1-celled; on rocks . . . .
. . . . *Thelomma mammosum*
1. Thallus distinctly lobed at the margin, verrucose-areolate in the center, dull, often pruinose; spores 2-celled; on wood and sometimes rocks. . . . .
. . . . *Thelomma californicum*

Thelomma californicum
Lobed nipple lichen

DESCRIPTION: Thallus very thick, lobed at the edge and verrucose-areolate in the center, gray to yellowish gray, coarsely pruinose to granular-scabrose, especially at the margin; black spore mass extruded from cone-shaped verrucae, often with a tattered white collar; fertile verrucae 1–3 mm in diameter; spores black or dark brown, 2-celled, constricted at the septum, 15–20 × 10–12 µm. CHEMISTRY: Cortex PD–, K–, KC+ rose-red, C–; medulla PD–, K–, KC–, C–, IKI–, UV+ blue-white (3-chlorodivaricatic acid); mazaedium PD–, K– (no lichen substances). HABITAT: On wood (e.g., fences and posts) and siliceous rocks in open areas near the coast. COMMENTS: Although restricted in distribution, this lichen can be very abundant and conspicuous. It is the only *Thelomma* with a lobed thallus. ***Thelomma occidentale*** and ***T. ocellatum***, also on wood, have much larger spores (22–28 × 14–15 µm). *Thelomma occidentale*, found from Califor-

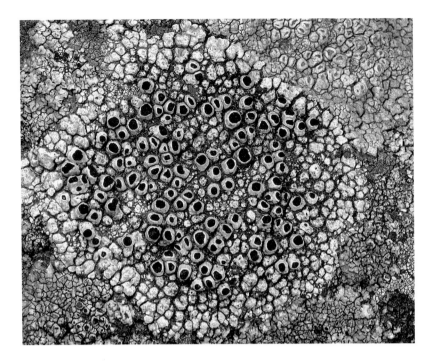

834. *Thelomma mammosum* Channel Islands, southern California ×2.5

nia to Alaska mainly along the coast, has a lumpy, areolate, brownish gray thallus with regular, round mazaedia, and its medulla is negative with IKI. The much rarer *T. ocellatum* has a gray thallus with irregular verrucae, with coarsely granular spore masses breaking through in irregular or sometimes round patterns. The medulla is IKI+ blue. It is known from Idaho.

Thelomma mammosum
Rock nipple lichen

DESCRIPTION: Thallus thick, pale gray to dark yellowish gray, sometimes light brown to pinkish brown, areolate to verrucose; fertile verrucae smooth, shaped like volcanoes or nipples, 0.8–1.6 mm in diameter; mazaedia often broad, with a thin white collar; spores black, 1-celled, rough-walled, spherical, 13–16 μm in diameter. CHEMISTRY: Cortex PD–, K–, KC+ pink, C–, UV+ white (3-chlorodivaricatic acid); mazaedium PD+ yellow, K+ red (norstictic and salazinic acids). HABITAT: On rocks along the coast and in the coastal mountains. COMMENTS: The similar but much rarer California lichen *T. santessonii* is KC– in the cortex but is UV+ blue-white (divaricatic acid) and has a rougher, dusky to dark yellow thallus with large fruiting verrucae up to 2.5 mm across. The distinct yellow is the best field character for distinguishing this species from *T. mammosum*.

Thelotrema (19 N. Am. species)
Barnacle lichens

DESCRIPTION: Crustose lichens with a thin or immersed thallus, shades of yellowish tan or pale brown. Photobiont green (*Trentepohlia*). Apothecia with a thin exciple distinct and generally separate from an outer thalloid margin, opening with a hole at the summit giving the fruiting body a perithecial appearance; exciple visible within the cavity as a frayed membrane; periphyses lining the wall of the exciple near the ostiole; paraphyses unbranched; asci thin-walled, K/I–; spores 2- to many-celled, sometimes muriform, colorless or brown, 1–8 per ascus. CHEMISTRY: Rich variety of compounds, especially β-orcinol depsidones and anthraquinones. HABITAT: On bark or, less frequently, wood or rock, in moist or tropical regions. COMMENTS: *Thelotrema* can superficially resemble species of *Pertusaria* or various perithecial lichens, but the ostioles of these lichens are not cavities. Most similar are two tropical genera, *Myriotrema* and *Ocellularia*. Both of these lack periphyses in the inner wall of the exciple and often have a sterile column of tissue (a columella) within the hymenium. In *Ocellularia*, the columella and exciple are black. *Coccotrema* species, although unrelated, have somewhat similar apothecia (also with periphyses) but with large, 1-celled spores.

KEY TO SPECIES (based on Harris, *More Florida Lichens*, 1995)

1. Thallus with sparse, thick isidia; spores brown, submuriform, 16–28 × 7–12 μm; coastal plain.............................. [*Thelotrema santense*]
1. Thallus not isidiate; spores colorless 2

2. Spores muriform, over 50 μm long; mainly along the northeastern and northwestern coasts, rare in Florida .... *Thelotrema lepadinum*
2. Spores only transversely septate, 8–15-celled, 25–45 × 6–8 μm 3

3. Apothecia 0.2–0.35 mm across; thallus well developed, with a thick medulla; coastal plain................... .... [*Thelotrema lathraeum*]
3. Apothecia 0.4–0.6 mm across; thallus thin, with an imperceptible medulla; Appalachian to coastal plain..... .... [*Thelotrema subtile*]

Thelotrema lepadinum
Bark barnacles

DESCRIPTION: Thallus within the bark tissue creating a brownish stain. Apothecia resembling small volcanoes or barnacles on the bark, 0.7–1.6 mm in diameter; spores colorless, broadly fusiform, muriform, (48–)67–100(–135) × (9–)12–20(–25) µm, (2–)4(–8) per ascus. CHEMISTRY: No lichen substances. HABITAT: On the bark of deciduous or coniferous trees in humid forests. COMMENTS: This is by far the most common *Thelotrema* on the continent. At least 16 species in Florida can be distinguished on the basis of spore septation and color, chemistry, apothecial characters, and presence or absence of isidia. A few of the more common species are included in the key. Two species ranging farther northward include *T. subtile* (from the Appalachians to northern Florida) and *T. petractoides* (on the British Columbia coast), both with transversely septate (i.e., not muriform), 8–15-celled, colorless spores.

Tholurna (1 species worldwide)

KEY TO SPECIES: See Key A.

Tholurna dissimilis
Urn lichen

DESCRIPTION: Thallus fruticose, greenish gray to brownish gray, consisting of an expanded, sometimes squamulose basal layer and radiating tufts of inflated, usually hollow, furrowed, unbranched podetia, 0.2–1 mm wide and 1–3 mm high, forming tufts 5–7 mm high, up to 2 cm across. Each podetium is terminated by a reddish brown mazaedial apothecium that looks like a little brown pot or urn with a slightly flaring lip, filled with black spores. Photobiont green (*Trebouxia*). Apothecia 0.2–0.7 mm in diameter; spores dark brown, 2-celled, distinctly constricted at the septum, with spiraled ridges on the wall, 16–21 × 8–10 µm. CHEMISTRY: All reactions negative (no lichen substances). HABITAT: On branches and twigs of exposed conifers, mostly in the subalpine zone of more humid

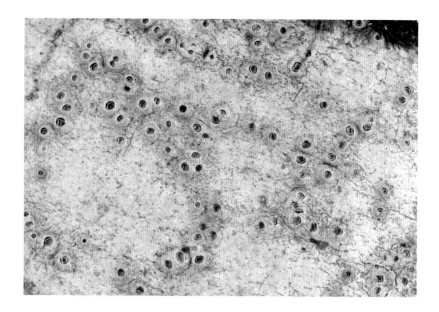

835. *Thelotrema lepadinum*   coastal Alaska ×5.8

836. *Tholurna dissimilis*   herbarium specimen (Trevor Goward) British Columbia ×7.3

837. *Thyrea confusa*
Ozarks, Missouri ×4.3

mountain ranges, but also on treetops at lower elevations. COMMENTS: This inconspicuous but fascinating lichen cannot be mistaken for any other.

Thyrea (2 N. Am. species)

KEY TO SPECIES: See Keys A and I.

Thyrea confusa
Jelly strap

DESCRIPTION: Fruticose to almost foliose or squamulose jelly lichens, black when dry, often with a bluish tint resulting from white pruina, gelatinous when wet; clumps attached at a single point with the flattened lobes or branches fanning out or almost erect; lobes deeply divided, strap-shaped to rounded, 0.3–1.5 (–3) mm wide, not thickened at the margins; upper surface generally covered with granular isidia. Photobiont blue-green (Chroococcales, e.g., *Gloeocapsa*), with small, irregular cells surrounded by thick walls or sheaths, more or less concentrated close to the lobe surfaces, leaving a thin, poorly defined photobiont-free zone. Apothecia uncommon and rarely noticed, immersed in the thallus, opening by a pore into a cavity and resembling a perithecium, expanded and disk-like when fully mature; spores colorless, 1-celled, 8 per ascus. CHEMISTRY: No lichen substances. HABITAT: On calcareous rock, especially on surfaces that are sometimes wet. DISTRIBUTION: Scattered localities throughout North America, but especially in the central states. It is too poorly known to be mapped at this time. COMMENTS: *Lichinella nigritella* is often a neighbor of *Thyrea confusa* and can be distinguished by its thinner lobes with slightly raised or thickened margins, and its lack of pruina. The fruiting bodies are also different, but because they are so inconspicuous, this is rarely a help in distinguishing the two genera. *Thyrea confusa* has been called **Th. pulvinata** in older literature. The taxonomy of this and related species is still unsettled. The thalli of *Thyrea* species are much stiffer than those of *Collema* species, the photobiont is different, and the photobiont is less uniformly distributed in the thallus. **Phylliscum demangeonii** is a closely related, rare lichen growing on siliceous rocks in alpine habitats. It is composed of small, round, umbilicate scales (mostly under 7 mm in diameter). Its apothecia are also buried in the thallus and opening by a tiny pore, and usually there are 16 spores per ascus.

Toninia (24 N. Am. species)
Blister lichens

DESCRIPTION: Crustose or squamulose lichens, most with thick thalli; lobes and verrucae flat to strongly convex, usually with a well-developed upper and sometimes lower cortex (a few species have poorly developed thalli or are parasitic on other lichens), pale gray to dark brown, frequently pruinose, at least in part. Photobiont green (unicellular). Apothecia lecideine to biatorine, with thick, radiating hyphae in the exciple; paraphyses predominantly unbranched except at the tips, with expanded, often pigmented tips. Apothecial tissues, especially the epihymenium and exciple, often reddish, brown, gray, or green, with various color changes with nitric acid or K; asci *Biatora*-type; spores colorless, 1–8-celled, ellipsoid to needle-shaped, 8 per ascus. CHEMISTRY: Mostly negative with spot tests, but a few species contain atranorin, zeorin, fatty acids, or a number of unidentified compounds. HABITAT: On soil or rock, usually containing some lime. Most species become established on lichen thalli as weak parasites before becoming free-living, and some remain as parasites. The preferred hosts are generally cyanobacterial lichens such as *Collema* and *Placynthium*, but some common *Toninia* species grow on green algal lichens. They are often in lichen communities that include calcareous soil and rock lichens such as *Endocarpon, Fulgensia, Placidium,* and *Peltula*. COMMENTS: The genus is best characterized as crustose lichens, with thick, often pruinose thalli, lecideine apothecia, and (usually) septate spores. It can have the aspect of a *Psora,* but that genus has 1-celled spores, asci of the *Porpidia* type, and the paraphyses are not expanded. Crustose (as opposed to squamulose) species can be mistaken for a *Catillaria,* a genus with 2-celled spores including some saxicolous species. The ascus tip in *Catillaria* has a uniformly staining tholus in K/I (Fig. 14c). Some species of *Buellia* (subgenus *Diplotomma*) or *Phaeorrhiza* may key out to *Toninia,* but they have ellipsoid, brown, 2-celled spores.

KEY TO SPECIES

1. Spores 4–8-celled, 23–42 μm long 2
1. Spores 1–2-celled, 12–24 μm long 4
2. Thallus heavily and coarsely pruinose; epihymenium gray, K+ violet; Rocky Mountains, Yukon to New Mexico . . . . *Toninia alutacea*
2. Thallus without pruina; epihymenium greenish to brown or red-brown, K– or K+ red; widespread in western U.S. and adjacent southern Canada. 3
3. Epihymenium reddish brown, K+ red; thallus dark olive-brown, rarely reddish brown . . [*Toninia ruginosa*]
3. Epihymenium green, K–; thallus pale to dark brown, not olive. . . . . [*Toninia squalida*]
4.(1) Thallus crustose, areolate to rimose-areolate, without pruina; epihymenium greenish brown or green, K–; Colorado. . . . . [*Toninia phillippea*]
4. Thallus squamulose, pruinose or not. 5
5. Pruina entirely absent from the thallus and apothecia; areoles and squamules flat to convex, with deep depressions and pores; epihymenium brown to green, K–. . . . . *Toninia tristis* ssp. *asiae-centralis*
5. Pruina present on the thallus and often the apothecia, sometimes sparse or patchy; epihymenium gray to violet, K+ violet. 6
6. Squamules very small, 0.4–1(–2) mm across, almost granular in appearance, often scattered, covered with coarse pruina; arctic and south into the Canadian Rockies . . . . [*Toninia arctica*]
6. Squamules medium to large, 1–4(–5) mm across 7
7. Pruina coarse and granular; southern Rockies . . . . [*Toninia subdiffracta*]
7. Pruina fine and powdery. 8
8. Squamules strongly convex and folded throughout; thallus not lobed at the margins or forming rosettes; pruina sometimes confined to most exposed surfaces of squamules and often lacking on the apothecia; on soil; widespread and common . . . . *Toninia sedifolia*
8. Squamules flattened and somewhat lobed at the margins, convex in the center; thallus often forming rosettes; pruina uniform and dense on the thallus as well as the apothecia; on calcareous rock; southern Rockies . . . . [*Toninia candida*]

Toninia alutacea
Frosted blister lichen

DESCRIPTION: Thallus with convex verrucae and areoles, lobed at the periphery, entirely covered with a heavy white pruina. Apothecia black, heavily pruinose, closely adnate, 1–2 mm in diameter; exciple and epihymenium gray or purplish, K+ violet, and violet in nitric acid; hypothecium colorless to

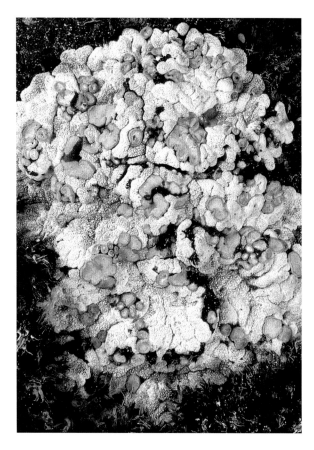

838. *Toninia alutacea* Rocky Mountains, Alberta ×4.1

839. *Toninia sedifolia* Kaibob Plateau, northern Arizona ×6.0

pale yellowish brown; spores narrowly fusiform, 4-celled, 20–34 × 2.8–4.8 μm. CHEMISTRY: All reactions negative (unknowns and triterpenes). HABITAT: On calcareous rocks, mostly arctic and alpine. COMMENTS: *Toninia alutacea* is similar to *T. candida* and sometimes even *T. sedifolia*, but these have exclusively 2-celled spores. *Toninia sedifolia* is not lobed at the margins and is less pruinose than either *T. candida* or *T. alutacea*.

Toninia sedifolia (syn. *T. caeruleonigricans*)
Earth-wrinkles, blue blister lichen

DESCRIPTION: Thallus heaped into folded and puckered irregular, convex warts and squamules, without radiating lobes at the edge, blue-gray color caused by a dense, fine white pruina on exposed surfaces, but dark olive to brown on surfaces without pruina. Apothecia black, sometimes lightly pruinose, flat to convex and contorted, at first with thin margins but finally losing them; hypothecium and inner part of exciple red-brown; epihymenium and outer parts of exciple gray to purplish black, K+ violet, and violet in nitric acid; spores fusiform, 2-celled, 12–24 × 3–5 μm. CHEMISTRY: All reactions negative (unidentified substances). HABITAT: On soil, and sometimes rock, from the Arctic to the desert, wherever the substrate contains abundant calcium and the habitat is relatively dry. COMMENTS: This is one of the most common species of the genus and is rather variable in thallus development and pruinosity.

Toninia tristis ssp. *asiae-centralis*
Pitted blister lichen

DESCRIPTION: Thallus brown, epruinose, with convex to flattened irregular squamules, some having deep depressions or pores (shown in the upper right corner of plate 840); black dots often cover the thallus. Apothecia black, without pruina, typically 0.8–1.5 mm in diameter; epihymenium dark brown, K+ reddish purple, dark red with nitric acid; hypothecium very pale yellowish to colorless, K+ reddish purple; exciple dark purplish brown at edge, colorless within, K–;

spores rod-shaped, 1–2-celled, 12.5–19 × 3.5–5 μm. CHEMISTRY: All reactions negative (unidentified compounds). HABITAT: Over soil and in rock crevices. COMMENTS: Only *T. bullata*, with 4–8-celled spores, has similar pores in the thallus.

Trapelia (6 N. Am. species)

DESCRIPTION: Crustose lichens with thin, pinkish white to pale greenish gray thalli; soredia present in a few species. Photobiont green (*Chlorella* or *Protococcus*). Apothecia yellowish brown or reddish brown to almost black, basically lecanorine, the outer thalloid margin either well developed and tattered or virtually absent, leaving a thin but distinct, pale to dark brown exciple; paraphyses abundantly branched and anastomosing, the tips barely expanded; hypothecium very pale, sometimes pinkish; exciple pale to dark brown; epihymenium brown; asci cylindrical with thin walls, entirely K/I– or very pale blue, with a low tholus that sometimes has an K/I+ blue cap; spores colorless, 1-celled, ellipsoid, often with a diffuse halo visible in ink preparations; 8 spores per ascus. CHEMISTRY: All species PD–, K–, KC+ red, C+ pink (gyrophoric acid) except *T. mooreana*. HABITAT: On siliceous rocks; one species on bark (*T. corticola*, a mostly sterile, sorediate species); in full sun or shade. COMMENTS: *Trapelia* is distinguished from *Trapeliopsis* in having thinner thalli, a lecanorine or double apothecial margin, and larger spores. It also tends to be saxicolous, rather than on wood or soil.

KEY TO GENUS: See Key F.

Trapelia involuta
Pebble lichen

DESCRIPTION: Thallus pale gray-green, sometimes pinkish brown in part, at first consisting of small, dispersed or contiguous areoles, often somewhat lobed at the edge, without soredia or isidia. Apothecia dark red-brown to almost black, 0.4–0.7 (–1) mm in diameter, marginless or often with a thin, thallus-colored, ragged margin, sometimes appearing as two margins, one inside the other, but some specimens with apothecia having thin to disappearing biatorine margins; spores 17–22 × 8–11 μm. HABITAT: On

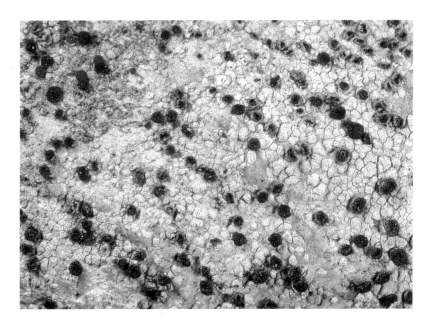

840. *Toninia tristis* ssp. *asiae-centralis* southwestern Yukon Territory ×5.8

841. *Trapelia involuta* coastal Massachusetts ×8.3

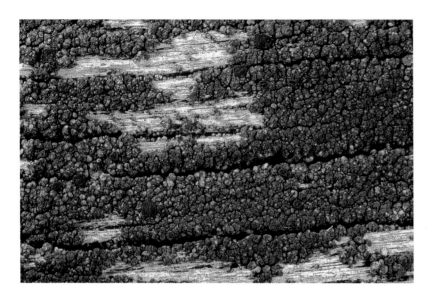

842. *Trapeliopsis flexuosa* Appalachians, Virginia ×7.0

cept for the lack of a double margin in the apothecia, *Trapeliopsis* is almost identical to *Trapelia* in its essential characteristics. A few species of *Trapeliopsis* have thin, membranous thalli, although this is anomalous for the genus. ***Trapeliopsis gelatinosa,*** for example, has a membranous to thin, granular, green or gray-green thallus, sometimes sorediate, on soil and peat. It has dark brown-black apothecia, 0.2–1 mm in diameter, and KC–, C– spot test reactions; it has an Appalachian–Great Lakes distribution.

KEY TO SPECIES

1. Thallus squamulose, thick, not producing soredia, pale gray; on rocks and soil from California to Oregon. . . . . *Trapeliopsis wallrothii*
1. Thallus areolate to verrucose, not squamulose; usually sorediate in part; widespread 2
2. Thallus pale gray to greenish white or pinkish; apothecia pink, brown, or dark gray, sometimes with all colors represented on the same thallus or even the same apothecium; on soil or wood. . . . . *Trapeliopsis granulosa*
2. Thallus dark gray to greenish black; apothecia almost always dark lead-gray, but occasionally pale; almost exclusively on wood. . . . . *Trapeliopsis flexuosa*

small stones and pebbles or boulders in open areas or shaded forests. COMMENTS: *Trapelia coarctata* is less common, with a thin, smooth or slightly cracked, continuous thallus. It is widespread but most common near the west and east coasts. Also, a fertile *Trapelia* grows on shaded forest boulders in the northeast, with a rather thick, almost continuous to cracked thallus, and large red-brown apothecia (up to 2 mm in diameter) with raised, often double, margins; this may represent a distinct species. ***Trapelia placodioides*** closely resembles *T. coarctata* but has irregular green soralia instead of apothecia and a slightly lobed thallus margin. It is a fairly common, shade-tolerant lichen on forest boulders in the northeast (rarer elsewhere). See also *Trapeliopsis*. *Trapelia involuta* seems to be very pollution-tolerant, occurring in residential areas of the Bronx and in Central Park in New York City.

Trapeliopsis (6 N. Am. species)

DESCRIPTION: Crustose lichens, most with thick, granular, areolate, verrucose, or squamulose thalli, dark olive-gray to very pale greenish gray, usually sorediate. Photobiont green (possibly *Chlorella* or *Pseudochlorella*). Apothecia biatorine, pinkish or yellowish brown to almost black, with prominent to disappearing margins; paraphyses branched and anastomosing, not expanded at the tips; asci cylindrical, slightly thickened at the apex, K/I– or weakly blue; spores colorless, 1-celled, ellipsoid, 8 per ascus. CHEMISTRY: Most species PD–, K–, KC+ red, C+ pink (gyrophoric acid). HABITAT: On rocks, soil, moss, or wood. COMMENTS: Ex-

Trapeliopsis flexuosa
Board lichen

DESCRIPTION: Thallus composed of dark, steel-gray to greenish black, round areoles or large granules, bursting open in maturity and giving rise to dark green or greenish yellow soredia. Frequently fertile with flat, lead-gray apothecia, 0.2–0.7 mm in diameter, having prominent, wavy (flexuose) margins (hence the scientific name); epihymenium and exciple olive-brown; hypothecium colorless; spores 7–10 × 3–5 μm. CHEMISTRY: Gyrophoric acid. HABITAT: On weathered wood, especially fences and boards, in full sun. COMMENTS: This easily overlooked lichen is actually very attractive when viewed under the hand lens. *Trapeliopsis granulosa* can sometimes be similar, but at least some of the apothecia on that species become convex and marginless, and tend to be variegated in color or entirely pinkish, although they can be dark gray. The areoles of *T. granulosa* are also larger (over 1 mm in diameter) and much paler.

Trapeliopsis granulosa
Mottled-disk lichen

DESCRIPTION: Thallus pale gray to greenish white, sometimes pinkish, with granular to hemispherical verrucae, some breaking into piles of coarse, green or almost white soredia. Apothecia pale pinkish brown to very dark brown or blackish gray, often mottled, 0.4–1.5 mm in diameter; margins usually persistent, but sometimes becoming very thin and disappearing; spores 9–13 × 4–6 μm. CHEMISTRY: Thallus PD–, K–, KC+ red, C+ pink (gyrophoric acid). HABITAT: On acid soil, peat, or rotting wood, including surfaces recently charred, rarely rock. COMMENTS: The C+ pink reaction of the thallus helps distinguish this species from most other white soil and wood crusts. Sterile species of *Ochrolechia* have continuous thalli. In ***T. pseudogranulosa***, the verrucae break down into an almost continuous granular or powdery crust, generally with orange spots (K+ deep red-purple). It grows in humid, shaded habitats in the Pacific Northwest. See also *T. flexuosa*. IMPORTANCE: Because this species is an important colonizer of bare soil and charred wood, forest managers in Quebec have disseminated this lichen over tracts of recently burned forest after fire has removed the litter from the forest floor. The white thalli reflect the sunlight, cooling the soil and allowing it to accumulate more moisture, and the lichen itself helps prevent erosion.

Trapeliopsis wallrothii
Scaly mottled-disk lichen

DESCRIPTION: Thallus squamulose, dull white to pale or pinkish gray, with indistinct radiating ridges; lobes 0.3–1 mm across, closely appressed, convex or ascending, almost like *Cladonia* squamules; often sterile. Apothecia broad, flat to convex, lead-gray to reddish brown or mottled, with a persistent or disappearing margin; spores 8–14 × 4–5 μm. CHEMISTRY: Thallus PD–, K–, KC+ red, C+ pink (gyrophoric acid). HABITAT: On rocks and soil, sometimes peat. COMMENTS: Apothecia resemble those of *T. granulosa*, but the squamulose rather than granular to verrucose thallus distinguishes them. The dull surface and C+ red medulla makes *T. wallrothii* easily distinguishable from any *Cladonia*.

843. *Trapeliopsis granulosa* Cascades, Oregon ×5.8

844. *Trapeliopsis wallrothii* Coast Range, northern California ×2.3

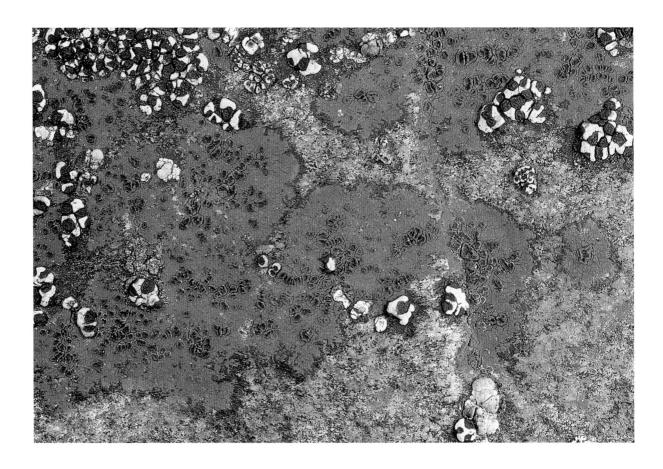

845. *Tremolecia atrata* interior Alaska ×4.9

Tremolecia (1 N. Am. species)

KEY TO SPECIES: See *Lecidea*.

Tremolecia atrata
Rusty-rock lichen

DESCRIPTION: Thallus dark rusty-orange, thin, smooth or cracked into irregular areoles, without soredia. Photobiont green (*Trebouxia?*). Apothecia pitch-black, concave, partly or entirely immersed in the thallus, mostly under 0.5 mm in diameter; margins black, thick; epihymenium greenish; paraphyses abundantly branched; hypothecium dark brown; exciple brown-black throughout; asci thickened at the apex but K/I– (*Tremolecia*-type); spores colorless, broadly ellipsoid, 11–15(–18) × 6–8(–10) μm, 1-celled, 8 per ascus. CHEMISTRY: Thallus and medulla PD–, K–, KC–, C– (no lichen substances or with traces of stictic acid). HABITAT: On siliceous rocks, especially those rich in iron, on exposed mountain ridges. COMMENTS: Many crustose lichens have rusty-orange thalli, and some have immersed apothecia as well. For example, ***Rhizocarpon oederi*** is very similar and grows in similar habitats. Its apothecia are convex, rough, and uneven, and its spores are pale brown and 4-celled or somewhat muriform. Several species of *Porpidia* also have rusty-orange thalli but have larger, flat to convex apothecia, halonate spores, and different asci. See also *Lecidea lapicida*.

Trypethelium (ca. 8 N. Am. species)
Speckled blister lichens

DESCRIPTION: Crustose lichens with thin thalli, usually embedded in bark and visible as a yellowish or brownish olive discoloration, but some species with a yellow- or orange-pigmented pruina or thin incrustation; with or without a black prothallus. Photobiont green (*Trentepohlia*). Perithecia solitary or grouped within warts (pseudostromata), partly emerging or entirely immersed with only the ostioles showing; walls of the warts or perithecia carbonized and black; paraphyses persistent, branched, anastomosing; spores

4–16-celled, walls unevenly thickened, producing lens-shaped locules. CHEMISTRY: Pigmented parts of thallus or fruiting warts sometimes K+ red-violet (anthraquinones); thallus sometimes UV+ yellow (lichexanthone). HABITAT: On bark, mostly in tropical or subtropical woodlands. COMMENTS: The species of *Trypethelium* that produce fruiting warts are easily distinguishable from most other lichens except **Bathelium**. The warts in that genus have brown walls composed of hyphae shaped like jigsaw puzzle pieces. Species of *Trypethelium* with solitary perithecia can resemble other perithecial lichens on bark, especially *Pyrenula*, which has similar (but brown) spores. Species of *Laurera* are similar and share the same habitats, but the perithecia are usually solitary or almost so, and the spores are muriform and very large (90–270 × 20–40 μm). See Comments under *T. tropicum*.

KEY TO SPECIES (including *Laurera*, *Astrothelium*, and *Pseudopyrenula*) (based largely on Harris, *More Florida Lichens*, 1995). With the exception of *T. virens* (East Temperate), all found in tropical or subtropical areas, especially southeastern coastal plain.

1. Perithecia buried in a wart-like pseudostroma or in thallus 2
1. Perithecia discrete, not buried in a pseudostroma or in thallus. 9
2. Spores thin-walled, muriform, 200–270 × 30–40 μm, 4 per ascus . . . . *Laurera megasperma*
2. Spores thick-walled, unevenly thickened, muriform or only transversely septate, mostly less than 75 × 30 μm; spores 8 per ascus. 3
3. Spores 6–14-celled 4
3. Spores 4-celled 5
4. Medulla of pseudostromata yellow to tan, K+ purple (anthraquinones); spores 10–14-celled, 40–50 × 9–12 μm . . . . [*Trypethelium eluteriae*]
4. Medulla of pseudostromata not pigmented, K–; spores 6–12-celled, 38–52 × 7–10 μm . . . . . . *Trypethelium virens*
5. Perithecia grouped two or three together sharing a single ostiole, pruinose (at least around the ostiole), dusted with yellow or orange anthraquinones (K+ purple); spores 28–35 × 10–13 μm . . . . . [*Astrothelium versicolor*]
5. Perithecia not sharing an ostiole and not pruinose; anthraquinones absent or present; spores 18–28 × 6–10 μm 6
6. Perithecia superficial; thallus whitish, without a cortex; hymenium containing many oil drops; spore locules with angular walls . . . . [*Pseudopyrenula diluta* (syn. *P. subnudata*)]
6. Perithecia entirely immersed in pseudostromata with only the ostioles visible; thallus dark brownish to olive, with a cortex or below the bark cells; spore locules lens-shaped. 7
7. Pseudostromata brown . . . . . [*Bathelium carolinianum*]
7. Pseudostromata thallus-colored or black 8
8. Thallus orange, K+ purple . . . . [*Trypethelium aeneum*]
8. Thallus olive to pale yellowish olive, K– . . . . [*Trypethelium variolosum* (syn. *T. ochroleucum*)]
9.(1) Perithecia conical; thallus light brown to pale orange; excipulum not distinguishable from involucrellum; perithecial cavity colorless; spore locules lens-shaped; spores with a gelatinous halo . . . *Trypethelium tropicum*
9. Perithecia hemispherical; thallus whitish; excipulum distinct from involucrellum, at least at base; perithecial cavity yellowish; spore locules with angular walls; spores without a halo . . . . [*Pseudopyrenula diluta*]

Trypethelium aeneum

DESCRIPTION: Thallus yellow-orange to orange, or predominantly green in shade forms (as shown in plate 846); perithecia almost entirely immersed, solitary and scattered or clustered; perithecial walls carbonaceous, forming a complete envelope; hymenium often inspersed with oil; spores 4-celled, 21–28 × 7.5–9 μm, halonate. CHEMISTRY: Thallus K+ purple, UV+ orange (anthraquinones). HABITAT: On hardwood bark in sand scrub. COMMENTS: *Trypethelium* species with pigmented, K+ purple material in the fruiting warts (not on the thallus surface) include *T. eluteriae* (10–14-celled spores, with yellow to tan powder in the warts) and *T. subeluteriae* (13–16-celled spores, with orange pigment in the warts). See also *Pyrenula cruenta*.

Trypethelium tropicum

DESCRIPTION: Thallus within the bark, creating a yellowish olive stain; perithecia black, not covered with thallus tissue, 0.3–0.5 mm in diameter, solitary or somewhat clustered and coalescing at the bases, but not collected into

846. *Trypethelium aeneum* central Florida ×5.9

847. *Trypethelium tropicum* Gulf Coast, Florida ×5.6

fruiting warts; ostioles sometimes whitish; hymenium filled with tiny oil drops; spores 4-celled, 20–27 × 6–8 µm, halonate. CHEMISTRY: Thallus UV–, K– (no lichen substances). HABITAT: On hardwood bark. COMMENTS: Most species of *Trypethelium* have perithecia clustered into warts that are at least partly covered with thallus. This species is the exception and therefore looks more like a *Pyrenula* than a *Trypethelium* but with colorless spores. **Bathelium carolinianum,** with its sometimes solitary perithecia, somewhat resembles *T. tropicum*. The perithecia of *Bathelium* species, however, have brown walls and are most often clustered, coalescing into low, brownish pseudostromata when mature. *Bathelium carolinianum* (long known under the name, **Trypethelium mastoideum,** which actually refers to a rare African species) is a common, coastal plain lichen with 4-celled spores, 18–28 × 6–9 µm.

Trypethelium virens

DESCRIPTION: Thallus within the bark, creating a yellowish brown or olive stain; perithecia clustered into low warts or blisters (pseudostromata), almost entirely covered with bark thallus tissue but showing through as aggregations of black dots (the ostioles); pseudostromata are usually rather shallow and spread out, but can be raised, as shown in plate 848, especially when growing on holly; spores 8–12-celled, 38–52 × 7–10 µm, IKI+ pale violet, halo absent or extremely thin. CHEMISTRY: Thallus UV–; medulla of pseudostroma K– (no lichen substances). HABITAT: On the bark of living trees with green inner bark tissues, such as holly, beech, young oaks, and tropical species, perhaps at least partially parasitic on the bark. COMMENTS: This is the most widely distributed species of *Trypethelium*. At the northern edge of its range, it usually lacks pseudostromata, having only scattered or clustered black pycnidia, but it still produces characteristic oval, olive to brownish patches on beech trunks. Some species of *Pyrenula* produce similar patches, but their perithecia are almost always scattered, and the spores are brown.

Tuckermannopsis (7 N. Am. species)
Wrinkle-lichens, ruffle lichens

DESCRIPTION: Medium-sized brown to olive foliose lichens, with lobes 1–4 mm wide, often ascending and ruffled, with sparse rhizines but sometimes with marginal cilia; pseudocyphellae sparse or absent. Photobiont green (*Trebouxia*). Apothecia lecanorine, with shiny brown disks, produced on the underside of reflexed lobe margins, or laminal in a few species; spores spherical, ca. 4–5 μm in diameter; pycnidia on the lobe margins, generally black and prominent; conidia dumbbell-shaped, 4–6 μm long. CHEMISTRY: Cortex usually lacking substances (K–), but atranorin present in a few species (K+ yellow); medulla often with orcinol depsides or depsidones (UV+ white, KC+ pink, C+ pink). HABITAT: Mostly on bark and branches, especially of conifers or birch, rarely on rock. COMMENTS: In size and color, species of *Tuckermannopsis* resemble *Melanelia*, but species of *Melanelia* are generally more closely appressed, have laminal apothecia, often have conspicuous pseudocyphellae, and produce ellipsoid spores. Until recently, *Tuckermannopsis* was included within *Cetraria*. Similar species of *Cetraria* differ in details of the asci and thallus anatomy, and often contain fatty acids.

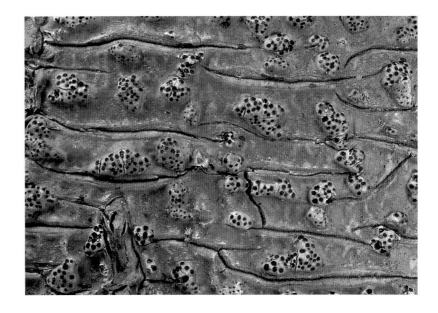

848. *Trypethelium virens* coastal plain, North Carolina ×3.0

KEY TO SPECIES

1. Soredia produced along the lobe margins............ .... *Tuckermannopsis chlorophylla*
1. Soredia absent 2

2. Lobes long and strap-shaped, flat; dichotomously branched, forming shrubby clumps; linear-elongate pseudocyphellae sometimes present along the margins .... *Tuckermannopsis subalpina*
2. Lobes rounded or somewhat elongated, flat to crinkled and crisped; branching usually irregular; pseudocyphellae present or absent, never linear-elongate 3

3. Isidia present on lobe margins and sometimes upper surface, cylindrical to spherical; apothecia absent; pycnidia sparse or very inconspicuous; lobes elongated ... . *Tuckermannopsis coralligera*
3. Isidia absent, although marginal lobules or warty tubercles sometimes produced; apothecia abundant; pycnidia abundant and conspicuous; lobes rounded 4

4. Lobes flat 5
4. Lobes undulating or crisped at the margins 6

5. Thallus brown, at least when dry, brown to olive when wet; pseudocyphellae sparse; apothecia on or close to the lobe margins .... *Tuckermannopsis sepincola*
5. Thallus olive when dry, grass-green when wet; pseudocyphellae abundant and conspicuous; apothecia on the lobe surface .... *Tuckermannopsis fendleri*

6. Warty tubercles and (or) lobules frequent on the lobe surface and margins; pseudocyphellae abundant and conspicuous, especially on the tubercles; medulla (especially in the apothecial margin) usually yellow or orange in spots .... *Tuckermannopsis platyphylla*
6. Warty tubercles absent; lobules present or absent; pseudocyphellae absent; medulla white throughout .. 7

7. Medulla C–, KC–, UV– (protolichesterinic acid present); lobules often present on the lobe margins .... *Tuckermannopsis orbata*
7. Medulla C+ pink or KC+ pink to red (protolichesterinic acid absent); lobules absent 8

8. Medulla C–, KC+ pink, UV+ blue-white (alectoronic acid) .... *Tuckermannopsis americana*
8. Medulla C+ pink, KC+ red, UV– (olivetoric acid) .... [*Tuckermannopsis ciliaris*]

849. *Tuckermannopsis americana* coastal Maine ×3.8

Tuckermannopsis americana
(syn. *Cetraria halei, C. ciliaris* var. *halei*)
Fringed wrinkle-lichen

DESCRIPTION: Lobes ascending, 3–6 mm across, dark brown to olive-brown, often (but not always) with rather long, brown, unbranched or forked cilia on the margins; lower surface pale to dark brown, coarsely wrinkled, with sparse, mostly undivided rhizines. Apothecia common, up to 7 mm in diameter; pycnidia prominent, barrel-shaped, black, predominantly marginal but in part laminal. CHEMISTRY: Cortex K– or K+ yellow (atranorin); medulla PD–, K–, KC+ pink, C–, UV+ blue-white (alectoronic acid). HABITAT: On twigs and branches, mostly of conifers and birch, in exposed sites. COMMENTS: Three chemically distinct but outwardly very similar species are separated in the key. *Tuckermannopsis ciliaris* contains olivetoric acid and is basically temperate in distribution, with an Appalachian–Great Lakes range except for a disjunction in northern Alaska. It is more abundant than *T. americana* in the southern part of its range. *Tuckermannopsis orbata* is also very similar to *T. americana*. It contains fatty acids, normally lacks cilia, and has a tendency to be more finely divided or lobulate. See also *T. platyphylla*.

Tuckermannopsis chlorophylla
(syn. *Cetraria chlorophylla*)
Powdered wrinkle-lichen

DESCRIPTION: Thallus chocolate-brown to greenish brown, paler in the shade; lobes divided and branched, concave and crisped, 0.7–2.5 mm wide; soredia produced in soralia all along the lobe margins and sometimes in irregular patches on the surface, fine powdery, white to greenish but flecked with black; lower surface somewhat paler than upper surface, shiny, with sparse, short, pale rhizines. Apothecia rarely present. CHEMISTRY: Medulla PD–, K–, KC–, C–, UV– (protolichesterinic acid). HABITAT: On bark and wood, rarely rock, in sun or shade. COMMENTS: This very common, brown, foliose lichen of the western conifer forests is the only *Tuckermannopsis* with soredia. Some particularly brown specimens of *Nephroma parile* can be confused with *T. chlorophylla*, but that species has blue-green cyanobacteria rather than grass-green algae, and the soredia are much coarser. The only camouflage lichens (*Melanelia* and *Neofuscelia*) with marginal soredia are *Melanelia albertana*, which has a C+ red medulla, and *M. culbersonii*, which is K+ red.

Tuckermannopsis coralligera
(syn. *Cetraria coralligera*)
Coral-edged wrinkle-lichen

DESCRIPTION: Thallus narrow-lobed (0.4–0.7 mm), flat and adnate, shiny olive or brown, with spherical to cylindrical isidia (becoming branched in well-developed specimens) along the margins and then on the lobe surface; lower surface dark olive to brown with predominantly marginal rhizines, almost like cilia. Without apothecia. CHEMISTRY: All reactions negative (protolichesterinic acid). HABITAT: Mostly on wood of conifers, occasionally on bark. COMMENTS: *Tuckermannopsis orbata* can sometimes appear to be marginally isidiate, but the isidia are more like tiny lobules, not cylindrical, and the lobes are more ascending and broader. Most similar are the isidiate species of *Melanelia*, such as *M. elegantula* or *M. exasperatula*. These have broader lobes, mostly over 1 mm wide, and the isidia are not initially marginal. *Tuckermannopsis fendleri* is the fertile, nonisidiate sister species of *T. coralligera*.

850. *Tuckermannopsis chlorophylla* Cascades, Oregon ×2.8

851. *Tuckermannopsis coralligera* (damp) Chiricahua Mountains, southeastern Arizona ×2.8

Tuckermannopsis fendleri (syn. *Cetraria fendleri*)
Dwarf wrinkle-lichen

DESCRIPTION: Thallus dull to shiny greenish brown or olive, becoming green when wet; lobes finely divided, 0.5–1.6 mm wide, sometimes producing tiny lobules along the margins, but without isidia or soredia; dot-like pseudocyphellae common along margins; lower surface pale brown or almost white, with scattered unbranched rhizines. Apothecia abundant, laminal, with prominent, warty or toothed margins; pycnidia usually along the lobe margins. CHEMISTRY: All reactions negative (protolichesterinic acid). HABITAT: On conifer bark, especially small branches, sometimes on wood. COMMENTS: *Tuckermannopsis sepincola* is similar in size and also produces apothecia profusely. Its thallus, however, is clearly yellow-brown to red-brown and does not turn green when wet. In addition, its lobes are broader (1–2 mm) and more ascending. Other species of *Tuckermannopsis* have marginal apothecia originating on the lower lobe surface. A similar

852. *Tuckermannopsis fendleri* (damp) Cape Cod, Massachusetts ×5.0

853. *Tuckermannopsis orbata* (western population) Coast Range, northern California ×2.0

southwestern species, *"Cetraria" weberi*, has somewhat broader, rounder lobes and has a C+ red medulla (olivetoric acid).

Tuckermannopsis orbata (syn. *Cetraria orbata*)
Variable wrinkle-lichen

DESCRIPTION: Thallus olive-brown to pale green (shade forms), or brown, with ascending or appressed, ruffled lobes, 1.5–3.5 mm wide, the margins bearing prominent black pycnidia, flattened and branched lobules, or brown to black cilia, or various combinations, including none of the above; soredia and true isidia absent, although the lobules, especially when branched and bearing pycnidia, often appear isidia-like; lower surface pale brown, strongly wrinkled, sometimes with scattered, pale rhizines. Apothecia very common and often abundant, originating on the lower or upper sides of the lobe margins, or on the lobe surface; pycnidia can be laminal as well as marginal. CHEMISTRY: Medulla PD–, K–, KC–, C– (protolichesterinic acid). HABITAT: On branches and twigs of conifers and birch, rarely other hardwoods. COMMENTS: The only character that does not seem to vary in this species is the chemistry. Specimens lacking lobules can be distinguished from *T. ciliaris* and *T. americana* only by chemistry. Eastern populations are more frequently ciliate or lobulate than western populations, but there are all intergradations. The eastern specimen shown in plate 854 is a particularly pale shade form. See also *T. platyphylla*.

Tuckermannopsis platyphylla (syn. *Cetraria platyphylla*)
Broad wrinkle-lichen

DESCRIPTION: Thallus dark reddish brown, rarely yellowish brown (never with olive caste when dry); lobes broad and rounded, (3–)5–10 mm wide; upper surface irregular and rough, often wrinkled, sometimes with short cylindrical or flattened, wart-like tubercles and lobules (also on the lobe margins), giving the thallus an isidiate appearance; tubercles often with white pseudocyphellae at the summit; cilia absent; medulla predominantly white but with scattered yellow to orange spots, especially in the apothecial margin; lower surface almost as dark as the upper surface, strongly wrinkled, with sharply raised, reticulate ridges; rhizines few, scattered. Apothecia frequently produced but not abundant, marginal to laminal, originating on the upper surface of the lobe margins; pycnidia marginal or laminal. CHEMISTRY: Medulla PD–, K–, KC–, C– in white areas, but K+ orange or yellow in spots, and KC+ orange where the medulla is yellow, most reliably revealed in the apothecial margin (pigment unknown; also contains lichesterinic and protolichesterinic acids and sometimes atranorin). HABITAT: Fairly common on conifers in western interior forests. COMMENTS: The lobes of *T. platyphylla* are considerably broader than those of other species in the genus. It has lobed tubercles like those of *T. orbata*, but they are on the upper surface rather than the margins, and the lobe surface is rougher and more wrinkled. The dark lower surface is unlike most other species of *Tuckermannopsis*, and the presence of a yellow medullary pigment is unique in the genus.

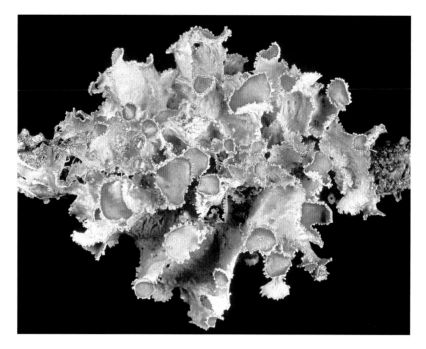

854. *Tuckermannopsis orbata* (eastern population, shade form) Appalachians, Tennessee ×3.5

855. *Tuckermannopsis platyphylla* Coast Range, Oregon ×4.2

the lobes, rarely conspicuous and sometimes absent; margins sometimes ciliate or toothed with black, tiny projections; lower surface paler than upper surface. Apothecia common, on the lobe margins. CHEMISTRY: All reactions negative (protolichesterinic and lichesterinic acids). HABITAT: On twigs and branches of subalpine shrubs and heath, occasionally on the ground among mosses. COMMENTS: Although classified as a species of *Tuckermannopsis* because of its spherical spores and asci, this lichen most closely resembles a species of *Cetraria* in its general appearance, differing mainly in the substrate preference, the flat rather than channeled lobes, and the paucity of conspicuous pseudocyphellae.

Umbilicaria (30 N. Am. species)
Rock tripes

DESCRIPTION: Umbilicate foliose lichens (and one rare crustose-squamulose species), very dark to pale brown or gray, shiny or dull, some species developing a crusty ("crystalline") white deposit of material resembling rock salt (actually consisting of dead cells mixed with gelatinous material sloughed off from the upper cortex); usually growing as discrete, rounded thalli, 1–30 cm in diameter, attached by a single, central holdfast showing as a bump on the upper surface (the umbo), but some species growing in crowded clumps (polyphyllous, literally "many-leaved"); rarely isidiate or sorediate. Photobiont green (*Trebouxia*). Lower surface smooth, wrinkled, with plates of tissue, or covered with granular to smooth, unbranched or branched rhizines or analogous outgrowths. Apothecia basically lecideine, but various species distinguished by sterile ridges and fissures growing through the hymenium, creating concentric or radiating patterns on the disks, or disks smooth, without such sterile tissues; spores 1-celled and colorless to brown and muriform, mostly 8 per ascus. CHEMISTRY: Medulla commonly containing gyrophoric acid (PD–, K–, KC+ red, C+ pink), rarely with other compounds including norstictic and stictic acids (PD+ yellow to orange, K+ red or yellow). HABITAT: On siliceous rocks such as granite. COMMENTS: The genus *Lasallia* is most similar to *Umbilicaria* but has pustulate thalli and only 1 or 2 very large, muriform spores per ascus. There are many other umbilicate lichens that are entirely unrelated to *Umbilicaria* and have very different fruiting bodies. For example, *Dermatocarpon* has perithecia, *Rhizoplaca* and *Omphalora* have lecanorine apothecia, and ***Glyphole-***

856. *Tuckermannopsis sepincola* Lake Superior region, Ontario ×3.8

Tuckermannopsis sepincola (syn. *Cetraria sepincola*)
Chestnut wrinkle-lichen

DESCRIPTION: Thallus yellow- to red-brown, shiny, forming rounded cushions mostly under 2 cm across; with rounded, ascending lobes, 0.7–2 mm wide, without soredia, isidia, or lobules, but occasionally sparsely ciliate; lower surface pale brown, wrinkled, with few rhizines. Apothecia profuse, formed on or close to the upper side of the lobe margins; disks shiny red-brown with thick or thin, smooth margins that usually have white pseudocyphellae. CHEMISTRY: All reactions negative (protolichesterinic acid). HABITAT: On twigs and branches of a variety of trees in open woodlands and wetlands. COMMENTS: The pure brown color (when dry) and apothecia on the upper lobe surface distinguish *T. sepincola* from most other species of *Tuckermannopsis*. See Comments under *T. fendleri*.

Tuckermannopsis subalpina (syn. *Cetraria subalpina*)
Arboreal Iceland lichen

DESCRIPTION: Thallus shades of yellowish, reddish, or greenish brown, usually pale, uniform, forming shrubby cushions of strap-shaped, dichotomously branching lobes, often curled and tangled, (0.5–)2–5 mm wide, elongate pseudocyphellae scattered along the edges of

857. *Tuckermannopsis subalpina* (wet) Cascades, Washington ×2.7

cia scabra, a rare lichen of the arid interior, has immersed apothecia and hundreds of tiny, spherical spores per ascus similar to those of other members of its family, the Acarosporaceae. Although many *Umbilicaria* species often occur without fruiting bodies, this is rarely the case for the other genera. In North America, isidia occur only in *U. deusta,* and there is only one sorediate species: *U. hirsuta.* The latter (described in the key) is rather rare, with scattered localities across North America including the Rocky Mountains, northern Alaska, southern Ontario, and especially the northern Appalachians. IMPORTANCE: The utility of species of *Umbilicaria* is evidenced by its many common names. Rock tripe, *tripes de roches,* and *iwatake* all refer to *Umbilicaria,* especially in its use as human food. Various species have been eaten by the Huran, Naskapi, Chipewyan, and Cree tribes, as well as by Inuit people, in times of emergency or famine. Stories of pilots surviving on rock tripe after crashing in the north woods are not uncommon. Members of the ill-fated Franklin expedition of 1820–21 survived the winter, in part, by eating *tripes de roche,* which they gathered each day as long as they had strength. We have mentioned earlier the use of a Japanese species,

U. esculenta (*iwatake*), as a commercial delicacy (see Importance under *U. muehlenbergii,* and Chapter 10). Because almost all species of *Umbilicaria* contain large amounts of gyrophoric acid, and because the lichen is fairly easy to collect, rock tripes have been favored by North American craftspersons as a source of fermentation dyes. It produces a very beautiful purple that, unfortunately, fades with exposure to light but can be set with mordants.

KEY TO SPECIES (including *Lasallia*)
1. Lobes pustulate or blistered, with depressions on the lower surface corresponding to pustules on the upper surface...2
1. Lobes not pustulate: smooth and even, wrinkled, bumpy or with a network of depressions and sharp ridges; if warty, the warts do not have corresponding depressions on the lower surface 4

2. Rhizines sparse; lobes usually crowded and overlapping, convex or undulating and crisped at the margins .....
..........................*Umbilicaria caroliniana*
2. Rhizines absent; lobes usually solitary and flat.......3

3. Lower surface black and coarsely papillate like very rough sandpaper; upper surface rather smooth, with low pustules having sloping sides . . . . *Lasallia pensylvanica*

3. Lower surface pale to dark brown, rather smooth or slightly roughened; upper surface very rough, with abundant pustules having almost vertical sides . . . . *Lasallia papulosa*

4.(1) Rhizines present, sparse or abundant 5

4. Rhizines absent 13

5. Lower surface black; upper surface smooth and even. 6

5. Lower surface pale to dark brown, pink, or gray; upper surface rough 10

6. Thallus with crowded, strongly folded, overlapping lobes; rhizines sparse, thick and irregular; southern Appalachians and Alaska . . . . . *Umbilicaria caroliniana*

6. Thallus relatively round and flat; rhizines abundant 7

7. Thallus upper surface dull, never shiny; rhizines developing out of a granular lower surface and forming a velvety mat 8

7. Thallus upper surface shiny; lower surface not granular; rhizines not forming a short, velvety mat 9

8. Thallus mostly brown, thin and membranous or moderately thick, rather fragile . . . *Umbilicaria mammulata*

8. Thallus mostly gray, thick and stiff like cardboard, rather tough . . . . *Umbilicaria americana*

9. Lower surface covered with irregular plates of tissue with short, highly branched rhizines developing from the plate margins; rhizines confined to lower surface; apothecia abundant, flat, usually angular, with concentric ridges; western montane . . [*Umbilicaria angulata*]

9. Lower surface rough, but plates confined to area immediately around attachment; rhizines unbranched or forked, developing from the lower cortex (not plates), often forming clumps on the upper surface; apothecia rare, convex, round, with radiating ridges; western North America and Newfoundland . . . . [*Umbilicaria polyrrhiza*]

10.(5) Thallus dull gray to gray-brown, 2–8 cm across, with coarse, granular soredia developing from the breakdown of the upper cortex close to the downturned margins; lower surface rough, with abundant, tapered, unbranched or forked rhizines; apothecia very rare; mainly northeastern, scattered elsewhere . . . . [*Umbilicaria hirsuta*]

10. Thallus gray or brown, surface without soredia, but covered with crystal-like deposits, at least close to the umbilicus; apothecia often present 11

11. Lower surface gray; rhizines unbranched, smooth; apothecial disks convex with thick, concentric ridges. . . . . *Umbilicaria proboscidea*

11. Lower surface pink to pale brown; rhizines dichotomously branched; apothecial disks more or less flat, with or without concentric ridges 12

12. Marginal cilia or cilia-like rhizines absent, but rhizines abundant on lower surface; apothecia broadly attached and adnate; apothecial disks with central buttons of sterile tissue or smooth and even; cortex KC+ red, C+ pink, although the reactions are sometimes faint and difficult to see (gyrophoric acid) . . . . *Umbilicaria virginis*

12. Marginal cilia or cilia-like rhizines common and abundant; nonmarginal rhizines sparse; apothecia constricted at the base and raised; apothecial disks with concentric ridges of sterile tissue; cortex KC–, C– (gyrophoric acid absent) . . . . *Umbilicaria cylindrica*

13.(4) Isidia present on thallus surface, mostly granular . . . . *Umbilicaria deusta*

13. Isidia absent 14

14. Upper surface of thallus pruinose, or with coarse, crystal-like deposits. 15

14. Upper surface of thallus entirely without pruina or deposits 18

15. Lower surface pruinose, at least close to margins . . . 16

15. Lower surface not pruinose 17

16. Apothecia adnate; apothecial disks with concentric ridges of sterile tissue; thallus thin to moderately thick . . . . *Umbilicaria proboscidea*

16. Apothecia raised; apothecial disks with central buttons of sterile tissue; thallus thick and stiff . . . . *Umbilicaria krascheninnikovii*

17. Upper surface bumpy but without a reticulate pattern of ridges; apothecia abundant, raised, with smooth and even disks; medulla KC–, C– . . . . . . *Umbilicaria rigida*

17. Upper surface with a network of depressions and sharp ridges, or wrinkled; apothecia not common, adnate, with disks having central buttons of sterile tissue; medulla KC+ red, C+ pink (gyrophoric acid) . . . . *Umbilicaria decussata*

18.(14) Lower surface with a network of rough and papillate membranes 19

18. Lower surface smooth, papillate, or tuberculate, without membranes. 20

19. Margins of the thallus perforated with irregular holes and forming finely divided lobes; apothecial disks with concentric ridges; medulla PD+ orange, K+ yellow (stictic acid) . . . . *Umbilicaria torrefacta*

19. Margins of the thallus not perforated and lobed; apothecial disks with radiating ridges; medulla PD–, K–. . . . . *Umbilicaria muehlenbergii*
20. Thallus round and flat; upper surface at least partly covered with convex warts or areoles over a smooth, black, basal layer. . . . . *Umbilicaria hyperborea*
20. Thallus forming cushions of crowded, overlapping lobes; upper surface smooth, not covered with warts or areoles 21
21. Lower surface black; lobe margins turned up; apothecia rare. . . . . *Umbilicaria polyphylla*
21. Lower surface pale or dark brown; lobe margins turned down; apothecia abundant . . . . *Umbilicaria phaea*

Umbilicaria americana
Frosted rock tripe

DESCRIPTION: Thallus pale gray or brownish gray, usually with a coarse white pruina over the surface, thick and rather stiff, 2–7 (–12) cm in diameter; lower surface covered with a velvet-like nap of closely packed, unbranched or forked, black rhizines, each one coated with a layer of black granules (Fig. 35b). Apothecia uncommon, convex, with disks having concentric ridges or, in old apothecia, a complex of concentric and radiating ridges. CHEMISTRY: Medulla PD–, K–, KC+ red, C+ red (gyrophoric acid). HABITAT: On granitic, steep rock faces, usually in relatively protected or partially shaded sites. COMMENTS: This is one of the thickest, most cardboard-like rock tripes. It has been only recently separated from *U. vellea*, a rarer, mainly arctic-alpine species, with pale, branched, naked rhizines scattered on the lower surface, interspersed with stubby rhizines with black granules (Fig. 35a). It is smaller than *U. americana* (mostly 3–8 cm across). Both species can produce thin plates of tissue on the lower surface, but they are more common and extensive in *U. vellea*. The black, microscopic granules covering the rhizines of *U. americana*, *U. vellea*, and a few related species can become dispersed and act as propagules for the lichen. *Umbilicaria hirsuta* also has a pale brownish gray thallus, but it produces soredia close to the margins and has a pale brown lower surface.

Umbilicaria caroliniana
Folded rock tripe

DESCRIPTION: Thallus smooth, shiny brown, 2.5–10 cm across, with folded, crowded, convex lobes mostly 20–40 mm in diameter, sometimes weakly pustulate, with the central holdfast sometimes hard to locate; lower surface coal-black, very rough (granular to minutely papillate), with scattered, long, black, smooth to wrinkled rhizines. Apothecia rare, with few concentric ridges on the disks. CHEMISTRY: Medulla PD–, K–, KC+ red, C+ red (gyrophoric acid). HABITAT: On exposed rocks. COMMENTS: The crowded, folded lobes and scattered black rhizines of this species distinguish it from *U. mammulata*. The geographic distribution of *U. caroliniana* is a classic example of a disjunction, with no known records between the southern Appalachian Mountains at high elevations and the region encompassing the northern Yukon Territory and the Brooks Range area of Alaska. These two areas are considered to be Ice Age refugia and are known for relict populations of species wiped out elsewhere in North America during the most recent continental glaciation.

Umbilicaria cylindrica
Fringed rock tripe

DESCRIPTION: Thallus light to dark brown beneath a crusty white deposit, making the dominant color gray to white, thick and stiff, usually under 4 cm in diameter, polyphyllous or deeply divided into lobes, the margins sometimes grading into black, branched cilia; lower surface smooth to minutely granular, pink to pale brown with sparse to abundant, pale brown to black, branched rhizines, almost all marginal, adding to the fringe of black cilia surrounding the thallus. Apothecia abundant, slightly stalked, with thick, concentric ridges on the disks. CHEMISTRY: PD–, K–, KC–, C– (no lichen substances). A chemical race containing norstictic acid (PD+ yellow, K+ red) has been reported from Norway. HABITAT: On exposed rocks. COMMENTS: This extraordinary little rock tripe with a ciliate margin bears some resemblance to *U. virginis* because of its pale lower surface and many long, branched rhizines. The

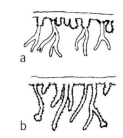

Fig. 35 The rhizines of (a) *U. vellea* compared with those of (b) *Umbilicaria americana*. Scale = 1 mm. (Drawing by Alexander Mikulin. Reprinted by permission from McCune and Goward, 1995, p. 170.)

858. *Umbilicaria americana* central Idaho ×2.4

apothecia of *U. virginis*, however, are flat and almost entirely unadorned, except for central sterile knobs or "buttons," and the thallus is not fringed with cilia, although rhizines can sometimes extend beyond the margin.

Umbilicaria decussata
Netted rock tripe

DESCRIPTION: Thallus more or less round, mostly 1–5 cm in diameter, thick and stiff, dark gray to brownish gray, with a coarse "crystalline" deposit; upper surface covered with a network of high, sharp ridges, continuing to the thallus edge (although lower there); lower surface smooth, sooty black, paler close to the thallus margin, sometimes with radiating folds close to the holdfast, but without rhizines. Apothecia rare, disks with a central knob or series of ridges. CHEMISTRY: Medulla PD–, K–, KC+ red, C+ red (gyrophoric acid). HABITAT: Arctic and alpine habitats on rock. COMMENTS: *Umbilicaria decussata* is very similar to *U. lyngei*, which has a thinner,

859. *Umbilicaria caroliniana* (damp) Appalachians, Tennessee ×1.2

860. *Umbilicaria cylindrica* interior Alaska ×4.2

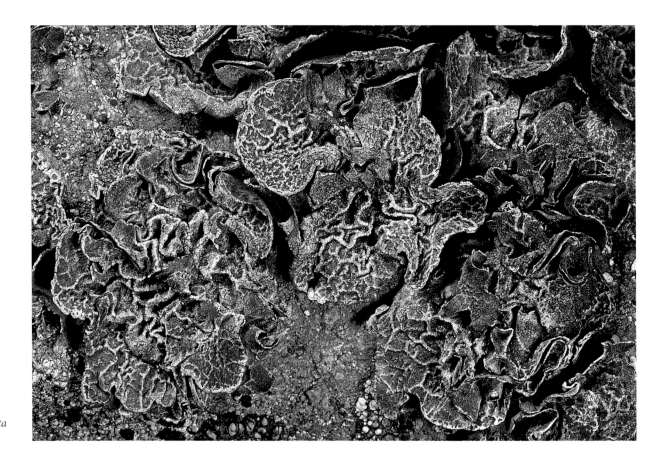

861. *Umbilicaria decussata* Coast Mountains, British Columbia ×3.0

862. *Umbilicaria deusta* coastal Maine ×1.9

more pliant thallus with reticulate ridges on the upper side that disappear toward the margin. The apothecia of *U. lyngei* have smooth disks. *Umbilicaria proboscidea* also has a network of pruinose ridges, but the lower surface is pale to dark brown, never sooty black. See also *U. krascheninnikovii*.

Umbilicaria deusta
Peppered rock tripe

DESCRIPTION: Thallus thin and fragile, single, round, or becoming irregularly lobed and even polyphyllous (with crowded, separate lobes), 1–5(–9) cm in diameter, dark brown to almost black, the upper surface more or less covered with tiny black, granular isidia; lower surface dark brown to black, smooth or sometimes dimpled with small pits, without rhizines. Apothecia extremely rare. CHEMISTRY: Medulla PD–, K–, KC+ red, C+ red (gyrophoric acid). HABITAT: On exposed boulders and outcrops. COMMENTS: The presence of granular isidia makes this species unique among the rock tripes. Young

863. *Umbilicaria hyperborea* mountains, northwestern British Columbia ×3.7

specimens that lack isidia, however, can resemble *U. polyphylla*. Both species can be polyphyllous and are smooth below.

Umbilicaria hyperborea
Blistered rock tripe

DESCRIPTION: Thallus medium to dark brown, margins even to deeply lobed, upper surface slightly uneven to very strongly verrucose, with convex, curved, or worm-like bumps crowded together or sometimes separated, leaving smooth, black areas between, not at all patterned; thallus margins and occasionally upper surface becoming lobulate, but not isidiate; lower surface smooth (or obscurely roughened), without depressions associated with the pustules or verrucae on the upper surface, pale to dark brown, or black, normally without rhizines. Apothecia common, flat to convex, closely appressed or slightly raised, with a system of very prominent, complex ridges on the disks. CHEMISTRY: Medulla PD–, K–, KC+ red, C+ red (gyrophoric and often umbilicaric acids). HABITAT: Very common on exposed rocks in arctic or alpine sites. COMMENTS: The most important characteristics of *U. hyperborea* are the smooth undersurface, lack of rhizines (although a variety with scattered rhizines is known), and very uneven, hilly (not sharply ridged) upper surface, with apothecial disks that are extremely complex. It is similar to the much rarer *U. arctica*, an arctic to northern boreal species, which has a predominantly gray to tan lower surface and higher, more pronounced hills and ridges on the upper surface. *Umbilicaria proboscidea*, often associated with *U. hyperborea*, has a radiating network of sharp ridges. The relatively rare but widespread *U. nylanderiana* has a sooty black lower surface with microscopic black granules.

Umbilicaria krascheninnikovii
Salty rock tripe

DESCRIPTION: Thallus black to dark gray-brown, thick and stiff, mostly 1.5–4 cm in diameter, heavily encrusted with a salt-like white deposit, with high, sharp ridges over most of the thallus; lower surface smooth, pale brown to pink, often grad-

864. *Umbilicaria krascheninnikovii* Olympics, Washington ×6.3

ing to mouse-gray and pruinose at the margins, with no rhizines. Apothecia slightly stalked, with prominent black margins, the disk smooth and flat except for one prominent, and other less prominent, irregular fissures or ridges in the center of the disk. CHEMISTRY: Medulla mostly PD–, K–, KC–, C–, but KC+ red, C+ red close to the algal layer and especially in and around the apothecia and pycnidia (gyrophoric acid). HABITAT: Exposed rock in alpine sites. COMMENTS: Compared with the similar *U. proboscidea*, *U. krascheninnikovii* is a smaller lichen (*U. proboscidea* can be 10 cm across), and the apothecia in *U. proboscidea* have disks that are entirely composed of concentric or intricately bent, deep fissures and ridges. *Umbilicaria lyngei* and *U. decussata* are also similar but have a sooty black undersurface. *Umbilicaria virginis* is a closely related species with abundant rhizines.

Umbilicaria mammulata
Smooth rock tripe

DESCRIPTION: Thallus smooth and even but not shiny, thin to moderately thick, sometimes folded close to the margin, typically 4–15(–30) cm in diameter, warm red-brown to grayish brown, rarely pruinose around the umbo; lower surface pitch-black, usually chinky to papillate, but occasionally fairly smooth between the rhizines, which almost cover the lower surface with a velvet-like nap; rhizines brown to black, generally branched, either very short and covered with black granules, or fairly long and tapering, largely free of granules; overlapping plates of tissue sometimes occur immediately around the central holdfast. Apothecia rarely seen. CHEMISTRY: Medulla PD–, K–, KC+ red, C+ red, often only close to the algal layer (gyrophoric acid). HABITAT: On boulders and steep rock walls in forests and sometimes in the open, frequently close to lakeshores. COMMENTS: The rare pruinose specimens can be confused with *U. americana*, with which it frequently occurs. The latter has a much thicker, cardboard-like thallus and is almost entirely gray. Another common neighbor, *U. muehlenbergii*, has a somewhat shiny thallus and an entirely different kind of lower surface, and it is frequently fertile. *Umbilicaria mammulata* is among the largest lichens in the world in terms of mass and diameter. A specimen from the Smoky Mountains of Tennessee measured 63 cm (over 2 ft) across! A cliff covered with dinner plate–sized thalli is a spectacular but not unusual sight where the air is humid and pollution levels are low.

Umbilicaria muehlenbergii
Plated rock tripe

DESCRIPTION: Thallus dark brown to grayish brown, surface satiny and almost shiny, never pruinose, smooth to verruculose, pitted with shallow depressions especially when fertile, 6–15(–20) cm in diameter, often folded and undulating; lower surface pale brown to black, roughened with sharp papillae, covered with a network of overlapping plates parallel with the surface, without true rhizines, although the plates often narrow into short, slender filaments at their edges. Apothecia common, without margins, 1–4(–6) mm in diameter, consisting of radiating, branched ridges split by fissures, usually sunken into depressions in the thallus. CHEMISTRY: Medulla PD–, K–, KC+ red, C+ red (gyrophoric acid). HABITAT: On boulders and steep rock walls in forests and in the open. COMMENTS: Sterile specimens can be distinguished from *U. mammulata* by the lower surface. *Umbilicaria polyrrhiza,* a western montane species,

has the same kind of marginless, radiating apothecia, but its lower surface has abundant, long, forked rhizines, some of which project out at the margins and even through holes in the upper thallus surface. A network of plates on the lower surface can be seen in a number of rock tripes. *Umbilicaria muehlenbergii* has only plates (no rhizines) below, and they are usually pale to dark brown, not black; ***U. cinereorufescens*** is a smaller, grayer species from Alaska and northern Yukon (rare in Arizona) with lobed and folded thalli, and short, distinctly ball-tipped, unbranched rhizines as well as thick and lumpy plates of tissue on the lower surface; *U. torrefacta* has plates that are restricted mainly to the area close to the central holdfast, and the lower surface is pale brown; ***U. angulata,*** a western montane species, has thick black plates mixed with branched black rhizines. IMPORTANCE: The Woods Cree (Nihitahawak) people of southeastern Saskatchewan added pieces of *U. muehlenbergii* to fish broth to make it into a thick soup. It was thought to be nutritious for sick people because it did not upset the stomach.

865. *Umbilicaria mammulata* central Massachusetts ×2.1

866. *Umbilicaria muehlenbergii* North Woods, Maine ×0.96

867. *Umbilicaria phaea*
southern Idaho ×4.6

Umbilicaria phaea
Emery rock tripe

DESCRIPTION: Thallus smooth, rather shiny, brown, single or lobed, rarely becoming polyphyllous, up to 6 cm across; lower surface pale, rough and papillate (like emery cloth), only rarely producing rhizines. Apothecia common, partially embedded, small, usually angular or even star-shaped, partially embedded, with disks having concentric ridges. CHEMISTRY: Medulla PD–, K–, KC+ red, C+ red (gyrophoric acid). HABITAT: On exposed rocks mainly in hot, arid sites. COMMENTS: This is one of the most common western rock tripes. It can approach *U. polyphylla* in form but never develops such a crowd of lobes and generally has a paler lower surface. A variety of *U. phaea* (**var. coccinea**), found from northern California to eastern Washington, becomes red-orange because of a heavy dusting of anthraquinone pigments (cortex K+ deep red-purple). *Umbilicaria hyperborea* has a more verrucose upper surface and rounder, superficial (not embedded) apothecia. *Umbilicaria angulata* is similar to *U. phaea* with regard to its upper surface and apothecia, but it has abundant rhizines. Its thallus is brown and satiny, smooth or cracked into broad areoles or sometimes folded. The lower surface of *U. angulata* has abundantly branched, relatively slender black rhizines mixed with thick black plates of tissue. It is an alpine to subalpine species occurring from California to coastal British Columbia.

Umbilicaria polyphylla
Petaled rock tripe

DESCRIPTION: Thallus very dark brown to almost black, 2–6 cm in diameter, polyphyllous, producing a pile of ascending, often narrow lobes (like flower petals); upper surface smooth, often with a satiny shine; lower surface uniformly sooty black, smooth or minutely papillate, without rhizines or plates. Apothecia very rare. CHEMISTRY: Medulla PD–, K–, KC+ red, C+ red (gyrophoric and umbilicaric acids). HABITAT: On exposed rocks, mostly arctic-alpine. COMMENTS: The smooth, polyphyllous thallus and black lower surface make this lichen very distinctive. See also *U. phaea*.

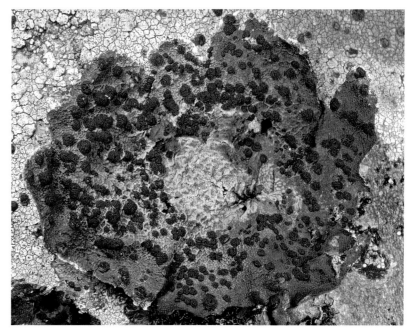

Umbilicaria proboscidea
Netted rock tripe

DESCRIPTION: Thallus gray-brown to very dark brown, usually round, thin and fragile, 3–6(–10) cm across; upper surface with a reticulate network of ridges, prominent and sharp close to the umbo and fading to low, convex ridges at the edges, often coarsely pruinose or with crystal-like deposits close to the umbo; lower surface pale tan in center, dark gray at edge, sometimes mottled or pruinose; rhizines extremely rare, and, if present, sparse. Apothecia common, round, becoming convex, with thick, concentric ridges on the disk. CHEMISTRY: Medulla PD–, K–, KC+ red, C+ red (gyrophoric acid); rarely PD+ yellow, K+ red, KC+ red, C+ red (with norstictic acid as well). HABITAT: On exposed rocks in alpine or boreal to arctic sites. COMMENTS: The most similar, common rock tripe is *U. krascheninnikovii*, a smaller species more uniformly pale on the lower surface, with a flat, smooth to "buttoned" apothecial disk. See also *U. decussata*.

Umbilicaria rigida
Roughened rock tripe

DESCRIPTION: Thallus dark gray or brownish black, single or divided into lobes, sometimes polyphyllous, 2–5 cm in diameter; stiff and rigid, but not especially thick; upper surface extremely rough, chinky-areolate, often with a thick deposit of crystal-like material, especially close to the umbo, where sharp ridges can develop, although not in a reticulate pattern; lower surface minutely textured or papillate, varying from pale tan to dark mouse-gray; rhizines absent. Apothecia very common, raised on short stalks, with flat, smooth disks and prominent margins. CHEMISTRY: Medulla PD–, K–, KC–, C– (usually with no lichen substances, but norstictic acid has been reported). HABITAT: Arctic-alpine on exposed rocks, especially those frequented by birds. COMMENTS: The very rough, unpatterned upper surface and stalked, smooth apothecia characterize this common western species. ***Umbilicaria lyngei*** has similar apothecia, but the upper surface is ridged in a dis-

868. *Umbilicaria polyphylla* Klamath Range, northwestern California ×2.9

869. *Umbilicaria proboscidea* interior Alaska ×2.0

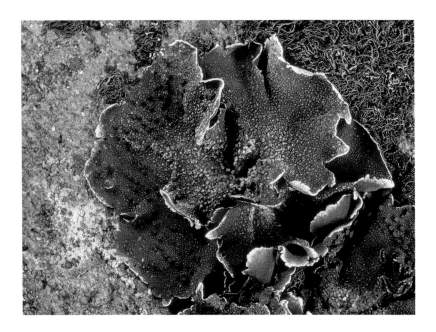

870. *Umbilicaria rigida* Olympics, Washington ×2.5

871. *Umbilicaria torrefacta* mountains, northwestern British Columbia ×2.5

tinctly reticulate pattern. It is basically an arctic species also known from the mountains of Oregon, Washington, and British Columbia.

Umbilicaria torrefacta
Punctured rock tripe

DESCRIPTION: Thallus medium to very dark brown, often shiny, ca. 2–6 cm in diameter, irregular in shape with thin areas breaking through and causing perforations at the periphery and a tattered appearance; upper surface cracked into areoles, some rather convex, with black, smooth areas between; lower surface pale brown, smooth to rough papillate, with a network of plates, some of which become fringed and resemble rhizines. Apothecia common, 0.5–2 mm in diameter, flat to convex, with concentric ridges on the disk. CHEMISTRY: Typically medulla PD+ orange, K+ yellow, KC–, C– (stictic acid); or, especially in the arctic, PD–, K–, KC+ red, C+ pink (gyrophoric acid). HABITAT: On exposed rocks in arctic and alpine sites. COMMENTS: *Umbilicaria torrefacta* is the only North American rock tripe that regularly produces β-orcinol depsidones (i.e., stictic acid), although norstictic acid apparently occurs in some European populations of *U. cylindrica*. The chemistry and perforate thalli, together with plates on the lower surface, characterize the species. *Umbilicaria hyperborea* has a similar upper surface and can even become slightly perforate, but its lower surface is mostly smooth. *Umbilicaria angulata,* which can have plates close to the holdfast, usually produces branched, cylindrical rhizines as well.

Umbilicaria virginis
Blushing rock tripe

DESCRIPTION: Thallus very dark gray to brownish gray, coarsely pruinose or encrusted with crystal-like material, at least close to the umbo, dull elsewhere, mostly 1.5–4 cm in diameter, but reported up to 10 cm, sharply ridged in the center fading to convex, irregular hills closer to the margins; lower surface smooth to minutely roughened, pale buff or pink except at the margins, which can be brown, with abundant, long, slender, pale to dark, sparsely branched rhizines especially close to the margins. Apothecia common, 0.6–2(–4) mm in diameter, flat, with prominent black margins and smooth disks having a single sterile ring or button (occasionally several) in the center. CHEMISTRY: Cortex PD–, K–, KC+ pink to red, C+ vaguely pink to red (difficult to discern in dark thalli); medulla PD–, K–, KC–, C–, or weakly KC+ pink, C+ pink (gyrophoric acid); rarely PD+ yellow, K+ red, KC–, C– (norstictic acid). HABITAT: On exposed rocks in the arctic tundra or on alpine ridges. COMMENTS: The pinkish lower surface and abundant rhizines are the distinguishing features of this very common, western montane species. *Umbilicaria krascheninnikovii* is superficially similar but has no rhizines. See also *U. cylindrica*.

Usnea (79 N. Am. species)
Beard lichens, old man's beard

DESCRIPTION: Pendent or shrubby, filamentous, fruticose lichens; yellowish green (usnic-yellow) or darkening, occasionally reddish because of pigment in the cortex; attachment area sometimes blackened for a millimeter or two at the base of the thallus. Branches round or angular in cross section and characterized by a central, cartilagenous cord (axis) of supporting tissue (Figs. 8b and 36); cortex thick, thin, or falling away, often fissured around the circumference of the branch (circular cracks), giving the thallus a segmented appearance (Fig. 37e); medulla thin or thick, loose and cottony (lax) or dense, white or pigmented yellow to dark red; axis slender or relatively broad, solid or rarely hollow (in some tropical species). Surface of branches smooth, or uneven because of tiny, translucent bumps called papillae (thickenings of the cortex), warts called tubercles (containing medullary tissue), or spiny perpendicular side branches called fibrils (containing an axis) (Fig. 37). Soralia common, with rounded mounds of soredia, or more or less depressed (excavate) with relatively sparse soredia; isidia very frequent, developing within soralia and mixed with the soredia, or occurring directly on the branch surface. (Tiny, isidia-like spinules called isidiomorphs are not distinguished from true isidia here.) Photobiont green (*Trebouxia?*). Apothecia lecanorine, with broad, flat, yellow to brown, often pruinose disks and thin, sometimes spiny margins; spores colorless, ellipsoid, small, 8 per ascus. CHEMISTRY: Cortex KC+ dark yellow (usnic acid) in all species; medulla with a varied

872. *Umbilicaria virginis* Rocky Mountains, Colorado ×3.9

873. *Usnea*-covered fir tree, north shore of Lake Superior, Ontario

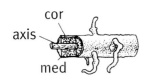
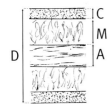

Fig. 36. (a) The structure of an *Usnea* filament showing the cortex (*cor*), medulla (*med*), and slightly stretched central axis (*axis*). (Reproduced courtesy of the Canadian Museum of Nature, from Brodo, 1988, fig. 76b.). (b) An *Usnea* filament in side view indicating how the cortex, medulla, and axis are measured. *A*, axis; *M*, medulla; *C*, cortex; *D*, diameter of entire filament. A/D ×100 = axis percent, M/D ×100 = medulla percent; C/D ×100 = cortex percent. (Drawing by I. Brodo.)

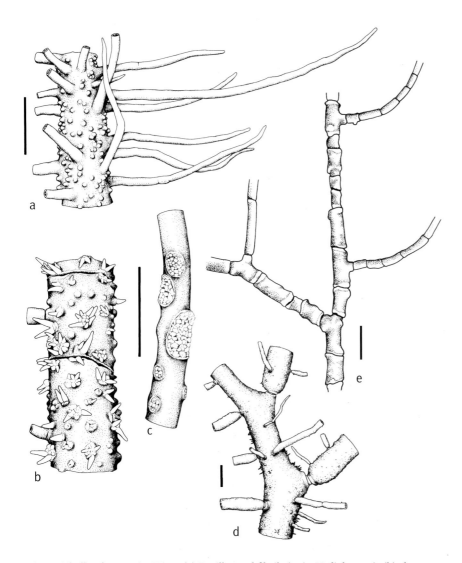

Fig. 37. Thallus features in *Usnea*. (a) Papillae and fibrils (as in *U. diplotypus*); (b) clusters of isidia arising from small soralia (*U. rubicunda*); (c) soralia (*U. glabrescens*); (d) inflated branches constricted at the base (as in *U. fragilescens*); (e) bone-like articulations (*U. trichodea*). (a–d reproduced courtesy of the British Lichen Society from Purvis et al., 1992, fig. 26.)

chemistry including many β-orcinol depsides and depsidones such as barbatic, diffractaic, squamatic, salazinic, norstictic, and stictic acids. HABITAT: On trees and shrubs, and a few species on rock. COMMENTS: *Alectoria*, *Ramalina*, and *Evernia* are also yellowish green, hair-like, bushy, or pendent lichen genera, but none of these have the central cord that characterizes *Usnea* (see Fig. 8). *Alectoria sarmentosa*, in particular, is often mistaken for species of *Usnea*. Naming specimens of *Usnea* is often difficult because of the morphological and chemical variation that occurs in many species. Papillae can be abundant or almost absent in some species; soralia are sometimes mixed with isidia or the isidia can be very sparse; normally pendent species are shrubby when they are young, and shrubby species can become somewhat pendent (subpendent) when exceptionally well developed. Lichen compounds are often so sparse as to give ambiguous spot test results. An important character for the purpose of species identification is the relative thickness of the cortex, medulla, and central axis, often expressed as a percentage of the branch diameter, as we have done in the descriptions that follow (see Fig. 36b). (Note that one must double the percentage of the cortex and medulla, then add these figures to the percentage of the axis, to arrive at 100 percent. This is not really necessary because, in practice, it is normal to compare the thickness of the cortex and medulla on only one side of the branch to the width of the entire axis.) The type of soralium is important. Is it mixed with isidia? If containing only soredia, does it form a round mound, or is it excavate, lying in a depression that is shallow or perhaps deep enough to reach the axis? *Usnea* taxonomy is still in a state of flux, and much remains to be done before we have a clear understanding of species limits. IMPORTANCE: *Usnea* is one of the most commercially important lichens because of its production of usnic acid, an effective antibiotic against certain bacteria. Species of *Usnea* are used in Chinese medicine, contemporary homeopathic medicine, and traditional medicine in the Pacific Islands, New Zealand, and every continent except Australia. Thousands of kilograms of *Usnea* are collected and imported each year by pharmaceutical firms, especially in central Europe, for extracting usnic acid. The extracts are used in ointments for superficial skin infections and are incorporated into cosmetics such as deodorants for the control of bacteria. On the other hand, because usnic acid causes serious allergic reactions in many people, its use in medicines has not been entirely without risk.

Some species of *Usnea* are used in dye-making, and in Mexico, it is used to brew beer. *Usnea* species are also important to wildlife as both food (especially for deer) and nesting material (northern parula warblers and other birds).

KEY TO SPECIES

1. Thallus pendent..................................2
1. Thallus growing in bushy tufts from a single point, or almost pendent..................................11

2. Surface of main branches rough and eroded, with the cortex thin and crumbling; main branches almost undivided, but with short to moderately long perpendicular side branches abundant all along the filaments ....*Usnea longissima*
2. Surface of main branches with a continuous cortex, with depressions and ridges, or with tubercles or papillae, not at all eroded..............................3

3. Axis pink to brown; medulla white or reddish, usually CK+ orange (diffractaic acid)....................4
3. Axis and medulla white 5

4. Medulla usually pink to red, dense or loose, 20–30%; older branches with abundant whitish warts; cortex very thick and hard..................*Usnea ceratina*
4. Medulla white, dense, 16–21%; whitish warts absent; cortex thin ....*Usnea trichodea*

5. Isidia present 6
5. Isidia absent 9

6. In Florida; isidia usually developing within soralia on branch tips; medulla K+ red (norstictic and galbinic acids).............................*Usnea dimorpha*
6. Boreal and western, rarely in the Appalachians; isidia not developing within soralia, although sometimes in small clusters; medulla K– or K+ red (but norstictic acid absent) 7

7. Branches smooth, without papillae, with circular, thickened cracks (Fig. 37e), at least at the base; medulla PD+ red, K–, KC+ pink (protocetraric acid)............... ....*Usnea hesperina*
7. Branches with papillae, lacking circular, thickened cracks; medulla PD– or PD+ yellow, K– or K+ red, KC– (lacking protocetraric acid) 8

8. Surface often dented or pitted; thallus more or less uniform in color; cortex very thin and fragile (cortex 7–10%, medulla 22–26%, axis 28–38%); medulla PD–, K–, or PD + yellow, K+ red (salazinic acid) ....*Usnea scabrata*

8. Surface more or less even, not pitted or dented; main branches dark, especially at the base; cortex moderately thick and tough (cortex 7–10%, medulla 14–21%, axis 36–43%); medulla PD+ yellow, K+ red (salazinic acid) ....*Usnea filipendula*

9.(5) Medulla loose and webby (lax). Branches dented and pitted, not divided into segments by circular cracks ... .*Usnea cavernosa*
9. Medulla thin and compact........................10

10. Branches strongly ridged, often broken into small segments, but not bone-like (with swollen, circular cracks); medulla PD+ yellow, K+ red, KC–, CK– (norstictic acid); rare, Appalachian–Great Lakes distribution.........................[*Usnea angulata*]
10. Branches terete, not dented, pitted, or ridged, with swollen, circular cracks and bone-like segments (Fig. 37e), at least at the base; medulla PD+ red, K–, KC+ pink, CK + orange (protocetraric and diffractaic acids) ....*Usnea hesperina*

11.(1) Thallus with apothecia 12
11. Thallus without apothecia 15

12. In central to eastern North America; medulla rather dense..13
12. In southwestern U.S.; medulla loose (lax), with white cottony hyphae 14

13. Medulla usually red, at least in part, or sometimes white; base of thallus not blackened; medulla K– or K+ red (norstictic acid); papillae, if present, wide at the base; very common lichen ....*Usnea strigosa*
13. Medulla white; base of thallus blackened; medulla K+ red (salazinic acid); papillae usually present, cylindrical; not a common lichen ....[*Usnea subfusca*]

14. Spiny fibrils abundant and crowded on the back of the apothecia; base of thallus narrowly or irregularly blackened or not blackened at all; medulla K–, PD– (salazinic acid absent) ....*Usnea cirrosa*
14. Spiny fibrils very sparse on the back of the apothecia; base of thallus extensively blackened; medulla usually K+ red, PD+ orange or yellow (salazinic acid) ....*Usnea arizonica*

15.(11) On rock 16
15. On trees or wood 17

16. Branches spotted with black bands, coalescing into more or less blackened tips; round soralia at the branch tips, lacking isidia; northern arctic with rare occurrences in the mountains of Washington and Oregon.. ..... [*Usnea sphacelata* (syn. *Neuropogon sulphureus*)]
16. Branches uniform in color, without black bands; isidia developing in soralia; temperate...*Usnea amblyoclada*

17. Branches clearly constricted at base and usually the axils, giving the branches an inflated appearance (Fig. 37d)..18
17. Branches not constricted at base, more or less uniform in diameter..................................20
18. Isidia absent; soredia in excavate soralia mainly at the branch tips....................... *Usnea glabrata*
18. Both isidia and soredia developing within soralia, concentrated at the branch tips or not 19
19. Soralia scattered over branches, remaining discrete; thallus black at the base for 1–2 mm . *Usnea fragilescens*
19. Soralia concentrated at branch tips, becoming confluent; thallus usually more or less uniform in color, but sometimes darkening at the base . . . *Usnea cornuta*
20.(17) Thallus cortex distinctly reddish to red-brown or orange, sometimes mottled......... *Usnea rubicunda*
20. Thallus cortex shades of yellowish green (usnic-yellow), not reddish 21
21. Axis and medulla pink or red (rarely white in *U. ceratina*)...22
21. Axis and medulla white 24
22. Axis broad and hollow; medulla PD+ yellow, K+ red (norstictic acid) .... *Usnea baileyi*
22. Axis solid throughout; medulla PD–, K– (norstictic acid absent)...................................23
23. Conspicuous whitish warts or tubercles present on branch surface; medulla thick, loose or dense; contains diffractaic acid .... *Usnea ceratina*
23. Whitish warts absent; medulla thin and dense; lacking diffractaic acid .... *Usnea mutabilis*
24.(21) Isidia absent, soralia abundant................25
24. Isidia present, either directly on the branches or in soralia 26
25. Soralia excavate, wider than half the branch diameter; medulla K– or K+ red (salazinic acid); base of thallus pale or blackened .... *Usnea lapponica*
25. Soralia level with branch surface or mounded, round and relatively small; medulla usually K+ yellow (stictic acid complex); base of thallus blackened............
.... [*Usnea glabrescens*]
26. Branches ridged, at least partly angular in cross section; isidia present, not arising from soralia; cortex thin and fragile............................ *Usnea hirta*
26. Branches not ridged, uniformly round in cross section; isidiate soralia present or absent; cortex thick or moderate, not fragile 27
27. Thallus not blackened at base; branches rather smooth, with few or inconspicuous papillae; cortex much thicker than medulla and very hard (cortex 18–22%, medulla 6–10%, axis 40–58%); medulla PD+ red, KC+ pink to red (protocetraric acid) .... *Usnea subscabrosa*
27. Thallus blackened at the base; main branches bumpy with abundant papillae; cortex thinner than medulla (cortex 7–14%, medulla 13–25%, axis 33–45%); medulla PD–, or PD+ orange or yellow, KC– (protocetraric acid absent)...28
28. Isidia scattered or clustered, abundant or sparse, not normally arising from soralia; branches relatively soft and pliable .... *Usnea filipendula*
28. Isidia usually clustered in round soralia (together with soredia); branches relatively stiff.................29
29. Main branches more or less equally dichotomous, tips not twisted or contorted; medulla usually K–, PD–, UV+ (squamatic acid), sometimes K+ dark yellow, PD+ orange, UV– (thamnolic acid), salazinic acid absent........................ [*Usnea subfloridana*]
29. Main branches in unequal dichotomies, tips twisted or contorted; medulla K+ red, PD+ yellow to orange (salazinic acid)................ *Usnea diplotypus* group

Usnea amblyoclada
Rock beard lichen

DESCRIPTION: Thallus usually shrubby and compact, 2–10 cm long; branches with many circular cracks, round or angular in section, sometimes dented or ridged, often with fibrils but without papillae or tubercles; soralia small and round, less than half the branch diameter, superficial, often containing both soredia and isidia, but isidia also occurring by themselves, scattered on the branch (not clustered); cortex thin (4.5–9 percent); medulla thick and dense (17–29 percent); axis thick (32–50 percent). Apothecia unknown. CHEMISTRY: Medulla PD+ yellow, K+ yellow turning red, KC–, C– (norstictic, salazinic, and galbinic acids). HABITAT: On siliceous rocks, rarely on bark. COMMENTS: The only other *Usnea* commonly collected on rock is the Appalachian species ***Usnea halei,*** a shrubby to almost pendent lichen with larger, rounded soralia containing abundant isidia. Although chemically similar, it lacks galbinic acid. Elements of both *U. amblyoclada* and *U. halei* have been called ***U. herrei*** in the past. The primarily corticolous ***U. dasaea,*** a species from the southern U.S. and the Appalachians, frequently grows on rock and is chemically identical to *U. amblyoclada*. It is distinguished by its dense, spinule-like fibrils, clustered, thinner isidia

never tipped with black, and larger, convex soralia that are broader than half the branch diameter.

Usnea arizonica
Western bushy beard

DESCRIPTION: Thallus bushy, stiff, 3–10 cm long, main branches covered with short, perpendicular, spiny side branches and fibrils of different lengths, strongly papillate, especially on the older parts of the branches, round in section or somewhat dented and ridged, without isidia or soredia; cortex thick and tough (11–13 percent); medulla lax (14–26 percent); axis fairly broad (30–45 percent). Apothecia abundant, at the tips of the branches, with broad yellowish, pruinose disks, 3–10 mm in diameter, and spiny margins; fibrils relatively sparse on the back of the apothecia. CHEMISTRY: Medulla PD+ yellow, K+ yellow slowly turning reddish (salazinic acid), occasionally PD–, K–; the reactions are often faint and slow. HABITAT: Common on trees. COMMENTS: This is one of a group of abundantly fruiting, spiny beard lichens and is the most common western representative. Another southwestern species in this group (also rarely in British Columbia) is *U. rigida,* which is PD+ red-orange, K– (protocetraric acid). *Usnea strigosa* (with a red or white medulla and usually norstictic, psoromic, or fumarprotocetraric acids) and *U. subfusca* (with a white, usually dense medulla and containing salazinic acid) are similar lichens in eastern North America. In the southeast, others occur that have a different chemistry. *Usnea cirrosa* is very similar, but the fibrils are about equal in length and the backs of the apothecia have more abundant, crowded, spiny fibrils. The medulla is entirely PD–, K– (usnic acid alone) in populations north of Mexico.

Usnea baileyi (syn. *U. antillarum, U. implicita*)
Hollow beard lichen

DESCRIPTION: Thallus shrubby to almost pendent, very stiff; branches strongly tapering, tuberculate-papillate, producing isidia or isidiate soralia in places; cor-

874. *Usnea amblyoclada* Sonoran Desert, southern Arizona ×1.3

875. *Usnea arizonica* Coast Range, California ×1.5

876. *Usnea baileyi* northern Florida ×1.2

877. *Usnea cavernosa* mountains, southeastern Arizona ×0.93

tex moderate (ca. 6 percent); medulla usually red, extremely thin and dense (ca. 5–6 percent); axis almost filling the branch (ca. 80 percent) and hollow, much like the stereome in *Cladonia*. CHEMISTRY: Medulla PD+ yellow, K+ red, KC–, C– (norstictic acid). HABITAT: On exposed trees and shrubs. COMMENTS: The only other beard lichen in North America with a hollow axis and thin, reddish medulla is ***U. perplectata**,* also a subtropical species found in Florida. It reacts K– and contains diffractaic acid. Its isidiate branches can be noticeably ridged much like those of *U. hirta,* with the isidia occurring along the ridges.

Usnea cavernosa
Pitted beard lichen

DESCRIPTION: Thallus long and pendent, 10–40 cm long, branching in dichotomies with few perpendicular side branches; main branches constricted at nodes and conspicuously dented and ridged; most branches under 0.3 mm wide, but some main branches reaching 1.6 mm; without any papillae, isidia, or soredia; cortex thin (ca. 7 percent); medulla broad and fairly lax (26–30 percent), about the same width as the axis (30–34 percent). CHEMISTRY: Medulla PD+ yellow or PD–, K+ yellow becoming red (sometimes faint) (salazinic acid) or K–, KC–, C– (usnic acid alone). HABITAT: Hanging from tree branches in well-lit boreal forests, especially on spruce. COMMENTS: Very few pendent species of *Usnea* have no papillae, soredia, or isidia. *Usnea trichodea* is most similar but has a brownish axis and, although its branches are somewhat angular in places, they do not have conspicuous dents and pits. Some forms of *U. scabrata* have ridges and shallow depressions, but they are always papillate and generally produce at least some isidia. See also *U. hesperina*. IMPORTANCE: This species was used by the Wylackie people of California for tanning leather. They would take "brains" and wrap it in the lichen to make a brick-like form that was rubbed and "crumbled into the hide."

Usnea ceratina
Warty beard lichen

DESCRIPTION: Thallus coarse, shrubby to scraggly and pendent, 6–20 cm long, with abundant wart-like tubercles covering the larger branches; tubercles becoming white at the top as they age, giving the lichen a white-spotted appearance, but they do not break down into true soralia; branches heavily papillate; frequently with clusters of isidia or isidiate soralia, or bearing thorn-like, pointed fibrils; cortex thick and tough, ((6–)9–13 percent); medulla thick, lax or dense (20–30 percent), usually pink to almost red, rarely white; axis broad (23–33 percent), also pinkish. Apothecia not known. CHEMISTRY: Medulla PD–, K–, KC–, C– (diffractaic acid). HABITAT: On conifers and shrubs in humid, open forests. COMMENTS: With its whitish tubercles and pink medulla, this is one of the more easily identified beard lichens. Specimens from west of the Rockies tend to be larger than the eastern ones. California populations, which have very small cells in the cortex, are sometimes called *U. californica*.

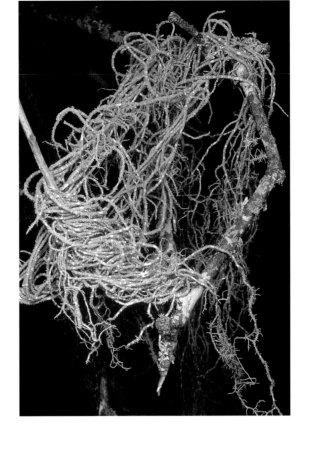

878. *Usnea ceratina* Coast Range, northern California ×1.0

Usnea cirrosa
Sundew beard lichen

DESCRIPTION: Thallus forming short, dense clumps, mostly under 4 cm long, branching from the base, each branch terminating in a large, pale apothecium fringed with spiny fibrils that continue on the back of the apothecium; branches with abundant stout, spine-like, strongly tapered fibrils that are approximately the same length all along the branch; branches papillate or smooth between the fibrils; cortex tough (7–9 percent); medulla white, lax or fairly dense (17–20 percent); axis broad (39–50 percent). CHEMISTRY: All reactions negative (usnic alone), although salazinic acid has been reported in a few Mexican specimens. HABITAT: On exposed trees. COMMENTS: This *Usnea* has the superficial appearance of *U. arizonica*, but it is even spinier with very abundant, tapering fibrils all along the branches. In *U. arizonica*, the backs of the apothecia have few fibrils, the stems are much more branched, and papillae are more abundant and prominent. *Usnea strigosa* and its chemical variants often have a red medulla and are strictly eastern.

Usnea cornuta
Inflated beard lichen

DESCRIPTION: Thallus small, shrubby, with a pale to slightly blackened base; branches constricted at the base, giving them an inflated appearance (Fig. 37d); surface normally papillate, occasionally dented; soralia common, especially near the tips of the branches, coalescing, containing abundant isidia as well as soredia; fibrils common or sparse; cortex thin and shiny (4–9 percent); medulla white, broad and lax (27–40 percent); axis narrow or wide (12–44 percent). Apothecia rarely seen. CHEMISTRY: Several chemical races: (1) the most common, medulla and axis PD+ orange, K+ yellow to red, KC–, C– (salazinic acid, sometimes with stictic acid); (2) less common, medulla PD+ red, K+ yellow, KC+ pink, C– (protocetraric and thamnolic acids); (3) rarely, medulla PD+ yellow, K– or K+ yellow, KC–, C– (psoromic and 2'-O-demethylpsoromic acids). HABITAT: On trees (especially conifers) and shrubs in open, humid forests.

879. *Usnea cirrosa* Chisos Mountains, southwestern Texas ×2.4

880. *Usnea cornuta* Appalachians, Tennessee ×1.8

COMMENTS: This species belongs to a group of beard lichens with inflated and constricted branches having a lax medulla and narrow central axis. Most similar is ***U. fragilescens* var. *mollis*,** which has smaller, more discrete soralia and produces them on secondary as well as terminal branches. It differs chemically (see Comments under *U. fragilescens*). ***Usnea wirthii*** is a smaller member of this group from the Pacific Northwest. It has a yellow medulla and axis and usually has red spots on the cortex. It contains norstictic, stictic, or psoromic acid. *Usnea glabrata*, also small, has large hemispherical soralia that contain no isidia. It usually has protocetraric acid, often with other compounds (but not salazinic acid). ***Usnea subfloridana*,** a common boreal species that has a similar habit and size, has a dense medulla and no basal constrictions, and it has a clearly blackened base.

881. *Usnea dimorpha* central Florida ×2.8

882. *Usnea diplotypus* (Well-developed papillae can be seen on the branches in the upper left, and clusters of divergent isidia arising from soralia occur on many branches. Short perpendicular branches [fibrils] are abundant everywhere.) Lake Superior region, Ontario ×3.0

Usnea dimorpha
Powder-tipped beard lichen

DESCRIPTION: Thallus robust, long, and pendent, up to 25 cm long, with many short, perpendicular fibrils; main branches somewhat dented and irregularly flattened in places, with many circular cracks that divide the long branches into bone-like segments (Fig. 37e); surface almost entirely smooth, with low and inconspicuous papillae confined mainly to the largest branches; abundant isidiate soralia on the branch tips, almost dissolving the cortex of the older fine branches into granules; cortex thin (7–8 percent); medulla white, broad, cottony (25–32 percent); axis slender (18–28 percent). CHEMISTRY: Medulla PD+ orange, K+ blood red, KC–, C– (norstictic and galbinic acids). HABITAT: On exposed trees, especially oaks. COMMENTS: *Usnea longissima*, a more northern species, has no soredia or isidia and reacts negatively with all reagents. *Usnea trichodea* has more bone-like articulations and lacks the little side branches. Neither species has soredia or isidia. ***Usnea angulata*** is a sparsely orediate pendent beard lichen with norstictic acid that may key out to *U. dimorpha*. It is rare, from the Appalachian–Great Lakes region. Its branches are prominently ridged and angular throughout, the medulla is very thin and dense, and the axis is broad and yellowish to brownish.

Usnea diplotypus
Ragged beard lichen

DESCRIPTION: Thallus shrubby to almost pendent, yellowish green, with or without a blackened base; main branches 1–1.5 mm wide, branching in unequal dichotomies, with branch tips often twisted and irregular; perpendicular side branches and fibrils fairly abundant; short or cylindrical papillae, and sometimes warts, prominent and abundant, especially on the main branches, often forming low ridges that twist around the branches; isidiate soralia common; cortex fairly thin, 9–11 percent; medulla white, lax or dense, 15–25 percent; axis broad, 33–45 percent. Apothecia rare. CHEMISTRY: Medulla PD+ yellow, K+ red, KC–, C– (salazinic acid). HABITAT: On trees, mostly conifers. DISTRIBUTION: Our knowledge of *U. diplotypus*

883. *Usnea filipendula*
Lake Superior region,
Ontario ×3.0

Usnea filipendula (syn. *U. dasypoga*)
Fishbone beard lichen

DESCRIPTION: Thallus pendent, commonly 20 cm or more, with several slender main branches, 0.3–1 mm thick (often much darker than the side branches, although this is not evident in plate 883), and many side branches and fibrils; distinctly blackened at the base; surface abundantly papillate with rather tall, cylindrical papillae; abundant isidia arising in clusters or singly, or within very small soralia; cortex moderately thick (7–14 percent); medulla rather lax (occasionally dense) (14–21 percent); axis moderately thick (36–43 percent). CHEMISTRY: Medulla PD+ yellow, K+ red, KC–, C– (salazinic acid). HABITAT: A forest species, growing mainly on spruce. COMMENTS: *Usnea scabrata* is similar but has a thinner, brittle cortex, no soralia, is more straw-colored without darkened main branches, and is often K– (lacking salazinic acid). In addition, it often has ridged and dented branches, and its medulla is looser, about the same width as the axis. *Usnea madeirensis* is a shrubby to almost pendent species common along the very humid west coast. It has superficial isidiate soralia and salazinic acid like *U. filipendula* but has many circular cracks on the oldest branches. As in *U. subfloridana*, the base is black. See also Comments under *U. diplotypus*. IMPORTANCE: On the island of Sakhalin, in the Russian Far East, *U. filipendula* has been used in powdered form to treat wounds, and modern tests for antibacterial activity have been positive.

in North America is currently too incomplete for mapping, but the species appears to be widespread in the coniferous forest region from coast to coast, perhaps extending southward in the eastern mountains. COMMENTS: *Usnea diplotypus* resembles *U. filipendula* but is shrubbier, usually has a thicker medulla, and has thicker main branches (1–1.5 mm). Both are (or can be) abundantly isidiate, and both contain salazinic acid. *Usnea diplotypus* is most easily mistaken for *Usnea subfloridana*, a very common boreal species that always has a black base, branches in equal dichotomies, and differs in chemistry. Most North American specimens contain squamatic acid (PD–, K–, KC–, C–, UV+ blue-white), but some have thamnolic acid (PD+ yellow-orange, K+ deep yellow) or, rarely, other compounds including barbatic or baeomycesic acid. *Usnea subfloridana* is shrubby to almost pendent and has a thick cortex and rather cottony medulla.

Usnea fragilescens
Inflated beard lichen

DESCRIPTION: Thallus usually shrubby, with very divergent branches, 3–6 cm long, and a conspicuously blackened, broad base; branches appearing inflated, constricted at the points of attachment (Fig. 37d), smooth or abundantly papillate, with many perpendicular fibrils, abundantly to sparsely sorediate, with excavate to mounded soralia sometimes containing isidia; cortex shiny, thin (4–8 percent); medulla very broad and lax (25–39 percent); axis narrow (14–35 percent). CHEMISTRY: Medulla (some-

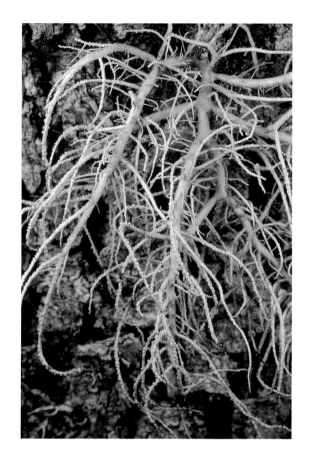
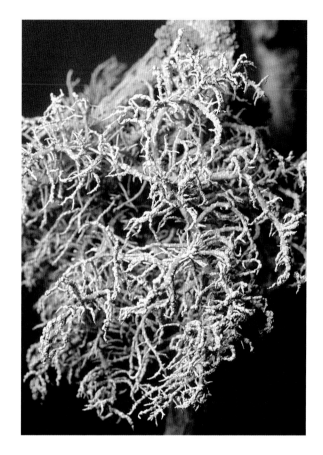

884. *Usnea fragilescens* northern California coast ×1.3

885. *Usnea glabrata* Coast Range, California ×3.6

times only close to the axis) PD+ yellow, K+ yellow to red, KC–, C– (stictic, norstictic, and constictic acids). HABITAT: On branches of exposed trees and shrubs in humid regions. COMMENTS: *Usnea subfloridana,* which also has a blackened base, has a narrower medulla and only slightly constricted branches. It usually contains squamatic acid in North America. *Usnea glabrata* is a purely sorediate species usually without papillae on the branches. It contains protocetraric acid (PD+ orange-red) or only usnic acid (PD–, K–). *Usnea cornuta* is particularly similar, but its soralia are concentrated on the branch tips, often coalescing, and it usually has salazinic acid. Almost all North American specimens of *U. fragilescens* represent **var. mollis,** which is more richly branched and more heavily isidiate than var. *fragilescens* and almost always occurs on tree bark; var. *fragilescens* is mostly on rock.

Usnea glabrata
Lustrous beard lichen

DESCRIPTION: Thallus forming short, shrubby tufts with a black base; branches inflated, constricted at the base, smooth (almost without papillae); without isi dia, but with abundant, tuberculate to excavate soralia (Fig. 37c); cortex thin, medulla lax and broad, and axis narrow. CHEMISTRY: Medulla PD+ red-orange, K–, KC– or KC+ pink, C– (fumarprotocetraric and (or) protocetraric acids), or negative (only usnic acid in the cortex). HABITAT: On branches and twigs of deciduous trees and conifers. DISTRIBUTION: Widely distributed in North America, boreal to temperate regions, but mapping the range is impossible at present. COMMENTS: Species in the *U. lapponica* and *U. glabrescens* groups also have shrubby thalli with true soralia (no isidia), but they do not have inflated or constricted branches and they generally react K+ yellow to red because of stictic, norstictic, or salazinic acid. *Usnea wirthii* is a northwestern beard lichen very much like *U. glabrata* in size and general appearance. It differs in the pale sulphur-yellow color of the central axis, the abundant papillae on the main branches, the occasional production of some isidia in the soralia, and the chemistry (PD+ yellow or orange, K+ yellow, usually turning red; norstictic acid, often accompanied or replaced by stictic acid, or rarely with psoromic acid). In addition, *U. wirthii* almost always is speckled with dark red spots on the branches.

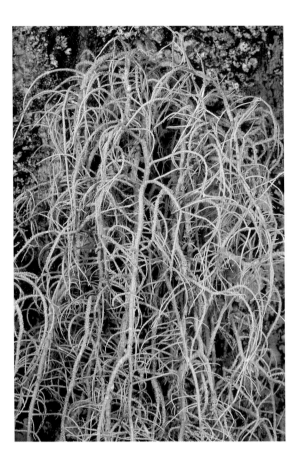

886. *Usnea hesperina* Appalachians, North Carolina ×1.8

887. *Usnea hirta* Lake Superior region, Ontario ×2.5

Usnea hesperina
Silken beard lichen

DESCRIPTION: Thallus pendent, up to 30 cm long, distinctly grayish green, not blackened at the base, with dichotomous branching mostly close to the base, and perpendicular side branches; branches smooth, with few or no papillae, but divided into bone-like segments by slightly swollen circular cracks, at least at the base; small isidiate soralia occasionally present, rarely abundant; cortex thick and shiny (8–17 percent); medulla white, thin, dense (6–16 percent); axis colorless, thick (43–64 percent). CHEMISTRY: Medulla PD+ red-orange, K–, KC+ pink, C– (protocetraric acid). HABITAT: On the branches of exposed hardwood and coniferous trees. COMMENTS: *Usnea trichodea* is similar in some respects (e.g., in the articulated branches and smooth surface), and the relatively rare form with a white axis might key out here; however, it is never isidiate and differs chemically. *Usnea chaetophora* and *U. merrillii,* which contain salazinic acid (PD+ yellow, K+ red, KC–), can also have articulated branches (especially *U. merrillii*). The medulla of both these species can be as thick as the axis and is always dense, and both species occasionally have isidiate soralia. In *U. chaetophora,* the basal main branches become dark reddish brown and often show black blotches. *Usnea chaetophora* is a northwestern species common along the coast, and *U. merrillii* is most common in the Appalachians, although some northwestern specimens closely resemble it. Other species with protocetraric acid such as *U. glabrata* and *U. cornuta* have shrubby, sorediate thalli; *U. rigida,* found in the southwestern United States and British Columbia, always has large apothecia.

Usnea hirta (syn. *U. variolosa*)
Bristly beard lichen, shaggy beard lichen

DESCRIPTION: Thallus forming short, densely branched, compact tufts rarely over 5 cm long, pale yellowish green, not blackened at the base; branches strongly ridged and angular in cross section, without papillae or soredia, but producing abundant isidia all along the branches; cortex thin (3–10 percent); medulla white, thick, and lax (23–30 percent); axis narrow (30–38 percent). CHEMISTRY: Medulla PD–, K–, KC–, C– (usnic acid alone or with fatty acids; rarely with diffractaic acid). Norstictic acid has been reported for the species, at least from Europe. HABITAT: *Usnea hirta* is most typically found on hard, weathered wood, but it also grows on coniferous trees and even rocks. COMMENTS: Among the small, shrubby species of *Usnea*, this is the only one that is completely isidiate and has ridged or angular branches.

888. *Usnea lapponica* (detail) Lake Superior region, Ontario ×5.6

Usnea lapponica (syn. *U. laricina*)
Powdered beard lichen

DESCRIPTION: Thallus forming small, shrubby tufts under 8 cm long, usually with few main branches and many perpendicular side branches and fibrils; branches round or dented, densely papillate, with deeply excavate soralia sometimes reaching the central axis, containing powdery soredia but no isidia; base usually pale; cortex thin or thick (4–11 percent); medulla white, cottony or dense, usually thick but variable (12–30 percent); axis also variable in thickness (27–60 percent). CHEMISTRY: Medulla PD–, K–, KC–, C– (usnic acid alone), or PD+ yellow, K+ red, KC–, C– (salazinic acid); rarely with other compounds. HABITAT: On branches of conifers, mainly in the boreal forest. COMMENTS: The deeply concave soralia and shrubby habit of this common beard lichen are its best identifying characteristics. *Usnea glabrescens* (including what has been called *U. fulvoreagens*) also has concave soralia, but the branching is more regularly dichotomous and the base is more frequently black. It contains various combinations of norstictic, salazinic, and stictic acids, with most specimens reacting K+ red. *Usnea substerilis* is very similar in habit, but the soralia are not as deeply excavate (rarely reaching the central cord) and often develop isidia. This species usually has salazinic acid, often with barbatic acid.

Usnea longissima
Methuselah's beard lichen

DESCRIPTION: Thallus pendent, extremely long (up to 3 m), consisting of slender, almost undivided main branches with many perpendicular side branches and fibrils of about equal length (3–40 mm), round to angular in section, often with circular cracks; cortex smooth, but disintegrating on the main stems, leaving rough patches of white medulla over the pinkish to brown central cord; soralia or isidia occasionally form on the side branches. CHEMISTRY: Medulla PD–, K–, KC–, C– (various β-orcinol depsides including evernic, barbatic, or diffractaic acid). HABITAT: On trees of all kinds in open or somewhat shaded, humid forests and often near lakes or streams. COMMENTS: The unbranched strands with perpendicular side branches are unique to this lichen, and the disintegrating cortex is a confirming character. Eastern specimens of *U. longissima* are generally much smaller than those found along the west coast. IMPORTANCE: *Usnea longissima* is one of the most pollution-sensitive lichens. Its presence can be used as an indication of pure air, just as its disappearance (as in most of Europe) indicates deteriorating air quality. In places where it grows (or grew) abundantly, such as the coniferous rain forests along the west coast, it has been used traditionally as material for diapers, feminine hygiene products, and bedding, and for straining medicines. It

889. *Usnea longissima*
Cascades, Oregon
~×1

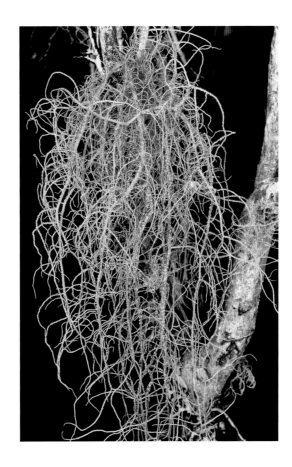

890. *Usnea mutabilis* coastal Massachusetts ×1.5

891. *Usnea rubicunda* California coast ×1.1

has been used in both China and India as an expectorant, and in Europe for strengthening hair (following the "doctrine of signatures"; see Chapter 10). *Usnea longissima* was probably the original "tinsel" on Christmas trees, indicating its past abundance in northern Europe, but its rarity and sensitivity to environmental disturbance should preclude such a use today, even in North America.

Usnea mutabilis
Bloody beard lichen

DESCRIPTION: Thallus shrubby to subpendent, rather dark grayish green; branching irregular or dichotomous, with few perpendicular fibrils; axis and medulla dull red; surface somewhat warty or papillate, or sometimes almost without papillae; isidia abundant, occurring singly or in clusters; cortex brittle, thin or thick (18–23 percent); medulla thin and dense (8–18 percent); axis thick (29–40 percent). Apothecia not seen. CHEMISTRY: Medulla PD–, K–, KC–, C– (no lichen substances). HABITAT: On trees in deciduous and pine forests. COMMENTS: A few other species of *Usnea* have a deeply pigmented medulla. *Usnea strigosa* also lacks conspicuous papillae, but it has no isidia, and it usually contains norstictic acid. *Usnea ceratina* is much more pendent and is covered with whitish tubercles. See also *U. baileyi*.

Usnea rubicunda
Red beard lichen

DESCRIPTION: Thallus shrubby to almost pendent, typically less than 8 cm long, usually distinctly red-orange because of a pigment in the cortex, but in some specimens the pigment is sparse or spotty at the surface, leaving the thallus mostly greenish; base pale, not blackened; branches often strongly papillate; fibrils and isidiate soralia usually abundant (Fig. 37b); cortex thick and glassy (10–18 percent); medulla white, dense, and thin (10–22 percent); axis colorless (35–44(–60) percent). CHEMISTRY: Medulla PD+ orange, K+ yellow, KC–, C– (stictic acid complex); or less frequently, PD+ yellow, K+ red (norstictic or salazinic acids). HABITAT: On branches and trunks of a variety of trees in open forests such as coastal pine

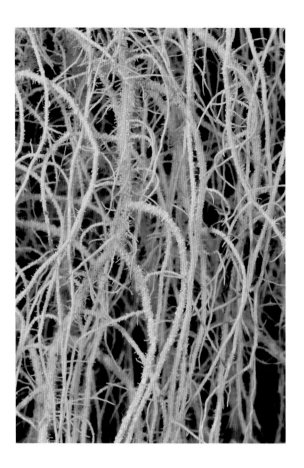

892. *Usnea scabrata* herbarium specimen (Trevor Goward), British Columbia ×4.2

forests. COMMENTS: This is by far the most common of the reddish species of *Usnea*. ***Usnea subcornuta*** is an uncommon species from California with a thicker, cottony medulla. The red pigment in this species is only on the inner portion of the cortex and in part of the medulla. Other species may have a red or pinkish medulla (e.g., *U. strigosa* and *U. mutabilis*) or axis (e.g., *U. trichodea* and *U. longissima*), but they are entirely yellowish green on the outside.

Usnea scabrata
Straw beard lichen

DESCRIPTION: Thallus very pale greenish yellow (straw-colored), not normally darkening, pendent, slender, 0.3–0.8 mm wide, rarely to 2 mm, branched at the base, short fibrils very sparse or common; tips of the branches abundantly branched and tapering, often sinuous; branches uneven because of irregular, sharp ridges, broad to deep depressions, and sometimes warts, giving some forms a very rough appearance; isidia usually abundant, as shown in plate 892, but sometimes sparse, arising singly or in clusters, rarely associated with soralia; papillae abundant, tall to short and rounded, sometimes on the same thallus; cortex thin (7–10 percent); medulla white and lax (22–26 percent); axis slender (28–38 percent). CHEMISTRY: Almost all inland specimens: medulla PD–, K–, KC–, C– (only usnic acid); elsewhere, especially along the humid west coast: PD+ deep yellow-orange, K+ yellow turning red, KC–, C– (salazinic acid). HABITAT: On conifers in forests or open habitats. COMMENTS: This species resembles *U. filipendula,* another pendent, boreal *Usnea* with isidia. *Usnea filipendula,* however, has a thin, dense medulla, and darkened main branches (especially close to the base), and always contains salazinic acid. Specimens having branches with only shallow depressions, consistently small, thin papillae, and isidiate soralia, apparently common in the southwest and California, can be called *U. scabiosa;* like *U. scabrata,* it can contain salazinic acid or only usnic acid.

Usnea strigosa
Bushy beard lichen

DESCRIPTION: Thallus dark yellowish green or gray-green, shrubby, 3–8 cm long, with divergent main branches that are abundantly covered with perpendicular side branches and fibrils, giving the whole lichen a bristly appearance; without isidia or soredia; papillae usually sparse or inconspicuous, but well developed in some populations; cortex thick and hard (10–16 percent); medulla dark red or white (or partly red), usually dense, variable in thickness (9–26 percent); axis usually broad, (24–)33–48 percent, white or pink; large apothecia very common at the tips of the branches, yellow to pale orange, pruinose, with fibrils on the margins. CHEMISTRY: There are five chemical races: (1) the most common throughout the range contains norstictic acid (ca. 60 percent of all collections), often with galbinic acid (PD+ yellow, K+ red, KC–, C–); (2) in the southeast, a common race contains psoromic acid (PD+ deep yellow, K–); (3) specimens containing only usnic acid (PD–, K–) are also found throughout the range. (4) Less common are specimens with fumarprotocetraric (PD+ red, K± brown) or (5) thamnolic acid (PD+ orange, K+ yellow). HABITAT: On deciduous trees and shrubs, mostly in well-lit habitats but sometimes in oak forests. COMMENTS: *Usnea arizonica,* a very similar species from the southwest, contains salazinic acid and its medulla

893. *Usnea strigosa* Cape Cod, Massachusetts ×2.0

894. *Usnea subscabrosa* Ouachita Mountains, Arkansas ×0.80

is always white. *Usnea subfusca* is a nonisidiate northeastern species with abundant fibrils, differing from *U. strigosa* in its conspicuous papillae, blackened base, and chemistry (salazinic acid). It has a thick cortex, dense, thin, white medulla, and broad axis, and it tends to be more pendent than *U. strigosa*. *Usnea mutabilis*, another eastern species with a red medulla, is shrubby to almost pendent and has abundant isidiate soralia but virtually no papillae. It does not produce apothecia and contains only usnic acid. See also Comments under *U. cirrosa*.

Usnea subscabrosa
Horny beard lichen

DESCRIPTION: Thallus shrubby to almost pendent, dark straw-colored, 6–20 cm long, with a pale base; branches rather smooth or slightly papillate close to the base, scattered (often abundant) small, round isidiate soralia; cortex very thick and hard (18–22 percent); medulla white, thin, dense (6–10 percent); axis broad, tough, white (40–58 percent). CHEMISTRY: PD+ red-orange, K–, KC+ pink, C– (proto-

895. *Usnea trichodea*
Cape Cod, Massachusetts × 2.7

cetraric acid). HABITAT: On trees of various kinds and occasionally on rocks. COMMENTS: The extremely thick cortex (often twice the width of the medulla) is the best character for identifying this species. *Usnea hesperina* also contains protocetraric acid and has smooth branches, but it does not have such a thick cortex and has conspicuous, somewhat swollen circular cracks that divide the branches into bony segments. *Usnea subscabrosa* also looks a bit like *U. mutabilis*, which has a red medulla (PD−).

Usnea trichodea
Bony beard lichen

DESCRIPTION: Thallus pendent, 10–30 cm long, branching in uneven dichotomies with some perpendicular side branches; main branches uniformly slender, under 0.4 mm in diameter, divided into articulating, bone-like segments by circular, somewhat swollen white cracks (Fig. 37e), but otherwise smooth, with no papillae, isidia, or soredia; cortex thin (9–12 percent); medulla white, dense, and almost powdery (16–21 percent); axis reddish to brown, rarely white (37–50 percent). CHEMISTRY: Medulla PD− or PD+ orange, K− or K+ yellow, KC+ pinkish orange, CK+ red-orange, C− (diffractaic acid or constictic acid, sometimes with barbatic acid). The branches are so slender and the medulla so minimal that spot tests made directly on the thallus are rarely reliable. HABITAT: On trees. COMMENTS: This species is very distinctive because of its smooth branches and reddish central cord. In the east, only *U. longissima* might be confused with it. The branching is quite different in the two species, however, and the cortex of *U. trichodea* does not disintegrate as it does in *U. longissima*. See also *U. hesperina* and *U. dimorpha*.

Verrucaria (75 N. Am. species)
Speck lichens

DESCRIPTION: Crustose lichens with white, olive, brown, or black, thin or thick thalli, often endolithic with only the fruiting bodies visible. Photobiont can be one of several genera of unicellular green algae. Ascomata are black perithecia variable in the color and development of the involucrellum and excipile, as well as how deeply or superficially the perithecia develop with respect to the thallus or rock (see Fig. 17); paraphyses disintegrating or not developing (not seen in mature perithecia); asci K/I−; spores colorless, 1-celled, ellipsoid to spherical, without a gelatinous halo, 8 per ascus. CHEMISTRY: All reactions negative (no lichen substances). HABITAT: On calcareous or siliceous rocks, in dry to aquatic situations. Aquatic species are specifically associated with either fresh or salt water. COMMENTS: The large number of species described in this genus, and their superficial similarity to one another, make identifying species of *Verrucaria* especially challenging. We have illustrated only 2 of the 75 species known for the continent because color photographs rarely help distinguish them, and careful sectioning techniques are necessary to reveal important differences in perithecial tissues. A number of common species have been added to the key. Closely related and very similar genera include **Thelidium**, with 2-celled spores, and **Polyblastia**, with muriform spores. Other lichens with perithecia usually have persistent, branched or unbranched paraphyses. See also *Staurothele*.

KEY TO SPECIES

1. On dry rocks 2
1. On at least periodically submerged rocks 6

2. Thallus dark brown to black; lower half of medulla black. Perithecia half to entirely immersed in thallus; spores 14–24 × 7–11 µm .... [*Verrucaria nigrescens*]
2. Thallis white, gray, or barely perceptible; medulla, if discernible, entirely white 3

3. Thallus very thick, composed of contiguous or scattered areoles, pale gray to bluish gray, surface rough and scabrose; in the arid southwest ....
   [*Verrucaria inficiens*; see Comments under *Placidium*]
3. Thallus very thin, never areolate; widespread 4

4. Thallus largely endolithic, forming only a whitish stain; perithecia tiny, 0.15–0.3 mm in diameter, immersed in tiny pits in the rock; spores (18–)23–31 × 9–14 µm .....
   .... *Verrucaria calciseda*
4. Thallus usually visible on the rock surface; perithecia 0.25–0.4 mm in diameter 5

5. Thallus white; perithecia not more than half immersed in thallus or rock; spores ellipsoid, 18–28 × 9–14 µm ...
   .... [*Verrucaria muralis*]
5. Thallus greenish gray; perithecia two-thirds immersed; spores narrowly ellipsoid, 16–24 × 8–10 µm ....
   .... [*Verrucaria calkinsiana*]

6.(1) In streams or on lakeshores. Thallus smooth to rimose; perithecia partially immersed; black involucrellum distinct from pale exciple; spores 25–30 × 10–12 µm; widespread .... [*Verrucaria aethiobola*]
6. In saltwater habitats in the intertidal zone 7

7. Thallus brown, very thin and membranous, smooth, without any black bumps or ridges; spores 8–11(–12) × 4.5–6.5 µm; on both coasts ....
   .... [*Verrucaria halizoa* (syn. *V. microspora*)]
7. Thallus black to dark olive-brown when dry, thick, smooth or rough 8

8. Thallus distinctly lobed at the margins, with radiating black ridges; perithecia entirely immersed and showing at the surface only as pale ostioles; spores almost spherical, 7–10 × 6.3–8 µm; frequent, from Vancouver Island to southeastern Alaska .... [*Verrucaria epimaura*]
8. Thallus not at all lobed at the margins; black ridges and bumps present or absent; perithecia immersed or somewhat prominent; spores ellipsoid 9

9. Thallus rimose-areolate when dry, black when wet, normally with a rough upper surface caused by abundant carbonized pegs or elongate ridges; perithecia immersed but creating bumps on the thallus surface; spores 10–20 × 6–10 µm; extremely common on both coasts .... *Verrucaria maura*

9. Thallus smooth or with inconspicuous cracks when dry, never areolate; green to greenish black when wet, lacking any carbonized pegs or ridges; perithecia entirely immersed, not creating bumps on the surface; spores 8–10(–12) × 4–8 µm; fairly common on both coasts....
   .... [*Verrucaria mucosa*]

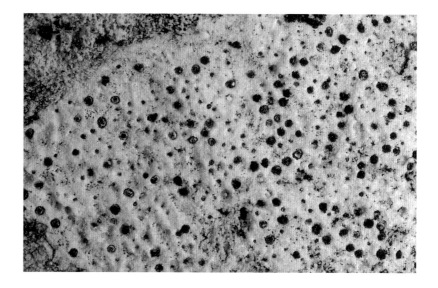

896. *Verrucaria calciseda* northern Arizona ~× 3.8

Verrucaria calciseda
Pitted stone lichen

DESCRIPTION: Thallus endolithic, leaving only a white stain and dot-like perithecia to reveal the presence of the lichen; perithecia 0.15–0.3 mm in diameter, eating away tiny depressions in the soft limestone substrate, giving the rock a pitted appearance, especially where the perithecia have rotted away; involucrellum fused to the exciple and not distinguishable, forming an unbroken, dark brown envelope around the hymenium (Fig. 17c); spores (18–)23–31 × 9–14 µm. HABITAT: On exposed limestone rock. DISTRIBUTION: Widely scattered in temperate to boreal regions rich in limestone. COMMENTS: *Verrucaria calciseda* can sometimes be confused with **V. muralis,** a common, widespread species that often has a thin white thallus visible on the rock surface. Its perithecia are larger (0.25–4 mm in diameter), only half-immersed, and have a black, spreading involucrellum and colorless exciple, making the base of the fruiting body pale rather than brown (Fig. 17f). The spores are about the same size (18–28 × 9–14 µm). *Verrucaria calkinsiana* is a common eastern species like *V. muralis* but with a dirty gray thallus and perithecia 1/3 immersed and with somewhat narrower spores, 16–24 × 8 10 µm. A number of *Verrucaria* spe-

897. *Verrucaria maura*
British Columbia coast
×5.8

cies on limestone have well-developed, dark thalli. The most common is *V. nigrescens,* with a thick areolate thallus, very dark brown when dry, and black perithecia showing as buried black dots, 0.15–0.3 mm in diameter. The medulla (below the algae) is black, and the spores are 14–24 × 7–11 μm.

Verrucaria maura
Sea tar, black seaside lichen

DESCRIPTION: Thallus dark brown to black, covering very large areas of rock, thick, cracked into irregular areoles; surface marked by numerous tiny black, raised points and ridges; medulla coal-black below the colorless alga-containing tissue; perithecia 0.3–0.5 mm in diameter, almost entirely buried in the thallus but producing low, rounded bumps on the thallus surface; spores 12–15(–19) × 6–10 μm. HABITAT: Mainly on siliceous, coastal rocks in the upper part of the intertidal zone and continuing into the salt-spray zone. COMMENTS: This species creates a conspicuous, often broad black band on coastal rocks along both east and west coasts (and is widespread elsewhere in the world). It can easily be mistaken (and has been) for the residue of oil spills along the Pacific coast, but it is an entirely natural phenomenon. There are a number of marine species of *Verrucaria*, some with thin olive-brown thalli and elongate black ridges (e.g., *V. degelii*), black dots (e.g., *V. erichsenii*), or no black dots or ridges except for the perithecia themselves (e.g., *V. halizoa*). *Verrucaria epimaura,* fairly common from Vancouver Island north to Alaska, has a thick, lobed thallus with radiating black ridges, and usually develops on top of *V. maura*. The spherical perithecia are entirely embedded in the thick black basal tissue of the thallus and contain spherical spores. *Verrucaria mucosa,* found on both coasts, has a thick, perfectly smooth, olive-green, somewhat gelatinous thallus (at least when wet) with entirely immersed perithecia. A black *Verrucaria* common on dry rock (limestone) is *V. nigrescens* (see Comments under *V. calciseda*).

Vestergrenopsis (2 N. Am. species)
Brownette lichens

DESCRIPTION: Small, appressed, foliose to almost crustose cyanobacterial lichens, olive to yellowish brown, with radiating lobes; lower surface pale with sparse, pale rhizines; tissues of the thallus resemble a pseudoparenchyma, especially close to the base, but there is no continuous upper cortex. Photobiont blue-green (*Scytonema*). Apothecia lecanorine, with colorless 1-celled, straight or bean-shaped spores, 12–16 per ascus. CHEMISTRY: All reactions negative (no lichen substances). HABITAT: On exposed siliceous rocks, dry or periodically wet. COMMENTS: *Vestergrenopsis* can resemble several other cyanobacterial lichens with lobed margins, especially *Placynthium*, **Koerberia**, and *Massalongia*. Species of *Placynthium* all have a blue-green lower surface and rhizine-like hyphae, except for *P. stenophyllum*, a limestone lichen that has extremely narrow (0.05–0.2 mm wide) lobes. *Massalongia* has broader lobes and contains *Nostoc* as the photobiont, and the spores are 2–4-celled. *Koerberia* is most similar but has narrower, 1–2-celled spores, often twisted in the ascus. All three genera have only 8 spores per ascus.

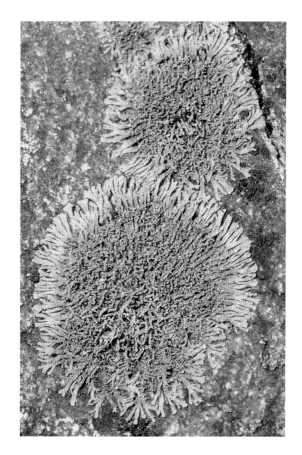

898. *Vestergrenopsis isidiata* mountains, coastal Alaska ×4.7

KEY TO SPECIES: See *Pannaria*.

Vestergrenopsis isidiata
Peppered brownette lichen

DESCRIPTION: Thallus with very narrow, longitudinally grooved lobes, 0.1–0.4 mm across, sometimes broadening to 0.8 mm at the tips, with very long, almost unbranched, slender (but sometimes flattened) isidia on older portions of the thallus. Apothecia very rare. HABITAT: On slightly moistened rocks; arctic and alpine habitats. COMMENTS: The olive, narrow, grooved lobes and long isidia characterize this uncommon arctic species. It reaches sea level at the foot of glaciers, where it is an early invader of exposed rocks. *Vestergrenopsis elaeina*, with a similar geographic distribution, is its fertile, nonisidiate counterpart. It has somewhat broader, radiating lobes, 0.3–0.7 mm wide over much of their length, but fanning out to as much as 1.3 mm wide at the tips. The center of the thallus is somewhat verrucose and is frequently covered with red-brown apothecia. *Koerberia biformis* is an inconspicuous olive lichen with laminal, cylindrical isidia very similar to those of *V. isidiata*, but it grows on bark in the southwest. It is commonly fertile with red-brown biatorine apothecia. *Koerberia sonomensis* has lobes 0.1–0.4 mm wide that fan out at the tips to 0.2–0.8 mm wide, with flattened lobules along the margins that sometimes resemble isidia. It grows on somewhat moist siliceous rocks in the Sierra Nevada and coastal ranges from California to Oregon, as well as in British Columbia, where it occurs on coastal rocks. See Comments under *Vestergrenopsis*.

Vulpicida (4 N. Am. species)
Sunshine lichens, yellow ruffle lichens

DESCRIPTION: Small to medium-sized, foliose to almost fruticose lichens, bright golden yellow to dark yellowish green, with flat, appressed to ascending or erect lobes, sometimes branched several times and contorted or twisted; upper surface usually prominently ridged, but lacking pseudocyphellae; medulla bright yellow. Photobiont green (*Trebouxia?*). Apothecia lecanorine, produced close to the lobe margins, with shiny red brown disks; spores colorless, very

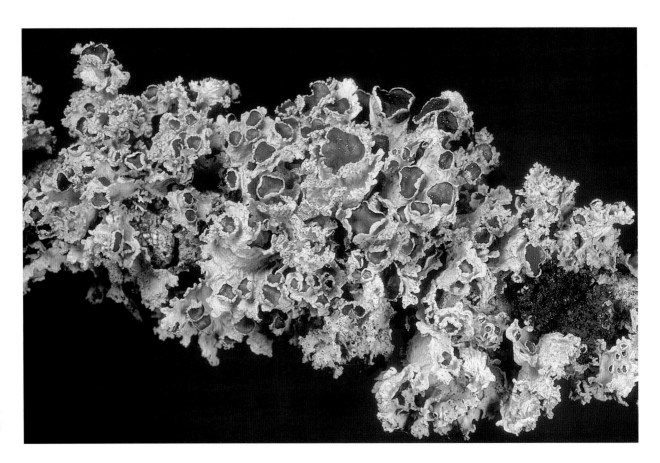

899. *Vulpicida canadensis* mountains, central Washington ×2.7

broadly ellipsoid, 1-celled, 8 per ascus; pycnidia marginal or laminal; conidia lemon- or dumbbell-shaped. CHEMISTRY: Three yellow pigments are present in all species: usnic acid in the cortex, and pinastric and vulpinic acids in the medulla. Zeorin occurs in easily detectable concentrations only in *V. pinastri*. Spot tests are negative or not discernible because of the yellow pigments. HABITAT: On rock, soil, bark, or wood. COMMENTS: The bright yellow color of *Vulpicida* species makes them very distinctive. Species of *Letharia* have the same color (and have some of the same pigments) but have round to angular branches and are clearly fruticose in habit. Lichens containing only usnic acid (e.g., *Ahtiana* and *Flavoparmelia*) are never more than pale greenish yellow and the medulla is white. Some species of *Pseudocyphellaria* have a bright yellow medulla, but the thalli themselves are generally brown, with a fuzzy lower surface that has conspicuous pseudocyphellae. IMPORTANCE: Like *Letharia*, most species of *Vulpicida* have been used as a source of bright yellow dyes. For example, the Gitksan people in British Columbia used *V. canadensis* for dying the wool of mountain goats.

KEY TO SPECIES

1. Soredia present on the lobe margins; pycnidia sparse or very inconspicuous . . . . *Vulpicida pinastri*
1. Soredia absent; pycnidia abundant and conspicuous. . 2
2. Growing on soil or moss; thallus lacking rhizines; apothecia rare. . . . . *Vulpicida tilesii*
2. Growing on bark or wood; thallus with sparse rhizines; apothecia abundant 3
3. Lobes smooth, with rounded depressions, or bumpy; pycnidia prominent; eastern U.S. . . . . . . *Vulpicida viridis*
3. Lobes with a network of depressions and sharp ridges or wrinkled; pycnidia entirely buried in thallus with just the ostiole visible; western montane . . . .
. . . . *Vulpicida canadensis*

Vulpicida canadensis
Brown-eyed sunshine lichen

DESCRIPTION: Thallus forming large tufts, with rounded, wrinkled lobes, 2–7 mm across; soredia absent, but with abundant red-brown apothecia up to 7 mm in diameter; lower surface yellow,

900. *Vulpicida pinastri* White Mountains, New Hampshire ×3.2

slightly paler than the upper surface, often with sharp ridges and wrinkles; rhizines sparse or absent, pale to dark. Pycnidia immersed in the thallus lobes, appearing as black dots. HABITAT: Common and conspicuous on bark and wood of conifers in open, relatively dry sites characteristic of pine forest east of the Cascades, often found with *Bryoria* species. COMMENTS: From afar, *Letharia columbiana* can look like *V. canadensis,* but a closer look reveals its shrubby, angular branches and white medulla. IMPORTANCE: See Importance under *Vulpicida*.

Vulpicida pinastri (syn. *Cetraria pinastri*)
Powdered sunshine lichen

DESCRIPTION: Thallus forming appressed or ascending, greenish yellow to bright yellow rosettes; lobes mostly 1.5–5 mm wide, the margins sometimes finely divided or ruffled, dissolving into brilliant yellow, farinose soredia; lower surface pale yellow to almost white; rhizines almost absent to fairly abundant, pale to dark brown. Apothecia uncommon. HABITAT: On bark, wood, and rock in open and shaded sites. COMMENTS: This pretty yellow lichen, with its ruffled, bright yellow, powdery margins, is a common sight in the boreal forest. It can hardly be mistaken for anything else, except perhaps yellow species of *Pseudocyphellaria* (see *Ps. aurata* and *Ps. crocata*), which have pseudocyphellae on the lower surface, or *Allocetraria oakesiana*, which has a white medulla.

Vulpicida tilesii (syn. *Cetraria tilesii*)
Goldtwist, limestone sunshine lichen

DESCRIPTION: Thallus consisting of round cushions of erect, flat lobes, 0.5–2 mm wide, irregularly branched, often finely divided with lobules on the margins; no soredia or isidia; lower surface of vertical lobes the same color as the upper surface, smooth, without rhizines. Apothecia extremely rare. HABITAT: On calcium-rich soil in alpine or arctic tundra. COMMENTS: The two species

901. *Vulpicida tilesii* southwestern Yukon ×3.3

902. *Vulpicida viridis* Ozarks, Arkansas ×2.6

of *Flavocetraria* occur in similar habitats. They are greenish yellow (usnic acid in the cortex), have a white medulla, and have more wrinkled or channeled lobes.

Vulpicida viridis (syn. *Cetraria viridis*)
Hidden sunshine lichen

DESCRIPTION: Thallus pale yellowish green, or sometimes dark grayish green, forming small, ascending rosettes with rounded lobes, 2–6 mm wide, with toothed to finely lobed margins; soredia absent; lower surface very pale yellowish to almost white, wrinkled and pitted, with scattered, forked, pale rhizines (sometimes almost absent). Apothecia very abundant, 2–5 mm in diameter, produced on the upper surface of the lobes, with shiny red-brown disks and slightly to abundantly lobulate margins; pycnidia black, spherical, also on the lobe surface, but most abundant close to the margins.
HABITAT: On bark, especially of mature oaks in the south, and Atlantic white cedar (*Chamaecyparis thyoides*) in bogs along the northeastern coastal plain.
COMMENTS: At first glance this species looks like a very

pale, yellowish *Tuckermannopsis*. It is greener and often darker than *V. canadensis* and has prominent black pycnidia rather than pycnidia immersed in the thallus. Both *V. viridis* and *V. canadensis* are found only in North America, although their distributions do not overlap.

Xanthoparmelia (51 N. Am. species)
Rock-shield lichens

DESCRIPTION: Small to medium-sized, yellowish green (usnic-yellow) foliose lichens, closely or loosely attached or, in some species, rolling freely over the ground as vagrants; lobes mostly 0.5–4 mm wide; often with isidia, either cylindrical (more or less uniform in diameter) or globose, constricted at the base and frequently breaking open; rarely with soredia; with lower surface pale brown to ebony black; rhizines unbranched or forked, usually abundant, but lacking in some soil species. Photobiont green (*Trebouxia*). Apothecia abundant in some species, lecanorine, with broad, brown disks and thallus-colored margins; spores colorless, ellipsoid, 6–13 × 4–8 μm, 8 per ascus; laminal pycnidia common, immersed in the thallus, visible as black spots; conidia mostly 5–8 μm long, bacilliform or with bulges at the tips or along the length. CHEMISTRY: Cortex K–, KC+ yellow (usnic acid); medulla containing a wide variety of lichen substances, especially β-orcinol depsidones such as salazinic, norstictic, and stictic acids, and depsides such as barbatic and diffractaic acids. HABITAT: On rocks, especially siliceous, noncalcareous types, and on mineral soil; most species in open, relatively dry sites, although a few species can tolerate shade and thrive on forest boulders; very rarely, on hard weathered wood. COMMENTS: The most important characters for naming species of *Xanthoparmelia* are degree of adnation to the substrate, color of the lower surface, presence of isidia of different types, and chemistry. Because some important chemical differences are not revealed by spot tests alone (e.g., norstictic acid and salazinic acid react PD+ yellow and K+ blood-red), some additional chemical analyses often are needed to make a positive identification. Other yellow-green foliose lichens on rock include *Arctoparmelia*, boreal to arctic species with a white to mouse-gray lower surface and containing alectoronic acid (medulla KC+ red, UV+ blue-white), and *Flavoparmelia*, with broad, round lobes and granular soredia or pustulate isidia. Some lobate species of *Lecanora*, such as *L. muralis* and *L. novo-mexicana*, and the umbilicate lichens *Rhizoplaca* and *Omphalora* can also resemble a *Xanthoparmelia*. *Xanthoparmelia mougeotii*, a small, narrow-lobed, mainly western species, is the only sorediate rock-shield lichen in North America; it contains stictic acid (PD+ orange, K+ yellow). See also *Allocetraria oakesiana*, which can sometimes grow on rock.

KEY TO SPECIES (including *Arctoparmelia*)

1. Soredia present on upper surface of thallus in hemispherical mounds 2
1. Soredia absent 3

2. Upper surface dull, grayish toward center of thallus; lower surface dull, dark gray to black; medulla KC+ red, UV+ white (alectoronic acid); arctic and northern boreal .... [*Arctoparmelia incurva*]
2. Upper surface shiny at lobe tips, brownish toward center of thallus; lower surface shiny black; medulla K+ yellow to red, PD+ yellow-orange (stictic and norstictic acids); southwestern montane habitats and coastal Nova Scotia; rare .... [*Xanthoparmelia mougeotii*]

3. Isidia present....................................... 4
3. Isidia absent....................................... 9

4. Lower surface and rhizines black; medulla PD+ orange, K+ yellow to red (norstictic and stictic acids) .... *Xanthoparmelia conspersa*
4. Lower surface and rhizines pale to dark brown 5

5. Medulla K+ yellow, orange, or red................. 6
5. Medulla K–....................................... 7

6. Isidia mainly cylindrical, often branched, not globular; widespread, especially in northeast, also southwest; medulla AI+ blue, PD+ orange, K+ dark yellow (stictic acid) .... *Xanthoparmelia plittii*
6. Isidia globular, at least when young, later cylindrical; western; medulla AI–, PD+ yellow-orange, K+ red (salazinic acid) .... *Xanthoparmelia mexicana*

7. Medulla PD–, hypoprotocetraric acid present .... *Xanthoparmelia weberi*
7. Medulla PD+ distinct yellow or red, hypoprotocetraric acid absent 8

8. Medulla PD+ bright yellow (psoromic acid); western.. .... *Xanthoparmelia lavicola*
8. Medulla PD+ red (fumarprotocetraric acid); common in southeast, especially the Ozarks................. .... [*Xanthoparmelia subramigera*]

9.(3) Lower surface dark gray to white or rarely pale brownish, dull and velvety; old thalli forming full or partial concentric rings; medulla KC+ red, PD–, UV+ white (alectoronic acid); mainly boreal to arctic ... 10

9. Lower surface brown or black, shiny, not velvety; old thalli not normally forming concentric rings; medulla KC–, PD+ yellow, orange, or red, UV– (alectoronic acid absent) 11

10. Lower surface gray .... *Arctoparmelia separata*
10. Lower surface white or pale brown .... *Arctoparmelia centrifuga*

11. Growing on soil 12
11. Growing on rock 13

12. Lobes 1–2 (–4) mm wide, convex, but margins not strongly inrolled; thallus attached to soil and pebbles .... *Xanthoparmelia wyomingica*
12. Lobes 1.5–5 mm wide, margins strongly inrolled forming tubes; thallus not attached (vagrant), on arid soils .... *Xanthoparmelia chlorochroa*

13. Lobes convex; thallus very loosely attached, ascending; pycnidia absent and apothecia rare .... *Xanthoparmelia wyomingica*
13. Lobes flat; thallus closely or somewhat loosely attached to substrate; pycnidia and apothecia abundant 14

14. Lower surface and rhizines black 15
14. Lower surface and rhizines pale tan or brown 17

15. Medulla K– or brownish, PD+ red (fumarprotocetraric acid) .... *Xanthoparmelia hypomelaena*
15. Medulla K+ yellow or red, PD+ yellow-orange 16

16. Thallus closely attached to rock and difficult to remove intact; lobes short and crowded, often overlapping and finely divided; contains stictic and norstictic acids ... . [*Xanthoparmelia angustiphylla* (previously called *X. hypopsila* in North America)]
16. Thallus loosely attached and easily removed; lobes strap-shaped; contains salazinic acid .... [*Xanthoparmelia tasmanica*]

17.(14) Medulla K–, PD+ red (fumarprotocetraric acid) .. .. *Xanthoparmelia novomexicana*
17. Medulla K+ yellow or red, PD+ distinct yellow or orange (fumarprotocetraric acid absent) 18

18. Thallus loosely attached to substrate over entire surface; lobes square to truncated or elongated, usually somewhat constricted just behind the tips; often spotted with maculae; medulla AI– (salazinic acid) .... *Xanthoparmelia somloënsis*
18. Thallus closely appressed to substrate, at least over most of the surface; lobes rounded, not constricted behind the tips; maculae absent 19

19. Medulla AI– (salazinic acid); thallus very closely attached to substrate; western ... *Xanthoparmelia lineola*
19. Medulla AI+ blue (stictic and norstictic acids); thallus closely to moderately attached to substrate; widespread in northeast and parts of the west .... *Xanthoparmelia cumberlandia*

Xanthoparmelia chlorochroa
Tumbleweed shield lichen

DESCRIPTION: Thallus vagrant, forming rounded cushions of dichotomously branched lobes; lobes 1–5 mm wide with the margins strongly curled inward, almost forming tubes; without isidia or soredia; lower surface mostly pale brown to smoky gray-brown, blackened in older parts of the thallus in some broader forms, with abundant, black, forked rhizines. Apothecia extremely rare. CHEMISTRY: Medulla PD+ orange, K+ red, KC–, C– (salazinic acid, with traces of norstictic acid). HABITAT: Growing loose over the soil, rolling around like tumbleweed; most abundant in high-elevation plains and semi-deserts. COMMENTS: This species intergrades with *X. wyomingica*, a western alpine species that usually remains attached to pebbles and has shorter, flatter, overlapping lobes. ***Xanthoparmelia neochlorochroa*** and ***X. norchlorochroa*** are two similar vagrant lichens that contain norstictic acid rather than salazinic acid. The former is almost identical to *X. chlorochroa* except for its chemistry and the fact that it more frequently has a blackened lower surface; *X. norchlorochroa* has a dull black lower surface, often with pronounced reticulate ridges and veins, without any rhizines, and with lobes rolled into tubes. IMPORTANCE: *Xanthoparmelia chlorochroa* has proved to be a reliable indicator of excellent proghorn antelope habitat and has been used as such by wildlife managers reintroducing antelope to certain areas in the Great Basin. The lichen is eaten by antelope and is especially important in early spring and under drought conditions. *Xanthoparmelia chlorochroa* is highly prized by the Ramah Navajo of New Mexico, and by craftspeople in other parts of the west, as a source of boiling-water dyes for reddish brown colors.

903. *Xanthoparmelia chlorochroa* northwestern New Mexico ×2.7

Xanthoparmelia conspersa
Peppered rock-shield

DESCRIPTION: Thallus closely or loosely attached to the rock, rather shiny, commonly browned at the margins; lobes fairly narrow, 1–3 mm wide, crowded and often overlapping, with sparse to dense globular to branched cylindrical isidia on the upper surface, sometimes also developing abundant lobules in the center of the thallus; lower surface pitch black except for a pale to dark brown area close to lobe tips. Apothecia rare, with isidiate margins. HABITAT: On siliceous rocks, especially granite, in sunny locations. CHEMISTRY: Medulla PD+ red-orange, K+ deep yellow or turning red, KC–, C– (stictic acid complex including cryptostictic acid, with varying amounts of norstictic acid). COMMENTS: This is the most common of the isidiate species that contain stictic and (or) norstictic acid and have a black lower surface. In the chemically identical *X. plittii,* the lower surface is pale to dark brown throughout. Similar but much rarer southeastern species include *X. isidiascens,* with a loosely attached thallus lacking cryptostictic acid, and *X. piedmontensis,* with fumarprotocetraric acid (PD+ red, K– or brownish). Specimens with few isidia can be mistaken for *X. angustiphylla,* a sister species that lacks isidia and has an Appalachian–Great Lakes and southwestern distribution. IMPORTANCE: This lichen reportedly has been used in southeastern and eastern Africa for the treatment of venereal disease and snakebite.

Xanthoparmelia cumberlandia
Cumberland rock-shield

DESCRIPTION: Thallus closely to loosely attached to the rock, often forming very large patches (up to 12 cm across); lobes rounded or somewhat toothed, 1.5–4 mm wide, frequently with a black edge; lower surface pale to medium brown, with pale, unbranched rhizines. Apothecia common, 2–8 mm in diameter, often with rolled, toothed margins; pycnidia very common, creating patches of black dots on the

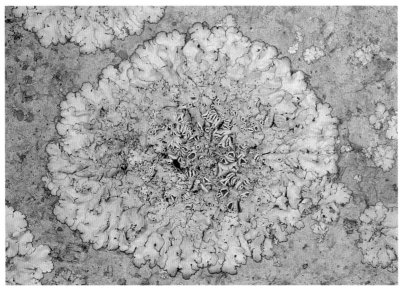

904. *Xanthoparmelia conspersa*   central Massachusetts ×1.9
905. *Xanthoparmelia cumberlandia*   mountains, western Texas ×0.87

thallus surface. CHEMISTRY: Medulla PD+ orange, K+ yellow, slowly darkening to orange or red, KC–, C–, AI+ blue (stictic, constictic, and norstictic acids). HABITAT: On exposed or somewhat shaded outcrops and boulders. COMMENTS: This is one of the most common and widespread rock-shield lichens in North America. *Xanthoparmelia somloënsis*, a common eastern species, is more loosely attached and has salazinic acid (K+ yellow changing rapidly to blood-red, AI–). *Xanthoparmelia angustiphylla* contains stictic acid but has a black lower surface and smaller lobes. See Comments under *X. conspersa*.

Xanthoparmelia hypomelaena
Ozark rock-shield

DESCRIPTION: Thallus closely attached to the rock, forming circular colonies, with lobes 0.7–2 (–3) mm across, without isidia or soredia, but very narrow lobules sometimes developing on old parts of the thallus; lower surface black except for a brown zone close to the margin. Apothecia common, 1–5 mm

906. *Xanthoparmelia hypomelaena* (two thalli, showing variation in lobe width) eastern Missouri ×2.3

907. *Xanthoparmelia lavicola* central Arizona ×1.6

in diameter; pycnidia common. CHEMISTRY: Medulla PD+ red, K± brown, KC–, C– (fumarprotocetraric and succinprotocetraric acids). HABITAT: On exposed rocks in openings and on ridges. COMMENTS: This species is almost entirely limited to the Ozark Plateau. Like *X. cumberlandia,* it has adnate, rounded lobe tips and lacks soredia or isidia, but it has a black lower surface and differs chemically. ***Xanthoparmelia angustiphylla*** also has a black lower surface but its medulla is PD+ orange, K+ yellow to red (stictic and norstictic acids). See also *X. novomexicana*.

Xanthoparmelia lavicola
(syn. *X. kurokawae*)
Salted rock-shield

DESCRIPTION: Thallus dull, almost pruinose in places, forming circular patches up to 7 cm across, with rounded, flat to undulating lobes, 1.5–5 mm wide; with globose to almost cylindrical isidia having brownish tips; lower surface pale brown with sparse, pale rhizines. Apothecia unknown and pycnidia rare. CHEMISTRY:

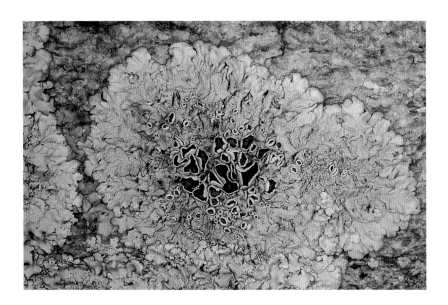

908. *Xanthoparmelia lineola* southern Arizona ×1.0

is most closely adnate; *X. coloradoensis* is loosely attached and has some overlapping lobes; *X. wyomingica* has even more loosely attached, shingled, somewhat rolled lobes; *X. chlorochroa* is entirely vagrant, growing unattached at maturity, with strongly inrolled lobes. See also the eastern *X. somloënsis*.

Xanthoparmelia mexicana
Salted rock-shield

DESCRIPTION: Thallus closely or loosely attached to rock, often shiny at the lobe tips; lobes rounded, 1.5–5 mm wide, with dense, globose isidia that commonly become cylindrical and branched; lower surface pale brown, or smoky gray-brown at margins. Apothecia and pycnidia rarely seen. CHEMISTRY: Medulla PD+ yellow-orange, K+ yellow turning red; KC–, C– (salazinic acid, varying amounts of norstictic acid, and sometimes consalazinic acid). HABITAT: On exposed rocks, rarely on hard wood, in oak or pine scrub. COMMENTS: *Xanthoparmelia mexicana* is the central and most widely distributed of a group of chemically distinct species with mainly globular isidia and pale lower surfaces, almost all of them from the arid or semiarid west: *X. ajoensis* is PD– (or somewhat orange), K–, C–, KC– or pale yellow, CK+ orange, UV+ blue (diffractaic and squamatic acids); *X. lavicola* is PD+ bright yellow, K– (psoromic acid); *X. maricopensis* is PD+ yellow, K+ red (norstictic and connorstictic acids); *X. schmidtii*, a rare species from Tulare County, California, and Arizona, has barbatic (and (or) diffractaic) acid in addition to salazinic and often norstictic acids; *X. subramigera*, common in the southern U.S., is PD+ red, K+ brownish, KC+ brown-pink, C– (fumarprotocetraric acid and (or) protocetraric acid); and *X. weberi* is PD–, K–, KC–, C– (hypoprotocetraric acid).

Medulla PD+ bright yellow, K–, KC–, C– (psoromic and 2'-*O*-demethylpsoromic acids). HABITAT: On rock. COMMENTS: This is one of a large number of species closely related to *X. mexicana*, all having globose to subglobose isidia and a pale lower surface, and common in the southwest. This species is unusual because it contains psoromic acid, as does *X. psoromifera*, a nonisidiate sister species relatively common in the arid, southwestern U.S. *Xanthoparmelia nigropsoromifera* from the southwest, is another nonisidiate species that contains psoromic acid, but it has a coal-black lower surface and broad lobes (2.5–5 mm across) with a rather wrinkled upper surface.

Xanthoparmelia lineola
Tight rock-shield

DESCRIPTION: Thallus dull, often darkening on older parts of the thallus, forming large, closely attached patches, with thick, rounded or irregular, contiguous lobes, 0.8–3(–5) mm wide, sometimes crowded and overlapping, without isidia or soredia; lower surface pale brown, often wrinkled and cracked, with pale rhizines. Apothecia common, somewhat raised, 2–5 mm in diameter; pycnidia generally abundant. CHEMISTRY: Medulla PD+ deep yellow, K+ blood-red, KC–, C– (salazinic acid and a trace of norstictic acid). HABITAT: On exposed rocks. COMMENTS: This lichen is one of several nonisidiate western species with salazinic acid forming a continuum based on the degree of adnation to the substrate: *Xanthoria lineola*

Xanthoparmelia novomexicana
(syn. *X. arseneana*)
New Mexican rock-shield

DESCRIPTION: Thallus closely attached, lobes 0.6–2.5 mm wide, with neither soredia nor isidia; lower surface pale. Apothecia round, even, 1–4 mm in diameter; disks tend to be very dark brown to almost black; pycnidia common. CHEMISTRY: Medulla PD+ red, K±

909. *Xanthoparmelia mexicana* Channel Islands, southern California ×2.5

910. *Xanthoparmelia novomexicana* Sonoran Desert, southern Arizona ×2.0

brownish, KC–, C– (fumarprotocetraric acid). HABITAT: In open, arid sites. COMMENTS: This is one of the few nonisidiate *Xanthoparmelia* species with fumarprotocetraric acid. Another is *X. monticola* in the Appalachians, which also has a pale lower surface but with lobes that are loosely attached and overlapping. See also *X. hypomelaena*.

Xanthoparmelia plittii
Plitt's rock-shield

DESCRIPTION and COMMENTS: This species is virtually identical to *X. conspersa* except for its pale to dark brown or mottled (never black) lower surface. It is the isidiate equivalent of *X. cumberlandia*. CHEMISTRY: Medulla PD+ red-orange, K+ deep yellow darkening to orange, KC–, C–, AI+ blue (stictic, constictic, and norstictic acids). HABITAT: On exposed or somewhat shaded rocks.

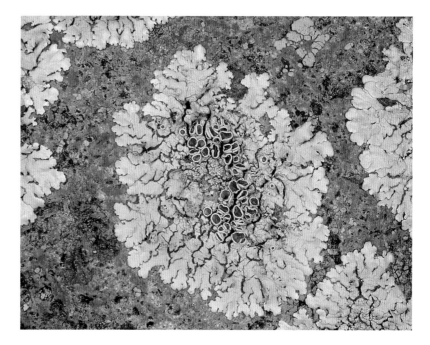

911. *Xanthoparmelia plittii* central Massachusetts ×1.4

912. *Xanthoparmelia somloënsis* Lake Superior region, Ontario ×1.0

Xanthoparmelia somloënsis
Shingled rock-shield

DESCRIPTION: Thallus often darkened and grayish in the old, central portions, usually shiny at the lobe tips; lobes narrow, 0.3–2.5 (–4) mm wide, often somewhat constricted just behind the tips, and almost always elongate or strap-shaped with square or angular tips, convoluted and overlapping, loosely attached, without isidia or soredia; with or without maculae (see Comments below); lower surface pale brown although the edges of the lobes are sometimes blackened; rhizines sparse, pale. Apothecia common, somewhat raised, 3–7(–15) mm in diameter; pycnidia very common. CHEMISTRY: Medulla PD+ yellow, K+ yellow turning to blood-red, KC–, C–, AI– (salazinic and consalazinic acids, sometimes with lobaric acid). HABITAT: On exposed boulders and outcrops. COMMENTS: *Xanthoparmelia somloënsis* is the most widely distributed of a group of closely related taxa characterized by lacking soredia or isidia, having a pale brown lower surface, containing salazinic acid, and growing directly on rock. It is said to be spotted with maculae, but if maculae are present, they are often so difficult to see that the character is unreliable. The loose attachment and strap-shaped lobes constricted near the tips are more useful for defining the species. Plate 912 shows a form with finely scalloped lobe tips, but this feature is not always present. In the absence of maculae, the species approaches *X. lineola* (closely adnate, with spreading or crowded and overlapping lobes) and *X. coloradoensis* (rounder lobes, more shingle-like). It is also related to *X. wyomingica* and *X. chlorochroa*, which are on soil and pebbles. (See Comments under these species.) *Xanthoparmelia cumberlandia* and *X. californica* can be superficially similar, but they are more closely attached and differ chemically: the former has stictic acid, and the latter contains norstictic acid.

Xanthoparmelia weberi
Salted rock-shield

DESCRIPTION: Thallus adnate but loosely attached to rock, dull; lobes rounded, rather broad, (2–)3–5(–8) mm wide, with abundant, rather fat isidia, globular at first, later cylindrical and

branched; lower surface pale brown, with short, pale rhizines. Apothecia and pycnidia rare. CHEMISTRY: Medulla PD–, K–, KC– and C– (hypoprotocetraric and 4-*O*-demethylnotatic acids). HABITAT: On exposed rocks, rarely on wood. COMMENTS: This is one of the very few *Xanthoparmelia* species with a K– medulla and a pale brown lower surface. It is essentially a chemical variant of *X. mexicana*. (See Comments under that species.)

Xanthoparmelia wyomingica
Shingled rock-shield, variable rockfrog

DESCRIPTION: Thallus often edged in black, loosely attached; lobes narrow, 1–2(–4) mm wide, overlapping each other in a pile; lower surface pale at center, but can be black close to lobe tips, with rather sparse rhizines. CHEMISTRY: Medulla PD+ yellow, K+ yellow turning red, KC–, C– (salazinic acid and some norstictic acid). HABITAT: On rock or mossy soil and pebbles. COMMENTS: This species resembles *X. chlorochroa* but re-

913. *Xanthoparmelia weberi* (with a species of *Dermatocarpon* and other lichens) Sonoran Desert, southern Arizona ×1.2

914. *Xanthoparmelia wyomingica* Rocky Mountains, Colorado ×1.5

mains attached to the rock, and its lobes are flatter, not as tightly inrolled. See Comments under *X. chlorochroa, X. lineola,* and *X. somloënsis.*

Xanthoria (13 N. Am. species)
Sunburst lichens, orange lichens

DESCRIPTION: Small to medium-sized foliose lichens, orange, yellow-orange, or reddish orange, usually most strongly pigmented where they are fully exposed to sun. Photobiont green (*Trebouxia*). Lobes narrow and erect to rather broad and appressed, overlapping and almost squamulose in some species; without pseudocyphellae, but frequently sorediate or producing granules or fragments called blastidia; lower surface white, sometimes partially yellow close to the margin; with or without rhizines (see Comments). Apothecia lecanorine, common in some species, with orange disks and margins; spores colorless, polarilocular, 8 per ascus; pycnidia frequent, embedded in the thallus and visible only as a dark orange dot, or superficial and dark like an inflamed pimple; conidia ellipsoid or bacilliform, 1.5–5 μm long. CHEMISTRY: Cortex K+ dark red-purple (various anthraquinone pigments, especially parietin). HABITAT: On a wide variety of substrates including rock (especially, but not exclusively, calcareous rock), bark, twigs, wood, and occasionally soil; usually in sunny locations. COMMENTS: The genus *Xanthoria* has a central position within the family Teloschistaceae, with the crustose genus *Caloplaca* on one side, and the fruticose genus *Teloschistes* on the other. The boundaries between these genera are not always distinct. Some closely attached species of *Xanthoria* are very close to *Caloplaca* (e.g., *X. elegans*) and others with ascending, partially terete lobes approach *Teloschistes* (e.g., *X. candelaria*). The majority, however, have clearly different upper and lower surfaces and are unambiguously foliose. Whether or not growth form is a good character to use in classification is another question. The rhizines in *Xanthoria* are slender, usually unbranched, but sometimes frayed at the ends. They can be attached to the substrate or free. Thalli can, however, be attached directly to the substrate by the lower surface or with broad, very short holdfasts, which are not the same as rhizines.

KEY TO SPECIES

1. Soredia absent; apothecia abundant............2
1. Soredia present; apothecia absent or occasional.....8
2. Thallus loosely attached over entire surface, forming round cushions of overlapping lobes..............3
2. Thallus closely appressed to substrate, forming round, flat rosettes......................................5
3. Rhizines absent (thallus sometimes attached by stout holdfasts); conidia ellipsoid; on rocks, bark, and wood ....*Xanthoria polycarpa*
3. Rhizines abundant, both attached to the substrate and unattached; conidia rod-shaped; on trees, rarely on rock..4
4. Spores mostly 13–15.5 × 5–7.5 μm, septum 1.5–4 μm wide; arid western interior.....[*Xanthoria montana*]
4. Spores mostly 15.5–18 × 7.5–9.5 μm, septum 5–8.5 μm wide; widespread, but not in western interior region.. ....*Xanthoria hasseana*
5.(2) Lobes usually wider than 1 mm, rounded, flat or concave, clearly foliose; lower surface smooth, with scattered long rhizines..............*Xanthoria parietina*
5. Lobes usually narrower than 1 mm, short or elongated to linear, appearing crustose, at least in center; rhizines absent or very rare............................6
6. Lower surface with a cortex, wrinkled; lobes convex; on rock..........................*Xanthoria elegans*
6. Lower surface without a distinct cortex and appearing crustose; lobes convex or flat....................7
7. On bark or wood. Lobes often pruinose, rather thick; apothecia mostly under 1.5 mm in diameter, concave when young, becoming flattened; common in California..........................[*Xanthoria tenax*]
7. On rock................lobed species of *Caloplaca*
8.(1) Thallus very closely appressed to substrate, almost crustose, without rhizines; soredia laminal, originating from the disintegration of pustules; on rock........ ....*Xanthoria sorediata*
8. Thallus loosely attached, ascending, clearly foliose or squamulose; rhizines present or absent; soredia rarely laminal; on bark, wood, or rock..................9
9. Rhizines abundant; pycnidia buried in the thallus, not conspicuous 10
9. Rhizines absent or very sparse; pycnidia conspicuous or not..12
10. Soredia mostly greenish yellow, within hood- or lip-like expansions of the lobe tips between the upper and lower cortices....................*Xanthoria fallax*
10. Soredia yellow to orange, on the lobe margins or sometimes laminal, or on the lower surface of hood-shaped

expansions of the lobe tips, not between the upper and lower cortices.................................11

11. Mostly north central to northeastern North America; soredia not produced in hood-shaped expansions; conidia always rod-shaped .... *Xanthoria ulophyllodes*
11. Western montane and California; soredia frequently produced in hood-shaped expansions; conidia ellipsoid or rod-shaped............. [*Xanthoria oregana*]

12.(9) Lobes flattened to almost cylindrical, branched, erect, sometimes almost fruticose in habit; soredia produced on or just under the lobe tips; pycnidia immersed and inconspicuous; conidia ellipsoid, 2–3.5 μm long .... *Xanthoria candelaria*
12. Lobes flat to convex, rounded or divided; soredia on the lower surface of the lobe tips, not forming hoods; pycnidia prominent, resembling orange pimples; conidia rod-shaped, 3–4 μm long.......... *Xanthoria fulva*

Xanthoria candelaria
Shrubby sunburst lichen

DESCRIPTION: Thallus foliose to almost fruticose, yellow to yellow-orange, forming small cushions up to approximately 3 cm across, or forming extensive, continuous colonies; lobes narrow, branched, flattened or almost terete, 0.2–0.5 mm wide, usually ascending and even erect, as shown in plate 915; fine to granular soredia or blastidia produced along the margins and on the lower surface of the lobe tips; lower surface white, or yellow on erect lobes, generally without rhizines or holdfasts. Apothecia not common, but they can be large and conspicuous (up to 4 mm in diameter); pycnidia immersed and inconspicuous; conidia ellipsoid, 2–3(–3.5) × 1–1.5 μm. HABITAT: On bark and branches, especially spruce, or on rock. This species is one of the lichens that is particularly characteristic of bird perches and other highly fertilized sites such as trees near farmyards, or places where it it subjected to mineral-rich dust or ocean spray. COMMENTS: The ascending, narrow lobes of *X. candelaria* often appear almost fruticose, especially on twigs, where it sometimes forms extremely dense and colorful colonies. Because of its variability, *X. candelaria* can be confused with several other sorediate, orange lichens, but it is the only one with ellipsoid conidia. *Xanthoria borealis* is a rare arctic species on rock and soil. It is dark red-orange and has broader, flatter lobes (1–2 mm across), with rhizines at the lobe bases. The granular soredia (technically, blastidia) are the same color as the thallus. *Xanthoria mendozae* is fairly common on granite and sandstone in the drier montane areas of the west from Colorado and California to southern British Columbia. It is a paler yellowish orange lichen, with round, greenish yellow, rather fuzzy soredia produced on the lower surface of the lobes. *Xanthoria oregana* is widespread in the west, especially in drier areas, and is commonly found on both bark and rock. Its soredia are marginal or close to the margin, or cover the lower surface of the lobes, which can curl inward and be somewhat helmet-shaped. The conidia are a curious mixture of ellipsoid and bacilliform types within the same pycnidium. The corticolous species are discussed under *X. fallax*.

915. *Xanthoria candelaria* Alaskan coast ×5.1

916. *Xanthoria elegans* (slightly pruinose form, especially on upper right) southwestern Yukon ×5.1

Xanthoria elegans
Elegant sunburst lichen

DESCRIPTION: Thallus foliose to almost crustose, very closely attached, with narrow, convex, radiating lobes, 0.4–1(–1.3) mm wide, extremely variable in color, from pale yellowish orange to dark red-orange, rarely pruinose; soredia and isidia absent, but some forms can produce papillate outgrowths on the lobe surface; lower surface white, wrinkled, without rhizines. Apothecia can be abundant, 1–3 mm in diameter, thallus-colored. HABITAT: On rocks of all kinds, in open sites typically rich in nutrients (such as bird perches), and on wood or bone, rarely on soil. COMMENTS: *Xanthoria elegans*, although a foliose lichen with a well-developed lower cortex, closely resembles some lobate species of *Caloplaca* (e.g., *C. saxicola*, *C. ignea*, or *C. trachyphylla*). All these lack a lower cortex and cannot be removed from the substrate intact. Specimens that produce granular soredia on the lobe surface are *X. sorediata*, a very similar species having about the same geographic range. IMPORTANCE: Because of the tendency of *X. elegans* to flourish on rocks fertilized by birds and mammals (see plate 25), the presence of the lichen has been used by Inuit hunters to locate the burrows of hoary marmots. Unfortunately, it is also used by poachers to find the nests of peregrine falcons.

Xanthoria fallax
Hooded sunburst lichen

DESCRIPTION: Thallus yellow-orange to dark orange, with rather short, narrow to broad lobes with rounded, somewhat divided tips, 0.8–2 mm wide, appressed or raised; soralia labriform on the lobe tips, with greenish yellow, powdery soredia (30–50 μm in diameter) that develop within a split between the upper and lower cortices, forming a kind of rounded hood; lower surface white, with white rhizines. Apothecia occasionally seen, mostly 0.7–1 mm in diameter; pycnidia immersed, not common or conspicuous; conidia bacilliform, 3–4.5 μm long. HABITAT: On the bark of a variety of trees, especially oaks,

917. *Xanthoria fallax* southern Ontario ×4.9

918. *Xanthoria fulva* Rocky Mountains, Colorado ×6.1

elms, and poplars, less commonly on wood, and rarely on rock. It is mainly a low-elevation species and is commonly seen on roadside trees or trees close to farms. COMMENTS: Among the sorediate species of *Xanthoria*, *X. fulva*, *X. ulophyllodes*, and **X. oregana** are most similar to *X. fallax*. *Xanthoria fulva* has narrower, more ascending lobes and lacks rhizines, and the granular soredia are mainly on the lower surface. *Xanthoria ulophyllodes*, also very common, has rhizines like *X. fallax*, but the thallus is more ascending and the blastidia-type soredia develop on the margins or on the upper surface, not between the cortices of the lobe tips. In *X. oregana*, a widespread western lichen, the lobes dissolve into blastidia along the margins (as well as having soredia on the lower surface), but the lobes can roll inward and resemble hoods as in *X. fallax*. It usually has rhizines. **Xanthoria mendozae** can also have inrolled lobes, but has larger, spherical, yellow-green soredia (45–85 µm in diameter). It is a western montane lichen always found on rock. See Comments under *X. candelaria*.

Xanthoria fulva
Bare-bottommed sunburst lichen

DESCRIPTION: Thallus dark red-orange to medium orange; lobes rounded or finely divided, 0.2–0.6 mm wide, flat to somewhat convex, ascending to erect (especially in the northwest) or mostly horizontal (especially in the southeast), with mealy soredia on the edge of the lower surface, not forming hoods; lower surface white, with few rhizines or holdfasts. Apothecia rarely seen; pycnidia common, resembling dark orange pimples; conidia bacilliform, 3–4.5 µm long. HABITAT: Mostly on bark, sometimes on wood, and rarely on rock. COMMENTS: See Comments under *X. fallax* and *X. ulophyllodes*.

919. *Xanthoria hasseana* Klamath Range, northwestern California ×5.1

Xanthoria hasseana
Poplar sunburst lichen

DESCRIPTION: Thallus yellow-orange to orange, in small, loosely attached rosettes with finely divided, overlapping lobes, 0.3–0.9 mm wide, without soredia or isidia; lower surface white, with abundant, relatively long rhizines. Apothecia almost always present, 0.6–3 mm in diameter, confined to the center of the thallus, with dark orange disks and thallus-colored margins that can develop white rhizines on the lower half; spores ellipsoid, (14.5–)15.5–18(–20) × (6–)7.5–9.5(–10) μm, with a broad central septum (mostly 5–8.5 μm); pycnidia frequent, immersed or protruding, dark orange; conidia bacilliform, 3–4 μm long. HABITAT: On bark, especially poplars, in open or partially shaded sites, rarely on wood or rock. COMMENTS: This is the most common and widespread of the nonsorediate, corticolous species of *Xanthoria*, and can be recognized in the field by its minutely lobed, loosely attached thallus and abundant rhizines. It should be compared with *X. polycarpa*, which has no true rhizines and has ellipsoid conidia. *Xanthoria montana*, a common species of the arid western interior, has smaller, cylindrical spores that have a narrow septum (see key).

Xanthoria parietina
Maritime sunburst lichen, wall lichen

DESCRIPTION: Thallus yellow-orange to orange, with shade forms that are gray-green with orange patches, forming large rosettes up to 10 cm in diameter; lobes broad, 0.7–3.2 mm across, flat to wrinkled and somewhat concave at the tips, without soredia or isidia; lower surface white, attached to the substrate by the lower cortex, with broad holdfasts, or with sparse, short rhizines. Apothecia almost always present, broad (1–8 mm in diameter), flat to concave, with dark orange disks and thallus-colored margins; pycnidia almost entirely immersed, dark orange; conidia ellipsoid, 2.5–4 × 1–1.5 μm. HABITAT: On bark (especially elms and poplars), rocks, or concrete in sunny sites, usually on or close to the seashore, rarely inland. COMMENTS: This is the only broad-lobed *Xanthoria* in North America. The sparseness of true rhizines distinguishes rare, narrow-lobed specimens from most other nonsorediate species, except for *X. polycarpa*, which forms small cushions rather than flat rosettes, and *X. elegans*, with more firmly attached, very convex lobes. *Xanthoria parietina* is very common along the northeast coast but is rare on the west coast. IMPORTANCE: Following the "doctrine of signatures," the orange color of *X. parietina* indicated to medieval herbalists the possible utility of this lichen in the treatment of jaundice.

Xanthoria polycarpa (syn. *X. ramulosa*, *X. alaskana*)
Pin-cushion sunburst lichen

DESCRIPTION: Thallus yellow-orange to orange, forming small cushions up to 2.5 cm across; lobes very narrow (0.2–0.7 mm wide), abundantly branched and irregular, without soredia or isidia; lower surface white, wrinkled, without true rhizines but sometimes with broad holdfasts. Apothecia abundant, crowding the center of the cushions, 1–4.5 mm in diameter, with dark orange disks and thallus-colored margins lacking a fringe of rhizines; pycnidia immersed and inconspicuous; conidia ellipsoid, 2–3 μm long. HABITAT: On bark, especially twigs and branches of spruce, oak, and fir, on rock, or wood. In the east, this is a largely coastal

920. *Xanthoria parietina* Maine coast ×3.1

921. *Xanthoria polycarpa* Willamette Valley, Oregon ×4.3

species, conspicuous on exposed trees and rocks close to the ocean. In the west, it is more tolerant of inland sites. COMMENTS: *Xanthoria hasseana* and *X. montana* are superficially similar but have abundant rhizines and bacilliform conidia. Specimens of *X. candelaria* that have very sparse soredia can be distinguished by their erect, almost terete lobes. *Xanthoria tenax,* a closely appressed, almost crustose, corticolous species with ellipsoid conidia, appears to be confined to California, where it can grow in mixed colonies with *X polycarpa*. It has virtually no lower cortex and is therefore very similar to the genus *Caloplaca*. See also *X. parietina*.

Xanthoria sorediata
Sugared sunburst lichen

DESCRIPTION and COMMENTS: *Xanthoria sorediata* closely resembles *X. elegans* in general habit and habitat, but the upper surface of the older parts of the lobes develops small, spherical pustules that dissolve into coarse, granular soredia, and apothecia are rare. Sometimes the soredia are not

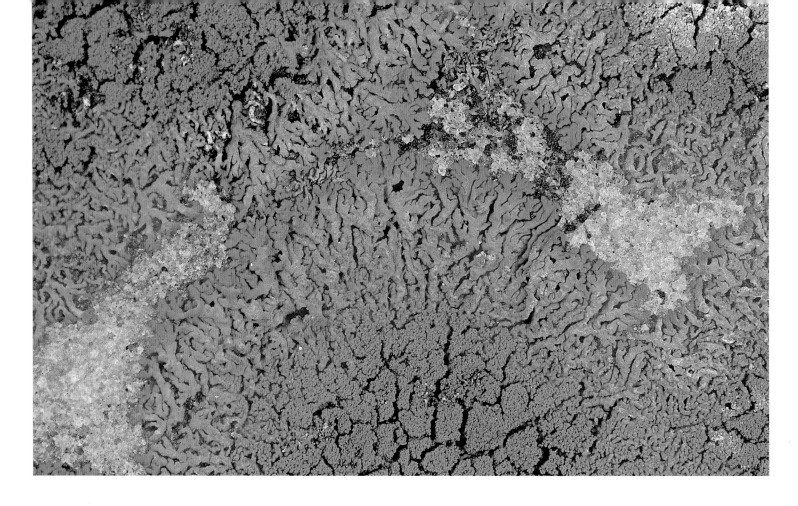

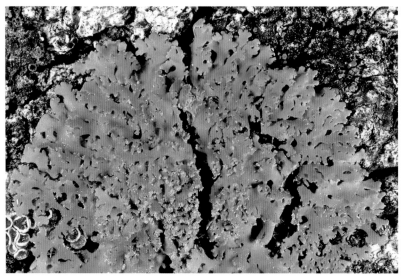

922. *Xanthoria sorediata* Lake Superior region, Ontario ×5.4

923. *Xanthoria ulophyllodes* Lake Superior region, Ontario ×5.4

abundant, but other times they cover most of the thallus. HABITAT: On exposed or somewhat shaded rocks, usually calcareous, or on fertilized substrates.

Xanthoria ulophyllodes
Powdery sunburst lichen

DESCRIPTION: Thallus yellow-orange to medium orange, with radiating, branched lobes, 0.3–1.4 mm wide; soralia marginal or sometimes laminal, delimited and crescent-shaped or coalescing over fairly large areas of the thallus; soredia yellow to orange; lower surface white, with abundant rhizines. Apothecia not common, up to 2.7 mm in diameter; pycnidia common, immersed and not conspicuous; conidia bacilliform to almost ellipsoid, 2.5–3.5 μm long. HABITAT: Mostly on bark, especially oaks and elms, rarely on rock. COMMENTS: Because the soralia can be crescent-shaped and rhizines are abundant, this species is often confused with *X. fallax* (see Comments under *X. fallax*). *Xanthoria fulva* and ***X. oregana*** have soredia on the lower surface of the lobes, and the pycnidia are more prominent.

Xylographa (6 N. Am. species)
Woodscript lichens

DESCRIPTION: Crustose lichens with whitish to gray thalli, often growing within the wood substrate and visible only as a stain. Photobiont green (unicellular). Fruiting bodies ellipsoid to elongate and branched lirellae, brown to black, usually aligned with the wood grain; exciple thin, brown; paraphyses usually unbranched; epihymenium brown; asci almost uniformly K/I+ light blue (*Trapelia*-type); spores colorless, 1-celled, ellipsoid, 8 per ascus; pycnidia brown to black, immersed, showing as dots; conidia rather long, slender, often bent. CHEMISTRY: Often containing stictic acid and (or) norstictic acid. HABITAT: All species generally on hard, weathered wood. COMMENTS: In other script lichens, the exciple (which forms the wall of the fruiting body) is at least partially black, and the spores have several cells. Species of *Xylographa* are distinguished by the size and shape of their fruiting bodies and spores, height of the hymenium, amount of thallus, and chemistry. *Agyrium rufum,* an unlichenized fungus, has orange, marginless apothecia that can become elongated and resemble those of *Xylographa,* and its spores are also ellipsoid, colorless, and 1-celled. It is widely distributed and quite common.

KEY TO GENUS: See Key E.

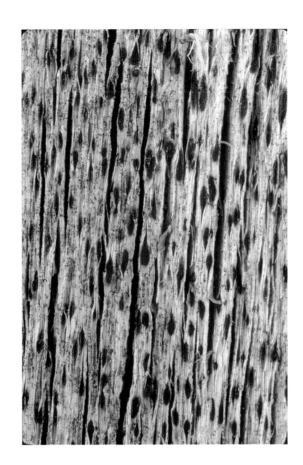

924. *Xylographa parallela* herbarium specimen (Canadian Museum of Nature), Ontario ×8.0

Xylographa parallela
(syn. *X. abietina*)
Black woodscript

DESCRIPTION: Thallus within the wood, producing only a gray stain; lirellae black or very dark brown, long and slender, 0.5–2 mm long, 0.1–0.25 mm wide, unbranched, following the wood grain, with a thin margin that disappears as the lirellae mature; spores 11–16 × 5–7.5(–8.5) μm. CHEMISTRY: Medulla directly under the lirellae, PD+ yellow-orange, K+ yellow, KC–, C– (stictic, constictic, and often traces of norstictic acids), or PD–, K– (no detectable substances because of the scarcity of thallus); IKI+ blue. (Test thick slices through the fruiting bodies including the wood and lichen tissue below.) COMMENTS: This is the most widespread of the woodscript lichens. It is very common, although it is infrequently collected. Also common is *X. vitiligo,* which produces elliptical, brown soralia, although the rest of the thallus is typically buried in the wood. The soralia react PD+ orange and K+ yellow (stictic acid). Its lirellae are brown and broadly elliptical. *Xylographa opegraphella* is a woodscript that occurs along the northeastern and northwestern seacoasts, often growing on beach logs. It has a relatively thick, greenish gray, verruculose thallus containing norstictic acid (PD+ yellow, K+ red) or rarely stictic acid. It has short, brown lirellae often branched into a small Y-shape, and it has narrow spores, 10–13 × 3–4.5 μm.

APPENDIX / CLASSIFICATION OF GENERA OF LICHEN FUNGI AND LICHEN-LIKE FUNGI MENTIONED IN THIS BOOK

This classification is based primarily on three sources—*Dictionary of the Fungi* (Hawksworth et al. 1995), the proposal by Tehler in Nash's *Lichen Biology* (1996), and *More Florida Lichens* (Harris 1995)—plus some recent publications on phylogeny. We have also benefited from comments made on early drafts by Richard C. Harris and François Lutzoni. The classification of the order Lecanorales into suborders is particularly tentative, owing to many changes that are continually suggested based on molecular studies. A query (?) following a taxon indicates that the taxon's classification is particularly problematic. Genera in square brackets are mentioned, but not pictured, in the text.

BASIDIOMYCETES (BASIDIOMYCOTA)

Agaricales
    Tricholomataceae: *Omphalina*
Cantharellales
    Clavariaceae: *Multiclavula*
Stereales
    Meruliaceae: [*Dictyonema*]

ASCOMYCETES (ASCOMYCOTA)

Euascomycetes
[Unitunicate ascohymenial orders: asci with single-layered walls; ascomata with paraphyses in the hymenium]

Agyriales
    Agyriaceae (including Rimulariaceae and Trapeliaceae): [*Agyrium*], [*Lithographa*], *Placopsis*, *Placynthiella*, [*Ptychographa*], [*Rimularia*], *Trapelia*, *Trapeliopsis*, *Xylographa*
    Schaereriaceae?: *Schaereria*
Gyalectales
    Gyalectaceae: [*Belonia*], *Coenogonium*, *Dimerella*, *Gyalecta*, [*Pachyphiale*]
Leotiales
    Baeomycetaceae: *Baeomyces*
    Mycocaliciaceae?: [*Chaenothecopsis*], [*Mycocalicium*], [*Phaeocalicium*], [*Stenocybe*]
    Sphinctrinaceae?: [*Pyrgidium*], [*Sphinctrina*]
Lichinales
    Heppiaceae: *Heppia*

Lichinaceae: *Ephebe*, [*Euopsis*?], [*Lempholemma*], *Lichinella*, [*Peccania*], [*Phylliscum*], [*Psorotichia*], *Pyrenopsis*, [*Synalissa*], [*Thermutis*], *Thyrea*

Peltulaceae: *Peltula*

Ostropales (including Graphidales)

Gomphillaceae: *Gomphillus*, [*Gyalideopsis*], *Sagiolechia*

Graphidaceae: *Glyphis*, *Graphina*, *Graphis*, *Phaeographina*, *Phaeographis*, *Sarcographa*

Solorinellaceae: [*Gyalidea*]

Stictidaceae: [*Absconditella*], *Conotrema*, [*Petractis*], [*Thelopsis*]

Thelotremataceae: *Diploschistes*, *Myriotrema*, *Ocellularia*, *Thelotrema*

[Bitunicate ascohymenial orders: asci with two-layered walls and special structures at the tip; ascomata with paraphyses in the hymenium]

Caliciales (unitunicate, but probably by reduction, and therefore close to Lecanorales)

Caliciaceae: [*Acrocyphus*], *Calicium*, *Cyphelium*, *Thelomma*, *Tholurna*

Coniocybaceae: *Chaenotheca*, [*Sclerophora*]

Lecanorales

Suborder Acarosporineae

Acarosporaceae: *Acarospora*, [*Glypholecia*], *Polysporina*, *Sarcogyne*, [*Sarcosagium*], *Sporastatia*, [*Thelocarpon*]

Biatorellaceae: [*Biatorella*]

Hymeneliaceae: *Aspicilia*, [*Eiglera*], [*Hymenelia*], *Ionaspis*, *Lobothallia*, [*Melanolecia*], *Tremolecia*

Suborder Cladoniineae

Cladoniaceae: *Cladina*, *Cladonia*, *Gymnoderma*, *Pycnothelia*

Crocyniaceae: *Crocynia*

Icmadophilaceae: *Dibaeis*, *Icmadophila*, *Siphula*?, *Thamnolia*?

Lecideaceae: *Lecidea*

Micareaceae: *Micarea*, *Psilolechia*

Pilocarpaceae: *Byssoloma*, [*Fellhanera*]

Porpidiaceae: *Amygdalaria*, *Bellemerea*, [*Clauzadea*], [*Immersaria*], [*Koerberiella*], *Porpidia*

Psoraceae: *Lecidoma*?, *Protoblastenia*, *Psora*, *Psorula*

Rhizocarpaceae: [*Catolechia*], *Rhizocarpon*

Sphaerophoraceae: [*Bunodophoron*], *Sphaerophorus*

Squamarinaceae?: *Squamarina*

Stereocaulaceae: *Pilophorus*, *Stereocaulon*

Suborder Lecanorineae

Biatoraceae: *Biatora*, *Cliostomum*, *Hypocenomyce*, *Phyllopsora*

Catillariaceae: [*Catillaria*], *Toninia*, [*Xanthopsorella*]

Gypsoplacaceae: *Gypsoplaca*

Lecanoraceae: *Bacidia*, [*Bacidina*], *Candelaria*, *Candelariella*, *Candelina*, [*Carbonea*], [*Catinaria*], [*Cladidium*], *Haematomma*, *Japewia*?, *Lecania*, *Lecanora*, *Lecidella*, [*Megalaria*], *Mycobilimbia*, *Pleopsidium*, *Pyrrhospora*, *Rhizoplaca*, [*Schadonia*], *Scoliciosporum*, [*Solenopsora*], *Speerschneidera*, [*Strangospora*], *Tephromela*, [*Waynea*]

Mycoblastaceae: *Mycoblastus*

Ophioparmaceae: *Loxospora*?, *Loxosporopsis*?, *Ophioparma*

Parmeliaceae: *Ahtiana*, *Alectoria*, *Allantoparmelia*, *Allocetraria*, *Anzia*, [*Arctocetraria*], *Arctoparmelia*, *Asahinea*, *Brodoa*, *Bryocaulon*, *Bryoria*, *Bulbothrix*, *Canomaculina*, *Canoparmelia*, *Cavernularia*, *Cetraria*, *Cetrariella*, *Cetrelia*, *Cornicularia*, *Dactylina*, *Esslingeriana*, *Evernia*, *Everniastrum*, *Flavocetraria*, *Flavoparmelia*, *Flavopunctelia*, *Hypogymnia*, *Hypotrachyna*, *Imshaugia*, *Kaernefeltia*, *Letharia*, *Masonhalea*, *Melanelia*, *Menegazzia*, *Myelochroa*, *Neofuscelia*, *Nodobryoria*, *Omphalora*, [*Paraparmelia*], *Parmelia*, *Parmelina*, *Parmelinopsis*, *Parmeliopsis*, *Parmotrema*, *Platismatia*, *Protoparmelia*, *Pseudephebe*, *Pseudevernia*, *Pseudoparmelia*, *Punctelia*, [*Relicina*], *Rimelia*, *Sulcaria*, *Tuckermannopsis*, *Usnea*, *Vulpicida*, *Xanthoparmelia*

Physciaceae: *Amandinea*, *Anaptychia*, *Buellia*, *Dimelaena*, *Diploicia*, *Dirinaria*, *Heterodermia*, *Hyperphyscia*, *Phaeophyscia*, *Phaeorrhiza*, *Physcia*, *Physciella*, *Physconia*, *Pyxine*, *Rinodina*

Ramalinaceae: *Niebla*, *Ramalina*

Suborder Peltigerineae

Coccocarpiaceae?: *Coccocarpia*, *Spilonema*

Collemataceae?: *Collema*, *Leptogium*

Lobariaceae: *Dendriscocaulon*, *Lobaria*, *Pseudocyphellaria*, *Sticta*

Nephromataceae: *Nephroma*

Pannariaceae: [*Degelia*], *Erioderma, Fuscopannaria, Leioderma, Leproloma, Pannaria, Parmeliella, Psoroma*

Peltigeraceae: *Hydrothyria, Massalongia?, Peltigera, Solorina*

Placynthiaceae?: [*Koerberia*], *Leptochidium, Placynthium, Polychidium, Vestergrenopsis*

Suborder Pertusariineae

Coccotremataceae: *Coccotrema*

Megasporaceae: *Megaspora*

Pertusariaceae: *Ochrolechia, Pertusaria,* [*Varicellaria*]

Suborder Teloschistineae

Brigantiaeaceae?: *Brigantiaea*

Fuscideaceae: *Fuscidea,* [*Maronea*], *Lopadium?, Orphniospora, Ropalospora*

Megalosporaceae?: [*Megalospora*]

Letrouitiaceae: *Letrouitia*

Teloschistaceae: *Caloplaca, Fulgensia, Teloschistes, Xanthoria*

Suborder Umbilicariineae?

Umbilicariaceae: *Lasallia, Umbilicaria*

Suborder uncertain

Phlyctidaceae: *Phlyctis*

Loculoascomycetes

[Orders with asci surrounded by a tissue of highly branched and anastomosing pseudoparaphyses rather than true paraphyses (which originate at the base and grow upward); asci typically double-walled, opening by a "jack-in-the-box" extension of the inner ascus (containing the spores) beyond the outer wall. Some orders have disk-like or script-like ascomata lacking a true exciple, although sometimes with carbonized walls; other orders have perithecium-like ascomata.]

Arthoniales (disk- or script-like ascomata)

Arthoniaceae: *Arthonia,* [*Arthothelium*], *Cryptothecia*

Chrysothricaceae: *Chrysothrix*

Roccellaceae: [*Combea*], [*Cresponea*], *Dendrographa, Dirina, Hubbsia, Lecanactis, Lecanographa, Opegrapha, Roccella, Roccellina,* [*Schismatomma*], *Schizopelte,* [*Sigridea*]

Dothideales (perithecium-like ascomata)

Dacampiaceae: [*Eopyrenula*]

Xanthopyreniaceae: [*Pyrenocollema*]

Patellariales? (disk-like ascomata)

Arthrorhaphidaceae: *Arthrorhaphis*

Pyrenulales (perithecium-like ascomata)

Monoblastiaceae?: [*Acrocordia*], [*Anisomeridium*], [*Monoblastia*]

Pyrenulaceae: [*Lithothelium*], *Pyrenula,* [*Pyrgillus*]

Strigulaceae?: [*Strigula*]

Trypetheliaceae: *Astrothelium,* [*Bathelium*], *Laurera,* [*Polymeridium*], [*Pseudopyrenula*], *Trypethelium*

Trichotheliales (perithecium-like ascomata)

Trichotheliaceae: *Porina,* [*Trichothelium*]

Verrucariales (perithecium-like ascomata)

Verrucariaceae: [*Catapyrenium*], *Dermatocarpon, Endocarpon, Normandina, Placidium,* [*Polyblastia*], *Staurothele, Thelidium, Verrucaria*

Order Uncertain

Thrombiaceae: [*Thrombium*]

Nonfertile Lichens of Uncertain Position

[*Cystocoleus*], *Lepraria, Leprocaulon, Racodium*

GLOSSARY

ADNATE. Tightly attached to the surface, like species of *Dirinaria* or *Bulbothrix*.

AI. A buffered iodine solution used for detecting stictic acid. Mix 1.5 ml of 20% Lugol's solution with 18.5 ml of pH 11.0 buffer.

ALGA (ALGAE). Green photosynthetic organism containing chloroplasts and nuclei and belonging to the kingdom Protoctista.

ALGAL LAYER. Layer of algal cells in a stratified lichen thallus (Fig. 4). Sometimes used synonymously with "photobiont layer" (a layer of either green algae or cyanobacteria).

AMPHITHECIUM. The portion of a lecanorine apothecium external to the exciple (Fig. 13c), usually containing algae, constituting the thalline margin.

AMYLOID. Containing carbohydrates that turn blue in an iodine solution (IKI).

ANASTOMOSING. Forming a net-like, interconnected growth (Fig. 16c).

ANISOTOMIC. Dividing in unequal dichotomies to produce a distinguishable main branch with side branches, as in *Cladina rangiferina*.

APICAL. At the tip or summit (i.e., the apex).

APOTHECIUM (APOTHECIA). A disk- or cup-shaped ascoma, usually with an exposed hymenium (Figs. 12, 13).

APPRESSED. Flattened and closely adnate, as in *Xanthoparmelia cumberlandia*.

AREOLATE. Broken up into areoles, often appearing tile-like (Fig. 2b; *Lecidea tessellata*).

AREOLE. (1) A small, irregular, often angular patch of thallus; (2) patch of vegetative tissue (containing algae and cortex) on the podetial surface of some *Cladonia* species.

ASCENDING. Lifting from the surface and becoming free from it, at least in part, like the lobes of species of *Tuckermannopsis*.

ASCOCARP. See *ascoma*.

ASCOHYMENIAL. Pertaining to a type of ascoma having a hymenium with true paraphyses rather than pseudoparaphyses or paraphysoids (Fig. 13a–e).

ASCOLOCULAR. Pertaining to a type of ascoma in which the asci arise within a uniform mass of fungal tissue and are separated in maturity, not by true paraphyses but by abundantly branched pseudoparaphyses or paraphysoids (Fig. 13f).

ASCOMA (ASCOMATA). The fruiting body of an Ascomycete; the structure that bears the asci, which in turn contain the ascospores (Figs. 12, 17). Apothecia and perithecia are types of ascomata.

ASCOMYCETE. A fungus that produces its sexual spores within an ascus.

ASCOSPORE. A spore produced in an ascus (Figs. 14, 15).

ASCUS (ASCI). The sac-like structure in Ascomycetes in which the ascospores are formed (Fig. 14). The sexual fusion of nuclei and reduction division occur within the ascus.

AXIAL BODY. Conical or roughly cylindrical, vertically oriented, nonamyloid structure in the tholus of certain types of asci such as the *Bacidia*- and *Biatora*-types (Fig. 14a,b); also called an apical cushion or masse axiale.

BACILLIFORM. Stick- or rod-shaped; cylindrical, usually straight.

BASIDIOMYCETE. A fungus that produces its sexual spores as external buds on a club-like basal cell (the basidium). Mushrooms, bracket fungi, and coral fungi, among others, belong to the Basidiomycetes, one of the main classes in the kingdom Fungi.

BASIDIUM. A club-shaped cell in the fruiting body of a Basidiomycete in which sexual fusion and reduction division occurs, with the subsequent budding off of 2–4 spores (basidiospores).

BIATORINE. Referring to a type of apothecium having a relatively soft, clear, or lightly pigmented (not carbonized) margin containing no photobiont cells (Fig. 13b) and usually resembling the disk in color.

BLASTIDIUM (BLASTIDIA). A granule-sized fragment of a lichen thallus that is formed by a budding off of the thallus margin (Fig. 19e).

BLUE-GREEN ALGAE. Cyanobacteria.

BOREAL. Refers to a northern region dominated by conifer forests; in North America, the region forms a belt from the maritime provinces of eastern Canada to Alaska.

BRYOPHYTES. Mosses, liverworts (hepatics), and hornworts.

C. Bleaching solution (sodium hypochlorite) or undiluted commercial bleach (e.g., Clorox™ or Javex™), used as a reagent in spot tests for revealing certain lichen substances (see Chapter 13, table 2).

CALCAREOUS. Containing lime or chalk (calcium carbonate), producing vigorous bubbling (CO_2) in the presence of a strong acid. Calcareous rocks include limestone, dolomite, and marble; some sandstones and soils can also be calcareous.

CAMPYLIDIUM (CAMPYLIDIA). Helmet-shaped conidia-bearing structure found in many tropical, foliicolous lichens.

CAPITATE. Referring to a type of rounded, almost hemispherical structure (usually a soralium); see *Melanelia sorediata*.

CAPITULUM. The tiny spherical or cup-shaped apothecium formed at the summit of a slender stalk; found in *Calicium, Chaenotheca*, and related genera (Fig. 12f).

CARBONACEOUS. Opaque black, usually brittle, referring to tissue such as the exciple. It is hard to distinguish individual cells in carbonaceous tissue.

CARTILAGINOUS. Tough, pliable cartilage- or sinew-like tissue. The term usually refers to supporting tissue (see Fig. 8).

CEPHALODIUM (CEPHALODIA). A small gall-like growth that contains cyanobacteria and occurs within the tissues or on the surface of some lichens with green algal photobionts (see plate 16; also, *Placopsis lambii*).

CHLOROCOCCALES. An order of green algae having taxa with individual spherical cells.

CHLOROPLAST. The structure in a green cell that contains chlorophyll, the substance responsible for photosynthesis.

CHROOCOCCUS. A genus of cyanobacteria with large cells in colonies of 2 to 4 in thin, uniform, layered or unlayered, colorless, gelatinous envelopes (Fig. 1b). Individual cells do not have a clearly defined gelatinous sheath as in *Gloeocapsa*.

CILIA. Hair-like appendages on the margins of the thallus or apothecia of many foliose and fruticose lichens (Fig. 10; plate 13).

COCCOMYXA. A genus of green algae with small, ellipsoid cells that contain a single chloroplast lying against one wall.

COLUMELLA. A vertical protuberance or column of sterile tissue found in some ascomata.

CONIDIUM (CONIDIA). An asexual spore usually formed in large numbers within special structures such as pycnidia and campylidia. Conidia sometimes serve as male sexual cells (spermatia) (Fig. 18).

CONSOREDIUM. A round cluster of tiny soredia that resembles a large soredium or granule.

CONTINUOUS. Unbroken, or broken only by cracks (Fig. 2c; *Lecanora caesiorubella*).

CORALLOID. Composed of, or having, minutely branched cylindrical outgrowths, as in *Caloplaca coralloides* or *Sphaerophorus globosus*.

CORTEX. The outer protective layers of a lichen thallus or apothecium, completely fungal in composition (except for some dead algal cells), often composed of hyphae with thick, gelatinized walls (Figs. 4, 13).

CORTICATE (CORTICAL). Having a cortex; pertaining to a cortex.

CORTICOLOUS. Growing on bark.

CRENULATE. Having a scalloped margin with rounded teeth or lobes (Fig. 3).

CRISPED. Having a ruffled, wavy, or twisted margin, as in many species of *Cetrelia* and *Parmotrema*.

CRUSTOSE. A thallus type that is generally in contact with the substratum at all points and lacks a lower cortex; cannot be removed intact from its substrate without removing a portion of the substrate as well (Fig. 2a–d; plate 10).

CRYPTOLECANORINE. (1) A kind of apothecium that is almost entirely sunken into the thallus and thereby lacks a prominent margin; (2) the partial, alga-containing margin in such apothecia (e.g., *Aspicilia* or *Ionaspis*).

CUDBEAR. Scottish name for some lichens used for fermentation dyes, especially the mainly European crustose species *Ochrolechia tartarea*.

CUFF-SHAPED. Referring to laminal soralia that burst through the upper cortex of a hollow lobe, leaving a hole in the center of the soralium (Fig. 20f), occurring in *Menegazzia terebrata*.

CYANOBACTERIA. Photosynthetic, chlorophyll-containing organisms related to bacteria (in the kingdom Monera), without organized nuclei or chloroplasts; sometimes called blue-green algae.

CYANOLICHEN. A lichen with cyanobacteria as the photobiont.

CYPHELLA (CYPHELLAE). A specialized, depressed pore in a lichen thallus, lined with small, loosely packed, spherical cells (Fig. 11c; plate 15); characteristic of the genus *Sticta*.

DECORTICATE. Having had a cortex that has now fallen away or decomposed.

DICHOTOMOUS. Branching into two equal parts, as in the letter "Y" (see, e.g., *Speerschneidera*).

DISJUNCT (DISJUNCTION). A population that is geographically remote from other occurrences of the same species (as in the distribution pattern of *Umbilicaria caroliniana*).

DORSIVENTRAL. With distinguishable upper and lower surfaces.

EFFIGURATE. Having a definite, usually somewhat lobed margin (Fig. 2d), as in *Dimelaena oreina*.

ELLIPSOID. Oval in outline; more or less football-shaped (Fig. 15a–d).

ENDEMIC. Found only in a certain, usually limited, region; in this book, used in reference to lichens found only in North America or smaller regions, for example, the Appalachian Mountains (see distribution map for *Hypotrachyna croceopustulata*).

ENDOLITHIC. Growing within the upper layers of a rock, that is, under and around the rock crystals, often with little or no thallus visible on the outer rock surface (plate 32).

ENDOPHLOEODAL. Growing largely within the upper layers of bark tissue (Fig. 9b), as in the thallus of *Graphis scripta*.

EPIHYMENIUM. The uppermost portion of the hymenium formed by the tips of the paraphyses, which are frequently expanded or branched; often pigmented and sometimes containing tiny granules (Fig. 13); considered here synonymous with *epithecium*.

EPISPORE. See *perispore*.

EPITHECIUM. See *epihymenium*.

EXCAVATE. Hollowed out or depressed; often referring to a type of soralium (Fig. 20h).

EXCIPLE. An area in an apothecium external to and below the hypothecium, forming the apothecial margin in lecideine or biatorine apothecia (Fig. 13a–b); much reduced in lecanorine apothecia (Fig. 13c).

EXCIPULUM. Synonymous with exciple; in this book, however, used with specific reference to the perithecial wall (Fig. 17).

F. form; a formal subdivision of a species, subspecies, or variety usually applied to a relatively minor morphological, chemical, or ecological variant.

FAMILY. A taxonomic category consisting of closely related genera.

FARINOSE SOREDIA. Very fine, powdery soredia, as on the podetial surface of *Cladonia deformis*.

FIBRIL. A short branch, usually perpendicular to the main filament, in *Usnea* (Fig. 37a; see *U. diplotypus* or *U. longissima*).

FILAMENTOUS. Hair-like.

FISSURAL. Resembling a gaping slit, often oval or fusiform in shape; usually referring to a type of soralium (Fig. 20d).

FIXED NITROGEN. Nitrogen from the air that has been chemically transformed into nitrogen-containing compounds such as nitrates that plants can use for growth.

FLEXUOSE. Wavy.

FLOCCULENT. Having a loose, cottony, fibrous texture, like the nubby surface of an old sweater; as seen, for example, on the surface of *Cladina portentosa*.

FOLIICOLOUS. Growing on leaves of vascular plants (especially in the tropics; e.g., *Calopadia fusca*).

FOLIOSE. Pertaining to a more or less "leafy" lichen thallus, distinctly dorsiventral, and varying in its attachment to the substrate from completely adnate to umbilicate (Fig. 2f–g; plate 9).

FRUITING BODY. The sexual reproductive structure of a lichen fungus (e.g., apothecium, perithecium, mushroom); in most lichens, the ascoma.

FRUTICOSE. Pertaining to a lichen thallus that is stalked, pendent, or shrubby, and normally with no clearly distinguishable upper and lower surfaces (Fig. 2h–j; plate 9).

FUSIFORM. Narrow, tapering at both ends, usually with pointed tips; spindle- or cigar-shaped (Fig. 15f).

GENUS. A group of closely related species, presumably with the same ancestor; constitutes the first word of the two-word name of every organism.

GLOEOCAPSA. Cyanobacteria consisting of groups of 2–8 spherical cells enclosed within a thick, gelatinous matrix. Individual cells have their own, often reddish, gelatinous sheaths that usually are K+ purple (Fig. 1c).

GRANULAR (GRANULOSE). (1) Having granules or granule-like particles (as in the thallus of *Bacidia rubella*). (2) Pertaining to soredia that are large enough to be easily distinguished under a dissecting microscope (as in *Cladonia chlorophaea*); see farinose.

GRANULE. A spherical or nearly spherical corticate particle.

GRAPHID. A lichen with a lirella-type apothecium, resembling *Graphis*.

HALO. A transparent gelatinous covering, often irregular in thickness, surrounding an ascospore (Fig. 15k); technically called a *perispore*, and referred to as an *epispore* by some lichenologists; revealed by an India ink preparation (see "Microscopic Study," Chapter 13).

HALONATE. Having a gelatinous perispore or "halo" (Fig. 15k).

HAUSTORIUM (HAUSTORIA). A special branch of a mycobiont that penetrates or otherwise attaches itself to the photobiont cell for the purpose of food absorption.

HEATH. A vegetation type consisting of extensive areas of low shrubs and few trees.

HEMIAMYLOID. Turning dark blue with iodine (IKI) when pretreated with KOH, but negative to greenish, changing to red-orange with IKI in tissues not pretreated with KOH.

HERBARIUM. A collection of dried plants (or lichens).

HETEROCOCCUS. A genus of yellow-green algae (Xanthophyceae) found in some species of *Verrucaria* as single-celled or short, filamentous units, but which in culture becomes filamentous and branched (as in Fig. 1i).

HETEROCYST. A specialized, thick-walled, colorless cell in certain cyanobacteria such as *Nostoc*, and the site of most nitrogen fixation (Fig. 1d).

HOLDFAST. The relatively thick and, in many cases, only attachment point of some lichens, especially *Usnea* and umbilicate lichens such as *Umbilicaria*.

HYMENIUM. The spore-bearing layer of an ascoma, consisting of asci and paraphyses, paraphysoids, or pseudoparaphyses (Fig. 13).

HYPHA (HYPHAE). The filamentous elements of a fungus, often modified and resembling round or angular cells (Fig. 5).

HYPHOPHORE. A stalked, often umbrella-shaped structure bearing thread-shaped conidia in certain lichens, especially those growing on leaves. See *Gomphillus*.

HYPOTHALLUS. A specialized tissue developing as a basal layer on certain foliose and squamulose lichens (e.g., *Anzia*, *Coccocarpia*; Fig. 6g); sometimes considered synonymous with *prothallus*, but not in this book.

HYPOTHECIUM. The tissue just below the hymenium (and subhymenium) but above the exciple (Fig. 13); often with a distinctive color or texture, as in *Buellia*, but sometimes merging with the exciple, as in *Biatora vernalis*.

IKI. A 1.5% solution of iodine in 10% potassium iodide; Lugol's solution (Chapter 13). Lower percentages of IKI are sometimes used for special types of staining procedures.

IMBRICATE. Overlapping in a shingle-like fashion; usually pertaining to scales or squamules, as in *Psora himalayana*.

INFLATED. Swollen and hollow, like the lobes of *Hypogymnia enteromorpha*.

INVOLUCRELLUM. The covering or cap external to the excipulum and present on many perithecia; usually black and carbonaceous (Fig. 17).

ISIDIUM (ISIDIA). A minute thalline outgrowth that is corticate and contains photobiont cells. Isidia are easily detached from the thallus and serve as vegetative reproductive units (Fig. 19d; plate 18).

ISOTOMIC. Dividing in regular dichotomies so that a main stem is not easily distinguishable, as in *Cladina stellaris*.

K. A 10% solution of potassium hydroxide (KOH) used in various microscopic preparations, or in spot tests for revealing certain lichen substances (see Chapter 13, table 2); can be substituted with household lye (sodium hydroxide, NaOH), 10 pellets per 20–30 ml (1–1.5 ounces) of water.

KC. A spot test for revealing certain lichen substances, performed by wetting the tested area with K followed by the application of C (see Chapter 13, table 2).

K/I. An analytical staining procedure for the hymenium or asci, performed by pretreating the tissue with K, removing the K by replacing it with water, and then replacing the water with 1.5% IKI. (See Chapter 13).

LABRIFORM. (1) Lip-shaped. (2) Pertaining to soralia, generally formed by an upturned thallus margin or a bursting hollow thallus lobe, sorediate on the lower or inside (i.e., exposed) surface, as in *Hypogymnia physodes* (Fig. 20e).

LAMINAL. On the upper surface of a thallus (laminal soralium: Fig. 20a).

LAX. Loose, not compact; usually referring to the medulla.

LECANORINE. Pertaining to an apothecium having a margin containing a photobiont, and usually resembling the thallus in color and texture, as in the genus *Lecanora* (Figs. 12b, 13c).

LECIDEINE. Pertaining to an apothecium with no photobiont cells in the margin, and in which the exciple is at least partially carbonized, forming a black apothecial margin, as in the genus *Lecidea* (Figs. 12a, 13a); see also *biatorine*.

LEPROSE. Composed entirely of soredia, referring to a thallus surface or the thallus itself (Fig. 2a; plate 24), as in *Lepraria*.

LICHEN. An association of a fungus and a photosynthetic symbiont (photobiont) resulting in a stable vegetative body with a specific structure, in which the fungus encloses the photobiont.

LICHENICOLOUS. Growing on or in a lichen, pertaining to fungi and lichens at various levels of parasitism with respect to the host lichen (plates 4, 5).

LICHENIZED. Pertaining to a fungus, alga, or cyanobacterium living within a lichen association.

LICHENOMETRY. A method of dating surfaces (usually rocks) using lichen growth rate and size. The size of the oldest lichen on a particular surface allows us to estimate how long that surface has been available for colonization (see Fig. 24; plates 26A,B).

LIGNICOLOUS. Growing on bare wood (lignum), as on a log or a wooden fence (plate 28).

LIRELLA (LIRELLAE). An elongated, sometimes branched apothecium, as in *Graphis* (Fig. 12e).

LOBE. A rounded or somewhat elongated division or projection of a thallus margin; measured at its widest point (Fig. 3).

LOBULE. A small, often scale-like lobe growing from a foliose thallus either along its margin or from the surface, sometimes also appearing along apothecial margins, generally of the same color and character as the parent thallus (Fig. 19f; plate 19). Lobules that are constricted at the base and function as propagules are often called *phyllidia*.

LOBULATE. Having many lobules.

LOCULE. The cell cavity in an ascospore; the locule sometimes develops a distinctive shape caused by unevenly thickened spore walls (Fig. 15c,d,i,j,l).

MACROLICHEN. A foliose or fruticose lichen.

MACULATE. Spotted or blotched by *maculae*, which are pale round or reticulate areas caused by gaps in the photobiont layer below the cortex (Fig. 4; plate 13).

MARGINAL. Along the thallus margins (Fig. 20b: marginal soralia).

MAZAEDIUM (MAZAEDIA). A dry, powdery mass of ascospores and paraphyses formed by the disintegration of the asci in the ascoma of some lichens, mainly in the order Caliciales, such as *Chaenotheca* and *Cyphelium* (Fig. 12f).

MEALY. Coarsely granular, like cornmeal.

MEIOSIS. A cell division in which the doubled chromosome number of a parent nucleus (the result of a sexual union) is reduced to a single set in each daughter nucleus, while providing an opportunity for the separating chromosomes to exchange parts, thereby increasing potential genetic diversity in the progeny.

MEDULLA. The internal layer in a thallus or lecanorine apothecium, generally composed of loosely packed fungal hyphae (Figs. 4, 13).

MICROLICHEN. A crustose or squamulose lichen. Dwarf fruticose lichens such as *Polychidium* and *Ephebe* are often considered to be microlichens.

MORPHOGENESIS. A change in form or shape over time, usually triggered by some external or internal event.

MORPHOLOGY. Physical characteristics, including external shape and internal anatomy.

MURIFORM. Spores divided into several cells by both longitudinal walls and crosswalls; the cells therefore look like stones or bricks in a wall (Fig. 15g,h).

MUSCICOLOUS. Growing over bryophytes, especially moss (e.g., like *Biatora vernalis*).

MUTUALISM. A type of symbiosis in which both components benefit from the association.

MYCOBIONT. The fungal symbiont or partner in a lichen.

MYRMECIA. A genus of green algae with round to slightly pear-shaped single cells containing a flat or (in lichen thalli) strongly folded and twisted chloroplast. Unlike *Trebouxia*, which it closely resembles, the chloroplast has no central pyrenoid.

NITROGEN FIXATION. The chemical transformation of atmospheric nitrogen, which cannot be used by plants, into substances such as nitrates and ammonium compounds that can be used by plants.

NONAMYLOID. Not turning blue with iodine (IKI), whether or not the tissue has been pretreated with K. (See also *hemiamyloid*.)

NOSTOC. A genus of cyanobacteria found in many lichens; producing bead-like chains or filaments of cells including thick-walled, colorless heterocysts (Fig. 1d); when lichenized, the filaments may be very short, consisting of only a few cells.

NUCLEUS (NUCLEI). A spherical structure within the cells of fungi, protoctists, and plants that contains genetic material, that is, the chromosomes and genes.

OCEANIC. Pertaining to a climate characterized by mild, wet winters, cool, moist summers, and frequent fogs.

OCULAR CHAMBER. A dome-shaped indentation of the lower edge of the tholus in certain types of ascus tips such as those of *Lecanora*-type asci (Fig. 14e).

ORCHIL. A reddish purple dye derived by "fermentation" methods from lichens such as *Roccella* (see Chapter 10, plate 69).

OSTIOLE. The small, usually round apical pore in pycnidia and various types of ascomata, especially perithecia, through which the ascospores escape (Fig. 17).

PAPILLAE. Small rounded or cylindrical bumps like "goose pimples," on the cortex of certain lichens, especially *Usnea* (Fig. 37a). Papillae do not contain medullary tissue.

PARAPHYSIS (PARAPHYSES). A sterile fungal filament, sometimes branched, attached at the base and free at the summit, associated with asci in the hymenium (Figs. 13, 16).

PARAPHYSOIDS. Sterile hymenial tissue between the asci, abundantly branched with frequent anastomoses, as in *Arthonia* (Fig. 13f).

PD. *Para*-phenylenediamine, a reagent used in spot tests for revealing certain lichen substances (see Chapter 13, table 2).

PELTATE. Attached at the center of the lower surface; umbrella-like.

PENDENT. Hanging straight down and usually soft and pliable (Fig. 2j; *Usnea longissima*); grades into *almost pendent* forms that are bushy and relatively stiff at the base but pendent over most of their length, as in *Usnea diplotypus*.

PERIPHYSES. Short, hair-like hyphae that sometimes line the inner walls of a perithecium (or an apothecium opening by a pore) near the ostiole (Fig. 17a).

PERISPORE. See *halo*.

PERITHECIUM (PERITHECIA). A flask-shaped ascoma opening by a pore at the summit (Fig. 17); may be prominent, but is more often partially or completely embedded in the thallus tissue; used here to include all similarly shaped ascomata regardless of their developmental origin (i.e., ascohymenial or ascolocular).

PHOTOBIONT. The photosynthetic component (symbiont) in a lichen thallus, either algae in the strict sense (e.g., green algae) or cyanobacteria (blue-green algae) (Fig. 1).

PHYLLIDIUM (PHYLLIDIA). See *lobule*.

PHYLLOCLADIUM (PHYLLOCLADIA). A minute, scale-like, lobed, cylindrical, or granular outgrowth on the stalks and branches of species of *Stereocaulon* (Fig. 8g).

PODETIUM (PODETIA). A stalk formed by a vertical extension of lower apothecial tissues (usually the hypothecium and stipe) and secondarily invested with an algal layer and sometimes a cortex, as in most specimens of *Cladonia*, or remaining almost free of vegetative tissue, as in *Dibaeis* (and, therefore, strictly speaking, are not true podetia). The fertile tissue or apothecial disks can be present (as in the red-fruited *Cladonia bellidiflora*) or commonly absent (as in *C. coniocraea*); podetia can be either short and unbranched (Fig. 2h) or quite tall and abundantly branched (Fig. 2i).

POLARILOCULAR. Pertaining to spores having two cell cavities (locules) separated by a relatively thick septum through which a narrow canal passes (Fig. 15d); characteristic of *Caloplaca* and related genera.

POLYPHYLLOUS. In species of *Umbilicaria* or *Dermatocarpon*: composed of numerous crowded lobes rather than a single peltate or umbilicate thallus (see *U. polyphylla*).

PRIMARY SQUAMULES. Small, scale-like lobes forming the basal or primary thallus of *Cladonia* species.

PRIMARY THALLUS. A squamulose or crustose thallus from which fruticose stalks or podetia arise as secondary components. Examples are found in *Cladonia*, *Pilophorus*, and *Baeomyces*.

PROPAGULE. A reproductive unit, either sexual (such as an ascospore) or vegetative (like a soredium or isidium).

PROSOPLECTENCHYMA. Fungal tissue consisting of coalesced, rather elongate hyphal cells often with thick, gelatinized walls (Fig. 5e,g). (Compare with pseudoparenchyma.)

PROTHALLUS. The purely fungal, white or darkly pigmented border of many crustose thalli, often visible as a fungal mat between the areoles or granules (Fig. 2b; see also *Lecanora thysanophora* and *Rhizocarpon geographicum*).

PRUINA. Powdery, frost-like deposit (usually white or gray, rarely yellow or reddish), typically composed of calcium oxalate or pigment crystals, dead cortical tissue, or some mixture of them; often occurs on a thallus (plate 11) or apothecial surface (see *Lecanora rupicola*).

PRUINOSE. Having a frosted appearance caused by a deposit of pruina.

PSEUDOCYPHELLA (PSEUDOCYPHELLAE). A tiny white dot or pore caused by a break in the cortex and the extension of medullary hyphae to the surface (Figs. 4, 11a,b).

PSEUDOPARAPHYSES. Branched, anastomosing hyphae between the asci in perithecium-like ascolocular ascomata, growing from the roof of the ascoma toward the base, as in *Anisomeridium*.

PSEUDOPARENCHYMA. Fungal tissue that appears cellular in section because of short, rounded to almost square cells of highly branched, irregularly oriented fungal filaments (Fig. 5d,f); sometimes called *paraplectenchyma*.

PSEUDOSTROMA (PSEUDOSTROMATA). Wart-like mass of vegetative fungal tissue containing some material from the substrate (usually bark) and supporting fruiting bodies of various kinds, as in

Trypethelium species and *Glyphis cicatricosa*. *Stromata*, which consist of fungal tissue alone, do not occur in lichens.

PUSTULE. A more or less hollow wart or verruca, small and knobby as in *Loxospora pustulata*, or broad and blister-like as in *Lasallia*.

PUSTULATE. Having many pustules.

PYCNIDIUM (PYCNIDIA). A small globular or flask-shaped body in which conidia are formed, embedded in a thallus or entirely superficial, often closely resembling a perithecium (Fig. 18).

PYRENOCARP (PYRENOCARPOUS). A lichen in which the fruiting bodies are perithecia; for example, *Verrucaria* or *Pyrenula*.

PYRENOID. A round, conspicuous body associated with the production and storage of starch and found within or attached to the chloroplast of some green algae. In *Trebouxia*, it is in the center of the chloroplast and is very conspicuous (often mistaken for the nucleus, which is almost invisible without staining and is in the colorless cytoplasm).

REFUGIUM. A geographic region that escaped destruction from continental glaciers or inland seas and thereby provided habitat for certain organisms during the ice ages or ancient periods of marine incursions; examples are the Yukon-Alaska corridor and the southern Appalachian Mountains.

RELICT. A remnant population far removed or disjunct from the main population center, usually surviving in a refugium.

RETICULATE. Net-like and interconnected, like the branches of *Ramalina menziesii*.

RHIZINE. A purely hyphal extension of the lower cortex that generally serves to attach a foliose thallus to its substrate; of various lengths, thicknesses, colors, and degrees of branching (Figs. 4, 6).

RIMOSE. Having a minutely cracked appearance (Fig. 2c).

SAPROPHYTE. An organism that lives on decaying organic matter.

SAXICOLOUS. Growing on rock, stone, pebbles, concrete, or brick.

SCABROSE. Having a minutely roughened, almost crusty surface, generally caused by an accumulation of dead cortical material (see *Peltigera scabrosa*); often intergrades with pruina, which is more powdery than crusty.

SCHIZIDIUM (SCHIZIDIA). A lichen fragment consisting of the upper layers of a thallus (with the cortex and photobiont) and serving as a vegetative reproductive unit; formed by a scaling off of the thallus surface (as in *Fulgensia*) or a breakdown of pustules (as in some species of *Hypotrachyna* or *Neofuscelia*) (Fig. 19g).

SCYTONEMA. A genus of cyanobacteria with a filamentous thallus (Fig. 1e) that branches by breaking through its gelatinous sheath (false branching).

SENSU LATO (S. LAT.) Meaning "in a broad sense"; usually used in reference to species or genera that are somewhat heterogeneous and may include other taxa.

SEPTUM (SEPTA). A cross wall in a fungal hypha or spore (Figs. 5, 15b–j).

SESSILE. Sitting on the surface, without a stalk of any kind, as with the apothecia of *Lecanora pacifica*.

SILICEOUS. Pertaining to a rock or soil rich in silica and lacking calcium; examples are granite, certain types of gneiss, syenite, and quartz sand.

SORALIUM (SORALIA). An area of a thallus in which the cortex has broken down or cracked and soredia are produced; can be in many forms (Fig. 20; plate 17); sometimes contains isidia as well as soredia, as in *Lobaria pulmonaria* and *Usnea diplotypus*.

SOREDIUM (SOREDIA). A vegetative propagule of a lichen consisting of a few algal cells entwined and surrounded by fungal filaments, and without a cortex; generally produced in localized masses called *soralia*, or covering large, diffuse areas on a thallus (Fig. 19h–j).

SP. Abbreviation of "species," generally used where the species is unknown or unspecified.

SPECIES. The basic evolutionary unit of an organism; named with two words, the first being the genus to which the species belongs, and the second the species' own name or "epithet." With respect to lichens, the name of a species refers to its fungal component. Lichen species are defined by discontinuities in various morphological, chemical, ecological, and geographic characteristics and, more and more frequently in recent years, by analysis of the actual genetic material (DNA) of the fungal component.

spore. A single- or multicelled reproductive body capable of giving rise to a new organism; as used here, refers specifically to an ascospore (Fig. 15).

spp. Several unspecified species.

squamule. A small, scale-like lobe or areole, lifting from the surface, at least at the edges (as in *Acarospora fuscata* or *Peltula* species), and sometimes strongly ascending and almost foliose (as in some species of *Cladonia*).

squamulose. Composed of or characterized by having squamules (Fig. 2e).

squarrose. With short, stiff, perpendicular branches; having the general appearance of a bottlebrush, as in certain types of rhizines (Fig. 6d–e).

ssp. Subspecies, a formal subdivision of a species, used for important morphologically, chemically, or ecologically distinct segregates, usually somewhat geographically isolated.

stereome. A tough, cartilaginous cylinder forming the supporting tissue for species of *Cladonia* and *Cladina* (Fig. 8f).

stichococcus. A small, unicellular green alga having short, cylindrical (rod-shaped) cells (Fig. 1g).

stigonema. A genus of filamentous cyanobacteria having "true branching" resulting from perpendicular division of cells within the filament (Fig. 1a). (Compare with *Scytonema*, Fig. 1e). Lichens with *Stigonema* (e.g., *Ephebe*) do not look very different from the free-living cyanobacteria.

stratified. Layered; in reference to lichen thalli having distinguishable layers of tissue including a cortex, photobiont layer, medulla, and often, a lower cortex (Fig. 4).

striate. With parallel fine longitudinal lines, ridges, or grooves.

stroma (stromata). See *pseudostroma*.

sub- (1) Partially. (2) Incompletely. (3) Approaching. (4) Under; as in *subfoliose* (not quite foliose), *submarginal* (close to the margin), *submuriform* (transversely septate spores having only a few longtudinal septa), *subpendent* (almost pendent), or *subhymenium* (the layer often distinguishable just below the hymenium and above the hypothecium).

substrate. The surface upon which a lichen grows; a nutritional relationship is not implied and rarely occurs in lichens.

superficial. Used here with reference to apothecia and other structures that sit on the surface of the thallus (i.e., sessile, not immersed or stalked).

symbiosis. A long-term, usually physical association between at least two dissimilar organisms. Mutualism and parasitism are types of symbiosis in which one or both organisms are changed in some way, either beneficially or detrimentally. In commensalism, neither of the associated organisms is affected.

taxon (taxa). A unit in a classification scheme; most commonly used with reference to a genus, species, or subdivision of a species (subspecies, variety, or form).

taxonomy. The study of taxa, especially with respect to their identification and classification, and the correct application of their names.

terete. Round in cross section (cylindrical) like the branches of *Bryoria* species.

terricolous. Growing on soil, sand, or peat.

thalline. Pertaining to the lichen thallus; similar to the thallus in appearance or structure.

thalline margin. See *amphithecium*.

thallus. In lichens, the vegetative body consisting of both algal and fungal components (Fig. 2).

thin-layer chromatography. A method used for the identification of chemical compounds, specifically lichen substances (see Chapter 13, plate 89).

tier. A platform-like expansion or flat cup on the podetia of some species of *Cladonia* (e.g., *Cladonia cervicornis* ssp. *verticillata*), often proliferating from the center or margins with one or more new branches.

tholus. The thickened tip of an ascus, frequently staining in iodine (IKI or K/I) in various ways (Fig. 14).

tlc. Thin-layer chromatography.

tomentose. Having tomentum; with a downy or woolly appearance.

tomentum. A covering of fine hair or fuzz usually caused by a superficial growth of colorless hyphae (plate 12).

TREBOUXIA. A genus of single-celled green algae with one distinctive, disk-shaped chloroplast almost filling the cell. The chloroplast has a lobed or scalloped margin and a single round to oval pyrenoid at the center. *Trebouxia* is the most common green photobiont in lichens. When followed by a query (?) in the text, it is used in the broad sense, including *Pseudotrebouxia* (Fig. 1f).

TRENTEPOHLIA. A genus of filamentous green algae found in many crustose lichens (Fig. 1j); when lichenized, the alga often produces very short filaments or is single-celled. The orange-red pigmented globules, common in the cells of unlichenized individuals, are infrequent or absent in lichenized individuals.

TUBERCLE. A wart-like protuberance that contains some medullary tissue; characteristic of *Usnea ceratina* and *Ramalina paludosa*.

TUBERCULATE. (1) Having the general form of a tubercle, usually referring to small, round soralia (as in *Bryoria fuscescens*); (2) having many tubercles, as in some species of *Usnea*.

UMBILICATE. Attached by a single, central holdfast (an *umbilicus*) on the lower surface of the thallus (Fig. 2g).

UMBILICUS. A short, thick, purely fungal, central attachment organ present on certain foliose lichens such as *Umbilicaria* and *Dermatocarpon*.

UMBO. The (1) central bump on the upper surface of an umbilicate lichen, corresponding to the position of the umbilicus; (2) a bump of sterile tissue in the center of an apothecial disk.

UV. Ultraviolet light. A number of lichen substances fluoresce and therefore can be detected in long-wave UV light (365 nm) (plate 90); shortwave UV light (254 nm) is used in thin-layer chromatography for analyzing plates made with gels containing certain fluorescent dyes.

VAGRANT. Growing unattached to rock or soil and therefore able to roll freely over the ground (plate 34).

VAR. Variety; a formal subdivision of a species or subspecies, usually used for recurring, genetically based variants in morphology, chemistry, or habitat.

VEIN. In lichens, broad or narrow ridges or thickenings, often pigmented, on the lower surface of some lichens such as *Peltigera* (e.g., *P. kristinssonii*), but not functioning as conducting tissue.

VERRUCA (VERRUCAE). A conspicuous, wart-like, thalline protuberance (e.g., *Pertusaria plittiana*).

VERRUCOSE. With a rough, warty surface (e.g., *Ochrolechia oregonensis*).

VERRUCULOSE. Minutely verrucose, like the thallus of *Lecanora circumborealis*.

WOOD. Lignum; trunks, logs, and stumps having no bark.

XANTHOCAPSA. Cyanobacteria consisting of groups of 2–8 spherical cells enclosed within a thick brownish to yellow-brown gelatinous matrix. Gelatinous sheaths of individual cells usually are K–.

FURTHER READING AND BIBLIOGRAPHY

The references listed below include those that were particularly important in the preparation of the introductory chapters of the book, as well as those that would be of interest to readers wanting more information about particular subjects. The organization of the sections only approximates the chapter headings in order to avoid needless duplication. Readers should be aware that many of the books in the "General" section provide the most up-to-date treatments of the topics listed separately. With few exceptions, we have listed only those papers and books that are in English, although there are many excellent works in foreign languages that are useful to those interested in North American lichen identification.

A great deal of up-to-date information on North American lichenology and lichenologists can be found on the Internet webpage of The American Bryological and Lichenological Society (ABLS) and the linkages provided there: *http://www.unomaha.edu/~abls/*.

A complete, searchable database of lichen references based on T. Esslinger's "Recent Literature on Lichens" published in *The Bryologist* is available on the Internet. See Culberson, Egan, and Esslinger (2000) below.

GENERAL, INCLUDING ANATOMY, MORPHOLOGY, PHYSIOLOGY, AND ECOPHYSIOLOGY

French or Spanish may be the first language of many of our readers. Pour un ouvrage général en français, consultez: Van Haluwyn et LeRond (1993); por una obra en español, puede consultar: Marcano (1994).

Ahmadjian, V. 1993. *The lichen symbiosis.* New York: John Wiley & Sons.

Ahmadjian, V., and M. E. Hale, eds. 1974. *The lichens.* New York: Academic Press.

Brown, D. H., D. L. Hawksworth, and R. H. Bailey. 1976. *Lichenology: Progress and problems.* New York: Academic Press.

Culberson, W. L., R. S. Egan, and T. L. Esslinger. 2000. Search recent literature on lichens. Online: *http://www.toyen.uio.no/botanisk/bot-mus/lav/sok_rll.htm*.

Galun, M., ed. 1988. *CRC handbook of lichenology*, 3 vols. Boca Raton, Fla.: CRC Press.

Hale, M. E., Jr. 1983. *The biology of lichens,* 3d ed. London: Edward Arnold.

Hawksworth, D. L., and D. J. Hill. 1984. *The lichen-forming fungi.* New York: Chapman and Hall.

Jahns, H. M. 1974. Anatomy, morphology, and development. In *The lichens*, edited by V. Ahmadjian and M. E. Hale. New York: Academic Press.

Kershaw, K. A. 1985. *Physiological ecology of lichens*. Cambridge: Cambridge University Press.

Lawrey, J. D. 1984. *Biology of lichenized fungi.* New York: Praeger.

Marcano, V. 1994. *Introducción al estudio de los líquenes y su clasificación.* Merida, Venezuela: Ed. Museo de Ciencia, Tecnología, Artes y Oficios. [A general textbook on lichens, in Spanish.]

Nash, T. H., III, ed. 1996. *Lichen biology.* Cambridge: Cambridge University Press.

Ozenda, P. 1963. *Lichens. Handbuch der Pflanzenanatomie 6(9), Abt.: Spez. Teil.* Berlin-Nikolassee: Gebrüder Borntraeger. [A general treatment of anatomy and morphology, in French.]

Purvis, O. W., B. J. Coppins, D. L. Hawksworth, P. W. James, and D. M. Moore. 1992. *The lichen flora of Great Britain and Ireland.* London: Natural History Museum.

Rikkinen, J. 1995. What's behind the pretty colours? A study on the photobiology of lichens. *Bryobrothera* 4:1–239.

Van Haluwyn, Ch., and M. Lerond 1993. *Guide des Lichens.* Paris: Éditions Lechevalier. [A general textbook on lichens, in French.]

ECOLOGY

Ahti, T. 1964. Macrolichens and their zonal distribution in boreal and arctic Ontario, Canada. *Annales Botanici Fennici* 1:1–35.

Barkman, J. J. 1958. *Phytosociology and ecology of cryptogamic epiphytes: Including a taxonomic survey and description of their vegetation units in Europe.* Assen, Netherlands: Van Gorcum.

Brodo, I. M. 1974. Substrate ecology. In *The lichens*, edited by V. Ahmadjian and M. E. Hale. New York: Academic Press.

Crittenden, P. D., and K. A. Kershaw. 1978. Discovering the role of lichens in the nitrogen cycle in Boreal-Arctic ecosystems. *Bryologist* 81:258–267.

Denison, W. C. 1973. Life in tall trees. *Scientific American* 228(5):75–80.

———. 1979. *Lobaria oregana*, a nitrogen-fixing lichen in old-growth Douglas fir forests. In *Symbiotic nitrogen fixation in the management of temperate forests*: proceedings of a workshop, April 2–5, 1979, edited by J. C. Gordon, C. T. Wheeler, and D. A. Perry. Corvallis, Ore.: Forest Research Laboratory, Oregon State University.

Ferry, B. W. 1982. Lichens. In *Experimental microbial ecology*, edited by R.G. Burns and J.H. Slater. Oxford: Blackwell Scientific Publications.

Forman, R. T. T., and D. L. Dowden. 1977. Nitrogen fixing lichen roles from desert to alpine in the Sangre de Cristo Mountains, New Mexico. *Bryologist* 80:561–570.

Gagnon, J.-D. 1966. Le lichen *Lecidea granulosa* constitue un milieu favorable à la germination de l'Epinette noire. *Naturaliste canadien* 93:89–98.

Kershaw, K. A. 1976. Studies on lichen-dominated systems. XX. An examination of some aspects of the northern boreal lichen woodlands in Canada. *Canadian Journal of Botany* 55:393–410.

Knops, J. M. H., T. H. Nash III, V. L. Boucher, and W. H. Schlesinger. 1991. Mineral cycling and epiphytic lichens: implications at the ecosystem level. *Lichenologist* 23:309–321.

Lawrey, J. D. 1986. Biological role of lichen substances. *Bryologist* 89:111–122.

Millbank, J. W. 1978. The contributions of nitrogen fixing lichens to the nitrogen status of their environment. In *Environmental role of nitrogen-fixing blue-green algae and asymbiotic bacteria*, edited by U. Granhall. Stockholm: Swedish Natural Science Research Council.

Seaward, M. R. D. ed. 1977. *Lichen ecology.* London: Academic Press.

———. 1988. Contributions of lichens to ecosystems. In *Handbook of lichenology*, vol. 2, edited by M. Galun. Boca Raton, Fla.: CRC Press.

Syers, J. K., and I. K. Iskandar. 1974. Pedogenetic significance of lichens. In *The lichens*, edited by V. Ahmadjian and M. E. Hale. New York: Academic Press.

ANIMALS AND LICHENS

For a more extensive bibliography see Sharnoff & Rosentreter (1998) below.

Ahti, T., and R.L. Hepburn. 1967. Preliminary studies on woodland caribou range, especially on lichen stands, in Ontario. *Ontario Department of Lands and Forests Research Report (Wildlife)* 74:1–134.

Armleder, H. M., S. K. Stevenson, and S. D. Walker. 1992. *Estimating the abundance of arboreal forage lichens.* Land Management Handbook, field guide insert 7. Victoria: British Columbia Ministry of Forests.

Bergerud, A. T. 1972. Food habits of Newfoundland caribou. *Journal of Wildlife Management* 36:913–923.

Edwards, R. Y., and R. W. Ritcey. 1960. Foods of caribou in Wells Gray Park, British Columbia. *Canadian Field-Naturalist* 74:3–7.

Fleischner, T. L. 1994. Ecological costs of livestock grazing in western North America. *Conservation Biology* 8: 629–644.

Fox, J. L., C. A. Smith, and J. W. Schoen. 1989. *Relationship between mountain goats and their habitat in Southeastern Alaska.* General Technical Report PNW-GTR-246. Portland, Ore.: USDA Forest Service, Pacific Northwest Research Station.

Gerson, U., and M. R. D. Seaward. 1977. Lichen-invertebrate associations. In *Lichen Ecology*, edited by M. R. D. Seaward. London: Academic Press.

Hayward, G. D., and R. Rosentreter. 1994. Lichens as nesting material for northern flying squirrels in the northern Rocky Mountains. *Journal of Mammalogy* 75: 663–673.

Hodgman, T. P., and R. T. Bowyer. 1985. Winter use of arboreal lichens, Ascomycetes, by white-tailed deer,

Odocoileus virginianus, in Maine. *Canadian Field-Naturalist* 99:313–316.

Jobin, L. 1973. L'arpenteuse de la pruche. *Feuillet d'Information CRFL* 4:1–6. Ste-Foy: Environnement Canada, Centre de Recherches Forestières des Laurentides.

Lawrey, J. D. 1987. Nutritional ecology of lichen/moss arthropods. In *Nutritional ecology of insects, mites, spiders, and related invertebrates*, edited by F. Slansky Jr. and J. G. Rodriguez. New York: John Wiley & Sons.

Longhurst, W. M., G. E. Connolly, B. M. Browning, and E. O. Garton. 1979. Food interrelationships of deer and sheep in parts of Mendocino and Lake Counties, California. *Hilgardia* 47:191–247.

Martell, A. M. 1981. Food habits of southern red-backed voles (*Clethrionomys gapperi*) in northern Ontario. *Canadian Field-Naturalist* 95:325–328.

Maser, Z., C. Maser, and J. M. Trappe. 1985. Food habits of the northern flying squirrel (*Glaucomys sabrinus*) in Oregon. *Canadian Journal of Zoology* 63:1084–1088.

Meehan, W. R., T. R. Merrell, and T. A. Hanley, eds. 1984. *Fish and wildlife relationships in old-growth forests*: proceedings of a symposium, Juneau, Alaska, 12–15 April 1982. [US]: American Institute of Fishery Research Biologists.

Richardson, D. H. S., and C. M. Young. 1977. Lichens and vertebrates. In *Lichen ecology*, edited by M. R. D. Seaward. London: Academic Press.

Scotter, G. W. 1961. Productivity of arboreal lichens and their possible importance to barren-ground caribou (*Rangifer arcticus*). *Archivum Societatis Zoologicae botanicae Fennicae 'Vanamo'* 16:155–161.

———. 1967. The winter diet of barren-ground caribou in northern Canada. *Canadian Field-Naturalist* 81: 33–39.

———. 1972. Reindeer ranching in Canada. *Journal of Range Management* 25:167–174.

Seyd, E. L., and M. R. D. Seaward. 1984. The association of oribatid mites with lichens. *Zoological Journal of the Linnean Society* 80:369–420.

Sharnoff, S., and R. Rosentreter. 1998. Lichen use by wildlife in North America. Online: *http://www.lichen.com/fauna.html*

Stevenson, S. K., and J. A. Rochelle. 1984. Lichen litterfall—its availability and utilization by black-tailed deer. In *Fish and wildlife relationships in old-growth forests*: proceedings of a symposium, Juneau, Alaska, 12–15 April 1982, edited by W. R. Meehan, T. R. Merrell, and T. A. Hanley, [US]: American Institute of Fishery Research Biologists.

Ure, D. C., and C. Maser. 1982. Mycophagy of red-backed voles in Oregon and Washington. *Canadian Journal of Zoology* 60:3307–3315.

GEOGRAPHY OF NORTH AMERICAN LICHENS

Only general references are listed here. Sixty-five other publications were consulted in the plotting of the distribution maps.

Barbour, M. G., and W. D. Billings. 1988. *North American Terrestrial Vegetation*. Cambridge: Cambridge University Press.

Brodo, I. M., and S. P. Gowan. 1983. Un aperçu de la répartition des lichens de l'est de l'Amérique du Nord. *Bulletin de la Société botanique du Québec* 5:13–30.

Daubenmire, R. 1978. *Plant geography, with special reference to North America*. New York: Academic Press.

Ecological Stratification Working Group. 1996. *A national ecological framework for Canada*. Ottawa: Agriculture and Agri-Food Canada, State of the Environment Directorate, Environment Canada.

Gowan, S. P. 1983. "The phytogeography of the lichens of Fundy National Park." Master's thesis, Carleton University.

Hale, M. E. 1961. *Lichen handbook: A guide to the lichens of eastern North America*. Washington, D.C.: Smithsonian Institution, Publication 4434.

Commission for Environmental Cooperation. 1997. *Ecological regions of North America: Towards a common perspective*. Montreal: Commission for Environmental Cooperation.

Vankat, J. L. 1979. *The natural vegetation of North America*. New York: John Wiley & Sons.

LICHENS AND PEOPLE: USES AND IMPACTS

For a more extensive bibliography see Sharnoff (1997) below.

Brough, S. G. 1984. Dye characteristics of British Columbia forest lichens. *Syesis* 17:81–94.

Casselman, K. D. 1996. *Lichen dyes: A source book*. Cheverie, N.S.: Studio Vista.

———. 1999. "Lichen dyes & dyeing: A critical bibliography of European and North American literature in a culturally marginalized field." Master's thesis, Saint Mary's University.

Casselman, K. L. 1993. *Craft of the dyer: Colour from plants and lichens*, 2d ed. New York: Dover Publications.

Herre, A. W. C. T. 1936. Our vanishing lichen flora. *Madroño* 3:198–200.

Innes, J. L. 1988. The use of lichens in dating. In *Handbook of lichenology*, vol. 3, edited by M. Galun. Boca Raton, Fla.: CRC Press.

Kok, A. 1966. A short history of orchil dyes. *Lichenologist* 3:248–272.

Llano, G. A. 1944. Lichens—their biological and economic significance. *Botanical Review* 10:1–65.

———. 1948. Economic uses of lichens. *Economic Botany* 2:15–45.

———. 1956. Utilization of lichens in the arctic and subarctic. *Economic Botany* 10:367–392.

McGrath, J. W. 1977. *Dyes from plants and lichens*. Toronto: Van Nostrand Reinhold.

Moerman, D. E. 1998. *Native American ethnobotany*. Portland, Ore.: Timber Press.

Moxham, T. H. 1981. Lichens in the perfume industry. *Dragoco Report* 2:31–39.

Richardson, D. H. S. 1975. *The vanishing lichens*. Newton Abbot, England: David & Charles.

———. 1988. Medicinal and other economic aspects of lichens. In *Handbook of lichenology*, vol. 3, edited by M. Galun. Boca Raton, Fla.: CRC Press.

Sharnoff, S. D. 1997. Lichens and people. Online: *http://www.lichen.com/people.html*

Turner, N. J. 1977. Economic importance of black tree lichen (*Bryoria fremontii*) to the Indians of western North America. *Economic Botany* 31:461–470.

Turner, N. J., J. Thomas, B. F. Carlson, and R. T. Ogilvie. 1983. Ethnobotany of the Nitinaht Indians of Vancouver Island. Victoria: *Occasional papers of the British Columbia Provincial Museum* No. 24.

ENVIRONMENTAL MONITORING

The USDA Forest Service Region 6 (Pacific Northwest) maintains a website on lichens and air quality. Online: *http://www.fs.fed.us/r6/aq/lichen/*

Ferry, B. W., M. S. Baddeley, and D. L. Hawksworth, eds. 1973. *Air pollution and lichens*. Toronto: University of Toronto Press.

Goward, T. 1994. Notes on oldgrowth-dependent epiphytic macrolichens in inland British Columbia, Canada. *Acta Botanica Fennica* 150:31–38.

Huckaby, L. S., ed. 1993. Lichens as bioindicators of air quality. General Technical Report RM-224. Fort Collins, Colo.: USDA Forest Service, Rocky Mountain Forest and Range Experiment Station.

LeBlanc, F., and J. DeSloover. 1970. Relation between industrialization and the distribution and growth of epiphytic lichens and mosses in Montreal. *Canadian Journal of Botany* 48:1485–1496.

Richardson, D. H. S. 1992. *Pollution monitoring with lichens*. Naturalists' Handbooks 19. Slough, England: Richmond Publishing.

Selva, S. B. 1994. Lichen diversity and stand continuity in the northern hardwoods and spruce-fir forests of northern New England and western New Brunswick. *Bryologist* 97:424–429.

Wetmore, C. M. 1988. Lichens and air quality in Indiana Dunes National Lakeshore. *Mycotaxon* 33:25–39.

———. 1988. Lichens of Sleeping Bear Dunes National Seashore. *Michigan Botanist* 27:111–118.

CHEMISTRY

Carlin, G. 1999. Chemistry. In *Nordic lichen flora,* vol. 1, edited by T. Ahti, et al., Bohuslän, Sweden: Nordic Lichen Society. [Includes a table of chemical spot tests for 60 lichen compounds.]

Culberson, C. F. 1969. *Chemical and botanical guide to lichen products*. Chapel Hill: University of North Carolina Press.

———. 1970. Supplement to "Chemical and botanical guide to lichen products." *Bryologist* 73:177–377.

———. 1972. Improved conditions and new data for the identification of lichen products by a standardized thin-layer chromatographic method. *Journal of Chromatography* 72:113–125. [The basic reference for TLC work.]

Culberson, C. F., and A. Johnson. 1982. Substitution of methyl *tert.*-butyl ether for diethyl ether in the standardized thin-layer chromatographic method for lichen products. *Journal of Chromatography* 238: 483–487.

Culberson, C. F., W. L. Culberson, and A. Johnson. 1977. *Second supplement to "Chemical and Botanical Guide to Lichen Products."* St. Louis, Mo.: American Bryological and Lichenological Society.

Elix, J. A. 1996. Biochemistry and secondary metabolites. In *Lichen biology*, edited by T. H. Nash III, Cambridge: Cambridge University Press.

Feige, G. B., H. T. Lumbsch, S. Huneck, and J. A. Elix. 1993. Identification of lichen substances by a standardized high-performance liquid chromatographic method. *Journal of Chromatography* 646:417–427.

Huneck, S., and I. Yoshimura. 1996. *Identification of lichen substances*. Berlin: Springer-Verlag.

Mietzsch, E., H. T. Lumbsch, and J. A. Elix. 1992. *Wintabolites (Mactabolites for Windows), users manual*. Essen, Germany: Universität Essen.

White, F. J., and P. W. James. 1985. A new guide to microchemical techniques for the identification of lichen substances. *British Lichen Society Bulletin* 57 (supplement):1–41.

THE TAXONOMY AND CLASSIFICATION OF LICHENS

Esslinger, T. L. 1999. A cumulative checklist for the lichen-forming, lichenicolous and allied fungi of the continental United States and Canada. North Dakota State University. Online: *http://www.ndsu.nodak.edu/instruct/esslinge/chcklst/chcklst7.htm*

Esslinger, T. L., and R. S. Egan. 1995. A sixth checklist for the lichen-forming, lichenicolous and allied fungi of the continental United States and Canada. *Bryologist* 98:467–549.

Hawksworth, D. L. 1974. *Mycologist's handbook: An introduction to the principles of taxonomy and nomenclature in the fungi and lichens*. Kew: Commonwealth Mycological Institute.

Hawksworth, D. L., P. M. Kirk, B. C. Sutton, and D. N. Pegler. 1995. *Dictionary of the fungi*, 8th ed. Egham, England: International Mycological Institute.

Harris, R. C. 1994. *A guide to the higher groups of New York State lichens.* New York: R. C. Harris.

———. 1995. *More Florida lichens, including the 10 cent tour of the pyrenolichens.* New York: R. C. Harris.

Tehler, A. 1996. Systematics, phylogeny and classification. In *Lichen biology*, edited by T. H. Nash III. Cambridge: Cambridge University Press.

IDENTIFICATION GUIDEBOOKS WITH KEYS

A searchable, downloadable list of all literature pertaining to the identification of North American lichens is available on the Internet. See May and Brodo (2000) below. Other books and papers are the most useful regional guides, most of which include keys.

Brodo, I. M. 1968. The lichens of Long Island, New York: A vegetational and floristic analysis. *New York State Museum & Science Service Bulletin* 410:1–330.

———. 1988. Lichens of the Ottawa region, 2d ed. *Ottawa Field-Naturalists' Club Special Publication* 3:1–115.

———. 1990. *Lichens de la région d'Ottawa*, 2d ed. Ottawa: Musée national des sciences naturelles. [A French edition of Brodo (1988).]

Dey, J. P. 1978. Fruticose and foliose lichens of the high-mountain areas of the southern Appalachians. *Bryologist* 81:1–93.

Fink, B. 1910. The lichens of Minnesota. *Contributions from the U.S. National Herbarium* 14(1):1–269.

———. 1935. *The lichen flora of the United States.* Ann Arbor: University of Michigan Press.

Flenniken, D. G. 1999. *The macrolichens in West Virginia.* Sugarcreek, Ohio: D. G. Flenniken.

Gowan, S. P., and I. M. Brodo. 1988. The lichens of Fundy National Park, New Brunswick, Canada. *Bryologist* 91:i, 255–325.

Goward, T. 1999. *The lichens of British Columbia, illustrated keys.* Part 2, *Fruticose species.* Victoria: British Columbia Ministry of Forests, Research Program, Special Report Series no. 9.

Goward, T., B. McCune, and D. Meidinger. 1994. *The lichens of British Columbia, illustrated keys.* Part 1, *Foliose and squamulose species.* Victoria: British Columbia Ministry of Forests, Research Program, Special Report Series no. 8.

Hale, M. E. 1979. *How to know the lichens*, 2d ed. Dubuque, Iowa: W. C. Brown.

Hale, M. E., and M. Cole. 1988. *Lichens of California.* Berkeley: University of California Press.

Harris, R. C. 1995. *More Florida lichens, including the 10 cent tour of the pyrenolichens.* New York: R. C. Harris.

Howard, G. E. 1950. *Lichens from the state of Washington.* Seattle: University of Washington Press.

May, P. F., and I. M. Brodo. 2000. Identifying North American Lichens—A Guide to the literature. Farlow Herbarium, Cambridge, Mass. Online: *http://www.herbaria.harvard.edu/Data/Farlow/lichenguide/index.html.*

McCune, B., and L. Geiser. 1997. *Macrolichens of the Pacific Northwest.* Corvallis: Oregon State University Press.

McCune, B., and T. Goward. 1995. *Macrolichens of the northern Rocky Mountains.* Eureka, Calif.: Mad River Press.

Medlin, J. J. 1996. *Michigan lichens.* Bloomfield Hills, Mich.: Cranbrook Institute of Science Bulletin 60.

Nearing, G. G. 1947. *The lichen book. Handbook of the lichens of Northeastern United States.* Ridgewood, N.J.: G. G. Nearing.

Noble, W. 1982. "The lichens of the coastal Douglas-Fir dry subzone." Ph.D. dissertation, University of British Columbia. [Reformatted edition available from Bruce McCune, 1840 NE Seavy Avenue, Corvallis, Oregon; *mccune@proaxis.com*]

St. Clair, L. L. 1999. *A color guidebook to the common Rocky Mountain lichens.* Provo, Utah: Bean Life Science Museum of Brigham Young University.

Thomson, J. W. 1979. *Lichens of the Alaskan Arctic slope.* Toronto: University of Toronto Press.

———. 1984. *American Arctic lichens.* vol. 1, *The macrolichens.* New York: Columbia University Press.

———. 1997. *American Arctic lichens.* vol. 2, *The microlichens.* Madison: The University of Wisconsin Press.

Tønsberg, T. 1992. The sorediate and isidiate, corticolous, crustose lichens in Norway. *Sommerfeltia* 14:1–331.

Vitt, D. H., J. E. Marsh, and R. B. Bovey. 1988. *Mosses, lichens & ferns of northwest North America.* Edmonton, Alberta: Lone Pine Publishing.

Wetmore, C. M. 1967. *Lichens of the Black Hills of South Dakota and Wyoming.* East Lansing: Publications of the Museum, Michigan State University, Biological Series, vol. 3, no. 4.

———. 1976. Macrolichens of Big Bend National Park, Texas. *Bryologist* 79:296–313.

INDEX OF NAMES

Note: Names in **boldface** indicate a main entry with an illustration; *italicized* names are synonyms; plain text is used for other species mentioned in the text or keys, as well as for English names. Page numbers in **boldface** indicate main entries; *italics* indicate entries in the keys; plain text is used for all other places the name occurs in the text. The author names and their abbreviations follow those in the North American lichen checklist (Esslinger 1999).

Absconditella Vězda, *128*, 302, 752
Acarospora A. Massal., 18, 55, 70, *128*, *132*, *134*, *136*, **145**, *145–146*, 167, 207, 210, 578, 586, 659, 752
　badiofusca (Nyl.) Th. Fr., 147
　chlorophana (Wahlenb.) A. Massal., 578
　contigua H. Magn., *134*, *145*, **146**, 148
　fuscata (Schrader) Arnold, 8, *145*, **146–147**, 763
　glaucocarpa (Ach.) Körber, *146*, **147**
　heppii (Nägeli *ex* Hepp) Nägeli *ex* Körber, *145*, 147
　macrospora (Hepp) Bagl., 147
　nodulosa (Dufour) Hue, 148
　schleicheri (Ach.) A. Massal., *145*, *146*, **147–148**, 166, 578
　smaragdula (Wahlenb.) A. Massal., 147
　stapfiana (Müll. Arg.) Hue, 7, 148
　strigata (Nyl.) Jatta, *146*, **148**
　thelococcoides (Nyl.) Zahlbr., 148
Acarosporaceae, 752
Acarosporineae, 752
Acrocordia A. Massal., 753
　cavata (Ach.) R. C. Harris, *126*
　conoidea (Fr.) Körber, *125*
Acrocyphus Léveillé, 752
　sphaerophoroides Léveillé, 657
Agaricales, 751
Agrestia J. W. Thomson, 170
　hispida (Mereschk.) Hale & Culb., 169
Agyriales, 751
Agyriaceae, 751
Agyrium Fr., 751
　rufum (Pers.) Fr., 749
Ahtiana Goward, 95, *139*, *148*, **148**, 156, 158, 479, 596, 730, 752
　aurescens (Tuck.) Thell & Randlane, *148*, **149**
　pallidula (Tuck. *ex* Riddle) Goward & Thell, 73, *148*, **149**
　sphaerosporella (Müll. Arg.) Goward, 65, *148*, 149, **150**, 316
Alectoria Ach., 16, 30, 46, 60, 79, 83, 95, *119*, **150**, *150–152*, 178, 179, 311, 460, 620, 672, 710, 752
　fallacina Mot., 154
　imshaugii Brodo & D. Hawksw., 72, *150*, **152**, 154
　"jubata," 78
　lata (Taylor) Lindsay, *150*, **152–153**, 154
　nigricans (Ach.) Nyl., *120*, *121*, *151*, **153**, 154, 184
　ochroleuca (Hoffm.) A. Massal., 64, 67, 79, *150*, **153–154**
　sarmentosa (Ach.) Ach., 60, 63, 76, *151*,153, **154**, 155, 625–626, 632, 710
　sarmentosa ssp. vexillifera (Nyl.) D. Hawksw., 154
　vancouverensis (Gyelnik) Gyelnik *ex* Brodo & D. Hawksw., 154
Allantoparmelia (Vainio) Essl., *140*, *141*, **154**, *154*, 178, 345, 422, 752
　almquistii (Vainio) Essl., 156
　alpicola (Th. Fr.) Essl., *154*, **154**, 156
Allocetraria Kurokawa & Lai, 14, 148, **156**, 231, 752
　madreporiformis (Ach.) Kärnef. & Thell, *120*, **156–157**, *293*, 295
　oakesiana (Tuck.) Randlane & Thell, 69, *138*, *139*, 148, 149, **157–158**, 316, 731, 733

almond lichens, 159
   powdery, 159
Amandinea M.Choisy *ex* Scheid. & H. Mayrh., *131*, **158**, 186, 644, *752*
   coniops (Wahlenb.) M.Choisy *ex* Scheid. & H. Mayrh., 158
   dakotensis (H. Magn.) P. May & Sheard, 158
   milliaria (Tuck.) P. May & Sheard, 158
   polyspora (Willey) E. Lay & P. May, 158
   punctata (Hoffm.) Coppins & Scheid., **158,** 159, *186*
Amygdalaria Norman, **159**, 167, *391*, 583, 633, 752
   panaeola (Ach.) Hertel & Brodo, 20, *123*, **159**
Anaptychia Körber, 14, **160**, 546, 560, 752
   bryorum Poelt, 160
   "*palmatula*" (as used by some authors), 160
   palmulata (Michaux) Vainio, 33, 69, *141*, **160**
   setifera Räsänen, *119*, *140*, *141*, **160–161,** *339*
   ulotrichoides (Vainio) Vainio, 541, 566
angel's hair, 630
Anisomeridium (Müll. Arg.) M. Choisy, 753
   nyssaegenum (Ellis & Everh.) R. C. Harris, *126*
   polypori (Ellis & Everh.) M. E. Barr, *126*
antler lichens, 588
   common, 591
   ghost, 589
   powder-tipped, 313
   western, 591
Anzia Stizenb., 14, **161**, 752, 759
   americana Yoshim. & Sharp, 161
   colpodes (Ach.) Stizenb., *15*, *69*, *141*, **161–162**
   ornata (Zahlbr.) Asah., 162
arctic tumbleweed, 427
Arctocetraria Kärnefelt & Thell, 214, 752
   andrejevii (Oksner) Kärnefelt & Thell, 213, *215*
   nigricascens (Nyl.) Kärnefelt & Thell, 213–214
Arctoparmelia Hale, 49, *139*, **162**, 479, 733, 752
   centrifuga (L.) Hale, 41, **162**, *163*, *734*
   incurva (Pers.) Hale, *139*, 162, *733*
   separata (Th. Fr.) Hale, 162, **163,** *734*
Arthonia Ach., 23, 28, 29, *126*, *130*, **163**, *163*, *326*, 443, 753
   caesia (Flotow) Körber, *163*, **164**
   cinnabarina (DC.) Wallr., *163*, **164**
   ilicina Taylor, 163
   patellulata Nyl., 24, *131*, *163*, 165
   phaeobaea Norman, *131*, *163*
   radiata (Pers.) Ach., 26, *163*, **165**
   rubella Fée) Nyl., 163

Arthoniaceae, 753
Arthoniales, 753
Arthopyrenia A. Massal., 23
Arthothelium A. Massal., *126*, *163*, 753
   ruanum (A. Massal.) Körber, *163*
   spectabile A. Massal., *163*
Arthrorhaphidaceae, 753
Arthrorhaphis Th. Fr., *123*, *130*, **165**, 753
   alpina (Schaerer) R. Sant., 166
   citrinella (Ach.) Poelt, *134*, **165**
Asahinea Culb. & C. Culb., **166**, 314, 451, 574, 752
   chrysantha (Tuck.) Culb. & C. Culb., 67, *138*, **166**, 453
   scholanderi (Llano) Culb. & C. Culb., *139*, 167
Ascomycetes, 4, 751
Ascomycota, 751
Aspicilia A. Massal., 41, 49, 85, *133*, 145, 159, **167**, *167*, 170, 174, 175, 300, 305, 384, 422, 429, 633, 752, 757
   alphoplaca (Wahlenb.) Poelt & Leuckert, 422
   caesiocinerea (Nyl. *ex* Malbr.) Arnold, 169
   calcarea (L.) Mudd, 169, 170
   candida (Anzi) Hue, *132*, *167*, **167**, 168, *374*
   cinerea (L.) Körber, *167*, **168–169**
   contorta (Hoffm.) Kremp., 51, 79, *136*, *167*, **169**
   esculenta (Pallas) Flagey, 79
   fruticulosa (Eversm.) Flagey, 169–170
   hispida Mereschk., 20, 51, 71, *121*, *167*, **169–170**, *385*
   myrinii (Fr.) Stein, 175
   praeradiosa (Nyl.) Poelt & Leuckert, 423
   verrucigera Hue, 168–169
asterisk lichen, 165
Astrothelium Eschw., **170**, 689, 753
   versicolor Müll. Arg., 46, **170**, 689
axil-bristle lichens, 447
   powdery, 447
   rock, 448
   smooth, 447

Bacidia De Not., 26, *130*, **170**, *171*, 176, 190, 192, 302, 443, 444, 546, 653, 752
   chlorococca (Stenh.) Lettau, 653
   circumspecta (Nyl. *ex* Vainio) Malme, *171*
   diffracta S. Ekman, 172
   fusca (A. Massal.) Du Rietz, 444
   heterochroa (Müll. Arg.) Zahlbr., *171*
   laurocerasi (Delise *ex* Duby) Zahlbr., *171*
   luteola "(Ach.) Mudd," 171
   obscurata (Sommerf.) Zahlbr., 444
   polychroa (Th. Fr.) Körber, *171*, *172*
   rubella (Hoffm.) A. Massal., *171*, **171–172,** 758

   sabuletorum (Schreber) Lettau, 27, 48, *171*, **172–173,** 446
   schweinitzii (Fr. *ex* E. Michener) A. Schneider, 27, 64, *171*, **173,** 535
   suffusa (Fr.) A. Schneider, *171*
Bacidina Vĕzda, *130*, 170, *171*, 176, 443, 444, 653, 752
   inundata (Fr.) Vĕzda, *171*, 363, 364
Baeomyces Pers., *121*, *127*, *136*, **173**, *173*, 231, 299, 300, 567, 610, 751, 761
   carneus Flörke, 174
   fungoides (Sw.) Ach., 299
   placophyllus Ach., *173*, **174**
   roseus Pers., 299
   rufus (Hudson) Rebent., *135*, *173*, **174,** 175
Baeomycetaceae, 751
bag lichen, brittle, 341
bark barnacles, 681
barnacle lichens, 680
Basidiomycetes, 4, 29, 751
Basidiomycota, 751
Bathelium Ach., 689, 690, 753
   carolinianum (Tuck.) R. C. Harris, 689, 690
   madreporiforme (Eschw.) Trevisan, 368
bear hair, 179
beard lichens, 16, 30, 83, 95, 709
   bloody, 723
   bony, 726
   bristly, 720
   bushy, 724
   chestnut, 460
   fishbone, 718
   hollow, 713
   horny, 725
   inflated, 715, 718
   lustrous, 719
   Methuselah's, 721
   pitted, 714
   powdered, 721
   powder-tipped, 717
   ragged, 717
   red, 723
   rock, 712
   shaggy, 720
   silken, 720
   straw, 724
   sundew, 715
   warty, 715
   western bushy, 713
Bellemerea Hafellner & Roux, *133*, 159, 167, **174,** 752
   alpina (Sommerf.) Clauzade & Roux, 108, **175–176**
   cinereorufescens (Ach.) Clauzade & Roux, 175
Belonia, 751

beret lichens, 173
   brown-, 174
   carpet, 174
Biatora Fr., 23, 26, 28, *129*, 170, **176**, 192, 279, 302, 344, 365, 444, 572, 752
   anthracophila (Nyl.) Hafellner, 344
   subduplex (Nyl.) Printzen, 176
   vernalis (L.) Fr., *131*, **176**, 759, 760
Biatoraceae, 752
Biatorella De Not., 614, 752
Biatorellaceae, 752
black-eye lichen, 676
black-foam lichens, 161
black moss, 79
black-on-black lichen, 474
black saddle lichen, 514
black seaside lichen, 728
black tree-lichen, 79, 180
black tree moss, 96
blister lichens, 683
      bark, 368
      blue, 684
      frosted, 683
      pitted, 684
      speckled, 688
bloodspot lichens, 96, 330, 473
    alpine, 473
    Pacific, 473
    rock, 332
    sunken, 332
    tree, 331
bloody-heart lichen, 446
board lichen, 686
bone lichens, 345
   seaside, 349
bootstrap lichen, 289
boulder lichens, 582
   concentric, 584
   orange, 584
   smoky-eye, 583
bread moss, 79
Brigantiaeceae, 753
Brigantiaea Trevisan, *127*, **177**, 423, 424, 753
   fuscolutea (Dicks.) R. Sant., 177
   leucoxantha (Sprengel) R. Sant., **177**, 414
British soldiers, 250
   Gritty, 254
   powder-foot, 257
brødmose, 79, 218
Brodoa Goward, *140*, *141*, *154*, *156*, **177**, 345, 752
   oroarctica (Krog) Goward, *139*, *154*, **177–178**
brook lichen, 297
   rusty, 363
brown lichens, 430
   lattice, 437
   northern, 438

brownette lichens, 729
   peppered, 729
brown-fuzz lichens, 544
Bryocaulon Kärnefelt, *150*, **178**, 179, 180, 460, 752
   divergens (Ach.) Kärnefelt, 20, 21, *120*, *151*, **179**, 180, 184, 215
   pseudosatoanum (Asah.) Kärnefelt, *152*, 179, 180
Bryoria Brodo & D. Hawksw., 14, 16, 30, 46, 60, 61, 78, 79, 80, 83, 95, 96, 105, *119*, *150*, 178, **179**, 460, 672, 731, 752, 763
   abbreviata (Müll. Arg.) Brodo & D. Hawksw., 460
   capillaris (Ach.) Brodo & D. Hawksw., 12, 107, *151*, **179–180**, 183
   fremontii (Tuck.) Brodo & D. Hawksw., 72, 78, 79, *119*, *151*, 179, **180**, 185, 461
   friabilis Brodo & D. Hawksw., 184
   furcellata (Fr.) Brodo & D. Hawksw., *151*, **181**
   fuscescens (Gyelnik) Brodo & D. Hawksw., *151*, **181–182**, 183, 185, 764
   glabra (Mot.) Brodo & D. Hawksw., 182, **182**, 183
   implexa (Hoffm.) Brodo & D. Hawksw., 184
   lanestris (Ach.) Brodo & D. Hawksw., 68, *151*, **182–183**
   nadvornikiana (Gyelnik) Brodo & D. Hawksw., 68, *151*, 179, **183**
   nitidula (Th. Fr.) Brodo & D. Hawksw., 41, *120*, *151*, 153, 179, **183–184**
   oregana (Tuck. *ex* Willey) Brodo & D. Hawksw., 461
   pikei Brodo & D. Hawksw., 179
   pseudocapillaris Brodo & D. Hawksw., 179, 673–674
   pseudofuscescens (Gyelnik) Brodo & D. Hawksw., *152*, 179, **184**
   salazinica Brodo & D. Hawksw., 184
   simplicior (Vainio) Brodo & D. Hawksw., 181, 183
   spiralifera Brodo & D. Hawksw., 184, 673–674
   subcana (Nyl. *ex* Stizenb.) Brodo & D. Hawksw., 182
   tortuosa (G. Merr.) Brodo & D. Hawksw., *151*, 179, 180, **184–185**
   trichodes (Michaux) Brodo & D. Hawksw., 58, 68, 107, *152*, 183, **185**
   trichodes ssp. americana (Mot.) Brodo & D. Hawksw., 185, 186
   trichodes ssp. trichodes, 185
Buellia De Not., *130*, *131*, 158, **186**, *186*–187, 300, 302, 394, 475, 544, 583, 633, 644, 683, 752, 759

   alboatra (Hoffm.) Th. Fr., *186*
   curtisii (Tuck.) Imshaug, 189
   disciformis (Fr.) Mudd, 12, *187*, **187**, 189
   erubescens Arnold, 187
   griseovirens (Turner & Borrer *ex* Sm.) Almb., *124*
   halonia (Ach.) Tuck., 34, 73, *186*, **187–188**
   ocellata (Flotow) Körber, 12
   penichra (Tuck.) Hasse, 186
   punctata (Hoffm.) A. Massal., 158
   retrovertens Tuck., 186
   spuria (Schaerer) Anzi, *167*, *186*, **188**, 384
   stigmaea Tuck., 188
   stillingiana J. Steiner, 17, 69, 108, *187*, **188**
Bulbothrix Hale, 18, 19, 138, *141*, *142*, **189**, *189*, 479, *480*, 752, 755
   confoederata (Culb.) Hale, 70, **189**, *189*
   coronata (Fée) Hale, 189
   goebelii (Zenker) Hale, *189*, **190**, 191, 487
   isidiza (Nyl.) Hale, *189*, 211
   laevigatula (Nyl.) Hale, *189*, **190**, 191
bull's-eye lichens, 571
   pink, 572
Bunodophoron A. Massal., 752
   melanocarpum (Sw.) Wedin, *119*, *120*, 657
burning bush, 630
button lichens, 158, 186
   boreal, 187
   common, 188
   lobed, 302
   seaside, 187
   sunken, 188
   tiny, 158
Byssoloma Trevisan, *128*, *130*, *131*, 170, **190**, **192**, 752
   leucoblepharum (Nyl.) Vainio, 192
   meadii (Tuck.) S. Ekman, **192**
   pubescens Vězda *ex* R. C. Harris, 192

Caliciaceae, 752
Caliciales, 752, 760
Calicium Pers., 5, 14, 28, 93, 99, *127*, **192**, 222, 752, 756
   abietinum Pers., 193
   adspersum Pers., 193
   glaucellum Ach., *193*
   lenticulare Ach., *193*
   salicinum Pers., 193, 194
   trabinellum (Ach.) Ach., *193*, **193**, 194
   viride Pers., 24, 192, **193–194**
Calopadia Vězda, *130*, **194**
   fusca (Müll Arg.) Vězda, 52, 194, 758
Caloplaca Th. Fr., 18, 34, 46, 70, 77, 102, *123*, *127*, *132*, *134*, 177, **194**, *195*–196, 205, 207, 319, 320, 331, 423, 578, 581, 586, 614, 616, *742*, *744*, *747*, 753, 761
   ahtii Søchting, 202

approximata (Lynge) H. Magn., 196, 199
arenaria (Pers.) Müll. Arg., *195*, **196–197**, 199
bolacina (Tuck.) Hcrrc, *196*, **197**
borealis (Vainio) Poelt, 202
brattiae W. A. Weber, 205
castellana (Räsänen) Poelt, 199
cerina (Ehrh. *ex* Hedwig) Th. Fr., *195*, **197–198**
chrysophthalma Degel., *196*, 201
cirrochroa (Ach.) Th. Fr., 198
citrina (Hoffm.) Th. Fr., *123*, *196*, **198**, 200
coralloides (Tuck.) Hulting, *77*, *118*, *195*, **198–199**, *385*, 757
decipiens (Arnold) Blomb. & Forss., 198
discolor (Willey) Fink, 201
epithallina Lynge, *7*, *195*, **199**
feracissima H. Magn., *48*, *195*, **199**, 200
ferruginea (Hudson) Th. Fr., *195*
flavogranulosa Arup, *17*, *77*, *196*, *197*, **200**, 203, 205
flavorubescens (Hudson) J. R. Laundon, *196*, **200–201**, 203
flavovirescens (Wulfen) Dalla Torre & Sarnth, *196*, *200*, **201**, 203
fraudens (Th. Fr.) H. Olivier, 196–197
holocarpa (Hoffm. *ex* Ach.) M. Wade, *9*, *27*, *195*, **201–202**
ignea Arup, *12*, *195*, **202**, 205, 744
inconspecta Arup, *195*, *196*, *200*, **202–203**
lithophila H. Magn., 202, 203
litoricola Brodo, 27
luteominia (Tuck.) Zahlbr., *195*, *196*, **203**
luteominia var. bolanderi (Tuck.) Arup, *127*, 203, 331
luteominia var. luteominia, 203
marina (Wedd.) Zahlbr., 197, 200
microphyllina (Tuck.) Hasse, *196*, 198
microthallina (Wedd.) Zahlbr., 200
pellodella (Nyl.) Hasse, *136*, *195*, **203–204**
rosei Hasse, 200
saxicola (Hoffm.) Nordin, 202, 204, 744
scopularis (Nyl.) Lettau, 205
sideritis (Tuck.) Zahlbr., *195*, 204
stantonii W. A. Weber *ex* Arup, 197
stillicidiorum (Vahl) Lynge, 198
trachyphylla (Tuck.) Zahlbr., *7*, *148*, *195*, *202*, **204–205**, 744
ulmorum (Fink) Fink, 197
velana (A. Massal.) Du Rietz, *196*, 201
verruculifera (Vainio) Zahlbr., *66*, *77*, *195*, *197*, **205**
camouflage lichens, 430, 449
abraded, 439
alpine, 439
Appalachian, 432
blistered, 450
brown-eyed, 440
California, 435
dimpled, 441
elegant, 433
erupted, 451
lustrous, 434
many-spored, 436
mealy, 433
northern, 438
powdered, 438
powder-rimmed, 431
rimmed, 435
shingled, 437
shiny, 434
spotted, 437
whiskered, 439
canary weed, 647
Candelaria A. Massal., *118*, *134*, *137*, **205**, *205*, 752
concolor (Dicks.) Stein, *64*, *69*, *205*, **205–206**
fibrosa (Fr.) Müll. Arg., *205*, **206**
Candelariella Müll. Arg., *128*, *195*, *198*, **206–207**, *207*, 210, 752
aurella (Hoffm.) Zahlbr., *207*, **207**, 209
efflorescens R. C. Harris & W. R. Buck, *123*, *206*, *207*, **207**, 208
placodizans (Nyl.) H. Magn., 208, 209
reflexa (Nyl.) Lettau, 206, 207
rosulans (Müll. Arg.) Zahlbr., *73*, *134*, *207*, **208**
spraguei (Tuck.) Zahlbr., 208
terrigena Räsänen, *207*, **208**, 209
vitellina (Hoffm.) Müll. Arg., *26*, *207*, **208–209**
xanthostigma (Ach.) Lettau, 207
Candelina Poelt, *128*, *207*, **209–210**, 752
mexicana (de Lesd.) Poelt, 210
submexicana (de Lesd.) Poelt, *74*, *137*, **210**
candleflame lichens, 205
fringed, 206
candlewax lichens, 148
eastern, 149
mountain, 150
pallid, 149
candy lichen, 360
can-of-worms lichen, 96, 288
Canomaculina Elix & Hale, **210–211**, *479*, *480*, *642*, 752
neotropica (Kurok.) Elix, 211, 501
subtinctoria (Zahlbr.) Elix, **211**, *494*, 501, *642*, 644
Canoparmelia Elix & Hale, **211**, *479*, *480*, 596, 752
amazonica (Nyl.) Elix & Hale, 211
caroliniana (Nyl.) Elix & Hale, **211–212**, *481*, 487
crozalsiana (de Lesd. *ex* Harm.) Elix & Hale, 212, *447*, *481*
cryptochlorophaea (Hale) Elix & Hale, 212
salacinifera (Hale) Elix & Hale, 211–212
texana (Tuck.) Elix & Hale, *142*, **212**, *481*, *491*
Cantharellales, 751
Carbonea (Hertel) Hertel, *391*, 752
vorticosa (Flörke) Hertel, *392*
caribou lichens, 59, 223
Catapyrenium Flotow, *50*, *296*, *307*, *569*, 753
cinereum (Pers.) Körber, 570
lacinulatum (Ach.) Breuss, 569
plumbeum (de Lesd.) J. W. Thomson, 570
psoromoides (Borrer) R. Sant., 571
tuckermanii (Rav. *ex* Mont.) J. W. Thomson, 570
Catillaria A. Massal., *26*, *428*, *443*, *633*, *636*, *683*, 752
chalybeia (Borrer) A. Massal., *131*
glauconigrans (Tuck.) Hasse, *131*
nigroclavata (Nyl.) Schuler, *131*
Catillariaceae, 752
Catinaria Vainio, *428*, 752
atropurpurea (Schaerer) Vězda & Poelt, *131*
laureri (Hepp *ex* Th. Fr.) Degel., *429*
Catolechia Flotow, *186*, 752
wahlenbergii (Ach.) Körber, *186*
Cavernularia Degel., *141*, **212–213**, *213*, *345*, *443*, 752
hultenii Degel., *75*, *213*, **213**
lophyrea (Ach.) Degel., *213*, **213**, 214
centipede lichens, 334
elegant, 338
powdered, 339
Cetraria Ach., *30*, *95*, *120*, *139*, *179*, **213–214**, *214–215*, *314*, *365*, *427*, *691*, *696*, 752
aculeata (Schreber) Fr., *20*, *121*, *151*, *179*, *184*, *213*, *214*, **215**, *580*
arenaria Kärnefelt, *214*, *215*, 216
aurescens Tuck., 149
californica Tuck., 365
chlorophylla (Willd.) Vainio, 692
ciliaris Ach. var. *halei* (Culb. & C. Culb.) Ahti, 692
coralligera (W. A. Weber) Hale, 692
cucullata (Bellardi) Ach., 314
culbersonii Hale, 432
ericetorum Opiz, *72*, *214*, **215–216**, 218
ericetorum ssp. reticulata (Räsänen) Kärnefelt, 215
fendleri (Nyl.) Tuck., 693
halei Culb. & C. Culb., 692
hepatizon (Ach.) Vainio, 435
idahoensis Essl., 310
islandica (L.) Ach., *13*, *79*, *83*, *213*, *214–215*, **216–218**

islandica ssp. crispiformis (Räsänen) Kärnefelt, 217–218
islandica ssp. islandica, 67, 218
laevigata Rass., *214, 216,* **218,** *219*
merrillii Du Rietz, 366
muricata (Ach.) Eckfeldt, 184, 215
nigricans Nyl., 216
nivalis (L.) Ach., 314
oakesiana Tuck., 158
orbata (Nyl.) Fink, 694
pallidula Tuck. *ex* Riddle, 149
pinastri (Scop.) Gray, 731
platyphylla Tuck., 694
sepincola (Ehrh.) Ach., 696
subalpina Imshaug, 696
tilesii Ach., 731
viridis Schwein., 732
weberi Essl., 693
Cetrariella Kärnefelt & Thell, *139, 214,* **219,** 752
delisei (Bory *ex* Schaerer) Kärnefelt & Thell, *120, 214,* **219–220**
fastigiata (Delise *ex* Nyl.) Kärnefelt & Thell, 220
Cetrelia Culb. & C. Culb., 20, 44, **220,** *220, 480, 574, 606,* 752, *757*
alaskana (C. Culb. & Culb.) Culb. & C. Culb., *139, 167, 220*
cetrarioides (Duby) Culb. & C. Culb., *220*
chicitae (Culb.) Culb. & C. Culb., *220,* **220,** 221
monachorum (Zahlbr.) Culb. & C. Culb., 220
olivetorum (Nyl.) Culb. & C. Culb., *220,* **220–221**
Chaenotheca Th. Fr., 5, 14, 93, 99, *192,* **221–222,** 752, *756, 760*
brunneola (Ach.) Müll. Arg., *193,* **222**
chrysocephala (Turner *ex* Ach.) Th. Fr., 222
ferruginea (Turner & Borrer) Mig., 222
furfuracea (L.) Tibell, *193,* **222,** *223, 596*
Chaenothecopsis Vainio, *192,* 222, 751
debilis (Turner & Borrer *ex* Sm.) Tibell, *192*
chalk-crust, California, 371
Chiodecton
montagnaei Tuck., 291
chocolate chip lichens, 654
common, 655
fringed, 656
orange, 655
christmas lichen, 291
Chrysothrichaceae, 753
Chrysothrix Mont., 38, 39, **223,** *396, 596,* 753
candelaris (L.) J. R. Laundon, *123,* **223**
chlorina (Ach.) J. R. Laundon, 223
cinder lichen, 168

Cladidium Hafellner, 752
bolanderi (Tuck.) B. D. Ryan, 385
Cladina Nyl., 55, 56, 78, 79, 83, 95, *121, 179,* **223–224,** 231, 255, 259, 263, 278, 610, 663, 752, *763*
arbuscula (Wallr.) Hale & Culb., **224,** 225, 230, 232
arbuscula ssp. arbuscula, 224
arbuscula ssp. beringeriana (Ahti) N. S. Golubk., 224
evansii (Abbayes) Hale & Culb., 70, **224,** 225, 227, 230, *231*
impexa de Lesd.
mitis (Sandst.) Hustich, 224, **225,** 226, 227, 230, 232
pacifica (Ahti) Ahti, 225
portentosa (Dufour) Follmann, 225, *231,* 758
portentosa ssp. pacifica (Ahti) Ahti, **225–227**
portentosa ssp. portentosa, 225
rangiferina (L.) Nyl., 60, 79, 95, 224, **227–228,** 229, 230, 232, 755
stellaris (Opiz) Brodo, 38, 57, 60, 83, 224, 232, 759
stygia (Fr.) Ahti, 228, **229,** 232
submitis (A. Evans) Hale & Culb., 225, **230,** *232,* 259
subtenuis (Abbayes) Hale & Culb., 224, **230–231,** *232*
terrae-novae (Ahti) Hale & Culb., 228, 230
Cladonia P. Browne, 14, 16, 18, 20, 44, 50, 55, 56, 95, 109, *121, 134, 135,* 224, **231,** *231–237, 293, 304, 344,* 462, 470, 566, 610, 687, 714, 752, *755,* 761, *763*
abbreviatula G. Merr., *236,* 251
albonigra Brodo & Ahti, 247
amaurocraea (Flörke) Schaerer, *232,* **237,** 278
anitae Culb. & C. Culb., *236,* 251
apodocarpa Robbins, *233,* **237–238,** *242*
artuata S. Hammer, 250, 255
asahinae J. W. Thomson, *234,* **238–239,** 254
atlantica A. Evans, 250
bacillaris Nyl., 259
balfourii Crombie, 273
beaumontii (Tuck.) Vainio, *236,* **239,** *271*
bellidiflora (Ach.) Schaerer, *234, 236,* **239,** 240, 251, *761*
borealis S. Stenroos, *234,* **239,** 241, 266
boryi Tuck., *233,* **241,** 245, 253, 263, 278
botrytes (K. Hagen) Willd., *236,* **242**
brevis (Sandst.) Sandst., 270
caespiticia (Pers.) Flörke, *135, 233, 236,* 238, **242**
calycantha Delise *ex* Nyl., 268
capitata (Michaux) Sprengel, 264

carassensis Vainio, 250
cariosa (Ach.) Sprengel, *237,* **242–243,** 265, 274–275
carneola (Fr.) Fr., 12, *233,* **243,** 244, 251, 266, 277, 388
caroliniana Tuck., *232, 241,* **243,** *245,* 253, 278
cenotea (Ach.) Schaerer, *233,* **245**
cervicornis (Ach.) Flotow, 261
cervicornis ssp. cervicornus, 245
cervicornis ssp. verticillata (Hoffm.) Ahti, *234,* **245–246,** 257, 268, *763*
chlorophaea (Flörke *ex* Sommerf.) Sprengel, 33, *233,* **247,** 248, 254, 266, 268, 758
clavulifera Vainio, 270
coccifera (L.) Willd., 239, 266
coniocraea (Flörke) Sprengel, 30, *235, 236,* **247, 249,** 262, 269, 273, *761*
conista A. Evans, 247
cornuta (L.) Hoffm., *235,* 247, 262, 269, 273
cornuta ssp. cornuta, 249
cornuta ssp. groenlandica (E. Dahl) Ahti, 249, 262
crispata (Ach.) Flotow, *234, 236,* **249–250,** 255, 271
cristatella Tuck., 34, 55, 94, *236,* 239, 242, **250–251,** 254, 257, 259, 269
cylindrica (A. Evans) A. Evans, 272
dahliana Kristinsson, 274
decorticata (Flörke) Sprengel, *235, 236,* 271
deformis (L.) Hoffm., 30, *233,* **251,** 266, 274, 758
didyma (Fée) Vainio, *235,* **251**
didyma var. didyma, 251
didyma var. vulcanica (Zoll. & Moritzi) Vainio, 251–252, 264
digitata (L.) Hoffm., *233,* **252,** 277
dimorphoclada Robbins, *232,* 245, **252–253,** 259, 263, 278
ecmocyna Leighton, *234, 235, 236,* **253–254,** 257, 261
ecmocyna ssp. ecmocyna, 254
ecmocyna ssp. intermedia (Robbins) Ahti, 253–254
ecmocyna ssp. occidentalis Ahti, 253–254
farinacea (Vainio) A. Evans, *235,* 249, 273
fimbriata (L.) Fr., *234,* 239, 247, **254,** 262
floerkeana (Fr.) Flörke, *235,* **254,** 255
floridana Vainio, *232,* 250, **254–255**
furcata (Hudson) Schrader, 12, 224, *232, 236,* **255–256,** 261, 270
glauca Flörke, 245, 273
gonecha (Ach.) Asah., 274
gracilis (L.) Willd., *237,* 254, **256–257,** 265, 269
gracilis ssp. elongata (Jacq.) Vainio, 256–257

INDEX OF NAMES 775

gracilis ssp. **gracilis,** *235, 236,* **256–257**
gracilis ssp. *nigripes* (Nyl.) Ahti, 256
gracilis ssp. **turbinata** (Ach.) Ahti, *235,* **257,** *261*
gracilis ssp. *vulnerata* Ahti, 256, 261
grayi G. Merr. *ex* Sandst., 51, 247
hypoxantha Tuck., *236,* 251
incrassata Flörke, *236,* 251, 252, **257–258**
kanewskii Oksner, 278
leporina Fr., *232,* **258–259,** *263, 264, 272*
luteoalba Wheldon & A. Wilson, 269
macilenta Hoffm., *235,* 251, 254, **259,** *264, 273, 277*
macrophylla (Schaerer) Stenh., *237*
macrophyllodes Nyl., *233, 234,* **260–261,** *276*
magyarica Vainio, 266, 268
major (K. Hagen) Sandst., 254
mateocyatha Robbins, *234,* **261,** *265, 271*
maxima (Asah.) Ahti, *235, 236, 257,* **261**
merochlorophaea Asah., 247
multiformis G. Merr., *232, 235, 236, 255,* **261–262,** *265*
norvegica Tønsberg & Holien, 259
ochrochlora Flörke, *234, 235, 247, 249,* **262–263,** *269, 273*
pachycladodes Vainio, *232, 245, 259,* **263,** *278*
parasitica (Hoffm.) Hoffm., *235, 236,* 251, **263–264**
perforata A. Evans, 51, 88, *232,* **264,** *330*
petrophila R. C. Harris, 238
peziziformis (With.) J. R. Laundon, *237, 238, 243,* **264,** *271*
phyllophora Hoffm., *234, 235, 257, 261,* **265,** *275*
piedmontensis G. Merr., 269
pityrea (Flörke) Fr., 273
pleurota (Flörke) Schaerer, *233, 239, 243,* 251, **265–266,** *277*
pocillum (Ach.) Grognot, *233, 234,* **266,** *267*
polycarpia G. Merr., 238, 267, 270
polycarpoides Nyl., 55, *237, 238, 243,* **266–267,** *269, 270*
polydactyla (Flörke) Sprengel, 252, 277
porocypha S. Hammer, 250, 255
prostrata A. Evans, 51, *233,* **267**
pseudorangiformis Asah., 232
pyxidata (L.) Hoffm., *235, 247, 266,* **267–268**
ramulosa (With.) J. R. Laundon, 262, 273
rappii A. Evans, *234, 245,* **268**
rappii var. exilior (Abbayes) Ahti, 268
rappii var. rappii, 268
ravenelii Tuck., *236,* 251
rei Schaerer, *234,* **269,** *273, 278*
robbinsii A. Evans, *134, 233, 236,* **269–270**

scabriuscula (Delise) Nyl., *232, 235, 249, 255,* **270**
schofieldii Ahti & Brodo, 260
singularis S. Hammer, 271
sobolescens Nyl. *ex* Vainio, *237, 238, 243, 265, 267,* **270–271,** *275*
squamosa Hoffm., *234, 236, 239,* **271,** *273*
squamosa var. *subsquamosa* (Nyl. *ex* Leighton) Vainio, 271
strepsilis (Ach.) Grognot, 51, 104, *233, 236, 261, 269,* **271–272**
stricta (Nyl.) Nyl., 275
subcervicornis (Vainio) Kernst., 260
subfurcata (Nyl.) Arnold, 250, 255
subradiata (Vainio) Sandst., *235, 249,* **272–273**
subrangiformis (as used in North America), 255
subsetacea Robbins *ex* A. Evans, 278
subulata (L.) F. H. Wigg., *234, 235, 249,* **273**
sulphurina (Michaux) Fr., *233,* 251, **274**
symphycarpia (Flörke) Fr., *233, 237, 238, 243, 270,* **274–275**
transcendens (Vainio) Vainio, 239, 277
trassii Ahti, *234,* **275**
turgida Hoffm., *236,* **275–276**
umbricola Tønsberg & Ahti, *233, 235, 252, 259,* **276–277**
uncialis (L.) F. H. Wigg., 224, *232, 237, 245, 253, 259, 263, 264, 276,* **277–278**
verruculosa (Vainio) Ahti, 72, *234,* **278**
verticillata (Hoffm.) Schaerer, 245
wainioi Savicz, 228, 232
cladonia, 57, 231
  bighorn, 249
  bramble, 254
  dragon, 271
  felt, 265
  fence-rail, 263
  fishnet, 241
  frosted, 253
  giant, 261
  granite thorn, 243
  large-leaved, 260
  many-forked, 255
  mealy forked, 270
  olive, 271
  perforate, 264
  prostrate thorn, 252
  resurrection, 267
  shaded, 276
  smooth, 256
  stalkless, 237
  stubby-stalked, 242
  thick-walled, 263
  thorn, 263, 277
  yellow tongue, 269

Cladoniaceae, 752
Cladoniineae, 752
clam lichens, 343
  common, 344
  small, 344
Clauzadea Hafellner & Bellem., 581, *583,* 752
  monticola (Ach. *ex* Schaerer) Hafellner & Bellem., *583*
Clavariaceae, 751
Cliostomum Fr., **279,** 302, 596, 597, 752
  griffithii (Sm.) Coppins, *131,* **279**
  leprosum (Räsänen) Holien & Tønsberg, 597
  vitellinum Gowan, 597
club-mushroom "lichens," 444
coal-dust lichens, 580
  common, 581
cobblestone lichens, 145
  brown, 146
  gold, 146, 578
  hoary, 148
  rimmed, 147
Coccocarpia Pers., *139, 143,* **279–280,** *475,* 752, *759*
  cronia (Tuck.) Vainio, 280
  domingensis Vainio, 281
  erythroxyli (Sprengel) Swinscow & Krog, **280,** 281, *476*
  palmicola (Sprengel) Arv. & D. J. Galloway, **280–281,** *475*
  stellata Tuck., 280
Coccocarpiaceae, 752
Coccotrema Müll. Arg., 77, **281,** 680, 753
  maritimum Brodo, 77, *124,* **281**
  pocillarium (Cummings) Brodo, 281
Coccotremataceae, 753
Coelocaulon Link
  aculeatum (Schreber) Link, 215
Coenogonium Ehrenb., *118, 121,* **281,** 751
  implexum Nyl., **282**
  interplexum Nyl., 282
  interpositum Nyl., 282
  linkii Ehrenb., 282
  moniliforme Tuck., 281
Collema F. H. Wigg., 5, 14, 15, 31, 55, 58, 99, *134, 137,* **282,** 282–283, 334, 399, 400, 403, 406, 414, 477, 682, 683, 752
  bachmanianum (Fink) Degel., 287
  coccophorum Tuck., *283,* **283–284,** *286, 287*
  conglomeratum Hoffm., *282,* 283
  crispum (Hudson) F. H. Wigg., 284
  cristatum (L.) F. H. Wigg, *283,* **284–285,** *285, 286*
  flaccidum (Ach.) Ach., *283,* 285
  furfuraceum (Arnold) Du Rietz, *283,* **285,** *286*

fuscovirens (With.) J. R. Laundon, 283, 288
nigrescens (Hudson) DC., 283, 285, **285,** 286
polycarpon Hoffm., 283, **286,** 287
pulcellum Ach., 283, **286**
pulcellum var. leucopeplum (Tuck.) Degel., 285, 286
pulcellum var. pulcellum, 283, 285, 286
pulcellum var. subnigrescens (Müll. Arg.) Degel., 285, 286
ryssoleum (Tuck.) A. Schneider, 283
subflaccidum Degel., 283, 285, **286–287**
tenax (Sw.) Ach., 56, 283, **287**
tuniforme (Ach.) Ach., 288
undulatum Laurer *ex* Flotow, 285, **287–288**
undulatum var. granulosum Degel., 283, 288
undulatum var. undulatum, 288
Collemataceae, 752
Combea De Not., 652, 753
californica (Th. Fr.) Follmann & M. Geyer, 652
comet-spored lichen, 649
comma lichens, 163
bloody, 164
frosted, 164
Coniocybaceae, 752
Coniocybe Ach., 221
Conotrema Tuck., **288,** 752
urceolatum (Ach.) Tuck., 46, 96, *125,* **288**
copper patch lichen, 660
coral lichens, 657, 658
fragile, 658
Coriscium Vainio
viride (Ach.) Vainio, 470
Cornicularia (Schreber) Hoffm., **289,** 365, 752
californica (Tuck.) Du Rietz, 365
divergens Ach., 179
normoerica (Gunn.) Du Rietz, *120,* **289**
cotton lichen, lobed, 289
cottonthread lichens, 397
cowpie lichen, 304
cracked lichens, 145
crater lichens, 303, 304, 670
desert, 303
crazy-scale lichen, 275
Cresponea Egea & Torrente, 369, 753
chlorocronia (Tuck.) Egea & Torrente, 369
crimson dot lichen, 614
northern, 614
southern, 615
sulphured, 615
Crocynia (Ach.) A. Massal., **289,** 752
pyxinoides Nyl., *123,* **289, 291**
Crocyniaceae, 752

crottle, 80, 483
smoky, 482
cryptic paw, 455
Cryptothecia Stirton, *123,* **291,** 753
rubrocincta (Ehrenb.:Fr.) Thor, 75, 290, **291**
striata Thor, **291**
cudbear, 81, 463
Cybebe Tibell, 222
Cyphelium Ach., 28, *127,* **291–292,** *292,* 679, 752, 760
chloroconium (Tuck.) Zahlbr., 292
inquinans (Sm.) Trevisan, *292,* **292**
karelicum (Vainio) Räsänen, 292
lucidum (Th. Fr.) Th. Fr., *292,* **292–293**
notarisii (Tul.) Blomb. & Forss., *127,* 292
pinicola Tibell, 292
tigillare (Ach.) Ach., 292
trachylioides (Nyl. ex Branth & Rostrup) Erichsen, 292
Cystocoleus Thwaites, *122,* 620, 753
ebeneus (Dillwyn) Thwaites, 619, 620

Dacampiaceae, 753
Dactylina Nyl., 156, 231, **293,** 752
arctica (Richardson) Nyl., *121, 293,* **293,** 294
arctica ssp. arctica, 293
arctica ssp. beringica (C. D. Bird & J. W. Thomson) Kärnefelt & Thell, 293
madreporiformis (Ach.) Tuck., 156
ramulosa (Hook.) Tuck., 67, *121,* 156, *293,* **293–295**
deer moss, 224
Degelia Arv. & D. J. Galloway, 279, 475, 753
plumbea (Lightf.) P.M. Jørg. & P. James, 280, 476
Dendriscocaulon Nyl., **295,** 579, 752
intricatulum (Nyl.) Henssen, *120, 121, 122,* **295**
Dendrographa Darbish., 73, *119,* **295,** 305, 648, 753
alectorioides Sundin & Tehler f. parva Sundin & Tehler, 297
leucophaea (Tuck.) Darbish., **297**
leucophaea f. leucophaea, 297
leucophaea f. minor (Darbish.) Sundin & Tehler, 296, 297
minor Darbish., 297
Dermatocarpon Eschw., 14, 29, *137,* **297,** *297,* 471, 696, 753, 761, 764
fluviatile (Weber) Th. Fr., 297
hepaticum (Ach.) Th. Fr., 570
intestiniforme (Körber) Hasse, 298, 299
lachneum (Ach.) A. L. Sm., 570
luridum (With.) J. R. Laundon, 35, *297,* **297–298,** 343
miniatum (L.) W. Mann, 51, *297,* **298–299**

moulinsii (Mont.) Zahlbr., *297,* 299
reticulatum H. Magn., *297,* **299**
rivulorum (Arnold) Dalla Torre & Sarnth., 298
tuckermanii (Rav. *ex* Mont.) Zahlbr., 570
Dibaeis Clem., *121, 127, 173,* 231, **299,** 567, 610, 752
absoluta (Tuck.) Kalb & Gierl, 300, 360
baeomyces (L. f.) Rambold & Hertel, 18, 50, 55, *122, 173, 174,* **299–300,** 360
Dictyonema C. Agardh., 470, 751
Dimelaena Norman, *132,* **300,** *300,* 644, 659, 660, 752
californica (H. Magn.) Sheard, 300, 301
oreina (Ach.) Norman, 41, 85, *300,* **300,** 301, 757
radiata (Tuck.) Müll. Arg., 11, 66, 74, *300,* **300–301**
suboreina (de Lesd.) Hale & Culb., 300
thysanota (Tuck.) Hale & Culb., 300, 301
Dimerella Trevisan, *128,* **301,** 329, 751
lutea (Dicks.) Trevisan, **302**
pineti (Ach.) Vězda, 302
dimple lichens, 301, 329
orange, 302
rock, 329
Diploicia A. Massal., *131, 186,* **302,** 305, 752
canescens (Dicks.) A. Massal., *186,* **302–303**
Diploschistes Norman, 23, 28, *125,* **303,** 449, 752
actinostomus (Ach.) Zahlbr., 303, 305
aeneus (Müll. Arg.) Lumbsch, 303
diacapsis (Ach.) Lumbsch, *303,* **303,** 304
muscorum (Scop.) R. Sant., 7, *303,* **304,** 305
scruposus (Schreber) Norman, 303, **304–305**
Diplotomma Flotow, *130, 186,* 683
Dirina Fr., *132,* **305,** 371, 648, 753
approximata Zahlbr., 648
catalinariae Hasse, *122,* **305**
catalinariae f. catalinariae, 305
catalinariae f. sorediata Tehler, 73, *123,* 305
paradoxa (Fée) Tehler, 305
Dirinaria (Tuck.) Clem., *141,* 302, **305,** 546, *546,* 558, 616, 752, 755
aegialita (Afz.) B. Moore, **306,** 307, 548
applanata (Fée) D. D. Awasthi, 307, 558
aspera (H. Magn.) D. D. Awasthi, 306
confluens (Fr.) D. D. Awasthi, 306
confusa D. D. Awasthi, 75, *144,* **306,** 307, *547*
frostii (Tuck.) Hale & Culb., 307, 548
picta (Sw.) Clem. & Shear, 31, 53, **306–307,** *548,* 551, 558, 617
purpurascens (Vainio) B. Moore, 306, **307,** *547*

INDEX OF NAMES 777

disk lichens, *118*, 390, 394
    gray-orange, 393
    sunken (see under "sunken disk lichens")
dog-lichens, 82, 503, 506
    alternating, 507
    field, 520
    membranous, 513
    pale-bellied, 518
    scaly, 518
Dothideales, 753
dot lichens, 165, 170, 176, 279, 364, 428, 443, 444
    city, 653
    frazzled, 325
    frosty-rimmed, 171
    golden, 165
    hidden, 364
    leaf, 194
    multicolored, 279
    orange, 586
    shadow, 443
dragon funnel, 271
dust lichens, 396
    fluffy, 396
    gold, 223
    lobed, 398
    sulphur
    zoned, 397

earth-crust, brown, 395
earthscales, 50, 71, 332, 569
    changing, 330
earth wrinkles, 684
easter lichen, 663, 666
Eiglera Hafellner, 363, 752
    flavida (Hepp.) Hafellner, 363
elegant centipede, 338
elf-ear lichen, 461
Endocarpon Hedwig, 12, **307–308,** 569, 661, 683, 753
    pulvinatum Th. Fr., *134,* **308**
    pusillum Hedwig, *135, 136,* **308–309,** 571
    tortuosum Herre, 308
Eopyrenula R. C. Harris, 289, 612, 753
    intermedia Coppins, *125,* 612
Ephebe Fr., 5, 14, *118,* **309,** 579, 619, 658, 752, 760, 763
    lanata (L.) Vainio, *122,* **309**
Erioderma Fée, 19, *143,* **309–310,** 476, 503, 753
    mollissimum (Samp.) Du Rietz, 310
    pedicellatum (Hue) P. M. Jørg., 310
    sorediatum D. J. Galloway & P. M. Jørg., *75, 87,* **310,** *395, 503*
Esslingeriana Hale & M. J. Lai, 214, **310,** *574,* 752
    idahoensis (Essl.) Hale & M. J. Lai, *142,* **310–311,** *574*

Euascomycetes, 751
Euopsis Nyl., 611, 752
Evernia Ach., 16, 58, 82, 96, *119,* 150, **311,** 458, 589, 620, 674, 710, 752
    divaricata (L.) Ach., 73, *120, 311,* **311–312**
    mesomorpha Nyl., 31, 82, *311,* **312,** 313
    perfragilis Llano, *311,* 312
    prunastri (L.) Ach., 57, 72, 82, *311,* **312–313,** 589, 591
Everniastrum Hale *ex* Sipman, **313,** 752
    catawbiense (Degel.) Hale *ex* Sipman, *119,* **313–314,** 360, 589
eyelash lichens, 189
    matted, 190
    rough, 190
    smooth, 189

false orchil, 295, 297
fan lichen, 521
Farnoldia Hertel
    jurana (Schaerer) Hertel, 583
Fellhanera Vězda, 752
    bouteillei (Desmaz.) Vězda, *131*
felt lichen, 504
finger lichens, 293
    arctic, 293
    fog, 652
    frosted, 293
    water, 653
firedot lichens, 194, 201
    bark sulphur-, 200
    brick-spored, 177
    coral, 198
    desert, 204
    flame, 202
    grainy seaside, 200
    granite, 196
    gray-rimmed, 197
    mealy, 198
    olive, 203
    parasitic, 199
    red, 203
    ringed, 205
    seaside, 202
    sidewalk, 199
    sulphur-, 201
    waxy, 197
fishnet, 625
Fistulariella Bowler & Rundel
    dilacerata [unpublished name], 623
    roesleri (Hochst. *ex* Schaerer) Bowler & Rundel, 629
five o'clock shadow, 544
Flavocetraria Kärnefelt & Thell, *120, 138,* 214, **314,** *314,* 732, 752
    cucullata (Bellardi) Kärnefelt & Thell, *314,* **314,** 315

    nivalis (L.) Kärnefelt & Thell, *314,* **314–315**
Flavoparmelia Hale, *118,* 148, **316,** 317, 479, 596, 730, 733, 752
    baltimorensis (Gyelnik & Fóriss) Hale, *139,* **316–317**
    caperata (L.) Hale, 13, 83, 91, *138, 139,* 150, 316–317, **317,** 318
    rutidota (Hook. f. & Taylor) Hale, *138,* 317, 318
Flavopunctelia (Krog) Hale, 316, **317,** 479, 606, 752
    flaventior (Stirton) Hale, *138,* **317–318,** 319
    praesignis (Nyl.) Hale, *138,* **318,** 607
    soredica (Nyl.) Hale, *138,* 317, 318, **319**
foam lichens, 663
    bony, 669
    eyed, 669
    finger-scale, 665
    grand, 665
    pixie, 666
    powdered, 664
    rock, 667
    snow, 667
    variegated, 670
    woolly, 667, 669
fog fingers, 652
fog lichens, 457
    armored, 458
    black-footed, 460
    bouquet, 458
    powdery, 457
foxhair lichen, 178
    heath, 179
    northern, 179
foxtail lichens, 460
    pendent, 461
    tufted, 460
freckle pelts
    common, 504
    flaky, 505
    ruffled, 513
fringe lichens, 160, 334
    Appalachian, 335
    coastal, 337
    cupped, 336, 338
    elegant, 338
    flowering, 336
    hanging, 160
    orange-bellied, 335
    orange-tinted, 339
    powdered, 339
    scaly, 341
    shaggy, 160
    white, 334
frost lichens, 560
    crescent, 565

ground, 565
hairy, 566
bottlebrush, 561, 564
elegant, 563
fancy, 561
yellow-edged, 564
Fulgensia A. Massal. & De Not., *122, 123, 127, 134, 148, 166,* **319**, *319*, *683*, *753*, *762*
   bracteata (Hoffm.) Räsänen, *33*, *50*, *319*, **319**, *320*
   desertorum (Tomin) Poelt, *319*, **319–320**
   fulgens (Sw.) Elenkin, *319*, *320*
   subbracteata (Nyl.) Poelt, *319*, *320*
funnel lichens, 273
   pale-fruited, 239
   powdered, 245
Fuscidea V. Wirth & Vězda, *129,* **320**, *391, 392, 649, 753*
   arboricola Coppins & Tønsberg, *124*, *321*
   intercincta (Nyl.) Poelt, *26*, *321*
   recensa (Stirton) Hertel, V. Wirth & Vězda, **321**
   thomsonii Brodo & V. Wirth, *321*
Fuscideaceae, 753
Fuscopannaria P. M. Jørg., *135,* **321**, *428*, *461, 475, 476, 477, 484, 753*
   ahlneri (P. M. Jørg.) P. M. Jørg., *321*, *476*
   leucophaea (Vahl.) P. M. Jørg., **321–322**, *476*
   leucosticta (Tuck.) P. M. Jørg., **322–323**, *476*, *478*
   leucostictoides (Ohlsson) P. M. Jørg., **323**, *476*, *485*
   maritima (P. M. Jørg.) P. M. Jørg., *322*
   mediterranea (Tav.) P. M. Jørg., *321*
   praetermissa (Nyl.) P. M. Jørg., *322*, **323–324**, *476*
   saubinetii (Mont.) P. M. Jørg., *322*, **324**, *476*
fuzzy-rim lichens, 190

Glyphis Ach., **324**, *752*
   cicatricosa Ach., *29*, *126*, **324**, *762*
Glypholecia Nyl., *752*
   scabra (Pers.) Müll. Arg., *137*, *696–697*
gnome fingers, 610
gold-eye lichen, 674
goldspeck lichens, 206
   common, 208
   hidden, 207
   powdery, 207
   sagebrush, 208
   tundra, 208
goldtwist, 731
Gomphillaceae, 752
Gomphillus Nyl., **325**, *752*, *759*
   americanus Essl., *122*, *127*, **325**

Gonohymenia J. Steiner, 414
   nigritella (Lettau) Henssen, 414
grain-spored lichens, 650
   frosted, 650
Graphidaceae, 752
Graphidales, 752
Graphina Müll. Arg., *27*, *126*, **325**, *325–326*, *535*, *752*
   abaphoides (Nyl.) Müll. Arg., *326*
   cypressi Müll. Arg. *325*
   incrustans (Fée) Müll. Arg., *325*
   peplophora M. Wirth & Hale, *326*, **326**
   xylophaga R. C. Harris, *325*
Graphis Adans., *18*, *23*, *24*, *27*, *28*, *126*, *163*, **326**, *326–327*, *471*, *535*, *536*, *752*, *758*, *759*
   afzelii Ach., *46*, *326*, **327**
   caesiella Vainio, *327*
   elegans (Borrer *ex* Sm.) Ach., *328*
   insidiosa (C. Knight & Mitten) Hook. f., *326*
   librata C. Knight, *327*
   lucifica R. C. Harris, *327*
   scripta (L.) Ach., *326*, *327*, **327–328**, *536*, *757*
   striatula (Ach.) Sprengel, *328*
   subelegans Nyl., *327*, **328**
greater thatch soldiers, 274
green light, 452
greenshield lichens, 316
   common, 317
   rock, 316
   speckled, 317
gut lichen, 349
Gyalecta Ach., *130*, *302*, *329*, *751*
   foveolaris (Ach.) Schaerer, *329*
   jenensis (Batsch) Zahlbr., **329**
   truncigena (Ach.) Hepp, *26*, *329*
Gyalectaceae, 751
Gyalectales, 751
Gyalidea Lettau *ex* Vězda, *329*, *752*
Gyalideopsis Vězda, *752*
   anastomosans P. James & Vězda, *325*
Gymnoderma Nyl., *329*, *752*
   lineare (A. Evans) Yoshim. & Sharp, *69*, *88*, *134*, *140*, *141*, **329–330**
Gypsoplaca Timdal, **330**, *752*
   macrophylla (Zahlbr.) Timdal, *136*, **330**
Gypsoplacaceae, 752

Haematomma A. Massal., *96*, *127*, **330–331**, *331*, *424*, *614*, *752*
   accolens (Stirton) Hillm., *331*, **331**, *332*
   americanum Kalb & Staiger, *331*
   californicum Sigal & D. Toren, *473*
   fenzlianum Massal., *17*, *70*, *331*, **332**
   flexuosum Hillm., *331*
   ochroleucum (Neck.) J. R. Laundon, *332*
   pacificum Hasse, *473*

   persoonii (Fée) Massal., *331*, **332**, *333*
   rufidulum (Fée) Massal., *331*
   subpuniceum (Fée) de Lesd., *332*
hair lichens, 16, 30, 46, 58
hairball lichen, rock, 659
heath lichens, 213
   spiny, 215
Heppia Nägeli, *308*, **332–333**, *522*, *597*, *751*
   adglutinata (Kremp.) A. Massal., *333*
   conchiloba Zahlbr., *135*, *139*, **333**, *522*, *525*
   lutosa (Ach.) Nyl., *333*
Heppiaceae, 751
Heterodermia Trevisan, *13*, *141*, *160*, **334**, *546*, *546*, *552*, *616*, *752*
   albicans (Pers.) Swinscow & Krog, **334–335**, *339*, *340*, *549*
   appalachensis (Kurok.) Culb., **335**, *339*, *548*
   casarettiana (A. Massal.) Trevisan, *142*, *340*, *341*, *548*
   crocea R. C. Harris, **335–336**, *547*
   diademata (Taylor) D. D. Awasthi, **336**, *338*, *547*
   echinata (Taylor) Culb., *18*, **336–337**, *339*, *547*
   erinacea (Ach.) W. A. Weber, *336*, **337**, *339*, *547*
   granulifera (Ach.) Culb., *335*, *361*, *547*
   hypoleuca (Muhl.) Trevisan, *35*, *336*, **338**, *547*
   leucomela (L.) Poelt, *119*, *141*, *161*, *335*, *336*, **338–339**, *548*
   leucomelaena (L.) Poelt, *338*
   leucomelos (L.) Poelt, *338*
   namaquana Brusse, *337*, *339*
   obscurata (Nyl.) Trevisan, *69*, *335*, *338*, **339**, *340*, *548–549*
   pseudospeciosa (Kurok.) Culb., *340*
   rugulosa (Kurok.) Wetmore, *336*, *338*, *547*
   speciosa (Wulfen) Trevisan, *19*, *335*, *338*, **339–340**, *548–549*
   squamulosa (Degel.) Culb., *142*, *338*, **341**, *547*
Heteroplacidium Breuss, 569
   acarosporoides (Zahlbr.) Breuss, 569
honeycomb lichens, 212
   fruiting, 213
   powdered, 213
horsehair lichens, 78, 96, 179, 185
   bay, 673
   brittle, 182
   burred, 181
   edible, 180
   gray, 179
   grooved, 672
   inedible, 184
   mountain, 184
   pale-footed, 181

red, 460
shiny, 182
spiny gray, 183
tundra, 183
yellow-twist, 184
Hubbsia W. A. Weber, **341**, 753
    californica (Räsänen) W. A. Weber, 342
    parishii (Hasse) Tehler, Lohtander, Myllys & Sundin, *121*, **341–342**, 653
Hydrothyria J. L. Russell, 38, **342**, 400, 753
    venosa J. R. Russell, *136*, *137*, 298, **342–343**, *400*
Hymenelia Kremp., *167*, *363*, 752
    epulotica (Ach.) Lutzoni, 363
    lacustris (With.) M. Choisy, 363
Hymeneliaceae, 752
Hyperphyscia Müll. Arg., *18*, *132*, *140*, *141*, 305, 306, **343**, 546, 616, 644, 752
    adglutinata (Flörke) H. Mayrh. & Poelt, *124*, **343**, *538*
    minor (Fée) D. D. Awasthi, 343
    syncolla (Tuck. *ex* Nyl.) Kalb, *13*, *70*, **343**, *344*, *538*, *540*
Hypocenomyce M. Choisy, *17*, **343–344**, 752
    anthracophila (Nyl.) P. James & Gotth. Schneider, *135*, **344**, 345
    castaneocinerea (Räsänen) Timdal, 345
    friesii (Ach.) P. James & Gotth. Schneider, 345
    leucococca R. Sant., 345
    scalaris (Ach.) M. Choisy, *135*, **344–345**
    xanthococca (Sommerf.) P. James & Gotth. Schneider, 345
Hypogymnia (Nyl.) Nyl., 48, *72*, *95*, *139*, *140*, *142*, *154*, *161*, *178*, *213*, **345**, *345–346*, 441, 443, 591, 752
    apinnata Goward & McCune, *346*, **346–347**, *349*, *350*, *352*, *354*
    austerodes (Nyl.) Räsänen, *346*, **347–348**, *355*
    bitteri (Lynge) Ahti, *346*, **348**, *355*
    duplicata (Ach.) Rass., *346*, **348–349**
    enteromorpha (Ach.) Nyl., *346*, **349**, *350*, *352*, *353*, *354*, *759*
    heterophylla L. Pike, *346*, **349–350**
    imshaugii Krog, *346*, *349*, **350**, *351*, *591*
    inactiva (Krog) Ohlsson, *346*, *349*, **350**, *352*, *591*
    krogiae Ohlsson, *69*, *161*, *346*, **350**, *352*, *354*, *356*
    metaphysodes (Asah.) Rass., *346*, *349*, **353**
    mollis L. Pike & Hale, 348
    occidentalis L. Pike, *346*, *349*, **353**, *354*
    oceanica Goward, 347, 356
    oroarctica Krog, 177
    physodes (L.) Nyl., *31*, *60*, *91*, *346*, *352*, **354**, *356*, *759*

pulverata (Nyl. *ex* Crombie) Elix, 178
    rugosa (G. Merr.) L. Pike, *346*, *353*, **354**, *355*
    subobscura (Vainio) Poelt, *346*, *348*, **354–355**
    tubulosa (Schaerer) Hav., *213*, *314*, *346*, *354*, **355–356**
    vittata (Ach.) Parrique, *346*, *354*, **356**
Hypotrachyna (Vainio) Hale, *15*, *109*, *142*, **356–357**, *357*, 420, 479, *480–481*, 486, 752, 762
    croceopustulata (Kurok.) Hale, *357*, **357**, *358*, *757*
    formosana (Zahlbr.) Hale, 358
    imbricatula (Zahlbr.) Hale, *357*
    laevigata (Sm.) Hale, 359, 360
    livida (Taylor) Hale, *357*, **358**, *359*, 448
    osseoalba (Vainio) Park & Hale, *33*, *357*, **358–359**, *360*, *487*
    prolongata (Kurok.) Hale, *357*
    pulvinata (Fée) Hale, *357*, *358*, **359**
    pustulifera (Hale) Skorepa, 358, 359
    revoluta (Flörke) Hale, *69*, *142*, *314*, *357*, **359–360**, *486*
    rockii (Zahlbr.) Hale, *357*, 360
    showmanii Hale, 359
    sinuosa (Sm.) Hale, *72*, *138*, 360
    taylorensis (M. E. Mitch.) Hale, *357*, 360
Hysterium Pers., 472

Iceland lichens, 79, 83, 213, 215
    arboreal, 696
    sand-loving, 215
    snow-bed, 219
    striped, 218
    true, 216
Icelandmoss, 213
Icmadophila Trevisan, *130*, *300*, **360**, 752
    ericetorum (L.) Zahlbr., 46, *300*, **360**
Icmadophilaceae, 752
Illosporium Mart.
    carneum Fr., 7
Immersaria Ramboldt & Pietschmann, *159*, 752
    carbonoidea (J. W. Thomson) Esnault & Roux, 159
Imshaugia S. F. Meyer, *141*, **360–361**, *361*, 481, 546, *546*, 752
    aleurites (Ach.) S. F. Meyer, *361*, **361–362**, 448, *547*
    placorodia (Ach.) S. F. Meyer, 46, 69, *149*, *361*, **362**, *547*
ink lichens, 573
Ionaspis Th. Fr., *133*, *167*, **362–363**, *363*, 752, *757*
    alba Lutzoni, 362
    lacustris (With.) Lutzoni, *128*, *363*, **363**, *364*

    lavata H. Magn., *363*, **363–364**
    odora (Ach.) Th. Fr., 364
iwatake, 79, 697

Japewia Tønsberg, **364**, 752
    carollii (Coppins & P. James) Tønsberg, 365
    tornoensis (Nyl.) Tønsberg, *128*, **364–365**
jelly flakes, 287
jelly lichens, 5, 14, 31, 99, *118*, *120*, 280, 282, 414
    blistered, 285, 286
    fingered, 284
    shaly, 286
    soil, 287
    tar, 283
    tree, 286
    whiskered, 399, 411
jelly strap, 682
jellyskin lichens, 400
    antlered, 403
    bearded, 411
    blistered, 404
    blue, 401, 404
    dimpled, 408
    dixie, 401
    edge-fruiting, 406
    four-spored, 408
    frilly, 407
    large-spored, 407
    petaled, 405
    rough bearded, 410
    ruffled, 402
    stretched, 406
    tattered, 404
jester lichen, 258
jewel lichens, 194

Kaernefeltia Thell & Goward, *119*, **365**, 752
    californica (Tuck.) Thell & Goward, *119*, *151*, **365**, *366*, 460
    merrillii (Du Rietz) Thell & Goward, *13*, *73*, *140*, *141*, *143*, *365*, **366**
kidney lichens, 96, 451
    alpine, 453
    arctic, 452
    cryptic, 455
    fringed, 454
    mustard, 455
    naked, 452
    peppered, 454
    pimpled, 457
    powdery, 456
Koerberia A. Massal., 573, 729, 753
    biformis A. Massal., 729
    sonomensis (Tuck.) Henssen, *475*, 729
Koerberiella, 752

lace lichen, 625
lace-scale lichens, 546
ladder lichen, 245
　　slender, 268
lakezone lichen, 662
Lasallia Mérat, 14, 41, 79, 81, *137*, **366**, 696, 697, 753, 762
　　papulosa (Ach.) Llano, 69, **366–367**, *698*
　　pensylvanica (Hoffm.) Llano, **367–368**, *698*
　　pustulata (L.) Mérat, 367
Laurera Rchb., **368**, 689, 753
　　megasperma (Mont.) Riddle, 29, **368**, *689*
　　subdisjuncta (Müll. Arg.) R. C. Harris, 368
leather lichen, 297
Lecanactis Körber, *130*, **368–369**, 371, 753
　　abietina (Ach.) Körber, **369**
　　chlorocronia Tuck., 369
　　megaspora (G. Merr.) Brodo, 369
　　nashii Egea & Torrente, 371
Lecania A. Massal., *132*, **369–370**, *370*, 372, 424, 657, 752
　　brunonis (Tuck.) Herre, 73, *136*, **370**, *370*
　　cyrtella (Ach.) Th. Fr., 370, 371
　　dimera (Nyl.) Th. Fr., 370
　　dubitans (Nyl.) A. L. Sm., 27, *370*, **370–371**
　　erysibe (Ach.) Mudd, 370
　　fuscella (Schaerer) Körber, 370
　　hassei (Zahlbr.) W. Noble, 370
　　nylanderiana A. Massal., 370
　　perproxima (Nyl.) Zahlbr., 370
Lecanographa Egea & Torrente, *126*, *132*, 369, **371**, 753
　　amylacea (Ehrh. *ex* Pers.) Egea & Torrente, 371
　　hypothallina (Zahlbr.) Egea & Torrente, **371–372**
Lecanora Ach., 14, 18, 26, 28, 44, *122*, *128*, *133*, *134*, 305, 343, **372**, *372–375*, 424, 463, 586, 615, 660, 676, 733, 752, 759
　　albella (Pers.) Ach., 377
　　albescens (Hoffm.) Branth & Rostrup, 380
　　allophana Nyl., 46, *373*, **375**, 389
　　argentea Oksner & Volkova, 377
　　argopholis (Ach.) Ach., *374*, **375–376**
　　beringii Nyl., 390
　　bipruinosa Fink, *374*, 383
　　caesiorubella Ach., *373*, **376–377**, *757*
　　caesiorubella ssp. caesiorubella Ach., 377
　　caesiorubella ssp. glaucomodes (Nyl.) Imshaug & Brodo, 377
　　caesiorubella ssp. merrillii Imshaug & Brodo, 376, 377
　　caesiorubella ssp. prolifera (Fink) R. C. Harris, 376, 377
　　caesiorubella ssp. saximontana Imshaug & Brodo, 377

californica Brodo, 377
cateilea (Ach.) A. Massal., 377
cenisia Ach., 24, *372*, *374*, *375*, 376, **377**, 378, 677
chlarotera Nyl., *372*, 382
cinereofusca H. Magn., 108, *373*, **377–378**
cinereofusca var. appalachensis Brodo, 378
circumborealis Brodo & Vitik., *372*, *373*, **379**, 764
confusa Almb., 385, 389
conizaeoides Nyl. *ex* Crombie, 388
crenulata Hook., 380
cupressi Tuck., 18, *128*, *373*, **379**, 389
dispersa (Pers.) Sommerf., 9, 48, *374*, **380**, 381
epibryon (Ach.) Ach., 375
expallens Ach., 615
fuscescens (Sommerf.) Nyl., *129*, 373
gangaleoides Nyl., 677
garovaglii (Körber) Zahlbr., *374*, **380**, 381, 383, 384
glabrata (Ach.) Malme, 85, 375
grantii H. Magn., 389
hagenii (Ach.) Ach., *372*, **380–381**, 390
horiza (Ach.) Lindsay, 375
hybocarpa (Tuck.) Brodo, 69, *372*, *373*, 379, **381–382**
intricata (Ach.) Ach., 387
lacustris (With.) Nyl., 363
leprosa Fée, 373
louisianae de Lesd., 373
marginata (Schaerer) Hertel & Rambold, *133*, *374*, **382**, 384, 392
mellea W. A. Weber, 374
mughicola Nyl., 387
muralis (Schaerer) Rabenh., 40, 90, 91, *134*, *136*, 300, *374*, 380, **383**, 384, 423, 733
novomexicana H. Magn., 73, *374*, 380, **384**, 640, 733
opiniconensis Brodo, 641
oreinoides (Körber) Hertel & Rambold, *133*, *167*, 188, 300, *375*, **384**
pacifica Tuck., 46, 72, *374*, **385**, 762
phaedrophthalma Poelt, 380, 383
phryganitis Tuck., *120*, **385–386**
pinguis Tuck., 77, *374*, **386**
piniperda Körber, *372*, *373*, **386–387**
polytropa (Hoffm.) Rabenh., *374*, *375*, **387**
pseudomellea B. D. Ryan, *374*, 380
pulicaris (Pers.) Ach., 27, *372*, 379, 382
rugosella Zahlbr., 374
rupicola (L.) Zahlbr., *133*, *375*, **387–388**, *761*
saligna (Schaerer) Zahlbr., 386
sambuci (Pers.) Nyl., 381
sierrae B. D. Ryan & T. Nash, 380, 383
straminea Ach., *374*, 383

strobilina (Sprengel) Kieffer, 27, *128*, *373*, *379*, **388**, 389
subfusca (L.) Ach., 376, 377, 382, 383, 425, 587
subimmergens Vainio, 374
subintricata (Nyl.) Th. Fr., 386
subpallens Zahlbr., 377
symmicta (Ach.) Ach., *128*, *129*, *372*, *373*, *379*, 385, **388–389**
thysanophora R. C. Harris, 17, 46, *123*, *164*, *373*, *374*, **389**, 396, 761
umbrina (Ach.) A. Massal., 381
valesiaca (Müll. Arg.) Stizenb., 383
varia (Hoffm.) Ach., 379
xylophila Hue, 77, *373*, 376, **389–390**
zosterae (Ach.) Nyl., *372*, 381, **390**
Lecanoraceae, 752
Lecanorales, 752
Lecanorineae, 752
Lecidea Ach., 26, 28, 55, *129*, *167*, 320, 321, 382, **390–391**, *391–392*, 394, 581, 583, 586, 633, 651, 676, 688, 752, 759
　　albofuscescens Nyl., 391
　　atrobrunnea (Ramond *ex* Lam. & DC.) Schaerer, 17, 41, *136*, *391*, **392–393**, 660, 676
　　auriculata Th. Fr., 391
　　brunneofusca H. Magn., 392
　　confluens (Weber) Ach., 394
　　fuscoatra (L.) Ach., 392
　　lapicida (Ach.) Ach., 17, *391*, 392, **393**, 394, 586, 688
　　oreinodes (Körber) W. A. Weber, 384
　　plana (J. Lahm) Nyl., 393
　　ramulosa Th. Fr., 385
　　syncarpa Zahlbr., 392
　　tesselina Tuck., 384
　　tessellata Flörke, *176*, *392*, **393–394**, 755
Lecideaceae, 752
Lecidella Körber, *187*, *391*, **394**, 614, 615, 752
　　asema (Nyl.) Knoph & Hertel, 187–188, 394
　　carpathica Körber, 392, 394
　　elaeochroma (Ach.) Hazsl., 395
　　elaeochromoides (Nyl.) Knoph & Hertel, 392, 394
　　euphorea (Flörke) Hertel, 391, 394
　　patavina (A. Massal.) Knoph & Leuckert, 392, 394
　　scabra (Taylor) Hertel & Leuckert, 615
　　spitsbergensis (Lynge) Hertel & Leuckert, 394
　　stigmatea (Ach.) Hertel & Leuckert, 392, **394–395**
　　wulfenii (Hepp) Körber, 391, 395
Lecidoma Gotth. Schneider & Hertel, **395**, 752
　　demissum (Rutstr.) Gotth. Schneider & Hertel, *129*, **395**

Leioderma Nyl., *143*, 309–310, **395**, 476, *503*, 753
    sorediatum D. J. Galloway & P. M. Jørg., 310, **395–396**, *504*
lemon lichen, 205
lemon-lime lichen, 595
Lempholemma Körber, *137*, 282, 752
    myriococcum (Ach.) Th. Fr., 282
    polyanthes (Bernh.) Malme, 282
Leotiales, 751
Lepraria Ach., 30, 38, *123*, 164, 291, 389, **396**, *396*, 398, 399, 753, 759
    caesioalba (de Lesd.) J. R. Laundon, 397
    finkii (de Lesd.) R. C. Harris, 396
    incana (L.) Ach., 397
    lobificans Nyl., 12, *123*, *396*, **396–397**
    neglecta (Nyl.) Erichsen, *396*, **397**
    zonata Brodo, 397
Leprocaulon Nyl. *ex* Lamy, *120*, *121*, **397–398**, 753
    albicans (Th. Fr.) Nyl. *ex* Hue, 398
    gracilescens (Nyl.) Lamb & Ward, *122*, **398**
    microscopicum (Vill.) Gams *ex* D. Hawksw., 398
    subalbicans (Lamb) Lamb & Ward, 398
Leproloma Nyl. *ex* Crombie, 290, *396*, **398**, 753
    membranaceum (Dicks.) Vainio, *123*, *396*, **398–399**
    vouauxii (Hue) J. R. Laundon, 399
Leptochidium M. Choisy, *137*, 282, **399**, *400*, *400*, 410, 411, 753
    albociliatum (Desmaz.) M. Choisy, 19, **399–400**, *400*
Leptogium (Ach.) Gray, 5, 14, 15, *134*, *137*, 279, 282, 342, 343, 399, **400**, *400–401*, 522, 752
    arsenei Sierk, *400*, 403
    austroamericanum (Malme) C. W. Dodge, *400*, **401**, 402, 405, 407
    azureum (Sw.) Mont., *401*, **401–402**, 404, 405
    burgessii (L.) Mont., *400*, 410
    burnetiae C. W. Dodge, *400*, *400*, 410, 411
    californicum Tuck., 405, 408
    chloromelum (Sw. *ex* Ach.) Nyl., *401*, **402–403**, 404, 407
    corniculatum (Hoffm.) Minks, *401*, **403**
    corticola (Taylor) Tuck., *401*, **404**, 405, 408
    cyanescens (Rabenh.) Körber, *400*, *401*, **404–405**
    dactylinum Tuck., 405
    denticulatum Tuck., 405
    floridanum Sierk, 402, 407
    furfuraceum (Harm.) Sierk, 408
    gelatinosum (With.) J. R. Laundon, *401*, **405**, 408
    isidiosellum (Riddle) Sierk, 407
    laceroides (de Lesd.) P. M. Jørg., *400*, 410
    lichenoides (L.) Zahlbr, *401*, 405, **405–406**, 580
    marginellum (Sw.) Gray, 75, *401*, **406**
    milligranum Sierk, *401*, 403, **406–407**
    palmatum (Hudson) Mont., 403
    phyllocarpum (Pers.) Mont., *401*, 402, 404, **407**
    platynum (Tuck.) Herre, *400*, *401*, **407–408**
    polycarpum P. M. Jørg. & Goward, *401*, **408**, 409
    pseudofurfuraceum P. M. Jørg., *400*, **408–410**, 411
    rivale Tuck., 343, *400*
    rugosum Sierk, *400*, 408, **410**, 411
    saturninum (Dicks.) Nyl., *400*, *400*, 410, **411**
    sinuatum (Hudson) A. Massal., 405
    subaridum P. M. Jørg. & Goward, 405
    tenuissimum (Dicks.) Körber, 406, 580
Letharia (Th. Fr.) Zahlbr., 16, 34, 73, 83, *118*, **411–412**, *412*, 457, 730, 752
    columbiana (Nutt.) J. W. Thomson, 73, *412*, **412**, 731
    vulpina (L.) Hue, 46, 80, 81, 83, *412*, **412**, 413
Letrouitia Hafellner & Bellem., *127*, *130*, 177, *412*, **414**, 423, 424, 753
    domingensis (Pers.) Hafellner & Bellem., 414
    parabola (Nyl.) R. Sant. & Hafellner, 177, **414**
    vulpina (Tuck.) Hafellner & Bellem., 414
Letrouitiaceae, 753
lettuce lichen, 417
lettuce lung, 417
Lichinaceae, 752
Lichinales, 751
Lichinella Nyl., 282, 289, **414**, 752
    cribellifera (Nyl.) P. Moreno & Egea, 415
    nigritella (Lettau) P. Moreno & Egea, *136*, *137*, **414–415**, 682
Lithographa Nyl., 751
    tesserata (DC.) Nyl., *126*
Lithothelium Müll. Arg., 612, 753
    hyalosporum (Nyl.) Aptroot, *126*, 612
    septemseptatum (R. C. Harris) Aptroot, 612
Lobaria (Schreber) Hoffm., 5, 6, 19, 46, 58, 59, 72, 89, 93, 95, *141*, *143*, 144, 295, **415**, *415*, 535, 591, 671, 752
    amplissima (Scop.) Forssell, 295
    hallii (Tuck.) Zahlbr., 19, *415*, **416**, 421
    linita (Ach.) Rabenh., *139*, *415*, **416**, 417
    oregana (Tuck.) Müll. Arg., 40, 59, 72, *137*, **417**, 418, 595
    pseudopulmonaria Gyelnik, 417
    pulmonaria (L.) Hoffm., 13, 15, 22, 30, 31, 60, 79, 81, 83, 87, *415*, 416, **417**, 419, 421, 593, 762
    quercizans Michaux, 69, *415*, **420**
    ravenelii (Tuck.) Yoshim., *415*, **420**, 421
    retigera (Bory) Trevisan, 83, 415, 417, 421, 593
    scrobiculata (Scop.) DC., *138*, *415*, 416, **421**, 455
    tenuis Vainio, 33, *415*, 420, **421–422**
Lobariaceae, 95, 752
lobed cotton lichen, 289
Lobothallia (Clauzade & Roux) Hafellner, *132*, *133*, *140*, *141*, *154*, 374, **422**, *422*, 752
    alphoplaca (Wahlenb.) Hafellner, *154*, *422*, **422–423**
    melanaspis (Ach.) Hafellner, 423
    praeradiosa (Nyl.) Hafellner, *154*, *422*, *422*, **423**, 424
Loculoascomycetes, 753
loop lichens, 356
    grainy, 358
    green, 360
    powdered, 359
    smooth, 359
    wrinkled, 358
    yellow-cored, 357
Lopadium Körber, *130*, 177, 194, **423**, 753
    disciforme (Flotow) Kullh., 173, 423, **424**
    fuscum Müll. Arg., 194
    pezizoideum (Ach.) Körber, 424
Loxospora A. Massal., *132*, **424**, *425*, 752
    cismonica (Beltr.) Hafellner, *425*, **425**
    elatina (Ach.) A. Massal., 426
    ochrophaea (Tuck.) R. C. Harris, *425*, **425**
    pustulata (Brodo & Culb.) R. C. Harris, 33, 69, *122*, *123*, *425*, **426**, 762
Loxosporopsis Henssen, **426**, 752
    corallifera Brodo, Henssen & Imshaug, 31, *122*, *123*, *132*, **426–427**
lung lichen, 415, 417
lungworts, 83, 415, 417
    cabbage, 416
    Dixie, 420
    gray, 416
    slender, 421
    smooth, 420
    textured, 421

map lichens, 84, 632
    crescent, 636
    dusky, 637
    false, 651
    lemon, 637
    single-spored, 634
    smooth, 636

snowy, 634
yellow, 635
mapledust lichen, 389
Maronea A. Massal., 753
    constans (Nyl.) Hepp, 133
Masonhalea Kärnefelt, 60, 214, **427,** 752
    richardsonii (Hook.) Kärnefelt, 64, 67, *120, 214,* **427**
Massalongia Körber, *140,* **427,** *475,* 573, 729, 753
    carnosa (Dicks.) Körber, 72, *139,* 324, **427–428,** *475*
    microphylliza (Nyl. *ex* Hasse) Henssen, 428
matchstick lichens, 566
    devil's, 567
    powdered, 568
    tapered, 568
medallion lichens, 305, 306
    grainy, 306
    powdery, 306
    purple-eyed, 307
Megalaria Hafellner, **428,** 752
    brodoana S. Ekman & Tønsberg, 429
    grossa (Pers. *ex* Nyl.) Hafellner, 429
    laureri (Th. Fr.) Hafellner, *131,* **429**
Megalospora Meyen, 753
    porphyrites (Tuck.) R. C. Harris, 614
Megalosporaceae, 753
Megaspora (Clauz. & Roux) Hafellner & V. Wirth, **429,** 753
    verrucosa (Ach.) Hafellner & V. Wirth, *133,* **429–430,** 530
Megasporaceae, 753
Melanelia Essl., *140, 143, 154, 214,* **430,** *430–431,* 449, 451, 563, 691, 752
    albertana (Ahti) Essl., 70, *430,* **431–432,** 692
    commixta (Nyl.) Thell, 435, 439
    culbersonii (Hale) Thell, 69, *430,* **432–433,** 438, 441, 692
    disjuncta (Erichsen) Essl., *430,* **433,** 438, 441
    elegantula (Zahlbr.) Essl., *430–431,* **433–434,** 440, 692
    exasperata (De Not.) Essl., *431,* 440–441
    exasperatula (Nyl.) Essl., *431,* **434,** 438, 692
    fuliginosa (Fr. *ex* Duby) Essl., *431,* **434–435,** 439
    glabra (Schaerer) Essl., 74, *431,* **435**
    glabratula (Lamy) Essl., 434
    glabroides (Essl.) Essl., 435
    halei (Ahti) Essl., 437
    hepatizon (Ach.) Thell, *141, 431,* 432, **435–436,** 450
    infumata (Nyl.) Essl., 433–434
    multispora (A. Schneider) Essl., *431,* **436,** 440

    olivacea (L.) Essl., *431,* **437,** 438, 439, 440
    panniformis (Nyl.) Essl., 156, 160, *431,* **437–438,** 482
    septentrionalis (Lynge) Essl., 64, 68, *431,* 437, **438**
    sorediata (Ach.) Goward & Ahti, *430,* 433, **438,** 441, 756
    stygia (L.) Essl., 156, *431,* 435, **439,** 450, 587
    subargentifera (Nyl.) Essl., *430, 431,* **439,** 440
    subaurifera (Nyl.) Essl., 31, *430–431,* 432, 434, **439–440**
    subelegantula (Essl.) Essl., 433, 434, 438
    subolivacea (Nyl.) Essl., 71, *431,* 436, 437, **440–441**
    tominii (Oksner) Essl., *430–431,* 433, 438, 439, **441**
    trabeculata (Ahti) Essl., 436, 438
Melanolecia Hertel, 583, 752
    transitoria (Arnold) Hertel *ex* Poelt, 583
Melanotheca Körber
    cruenta (Mont.) Müll Arg., 612
Menegazzia A. Massal., *140,* 213, 345, **441,** 752
    terebrata (Hoffm.) A. Massal., 69, *345,* **441–443,** 757
Meruliaceae, 751
Micarea Fr., *131,* 176, 364, 365, **443,** 572, 653, 752
    erratica (Körber) Hertel, Rambold, & Pietschmann, 392
    melaena (Nyl.) Hedl., 443
    peliocarpa (Anzi) Coppins & R. Sant., **443–444**
    prasina Fr., *131,* 444
Micareaceae, 752
monk's-hood lichen, 354
    brownish, 356
Monoblastia Riddle, *125,* 753
Monoblastiaceae, 753
moon lichens, 670
    fringed, 671
    peppered, 672
    powdered, 672
moonglow lichens, 300
    golden, 300
    silver, 300
moss-dots
    four-celled, 444
    six-celled, 172
moss-shingle,
    brown-gray, 477
    green, 604
moss-thorns, 580
mottled-disk lichen, 687
    scaly, 687
mountain pink lichen, 363
mouse ears, 309, 310, 395
mouse lichens, 321, 475

Multiclavula R. Petersen, 8, 30, *117,* **444,** 470, 751
    corynoides (Peck) R. Petersen, **444**
    muscida (Fr.) R. Petersen, 444
    vernalis (Schwein.) R. Petersen, 444
mushroom lichens, 470
    greenpea, 470
mustard lichen, 617
Mycobilimbia Rehm, *130,* 170, 176, 365, 395, 443, **444,** 752
    berengeriana (A. Massal.) Hafellner & V. Wirth, *129,* 172–173, 444
    fusca (A. Massal.) Hafellner & V. Wirth, 444
    hypnorum (Lib.) Kalb & Hafellner, *129,* 173, 444
    lobulata (Sommerf.) Hafellner, 395
    obscurata (Sommerf.) Rehm, 444
    sabuletorum (Schreber) Hafellner, 172
    tetramera (De Not.) W. Brunnbauer, 172, **444, 446**
Mycoblastaceae, 752
Mycoblastus Norman, *128,* 364, **446,** 752
    affinis (Schaerer) Schauer, 447
    alpinus (Fr.) Kernst., 447
    caesius (Coppins & P. James) Tønsberg, *124,* 446
    fucatus (Stirton) Zahlbr., 446
    sanguinarius (L.) Norman, **446–447**
Mycocaliciaceae, 751
Mycocalicium Vainio, *192,* 222, 751
    subtile (Pers.) Szat., *192*
Myelochroa (Asah.) Elix & Hale, *142,* **447,** 480, 486, 752
    aurulenta (Tuck.) Elix & Hale, 69, *142,* 212, **447,** 480–481, 619
    galbina (Ach.) Elix & Hale, 358, **447–448,** 480–481
    obsessa (Ach.) Elix & Hale, **448,** 480–481, 487
Myriotrema Fée, *125,* **449,** 449, 462, 680, 752
    glaucescens (Nyl.) Hale, 449
    rugiferum (Harm.) Hale, **449,** 449
    subcompunctum (Nyl.) Hale, 449
    wightii (Taylor) Hale, 449

Neofuscelia Essl., *140, 143, 430, 430,* **449,** 692, 752, 762
    ahtii (Essl.) Essl., 450
    atticoides (Essl.) Essl., *141, 431,* **450**
    brunella (Essl.) Essl., 450
    infrapallida (Essl.) Essl., 450
    loxodes (Nyl.) Essl., *431,* 433, **450,** 451
    occidentalis (Essl.) Essl., 450
    subhosseana (Essl.) Essl., *431,* **450, 451**
    verruculifera (Nyl.) Essl., 450, 451
Nephroma Ach., 6, 37, 58, 72, 89, 96, *143,* 295, 415, **451,** *451,* 503, 592, 654, 752

arcticum (L.) Torss., 22, *138*, 167, *451*, **452,** 453
bellum (Sprengel) Tuck., *451*, **452–453**
expallidum (Nyl.) Nyl., *140*, 167, *451*, 452, **453**
helveticum Ach., *451*, 452, **453,** 457, *475*
helveticum var. helveticum, 454
helveticum var. sipeanum (Gyelnik) Goward & Ahti, 454
isidiosum (Nyl.) Gyelnik, *451*, **454–455,** 456
laevigatum Ach., 66, 75, *451*, 452, 454, **455**
occultum Wetmore, 93, *137*, *451*, **455–456**
parile (Ach.) Ach., *451*, 454, 455, **456–457,** 507, 672, 692
resupinatum (L.) Ach., *451*, 454, **457**
Nephromataceae, 752
Neuropogon Nees & Flotow
  sulphureus (J. Körnig) Hellbom, 711
Niebla Rundel & Bowler, 16, 73, *119*, 295, 305, 311, **457,** *620,* 620, 652, 752
  cephalota (Tuck.) Rundel & Bowler, **457–458,** *620*
  ceruchis (Ach.) Rundel & Bowler, 458
  ceruchoides Rundel & Bowler, 458
  combeoides (Nyl.) Rundel & Bowler, **458,** 459, *621*
  homalea (Ach.) Rundel & Bowler, 73, **458,** 460, *621*
  laevigata Bowler & Rundel, **460**
  procera Rundel & Bowler, 458
nipple lichens, 610, 678
  lobed, 679
  rock, 680
Nodobryoria Common & Brodo, *150*, 178, 179, 180, 365, 366, **460,** 752
  abbreviata (Müll. Arg.) Common & Brodo, *119*, *151*, 365, **460–461**
  oregana (Tuck.) Common & Brodo, *119*, *151*, 460, **461**
Normandina Nyl., **461,** 470, 753
  pulchella (Borrer) Nyl., *135*, **461–462**

oakmoss lichen, 82, 311, 312, 313
  boreal, 312
  mountain, 311
Ocellularia G. Meyer, *125*, 449, **462,** 680, 752
  americana Hale, 462
  granulosa (Tuck.) Zahlbr., **462–463**
  sandfordiana (Zahlbr.) Hale, 463
Ochrolechia A. Massal., 28, 44, 108, 109, *128*, *133*, 281, **463,** 463–464, 526, 534, 687, 753
  africana Vainio, *464,* **464–465,** 468
  androgyna (Hoffm.) Arnold, *124*, *464,* **465,** 466
  arborea (Kreyer) Almb., 465
  farinacea Howard, 468
frigida (Sw.) Lynge, 385, *464,* **465–466**
juvenalis Brodo, 72, 104, *464,* **466**
laevigata (Räsänen) Vers. *ex* Brodo, 72, *464,* **466–467**
mexicana Vainio, 109, *464,* 465, 468
oregonensis H. Magn., 65, 72, 81, 104, 463, *464,* **467,** 764
pseudopallescens Brodo, *464,* 466
subathallina H. Magn., 466
subisidiata Brodo, 470
subpallescens Vers., *464,* 467
subplicans (Nyl.) Brodo, 124
szatalaënsis Vers., *464,* 468
tartarea (L.) A. Massal., 81, 463, 757
trochophora (Vainio) Oshio, *464,* 465, **468,** 470, 534
upsaliensis (L.) A. Massal., 18, 48, *464,* **468,** 469
yasudae Vainio, *464,* **468–470**
oilskin
  blue, 404
  peacock, 408
  wrinkled, 407
old man's beard, 709
old-wood lichen, 368, 369
olive-thorn lichen, 295
Omphalina Quélet, 29, *117,* **470,** 751
  ericetorum (Pers.:Fr.) M. T. Lange, 470
  hudsoniana (H. S. Jenn.) H. E. Bigelow, *135,* 470
  umbellifera (L.:Fr.) Quélet, **470**
Omphalodium Meyen & Flotow
  arizonicum (Tuck. *ex* Willey) Tuck., 471
Omphalora T. Nash & Hafellner, 14, **470,** 696, 733, 752
  arizonica (Tuck. *ex* Willey) T. Nash & Hafellner, 72, *136,* **471,** 638
Opegrapha Ach., 23, 24, *126,* 326, 369, 371, **471–472,** *472,* 753
  atra Pers., 472
  cinerea Chevall., 472
  herbarum Mont., 472
  hypothallina (Zahlbr.) Tehler, 371
  lichenoides Pers., 472
  longissima Müll. Arg., 472
  pulicaris (as commonly used), 472
  rufescens Pers., 472
  rupestris Pers., 472
  varia Pers., *472,* **472**
  viridis (Pers. *ex* Ach.) Behlen & Desberger, 472
  vulgata Ach., 472
Ophioparma Norman, *127,* 331, **473,** *473,* 752
  herrei (Zahlbr.) Kalb & Staiger, 473
  lapponica (Räsänen) Hafellner & W. Rogers, *473,* 474

rubricosa (Müll. Arg.) S. Ekman, *473,* **473,** 474
ventosa (L.) Norman, *473,* **473–474**
Ophioparmaceae, 752
orange bush lichens, 674
  powdered, 675
  slender, 675
orange lichens, 742
orchil lichens, 81, 463, 647
organ-pipe lichen, 249
Orphniospora Körber, **474,** 753
  moriopsis (A. Massal.) D. Hawksw., *129,* **474–475**
Ostropales, 752
owl lichens, 654

Pachyospora A. Massal.
  verrucosa (Ach.) A. Massal., 429
Pachyphiale Lönnr., 329, 751
  fagicola (Hepp) Zwackh, *128,* 302
palmetto lichen, 622
Pannaria Delise, 89, 108, *135,* *140,* *141,* *143,* 279, 309, 321, 427, 428, 461, **475,** 475–476, 484, 597, 729, 753
  conoplea (Ach.) Bory, *135,* *143,* 321, 475, **476,** 477, 479
  leucophaea (Vahl) P. M. Jørg., 321
  leucosticta (Tuck.) Tuck. *ex* Nyl., 322
  leucostictoides Ohlsson, 323
  lurida (Mont.) Nyl., 87, 476, **476–477,** 478
  pezizoides (Weber) Trevisan, *133,* 321, 476, **477–478,** 604
  rubiginosa (Ach.) Bory, 322, 323, 476, 477, **478**
  tavaresii P. M. Jørg., 475, 476, 478, **479**
Pannariaceae, 753
Paraparmelia Elix & J. Johnston, *480,* 752
  alabamensis (Hale & McCull.) Elix & J. Johnston, *142,* 212, 481
Parmelia Ach., 20, 55, 95, 104, *140,* *142,* *143,* *144,* 166, 420, 430, **479,** *480*–481, 486, 595, 752
  albertana Ahti, 431
  appalachensis Culb., 606
  bolliana Müll. Arg., 606
  disjuncta Erichsen, 433
  elegantula (Zahlbr.) Szat., 433
  exasperatula Nyl., 434
  fertilis Müll. Arg., 484
  flaventior Stirton, 317
  fraudens (Nyl.) Nyl., 479, *481,* 484
  glabra (Schaerer) Nyl., 435
  glabratula (Lamy) Nyl., 434
  hygrophila Goward & Ahti, 31, *481,* **481–482,** 483, 484
  kerguelensis A. Wilson, 483
  multispora A. Schneider, 436

olivacea (L.) Ach., 437
omphalodes (L.) Ach., 80, 81, *481*, **482–483**, 565
panniformis (Nyl.) Vainio, 437
praesignis Nyl., 318
pseudosulcata Gyelnik, 483
rudecta Ach., 608
saxatilis (L.) Ach., 15, 31, 80, 83, 167, 479, *481*, 482–483, **483**, 484
septentrionalis (Lynge) Ahti, 438
sorediosa Almb., 431, 438
sphaerosporella Müll. Arg., 150
squarrosa Hale, 15, *481*, 483, **484**
stictica (Duby) Nyl., 608
stygia (L.) Ach., 439
subargentifera Nyl., 439
subaurifera Nyl., 439
subolivacea Nyl., 440
subrudecta Nyl., 609
substygia Räsänen, 441
sulcata Taylor, 30, 61, 104, *481*, **484**, 485
ulophyllodes (Vainio) Savicz, 319
Parmeliaceae, 95, 752
Parmeliella Müll. Arg., 135, 140, 279, 321, 428, 475, 476, **484**, 546, 597, 753
pannosa (Sw.) Müll. Arg., 475
plumbea (Lightf.) Vainio, 280
triptophylla (Ach.) Müll. Arg., 323, 475–476, **485**
Parmelina Hale, *142*, 421, 447, 479, *480*, **486**, 752
aurulenta (Tuck.) Hale, 447
dissecta (Nyl.) Hale, 487
galbina (Ach.) Hale, 447
horrescens (Taylor) Hale, 486
obsessa (Ach.) Hale, 448
quercina (Willd.) Hale, 74, 448, *480*, **486**
Parmelinopsis Elix & Hale, *142*, 447, 479, *480*, **486**, 752
cryptochlora (Vainio) Elix & Hale, 487
horrescens (Taylor) Elix & Hale, *480*, **486–487**, 644
minarum (Vainio) Elix & Hale, *480*, **487**, 488
spumosa (Asah.) Elix & Hale, *480*, **487–488**
Parmeliopsis Nyl., **488**, 752
aleurites (Ach.) Nyl., 361
ambigua (Wulfen) Nyl., 46, *138*, 360, **489**
capitata R. C. Harris, 489
hyperopta (Ach.) Arnold, *142*, 361, **489**, 490, 546
placorodia (Ach.) Nyl., 362
subambigua Gyelnik, 489
Parmotrema A. Massal., 99, 103, 211, 220, 316, 479, *480*, **489–490**, 490–491, 574, 642, 752, 757
arnoldii (Du Reitz) Hale, 31, *491*, **492**, 493, 499

austrosinense (Zahlbr.) Hale, *491*, **492–493**
cetratum (Ach.) Hale, 642
chinense (Osbeck) Hale & Ahti, 31, 83, *491*, *492*, **493–494**, 497, 499
crinitum (Ach.) M. Choisy, 211, *491*, **494**, 501, 644
cristiferum (Taylor) Hale, *491*, **494–495**
diffractaicum (Essl.) Hale, 643
dilatatum (Vainio) Hale, *491*, **495**, 496, 499
dominicanum (Vainio) Hale, 495
endosulphureum (Hillm.) Hale, *491*, **495–496**, 501
eurysacum (Hue) Hale, 491
gardeneri (C. W. Dodge) Sérus., 495, 499
hypoleucinum (Steiner) Hale, 493, 496
hypotropum (Nyl.) Hale, 11, *491*, *492*, 493, **496–497**, 643
internexum (Nyl.) Hale, 494
louisianae (Hale) Hale, 496
madagascariaceum (Hue) Hale, 502
margaritatum (Hue) Hale, 492, 494, 501, 643
mellissii (C. W. Dodge) Hale, *491*, 494, 501
michauxianum (Zahlbr.) Hale, *480*, *491*, **497**, 499
perforatum (Jacq.) A. Massal., 18, 19, 83, *491*, 494, 497, **498–499**
perlatum (Hudson) M. Choisy, 493
praesorediosum (Nyl.) Hale, *491*, 495, **499**
preperforatum (Culb.) Hale, 499
rampoddense (Nyl.) Hale, *491*, 495, **499–500**
reticulatum (Taylor) M. Choisy, 643
rigidum (Lynge) Hale, 499
rubifaciens (Hale) Hale, 495
simulans (Hale) Hale, 643
stuppeum (Taylor) Hale, 30, *491*, *492*, 496, **500–501**
subisidiosum (Müll. Arg.) Hale, 643
subtinctorium (Zahlbr.) Hale, 211
sulphuratum (Nees & Flotow) Hale, *138*, *490*, 495, **501**, 502
tinctorum (Delise ex Nyl.) Hale, 70, 104, *491*, 495, **501**, 502
ultralucens (Krog) Hale, 70, *491*, 494, **501–502**
xanthinum (Müll. Arg.) Hale, *138*, *139*, *490*, **502–503**
Patellariales, 753
paw lichens, 451
pebble lichen, 685
pebblehorn lichen, 278
Peccania A. Massal. ex Arnold, 611, 752
peg lichen, 265, 266, 270
 powdery, 272
pelt lichens, 503

black-and-white, 510
born-again, 518
carpet, 516
common freckle, 504
concentric, 509
dark-veined, 510
flaky freckle, 505
flat-fruited, 510
fringed, 517
many-fruited, 517
peppered, 509
ruffled freckle, 513
scabby, 521
scaly, 511
tree, 507
veinless, 513
Peltigera Willd., 5, 6, 13, 14, 20, 37, 38, 40, 58, 59, *139*, *143*, *144*, 310, 395, 415, 451, 461, **503**, 503–504, 654, 753, 764
aphthosa (L.) Willd., 83, *503*, **504–505**, 507, 513, 521
britannica (Gyelnik) Holt.-Hartw. & Tønsberg, 6, 22, *503*, **505–506**, 513, 522
canina (L.) Willd., 15, 35, 83, *504*, **506–507**, 513, 520, 521
cinnamomea Goward, 520
collina (Ach.) Schrader, *143*, 457, *503*, **507**, 508, 521, 672
degenii Gyelnik, 517–518
didactyla (With.) J. R. Laundon, 7, *503*, **507–509**, 511, 521
didactyla var. *didactyla*, 508–509
didactyla var. *extenuata* (Nyl. ex Vainio) Goffinet & Hastings, 508–509, 521
elisabethae Gyelnik, 33, *504*, **509**, 510, 517, 520
evansiana Gyelnik, *504*, 509, 511
horizontalis (Hudson) Baumg., *504*, 509, **510**, 518
hymenina (Ach.) Delise, *504*, 513, 518
kristinssonii Vitik., 73, *504*, **510–511**, 521, 764
lepidophora (Vainio) Bitter, *504*, 508, 509, **511**
leucophlebia (Nyl.) Gyelnik, *503*, 505, **512–513**, 521
malacea (Ach.) Funck, *504*, 505, 506, **513**, 514
membranacea (Ach.) Nyl., *504*, 506–507, **513**, 515
neckeri Hepp ex Müll. Arg., *504*, **514–515**
neopolydactyla (Gyelnik) Gyelnik, *504*, **516**, 518, 521
pacifica Vitik., 33, *504*, **517**, 518, 520
polydactylon (Necker) Hoffm., *504*, 510, 513, 514, 516, **517–518**
ponojensis Gyelnik, *504*, **518**, 519, 521

praetextata (Flörke *ex* Sommerf.) Zopf, 504, 507, 509, 517, **518–520**
retifoveata Vitik., 521
rufescens (Weiss) Humb., 19, 20, 35, 55, 504, 506, 508–509, 510, 518, **520–521**
scabrosa Th. Fr., 18, 504, 507, 510, **521**, 762
scabrosella Holt.-Hartw., 521
spuria (Ach.) DC., 507
venosa (L.) Hoffm., 136, 445, 503, 505, **521–522**
Peltigeraceae, 753
Peltigerineae, 752
Peltula Nyl., 35, 50, 56, 58, 74, 135, 308, 333, **522**, 522, 597, 683, 752, 763
bolanderi (Tuck.) Wetmore, 523
clavata (Kremp.) Wetmore, 523
cylindrica Wetmore, 121, 134, 308, 522, **523**, 525
euploca (Ach.) Poelt, 135, 522, **523**, 524, 525
michoacanensis (de Lesd.) Wetmore, 524
obscurans (Nyl.) Gyelnik, 522, **523–524**
obscurans var. deserticola (Zahlbr.) Wetmore, 524
obscurans var. hassei (Zahlbr.) Wetmore, 524
obscurans var. obscurans, 524
patellata (Bagl.) Swinscow & Krog, 522, **524–525**
polyspora (Tuck.) Wetmore, 524
richardsii (Herre) Wetmore, 66, 522, 524, **525**
tortuosa (Nees) Wetmore, 523, 525
zahlbruckneri (Hasse) Wetmore, 523
Peltulaceae, 752
pepper-spore lichens, 644
tundra, 647
creamy, 645
poplar, 646
scaly, 645
Pertusaria DC., 23, 27, 61, 124, 127, 128, 133, 281, 424, 430, 462, 463, **525–526**, 526–527, 680, 753
amara (Ach.) Nyl., 63, 69, 123, 124, 527, **527–528**, 535
arizonica Dibben, 528
borealis Erichsen, 614
consocians Dibben, 526, 533
copiosa Erichsen, 527
dactylina (Ach.) Nyl., 122, 123, 127, 527, **528**, 530
epixantha R. C. Harris, 533
flavicunda Tuck., 526, 527, **528–529**
globularis (Ach.) Tuck., 470
hypothamnolica Dibben, 527
leioplaca DC., 526
leucostoma A. Massal., 526
macounii (Lamb) Dibben, 24, 526, **529**, 533

multipunctoides Dibben, 527, 530
neoscotica Lamb, 532, 533
ophthalmiza (Nyl.) Nyl., 527, **530**, 532, 533
ostiolata Dibben, 526
panyrga (Ach.) A. Massal., 527, 528, **530**, 531
paratuberculifera Dibben, 526, **530**, 531
plittiana Erichsen, 526, **530–532**, 533, 764
pupillaris (Nyl.) Th. Fr., 614
pustulata (Ach.) Duby, 526, 535
rubefacta Erichsen, 526
sinusmexicani Dibben, 526
subambigens Dibben, 72, 527, 530, **532**
subobducens Nyl., 526
subpertusa Brodo, 118, 526, 529, **532–533**
tetrathalamia (Fée) Nyl., 526, 530
texana Müll. Arg., 526, **533**, 535
trachythallina Erichsen, 124, 527, 530, **533–534**
velata (Turner) Nyl., 527, **534–535**
xanthodes Müll. Arg., 65, 70, 118, 526, 533, **535**
Pertusariaceae, 753
Pertusariineae, 753
Petractis Fr., 752
farlowii (Tuck. *ex* Nyl) Vězda, 329
Phaeocalicium A. F. W. Schmidt, 192, 222, 751
populneum (Brond *ex* Duby) A. F. W. Schmidt, 192
Phaeographina Müll. Arg., 126, **535**, 535, 752
caesiopruinosa (Fée) Müll. Arg., 535, **535**, 536
quassiaecola (Fée) Müll. Arg., 535, **535–536**
Phaeographis Müll. Arg., 126, 163, 326, 471, 535, **536**, 649, 752
inusta (Ach.) Müll. Arg., 327, **536–537**
Phaeophyscia Moberg, 46, 55, 140, 141, 142, 143, 160, 343, 430, **538**, 538, 546, 558, 560, 566, 616, 657, 752
adiastola (Essl.) Essl., **538–539**, 539, 543
cernohorskyi (Nádv.) Essl., 19, 70, 538, **539–540**
ciliata (Hoffm.) Moberg, 69, 538, **540**, 542
constipata (Norrlin & Nyl.) Moberg, 139, 538, **540–541**
decolor (Kashiw.) Essl., 538, 540, **541–542**, 544
endococcinodes (Poelt) Essl., 542
hirsuta (Mereschk.) Essl., 539
hirtella Essl., 539, 540
hispidula (Ach.) Essl., 144, 538, 539, **542**, 547–548
hispidula ssp. hispidula, 542
hispidula ssp. limbata Poelt, 542
imbricata (Vainio) Essl., 542
insignis (Mereschk.) Moberg, 542
kairamoi (Vainio) Moberg, 540, 544

orbicularis (Necker) Moberg, 539, **542–543**, 543, 560
pusilloides (Zahlbr.) Essl., 538, **543**
rubropulchra (Degel.) Essl., 69, 538, 539, **543–544**, 619
sciastra (Ach.) Moberg, 140, 538, **544**
squarrosa Kashiw., 538, 542
Phaeorrhiza H. Mayrh. & Poelt, 132, **544**, 644, 683, 752
nimbosa (Fr.) H. Mayrh. & Poelt, 136, 544, **545**, 644, 647
sareptana (Tomin) H. Mayrh., 544, 545
Phlyctidaceae, 753
Phlyctis Wallr., 132, 526, **545**, 753
agelaea (Ach.) Flotow, 546
argena (Sprengel) Flotow, 123, **545–546**
Phylliscum Nyl., 752
demangeonii (Moug. & Mont.) Nyl., 682
Phyllopsora Müll. Arg., 176, **546**, 752
parvifolia (Pers.) Müll. Arg., 129, 135, **546**
Physcia (Schreber) Michaux, 27, 28, 46, 141, 142, 302, 305, 306, 334, 361, 362, 488, 489, 538, **546**, 546–549, 560, 561, 616, 657, 752
adscendens (Fr.) H. Olivier, 18, 548, **549**, 550, 560
aipolia (Ehrh. *ex* Humb.) Fürnr., 69, 547, **549–550**, 556, 558
americana G. Merr., 306, 548, **550–551**, 556, 558
atrostriata Moberg, 30, 70, 549, 551, **551–552**, 558
biziana (A. Massal.) Zahlbr., 73, 144, 547, **552–553**
caesia (Hoffm.) Fürnr., 40, 548, **553**
callosa Nyl., 73, 549, **553–554**
cascadensis H. Magn., 552, 556
chloantha (Ach.) Vainio, 560
clementii (Turner) Maas Geest., 551
crispa Nyl., 552
dimidiata (Arnold) Nyl., 552
dubia (Hoffm.) Lettau, 538, 549, 553, **554**, 555, 559, 560
halei J. W. Thomson, 547, **555**, 559
leptalea (Ach.) DC., 549, 558
melanchra Hue, 560
mexicana de Lesd., 552
millegrana Degel., 33, 549, 553, **555**, 558
neogaea R. C. Harris, 547, 550, 551, **556**
phaea (Tuck.) J. W. Thomson, 547, 553, **556**, 557
pseudospeciosa J. W. Thomson, 340, 549
pumilior R. C. Harris, 547, 550, **556**, 557
semipinnata (J. F. Gemlin) Moberg, 549
sorediosa (Vainio) Lynge, 142, 548, 552, **556–558**
stellaris (L.) Nyl., 547, 549, 553, **558**
subtilis Degel., 549, 554, 555, **558–559**

tenella (Scop.) DC., *548, 549*, **559–560**
undulata Moberg, *552*
Physciaceae, 752
Physciella Essl., *141*, 538, 546, **560**, 752
   chloantha (Ach.) Essl., *538, 548*, 554, **560**
   melanchra (Hue) Essl., *542*, 560
Physconia Poelt, 18, 35, 46, *140, 142*, 160, 430, 538, 546, 552, **560–561**, *561*, 616, 752
   americana Essl., 33, 160, *561*, **561**, 562, 565
   detersa (Nyl.) Poelt, 15, 19, *561*, **561**, 563, 564, 565, 566
   elegantula Essl., *561*, **563**
   enteroxantha (Nyl.) Poelt, 65, *561*, **564**, 565
   grisea (Lam.) Poelt, 566
   isidiigera (Zahlbr.) Essl., 74, *561*, 563, **564–565**
   leucoleiptes (Tuck.) Essl., 561, 564, 565
   muscigena (Ach.) Poelt, 12, *139, 561*, **565**
   perisidiosa (Erichsen) Moberg, *561*, 564, **565–566**
   subpallida Essl., 160
   thomsonii Essl., *548, 561*, **566**
pillow lichens, 345
Pilocarpaceae, 752
Pilophoron Th. Fr., 566
   aciculare (Ach.) Th. Fr., 567
   cereolus (Ach.) Th. Fr., 568
   hallii (Tuck.) Vainio, 568
Pilophorus Th. Fr., 14, 22, *121*, 231, **566–567**, *567*, 657, 663, 761
   acicularis (Ach.) Th. Fr., 55, *121, 567*, **567–568**
   cereolus (Ach.) Th. Fr., *567*, **568**, 663, 666
   clavatus Th. Fr., *567*, **568**
   fibula (Tuck.) Th. Fr., 568
   nigricaulis Satô, 567
   robustus Th. Fr., 567
   vegae Krog, 567
pimple lichens, 582
pin lichens, 221, 259
pink earth lichen, 50, 299, 360
pitted stone lichen, 727
pixie-cup lichen, 238, 243, 247
   boreal, 239
   carpet, 266
   crowned, 243
   finger, 252
   mealy, 247
   pebbled, 267
   red-fruited, 265
   rosette, 266
   mixed-up, 261
pixie-hair lichens, 281, 282
Placidium A. Massal., 50, 56, *136*, 309, 333, 522, **569**, 597, 683, 727, 753
   lachneum (Ach.) Breuss, 570
   lacinulatum (Ach.) Breuss, **569–570**

   squamulosum (Ach.) Breuss, 570
   tuckermanii (Ravenel *ex* Mont.) Breuss, *135*, **570–571**
Placopsis (Nyl.) Lindsay, 22, *132*, 567, **570**, 751
   cribillens (Nyl.) Räsänen, 571
   gelida (L.) Lindsay, 55, 571, 572
   lambii Hertel & V. Wirth, 22, *123*, 571, **572**, 756
   roseonigra Brodo, 571
Placynthiaceae, 753
Placynthiella Elenkin, 57, *129, 391*, **572**, 751
   icmalea (Ach.) Coppins & P. James, *122*, 573
   oligotropha (J. R. Laundon) Coppins & P. James, 50, 55, *122*, **572–573**
   uliginosa (Schrader) Coppins & P. James, 572
Placynthium (Ach.) Gray, *130, 131*, **573**, 683, 729, 753
   asperellum (Ach.) Trevisan, 573
   flabellosum (Tuck.) Zahlbr., 573
   nigrum (Hudson) Gray, 48, *135*, **573**
   stenophyllum (Tuck.) Fink, 573, 729
Plagiocarpa R. C. Harris, 612
Platismatia Culb. & C. Culb., *141, 142*, 166, 167, 310, 451, 480, 491, **574**, *574*, 589, 610, 752
   glauca (L.) Culb. & C. Culb., 220, 310, *491, 574*, **574**, 575, 577
   herrei (Imshaug) Culb. & C. Culb., 32, *574*, **574–576**
   lacunosa (Ach.) Culb. & C. Culb., 72, *574*, **576**
   norvegica (Lynge) Culb. & C. Culb., 75, *574*, **576**, 577
   stenophylla (Tuck.) Culb. & C. Culb., *574*, **576–577**
   tuckermanii (Oakes) Culb. & C. Culb., *574*, 576, **577–578**
Pleopsidium Körber, 11, 34, *132, 145*, 207, 210, **578**, 752
   chlorophanum (Wahlenb.) Zopf, 578
   flavum (Bellardi) Körber, *145*, **578–579**
   oxytonum (Ach.) Rabenh., 578
Polyblastia A. Massal., *125*, 661, 677, 726, 753
Polychidium (Ach.) Gray, 5, 14, *118*, 460, **579**, 658, 753, 760
   contortum Henssen, 75, *122*, **579–580**
   dendriscum (Nyl.) Henssen, 579
   muscicola (Sw.) Gray, 48, *122*, **580**
Polymeridium (Müll. Arg.) R. C. Harris, 753
   quinqueseptatum (Nyl.) R. C. Harris, *125*
Polysporina Vězda, 50, 55, *129, 145*, **580–581**, 650, 659, 752
   simplex (Davies) Vězda, **581**
   urceolata (Anzi) Brodo, 581
Porina Mull. Arg., 23, 29, *125*, **582**, 753

   guaranitica Malme, 582
   heterospora (Fink) R. C. Harris, *126*, **582**
   scabrida R. C. Harris, *126*
Porpidia Körber, 26, 27, 50, 55, 102, 159, 392, 394, 581, **582–583**, *583*, 586, 633, 688, 752
   albocaerulescens (Wulfen) Hertel & Knoph, *583*, **583–584**
   carlottiana Gowan, 583
   crustulata (Ach.) Hertel & Knoph, *583*, **584**, 585
   flavocaerulescens (Hornem.) Hertel & A. J. Schwab, *391, 393, 583*, **584–586**
   macrocarpa (DC.) Hertel & A. J. Schwab, 24, *393*, 583, 584, 586
   melinodes (Körber) Gowan & Ahti, 586
   speirea (Ach.) Kremp., 583, 634
   tahawasiana Gowan, *583*, 584
   thomsonii Gowan, 584
Porpidiaceae, 752
potato-chip lichen, 471
powderhorn,
   antlered, 273
   common, 247
   lipstick, 259
   smooth-footed, 262
powder-puff lichen, 224
pox lichens, 611
   eastern, 613
   yellow, 613
Protoblastenia (Zahlbr.) J. Steiner, *127*, 583, **586**, 614, 752
   incrustans (DC.) J. Steiner, 586
   rupestris (Scop.) J. Steiner, **586**
Protoparmelia M. Choisy, *133*, 156, 372, 374, 422, **586**, 752
   badia (Hoffm.) Hafellner, **586–587**
   ochrococca (Nyl.) P.M. Jørg., Rambold & Hertel, 586
Pseudephebe M. Choisy, 48, *120*, 179, 289, **587**, *587*, 619, 752
   minuscula (Nyl. *ex* Arnold) Brodo & D. Hawksw., 41, *120*, 156, 289, 439, 580, *587*, **587**, 588
   pubescens (L.) M. Choisy, 580, *587*, **587–588**
Pseudevernia Zopf, *119, 141, 142*, 313–314, **588–589**, *589*, 752
   cladonia (Tuck.) Hale Culb., 69, 82, *589*, **589**, 591
   consocians (Vainio) Hale & Culb., 64, 69, 82, *589*, **590–591**
   furfuracea (L.) Zopf, 82, 591
   intensa (Nyl.) Hale & Culb., 74, 82, *589*, **591**
Pseudocyphellaria Vainio, 5, 19, 20, 21, 46, 58, 72, 95, *143*, 295, 415, 455, 457, **591–592**, *592*, 671, 730, 731, 752
   anomala Brodo & Ahti, 31, 72, *117, 592*, **592–593**, 672
   anthraspis (Ach.) H. Magn., 21, 417, *592*, **593**

INDEX OF NAMES   787

aurata (Ach.) Vainio, *592*, **593–594**, *731*
crocata (L.) Vainio, *75*, *592*, **594–595**, *731*
rainierensis Imshaug, *93*, *592*, **595**
Pseudoparmelia Lynge, *480*, **595**, *752*
  alabamensis (Hale & McCull.) Hale, 212
  amazonica (Nyl.) Hale, 211
  baltimorensis (Gyelnik & Fóriss) Hale, 316
  caperata (L.) Hale, 317
  caroliniana (Nyl.) Hale, 211
  cryptochlorophaea (Hale) Hale, 212
  cubensis (Nyl.) Elix & Nash, 596
  floridensis Elix & Nash, 596
  rutidota (Hook. f. & Taylor) Hale, 317
  salacinifera (Hale) Hale, 211
  sphaerospora (Nyl.) Hale, 596
  texana (Tuck.) Hale, 212
  uleana (Müll. Arg.) Elix & Nash, *138*, *481*, **595–596**
Pseudopyrenula Müll. Arg., *689*, 753
  diluta (Fée) Müll. Arg., *689*
  subnudata Müll. Arg., 689
Psilolechia A. Massal., **596**, *752*
  lucida (Ach.) M. Choisy, *128*, *223*, *279*, **596–597**
Psora Hoffm., *17*, *50*, *56*, *74*, *136*, *308*, *330*, *395*, *522*, *597*, 597, *604*, *605*, *683*, *752*
  anthracophila (Nyl.) Arnold, 344
  californica Timdal, *598*, 599, 601
  cerebriformis W. A. Weber, *71*, *597*, **598**, 599
  crenata (Taylor) Reinke, *597*, 598, **599**
  decipiens (Hedwig) Hoffm., *50*, *597*, *598*, **599**, 600
  friesii (Ach.) Hellbom, 345
  globifera (Ach.) A. Massal., *71*, *598*, **599**, 600
  himalayana (C. Bab.) Timdal, *597*, **599–601**, *759*
  icterica (Mont.) Müll. Arg., *65*, *71*, *134*, *597*, **601**
  luridella (Tuck.) Fink, 599, 602
  montana Timdal, *597*, 602
  nipponica (Zahlbr.) Gotth. Schneider, *71*, *135*, *598*, **601–602**
  novomexicana de Lesd., 601
  pacifica Timdal, *597*, *599*, **602**, 603
  pseudorussellii Timdal, *598*, **602–603**, 604
  rubiformis (Ach.) Hook., 601, 602
  rufonigra (Tuck.) A. Schneider, 604
  russellii (Tuck.) A. Schneider, *598*, 603
  scalaris (Ach. *ex* Lilj.) Hook., 344
  tuckermanii R. Anderson *ex* Timdal, *598*, 601, **603–604**
Psoraceae, 752
Psoroma Michaux, *475*, **604**, *753*
  hypnorum (Vahl) Gray, *133*, *135*, *136*, *445*, *476*, *477*, **604**, *605*
  tenue Henssen, 604

Psorotichia A. Massal., 752
  schaereri (A. Massal.) Arnold, 611
Psorula Gotth. Schneider, *136*, *522*, 597, **604**, *752*
  rufonigra (Tuck.) Gotth. Schneider, *597*, **604–605**, *659*
Ptychographa Nyl., 751
  xylographoides Nyl., *126*
puffed lichen, 354
Punctelia Krog, *20*, *142*, *144*, *221*, *317*, *479*, *480*, *596*, **605–606**, *606*, *752*
  appalachensis (Culb.) Krog, *606*, **606**
  bolliana (Müll. Arg.) Krog, *70*, *606*, **606–607**
  borreri (Sm.) Krog, 610
  hypoleucites (Nyl.) Krog, *606*, **607–608**
  missouriensis G. Wilh. & Ladd, 609
  perreticulata (Räsänen) G. Wilh. & Ladd, 610
  punctilla (Hale) Krog, 609
  rudecta (Ach.) Krog, *81*, *211*, *606*, **608**
  semansiana (Culb. & C. Culb.) Krog, *606*, 607, 608
  stictica (Duby) Krog, *140*, *143*, *606*, **608–609**
  subpraesignis (Nyl.) Krog, *606*, 607
  subrudecta (Nyl.) Krog, *143*, *318*, *606*, **609–610**
pustule crust lichen, 426
Pycnothelia Dufour, **610**, *752*
  papillaria Dufour, *121*, **610**
Pyrenocollema Reinke, *52*, 753
  halodytes (Nyl.) R. C. Harris, *125*
Pyrenopsis (Nyl.) Nyl., *5*, *132*, *137*, **611**, *752*
  polycocca (Nyl.) Tuck., *124*, **611**
Pyrenula A. Massal., *27*, *29*, *125*, **611–612**, *612*, *689*, *690*, *753*, *762*
  anomala (Ach.) Vainio, *612*
  cerina Eschw., 613
  cinerea Zahlbr., 612
  concatervans (Nyl.) R. C. Harris, *612*
  cruenta (Mont.) Vainio, *612*, **612–613**, *689*
  cruentata (Müll. Arg.) R. C. Harris, 613
  falsaria (Zahlbr.) R. C. Harris, *612*
  leucostoma Ach., *612*
  mamillana (Ach.) Trevisan, *612*
  marginata Hook., *612*
  microcarpa Müll. Arg., *612*
  occidentalis (R. C. Harris) R. C. Harris, *612*, 614
  ochraceoflava (Nyl.) R. C. Harris, 613
  ochraceoflavens (Nyl.) R. C. Harris, *612*, **613**
  pseudobufonia (Rehm) R. C. Harris, *46*, *612*, **613–614**
  punctella (Nyl.) Trevisan, *612*
  ravenellii (Tuck.) R. C. Harris, *612*
  thelomorpha Tuck., *612*

Pyrenulaceae, 753
Pyrenulales, 753
Pyrgidium, 751
Pyrgillus Nyl., 753
  javanicus (Mont. & Bosch) Nyl., *127*
Pyrrhospora Körber, *331*, *586*, **614**, *752*
  cinnabarina (Sommerf.) M. Choisy, *124*, *127*, **614–615**, 616
  elabens (Fr.) Hafellner, *391*, 614
  quernea (Dicks.) Körber, *129*, *396*, 614, **615**
  russula (Ach.) Hafellner, *127*, **615–616**
  varians (Ach.) R. C. Harris, *129*, 614, 615
Pyxine Fr., *109*, *142*, *302–303*, *546*, *558*, *560–561*, **616**, *752*
  albovirens (G. Meyer) Aptroot, *548*, 616
  berteriana (Fée) Imshaug, *66*, *74*, *547*, **616**, 617
  caesiopruinosa (Nyl.) Imshaug, 616
  cocoes (Sw.) Nyl., *548*, **616–617**, 618
  eschweileri (Tuck.) Vainio, *142*, *548*, *558*, **617**, 619
  sorediata (Ach.) Mont., *548*, 616, **617**, 619
  subcinerea Stirton, 617

qelquaq, 421
quill lichen, 237
quilt lichens, 320

Racodium Pers.: Fr., *122*, **619**, *753*
  rupestre Pers., *281*, **619–620**
rag lichens, 574
  arctic, 166
  crinkled, 576
  crumpled, 577
  ground, 166
  oldgrowth, 576
  ribbon, 576
  tattered, 574
  tinted, 310
  varied, 574
ragbag, 574
ragged-rim lichens, 424
  eastern, 425
  frosted, 425
Ramalina Ach., *14*, *16*, *37*, *58*, *89*, *119*, *150*, *311*, *312*, *457*, *458*, **620**, *620–621*, *674*, *710*, *752*
  americana Hale, *621*, **621–622**, *623*, *624*, *630*
  canariensis J. Steiner, 627
  culbersoniorum La Greca, 622
  celastri (Sprengel) Krog & Swinscow, *70*, *621*, **622–623**, *625*
  complanata (Sw.) Ach., *621*, **623**, *628*
  dendriscoides Nyl., 624, 629
  denticulata Nyl., 623
  dilacerata (Hoffm.) Hoffm., *621*, **623–624**, 629
  duriaei (De Not.) Bagl., 628

ecklonii (Sprengel) G. Meyer & Flotow, 623
evernioides Nyl., 628
farinacea (L.) Ach., 32, 312, *621*, **624**, 629, 630
geniculata Hooker f. & Taylor, 624
intermedia (Delise *ex* Nyl.) Nyl., 69, *621*, **624–625**, *627*, 629
lacera (With.) J. R. Laundon, 628
leptocarpha Tuck., *620*, *621*, **625**, *626*, 630
maciformis (Delise) Bory, 38
menziesii Taylor, 40, 57, 61, 74, *620*, **625–626**, *632*, 762
montagnei De Not., *621*, **626–627**, *628*, 630, 632
obtusata (Arnold) Bitter, *621*, *624*, **627–628**, 629
paludosa B. Moore, 75, *621*, *622*, *627*, **628**, 764
peruviana Ach., *621*, *624*, **628–629**
petrina Bowler & Rundel, 624
pollinaria (Westr.) Ach., *621*, *624*, *627*, **629**
roesleri (Hochst. *ex* Schaerer) Hue, *621*, **629**
sinensis Jatta, *136*, *621*, *622*, **630**
stenospora Müll. Arg., *621*, *626*, **630**, *631*
subleptocarpha Rundel & Bowler, *620*, *621*, *624*, *625*, **630**, *632*
thrausta (Ach.) Nyl., 150, *620*, **630**, *632*
usnea (L.) R. Howe, *620*, *625*, *626*
willeyi R. Howe, 69, *621*, *626*, *628*, **632**, *633*
ramalina, 620
  bumpy, 623
  chalky, 629
  dotted, 624
  fan, 630
  frayed, 629
  hooded, 627
  punctured, 623
  rock, 624
  sinewed, 621
  slit-rimmed, 630
  striped, 626
  thorny, 632
  warty, 628
Ramalinaceae, 752
rash lichens, 611
  bark, 612
reindeer lichens, 38, 39, 56, 57, 60, 83, 95, 99, 179, 223, 224, 255, 278, 663
  black-footed, 229
  Dixie, 230
  dune, 230
  green, 225
  gray, 228
  maritime, 225
  star-tipped, 228
reindeer moss, 3, 59, 223

Reinkella Darbish.
  parishii Hasse, 341
Relicina (Hale & Kurok.) Hale, *138*, *142*, *189*, 752
  abstrusa (Vainio) Hale, 189
  eximbricata (Gyelnik) Hale, 189
Rhizocarpaceae, 752
Rhizocarpon Ramond *ex* DC., 26, 27, 55, 102, *130*, *131*, 300, 428, 475, 578, 583, **632–633**, *633–634*, 652, 752
  badioatrum (Flörke *ex* Sprengel) Th. Fr., 633
  bolanderi (Tuck.) Herre, 633
  chioneum (Norman) Th. Fr., *633*, **634**
  cinereovirens (Müll. Arg.) Vainio, 636
  concentricum (Davies) Beltr., *584*, *633*
  disporum (Nägeli *ex* Hepp) Müll. Arg., *633*, **634–635**
  eupetraeoides (Nyl.) Blomb. & Forss., *634*, 635
  eupetraeum (Nyl.) Arnold, 634
  geminatum Körber, *633*, 634–635
  geographicum (L.) DC., 17, 41, 84, *634*, **635**, *637*, 761
  grande (Flörke *ex* Flotow) Arnold, *634*, 635
  hochstetteri (Körber) Vainio, *131*, 428, *633*, **636**
  inarense (Vainio) Vainio, 635
  lavatum Hazsl., *363*, 638
  lecanorinum Anders, *634*, *635*, **636**
  macrosporum Räsänen, *634*, *635*, **637**
  obscuratum (Ach.) A. Massal., *633*, **637–638**
  oederi (Weber) Körber, *393*, *633*, 688
  plicatile (as commonly used), 633
  rubescens Th. Fr., 633
  superficiale (Schaerer) Vainio, 635
  umbilicatum (Ramond) Flagey, 634
Rhizoplaca Zopf, 14, 51, 59, 71, *133*, *137*, *170*, 372, 471, **638**, *638*, *696*, *733*, 752
  chrysoleuca (Sm.) Zopf, *638*, **639**, *640*, *641*
  glaucophana (Nyl. *ex* Hasse) W. A. Weber, 638
  haydenii (Tuck.) W. A. Weber, 120, 170, *638*, **639–640**, *641*
  marginalis (Hasse) W. A. Weber, 638
  melanophthalma (DC.) Leuckert & Poelt, 51, 384, *638*, *639–640*, **640**, *641*
  peltata (Ramond) Leuckert & Poelt, 65, 71, *638*, **640–641**
  subdiscrepans (Nyl.) R. Sant., *134*, *375*, *638*, *639*, **641**
ribbon lichen, yellow, 158
Rimelia Hale & A. Fletcher, 20, 479, *480*, **642**, *642*, 752
  cetrata (Ach.) Hale & A. Fletcher, *642*, **642**
  diffractaica (Essl.) Hale & A. Fletcher, 643

reticulata (Taylor) Hale & A. Fletcher, 20, 87, *491*, *494*, 500, *642*, **642–643**
  simulans (Hale) Hale, 643
  subisidiosa (Müll. Arg.) Hale & A. Fletcher, 211, *642*, **643–644**
Rimeliella Kurok., 210
rim-lichens, 369, 372
  beaded, 377
  bean-spored, 370
  black-eyed, 379
  brown-eyed, 375
  bumpy, 381
  chocolate, 586
  cypress, 379
  driftwood, 389
  flat-fruited, 390
  frosted, 376
  fused, 388
  fuzzy-, 190
  granite-speck, 387
  Hagen's, 380
  mealy, 388
  mortar, 380
  multicolored, 385
  New Mexico, 384
  peg-legged, 648
  sagebrush, 380
  seaside sulphur-, 386
  shrubby, 385
  smoky, 377
  stonewall, 383
  sunken, 384
  varying, 375
  white, 387
  white-, 661
  wood-spot, 386
Rimularia Nyl., 751
  insularis (Nyl.) Rambold & Hertel, 387, 388
Rimulariaceae, 751
ring lichens, 162
  concentric, 162
  rippled, 163
Rinodina (Ach.) Gray, *132*, *158*, 300, *343*, 372, 544, **644**, *644–645*, 752
  adirondacki H. Magn., 645
  applanata H. Magn., 645
  archaea (Ach.) Arnold, *644–645*
  ascociscana Tuck., *644–645*, **645**, *646*
  bischoffii (Hepp) A. Massal., 644
  bolanderi H. Magn., 74, *645*, **645–646**
  colobina (Ach.) Th. Fr., 645
  confragosa (Ach.) Körber, *644*, 646
  conradii Körber, 644
  dakotensis H. Magn., 158, 645
  gennarii Bagl., 644
  glauca Ropin, 645
  marysvillensis H. Magn., 645

milliaria Tuck., 158, 645
mniaraea (Ach.) Körber, 647
polyspora Th. Fr., 647
populicola H. Magn., 70, *645,* **646–647**
roscida (Sommerf.) Arnold, 646, 647
subminuta H. Magn., 27, *645*
tephraspis (Tuck.) Herre, *644*
turfacea (Wahlenb.) Körber, *67,* **644–645,** **647**
Roccella DC., 13, 14, 73, 81, *119,* 295, 463, **647,** 653, 753, 760
babingtonii Mont., 81, 88, **647–648**
fimbriata Darbish., 648
Roccellaceae, 753
Roccellina Darbish., 305, 371, **648,** 753
conformis Tehler, *132,* **648**
franciscana (Zahlbr. *ex* Herre) Follmann, *132,* 648
rockbright, 638
rockfrog, variable, 741
rock gnome, 329
rock grub lichens, 154
rock grubs, 154
rock hair, 619
rock licorice, 414
rock-olive lichens, 522
common, 523
cylindrical, 523
giant, 525
powdery, 523
stuffed, 524
rock pimples, 661
rock-posy lichens, 638
brown, 640
green, 640
orange, 639
scattered, 641
rockshag lichen, 309
rock-shield lichens, 733
Cumberland, 735
New Mexican, 738
Ozark, 736
peppered, 735
Plitt's, 739
salted, 737, 738, 740
shingled, 740, 741
tight, 738
rock tripe, 41, 79, 81, 83, 296, 366, 463, 696
blistered, 703
blushing, 708
emery, 706
folded, 699
fringed, 699
frosted, 699
netted, 701, 707
peppered, 702
petaled, 706

plated, 704
punctured, 708
roughened, 707
salty, 703
smooth, 704
rockwool, 587
coarse, 587
fine, 587
pale, 657
Ropalospora A. Massal., 170, 320, **649,** 753
chlorantha (Tuck.) S. Ekman, *130,* **649**
lugubris (Sommerf.) Poelt, 649
viridis (Tønsberg) Tønsberg, *124,* 649
rosette lichens, 361, 546
beaded, 553
black-bottomed, 556
black-eyed, 556
blue-gray, 553
buttoned, 616
cryptic, 560
dwarf, 556
fringed, 559
frosted, 552
granite, 555
hoary, 549
hooded, 549
mealy, 555
powderback, 553
powder-tipped, 554
powdery, 550
rockmoss, 427
slender, 558
spotted, 556
star, 558
streaked, 551
ruffle lichens, 489, 691
cracked, 495, 642
green, 502
long whiskered, 499
mottled, 210, 211
palm, 501
perforated, 499
powder-crown, 499
powdered, 492, 493, 496
powder-edged, 500
salted, 494
spotted, 501
sulphur, 501
unperforated, 497
unwhiskered, 492, 494
yellow, 729
yellow-cored, 495
rusty-rock lichen, 688

Sagiolechia A. Massal., 752
rhexoblephara (Nyl.) Zahlbr., *130*
saguaro lichens, 212

Sarcographa Fée, *126,* 325, 536, **649,** 752
intricans (Nyl.) Müll. Arg., 650
labyrinthica (Ach.) Müll. Arg., 74, **650**
medusulina (Nyl.) Müll. Arg., 650
tricosa (Ach.) Müll. Arg., 650
Sarcogyne Flotow, 55, *130,* 145, 581, 583, **650,** 659, 752
bicolor H. Magn., 581, 650
clavus (DC.) Kremp., 650–651
privigna (Ach.) A. Massal., 581, 651
regularis Körber, 147, 581, **650–651**
similis H. Magn., 650–651
Sarcosagium A. Massal., 752
campestre (Fr.) Poetsch & Schiedem., *129*
saucer lichens, 463
arctic, 465
coral, 468
double-rim, 467
frosty, 464
juvenile, 466
powdery, 465
rosy, 468
smooth, 466
tundra, 468
sausage lichen, mountain, 177
scale lichens, 308, 597
blackberry, 599
blue-edged, 604
blushing, 599
bordered, 602
brain, 598
brick, 599
brown-eyed, 603
butterfly, 601
mountain, 599
Pacific, 602
yellow, 601
scatter-rug lichens, 489
Schadonia Körber, 423, 424, 752
alpina Körber, 424
fecunda (Th. Fr.) Vězda & Poelt, 424
Schaereria Körber, 26, *391,* **651,** 751
cinereorufa (Schaerer) Th. Fr., 652
fuscocinerea (Nyl.) Clauzade & Roux, *392,* **651–652**
tenebrosa (Flotow) Hertel & Poelt, 651
Schaereriaceae, 751
Schismatomma Flotow & Körber *ex* A. Massal., *126,* 305, 369, 753
californicum (Tuck.) Zahlbr., *132*
hypothallinum (Zahlbr.) Hasse, 371
Schizopelte Th. Fr., 73, **652,** 753
californica Th. Fr., *121, 342,* **652–653**
Sclerophora Chevall., 222, 752
Scoliciosporum A. Massal., *130,* 170, 171, **653,** 752
chlorococcum (Stenh.) Vězda, *171,* **653**

sarcothamni (Vainio) Vĕzda, 653
umbrinum (Ach.) Arnold, *171*
scribble lichens, 471, 472
script lichens, *118*, 325, 326
blistered, 324–325
brick-spored, 535
common, 327
dark-spored, 536
fluted, 328
pastry, 326
powdered, 327
warty, 649
seaside lichen, black, 728
sea tar, 728
sea-storm lichens, 220
shadow lichens, 538
dark, 544
hairy, 539
mealy, 542
orange-cored, 543
pincushion, 540
pompon, 543
powder-tipped, 539
smooth, 540
starburst, 541
whiskered, 542
shadow-crust lichens, 343
grainy, 343
smooth, 343
shag lichens, 309
shell lichens, 279
fruiting, 280
salted, 280
shield lichens, 211, 479, 486
bottlebrush, 484
Carolina, 211
fringed, 486
hairless-spined, 487
hairy-spined, 486
hammered, 484
pustuled, 487
salted, 483
Texas, 212
tumbleweed, 734
western, 481
shingle lichens, 475, 484
black-bordered, 485
brown, 321
brown-eyed, 478
coral-rimmed, 479
mealy-rimmed, 476
moss, 323
petaled, 323
pink-eyed, 324
rimmed, 322
rock, 321
veined, 476

sieve lichen, 261
Sigridea Tehler, 305, 753
californica (Tuck.) Tehler, *132*
Siphula Fr., **653**, 752
ceratites (Wahlenb.) Fr., 38, *121*, **653–654**
snow lichens, 314
crinkled, 314
curled, 314
soil jelly lichen, 287
soil paint lichen, 147
soil ruby lichens, 332
common, 333
Solenopsora A. Massal., 370, 752
candicans (Dicks.) J. Steiner, *370*
Solorina Ach., 5, *140, 142, 143*, 333, **654,** *654–655,* 753
bispora Nyl., 654, 656
crocea (L.) Ach., 22, *655,* **655**
octospora (Arnold) Arnold, 654, 656
saccata (L.) Ach., 50, 654, *655,* **655–656**
simensis Hochst., 656
spongiosa (Ach.) Anzi, *131*, 654, **656**
Solorinellaceae, 752
soot lichens, 291
cupped, 292
yellow, 292
speck lichens, 677, 726
southern soldiers, 251
speckleback lichens, 605
forest, 609
green, 317
seaside, 608
specklebelly lichens, 591
dimpled, 593
green, 593
netted, 592
oldgrowth, 595
yellow, 594
speckled greenshield, 317
fruiting, 318
powder-edged, 319
speckled shield lichens, 605
Appalachian, 606
eastern, 606
powdered, 609
rock, 608
rough, 608
southwestern, 607
Speerschneidera Trevisan, **657**, 752, 757
euploca (Tuck.) Trevisan, 74, *120, 141*, 541, **657**
Sphaerophoraceae, 752
Sphaerophorus Pers., 16, *120, 121*, **657,** *657,* 752
fragilis (L.) Pers., *657,* **658**
globosus (Hudson) Vainio, *119*, 657, **658,** 659, 757

Sphinctrina Fr., *192*, 751
turbinata (Pers. ex Fr.) de Not, *193*
Sphinctrinaceae, 751
Spilonema Bornet, 309, 579, 597, **658,** 752
paradoxum Bornet, 659
revertens Nyl., *122*, 605, **659**
spiral-spored lichens, 412
split-peg lichen, 242, 274
split-peg soldiers, 242
Sporastatia A. Massal., *129*, 300, **659,** 752
testudinea (Ach.) A. Massal., **660**
polyspora (Nyl.) Grummann, 660
spotted black-foot, 275
spraypaint, 360
Squamarina Poelt, *133, 174,* **660,** 752
cartilaginea (With.) P. James, 661
degelii Frey & Poelt, 661
lentigera (Weber) Poelt, 50, *136,* **661**
Squamarinaceae, 752
starburst lichens, 360, 488
American, 362
gray, 489
green, 489
salted, 361
Staurothele Norman, *125*, 308, 309, 363, 570, **661,** *661,* 726, 753
areolata (Ach.) Lettau, 570, 661, 662
clopima (Wahlenb.) Th. Fr., 662
clopimoides (Arnold) J. Stein, 662
diffractella (Nyl.) Tuck., *661,* **661–662**
drummondii (Tuck.) Tuck., *661,* **662**
fissa (Taylor) Zwackh, *661,* **662–663**
fuscocuprea (Nyl.) Zschacke, 662
Stenocybe (Nyl.) Körber, *192*, 222, 751
major (Nyl.) Körber, *192*
Stereales, 751
Stereocaulaceae, 752
Stereocaulon Hoffm., 16, 22, 49, 55, 56, 58, 60, *121*, 231, 398, 566, 567, 657, **663,** *663–664,* 752, 761
alpinum Laurer *ex* Funck, 665–666
arenarium (Savicz) Lamb, 670
botryosum Ach., 664
condensatum Hoffm., 663, 664
coniophyllum Lamb, *663,* **664–665**
dactylophyllum Flörke, 68, *664,* **665,** 667, 669
glareosum (Savicz) H. Magn., 663, 664
grande (H. Magn.) H. Magn., *663–664,* **665–666**
groenlandicum (E. Dahl) Lamb, 670
incrustatum Flörke, 667
intermedium (Savicz) H. Magn., 667, 669
paschale (L.) Hoffm., *664,* **666,** 669
pileatum Ach., 48, *121*, 568, 663, 664, **666**
rivulorum H. Magn., *664,* **667**
sasakii Zahlbr., 664, 666, **667,** 668, 669

saxatile H. Magn., *664, 665,* **667, 669**
spathuliferum Vainio, *664*
sterile (Savicz) Lamb *ex* Krog, *667, 669*
subcoralloides (Nyl.) Nyl., *669*
tennesseense H. Magn. *ex* Degel., *664, 665,* **669**
tomentosum Fr., *50, 664, 666, 667,* **669–670**
vesuvianum Pers., *55, 663,* **670**
Sticta (Schreber) Ach., *6, 19, 20, 21, 72, 95, 143, 295, 415, 455, 457, 591,* **670–671,** *671, 752, 757*
arctica Degel., *139, 670*
beauvoisii Delise, *671,* **671–672**
fuliginosa (Hoffm.) Ach., *671,* **672**
limbata (Sm.) Ach., *671,* **672,** *673*
oroborealis Goward & Tønsberg, *295*
sylvatica (Hudson) Ach., *672*
weigelii (Ach.) Vainio, *455, 671, 672*
wrightii Tuck., *671*
Stictidaceae, *752*
stippleback lichens, *297*
  common, *298*
  northwest, *299*
  sandpaper, *299*
  streamside, *297*
stippled lichens, *307*
  inflated, *308*
  scaly, *308*
stipplescale lichens, *569*
  brown, *569*
  tree, *570*
Strangospora Körber, *752*
  microhaema (Norman) R. Anderson, *127*
  moriformis (Ach.) Stein, *129*
strap lichens
  southern, *630*
  western, *625*
Strigula Fr., *125, 582, 753*
  stigmatella (Ach.) R. C. Harris, *126, 582*
Strigulaceae, *753*
stubble lichens, *5, 14, 92, 99, 192, 221, 596*
  brown-head, *222*
  green, *193*
  sulphur, *222*
  yellow collar, *193*
Sulcaria Bystrek, *150, 179,* **672,** *752*
  badia Brodo & D. Hawksw., *119, 151,* **673**
  isidiifera Brodo, *674*
sulphur lichens, *319*
  desert, *319*
  tundra, *319*
sulphur-cups
  greater, *274*
  lesser, *251*
sulphur-rim lichen, seaside, *386*

sunburst lichens, *742*
  bare-bottomed, *745*
  elegant, *744*
  hooded, *744*
  maritime, *746*
  pin-cushion, *746*
  poplar, *746*
  powdery, *748*
  shrubby, *743*
  sugared, *747*
sunken disk lichens, *167*
  brown, *174*
  chalky, *167*
  chiseled, *169*
  false, *429*
  puffed, *422*
sunshine lichens, *729*
  brown-eyed, *730*
  hidden, *732*
  limestone, *731*
  powdered, *731*
surprise lichen, *173*
Synalissa Fr., *611, 752*
tar-jelly, *287*
tarpaper lichens, *282*
tar-spot lichens, *572*
Teloschistaceae, *753*
Teloschistes Norman, *26, 89, 118, 199,* **674,** *674, 742, 753*
  arcticus Zahlbr., *67, 674*
  californicus Sipman, *161, 674*
  chrysophthalmus (L.) Th. Fr., *18,* **674,** *674, 675*
  contortuplicatus (Ach.) Clauzade & Rondon *ex* Vězda, *674*
  exilis (Michaux) Vainio, *11,* **674,** *675*
  flavicans (Sw.) Norman, *674,* **675,** *676*
Teloschistineae, *753*
Tephromela M. Choisy, *372,* **675–676,** *752*
  armeniaca (DC.) Hertel & Rambold, *391, 393,* **676**
  atra (Hudson) Hafellner, *133,* **676–677**
Thamnolia Ach. *ex* Schaerer, *16, 109, 654,* **677,** *752*
  vermicularis (Sw.) Ach. *ex* Schaerer, *121,* **677,** *678*
  vermicularis var. subuliformis Asah., *677*
  vermicularis var. vermicularis, *677*
the dotted line, *624*
Thelidium A. Massal., *125,* **677,** *726, 753*
  pyrenophorum (Ach.) Mudd, **678**
Thelocarpon Nyl. *ex* Hue, *124, 125, 752*
  epibolum Nyl., *26*
Thelomma A. Massal., *28, 127, 292,* **678–679,** *679, 752*

  californicum (Tuck.) Tibell, *48, 302, 679,* **679–680**
  mammosum (Hepp) A. Massal., *679,* **680**
  occidentale (Herre) Tibell, *679*
  ocellatum (Körber) Tibell, *679–680*
  santessonii Tibell, *680*
Thelopsis Nyl., *124, 125, 752*
Thelotrema Ach., *23, 28, 125, 449,* **680,** *680, 752*
  lathraeum Tuck., *680*
  lepadinum (Ach.) Ach., *24,* **680,** *681*
  petractoides P. M. Jørg. & Brodo, *681*
  santense Tuck., *680*
  subtile Tuck., *680, 681*
Thelotremataceae, *752*
Thermutis Fr., *752*
  velutina (Ach.) Flotow, *658*
Tholurna Norman, **681,** *752*
  dissimilis (Norman) Norman, *121, 657,* **681–682**
thornbush lichens, *365*
  coastal, *365*
  flattened, *366*
Thrombiaceae, *753*
Thrombium Wallr., *753*
  epigaeum (Pers.) Wallr., *124*
Thyrea A. Massal., *5, 414,* **682,** *752*
  confusa Henssen, *137, 415,* **682–683**
  pulvinata (Schaerer) A. Massal., *682*
ticker-tape lichen, *348*
tile lichen, *390, 393*
  brown, *392*
toadskin lichens, *366*
  blackened, *367*
  common, *366*
Toninia A. Massal., *130, 131, 136, 308, 395, 597,* **683,** *683, 752*
  alutacea (Anzi) Jatta, *66, 683,* **683–684**
  arctica Timdal, *683*
  bullata (Meyen & Flotow) Zahlbr., *685*
  caeruleonigricans (as commonly used), *684*
  candida (Weber) Th. Fr., *683, 684*
  lobulata (Sommerf.) Lynge, *395*
  philippea (Mont.) Timdal, *683*
  ruginosa (Tuck.) Herre, *683*
  sedifolia (Scop.) Timdal, *683,* **684**
  squalida (Ach.) A. Massal., *683*
  subdiffracta Timdal, *683*
  tristis (Th. Fr.) Th. Fr. **ssp. asiae-centralis** (H. Magn.) Timdal, *683,* **684–685**
toy soldiers, *239*
Trapelia M. Choisy, *128, 133,* **685,** *686, 751*
  coarctata (Sm.) M. Choisy, *26, 686*
  corticola Coppins & P. James, *685*
  involuta (Taylor) Hertel, **685–686**
  mooreana (Carrol) P. James, *685*
  placodioides Coppins & P. James, *686*

Trapeliaceae, 751
Trapeliopsis Hertel & Gotth. Schneider, 124, 129, 685, **686**, 686, 751
    flexuosa (Fr.) Coppins & P. James, 122, 573, 686, **686**, 687
    gelatinosa (Flörke) Coppins & P. James, 686, 686
    granulosa (Hoffm.) Lumbsch, 55, 56, 122, **687**
    pseudogranulosa Coppins & P. James, 687
    wallrothii (Flörke) Hertel & Gotth. Schneider, 135, 136, 598, 686, **687**
tree coral, 657
tree-coral lichen, tiny, 426
treeflute, 441
tree-hair lichen, 79, 96, 179, 180
tree lichen, black
tree moss, 82
treepelt lichen, 395
Tremolecia M. Choisy, 26, *391*, **688**, 752
    atrata (Ach.) Hertel, 48, *391*, 393, 586, **688**
Tricholomataceae, 751
Trichotheliaceae, 753
Trichotheliales, 753
Trichothelium Müll. Arg., 753
    cestrense (Michener) R. C. Harris, 126
tripes de roches, 79, 697
trumpet lichen, 247, 254
Trypetheliaceae, 753
Trypethelium Sprengel, 125, 170, 368, 612, **688–689**, 753, 762
    aeneum (Eschw.) Zahlbr., 689, **689**, 690
    eluteriae Sprengel, 689
    mastoideum (Ach.) Ach., 690
    ochroleucum (Eshw.) Nyl., 689
    subeluteriae Makhija & Patwardhan, 689
    tropicum (Ach.) Müll. Arg., 689, **689–690**
    variolosum Ach., 689
    virens Tuck. *ex* Michener, 29, 46, 689, **690–691**
tube lichens, 345
    beaded, 346
    budding, 349
    deflated, 353
    forked, 350
    freckled, 350
    heath, 354
    hooded, 354
    lattice, 353
    mottled, 350
    powdered, 348
    powder-headed, 355
    seaside, 349
    varnished, 347
    wrinkled, 354

Tuckermannopsis Gyelnik, 30, *141*, *143*, *148*, *149*, *214*, *430*, 577, **691**, *691*, 733, 752, 755
    americana (Sprengel) Hale, 691, **692**, 694
    chlorophylla (Willd.) Hale, *140*, *158*, *430*, *431*, 691, **692**, 693
    ciliaris (Ach.) Gyelnik, 60, 691, **692**, 694
    coralligera (W. A. Weber) W. A. Weber, *140*, *430*, 691, **692–693**
    fendleri (Nyl.) Hale, 365, *431*, 691, 692, **693–694**, 696
    orbata (Nyl.) M. J. Lai, 366, 691, 692, **694**, 695
    platyphylla (Tuck.) Hale, 73, 691, 692, **694–695**
    sepincola (Ehrh.) Hale, 691, 693, **696**
    subalpina (Imshaug) Kärnefelt, *140*, *214*, 691, **696**, 697
tumbleweed, arctic, 427
turban lichen, 264

Umbilicaria Hoffm., 14, 41, 48, 79, 81, *137*, 296, 366, 463, 471, **696–697**, *697–699*, 753, 758, 761, 764
    americana Poelt & T. Nash, 698, 699, 700, 704
    angulata Tuck., 698, 705, 706, 708
    arctica (Ach.) Nyl., 703
    caroliniana Tuck., **697–698**, 699, 701, 757
    cinereorufescens (Schaerer) Frey, 705
    cylindrica (L.) Delise *ex* Duby, 698, **699**, **701**, 708, 709
    decussata (Vill.) Zahlbr., 698, **701–702**, 704, 707
    deusta (L.) Baumg., 697, 698, **702–703**
    esculenta (Miyoshi) Minks, 83, 697
    hirsuta (Sw. *ex* Westr.) Hoffm., 697, 698, 699
    hyperborea (Ach.) Hoffm., 24, 699, **703**, 706, 708
    krascheninnikovii (Savicz) Zahlbr., 698, 702, **703–704**, 707, 709
    lyngei Schol., 701–702, 704, 707
    mammulata (Ach.) Tuck., 69, 698, 699, **704**, 705
    muehlenbergii (Ach.) Tuck., 79, 697, 699, **704–705**
    nylanderiana (Zahlbr) H. Magn., 703
    phaea Tuck., 699, **706**
    phaea var. coccinea Llano, 706
    polyphylla (L.) Baumg., 699, 703, **706–707**, 761
    polyrrhiza (L.) Fr., 698, 704
    proboscidea (L.) Schrader, 12, 698, 702, 703, 704, **707**
    rigida (Du Rietz) Frey, 698, **707–708**
    torrefacta (Lightf.) Schrader, 698, 705, **708**

    vellea (L.) Hoffm., 699
    virginis Schaerer, 698, 699, 701, 704, **708–709**
Umbilicariaceae, 753
Umbilicariineae, 753
urn lichen, 681
urn-disk lichens, 423
Usnea Dill. *ex* Adans., 16, 30, 34, 37, 48, 58, 60, 74, 79, 82, 83, 84, 86, 89, 90, 93, 94, *119*, *150*, *152*, 311, 312, 620, **709–711**, *711–712*, 752, 758, 760, 764
    amblyoclada (Müll. Arg.) Zahlbr., 711, **712–713**
    angulata Ach., 711, 717
    antillarum (Vainio) Zahlbr., 713
    arizonica Mot., 711, **713**, 715, 724
    baileyi (Stirton) Zahlbr., 75, 94, *712*, **713–714**, 723
    californica Herre, 715
    cavernosa Tuck., 711, **714**
    ceratina Ach., 711, 712, **715**, 723, 764
    chaetophora Stirton, 720
    cirrosa Mot., 11, 711, 713, **715**, 716, 725
    cornuta Körber, 712, **715–716**, 719, 720
    dasaea Stirton, 712
    dasypoga (as commonly used), 718
    dimorpha (Müll. Arg.) Mot., 711, **717**, 726
    diplotypus Vainio, 710, 712, **717–718**, 758, 761, 762
    filipendula Stirton, 711–712, **718**, 724
    fragilescens Hav. *ex* Lynge, 710, 712, 716, **718–719**
    fragilescens var. fragilescens, 719
    fragilescens var. mollis (Vainio) P. Clerc, 716, 719
    fulvoreagens (Räsänen) Räsänen, 721
    glabrata (Ach.) Vainio, 712, 716, **719**, 720
    glabrescens (Nyl. *ex* Vainio) Vainio, 710, 712, 719, 721
    halei P. Clerc, 712
    herrei Hale, 712
    hesperina Mot., 711, 714, **720**, 726
    hirta (L.) F. H. Wigg., 712, 714, **720–721**
    implicita (Stirton) Zahlbr., 713
    lapponica Vainio, 712, 719, **721**
    laricina Vainio *ex* Räsänen, 721
    longissima Ach., 10, 89, 94, 711, 717, **721–723**, 724, 726, 758, 761
    madeirensis Mot., 718
    merrillii Mot., 720
    mutabilis Stirton, 712, **723**, 724, 725, 726
    perplectata Mot., 714
    rigida (Ach.) Mot., 713, 720
    rubicunda Stirton, 710, 712, **723–724**
    scabiosa Mot., 724
    scabrata Nyl., 711, 714, 718, **724**

sphacelata R. Br., *711*
strigosa (Ach.) Eaton, *711, 713, 715, 723,* **724–725**
subcornuta Stirton, *724*
subfloridana Stirton, *712, 716, 718, 719*
subfusca Stirton, *711, 713, 725*
subscabrosa Mot., *94, 712,* **725–726**
substerilis Mot., *721*
trichodea Ach., *710, 711, 714, 717, 720, 724,* **726**
variolosa Mot., *720*
wirthii Clerc, *716, 719*

vagabond lichen, 169
Varicellaria Nyl., *526, 753*
    rhodocarpa (Körber) Th. Fr., *127*
Vermilacinia Spjut & Hale, *457*
    cephalota (Tuck.) Spjut & Hale, *457*
    ceruchoides (Rundel & Bowler) Spjut, *458*
    combeoides (Nyl.) Spjut & Hale, *458*
    laevigata (Bowler & Rundel) Spjut, *460*
    procera (Rundel & Bowler) Spjut, *458*
Verrucaria Schrader, *5, 29, 124, 363, 661, 662, 677,* **726,** *727, 753, 758, 762*
    aethiobola Wahlenb., *727*
    calciseda DC., *50, 727,* **727–728**
    calkinsiana Servit, *727*
    degelii R. Sant., *728*
    epimaura Brodo, *727, 728*
    erichsenii Zschacke, *728*
    halizoa Leighton, *727, 728*
    inficiens Breuss, *570, 727*
    maura Wahlenb., *77, 727,* **728**
    microspora (as commonly used), *727*
    mucosa Wahlenb., *727, 728*
    muralis Ach., *727*
    nigrescens Pers., *727, 728*
Verrucariaceae, *753*
Verrucariales, *753*
Vestergrenopsis Gyelnik, *140, 475, 573,* **729,** *753*
    elaeina (Wahlenb.) Gyelnik, *729*
    isidiata (Degel.) E. Dahl, *122, 475,* **729**
V-fingers, 156
vinyl lichens, 400
volcano lichen, 281, 449, 462
    frosted, 462
Vulpicida J.-E. Mattsson & M. J. Lai, *34, 83, 137,* **729–730,** *730, 752*
    canadensis (Räsänen) J.-E. Mattsson & M. J. Lai, *72, 411, 730,* **730–731,** *733*
    pinastri (Scop.) J.-E. Mattsson & M. J. Lai, *730,* **731**
    tilesii (Ach.) J.-E. Mattsson & M. J. Lai, *67, 118, 314, 730,* **731–732**
    viridis (Schwein.) J.-E. Mattsson & M. J. Lai, *138, 730,* **732–733**

wall lichen, 743
wand lichen, 269
    western, 278
wanderlust lichen, 639
wart lichens, 525
    arctic, 530
    bitter, 527
    finger, 528
    frosted, 532
    Macoun's, 529
    mesa, 532
    powdered, 533
    ragged, 530
    rimmed, 534
    rock, 530
    speckled, 170
    spotted, 530
    sulphur, 528
    Texas, 533
    volcano, 535
watercolor lichens, 362
water-crust, red, 611
waterfan, 298, 342
waterfingers, 653
Waynea Moberg, *752*
    californica Moberg, *135*
we'ia, 79, 96
whisker lichens, 221
white-rim lichen, 661
whitewash lichens, 545
whiteworm lichen, 677
witch's hair, 96, 150, 154
    flowering, 152
    gray, 153
    green, 153
    spiny, 152
wolf lichens, 16, 73, 83, 411, 412
    brown-eyed, 412
wooden soldiers, 242
woodscript lichens, 749
    black, 749
woollybear lichens, 579
wrinkle-lichens, 691
    broad, 694
    chestnut, 696
    coral-edged, 692
    dwarf, 693
    fringed, 692
    powdered, 692
    variable, 694

Xanthoparmelia (Vainio) Hale, *34, 48, 50, 51, 55, 59, 71, 74, 103, 139, 162, 170, 212, 316, 317, 479, 502, 609, 638,* **733,** *733–734, 752*
    ajoensis (T. Nash) Egan, *738*
    angustiphylla (Gyelnik) Hale, *734, 735, 736, 737*
    arseneana (Gyelnik) Hale, *738*
    californica Hale, *740*
    centrifuga (L.) Hale, *162*
    chlorochroa (Tuck.) Hale, *51, 81, 734,* **734–735,** *738, 740, 742*
    coloradoensis (Gyelnik) Hale, *738, 740*
    conspersa (Ehrh. *ex* Ach.) Hale, *733,* **735,** *736*
    cumberlandia (Gyelnik) Hale, *734,* **735–736,** *737, 739, 740, 755*
    hypomelaena (Hale) Hale, *734,* **736–737,** *739*
    hypopsila (Müll. Arg.) Hale, *734*
    isidiascens Hale, *735*
    kurokawae (Hale) Hale, *737*
    lavicola (Gyelnik) Hale, *733,* **737–738**
    lineola (E. C. Berry) Hale, *734,* **738,** *740, 742*
    maricopensis T. Nash & Elix, *738*
    mexicana (Gyelnik) Hale, *31, 733,* **738,** *739, 741*
    monticola (J. P. Dey) Hale, *739*
    mougeotii (Schaerer) Hale, *733*
    neochlorochroa Hale, *734*
    nigropsoromifera (T. Nash) Egan, *738*
    norchlorochroa Hale, *734*
    novomexicana (Gyelnik) Hale, *734, 737,* **738–739**
    piedmontensis (Hale) Hale, *735*
    plittii (Gyelnik) Hale, *733, 735,* **739–740**
    psoromifera (Kurok.) Hale, *738*
    schmidtii Hale, *738*
    separata (Th. Fr.) Hale, *163*
    somloënsis (Gyelnik) Hale, *734, 736, 738,* **740,** *742*
    subramigera (Gyelnik) Hale, *733, 738*
    tasmanica (Hook. f. & Taylor) Hale, *734*
    weberi (Hale) Hale, *733, 738,* **740–741**
    wyomingica (Gyelnik) Hale, *734, 738, 740,* **741–742**
Xanthopsorella Kalb & Hafellner, *752*
    texana (W. A. Weber) Kalb & Hafellner, *134*
Xanthopyreniaceae, *753*
Xanthoria (Fr.) Th. Fr., *34, 46, 48, 52, 77, 118, 134, 137, 195, 205,* **742,** *742–743, 753*
    alaskana J. W. Thomson, *746*
    borealis R. Sant. & Poelt, *743*
    candelaria (L.) Th. Fr., *40, 206, 742, 743,* **743,** *745, 747*
    elegans (Link) Th. Fr., *13, 18, 40, 71, 195, 202, 204, 742,* **744,** *746, 747*
    fallax (Hepp) Arnold, *31, 46, 85, 206, 742, 743,* **744–745,** *748*
    fulva (Hoffm.) Poelt & Petutschnig, *743,* **745,** *748*
    hasseana Räsänen, *206, 742,* **746,** *747*

mendozae Räsänen, *743, 745*
montana L. Lindblom, *742, 746, 747*
oregana Gyelnik, *743, 743, 745, 748*
parietina (L.) Th. Fr., 26, 83, 86, *742*, **746, 747**
polycarpa (Hoffm.) Rieber, *742*, **746–747**
ramulosa (Tuck.) Herre, 746

sorediata (Vainio) Poelt, *123, 742, 744*, **747–748**
tenax L. Lindblom, *742, 747*
ulophyllodes Räsänen, *743, 745*, **748**
Xylographa Th. Fr., 46, *126*, 471, **749,** 751
  abietina (Pers.) Zahlbr, 749
  opegraphella Nyl. *ex* Rothr., 749

parallela (Ach.:Fr.) Behlen & Desberger, **749**
vitiligo (Ach.) J. R. Laundon, 749

yellow ribbon lichen, 158
yolk lichens, 206, 209
  Mexican, 210